Principles of Plant Genetics and Breeding

Principles of Plant Genetics and Breeding

Second Edition

George Acquaah
Bowie State University, Maryland, USA

WILEY-BLACKWELL

A John Wiley & Sons, Ltd., Publication

This edition first published 2012 © 2012 by John Wiley & Sons, Ltd

Wiley-Blackwell is an imprint of John Wiley & Sons, formed by the merger of Wiley's global Scientific, Technical and Medical business with Blackwell Publishing.

Registered office: John Wiley & Sons, Ltd, The Atrium, Southern Gate, Chichester, West Sussex, PO19 8SQ, UK

Editorial offices: 9600 Garsington Road, Oxford, OX4 2DQ, UK
The Atrium, Southern Gate, Chichester, West Sussex, PO19 8SQ, UK
111 River Street, Hoboken, NJ 07030-5774, USA

For details of our global editorial offices, for customer services and for information about how to apply for permission to reuse the copyright material in this book please see our website at www.wiley.com/wiley-blackwell.

The right of the author to be identified as the author of this work has been asserted in accordance with the UK Copyright, Designs and Patents Act 1988.

Library of Congress Cataloging-in-Publication Data
Acquaah, George.
  Principles of plant genetics and breeding / George Acquaah. — 2nd ed.
    p. cm.
  Includes bibliographical references and index.
  ISBN 978-0-470-66476-6 (cloth) — ISBN 978-0-470-66475-9 (pbk.) 1. Plant
breeding. 2. Plant genetics. I. Title.
  SB123.A334 2012
  631.5'2—dc23   2012010941

A catalogue record for this book is available from the British Library.

Wiley also publishes its books in a variety of electronic formats. Some content that appears in print may not be available in electronic books.

Cover design by Dan Jubb.

Set in 10/12pt, Galliard-Roman by Thomson Digital, Noida, India
Printed and bound in Singapore by Markono Print Media Pte Ltd

First Impression 2012

Dedication

To my wife, Theresa, with love and appreciation for uncommon character.

Contents

Companion website

The book is accompanied by a companion resources site:

www.wiley.com/go/acquaah/plantgeneticsandbreeding

With figures and tables from the book.

Preface

The second edition of *Principles of Plant Genetics and Breeding* represents a thoroughly overhauled version of the preceding edition, following recommendations and suggestions from users and reviewers. The major changes in the new edition include restructuring and reordering the chapters to follow more closely with how plant breeding is done in practice, and expanding the molecular genetics component. Also, the basic science information has been reduced. Two of the chapters in the first edition have been transferred to the back of the textbook as supplementary material, so it may be referred to by users only as needed. In this way, students and users who already have a background in genetics will not feel obligated to study those chapters before advancing to more plant breeding related topics. A feature of the first edition that is retained and expanded in the second edition is the inclusion of contributions on selected topics by industry professionals. The book is copiously illustrated to facilitate teaching and learning of the topics.

The book is organized into nine sections. Section I is an overview and historical perspective of plant breeding. Chapter 1 provides an introduction to the field of plant breeding, describing its importance to society while Chapter 2 provides historical perspectives, highlighting the contributions by researchers to knowledge in the field. The two chapters in Section II are devoted to discussing pertinent population and quantitative genetic concepts, to assist the reader in better understanding the practices of plant breeders.

Section III, reproductive systems, comprises four chapters. Chapter 5, autogamy, and Chapter 6, allogamy, focus on reproductive and genetic issues as they pertain to self-pollinated and cross-pollinated species, respectively. Chapter 7 is devoted to discussing the genetic issues associated with crossing plants to reorganize the genetic matrix, while Chapter 8 ends the section with a discussion of issues associated with clonal propagation.

Section IV, deals with germplasm for breeding. It is impossible to conduct plant breeding without the proper germplasm. Chapter 9 in this section focuses on variation and its genetic basis, while Chapter 10 focuses on domestication of plant species. The discussion includes the dependence of plant breeding on heritable variation. Finally, Chapter 11 addresses the matter of plant genetic resources used in plant breeding. It includes a discussion of how germplasm is collected and managed for long term use by breeders.

Section V is devoted to discussing common breeding objectives pursued by plant breeders. The discussions include the genetic basis of those traits and the implication in their breeding. Chapter 12 focuses on breeding for increased yield and improving morphological traits that enhance crop productivity. In the ensuing chapter, 13, breeding for selected quality traits is the focus of discussion. Breeding for disease and pest resistance is a major breeding objective in most crops. This is the subject of Chapter 14, while Chapter 15 is devoted to issues pertaining to breeding for resistance or tolerance to selected abiotic factors, such as salt tolerance.

The topics of Section VI focus on selection or breeding methods. In this section, breeding methods for autogamous species is the subject of Chapter 16, while Chapter 17 is devoted to breeding allogamous species. Chapter 18 concerns the selection methods used for breeding hybrid cultivars while the last chapter in the section, 19, is devoted to discussing the breeding methods used for clonally propagated species. The discussions in these chapters provide the advantages and disadvantages of each method, and include alternative approaches.

Molecular breeding is the subject of Section VII, which received the most overhaul. The concept of markers and various commonly used molecular markers in plant breeding are discussed in detail in Chapter 20, including their advantages and disadvantages, as well the cost and ease of application in breeding. Chapter 21 is devoted to discussing the mapping of genes and the importance of such maps in plant breeding. Marker assisted selection (MAS) as a method of facilitating plant breeding is the subject of

Chapter 22. Chapter 23 focuses on the use of mutagenesis for inducing variability for crop improvement. The discussions include the types of mutagens commonly used in crop improvement, and the success of this approach to breeding. Many important crop species are polyploids. The methods used for improving polyploids are discussed in Chapter 24. The last chapter in this section, 25, addresses the subject of molecular genetic modifications, including the role of genetic engineering in plant improvement. Also, in this chapter, the contemporary subject of genome-wide genetics is introduced.

Section VIII is titled marketing and societal issues in plant breeding. In Chapter 26, the reader is exposed to the process of preparing a cultivar for release to farmers for use. Commercial seed producers ensure the quality of their products through the seed certification process, as described in Chapter 27. Plant breeders protect their products through securing legal protection, the subject of Chapter 28. The last two chapters, 29 and 30, end the section with discussions on social concerns that arise from the applications of biotechnological tools, and issues confronting breeders on the international plant breeding scene.

The last section, IX, is devoted to discussion of the breeding of selected crops. This section includes discussions on the genetics of selected crop plants, germplasm used, and breeding methods used for their improvement. Professional highlights are provided for these chapters.

An effort has been made to organize this book such that the sequence of discussion of topics follows closely the sequence in conducting a plant breeding project. A plant breeding course, at the minimum, is usually an upper level course at the undergraduate level. It is assumed that a student taking a plant breeding course would have received prior instruction in the basic biology, including genetics, botany, and physiology. A review of basic genetic principles is helpful to better understanding the material in this book and a plant breeding course in general. Sometimes, some of this basic material is reviewed as appropriate. In addition, some of the underlying science is presented in the supplementary chapters of the book.

Acknowledgements

The author extends special gratitude to Drs Herman van Eck, Rients Niks, Marieke Jeuken, Gerard van der Linden, Yuling Bai, Paul Arens, Luisa Trindade, Chris Maliepaard and Jaap van Tuyl of the academic staff of the Laboratory of Plant Breeding, Wageningen University and Research Center, in The Netherlands, for their outstanding contribution to this edition. Specifically, Drs Van Tuyl and Arens reviewed and edited Chapter 7 (Hybridization), while Dr van der Linden reviewed and edited Chapter 15 (Breeding for resistance to abiotic stress). Dr Jeuken wrote a boxed reading article on lettuce BILs, while Dr Bai contributed two articles, a supplement to Chapter 39 (Breeding tomatoes) as well as a paper on the introgression breeding of tomatoes as part of the industry highlights featured in the book. Chapter 13 (Breeding for quality) was reviewed by Dr Trindale. Dr Miliepaard deserves special mention for reviewing almost the entire first edition of the book and for making suggestions for accuracy and general improvement of the second edition.

Of the Wageningen team members, the author reserves his profound and deepest appreciation for the invaluable contributions of Dr Herman van Eck who made the initial contact with a proposal to assist with reorganizing the second edition, putting the team together and reviewing Chapter 23 (Mutagenesis in plant breeding) as well. Dr van Eck and Dr Niks collaborated with the author to reorder and restructure the chapters of the first edition to make the contents of the second edition flow more meaningfully. They also suggested additional chapters and topics for inclusion in the new edition. Dr van Eck provided the author with a collection of published literature and personal notes to assist with writing new chapters and updating others. Dr. Niks' additional role included critically reviewing and editing several chapters, including 5 (Autogamy), 6 (Allogamy), 7 (Hybridization), 8 (Clonal propagation) 10 (Domestication), 11 (Plant genetic resources) and, especially, chapter 14 (Breeding for disease resistance), which was overhauled according to his recommendations. The second edition is clearer and more accurate because of his thorough review and insightful critique of the chapters he reviewed.

Notwithstanding the tremendous contribution of the Wageningen team, the final content of the book remains entirely the responsibility of the author. The author also acknowledges with deep fondness the support of Dr Theresa Acquaah, his wife, for her moral support during the preparation of this edition. The final and ultimate appreciation is reserved for the author's mentor and source of inspiration, Dr J.C. El Shaddai.

Industry highlights boxes

Industry highlights boxes: authors

Acquaah, G., Bowie State University, Computer Science Building, Bowie, MD 20715, USA

Afriyie, T., International Maize and Wheat Improvement Center (CIMMYT), PO Box 5689, Addis Ababa, Ethiopia

Anderson, J.A., University of Minnesota, Twin Cities, Department of Agronomy and Plant Genetics, 411 Borlaug Hall, St Paul, MN 55108-6026, USA

Baenziger, P.S., University of Nebraska-Lincoln, Department of Agronomy & Horticulture, 330 K, Lincoln NE 68583-0915, USA

Bai, Y., Wageningen UR Plant Breeding, Droevendaalsesteeg 1, 6708 PB Wageningen, The Netherlands

Betrán, F.J., Texas A&M University, College Station, TX 77843, USA

Borrell, A., DPI&F, Hermitage Research Station, Warwick, QLD 4370, Australia

Bradshaw, J.E., James Hutton Institute, Invergowrie, Dundee DD2 5DA, UK

Brown-Guedira, G., USDA-ARS Plant Science Research Unit, Dept. of Crop Science, North Carolina State University, 840 Main Campus Drive, Box 7258, Raleigh, NC 27606, USA

Burton, J.W., USDA Plant Science Building, 3127 Ligon Street, Raleigh, NC 27607, USA

Ceccarelli, S., The International Center for Agricultural Research in the Dry Areas (ICARDA), PO Box 5466 Aleppo, Syria

Chalyk, S., Institute of Genetics, Chisinau, Moldova

Chen, X., USDA-ARS Wheat Genetics, Quality, Physiology & Disease Research Unit, Washington State University, 209 Johnson Hall, Pullman, WA 99164-6420, USA

Deerfield, D.W., II, Pittsburgh Supercomputing Center, 4400 Fifth Avenue, Pittsburgh, PA 15213, USA

Dubcovsky, J., University of California at Davis, Department of Agronomy and Range Science, 281 Hunt Hall, Davis, CA 95616-8515, USA

Elias, E., North Dakota State University, Department of Plant Sciences, 470G Loftsgard Hall, Fargo, ND 58105, USA

van Eck, H.J., Laboratory for Plant Breeding, Wageningen University, Droevendaalsesteeg 1, 6708PB Wageningen, The Netherlands

Fritz, A., Kansas State University, Department of Agronomy, 4012 Throckmorton Hall, Manhattan, KS 66506, USA

Garland-Campbell, K.A., USDA-ARS Wheat Genetics, Quality, Physiology & Disease Research Unit, Washington State University, 209 Johnson Hall, Pullman, WA 99164-6420, USA

Gaur, P.M., International Crops Research Institute for the Semi-Arid Tropics, Patancheru 502 324, AP, India

Gill, B.S., Kansas State University, Wheat Genetics Resource Center, Department of Plant Pathology, 4024 Throckmorton Hall, Manhattan, KS 66506, USA

Gill, K.S., Washington State University, Department of Crop and Soil Sciences, Johnson 277, P.O. BOX 646420, Pullman, WA 99164-6420, USA

Gowda, C.L.L., International Crops Research Institute for the Semi-Arid Tropics, Patancheru 502 324, AP, India

Grando, S., The International Center for Agricultural Research in the Dry Areas (ICARDA), PO Box 5466 Aleppo, Syria

Haley, S., Colorado State University, Department of Soil and Crop Sciences, C101 Plant Sciences, Fort Collins, CO 80526, USA

Hammer, G., University of Queensland, School of Land and Food, QLD 4072 Australia

Hammer, K., Institute of Crop Science, Agrobiodiversity Department, University Kassel, Steinstr. 19, D-37213 Witzenhausen, Germany

Hammer-Spahillari, M., Dr. Junghanns GmbH, D-06449 Aschersleben, Germany

Harrington, J., Pioneer Hi-Bred, a DuPont Business, Des Moines, IA, USA

Henzell, B., DPI&F, Hermitage Research Station, Warwick, QLD 4370, Australia

Heuser, F., Institute of Crop Science, Agrobiodiversity Department, University Kassel, Steinstr. 19, D-37213 Witzenhausen, Germany

Irish, B.M., USDA-ARS, Tropical Agriculture Research Station (TARS) Mayaguez, 2200 Pedro Albizu Campos Ave. Suite. 201, Mayaguez, Puerto Rico 00680-5470

Jaradat, A.A., Agricultural Research Service, USDA, 803 Iowa Ave., Morris 56267 MN, USA

Jordan, D., DPI&F, Hermitage Research Station, Warwick, QLD 4370, Australia

Jueken, M., Wageningen UR Plant Breeding, Droevendaalsesteeg 1, 6708 PB Wageningen, The Netherlands

Keim, D.L., Delta and Pine Land Company, One Cotton Row, PO Box 157, Scott, MS 38772, USA

Khoshbakht, K., Institute of Crop Science, Agrobiodiversity Department, University Kassel, Steinstr. 19, D-37213 Witzenhausen, Germany

Kianian, S.F., North Dakota State University, Department of Plant Sciences, 470G Loftsgard Hall, Fargo, ND 58105, USA

Kidwell,K.K., Washington State University, Department of Crop and Soil Sciences, Johnson 277, P.O. BOX 646420, Pullman, WA 99164-6420, USA

Kindiger, B., USDA-ARS Grazinglands Research Laboratory, El Reno, OK 73036, USA

Klein, P., Texas A&M University, Institute for Plant Genomics & Biotechnology, College Station, USA

Klein, R., USDA-ARS, Southern Agricultural Research Station, College Station, USA

Lapitan, N., Colorado State University, Department of Soil and Crop Sciences, C101 Plant Sciences, Fort Collins, CO 80526, USA

McClung, A.M., USDA-ARS, Rice Research Unit, 1509 Aggie Dr., Beaumont, TX 77713, USA

Metz, S., Monsanto Corporation, 800 North Lindbergh Blvd, St Louis, MO 63167, USA

Mullet, J., Texas A&M University, Institute for Plant Genomics & Biotechnology, College Station, USA

Nguyen, H., University of Missouri, Plant Sciences Unit and National Center for Soybean Biotechnology, Columbia, MO 65211, USA (previously Texas Tech University, Lubbock, USA)

Nicholas, H.B., Jr., Pittsburgh Supercomputing Center, 4400 Fifth Avenue, Pittsburgh, PA 15213, USA

Nigam, S.N., International Crops Research Institute for the Semi-Arid Tropics, Patancheru 502 324, AP, India

Novy, R., USDA-Agricultural Research Service, Aberdeen, ID 83210, USA

Ohm, H., Purdue University, Department of Agronomy, 1150 Lilly Hall, West Lafayette, IN 47907-1150, USA

Rai, K.N., International Crops Research Institute for the Semi-Arid Tropics, Patancheru 502 324, AP, India

Reddy, B.V.S., International Crops Research Institute for the Semi-Arid Tropics, Patancheru 502 324, AP, India

Rooney, W., Texas A&M University, College Station, TX 77843, USA

Ropelewski, A.J., Pittsburgh Supercomputing Center, 4400 Fifth Avenue, Pittsburgh, PA 15213, USA

Rosenow, D., Texas A&M Agricultural Research & Extension Center, Lubbock, TX 79403-9803, USA

Santra, D., Washington State University, Department of Crop and Soil Sciences, Johnson 277, P.O. BOX 646420, Pullman, WA 99164-6420, USA

Saxena, K.B., International Crops Research Institute for the Semi-Arid Tropics, Patancheru 502 324, AP, India

Simpson, C., Texas A&M University, College Station, TX 77843, USA

Smith, J.R., USDA-ARS, Crop Genetics and Production Research Unit, Stoneville, MS 38776, USA

Soria, M., University of California at Davis, Department of Agronomy and Range Science, 281 Hunt Hall, Davis, CA 95616-8515, USA

Sorrells, M., Cornell University, Department of Plant Breeding, 252 Emerson Hall, Ithaca, NY 14853-1902, USA

Souza, E., University of Idaho, Aberdeen Research and Extension Center, 1693 South 2700 West, Aberdeen, ID 83210, USA

Talbert, L., Montana State University, Bozeman, Department of Plant Sciences and Plant Pathology, 406 Leon Johnson Hall, Bozeman, MT 59717-3150, USA

Teklu, Y., Institute of Crop Science, Agrobiodiversity Department, University Kassel, Steinstr. 19, D-37213 Witzenhausen, Germany

Thomas, W.T.B., Scottish Crop Research Institute, Invergowrie, Dundee DD2 5DA, UK

Ude, G.N., Natural Science Department, Bowie State University, Bowie, MD 20715, USA

Upadhyaya, H.D., International Crops Research Institute for the Semi-Arid Tropics, Patancheru 502 324, AP, India

Section 1

Overview and historical perspectives

Chapter 1 Introduction
Chapter 2 History of plant breeding

It is informative for students of plant breeding to have an historical perspective of the discipline. Understanding the time line of advances in one's discipline is instructive in itself. It helps the student to put current advances in plant breeding in the proper perspective, appreciating the challenges and opportunities along the way as professionals contributed to knowledge in the discipline.

1

Introduction

Purpose and expected outcomes

Agriculture is the deliberate planting and harvesting of plants and herding animals. This human invention has and continues to impact society and the environment. Plant breeding is a branch of agriculture that focuses on manipulating plant heredity to develop new and improved plant types for use by society. People in society are aware and appreciative of the enormous diversity in plants and plant products. They have preferences for certain varieties of flowers and food crops. They are aware that whereas some of this variation is natural, humans with special expertise (plant breeders) create some of it. Generally, also, there is a perception that such creations derive from crossing different plants. This introductory chapter is devoted to presenting a brief overview of plant breeding, including its benefits to society and some historical perspectives. After completing this chapter, the student should have a general understanding of:

1 The need and importance of plant breeding to society.
2 The goals of plant breeding.
3 The art and science of plant breeding.
4 Trends in plant breeding as an industry.
5 Selected milestones and accomplishments of plant breeders.
6 The future of plant breeding in society.

1.1 What is plant breeding?

Plant breeding is a deliberate effort by humans to nudge nature, with respect to the heredity of plants, to an advantage. The changes made in plants are permanent and heritable. The professionals who conduct this task are called **plant breeders**. This effort at adjusting the *status quo* is instigated by a desire of humans to improve certain aspects of plants to perform new roles or enhance existing ones. Consequently, the term "plant breeding" is often used synonymously with "plant improvement" in modern society. It needs to be emphasized that the goals of plant breeding are focused and purposeful. Even though the phrase "to breed plants" often connotes the involvement of the sexual process in effecting a desired change, modern plant breeding also includes the manipulation of asexually reproducing

Principles of Plant Genetics and Breeding, Second Edition. George Acquaah.
© 2012 John Wiley & Sons, Ltd. Published 2012 by John Wiley & Sons, Ltd.

plants (plants that do not reproduce through the sexual process). Breeding is hence about manipulating plant attributes, structure and composition, to make them more useful to humans. It should be mentioned at the onset that it is not every plant character or trait that is readily amenable to manipulation by breeders. However, as technology advances, plant breeders are increasingly able to accomplish astonishing plant manipulations, needless to say not without controversy, as is the case involving the development and application of **biotechnology** to plant genetic manipulation. One of the most controversial of these modern technologies is **transgenesis**, the technology by which gene transfer is made across natural biological barriers.

Plant breeders specialize in breeding different groups of plants. Some focus on field crops (e.g., soybean, cotton), horticultural food crops (e.g., vegetables), ornamentals (e.g., roses, pine trees), fruit trees (e.g., citrus, apple), forage crops (e.g., alfalfa, grasses), or turf species. (e.g., Bluegrass, fescue) More importantly, breeders tend to specialize in or focus on specific species in these groups (e.g., corn breeder, potato breeder). This way, they develop the expertise that enables them to be most effective in improving the species of their choice. The principles and concepts discussed in this book are generally applicable to breeding all plant species.

1.2 The goals of plant breeding

The plant breeder uses various technologies and methodologies to achieve targeted and directional changes in the nature of plants. As science and technology advance, new tools are developed while old ones are refined for use by breeders. Before initiating a breeding project, clear breeding objectives are defined based on factors such as producer needs, consumer preferences and needs, and environmental impact. Breeders aim to make the crop producer's job easier and more effective in various ways. They may modify plant structure, so it will resist lodging and thereby facilitate mechanical harvesting. They may develop plants that resist pests, so that the farmer does not have to apply pesticides, or applies smaller amounts of these chemicals. Not applying pesticides in crop production means less environmental pollution from agricultural sources. Breeders may also develop high yielding varieties (or **cultivars**), so the

farmer can produce more for the market to meet consumer demands while improving his or her income. The term cultivar is reserved for variants deliberately created by plant breeders and will be introduced more formally later in the book. It will be the term of choice in this book.

When breeders think of consumers, they may, for example, develop foods with higher nutritional value and that are more flavorful. Higher nutritional value means reduced illnesses in society (e.g., nutritionally related ones such as blindness, rickettsia) caused by the consumption of nutrient-deficient foods, as pertains in many developing regions where staple foods (e.g., rice, cassava) often lack certain essential amino acids or nutrients. Plant breeders may also target traits of industrial value. For example, fiber characteristics (e.g., strength) of fiber crops such as cotton can be improved, while oil crops can be improved to yield high amounts of specific fatty acids (e.g., high oleic content sunflower seed). The latest advances in technology, specifically genetic engineering technologies, are being applied to enable plants to be used as bioreactors to produce certain pharmaceuticals (called **biopharming** or simply **pharming**).

The technological capabilities and needs of societies in the past restricted plant breeders to achieving modest objectives (e.g., product appeal, adaptation to production environment). It should be pointed out that these "older" breeding objectives are still important today. However, with the availability of sophisticated tools, plant breeders are now able to accomplish these genetic alterations in novel ways that are sometimes the only option, or are more precise and more effective. Furthermore, as previously indicated, plant breeders are able to undertake more dramatic alterations that were impossible to attain in the past (e.g., transferring a desirable gene from a bacterium to a plant!). Some of the reasons why plant breeding is important to society are summarized next.

1.3 The concept of genetic manipulation of plant attributes

The work of Gregor Mendel and further advances in science that followed his discoveries established that plant traits are controlled by hereditary factors or **genes** that consist of DNA (deoxyribose nucleic acid, the hereditary material). These genes are expressed in an environment to produce a trait. It follows, then, that in order to change a trait or its expression, one may change the *nature* or its genotype, and/or modify the *nurture* (environment in which it is expressed). Changing the environment essentially entails modifying the growing or production conditions. This may be achieved through an agronomic approach; for example, the application of production inputs (e.g., fertilizers, irrigation). While this approach is effective in enhancing certain traits, the fact remains that once these supplemental environmental factors are removed, the expression of the plant trait reverts to the *status quo*. On the other hand, plant breeders seek to modify plants with respect to the expression of certain selected attributes by modifying the genotype (in a desired way by targeting specific genes). Such an approach produces an alteration that is permanent (i.e., transferable from one generation to the next).

1.4 Why breed plants?

The reasons for manipulating plant attributes or performance change according to the needs of society. Plants provide food, feed, fiber, pharmaceuticals, and shelter for humans. Furthermore, plants are used for aesthetic and other functional purposes in the landscape and indoors.

1.4.1 Addressing world food and feed quality needs

Food is the most basic of human needs. Plants are the primary producers in the **ecosystem** (a community of living organisms including all the nonliving factors in the environment). Without them, life on earth for higher organisms would be impossible. Most of the crops that feed the world are cereals (Table 1.1).

Table 1.1 Twenty five major food crops of the world.

1 Wheat	11 Sorghum	21 Apples
2 Rice	12 Sugarcane	22 Yam
3 Corn	13 Millets	23 Peanut
4 Potato	14 Banana	24 Watermelon
5 Barley	15 Tomato	25 Cabbage
6 Sweet potato	16 Sugar beet	
7 Cassava	17 Rye	
8 Grapes	18 Oranges	
9 Soybean	19 Coconut	
10 Oats	20 Cottonseed oil	

The ranking is according to total tonnage produced annually. (Source: Harlan, 1976)

Plant breeding is needed to enhance the value of food crops, by improving their yield and the nutritional quality of their products, for healthy living of humans. Certain plant foods are deficient in certain essential nutrients to the extent that where these foods constitute the bulk of a staple diet, diseases associated with nutritional deficiency are often common. Cereals tend to be low in lysine and threonine, while legumes tend to be low in cysteine and methionine (both sulfur-containing amino acids). Breeding is needed to augment the nutritional quality of food crops. Rice, a major world food, lacks pro-vitamin A (the precursor of vitamin A). The Golden Rice project currently underway at the International Rice Research Institute (IRRI) in the Philippines and other parts of the world, is geared towards developing, for the first time ever, a rice cultivar with the capacity to produce pro-vitamin A (Golden rice 2, with a 20-fold increase in pro-vitamin A, has been developed by Syngenta's Jealott's Hill International Research Centre in Berkshire, UK). An estimated 800 million people in the world, including 200 million children, suffer chronic under-nutrition, with its attendant health issues. Malnutrition is especially prevalent in developing countries.

Breeding is also needed to make some plant products more digestible and safer to eat, by reducing their toxic components and improving their texture and other qualities. A high lignin content of the plant material reduces its value for animal feed. Toxic substances occur in major food crops, such as alkaloids in yam, cynogenic glucosides in cassava, trypsin inhibitors in pulses, and steroidal alkaloids in potatoes. Forage breeders are interested, amongst other things, in improving feed quality (high digestibility, high nutritional profile) for livestock.

1.4.2 Addressing food supply needs for a growing world population

In spite of a doubling of the world population in the last three decades, agricultural production rose at an adequate rate to meet world food needs. However, an additional three billion people will be added to the world population in the next three decades, requiring an expansion in world food supplies to meet the projected needs. As the world population increases, there would be a need for an agricultural production system that is aligned with population growth. Unfortunately, land for farming is scarce. Farmers have expanded their enterprise onto new lands. Further expansion is a challenge because land that can be used for farming is now being used for commercial and residential purposes to meet the demands of a growing population. Consequently, more food will have to be produced on less land. This calls for improved and high yielding cultivars to be developed by plant breeders. With the aid of plant breeding, the yields of major crops have dramatically changed over the years. Another major concern is the fact that most of the population growth will occur in developing countries, where food needs are currently most serious and where resources for feeding the people are already most severely strained, because of natural or human-made disasters, or ineffective political systems.

1.4.3 Need to adapt plants to environmental stresses

The phenomenon of global climatic change that is occurring is partly responsible for modifying the crop production environment (e.g., some regions of the world are getting drier and others saltier). This means that new cultivars of crops need to be bred for new production environments. Whereas developed economies may be able to counter the effects of unseasonable weather by supplementing the production environment (e.g., by irrigating crops), poorer countries are easily devastated by even brief episodes of adverse weather conditions. For example, development and use of drought resistant cultivars is beneficial to crop production in areas of marginal or erratic rainfall regimes. Breeders also need to develop new plant types that can resist various biotic (diseases and insect pests) and other abiotic (e.g., salt, drought, heat, cold) stresses in the production environment. Crop distribution can be expanded by adapting crops to new production environments (e.g., adapting tropical plants to temperate regions). Development of photoperiod insensitive crop cultivars would allow an expansion in production of previously photoperiod sensitive species.

1.4.4 Need to adapt crops to specific production systems

Breeders need to produce plant cultivars for different production systems to facilitate crop production and optimize crop productivity. For example, crop cultivars must be developed for rain-fed or irrigated production, and for mechanized or non-mechanized production. In the case of rice, separate sets of cultivars are needed for upland production and for paddy production. In organic production systems where pesticide use is highly restricted, producers need insect and disease resistant cultivars in crop production.

1.4.5 Developing new horticultural plant varieties

The ornamental horticultural production industry thrives on the development of new varieties through plant breeding. Aesthetics is of major importance to horticulture. Periodically, ornamental plant breeders release new varieties that exhibit new colors and other morphological features (e.g., height, size, shape). Also, breeders develop new varieties of vegetables and fruits with superior yield, nutritional qualities, adaptation, and general appeal.

1.4.6 Satisfying industrial and other end-use requirements

Processed foods are a major item in the world food supply system. Quality requirements for fresh produce meant for the table are different from those for the food processing industry. For example, there are table grapes and grapes bred for wine production. One of the reasons why the first **genetically modified** (GM) crop (produced by using genetic engineering tools to incorporate foreign DNA) approved for food, the "FlavrSavr^{TM}" tomato, did not succeed was because the product was marketed as table or fresh tomato, when in fact the gene of interest was placed in a genetic background for developing a processing tomato variety. Other factors contributed to the demise of this historic product. Different markets have different needs that plant breeders can address in their undertakings. For example, potato is a versatile crop used for food and industrial products. Different varieties are being developed by breeders for baking, cooking, fries (frozen), chipping, and starch. These cultivars differ in size, specific gravity, and sugar content, among other properties. High sugar content is undesirable for frying or chipping because the sugar caramelizes under high heat to produce undesirable browning of fries and chips.

1.5 Overview of the basic steps in plant breeding

Plant breeding has come a long way, from the cynical view of "crossing the best with best and hoping for the best" to carefully planned and thought-out strategies to develop high performance cultivars. Plant breeding methods and tools keep changing as technology advances. Consequently, plant breeding approaches may be categorized into two general types: **conventional** and **unconventional**. (This categorization is only for convenience.)

- **Conventional approach.** Conventional breeding is also referred to as **traditional** or **classical breeding**. This approach entails the use of tried, proven, and older tools. Crossing two plants (hybridization) is the primary technique for creating variability in flowering species. Various breeding methods are then used to discriminate among the variability (selection) to identify the most desirable recombinant. The selected genotype is increased and evaluated for performance before release to producers. Plant traits controlled by many genes (quantitative traits) are more difficult to breed. Age notwithstanding, the conventional approach remains the workhorse of the plant breeding industry. It is readily accessible to the average breeder and is relatively easy to conduct compared to the unconventional approach.

- **Unconventional approach.** The unconventional approach to breeding entails the use of cutting edge technologies for creating new variability that it is sometimes impossible to achieve with conventional methods. However, this approach is more involved, requiring special technical skills and knowledge. It is also expensive to conduct. The advent of recombinant DNA (rDNA) technology gave breeders a new set of powerful tools for genetic analysis and manipulation. Gene transfer can now be made across natural biological barriers, circumventing the sexual process (e.g., the *Bt* products that consist of bacterial genes transferred into crops to confer resistance to the European corn borer). Molecular markers are available to aid the selection process to make the process more efficient and effective.

Even though two basic breeding approaches have been described, it should be pointed out that they are best considered as complementary rather than independent approaches. Usually, the molecular tools are used to generate variability for selection, or to facilitate the selection process. After genetically modifying plants using molecular tools, it may be used as a parent in subsequent crosses, using conventional tools, to transfer the desirable genes into adapted and commercially desirable genetic backgrounds. Whether developed by conventional or molecular approaches, the genotypes are evaluated in the field by conventional methods and then advanced through the standard seed certification process before the farmer can have access to it for planting a crop. The unconventional approach to breeding tends to receive more attention from funding agencies than the conventional approach, partly because of its novelty and advertised potential, as well as the glamour of the technologies involved.

Regardless of the approach, a breeder follows certain general steps in conducting a breeding project. A breeder should have a comprehensive plan for a breeding project that addresses:

- **Objectives.** The breeder should first define a clear objective (or set of objectives) for initiating the breeding program. In selecting breeding objectives, breeders need to consider:
  (a) The producer (grower) from the point of view of growing the cultivar profitably (e.g., need for high yield, disease resistance, early maturity, lodging resistance).
  (b) The processor (industrial user) as it relates to efficiently and economically using the cultivar as raw materials for producing new product (e.g., canning qualities, fiber strength).
  (c) The consumer (household user) preference (e.g., taste, high nutritional quality, shelf life).

The tomato will be used to show how different breeding objectives can be formulated for a single crop. Tomato is a very popular fruit with a wide array of uses, each calling for certain qualities. For salads, tomato is used whole, and hence the small size is preferred; for hamburgers, tomato is sliced, round large fruits being preferred. Tomato for canning (e.g., puree) requires certain pulp qualities. Being a popular garden species, gardeners prefer a tomato cultivar that ripens over time so harvesting can be spaced. However, for industrial use, as in the case of canning, the fruits on the commercial cultivar must ripen together, so the field can be mechanically harvested. Furthermore, whereas appearance of the fruit is not top priority for a processor who will be making tomato juice, the appearance of fruits is critical in marketing the fruit for table use.

- **Germplasm.** It is impossible to improve plants or develop new cultivars without genetic variability. Once the objectives have been determined, the breeder then assembles the germplasm to be used to initiate the breeding program. Sometimes, new variability is created through crossing of selected parents, inducing mutations, or using biotechnological techniques. Whether used as such or recombined through crossing, the base population used to initiate a breeding program must of necessity include the gene(s) of interest. That is, you cannot breed for disease resistance, if the gene conferring resistance to the disease of interest does not occur in the base population.

- **Selection.** After creating or assembling variability, the next task is to discriminate among the variability to identify and select individuals with the desirable genotype to advance and increase in order to develop potential new cultivars. This calls for using standard selection or breeding methods suitable for the species and the breeding objective(s).

- **Evaluation.** Even though breeders follow basic steps in their work, the product reaches the consumer only after it has been evaluated. Agronomists may participate in this stage of plant breeding. In a way, evaluation is also a selection process, for it entails comparing a set of superior candidate genotypes to select one for release as a cultivar. The potential cultivars are evaluated in the field, sometimes at different locations and over several years, to identify the most promising one for release as a commercial cultivar.

- **Certification and cultivar release.** Before a cultivar is released, it is processed through a series of steps, called the seed certification process, to increase the experimental seed and to obtain approval for release from the designated crop certifying agency in the state or country. These steps in plant breeding are discussed in detail in this book.

1.6 How have plant breeding objectives changed over the years?

In a review of plant breeding over the past 50 years, Baenzinger and colleagues in 2006 revealed that while some aspects of how breeders conduct their operations have dramatically changed, others have stubbornly remained the same, being variations on a theme at best.

Breeding objectives in the 1950s and 1960s, and before, appeared to focus on increasing crop productivity. Breeders concentrated on yield and adapting crops to their production environment. Resistance to diseases and pests was also priority. Quality traits for major field crops, such as improved fiber strength for cotton and milling and baking quality in wheat, were important in the early breeding years. Attention was given to resistance to abiotic stresses such as winter hardiness and traits like lodging resistance, uniform ripening, and seed oil content of some species. Crop yield continued to be important throughout the 1990s. However, as analytical instrumentation that allowed high throughput, low cost, ease of analysis and repeatability of results became more readily available, plant breeders began to include nutritional quality traits into their breeding objectives. These included forage quality traits, such as digestibility and neutral detergent fiber.

More importantly, with advanced technology, quality traits are becoming more narrowly defined in breeding objectives. Rather than high protein or high oil, breeders are breeding for specifics, such as low linolenic acid content, to meet consumer preferences for eating healthful foods (low linolenic acid in oil provides it with stability and enhanced flavor, and reduces the need for partial hydrogenation of the oil and production of trans fatty acids). Also, a specific quality trait such as low phytate phosphorus in grains (e.g., corn, soybean) would increase feed efficiency and reduce phosphorus in animal waste, a major source of the environmental degradation of lakes.

Perhaps no single technology has impacted breeding objectives more in recent times than biotechnology (actually, a collection of biological technologies). The subject is discussed in detail in later chapters. Biotechnology has enabled breeders to develop a new generation of cultivars with genes included from genetically unrelated species (transgenic or GM cultivars). The most successful transgenic input traits to date have been herbicide resistance and insect resistance, which have been incorporated into major crops species like corn, cotton, soybean, and tobacco. According to a 2010 International Service for the Acquisition of Agri-Biotech Crops (ISAAA) report, GM is far from being a global industry, with only six countries (USA, Brazil, Argentina, India, Canada and China) growing about 95% of the total global acreage (use leads with about 50%). Some argue that biotechnology has become the tail that wags the plant breeding industry. Improvement in plant genetic manipulation technology has also encouraged the practice of gene stacking in plant breeding. Another significant contribution of biotechnology to changing breeding objectives is the creation of the "universal gene pool", whereby breeders, in theory, have limitless sources of diversity, and hence can be more creative and audacious in formulating breeding objectives.

In the push to reduce our carbon footprint and reduce environmental pollution, there is a drive towards the discovery and use of alternative fuel sources. Some traditional improvement of some crop species (e.g., corn) for food and feed is being changed to focus some attention on their industrial use, through increasing biomass for biofuel production, and as bioreactors for production of polymers and pharmaceuticals. In terms of reducing adverse environmental impact, one of the goals of modern breeding is to reduce the use of agrochemicals.

1.7 The art and science of plant breeding

The early domesticators relied solely on experience and intuition to select and advance plants they thought had superior qualities. As knowledge abounds and technology advances, modern breeders are increasingly depending on science to take the guesswork out of the selection process, or at least reduce it. At the minimum, a plant breeder should have a good understanding of genetics and the principles and concepts of plant breeding, hence the emphasis of both disciplines in this book. Students taking a course in plant breeding are expected to have taken at least an introductory course in genetics. Nonetheless, two supplementary chapters have been provided in this book; they review some pertinent genetic concepts that will aid the student in understanding plant breeding. By placing these fundamental concepts in the back of the book, users will not feel obligated to study them but can use them on as needed basis.

1.7.1 Art and the concept of the "breeder's eye"

Plant breeding is an applied science. Just like other non-exact science disciplines or fields, art is important to the success achieved by a plant breeder. Early plant breeders depended primarily on intuition, skill, and judgment in their work. These attributes are still desirable in modern day plant breeding. This book discusses the various tools available to plant breeders. Plant breeders may use different tools to tackle the same problem, the results being the arbiter of the wisdom in the choices made. In fact, it is possible for different breeders to use the same set of tools to address the same kind of problem with different results, due in part to the differences in their skill and experience. As is discussed later in the book, some breeding methods depend on phenotypic selection (based on appearance; visible traits). This calls for the proper design of the fieldwork to minimize the misleading effect of a variable environment on the expression of plant traits. Selection may be likened to a process of informed "eye-balling" to discriminate among variability.

A good breeder should have a keen sense of observation. Several outstanding discoveries were made just because the scientists who were responsible for these events were observant enough to spot unique and unexpected events. Luther Burbank selected one of the most successful cultivars of potato, the "Burbank potato", from among a pool of variability. He observed a seed ball on a vine of the "Early Rose" cultivar in his garden. The ball contained 23 seeds, which he planted directly in the field. At harvest time the following fall, he dug up and kept the tubers from the plants separately. Examining them, he found two vines that were unique, bearing large smooth and white potatoes. Still, one was superior to the others. The superior one was sold to a producer who named it Burbank. The Russet Burbank potato is produced on about 50% of all lands devoted to potato production in the United States.

Breeders often have to discriminate among hundreds and even tens of thousands of plants in a segregating population to select only a small fraction of promising plants to advance in the program. Visual selection is an art, but it can be facilitated by selection aids such as **genetic markers** (simply inherited and readily identified traits that are linked to desirable traits that are often difficult to identify). Morphological markers (not biochemical markers) are useful when visual selection is conducted. A keen eye is advantageous even when markers are involved in the selection process. As is emphasized later in this book, the breeder ultimately adopts a holistic approach to selection, evaluating the overall worth or desirability of the genotype, not just the trait targeted in the breeding program.

1.7.2 The scientific disciplines and technologies of plant breeding

The science and technology component of modern plant breeding is rapidly expanding. While a large number of science disciplines directly impact plant breeding, several are closely associated with it. These are plant breeding, genetics, agronomy, cytogenetics, molecular genetics, botany, plant physiology, biochemistry, plant pathology, entomology, statistics, and tissue culture. Knowledge of the first three disciplines is applied in all breeding programs. The technologies used in modern plant breeding are summarized in Table 1.2. These technologies are discussed in varying degrees in this book. The categorization is only approximate and generalized. Some of these tools are used to either generate variability directly or to transfer genes from one genetic background to another to create variability for breeding. Some technologies facilitate the breeding process through, for example, identifying individuals with the gene(s) of interest.

- **Genetics.** Genetics is the principal scientific basis of modern plant breeding. As previously indicated, plant breeding is about targeted genetic modification of plants. The science of genetics enables plant breeders to predict, to varying extents, the outcome of genetic manipulation of plants. The techniques and methods employed in breeding are determined based on the genetics of the trait of interest, regarding, for example, the number of genes coding for it and gene action. For example, the size of the segregating population to generate in order to have a

Table 1.2 An operational classification of technologies of plant breeding.

Classical/traditional tools	Common use of the technology/tool
Emasculation	making a completer flower female; preparation for crossing
Hybridization	crossing un-identical plants to transfer genes or achieve recombination
Wide crossing	crossing of distantly related plants
Selection	the primary tool for discriminating among variability
Chromosome counting	determination of ploidy characteristics
Chromosome doubling	manipulating ploidy for fertility
Male sterility	to eliminate need for emasculation in hybridization
Triploidy	to achieve seedlessness
Linkage analysis	for determining association between genes
Statistical tools	for evaluation of germplasm
Relatively advanced tools	
Mutagenesis	to induce mutations to create new variability
Tissue culture	for manipulating plants at the cellular or tissue level
Haploidy	used to create extremely homozygous diploid
Isozyme markers	to facilitate the selection process
In situ hybridization	detect successful interspecific crossing
More sophisticated tools	
DNA markers	
– RFLP	more effective than protein markers (isozymes)
– RAPD	PCR-based molecular marker
Advanced technology	
Molecular markers	SSR, SNPs, etc.
Marker assisted selection	facilitate the selection process
DNA sequencing	ultimate physical map of an organism
Plant genomic analysis	studying the totality of the genes of an organism
Bioinformatics	computer-based technology for prediction of biological function from DNA sequence data
Microarray analysis	to understand gene expression and for sequence identification
Primer design	for molecular analysis of plant genome
Plant transformation	for recombinant DNA work

chance of observing that unique plant with the desired combination of genes, depends on the number of genes involved in the expression of the desired trait.

- **Botany.** Plant breeders need to understand the reproductive biology of their plants as well as their taxonomic attributes. They need to know if the plants to be hybridized are cross-compatible, as well as to know in fine detail about flowering habits, in order to design the most effective crossing program.
- **Plant physiology.** Physiological processes underlie the various phenotypes observed in plants. Genetic manipulation alters plant physiological performance, which in turn impacts the plant performance in terms of the desired economic product. Plant breeders manipulate plants for optimal physiological efficiency, so that dry matter is effectively partitioned in favor of the economic yield. Plants respond to environmental factors, biotic (e.g., pathogens) and abiotic (e.g., temperature, moisture). These factors are sources of physiological stress when they occur at unfavorable levels. Plant breeders need to understand these stress relationships in order to develop cultivars that can resist them for enhanced productivity.
- **Agronomy.** Plant breeders conduct their work in both controlled (greenhouse) and field environments. An understanding of agronomy (the art and science of producing crops and managing soils) will help the breeder to provide the appropriate cultural conditions for optimal plant growth and development for successful hybridization and selection in the field. An improved cultivar is only as good as its cultural environment. Without the proper nurturing, the genetic potential of an improved cultivar would not be realized. Sometimes, breeders need to modify the plant growing environment to identify individuals to advance in a breeding program to achieve an objective (e.g., withholding water in breeding for drought resistance).
- **Pathology and entomology.** Disease resistance breeding is a major plant breeding objective. Plant breeders need to understand the biology of the insect pest or pathogen against which resistance is being sought. The kind of cultivar to breed, the methods to use in breeding and evaluation all depend on the kind of pest or pathogen (e.g., its races or variability, pattern of spread, life cycle, and most suitable environment).
- **Statistics.** Plant breeders need to understand the principles of research design and analysis. This knowledge is essential for effectively designing field and laboratory studies (e.g., for heritability, inheritance of a trait, combining ability) and for evaluating genotypes for cultivar release at the end of the breeding program. Familiarity with computers is important for record keeping and data manipulation. Statistics is indispensable to plant breeding programs. This is because the breeder often encounters situations in which predictions about outcomes, comparison of results, estimation of response to a treatment, and many more, need to be made. Genes are not expressed in a vacuum but in an environment with which they interact. Such interactions may cause certain outcomes to deviate from the expected. Statistics is needed to analyze the variance within a population to separate real genetic effects from environmental effects. Application of statistics in plant breeding can be as simple as finding the mean of a set of data, to complex estimates of variance and multivariate analysis.
- **Biochemistry.** In this era of biotechnology, plant breeders need to be familiar with the molecular basis of heredity. They need to be familiar with the procedures of plant genetic manipulation at the molecular level, including the development and use of molecular markers and gene transfer techniques.

While the training of a modern plant breeder includes these courses and practical experiences in these and other disciplines, it is obvious that the breeder cannot be an expert in all of them. Modern plant breeding is more *team work* than solo effort. A plant breeding team will usually have experts in all these key disciplines, each one contributing to the development and release of a successful cultivar.

1.8 Achievements of modern plant breeders

The achievements of plant breeders are numerous, but may be grouped into several major areas of impact – yield increase, enhancement of compositional traits, crop adaptation, and the impact on crop production systems.

1.8.1 Yield increase

Yield increase in crops has been accomplished in a variety of ways, including targeting yield *per se* or its

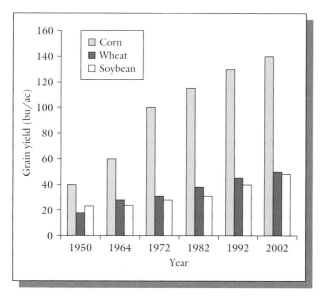

Figure 1.1 The yield of major world food crops is steadily rising, as indicated by the increasing levels of crops produced in the US agricultural system. A significant portion of this rise is attributable to the use of improved crop cultivars by crop producers. (Source: Drawn with data from the USDA.)

components, or making plants resistant to economic diseases and insect pests, and breeding for plants that are responsive to the production environment. Yields of major crops (e.g., corn, rice, sorghum, wheat, and soybean) have significantly increased in the USA over the years (Figure 1.1). For example, the yield of corn rose from about 2000 kg/ha in the 1940s to about 7000 kg/ha in the 1990s. In England, it took only 40 years for wheat yields to rise from 2 metric tons/ ha to 6 metric tons/ha. Food and Agriculture Organization (FAO) data comparing crop yield increases between 1961 and 2000 show dramatic changes for different crops in different regions of the of the world. For example, wheat yield increased by 681% in China, 301% in India, 299% in Europe, 235% in Africa, 209% in South America, and 175% in the USA. These yield increases are not totally due to the genetic potential of the new crop cultivars (about 50% is attributed to plant breeding) but are also due to the improved agronomic practices (e.g., application of fertilizer, irrigation). Crops have been armed with disease resistance to reduce yield loss. Lodging resistance also reduces yield loss resulting from harvest losses.

1.8.2 Enhancement of compositional traits

Breeding for plant compositional traits to enhance nutritional quality or meet an industrial need are major plant breeding goals. High protein crop varieties (e.g., high lysine or quality protein maize) have been produced for use in various parts of the world. Different kinds of wheat are needed for different kinds of products (e.g., bread, pasta, cookies, semolina). Breeders have identified the quality traits associated with these uses and have produced cultivars with enhanced expression of these traits. Genetic engineering technology has been used to produce high oleic sunflower for industrial use; it is also being used to enhance the nutritional value of crops (e.g., pro-vitamin A golden rice). The shelf life of fruits (e.g., tomato) has been extended through the use of genetic engineering techniques to reduce the expression of compounds associated with fruit deterioration.

1.8.3 Crop adaptation

Crop plants are being produced in regions to which they are not native, because breeders have developed cultivars with modified physiology to cope with variations in the duration of day length (photoperiod). Photoperiod insensitive cultivars will flower and produce seed under any day length conditions. The duration of the growing period varies from one region of the world to another. Early maturing cultivars of crop plants enable growers to produce a crop during a short window of opportunity, or even to produce two crops in one season. Furthermore, early maturing cultivars can be used to produce a full season crop in areas where adverse conditions are prevalent towards the end of the normal growing season. Soils formed under arid conditions tend to accumulate large amounts of salts; to use these lands for crop production, salt tolerant (saline and aluminum tolerance) crop cultivars have been developed for certain species. In crops such as barley and tomato there are commercial cultivars in use with drought, cold, and frost tolerance.

1.8.4 Impact on crop production systems

Crop productivity is a function of the genotype (genetic potential of the cultivar) and the cultural environment. The **Green Revolution** is an example of an outstanding outcome of the combination of plant breeding efforts and production technology to increase food productivity. A chemically intensive production system (use of agrochemicals-like fertilizers) calls for crop cultivars that are responsive to such high input growing conditions. Plant breeders have developed cultivars with the architecture for such environments. Through the use of genetic engineering technology, breeders have reduced the need for pesticides in the production of major crops (e.g., corn, tobacco, soybean) with the development of GM pest resistant cultivars, thereby reducing environmental damage from agriculture. Cultivars have been developed for mechanized production systems.

Industry highlights

Norman Ernest Borlaug: The man and his passion

George Acquaah

Bowie State University, Computer Science Building, Bowie, MD 20715, USA

"For more than half a century, I have worked with the production of more and better wheat for feeding the hungry world, but wheat is merely a catalyst, a part of the picture. I am interested in the total development of human beings. Only by attacking the whole problem can we raise the standard of living for all people, in all communities, so that they will be able to live decent lives. This is something we want for all people on this planet".

Norman E. Borlaug.

Dr Norman E. Borlaug has been described in the literature in many ways, including as "the father of the Green Revolution", "the forgotten benefactor of humanity", "one of the greatest benefactors of human race in modern times", and "a distinguished scientist-philosopher". He has been presented before world leaders and received numerous prestigious academic honors from all over the world. He belongs to an exclusive league, with the likes of Henry Kissinger, Elie Wiesel, and President Jimmy Carter – all Nobel Peace laureates. Yet, Dr Borlaug is hardly a household name in the United States. But, this is not a case of a prophet being without honor in his country. It might be more because this outstanding human being chooses to direct the spot light on his passion, rather than his person. As previously stated in his own words, Dr Borlaug has a passion for helping to achieve of a decent living status for the people of the world, starting with the alleviation of hunger. To this end, his theatre of operation is the developing countries, which are characterized by poverty, political instability, chronic food shortages, malnutrition, and prevalence of preventable diseases. These places are hardly priority sources for news for the first world media, unless an epidemic or catastrophe occurs.

Dr Borlaug was born on March 25,1914, to Henry and Clara Borlaug, Norwegian immigrants in the city of Saude, near Cresco, Iowa. He holds a BS degree in Forestry, which he earned in 1937. He pursued an MS in Forest Pathology, and later earned a PhD in Pathology and Genetics in 1942 from the University of Minnesota. After a brief stint with the E.I. du Pont de Nemours in Delaware, Dr Borlaug joined the Rockefeller Foundation team in Mexico in 1944, a move that would set him on course to accomplish one of the most notable accomplishments in history. He became the director of the Cooperative Wheat Research and Production Program in 1944, a program initiated to develop highyielding cultivars of wheat for producers in the area.

In 1965, the Centro Internationale de Mejoramiento de Maiz y Trigo (CIMMYT) was established in Mexico, as the second of the currently 16 International Agricultural Research Centers (IARC) by the Consultative Group on International Agricultural Research (CGIAR). The purpose of the center was to undertake wheat and maize research to meet the production needs of developing countries. Dr Borlaug served as the director of the Wheat Program at CIMMYT until 1979 when he retired from active research, but not until he had accomplished his landmark achievement, dubbed the Green Revolution. The key technological strategies employed by Dr Borlaug and his team were to develop high yielding varieties of wheat, and an appropriate agronomic package (fertilizer, irrigation, tillage, pest control) for optimizing the yield potential of the varieties. Adopting an interdisciplinary approach, the team assembled germplasm of wheat from all over the world. Key contributors to the efforts included Dr Burton Bayles and Dr Orville Vogel, both of the USDA, who provided the critical genotypes used in the breeding program. These genotypes were crossed with Mexican genotypes to develop lodging-resistant, semi-dwarf wheat varieties that were adapted to the Mexican production region (Figure B1.1). Using the improved varieties and appropriate agronomic package, wheat production in Mexico increased dramatically from its low 750 kg/ha to about 3200 kg/ha. The successful cultivars were introduced

Figure B1.1 Dr Norman Borlaug working in a wheat crossing block.

into other part of the world, including Pakistan, India, and Turkey in 1966, with equally dramatic results. So successful was the effort in wheat that the model was duplicated in rice in the Philippines in 1960. In 1970, Dr Norman Borlaug was honored with the Nobel Peace Prize for contributing to curbing hunger in Asia and other parts of the world where his improved wheat varieties were introduced (Figure B1.2).

Whereas the Green Revolution was a life-saver for countries in Asia and some Latin-American countries, another part of the world that is plagued by periodic food shortages, the sub-Saharan Africa, did not benefit from this event. After retiring from CIMMYT in 1979, Dr Borlaug focused his energies on alleviating hunger and promoting the general well-being of the people on the continent of Africa. Unfortunately, this time around, he had to go without the support of these traditional allies, the Ford Foundation, the Rockefeller Foundation, and the World Bank. It appeared the activism of powerful environmental groups in the developed world had managed to persuade these donors from supporting what, in their view, was an environmentally intrusive practice advocated by people such as Dr Borlaug. These environmentalists promoted the notion that high yield agriculture for Africa, whereby the agronomic package included inorganic fertilizers, would be ecologically disastrous.

Figure B1.2 A copy of the actual certificate presented to Dr Norman Borlaug as part of the 1970 Nobel Peace Prize Award he received.

Figure B1.3 Dr Twumasi Afriyie, CIMMYT Highland Maize Breeder and a native of Ghana, discusses the quality protein maize he was evaluating in a farmer's field in Ghana with Dr Borlaug.

Incensed by the distractions of "green politics", which sometimes is conducted in an elitist fashion, Dr Borlaug decided to press on undeterred with his passion to help African Farmers. At about the same time, President Jimmy Carter was collaborating with the late Japanese industrialist, Ryoichi Sasakawa, to address some of the same agricultural issues dear to Dr Borlaug. In 1984, Mr Sasakawa persuaded Dr Borlaug to come out of retirement to join them to vigorously pursue food production in Africa. This alliance gave birth to the Sasakawa Africa Association, presided over by Dr Borlaug. In conjunction with the Global 2000 of The Carter Center, the Sasakawa-Global 2000 was born, with a mission to help small-scale farmers to improve agricultural productivity and crop quality in Africa. Without wasting time, Dr Borlaug selected an initial set of countries in which to run projects. These included Ethiopia, Ghana, Nigeria, Sudan, Tanzania, and Benin (Figure B1.3). The crops targeted included popular staples such as corn, cassava, sorghum, cowpeas, as well as wheat. The most spectacular success was realized in Ethiopia, where the country recorded its highest ever yield of major crops in the 1995–1996 growing season.

The Sasakawa-Global 2000 operates in some 12 African nations. Dr. Borlaug was associated with CIMMYT and also held a faculty position at Texas A&M University, where he taught international agriculture until his death on September 12, 2009. On March 29, 2004, in commemoration of his ninetieth birthday, Dr. Borlaug was honored by the USDA with the establishment of the Norman E. Borlaug International Science and Technology Fellowship Program. The fellowship is designed to bring junior and mid-ranking scientists and policymakers from African, Asian, and Latin American countries to the United States to learn from their US counterparts.

References

Byerlee, D., and Moya, P. (1993). Impacts of international wheat breeding research in the developing world. Mexico, D.F., Mexico CIMMYT.

Borlaug, N.E. (1958). The impact of agricultural research on Mexican wheat production. *Transactions of the New York Academy of Science*, **20**:278–295.

Borlaug, N.E. (1965). Wheat, rust, and people. *Phytopathology*, **55**:1088–1098.

Borlaug, N.E. (1968). Wheat breeding and its impact on world food supply. Public lecture at the 3rd International Wheat Genetics Symposium, August 5–9, 1968, Canberra, Australia. Australian Academy of Science.

Brown, L.R. (1970). *Seeds of change: The Green Revolution and development in the 1970s.* Praeger, New York.

Dalrymple, D.G. (1986). Development and spread of high-yielding rice varieties in developing countries. Agency for International Development, Washington, DC.

Haberman, F.W. (1972). *Nobel Lectures. 1951–1970. Nobel Lectures, Peace.* Elsevier Publishing Company, Amsterdam, The Netherlands.

Wharton, C.R. Jr., (1969). The Green Revolution: cornucopia or Pandora's box? *Foreign Affairs*, **47**:464–476.

1.9 The plant breeding industry

Commercial plant breeding is undertaken in both the private and public sectors. Breeding in the private sector is primarily for profit. A sample of the major plant breeding companies in the world is presented in Table 1.3. It should be pointed out these companies operate under the umbrella of giant multinational corporations, such as Monsanto, Pioneer/Dupont, Novartis/Syngenta, and Advanta Seed Group, through mergers and acquisitions. Products from private seed companies are proprietary. In the United

Table 1.3 Selected seed companies in various parts of the world.

Holland
Bakker Brothers
Bejo Zaden BV
De Ruiter Seeds
Pinnar Seed BV
Van Dijke Zaden
Enza Zaden BV
Nikerson-Zwaan BV

Germany
Hild Samen GmbH
Nikerson-Zwaan GmbH

United Kingdom
CN Seed Ltd.
Tozer Seeds
Nikerson-Zwaan GmbH

France
Clause
Panam Semences
Vilmorin
Technisem
Gautier Semences

South Africa
Capstone Seeds
Gellman Seeds Pty. Ltd.
JW Seeds
Pinnar Seed
Sensako

Kenya
Pinnar Seed

United States of America
Seminis
Syntenta Seeds
Monsanto
DeKalb
Danson Seeds
Campbell's Seed
Burpee
Pioneer

States, an estimated 65–75% of all plant breeders are employed in the private sector. More importantly, crop species that are self-pollinated (e.g., wheat), and hence allow farmers to save seed to plant the next season's crop, are of less interest to commercial seed breeders in the private sector. An estimated 80% of wheat breeders in the United States are in the public sector while only about 7% of corn (cross-pollinated, readily amenable to hybrid production) breeders are

in the public sector. Most germplasm enhancement efforts (pre-breeding, introduction of exotic genes into cultivated germplasm) occur mainly in the public sector. Funds for public breeding in wheat come from contributions from the Wheat Growers Association.

The private sector dominates corn breeding throughout the industrial world. However, the roles of the public and private sectors differ markedly in Western Europe, different regions of the USA, Canada, and Australia, as outlined in the next sections.

1.9.1 Private sector plant breeding

Four factors are deemed by experts to be critical in determining the trends in investment in plant breeding by the private sector.

- **Cost of research innovation.** Modern plant breeding technologies are generally expensive to acquire and use. Consequently, the cost of research and development of new cultivars by these technologies are exorbitant. However, some of these innovations result in increased product quality and yield, and sometimes facilitate the production of the crop by the producer. Also, some innovations eventually reduce the duration of the cumulative research process.
- **Market structure.** Private companies are more likely to invest in plant breeding where the potential size of the seed market is large and profitable. Further, the attraction to enter into plant breeding will be greater if there are fixed costs in marketing the new cultivars to be developed.
- **Market organization of the seed industry.** Conventional wisdom suggests that the more concentrated a seed market, the greater the potential profitability a seed production enterprise would have. However, contemporary thought on industrial organization suggests that the ease of entry into an existing market would depend on the contestability of the specific market, and would subsequently decide the profitability to the company. Plant breeding is increasingly becoming a technology-driven industry. Through research and development, a breakthrough may grant a market monopoly to an inventor of a technology or product, until another breakthrough occurs that grants a new monopoly in a related market. For example, Monsanto, the developer of Roundup Ready® technology, is also the developer of the Roundup® herbicide, which is required for the technology to work.

• **Ability to appropriate the returns to research and distribution of benefits.** The degree to which a seed company can appropriate returns to its plant breeding inventions is a key factor in the decision to enter the market. Traditionally, cross-pollinated species (e.g., corn) that are amenable to hybrid breeding and high profitability have been most attractive to private investors. Public sector breeding develops most of the new cultivars in self-pollinated species (e.g., wheat, soybean). However, the private sector interest in self-pollinated species is growing. This shift is occurring for a variety of reasons. Certain crops are associated in certain cropping systems. For example, corn–soybean rotations are widely practiced. Consequently, producers who purchase improved corn are likely to purchase improved soybean seed. In the case of cotton, the shift is for a more practical reason. Processing cotton to obtain seed entails ginning and delinting, which are more readily done by seed companies than farmers.

Another significant point that needs to be made is that the for-profit private breeding sector is obligated not to focus only on profitability of a product to the company, but it must also price its products such that the farmer can use them profitably. Farmers are not likely to adopt a technology that does not significantly increase their income!

1.9.2 Public sector plant breeding

The USA experience

Public sector breeding in the USA is conducted primarily by land grant institutions and researchers in the federal system (i.e., the United States Department of Agriculture (USDA)). The traditional land grant institutional program is centered on agriculture and is funded by the federal government and the various states, often with support from local commodity groups. The plant research in these institutions is primarily geared towards improving field crops and horticultural and forest species of major economic importance to a state's agriculture. For example, the Oklahoma State University, an Oklahoma land grant university, conducts research on wheat, the most important crop in the state. A fee is levied on produce presented for sale at the elevator by producers and used to support agricultural research pertaining to wheat.

In addition to its in-house research unit, the Agricultural Research Service (ARS), the USDA often has scientists attached to land grant institutions to conduct research of benefit to a specific state as well as the general region. For example, the Grazinglands Research Laboratory at El Reno, Oklahoma, is engaged in forage research for the benefit of the Great Plains. Research output from land grant programs and the USDA is often public domain and often accessible to the public. However, just like the private sector, inventions may be protected by obtaining plant variety protection or a patent.

The UK experience[1]

The equivalent of a land grant system does not operate in the United Kingdom but, up to the 1980s, there were a number of public sector breeding programs at research institutes such as the Plant Breeding Institute (PBI) (now part of the John Innes Centre), Scottish Crop Research Institute (SCRI), Welsh Plant Breeding Station (now the Institute of Grassland and Environmental Research (IGER)) and National Vegetable Research Station (now Horticultural Research International (HRI)), with the products being marketed through the National Seed Development Organisation (NSDO). In addition, there were several commercial breeding programs producing successful finished cultivars, especially for the major crops. Following a review of "Near Market Research", the plant breeding program at the PBI and the whole portfolio of the NSDO were sold to Unilever and traded under the brand PBI Cambridge, later to become PBI Seeds. The review effectively curtailed the breeding activities in the public sector, especially of the major crops. Plant breeding in the public sector did continue at the IGER, HRI and SCRI but was reliant on funding from the private sector for a substantial part of the program. Two recent reviews of crop science research in the United Kingdom have highlighted the poor connection between much public sector research and the needs of the plant breeding and end-user communities. The need for public good plant breeding was recognized in the Biotechnology and Biological Sciences Research Council (BBSRC) Crop Science Review to translate fundamental research into deliverables for the end-user and is likely to stimulate pre-breeding activity at the very least in the public sector.

---

[1] The information regarding the United Kingdom experience is through personal communication with W.T.B. Thomas of the Scottish Crop Research Institute, Invergowrie, UK.

Crop research and development in European Community (EC) countries

Unlike the United States, the private sector is responsible for cultivar development of established crops, while the public sector focuses on research. However, in the case of new crops where risk investment is high, the public sector (governmental institutes) engages in both research and cultivar development. Several research and development arrangements occur in Europe.

- Agro-industrial programs. These programs involve partnership between two or more countries and may include private sector in some cases. Their activities include development of new potential crops (*Cuphea*, jojoba, castor bean, lupines, Jerusalem artichoke), industrial processing, primary production, transformations and utilization of biological feed stocks.
- Bilateral programs. These are informal partnerships between countries that may include germplasm and information exchange.
- National program. Universities and agricultural research institutes work on new crops.
- Industrial programs. These are conducted by the private sector and may include the search for new crops of pharmaceutical value, as well as bioactive compounds. Sometimes, public sector institutes may be engaged.

International plant breeding

There are other private sector efforts that are supported by foundations and world institutions, such as the Food and Agriculture Organization (FAO), Ford Foundation, and Rockefeller Foundation. These entities tend to address issues of global importance and also support the improvement of the so-called "orphaned crops" (crops that are of importance to developing countries, but not of sufficient economic value to attract investment by multinational corporations). Developing countries vary in their capabilities for modern plant breeding research. Some countries, such as China, India, Brazil, and South Africa, have advanced plant breeding research programs. Other countries have national research stations that devote efforts to the breeding of major national crops or plants, such as the Crops Research Institute in Ghana, where significant efforts have led to the country being a leading adopter of quality protein maize (QPM) in the world.

1.9.3 Public sector versus private sector breeding

Public sector breeding is disadvantaged in an increasingly privatized world. The issues of intellectual property protection, globalization and the constraints on public budgets in both developed and developing economies are responsible for the shift in the balance of plant breeding undertakings from the public to the private sector. This shift in balance has occurred over a period and differs from one country to another, as well as one crop to another. The shift is driven primarily by economic factors. For example, corn breeding in developed economies is dominated by the private sector. However, the trends in wheat breeding are variable in different parts of the world and even within regions in the same country. Public sector plant breeding focuses on problems that are of great social concern, even though they may not be of tremendous economic value (having poor market structure), whereas private sector breeding focuses on problems of high economic return. Public sector breeders can afford to tackle long term research while the private sector, for economic reasons, prefers to have quicker returns on investment. Public sector breeders also engage in minor crops in addition to the principal crops of importance to various states (in the case of the land grant system of the United States). A great contribution of public sector research is the training of plant breeders who work in both the public and private sectors. Also, the public sector is primarily responsible for germplasm conservation and preservation. Hence, private sector breeding benefits tremendously from public sector efforts.

It has been suggested by some that whereas scientific advances and cost of research are relevant factors in the public sector breeding programs, plant breeding investment decisions are not usually significantly directly impacted by the market structure and the organization of the seed industry.

A major way in which private and public breeding efforts differ is on the returns to research. Public sector breeders are primarily not profit oriented and can afford to exchange and share some of their inventions more freely. However, it must be pointed out that access to some public germplasm and technologies is now highly restricted, requiring significant protocol and fees to be paid for their use. The public sector plays a critical role in important activities such as the education and training of plant breeders, development of new methods of breeding, and germplasm preservation and enhancement. These activities are generally long term and less profitable, at least in the short run, and hence less attractive to the private sector.

1.10 Duration and cost of plant breeding programs

It is estimated that it takes about 7–12 years (or even longer) to complete (cultivar release) a breeding program for annual cultivars such as corn, wheat, and soybeans, and much longer for tree crops. The use of molecular techniques to facilitate the selection process may reduce the time for plant breeding in some cases. The use of tissue culture can reduce the length of breeding programs of perennial species. Nonetheless, the development of new cultivars may cost from hundreds of thousands of dollars to even several million dollars. The cost of cultivar development can be much higher if proprietary material is involved. Genetically engineered parental stock attracts a steep fee to use because of the costs involved in its creation. The cost of breeding also depends on where and by whom the activity is being conducted. Because of high overheads, similar products can be produced by breeders in developed and developing economies, but for dramatically higher cost in the former. Cheaper labor in developing countries can allow breeders to produce hybrids of some self-pollinated species less expensively, because they can afford to pay for hand pollination (e.g., cotton in India).

1.11 The future of plant breeding in society

For as long as the world population is expected to continue to increase, there will continue to be a demand for more food. However, with an increasing population comes an increasing demand for land for residential, commercial, and recreational uses. Sometimes, farmlands are converted to other uses. Increased food production may be achieved by increasing production per unit area or bringing new lands into cultivation. Some of the ways in which society will affect and be affected by plant breeding in the future are:

- **New roles of plant breeding.** The traditional roles of plant breeding (food, feed, fiber, and ornamentals) will continue to be important. However, new roles are gradually emerging for plants. The technology for using plants as bioreactors to produce pharmaceuticals will advance. The technology has been around for over a decade. Strategies are being perfected for the use of plants to generate pharmaceutical antibodies, engineering antibody-mediated pathogen resistance, and altering plant phenotype by immunomodulation. Successes that have been achieved include the incorporation of streptococcus surface antigen in tobacco and the herpes simplex virus in soybean and rice.
- **New tools for plant breeding.** New tools will be developed for plant breeders, especially in the areas of the application of biotechnology to plant breeding. New marker technologies continue to be developed and older ones advanced. Tools that will assist breeders to more effectively manipulate quantitative traits will be enhanced. Genomics and bioinformatics will continue to be influential in the approach of researchers to crop improvement. Marker assisted selection (MAS) will be important in plant breeding in the twenty-first century.
- **Training of plant breeders.** As discussed elsewhere in the book, plant breeding programs have experienced a slight decline in the number of graduates entering the field in the recent past. Because of the increasing role of biotechnology in plant genetic manipulation, graduates who combine skills and knowledge in both conventional and molecular technologies are in high demand. It has been observed that some commercial plant breeding companies prefer to hire graduates with training in molecular genetics, then provide them the needed plant breeding skills on the job.
- **The key players in plant breeding industry.** The last decade saw a fierce race by multinational pharmaceutical corporations to acquire seed companies. There were several key mergers as well. The modern technologies of plant breeding are concentrated in the hands of a few of these giant companies. The trend of acquisition and mergers is likely to continue in the future. Publically-supported breeding efforts will decline in favor of for-profit programs.
- **Yield gains of crops.** With the dwindling of arable land and the increasing policing of the environment by activists, there is an increasing need to produce more food or other crop products on the same piece of land in a more efficient and environmentally safer manner. High yield cultivars will continue to be developed, especially in crops that have received less attention from plant breeders. Breeding for

adaptation to environmental stresses (e.g., drought, salt) will continue to be important and will enable more food to be produced on marginal lands.

- **The biotechnology debate.** It is often said that these modern technologies for plant genetic manipulation benefit the developing countries the most because they are in dire need of food, both in quantity and nutritional value. On the other hand, the intellectual property that covers those technologies is owned by the giant multinational corporations. Efforts will continue to be made to negotiate fair use of these technologies. Appropriate technology transfer and support to these poorer developing nations will continue, to enable them develop capacity for the exploitation of these modern technologies.

Key references and suggested reading

Baezinger, P.S. (2006). Plant breeding training in the U. S. A. *HortScience*, **41**:28–29.

Baezinger, P.S., Russell, W.K., Graef, G.L., and Campbell, B. T., (2006). 50 years of crop breeding, genetics, and cytology. *Crop Science*, **46**:2230–2244.

Bliss, F. (2006). Plant breeding in the private sector of North America. *HortScience*, **41**:45–47.

Bliss, F.A. (2007). Education and preparation of plant breeders for careers in global crop improvement. *Crop Science*, **47**:S250–S261.

Charles, D., and Wilcox, B. (2002). *Lords of the harvest: Biotechnology, big money and the future of food*. Perseus Publishing.

Duvick, D.N. (1986). Plant breeding: Past achievements and expectations for the future. *Econ. Bot.*, **40**:289–297.

Frey, K.J. (1971). Improving crop yields through plant breeding, in *Moving off the Yield Plateau* (eds J.D. Eastin and R. D. Munson), ASA Spec. Publ. 20, ASA, CSSA, and SSSA, Madison, WI, pp. 15–58.

Frey, K. (1996). National plant breeding study: I. Human and financial resources devoted to plant breeding research and development in the United States in 1994, Spec. Rep. No. 98, Iowa Agric. Home Economics Exp. Stn., Ames, IA.

Gepts, P., and Hancock, J. (2006). The Future of Plant Breeding. *Crop Science*, **46**:1630–1634.

Hancock, J.F. (2006). Introduction to plant breeding and the public sector: Who will train plant breeders in the U.S. and around the world? *HortScience*, **41**:28–29.

Harlan, J.R. (1976). Plants and animals that nourish man, in *Food and agriculture* (A Scientific American Book) W.H. Freeman and Company, San Francisco, CA.

International Food Policy Research, Institute. (2002). Green Revolution – Curse or blessing? Washington, DC.

Khush, G.S. (2006). Plant breeding training in the international sector. *HortScience*, **41**:48–49.

Solheim, W.G.II. (1972). An earlier agricultural revolution. *Scientific American*, **226**(4):34–41.

Traxler, G., Acquaye, A.K.A., Frey, K., and Thro, A.M. (2005). Public sector plant breeding resources in the US: Study results for the year 2001 (http://www.csrees.usda.gov/nea/plants/in_focus/ptbreeding_if_study.html).

Internet resources for reference

http://www.foodfirst.org/media/opeds/2000/4-green-rev.html – Lessons from the Green Revolution (accessed March 28, 2012).

http://www.arches.uga.edu/~wparks/ppt/green/– Biotechnology and the Green Revolution, Interview with Norman Borlaug (accessed March 28, 2012).

Outcomes assessment

Part A

Please answer the following questions true or false.

1 Rice varieties were the first products of the experiments leading to the Green Revolution.
2 Rice is high in pro-vitamin A.
3 The IR8 was the rice variety released as part of the Green Revolution.
4 Wilhelm Johannsen developed the pure line theory.

Part B

Please answer the following questions.

1 .... won the Nobel Peace Prize in .... for being the chief architect of the Green Revolution.
2 Define plant breeding.
3 Give three common objectives of plant breeding.
4 Discuss plant breeding before Mendel's work was discovered.
5 Give the first two major wheat cultivars to come out of the Mexican Agricultural Program initiated in 1943.

Part C

Please discuss in the following questions in detail.

1 Plant breeding is an art and a science. Discuss.
2 Discuss the importance of plant breeding to society.
3 Discuss how plant breeding has changed through the ages.
4 Discuss the role of plant breeding in the Green Revolution.
5 Discuss the impact of plant breeding on crop yield.
6 Plant breeding is critical to the survival of modern society. Discuss.
7 Discuss the concept of breeder's eye.
8 Discuss the general steps in a plant breeding program.
9 Discuss the qualifications of a plant breeder.
10 Distinguish between public sector and private sector plant breeding.
11 Discuss the molecular and classical plant breeding approaches as complementary approaches in modern plant breeding.

2

History of plant breeding

Purpose and expected outcomes

Agriculture is a human invention that continues to impact society and the environment. The players on this stage advanced the industry with innovation, technology, and knowledge available to them during their era. The tools and methods used by plant breeders have been developed and advanced through the years. There are milestones in plant breeding technology as well as accomplishments by plant breeders over the years. In this chapter, individuals (or groups of people) whose contributions to knowledge, theoretical or practical, have impacted on what has become known in the modern era as plant breeding will be spotlighted. After studying this chapter, the student should be able to:

1 List and describe the contributions of some of the people through history whose discoveries laid the foundation for modern plant breeding.
2 Describe the contributions of Mendel to modern plant breeding.
3 Discuss the advances in plant breeding technologies.

2.1 Origins of agriculture and plant breeding

In its primitive form, plant breeding started after the invention of agriculture, when people of primitive cultures switched from a lifestyle of hunter–gatherers to sedentary producers of selected plants and animals. Views of agricultural origins range from the mythological to ecological. The Fertile Crescent in the Middle East is believed to be the cradle of agriculture, where deliberate tilling of the soil, seeding and harvesting occurred over 10 000 years ago. This lifestyle change did not occur overnight but was a gradual process during which plants were transformed from being independent, wild progenitors, to fully dependent (on humans) and domesticated varieties. Agriculture is generally viewed as an invention and discovery. During this period humans also discovered the time-honored and most basic plant breeding technique – **selection**, the art of discriminating among biological variation in a population to identify and pick desirable variants. Selection implies the existence of variability.

In the beginnings of plant breeding, the variability exploited was the naturally occurring variants and wild relatives of crop species. Furthermore, selection

Principles of Plant Genetics and Breeding, Second Edition. George Acquaah.
© 2012 John Wiley & Sons, Ltd. Published 2012 by John Wiley & Sons, Ltd.

was based solely on the intuition, skill, and judgment of the operator. Needless to say that this form of selection is practiced to date by farmers in poorer economies, where they save seed from the best-looking plants or the most desirable fruit for planting the next season. These days, scientific techniques are used in addition to the aforementioned qualities to make the selection process more precise and efficient. Even though the activities described in this section are akin to some of those practiced by modern plant breeders, it is not being suggested that primitive crop producers were necessarily conscious of the fact that they were manipulating the genetics of their crops.

2.2 The "Unknown Breeder"

Two distinct kinds or groups of people continue to impact plant improvement in significant ways, but with recognition that cannot be personalized.

2.2.1 The "farmer-breeder"

The term "breeder" is a modern day reference to professionals who knowingly manipulate the nature of plants to improve their appearance and performance in predetermined ways. These professionals operate with formal knowledge from the discipline of plant biology and allied disciplines. They are preceded by people who unknowingly and indirectly manipulated the nature of plants to their advantage. This category of "breeders" (to use the term very loosely), or "farmer-breeders", continues to impact world crop production today. Of course, the image of the farmer today is variable from one part of the world to another. In developing countries, many farmers still produce crops with primitive technologies, while high-tech defines the farmer of today in technologically advanced countries.

The age-old practice is for farmers to save seed from the current year's crop to plant the next season's crop. In doing so, farmers depend on their instincts, intuition, experience and keen observation to save seed from selected plants for planting the next crop. Performance and appeal are two key factors in the decision making process. For example, seeds from a plant without blemish among a plot of others with disease symptoms would be saved because it obviously had "something" that makes it ward off diseases. This may be described as primitive or rudimentary "breeding" for disease resistance. Similarly, farmers may save seed on the basis of other agronomic features of their preference, such as seed or fruit size, seed or fruit color, plant stature, and maturity, and in the process manipulate plant genetics without knowing it. I call this "unconscious breeding".

Over time, farmers create varieties of crops that are adapted to their cultural environments, the sole technique being the art of discrimination among variability, or selection as it is called in modern crop improvement. These creations are called **farmer-selected** varieties and sometimes **landraces**. The practice prevails in areas of subsistence agriculture, which represent many parts of the developing world. These varieties are highly adapted to local regions and can be depended upon by farmers who produce their crops with limited resources. Farmer-selected seed continues to sustain agricultural production in these parts of the world while the commercial seed supply system is being developed.

Farmer-selected varieties or landraces are an important source of breeding material for modern breeders. This primitive or exotic germplasm is heterogeneous and is useful for initiating some plant breeding programs.

2.2.2 The "no name" breeder

One of the common practices or traditions in modern plant breeding is to refer to germplasm whose source, name or breeding history is unknown as simply "No Name". This casual acknowledgement appears to absolve the breeder of any deliberate and willful infringement on intellectual property. These nameless products are modern day examples of cultivars that have fallen victim to improper record keeping.

2.3 Plant manipulation efforts by early civilizations

Archeological and historical records from early civilizations indicate that some of these communities engaged in rudimentary plant manipulations, albeit in the dark, without knowledge of plant heredity. Whereas it would not be far-fetched to assume that, just like farmers of the early civilizations who domesticated crops species would have also continued their selection practices to produce farmer-selected varieties, evidence of deliberate plant manipulation for the purposes of improvement are few. Archeological findings occasionally reveal some ancient practices which indicate that plant manipulation beyond phenotypic selection among natural variability occurred. Babylonians are said to have perceived the role of pollen in successful fruit production and applied it to the pistils of female date palms to produce fruit. The Assyrians did likewise in about 870 BC, artificially pollinating date palms.

2.4 Early pioneers of the theories and practices of modern plant breeding

Plant breeding as we know it today began in earnest in the nineteenth century. Prior to this era, a number of groundbreaking discoveries and innovations paved the way for scientific plant manipulation. Some of the early pioneers of plant breeding include the following:

Rudolph Camerarius (1665–1721). Rudolph Camerarius was a professor of philosophy at the University of Tubingen in Germany. He conducted research that contributed to establishing sexual differentiation, defining the male and female reproductive parts of the plant. His seminal work, *De sexu plantarum* (On the sex of plants), was published in 1694 in a letter to a colleague. Camerarius's work also described the functions of the reproductive parts in fertilization and showed that pollen is required for this key process in heredity.

Joseph Gottlieb Koelreuter (1733–1806). German botanist, J.G. Koelreuter became professor of natural history and director of the botanical gardens in Karlsruhe in 1764. He was a pioneer in the application of the discovery of sex in plants as a vehicle for their genetic manipulation. He observed that the hybrid offspring generally resembled the parent that supplied the pollen as closely as the parent on which seed was borne. Koelreuter conducted the first systematic experiments in plant hybridization, using the tobacco plant as subject. He recognized the role of insects and wind in pollination of flowers, and also conducted experiments to study artificial fertilization and development in tobacco plants. The golden rain tree genus (*Koelreuteria*) is named in his honor.

Louis de Vilmorin (1902–1969). Louis de Vilmorin was a noted French seedsman. His experiments in heredity contributed to our understanding of the cause of variation. Vilmorin conducted studies in plant improvement in vegetables using a method called genealogical selection, which is the modern breeding equivalent of progeny testing. He recognized that new varieties of plants could be developed by selecting certain characteristics, which would then be transmitted through genealogy to the progeny. In 1856, he published his "Note on the Creation of a New Race of Beetroot and Considerations on Heredity in Plants", which laid the theoretical groundwork for the modern seed breeding industry. The modern day company Vilmorin is a major player in the global seed industry; along with its international subsidiaries it is ranked among the top five largest seed companies in the world. The company is also credited with producing the first seed catalog for farmers and academics, among other significant publications.

Thomas Andrew Knight (1759–1838). This British horticulturalist and botanist conducted basic research in plant physiology that led to the discovery of the phenomenon of geotropism, the effects of gravity on seedlings. He also showed how decay in fruit trees was transmitted through grafting. In terms of practical crop improvement, Knight conducted research in the breeding of horticultural plants, including strawberries, cabbages, peas, apples and pears. The "Downton" strawberry that he developed is noted in the pedigree of most of the important modern strawberries. He is credited with pioneering work in the science of fruit breeding. In 1797 he published a *Treatise on the culture of apple and pear.* Knight is also said to have demonstrated segregation for seed characters of the garden pea but, unfortunately did not offer an explanation for the event as Mendel eventually did.

Carl Linnaeus (1707–1778). A Swedish botanist, physician, and zoologist, Carl Linnaeus is most noted for his work in plant taxonomy, which led to the development of his enduring conventions for naming living organisms, the universally accepted **binomial nomenclature**, also called Linnaean taxonomy or the scientific classification of organisms. The binomial nomenclature classifies nature within a hierarchy, assigning a two-part name to an individual, a *genus* and a *species* (specific epithet). His work was published in his most noted publication *Species Plantarum.* There are specific rules and guidelines for writing scientific names, which are in Latin, the genus beginning with a capital letter while the species does not; being non-English, the name is italicized (or underlined), for example, *Zea mays* (corn). Further, the genus can stand alone, but not the species (e.g., *Zea, Zea sp, or Z. mays*).

Charles Darwin (1809–1882). Charles Robert Darwin was an English naturalist with one of the most recognizable names of all times, because of his work that led to one of the most enduring theories ever, the **theory of evolution**. He proposed what is sometimes called the unifying theory of life sciences that all species of life have evolved over time from a common ancestor. The process of evolution is extremely slow, requiring thousands or even millions of years to bring about the gradual changes which incrementally result in the divergence or diversity of life that is now seen. The primary mechanism of evolution, he reckoned, is natural selection, the arbiter in deciding which individuals survive to contribute to the subsequent generations (survival of the fittest). Genetic mutations are the ultimate source of variation, but natural selection decides which modifications are advantageous and contribute to the survival of individuals. The survival or extinction of an organism depends on its ability to adapt to its changing environment. He published his seminal work in his 1859 book, *On the origin of species.*

For all intents and purposes, modern plant breeding is evolution happening in real time. Instead of thousands or millions of years to bring about a new variety, plant breeders achieve their goal in about ten years, depending upon the method used, among other factors. Random mutations may be used to create variation, but other more efficient methods are preferred today. Once generated, breeders use artificial selection (not natural selection) to discriminate among the variability to decide which individual plants to advance to the next step in the breeding program.

Gregor Mendel (1822–1844). Born in 1822, Gregor Mendel, an Augustinian monk, is known for his scientific research that led to the foundations of modern transmission genetics. Of German ethnicity, his nationality was Austrian–Hungarian. Even though several researchers in his time and prior to that time had conducted research or made observations similar to what he did, it was Mendel who was credited with being first to provide empirical evidence about the nature of heredity, the underpinnings of traits and how genes that condition them are transmitted from parents to offspring. He made his ground-breaking findings from making and studying *Pisum* (pea) hybrids. His paper *Experiments with hybrid plants* was published in 1866 to reveal what became known as the **laws of Mendel** – the laws of dominance, segregation, and independent assortment, which are the foundations of modern genetics. In fact, Mendel is often referred to as the father of modern genetics. In addition to the laws he established, Mendel also made two other significant contributions to the field of genetics – the development of pure lines, and good record keeping for use in statistical analysis that led to his discoveries (he counted plant variants).

Luther Burbank (1849–1926). An American botanist and horticulturalist, Burbank is known to have developed numerous varieties of fruits, flowers, grains, grasses and vegetables. One of his most remarkable creations is the **Russet Burbank potato**, which has a russet-colored skin and which is used worldwide today. This natural variant was isolated and propagated by Burbank.

It is significant to note that some of the most widely used plant breeding methods of selection were developed prior to the nineteenth century, preceding Mendel! These methods include mass selection, pedigree selection, and bulk breeding.

2.5 Later pioneers and trailblazers

Since the beginning of the nineteenth century, there has been an explosion of knowledge in plant breeding and its allied disciplines. Discussing each one would simply overwhelm this chapter. Consequently, only a sample of the key innovations or discoveries with direct and significant implication on plant breeding is discussed briefly. Some of these innovations or discoveries pertain to breeding schemes or methods and other applications that are discussed in detail later in the book and therefore are only introduced briefly.

M.M Rhoades and D.N. Duvick,. Cytoplasmic male sterility(CMS) was discovered as a breeding technique by Marcus Rhoades in 1933. Duvick was a major player in the discovery of various aspects of this technology. In 1965, he published a summary of work done in this area.

Nikolai I. Vavilov. Vavilov identified eight areas of the world which he designated **centers of diversity**

of crop species or centers of origin of crops. He distinguished between primary centers, where the crop was first domesticated, and secondary centers, which developed from plants migrating from the primary center. He also established the **law of homologous series** in heritable variation, showing the existence of parallelism in variability among related species. This law allows plant explorers to predict, within limits, forms that are yet to be described. Germplasm banks explore and collect germplasm from these centers to be classified and preserved for use by researchers.

E.R. Sears and C.M. Ricks. Sears and Ricks were first to apply their knowledge of cytogenetics to plant breeding of wheat and tomato, respectively. Their efforts showed how researchers could transfer genes and chromosomes from alien species to cultivated crop species. This achievement aided the use of cytogenetics in the evolutionary study of plant species.

H.J. Muller. The pioneering experiments by Muller (1927) showed that it is possible to alter the effect of genes. Using X-rays, he demonstrated that the physiology and genetics of an organism could be altered upon exposure to this radiation. Mutagenesis or mutation breeding became possible because of this discovery. In 1928, Stadler described the mutagenic effects of X-rays on barley.

Wilhelm Johannsen. The work of Johannsen pioneered the **single plant selection** method. He was the first to distinguish between **genotype** and **phenotype**. Working with the field bean, a self-pollinated species, he selected extreme individuals each generation and observed that improvement only occurred in the first generation (i.e., heritable variation did not extend beyond the first generation). Variation observed in the second and subsequent generations was environmental (not heritable). Repeated selfing, after some time, is unresponsive to selection because of lack of genetic variation. Prolonged selfing leads to an individual with extreme homozygosity. He called such products pure lines. This became the **pure line theory** in 1903.

H.H. Hardy and W. Weinberg. The work in 1908 of Hardy, an Englishman, and Weinberg, a German, laid the foundation for modern day breeding of cross-pollinated species. They independently demonstrated that in a large random-mating population, both gene and genotypic frequencies remained unchanged from one generation to the next, in the absence of change agents like mutation,

migration and selection. This later became known as the **Hardy–Weinberg equilibrium** or law. This concept is foundational to the breeding strategies employed for breeding cross-pollinated species.

Nilsson-Ehle. Nilsson-Ehle is credited with being the leader of the first scientific wheat breeding program, which was started by the Swedish Seed Association at Svalof. It was there, in 1912, that he developed the method of plant breeding called **bulk breeding** to cope with the large number of crosses, generations, and plants involved is his breeding program. His breeding program centered on the winter hardiness of wheat. He space-planted the F_1 and bulk-harvested the F_2.

H.V Harlan and M.N. Pope. Harlan and Pope first applied the **backcross breeding** scheme to plants in 1922, after observing its success with animal breeding. Unable to observe desired recombinants in the segregating population of a cross between the commercial cultivar, "Manchuria", a rough-awned wheat, and a smooth-awned exotic parent (donor parent), they resorted to a repeated crossing of the F_1 to the commercial or adapted parent (recurrent parent).

C.H. Goulden. Goulden developed the **single seed decent** (rapid generation advance) selection scheme in 1941 as a means of speeding up the attainment of homozygosity. This was later modified by Brim in 1966.

E.M. East and D.F. Jones. The concept of **recurrent selection** was independently proposed by Hayes and Garber in 1919, and East and Jones in 1920. Hayes and Garber also proposed the method of **synthetic breeding** in 1919.

F.H. Hull. Hull coined the term **recurrent selection** in 1945. His work included recurrent selection for combining ability.

F.E Comstock, H.F. Robinson, and P.H. Harvey. These breeders proposed the method of **reciprocal recurrent selection** in 1949.

C.M. Donald. An Australian biologist, Donald proposed the **ideotype breeding** concept as a way of managing plant breeding programs by modeling plant architecture. Breeding based on a plant model (archetype) meant that breeders paid more attention to their breeding goals and strategies. They could introgress exotic germplasm and expand genetic diversity in their program, following judicious strategies. Even though it did not attain prominence in plant breeding, notable applications were made by Wayne Adams (the major graduate advisor of the author of this book) at Michigan

State University, and by Rasmussen at the University of Minnesota.

H.H. Flor. Flor proposed the **gene-for-gene hypothesis** in 1956 to postulate that both host and parasite genetics were significant in determining whether or not a disease resistance reaction would be observed. The expression of resistance by the host was dominant while the expression of avirulence by the parasite was dominant. In other words, there was a single gene in the host that interacted with a single gene for the parasite.

G.H. Shull. George Shull coined the term "**heterosis**" for the phenomenon of **hybrid vigor**. His research on crossing corn, an open pollinated species, led to the observation of hybrid vigor. This observation had also been made by East and Yates and other researchers, but it was Shull who gave the correct interpretation of heterosis in 1908. Hybrid vigor is the reason why hybrid seed is a huge commercial success.

W.J. Beal. Beal was one of the pioneers in the development of hybrid corn. He is also noted for the oldest and continuously operated botanical garden (The W.J. Beal Botanical Garden) in the United States, located at Michigan State University. His noted publications include the *The New Botany*, *Grasses of North America*, and *History of Michigan Agricultural College*. In 1879, Beal started one of the longest running experiments in botany, designed to determine how long seed can remain viable. The experiment, which includes periodic retrieval and germination testing of the buried seeds, is scheduled to be completed in 2100.

Ronald Fisher. Though not a plant breeder, this biologist made major contributions to the field of statistics and genetics. He introduced the concept of **randomization** and the **analysis of variance** procedure that are indispensable to plant breeding research and evaluation. The concept of **likelihood** (maximum likelihood) is his original idea. His contributions to quantitative genetics aided breeders in the understanding and manipulation of quantitative traits.

C.C. Cockerham. Cockerham's contribution to the role of statistics in plant breeding was summarized in his seminal paper of 1961. It connected statistics to genetics by shedding light on sources of variation and variance components, and covariance among relatives in genetic analysis. There are other names that are associated with this effort, including Mather and Jinks, and Eberhardt and Comstock.

Murashige and Skoog. Tissue culture technology is vital to plant breeding. Many applications, such as embryo rescue, anther culture, micropropagation, *in vitro* selection, and somaclonal variation, depend on tissue culture. The development in 1962 of the Murashige–Skoog media (MS media). Modern methods of genetic engineering depend on tissue culture systems for key steps such as transformation and regeneration.

Watson and Crick. The understanding of heredity that underlies the ability of plant breeders to effectively manipulate plants at the molecular level to develop new cultivars, depends on the seminal work of Watson and Crick. Their discovery of the **double helical structure of the DNA molecule** laid the foundation for the understanding of the chemical basis of heredity.

Norman Borlaug. In the modern era of agriculture, Norman Borlaug deserves mention, not so much for his contribution to science as much as application of scientific principles to address world food and hunger, according to a methodology driven by his personal philosophy. This philosophy, dubbed the "Borlaug Hypothesis" by some economists, proposes to increase the productivity of agriculture on the best farmland to help curb deforestation by reducing demand for new farmland. His signature accomplishment, for which his name is synonymous, and for which he received the prestigious Nobel Prize (for Peace) in 1970 – the first agriculturalist to be so recognized – was the **Green Revolution**. While the award signified an acknowledgment of the positive impact of this work, the Green Revolution received criticism from a broad spectrum of sources. Undeterred by his detractors, Borlaug continued his advocacy for the poor and those plagued by perpetual hunger, working hard until his death in 2009 to alleviate world hunger.

Herb Boyer, Stanley Cohen, and Paul Berg. In 1973, Herb Boyer, Stanley Cohen, and Paul Berg lead the way into the brave new world of genetic manipulation in which DNA from one organism could be transferred into another, by achieving the feat with bacteria. Called **recombinant DNA technology,** the researchers successfully transferred foreign DNA into a bacterium cell. This began the era of **genetic engineering**. Currently, this is one of the major technologies in modern plant breeding, albeit controversial.

Industry highlights
Barley breeding in the United Kingdom

W.T.B. Thomas

Scottish Crop Research Institute, Invergowrie, Dundee DD2 5DA, UK

Targets

Barley breeding in the United Kingdom aims to produce new cultivars that offer an improvement in one or more of the key traits for the region (Table B2.1). New cultivars must have a good yield, preferably in excess of the currently established cultivars if targeted solely at the feed market. To be accepted for malting use, a new cultivar must offer improvement in one or more key facets of malting quality, primarily hot water extract, with no major defects in, for example, processability traits. Additionally, new cultivars must have minimum levels of disease resistance, which equates to being no worse than moderately susceptible, to the key diseases listed in Table B2.1.

Crossing to commercialization

Barley breeders therefore design crosses in which the parents complement each other for these target traits and attempt to select out recombinants that offer a better balanced overall phenotype. Whilst a wide cross may offer a better chance of producing superior recombinants, most barley breeders in the United Kingdom concentrate on narrow crosses between elite cultivars. The main reason for doing so is that a narrow cross between elite lines is more likely to produce a high mid-parental value for any one trait, so the proportion of desirable recombinants is thus far greater in the narrow cross than in the wide (Figure B2.1). Thus, the chances of finding a desirable recombinant for a complex trait such as yield in the wide cross is low and the chances of combining it with optimum expression for all the other traits is remote. As breeders are still making progress using such a narrow crossing strategy, it is possible that there is still an adequate level of genetic diversity with the elite barley gene pool in the United Kingdom. A similar phenomenon has been observed in barley breeding in the USA, where progress has been maintained despite a narrow crossing strategy (Rasmusson and Phillips, 1997). Rae *et al.* (2005) genotyped three spring barley cultivars (Cocktail, Doyen and Troon) on the 2005 United Kingdom recommended list with 35 Simple Sequence Repeat (SSR) markers and found sufficient allelic diversity to produce over 21 million different genotypes. It would appear, therefore, that the breeding challenge is not so much to generate variation as to identify the best recombinants.

The progress of a potential new barley cultivar in the United Kingdom, in common with that of the other cereals, proceeds through a series of filtration tests (Figure B2.2) and the time taken to pass through all but the first is strictly defined. The opportunity to reduce the time taken for breeders' selections is fairly limited given that multiplication of material for and conducting single and multisite trials takes at least three years, irrespective of whether out of season nurseries are used for shuttle breeding for the spring crop or doubled haploidy (DH) or Single Seed Descent (SSD) for the winter crop. The length of the breeding cycle is thus fairly well defined, with occasional reduction by a year when a cultivar from a highly promising cross is speculatively advanced by a breeder. A breeder may also delay submitting a line for official trials for an extra season's data but breeders now aim to submit the majority of their lines to official trials within 4–5 years of making a cross.

Table B2.1 Traits listed in the current UK recommended lists of barley (www.hgca.com).

Trait	Spring Barley	Winter Barley
Yield (overall and regional with fungicide)	Yes	Yes
Yield without fungicide	Yes	Yes
Height	Yes	Yes
Lodging resistance	Yes	Yes
Brackling resistance	Yes	
Maturity	Yes	Yes
Winter hardiness		Yes
Powdery Mildew resistance	Yes	Yes
Rhynchosporium resistance	Yes	Yes
Yellow rust resistance	Yes	Yes
Brown rust resistance	Yes	Yes
Net blotch resistance		Yes
BaYMV complex resistance		Yes
BYDV resistance	Yes	
Grain nitrogen content	Yes	Yes
Hot water extract	Yes	Yes
Screenings (2.25 and 2.5 mm)	Yes	Yes
Specific weight	Yes	Yes

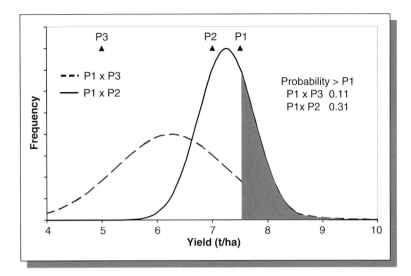

Figure B2.1 Frequency distribution of two crosses with a common parent (P1) and alternative second parents (P2 and P3). P2 is a slightly lower yielding parent, thus progeny from the cross will have a high mid-parent and small variation. P3 is a comparatively high yielding unadapted parent and the cross has a lower mid-parent but much greater variance. Areas under the shaded portion of both curves represent the fraction selected for high yield potential (>P1). Thus, whilst the extreme recombinant of P1 × P3 has a greater yield potential than that of P1 × P2, the probability of identifying superior lines for just this one trait is far greater for the latter. Figure courtesy of W.T.B Thomas.

Given that many breeders would have begun re-crossing such selections by this stage of their development, the approximate time for the breeding cycle in the United Kingdom is four years.

During the two years of National List Trials (NLT), potential cultivars are tested for Distinctness, Uniformity and Stability (DUS) using established botanical descriptors. A submission therefore has to be distinct from any other line on the National List and not have more than a permitted level of "off-types", currently equivalent to a maximum of 3 in 100 ear rows. Lines are tested over more than one year to ensure that they are genetically stable and do not segregate in a subsequent generation. DUS tests are carried out by detailed examination of 100 ear rows and three bulk plots (approximately 400 plants in total) submitted by the breeder. Thirty-three traits are examined routinely and there are three special and 59 approved additional traits. At the same time plot trials are carried out to establish whether the submission has Value for Cultivation and Use (VCU), and the VCU and DUS submissions are checked to verify that they are the same. Occasionally, a submission may fail DUS in NLT1, in which case the breeder has the option of submitting a new stock for a further two years of testing. Generally, the VCU results are allowed to stand and a cultivar can be entered into RLT before it has passed DUS in the anticipation that it will have succeeded by the time a recommendation decision has to be made. Full details can be obtained from www.defra.gov.uk/planth/pvs/VCU_DUS.htm.

The UK barley breeding community

The Plant Varieties and Seeds act of 1964, which enabled plant breeders to earn royalties on the certified seed produced for their cultivars, led to a dramatic increase in breeding activity in the United Kingdom. Formerly, it was largely the province of state funded improvement programs, such as that of the Plant Breeding Institute (PBI), of Cambridge, UK, that had produced the highly successful spring cultivar Proctor. The increase in breeding activity in the 1970s and early 1980s was largely as a result of dramatic expansion in the commercial sector, initially led by Miln Marsters, of Chester, UK, who produced Golden Promise, which dominated Scottish spring barley production for almost two decades. The two sectors co-existed until the privatization of the breeding activity at PBI and the state marketing arm, the National Seed Development Organisation, together with a change in government policy led to the withdrawal of the public sector from barley breeding. Barley breeding in the commercial sector in the United Kingdom is highly competitive with currently five UK-based crossing and selection programs. A number of other companies have their own selection programs based in the United Kingdom and many continental breeders have agency agreements for the testing and potential marketing of their products. For example, 41 spring and 34 winter barley lines were submitted for NLT1 for harvest 2004 and these were derived from 16 different breeders.

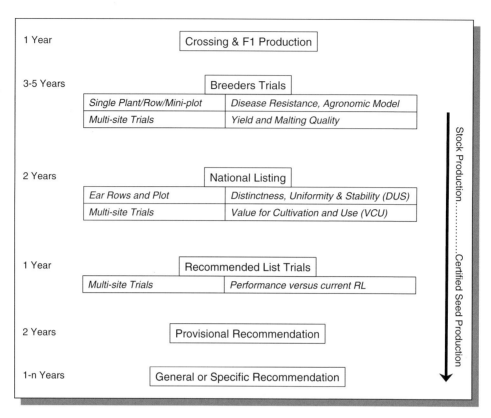

Figure B2.2 The phases in the development of a successful new cultivar from crossing to commercialization with the timescale for each. The exact nature of the scheme adopted in breeder's trials varies according to breeder and crop type but is either based upon a version of the pedigree or doubled haploid system. A cultivar may persist on the recommended list for *n* years, where *n* is the number of years where there is a significant demand for it. Figure courtesy of W.T.B Thomas.

The amount of certified seed produced for each cereal variety in the United Kingdom is published by the National Institute of Agricultural Botany. The total annual production of certified barley seed has been in decline since its peak of over 250 000 t in 1987 and has declined by 43% since 1995, with most due to a reduction in winter barley seed (Figure B2.3). There are a number of potential reasons, such as an increase in farm-saved seed, but the principal feature has been a marked decrease in winter barley cropping over the period whereas spring barley has remained fairly static and winter wheat has increased. Over this period, certified seed production has exceeded 100 000 metric tons for two spring (Optic and Chariot) and two winter (Regina and Pearl) barley cultivars; these can be considered notable market successes. There has been substantial production for a number of others but total production exceeded 25 000 tonnes for only six and seven spring and winter barley cultivars, respectively. When it is considered that over 830 lines were submitted for NLT over this period, the overall success rate is 1.6%. Nevertheless, real breeding progress is being made. Using yield data from the recommended list trials from 1993 to 2004 to estimate the mean yield of each recommended cultivar and then regressing that data against the year that it was first recommended revealed that genetic progress was in the order of 1% per annum (Rae *et al.*, 2005).

The impact of molecular markers

The first whole genome molecular maps of barley were published in 1991 (Graner *et al.*, 1991; Heun *et al.*, 1991) and were closely followed by QTL maps in 1992 (Heun, 1992) and 1993 (Hayes *et al.*, 1993) with well over 40 barley mapping studies now in the public domain. Despite this apparent wealth of information, barley breeders in the United Kingdom are largely relying on conventional Phenotypic Selection (PS) to maintain this progress. This is in marked contrast to the highly successful of Marker Assisted Selection (MAS) in the Australian Barley program (Langridge and

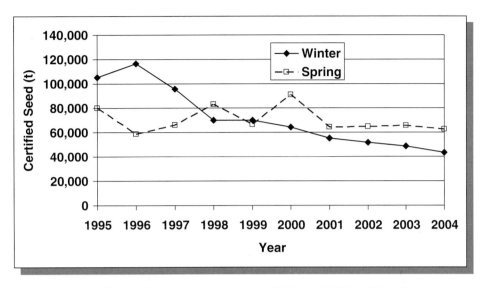

Figure B2.3 Tonnes of certified barley seed produced in the UK from 1995 to 2004. Figure courtesy of W.T.B Thomas.

Barr, 2003), which is probably a reflection of the different breeding strategies in the two countries. In the United Kingdom, improvement is being achieved in the elite gene pool, as noted above, whereas MAS has been deployed in an introgression breeding strategy in Australia. Given that most barley mapping studies have concentrated on diverse crosses to maximize polymorphism and facilitate map construction, there are very few published QTL studies that are relevant to current United Kingdom barley breeding strategies. Surveying results from eight different barley mapping populations Thomas (2003) found that there were very few instances where QTLs were co-located for three or more crosses for important traits such as yield and hot water extract.

Major gene targets

Markers have been developed for a number of known major genes and could potentially be deployed in MAS by United Kingdom breeders. Many of these major gene targets are, however, disease resistances, many of which have been defeated by matching virulence in the corresponding pathogen population. United Kingdom barley breeders have been required to select for at least some resistance to the key foliar pathogens listed in Table 2.11 since the introduction of minimum standards and have, accordingly, developed efficient phenotypic screens. There are exceptions, most notably the Barley Yellow Mosaic Virus complex, which is transmitted by infection of the roots with the soil borne fungus vector *Polymixa graminis*. A phenotypic screen therefore requires an infected site and the appropriate environment for infection and expression. Phenotypic screening can be expensive if a breeder is distant from an infected site and is subject to potential mis-classification.

Resistance due to the *rym4* allele was initially found in Ragusa and was effective against BaYMV strain 1 and a number of cultivars carrying this allele have been developed, initially by phenotypic screening. Markers to select for this resistance have also been developed, beginning with the RFLP probe MWG838 (Graner and Bauer, 1993), later converted to an STS (Bauer and Graner, 1995), and were used in some breeding programs in the United Kingdom and Europe. BaYMV strain 2, which became more frequent in the 1990s, could overcome the *rym4* resistance but another resistance, *rym5*, was identified in Mokusekko 3 as being effective against both strains. This resistance was co-located with *rym4* and the Simple Sequence Repeat (SSR) marker Bmac29 was found to be linked to it (Graner *et al.*, 1999). Bmac29 could not only distinguish between resistant and susceptible alleles but also between the *rym4* and *rym5* alleles derived from Ragusa and Mokusekko 3, respectively, but as it is 1.3 cM from the gene locus it is not effective in a wide germplasm pool, as *Hordeum spontaneum* lines predicted to be resistant by the marker were found to be susceptible (R.P. Ellis, unpublished data). Bmac29 has, however, proved to be particularly effective for United Kingdom, and European, barley breeders as they are working with a narrow genetic base and just the two sources of resistance. Other resistance loci have been identified together with suitable markers to deploy in a pyramiding strategy in an attempt to provide durable resistance (Ordon *et al.*, 2003) and a clear example of how the use of markers in MAS have evolved together with the pathogen.

Another example relates to a particular requirement of the Scotch whisky distilling industry. In grain and certain malt whisky distilleries, a breakdown product of the cyanogenic glycoside epiheterodendrin can react with copper in the still to form the carcinogen ethyl carbamate, which can be carried over into the final spirit in distilling, leading to a demand for barley cultivars that do not produce epiheterodendrin. The trait is controlled by a single gene with the non-producing allele originating in the mildew resistance donor "Arabische" used in the derivation of the cultivar Emir. The phenotypic assay for the trait involves the use of hazardous chemicals and the finding of a linked SSR marker (Bmac213) offered a simpler and safer alternative (Swanston *et al.*, 1999). The distance between the gene locus and the marker (6cM) meant that, in contrast to Bmac29, Bmac213 was not reliable in the cultivated gene pool. For instance, the cultivar Cooper and its derivatives possessed the non-producing allele yet were producers. However, the marker could still be used when the parents of a cross were polymorphic for both the phenotype and the marker. Recently, a candidate gene has been identified and markers used for reliable identification of non-producers developed (P. Hedley, personal communication).

QTL targets

Currently, United Kingdom barley breeders do not use MAS for any other malting quality targets. A QTL for fermentability was detected in a cross between elite United Kingdom genotypes (Swanston *et al.*, 1999) but the increasing allele was derived from the parent with relatively poor malting quality. When this QTL was transferred into a good malting quality cultivar, the results were inconclusive (Meyer *et al.*, 2004), probably because the effect of the gene was more marked in a poor quality background and any extra activity due to it was superfluous in a good quality background. This highlights one of the problems in developing MAS for complex traits such as yield and malting quality. Results from an inappropriate gene pool may well not translate to a target gene pool; it is therefore essential that QTL studies are carried out in the appropriate genetic background.

Future prospects

The genotyping of entries from Danish registration trials coupled with associations of markers with yield and yield stability phenotypes demonstrated that QTLs can be detected in the elite gene pool (Kraakman *et al.*, 2004) but the findings need validation before the markers can be used in MAS. At the Scottish Crop Research Institute (SCRI), extensive genotyping of UK RLT entries over the past 12 years will be undertaken in collaboration with the University of Birmingham, National Institute of Agricultural Botany, Home Grown Cereals Authority, barley breeders and representatives of the malting, brewing and distilling industries in a project funded by the Defra Sustainable Arable LINK scheme. The RLT phenotypic data set represents an extensive resource that can discriminate between the fine differences in elite cultivars and will facilitate the identification of meaningful associations within the project for validation and potential use in MAS. How MAS is then used by commercial breeders in the United Kingdom might well vary but could range from early generation selection to an enriched germplasm pool upon which phenotypic selection can be concentrated to identification of candidate submission lines carrying target traits.

Acknowledgements

W.T.B. Thomas is funded by the Scottish Executive Environmental and Rural Affairs Department.

References

Bauer, E., and Graner, A. (1995). Basic and applied aspects of the genetic analysis of the *ym4* virus resistance locus in barley. *Agronomie*, **15**:469–473.

Graner, A., and Bauer, E. (1993). RFLP Mapping of the *ym4* virus-resistance gene in barley. *Theoretical and Applied Genetics*, **86**:689–693.

Graner, A., Jahoor, A., Schondelmaier, J., *et al.* (1991). Construction of an RFLP map of barley. *Theoretical and Applied Genetics*, **83**:250–256.

Graner, A., Streng, S., Kellermann, A., *et al.* (1999). Molecular mapping and genetic fine-structure of the *rym5* locus encoding resistance to different strains of the Barley Yellow Mosaic Virus Complex. *Theoretical and Applied Genetics*, **98**:285–290.

Hayes, P.M., Liu, B.H., Knapp, S.J., *et al.* (1993). Quantitative trait locus effects and environmental interaction in a sample of North-American barley germ plasm. *Theoretical and Applied Genetics*, **87**:392–401.

Heun, M. (1992). Mapping quantitative powdery mildew resistance of barley using a Restriction-Fragment-Length-Polymorphism map. *Genome*, **35**:1019–1025.

Heun, M., Kennedy, A.E., Anderson, J.A., *et al.* (1991). Construction of a Restriction-Fragment-Length-Polymorphism map for barley (*Hordeum vulgare*). *Genome*, **34**:437–447.

Kraakman, A.T.W., Niks, R.E., Van den Berg, P.M.M.M., Stam, P., and van Eeuwijk, F.A. (2004). Linkage disequilibrium mapping of yield and yield stability in modern spring barley cultivars. *Genetics*, **168**:435–446.

Langridge, P., and Barr, A.R. (2003). Better barley faster: the role of marker assisted selection – Preface. *Australian Journal of Agricultural Research*, **54**:i–iv.

Meyer, R.C., Swanston, J.S., Brosnan, J., *et al.* (2004). Can Anonymous QTLs be Introgressed Successfully Into Another Genetic Background? Results From A Barley Malting Quality Parameter, in *Barley Genetics IX, Proceedings of the Ninth International Barley Genetics Symposium* (eds J. Spunar and J. Janikova), II. Agricultural Research Institute, Kromeriz, Czech Republic, pp. 461–467.

Ordon, F., Werner, K., Pellio, B., Schiemann, A., Friedt, W., and Graner, A. (2003). Molecular breeding for resistance to soil-borne viruses (BaMMV, BaYMV, BaYMV-2) of barley (*Hordeum vulgare L.*). *Journal of Plant Diseases and Protection*, **110**:287–295.

Rae, S.J., Macaulay, M., Ramsay, L., *et al.* (2005). Molecular breeding for resistance to soil-borne viruses (BaMMV, BaYMV, BaYMV-2) of barley (*Hordeum vulgare L.*). *Journal of Plant Diseases and Protection*, **110**:287–295.

Rasmusson, D.C., and Phillips, R.L. (1997). Plant breeding progress and genetic diversity from *de novo* variation and elevated epistasis. *Crop Science*, **37**:303–310.

Swanston, J.S., Thomas, W.T.B., Powell, W., *et al.* (1999). Using molecular markers to determine barleys most suitable for malt whisky distilling. *Molecular Breeding*, **5**:103–109.

Thomas, W.T.B. (2003). Prospects for molecular breeding of barley. *Annals of Applied Biology*, **142**:1–12.

2.6 History of plant breeding technologies/techniques

Modern plant breeding is an art and a science. The two key activities in plant breeding are the creation (or assembling) of variation and discriminating (selecting) among the available variability to identify and advance individuals that meet the breeding objectives. Consequently, advances in plant breeding technologies and techniques focus on facilitating and making these two distinct activities more efficient and cost effective.

2.6.1 Technologies/techniques associated with creation of variation

Plant breeders depend on variation for plant improvement. Variation may be natural in origin or it may be artificially generated in a variety of ways. Through the years, breeders have used various technologies and techniques in the quest for desired variation.

Artificial pollination

Artificial pollination, the deliberate transfer by humans of pollen from the flower (anther) of one plant to the flower (stigma) of another plant is an ancient practice, as previously noted. Babylonians and Assyrians were known to have conducted it on date palms. These ancient cultures did this without the benefit of knowing the underlying science of pollination and fertilization. These ancient efforts were not geared toward creating variation; they were primarily for fertilization for fruit production. Science-based artificial pollination started after the discovery of sex in plants by Camerarius and the ensuing work of Koelreuter. Artificial pollination (controlled pollination) is used in a variety of ways in modern plant breeding. Naturally cross-pollinating species can be artificially self-pollinated to create variability for selection or to generate special parental breeding stock for experimentation or development of new cultivars. Experiments in heredity (e.g., Mendel's) depend on controlled pollination. These applications are discussed in detail elsewhere in this book.

Hybridization

One of the commonly used techniques in modern plant breeding to create variation is hybridization (crossing) of genetically different plants. It is commonly used to generate the initial population in which selection is practiced in a breeding program. The F_2 is the most variable generation in which selection is often initiated. Breeders working in the field often have crossing blocks where controlled hybridization is conducted. Depending on the species and breeding objective, pollination may be done manually, or with the aid of natural agents (wind, insects). Whereas hybridization for the creation of variation may entail just two parents, there are various sophisticated hybridization schemes in modern plant breeding in which a number of parents are included (e.g., diallele crosses).

Hybridization is commonly conducted with parents that are crossable or genetically compatible. However, there are occasions in plant breeding where it is desirable or even necessary to seek to introduce genes into the breeding program from genetically distant sources. Wild germplasm is considered a rich source of genes for modern crop improvement. The term "wide cross" is used to refer to hybridization that involves plant materials from outside the pool for cultivated species. Some wide crosses involve two species (interspecific cross), or even genera (intergeneric cross). The more distant the parents used in hybridization, the higher the incidence of genetic complications pertaining to meiosis, and the lesser the chances of success. Breeders use certain techniques and technologies to boost the success of wide crosses.

Tissue culture/embryo culture

Tissue culture entails growing plants or parts of plants *in vitro* under an aseptic environment. It has various applications in modern plant breeding. Regarding the generation of variation, the specific application of tissue culture is in rescuing embryos produced from wide crosses. Due to genetic incompatibility arising from the genetic distance between parents in wide crosses, the hybrid embryo often does not develop adequately to produce a viable seed. The technique of **embryo culture** enables breeders to aseptically extract the immature embryo and culture it into a full grown plant that can bear seed.

Chromosome doubling

To circumvent a major barrier to interspecific crossing, breeders use the chromosome doubling technique to double the chromosomes in the hybrid created (which is reproductively sterile due to meiotic incompatibility) in order to provide paring partners for successful meiosis and restoration of fertility. Chromosome doubling is achieved through the application of the chemical colchicine.

Bridge cross

The bridge cross is another technique developed to facilitate wide crossing. This technique provides an indirect way of crossing two parents that differ in ploidy level (different number of chromosomes) through a transitional or intermediate cross. This intermediate cross is reproductively sterile and is subjected to chromosome doubling to restore fertility.

Protoplast fusion

Cell fusion or specifically protoplast (excluding cell wall) fusion is a technique used by breeders to effect *in vitro* hybridization in situations where normal hybridization is challenging. It can be used to overcome barriers to fertilization associated with interspecific crossing. The first successful application of this techniques occurred in 1975.

Hybrid seed technology/technique

Hybridization may be used as a means of generating variation for selection in a breeding program. It may also be done to create the end product of a breeding program. The discovery of the phenomenon of heterosis laid the foundation for hybrid seed technology. Breeders spend resources to design and develop special genotypes to be used as parents in producing hybrid seeds. Hybrid seed is expensive to produce and hence costs more than non-hybrid seed. In the 1990s, the **genetic use restriction technology** (**GURT**), colloquially, **terminator technology**, was introduced as a means of deterring the unlawful use of hybrid seed. This technology causes second generation seed from a hybrid crop to be reproductively sterile (i.e., a farmer cannot harvest a crop by saving seed from the current year's crop to plant the next season's crop). Allied techniques that drive the hybrid seed industry include male sterility and self-incompatibility, techniques used to manage pollination and fertility in the hybrid breeding industry.

Seedlessness technique

Whereas fertility is desired in a seed-bearing cultivar, sometimes seedless fruits are preferred by consumers. The observation that triploidy (or odd chromosome number set) results in hybrid sterility led to the application of this knowledge as a breeding technique. Crossing a diploid ($2n$) with a tetraploid ($4n$) yields a triploid ($3n$), which is sterile and hence produces no seed.

Mutagenesis

Evolution is driven by mutations that arise spontaneously in the population. Since the discovery in 1928 by H. Muller of the mutagenetic effects of X-rays on the fruit fly, the application of mutagens (physical and chemical) have been exploited by plant breeders to induce new variation. Mutation breeding is a recognized scheme of plant breeding that has yielded numerous successful commercial cultivars, in addition to being a source of variation.

DNA technology

The advent of the recombinant DNA technology in 1985 revolutionized the field of biology and enabled researchers to directly manipulate an organism directly at the DNA level. The most astonishing capacity of this technology is the ability of researchers to move DNA around without regard to genetic boundaries. Simply put, DNA (or gene) from an animal may be transferred into a plant. The DNA technology also allows researchers to isolate and clone genes and pieces of DNA for various purposes. This precise gene transfer is advantageous in plant improvement. Mutagenesis can now be targeted and precise instead of random, as in the use of mutagens in conventional applications.

A new category of cultivars, GM cultivars, has been developed using recombinant DNA technology. DNA technologies and techniques are exploding at a terrific rate, with new ones being regularly added while existing ones are refined and made more efficient and cost effective. One of the most useful applications of DNA technology in plant breeding is in molecular markers.

Important modern milestones associated with the creation of variation

- **Plant Variety Protection Act.** Enacted in 1970 and amended in 1994, the US Plant Variety Protection Act gave intellectual property rights to innovators who developed new crop varieties of sexually reproducing species and tuber-propagated species. The commercial seed industry is thriving because companies can reap benefits from their investments in the often expensive cultivar development ventures.
- **First commercial GM crop.** The **FlavrSavr tomato** was the first commercially approved and grown genetically modified (GM) crop for human consumption. It was developed in 1992 by the biotech company Calgene, using the antisense gene technology to down-regulate the production of the enzyme polygalacturonase that degrades pectin in fruit cell walls, resulting in fruit softening. FlavrSavr tomato hence ripens slowly and stays fresher on the shelf for a longer time. In 1995, *Bt* **corn**, engineered to resist the European corn borer was produced by the Pioneer Hibred company, while **RR (Roundup ready)** soybean, a Monsanto product, was introduced in 1996.

2.6.2 Technologies/techniques for selection

Selection or the discrimination among variability is the most fundamental of techniques used by plant breeders throughout the ages. In some cases, individual plants are the units of selection; in other cases, a large number of plants are chosen and advanced in the breeding program. With time, various strategies (breeding schemes) have been developed for selection in breeding programs.

Selection (breeding) schemes

Breeding schemes are discussed in detail in Chapters 15–18. They are distinguished by the nature and source of the population used to initiate the breeding program, as well as by the nature of the product. The most basic of these schemes is mass selection; others are recurrent selection, pedigree selection, and bulk population strategy.

Molecular marker technology

Marker technique is essentially selection by proxy. Selection is generally conducted by visually discriminating among variability, in the hope that the variation on hand is caused by differences in genotype and not by variation in the environment. Markers are phenotypes that are linked to genotypes (or precisely genes of interest). Markers are discussed in detail in Chapter 20. They are useful in facilitating the selection process and making it more efficient and cost effective. Molecular (DNA-based) markers have superseded morphological markers in scale of use in plant breeding. Marker assisted selection (MAS) is used to facilitate plant breeding (Chapter 21).

Gene mapping

Gene mapping entails a graphic representation of the arrangement of a gene or a DNA sequence on a chromosome. It can be used to locate and identify the gene (or group of genes) that conditions a trait of interest. It depends on availability of markers. The availability of molecular markers has greatly facilitated gene mapping. Furthermore, genomic DNA sequencing produces the most complete maps for species. Now, quantitative trait loci (QTLs) mapping is becoming more widespread. Modern plant breeding is greatly facilitated by genetic maps.

2.7 Genome-wide approaches to crop improvement

An organism's complete set of DNA is called its **genome**. The concept of genomics began with the successful sequencing of the genomes of a virus and a mitochondrion by Fred Sanger and his colleagues starting in the 1970s. Previously, researchers were limited to understanding plant structure and function piecemeal (gene-by-gene). With the advances in technology, whole genomes of certain species have been sequenced, thereby making all the genes they contain accessible to researchers. Because of the cost of such undertakings, whole genome sequences have so far been limited to the so-called model organisms, including *Arabidopsis*, rice, and corn. Through comparative genome analysis, researchers seek to establish correspondence between genes or other genomic features in different organisms, without the need to have whole genome maps of all organisms. In sum, the goal of plant genomics is to understand the genetic and molecular basis of all the relevant biological processes that pertain to a plant species, so that they can be exploited more effectively and efficiently for improving the species. Genomics is hence important in modern plant breeding efforts. Two of the major tools employed in genomics research are microarrays and bioinformatics.

2.8 Bioinformatics in crop improvement

Genomics programs generate large volumes of data or information that need to be organized and interpreted to increase our understanding of biological processes. Bioinformatics is the discipline that combines mathematical and computational approaches to understand biological processes. Researchers in this area engage in activities that include mapping and analyzing DNA and protein sequences, aligning different DNA and protein sequences for the purpose of comparison, gene finding, protein structure prediction, and prediction of gene expression. Bioinformatics will continue to have a major impact on how modern plant breeding is conducted.

2.9 Plant breeding in the last half century

The foregoing brief review has revealed that plant breeding as a discipline and practice has changed significantly over the years.

2.9.1 Changes in the science of breeding

It has been said several times previously that plant breeding is a science and an art. Over the last decade, it has become clear that science is what is going to drive the achievements in plant breeding. More importantly, is it clear that a successful plant breeding program has an interdisciplinary approach, for recent strides in plant breeding have come about because of recent advances in allied disciplines. High-tech cultivars need appropriate cultural environment for the desired productivity. Advances in agronomy (tillage systems, irrigation technology, and herbicide technology) have contributed to the expansion of crop production acreage. In other words, plant breeders do not focus on crop improvement in isolation but consider the importance of the ecosystem and its improvement to their success. Whereas most of the traditional plant breeding schemes and technologies previously discussed are still in use, the tools of biotechnology have been the dominant influence in the science of plant breeding.

2.9.2 Changes in laws and policies

In the United States, land grant institutions were established to promote and advance agricultural growth and productivity of the states, among other roles. Much of the effort of researchers is put in the public domain for free access. The Plant Variety Protection Act of 1970 that provided intellectual property rights to plant breeders was the major impetus for the proliferation of for-profit private seed companies, and their domination of the more profitable aspects of the seed market where legal protection and enforcement were clearer and more enforceable (e.g., hybrid seed). Plant breeders' rights legislation was implemented in the 1960s and 1970s in most of Western Europe. Australia and Canada adopted similar legislation much later, around 1990. The US Supreme Court ruled in 1980 to allow utility patent protection to be applied to living things. This protection was extended to plants in 1985. The European Patent Office granted such protection to GM cultivars in 1999.

2.9.3 Changes in breeding objectives

Breeding objectives depend on the species and the intended use of the cultivar to be developed. Over the years, new (alternative) species have been identified to address some traditional needs in some parts of the world. By the same token, the traditional uses of some species have been modified. For example, whereas corn continues to be used for food and feed in many parts of the world, corn has an increasingly industrial role in some industrialized countries (e.g., ethanol production for biofuel). Yield or productivity, adaptation to a production environment, and resistance to biotic and abiotic stresses will always be important. However, with time, as they are resolved, breeders shift their emphasis to other quality traits (e.g., oil content or more specific consumer needs, such as low linolenic acid content). Advances in technology (high throughput, low cost, precision, repeatability) have allowed breeders to pursue some of the challenging objectives that once were impractical to do. Biotechnology, especially recombinant DNA technology, has expanded the source of genes for plant breeding in the last half decade. Also, the increasing need to protect the environment from degradation has focused breeders' attention on addressing the perennial problem of agricultural sources of pollution.

2.9.4 Changes in the creation of variability

The primary way of creating variability for breeding has been through artificial crossing (hybridization) or mutagenesis (induced mutations). Hybridization is best done between crossable parents. However,

sometimes, breeders attempt to cross genetically distant parents, with genetic consequences. There are traditional schemes and techniques to address some of these consequences (e.g., wide cross, embryo rescue). The success of hybridization depends on the ability to select and use the best parents in the cross. Breeders have access to elite lines for use as parents. Furthermore, biotechnology tools are now available to assist in identifying suitable parents for a cross and also to assist in introgressing genes from exotic sources into adapted lines. Transgenesis (genetic engineering involving gene transfer across natural biological boundaries) and, more recently, cisgenesis (genetic engineering involving gene transfer among related and crossable species) can be used to assist breeders in creating useful variability for breeding. In the case of mutagenesis, advances in technology have enabled breeders to be more efficient in screening mutants (e.g., by TILLING). Products from mutation breeding, not being transgenic, are more acceptable to consumers who are unfavorably disposed to GM crops.

2.9.5 Changes in identifying and evaluating genetic variability

Identifying and measuring quantitative variability continues to be challenging, even though some progress has been made (e.g., QTLs – quantitative trait loci analysis and mapping). This has been possible because of the new kinds of molecular markers that have been developed and the accompanying throughput technologies. QTLs are more precisely mapped, in addition to the increased precision of linkage maps (marker dense). The abundance of molecular markers and availability of more accessible genomic tools has made it easier for researchers to readily characterize biodiversity.

2.9.6 Selecting and evaluating superior genotypes

Selection schemes have remained relatively the same for a long time. Here, too, the most significant change over the last half century has been driven by molecular technology. The use of molecular markers in selection (MAS – marker assisted selection) gained significant attention over the period. Most traits of interest to breeders are quantitatively inherited. The continuing challenge with this approach is the lack of precision (the need for more high resolution QTL maps) and higher throughput marker technology, amongst others. Selected genotypes are evaluated across time and space in the same old fashioned way.

Key references and suggested reading

Baezinger, P.S., Russell, W.K., Graef, G.L., and Campbell, B.T. (2006). 50 years of crop breeding, genetics, and cytology. *Crop Science*, **46**:2230–2244.

Borém, A., Guimarães, P.E., Federizzi, L.C., and Ferraz, J.F. (2002). From Mendel to genomics, plant breeding milestones: a review. *Crop Breeding and Applied Biotechnology*, **2**(4): 649–658.

Bhat, S.R., and Srinivasan. (2002). Molecular and genetic analyses of transgenic plants: considerations and approaches. *Nature*, **349**:726.

Brim, C.A. (1966). A modified pedigree method of selection in soybeans. *Crop Science*, **6**:220.

Crow, J.F. (1998). 90 Years ago: the beginning of hybrid maize. *Genetics*, **148**:923–928.

Donald, C.M. (1968). The breeding of crop ideotypes. *Euphytica*, **17**:385–403.

Duvick, D.N. (1996). Plant breeding: an evolutionary concept. *Crop Science*, **36**:539–548.

Duvick, D.N. (2001). Biotechnology in the 1930s: The development of hybrid maize. *Nature Reviews Genetics*, **2**:69.

Fernandez-Cornejo, J., *et al.* (2004). The Seed Industry in US Agriculture: An Exploration of Data and Information on Crop Seed Markets, Regulation, Industry Structure, and Research and Development. Agriculture Information Bulletin Number 786, Resource Economics Division, Economic Research Service, US Department of Agriculture, Washington, DC. (http://www.ers.usda.gov/publications/AIB786/)

Kasha, K.J. (1999). Biotechnology and world food supply. *Genome*, **42**:642–645.

Khush, G.S. (2001). Green revolution: the way forward. *Nature Reviews Genetics*, **2**:815–822.

Knight, J. (2003). A Dying Breed. *Nature*, **421**:568–570. (http://www.cambia.org/daisy/cambia/319)

Lindner, R. (2004). Economic Issues for Plant Breeding – Public Funding and Private Ownership. *Australasian Agribusiness Review*, **12**: Paper 6. (http://www.agrifood.info/review/2004/Lindner.html)

Rasmusson, D.C. (1991). A plant breeder's experience with ideotype breeding. *Field Crop Research*, **26**:191–200.

Wehner, T.C. (2005). History of Plant Breeding. Plant Breeding Methods, North Carolina State University, Raleigh, NC. (http://cuke.hort.ncsu.edu/cucurbit/wehner/741/hs741hist.html)

Xing, T. (1998). Bioinformatics and its impact on plant science. *Trends in Plant Science*, **3**:450.

Outcomes assessment

Part A

Please answer the following questions true or false.

1 JH Muller is associated with the discovery of the possible effect of X-rays on genetic material.
2 The term "heterosis" was coined by GH Shull.
3 Gregor Mendel is the author of the book *On the origin of species*.

Part B

Please answer the following questions.

1 The term "recurrent selection" was coined by ....
2 For what contribution to tissue culture are Murashige and Skoog known?
3 Who was Norman Borlaug?

Part C

Please discuss in the following questions in detail.

1 How is the farmer in a developing country like a plant breeder
2 Describe the contribution made by each of the following persons to modern plant breeding – Luther Burbank, Louis de Vilmorin, Joseph Koelreuter.
3 Briefly discuss the changes in the laws and policies that have impacted plant breeding over the years.
4 How has the science of plant breeding changed over the years?
5 Discuss the impact of DNA technologies on plant breeding.

Section 2

Population and quantitative genetic principles

Chapter 3 Introduction to concepts of population genetics
Chapter 4 Introduction to quantitative genetics

Plant breeders develop new cultivars by modifying the genetic structure of the base population used to start the breeding program. Students need to have an appreciation for population and quantitative genetics in order to understand the principles and concepts of practical plant breeding. In fact, there is what some call the breeders' equation, a mathematical presentation of a fundamental concept that all breeders must thoroughly understand. This section will help the student understand this and other basic breeding concepts.

3

Introduction to concepts of population genetics

Purpose and expected outcomes

Plant breeders manipulate plants based on the modes of their reproduction (i.e., self- or cross-pollinated). Self-pollinated plants are pollinated predominantly by pollen grains from their own flowers, whereas cross-pollinated plants are predominantly pollinated by pollen from other plants. These different reproductive behaviors have implications in the genetic structure of plant populations. In addition to understanding Mendelian genetics, plant breeders need to understand changes in gene frequencies in populations. After all, selection alters the gene frequencies of breeding populations. After studying this chapter, the student should be able to:

1 Define a population.
2 Discuss the concept of a gene pool.
3 Discuss the concept of gene frequency.
4 Discuss the Hardy–Weinberg law.
5 Discuss the implications of the population concept in breeding.
6 Discuss the concept of inbreeding and its implications in breeding.
7 Discuss the concept of combining ability.

3.1 Concepts of a population and gene pool

Some breeding methods focus on individual plant improvement, whereas others focus on improving plant populations. Plant populations have certain dynamics, which impact their genetic structure.

The genetic structure of a population determines its capacity to be changed by selection (i.e., improved by plant breeding). Understanding population structure is key to deciding the plant breeding options and selection strategies to use in a breeding program.

Principles of Plant Genetics and Breeding, Second Edition. George Acquaah.
© 2012 John Wiley & Sons, Ltd. Published 2012 by John Wiley & Sons, Ltd.

3.1.1 Definitions

A **population** is a group of sexually interbreeding individuals. The capacity to interbreed implies that every gene within the group is accessible to all members through the sexual process. A **gene pool** is the total number and variety of genes and alleles in a sexually reproducing population that are available for transmission to the next generation. Rather than the inheritance of traits, **population genetics** is concerned with how the frequencies of alleles in a gene pool change over time. Understanding population structure is important to breeding by either conventional or unconventional methods. It should be pointed out that the use of recombinant DNA technology, as previously indicated, has the potential to allow gene transfer across all biological boundaries to be made. Breeding of cross-pollinated species tends to focus on improving populations rather than individual plants, as is the case in breeding self-pollinated species. To understand population structure and its importance to plant breeding, it is important to understand the type of variability present, and its underlying genetic control, in addition to the mode of selection for changing the genetic structure.

3.1.2 Mathematical model of a gene pool

As previously stated, gene frequency is the basic concept in population genetics, which is concerned with both the genetic composition of the population as well as the transmission of genetic material to the next generation. The genetic constitution of a population is described by an array of gene frequencies. The genetic properties of a population are influenced in the process of transmission of genes from one generation to the next by four major factors – **population size, differences in fertility and viability, migration and mutation,** and **the mating system**. Genetic frequencies are subject to sample variation between successive generations. A plant breeder directs the evolution of the breeding population through the kinds of parents used to start the base population in a breeding program, how the parents are mated, and artificial selection.

The genetic constitution of individuals in a population is reconstituted for each subsequent generation. Whereas the genes carried by the population have continuity from one generation to the next, there is no such continuity in the genotypes in which these genes occur. Plant breeders often work with genetic phenomena in populations that exhibit no apparent Mendelian segregation, even though in actuality, they obey Mendelian laws. Mendel worked with genes whose effects were categorical (kinds) and were readily classifiable (ratios) into kinds in the progeny of crosses. Breeders, on the other hand, are usually concerned about differences in populations measured in degrees rather than kinds. Population genetics uses mathematical models to attempt to describe population phenomena. To accomplish this, it is necessary to make assumptions about the population and its environment.

Calculating gene frequency

To understand the genetic structure of a population, consider a large population in which random mating occurs, with no mutation or gene flow between this population and others, no selective advantage for any genotype, and normal meiosis. Consider also one locus, A, with two alleles, A and a. The frequency of allele A_1 in the gene pool is p, while the frequency of allele A_2 is q. Also, $p + q = 1$ (or 100% of the gene pool). Assume a population of N diploids in which two alleles (A, a) occur at one locus. Assuming dominance at the locus, three genotypes – AA, Aa, and aa – are possible in an F_2 segregating population. Assume the genotypic frequencies are D (for AA), H (for Aa) and Q (for aa). Since the population is diploid, there will be 2N alleles in it. The genotype AA has two A alleles. Hence, the total number of A alleles in the population is calculated as $2D + H$. The proportion or frequency of A alleles (designated as p) in the population is obtained as follows:

$$2D + H/2N = [D + {}^1/_2 H]/N = p$$

The same can be done for allele a, and designated q. Further, $p + q = 1$ and hence $p = 1 - q$. If $N = 80$, $D = 4$, and $H = 24$:

$$p = [D + {}^1/_2 H]/N = (4 + 12)/80 = 16/80 = 0.2$$

Since $p + q = 1$, $q = 1 - p$, and hence $q = 1 - 0.2 = 0.8$.

Hardy–Weinberg equilibrium

Consider a random mating population (each male gamete has an equal chance of mating with any female gamete). Random mating involving the previous locus (A/a) will yield the following genotypes: *AA*, *Aa*, and *aa*, with the corresponding frequencies of p^2, $2pq$, and q^2, respectively. The gene frequencies must add up to unity. Consequently, $p^2 + 2pq + q^2 = 1$. This mathematical relationship is called the **Hardy–Weinberg equilibrium**. Hardy, from England, and Weinberg, from Germany, discovered that equilibrium between genes and genotypes is achieved in large populations. They showed that the frequency of genotypes in a population depends on the frequency of genes in the preceding generation, not on the frequency of the genotypes.

Considering the previous example, the genotypic frequencies for the next generation following random mating can be calculated as follows:

$$
\begin{aligned}
AA &= p^2 &&= 0.2^2 &&= 0.04 \\
Aa &= 2pq &&= 2(0.2 \times 0.8) &&= 0.32 \\
aa &= q^2 &&= 0.82 &&= \underline{0.64} \\
\text{Total} & && &&= 1.00
\end{aligned}
$$

The Hardy–Weinberg equilibrium is hence summarized as:

$$p^2\,AA + 2pq\,Aa + q^2\,aa = 1\,(\text{or } 100\%)$$

This means that in a population of 80 plants as before, about three plants will have a genotype of *AA*, 26 will be *Aa*, and 51, *aa*. Using the previous formula, the frequencies of the genes in the next generation may be calculated as:

$$p = [\mathrm{D} + {}^1/_2\mathrm{H}]/\mathrm{N} = (3 + 13)/80 = 0.2$$

and $q = 1 - p = 0.8$

The allele frequencies have remained unchanged, while the genotypic frequencies have changed from 4, 24, and 52, to 3, 26, and 51, for *AA, Aa,* and *aa*, respectively. However, in subsequent generations, both the genotype and gene frequencies will remain unchanged, provided that:

1 random mating occurs in a very large diploid population;
2 allele *A* and allele *a* are equally fit (one does not confer a superior trait than the other);
3 there is no differential migration of one allele into or out of the population;
4 the mutation rate of allele *A* is equal to that of allele *a*.

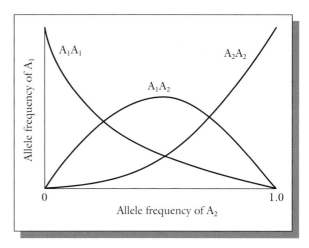

Figure 3.1 The relationship between allele frequencies and genotype frequencies in a population in Hardy–Weinberg equilibrium for two alleles. The frequency of the heterozygotes cannot be more than 50%, and this maximum occurs when the allele frequencies are $p = q = 0.5$. Further, when the frequency of an allele is low, the rare allele occurs predominantly in heterozygotes and there are very few homozygotes. (Adapted from Falconer, 1981.)

In other words, the variability does not change from one generation to another in a random mating population. The maximum frequency of the heterozygote (H) cannot exceed 0.5 (Figure 3.1). The Hardy–Weinberg law states that equilibrium is established at any locus after one generation of random mating. From the standpoint of plant breeding, two states of variability are present – two homozygotes (*AA, aa*), called "free variability", that can be fixed by selection and the intermediate heterozygous (*Aa*), called "hidden or potential variability", that can generate new variability through segregation. In outcrossing species, the homozygotes can hybridize to generate more heterozygotic variability. Under random mating and no selection, the rate of crossing and segregation will be balanced to maintain the proportion of free and potential variability at 50% : 50%. In other words, the population structure is maintained as a dynamic flow of crossing and segregation. However, with two loci under consideration, equilibrium will be attained slowly over many generations. If genetic linkage is strong, the rate of attainment of equilibrium will even be much slower.

Most of the important variation displayed by nearly all plant traits affecting growth, development and reproduction, is quantitative (continuous or polygenic variation; controlled by many genes). Polygenes demonstrate the same properties in terms of dominance, epistasis, and linkage as classical Mendelian genes. The Hardy–Weinberg equilibrium is applicable to these traits. However, it is more complex to demonstrate.

Another state of variability is observed when more than one gene affects the same polygenic trait. Consider two independent loci with two alleles each: *A, a* and *B, b*. Assume also the absence of dominance or epistasis. It can be shown that nine genotypes (*AABB, AABb, AaBb, Aabb, AaBB, AAbb, aaBb, aaBB, aabb*) and five phenotypes ([*AABB*, 2*AaBB*] + [2*AABb; AAbb, aaBB*] + [4*AaBb*, 2*aaBb*] + [2*Aabb*] + [*aabb*]), in a frequency of 1 : 4 : 6 : 4 : 1, will be produced following random mating. Again, the extreme genotypes (*AABB, aabb*) are the source of completely free variability. But *AAbb* and *aaBB*, phenotypically similar but contrasting genotypes, also contain latent variability. Termed homozygotic potential variability, it will be expressed in free state only when, through crossing, a heterozyote (*AaBb*) is produced, followed by segregation in the F_2. In other words, two generations will be required to release this potential variability in free state. Further, unlike the 50% : 50% ratio in the single locus example, only 1/8 of the variability is available for selection in the free state, the remainder existing as hidden in the heterozygotic or homozygotic potential states. A general mathematical relationship may be derived for any number (n) of genes as 1:n:n−1 of free:heterozygotic potential:homozygotic potential.

Another level of complexity may be factored in by considering dominance and non-allelic interactions (*AA = Aa = BB = Bb*). If this is so, the nine genotypes previously observed will produce only three phenotypic classes, [*AABB, 4AaBb, 2AaBB, 2AABb*] + [2*Aabb, 2aaBb, AAbb, aaBB*] + [*aabb*], in a frequency of 9 : 6 : 1. A key difference is that 50% of the visible variability is now in the heterozygous potential state that cannot be fixed by selection. The heterozygotes now contribute to the visible instead of the cryptic variability. From the plant breeding standpoint, its effect is to reduce the rate of response to phenotypic selection at least in the same direction as the dominance effect. This is because the fixable homozygotes are indistinguishable from the heterozygotes without a further breeding test (e.g., progeny row). Also, the classifications are skewed (9 : 6 : 1) in the positive (or negative) direction.

Key plant breeding information to be gained from the above discussion is that, in outbreeding populations, polygenic systems are capable of storing large amounts of cryptic variability. This can be gradually released for selection to act on through crossing, segregation, and recombination. The flow of this cryptic variability to the free state depends on the rate of recombination (which also depends on the linkage of genes on the chromosomes and the breeding system).

Given a recombination value of r between two linked genes, the segregation in the second generation depends on the initial cross, as M.D. Haywood and E.L. Breese demonstrated as follows:

Initial cross	Free	Homozygous potential
1. *AABB* × *aabb*	$1(1-r)$	$2r$
2. *AAbb* × *aaBB*	$2r$	$2(r-1)$

The second cross shows genes linked in the repulsion phase. The flow of variability from the homozygous potential to the free state depends on how tight a linkage exists between the genes. It will be at its maximum when $r = 0.5$ and recombination is free, and diminish with diminishing r. This illustration shows that with more than two closely linked loci on the same chromosome, the flow of variability would be greatly restricted. In species where selfing is the norm (or when a breeder enforces complete inbreeding) the proportion of heterozygotes will be reduced by 50% in each generation, dwindling to near zero by the eighth or ninth generation.

The open system of pollination in cross-pollinated species allows each plant in the gene pool to have both homozygous and heterozygous loci. Plant breeders exploit this heterozygous genetic structure of individuals in population improvement programs. In a natural environment, the four factors of genetic change mentioned previously are operational. Fitness or adaptive genes will be favored over non-adaptive ones. Plant breeders impose additional selection pressure to hasten the shift in the population genetic structure toward adaptiveness as well as increase the frequencies of other desirable genes.

An example of a breeding application of Hardy–Weinberg equilibrium

In disease resistance breeding, plant breeders cross an elite susceptible cultivar with one that has disease resistance. Consider a cross between two populations, susceptible x resistant. If the gene frequencies of an allele A in the two populations are represented by P_1 and P_2, the gene frequency in the $F_1 = (P_1 + P_2)/2 = p$. Assuming the frequency of the resistance gene in the resistant cultivar is $P_1 = 0.7$ and that in the susceptible elite cultivar is $P_2 = 0.05$, the gene frequency in the progeny of the cross p would be obtained as follows:

$$p = (P_1 + P_2)/2 = (0.7 + 0.05)/2 = 0.375.$$

Consequently, the gene frequency for the resistant trait is reduced by about 50% (from 0.7 to 0.375).

3.2 Issues arising from the Hardy–Weinberg equilibrium

For the Hardy–Weinberg equilibrium to be true, several conditions must be met. However, some situations provide approximate conditions to satisfy the requirements.

3.2.1 Issue of population size

The Hardy–Weinberg equilibrium requires a large random mating population (among other factors as previously indicated) to be true. However, in practice, the law has been found to be approximately true for most of the genes in most cross-pollinated species, except when non-random mating (e.g., inbreeding and assortative mating) occur. Whereas inbreeding is a natural feature of self-pollinated species, assortative mating can occur when cross-pollinated species are closely spaced in the field.

3.2.2 Issue of multiple loci

Research has shown that it is possible for alleles at two loci to be in random mating frequencies and yet not in equilibrium with respect to each other. Furthermore, equilibrium between two loci is not attained after one generation of random mating as the Hardy–Weinberg law concluded but is attained slowly over many generations. Also, the presence of genetic

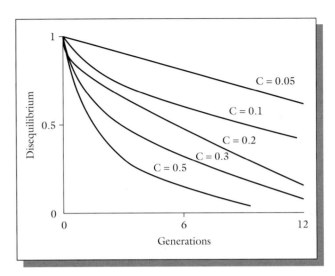

Figure 3.2 The approach to linkage equilibrium under random mating of two loci considered together. The value of c gives the linkage frequency between two loci. The effect of linkage is to slow down the rate of approach; the closer the linkage, the slower the rate. For $c = 0.5$, there is no linkage. The equilibrium value is approached slowly and is theoretically unattainable.

linkage will further slow down the rate of attainment of equilibrium (Figure 3.2). If there is no linkage ($c = 0.5$), the differential between actual frequency and the equilibrium frequency is reduced by 50% in each generation. At this rate, it would take about seven generations to reach approximate equilibrium. However, at $c = 0.01$ and $c = 0.001$, it would take about 69 and 693 generations, respectively, to reach equilibrium. A composite gene frequency can be calculated for genes at the two loci. For example, if the frequency at locus $Aa = 0.2$ and that for locus $bb = 0.7$, the composite frequency of a genotype $Aabb = 0.2 \times 0.7 = 0.14$.

3.3 Factors affecting changes in gene frequency

Gene frequency in a population may be changed by one of two primary types of processes – **systematic** or **dispersive**. A systematic process causes a change in gene frequency that is predictable in both direction and amount. A dispersive process, associated with small populations, is predictable only in amount, not direction. D.S. Falconer listed the systematic processes as **selection, migration,** and **mutation**.

3.3.1 Migration

Migration is important in small populations. It entails the entry of individuals into an existing population from outside. Because plants are sedentary, migration, when it occurs naturally, is via pollen transfer (gamete migration). The impact that this immigration will have on the recipient population will depend on the immigration rate and the difference in gene frequency between the immigrants and natives. Mathematically, $\Delta q = m(q_m - q_o)$, where $\Delta q =$ the changes in the frequency of genes in the new mixed population, $m =$ the number of immigrants, $q_m =$ the gene frequency of the immigrants, and $q_o =$ gene frequency of the host. Plant breeders employ this process to change frequencies when they undertake introgression of genes into their breeding populations. The breeding implication is that for open-pollinated (outbreeding) species, the frequency of the immigrant gene may be low, but its effect on the host gene and genotypes could be significant.

3.3.2 Mutation

Natural mutations are generally rare. A unique mutation (non-recurrent mutation) would have little impact on gene frequencies. Mutations are generally recessive in gene action, but the dominant condition may also be observed. Recurrent mutation (occurs repeatedly at a constant frequency) may impact gene frequency of the population. Natural mutations are of little importance to practical plant breeding. However, breeders may artificially induce mutation to generate new variability for plant breeding.

3.3.3 Selection

Selection is the most important process by which plant breeders alter population gene frequencies. Its effect is to change the mean value of the progeny population from that of the parental population. This change may be greater or lesser than the population mean, depending on the trait of interest. For example, breeders aim for higher yield but may accept and select for less of a chemical factor in the plant that may be toxic in addition to the high yield. For selection to succeed:

1 there must be phenotypic variation for the trait to allow differences between genotypes to be observed;
2 the phenotypic variation must at least be partly genetic.

3.4 Frequency dependent selection

Selection basically concerns the differential rate of reproduction by different genotypes in a population. The concept of fitness describes the absolute or relative reproductive rate of genotypes. The contribution of genotypes to the next generation is called the **fitness** (or **adaptive value** or **selective value**). The relative fitness of genotypes in a population may depend on its frequency relative to others. Selection occurs at different levels in the plant – phenotype, genotype, zygote, and gamete – making it possible to distinguish between haploid and diploid selections. The **coefficient of selection** is designated s, and has values between zero and one. Generally, the contribution of a favorable genotype is given a score of one, while a less favorable (less fit) genotype is scored $1 - s$.

An $s = 0.1$ means that for every 100 zygotes produced with the favorable genotype there will be 90 individuals with the unfavorable genotype. Fitness can exhibit complete dominance, partial dominance, no dominance, or overdominance. Consider a case of complete dominance of the A allele. The relative fitness of genotypes will be:

Genotypes	AA	Aa	aa	Total
Initial frequency	p^2	$2pq$	q^2	1
Relative fitness	1	1	$1 - s$	
After selecting	p^2	$2pq$	$q^2(1 - s)$	$1 - sq^2$

The total after selection is given by:

$$p^2 + 2pq + q^2(1 - s)$$
$$= (1 - q)(1 - q) + 2(1 - q)q + q^2 - sq^2$$
$$= 1 - 2q + q^2 + 2q - 2q^2 + q^2 - sq^2 + 1 - sq^2$$

To obtain the gene frequency in the next generation, use:

$$Q = (^1/_2 H + Q)/N$$
$$= [pq + q^2(1 - s)]/1 - sq^2$$

where $p = 1 - q$, and multiply $(1 - s)$ by q^2:

$$q_1 = [q(1 - q) + q^2 - sq^2]/1 - sq$$
$$= [q - q^2 + q^2 - sq^2]/1 - sq^2$$
$$= (q - sq^2)/1 - sq^2$$
$$= [q(1 - sq)]/1 - sq^2$$

The relationship between any two generations may be generalized as:

$$q(n + 1) = [q_n(1 - sq_n)]/1 - sq_n^2$$

Similarly, the difference in gene frequency, Δq, between any two generations can be shown to be:

$$\Delta q = q_1 - q$$
$$= [sq^2(1 - q)]/1 - sq^2$$

Other scenarios of change in gene frequency are possible.

Plant breeders use artificial selection to impose new fitness values on genes that control traits of interest in a breeding program.

3.5 Summary of key plant breeding applications

- Selection is most effective at intermediate gene frequency ($q = 0.5$) and least effective at very large or very small frequencies ($q = 0.99$ or $q = 0.01$). Furthermore, selection for or against a rare allele is ineffective. This is so because a rare allele in a population will invariably occur in the heterozygote and be protected (heterozygote advantage).
- Migration increases variation of a population. Variation of a population can be expanded in a breeding program through introductions (impact of germplasm). Migration also minimizes the effects of inbreeding.
- In the absence of the other factors or processes, any one of the frequency altering forces will eventually lead to fixation of one allele or the other.
- The forces that alter gene frequencies are usually balanced against each other (e.g., mutation to a deleterious allele is balanced by selection).
- Gene frequencies attain stable values called equilibrium points.
- In both natural and breeding populations, there appears to be a selective advantage for the heterozygote (hybrid). Alleles with low selection pressure may persist in the population in the heterozygote state for many generations.
- As population size decreases, the effect of random drift increases. This effect is of importance in germplasm collection and maintenance. The original collection can be genetically changed if a small sample is taken for growing to maintain the accession.

3.6 Modes of selection

There are three basic forms of selection – **stabilizing, disruptive,** and **directional** – the last form being the one of most concern to plant breeders. These forms of selection operate to varying degrees under both natural and artificial selection. A key difference lies in the goal. In natural selection, the goal is to increase the fitness of the species, whereas in plant breeding, breeders impose artificial selection usually to direct the population toward a specific goal (not necessarily the fittest).

3.6.1 Stabilizing selection

Selection as a process is ongoing in nature. Regarding traits that directly affect the fitness of a plant (i.e., viability, fertility), selection will always be directionally toward optimal phenotype for a given habitat. However, for other traits, once optimal phenotype has been attained, selection will act to perpetuate it as long as the habitat remains stable. Selection will be for the population mean and against either extreme expression of the phenotype. This mode of selection is called **stabilizing selection** (or also called balancing or optimum selection). Taking flowering as an example, stabilizing selection will favor neither early flowering nor late flowering. In terms of genetic architecture, dominance will be low or absent or ambidirectional, whereas epistasis will not generally be present. Stabilizing selection promotes additive variation.

3.6.2 Disruptive selection

Natural habitats are generally not homogeneous but consist of a number of "ecological niches" that are distinguishable in time (seasonal or long term cycles), space (microniches), or function. These diverse ecological conditions favor diverse phenotypic optima in form and function. Disruptive selection is a mode of selection in which extreme variants have higher adaptive value than those around the average mean value. Hence, it promotes diversity (polymorphism). The question then is how the different optima relate (dependent or independent) for maintenance and functioning. Also, at what rate does gene exchange occur between the differentially selected genotypes? These two factors (functional relationship and rate of gene exchange) determine the effect of genetic structure of a population. In humans, for example, a polymorphism that occurs is sex (female and male). The two sexes are 100% interdependent in reproduction (gene exchange is 100%). In plants, self-incompatibility is an example of such genetically controlled polymorphism. The rarer the self-incompatibility allele at a locus, the higher the chance of compatible mating (and vice versa). Such frequency dependent selection is capable of building up a large number of self-incompatibility alleles in a population. As previously indicated, several hundreds of alleles have been found in some species.

3.6.3 Directional selection

Plant breeders, as previously stated, impose **directional selection** to change existing populations or varieties (or other genotypes) in a predetermined way. Artificial selection is imposed on the targeted trait(s) to achieve maximal or optimal expression. To achieve this, the breeder employs techniques (crossing) to reorganize the genes form the parents in a new genetic matrix (by recombination), assembling "co-adapted" gene complexes to produce a fully balanced phenotype, which is then protected from further change by genetic linkage. The breeding system will determine whether the newly constituted gene combinations will be maintained. Whereas inbreeding (e.g., selfing) would produce a homozygous population that will resist further change (until crossed), outbreeding tends to produce heterozygous combinations. In heterozygous populations, alleles that exhibit dominance in the direction of expression targeted by the breeder will be favored over other alleles. Hence, directional selection leads to the establishment of dominance and/or genic interaction (episitasis).

3.7 Effect of mating system on selection

Four mating systems are generally recognized. They may be grouped into two broad categories as **random mating** and **non-random mating** (comprising **genetic assortative mating, phenotypic assortative mating,** and **disassortative mating**).

3.7.1 Random mating

In plants, **random mating** occurs when each female gamete has an equal chance of being fertilized by any male gamete of the same plant or with any other plant of the population and, furthermore, there is an equal chance for seed production. As can be seen from the previous statement, it is not possible to achieve true random mating in plant breeding because selection is involved. Consequently, it is more realistic to describe the system of mating as random mating with selection. Whereas true random mating does not change gene frequencies, existing variability in the population, or genetic correlation between close relatives, random mating with selection changes gene frequencies and the mean of the population, with little or no effect on homozygosity, population variance, or genetic correlation between close relatives in a large population. Small populations are prone to random fluctuation in gene frequency (genetic drift) and inbreeding, factors that reduce heterozygosity in a population. Random mating does not fix genes, with or without selection. If the goal of the breeder is to preserve desirable alleles (e.g., in germplasm composites), random mating will be an effective method of breeding.

3.7.2 Non-random mating

Non-random mating has two basic forms: (i) mating occurs between individuals that are related to each other by ancestral descent (promotes an increase in homozygosity at all loci), and (ii) individuals mate preferentially with respect to their genotypes at any particular locus of interest. If mating occurs such that the mating pair has the same phenotype more often than would occur by chance, it is said to be **assortative mating**. The reverse is true in **disassortative mating**, which occurs in species with self-incompatibility or sterility problems, promoting heterozygosity.

Genetic assortative mating

Genetic assortative mating or inbreeding entails mating individuals that are closely related by ancestry, the closest being selfing (self-fertilization). A genetic consequence of genetic assortative mating is the exposure of cryptic genetic variability that was inaccessible to selection and was being protected by heterozygosity (i.e., heterozygous advantage). Also, repeated selfing results in homozygosity and brings about fixation of types. This mating system is effective if the goal of the breeder is to develop homozygous lines (e.g., developing inbred lines for hybrid seed breeding or development of synthetics).

Pheotypic assortative mating

Mating may also be done on the basis of phenotypic resemblance. Called **phenotypic assortative mating**, the breeder selects and mates individuals on the basis of their resemblance to each other compared to the rest of the population. The effect of this action is the development of two extreme phenotypes. A breeder may choose this mating system if the goal is to develop an extreme phenotype.

Disassortative mating

Disassortative mating may be genetic or phenotypic. Genetic disassortative mating entails mating individuals that are less closely related than they would under random mating. A breeder may use this system to cross different strains. In phenotypic disassortative mating, the breeder may select individuals with contrasting phenotypes for mating. Phenotypic disassortative mating is a conservative mating system that may be used to maintain genetic diversity in the germplasm from which the breeder may obtain desirable genes for breeding as needed. It maintains heterozygosity in the population and reduces genetic correlation between relatives.

3.8 The concept of inbreeding

As previously indicated, plant breeding is a special case of evolution, whereby a mixture of natural and, especially, artificial selection operates rather than natural selection alone. The Hardy–Weinberg equilibrium is not satisfied in plant breeding because of factors including non-random mating. Outcrossing promotes random mating, but breeding methods impose certain mating schemes that encourage non-random mating, especially inbreeding. Inbreeding is measured by the **coefficient of inbreeding (F)**, which is the probability of identity of alleles by descent. The range of F is zero (no inbreeding; random mating) to one (prolonged selfing). It can be shown mathematically that:

$$[p^2(1 - F) + Fp] : [2pq(1 - F)] : [q^2(1 - F) + Fq]$$

If $F = 0$, then the equation reduces to the familiar $p^2 + 2pq + q^2$. However, if $F = 1$, it becomes $p : 0 : q$. The results show that any inbreeding leads to **homozygosis** (all or nearly all loci homozygous), extreme inbreeding leading to a complete absence of **heterozygosis** (all or nearly all loci heterozygous).

Differential fitness is a factor that mitigates against the realization of the Hardy–Weinberg equilibrium. According to Darwin, the more progeny left, on average, by a genotype in relation to the progeny left by other genotypes, the fitter it is. It can be shown that the persistence of alleles in the population depends on whether they are dominant, intermediate or recessive in gene action. An unfit (deleterious) recessive allele is fairly quickly reduced in frequency but declines slowly thereafter. On the other hand, an unfit dominant allele is rapidly eliminated from the population, while an intermediate allele is reduced more rapidly than a recessive allele because the former is open to selection in the heterozygote. The consequence of these outcomes is that unfit dominant or intermediate alleles are rare in cross-breeding populations, while unfit recessive alleles persist because they are protected by their recessiveness. The point that will be made later but is worth noting here is that inbreeding exposes unfit recessive alleles (they become homozygous and are expressed) to selection and potential elimination from the population. It follows that inbreeding will expose any unfit allele, dominant or recessive. Consequently, species that are inbreeding would have opportunity to purge out unfit alleles and hence carry less genetic unfitness load (i.e., have more allele fitness) than outcrossing species. Furthermore, inbreeders (self-pollinated species) are more tolerant of inbreeding while outcrossing species are intolerant of inbreeding.

Whereas outcrosing species have more heterozygous loci and carry more unfitness load, there are cases in which the heterozygote is fitter than either homozygote. Called **overdominance**, this phenomenon is exploited in hybrid breeding (Chapter 18).

3.9 Inbreeding and its implications in plant breeding

The point has already been made that the methods used by plant breeders depend on the natural means of reproduction of the species. This is because each method of reproduction has certain genetic consequences. In Figure 3.3a there is no inbreeding because there is no common ancestral pathway to the individual, A (i.e., all parents are different). However, in Figure 3.3b inbreeding exists because B and C have common parents (D and E), that is, they are full sibs. To calculate the amount of inbreeding, the standard pedigree is converted to an arrow diagram (Figure 3.3c). Each individual contributes $\frac{1}{2}$ of its genotype to its offspring. The coefficient of relationship (R) is calculated by summing up all the pathways between two individuals through a common ancestor as: $R_{BC} = \Sigma(\frac{1}{2})^s$, where s is the number of steps (arrows) from B to the common ancestor and back to C. For example, B and C probably inherited $(\frac{1}{2})(\frac{1}{2}) = \frac{1}{4}$ of their genes in common through ancestor D. Similarly, B and C probably inherited $\frac{1}{4}$ of their genes in common through ancestor E. The coefficient of relationship between B and C, as a result of common ancestry, is hence $R_{BC} = \frac{1}{4} + \frac{1}{4} = \frac{1}{2} = 50\%$. Other more complex pedigrees are shown in Figure 3.4.

As previously indicated, prolonged selfing is the most extreme form of inbreeding. With each selfing, the heterozygosity decreases by 50%, whereas the homozygosity increases by 50% from the previous generation. The approach to homozygosity depends on the intensity of inbreeding as illustrated in Figure 3.5. The more distant the relationship between parents, the slower is the approach to homozygosity. The coefficient of inbreeding (F), previously discussed, measures the probability of identity of alleles by descent. This can be measured at both the individual level as well as the population level. At the individual level, F measures the probability that any two alleles at any locus are identical by descent (i.e., they are both products of a gene present in a common ancestor) At the population level, F measures the percentage of all loci which were heterozygous in the base population but have now probably become homozygous due to the effects of inbreeding. There are several methods used for calculating F. The coefficient of inbreeding (Fx) of an individual may be obtained by counting the number of arrows (n) that connect the individual through one parent back to the common ancestor and back again to the other parent, and using the mathematical expression:

$$Fx = \sum (^1/_2)^n (1 + F_A)$$

3.9.1 Consequences

The genetic consequences of inbreeding were alluded to in a previous section. The tendency towards homozygosity with inbreeding provides an opportunity for recessive alleles to be homozygous and, hence, expressed. Whereas inbreeding generally has little or no adverse effect in inbred species, cross-bred species suffer adverse consequences when the recessive alleles are less favorable than the dominant alleles. Called **inbreeding depression**, it is manifested as a reduction in performance, because of the expression of less fit or deleterious alleles. The severity of inbreeding depression varies among species, being extreme in species such as alfalfa in which inbreeding produces homozygous plants that fail to survive. Furthermore, the effect of inbreeding is most significant in the first 5–8 generations and negligible after the eighth generation in many cases.

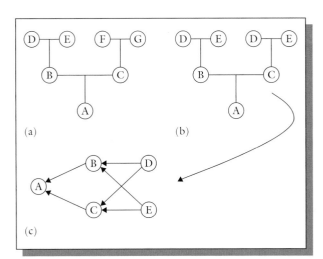

Figure 3.3 Pedigree diagrams can be drawn in the standard form (a, b) or converted to into an arrow diagram (c).

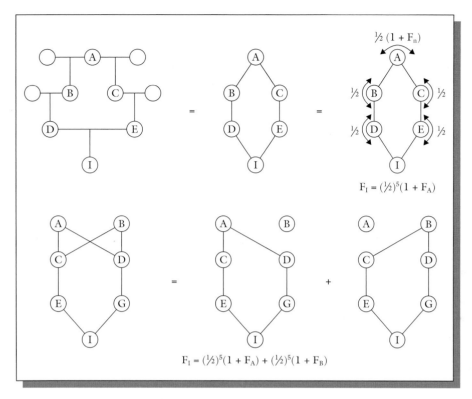

Figure 3.4 The inbreeding coefficient (F) may be calculated by counting the number of arrows that connect the individual through one parent back to the common ancestor and back again to the other parent and applying the formula shown in the figure.

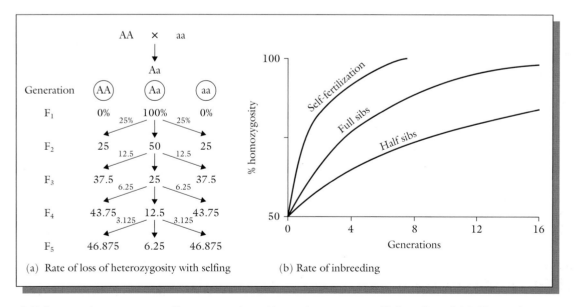

(a) Rate of loss of heterozygosity with selfing   (b) Rate of inbreeding

Figure 3.5 Increase in percentage of homozygosity under various systems of inbreeding. (a) Selfing reduces heterozygosity by 50% of what existed at the previous generation. (b) The approach to homozygosity is most rapid under self-fertilization.

3.9.2 Applications

Inbreeding is desirable in some breeding programs. Inbred cultivars of self-pollinated species retain their genotype through years of production. In cross-pollinated species, inbred lines are deliberately developed for use as parents in hybrid seed production. Similarly, partially inbred lines are used as parents in the breeding of synthetic cultivars and vegetatively propagated species by reducing the genetic load. Another advantage of inbreeding is that it increases the genetic diversity among individuals in a population, thereby facilitating the selection process in a breeding program.

3.9.3 Mating systems that promote inbreeding

Mating is a way by which plant breeders impact the gene frequencies in a population. Four **mating systems** are commonly used to effect inbreeding – self-fertilization, full-sib mating, half-sib mating, and backcrossing. Self-fertilization is the union of male and female gametes; full-sib mating involves the crossing of pairs of plants from a population. In half-sib mating the pollen source is random from the population, but the female plants are identifiable. In a backcross the F_1 is repeatedly crossed to one of the parents. Self-fertilization and backcrossing are the most extreme forms of inbreeding attaining a coefficient of inbreeding (F) of 15/16 after four generations of mating. Autopolyploids have multiple alleles and hence can accumulate more deleterious alleles that remain masked. Inbreeding depression is usually more severe in autopolyploids than diploid species. However, the progression to homozygosity is much slower in autopolyploids than in diploids.

3.10 Concept of population improvement

The general goal of improving open or cross-pollinated species is to change the gene frequencies in the population towards fixation of favorable alleles while maintaining a high degree of heterozygosity. Unlike self-pollinated species, in which individuals are the focus and homozygosity and homogeneity are desired outcomes of breeding, population improvement focuses on the whole group, not individual plants. Consequently, open-pollinated populations are not homogeneous.

3.11 Types

The population can be changed by one of two general strategies (i.e., there are two basic types of open-pollinated populations in plant breeding) – by **population improvement** and by development of **synthetic cultivars**. To develop cultivars by population improvement entails changing the population *en masse* by implementing a specific selection tactic. A cultivar developed this way is sustainable in a sense, maintaining its identity indefinitely through random mating within itself in isolation. The terminology "synthetic" is used to denote an open-pollinated cultivar developed from combining inbred or clonal parental lines. However, the cultivar is not sustainable and must be reconstituted from parental stock. Other usage of the term occurs in the literature.

3.11.1 Methods of population improvement

Some form of evaluation precedes selection. A breeding material is selected after evaluating the variability available. Similarly, advancing plants from one generation to the next is preceded by an evaluation to determine individuals to select. In self-pollinated species, individuals are homozygous and when used in a cross their genotype is precisely reproduced in their progeny. Hence, a **progeny test** is adequate for evaluating an individual's performance. However, open-pollinated species are heterozygous plants and are further pollinated by other heterozygous plants growing with them in the field. Progeny testing is thus not adequately evaluative of the performance of individual plants of such species. A more accurate evaluation of performance may be achieved by using pollen (preferably from a homozygous source – inbred line) to pollinate the plants. As previously described, the method of evaluating the performance of different mother plants in a comparative way using a common pollen source (tester line) is called a test cross. The objective of such a test is to evaluate the performance of a parent in a cross, a concept called combining ability.

The methods used by plant breeders in population improvement may be categorized into two groups, based on the process for evaluating performance. One group of methods is based solely on phenotypic selection and the other on progeny testing (genotypic selection). The specific methods include mass selection, half-sib, full-sib, recurrent selection, and synthetics.

Industry highlights
Introgression breeding on tomatoes for resistance to powdery mildew

Yuling Bai

Wageningen UR Plant Breeding, Droevendaalsesteeg 1, 6708 PB Wageningen, The Netherlands

Tomato and its wild relatives

Tomato (*Solanum lycopersicum*) is a very important vegetable both for the fresh market and for the processed food industry. Although cultivated as an annual, tomato grows as a perennial in its original habitat in Peru (Picken *et al.*, 1985). The original site of domestication of tomato is likely in Mexico (Taylor, 1986).

According the recent classification, tomato belongs to section *Lycopersicon* and has 12 wild relatives (Table B3.1). Of these 12 relatives, nine (numbers 1 to 9 in Table B3.1) are previously defined in the genus of *Lycopersicon* (referred to as

Table B3.1 Old and new names of tomato and its wild relatives.

No.	New *Solanum* name	*Lycopersicon* equivalent	Fruit color	Self-compatibility	Ability to be crossed with other *Solanum* species	Section name within *Solanum*
1	*Solanum lycopersicum*	*Lycopersicon esculentum*	Red	Self-compatible	Old "*esculentum*" group, crossable among these species, although it is sometimes only possible to make crosses in one direction.	Section *Lycopersicon*
2	*S. pimpinellifolium*	*L. pimpinellifolium*	Red			
3	*S. cheesmaniae*	*L. cheesmaniae*	Yellow			
4	*S. chmielewskii*	*L. chmeilewskii*	Green			
5	*S. neorickii*	*L. parviflorum*	Green			
6	*S. habrochaites*	*L. hirsutum*	Green	Self-incompatible		
7	*S. pennellii*	*L. pennellii*	Green			
8	*S. chilense*	*L. chilense*	Green		Old "*peruvianum*" group, crossable between these two species, but difficult to cross them with cultivated tomato and embryo rescue is often needed.	
9	*S. peruvianum*	*L. peruvianum*	Green			
10	*S. lycopersicoides*	*L. lycopersicoides*	Green-black		Most closely related to old *Lycopersicon* species and crossable to *S. lycopersicum*, *S. cheesmaniae*, *S. pimpinellifolium* and *S. pennellii*.	
11	*S. sitiens*	*L. sitiens*	Green		Also known as *S. rickii*, crossable with *S. lycopersicoides*	
12	*S. ochranthum*	*L. ochranthum*	Green		Unknown crossability with other *Solanum* species.	Section *Juglandifolium*
13	*S. juglandifolium*	*L. juglandifolium*	Green		Unknown crossability with other *Solanum* species.	

old *Lycopersicon* species). Accessions of nearly all these nine species have been successfully used to introduce valuable traits for crop improvement, especially monogenic sources conferring resistance to fungal, nematode, bacterial and viral diseases. The phelogenic relation of these old *Lycopersicon* species with cultivated tomato has been extensively studied, based on comparative analysis of morphology, self-compatibility, crossability, and molecular markers. The classical taxonomic traits which have been used to divide old *Lycopersicon* species are fruit color and self-compatibility. In the phylogeny generated with molecular markers, different patterns of species relationships have been obtained, some are congruent with results of classical taxonomy and others add resolution to new divisions that are not always in agreement. In general, we could conclude the following (i) species (*S. lycopersicum*, *S. lycopersicum* var *cerasiforme*, *S. cheesmaniae*, *S. pimpinellifolium*) with self-compatibility and red fruits are most closely related, (ii) *S. peruvianum* and *S. chilense* (green fruits and self-incompatible) are closely related species, and (iii) species with green fruits, including *S. chmielewskii*, *S. neorikii*, *S. habrochaites* and *S. pennellii*, have varied relationships with the rest depending on markers used for phylogeny.

Introgression breeding

Wild tomatoes have large genetic diversity, especially within the self-incompatible species like *S. chilense* and *S. peruvianum* (Rick, 1986). Tremendous variation has been revealed by molecular markers and it is striking that more genetic variation was observed within a single accession of the self-incompatible species than in all accessions of any of the self-compatible species (Egashira *et al.*, 2000). Compared to the rich reservoir in wild species, the cultivated tomato is genetically poor due to the inbreeding during tomato domestication. It is estimated that the genomes of tomato cultivars contain less than 5% of the genetic variation of their wild relatives (Miller and Tanksley, 1990). The lack of diversity in the cultivated tomato can be visualized using DNA technologies. Very few polymorphisms within the cultivated tomato gene pool are identified. Tomato domestication experienced severe genetic bottleneck as the crop was carried from the Andes to Central America and from there to Europe. The initial domestication process was, in part, reached by selecting preferred genotypes in the existing germplasm. In a predominantly inbreeding species, genetic variation tends to decrease, even without selection. As a consequence, genetic drift is a major process that reduces genetic variation.

It is most likely that no exchange of genetic information with the wild germplasm took place until the 1940. By then, the renowned geneticist and plant breeder Charlie Rick (University of California, Davis) observed that crosses between wild and cultivated species generated a wild array of novel genetic variation in the offspring. Since then, breeding from wild species via interspecific crosses followed by many backcrosses to cultivated tomatoes (so called introgression breeding) has led to the transfer of many favorable attributes in the cultivated tomato. Breeding barriers are sometimes expected in interspecific crosses, which include unilateral incompatibility, hybrid inviability, sterility, reduced recombination and linkage drag.

An example of introgression breeding

One of the common breeding objectives in tomato is breeding for resistance to the most destructive pests and pathogens. Tomato hosts more than 200 species of a wide variety of pests and pathogens that can cause significant economic losses. Tomato powdery mildew caused by *Oidium neolycopersici* occurred for the first time in 1986 in The Netherlands (Paternotte, 1988). It has since then spread within 10 years to all European countries and is nowadays a worldwide disease on tomato, except for Australia where another species (*O. lycopersici*) occurs (Kiss *et al.*, 2001). Upon the outbreak of *O. neolycopersici*, all tomato cultivars were susceptible and this fungus was the only one to be controlled by fungicides in greenhouse tomato production in Northwest Europe (Huang *et al.*, 2000). By 1996, our group was invited by Dutch vegetable seed companies to search for resistance genes against *O. neolycopersici*. Here, our practice on breeding tomatoes with resistance to powdery mildew is used as an example for introgression breeding.

1. *Search for resistance in wild relatives of tomato.* As the consequence of inbreeding during tomato domestication, the genetic diversity in cultivated tomato is now very narrow. However, large variation is present and exploitable in the wild *Solanum* species. Thus, the first step is to find wild tomato accessions with resistance to tomato powdery mildew.

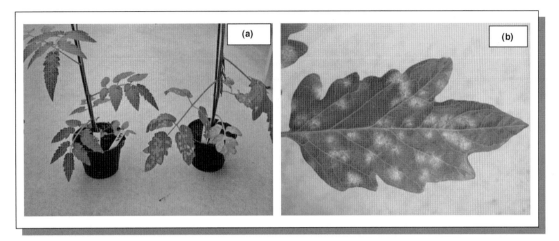

Figure B3.1 Tomato plants inocluated with tomato powdery mildew *(Oidium neolycopersici)*. (a) The plant on the left is from tomato wild species *Solanum pervianum* LA2172, showing no powdery midew infection; the plant on the right is from *S. lycoerpsicum* cv. Moneymaker, showing fungal clonies growing on infected leaves. (b) A closer look at the colonization of tomato powdery mildew growing on the upper side of the leaf. Pictures were taken 15 days post inoculation. Figure courtesy of Yuling Bai.

In the Tomato Genetics Resource Center (TGRC) in Davis, California (http://tgrc.ucdavis.edu/) and the Botanical and Experimental Garden (http://www.bgard.science.ru.nl/) in the Netherlands, thousands of accessions of the wild *Solanum* species have been collected and maintained. From these collections, we selected and tested some *Solanum* species with tomato powdery mildew. As expected, many wild accessions showed resistance (Figure B3.1).

2. *Inheritance of resistance.* Monogenic resistance is most exploited in tomato breeding programs. Modern tomato cultivars may harbor resistance to more than 10 pathogens. Thus, the second step is to study the inheritance of the resistance identified in the wild tomato species. For this purpose, resistant plants were selected and crossed to a susceptible cultivar, *S. lycopersicum* cv. Moneymaker to produce populations (usually F_2 populations) for inheritance study. Crosses between *S. lycopersicum* and wild tomato species can be easy but sometimes require strategies such as embryo rescue, especially for the self-incompatible species like *S. peruvianum*.

By using F_2 populations, inheritance of resistance identified in several wild species was characterized. Monogenic resistance to *O. neolycopersici* was found in *S. peruvianum* LA2172, *S. habrochaites* G1.1560 and G1.1290, and polygenic resistance in *S. neorikii* G.16101. Furthermore, by screening these F_2 plants with molecular markers, such as RAPD, AFLP and CAPS, the resistance in these species was mapped onto specific chromosomes. The resistance loci in *S. peruvianum* LA2172, *S. habrochaites* G1.1560 and G1.1290, named *Ol-4*, *Ol-1* and *Ol-3*, respectively, are all located on tomato chromosome 6. The *Ol-4* locus is on the short arm, while, *Ol-1* and *Ol-3* are on the long arm and closely linked if not allelic (Figure B3.2). In addition to these monogenic *Ol*-genes, three quantitative trait loci (QTLs) were identified governing the resistance in *S. neorickii* G1.1601. The *Ol-qtl1* interval overlaps with *Ol-1* and *Ol-3*, while the other two linked *Ol-qtls* are located on chromosome 12 in the vicinity of the *Lv* locus that confers resistance to another powdery mildew species, *Leveillula taurica*. Markers with close linkage to these loci were generated and can be applied in marker assisted selection in breeding programs.

3. *Generation of near isogenic lines.* Near-isogenic lines (NILs) that carry small introgressed chromosome fragments from related wild species in a cultivated tomato background are most useful pre-bred in a breeding program. To develop NILs that only differ in *Ol* genes for resistance to *O. neolycopersici*, resistant donor accessions were crossed with susceptible *S. lycopersicum* cv. Moneymaker (MM). BC crosses were made starting from crossing F_1 plants back to MM (Figure B3.3). During the backcrossing, selection of resistant

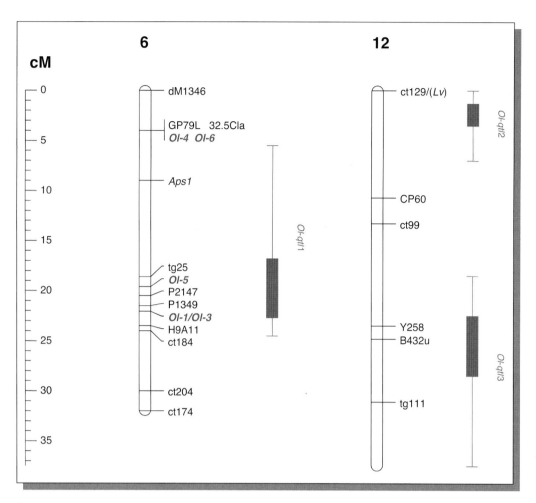

Figure B3.2 The chromosome locations of tomato loci for resistance to tomato powdery mildew caused by *Oidium neolycopersici*. On the left, genetic distance in cM is shown. On the right, map positions of markers and resistance loci are shown on tomato chromosome 6 and 12, respectively. The donors for *Ol-1*, *Ol-3*, *Ol-5* are *Solanum habrochaites* G1. 1560, G1. 1290, PI247087, respectively; for *Ol-4* is *S. peruvianum* LA2172 and for *Ol-qtls* are *S. neorickii* G1.1601. *Ol-6* is identified from an advanced breeding line with unknown source. As to *Ol-qtls*, bars indicate the QTL interval for which the inner bar shows a one-LOD support and the outer one shows a two-LOD support interval. Figure courtesy of Yuling Bai.

BC plants can be performed in two ways. One is by testing BC plants with powdery mildew and the other is to genotype them with markers linked to individual resistant loci. Since we have markers linked to each *Ol*-genes, we could apply marker assisted selection (MAS). We carried out the disease test because (i) it is a relatively easy disease assay which can be carried out at seedling stage and (ii) the resistance phenotype is clear to be scored. In the case that the disease assay is not easy to perform, for example due to (i) quarantine

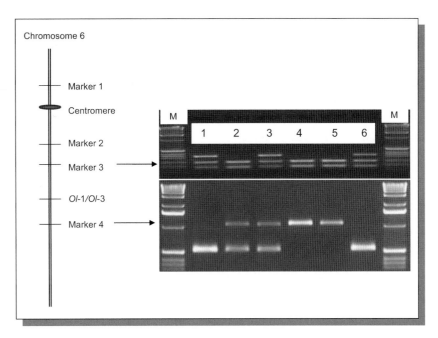

Figure B3.3 Illustration of marker assisted selection (MAS). On the left, a genetic linkage map of tomato chromosome 6 showing that the *Ol-1* and *Ol-3* genes, conferring resistance to tomato powdery mildew, are located at the same locus and are flanked by Markers 3 and 4. On the right, electrophoretic patterns of PCR markers showing marker genotypes of six plants; the upper panel for Marker 3 and the lower panel for Marker 4. Plants 1–4 are either BC_3 plants (for Marker 3) and BC_3S_1 plants (for Marker 4). Plants 5 and 6 are parental plants that are susceptible and resistant to tomato powdery mildew, respectively. M indicates DNA size marker of 1kb ladder. For MAS, marker flanking the target gene is often used. For Marker 3, BC_3 plants 1 and 3 are selected and expected to be resistant to powdery mildew since they have the marker allele of the resistant parent (plant 6). For Marker 4, plants 1–3 are selected and expected to be resistant since they have the resistant marker allele as the resistant parental plant 6 (plant 1 is homozygous and plants 2 and 3 are heterozygous). Figure courtesy of Yuling Bai.

pathogens or (ii) if the disease test has to be performed at the late developmental stage, MAS would be a convenient way to select resistant plants (Figure B3.4).

　　After several backcrossing generations, homozygous BC_nS_1 resistant plants of these crosses were selected (Figure B3.3). Since we have facilities for genome-wide analysis, we genotyped all selected plants with an AFLP marker to compare their genetic background with the recurrent parent MM. BC_nS_1 resistant plants that were genetically most similar to MM were maintained as NILs.

4.　*Releasing NILs to companies for producing of resistant cultivars.* The NILs harboring dominant *Ol* genes are valuable advanced breeding lines and have been used by seed companies for breeding tomato cultivars with resistance to tomato powdery mildew; they are nowadays available in the market. The NILs for the *Ol-qtls* are still being developed via marker assisted selection.

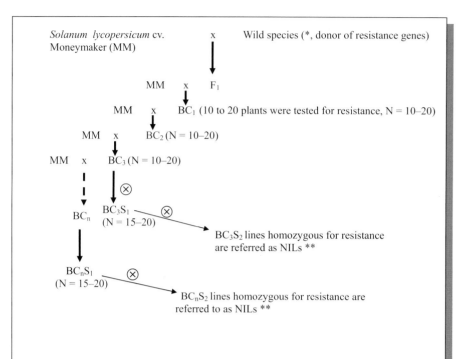

Solanum lycopersicum cv. Moneymaker (MM)   x   Wild species (*, donor of resistance genes)

MM x F_1

MM x BC_1 (10 to 20 plants were tested for resistance, N = 10–20)

MM x BC_2 (N = 10–20)

MM x BC_3 (N = 10–20)

BC_n BC_3S_1 (N = 15–20)

BC_3S_2 lines homozygous for resistance are referred as NILs **

BC_nS_1 (N = 15–20)

BC_nS_2 lines homozygous for resistance are referred to as NILs **

*, ** Usually it takes many generations to remove the deleterious genes that go along with the introduced genes due to linkage drag. Therefore, it is useful to start with advanced breeding lines having introgression from the wild species in order to shorten the backcrossing procedure. In our practice, we used advanced breeding lines derived from *S. habrochaites* G1.1560 (donor of the *Ol-1* gene) to produce the F_1. We checked BC_3S_1 plants for the uniformity of their genetic background by genotyping these plants with 12 AFLP primer combinations that produce genome-wide markers. Of the 30 AFLP marker alleles specific for *S. habrochaites* G1.1560, there are only present in the BC_3S_1 plants two that are fully cosegregating with the *Ol-1* gene, suggesting that the genetic background of these BC_3S_1 plants is genetically similar to MM. For the *Ol-4* genes derived from *S. peruvianum* LA2172, we started with the wild accession. With 12 AFLP primer combinations, 48 AFLP marker alleles were identified from *S. peruvianum* LA2172 and 11 of these alleles still segregated in the tested BC_3S_1 plants. Thus, when the backcross is started from wild accessions, more backcross generations are needed to make NILs.

Figure B3.4 Cross-pollinating scheme on the generation of near-isogenic lines (NIL) harbouring dominant resistance genes to tomato powdery mildew. Via backcrosses, new traits are introduced from wild tomato relatives. During the backcrosses, selection of resistant plants can be performed via (i) disease test and/or (ii) marker assisted selection (MAS, Figure B3.4). In addition, selected plants should have a similar morphology to the recurrent parent. Figure courtesy of Yuling Bai.

References

Egashira, H., Ishihara, H., Takshina, T., and Imanishi, S. (2000). Genetic diversity of the 'peruvianum-complex' (*Lycopersicon peruvianum* (L.) Mill. and *L. chilense* Dun.) revealed by RAPD analysis. *Euphytica*, **116**:23–31.

Huang, C.C., Van de Putte, P.M., Haanstra-van der Meer, J.G., Meijer-Dekens, F., and Lindhout, P. (2000). Characterization and mapping of resistance to *Oidium lycopersicum* in two *Lycopersicon hirsutum* accessions: Evidence for close linkage of two *Ol*-genes on chromosome 6. *Heredity*, **85**:511–520.

Kiss, L., Cook, R.T.A., Saenz, G.S., *et al.* (2001). Identification of two powdery mildew fungi, *Oidium neolycopersici* sp. nov. and *O. lycopersici*, infecting tomato in different parts of the world. *Mycological Research*, **105**:684–697.

Miller, J.C., and Tanksley, S.D. (1990). RFLP analysis of phylogenetic relationships and genetic variation in the genus *Lycopersicon*. *Theoretical and Applied Genetics*, **80**:437–448.

Paternotte, S.J. (1988). Echte meeldauw in tomaat geen echte bedreiging. *Groenten en Fruit*, **43**:30–31.

Picken, A.J., Hurd, R.G., and Vince-Prue, D. (1985). *Lycopersicon esculentum*, in *Handbook of flowering* (ed. A.H. Halevy). CRC Press, Boca Raton, FL, pp. 330–346.

Rick, C.M. (1986). Reproductive isolation in the *Lycopersicon peruvianum* complex, in *Solanaceae, Biology and Systematics* (ed. W.G. D'Arcy). Columbia University Press, New York, pp. 477–495.

Taylor, I.B. (1986). Biosystematics of the tomato, in *The tomato crop: a scientific basis for improvement* (eds J.G. Atherton and J. Rudich). Chapman and Hall, London, pp. 1–34.

Key references and suggested reading

Ayala, F.J., and Campbell, C.A. (1974). Frequency-dependent selection. *Ann. Ecology and Systematics*, **5**:115–138.

Cornelius, P.L., and Dudley, J.W. (1974). Effects of inbreeding by selfing and full-sib mating in a maize population. *Crop Science*, **14**:815–819.

Crow, J.F., and Kimura, M. (1970). *An introduction to population genetics theory*. Harper and Row, New York.

Falconer, D.S. (1981). *Introduction to quantitative genetics*, 2nd edn. Longman.

Hayward, M.D., and Breese, E.L. (1993). Population structure and variability, in *Plant Breeding: Principles and Practices* (eds M.D. Hayward, N.O. Bosemark, and I. Ramagosa). Chapman and Hall, London.

Li, C.C. (1976). *A first course in population genetics*. Box Wood, Pacific Grove, CA.

Internet resource

http://users.rcn.com/jkimball.ma.ultranet/BiologyPages/H/Hardy_Weinberg.html – Excellent discussion of population genetics (accessed March 28, 2012).

Outcomes assessment

Part A

Please answer these questions true or false.

1 Inbreeding promotes heterozygosity.
2 Naturally cross-breeding species are more susceptible to inbreeding than naturally self-pollinated species.
3 In the Hardy–Weinberg equilibrium gene frequencies add up to unity.
4 Open-pollinated species can be improved by mass selection.

Part B

Please answer the following questions.

1 Define the terms (i) population and (ii) gene pool.
2 Give three major factors that influence the genetic structure of a population during the processes of transmission of genes from one generation to another.
3 Explain the phenomenon of inbreeding depression.
4 Distinguish between assortative and disassortative matings.
5 Discuss the main types of mating systems used by plant breeders to effect inbreeding.

Part C

Please write a brief essay on each of the following topics.

1 Discuss the Hardy–Weinberg equilibrium and its importance in breeding cross-pollinated species.
2 Discuss the consequences of inbreeding.
3 Discuss the concept of combining ability.
4 Discuss the application of inbreeding in plant breeding.

4

Introduction to quantitative genetics

Purpose and expected outcomes

Most of the traits that plant breeders are interested in are quantitatively inherited. It is important to understand the genetics that underlie the behavior of these traits in order to develop effective approaches for manipulating them. After studying this chapter, the student should be able to:

1 Define quantitative genetics and distinguish it from population genetics.
2 Distinguish between qualitative traits and quantitative traits.
3 Discuss polygenic inheritance.
4 Discuss gene action.
5 Discuss the variance components of quantitative traits.
6 Discuss the concept of heritability of traits.
7 Discuss selection and define the "breeders' equation".
8 Discuss the concept of general worth of a plant.
9 Discuss combining ability

4.1 What is quantitative genetics?

Population genetics and quantitative genetics are closely related fields, both dealing with the genetic basis of phenotypic variation among the individuals in a population. Population genetics traditionally focuses on frequencies of alleles and genotypes, whereas quantitative genetics focuses on linking phenotypic variation of complex traits to its underlying genetic basis to enable researchers better understand and predict genetic architecture and long term change in populations (to predict the response to selection given data on the phenotype and relationships of individuals in the population). Historically, quantitative genetics has its roots in statistical abstractions of genetic effects, first described by Karl Pearson and Ronald Fisher in the early 1900s. The foregoing represents the classical view of quantitative genetics.

The modern molecular view of quantitative genetics focuses on the use of molecular genetics tools

---

Principles of Plant Genetics and Breeding, Second Edition. George Acquaah.
© 2012 John Wiley & Sons, Ltd. Published 2012 by John Wiley & Sons, Ltd.

(genomics, bioinformatics, computational biology, etc.) to reveal links between genes and complex phenotypes (quantitative traits). Genes that control quantitative traits are called quantitative trait loci (QTLs). Molecular-based QTL analyses are being used to evaluate the coupling associations of the polymorphic DNA sites with phenotypic variations of quantitative and complex traits and analyzing their genetic architecture. There is evidence of a paradigm shift in the field of quantitative genetics. In this chapter, both classical and molecular quantitative genetics are discussed.

4.2 Classical quantitative genetics

Discussion in this section includes genetic and environmental variances, relationships and genetic diversity, linkage and epistatic issues in populations.

4.2.1 Quantitative trait

Most traits encountered in plant breeding are quantitatively inherited. Many genes control such traits, each contributing a small effect to the overall phenotypic expression of a trait. Variation in quantitative trait expression is without natural discontinuities (i.e., the variation is continuous). The traits that exhibit continuous variations are also called **metric traits**. Any attempt to classify such traits into distinct groups is only arbitrary. For example, height is a quantitative trait. If plants are grouped into tall versus short plants, relatively tall plants could be found in the short group and, similarly, short plants could be found in the tall group.

4.2.2 Qualitative genetics versus quantitative genetics

The major way in which qualitative genetics and quantitative genetics differ may be summarized as follows:

- **Nature of traits.** Qualitative genetics is concerned with traits that have Mendelian inheritance and can be described according to kind and, as previously discussed, can be unambiguously categorized. Quantitative genetic traits are described in terms of the degree of expression of the trait, rather than the kind.
- **Scale of variability.** Qualitative genetic traits provide discrete (discontinuous) phenotypic variation, whereas quantitative genetic traits produce phenotypic variation that spans the full spectrum (continuous).
- **Number of genes.** In qualitative genetics, the effects of single genes are readily detectable, while in quantitative genetics single gene effects are not discernible. Rather, traits are under polygenic control (genes with small indistinguishable effects).
- **Mating pattern.** Qualitative genetics is concerned with individual matings and their progenies. Quantitative genetics is concerned with a population of individuals that may comprise of a diversity of mating kinds.
- **Statistical analysis.** Qualitative genetic analysis is quite straight forward; it is based on counts and ratios. On the other hand, quantitative analysis provides estimates of population parameters (attributes of the population from which the sample was obtained).

4.2.3 The environment and quantitative variation

All genes are expressed in an environment (phenotype = genotype + environmental effect). However, quantitative traits tend to be influenced to a greater degree than qualitative traits. Under significantly large environmental effects, qualitative traits (controlled by one or a few major genes) can exhibit quantitative trait inheritance pattern. A strong environmental influence causes the otherwise distinct classes to overlap (Figure 4.1).

4.2.4 Polygenes and polygenic inheritance

Quantitative traits are controlled by multiple genes or polygenes.

What are polygenes?

Polygenes are genes with effects that are too small to be individually distinguished. They are sometimes called **minor genes**. In polygenic inheritance, segregation occurs at a large number of loci affecting a trait. The phenotypic expression of polygenic traits is susceptible to significant modification by the variation in environmental factors to which plants in the population are subjected. Polygenic variation cannot be classified into discrete groups (i.e., variation is continuous). This is because of the large number of segregating loci, each with effects so small that it is not possible to identify individual gene effects in the

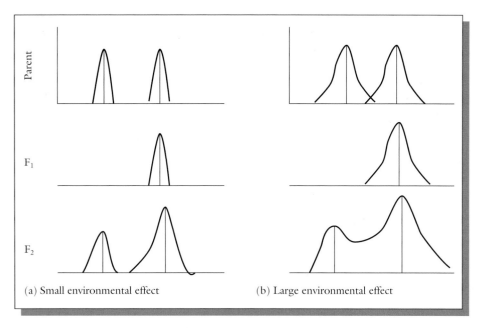

Figure 4.1 Environmental effect on gene expression. The phenotype = genotype + environment. Some traits are influenced a lot more than others by the environment. In cross (a) the environmental influence is small, such that the phenotypes are distinguishable in the F_2; in cross (b) the environmental influence is strong, resulting in more blurring of the differences among phenotypes in the segregating population.

segregating population or meaningfully describe individual genotypes. Instead, biometrics is used to describe the population in terms of means and variances. Continuous variation is caused by environmental variation and genetic variation due to the simultaneous segregation of many genes affecting the trait. These effects convert the intrinsically discrete variation to a continuous one. Biometrical genetics is used to distinguish between the two factors that cause continuous variability to occur.

Another aspect of polygenic inheritance is that different combinations of polygenes can produce a particular phenotypic expression. Furthermore, it is difficult to measure the role of environment on trait expression because it is very difficult to measure the environmental effect on a plant basis. Consequently, a breeder attempting to breed a polygenic trait should evaluate the cultivar in an environment that is similar to that prevailing in the production region. It is beneficial to plant breeding if a tight linkage of polygenes (called **polygenic block**; **linkage block**) that has favorable effects on traits of interest to the breeder is discovered.

In 1910, a Swedish geneticist, Nilsson-Ehle, provided a classic demonstration of polygenic inheritance.

In the process he helped to bridge the gap between our understanding of the essence of quantitative and qualitative traits. Polygenic inheritance may be explained by making three basic assumptions:

1 that many genes determine the quantitative trait;
2 these genes lack dominance;
3 the action of the genes are additive.

Nilsson-Ehle crossed two varieties of wheat, one with deep red grain of genotype $R_1R_1R_2R_2$, and the other white grain of genotype $r_1r_1r_2r_2$. The results are summarized in Table 4.1. He observed

Table 4.1 Transgressive segregation.

P_1	$R_1R_1R_2R_2$	×	$r_1r_1r_2r_2$
	(dark red)		(white)
F_1	$R_1r_1R_2r_2$		
F_2	1/16	=	$R_1R_1R_2R_2$
	4/16	=	$R_1R_1R_2r_2, R_1r_1R_2R_2$
	6/16	=	$R_1R_1r_2r_2, R_1r_1R_2r_2, r_1r_1R_2R_2$
	4/16	=	$R_1r_1r_2r_2, r_1r_1R_2r_2$
	1/16	=	$r_1r_1r_2r_2$

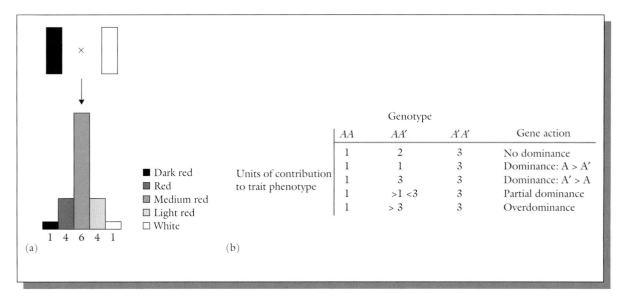

Figure 4.2 (a) Nilsson-Ehle's classic work involving wheat color provided the first formal evidence of genes with cumulative effect. (b) An illustration of gene action using numeric values.

that all the seed of the F_1 was medium red. The F_2 showed about 1/16 dark red and 1/16 white seed, the remainder being intermediate. The intermediates could be classified into 6/16 medium red (like the F_1), 4/16 red, and 4/16 light red. The F_2 distribution of phenotypes may be obtained as an expansion of the bionomial $(a + b)^4$, where $a = b = \frac{1}{2}$ (a binomial coefficient is the number of combinations of r items that can be selected from a set of n items).

His interpretation was that the two genes each had a pair of alleles that exhibited cumulative effects. In other words, the genes lacked dominance and their action was additive. Each allele R_1 or R_2 added some red to the phenotype, so that the genotypes of white contained neither of these alleles, while the dark red genotype contained only R_1 and R_2. The phenotypic frequency ratio resulting from the F_2 was $1:4:6:4:1$ (i.e., 9 genotypes and five classes) (Figure 4.2).

The study involved only two loci. However, most polygenic traits are conditioned by genes at many loci. The number of genotypes that may be observed in the F_2 is calculated as 3^n, where $n =$ number of loci (each with two alleles). Hence, for three loci the number of genotypes $= 27$, and for

10 loci it will be $3^{10} = 59\,049$. Many different genotypes can have the same phenotype. Consequently, there is no strict one-to-one relationship between genotype. For n loci, there are 3^n genotypes and $2n + 1$ phenotypes. Many complex traits such as yield may have dozens and conceivably even hundreds of loci.

Other difficulties associated with studying the genetics of quantitative traits are dominance, environmental variation, and epistasis. Not only can dominance obscure the true genotype, but both the amount and direction can vary from one gene to another. For example, allele A may be dominant to a, but b may be dominant to B. It has previously been mentioned that environmental effects can significantly obscure genetic effects. Non-allelic interaction is a clear possibility when many genes are acting together.

Number of genes controlling a quantitative trait

Polygenic inheritance is characterized by segregation at a large number of loci, affecting a trait as previously discussed. Biometrical procedures have been proposed to estimate the number of genes involved in a quantitative trait expression. However, such estimates, apart from not being reliable, have limited practical use. Genes may differ in the magnitude of their effects on traits, not to mention the possibility of modifying gene effects on certain genes.

Modifying genes

One gene may have a major effect on one trait and a minor effect on another. There are many genes in plants without any known effects besides the fact that they modify the expression of a major gene by either enhancing or diminishing it. The effect of modifier genes may be subtle, such as slight variations in traits like shape and shades of color of flowers or, in fruits, variation in aroma and taste. Those trait modifications are of concern to plant breeders as they conduct breeding programs to improve quantitative traits involving many major traits of interest.

4.2.5 Decision making in breeding based on biometrical genetics

Biometrical genetics is concerned with the inheritance of quantitative traits. As previously stated, most of the genes of interest to plant breeders are controlled by many genes. In order to effectively manipulate quantitative traits, the breeder needs to understand the nature and extent of their genetic and environmental control. M.J. Kearsey summarized the salient questions that need to be answered by a breeder who is focusing on improving quantitative (and also qualitative) traits, into four:

1　Is the trait inherited?
2　How much variation in the germplasm is genetic?
3　What is the nature of the genetic variation?
4　How is the genetic variation organized?

By having answers to these basic genetic questions, the breeder will be in apposition to apply the knowledge to address certain fundamental questions in plant breeding.

What is the best cultivar to breed?

As is discussed later in the book, there are several distinct types of cultivars that plant breeders develop – pure lines, hybrids, synthetics, multilines, composites, and so on. The type of cultivar is closely related to the breeding system of the species (self- or cross-pollinated) but more importantly on the genetic control of the traits targeted for manipulation. As breeders have more understanding of and control over plant reproduction, the traditional grouping between types of cultivars to breed and the methods used along the lines of the breeding system have diminished. The fact is that the breeding system can be artificially altered (i.e., self-pollinated species can be forced to outbreed, and vice versa). However, the genetic control of the trait of interest cannot be changed. The action and interaction of polygenes are difficult to alter. As Kearsey noted, breeders should make decisions of the type of cultivar to breed based on the genetic architecture of the trait, especially, the nature and extent of dominance and gene interaction (Section 4.2.6), more so than the breeding system of the species.

Generally, where additive variance and additive × additive interaction predominate, it is appropriate to develop pure lines and inbred cultivars. However, where dominance variance and dominance × dominance interaction suggest overdominance predominates, hybrids would be successful cultivars. Open-pollinated cultivars are suitable where a mixture of the above genetic architecture occurs.

What selection method would be most effective for improvement of the trait?

The kinds of selection methods used in plant breeding are discussed in Chapters 15–18. The genetic control of the trait of interest determines the most effective selection method to use. The breeder should pay attention to the relative contribution of the components of genetic variance (additive, dominance, epistasis) and environmental variance in choosing the best selection method. Additive genetic variance can be exploited for long term genetic gains by concentrating desirable genes in the homozygous state in a genotype. The breeder can make rapid progress where heritability is high by using selection methods that are dependent solely on phenotype (e.g., mass selection). However, where heritability is low, the methods of selection based on families and progeny testing are more effective and efficient. When overdominance predominates, the breeder can exploit short term genetic gain very quickly by developing hybrid cultivars for the crop.

It should be pointed out that as self-fertilizing species attain homozygosity following a cross, they become less responsive to selection. However, additive genetic variance can be exploited for a longer time in open-pollinated populations because relatively more genetic variation is regularly being generated through the ongoing intermating.

Should selection be on single traits or multiple traits?

Plant breeders are often interested in more than one trait in a breeding program, which they seek to improve simultaneously. The breeder is not interested in achieving disease resistance only but, in addition, high yield and other agronomic traits. The problem with simultaneous trait selection is that the traits could be correlated such that modifying one affects the other. The concept of correlated traits is discussed next. Biometrical procedures have been developed to provide a statistical tool for the breeder to use. These tools are also discussed in this section.

4.2.6 Gene action

Additional information on gene action can be found in the supplementary chapters at the end of the book. There are four types of gene action: **additive, dominance, epistasis**, and **overdominance**. Because gene effects do not always fall into clear-cut categories, and quantitative traits are governed by genes with small individual effects, they are often described by their gene action rather than by the number of genes by which they are encoded. It should be pointed out that gene action is conceptually the same for major genes as well as minor genes, the essential difference being that the action of a minor gene is small and significantly influenced by the environment. A general way of distinguishing between these types of gene action based on interaction among alleles is as follows:

	No allelic interaction	Allelic interaction
Within locus interaction	**Additive action**	Dominance action
Between loci interaction	**Additive action**	Epistatsis

Additive gene action

The effect of a gene is said to be additive when each additional gene enhances the expression of the trait by equal increments. Consequently, if one gene adds one unit to a trait, the effect of $aabb = 0$, $Aabb = 1$, $AABb = 3$, and $AABB = 4$. In the case of a single locus (A, a) the heterozygote would be exactly intermediate between the parents (i.e., $AA = 2$, $Aa = 1$, $aa = 0$). That is, the performance of an allele is the same irrespective of other alleles at the same locus. This means that the phenotype reflects the genotype in additive action, assuming the absence of environmental effect. Additive effects apply to the allelic relationship at the same locus. Furthermore, a superior phenotype will breed true in the next generation, making selection for the trait more effective to conduct. Selection is most effective for additive variance; it can be fixed in plant breeding (i.e., develop a cultivar that is homozygous).

- **Additive effect.** Consider a gene with two alleles (A, a). Whenever A replaces a, it adds a constant value to the genotype:

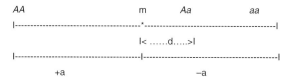

Replacing a by A in the genotype aa causes a change of a units. When both aa are replaced, the genotype is $2a$ units away from aa. The midparent value (the average score) between the two homozygous parents is given by m (representing a combined effect of both genes for which the parents have similar alleles and environmental factors). This also serves as the reference point for measuring deviations of genotypes. Consequently, $AA = m + a_A$, $aa = m - a$, and $Aa = m + d_A$, where a_A is the **additive effect** of allele A. This effect remains the same regardless of the allele with which it is combined.

- **Average effect.** In a random mating population, the term **average effect** of alleles is used because there are no homozygous lines. Instead, alleles of one plant combine with alleles from pollen from a random mating source in the population through hybridization to generate progenies. In effect, the allele of interest replaces its alternative form in a number of randomly selected individuals in the population. The change in the population as a result of this replacement constitutes the average effect of the allele. In other words, the average effect of a gene is the mean deviation from the population mean of individuals that received a gene from one parent, the gene from the other parent having come at random from the population.

- **Breeding value.** The average effects of genes of the parents determine the mean genotypic value of the progeny. Furthermore, the value of an individual judged by the mean value of its progeny is called the **breeding value** of the individual. This is the value that is transferred from an individual to its progeny. This is a measurable effect, unlike the average effect of a gene. However, the breeding value must always be with reference to the population to which an individual is to be mated. From a practical breeding point of view, the additive gene effect is of most interest to breeders because its exploitation is predictable, producing improvements that increase linearly with the number of favorable alleles in the population.

Dominance gene action

Dominance action describes the relationship of alleles at the same locus. Dominance variance has two components – variance due to homozygous alleles (which is additive) and variance due to heterozygous genotypic values. Dominance effects are deviations from additivity that make the heterozygote resemble one parent more than the other. When dominance is complete, the heterozygote is equal to the homozygote in effects (i.e., $Aa = AA$). The breeding implication is that the breeder cannot distinguish between the heterozygous and homozygous phenotypes. Consequently, both kinds of plants will be selected, the homozygotes breeding true while the heterozygotes will not breed true in the next generation (i.e., fixing superior genes will be less effective with dominance gene action).

- **Dominance effect.** Using the previous figure for additive effect, the extent of dominance (d_A) is calculated as the deviation of the heterozygote, Aa, from the mean of the two homozygotes (AA, aa). Also, $d_A = 0$ when there is no dominance while d is positive if A is dominant and negative if a_A is dominant. Furthermore, if dominance is complete $d_A = a_A$, whereas $d_A < a_A$ for incomplete (partial) dominance and $d_A > a_A$ for overdominace. For a single locus, $m = \frac{1}{2}(AA + aa)$, $a_A = \frac{1}{2}(AA - aa)$, while $d_A = Aa - \frac{1}{2}(AA + aa)$.

Overdominance gene action

Overdominance gene action exists when each allele at a locus produces a separate effect on the phenotype and their combined effect exceeds the independent effect of the alleles. From the breeding standpoint, the breeder can fix overdominance effects only in the first generation (i.e., F_1 hybrid cultivars) through apomixis.

Epistatic gene action

Epistasis is the interaction of alleles at different loci. It complicates gene action in that the value of a genotype or allele at one locus depends on the genotype at other epistatically interacting loci. In other words, the allelic effects at one locus depend on the genotype at a second locus. An effect of epistasis is that an allele may be deemed "favorable" at one locus and then deemed "unfavorable" under a different genetic background. In the absence of epistasis, the total genetic value of an individual is simply the sum total of the individual genotype values, because the loci are independent. Epistasis is sometimes described as the masking effect of the expression of one gene by another at a different locus.

Estimation of gene action or genetic variance requires the use of large populations and a mating design. The effect of the environment on polygenes makes estimations more challenging. As N.W. Simmonds observed, at the end of the day, what qualitative genetic analysis allows the breeder to conclude from partitioning variance in an experiment is to say that a portion of the variance behaves as though it could be attributed to additive gene action or dominance effect, and so forth.

4.2.7 Gene action and plant breeding

Understanding gene action is critical to the success of plant breeding. It is used by breeders several ways:

- in the selection of parents used in crosses to create segregating populations in which selection is practiced;

- in the choice of the method of breeding used in crop improvement;
- in research applications to gain understanding of the breeding material by estimating genetic parameters.

Gene action and methods of breeding

Breeding methods are discussed in detail in Chapters 15–18. The methods are grouped according to modes of pollination – self-pollinated or cross-pollinated.

Self-pollinated species. When additive gene action predominates in a self-pollinated species, breeders should consider using selection methods such as pure line selection, mass selection, progeny selection and hybridization. However, when non-additive gene action predominates, effective methods of breeding are the exploitation of heterosis in breeding hybrid cultivars.

Cross-pollinated species. When additive gene action predominates in a cross-pollinated species, recurrent selection may be used to achieve general combining ability (GCA). Specific breeding products to pursue include synthetic varieties and composites. In the case of non-additive gene action, heterosis breeding, just like in self-pollinated species, is recommended for breeding hybrid cultivars. Alternatively, breeders may consider recurrent selection for specific combining ability (SCA) for population improvement. Where both additive and non-additive gene action occur together, reciprocal recurrent selection may be used for population improvement.

Impact of breeding method on genetic variance

Additive genetic variance is known to decrease proportionally to the improvement following selection. In pure line selection, genetic variance is completely depleted with time, until further improvement is impossible. However, mutational events as well as genetic recombination can replenish some of the lost additive genetic variance. On the contrary, additive genetic variance cannot be depleted in intermating populations because auto conversion (self conversion) of non-additive genetic variance to additive genetic variance occurs. This conversion occurs because heterozygotes become fixed into homozygotes.

Estimating gene action

Gene action may be estimated by creating various crosses (e.g., diallele, partial diallele, line x tester cross, biparental cross, etc.) and applying various biometrical analyses to estimate components of genetic variance. Additive genetic variance is very important to breeders because it is the only genetic variance that responds to selection. In addition to the components of genetic variance, combining ability variances may also be used to measure gene action.

Factors affecting gene action

Gene action is affected by several factors, the key ones being the type of genetic material, mode of pollination, mode of inheritance, presence of linkage, as well as biometrical parameters (e.g., sample size, sampling method, and method of calculation). Alleles with a dominant, additive, or deleterious phenotypic effect affect heritability differently depending on whether they are in homozygous or heterozygous condition. Knowledge of the way genes act and interact will determine which breeding system optimizes gene action more efficiently and will elucidate the role of breeding systems in the evolution of crop plants.

Self-pollinated materials (e.g., mass selected cultivar, multiline, varietal blends) express additive and additive epistasis. A pure line cultivar will have additive gene action but without genetic variation. On the other hand, products derived from cross-pollinated species (e.g., composite variety, synthetic variety) will display additive, dominance and epistatic gene action. F_1 hybrid material will have no additive gene action and no genetic variation.

In terms of pollination, self-pollinated species exhibit additive gene action because this gene action is associated with homozygosity. On the contrary, non-additive gene action is associated with heterozygosity, and hence is more prevalent in cross-pollinated species than self-pollinated ones. Simply inherited (qualitative, oligogenic) traits predominantly exhibit non-additive and epistatic gene action, while polygenic traits are governed predominantly by additive gene action.

Gene action estimates are affected by the presence of genetic linkage. Estimates of additive gene action and dominance gene action can be biased up when the genes of interest are in the coupling phase (AB/ab). In the repulsion phase (Ab/aB), genes can cause estimates of dominance gene action to be biased upwards and additive action downwards.

4.2.8 Variance components of a quantitative trait

The genetics of a quantitative trait centers on the study of its variation. As D.S Falconer stated, it is in terms of variation that the primary genetic questions are formulated. Furthermore, the researcher is interested in partitioning variance into its components that are attributed to different causes or sources. The genetic properties of a population are determined by the relative magnitudes of the components of variance. In addition, by knowing the components of variance, the relative importance of the various determinants of phenotype may be estimated.

K. Mather expressed the phenotypic value of quantitative traits in this commonly used expression:

P(phenotype) = G(genotype) + E(environment)

Individuals differ in phenotypic value. When the phenotypes of a quantitative trait are measured, the observed value represents the phenotypic value of the individual. The phenotypic value is variable because it depends on genetic differences among individuals, as well as environmental factors and the interaction between genotypes and the environment (called G × E interaction). A third factor (GE) is therefore added to the previous conceptual equation, so that the total variance of a quantitative trait may be mathematically expressed as follows:

$$V_P = V_G + V_E + V_{GE}$$

where V_P = total **phenotypic variance** of the segregating population; V_G = **genetic variance**; V_E = **environmental variance**; and V_{GE} = variance associated with the genetic and environmental interaction.

The genetic component of variance may be further partitioned into three components as follows:

$$V_G = V_A + V_D + V_I$$

where V_A = **additive variance** (variance from additive gene effects), V_D = **dominance variance** (variance from dominance gene action), and V_I = **interaction** (variance from interaction between genes). Additive genetic variance (or simply additive variance) is the variance of breeding values and is the primary cause of

resemblance between relatives. Hence V_A is the primary determinant of the observable genetic properties of the population, and of the response to the population to selection. Further, V_A is the only component that the researcher can most readily estimate from observations made on the population. Consequently, it is common to partition genetic variance into two – additive versus all other kinds of variance. This ratio, V_A/V_P gives what is called the **heritability** of a trait, an estimate that is of practical importance in plant breeding (Section 4.2.9).

The total phenotypic variance may then be rewritten as follows:

$$V_P = V_A + V_D + V_I + V_E + V_{GE}$$

To estimate these variance components, the researcher uses carefully designed experiments and analytical methods. To obtain environmental variance, individuals from the same genotype or replicates are used.

An inbred line (essentially homozygous) consists of individuals with the same genotype. An F_1 generation from a cross of two inbred lines will be heterozygous but genetically uniform. The variance from the parents and the F_1 may be used as a measure of environmental variance (V_E). K. Mather provided procedures for obtaining genotypic variance from F_2 and backcross data. In sum, variances from additive, dominant and environmental effects may be obtained as follows:

$$V_{P1} = E; \; V_{P2} = E; \; V_{F1} = E$$
$$V_{F2} = \tfrac{1}{2}A + \tfrac{1}{4}D + E$$
$$V_{B1} = \tfrac{1}{4}A + \tfrac{1}{4}D + E$$
$$V_{B2} = \tfrac{1}{4}A + \tfrac{1}{4}D + E$$
$$V_{B1} + V_{B2} = \tfrac{1}{2}A + \tfrac{1}{2}D + 2E$$

where V_{P1} and V_{P2} are variances for the parents in a cross; VF_1 is the variance of the resulting hybrid; F_2 is the variance of the F_2 population; A and D are additive and dominant effects, respectively; E is the environmental effect; V_{B1} and V_{B2} are backcross variances. This represents the most basic procedure for obtaining components of genetic variance because it omits the variances due to epistasis, which are common with quantitative traits. More rigorous biometric procedures are needed to consider the effects of interlocular interaction.

It should be pointed out that additive variance and dominance variance are statistical abstractions rather than genetical estimates of these effects. Consequently, the concept of additive variance does not connote perfect additivity of dominance or epistasis. To exclude the presence of dominance or epistasis, all the genotypic variance must be additive.

4.2.9 The concept of heritability

Genes are not expressed in a vacuum but in an environment. A phenotype observed is an interaction between genes that encode it and the environment in which the genes are being expressed. Plant breeders typically select plants based on the phenotype of the desired trait, according to the breeding objective. Sometimes, a genetically inferior plant may appear superior to other plants only because it is located in a more favorable region of the soil. This may mislead the breeder. In other words, the selected phenotype will not give rise to the same progeny. If the genetic variance is high and the environmental variance is low, the progeny will be like the selected phenotype. The converse is also true. If such a plant is selected for advancing the breeding program, the expected genetic gain will not materialize. Quantitative traits are more difficult to select in a breeding program because they are influenced to a greater degree by the environment than are qualitative traits. If two plants are selected randomly from a mixed population, the observed difference in a specific trait may be due to the average effects of genes (hereditary differences), or differences in the environments in which the plants grew up, or both. The average effects of genes is what determines the degree of resemblance between relatives (parents and offspring), and hence is what is transmitted to the progenies of the selected plants.

Definition

The concept of the reliability of the phenotypic value of a plant as a guide to the breeding value (additive genotype) is called the **heritability** of the metrical trait. As previously indicated, plant breeders are able to measure phenotypic values directly but it is the breeding value of individuals that determines their influence on the progeny. Heritability is the proportion of the observed variation in a progeny that is inherited. The bottom line is that if a plant breeder selects plants on the basis of phenotypic values to be used as parents in a cross, the success of such an action in changing the traits in a desired direction is predictable only by knowing the degree of correspondence (genetic determination) between phenotypic values and breeding values. Heritability measures this degree of correspondence. It does not measure genetic control but rather how this control can vary.

Genetic determination is a matter of what causes a characteristic or trait; heritability, by contrast, is a scientific concept of what causes differences in a characteristic or trait. Heritability is, therefore, defined as a fraction: it is the ratio of genetically caused variation to total variation (including both environmental and genetic variation). Genetic determination, by contrast, is an informal and intuitive notion. It lacks quantitative definition and depends on the idea of a normal environment. A trait may be described as genetically determined if it is coded in and caused by the genes, and bound to develop in a normal environment. It makes sense to talk about genetic determination in a single individual but heritability makes sense only relative to a population in which individuals differ from one another.

Types of heritability

Heritability is a property of the trait, the population, and the environment. Changing any of these factors will result in a different estimate of heritability. There are two different estimates of heritability.

1 **Broad sense heritability.** Heritability estimated using the total genetic variance (V_G) is called **broad sense heritability.** It is expressed mathematically as:

$$H = V_G/V_P$$

   It tends to yield a high value (Table 4.2). Some use the symbol H^2 instead of H.

2 **Narrow sense heritability.** Because the additive component of genetic variance determines the response to selection, the **narrow sense heritability** estimate is more useful to plant breeders than the broad sense estimate. It is estimated as follows:

$$h^2 = V_A/V_P$$

Table 4.2 Heritability estimates of some plant architectural traits.

Trait	Heritability
Plant height	45
Hypocotyl diameter	38
Number of branches/plant	56
Nodes in lower third	36
Nodes in mid section	45
Nodes in upper third	46
Pods in lower third	62
Pods in mid section	85
Pods in upper third	80
Pod width	81
Pod length	67
Seed number per pod	30
100 seed weight	77

However, when breeding clonally propagated species (e.g., sugarcane, banana), in which both additive and non-additive gene action are fixed and transferred from parent to offspring, broad sense heritability is also useful. The magnitude of narrow sense heritability cannot exceed, and is usually less than, the corresponding broad sense heritability estimate.

Heritabilities are seldom precise estimates because of large standard errors. Traits that are closely related to reproductive fitness tend to have low heritability estimates. The estimates are expressed as a fraction, but may also be reported as a percentage by multiplying by 100. A heritability estimate may be unity (1) or less.

Factors affecting heritability estimates

The magnitude of heritability estimates depends on the genetic population used, sample size, and the method of estimation.

- **Genetic population.** When heritability is defined as $h^2 = V_A/V_P$ (i.e., in the narrow sense), the variances are those of individuals in the population. However, in plant breeding, certain traits such as yield are usually measured on a plot basis (not on individual plants). The amount of genotypic variance present for a trait in a population influences estimates of heritability. Parents are responsible for the genetic structure of populations they produce. More divergent parents yield a population that is more genetically variable. Inbreeding tends to increase the magnitude of genetic variance among individuals in the population. This means that estimates derived from F_2 will differ from, say, those from F_6.

- **Sample size.** Because it is impractical to measure all individuals in a large population, heritabilities are estimated from sample data. To obtain the true genetic variance for a valid estimate of the true heritability of the trait, the sampling should be random. A weakness in heritability estimates stems from bias and lack of statistical precision.

- **Methods of computation.** Heritabilities are estimated by several methods that use different genetic populations and produce estimates that may vary. Common methods include parent–offspring regression and variance component method. Mating schemes are carefully designed to enable the total genetic variance to be partitioned.

Methods of computation

The methods of estimating heritabilities have strengths and weaknesses.

- **Variance components.** The variance component method of estimating heritability uses the statistical procedure of **analysis of variance**. Variance estimates depends on the types of populations in the experiment. Estimating genetic components suffer from certain statistical weaknesses. Variances are less accurately estimated than means. Also, variances are not robust and are sensitive to departure from normality. An example of heritability estimate using F_2, and backcross populations is as follows:

$$V_{F2} = V_A + V_D + V_E$$
$$V_{B1} + V_{B2} = V_A + 2V_D + 2V_E$$
$$V_E = [V_{P1} + V_{P2} + V_{F1}]/3$$
$$H = (V_A + V_D)/(V_A + V_D + V_E) = V_G/V_P$$
$$h^2 = (V_A)/(V_A + V_D + V_E) = V_A/V_P$$

Example

Using the data in the following table:

	P$_1$	P$_2$	F$_1$	F$_2$	BC$_1$	BC$_2$
Mean	20.5	40.2	28.9	32.1	25.2	35.4
Variance	10.1	13.2	7.0	52.3	35.1	56.5

$V_E = [V_{P1} + V_{P2} + V_{F1}]/3$
$\quad = [10.1 + 13.2 + 7]$
$\quad = 30.3/3$
$\quad = 10.1$

$V_A = 2V_{F2} - (V_{B1} + V_{B2})$
$\quad = 2(52.3) - (35.1 + 56.5)$
$\quad = 104.6 - 91.6$
$\quad = 13.0$

$V_D = [(V_{B1} + V_{B2}) - F_2 - (V_{P1} + V_{P2} + F_1)]/3$
$\quad = [(35.1 + 56.5) - 52.3 - (10.1 + 13.2 + 7.0)]/3$
$\quad = [91.6 - 52.3 - 30.3]/3$
$\quad = 3.0$

Broad sense heritability

$H = [13.0 + 3.0]/[13.0 + 3.0 + 10.1]$
$\quad = 16.0/26.1$
$\quad = 0.6130$
$\quad = 61.30\%$

Narrow sense heritability

$h^2 = 13.0/[13.0 + 3.0 + 10.1]$
$\quad = 13.0/26.1$
$\quad = 0.4980$
$\quad = 49.80\%$

Other methods of estimation

$H = [V_{F2} - \tfrac{1}{2}(V_{P1} + V_{P2})]/V_{F2}$
$\quad = [52.3 - \tfrac{1}{2}(10.1 + 13.2)]/52.3$
$\quad = 40.65/52.3$
$\quad = 0.7772$
$\quad = 77.72\%$

(The estimate is very close to that obtained by using the previous formula.)

- **Parent–offspring regression.** The type of offspring determines if the estimate would be broad sense or narrow sense. This method is based on several assumptions: the trait of interest has diploid Mendelian inheritance; the population from which the parents originated is randomly mated; the population is in linkage equilibrium (or no linkage among loci controlling the trait); parents are non-inbred; no environmental correlation between the performance of parents and offspring.

The parent–offspring method of heritability is relatively straightforward. Firstly, the parent and offspring means are obtained. Cross products of the paired values are used to compute the covariance. A regression of offspring on mid-parent is then calculated. Heritability in the narrow sense is obtained as follows:

$$h^2 = b_{op} = V_A/V_P$$

where b_{op} is the regression of offspring on mid-parent, and V_A and V_P = additive variance and phenotypic variance, respectively.

However, if only one parent is known or relevant (e.g., a polycross):

$$b = \tfrac{1}{2}(V_A/V_P)$$

and

$$h^2 = 2b_{op}$$

Applications of heritability

Heritability estimates are useful for breeding quantitative traits. The major applications of heritability are:

1 To determine whether a trait would benefit from breeding. If especially the narrow sense heritability for a trait is high, it indicates that the use of plant breeding methods will likely be successful in improving the trait of interest.
2 To determine the most effective selection strategy to use in a breeding program. Breeding methods that use selection based on phenotype are effective when heritability is high for the trait of interest.
3 To predict gain from selection. Response to selection depends on heritability. A high heritability would likely result in high response to selection to advance the population in the desired direction of change.

Evaluating parental germplasm

A useful application of heritability is in evaluating the germplasm assembled for a breeding project to determine if there is sufficient genetic variation for successful improvement to be pursued. A replicated trial of the available germplasm is conducted and analyzed by ANOVA as follows:

Source	df	EMS
Replication	$r - 1$	
Genotypes	$g - 1$	$\sigma^2 + r\sigma_g^2$
Error	$(r - 1)(g - 1)$	σ^2

From the analysis, heritability may be calculated as:

$$H \text{ or } h^2 = [\sigma_g^2]/[\sigma_g^2 + \sigma_e^2]$$

Whether the estimate is heritability in the narrow or broad sense depends on the nature of the genotypes. Pure lines or inbred lines would yield additive type of variance, making the estimate narrow sense. Segregating population would make the estimate broad sense.

4.2.10 Response to selection in breeding

The focus of this section is on the **response to selection** (**genetic gain** or **genetic advance**). After generating variability, the next task for the breeder is the critical one of advancing the population through selection.

Selection, in essence, entails discriminating among genetic variation (heterogeneous population) to identify and choose a number of individuals to establish the next generation. The consequence of this is differential reproduction of genotypes in the population such that gene frequencies are altered and, subsequently, the genotypic and phenotypic values of the targeted traits. Even though artificial selection is essentially directional, the concept of "complete" or "pure" artificial selection is an abstraction because, invariably, before the breeder gets a chance to select plants of interest, some amount of natural selection would have already been imposed.

The breeder hopes, by selecting from a mixed population, that superior individuals (with high genetic potential) will be advanced, and consequently change the population mean of the trait in a positive way in the next generation. The breeder needs to have a clear objective. The trait to be improved needs to be clearly defined. Traits controlled by major genes are usually easy to select. However, polygenic traits, being genetically and biologically complex, present a considerable challenge to the breeder.

The **response to selection (R)** is the difference between the mean phenotypic value of the offspring of the selected parents and the whole of the parental generation before selection. The response to selection is simply the change of population mean between generations following selection. Similarly, the **selection differential (S)** is the mean phenotypic value of the individuals selected as parents expressed as a deviation from the population mean (i.e., from the mean phenotypic value of all the individuals in the parental generation before selection). Response to selection is related to heritability by the following equation:

$$R = h^2 S$$

Prediction of response in one generation – genetic advance due to selection

The genetic advance achieved through selection depends on three factors:

1 The total variation (phenotypic) in the population in which selection will be conducted.
2 Heritability of the target trait.
3 Selection pressure to be imposed by the plant breeder (i.e., the proportion of the population that is selected for the next generation).

A large phenotypic variance would provide the breeder with a wide range of variability from which to select. Even when the heritability of the trait of interest is very high, genetic advance would be small without a large amount of phenotypic variation (Figure 4.3). When the heritability is high, selecting and advancing only the top few performers is likely to produce a greater genetic advance than selecting many moderate performers. However, such a high selection pressure would occur at the expense of a rapid loss in variation. When heritability is low, the breeder should impose a lower selection pressure in order to advance as many high potential genotypes as possible.

In principle, the prediction of response is valid for only one generation of selection. This is because a response to selection depends on the heritability of the trait estimated in the generation from which parents are selected. To predict the response in subsequent generations, heritabilities must be determined in each generation. Heritabilities are expected to change from one generation to the next because, if there is a response, it must be accompanied by change in gene frequencies on which heritability depends. Also, selection of parents reduces the variance and the heritability, especially in the early generations. It should be pointed out that heritability changes are not usually large.

If heritability is unity ($V_A = V_P$; no environmental variance), progress in a breeding program would be perfect and the mean of offspring would equal the mean of the selected parents. On the other hand, if

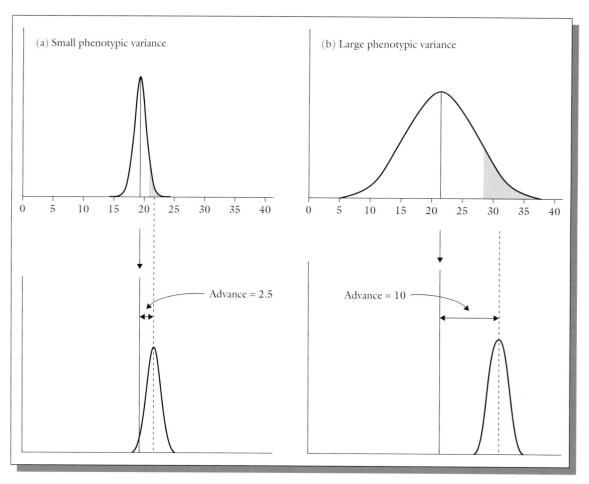

Figure 4.3 The effect of phenotypic variance on genetic advance. If the phenotypic variance is too small, the genetic variability from which to select will be limited, resulting in a smaller genetic gain. The reverse is true when the phenotypic variance is large.

heritability is zero, there would be no progress at all (R = 0).

The response in one generation may be expressed mathematically as

$$\bar{X}o - \bar{X}p = R = ih^2\sigma \,(\text{or } \Delta G = ih^2\sigma)$$

where $\bar{X}o$ is mean phenotype of the offspring of selected parents, $\bar{X}p$ is the mean phenotype of the whole parental generation, R is the advance in one generation of selection, h^2 is heritability, σ is phenotypic standard deviation of the parental population, i is intensity of selection, and ΔG is genetic gain or genetic advance.

This equation has been suggested by some to be one of the fundamental equations of plant breeding that must be understood by all breeders, hence it is called the **breeders' equation**. The equation is graphically illustrated in Figure 4.4. The factor "i", the intensity of selection (standardized selection differential), is a statistical factor that depends on the fraction of the current population retained for use as parents for the next generation. If the entire population is used, the selection intensity is zero. The breeder may consult statistical tables for specific values (e.g., at 1% I = 2.668; at 5% i = 2.06; at 10% i = 1.755). The breeder must decide the selection intensity to impose to achieve a desired objective.

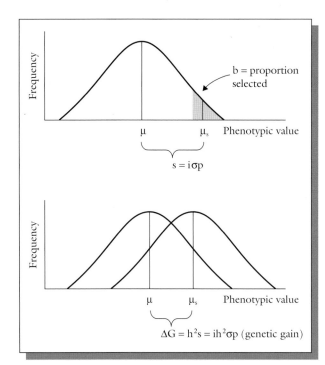

Figure 4.4 Genetic gain or genetic advance from selection indicates the progress plant breeders make from one generation to another based on the selection decisions they make.

The selection differential can be predicted if the phenotypic values of the trait of interest are normally distributed and the selection is by truncation (i.e., the individuals are selected solely in order of merit according to their phenotypic value – no individual being selected is less good than any of those rejected).

The response equation is effective in predicting response to selection, provided that the heritability estimate (h^2) is fairly accurate. In terms of practical breeding, the parameters for the response equation are seldom available, and hence not widely used. It is instructive to state that *predicted* response (theoretical estimate based on heritability and tabulated selection intensity) is different from *realized* response (what the breeder actually observes in the next generation following selection). Over the long haul, repeated selection tends to fix favorable genes, resulting in a decline in both heritability and phenotypic standard deviation. Once genes have been fixed, there will be no further response to selection.

Example

Using the data in the table below:

	X	σp	V_P	V_A	V_E
Parents	15	2	6	4	3
Offspring	20.2	15	4.3	2.5	3

$$R = ih^2\sigma$$

Parents

$$h^2 = V_A/V_P$$
$$= 4/6$$
$$= 0.67$$

for i at $p = 10\% = 1.755$ (read from tables and assuming a very large population):

$$R = 1.755 \times 0.67 \times 2$$
$$= 2.35$$

Offspring

$$h^2 = V_A/Vp$$
$$= 2.5/4.3$$
$$= 0.58$$
$$R = 1.755 \times 0.58 \times 1.5$$
$$= 1.53$$

Generally, as selection advances to higher generations, genetic variance and heritability decline. Similarly, the advance from one generation to the next declines, while the mean value of the trait being improved increases.

4.2.11 Concept of correlated response

Correlation is a measure of the degree of association between traits as previously discussed. This association may be on the basis of genetics or may be non-genetic. In terms of response to selection, genetic correlation is what is useful. When it exists, selection for one trait will cause a corresponding change in other traits that are correlated. This response to change by genetic association is called **correlated response**. Correlated response may be caused by pleiotropism or linkage disequilibrium. Pleiotropism is the multiple effect of a single gene (i.e., a single simultaneously affects several physiological pathways). In a random mating population, the role of linkage disequilibrium in correlated response is only important if the traits of interest are closely linked.

In calculating correlated response, genetic correlations should be used. However, the breeder often has access to phenotypic correlation and can use them if they were estimated from values averaged over several environments. Such data tend to be in agreement with genetic correlation. In a breeding program the breeder, even while selecting simultaneously for multiple traits, has a primary trait of interest and secondary traits. The correlated response (CRy) to selection in the primary trait (y) for a secondary trait (x) is given by

$$CRy = i_x h_x h_y \rho_g \sqrt{V_{py}}$$

where h_x and h_y are square roots of the heritabilities of the two respective traits and ρg is the genetic correlation between traits. This relationship may be reduced to:

$$CRy = i_x \rho_g h_x \sqrt{V_{Gy}}$$

since $h_y = \sqrt{(V_{Gy}/V_{py})}$.

It is clear that the effectiveness of indirect selection depends on the magnitude of genetic correlation and the heritability of the secondary traits being selected. Furthermore, given the same selection intensity and a high genetic correlation between the traits, indirect selection for the primary trait will be more effective than directional selection, if heritability of the secondary trait is high ($\rho_g h_x > h_y$). Such a scenario would occur when the secondary trait is less sensitive to environmental change (or can be measured under controlled conditions). Also, when the secondary trait is easier and less costly to measure, the breeder may apply a higher selection pressure to it.

Correlated response has wider breeding application in homozygous, self-fertilizing species and apomicts. Additive genetic correlation is important in selection in plant breeding. As previously discussed, the additive breeding value is what is transferred to offspring and can be changed by selection. Hence, where traits are additively genetically correlated, selection for one trait will produce a correlated response in another.

4.2.12 Selection for multiple traits

Plant breeders may use one of three basic strategies to simultaneously select multiple traits – **tandem selection, independent curling,** and **selection index**. Plant breeders often handle very large numbers of plants in a segregating population using limited resources (time, space, labor, money, etc.). Along with the large number of individuals are the many breeding traits often considered in a breeding program. The sooner they can reduce the numbers of plants to the barest minimum and, more importantly, to the most desirable and promising individuals, the better. Highly heritable and readily scorable traits are easier to select for in the initial stages of a breeding program.

Tandem selection

In this mode of selection, the breeder focuses on one trait at a time (serial improvement). One trait is selected for several generations, then another trait is focused on for the next period. The questions of how long each trait is selected before a switch and at what selection intensity are major considerations for the breeder. It is effective when genetic correlation does not exist between the traits of interest or when the relative importance of each trait changes throughout the years.

Independent curling

Also called truncation selection, independent curling entails selecting for multiple traits in one generation. For example, for three traits, A, B and C, the breeder may select 50% plants per family for A on phenotypic basis, and from that select 40% plants per family based on trait B; finally, from that subset, 50% of the plants per family is selected for trait C, giving a total of 10% selection intensity ($0.5 \times 0.4 \times 0.5$).

Index selection

A breeder has a specific objective for conducting a breeding project. However, selection is seldom made on the basis of one trait alone. For example, if the breeding project is for disease resistance, the objective will be to select a genotype that combines disease resistance with the qualities of the elite adapted cultivar. Invariably, breeders usually practice

Industry highlights
Recurrent selection with soybean

Joseph W. Burton

USDA Plant Science Building, 3127 Ligon Street, Raleigh, NC 27607, USA

Selection using a restricted index

Two commodities, protein meal and oil, are produced from soybean (*Glycine max* (L.) merr.) and give the crop its value. Soybean seeds are crushed, oil is extracted, and protein meal is what remains. On a dry-weight basis, soybeans are approximately 20% oil and 40% protein. Concentration of protein in the meal is dependent on protein concentration in soybean seeds. Protein meal is traded either as 44% protein or 48% protein. The 48% protein meal is more valuable, so increasing or maintaining protein concentration in soybean seeds has been a breeding objective. Protein is negatively associated with oil in seeds and in many breeding populations it is negatively associated with seed yield (Brim and Burton, 1979).

The negative association between yield and protein could be due to genetic linkage as well as physiological processes (Carter *et al.*, 1982). Thus, a breeding strategy is needed which permits simultaneous selection of both protein and yield. Increased genetic recombination should also be helpful in breaking unfavorable linkages between genes that contribute to the negative yield and protein relation. We devised a recurrent S_1 family selection program to satisfy the second objective and applied a restricted index to family performance to achieve the first objective.

Selection procedure

A population designated RS 4 was developed using both high yielding and high protein parents. The high yielding parents were the cultivars, "Bragg", "Ransom", and "Davis" (Figure B4.1). The high protein parents were 10 F_3 lines from cycle 7 of another recurrent selection population designated IA (Brim and Burton, 1979). In that population, selection had been solely for protein. Average protein concentration of the 10 parental F_3 lines was 48.0%. The base or C_0 population was developed by making seven or eight matings between each high protein line and the three cultivars, resulting in 234 hybrids. The S_0 generation was advanced at the USDA Winter Soybean Nursery in Puerto Rico, resulting in 234 S_1 families. These were tested in two replications at two locations. Both seed yield and protein concentration were determined for each family. Average protein concentration of the initial population was 45.6%. As this was an acceptable increase in

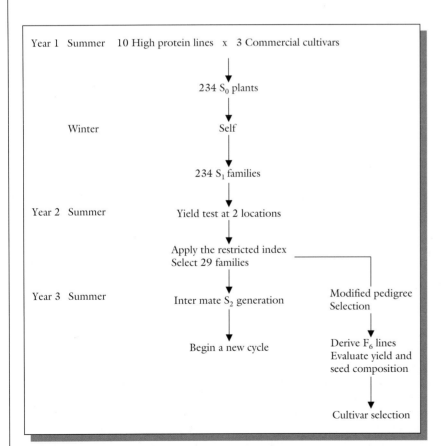

Figure B4.1 Recurrent S_1 family for yield and protein using a restricted index.

protein, a restricted selection index was applied aimed at increasing yield and holding protein constant. This index was the following:

$$I = \text{Yield} - (\sigma_{Gyp}/\sigma^2_{Gp}) \times \text{protein}$$

where σ_{Gp} is the estimated genetic covariance between yield and protein and σ^2_{Gp} is the estimated genetic variance of protein (Holbrook *et al.*, 1989). Using this index, 29 families were selected.

The following summer, these 29 families (now in the S_2 generation) were randomly intermated. To do this, the following procedure was used. Each day of the week, flowers for pollen were collected from 24 of the families and used to pollinate the remaining five families. A different set of 24 and 5 families were used as males and females, respectively, each day. This process was followed until each family had at least seven successful pollinations on seven different plants within each family. These were advanced in the winter nursery to generate the S_1 families for the next cycle of selection.

Development of "Prolina" soybean

Modified pedigree selection was applied to the S_1 families chosen in the first restricted index selection cycle. F_6 lines were tested in replicated yield tests. One of those lines, N87-984, had good yielding ability and 45% seed protein concentration. Because of heterogeneity for plant height within the line, F_9 lines were derived from N87-984 using single seed descent. These were yield tested in multiple North Carolina locations. The two lines most desirable in terms of uniformity, protein concentration, and seed yield were bulked for further testing and eventual release as the cultivar, Prolina (Burton *et al.*, 1999). At its release, Prolina had 45% protein compared with 42.7% for the check cultivar, Centennial, and similar yielding ability.

Recurrent selection using male sterility

In the previous example, intermating the selections was done using hand pollination. Hand pollination with soybean is time consuming and difficult. The average success rate in our program during the August pollinating season was 35%. Thus, a more efficient method for recombination would be helpful in a recurrent selection program which depends on good random mating among selected progeny for genetic recombination and reselection.

Genetic (nuclear) male sterility has been used for that purpose. Several nuclear male sterile alleles have been identified (Palmer *et al.*, 2004). The first male sterile allele to be discovered (*ms1*) is completely recessive (Brim and Young, 1971) to male fertility (*Ms$_1$*). Brim and Stuber (1973) described ways that it could be used in recurrent selection programs. Plants which are homozygous for the *ms$_1$* allele are completely male sterile. All seeds produced on male sterile plants result from pollen contributed by a male-fertile plant (*Ms$_1$Ms$_1$* or *Ms$_1$ms$_1$*) via an insect pollen vector. The ms1ms1 male sterile plants are also partially female sterile, so that seed set on male sterile plants is low, averaging about 35 seeds per plant. In addition, most pods have only one seed and that seed is larger (30–40% larger) than seeds which would develop on a fertile plant with similar genetic background. The *ms$_1$* allele is maintained in a line that is 50% *ms$_1$ms$_1$* and 50% *Ms$_1$ms$_1$*. This line is planted in an isolation block. One-half of the pollen from male-fertile plants carries the *Ms$_1$* fertile allele and one-half carries the *ms$_1$* sterile allele. Male sterile plants are pollinated by insect vectors, usually various bee species. At maturity, only seeds of male sterile plants were harvested. These occur in the expected genotypic ratio of 50% *Ms$_1$ms$_1$* : 50% *ms$_1$ms$_1$*.

One of the phenotypic consequences of ms1 male sterility and low seed set is incomplete senescence. At maturity, soybeans normally turn yellow, leaves abscise, and pods and seeds dry. Seed and pods on male sterile plants mature and dry normally, but the plants remain green and leaves do not abscise. Thus, they are easily distinguished from male-fertile plants.

To use nuclear male sterility in a recurrent selection experiment, a population is developed for improvement which segregates for one of the recessive male sterile alleles. This can be accomplished in a number of ways depending on breeding objectives. Usually, a group of parents with desirable genes are mated to male sterile genotypes. This can be followed by one or more backcrosses. Eventually, an F_2 generation which segregates for male sterility is allowed to randomly intermate. Seeds are harvested from male sterile plants. Then several different selection units are possible. These include the male sterile plant itself (Tinius, *et al.*, 1991); the seeds (plants) from a single male sterile plant (a half-sib family) (Burton and Carver, 1993); and selfed seeds (plants) of an individual from a male sterile plant (S_1 family) (Burton *et al.*, 1990). Selection can be among and/or within the families (Carver *et al.*, 1986). If appropriate markers are employed, half-sib selection using a tester is also possible (Feng *et al.*, 2004). As with all recurrent selection schemes selected individuals are intermated. These can be either remnant seed of the selection unit or progeny of the selection unit. The male sterile alleles segregate in both because both were derived in some manner from a single male sterile plant.

Recurrent mass selection for seed size

Because seed set on male sterile plants is generally low, we hypothesized that the size of the seed was not limited by source (photosynthate) inputs. Thus, selecting male sterile plants with the largest seeds would be selecting plants with the most genetic potential for producing large seeds. If so, this would mean that male-fertile plants derived from those selections would also produce larger seeds and perhaps have increased potential for overall seed yield.

To test this hypothesis, we conducted recurrent mass selection for seed size (mg/seed) in a population, N80-1500, that segregated for the *ms1* male sterile allele and had been derived from adapted high yielding cultivar and breeding lines (Burton and Brim, 1981). The population was planted in an isolation block. Intermating between male sterile and male-fertile plants occurred at random. In North Carolina, there are numerous wild insect pollen vectors so introduction of domestic bees was not needed. If needed, bee hives can be placed in or near the isolation block. At maturity, seeds were harvested from approximately 200 male-sterile plants. To make sure that the entire population was sampled, the block was divided into sections and equal numbers of plants were sampled from each section. Seeds from each plant were counted and weighed. The twenty plants with largest seeds (greatest mass) were selected. These 20 selections were grown in a winter nursery and bulk selfed to increase seed numbers. Equal numbers of seeds from the 20 selfed selections were combined and planted in another isolation block the following summer to begin another selection cycle (Figure B4.2).

With this method, one cycle of selection is completed each year. This is mass selection where only the female parent is selected. Additionally, the female parents all have an inbreeding coefficient of 0.5 because of the selfing seed increase during the winter. Thus, the expected genetic gain (Δ_G) for this selection scheme is:

$$\Delta_G = S(0.75)\sigma_A^2(\sigma_P^2)^{-1}$$

where S is the selection differential, σ_A^2 is the additive genetic variance and σ_P^2 is the phenotypic variance. This method was also used to increase oleic acid concentration in seed lipids (Carver *et al.*, 1986).

At the end of cycle 4 and cycle 7, selected materials from each cycle were evaluated in replicated field

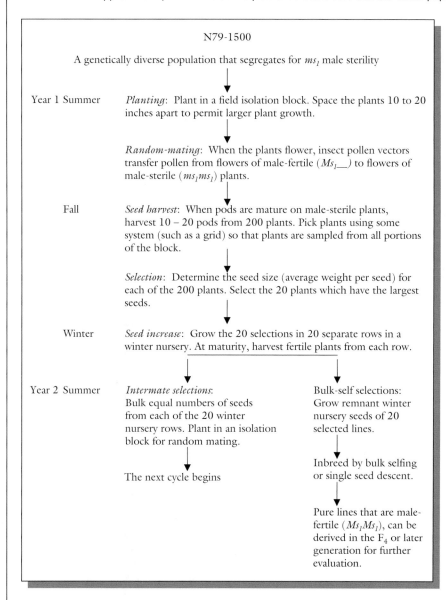

Figure B4.2 Recurrent mass selection for seed size in soybean using nuclear male sterility to inter mate selections.

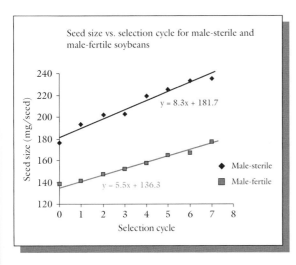

Figure B4.3 Seed size changes with each selection for male-sterile and male-fertile soybeans.

Figure B4.4 Distribution of seed diameters initially and after four cycles of selections for larger seeds.

trials. Results of those trials showed that the method had successfully increased both seed size and yield in the population. In seven cycles of selection, seed size of the male sterile plants increased linearly from 196 mg/seed to 235 mg/seed. Male-fertile seed size also increased linearly from 138 to 177 mg/seed (Figure B4.3). Not only did the mass increase, but the physical size of the seeds increased as well. The range in seed diameter initially was 4.8–7.1 mm. After four cycles of selection, diameter range had shifted and was 5.2–7.5 mm (Figure B4.4). Yield increased at an average rate of 63.5 kg/ha each cycle (Figure B4.5), about 15% overall. There was some indication that after cycle 5 changes in yield were leveling off as yields of selections from cycle 5 and cycle 7 were very similar.

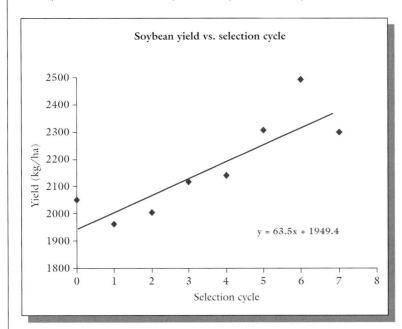

Figure B4.5 Correlated changes in seed yield with selection for increased seed size.

This method is relatively inexpensive. Little field space is required and only a balance is needed to determine which individual should be selected. The ability to complete one cycle each year also makes it efficient. The largest expense is probably that needed to increase the seeds from selected male sterile plants in a winter greenhouse or nursery. The method may be quite useful for introgressing unadapted germplasm into an adapted breeding population, followed by rapid improvement of productivity. The population could be sampled in any cycle using single seed descent. Pure lines developed from these populations would be handled exactly as those developed from single crosses in typical modified pedigree selection programs.

References

Brim, C.A., and Burton, J.W. (1979). Recurrent selection in soybeans: II. Selection for increased protein in seeds. *Crop Sci.*, **19**:494–498.

Brim, C.A., and Stuber, C.W. (1973). Application of genetic male sterility to recurrent selection schemes in soybeans. *Crop Sci.*, **13**:528–530.

Brim, C.A., and Young, M.F. (1971). Inheritance of a male-sterile character in soybeans. *Crop Sci.*, **11**:564–566.

Burton, J.W., and Brim, C.A. (1981). Registration of two soybean germplasm populations. *Crop Sci.*, **21**:801.

Burton, J.W., and Carver, B.F. (1993). Selection among S1 families vs. selfed half-sib and full-sib families in autogamous crops. *Crop Sci.*, **33**:21–28.

Burton, J.W., Koinange, E.M.K., and Brim, C.A. (1990). Recurrent selfed progeny selection for yield in soybean using genetic male sterility. *Crop Sci.*, **30**:1222–1226.

Burton, J.W., Carter, T.E.Jr., and Wilson, R.F. (1999). Registration of 'Prolina' soybean. *Crop Sci.*, **39**:294–295.

Carter, T.E.Jr., Burton, J.W., and Brim, C.A. (1982). Recurrent selection for percent protein in soybean seed- indirect effects on plant N accumulation and distribution. *Crop Sci.*, **22**:513–519.

Carver, B.F., Burton, J.W. Carter, T.E.Jr., and Wilson, R.F. (1986). Cumulative response to various recurrent selection schemes in soybean oil quality and correlate agronomic traits. *Crop Sci.*, **26**:853–858.

Feng, L., Burton, J.W., Carter, T.E.Jr., and Pantalone, V.R. (2004). Recurrent half-sib selection with test cross evaluation for increased oil content in soybean. *Crop Sci.*, **44**:63–69.

Holbrook, C.C., Burton, J.W., and Carter, T.E.Jr. (1989). Evaluation of recurrent restricted index selection for increasing yield while holding seed protein constant in soybean. *Crop Sci.*, **29**:324–329.

Palmer, R.G., Pfeiffer, T.W., Buss, G.R., Kilen, T.C. (2004). Quantitative genetics. P. 137–233, in (eds H.R. Boerma and J.E. Specht) *Soybeans, improvement, production, and uses.* Agronomy Monograph 16, 3rd edn, American Society of Agronomy, Crop Science Society of America, and Soil Science Society of America, Madison, WI.

Tinius, C.N., Burton, J.W., and Carter, T.E.Jr. (1991). Recurrent selection for seed size in soybeans. I. Response to selection in replicate populations. *Crop Sci.*, **31**:1137–1141.

selection on several traits simultaneously. The problem with this approach is that as more traits are selected for, the less the selection pressure that can be exerted on any one trait. Therefore, the breeder should select on the basis of two or three traits of the highest economic value. It is conceivable that a trait of high merit may be associated with other traits of less economic value. Hence, using the concept of selection on total merit, the breeder would make certain compromises, selecting individuals that may not have been selected were it based on a single trait.

In selecting on a multivariate phenotype, the breeder explicitly or implicitly assigns a weighting scheme to each trait, resulting in the creation of a univariate trait (an **index**) that is then selected. The index is the best linear prediction of an individual's breeding value. It takes the form of a multiple regression of breeding values on all the sources of information available for the population.

The methods used for constructing an index usually include heritability estimates, relative economic importance of each trait, and genetic and phenotypic correlation between the traits. The most common index is a linear combination that is mathematically expressed as:

$$I = \sum_{i=1}^{m} b_j z_j = b^I z$$

where z is the vector of phenotypic values in an individual and b is a vector of weights. For three traits, the form may be:

$$I = aA^1 + bB^1 + cC^1$$

where a, b, and c are coefficients correcting for relative heritability and the relative economic importance of traits A, B, and C, respectively, and A^1, B^1, and C^1 are the numerical values of traits A, B, and C expressed in standardized form. A standardized variable (X^1) is calculated as:

$$X^1 = (X - \bar{X})/\sigma_x$$

where, X is the record of performance made by an individual; $\bar{X}$ is the average performance of the population; and σ_x is the standard deviation of the trait.

The classical selection index has the following form:

$$I = b_1 x_1 + b_2 x_2 + b_3 x_3 + \ldots + b_m x_n$$

where x_1, x_2, x_3, x_n are the phenotypic performance of the traits of interest, and b_1, b_2, and b_3 are the relative weights attached to the respective traits. The weights could be simply the respective relative economic importance of each trait; the resulting index is called the **basic index** and may be used in cultivar assessment in official registration trials.

An index by itself is meaningless, unless it is used in comparing several individuals on a relative basis. Furthermore, in comparing different traits, the breeder is faced with the fact that the mean and variability of each trait is different and, frequently, the traits are measured in different units. Standardization of variables resolves this problem.

4.2.13 Concept of intuitive index

Plant breeding was described in Chapter 2 as both a science and an art. Experience (with the crop, the methods of breeding, breeding issues concerning the crop) is advantageous in having success in solving plant breeding problems. Plant breeders, as previously indicated, often must evaluate many plant traits in a breeding program. Whereas one or a few would be identified as key traits and focused on in a breeding program, breeders are concerned about the overall performance of the cultivar. During selection, breeders formulate a mental picture of the product desired from the project and balance good qualities against moderate defects as they make final judgments in the selection process.

Explicit indices are laborious, requiring the breeder to commit to extensive record keeping and statistical analysis. Most breeders use a combination of truncation selection and intuitive selection index in their programs.

4.2.14 The concept of general worth

For each crop, there is a number of traits which, considered together, define the overall desirability of the cultivar from the combined perspectives of the producer and the consumer. These traits may range between about a dozen to several dozens, depending on the crop, and constitute the primary pool of traits that the breeder may target for improvement. These traits differ in importance (economic and agronomic) as well as ease with which they can be manipulated through breeding. Plant breeders typically target one or few of these traits for direct improvement in a breeding program. That is, the breeder draws up a working list of traits to address the needs embodied in the stated objectives. Yield of the economic product is almost universally the top priority in a plant breeding program. Disease resistance is more of a local issue because what may be economically important in one region may not be important in another area. Even though a plant breeder may focus on one or a few traits at a time, the ultimate objective is the improvement of the totality of the key traits that impact the overall desirability or general worth of the crop. In other words, breeders ultimately have a holistic approach to selection in a breeding program. The final judgments are made on a balanced view of the essential trait of the crop.

4.2.15 Nature of breeding traits and their levels of expression

Apart from relative importance, the traits the plant breeder targets vary in other ways. Some are readily evaluated by visual examination (e.g., shape color, size), whereas others require a laboratory assay (e.g., oil content) or mechanical measurement (e.g., fiber traits of cotton). Special provisions (e.g., greenhouse, isolation block) may be required in disease breeding, whereas yield evaluations are most reliable when conducted over seasons and locations in the field.

In addition to choosing the target traits, the breeder will have to decide on the level of expression of each one, below which a plant material would be declared worthless. The acceptability level of expression of a trait may be narrowly defined (stringent selection) or broadly defined (loose selection). In industrial crops (e.g., cotton), the product quality may be strictly defined (e.g., a certain specific gravity, optimum length). In disease resistance breeding, there may not be a significant advantage in selecting for extreme resistance over selecting for less than complete resistance. On the other hand, in breeding nutritional quality, there may be legal guidelines as to threshold expression for toxic substances.

4.2.16 Early generation testing

Early generation testing is a selection procedure in which the breeder initiates testing of genetically heterogeneous lines or families in an earlier than normal generation. For example, recurrent selection with testers can be used to evaluate materials in early generations (Chapter 15). A major consideration of the breeder in selecting a particular breeding method is to maximize genetic gain per year. Testing early, if effective, helps to identify and select potential cultivars from superior families in the early phase of the breeding program. The early generation selection method has been favorably compared with other methods, such as pedigree selection, single seed descent, and bulk breeding. The question of how early the test is implemented often arises. Should it be in the F_1-, F_2- or F_3-derived families? Factors to consider in deciding on the generation in which selection is done include the trait being improved and the availability of off-season nurseries to use in producing additional generations per year (in lieu of selecting early).

4.2.17 Concept of combining ability

Over the years, plant breeders have sought ways of facilitating plant breeding through efficient selection of parents for a cross, effective and efficient selection within a segregating population, prediction of response to selection, amongst other needs. Quantitative assessment of the role of genetics in plant breeding entails the use of statistical genetics approaches to estimate variances and partition them into components. Because variance estimates are neither robust nor accurate, the direct benefits of statistical genetics to the breeder have been limited.

In 1942, Sprague and Tatum proposed a method of evaluating inbred lines to be used in corn hybrid production that was free of the genetic assumptions that accompany variance estimates. Called **combining ability**, the procedure entails the evaluation of a set of crosses among selected parents to ascertain the extent to which variances among crosses are attributable to statistically additive characteristics of the parents, and what could be considered the effect of residual interactions. Crossing each line with several other lines produces an additional measure in the mean performance of each line in all crosses. This mean performance of a line, when expressed as a deviation from the mean of all crosses, gives what Sprague and Tatum called the **general combining ability (GCA)** of the lines.

The GCA is calculated as the average of all F_1s having this particular line as one parent, the value being expressed as a deviation from the overall mean of crosses. Each cross has an expected value (the sum of GCAs of its two parental lines). However, each cross may deviate from the expected value to a greater or lesser extent, the deviation being the **specific combining ability (SCA)** of the two lines in combination. The differences of GCA are due to the additive and additive x additive interactions in the base population. The differences in SCA are attributable to non-additive genetic variance. Furthermore, the SCA is expected to increase in variance more rapidly as inbreeding in the population reaches high levels. GCA is the average performance of a plant in a cross with different tester lines, while SCA measures the performance of a plant in a specific combination in comparison with other cross combinations.

The mathematical representation of this relationship for each cross is as:

$$X_{AB} = \bar{X} + G_A + G_B + S_{AB}$$

where $\bar{X}$ is the general mean, G_A and G_B are the general combining ability estimates of the parents, and S_{AB} is the statistically unaccounted for residual or specific combining ability (SCA). The types of interactions that can be obtained depend upon the mating scheme used to produce the crosses, the most common being the diallel mating design (full or partial diallel).

Plant breeders may use a variety of methods for estimating combining abilities, including the polycross and top-crossing methods. However, the diallel cross (each line is mated with every other line) developed by B. Griffing in 1956 is perhaps the most commonly used method. The GCA of each line is calculated as follows:

$$Gx = [Tx/(n-2)] - [\Sigma T/n(n-2)]$$

where x represents a specific line. Using the data in Table 4.3. G_A can be calculated as:

$$
\begin{aligned}
G_A &= [T_A/n - 2] - [\Sigma T/n(n-2)] \\
&= [226/8] - [2024/10(10-2)] \\
&= 28.25 - 25.3 \\
&= 2.95
\end{aligned}
$$

The others may be calculated as for line A. The next step is to calculate the expected value of each cross. Using the cross CD as an example, the expected value is calculated as follows:

$$E(X_{CD}) = -4.18 + 5.33 + 22.49 = 23.64$$

SCA is calculated as follows:

$$
\begin{aligned}
SCA_{CD} &= 26 - 23.64 \\
&= 2.36
\end{aligned}
$$

Table 4.3 Calculating general and specific combining abilities.

	B	C	D	E	F	G	H	I	J	Total	GCA
A	26	24	29	28	22	21	27	21	28	226	2.98
B		21	35	30	26	22	29	14	19	222	2.45
C			26	21	10	14	13	17	23	169	−4.18
D				25	31	32	28	21	18	245	5.33
E					13	23	15	15	14	184	−2.3
F						20	31	17	15	185	−2.18
G							32	14	12	190	−1.55
H								35	38	248	5.7
I									17	171	−3.93
J										184	−2.3
										2024	0

This is done for each combination and a plot of observed values versus expected values plotted. Because the values of SCA are subject to sampling error, the points on the plot do not lie on the diagonal. The distance from each point to the diagonal represents the SCA plus sampling error of the cross. Additional error would occur if the lines used in the cross are not highly inbred (error due to the sampling of genotypes from the lines).

Combining ability calculations are statistically robust, being based on first degree statistics (totals, means). No genetic assumptions are made about individuals. The concept is applicable to both self-pollinated and cross-pollinated species, for identifying desirable cross combinations of inbred lines to include in a hybrid program or for developing synthetic cultivars. It is used to predict the performance of hybrid populations of cross-pollinated species, usually via a test cross or polycross. It should be pointed out that combining ability calculations are properly applied only in the context in which they were calculated. This is because GCA values are relative and depend upon the mean of the chosen parent materials in the crosses.

A typical ANOVA for combining ability analysis is as follows:

Source	df	SS	MS	EMS
GCA	$p-1$	S_G	M_G	$\sigma_E^2 + \sigma_{SCA}^2 + \sigma_{GCA}^2$
SCA	$p(p-1)/2$	S_S	M_S	$\sigma_E^2 + \sigma_{SCA}^2$
Error	m	S_E	M_E	σ_E^2

The method used of a combining ability analysis depends on available data: The method depends on available data:

- Parents + F_1 or F_2 and reciprocal crosses (i.e., p^2 combination).
- Parents + F_1 or F_2, without reciprocals (i.e., $\frac{1}{2} p (p + 1)$ combinations).
- $F_1 + F_2$ + reciprocals, without parents and reciprocals (i.e., $\frac{1}{2} p(p - 1)$ combination.
- Only F_1 generations without parents, reciprocals (i.e., $\frac{1}{2} p(p - 1)$ combinations.

4.2.18 Mating designs

Artificial crossing or mating is a common activity in plant breeding programs for generating various levels of relatedness among the progenies that are produced. Mating in breeding has two primary purposes:

1 To generate information for the breeder to understand the genetic control or behavior of the trait of interest.
2 To generate a base population to initiate a breeding program.

The breeder influences the outcome of a mating by the choice of parents, the control over the frequency each parent is involved in mating, and the number of offspring per mating, amongst other ways. A mating may be as simple as a cross between two parents, to the more complex diallel mating.

Hybrid crosses

These are reviewed here to give the student a basis for comparison with the random mating schemes to be presented.

- Single cross $= A \times B \rightarrow F_1 (AB)$
- Three-way cross $= (A \times B) \rightarrow F_1 \times C \rightarrow (ABC)$
- Backcross $= (A \times B) \rightarrow F_1 \times A \rightarrow (BC1)$
- Double cross $= (A \times B) \rightarrow F_{AB}; (C \times D) \rightarrow F_{CD};$ $F_{AB} \times F_{CD} \rightarrow (ABCD)$

These crosses are relatively easy to genetically analyze. The breeder exercises significant control over the mating structure.

Mating designs for random mating populations

The term **mating design** is usually applied to schemes used by breeders and geneticists to impose random mating for a specific purpose. To use these designs, certain assumptions are made by the breeder:

- The materials in the population have diploid behavior. However, polyploids that can exhibit disomic inheritance (alloploids) can be studied.
- The genes controlling the trait of interest are independently distributed among the parents (i.e., uncorrelated gene distribution).
- The absence of: non-allelic interactions, reciprocal differences, multiple alleles at the loci controlling the trait, and G × E interactions.

Biparental mating (or pair crosses) In this design, the breeder selects a large number of plants (n) at random and crosses them in pairs to produce $\frac{1}{2}n$ full-sib families. The biparental (BIP) is the simplest of the mating designs. If r plants per progeny family are evaluated, the variation within and between families may be analyzed as follows:

Source	df	MS	EMS
Between families	$(\frac{1}{2}n) - 1$	MS_1	$\sigma^2 w + r\sigma^2 b$
Within families	$\frac{1}{2}n(r - 1)$	MS_2	$\sigma^2 w$

where $\sigma^2 b$ is the covariance of full-sibs ($= \frac{1}{2}V_A + \frac{1}{4} V_D + V_{EC} = 1/r (MS_1 - MS_2)$ and $\sigma^2 w = \frac{1}{2}V_A + \frac{3}{4} V_D + V_{EW} = MS_2$)

The limitation of this otherwise simple to implement design is its inability to provide the needed information to estimate all the parameters required by the model. The progeny from the design comprise full-sibs or unrelated individuals. There is no further relatedness among individuals in the progeny. The breeder must make unjustifiable assumptions in order to estimate the genetic and environmental variance.

Polycross This design is for intermating a group of cultivars by natural crossing in an isolated block. It is most suited to species that are obligate cross-pollinated (e.g., forage grasses and legumes, sugarcane, sweet potato), but especially those that can be vegetatively propagated. It provides an equal opportunity for each entry to be crossed with every other entry. It is critical that the entries be equally represented and randomly arranged in the crossing block. If 10 or less genotypes are involved, the Latin square design may be used. For a large number of entries, the completely randomized block design may be used. In both cases, about 20–30 replications are included in the crossing block. The ideal requirements are hard to meet in practice because of several problems, placing the system in jeopardy of deviating from random mating. If all the entries do not flower together, mating will not be random. To avoid this, the breeder may plant late flowering entries earlier.

Pollen may not be dispersed randomly, resulting in concentrations of common pollen in the crossing block. Half-sibs are generated in a polycross because progeny from each entry has a common parent. The design is used in breeding to produce synthetic cultivars, recombining selected entries of families in recurrent selection breeding programs, or for evaluating the GCA of entries.

North carolina design I Design I is a very popular multipurpose design for both theoretical and practical plant breeding applications (Figure 4.5). It is commonly used to estimate additive and dominance variances as well as for evaluation of full- and half-sib recurrent selection. It requires sufficient seed for replicated evaluation trials, and hence is not of practical application in breeding species that are not capable of producing large amounts of seed. It is applicable to both self- and cross-pollinated species that meet this criterion. As a nested design, each member of a group of parents used as males is mated to a different group of parents. NC design I is a hierarchical design with non-common parents nested in common parents.

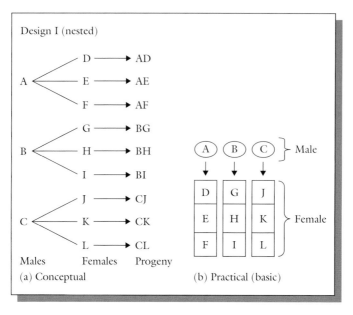

Figure 4.5 North Carolina Design I. (a) This design is a nested arrangement of genotypes for crossing in which no male is involved in more than one cross. (b) A practical layout in the field.

The total variance is partitioned as follows:

Source	df	MS	EMS
Males	$n - 1$	MS_1	$\sigma^2 w + r\sigma^2_{mf} + rf\sigma^2_m$
Females	$n_1(n_2 - 1)$	MS_2	$\sigma^2 w + r\sigma^2_{mf}$
Within progenies	$n_1 n_2(r - 1)$	MS_3	$\sigma^2 w$

$$\sigma^2_m = [MS_1 - MS_2]/rn_2 = {}^1\!/_4 V_A$$
$$r\sigma^2_{flm} = [MS_2 - MS_3]/r = ({}^1\!/_4)V_A + ({}^1\!/_4)V_D$$
$$\sigma^2 w = MS_3 = ({}^1\!/_2)V_A + ({}^3\!/_4)V_D + E$$

This design is most widely used in animal studies. In plants, it has been extensively used in maize breeding for estimating genetic variances.

North carolina design II In this design, each member of a group of parents used as males is mated to each member of another group of parents used as females. **Design II** is a factorial mating scheme similar to Design I (Figure 4.6). It is used to evaluate inbred lines for combining ability. The design is most adapted to plants that have multiple flowers, so that each plant can be used repeatedly as both male and female. Blocking is used in this design to allow all the mating involving a single group of males to a single group of females to be kept intact as a unit. The design is essentially a two-way ANOVA in which the variation may be partitioned into difference between males and females and their interaction. The ANOVA is as follows:

Source	df	MS	EMS
Males	$n_1 - 1$	MS_1	$\sigma^2 w + r\sigma^2_{mf} + rn\sigma^2_m$
Females	$n_2 - 1$	MS_2	$\sigma^2 w + r\sigma^2_{mf} + rn_1\sigma^2_f$
Males × females	$(n_1 - 1)(n_2 - 1)$	MS_3	$\sigma^2 w + r\sigma^2_{mf}$
Within progenies	$n_1 n_2(r - 1)$	MS_4	$\sigma^2 w$

$$\sigma^2_m = [MS_1 - MS_3]/rn_2 = ({}^1\!/_4)V_A$$
$$r\sigma^2_f = [MS_2 - MS_3]/rn_1 = ({}^1\!/_4)V_A$$
$$r\sigma^2_{mf} = [MS_3 - MS_4]/r = ({}^1\!/_4)V_D$$
$$\sigma^2 w = MS_4 = ({}^1\!/_2)V_A + ({}^3\!/_4)V_D + E$$

The design also allows the breeder to measure not only GCA but also SCA.

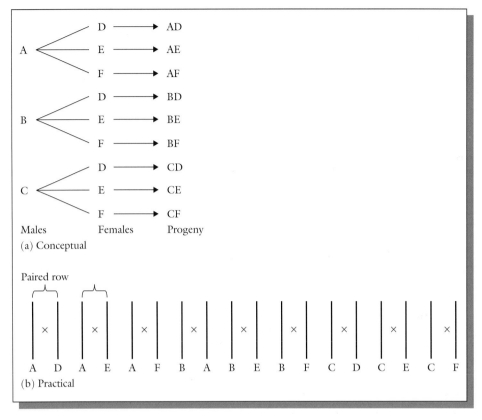

Figure 4.6 North Carolina Design II. (a) This is a factorial design. (b) Paired rows may be used in the nursery for factorial mating of plants.

North carolina design III　In this design, a random sample of F_2 plants is backcrossed to the two inbred lines from which the F_2 was descended. It is considered the most powerful of all the three North Carolina designs. However, it was made more powerful by the modifications made by Kearsey and Jinks that add a third tester (not just the two inbreds) (Figure 4.7). The modification is called the **triple test cross** and is capable of testing for non-allelic (epistatic) interactions that the other designs cannot, and also estimate additive and dominance variance.

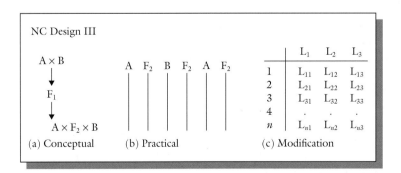

Figure 4.7 North Carolina Design III. The conventional form (a), the practical layout (b), and the modification (c) are shown.

Diallel cross A **complete diallel mating** design is one that allows the parents to be crossed in all possible combinations, including selfs and reciprocals. This is the kind of mating scheme required to achieve Hardy–Weinberg equilibrium in a population. However, in practice, a diallel with selfs and reciprocals is neither practical nor useful for several reasons. Selfing does not contribute to recombination of genes between parents. Furthermore, recombination is achieved by crossing in one direction making reciprocals unnecessary. Because of the extensive mating patterns, the number of parents that can be mated this way is limited. For p entries, a complete diallel will generate p^2 crosses. Without selfs and reciprocals, the number is $p(p-1)/2$ crosses.

When the number of entries is large, a **partial diallel mating** design, which allows all parents to be mated to some but not all other parents in the set, is used. A **diallel design** is most commonly used to estimate combining abilities (both general and specific). It is also widely used for developing breeding populations for recurrent selection.

Nursery arrangements for application of complete and partial diallel are varied. Because a large number of crosses are made, diallel mating takes a large amount of space, seed, labor, and time to conduct. Because all possible pairs are contained in one half of a symmetric Latin square, this design may be used to address some of the space needs.

There are four basic methods developed by Griffing that vary in either the omission of parents or the omission of reciprocals in the crosses. The number of progeny families (pf) for methods 1 through 4 are: $pf = n^2$, $pf = \frac{1}{2} n(n+1)$, $pf = n(n-1)$, and $pf = \frac{1}{2}n(n-1)$, respectively. The ANOVA for method 4, for example, is:

Source	df	EMS
GCA	$n_1 - 1$	$\sigma^2 e + r\sigma_g^2$ $+ r(n-2)\sigma^2$
SCA	$[n(n-3)]/2$	$\sigma^2 e + r\sigma_g^2$
Reps x Crosses	$(r-1)\{[n(n-1)/2]-1\}$	$\sigma^2 e$

Comparative evaluation of mating designs Hill, Becker and Tigerstedt roughly summarized these mating designs in two ways:

1 In terms of coverage of the population: BIPs > NCM-I > Polycross > NCM-III > NCM-II > diallel, in that order of decreasing effectiveness.
2 In terms of amount of information: Diallel > NCM-II > NCM-II > NCM-I > BIPs.

The diallel mating design is the most important for GCA and SCA. These researchers emphasized that it is not the mating design *per se*, but rather the breeder who breeds a new cultivar. The implication is that the proper choice and use of a mating design will provide the most valuable information for breeding.

4.3 Molecular quantitative genetics

Molecular quantitative genetics mainly focuses on evaluating the coupling association of the polymorphic DNA sites with the phenotypic variations of quantitative and complex traits. In addition, whereas classical quantitative genetics deals with the holistic status of all genes, molecular quantitative genetics dissects the genetic architectures of quantitative genes (concerned with the analytical status of the major genes and holistic status of the minor genes).

4.3.1 The genetic architecture of quantitative traits

The genetic architecture of quantitative traits entails the number of QTLs that influence a quantitative trait, the number of alleles that each QTL possesses, the frequencies of the alleles in the population, and the influence of each QTL and its alleles on the quantitative trait. Identifying and characterizing QTLs will provide a basis for selecting and improving plant species. The summation of QTL studies indicates that QTL alleles with large effects are rare; most quantitative traits are controlled by many loci with small effects.

Researchers commonly use one of two fundamental approaches to design and study quantitative traits. In what is called the top-down approach, they start with the trait of interest and then attempt to draw inferences about the underlying genetics from examining the degree of trait resemblance among related subjects. It is usually the first step taken to determine if there is any evidence for a genetic component. It is also described as the unmeasured genotype approach because it focuses on the inheritance pattern without measuring any genetic variations. Typical statistical analyses employed in this approach are **heritability** and **segregational analysis**.

In the second approach, the bottom-up (measured approach), researchers actually measure QTLs and use the information to draw inferences about which genes might have a role in the genetic architecture of a quantitative trait. Typical statistical analyses employed in this approach are **linkage analysis** and **association analysis.** The second approach is becoming more assessable with the advent of newer, more efficient and less expensive technologies to measure QTLs. These technologies include DNA microarrays and protein mass spectrometry. They allow researchers to quantitatively measure the expression levels of thousands of gene simultaneously, and thereby study gene expression at the both the RNA and protein levels as a quantitative trait.

4.3.2 Effects of QTL on phenotype

QTLs influence quantitative trait phenotype in various ways. They can influence quantitative trait levels (quantitative trait means can be different among different genotypes). Most of the statistical methods used for studying quantitative traits are based on genotypic means. The variation in phenotypic values may also vary among genotypes. Furthermore, QTLs also may affect the correlation among quantitative traits as well as the dynamics of traits (the change in phenotype over a period may be due to variations in a QTL).

QTL alleles have context-dependent effects. Their effects may differ in magnitude or direction in different genetic backgrounds, different environments or between male and females (i.e., genotype x genotype interaction – epistasis; genotype x environment interactions; genotype x sex interaction). These context-dependent effects are very common and play a significant role in genetic architecture, but they are very difficult to detect and characterize. In addition to meaning the masking of genotypic effects at one locus by genotypes of another locus, epistasis in quantitative genetics also refers to any statistical interaction between the genotypes at two or more loci. It is common between mutations that affect the same quantitative trait. Epistatic effects can be as large as main QTL effects; they can also occur in opposite directions between

different pairs of interacting loci and between loci with having significant main effects on the trait of interest. Epistasis has been found between closely linked QTLs and also polymorphisms at a single locus.

Pleiotropy, the effect of a gene on more than one phenotype, is important in the genetics of QTLs. In a narrow sense, pleiotropy can mean the effect of a particular allele on more than one phenotype and is the reason for the stable genetic correlations between quantitative traits, if the effects at multiple loci affecting the same trait are in the same direction. Understanding of pleiotropic connections between quantitative traits helps in predicting the correlated responses to artificial selection and assessing the contribution of new mutations to standing variation for quantitative traits. Pleiotropy is known to occur even between traits that are not functionally related. Consequently, the pleiotropic effects of different genes that affect pairs of traits are usually not in the same direction and do not result in significant genetic correlations between traits.

4.3.3 Molecular basis of quantitative variation

The causal molecular variants that affect quantitative trait loci are called **quantitative trait nucleotides(QTNs).** The distribution of QTN allele frequencies can indicate the nature of the selective forces operating on the trait. Inference of QTN allele frequencies is limited to association mapping (Chapter 20) designs in which all the variants in a candidate gene or gene region have been identified. QTNs allow researchers to map phenotype to genotype in the absence of biological context. QTNs consist of two components – **eQTLs (expression quantitative trait loci)** and **QTT (quantitative trait transcripts)** (Figure 4.8).

The eQTL is a region of the genome containing one or more genes that affect variation in gene expression, which is identified by linkage to polymorphic marker loci. It is technically a marriage of high-throughput expression profiling technology and QTL analysis. QTT is a transcript for which variation in its expression is correlated with variation in an organismal level quantitative trait phenotype. Numerous (even several hundreds) of QTT are believed to be associated with any single quantitative trait phenotype. Furthermore, these QTT are genetically correlated.

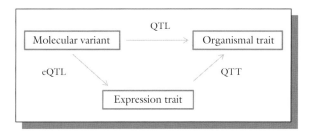

Figure 4.8 Systems genetics of complex traits. An integrative framework showing the relationship between DNA sequence variation and quantitative variation for gene expression and an organismal phenotype. QTLs allow researchers to map phenotype to genotype in the absence of biological context. To gain this context, they need to describe the flow of information from DNA to the organismal phenotype through RNA intermediates, proteins and other molecular endophenotypes.

4.4 Systems genetics

Systems genetics (also called genetical genomics) is considered to be the third in the developmental stages of quantitative genetics, after classical and molecular quantitative genetics in that order of development. This new paradigm is instigated by the fact that associating DNA sequence variation with variation of the phenotype of an organism skips all of the intermediate steps (e.g., the intermediate molecular phenotypes such as transcript abundance which are quantitative traits and vary genetically in populations) in the chain of the causation from genetic perturbation to phenotypic variation. Systems genetics detects variation in phenotypic traits and integrates them with underlying genetic variation. According to Mackay and colleagues, this approach to QTL analysis integrates DNA sequence variation, variation in transcript abundance and other molecular phenotypes, and variation in the phenotypes of the organism in linkage or association mapping, thereby allowing the researcher to interpret quantitative genetic variation in terms of biologically meaningful causal networks of correlated transcripts. In other words, it can be used to define biological networks and to predict molecular interactions by analyzing transcripts with expressions that co-vary within genetic populations. The approach can be used to analyze effects of genome-wide genetic variants on transcriptome-wide variation in gene expression.

4.5 Predicting breeding value

Breeding value (or genetic merit) of an individual as a genetic parent is the sum of gene effects of the individual as measured by the performance of its progeny. Statistically, it is measured as twice the deviation of the offspring from the population mean (since the individual only contributes half of the alleles to its offspring). This estimate measures the ability of an individual to produce superior offspring. This is the part of an individual's genotypic value that is due to independent gene effects, and hence can be transmitted. The mean breeding value becomes zero with random mating. This estimate is of importance to breeders because it assists them in selecting the best parents to use in their programs.

The **Best Linear Unbiased Prediction(BLUP)** is a common statistical method for estimating breeding values. It is unbiased because as more data are accumulated, the predicted breeding values approach the true values. BLUP is a method of estimating random effects. The context of this statistical method is the linear model:

$$y = XB + Zu + e$$

where y is is a vector of n observable random variables; B is vector of p unknown parameters with fixed value or effects; X and Z are known matrices; u and e are vectors of q and n, respectively, unobservable random variables (random effects).

To apply this technique, numerical scores are assigned to traits and compiled as predictions of the future. Simple traits can be most accurately and objectively measured and possibly predicted. Only one trait may be predicted in a model. This trait has to be objectively measurable with high accuracy. Furthermore, it has to be heritable.

4.6 Genomic selection (genome-wide selection)

Selection in conventional plant breeding generally relies on breeding values estimated from pedigree-based mixed models that cannot account for Mendelian segregation and, in the absence of inbreeding, can only explain one half of the genetic variability (individual contributes only half of their alleles to the next generation). Molecular markers have the capacity to track Mendelian segregation as several positions of the genome of the organism, thereby increasing the accuracy of estimates of genetic values (and the genetic progress achievable when the predictions are used for selection in breeding). Even though marker-assisted selection (MAS) (Chapter 22) has achieved some success, its application to improving quantitative traits is hampered by various factors. The biparental mating designs used for detection of loci affecting quantitative traits and statistical methods used are not well suited to traits that are under polygenic control (MAS uses molecular markers in linkage disequilibrium with QTL).

Genomic selection (or **genome-wide selection**) is proposed as a more effective approach to improving quantitative traits. It uses all the available molecular markers across the entire genome (there are thousands of genome-wide molecular markers) to estimate genetic or breeding values. Using high-density marker scores in the prediction model and high throughput genotyping, genomic selection avoids biased marker effect estimates and captures more of the variation due to the small-effect QTL.

Genomic selection method uses two types of data set – a training set and a validation set. It is applied in a population that differs from the reference population in which the marker effects were estimated. The training set is the reference population and has three components – (i) phenotypic information from the relevant breeding germplasm evaluated over a range of environmental conditions, (ii) molecular marker scores, and (iii) pedigree or kinship information. Marker effects are estimated on the training set, and the breeding values of new genotypes are predicted based solely on the marker effect. The validation set contains the selection candidates that have been genotyped but not phenotyped and selected on the basis of marker effects estimated in the training set.

Genomic selection has advantages. It can accelerate the selection cycles and increase the selection gains per unit time. This potential notwithstanding, the method needs to empirically validated in plant applications. It needs to go beyond computer simulations and become incorporated into breeding schemes. As the cost of marker technology continues to decrease, genotyping would continue to become more cost effective than phenotyping in a breeding program.

4.7 Mapping quantitative traits

The subject of mapping is treated in detail in Chapter 21.

Key references and suggested reading

Ali, A., and Johnson, D.L. (2000). Heritability estimates for winter hardiness in lentil under natural and controlled conditions. *Plant Breed.* **119**:283–285.

Bhatnagar, S., Betran, F.J., and Rooney, L.W. (2004). Combining abilities of quality protein maize inbreds. *Crop Sci.*, **44**:1997–2005.

Bohren, B.B., McKean, H.E., and Yamada, Y. (1961). Relative efficiencies of heritability estimates based on regression of offspring on parent. *Biometrics*, **17**:481–491.

Cockerham, R.E. H.F., and Harvey, P.H. (1949). A breeding procedure designed to make maximum use of both general and specific combining ability. *J. Amer. Soc. Agron.*, **41**:360–367.

Edwards, J.W., and Lamkey, K.R. (2002). Quantitative genetics of inbreeding in a synthetic maize population. *Crop Sci.*, **42**:1094–1104.

Falconer, D.S. (1981). *Introduction to quantitative genetics*. Longman Group Ltd., New York.

Falconer, D.S., and Mackay, T.F.C. (1996). *Introduction to quantitative genetics*, 4th edn. Longman, Harlow, UK.

Gardner, C.O. (1977). Quantitative genetic studies and population improvement in maize and sorghum, in *Proc. Int. Conf. Quantitative Genetics* (eds E. Pollak, O. Kempthorne, and T.B. Bailey). Iowa State University, Ames, IA.

Glover, M.A., Willmot, D.B., Darrah, L.L., Hibbard, B.E., and Zhu, X. (2005). Diallel analysis of agronomic traits using Chinese and US maize germplasm. *Crop Sci.*, **45**:1096–1102.

Griffing, B. (1956). A generalized treatment of the use of diallel crosses in quantitative inheritance. *Heredity*, **10**:31–50.

Griffing, B. (1956). Concept of general and specific combining ability in relation to a diallel crossing system. *Aust. J. Biol. Sci.*, **9**:463–493.

Heffner, E.L., Sorrells, M.E., and Jannink, J. (2009). Genomic selection for crop improvement. *Crop Sci.*, **49**:1–12 (2009).

Henderson, C.R. (1963). Selection index and expected genetic advance, in *Statistical Genetics and Plant breeding* (eds W.D. Hanson and H.F. Robinson). Publication No. 982, National Academy of Sciences and National Research Council, Washington, DC.

Hill, J., Becker, H.C., and Tigerstedt, P.M.A. (1998). *Quantitative and ecological aspects of plant breeding.* Chapman and Hall, London.

Holland, J.B. (2001). Epistasis and plant breeding. *Plant Breed. Rev.*, **21**:27–92.

Lin, C.Y. (1978). Index selection for genetic improvement of quantitative characters. *Theor. Appl. Genet.*, **52**:49–56.

Mackay, T.F.C., Stone, E.A., and Ayroles, J.F. (2009). The genetics of quantitative traits: challenges and prospects. *Nature Reviews Genetics*, **10**:565–577.

Meuwissen, T.H.E., Hayes, B.J., and Goddard, M.E. (2001). Prediction of total genetic value using genome-wide dense markermaps. *Genetics*, **157**:1819–1829.

Wricke, G., and Weber, W.E. (1986). *Quantitative genetics and selection in plant breeding.* Walter de Gruyter, Berlin.

Zhu, M., Yu, M., and Zhao, S. Understanding Quantitative Genetics in the Systems Biology Era. *Int. J. Biol. Sci.*, **5**:161–170.

Outcomes assessment

Part A

Please answer the following questions true or false.

1 Heritability is a population phenomenon.
2 Specific combining ability of a trait depends on additive gene action.
3 Polygenes have distinct and distinguishable effects.
4 Quantitative variation deals with discrete phenotypic variation.
5 Quantitative traits are also called metrical traits.
6 Quantitative traits are more influenced by the environment than qualitative traits.
7 Quantitative traits are controlled by polygenes.

Part B

Please answer the following questions.

1 What is quantitative genetics and how does it differ from qualitative genetics?
2 Give two specific assumptions of quantitative genetic analysis.
3 Describe additive gene action.
4 What is heritability of a trait?
5 What is the breeders' equation?

Part C

Please write a brief essay on each of the following topics.

1 Discuss the role of the environment in quantitative trait expression.
2 Discuss the concept of general worth of a plant.
3 Discuss the concept of intuitive selection.
4 Discuss the application of combining ability analysis in plant breeding.
5 Discuss a method of estimating heritability of a trait.

Section 3

Reproductive systems

Chapter 5 Introduction to reproduction and autogamy
Chapter 6 Allogamy
Chapter 7 Hybridization
Chapter 8 Clonal propagation and *in vitro* culture

Reproduction is the process by which plants multiply themselves. It is not only important to plant producers but also to plant breeders. The mode of reproduction determines the method of breeding the species and how the product of breeding is maintained for product identity preservation. It is important to add that, whereas in the past plant breeding methods were fairly distinct for self-pollinated species and for cross-pollinated species, such a clear distinction does not currently exist. Rather, the methods of plant breeding for the two groups tend to overlap.

5

Introduction to reproduction and autogamy

Purpose and expected outcomes

Rudolph Camerarius is credited with establishing sexual differentiation, noting that male and female sex organs exist in the Plant Kingdom. Some species produce flowers while others do not. In flowering species, reproduction involves the union of gametes following pollination. Plant breeders need to understand the mode of reproduction in order to manipulate plants effectively to develop new and improved ones for crop production. After studying this chapter, the student should be able to:

1 Discuss the importance of the mode of reproduction to plant breeding.
2 Distinguish between self-pollination and cross-pollination.
3 Discuss the natural barriers that favor or hinder each of the modes of reproduction.
4 Discuss the implications of mode of reproduction in schemes and strategies employed in plant breeding.
5 Discuss the use of male sterility and self-incompatibility in breeding.

5.1 Importance of mode of reproduction to plant breeding

Plant breeders need to understand the reproductive systems of plants for the following key reasons:

(i) The genetic structure of plants depends on their mode of reproduction. Methods of breeding are generally selected such that the natural genetic structure of the species is retained in the cultivar. Otherwise, special efforts will be needed to maintain the newly developed cultivar in cultivation.

(ii) In flowering species, artificial hybridization is needed to conduct genetic studies to understand the inheritance of traits of interest and for transfer of genes of interest from one parent to another. To accomplish this, the breeder needs

---

Principles of Plant Genetics and Breeding, Second Edition. George Acquaah.
© 2012 John Wiley & Sons, Ltd. Published 2012 by John Wiley & Sons, Ltd.

to understand thoroughly the floral biology and other factors associated with flowering in the species.

(iii) Artificial hybridization requires an effective control of pollination so that only the desired pollen is allowed to be involved in the cross. To this end, the breeder needs to understand the reproductive behavior of the species. Pollination control is critical to the hybrid seed industry.

(iv) The mode of reproduction also determines the procedures for multiplication and maintenance of cultivars developed by plant breeders.

5.2 Overview of reproductive options in plants

Four broad and contrasting pairs of reproductive mechanisms or options occur in plants.

(i) **Hermaphrodity versus unisexuality.** Hermaphrodites have both male and female sexual organs, and hence may be capable of self-fertilization. On the other hand, unisexuals, having one kind of sexual organ, are compelled to cross-fertilize. Each mode of reproduction has genetic consequences. Hermaphrodity promotes a reduction in genetic variability, whereas unisexuality, through cross-fertilization, promotes genetic variability.

(ii) **Self-pollination versus cross-pollination.** Hermaphrodites that are self-fertile may be self-pollinated or cross-pollinated. In terms of pollen donation, a species may be **autogamous** (pollen comes from the same flower – selfing) or **allogamous** (pollen comes from a different flower). There are finer differences in these types. For example, there may be differences between the time of pollen shed and stigma receptivity.

(iii) **Self-fertilization versus cross-fertilization.** Just because a flower is successfully pollinated does not necessarily mean fertilization would occur. The mechanism of self-incompatibility causes some species to reject pollen from their own flowers, thereby promoting outcrossing.

(iv) **Sexuality versus asexuality.** Sexually reproducing species are capable of providing seed through sexual means. Asexuality manifests in one of two ways – vegetative reproduction (in which no seed is produced) or agamospermy (in which seed is produced).

5.3 Types of reproduction

Plants are generally classified into two groups, based on the mode of reproduction, as either **sexually reproducing** or **asexually reproducing**. Sexually reproducing plants produce seed as the primary propagule. Seed is produced after sexual union (fertilization) involving the fusion of sex cells or **gametes**. Gametes are products of meiosis and, consequently, seeds are genetically variable. Asexual or vegetative reproduction mode entails the use of any vegetative part of the plant for propagation. Some plants produce modified parts, such as creeping stems (stolons or rhizomes), bulbs or corms, which are used for their propagation. Asexual reproduction is also applied to the condition whereby seed is produced without fusion of gametes (called apomixis). It should be pointed out that some plants could be reproduced by either the sexual or asexual mode. However, for either ease of propagation or for product quality, one mode of reproduction, often the vegetative mode, is preferred. Such is the case in flowering species such as potato (propagated by tubers or stem cuttings) and sugarcane (propagated by stem cuttings).

5.4 Sexual reproduction

Sexual reproduction increases genetic diversity through the involvement of meiosis. Flowering plants dominate the terrestrial species. Flowering plants may reproduce sexually or asexually.

5.4.1 Sexual lifecycle of a plant (alternation of generation)

The normal sexual lifecycle of a flowering plant may be simply described as consisting of events from first establishment to death (from seed to seed in seed-bearing species). A flowering plant goes through two basic growth phases – **vegetative** and **reproductive**, the former preceding the latter. In the vegetative phase, the plant produces vegetative growth only (stem, branches, leaves, etc., as applicable). In the reproductive phase, flowers are produced. In some species, exposure to a certain environmental factor (e.g., temperature, photoperiod) is required to switch from the vegetative to the reproductive phase. The duration between phases varies among species and

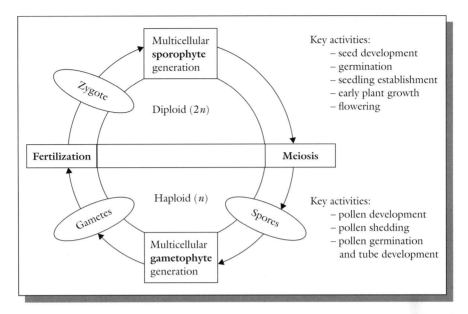

Figure 5.1 Schematic representation of the alternation of generations in flowering plants. The sporophyte generation is diploid and often the more conspicuous phase of the plant lifecycle. The gametophyte is haploid.

can be manipulated by modifying the growing environment.

For sexual reproduction to occur, two processes must occur in sexually reproducing species. The first process, meiosis, reduces the chromosome number of the diploid ($2n$) cell to the haploid (n) number. The second process, **fertilization,** unites the nuclei of two gametes, each with the haploid number of chromosomes, to form a diploid. In most plants, these processes divide the lifecycle of the plant into two distinct phases or generations, between which the plant alternates (called **alternation of generation**) (Figure 5.1). The first phase or generation, called the **gametophyte generation**, begins with a haploid spore produced by meiosis. Cells derived from the gametophyte by mitosis are haploid. The multicellular gametophyte produces gametes by mitosis. The sexual reproductive process unites the gametes to produce a **zygote** that begins the diploid **sporophyte generation** phase.

In lower plants (mosses, liverworts), the sporophyte is small and dependent upon the gametophyte. However, in higher plants (ferns, gymnosperms, angiosperms), the male gametophyte generation is reduced to a tiny pollen tube and three haploid nuclei (called the microgametophyte). The female gametophyte (called the megagametophyte) is a single multinucleated cell, also called the **embryo sac**. The genotype of the gametophyte or sporophyte influences sexual reproduction in species with self-incompatibility problems. This has implications in the breeding of certain plants; this is discussed further later in this chapter.

5.4.2 Duration of plant growth cycles

The plant breeder should know the lifecycle of the plant to be manipulated. The strategies for breeding are influenced by the duration of the plant growth cycle. Angiosperms (flowering plants) may be classified into four categories based on the duration of their growth cycle (Figure 5.2):

(i) **Annual.** Annual plants (or annuals) complete their lifecycle in one growing season. Examples of such plants include corn, wheat, and sorghum. Annuals may be further categorized into **winter annuals** or **summer annuals**. Winter annuals (e.g. wheat) utilize parts of two seasons. They are planted in the fall (autumn) and undergo a critical physiological inductive change called **vernalization**, which is required for flowering and fruiting in spring. In cultivation, certain non-annuals (e.g. cotton) are produced as though they were annuals.

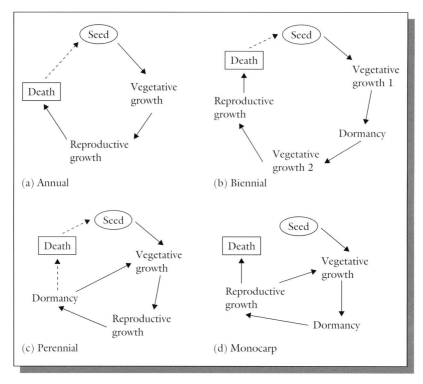

Figure 5.2 Flowering plants have one of four lifecycles – annual, biennial, perennial, and monocarp. Variations occur within each of these categories, partly because of the work of plant breeders.

(ii) **Biennial.** A biennial completes its lifecycle in two growing seasons. In the first season it produces basal roots and leaves; then it grows a stem, produces flowers and fruits, and dies in the second season. The plant usually requires a special environmental condition or treatment (e.g., vernalization) to be induced to enter the reproductive phase. For example, sugar beet grows vegetatively in the first season. In winter, it becomes vernalized and starts reproductive growth in spring.

(iii) **Perennial.** Perennials are plants that have the ability to repeat their lifecycles indefinitely by circumventing the death stage. They may be herbaceous, as in species with underground vegetative structures called **rhizomes** (e.g. indiangrass), or aboveground structures called **stolons** (e.g. buffalograss). They may also be woody, as in shrubs, vines (grape), and trees (orange).

(iv) **Monocarp.** Monocarps are annuals or biennials, but some persist in vegetative development for very long periods (e.g., the so-called "century plant") before they flower and set seed (e.g.,

bamboo and agave). Once flowering occurs, the plant dies. That is, monocarps are plants that flower only once. Other examples are bromeliads. The top part dies, so that new plants arise from the root system of the old plant.

It should be pointed out that certain plants that may be natural biennials or perennials are cultivated by producers as annuals. For example, sugar beet, a biennial, is commercially produced as an annual for its roots. For breeding purposes it is allowed to bolt to produce flowers for crossing, and subsequently to produce seed.

5.4.3 The flower structure

Genetic manipulation of flowering plants by conventional tools is accomplished using the technique of crossing, which involves flowers. To be successful, the plant breeder should be familiar with the flower structure, regarding the parts and their arrangement. Flower structure affects the way flowers are

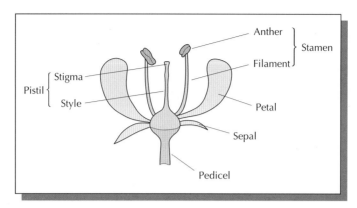

Figure 5.3 The typical flower has four basic parts – petals, sepals, pistil, and stamen. The shape, size, color and other aspects of these floral parts differ widely among species.

emasculated (prepared for crossing by removing the male parts to make the flower female). The size of the flower affects the kinds of tools and techniques that can be used for crossing.

5.4.4 General reproductive morphology

Four major parts of a flower are generally recognized – **petal, sepal, stamen,** and **pistil**. These form the basis of flower variation. Flowers vary in the color, size, numbers and arrangement of these parts. Typically, a flower has a receptacle to which these parts are attached (Figure 5.3). The male part of the flower (the **stamen**) comprises a stalk, called a **filament**, to which is attached a structure consisting of four pollen-containing chambers that are fused together (**anther**). The stamens are collectively called the **androecium**. The center of the flower is occupied by a pistil, which consists of the style, stigma and ovary (contains carpels). The pistil is also called the **gynoecium**. Sepals are often leaf-like structures that enclose the flower in its bud stage. Collectively, sepals are called the **calyx**. The showiest parts of the flower are the petals, collectively called the **corolla**.

5.4.5 Types of flowers

When a flower has all the four major parts, it is said to be a **complete flower** (e.g., soybean, tomato, cotton, tobacco). However, if a flower lacks certain parts (often petals or sepals), as is the case in many grasses (e.g., rice, corn, wheat), it is said to be an **incomplete flower**. Some flowers either have only stamens or a

pistil, but not both. When both stamens and a pistil occur in the same flower, the flower is said to be a **perfect flower** (bisexual), as in wheat, tomato and soybean. Some flowers are unisexual (either stamens or pistil may be absent) and are called **imperfect flowers**. If imperfect flowers have stamens they are called **staminate flowers**. When only a pistil occurs, the flower is a **pistilate flower**. A plant such as corn bears both staminate (tassel) and pistillate (silk) flowers on the same plant and is said to be a **monoecious plant**. However, in species such as asparagus and papaya, plants may either be pistilliate (female plant) or staminate (male plant) and are said to be **dioecious plants**. Flowers may either be **solitary** (occur singly or alone) or may be grouped together to form an **inflorescence**. An inflorescence has a primary stalk (peduncle) and numerous secondary smaller stalks (pedicels). The most common inflorescence types in crop plants are the cyme and raceme. A branched raceme is called a **panicle** (e.g., oats) while a raceme with sessile (short pedicels) is called a **spike** (e.g., wheat). From the foregoing it is clear that a plant breeder should know the specific characteristics of the flower in order to select the appropriate techniques for crossing.

5.4.6 Gametogenesis

Sexual reproduction entails the transfer of gametes to specific female structures where they unite and are then transformed into an embryo, a miniature plant. Gametes are formed by the process of gametogenesis. They are produced from specialized diploid cells

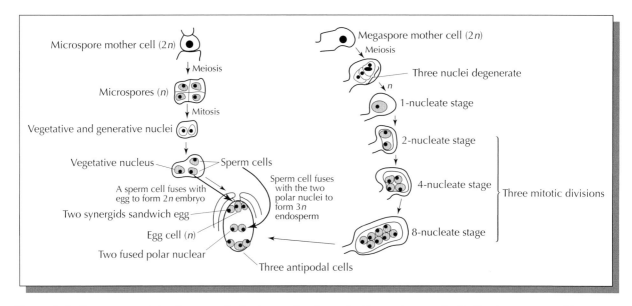

Figure 5.4 Gametogenesis in plants results in the production of pollen and egg cells. Pollen is transported by agents to the stigma of the female flower, from which it travels to the egg cell to unite with it.

called **microspore mother cells** in anthers and **megaspore mother cells** in the ovary (Figure 5.4). Microspores derived from the mother cells are haploid cells, each dividing by mitosis to produce an immature **male gametophyte** (pollen grain). Most pollen is shed in the two-cell stage, even though sometimes, as in grasses, one of the cells later divides again to produce two sperm cells. In the ovule, four megaspores are similarly produced by meiosis. The nucleus of the functional megaspore divides three times by mitosis to produce eight nuclei, one of which eventually becomes the egg. The **female gametophyte** is the seven-celled, eight-nucleate structure. This structure is also called the **embryo sac**. Two free nuclei remain in the sac. These are called polar nuclei because they originate from opposite ends of the embryo sac.

5.4.7 Pollination and fertilization

Pollination is the transfer of pollen grains from the anther to the stigma of a flower. This transfer is achieved through a vector or pollination agent. The common pollination vectors are wind, insect, mammals, and birds. Flowers have certain features that suit the various pollination mechanisms (Table 5.1). Insect-pollinated flowers tend to be showy and exude

Table 5.1 Pollination mechanisms in plants.

Pollination vector	Flower characteristics
Wind	Tiny flowers (e.g., grasses); dioecious species
Insects	
Bees	Bright and showy (blue, yellow); sweet scent; unique patterns; corolla provides landing pad for bees
Moths	White or pale color for visibility at night time; strong penetrating odor emitted after sunset
Beetles	White or dull color; large flowers; solitary or inflorescence
Flies	Dull or brownish color
Butterflies	Bright colors (often orange, red); nectar located at base of long slender corolla tube
Bats	Large flower with strong fruity pedicels; dull or pale colors; strong fruity or musty scents, flowers produce copious, thick nectar
Birds	Bright colors (red, yellow); odorless; thick copious nectar

strong fragrances. Birds are attracted to red and yellow flowers. When compatible pollen falls on a receptive stigma, a pollen tube grows down the style to the micropylar end of the embryo sac, carrying two sperms or male gametes. The tube penetrates the sac through the micropyle. One of the sperms unites with the egg cell, a process called **fertilization**. The other sperm cell unites with the two polar nuclei (called triple fusion). The simultaneous occurrence of two fusion events in the embryo sac is called **double fertilization**.

On the basis of pollination mechanisms, plants may be grouped into two **mating systems** – **self-pollinated** or **cross-pollinated**. Self-pollinated species accept pollen primarily from the anthers of the same flower (autogamy). The flowers, of necessity, must be bisexual. Cross-pollinated species accept pollen from different sources. In actuality, species express varying degrees of cross-pollination, ranging from lack of cross-pollination to complete cross-pollination.

Table 5.2 Examples of predominantly self-pollinated species.

Common name	Scientific name
Barley	*Hordeum vulgare*
Chickpea	*Cicer arietinum*
Clover	*Trifolium* spp.
Common bean	*Phaseolus vulgaris*
Cotton	*Gossypium* spp.
Cowpea	*Vigna unguiculata*
Eggplant	*Solanum melongena*
Flax	*Linum usitatissimum*
Jute	*Corchorus espularis*
Lettuce	*Letuca sativa.*
Oat	*Avena sativa*
Pea	*Pisum sativum*
Peach	*Prunus persica*
Peanut	*Arachis hypogaea*
Rice	*Oryza sativa*
Sorghum	*Sorghum bicolor*
Soybean	*Glycine max*
Tobacco	*Nicotiana tabacum*
Tomato	*Solanum lycopersicum*
Wheat	*Triticum aestivum*

5.5 Autogamy

Self-pollination or **autogamy** occurs in a wide variety of plant species – vegetables (lettuce, tomatoes, snap beans, endive), legumes (soybean, peas, lima beans) and grasses (barley, wheat, oats). Certain natural mechanisms promote or ensure self-pollination, specifically cleistogamy and chasmogamy, while other mechanisms prevent self-pollination (e.g., self-incompatibility, male sterility).

5.5.1 Mechanisms that promote autogamy

Cleistogamy is the condition in which the flower fails to open. The term is sometimes extended to mean a condition in which the flower opens only after it has been pollinated (as occurs in wheat, barley, lettuce), a condition called **chasmogamy**. Some floral structures, such as those found in legumes, favor self-pollination. Sometimes, the stigma of the flower is closely surrounded by anthers, making it prone to selfing.

Very few species are completely self-pollinated. The level of self-pollination is affected by factors including the nature and amount of insect pollination, air current, and temperature. In certain species, pollen may become sterilized when the temperature

dips below freezing. Any flower that opens prior to self-pollination is susceptible to some cross-pollination. A list of predominantly self-pollinated species is presented in Table 5.2.

5.5.2 Mechanisms that prevent autogamy

There are several mechanisms in nature that work to prevent self-pollination in species that otherwise would be self-pollinated. These include self-incompatibility, male sterility and dichogamy.

Self-incompatibility

Self-Incompatibility (or lack of self-fruitfulness) is a condition in which the pollen from a flower is not receptive on the stigma of the same flower, and hence incapable of setting seed. This happens in spite of the fact that both pollen and ovule development are normal and viable. It is caused by a genetically controlled physiological hindrance to self-fertilization. Self-incompatibility is widespread in nature, occurring in families such as Poaceae, Cruciferae, Compositae, and Rosaceae. The incompatibility reaction is genetically conditioned by a locus designated *S*, with multiple alleles that can number over 100 in some species such as *Trifolium pretense*. However, unlike monoecy and dioecy, all plants produce seed in self-incompatible species.

Self-incompatibility systems

Self-incompatibility systems may be classified into two basic types – **heteromorphic** and **homomorphic**.

- **Heteromorphic incompatibility.** This is caused by differences in the lengths of stamens and style (called **heterostyly**) (Figure 5.5). In one flower type called the **pin**, the styles are long while the anthers are short. In the other flower type, **thrum**, the reverse is true (e.g., in *Primula*). The pin trait is conditioned by the genotype *ss* while thrum is conditioned by the genotype Ss. Crosses of pin (*ss*) x pin (*ss*) as well as thrum (*Ss*) x thrum (*Ss*) are incompatible. However, pin (*ss*) x thrum (*Ss*) or vice versa is compatible. The condition described is **distyly** because of the two different types of style lengths of the flowers. In *Lythrum* three different relative positions occur (called **tristyly**).
- **Homomorphic incompatibility.** There are two kinds of homomorphic incompatibility – **gametophytic** and **sporophytic** (Figure 5.6).
  - (i) **Gametophytic incompatibility.** In gametophytic incompatibility (originally called the oppositional factor system), the ability of the

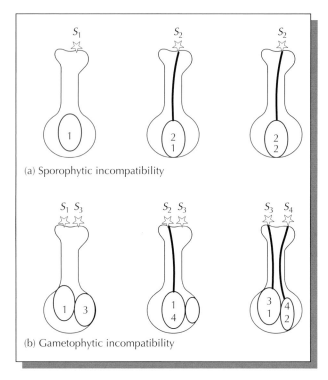

(a) Sporophytic incompatibility

(b) Gametophytic incompatibility

Figure 5.6 Types of self-incompatibility: (a) sporophytic and (b) gametophytic. Sporophytic incompatibility occurs in families such as Compositae and Cruciferae. It is associated with pollen grains with two generative nuclei, whereas gametophytic incompatibility is associated with pollen with one generative nucleus in the pollen tube as occurs in various kinds of clover.

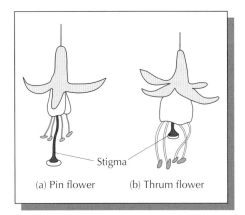

(a) Pin flower   (b) Thrum flower

Stigma

Figure 5.5 Heteromorphic incompatibility showing floral modifications in which anthers and pistils are of different lengths in different plants (heterostyly). This type of incompatibility is believed to be always of the sporophytic type. Pin and thrum flowers occurs in flowers such as *Primula*, *Forsythia*, *Oxalis*, and *Silia*.

pollen to function is determined by its own genotype and not the plant that produces it. Gametophytic incompatibility is more widespread than sporophytic incompatibility. Gametophytic incompatibility occurs in species such as red clover, white clover, and yellow sweet clover. Homomorphic incompatibility is controlled by a series of alleles at a single locus (S_1, S_2, ... S_n) or alleles at two loci in some species. The system is called homomorphic because the flowering structures in both the seed-bearing (female) and pollen-bearing (male) plants are similar. The alleles of the incompatibility gene(s) act individually in the style. They exhibit no dominance. The incompatible pollen is inhibited in the style. The pistil is diploid and hence contains two incompatibility alleles (e.g., S_1S_3, S_3S_4). Reactions occur if identical alleles in both pollen and style are encountered. Only heterozygotes for S alleles are produced in this system.

(ii) **Sporophytic incompatibility.** In sporophytic incompatibility, the incompatibility characteristics of the pollen are determined by the plant (sporophyte) that produces it. It occurs in species such as broccoli, radish, and kale. The sporophytic system differs from the gametophytic system in that the S allele exhibits dominance. Also, it may have individual action in both pollen and style, making this incompatibility system complex. The dominance is determined by the pollen parent. Incompatible pollen may be inhibited on the stigma surface. For example, a plant with genotype S_1S_2 where S_1 is dominant to S_2, will produce pollen that will function like S_1. Furthermore, S_1 pollen will be rejected by an S_1 style but received by an S_2 style. Hence, homozygotes of S alleles are possible.

Incompatibility is expressed in one of three general ways, depending on the species. The germination of the pollen may be decreased (e.g., in broccoli). Sometimes, removing the stigma allows normal pollen germination. In the second way, pollen germination is normal but pollen tube growth is inhibited in the style (e.g., tobacco). In the third scenario, the incompatibility reaction occurs after fertilization (e.g., in *Gesteria*). This third mechanism is rare.

Changing the incompatibility reaction

Mutagens (agents of mutation) such as X-rays, radioactive sources such as ^{32}P and certain chemicals have been used to make a self-infertile genotype self-fertile. Such a change is easier to achieve in gametophytic systems than sporophytic systems. Furthermore, doubling the chromosome number of species with the sporophytic system of incompatibility does not significantly alter the incompatibility reaction. This is because two different alleles already exist in a diploid that may interact to produce the incompatibility effect. Polyploidy only makes more of such alleles available. On the other hand, doubling the chromosome in a gametophytic system would allow the pollen grain to carry two different alleles (instead of one). The allelic interaction could cancel any

incompatibility effect to allow selfing to be possible. For example, diploid pear is self-incompatible whereas autotetraploid pear is self-fruitful.

Plant breeding implications of self-incompatibility

Infertility of any kind hinders plant breeding. However, this handicap may be used as a tool to facilitate breeding by certain methods. Self-incompatibility may be temporarily overcome by techniques or strategies such as the removal of the stigma surface (Figure 5.7) (or application of electric shock), early pollination (before inhibitory proteins form), or lowering the temperature (to slow down the development of the inhibitory substance). Self-incompatibility promotes heterozygosity. Consequently, selfing self-incompatible plants can create significant variability from which a breeder can select superior recombinants. Self-incompatibility may be used in plant breeding (for F_1 hybrids, synthetics, triploids), but first homozygous lines must be developed.

Self-incompatibility systems for hybrid seed production have been established for certain crops (e.g., cabbage, kale) that exhibit sporophytic incompatibility (Figure 5.8). Inbred lines (compatible inbreds) are used as parents. These systems generally are used to manage pollinations for commercial production of hybrid seed. Gametophytic incompatibility occurs in vegetatively propagated species. The clones to be hybridized are planted in adjacent rows.

Male sterility

Male sterility is a condition in plants whereby the anthers or pollen are non-functional. The condition may manifest most commonly as absence of or extreme scarcity of pollen, severe malformation or absence of flowers or stamens, or failure of pollen to dehisce. Just like self-incompatibility, male sterility enforces cross-pollination. Similarly, it can be exploited as a tool to eliminate the need for emasculation for producing hybrid seed. There are three basic kinds of male sterility based on the origin of the abnormality:

(i) **True male sterility** – This is due to unisexual flowers that lack male sex organs (dioecy and monoecy), or bisexual flowers with abnormal or non-functional microspores (leading to pollen abortion).

(a) **Overcoming pollination barriers:** Cut-style method in lily. The method involves cutting the style (just above the ovary) to circumvent stylar barriers, which normally inhibit pollen tube growth in interspecific crosses. After cutting 90% of the style, stigmatic fluid and pollen are applied on the stylar surface. Subsequently the stylar surface is capped with a piece of aluminum foil to prevent unwanted cross pollination and drying out.

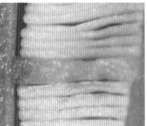

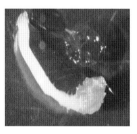

(b) **Overcoming post-fertilization reproductive barriers:** Embryo rescue in lily. In intersectional crosses, embryo's abort before maturation but can be rescued by means of tissue culturing. Swollen seed pods that start discoloring and/or softening are harvested and opened up under sterile conditions. Swollen (pollinated) ovules are identified and from these the still viable embryo is isolated and transferred to a germination medium. The embryo is kept in the dark to germinate after which it can be propagated to a bulblet that can be planted into soil.

Pictures and text with courtesy of JM van Tuyl/P Arens Wageningen UR – Plant Breeding

Figure 5.7 Overcoming reproductive barriers: (a) pollination barriers; (b) post-fertilization reproductive barriers.

(ii) **Functional male sterility** – The anthers fail to release their contents even though the pollen is fertile.

(iii) **Induced male sterility** – Plant breeders may use chemicals to induce sterility.

True male sterility

There are three kinds of pollen sterility – nuclear, cytoplasmic, cytoplasmic-genetic.

• **Genetic male sterility.** Genetic (nuclear, genic) **male sterility** is widespread in plants. The gene for sterility has been found in species including barley, cotton, soybean, tomato, potato, and lima bean. It is believed that nearly all diploid and polyploidy plant species have at least one male sterility locus. Genetic male sterility may be manifested as pollen abortion (pistillody) or abnormal anther development. Genetic male sterility is often conditioned by a single recessive nuclear gene, *ms*, the dominant allele, *Ms*, conditioning normal anther and pollen development. However, male sterility in alfalfa has been reported to be under the control of two independently inherited genes. The expression of the gene may vary with the environment. To be useful for application in plant breeding, the male sterility system should be stable in a wide range of environments and inhibit virtually all seed production. The breeder cannot produce and maintain a pure population of male sterile plants. The genetically male sterile types (*msms*) can be propagated by crossing them with a heterozygous pollen source (*Msms*). This cross will produce a progeny in which 50% of the plants will be male sterile (*msms*) and 50% male fertile (*Msms*). If the crossing block is

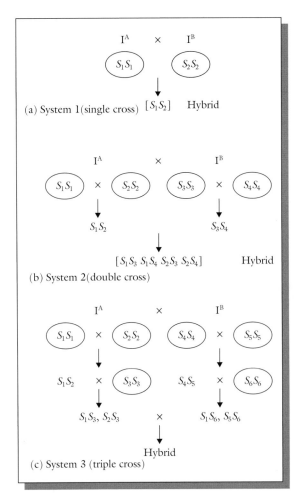

Figure 5.8 shown above contains:

(a) System 1 (single cross)

I^A × I^B

S_1S_1 S_2S_2

$[S_1S_2]$ Hybrid

(b) System 2 (double cross)

I^A × I^B

S_1S_1 × S_2S_2 S_3S_3 × S_4S_4

S_1S_2 S_3S_4

$[S_1S_3\ S_1S_4\ S_2S_3\ S_2S_4]$ Hybrid

(c) System 3 (triple cross)

I^A × I^B

S_1S_1 × S_2S_2 S_4S_4 × S_5S_5

S_1S_2 × S_3S_3 S_4S_5 × S_6S_6

S_1S_3, S_2S_3 × S_1S_6, S_5S_6

Hybrid

Figure 5.8 Application of self-incompatibility in practical plant breeding. Sporophytic incompatibility is widely used in breeding of cabbage and other Brassica species. The single cross hybrids are more uniforms and easier to produce. The top cross is commonly used. A single self-incompatible parent is used as female and is open-pollinated by a desirable cultivar as pollen source.

isolated, breeders will always harvest 50% male sterile plants by harvesting only the male sterile plants. The use of this system in commercial hybrid production is outlined Figure 5.9.

Markers linked to genetic male sterility have been identified in some crops (e.g., bright green hypocotyls in broccoli and potato leaf shape and green stem in tomato). Molecular markers (including SCAR, STS, RAPD) associated with male sterility have also been found in some plant species.

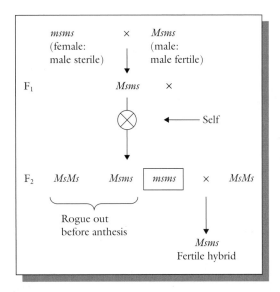

$msms$ (female: male sterile) × $Msms$ (male: male fertile)

F_1 $Msms$ ×

$\otimes$ ← Self

F_2 $MsMs$ $Msms$ $msms$ × $MsMs$

Rogue out before anthesis

$Msms$ Fertile hybrid

Figure 5.9 Genetic male sterility as used in practical breeding.

Male sterility may chemically be induced by applying a variety of agents. This is useful where cytoplasmic male sterility (CMS) genes have not been found. However, this chemical technique has not been routinely applied in commercial plant breeding; it needs further refinement.

- **Cytoplasmic male sterility.** Sometimes, male sterility is controlled by the cytoplasm (mitochondrial gene) but may be influenced by nuclear genes. A cytoplasm without sterility genes is described as normal (N) cytoplasm, while a cytoplasm that causes male sterility is called a sterile (s) cytoplasm or said to have **cytoplasmic male sterility (CMS)**. CMS is transmitted through the egg only (maternal factor). The condition has been induced in species such as sorghum by transferring nuclear chromosomes into a foreign cytoplasm (in this example, a milo plant was pollinated with kafir pollen and backcrossed to kafir). CMS has been found in species including corn, sorghum, sugar beet, carrot, and flax. This system has real advantages in breeding ornamental species because all the offspring is male sterile, hence allowing them to remain fruitless (Figure 5.10). By not fruiting, the plant remains fresh and in bloom for a longer time
- **Cytoplasmic-genetic male sterility.** Cytoplasmic male sterility may be modified by the presence of fertility-restoring genes in the nucleus. CMS is rendered ineffective when the dominant allele for the fertility-restoring gene (*Rf*) occurs, making

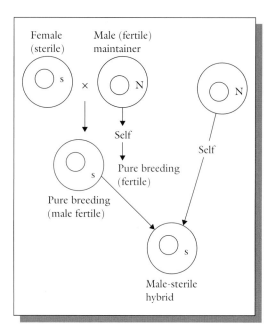

Figure 5.10 Cytoplasmic male sterility as applied in plant breeding (N, normal cytoplasm; s, sterile cytoplasm).

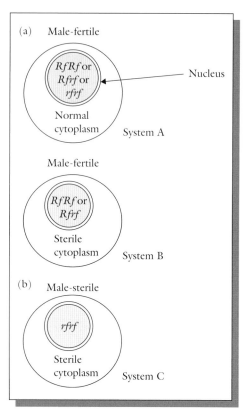

Figure 5.11 The three systems of cytoplasmic genetic male sterility. The three factors involved in CMS are the normal cytoplasm (N), the male sterile cytoplasm (S), and the fertility restorer (*Rf, rf*).

the anthers able to produce normal pollen (Figure 5.11). As previously stated, CMS is transmitted only through the egg, but fertility can be restored by *Rf* genes in the nucleus. Three kinds of progeny are possible following a cross, depending on the genotype of the pollen source. The resulting progenies assume that the fertility gene will be responsible for fertility restoration.

Exploiting male sterility in breeding

Male sterility is used primarily as a tool in plant breeding to eliminate emasculation in hybridization. Hybrid breeding of self-pollinated species is tedious and time consuming. Plant breeders use male sterile cultivars as female parents in a cross without emasculation. Male sterile lines can be developed by backcrossing.

Using genetic male sterility in plant breeding is problematic because it is not possible to produce a pure population of male sterile plants using conventional methods. It is difficult to eliminate the female population before either harvesting or sorting harvested seed. Consequently, this system of pollination control is not widely used for commercial hybrid seed production. On the contrary, CMS is used routinely in hybrid seed production in corn, sorghum, sunflower, and sugar beet. The application of male sterility in commercial plant hybridization is discussed in Chapter 17.

Dichogamy

Dichogamy is the maturing of pistils and stamens of a flower at different times. When this occurs in a self-pollinated species, opportunities for self-pollination are drastically reduced or eliminated altogether, thus making the plant practically cross-pollinated. There are two forms of dichogamy – **protogyny** (stigma is receptive before the anther is mature to release the pollen) and **protandry** (pollen is released from the anther before the female is receptive).

5.5.3 Genetic and breeding implications of autogamy

Self-pollination is considered the highest degree of inbreeding a plant can achieve. It promotes homozygosity of all gene loci and traits of the sporophyte. Consequently, should there be cross-pollination the resulting heterozygosity is rapidly eliminated. To be classified as self-pollinated, cross-pollination should not exceed 4%. The genotypes of gametes of a single plant are all the same. Furthermore, the progeny of a single plant is homogeneous. A population of self-pollinated species, in effect, comprises a mixture of homozygous lines. Self-pollination restricts the creation of new gene combinations (no introgression of new genes through hybridization). New genes may arise through mutation but such a change is restricted to individual lines or the progenies of the mutated plant. The proportions of different genotypes, not the presence of newly introduced types, define the variability in a self-pollinated species. Another genetic consequence of self-pollination is that mutations (which are usually recessive) are readily exposed through homozygosity for the breeder or nature to apply the appropriate selection pressure on.

Repeated selfing has no genetic consequence in self-pollinated species (no **inbreeding depression** or loss of vigor following selfing). Similarly, self-incompatibility does not occur. Because a self-pollinated cultivar is generally one single genotype reproducing itself, breeding self-pollinated species usually entails identifying one superior genotype (or a few) and multiplying it. Specific breeding methods commonly used for self-pollinated species are pure line selection, and also pedigree breeding, bulk populations, and backcross breeding.

5.6 Genotype conversion programs

To facilitate breeding of certain major crops, projects have been undertaken by certain breeders to create breeding stock of male sterile lines that plant breeders can readily obtain. In barley, over 100 spring and winter wheat cultivars have been converted to male sterile lines by USDA researchers. In the case of CMS, transferring chromosomes into foreign cytoplasm is a method of creating CMS lines. This approach has been used to create male sterility in wheat and sorghum. In sorghum, kafir chromosomes were transferred into milo cytoplasm by pollinating milo with kafir and backcrossing the product to kafir to recover all the kafir chromosomes as previously indicated.

5.7 Artificial pollination control techniques

As previously indicated, crossing is a major procedure employed in the transfer of genes from one parent to another in the breeding of sexual species. A critical aspect of crossing is pollination control to ensure that only the desired pollen is involved in the cross. In hybrid seed production, success depends on the presence of an efficient, reliable, practical, and economic pollination control system for large scale pollination. Pollination control may be accomplished in three general ways:

(i) **Mechanical control.** This approach entails manually removing anthers from bisexual flowers to prevent pollination, a technique called emasculation, or removing one sexual part (e.g., detasselling in corn), or excluding unwanted pollen by covering the female part. These methods are time consuming, expensive, and tedious, limiting the number of plants that can be crossed. It should be mentioned that in crops such as corn, mechanical detasselling is widely used in the industry to produce hybrid seed.

(ii) **Chemical control.** A variety of chemicals called chemical hybridizing agents (or other names, e.g., male gametocides, male sterilants, pollenocides, androcides) are used to temporally induce male sterility in some species. Examples of such chemicals include Dalapon®, Estrone®, Ethephon®, Hybrex®, and Generis®. The application of these agents induces male sterility in plants, thereby enforcing cross-pollination. The effectiveness is variable among products.

(iii) **Genetical control.** Certain genes are known to impose constraints on sexual biology by incapacitating the sexual organ (as in male sterility) or inhibiting the union of normal gametes (as in self-incompatibility). These genetic mechanisms are now discussed further.

5.8 Mendelian concepts in plant breeding

As previously stated, genetics is the principal science that underlies plant breeding. Mendel derived several postulates or principles of inheritance, which are often

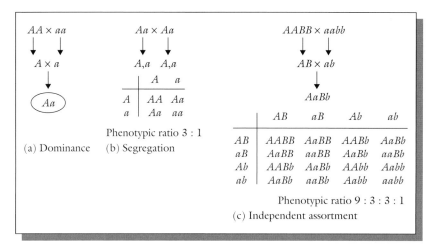

Figure 5.12 Mendel's postulates: (a) dominance, (b) segregation, and (c) independent assortment.

couched as Mendel's Laws of inheritance. Genes are transferred from parents to offspring following mating or crossing. Mendel's principles are best illustrated using self-pollinated species, as was the case in his original experiments.

5.8.1 Mendelian postulates

Because plant breeders transfer genes from one source to another, an understanding of **transmission genetics** is crucial to a successful breeding effort. The method of breeding used depends upon the heredity of the trait being manipulated, amongst other factors. According to Mendel's results from his hybridization studies in pea, traits are controlled by heritable factors that are passed from parents to offspring through the reproductive cells. Each of these unit factors occurs in pairs in each cell (except reproductive cells or gametes).

In his experiments, Mendel discovered that in a cross between parents displaying two contrasting traits, the hybrid (F_1) expressed one of the traits to the exclusion of the other. He called the expressed trait **dominant** and the suppressed trait **recessive**. This is the phenomenon of **dominance** and **recessivity**. When the hybrid seed was planted and self-pollinated, he observed that both traits appeared in the second generation (F_2) (i.e., the recessive trait reappeared) in a ratio of 3:1 dominant:recessive individuals (Figure 5.12). Mendel concluded that the two factors that control each trait do not blend but remain distant throughout the life of the individual and

segregate in the formation of gametes. This is called the **law of segregation**. In further studies in which he considered two traits simultaneously, he observed that the genes for different traits are inherited independently of each other. This is called the **law of independent assortment**. In summary, the two key laws are as follows:

(i) **Law I: Law of segregation**: Paired factors segregate during the formation of gametes in a random fashion such that each gamete receives one form or the other.

(ii) **Law II: Law of independent assortment:** When two or more pairs of traits are considered simultaneously, the factors for each pair of traits assort independently to the gametes.

Mendel's pair of factors is now known as **genes**, while each factor of a pair (e.g., *HH* or *hh*) is called an **allele** (i.e. the alternative form of a gene; *H* or *h*). The specific location on the chromosome where a gene resides is called a **gene locus** or simply **locus** (**loci** for plural).

5.8.2 Concept of genotype and phenotype

The term **genotype** is used to describe the totality of the genes of an individual. Because the totality of an individual's genes is not known, the term, in practice, is usually used to describe a very small subset of genes of interest in a breeding program or research. Conventionally, a genotype is written with an uppercase letter (*H*, *G*) indicating the **dominant allele** (expressed over the alternative allele), while a lower case letter (*h*, *g*) indicates the **recessive allele**. A plant that has two identical alleles for genes is **homozygous** at that locus (e.g., *AA*, *aa*, *GG*, *gg*) and is called a **homozygote**. If it has different alleles for a gene, it is **heterozygous** at these loci (e.g., *Aa*, *Gg*) and is called a **heterozygote**. Certain plant breeding methods are designed to produce products that are homozygous (breed true – most or all of the loci are homozygous) whereas others (e.g., hybrids) depend on heterozygosity for success.

The term **phenotype** refers to the observable effect of a **genotype** (the genetic makeup of an individual). Because genes are expressed in an environment, a phenotype is the result of the interaction between a genotype and its environment (i.e., phenotype = genotype + environment, or symbolically, $P = G + E$). Later in this book a more complete form of this equation is introduced as $P = G + E + GE + error$, where GE represents the interaction between the environment and the genotype. This interaction effect helps plant breeders in the cultivar release decision making process (Chapter 23).

5.8.3 Predicting genotype and phenotype

Based upon Mendel's laws of inheritance, statistical probability analysis can be applied to determine the outcome of a cross, given the genotype of the parents and gene action (dominance/recessivity). A genetic grid called a **Punnett square** (Figure 5.13) facilitates the analysis. For example, a monohybrid cross in which the genotypes of interest are $AA \times aa$, where *A* is dominant over *a*, will produce a hybrid genotype *Aa* in the F_1 (first filial generation) with an AA phenotype. However, in the F_2 ($F_1 \times F_1$), the Punnett square shows a genotypic ratio of *1AA:2Aa:1aa* and a phenotypic ratio of 3:1 because of dominance. A **dihybrid cross** (involving simultaneous analysis of two different genes) is more complex but conceptually like a **monohybrid cross** (only one gene of interest) analysis. An analysis of a dihybrid cross $AABB \times aabb$, using the Punnett square is illustrated in Figure 5.13. An alternative method of genetic analysis of a cross is by the **branch diagram** or forked line method (Figure 5.14).

Predicting the outcome of a cross is important to plant breeders. One of the critical steps in a hybrid program is to authenticate the F_1 product. The

Egg	Pollen	
	$\frac{1}{2}A$	$\frac{1}{2}a$
$\frac{1}{2}A$	$\frac{1}{4}AA$	$\frac{1}{4}Aa$
$\frac{1}{2}a$	$\frac{1}{4}Aa$	$\frac{1}{4}aa$

Phenotypic ratio of 3 : 1 *A*– : *aa*
(a) Monohybrid cross

Egg	Pollen			
	$\frac{1}{4}AB$	$\frac{1}{4}Ab$	$\frac{1}{4}Ab$	$\frac{1}{4}ab$
$\frac{1}{4}AB$	1/16 AABB	1/16 AABb	1/16 AaBB	1/16 AaBb
$\frac{1}{4}Ab$	1/16 AABb	1/16 AAbb	1/16 AaBB	1/16 Aabb
$\frac{1}{4}Ab$	1/16 aABB	1/16 AaBb	1/16 aaBB	1/16 aaBb
$\frac{1}{4}ab$	1/16 AaBb	1/16 Aabb	1/16 aaBb	1/16 aabb

Phenotypic ratio of 9 : 3 : 3 : 1   *A–B*– : *A–ab* : *aaB*– : *aabb*
(b) Dihybrid cross

Figure 5.13 The Punnett square procedure may be used to demonstrate the events that occur during hybridization and selfing: (a) a monohybrid cross and (b) a dihybrid cross show the proportions of genotypes in the F_2 population and the corresponding Mendelian phenotypic and genotypic ratios.

breeder must be certain that the F_1 truly is a successful cross and not a product of selfing. If a selfed product is advanced, the breeding program will be a total waste of resources. To facilitate the process, breeders may include a genetic marker in their program. If two plants are crossed, for example, one with purple flowers and the other white flowers, the F_1 plant is expected to have purple flowers because of dominance of purple over white flowers. If the F_1 plant has white flowers, it is proof that the cross was unsuccessful (i.e., the product of the "cross" is actually from selfing).

5.8.4 Distinguishing between heterozygous and homozygous individuals

In a segregating population where genotypes *PP* and *Pp* produce the same phenotype (because of dominance), it is necessary sometimes to know the exact genotype of a plant. There are two procedures that are commonly used to accomplish this task.

(i) **Test cross.** Developed by Mendel, a **test cross** entails crossing the plant with the dominant allele but unknown genotype with a homozygous

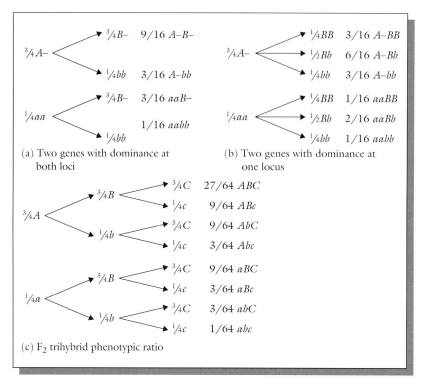

(a) Two genes with dominance at both loci

(b) Two genes with dominance at one locus

(c) F$_2$ trihybrid phenotypic ratio

Figure 5.14 The branch diagram method may be used to predict the phenotypic and genotypic ratios in the F$_2$ population.

recessive individual (Figure 5.15). If the unknown genotype is *PP*, crossing it with the genotype pp will produce all *Pp* offspring. However, if the unknown is *Pp* then a test cross will produce offspring segregating 50:50 for *Pp:pp*. The test cross also supports Mendel's postulate that separate genes control purple and white flowers.

(ii) **Progeny Test.** Unlike a test cross, a **progeny test** does not include a cross with a special parent

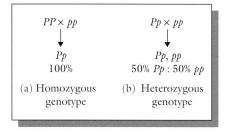

(a) Homozygous genotype

(b) Heterozygous genotype

Figure 5.15 The test cross. Crossing a homozygous dominant genotype with a homozygous recessive genotype always produces all heterozygotes (a). However, crossing a heterozygote with a homozygous recessive produces both homozygotes and heterozygotes (b).

but selfing of the F$_2$. Each F$_2$ plant is harvested and separately bagged and then, subsequently, planted. In the F$_3$, plants that are homozygous dominant will produce progeny that is uniform for the trait, while plants that are heterozygous will produce a segregating progeny row.

Plant breeders use the progeny test for a number of purposes. In breeding methodologies in which selection is based on phenotype, a progeny test will allow a breeder to select superior plants from among a genetically mixed population. Following an environmental stress, biotic or abiotic, a breeder may use a progeny test to identify superior individuals and further ascertain if the phenotypic variation is due to genetic effects or just caused by environment factors.

5.9 Complex inheritance

Just how lucky was Mendel in his experiments that yielded his landmark results? This question has been widely discussed among scientists over the years. Mendel selected traits whose inheritance patterns enabled him to avoid certain complex inheritance patterns that would have made his results and interpretations more challenging. The inheritance of traits such as those studied by Mendel is described as **simple (simply inherited) traits,** or having **Mendelian inheritance**. There are numerous other traits that have complex inheritance patterns that cannot be predicted by Mendelian ratios. Several factors are responsible for the observation of non-Mendelian ratios, as discussed next.

5.9.1 Incomplete dominance and codominance

Mendel worked with traits that exhibited complete dominance. Post-Mendelian studies revealed that frequently the masking of one trait by another is only partial (called **incomplete dominance** or **partial dominance**). A cross between a red-flowered (RR) and white-flowered (rr) snapdragon produces pink-flowered plants (Rr). The genotypic ratio remains 1:2:1 but a lack of complete dominance also makes the phenotypic ratio 1:2:1 (instead of the 3:1 expected for complete dominance).

Another situation in which there is no dominance occurs when both alleles of a heterozygote are expressed to equal degrees. The two alleles code for two equally functional and detectable gene products. Commonly observed and useful examples to plant breeding technology are **allozymes,** the production of different forms of the same enzyme by different alleles at the same locus. Allozymes catalyze the same reaction. This pattern of inheritance is called **codominant inheritance** and the gene action **codominance.** Some molecular markers are codominant. Whereas incomplete dominance produces blended phenotype, codominance produces distinct and separate phenotypes.

5.9.2 Multiple alleles of the same gene

The concept of **multiple alleles** can be studied only in a population. Any individual diploid organism can, as previously stated, have at most two homologous gene loci that can be occupied by different alleles of the same gene. However, in a population, members of a species can have many alternative forms of the same gene. A diploid, by definition can have only two alleles at each locus (e.g., C^1C^1, C^7C^{10}, C^4C^6). However, mutations may cause additional alleles to be created in a population. Multiple alleles of allozymes are known to occur. The mode of inheritance by which individuals have access to three or more alleles in the population is called **multiple allelism** (the set of alleles is called an **allelic series**). A more common example of multiple allelism, which may help the reader better understand the concept, is the **ABO** blood group system in humans. An allelic series of importance in plant breeding is the **S alleles** that condition self-incompatibility (inability of a flower to be fertilized by its own pollen). Self-incompatibility is a constraint to sexual biology and can be used as a tool in plant breeding, as previously discussed in detail in this chapter.

5.9.3 Multiple genes

Just as a single gene may have multiple alleles that produce different forms of one enzyme, there can be more than one gene for the same enzyme. The same enzymes produced by different genes are called **isozymes**. Isozymes are common in plants. For example, the enzyme phosphoglucomutase in *Helianthus debilis* is controlled by two nuclear genes and two chloroplast genes. Isozymes and allozymes were the first molecular markers developed for use in plant and animal genetic research.

5.9.4 Polygenic inheritance

Mendelian genes are also called **major genes** (or **oligogenes**). Their effects are easily categorized into several or many non-overlapping groups. The variation is said to be discrete. Some traits are controlled by several or many genes that have effects too small to be individually distinguished. These traits are called **polygenes** or **minor genes** and are characterized by non-discrete (or continuous) variation, because the effects of the environment on these genes allow their otherwise discrete segregation to be readily observed. Scientists use statistical genetics to distinguish between genetic variation due to the segregation of polygenes and environmental variation. Many genes of interest to plant breeders exhibit polygenic inheritance.

	AB	Ab	aB	ab
AB	AABB	AABb	AaBB	AaBb
Ab	AABb	AAbb	AaBb	Aabb
aB	AaBB	AaBb	aaBB	aaBb
ab	AaBb	Aabb	aaBb	aabb

(a) Complementary genes 9 : 7

	AB	Ab	aB	ab
AB	AABB	AABb	AaBB	AaBb
Ab	AABb	AAbb	AaBb	Aabb
aB	AaBB	AaBb	aaBB	aaBb
ab	AaBb	Aabb	aaBb	aabb

(b) Additive genes 9 : 6 : 1

	AB	Ab	aB	ab
AB	AABB	AABb	AaBB	AaBb
Ab	AABb	AAbb	AaBb	Aabb
aB	AaBB	AaBb	aaBB	aaBb
ab	AaBb	Aabb	aaBb	aabb

(c) Duplicate genes 15 : 1

	AB	Ab	aB	ab
AB	AABB	AABb	AaBB	AaBb
Ab	AABb	AAbb	AaBb	Aabb
aB	AaBB	AaBb	aaBB	aaBb
ab	AaBb	Aabb	aaBb	aabb

(d) Suppressor genes 13 : 3

	AB	Ab	aB	ab
AB	AABB	AABb	AaBB	AaBb
Ab	AABb	AAbb	AaBb	Aabb
aB	AaBB	AaBb	aaBB	aaBb
ab	AaBb	Aabb	aaBb	aabb

(e) Dominant epistasis 12 : 3 : 1

	AB	Ab	aB	ab
AB	AABB	AABb	AaBB	AaBb
Ab	AABb	AAbb	AaBb	Aabb
aB	AaBB	AaBb	aaBB	aaBb
ab	AaBb	Aabb	aaBb	aabb

(f) Recessive epistasis 9 : 3 : 4

Figure 5.16 Epistasis or non-Mendelian inheritance is manifested in a variety of ways, according to the kinds of interaction. Some genes work together while other genes prevent the expression of others.

5.9.5 Concept of gene interaction and modified mendelian ratios

Mendel's results primarily described discrete (discontinuous) variation, even though he observed continuous variation in flower color. Later studies established that the genetic influence on the phenotype is complex, involving the interactions of many genes and their products. It should be pointed out that genes do not necessarily interact directly to influence a phenotype but, rather, the cellular function of numerous gene products work together in concert to produce the phenotype.

Mendel's observation of dominance/recessivity is an example of an interaction between alleles of the same gene. However, interactions involving non-allelic genes do occur, a phenomenon called **epistasis**. There are several kinds of epistatic interactions, each modifying the expected Mendelian ratio in a characteristic way. Instead of the 9 : 3 : 3 : 1 dihybrid ratio for dominance at two loci, modifications of the ratio includes 9 : 7 (complementary genes), 9 : 6 : 1 (additive genes), 15 : 1 (duplicate genes), 13 : 3 (suppressor gene), 12 : 3 : 1 (dominant epistasis), and 9 : 3 : 4 (recessive epistasis) (Figure 5.16). Other possible ratios are 6 : 3 : 3 : 4 and 10 : 3 : 3. To arrive at these conclusions, researchers test data from a cross against various models, using the chi square statistical method. Genetic linkage, cytoplasmic inheritance, mutations, and transposable elements are considered the most common causes of non-Mendelian inheritance.

5.9.6 Pleiotropy

Sometimes, one gene can affect multiple traits, a condition called **pleiotropy**. It is not hard to accept this fact when one understands the complex process of the development of an organism in which the event of one stage is linked to those before (i.e., correlated traits). That is, genes that are expressed early in the development of a trait are likely to affect the outcome of the developmental process. In sorghum, the gene *hl* causes the high lysine content of seed storage proteins to increase as well as cause the endosperm to be shrunken. Declaring genes to be pleiotropic is often not clear cut, because closely associated or closely linked genes can behave this way. Conducting a large number of crosses may produce a recombinant, thereby establishing that linkage, rather than pleiotropy, exists.

Industry highlights

Haploids and doubled haploids: Their generation and application in plant breeding

Sergey Chalyk

Institute of Genetics, Chisinau, Moldova

Haploid plants are intensively utilized for investigation and improvement of many agricultural crops. Haploids are unique plants and can provide researchers with genetic information not possible with normal diploid individuals. The methods of obtaining and some advantages of using haploids in plant breeding are discussed here.

What are haploids?

The term "haploid" refers to a plant or an embryo that contains a gametic chromosome set. Spontaneous haploid plants are found to occur in many crop species, such as cotton, tomato, potato, soybean, tobacco, maize, barley wheat, rice, rye, and so on.

In general, haploids can be divided in three types. The first type is called a maternal haploid. These haploids contain the only nuclear material and cytoplasm from the maternal parent. They result either from the elimination of the chromosomes provided by the paternal parent during embryo development or by paternal sperm nuclei that are incapable of fertilization.

The second type is called an *in vitro* androgenic haploid. They can be obtained through the anther or by microspore culture and contain both the cytoplasm and nucleus of the developing microsporocyte.

The third type of haploid is called an *in vivo* androgenic haploid, since it arises by *in vivo* embryogenesis. This class of haploids develops from an egg cell or any other cell of the embryo sac by having the chromosomes of the maternal parent being lost during embryogenesis. Such haploids contain the cytoplasm of maternal plant and only the chromosomes of the paternal parent.

Advantages in the utilization of haploids

Haploid plants contain only one set of chromosomes. All their genes are hemizygous and each gene has only one allele. This particular feature of haploid plants allows them to be utilized in unique ways for breeding or genetic studies.

Firstly, haploid plants can be utilized for the accelerated development of homozygous lines and pure cultivars. For this purpose it is essential to double the chromosome number after a haploid individual is generated. Following diploidization, two identical sets of chromosomes are present in the doubled haploid individual. In these instances, each gene is now represented in two exact copies or two identical alleles. By utilizing this approach, breeders can obtain homozygous lines and pure cultivars two to three times faster than by utilizing conventional methods of breeding.

Secondly, haploids can also be utilized for the selection of genotypes that contain favorable genes. Since haploids possess only a single dose of their respective genomes, there is no possibility for intra-allelic genetic interactions. Each gene is expressed in a single dose. This significantly facilitates the search and selection of favorable genes and the development of superior breeding genotypes. In addition, since haploids possess only a single dose of each chromosome, the possible number of gene segregation products is significantly reduced. This allows the breeder to identify a favorable combination of genes with higher probability. This approach has special value for breeders or geneticists interested in developing an understanding of the inheritance and expression of quantitative traits. The enhanced probability of finding a favorable genotype is identified in the following example. Assume that the progeny of a hybrid plant is segregating for 10 genes. In a normal diploid, it would be necessary to grow 1 048 576 plants to obtain all possible combinations of these genes. For haploids, it would take only 1024 haploid plants to generate all the possible combinations at least once. This example clearly shows that a desired combination of genes can be found with far fewer individuals when haploids are utilized.

Thirdly, natural selection on haploids can be utilized as a genetic filter to identify or remove harmful mutant genes. Normally, a certain "genetic load" exists in any line, cultivar or population as a result of spontaneous mutation over time. In diploid plants this is hidden by homologous alleles and can weaken the plants. By utilizing haploids, all genes are expressed in a single dose, including mutant or deleterious genes. Consequently, haploids containing harmful, lethal and semi-lethal mutations either perish or are completely sterile. This approach leads to the natural cleansing of breeding material of genes that can reduce plant viability and productivity.

Generation of haploids and doubled haploids

Chromosome elimination

It was discovered that haploid plants of *Hordeum vulgare* could be obtained on a large scale following the hybridization of *Hordeum vulgare* with *Hordeum bulbosum* (Kasha and Kao, 1970). When *H. vulgare* and *H. bulbosum* are crossed, a normal double fertilization event occurs. However, during seed development, chromosomes of *H. bulbosum* are eliminated in both the embryo and endosperm. At approximately 10 days post-fertilization, most dividing cells in the embryo are haploid. Seed possessing a haploid embryo are removed from the spikes and placed on an embryo rescue nutrient agar culture. Approximately 50–60% of the cultured embryos develop into mature haploid plants. Colchicine, a mitotic inhibitor is applied to the haploid seedlings generating fertile spikelet/seed sectors with double the chromosome number. Haploids with these fertile sectors generate seed that have a normal diploid chromosome number.

Chromosome elimination is an alternative method for producing haploids commonly utilized in wheat. In this approach, pollen from either *H. bulbosum* or maize pollen is applied to the silks of an emasculated wheat spike. The application of maize pollen has proven to be the most successful approach by providing the highest frequency of haploids.

In vitro androgenesis

In vitro androgenesis refers to the culturing of the male gamete either in the form of an anther or as isolated microspores onto an appropriate culture media. For most crop species appropriate *in vitro* androgenesis culture media has been developed. The most successful culture media will be useful for a wide range of genotypes, such as that developed for barley (Kasha *et al.*, 2001). In their experiments more than 30 different barley genotypes have been examined and, in general, between 5000 and 15 000 embryos are produced per plate. Regeneration ability of the embryos ranged from 36 to 97%. About 70% of plants obtained by the method of isolated microspore cultures double chromosome number spontaneously, eliminating the necessity for the use of chromosome doubling procedures. This method can be used for mass production of haploids from any genotype of barley and with minor modification, genotypes of wheat.

In vivo androgenesis

Kermicle (1969) reported on the possibility of obtaining androgenic haploids in maize. He found that pollination of plants containing the homozygous gene *ig1* (*indeterminate gametophyte 1*) results in the development of 1–3% of seed with an androgenic haploid embryo. Additional research has identified that the *ig1* gene causes an increased number of mitotic divisions during the formation of the megaspore mother cell. The extra divisions of the egg cells lead to a loss of a normal fertilization event. Sperm usually penetrates egg cell, but sperm and egg cell fail to fuse. In this event, the developing embryo possesses only the chromosomes from the sperm nuclei.

Androgenic haploids are mostly used for the accelerated development of lines containing male sterile cytoplasm. For this purpose a series of *ig1 ig1* B-3Ld *Ig1* trisomics were developed that contain following types of sterile cytoplasm: C, S, SD, Vg, ME, MY, CA, L, Q (Kindiger, 1993; Kindiger and Hamann, 1994).

Induction of maternal haploids in maize

In maize, maternal haploids can occur spontaneously. Their rate is usually about one haploid per one or two thousand of normal diploid plants. Extensive research investigations by Chase (1969) suggested their use in inbred line development. However, the low frequency of natural haploid generation prevents an efficient use of this approach in breeding programs. An alternative approach to obtain and investigate maternal haploids in maize was reported following the discovery of a line called "Stock 6" (Coe, 1959). In this report, it was observed that pollen of Stock 6 induced the generation of haploidy. The discovery of a maize pollen source that contains a haploid-inducing factor, simplified and facilitated the ability to obtain haploids from a wide array of different genotypes.

Stock 6 has since been utilized for the development of many new haploid-inducing lines that possess dominant marker genes for isolation of haploids. Typically, dominant marker genes conferring anthocyanin production are utilized. Such genes cause the development or a deep red or purple pigment in the seeds, seedlings and/or plants. One such marker is called *R1-nj (R-navajo)*. This gene expresses anthocyanin in both the aleurone layer of the crown of the seed as well as the embryo. Following pollination of a breeding material by a haploid-inducing line that possesses marker gene *R1-nj*, seeds that develop with pigment in the aleurone layer but no pigment in the embryo will provide haploid plants (Figure B5.1). Seeds with pigment in both the aleurone and the embryo are hybrid.

Figure B5.1 Ear developed following pollination by haploid inducer MHI that contains marker gene, *R1-nj*. The arrow shows the kernel containing haploid embryo. Figure courtesy of Sergey Chalyk.

Generating doubled haploids

Doubling the chromosome number in haploids is often conducted through the use of colchicine or other mitotic inhibitors, such as nitrous oxide gas and some herbicides. Following treatment of the apical meristem by a mitotic inhibitor, chromosome numbers are doubled in small sectors of the haploid plant, including some sectors of the spike, ear or the tassel. Normally the doubled sectors produce seeds. These seeds are doubled haploids, pure line cultivars.

In our research, a method of colchicine treatment was evaluated on maize haploids. In the method, haploid seedlings were treated with colchicines at the stage when the length of the coleoptile was 1 cm or longer. Initially, the seeds were germinated at 26 °C for 4–5 days. Then a small tip of the coleoptile was removed and the seedlings were placed into 0.06% colchicine solution with 0.5% dimethyl sulfoxide (DMSO) for 12 h. Thereafter, the seedlings were washed and planted in the field. Following the treatment, approximately half of the haploids produced fertile pollen (49.4%). Most were selfed and 27.3% of the haploid individuals produced seeds. These seeds were doubled haploids since they contained normal diploid embryo. The seeds were quite viable and generated normal viable seedlings following planting in the field or in greenhouse.

Application of haploids and doubled haploids in plant breeding

Haploids have two primary uses in plant breeding. The first is the accelerated production of homozygous lines and pure cultivars. As we have discussed earlier, the doubling of haploids is the most rapid route toward the development of pure cultivars in self-pollinated seed crops. It can be achieved in a single generation and can be performed at any generation in a breeding program. For cross-pollinated crops, haploids are used primarily for the production of homozygous lines, which are in themselves utilized in the production of hybrid seed. At present, more than 200 varieties have been developed by utilizing a doubled haploid approach (Thomas *et al.*, 2003).

The second primary is that haploids provide a possibility of screening breeding material for the presence of advantageous genes. In both haploids and doubled haploids, all alleles are expressed. This facilitates selection of genotypes that are important for breeders. Selected haploids can be used for the improvement of any breeding material, including increasing the frequency of favorable genes in populations (i.e. in recurrent selection). As one example of haploid utility in a breeding program, the method of Haploid Sib Recurrent Selection can be presented (Eder and Chalyk, 2002). In this approach, the selection of favorable genotypes is performed on haploid plants. Every cycle of selection consists of two steps. The first step is to obtain haploids from a synthetic population. In our experiment haploids were obtained in a space-isolated nursery following pollination with a haploid-inducing line. The second step is growing haploid plants, the selection of haploid plants and pollination with a mixture of pollen collected from diploid plants of the same synthetic population at the same cycle of selection. Seeds from the haploids are obtained by applying pollen collected from diploid plants of the same synthetic population at the same cycle of selection. This step requires fertility in the haploid ears. Typically following pollination, nearly every ear possesses seed with a normal fertilized diploid embryo. This unique tendency of maize haploid ears allows the breeder to move forward in the breeding process without doubling the chromosome number of the haploid individuals. This makes the utilization of haploids simple and inexpensive.

In our experiments, selection was carried out for ear size. Each season, 1000–2000 haploid plants from an improved synthetic population were planted in the field. Of these, 200–300 haploids were pollinated by diploid representatives of the same synthetic population. At harvest time, about 20–30 haploids having the largest ears were selected for the next cycle of selection. Three cycles of Haploid Sib Recurrent Selection were completed for two synthetic populations, SP and SA. Initial and improved synthetics were evaluated in the field for four years. The performance results of synthetic SA are

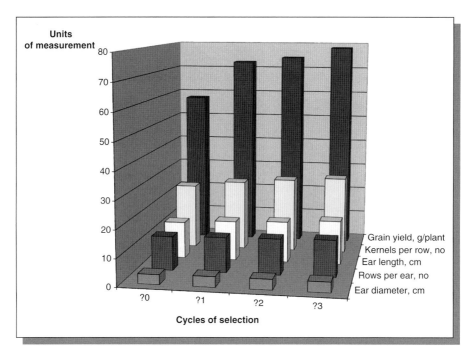

Figure B5.2 Grain yield and ear traits of synthetic population SA after three cycles of Haploid Sib Recurrent Selection, average over four years of testing. Figure courtesy of Sergey Chalyk.

presented in Figure B5.2. The data indicate that selection for ear size, utilizing haploids, can result in a significant increase of grain yield.

An important method that effectively estimates the efficiency of a recurrent selection program is the determination of gain per cycle. This parameter is used for comparing the efficiency of different recurrent selection schemes. Normally gain per cycle for grain yield in a recurrent selection scheme in maize approximates 2–5% and seldom exceeds 7% (Gardner, 1977; Weyhrich et al., 1998). The results obtained by Haploid Sib Recurrent Selection are presented in the Table B5.1. For synthetic population SA gain per cycle was 12.0%, and for synthetic SP it was 13.1%. Gain per cycle was distinctly higher than that observed when utilizing conventional recurrent selection methods. The conclusion drawn from the experiment is that the utilization of haploid plants for the selection of favorable genotypes greatly increases the efficiency of recurrent selection.

Table B5.1 Estimated gain per cycle for grain yield, ear and plant traits of SP and SA synthetic populations on average for four years of testing.

Traits	Gain per cycle (%)									
	SP					SA				
	1998	1999	2000	2001	Average	1998	1999	2000	2001	Average
Grain yield	11.0	16.4	1.7	17.8	13.1	16.7	21.0	10.3	8.4	12.0
Ear length	9.2	6.6	4.4	7.5	7.8	4.5	9.2	4.4	5.8	6.2
Seeds per row	2.6	6.7	1.1	9.6	6.2	7.5	11.4	4.8	4.8	6.3
Rows number	7.3	6.2	3.0	5.2	2.9	1.4	4.7	3.0	4.4	6.1
Ear diameter	5.7	4.8	0.6	4.1	4.1	3.5	3.7	2.2	2.6	2.0
Plant height	9.3	9.6	10.5	7.7	10.0	12.1	3.7	7.3	4.8	4.8
Ear height	10.0	7.6	10.0	11.8	10.9	16.6	8.4	13.8	11.1	6.7
Leaf length	8.5	6.2	3.8	9.0	7.8	3.3	2.8	1.5	3.2	2.6

Overall, numerous experiments have indicated that haploid generation can be successfully applied to several species on a large scale. The methods and citations given above provide only a few examples of this useful and efficient method. Utilization of haploids and doubled haploids can simplify the identification of genotypes that can provide a significant improvement in a variety of agronomic traits. In addition, haploids and doubled haploids can accelerate the generation of homozygous lines and pure cultivars. Therefore, haploid inducement technologies have a bright future in plant breeding.

References

Chase, S.S. (1969). Monoploids and monoploid-derivatives of maize (*Zea mays* L.). *The Botanical Review.* **35**:117–167.

Coe, E.H. (1959). A line of maize with high haploid frequency. *Amer. Nat.*, **93**:381–382.

Eder, J., and Chalyk, S. (2002). *In vivo* haploid induction in maize. *Theor. Appl. Genet.*, **104**:703–708.

Gardner, C.O. (1977). Population improvement in maize, in *Maize breeding and genetics* (ed. D.B. Walden). John Wiley & Sons, Inc., New York, p. 207–228.

Kasha, K.J., and Kao, K.N. (1970). High frequency haploid production in barley (*Hordeum vulgare* L.). *Nature*, **225**:874–876.

Kasha, K.J., Simion, E., Oro, R., Yao, Q.A., Hu, T.C., and Carlson, A.R. (2001). An improved *in vitro* technique for isolated microspore culture of barley. *Euphytica*, **120**:379–385.

Kermicle, J.L. (1969). Androgenesis conditioned by a mutation in maize. *Science*, **116**:1422–1424.

Kindiger, B. (1994). Registration of 10 genetic stocks of maize for the transfer of cytoplasmic male sterility. *Crop Sci.*, **34**:321–322.

Kindiger, B., and Hamann, S. (1993). Generation of haploids in maize: a modification of the *indeterminate gametophyte* (*ig*) system. *Crop Sci.*, **33**:342–344.

Thomas,B., W.T., Forster, B.P., and Gertsson, B. (2003). Doubled haploids in breeding, in *Doubled haploid production in crop plants* (eds M. Maluszynski, K. Kasha, B. P. Forster and I. Szarejko) Klewer Academic Publishers, pp. 337–350.

Weyhrich, R.A., Lamkey, K.R., and Hallauer, A.R. (1998). Responses to seven methods of recurrent selection in the BS11 maize population. *Crop Sci.*, **38**:308–321.

Key references and suggested reading

Acquaah, G. (2004). *Horticulture: Principles and practices*, 3rd edn. Prentice Hall, Upper Saddle River, NJ.

Kiesselbach, T.A. (1999). *The structure and reproduction of corn. 50th anniversary edition*. Cold Spring Habor Laboratory Press, Cold Spring Habor, New York.

Stern, K.R. (1997). *Introductory plant biology*, 7th edn. Wm. C. Brown Publishers, Dubuque, IA.

Outcomes assessment

Part A

Please answer the following questions true or false.

1 Biennial plants complete their lifecycle in two growing seasons.
2 A staminate flower is a complete flower.
3 Self-pollination promotes heterozygosity of the sporophyte.
4 The union of egg and sperm is called fertilization.
5 A branched raceme is called a panicle.
6 The carpel is also called the androecium.

Part B

Please answer the following questions.

   1 Plants reproduce by one of two modes, ...

   2 Distinguish between monoecy and dioecy.

   3 .... is the transfer of pollen grain from the anther to the stigma of a flower.

   4 What is self-incompatibility?

   5 Distinguish between heteromorphic self-incompatibility and homomorphic self-incompatibility.

   6 What is apomixis?

   7 Distinguish between apospory and displospory as mechanisms of apomixis

Part C

Please write a brief essay on the following topics.

   1 Discuss the genetic and breeding implications of self-pollination.

   2 Discuss the genetic and breeding implications of cross-pollination.

   3 Fertilization does not always follow pollination. Explain.

   4 Discuss the constraints of sexual biology in plant breeding.

   5 Discuss how cytoplasmic male sterility (CMS) is used in a breeding program.

   6 Discuss how genetic male sterility is used in a breeding program

6

Allogamy

Purpose and expected outcomes

There are major world crops that reproduce via allogamy. The breeding methods for autogamy are different from those for allogamy because the mode of reproduction has such profoundly different genetic consequences. After studying this chapter, the student should be able to:

1 Discuss the natural mechanisms that favor allogamy.
2 Discuss the genetic consequences of allogamy.
3 Discuss the implications of allogamy in crop improvement.

6.1 What is allogamy?

Allogamy occurs when fertilization of the flower of a plant is effected by pollen donated by a different plant within the same species. This is synonymous with **cross-pollination**, **cross-fertilization** or **outbreeding**, involving the actual fusion of gametes (sperm and ovum). An (incomplete) list of allogamous species is presented in Table 6.1.

6.2 Mechanisms that favor allogamy

Allogamous species depend on agents of pollination, especially wind and insects, and hence tend to produce large amounts of pollen and have large, bright-colored fragrant flowers to attract insects. They commonly have taller stamens than carpels or use other

mechanisms to better ensure the dispersal of pollen to other plants flowers. Other provisions that promote cross-fertilization are mechanisms that control the timing of the receptiveness of the stigma and shedding of pollen and, thereby, prevent autogamy within the same flower. In **protandry**, the anthers release their pollen before the stigma of the same flower is receptive (protandrous flower). In **protogyny**, the stigma is receptive before the pollen is shed from the anthers of the same flower (protogynous flower). Several mechanisms occur in nature by which cross-pollination is ensured, the most effective being dioecy, monoecy, dichogamy, and self-incompatibility. Some mechanisms are stringent in enforcing cross-pollination (e.g., dioecy), while others are less so (e.g., monoecy). These mechanisms are exploited by plant breeders during the controlled pollination phase of their breeding programs, so that

---

Principles of Plant Genetics and Breeding, Second Edition. George Acquaah.
© 2012 John Wiley & Sons, Ltd. Published 2012 by John Wiley & Sons, Ltd.

Table 6.1 Examples of predominantly cross-pollinated species.

Common name	Scientific name
Alfalfa	*Medicago sativa*
Annual ryegrass	*Lolium multiflorum*
Banana	*Musa* spp.
Birdsfoot trefoil	*Lotus corniculatus*
Cabbage	*Brassica oleracea*
Carrot	*Daucus carota*
Cassava	*Manihot esculentum*
Cucumber	*Cucumis sativa*
Fescue	*Festuca* spp.
Kentucky bluegrass	*Poa pratense*
Maize	*Zea mays*
Muskmelon	*Cucumis melo*
Onion	*Allium* spp.
Potato	*Solanum tuberosum*
Radish	*Raphanus sativus*
Rye	*Secale cereale*
Sugarbeet	*Beta vulgaris*
Sunflower	*Helianthus annuus*
Sweet potato	*Ipomoea batatus*
Watermelon	*Citrullus lanatas*

Though predominantly pollinated, some of these species may have another reproductive mechanism in breeding and crop cultural systems. For example, banana is vegetatively propagated (and not grown from seed), as are cassava and sweet potato; cabbage and maize are produced as hybrids.

only desired pollen sources participate in siring the next plant generation.

6.2.1 Monoecy

Some flowers are complete (possess all the four basic parts) while others are incomplete (are missing one or more of the four basic floral parts). Furthermore, in some species, the sexes are separate. When separate male and female flowers occur on the same plant, the condition is called **monoecy.** Sometimes, the male and female flowers occur in different kinds of inflorescence (different locations, as in corn). Other examples of monoecious plant include most figs, birch, and pine trees. It is easier and more convenient to self-pollinate plants when the sexes occur in the same inflorescence. In terms of seed production, monoecy and dioecy may appear to be inefficient because not all flowers produce seed. Some flowers produce only pollen.

6.2.2 Dioecy

When the sexes occur on different plants (i.e., there are female plants and male plants), the condition is called **dioecy.** Examples of dioecious crop species include date, hops, asparagus, spinach, holly, and hemp. The separation of the sexes means that, by necessity, all seed from dioecious species are hybrid in composition. Where the economic product is the seed or fruit, it is imperative to have female and male plants in the field in an appropriate ratio. In dioecious fruit orchards (e.g., date, persimmons), 3–4 males per 100 females may be adequate. In hops, the commercial product is the female inflorescence. Unfertilized flowers have the highest quality. Consequently, it is not desirable to grow pollinators in the same field when growing hops. Dioecious crops propagated by seed may be improved by mass selection or controlled hybridization.

6.3 Genetic and breeding implications of allogamy

The genotype of the sporophytic generation is highly heterozygous while the genotypes of gametes of a single plant are all different. The genetic structure of a cross-pollinated species is characterized by a high level of heterozygosity. However, this is not to say that at each locus heterozygosity occurs. Especially when the allele frequency of certain genes is high (Chapter 3), a plant may very well be homozygous for that locus. Another source of some homozygosity may be due to occasional selfing in a plant. Unlike allogamous species in which formation of new gene combinations are discouraged, cross-pollinated species share a wide gene pool from which new combinations are created to form the next generation.

It is instructive to state that in autogamous crops in principle the whole genotype is transmitted through the generations (i.e., they are "immortal"). Homozygous plants reproduce genetically identically. Consequently, the unit of selection in a mixture of homozygous lines is genotype. In contrast, in allogamous crops the unit of selection is the single gene. The gene in this case is "immortal." Genotypes perish (lose their identity) at each round of sexual reproduction. The only way the genotype can become immortal and be the unit of selection in allogamous crops is when they are clonally propagated, as is the case in potato.

Allogamous species may undergo self-fertilization to a varying extent. In that case the progeny usually suffers from **inbreeding depression**. Deleterious recessive alleles that were suppressed because of heterozygous advantage have opportunities to become homozygous, and therefore become expressed. However, such depression is reversed upon cross-pollination. **Hybrid vigor** (the increase in vigor of the hybrid over its partially homozygous and distinct parents) is exploited in hybrid seed production (Chapter 18). In addition to hybrid breeding, population-based improvement methods (e.g., mass selection, recurrent selection, and synthetic cultivars) are common methods of breeding cross-pollinated species.

segregational load (due to heterozygous advantage), and substitutional or frequency-dependent load (occurs during transient polymorphism; it arises in a population in which natural selection is acting to substitute one allele for another). Genetic load generally lowers the viability of a population.

Because of selection, the frequency of deleterious recessive alleles in a population is expected to decrease rapidly with higher levels of inbreeding. Eventually, these alleles may be lost from the population, a process sometimes referred to as purging populations of their genetic load. Populations that have experienced long periods of inbreeding are expected to show less inbreeding consequences.

6.4 Inbreeding depression

As previously stated, inbreeding or crossing closely related parents results in reduced fitness or vigor of individuals in the progenitor population, a condition called inbreeding depression. Reduction in fitness usually manifests itself as a reduction in vigor, fertility, and productivity, and is seen as lower biomass per plant, lower fecundity, malformation of organs and lower germination of seeds. The effect of inbreeding is more severe in the early generations (5–8) than in later generations. Just like heterosis, inbreeding depression is not uniformly manifested in plants. Plants including onions, sunflower, cucurbits, maize and rye are rather tolerant of inbreeding with low or no inbreeding depression. On the other hand, crops such as alfalfa and carrot are highly intolerant of inbreeding.

6.4.1 The concept of genetic load

Genetic load (or genetic burden) may be defined as the decrease in fitness of the average individual in a population due to the presence of deleterious genes or genotypes in the gene pool. In other words, it is the reduction in selective value for a population compared to what it would otherwise have if all the individuals had the most favored genotype. Statistically, its value ranges between zero (no load) and one. It is generally believed that most species carry a genetic load of 3–5 recessive lethal genes. The genes are mostly hidden (enjoy heterozygous advantage). Inbreeding usually causes the genetic load to increase. Genetic load has three components – mutational load (due to harmful mutations),

6.5 Hybrid vigor

Hybrid vigor or **heterosis** is opposite and complementary to inbreeding depression (reduction in fitness as a direct result of inbreeding). In theory, the heterosis observed after crossing is expected to be equal to the depression upon inbreeding, considering a large number of crosses between lines derived from a single base population. In practice, plant breeders are interested in heterosis expressed by specific crosses between selected parents, or between populations that have no known common recent origin. Furthermore, because heterosis is subject to the interactions between genotype and environment, it is desirable to describe the heterosis of a particular hybrid line for a specific trait at a specific location or under specified environmental conditions.

Hybrid vigor may be defined as the increase in size, vigor, fertility, and overall productivity of a hybrid plant, F_1, over the mid-parent value (average performance of the two parents P_1 and P_2). It is calculated as the difference between the cross-bred and inbred means.

$$\text{Hybrid vigor} = \{[F_1 - (P_1 + P_2)/2]/[(P_1 + P_2)/2]\}$$

The estimate is usually calculated as a percentage (i.e., $\times 100$).

The synonymous term, **heterosis**, was coined by G.H. Shull. Heterosis is of little commercial value (and hencevalue to the farmer) if a hybrid will only exceed the mid-parent in performance. Hence, the *practical* definition of heterosis is hybrid vigor that greatly exceeds the better or higher parent in a cross. Such advantageous hybrid vigor is observed, in particular, when breeders cross parents that are genetically diverse. Heterosis occurs when two inbred lines of outbred species are crossed.

In theory, heterosis may be "positive" or "negative". This is largely an artificial distinction. Positive heterosis is generally desired for traits like yield, while negative heterosis is desired for traits such as early maturity. Three kinds of heterosis may be distinguished as – mid-parent, standard variety, and better parent (also called heterobeltiosis). Standard variety (or check) heterosis is measured by comparing the hybrid to existing high yield commercial variety. Considering the fact that breeders aim to develop cultivars that excel in performance to existing commercial ones, standard variety heterosis is perhaps most desirable to breeders.

Heterosis, though widespread in the plant kingdom, is not uniformly manifested in all species and for all traits. It is manifested at a higher intensity in traits that have fitness value, and also more frequently and at higher levels among cross-pollinated species than self-pollinated species. All breeding methods that are preceded by crossing make use of heterosis to some extent. However, it is only in hybrid cultivar breeding and the breeding of clonally propagated varieties that the breeder has the opportunity to exploit the phenomenon to full advantage.

Hybrids may have dramatically increased yields compared to open-pollinated cultivars. By the early 1930s (before extensive use of hybrids), maize yield in the United States averaged 1250 kg/ha. By the early 1970s (following the adoption of hybrids), maize yields quadrupled to 4850 kg/ha. The contribution of hybrids (genotype) to this increase was estimated at about 60%, the remainder being attributed to production practices.

6.5.1 Genetic basis of heterosis

Three schools of thought have been advanced to explain the genetic basis for why fitness lost on inbreeding tends to be restored upon crossing. The two most commonly known are the **dominance theory**, first proposed by C.G. Davenport in 1908 and later by I.M. Lerner, and the **overdominance theory,** first proposed by G. H. Shull in 1908 and later by K. Mather and J.L. Jinks. A third theory, the mechanism of epistasis (non-allelic gene interactions), has also been proposed by researchers (such as A.C. Fasoulas and R.W. Allard in 1962). Any viable theory should

account for both inbreeding depression in cross-pollinated species upon selfing and increased vigor in F_1, upon hybridization. It should be pointed out that the proposed mechanisms do not occur in exclusion to one another but indeed could operate simultaneously, each in different genes. Further, even though the dominance theory is the most favored by most scientists, none of the theories is completely satisfactory.

Dominance theory

The dominance theory assumes that vigor in plants is conditioned by dominant (functional) alleles, recessive alleles being deleterious or neutral in effect, mostly representing loss-of-function versions of the original dominant gene. It follows then that a genotype with more dominant alleles will be more vigorous than one with few dominant alleles. Consequently, inbreeding parents that are homozygous dominant or heterozygous at most loci will be vigorous but upon inbreeding heterozygous loci may result in progeny that is homozygous for recessive non-functional alleles at several or many loci, resulting in inbreeding depression. If such inbreeding is done on two parents that are of distinct origin, chances are low that they will carry deleterious alleles for the same loci. Therefore, crossing two such largely homozygous parents with complementary dominant and recessive alleles will concentrate more favorable alleles in the hybrid than either inbred parent. In practice, linkage and the large number of genes to be taken care of prevent the breeder from developing inbred lines that contain all dominant alleles in homozygous state. Inbreeding depression occurs upon selfing because the deleterious recessive alleles that are protected in the heterozygous condition (heterozygous advantage) become homozygous and are expressed. In corn, inbred lines have been developed with a limited number and limited deleteriousness of homozygous recessive alleles, resulting in only limited inbreeding depression. These inbred lines are sufficiently fit to produce enough seeds to serve as parents for hybrid cultivar seed production.

To illustrate this theory, assume a quantitative trait like seed yield is conditioned by four loci. Assume that each allele in the dominant homozygous or heterozygous state contributes two units to the phenotype, while a recessive homozygous genotype contributes one unit. A cross between two inbred parents produces the following outcome:

	P_1 ($AAbbCCdd$)	x	P_2 ($aaBBccDD$)
Phenotypic value	$2+1+2+1=6$	$\downarrow$	$1+2+1+2=6$
	F_1	($AaBbCcDd$) $2+2+2+2=8$	

because the homozygous and heterozygous dominant state will both contribute two units to the phenotype. The result is that the F_1 would be more productive than either parent.

D.L. Falconer developed a mathematical expression for the relationship between the parents in a cross that leads to heterosis as follows:

$$HF_1 = \Sigma dy^2$$

where HF_1 is the deviation of the hybrid from the mid-parent value, d is the degree of dominance, and y

gene (e.g., A, a) are contrasting but each has a different favorable effect in the plant. In this view allele a is not supposed to have a loss of function. Consequently, a heterozygous locus would have greater positive effect than either homozygous locus and, by extrapolation, a genotype with more heterozygous loci would be more vigorous than one with less heterozygous loci.

To illustrate this phenomenon, consider a quantitative trait conditioned by four loci. Assume that recessive, heterozygote, and homozygote dominants contribute 1, 2, and $1\frac{1}{2}$ units to the phenotypic value, respectively:

	P_1 ($AAbbCCdd$)	x	P_2 ($aaBBccDD$)
Phenotypic value	$1\frac{1}{2}+1+1\frac{1}{2}+1=5$	$\downarrow$	$1+1\frac{1}{2}+1+1\frac{1}{2}=5$
		($AaBbcCdD$) $2+2+2+2=8$	

is the difference in gene frequency in the parents of the cross. From the expression, maximum mid-parent heterosis (HF_1) will occur when the values of the two factors (d, y) are each unity. That is, the populations to be crossed are fixed for opposite alleles ($y = 1.0$) and there is complete dominance ($d = 1.0$).

Overdominance theory

The phenomenon of a heterozygote being superior to the best performing homozygote is called **overdominance** (i.e., heterozygosity *per se* is assumed to be responsible for heterosis). A possible explanation for this could be the fact that genes normally have pleiotropic effects and, thereby, contribute simultaneously to many measurable traits of the plant. The overdominance theory assumes that the alleles of a

Heterozygosity leads to the highest trait values of the three genotypes.

Where the dominance theory applies, heterosis, theoretically, can be fixed in a pure line; however, where overdominance applies, this cannot occur. Of course, both theories are not exclusive. Some types of gene may contribute to heterosis because of the dominance effect, others because of the overdominance effect.

6.5.2 Biometrics of heterosis

Heterosis may also be defined in two basic ways:

(i) **Better-parent heterosis**. This is calculated as the degree by which the F_1 mean exceeds the better parent in the cross.

(ii) **Mid-parent heterosis**. Calculated as the degree by which the F_1 mean exceeds the mean of the parents in the cross.

For breeding purposes, the breeder is most interested to know whether heterosis can be exploited (e.g., fixed) for crop improvement. To do this, the breeder needs to understand the types of gene action involved in the phenomenon as it operates in the breeding population of interest. As Falconer indicated, in order for heterosis to manifest for the breeder to exploit, some level of dominance gene action must be present in addition to the presence of relative difference in gene frequency in the two parents. Below it is derived that the degree of heterosis will depend on the number of loci that have contrasting alleles in the two parental populations or lines, as well as on the level of dominance at each locus.

Given two populations (A, B) in Hardy–Weinberg equilibrium, with genotypic values and frequencies for one locus with two alleles (A_1 and A_2) occurring in frequencies p and q (in which $q = 1 - p$) respectively for population A, and in frequencies r and s (in which $s = 1 - r$) for population B as follows:

$$P_A = (p - q)a + 2pqd$$
$$= (2p - 1)a + 2(p - q^2)d$$
$$P_B = (r - s)a + 2rsd$$
$$= (2r - 1) + 2(r - r^2)d$$
$$F = (pr - qs)a + (ps + qr)d$$
$$= (p + r - 1)a + (p + r - 2pr)d$$

Calculating heterosis (H_{MP}) as a deviation from the mid-parent values is as follows:

$$\begin{aligned}H_{MP} &= F_1 - (P_1 + P_2)/2 \\ &= [(p + r - 1)a + (p + r - 2pr)d] \\ &\quad - 1/2[(2p - 1)a + 2(p - q^2)d \\ &\quad + (2r - 1)a + 2(r - 1)a + 2(r - r^2)d] \\ &= (p - r)^2 d\end{aligned}$$

From the foregoing, if d = 0 (no dominance), heterosis = 0 (i.e., F = MP, the mean of mid-parents). On the other hand, if in population A $p = 0$ or 1 and by the same token in population B $r = 1$ or 0 for the same locus, depending on whether the allele is in homozygous recessive or dominant state there will be a heterotic response. In the first generation, the

Genotypes	Gene frequency		Genotypic values	Number of A_1 alleles
	Population A	Population B		
A_1A_1	p^2	r^2	$+a$	2
A_1A_2	$2pq$	$2rs$	d	1
A_2A_2	q^2	s^2	$-a$	0

After a cross (A × B) between the populations in Hardy–Weinberg equilibrium and genotypic values and frequencies in the cross as follows:

Genotypes	Frequencies	Genotypic values
A_1A_1	pr	$+d$
A_1A_2	$ps+qr$	d
A_2A_2	qs	$-d$

where p and r are the frequencies of allele A_1 and q and s are frequencies of alleles A_2 in the two populations. The mean genotypic values of the populations are P_A and P_B:

heterotic response will be due to the loci where $p = 1$ and $r = 0$, or vice versa. The highest heterosis will occur when one allele is fixed in one population and the other allele in the other. If gene action is completely additive, the average response would be equal to the mid-parent, and hence heterosis will be zero. On the other hand, if there is dominance and/or epistasis, heterosis will occur.

Plant breeders develop cultivars that are homozygous (in autogamous crops). When there is complete or partial dominance, the best genotypes to develop are homozygotes or heterozygotes, where there could be opportunities to discover transgressive segregates. On the other hand, when non-allelic interaction is significant, the best genotype to breed would be a heterozygote.

Some recent views on heterosis have been published. Some maize researchers have provided evidence to the effect that the genetic basis of heterosis is partial dominance to complete dominance. A number of research data supporting overdominance suggest that it resulted from pseudo-overdominance arising from dominant alleles in repulsion phase linkage. Yet, still, some workers in maize research have suggested epistasis between linked loci to explain the heterosis.

6.5.3 Factors to consider in using heterosis in breeding

Springer and Stupar (2007) summarized four factors to note when considering the application of heterosis in crop improvement:

(i) The magnitude of heterosis is variable among species. The effects of heterosis are stronger and more ubiquitous in corn (allogamous species) than in, say, *Arabidopsis (autogamous species).*

(ii) The level of heterosis for specific traits varies and is not correlated in different hybrids of the same species. This indicates that the phenomenon of heterosis is not conditioned by the action of a single locus, nor does it simply represent the overall extent of heterozygosity between parents.

(iii) Generally, heterosis increases as the genetic distance between the parental inbreds increases. However, there is a threshold that when genetic distance between the parents is exceeded, heterosis decreases. Whereas this appears to suggest a relationship between genetic diversity and heterosis, this relationship is not strong enough to make it a predictive tool.

(iv) The allelic variation that produces heterosis does not represent the totality of variation that occurs. Not all allelic variants in a species' population will be fixed in inbred lines because variants with strong deleterious phenotypes will be selected against by breeders. Consequently, the range of allelic variation in inbred lines that can contribute to heterosis is limited to only the variation with beneficial effects for a specific trait or that which has limited deleterious effects. In other words, not all allelic variation between parental pairs contributes to heterosis. Some allelic variation will not be fixed, that is, when the homozygous state of a certain allele is deleterious to the genotype.

6.6 Concept of heterotic relationship

Genetic diversity in the germplasm used in a breeding program affects the potential genetic gain that can be achieved through selection. The most costly and time consuming phase in a hybrid program is the identification of parental lines that would produce superior hybrids when crossed. Hybrid production exploits the phenomenon of heterosis, as already indicated. Genetic distance between parents plays a role in heterosis.

In general, heterosis is considered an expression of the genetic divergence among cultivars. When heterosis is significant for certain trait(s), it may be concluded that there is genetic divergence among the parental cultivars for that/those trait(s). Information on the genetic diversity and distance among the breeding lines, and the correlation between genetic distance and hybrid performance, are important for determining breeding strategies, classifying the parental lines, defining heterotic groups, and predicting future hybrid performance.

6.6.1 Definition

Plant breeders seek ways of facilitating the use of available germplasm effectively for plant improvement. One such way is to classify inbred lines into **heterotic groups** for creation of predictable hybrids. Crosses between inbreds from different heterotic groups result in vigorous F_1 hybrids, with significantly more heterosis than F_1 hybrids from inbreds within the same heterotic group or pattern. When two parental lines result in progeny with high heterosis, they are said to have high (or favorable) **combining ability**. A **heterotic group** may hence be defined as a group of related or unrelated genotypes from the same or different populations, which display combining ability when crossed with genotypes from other germplasm groups. A **heterotic pattern,** on the other hand, involves specific pairs of heterotic groups, which may be populations or lines, that express high heterosis and, consequently, high hybrid performance in their crosses. Such a pattern has been established in the US Corn Belt germplasm for Reid Stiff Stalk vs. Lancaster. Heterotic patterns can be revealed through combining ability tests. It should be pointed out that heterotic patterns can change, based on the environment under which the evaluation is conducted. Stability of heterotic patterns is useful information for formulating an effective breeding strategy for a region with variable growing conditions.

Knowledge of the heterotic groups and patterns is helpful in plant breeding. It helps breeders to use their germplasm in an efficient and consistent manner through exploitation of complementary lines to maximize the outcomes of a hybrid breeding program. Breeders may use heterotic group information for cataloging diversity and directing the introgression of traits and creation of new heterotic groups.

The concept of heterotic groups was first developed by maize researchers who observed that inbred lines selected out of certain populations tended to produce, in particular, superior performing hybrids when hybridized with inbreds from a certain other group. The existence of heterotic groups has been attributed to the possibility that populations of divergent backgrounds might have unique allelic diversity that could have originated from founder effects, genetic drift, or accumulation of unique diversity by mutation or selection. Interallelic interaction (overdominance) or repulsion phase linkage among loci showing dominance (pseudo-overdominance) could explain the significantly greater heterosis observed following a cross between genetically divergent populations. Experimental evidence supports the concept of heterotic patterns. Such research has demonstrated that intergroup hybrids significantly out-yielded intragroup hybrids. In maize, one study showed that intergroup hybrids between Reid Yellow Dent × Lancaster Sure Crop out-yielded intragroup hybrids by 21%.

D. Melchinger and R.R.Gumber (1998) noted that heterotic groups are the backbone of successful hybrid breeding and, hence, a decision about them should be made at the beginning of a hybrid crop improvement program. They further commented that once established and improved over a number of selection cycles, it is extremely difficult to develop new and competitive heterotic groups. This is because, at an advanced stage, the gap in performance between improved breeding materials and unimproved source materials is often too large. However, the chance to develop new heterotic groups could be enhanced with a change in breeding objectives. Once developed, heterotic groups should be broadened continuously by introgressing unique germplasm in order to sustain medium and long term gains from selection.

6.6.2 Methods for developing heterotic groups

A number of procedures may be used by breeders to establish heterotic groups and patterns. These include pedigree analysis, geographic isolation inference, measurement of heterosis, and combining ability analysis. Some have used diallel analysis to obtain preliminary information on heterotic patterns. The procedure is recommended for use with small populations. The technology of molecular markers may be used to refine existing groups and patterns or to expedite the establishment of new ones through the determination of genetic distances.

To establish a heterotic group and pattern, breeders make crosses between or within populations. Intergroup hybrids have been shown to be superior over intragroup hybrids in establishing heterotic relationships. In practice, most of the primary heterotic groups were not developed systematically but rather by relating the observed heterosis and hybrid performance with the origin of parents included in the crosses. One of the earliest contributions to knowledge in the area of developing heterotic patterns was made in 1922. Comparing heterosis for yield in a large number of intervarietal crosses of maize, it was discovered that hybrids between varieties of different endosperm types (flint vs. dent) produced a higher performance than among varieties with the same endosperm type. This discovery, by F.D. Richey, suggested that crosses between geographically or genetically distant parents expressed higher performance and, hence, increased heterosis. This information led to the development of the most widely used heterotic pattern in the US Corn Belt – the Reid Yellow Dent × Lancaster Sure Crop.

6.6.3 Heterotic groups and patterns in crops

Heterotic patterns have been studied in various species. For certain crops breeders have defined standard patterns that guide in the production of hybrids. As previously indicated in maize, for example, a widely used scheme for hybrid development in temperate maize is the Reid × Lancaster heterotic pattern. These heterotic populations were discovered from pedigree and geographic analysis of inbred lines used in the Corn Belt of the United States. In Europe, a common pattern for maize is the European flint $\times$ Corn Belt Dent, identified based on endosperm types. In France, $F_2 \times F_6$ heterotic pattern derived from the same open pollinated cultivars was reported. Other patterns include ETO-composite $\times$ Tuxpeno and Suwan 1 $\times$ Tuxpeno in tropical regions. Alternative heterotic patterns continue to be sought.

In rice, some research suggests two heterotic groups within *O. indica*, one including strains from southeast China and another containing strains from southeast Asia. In rye, the two most widely used germplasm groups are the Petkus and Carsten, while in faba bean three major germplasm pools are available, namely, Minor, Major, and Mediterranean.

Even though various approaches are used for the identification of heterotic patterns, they generally follow certain principles. The first step is to assemble a large number of germplasm sources and then make populations of crosses from among which the highest performing hybrids are selected as potentially profitable heterotic groups and patterns. If established heterotic patterns already exist, the performance of the putative patterns with the established ones is compared. Where the number of inbred lines in a breeding program is too large to permit the practical use of a diallel cross, the germplasm may first be grouped based on genetic similarity. For these groups, representatives are selected for evaluation in a diallel cross.

According to Melchinger, the choice of a heterotic group or pattern in a breeding program should be based on the following criteria:

1 high mean performance and genetic variance in the hybrid population;
2 high *per se* performance and good adaptation of parent population to the target region; and
3 low inbreeding depression of inbreds.

6.6.4 Estimation of heterotic effects

Consider a cross between two inbred lines, A and B, with population means of $\bar{X}_{P1}$ and $\bar{X}_{P2}$, respectively. The phenotypic variability of the F_1 is generally less than the variability of either parent. This could indicate that the heterozygotes are less subject to environmental influences than the homozygotes. The heterotic effect resulting from the crossing is roughly estimated as:

$$H_{F1} = \bar{X}_{F1} - 1/2(\bar{X}_{P1} + \bar{X}_{P2})$$

This equation indicates the average excess in vigor exhibited by F_1 hybrids over the midpoint (midparent) between the means of the inbred parents. K.R. Lamkey and J.W. Edward coined the term **panmitic mid-parent heterosis** to describe the deviation in performance between a population cross and its two parent populations in Hardy–Weinberg equilibrium. Heterosis in the F_2 is 50% less than that manifested in the F_1.

6.6.5 Breeding implications

Perhaps the most obvious genetic implication in breeding cross-pollinated species is their tendency to be heterozygous because of lack of restriction on pollination. Unlike self-pollinated cultivars that are stable and do not segregate (unless cross-fertilization took place in a recent generation) upon selfing and, hence, require no maintenance to keep their genetic identity; open-pollinated cultivars of cross-pollinated species are less stable, changing the genetic identity of all the constituting plants from one generation to the next. From generation to generation certain genes may be selected against or for, changing the allele frequencies. For example, after a cold period plants with low frost tolerance may die or suffer damage, while plants possessing one or more alleles for cold tolerance will have normal reproduction. Such an event will cause an increased allele frequency for frost tolerance. The next generation may suffer some other environmental restriction, leading to the shift in allele frequency of another gene. This trait is of more importance to regions where a commercial seed production system is lacking, leaving farmers to save seed from the current season's harvest to plant the next season's crop.

Plant breeders employ various breeding schemes to develop cultivars that significantly resist changes in genetic structure or composition even under an open-pollination production environment. By the same token, if a cultivar is produced by a process in which controlled cross-pollination is enforced, the only way to prevent the cultivar from returning to the natural way of being susceptible to cross-fertilization is to continue to enforce restricted pollination in its maintenance. In the case of a hybrid cultivar (Chapter 17), the producer can reap the optimum benefits of this cultivar only in one production season.

Key References and Suggested Reading

Melchinger, A.E., and Gumbler, R.K. (1998). Overview of heterosis and heterotic groups in agronomic crops, in *Concept and breeding of heterosis in crop plants* (eds K.R. Lamkey and J.E. Staub). Special Publication No. 25, CCSSA, Madison, WI, pp. 29–44.

Smith, J.S.C., Chin, E.C.L., Shu, H., *et al.* (1997). An evaluation of the utility of SSR loci as molecular markers in maize (*Zea mays L.*): Comparisons with data from RFLPs and pedigree. *Theor. Appl. Genet.,* **95**:163–173.

Springer, N.M., and Stupar, R.M. (2007). *Allelic variation and heterosis in maize; How do two halves make more than a whole?* Cold Spring Harbor Laboratory Press, Cold Spring Harbor, New York.

Outcomes assessment

Part A

Please answer the following questions true or false.

1   Hybrid vigor is highest in a cross between two identical parents.
2   CMS may be used in hybrid breeding to eliminate emasculation.
3   The A inbred line is male sterile.
4   G.H. Shull proposed the dominance theory of heterosis.
5   A hybrid cultivar is the F_1 offspring of a cross between inbred lines.

Part B

Please answer the following questions.

1   Define a hybrid cultivar.
2   What is hybrid vigor and what is its importance in hybrid breeding.
3   What is an inbred line?
4   What is a heterotic group?
5   Explain the dominance of single cross hybrids in modern corn hybrid production.

Part C

Please write a brief essay on each of the following topics.

1   Discuss the dominance theory of heterosis.
2   Discuss the importance of synchronization of flowering in hybrid breeding.
3   Discuss inbred lines and their use in hybrid breeding.
4   Discuss the contributions of G.H. Shull and D.F. Jones in hybrid breeding.

7

Hybridization

Purpose and expected outcomes

One of the principal techniques of plant breeding is artificial mating (crossing) of selected parents to produce new individuals that combine the desirable characteristics of the parents. This technology is restricted to sexually reproducing species that are compatible. However, in the quest for new desirable genes, plant breeders sometimes attempt to mate individuals that are biologically distantly related. In such interspecific crosses pre- and post-fertilization barriers may occur. It is important for the breeder to understand the problems associated with making a cross and how barriers to crossing, where they exist, can be overcome. After studying this chapter, the student should be able to:

1 Define sexual hybridization and discuss its genetic consequences.
2 Define a wide cross and discuss its objectives and consequences.
3 Discuss the challenges to wide crosses and techniques for overcoming them.

7.1 Concept of gene transfer and hybridization

Crop improvement typically involves the transfer of genes from one source or genetic background to another, or combining genes from different sources that complement each other, with the hope that the new cultivar will combine the best of both parents, while being distinct from both. When a plant breeder has decided on the combination of traits that are to be incorporated in the new cultivar to be developed, the next crucial step is to find one or more sources of the appropriate gene(s) for such characters. In flowering species, the conventional method of gene transfer or gene combination is by crossing or **sexual**

hybridization. This procedure causes genes from the two parents to be assembled into a new genetic matrix. It follows that if parents are not genetically compatible, gene transfer by sexual means cannot occur at all or, at best, may be fraught with complications. The product of hybridization is called a **hybrid**.

Sexual hybridization can occur naturally through agents of pollination. Even though self-pollinating species may be casually viewed as "self-hybridizing", the term hybridization is reserved for crossing between unidentical parents (the degree of divergence is variable). Artificial sexual hybridization is the most common conventional method of generating a segregating population for selection in breeding flowering species. In some breeding programs, the hybrid (F_1)

Principles of Plant Genetics and Breeding, Second Edition. George Acquaah.
© 2012 John Wiley & Sons, Ltd. Published 2012 by John Wiley & Sons, Ltd.

is the final product of plant breeding (see hybrid breeding in Chapter 18). However, in most situations the F_1 is selfed (to give an F_2) to generate recombinants (as a result of recombination of the parental genomes) or a segregating population, in which selection is practiced. In clonally propagated crops, the F_1 usually segregates sufficiently and its clonally produced descendants will be submitted to selection without further crossing or selfing.

The tools of modern biotechnology now enable the breeder to transfer genes by circumventing the sexual process (i.e., without crossing). More significantly, gene transfer can transcend natural reproductive or genetic barriers. Transfers can occur between unrelated plants and even between plants and animals (by genetic transformation; Chapter 24).

7.2 Applications of crossing in plant breeding

Sometimes, crossing is done for specific purposes, within the general framework of generating variability. Hybridization precedes certain methods of selection in plant breeding to generate general variability.

- **Gene transfer.** Sometimes, only a specific gene (or a few) needs to be incorporated into an adapted cultivar. Crossing is used for the gene transfer process, followed by additional strategic crossing to retrieve the desirable genes of the adapted cultivar (backcrossing; Chapter 15).
- **Recombination.** Genetically diverse parents may be crossed in order to recombine their desirable traits. The goal of recombination, which is a key basis of plant breeding, is to forge desirable linkage blocks.
- **Break undesirable linkages.** Whereas forging desirable linkage blocks is a primary goal of plant breeding, sometimes crossing is applied to provide opportunities for undesirable linkages to be broken.
- **For heterosis.** Hybrid vigor (heterosis) is the basis of hybrid seed development. Specially developed parents are crossed in a predetermined fashion to capitalize on the phenomenon of heterosis for cultivar development.
- **For maintenance of parental lines.** In hybrid seed development programs, crossing is needed to maintain special parents used in the breeding program (e.g., CMS lines, maintainer lines).
- **For maintenance of diversity in a gene pool.** Plant breeders may use a strategy of introgression

(crossing and backcrossing selected entries with desired traits into adapted stocks) and incorporation to develop dynamic gene pools from which they can draw materials for crop improvement.
- **For evaluation of parental lines.** Inbred lines for hybrid seed development are evaluated by conducting planned crosses to estimate combining abilities in order to select appropriate parents for used in hybrid seed development.
- **For genetic analysis.** Geneticists make planned crosses to study the inheritance and genetic behavior of traits of interest.

7.3 Artificial hybridization

Artificial hybridization is the deliberate crossing of selected parents (controlled pollination). There are specific methods for crossing that depend on the species in which the cross is being made, which differ according to factors including floral morphology, floral biology, possible genetic barriers, and environmental factors. Methods for selected species are described later in this book. However, there are certain basic factors to consider in preparation for hybridization:

- Parents should belong to the same or closely related plant species. In the case that they belong to different (related) plant species, all kinds of techniques may be required to obtain hybrid progeny.
- The parents, obviously, together should supply the critical genes needed to accomplish the breeding objective.
- One parent is usually designated as female. Whereas some breeding methods may not require this designation, breeders usually select one parent to be a female and the other a male (pollen source). This is especially so when hybridizing self-pollinated species. Whenever genetic markers are available (e.g., white flowers, white seeds), the female exhibits the recessive morphological trait. In some cases, selected parents of cross-pollinated species may be isolated and allowed to randomly cross-pollinate each other.
- The female parent usually needs some special preparation. In complete flowers (those having both male and female organs), the flowers of the parent selected to be female are prepared for hybridization by removing the anthers, a tedious procedure called **emasculation** (discussed next). Emasculation is

eliminated in some crossing programs by taking advantage of male sterility (renders pollen sterile) when it occurs in the species.

- Pollen is often physically or manually transferred. Artificial hybridization often includes artificial pollination, whereby the breeder physically deposits pollen from the male parent onto the female stigma. However, when hybridization is conducted on large scale (e.g., commercial hybrid seed development), hand pollination is rarely a feasible option.

7.4 Artificial pollination control techniques

As previously indicated, crossing is a major procedure employed in the transfer of genes from one parent to another in the breeding of sexual species. A critical aspect of crossing is pollination control to ensure that only the desired pollen is involved in the cross. In hybrid seed production, success depends on the presence of an efficient, reliable, practical, and economic pollination control system for large-scale pollination. Pollination control may be accomplished in three general ways:

(i) **Mechanical control.** This approach entails manually removing anthers from bisexual flowers to prevent pollination, a technique called emasculation, or removing one sexual part (e.g., detasselling in corn), or excluding unwanted pollen by covering the female part. These methods are time consuming, expensive, and tedious, limiting the number of plants that can be crossed. It should be mentioned that in crops such as corn, mechanical detasselling is widely used in the industry to produce hybrid seed.

(ii) **Chemical control.** A variety of chemicals called chemical hybridizing agents, or by other names (e.g., male gametocides, male sterilants, pollenocides, androcides), are used to temporally induce male sterility in some species. Examples of such chemicals include Dalapon®, Estrone®, Ethephon®, Hybrex®, and Generis®. The application of these agents induces male sterility in plants, thereby enforcing cross pollination. The effectiveness is variable among products.

(iii) **Genetical control.** Certain genes are known to impose constraints on sexual biology by incapacitating the sexual organ (as in male sterility) or inhibiting the union of normal gametes (as in self-incompatibility). These genetic mechanisms are discussed further Chapter 5.

7.5 Flower and flowering issues in hybridization

The flower has a central role in hybridization. The success of a crossing program depends on the condition of the flower regarding its overall health, readiness or receptiveness to pollination, maturity, and other factors. The actual technique of crossing depends on floral biology (time of pollen shedding, complete or incomplete flower, self- or cross-pollinated, size and shape of individual flowers and of the inflorescence).

7.5.1 Flower health and induction

It is important that plants in a crossing block (or to be crossed) be in excellent health and be properly developed. This is especially so when flowers are to be manually emasculated. Once successfully crossed, an adequate amount of seed should be obtained for planting the first generation. The parents to be mated should receive proper lighting, moisture supply, temperature, nutrition, and protection from pests. Parents should be fertilized with the proper amounts of nitrogen, phosphorus, and potassium for vigorous plant growth to develop an adequate number of healthy flowers.

Plants growing in the greenhouse should be provided with the proper intensity and duration of light. If the species is photoperiod sensitive, the lighting should be adjusted accordingly. Proper temperature is required for proper plant growth and development. In some species, a special temperature treatment (vernalization, usually some period of low temperature) is required for flower induction. Furthermore, temperature affects pollen shed in flowers. Consequently, extreme temperatures may cause inadequate amounts of pollen to be shed for successful artificial pollination. Pollen quantity and quality are influenced by the relative humidity of the growing environment. Extreme moisture conditions should be avoided.

7.5.2 Synchronization of flowering

In artificial pollination, the breeder should be familiar with the species to know its flowering habits regarding time from planting to flowering, duration of flowering, mechanisms and timing of natural anther dehiscence and fertilization, and time of peak pollen production, in order to take advantage of the window of opportunity of anthesis (pollen shed) for best crossing outcomes. To ensure that parents in a crossing program will have flowers at the same time, the practice of staggered planting – to plant sets of parents at different times – is recommended. This way, a late-planted early flowering genotype may be pollinated by an early-planted late flowering genotype. When depending on natural pollination, interspersed planting on different dates will favor even pollen distribution.

Photoperiod may be manipulated in photoperiod sensitive species to delay or advance flowering as appropriate, in order to synchronize flowering of the parents in a cross. Other techniques that have been used in specific cases include manipulation of temperature and planting density, removal of older flowers to induce new flushes of flowers, and pinching (e.g., removal of plant apex to induce tillering or branching for additional flowers). In corn, the silk of an early flowering inbred parent may be cut back to delay the time to readiness for pollination.

7.5.3 Selecting female parents and suitable flowers

After selecting lines to be parents in a cross, it is necessary in artificial crosses to designate one parent as female (as previously stated), as well as identify which type of flowers on the parent would be most desirable to cross. In crossing programs in which the CMS system is being used, it is critical to know which plants to use as females (these would be the male sterile genotypes, or A and B lines; Chapter 17). Because the pollen or male gamete is practically without cytoplasm, and because certain genes occur in the extranuclear genome (such as CMS), it is critical that parents selected as female plants be selected judiciously.

Markers are important to plant breeding as was discussed previously. Some markers may be used to distinguish between selfed and hybrid seed on the female plant. For example, in sorghum, waxy endosperm is conditioned by a recessive allele while normal endosperm is under the control of the dominant allele. If a waxy female is crossed with a normal male, all F_1 seed with waxy endosperm would be products of selfing (undesirable) while normal seed would indicate a successful hybrid. Other markers, molecular and morphological, may be strategically included in a crossing program to allow the authentication of hybridity. In terms of flower characteristics, bigger flowers are easier to handle than tiny ones. Whenever possible, the parent with bigger flowers should be used as female.

Another critical aspect of flower physiology is the age of the flower when it is most receptive to pollination. The breeder usually determines the optimal stage of flower maturity by examining its physical appearance. Tell-tale signs depend on species. Usually, fully opened flowers would have already been pollinated by undesirable pollen. In most plant species flowers are emasculated in the bud stage just as the petals begin to show through the bud. Rice is ready in the boot stage, whereas wheat is best emasculated when florets are light green with well-developed but still green anthers and feathery stigmas that extend about a quarter of the length of the florets. Furthermore, flowers in the same inflorescence usually have different maturity levels. In species such as the broad bean (*Vicia faba*), the first inflorescence is more suitable for crossing than later ones. Also, flowers at the base and middle of inflorescences give better results than those at the top. Flowers in the inflorescence that are not used for crossing may be removed, while the ones that are used in crossing may be marked with a label or small clip or peg.

7.6 Emasculation

The process of making a bisexual flower female by removing the male parts or incapacitating them is called **emasculation**. It should be pointed out right away that emasculation is not a universal requirement for artificial crossing of plants. Species with fertility-regulating mechanisms (e.g., male sterility, self-incompatibility, protogyny, monoecy, dioecy) may be crossed without the often tedious and time consuming process of emasculation.

7.6.1 Factors to consider for success

Apart from picking the right flowers, it is critical to know the duration of stigma receptivity and pollen viability. The maximum time between emasculation and pollination that can be tolerated varies among species. Since the anthers were removed before they were mature, the female parts often are not yet receptive at the moment of emasculation. This makes it necessary to pollinate at a later time, either during the same day or even later. The caution to observe is that prolonged delay between the two operations increases the chance of contamination from undesirable pollen. To reduce this risk, emasculated flowers may be covered with bags (e.g., glassine, paper or cloth bag).

Pollen quality and quantity vary with the weather and time of day. For example, in chickpea, some breeders prefer to emasculate in the evening and pollinate in the morning. Because emasculation is done before anthers are mature in species such as wheat and barley, pollination is done 2–3 days later when the stigma is receptive. In extreme cases, such as in sugar beet, pollination may immediately follow emasculation or be delayed for up to 12 days.

7.6.2 Methods of emasculation

There are several techniques of emasculation used by plant breeders that include the use of instruments or chemicals. A pair of forceps or tweezers is one of the most widely used instruments in the emasculation of flowers. Different shapes and sizes are used according to the size and structure of the flower. The methods of emasculation may be classified as direct or indirect.

Direct anther emasculation

The technique of removing anthers from selected flowers is the most common procedure for emasculation of flowers (usually using a pair of forceps). When handling plants with inflorescence, it is important to firstly thin out the bunch by removing immature flowers as well as old ones. This improves the survival of the emasculated flowers. Breeders of various crops have developed convenient ways of removing the anthers. Sometimes, the sepals are first removed, followed by the petals before access is gained to the anthers. In soybean and sesame, a skilled person may be able to remove the petals and anthers in one attempt. In flowers such as soybean, the pedicel is easily broken as a result of physical handling of the delicate flower during emasculation. In wheat and barley, the florets are clipped with

scissors. Specific techniques for specific crops are discussed in Part II of this book.

Indirect anther emasculation

In these methods, the anthers are incapacitated without being removed from the flower. Incapacitation is achieved in several ways:

- **Thermal inactivation.** The inflorescence is first thinned out to leave only flowers at the proper stage for emasculation. It is then immersed in hot water (e.g., held in a thermos bottle) to kill the pollen without injuring the pistil. The temperature and time of immersion are variable (e.g., 43 °C for 5 minutes in rice; 47–48 °C for 10 minutes in sorghum). The inflorescence is allowed to dry before pollinating about 30–60 minutes later.
- **Alcohol emasculation.** In species such as alfalfa the raceme is immersed in 57% ethanol for 10 seconds and then rinsed in water for a few seconds.
- **Commercial gametocides.** These are chemicals designed to kill the anthers (e.g., sodium methyl arsenate).

If pollination is not to follow emasculation immediately, the flowers should be covered to exclude contaminating pollen from elsewhere. Once properly pollinated, the flower should be tagged for identification.

7.7 Pollination

Successful pollination depends on pollen maturity, quality (freshness), and timing of pollination, among other factors.

7.7.1 Collection and storage

In some species (e.g., soybean) pollination immediately follows emasculation. In this case, there is no need for storage. Fresh pollen gives the best success of crossing. Good pollen flowers may be picked and placed in a Petri dish or some suitable container for use. In some species, mechanical vibrations may be used to collect pollen. Pollen is most copious at peak anthesis. Generally, pollen loses viability quickly. However, in some species, pollen may be stored at a cool temperature and an appropriate humidity for the species for an extended period.

7.7.2 Application of pollen

Commonly, pollen is applied directly to the stigma by using a fine brush or dusting off the pollen onto the stigma directly from the flower of the pollen source (e.g., the staminal column may be used as brush). Sometimes, an object such as a cotton bulb or a tooth pick is used to deposit pollen on the stigma. In some flowers, pollen deposition is made without direct contact with the stigma. Instead, pollen may be injected or dusted into a sack covering the emasculated inflorescence and agitated to distribute the pollen over the inflorescence. A key precaution against contamination during pollination is for the operators to disinfect their hands and tools between pollinations when different varieties are involved. It is critical to tag the pollinated flower for identification at the time of harvesting.

7.7.3 Tagging after pollination

After depositing the desired pollen, it is critical to identify the flowers that were pollinated with an appropriate tag or label. The information on the label should include the date of emasculation, date of pollination, name of seed parent, and the name of pollen parent. The tag should be attached to the pedicel of the emasculated flower not the branch.

7.8 Number of F$_1$ crosses to make

There are practical factors to consider in deciding on the number of crosses to make for a breeding project. These include the ease of making the crosses from the standpoint of floral biology and the constraints of resources (labor, equipment, facilities, and funds). It will be easier to make more crosses in species in which emasculation is not needed (e.g., monoecious and dioecious species) than in bisexual species. Some breeders make a small number of carefully planned crosses, while others make thousands of cross combinations.

Generally, a few hundred cross combinations per crop per year would be adequate for most purposes for species in which the F$_1$ is not the commercial product. More crosses may be needed for species in which hybrids are commonly produced, for the purpose of discovering heterotic combinations. As is discussed next, breeding programs that go beyond

the F$_1$ usually require very large F$_2$ populations. Regarding the number of flowers per cross combination, there is variation according to fecundity. Species such as tomato may need only one or two crosses, since each fruit contains over 100 seeds. Plants that tiller also produce large numbers of seed. Each crop species has its own reproduction rate, which may be huge (e.g., tobacco: 1000s of seeds produced per plant, 100s per bowl) or relatively small (e.g., pea: about 100 per plant, about 2–5 per pod).

7.9 Genetic issues in hybridization

Because hybridization involves combining two sets of genes in a new genetic matrix through the meiotic process, it is accompanied by a variety of genetic based effects.

7.9.1 Immediate effect

The immediate effect of hybridization is the assembly of two different genomes into a newly created individual. Several genetic consequences may result from such a union of diverse genomes, some of which may be desirable, some of which may not be desirable. The key ones are:

- **Expression of recessive lethal gene.** Crossing may bring together recessive lethal genes (that were in the heterozygous state) into the expressible homozygous state. The resulting hybrid may die or loose vigor. By the same token, hybridization can also mask the expression of a recessive allele by creating a heterozygous locus. Individuals carry a certain genetic load (or genetic burden), representing the average number of recessive lethal genes carried in the heterozygous condition by an individual in a population. Selfing or inbreeding predisposes an individual to having deleterious recessive alleles that were protected in the heterozygous state to becoming expressed in the homozygous recessive form
- **Hybrid necrosis.** The crossing of parents that are somewhat distantly related (but still the same crop species) may result, especially, in the phenomenon of hybrid necrosis. Interactions between pairs of genes in both parents may work out unfavorably to the physiology of the plant. This phenomenon has been reported not only in wheat and rye but also in *Arabidopsis*.

- **Heterosis.** Genes in the newly constituted hybrid may complement each other to enhance the vigor of the hybrid. The phenomenon of hybrid vigor (heterosis) is exploited in hybrid seed development (Chapter 17).
- **Transgressive segregation.** Hybrids have features that may represent an average of the parental features, or a bias toward the features of one parent, or even new features that are unlike either parent (transgressive segregates). When the parents "nick" in a cross, transgressive segregates with performance superseding either parent is likely to occur in the segregating population.
- **Genome-plastome incompatibility.** Plastomes (the genetic material found in plastids such as in chloroplasts) and genomes in most genera function to form normal plants, regardless of the taxonomical distances between the plastid and nuclear genomes. However, in some genera, plastomes and genomes, having co-evolved to a significant degree, are only compatible within a specific combination.

7.9.2 Subsequent effect

The subsequent effect of hybridization, which is often the reason for hybridizing parents by breeders, occurs in the F_2 and later generations. By selfing the F_1 hybrid, the parental genes are reorganized into new genetic matrices in the offspring. This occurs through the process of meiosis, a nuclear division process that occurs in flowering plants. Contrasting alleles segregate and subsequently recombine in the next generation to generate new variability. Furthermore, the phenomenon of crossing over that leads to the physical exchange of parts of chromatids from homologous chromosomes provides an opportunity for recombination of linked genes, also leading to the generation of new variation.

7.9.3 Gene recombination in the F_2

The goal of crossing for generating variability for selection is to produce a large number of gene recombinations from the parents used in the cross. In hybrid seed programs, the F_1 is the end product for commercial use. However, in other crosses, the F_2 and subsequent generations are evaluated to select genotypes that represent the most desirable recombination of parental genes. The F_2 generation has the largest number of different gene combinations of any generation following a cross. The critical question in plant breeding is what size of F_2 population to generate in order to have the chance of including that ideal recombinant this is homozygous for all the desirable genes in the parent. Three factors determine the number of gene recombinations that would be observed in an F_2 population:

(i) The number of gene loci for which the parents in a cross differ.
(ii) The number of alleles at each locus.
(iii) The linkage of the gene loci.

Plant breeders are often said to play the numbers game. Table 7.1 summarizes the challenges of breeding in terms of the size of the F_2 population to grow. If the parents differ by only one pair of allelic genes, the breeder needs to grow at least 16 plants in the F_2 to have the chance to observe all the possible gene combinations (according to Mendel's laws). On the other hand, if the parents differ in 10 allelic pairs, the F_2 population size needed is 59 049 (obtained by the formula 3^n, where n is the number of loci). The frequencies illustrate how daunting a task it is to select for quantitative traits.

The total possible genotypes in the F_2 based on the number of alleles per locus is given by the relationship

Table 7.1 The variability in an F_2 population as affected by the number of genes that are different between the two parents.

Number of heterozygous loci	Number of heterozygous in the F_2	Number of different genotypes in the F_2	Minimum population size for a chance to include each genotype
n	2^n	3^n	4^n
1	2	3	4
2	4	9	16
6	64	729	4096
10	1024	59 049	1 048 576
15	32 768	14 348 907	1 076 741 824

$[k(k + 1)/2]^n$, where k is the number of alleles at each locus and n is the number of heterozygous loci. With one heterozygote and two alleles, there will be only three kinds of genotypes in the F_2, while with one heterozygote and four alleles, there will be ten. The effect on gene recombination by linkage is more important than for the number of alleles. Linkage may be desirable or undesirable. Linkage reduces the frequency of gene recombination (it increases parental types). The magnitude of reduction depends on the phase (**coupling phase** – with both dominant gene loci in one parent, e.g., *AB/ab* – and **repulsion phase** – with one dominant and one recessive loci in one parent, e.g., *Ab/aB*). The effect of linkage in the F_2 may be calculated as $\frac{1}{4}(1 - P)^2 \times 100$ for the coupling phase, and $\frac{1}{4}P^2 \times 100$ for the repulsion phase, for the proportion of *AB/AB* or *ab/ab* genotypes in the F_2 from a cross between *AB/ab* × *Ab/aB*. Given, for example, a crossing over value of 0.10, the percentage of the homozygotes will be 20.25% in the coupling versus only 0.25% in the repulsion phase. If two genes were independent (crossing over value = 0.50), only 6.25% homozygotes would occur. The message here is that the F_2 population should be as large as possible.

With every advance in generation, the heterozygosity in the segregating population decreases by 50%. The chance of finding a plant that combines all the desirable alleles decreases as the generations advance, making it practically impossible to find such a plant in advanced generations. Some calculations by J. Sneep will help clarify this point. Assuming 21 independent gene pairs in wheat, he calculated that the chance of having a plant with all desirable alleles (either homozygous or heterozygous) are 1 in 421 in the F_2, 1 in 49 343 in the F_3, and 1 in 176 778 in the F_4, and so on. However, to be certain of finding such a plant, he recommended that the breeder grow four times as many plants.

Another genetic consequence of hybridization is the issue of linkage drag. As previously noted, genes that occur in the same chromosome constitute a linkage block. However, the phenomenon of crossing over provides an opportunity for linked genes to be separated and not inherited together. Sometimes, a number of genes are so tightly linked they are resistant to the effect of recombination. Gene transfer by hybridization is subject to the phenomenon of **linkage drag**, the unplanned transfer of other genes associated with those targeted. If a desired gene is strongly linked with other undesirable genes, a cross to transfer the desired gene will invariably be accompanied by the linked undesirable genes.

7.10 Types of populations generated through hybridization

A breeding program starts with an initial population that is obtained from previous programs and existing variable populations (e.g., landraces), or is created through a planned cross. Hybridization may be used to generate a wide variety of populations in plant breeding, ranging from the very basic two-parent cross (single cross) to very complex populations in which hundreds of parents could be involved. Single crosses are the most widely used in breeding. Commercial hybrids are mostly produced by single crosses. Complex crosses are important in breeding programs where the goal is population improvement. Hybridization may be used to introgress new alleles from wild relatives into breeding lines. Because the initial population is critical to the success of the breeding program, it cannot be emphasized enough that it be generated with much planning and thoughtfulness.

Various mating designs and arrangements are used by breeders and geneticists to generate plant populations. These designs require some type of cross to be made. Factors that affect the choice of a mating design, include the predominate type of pollination (self- or cross-pollinated), the type of crossing used (artificial or natural), the type of pollen dissemination (wind or insect), the presence of male sterility system, the purpose of the project (for breeding or genetic studies), and the size of the population required. In addition, the breeder should be familiar with how to analyze and interpret or use the data generated from the mating.

The primary purpose of crossing is to expand genetic variability by combining genes from the parents involved in the cross to produce offspring that contain genes they never had before. Sometimes, multiple crosses are conducted to generate the variability in the base population to begin the selection process in the program. Based on how the crosses are made and their effects on the genetic structure of the plants or the population, methods of crossing may be described as either divergent or convergent.

7.10.1 Divergent crossing

Genetically divergent parents are crossed for recombination of their desirable genes. To optimize results, parents should be carefully selected to have maximum number of positive traits and a minimum number of negative traits with no negative traits in common (i.e., elite × elite cross). This way, recombinants that possess both sets of desirable traits will occur in significant numbers in the F_2. The F_1 contains the maximum number of desirable genes from both parents. There are several ways to conduct divergent crosses (Figure 7.1).

- **Single cross.** If two elite lines are available that together possess all desired traits at adequate levels, one cross (single cross [$A \times B$]) may be all that is needed by in the breeding program.
- **Three-way cross.** Sometimes, for combining all desirable traits several cultivars or elite germplasm are required, since each pair may have some negative traits in common. In this case, multiple crosses may

be required in order to have the opportunity of obtaining recombinants that combine all the desirable traits. The method of three-way crosses ($[A \times B] \times C$) may be used. If a three-way cross product will be the cultivar, it is especially important that the third parent (C) be adapted to the region of intended use, since it contributes more genes than each of the A and B parents.

- **Double cross.** A double cross is a cross of two single crosses $\{[(A \times B)] \times [(C \times D)]\}$. The method of successive crosses is time consuming. Further, complex crosses such as double cross have a low frequency of yielding recombinants in the F_2 that possess a significant number of desirable parental genes. When this method is selected, the number of targeted desirable traits should be small (at most about 10). The double cross hybrid is genetically more broad-based than the single cross hybrid but is more time consuming to make.
- **Diallel cross.** A diallel cross is one in which each parent is crossed with every other parent in the set (complete diallel), yielding n − (n − 1)/2 different

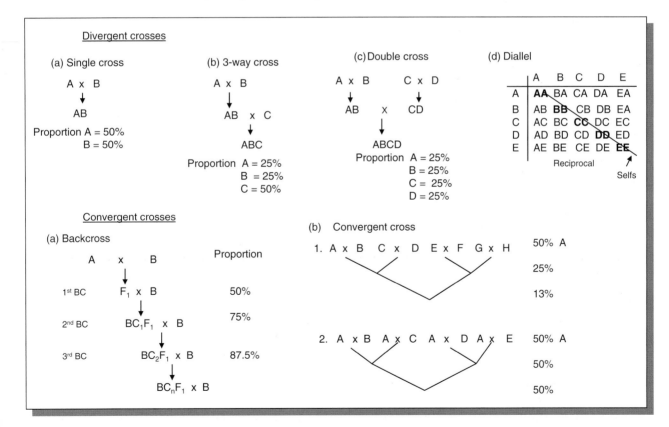

Figure 7.1 The basic types of crosses used by plant breeders. Some crosses are divergent (a) while others are convergent (b).

combinations (where n is the number of entries). This method entails making a large number of crosses. Sometimes, the partial diallel is used, in which only certain parent combinations are made. The method is tedious to apply to self-pollinated species. Generally, it is a crossing method for genetic studies and less for the purpose of creating populations for breeding.

7.10.2 Convergent crosses

These are conservative methods of crossing plants. The primary goal of convergent crossing is to incorporate a specific trait into an existing cultivar without losing any of the existing desirable traits. Hence, one (or several) parents serve as a **donor** of specific genes and is usually involved in the cross only once. Subsequent crosses entail crossing the desirable parent (recurrent parent) repeatedly to the F_1 in order to retrieve all the desirable traits. A commonly used convergent cross is the **backcross** (Chapter 15).

7.11 Wide crosses

The first choice of parents for use in a breeding program is cultivars and experimental materials with desirable traits of interest. Most of the time, plant breeders make elite × elite crosses (they use adapted and improved materials). Even though genetic gains from such crosses may not always be dramatic, they are, nonetheless, significant enough to warrant the practice. After exhausting the variability in the elite germplasm as well as in the cultivated species, the breeder may look elsewhere, following the recommendation by Harlan and de Wet. These researchers proposed that the search for desired genes should start from among materials in the primary gene pool (related species), then proceed to the secondary gene pool and, if necessary, the tertiary gene pool. Crossing involving materials outside the cultivated species is collectively described as **wide crosses**. When the wide cross involves another species, it is called an **interspecific cross** (e.g., kale). When it involves a plant from another genus, it is called an **intergeneric cross** (e.g., wheat). Crosses between crops with their wild progenitor species should not be considered wide crosses, since, despite the sometimes used different scientific names (barley *Hordeum vulgare* was derived from *H. spontaneum*; lettuce *Lactuca sativa* was derived from *L. serriola*). Genetically such "species" are fully compatible and behave genetically as an **intraspecific** cross (i.e. cross within the same species).

7.11.1 Objectives of wide crosses

Wide crosses may be undertaken for practical and economic reasons, research purposes, or to satisfy curiosity. Specific reasons for wide crosses include:

- **Economic crop improvement.** The primary purpose of wide crosses is to improve a species for economic production by transferring one or a few genes, or a segment of chromosomes or whole chromosomes from a donor across interspecific or intergeneric boundaries. The genes may condition a specific disease or pest resistance, or may be a product quality trait, or flower shape or color novelty in ornamentals, among other traits. In some species, such as sugarcane, cotton, sorghum, and potato, hybrid vigor is known to have accompanied certain crosses.
- **New character expression.** Novelty is highly desirable in the ornamental industry. Combining genomes from diverse backgrounds may trigger a complementary gene action or even introduce a few genes that could produce previously unobserved phenotypes that may be superior to the parental expression of both qualitative and quantitative traits.
- **Creation of new alloploids.** Wide crosses often produce sterile hybrids. The genome of such hybrids can be doubled to create a new fertile alloploid species (a polyploid with the genomes of different species), such as triticale, which is a synthetic species that consists of the genomes of (mostly) tetraploid wheat and rye.
- **Scientific studies.** Cytogenetic studies following a wide cross may be used to understand the phylogenic relationships between the species involved.
- **Curiosity and aesthetic value.** Wide crosses may produce unique products of ornamental value, which can be useful to the horticultural industry. Sometimes just being curious is a good enough reason to try new things.

7.11.2 Selected success with wide crosses

Developing commercial cultivars with genes introduced from the wild can be an expensive and long process (see pre-breeding in Chapter 8). Some linkages with genes of the wild donor need to be broken.

In tomato, it took 12 years to break the linkage between nematode resistance and undesirable fruit characteristics. Nonetheless, some significant successes have been accomplished through wide crosses.

- **Natural wide crosses.** Natural wide crosses have been determined by scientists to be the origin of numerous modern day plants of economic importance. Ornamentals such as irises, cannas, dahlias, roses, and violets, are among the list of such species. In tree crops, apples, cherries, and grapes are believed to have originated as natural wide crosses, and so are field crops such as wheat, tobacco and cotton, as well as horticultural crops like strawberry and sweet potatoes. Most natural wide cross products of economic value to modern society are used as ornamentals and are usually propagated vegetatively. This led G.L. Stebbins to observe that wide crosses may be more valuable in vegetatively propagated species than in seed propagated species.
- **Synthetic (artificial) wide crosses.** Apart from natural occurrences, plant breeders over the years have introgressed desirable genes into adapted cultivars from sources as close as wild progenitors to distant ones such as different genera. Practical applications of wide crosses may be grouped into three categories:
  - (i) **Gene transfer between species with the same chromosome number.** Wide crosses between two tomato species, *Lycopersicon pimpinellifolium* × *L. esculentum*, have been conducted to transfer genes resistant to diseases such as leaf mold and *Fusarium* wilt. Gene transfers in which both parents have identical chromosome numbers is often without complications beyond minor ones (e.g., about 10% reduction in pollen fertility). It is estimated that nearly all commercially produced tomatoes anywhere in the world carry resistance to *Fusarium* that derived from a wild source.
  - (ii) **Gene transfer between species with different number of chromosomes.** Common wheat is a polyploid (an allohexaploid) with a genomic formula of *AABBDD*). It has 21 pairs of chromosomes. There is diploid wheat, einkorn (*Triticum monococcum*), with seven pairs of chromosomes and a genomic formula of *AA*. It should be pointed out that later studies of the origin of the A genome showed that the diploid component of the *Triticum* genus is comprised of two distinct biological species, *T. monococcum* and *T. urartu*. The *A* in bread wheat is believed to be from *T. urartu*. There are several tetraploid wheats (*AABB*) such as emmer wheat (*T. dicoccum*). Transfer of genes from species of lower ploidy to common wheat is possible (but not always the reverse). Stem rust resistance is one such gene transfer that was successful.
  - (iii) **Gene transfer between two genera.** Common wheat comprises three genomes of which one (DD) is from the genus *Aegilops* (*A. tauschii*). Consequently, gene transfers have been conducted between *Triticum* and *Aegilops* (e.g., for genes that confer resistance to leaf rust). Other important donors of resistance are *Secale cereal* (rye) and *Agropyron* sp.
- **Developing new species via wide crossing.** A species is defined as a population of individuals capable of interbreeding freely with one another but which, because of geographic, reproductive, or other barriers, does not in nature interbreed with members of other species. One of the long-term "collaborative" breeding efforts is the development of the triticale (X *Triticosecale* Wittmack). The first successful cross, albeit sterile, is traced back to 1876; the first fertile triticale was produced in 1891. The development of this new species occurred over a century, during which numerous scientists modified the procedure to reach its current status where the crop is commercially viable. Triticale is a wide cross between *Triticum* (wheat) and *Secale* (rye), hence triticale (a contraction of the two names). It is predominantly a self-fertilizing crop. The breeding of triticale is discussed in Chapter 24.

7.12 Issue of reproductive isolation barriers

Hybridization is often conducted routinely without any problems when individuals from the same species are involved, provided there is no fertility regulating mechanisms operating. Even when such mechanisms exist, hybridization can be successfully conducted by providing appropriate pollen sources. Sometimes, plant breeders are compelled to introduce desired genes from distant relatives or other more-or-less related species. Crossing plants from two different species or sometimes even plants from two genera is challenging and has limited success. Often, the breeder needs to use additional techniques (e.g.,

Table 7.2 A summary of the reproductive isolation barriers in plants as first described by G.L. Stebbins.

---

1. External barriers

(a) Spatial isolation mechanisms: associated with geographic distances between two species.

(b) Pre-fertilization reproductive barriers: prevents union of gametes. Includes ecological isolation (e.g., spring and winter varieties), mechanical isolation (differences in floral structures), and gametic incompatibility.

2. Internal barriers

(c) Post-fertilization reproductive barriers: leads to abnormalities following fertilization (hybrid inviability or weakness and sterility of plants).

---

embryo rescue) to intervene at some point in the process in order to obtain a mature hybrid plant.

Reproductive isolation barriers may be classified into three categories, as suggested by researchers such as G.L. Stebbins, T. Dobzhansky and D. Zohary (Table 7.2). These barriers maintain the genetic integrity of the species by excluding gene transfer from outside species. Some barriers occur before fertilization, some after fertilization. These barriers vary in the degree of difficulty by which they may be overcome through breeding manipulations.

- **Spatial isolation. Spatial isolation** mechanisms are usually easy to overcome. Plants that have been geographically isolated may differ only in photoperiod response. In which case, the breeder can cross the plants under controlled environment (e.g., greenhouse) by manipulating the growing environment to provide the proper duration of day length needed to induce flowering.
- **Pre-fertilization reproductive barrier.** These barriers occur between parents in a cross. Crops such as wheat have different types that are **ecologically isolated**. There are spring wheat types and winter wheat types. Flowering can be synchronized between the two groups by, for example, vernalization (a cold temperature treatment that exposes plants to about 3–4 °C) of the winter wheat to induce flowering (normally accomplished by exposure to the winter conditions). **Mechanical isolation** may take the form of differences in floral morphology that prohibits the same pollinating agent (insect) from pollinating different species. A more serious barrier to gene transfer is **gametic**

incompatibility whereby fertilization is prevented. This mechanism is a kind of self-incompatibility (Chapter 4). The mechanism is controlled by a complex of multiple allelic system of S-genes that prohibit gametic union. The breeder has no control over this barrier.

- **Post-fertilization reproductive barriers.** These barriers occur between hybrids. After fertilization various hindrances to proper development of the embryo (hybrid) may arise, sometimes resulting in abortion of the embryo, or even formation of a haploid (rather than a diploid). The breeder may use embryo rescue techniques to remove the embryo and culture it to a full plant. Should the embryo develop naturally, the resulting plant may be unusable as a parent in future breeding endeavors because of a condition called **hybrid weakness**. This condition is caused by factors such as disharmony between the united genomes. Some hybrid plants may fail to flower because of **hybrid sterility** (F_1 sterility) resulting from meiotic abnormalities. On some occasions, the hybrid weakness and infertility manifest in the F_2 and later generations (called **hybrid breakdown**).

7.13 Overcoming challenges of reproductive barriers

The reproductive barriers previously discussed confront plant breeders who attempt gene transfer between distant genotypes via hybridization. The primary challenge of wide crosses is obtaining fertile F_1 hybrids, because of the mechanisms that promote, especially, gametic incompatibility. As previously indicated, this mechanism acts to prevent (a) the pollen from reaching the stigma of the other species, (b) germination of the pollen and inhibition of growth of the pollen tube down the style, or the union of male gamete and the egg if the pollen tube reaches the ovary, and (c) the development of the zygote into a seed and the seed into a mature plant. Gametic incompatibility ends where fertilization occurs. However, thereafter, there are additional obstacles to overcome. Gametic incompatibility and hybrid breakdown are considered to be barriers to hybridization that are outside the control of the breeder.

Several techniques have been developed to increase the chance of recovering viable seed and plants from a wide cross. These techniques are based on the nature of the barrier. All techniques are not applicable to all species.

- **Overcoming barriers to fertilization.**
  (i) **Conduct reciprocal crosses.** Generally, it is recommended to use the parent with the larger chromosome number as the female in a wide cross for higher success. This is because some crosses are successful only in one direction. Hence, where there is no previous information about crossing behavior, it is best to cross in both directions.
  (ii) **Shorten the length of the style.** The pollen tube of a short-styled species may not be able to grow through a long style to reach the ovary. Thus, shortening a long style may improve the chance of a short pollen tube reaching the ovary. This technique has been successfully tried in corn and in lily.
  (iii) **Apply growth regulators.** Chemical treatment of the pistil with growth-promoting substances (e.g., NAA, GA) tends to promote rapid pollen tube growth or extend the period over which the pistil remains viable.
  (iv) **Modify ploidy level.** A diploid species may be converted to a tetraploid to be crossed to another species. For example, narrow leaf trefoil, (*Lotus tenuis*, $2n = 12$) was successfully crossed with broadleaf bird's foot trefoil (*L. corniculatus*, $2n = 24$) after chromosome doubling of the *L. tenuis* accession.
  (v) **Use mixed pollen.** Mixing pollen from a compatible species with pollen from an incompatible parent makes it possible to avoid the unfavorable interaction associated with cross-incompatibility.
  (vi) **Remove stigma.** In potato, wide crosses were accomplished by removing the stigma before pollination and substituting it with a small block of agar fortified with sugar and gelatin.
  (vii) **Grafting.** Grafting the female parent to the male species has been reported in some crops to promote pollen tube growth and subsequent fertilization.
  (viii) **Protoplast fusion.** A **protoplast** is all the cellular component of a cell excluding the cell wall. Protoplasts may be isolated by either mechanical or enzymatic procedures. Mechanical isolation involves slicing or chopping of the plant tissue to allow the protoplast to slip out through a cut in the cell wall. This method yields low numbers of protoplasts.

The preferred method is the use of hydrolytic enzymes to degrade the cell wall. A combination of three enzymes – cellulase, hemicellulase, and pectinase – is used in the hydrolysis. The tissue used should be from a source that would provide stable and metabolically active protoplasts. This calls for monitoring plant nutrition, humidity, day length, and other growth factors. Often, protoplasts are extracted from leaf mesophyll or plants grown in cell culture. The isolated protoplast is then purified, usually by the method of flotation. This method entails first centrifuging the mixture from hydrolysis at about $50 \times g$, and then re-suspending the protoplasts in high concentration of fructose. Clean, intact protoplasts float and can be retrieved by pipetting. Protoplasts can also be used to create hybrids *in vitro* (as opposed to crossing mature plants in conventional plant breeding).

- **Overcoming the problem of inadequate hybrid seed development.** Abnormal embryo or endosperm development following a wide cross may be overcome by using proper parent selection and reciprocal crossing as previously described. In addition, the technique of embryo rescue is an effective and common technique. The embryo is aseptically extracted and nurtured into a full plant under tissue culture conditions.
- **Overcoming lack of vigor of the hybrid.** Hybrids may lack the vigor to grow properly to flower and produce seed. Techniques such as proper parent selection, reciprocal crossing, and grafting the hybrid onto one of the parents may help.
- **Overcoming hybrid sterility.** Sterility in hybrids often stems from meiotic complications due to lack of appropriate pairing partners. Sterility may be overcome by doubling the chromosome number of the hybrid to create pairing mates for all chromosomes, and hence produce viable gametes.

7.14 Bridge crosses

Bridge crossing is a technique of indirectly crossing two parents that differ in ploidy levels through a transitional or an intermediate cross (Figure 7.2). For example R.C. Buckner and his colleagues succeeded in crossing the diploid Italian ryegrass (*Lolium multiflorum*, $2n = 2 \times = 14$) with the hexaploid tall fescue (*Festuca arundinacea*, $2n = 6 \times = 42$) via the bridge cross technique. The

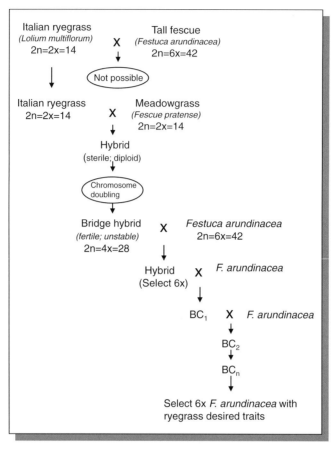

Figure 7.2 An example of a bridge cross. To hybridize Italian ryegrass and tall fescue, the breeder may firstly make an intermediary cross with meadowgrass, followed chromosome doubling.

intermediate cross was between *L. multiflorum* and diploid meadow fescue (*Festuca pratensis*, $2n = 2 \times = 14$). The resulting embryo was rescued and the chromosome number doubled to produce a fertile but genetically unstable tetraploid hybrid (ryegrass-meadow fescue). Using tall fescue as recipient, the *L. multiflorum* × *F. pratensis product* was backcrossed to tall fescue, resulting in the transfer of genes from *L. multiflorum* to *F. arundinacea*. A 42-chromosome cultivar of tall fescue with certain Italian ryegrass traits was eventually recovered and stabilized. Another example of a successful bridge cross is *Allium cepa* receiving genes from *A. fistulosum* through the *A. roylei* bridge.

Key references and suggested reading

Chandler, J.M., and Beard, B.H. (1983). Embryo culture of *Helianthus* hybrids. *Crop Science*, **23**: 1004–1007.

Forsberg, R.A., (ed.) (1985). *Triticale*. Crop Science of America Special Publication No. 9, American Society of Agronomy, Madison, WI.

Morrison, L.A., Riera-Lizaraza, O., Cremieux, L., and Mallory-Smith, C.A. (2002). Jointed goatgrass (*Aegilops cylindrica* Host) × wheat (*Triticum aestivum* L.) hybrids: Hybridization dynamics in Oregon wheat fields. *Crop Science*, **42**: 1863–1872.

Singh, A.K., Moss, J.P., and Smartt, J. (1990). Ploidy manipulations for interspecific gene transfer. *Advances in Agronomy*, **43**: 199–240.

Stalker, H.T. (1980). Utilization of wild species for crop improvement. *Advances in Agronomy*, **33**: 111–147.

Stoskopf, N.C. (1993). *Plant breeding: theory and practice.* Westview Press. Boulder, CO.

Zohary, D. (1973). Gene-pools for plant breeding, in *Agricultural Genetics* (ed. R. Moav). John Wiley and Sons, Inc., New York.

Internet resource

http://www.actahort.org/books/200/200_3.htm – Application of wide crosses in tomato improvement (accessed February 29, 2012).

Outcomes assessment

Part A

Please answer the following questions true or false.

1 A hybrid is a product of unidentical parents.
2 Emasculation is undertaken to make a flower female.
3 An integeneric cross occurs between two species.
4 Wheat is a product of a wide cross.
5 Bridge crosses are used to facilitate crosses between two parents of identical ploidy levels.

Part B

Please answer the following questions.

1 What is hybridization?
2 What are wide crosses?
3 Give three specific reasons why wide crosses may be undertaken.
4 Explain the phenomenon of linkage drag.
5 Give examples of major crops that arose by wide crosses.

Part C

Please write a brief essay on each of the following topics.

1 Discuss the basic steps in artificial hybridization.
2 Discuss the challenges of wide crosses.
3 Discuss the techniques used for overcoming the challenges to wide crosses.
4 Discuss the technique of bridge crossing.
5 Discuss the genetic issues in hybridization.

Clonal propagation and *in vitro* culture

Purpose and expected outcomes

Conventional plant breeding entails sexual recombination between plants to obtain novel and useful combinations of traits in the resulting new product that can be reproduced more or less "true to type". Clonal propagation, on the other hand, does not entail recombination but rather the asexual multiplication of plants such that uniformity and identity are preserved intact. Plant propagators and horticulturists commonly use clonal propagation in their operations. Sometimes, plant breeders need a large number of clones for genetic studies. Modern plant breeding tools such as biotechnology include the use of techniques for generating clones and for genetic modification under aseptic conditions.

After completing this chapter, the student should be able to discuss:

1 Micropropagation and its applications in breeding.
2 The importance of cell and tissue culture in plant breeding.
3 The characteristics of asexual propagation that have breeding implications.
4 Apomixis and the breeding of apomictic species.
5 The advantages and limitations of clonal propagation.

8.1 What is a clone?

In biology, a clone is an organism whose genetic information is identical to that of the parent organism (progenitor) from which it was created. In plants, a clone means a genetically uniform plant material derived from a single individual and propagated exclusively by vegetative (non-seed) methods. Clones are not produced via recombination and meiosis but from replication and mitosis. Plant breeders generally make a cross to obtain from true seeds a segregating progeny that will be followed by cycles of clonal reproduction, during which the superior types will be maintained and the inferior types discarded. There is, therefore, no further genetic segregation during these cycles of clonal reproduction (in contrast to sexual progeny obtained from crosses between homozygous autogamous parents and allogamous populations).

Principles of Plant Genetics and Breeding, Second Edition. George Acquaah.
© 2012 John Wiley & Sons, Ltd. Published 2012 by John Wiley & Sons, Ltd.

8.2 Clones, inbred lines, and pure lines

As previously discussed, plants may be naturally sexually or asexually propagated. Further, sexually propagated species may be self-fertilized or cross-fertilized. These natural modes of reproduction have implications in the genetic structure and constitution of plants and breeding implications as already discussed. Plant breeders are able to manipulate the natural reproductive systems of species to develop plants that have atypical genetic constitution. The terms **pure line, inbred line**, and **clone** are applied to materials developed by plant breeders to connote sameness of genetic constitution in some fashion. However, there are some significant distinctions among them.

- **Pure lines.** These genotypes are developed as cultivars of self-pollinated crops for direct use by farmers. As products of repeated selfing of single plants, pure lines are homogeneous and homozygous and can be naturally maintained by selfing.
- **Inbred lines.** These are genotypes that are developed to be used as parents in the production of hybrid cultivars and synthetic cultivars in the breeding of cross-pollinated species. They are not meant for direct release for use by farmers. They are homogenous and homozygous, just like pure lines. However, unlike pure lines, they need to be artificially maintained because they are produced by forced selfing (not natural selfing) of naturally cross-pollinated species.
- **Clones.** Clones are identical copies of a genotype. Together, they are phenotypically homogeneous. However, individually, they are highly heterozygous. Asexually or clonally propagated plants produce genetically identical progeny.

8.3 Categories of clonally propagated species based on economic use

Clonally propagated species may be divided into several broad categories on the basis of economic use:

- **Those cultivated for a vegetative product** – Important species vegetatively cultivated for a vegetative product include sweet potatoes, yams, cassava, sugarcane, and Irish potatoes. These species tend to exhibit certain reproductive abnormalities. For example, flowering is reduced and so is fertility. Some species such as potatoes have cytoplasmic male sterility. Sometimes, flowering is retarded (e.g., by chemicals) in production (e.g., in sugarcane).
- **Those cultivated for a fruit** – Plants in this category include fruit trees and cane fruits. Examples include apple, pear, grape, strawberry, and banana.
- **Those cultivated for floral products** – Plant in this category include tulips and many cut flower species.

8.4 Categories of clonally propagated species for breeding purposes

For breeding purposes, vegetatively propagated crops may be grouped into one of four groups based on flowering behavior:

(i) **Those with normal flowering and seed set.** Species in this category produce normal flowers and are capable of sexual reproduction (to varying extents) without artificial intervention (e.g., sugar cane). However, in crop production, the preference is to propagate them sexually. Such species enjoy the advantages of both sexual and asexual reproduction. Hybridization is used to generate recombinants (through meiosis) and introduce new genes into the adapted cultivar, while vegetative propagation is used to maintain indefinitely, the advantages of the heterozygosity arising from hybridization.

(ii) **Those with normal flowers but have poor seed set.** Some plant species produce normal looking flowers that have poor seed set, or set seed only under certain conditions but not under others. These restrictions on reproduction make it unattractive to use seed as a means of propagation. However, the opportunity for hybridization may be exploited to transfer genes into adapted cultivars.

(iii) **Produce seed by apomixis.** The phenomenon of apomixis results in the production of seed without fertilization, as was first discussed in Chapter 4. Over 100 species of perennial grasses have this reproductive mechanism.

(iv) **Non-flowering species.** These species may be described as "obligate asexually propagated species" because they have no other choice. Without flowers (or with sterile flowers) these species can only be improved by asexual means. Genetic diversity is not obtained via recombination but by other sources (e.g., mutation).

8.5 Types of clonal propagation

Clonal propagation can be a natural process or artificially conducted.

- **Natural clonal propagation.** Clones are common in nature. Clonal propagation is important for many herbaceous species as well as woody perennial plants throughout the world. The genera *Prunus* and *Populus* produce clones by suckering, while *Betula, Carpinus, Corylus, Quercus, Salix* and *Tilia* are all genera with the ability to self-coppice.
- **Artificial clonal propagation.** Some crops are commercially produced by clonal propagation using various parts (e.g., tubers, corms, bulbs, stolons, etc.). Some species are obligated to this mode of propagation because they have lost their capacity to flower (e.g., leek, some potato cultivars). However, some of these species that are produced clonally as a preference may also have viable sexual reproduction as an option (e.g., potato, strawberry). In fact, some species with flowering capabilities have be traditionally reproduced clonally for a long time (e.g., apple, roses, ornamental trees and shrubs). Plants derived from true seeds of those same species often have a long juvenile stage and a take long time to reach commercially interesting size (orchids, tulips, chrysanthemum, potato). Artificial methods of clonal propagation include cuttings, grafting, and the more sophisticated laboratory technique of tissue culture. For orchids, *in vitro* clonal propagation is the only commercially viable method of micropropagation.

  Clonal multiplication of the cultivar is very important in horticulture and silviculture (tree production). The first step in clonal propagation is to identify and select a genetically superior (elite) plant. The part of the plant used as a propagule varies among species and includes stem, roots, bulbs, and stolons as stated previously.

8.6 Importance of clonal propagation in plant breeding

Clonal propagation has several significant applications in plant breeding.

- **Quick production of quality breeding and planting stock.** In tree breeding, somatic propagation of trees is advantageous over the use of seeds to raise stock. It is fast and economical. Breeding time is shortened by using micropropagation.
- **Early flower induction.** Clonal propagation is known to cause earlier flowering in plant than seed produced plants. This helps to speed up the reproductive cycle of the species and hence accelerate the breeding and testing time in a breeding program.
- **Germplasm maintenance.** Genotypes in clonal banks are maintained obtained by clonal propagation. Clonal testing can be conducted to evaluate the accessions and their interaction with the environment in a greenhouse setting.
- **Maintenance of genetic uniformity.** Cross-pollinated species are naturally highly heterozygous, resulting in the progenies from such genotypes not only being true to type, but also missing some of the parental qualities. Clonal propagation is used to maintain the genetic characteristics of species.
- **Production of disease-free plants.** Vegetative propagation techniques such as grafting the *in vitro* culture of specific tissues can produce healthy clones from disease susceptible species.
- **Propagating problematic species.** Some species produce few or no viable seeds, or may have seed dormancy issues, while others have seeds with poor germination capacity. Sometimes, germination is very slow, causing such species to take a long time to produce marketable size trees. The use of clonal propagation can usually be a more rapid and economical technique than seeds in multiplying such species.
- **Multiplication of sexually derived sterile hybrids.** Interspecific hybridization in plants commonly results in products that are sterile. Clonal propagation may be used to multiply such hybrids.
- **Maintenance of genetic gain.** After a breeder makes a cross, the next step is to conduct repeated cycles of clonal reproduction to identify and maintain superior plants. Clonal propagation helps capture and maintain maximum genetic gains

8.7 Breeding implications of clonal propagation

There are certain characteristics of clonal propagation that have breeding implications.

- Clonal species with viable seed and high pollen fertility can be improved by crosses.
- Unlike crossing in sexually reproducing species, which often requires additional steps to fix the

genetic variability in a genotype for release as a cultivar (except for hybrid cultivars), clonal cultivars can be released immediately following a cross, provided a desirable genotype combination has been achieved.

- When improving species whose economic parts are vegetative products, it is not important for the product of crossing to be fertile.
- Because of the capacity to multiply from vegetative material (either through methods such as cuttings or micropropagation), the breeder only needs to obtain a single desirable plant to be used as stock.
- Heterosis, (hybrid vigor) if it occurs, is fixed in the hybrid product. That is, unlike hybrid cultivars in seed-producing species, there is no need to reconstitute the hybrid. Once bred, heterozygosity is maintained indefinitely.
- It is more difficult to obtain large quantities of planting material from clones in a short time.
- Plant species that are vegetatively parthenocarpic (e.g., banana) cannot be improved by the method of crossing parents.
- Species such as mango and citrus produce polyembryonic seeds. This reproductive irregularity complicates breeding because clones of the parent are mixed with hybrid progeny.
- Many clonal species are perennial outcrossers and intolerant of inbreeding. These are highly heterozygous.
- Unlike sexual crop breeding in which the genotype of the cultivar is determined at the end of the breeding process (because it changes with inbreeding), the genotype of a clone is fixed and determined at the outset.
- Both general combining ability (GCA) and specific combining ability (SCA), that is performance in crosses, can be fully exploited with appropriate breeding approaches.

8.8 Genetic issues in clonal breeding

- **Genetic make-up.** All the progeny from an individual propagated asexually are genetically identical (clones) and uniform. Clones are products of mitosis. Any variation occurring among them is environmental in origin.
- **Heterozygosity and heterosis.** Many species that are asexually propagated are highly heterozygous. They are highly heterotic. Consequently, they are susceptible to inbreeding depression. For those

species that can be hybridized without problems, any advantage of asexual propagation is that heterosis, where it occurs, is fixed for as long as the cultivar is propagated asexually.
- **Ploidy.** Many known species that are asexually propagated are interspecific hybrids or have high ploidy.
- **Chimerism.** Clones are stable over many generations of multiplication. The only source of natural variation, albeit rare, is somatic mutation in the bud. Plant breeders may generate variability by the method of mutagenesis. Whether natural or artificial, somatic mutations are characterized by tissue mosaicism, a phenomenon called chimerism.

  A chimera or chimeric change occurs when an individual consists of two or more genetically different types of cells. Though heritable changes, these mosaics can only be maintained by vegetative propagation (not transferable to progenies by sexual means).

There are four basic types of chimeras.

(i) **Sectorial.** This chimera is observed in a growing shoot as two different tissues located side-by-side. The effect of this modification is that the stem develops with two distinct tissues on each half.

(ii) **Periclinal.** This type of chimera consists of two thin layers of different genetic makeup, one over the other.

(iii) **Mericlinal.** When an outer layer of different genetic tissue does no completely extend over the layer below, the chimera is mericlinal.

(iv) **Graft chimeras.** Unlike the first three chimeras that have genetic origin, a graft chimera is a nonheritable mixture of tissues that may occur after grafting is made.

Additional information on chimeras is found in Chapter 12 (Section 12.8). Whereas they are undesirable in crop plants, chimeras may be successfully exploited in horticulture.

8.9 Breeding approaches used in clonal species

Several breeding approaches are used in the breeding of clonally propagated species.

8.9.1 Planned introduction

Just like seed, vegetative material (whole plants or parts) may be introduced into a new production area for evaluation and adaptation to the new area. Seedlings or cuttings may be introduced. However, the technology of tissue culture allows a large variety of small samples to be introduced in sterile condition. These disease-free samples are easier to process through plant quarantine.

8.9.2 Clonal selection

Clonal selection has two primary goals – to maintain disease-free and genetically pure clones, and the development of new cultivars.

Purifying an infected cultivar

Clonal cultivars may become infected by pathogens, some of which may be systemic (e.g., viruses). Two general approaches may be used to purify a cultivar to restore it to its disease-free original genetic purity.

(i) **Screening for disease-free material.** Plant materials may be visually inspected for the presence of pathogens. However, because some pathogens may be latent, a variety of serological and histological techniques are used to detect the presence of specific pathogens. Called **indexing**, these techniques can detect latent viruses (**viral indexing**) as well as other pathogens. A negative test may not always be proof of the absence of pathogens. It could be that the particular assay is not effective. The clean clonal material is then used as starting material for multiplication for propagation.

(ii) **Elimination of pathogen.** A positive test from indexing indicates the presence of a pathogen. Should this be the only source of planting material, the breeder has no choice but to eliminate the pathogen from the plant tissue by one of several methods.
- **Tissue culture.** Even when the pathogen is systemic, it is known that tissue from the terminal growing points is often pathogen-free. Tissue from these points may be ascetically removed and cultured under tissue culture conditions to produce disease-free plantlets. Through micropropagation, numerous disease-free plants can be obtained.

- **Heat treatment.** This may be of short or long duration. Short-duration heat treatment is administered to the plant material for about 30 min to four hours at 43–57 °C. This could be in the form of hot air treatment or by soaking the material in hot water. This works well for fungal, bacterial, and nematode infection. For viruses, a longer treatment of several weeks (2–4 weeks) is used. Potted plants are held at 37 °C in a controlled environment for the duration of the treatment. Cuttings from the treated plants may be used as scions in grafts, or rooted into a seedling.
- **Chemical treatment.** This surface sterilization treatment is suitable for elimination of pathogens that are external to the plant material (e.g., in tubers).
- **Use of apomictic seed.** Viral infections are generally not transmitted through seed in cultivars that are capable of apomixis (e.g., citrus).

Clonal selection for cultivar development

This procedure is effective if variability exists in the natural clonal population.

Year 1.	Assemble clonal population. Plant and expose to diseases of interest. Select resistant clones with other superior traits and harvest individually.
Year 2.	Grow progenies of selected clones and evaluate as in year 1. Select superior clones.
Year 3.	Conduct preliminary yield trials. Select superior clones.
Years 4–6.	Conduct advanced yield trials at multilocations for cultivar release.

8.9.3 Crossing with clonal selection

This procedure is applicable to species that are capable of producing seed in appreciable quantities. Because heterosis can be fixed in clonal populations, the breeder may conduct combining ability analysis to determine the best combiners to be used in crosses. The steps below are not applicable to trees in which much longer times are needed.

Year 1.	Cross selected parents. Harvest F_1 seed.
Year 2.	Plant and evaluate F_1s. Select vigorous and healthy plants.

Year 3. Space plant clonal progeny rows of selected plants. Select about 100–200 superior plant progenies.

Year 4. Conduct preliminary yield trials.

Years 5–7. Conduct advanced yield trials for cultivar release.

Other techniques that are applicable include backcrossing to transfer specific traits and wide crossing. The challenges with backcrossing are several. As previously indicated, clonal species are very heterozygous and prone to inbreeding depression. Backcrossing to one parent (the recurrent parent) provides opportunity for homozygosity and consequently inbreeding depression. To prevent this, breeders may cross the backcross to another clone instead of the recurrent parent, followed selection to identify superior plants. The process is repeated as needed.

8.9.4 Mutation breeding

The subject is discussed in detail in Chapter 22. Inducing variability via mutagenesis is challenging for two key reasons. Being rare events, a large population of M_1V_2 is needed to have a good chance of observing desired mutants. Obtaining a large number of vegetative propagules is difficult. Also, mutations occur in individual cells. Without the benefit of meiosis, the mutated clonal material develops chimeras. Using adventitious buds as starting material reduces the chance of chimeras. A mutation in the epidermal cell (usually there is one) would result in an adventitious shoot that originated from a single mutant cell. This technique is not universally applicable.

8.9.5 Breeding implications, advantages and limitations of clonal propagation

There several advantages and limitations of breeding clonally propagated species:

Advantages
- Sterility is not a factor in clonal propagation because seed is not involved.
- Because clonal plants are homogeneous, the commercial product is uniform.
- Micropropagation can be used to rapidly multiply planting material.
- Heterozygosity and heterosis are fixed in clonal populations.

Disadvantages
- Clonal propagules are often bulky to handle (e.g., stems, bulbs).
- Clones are susceptible to devastation by an epidemic. Because all plants in the clonal population are identical, they are susceptible to the same strain of pathogen.
- Clonal propagules are difficult to store for a long time, because they are generally fresh and succulent materials.

8.10 Natural propagation

Some crops rely on clonal propagation: tubers, corms, cuttings, bulbs, stolons, and so on. Such crop species may have lost the capacity of flowering (leek, some potato cultivars). Their progeny is genetically identical to the plant from which it was derived (except if the primordial cell contained some mutation (**chimerism**)). Normally, such species may also have sexual reproduction as a natural option. Potato, for example, may form berries with true seeds, strawberry produces fruits with seeds. Such seeds produce genetically heterogeneous progeny because of segregation, since most clonally reproducing species have a high level of heterozygosity. Plants from natural clonal tissues are usually vigorous and can produce flowers and fruit in the same or next season. Plants derived from true seeds of those same species often have a long juvenile stage and take a long time to reach a commercially interesting size (orchids, tulips, chrysanthemum, potato). The same is true for species that naturally do not reproduce clonally, but as crops have been reproduced that way for a long time. Examples are apple, rose and ornamental trees and shrubs, which are reproduced by grafting or cutting.

8.11 *In vitro* culture

In vitro culture or **tissue culture** of cells, tissues, organs, and protoplasts is used as a technique by plant breeders and growers for propagation. It is critical in some modern plant breeding approaches, specifically biotechnology, in which genetic alterations are conducted under aseptic conditions. The cell is the fundamental unit of structure and function of a plant, containing all the genetic information. Tissues and even single cells can be nurtured to develop into full plants. In biotechnology, it is critical to be able to nurture a single cell into a full plant in order to apply some of the more sophisticated techniques, such as gene transfer or transformation. The technique of tissue culture may be used to assist plant breeders who realize wide crosses to be able to nurture young embryos into full plants. Plant germplasm of vegetatively propagated species may be maintained in germplasm banks using the tissue culture technique.

The most critical aspect of *in vitro* culture is the provision of a sterile environment. A plant has certain natural defenses against pathogens and the abiotic environment in which it grows. Cells and tissues lack such protection once extracted from the parent plant. The environment for growing plants in the soil under natural conditions should provide adequate moisture, nutrients, light, temperature, and air. Plant performance can be enhanced by supplementing the growth environment (e.g., by fertilization, irrigation). In tissue and cell culture, plant materials are grown in a totally artificial environment in which nutrients, plus additional factors (e.g., growth regulators) and sometimes antibacterial substances, are supplied. The cultural environment in tissue culture may be adjusted by the researcher to control the growth and development of the cultured material. For example, the researcher may modify the hormonal balance in the culture medium to favor only root or only shoot development. The components of a tissue culture medium may be categorized into four groups: mineral elements, organic compounds, growth regulators, and a physical support system.

8.12 Micropropagation

Seed is the preferred propagule for use in the propagation and cultivation of most agronomic species. This is because they are easy to handle before and during the production of the plant, and seeds of most species can be stored for many years or decades. However, a number of major food crops and horticultural species are vegetatively (asexually) propagated as a preferred method because of biological reasons (e.g., self-incompatibility) and the lack of uniformity in seed progeny. **Micropropation** is the *in vitro* clonal propagation or reproduction of plants. It is used more commonly for commercial propagation of ornamentals and other high-priced horticultural species than for field crop species. Micropropagation

can utilize pre-existing meristems (specific regions where cells are undifferentiated or have no specific assigned roles or function) or non-meristematic tissue. The method of micropropagation commonly used may be divided into three categories: (i) **axillary shoot production,** (ii) **adventitious shoot production,** and (iii) **somatic embryogenesis.**

Micropropagation may be summarized in five general steps:

1 **Selection of explant.** The plant part (e.g., meristem, leaf, stem tissue, buds) to initiate tissue culture is called the **explant**. It must be in good physiological condition and disease-free. Factors that affect the success of the explant include its location on the plant, age, or developmental phase. Explants that contain shoot primordia (e.g., meristems, node buds, shoot apices) are preferred. Also, explants from younger (juvenile) plants are more successfully used in micropropagation.

2 **Initiation and aseptic culture establishment.** The explant is surface sterilized (e.g., with Chlorox, alcohol) before being placed on the medium. Small amounts of plant growth regulators may be added to the medium for quick establishment of the explant.

3 **Proliferation of axillary shoots.** Axillary shoot proliferation is induced by adding cytokinin to the shoot culture medium. Cytokinin to auxin ratio of about 50:1 produces shoots with minimum callus formation. New shoots may be subcultured at an interval of about four weeks.

4 **Rooting.** Addition of auxin to the medium induces root formation. Roots must be induced on the shoot to produce plantlets for transfer into the soil. It is possible to root the shoot directly in the soil.

5 **Transfer to natural environment.** Before transferring into the field, seedlings are gradually moved from ideal lab conditions to more natural climate room or greenhouse conditions by reducing the relative humidity and increasing light intensity, a process called **hardening off.**

8.12.1 Axillary shoot production

Pre-existing meristems are used to initiate **shoot culture** (or shoot-tip culture). The size of the shoot tip ranges between 1 and 10 mm in length. Cytokinin is used to promote axillary shoot proliferation. Some species (e.g., sweet potato) do not respond well to this treatment. Instead, shoots consisting of single or multiple nodes per segment are used. These explants are placed horizontally on the medium and from them single unbranched shoots arise that may be induced to root to produce plantlets.

Shoot tips are easy to excise from the plant and are genetically stable. They contain pre-formed incipient shoot and are phenotypically homogeneous. These explants have high survival and growth rates. Axillary and terminal buds have the advantages of shoot tips but they are more difficult to disinfect. On the other hand, meristem tips contain pre-formed meristems and are genetically stable and phenotypically homogeneous but are more difficult to extract from the plant. Furthermore, they have low survival rates. Meristems in general tend to be free of virus infection, even if the rest of the plant is infected. Meristems may therefore be the ideal explant to cure virus infected valuable clones.

8.12.2 Adventitous shoot production

Adventitious shoots originate from adventitious meristems. Non-meristematic tissue can be induced to form plant organs (e.g., embryos, flowers, leaves, shoots, roots). Differentiated plant cells (with specific functional roles) can be induced to dedifferentiate from their current structural and functional state, and then embark upon a new developmental path to produce new structures. Adventitious shoot production through organogenesis occurs by one of two pathways – **indirect** or **direct**.

(i) **Indirect organogenesis.** The indirect organogenetic pathway goes through a stage in which a mass of dedifferentiated cells (**callus**) forms (i.e., the explant forms a callus from which adventitious meristems are induced and from which plant regeneration is initiated). The callus consists of an aggregation of meristem-like cells that are developmentally plastic (can be manipulated to redirect morphogenic end point). The negative side of this method is that the callus phase sometimes introduces mutations (somaclonal variation, making this not always a 100% clonal procedure). The callus phase also makes it more technically challenging than shoot tip micropropagtion.

(ii) **Direct organogenesis.** Direct organogenesis bypasses a callus stage in forming plant organs. The cells in the explant act as direct precursors of a new primordium. This pathway is less common than the callus mediated pathway.

8.12.3 Somatic adventitious embryogenesis

A zygote is formed after an egg has been fertilized by a sperm. The zygote then develops into an embryo (**zygotic embryo**). *In vitro* tissue culture techniques may be used to induce the formation of embryos from somatic tissue (**non-zygotic embryo** or **somatic embryogenesis**) using growth regulators. Somatic embryos arise from a single cell rather than budding from a cell mass as in zygotic embryos. This option is very important in biotechnology since transgenesis in plants may involve the manipulation of single somatic cells. However, without successful regeneration, plant transformation cannot be undertaken. Somatic embryogenesis has been extensively studied in Apiaceae, Fabaceae, and Solanaceae.

8.13 Concept of totipotency

Plants reproduce sexually or asexually. **Clones** are identical copies of a genotype, derived from somatic tissue or cells of the source plant. Together, they are phenotypically homogeneous, since they all originate from the same source plant either in one or more clonal generations of reproduction. However, individually, they are highly heterozygous. Clonally propagated plants produce genetically identical progeny. Pieces of plant parts (leaf, stem, roots, tubers) can be used to grow full plants in the soil. *In vitro* (growing plants under sterile conditions) plant culture was first proposed in the early 1900s. By 1930s, cell culture had been accomplished. Each cell in a multicelluar organism is theoretically **totipotent** (i.e., endowed with the full complement of genes to direct the development of the cell into a full organism). In theory, a cell can be taken from a root, leaf, or stem and cultured *in vitro* into a complete plant.

8.14 Somaclonal variation

Clones, as previously stated, are exact replicas of the genotype from which its source tissue had been derived. Commonly, however, clonal propagation, occurring under a tissue culture environment, produces materials that are not exact replicas of the original material used to initiate the culture. Such variation, which results not from meiosis but from the culture of somatic tissue, is referred to as **somaclonal**

Table 8.1 Crops where desirable and heritable somaclonal variations have been reported.

Species	Characteristics modified
A. Monocotyledons	
1. *Allium sativa*	Bulb size and shape; clove no.; aerial bulbit
2. *Avena sativa*	Plant height; heading date; awns
3. *Hordeum* spp.	plant height; tillering
4. *Lolium* hybrids	Leaf size; flower, vigor; survival
5. *Oryza sativa*	Plant height; heading date; seed fertility: grain no and weight
6. *Saccharum officinarum*	Diseases (eye spot, fiji virus, downy mildew, leaf scald)
7. *Triticum aestivum*	Plant and ear morphology; awns; gliadins; amylase; grain weight, yield
8. *Zea mays*	T toxin resistance; male fertility; mtDNA
B. Dicotyledons	
9. *Lactuca sativa*	Leaf weight, length, width, flatness and color
10. *Solanum lycopersicum*	Leaf morphology; branching habit; fruit color; pedicel; male fertility; growth
11. *Medicago sativa*	Multifoliate leaves; elongated petioles; growth; branch no.; plant height; dry matter yield
12. *Solanum tuberosum*	Tuber shape; maturity date; plant morphology; resistance for early and late blight; photoperiod; leaf color; vigor; height.; skin color

variation, with the variants referred to as **soma-clones**. The variation observed may be transient (epigenetic) or heritable (genetic in origin). It is important to authenticate the presence of a true mutational event before using the somaclone in a breeding program as a valuable source of variation. Somaclonal variants can be recovered in tissue culture with selection pressure (e.g., deliberate inclusion of a toxic agent in the culture medium) or without selection pressure (the basic cultural medium).

A variety of mechanisms have been implicated in this phenomenon. Chromosomal changes, both polyploidy and aneuploidy, have been observed in potato, wheat, and ryegrass. Some research suggests mitotic crossovers to be involved whereas cytoplasmic factors (mitochondrial genes) have been implicated by others. Further, point mutation, transposable elements, DNA methylation and gene amplification are other postulated mechanisms for causing somaclonal variation. One more trivial source of variation in plants derived from tissue culture is that they are derived from a mutated section of the explants. Somatic cells may have undergone mutations (leading to chimeras; Chapter 23). Tissue from chimeric plants may lead to genetically different progeny.

As a breeding tool, breeders may deliberately plan and seek these variants by observing certain factors in tissue culture. Certain genotypes are more prone to genetic changes in tissue culture, polyploids generally being more so than diploids. Also, holding the callus in an undifferentiated state for prolonged periods enhances the chances of somaclonal variation occurring. Not unexpectedly, the tissue culture environment (medium components) may determine the chance for heritable changes in the callus. The inclusion of auxin 2, 4-*D* enhances the chances of somaclonal variation.

The value of somaclonal variation as a breeding tool is evidenced by the successes in various species (Table 8.1). These include disease resistance (e.g., *Helminthosporium sacchari* in sugarcane and *Fusarium* in *Apium graveolens*), resistance to various abiotic stresses.

8.15 Apomixis

Seed production in higher plants that are sexually propagated species normally occurs after a sexual union in which male and female gametes fuse to form a zygote, which then develops into an embryo. However, some species have the natural ability to develop seed without fertilization, a phenomenon called **apomixis**. The consequence of this event is that apomictically produced seeds are clones of the mother plant. That is, apomixis is the asexual production of seed. Unlike sexual reproduction, there is no opportunity in apomixis for new recombination to occur to produce diversity in the offspring. However, the seeds have the same advantages as non-apomictic seeds: they can be stored and sown, and have less chance of being infected by virus than whole plants or plant organs.

8.15.1 Occurrence in nature

Apomixis is widespread in nature and occurs in several unrelated plant families. About 10% of the estimated 400 plant families and a mere 1% of the estimated 40 000 plant species they comprise exhibit apomixis. The plant families with the highest frequency of apomixis are Gramineae (Poaceae), Compositae, Rosaceae, and Asteraceae. Many species of *Citrus*, mango, perennial forage grasses, and guayule reproduce apomictically.

Some species can produce both sexual and apomictic seeds and are called **partial apomicts** (e.g., bluegrass, *Poa pratensis*). Species such as bahiagrass (*Paspalum notatum*) reproduce exclusively, or nearly so, by apomixis and are called **complete apomicts**. There are several indicators of apomixis. When the progeny from a cross in a cross-pollinated species fails to segregate, appearing uniform and identical to the mother plant, this could indicate apomixis. Similarly, when plants expected to exhibit high sterility (e.g., aneuploids, triploids) instead show significantly high fertility, apomixis could be the cause. Obligate apomicts may display multiple floral features (e.g., multiple stigmas and ovules per floret, double or fused ovaries), or multiple seedlings per seed. Facultative apomixis may be suspected if the progeny of a cross shows an unusually high number of identical homozygous individuals that resemble the mother plant in addition to the presence of individuals that are clearly different (hybrid products).

The indicators suggested are by no means conclusive evidence of apomixis. To confirm the occurrence of apomixis and to discover its mechanisms require additional progeny test as well as cytological tests of megasporogenesis and embryo sac development.

8.15.2 Benefits of apomixis

The benefits of apomixis may be examined from the perspectives of the plant breeder and the crop producer, as well as impact on the environment.

Benefits to the plant breeder

Apomixis, essentially, is a natural mechanism of cloning plants through seed. If it could be developed to become a breeding tool, it would allow plant breeders to develop hybrids that can retain their original genetic properties indefinitely with repeated use, without a need to reconstitute them. In other words, hybrid seed could be produced from hybrid seed. The plant breeder would not need to make crosses each year to produce the hybrid. This advantage would accelerate breeding programs and reduce development cost of hybrid cultivars. Apomixis would be greatly beneficial when uniformity of product is desired. Breeders could use this tool to quickly fix superior gene combinations. That is, vigor could be duplicated, generation after generation without decline. Furthermore, commercial hybrid production could be implemented for species without fertility control mechanisms (e.g., male-sterility system), neither would there be a need for isolation in F_1 hybrid seed production. There would be no need to maintain and increase parental genotypes. Cultivar evaluation could proceed immediately following a cross. This advantage is applicable to clonal propagation in general.

In addition to these obvious benefits, it is anticipated that plant breeders would divert the resources saved (time, money) into other creative breeding ventures. For example, cultivars could be developed for smaller and more specific production environments. Also, more parental stock could be developed to reduce the risk of genetic vulnerability through the use of a few elite genetic stocks as parents in hybrid development.

There are some plant breeding concerns associated with apomixis. Species that exhibit partial apomixis are more challenging to breed because they produce both sexual and apomitic plants in the progeny. Complete apomicts are easier to breed by conventional methods.

Benefits to the producer

The most obvious benefit of apomitic cultivars to crop producers is the ability to save seed from their field harvest of hybrid cultivars to plant the next season. Because apomixis fixes hybrid vigor, the farmer does not need to purchase fresh hybrid seed each season. This would especially benefit the producer in poor economies, who often cannot afford the high price of hybrid seed. Apomixis, as previously indicated, could accelerate plant breeding. This could translate into less expensive commercial seed for all producers. Realistically, such benefits will materialize only if commercial breeders can make acceptable profit from using the technology.

Impact on the environment

Some speculate that apomixis has the potential to reduce biodiversity because it produces clonal cultivars and, hence, uniform populations that are susceptible to disease epidemics. However, others caution that the suspected reduction in biodiversity would be minimal, since apomixis occurs naturally in polyploids, which occurs less frequently than diploids.

8.15.3 Mechanisms of apomixis

Apomixis arises by a number of mechanisms, of which four major ones that differ according to origin (cell that undergoes mitosis to produce the embryo) are discussed here. Seed formation without sexual union is called **agamospermy**, the mechanism may be divided into two categories – **gametophytic apomixis** and **adventitious apomixis**. There are two types of gametophytic apomixis – **apospory** and **diplospory**.

Apospory

This is the most common mechanism of apomixis in higher plants. It is a type of agamospermy that involves the nucellar. The somatic cells of the ovule divide mitotically to form unreduced ($2n$) embryonic sac. The megaspore or young embryo sac aborts, as occurs in species such as Kentucky bluegrass (*Poa pratensis*).

Diplospory

The unreduced megaspore mother cell produces the embryo sac following mitosis instead of meiosis. This cytological event occurs in species such as *Tripsacum*.

Adventitious embryo

Unlike apospory and diplospory, in which an embryo sac is formed, no embryo sac is formed in adventitious embryony. Instead, the source of the embryo could be somatic cells of the ovule, integuments, or ovary wall. This mechanism occurs commonly in *Citrus* but rarely in other higher plants.

Parthenogenesis

This mechanism produces an outcome that is essentially equivalent to haploidy. The reduced (n) egg nucleus in a sexual embryo sac develops into a haploid embryo without fertilization by the sperm nucleus. In this case, the embryo is not an identical genotype as the plant on which it is formed.

Other less common mechanisms of apomixis are **androgenesis** (development of a seed embryo from the sperm nucleus upon entering the embryo sac) and **semigamy** (sperm nucleus and egg nucleus develop independently without uniting, leading to one haploid embryo). The resulting haploid plants contain sectors of material from both maternal and paternal origin, and are therefore chimeric.

8.16 Other tissue culture applications

There are other tissue culture based applications of interest to plant breeders besides micropropagation.

8.16.1 Synthetic seed

Somatic embryogenesis has potential commercial applications, one of which is in **synthetic seed** technology (production of artificial seeds). A synthetic seed consists of somatic embryos enclosed in protective coating. There are two types currently being developed:

(i) **Hydrated synthetic seed** – seed is encased in hydrated gel (e.g., calcium alginate).
(ii) **Desiccated syntethic seed** – seed is coated with water soluble resin (e.g., polyoxethylene).

To develop synthetic seed, it is critical to achieve a quiescent phase, which is typically lacking in somatic embryogenesis (i.e., without quiescence, there is continuous growth, germination, and eventually death, but no stationary stage as in embryos in mature seeds). The application will depend on the crop. Lucerne (*Medicago sativa*) and orchardgrass (*Dactylis glomerata*) are among the species that have received much attention in artificial seed development. Potential application of artificial seed is in species that are highly heterozygous and in which conventional breeding is time consuming. Trees can be cloned more readily by this method. In some tropical species that are seed propagated but in which seeds have short period of viability, artificial seed production could be economical because of the high economic value of these crops (e.g., cocao, coconut, oil palm, coffee). Also, hybrid synthetic seed could be produced in species in which commercial hybrid production is problematic (e.g., cotton, soybean).

8.16.2 Limitations to commercialization of synthetic seed technology

Whereas the prospect of commercial synthetic seed is appealing, several factors make this impractical at present. Problems may occur at maturation, germination, rooting, shot apex formation, or acclimatization.

- Large-scale production of high quality viable propagules remains a key challenge.
- A major limitation is the poor conversion of apparently normal propagules into normal plantlets.
- Improper development and maturation of somatic embryos causes poor germination and conversion problems.
- Poor storage of synthetic seeds due to lack of dormancy and stress tolerance in the somatic embryos.
- Mechanical damage, lack of oxygen supply, invasion by microbes, and lack of nutrients all contribute to poor germination of synthetic seeds.

Strategies are available for addressing some of these challenges. They vary among species and include desiccation, a process that can damage the embryo.

8.16.3 Production of virus-free plants

Viral infections are systemic, being pervasive in the entire affected plant. Heat therapy is a procedure that is used for ridding infected plants of viral infections. After heat treatment, subsequent new growth may be free of viruses. More precisely, meristems dissected from leaf and shoot primordia are often free of viruses even when the plant is infected. Tissue culture technology is used to nurture the excised meristematic tissue into full plants that are free from viruses.

The process starts with detection (e.g., by ELISA) of the presence of a viral infection in the plant. Once confirmed, the meristems on the shoots are aseptically removed and sterilized (dipped in 75–99% ethanol or 0.1–0.5% sodium hypochlorite or household bleach for a few seconds or minutes). The explant is submitted to tissue culture as previously described. Sometimes, to increase the success of viral elimination, researchers may include chemicals (e.g., Ribavirin, Virazole) in the tissue culture medium. The plants produced must be tested to confirm virus-free status.

The virus-free plants are used to produce more materials (by mircropropagation) for planting a virus-free crop. It should be pointed out that virus elimination from plants does not make them virus resistant. The producer should adopt appropriate measures to protect the crop from infection.

8.16.4 Applications in wide crosses

Wide cross production is discussed in Chapter 7.

Embryo rescue

Sometimes, especially in crosses between different plant species, the embryo formed after fertilization in wide crosses fails to develop any further. The breeder may dissect the flower to remove the immature embryo. The embryo is then nurtured into a full plant by using tissue culture technology. This technique is called **embryo rescue**. The fertilized ovary is excised within several days of fertilization to avoid an abortion (due to, e.g., abnormal endosperm development). Normal embryogenesis ends at seed maturation. The development of the embryo goes through several stages with certain distinct features. The globular stage is undifferentiated, while the heart stage is differentiated and capable of independent growth. The torpedo stage and cotyledonary stage of embryo development follow these early stages. Prior to differentiation, the developing embryo is heterotrophic and dependent on the endosperm for nutrients. Excising the embryo prematurely gives it less a chance of surviving the embryo rescue process. Just like all tissue culture work, embryo rescue is conducted aseptically and cultured on the medium appropriate for the species.

Somatic hybridization

Somatic hybridization was discussed in Chapter 5.

8.17 Production of haploids

Haploids contain half the chromosome number of somatic cells. Anthers contain immature microspores or pollen grains with the haploid (*n*) chromosome number. If successfully cultured (anther culture), the plantlets resulting will have a haploid genotype. Haploid plantlets may arise directly from embryos or indirectly via callus, as previously discussed. To have maximum genetic variability in the plantlets, breeders usually use anthers from F_1 or F_2 plants. Usually, the haploid plant is not the goal of anther culture. Rather, the plantlets are diplodized (to produce diploid plants) by using colchicine for chromosome doubling. This strategy yields a highly inbred line that is homozygous at all loci, after just one generation.

Methods used for breeding self-pollinated species generally aim to maintain their characteristic narrow genetic base through repeated selfing over several generations for homozygosity. The idea of using haploids to produce instant homozygotes by artificial doubling has received attention. Haploids may be produced by one of several methods:

- Anther culture to induce androgenesis.
- Ovary culture to induce gynogenesis.
- Embryo rescue from wide crosses.

8.17.1 Anther culture

Flower buds are picked from healthy plants. After surface sterilization, the anthers are excised from the buds and cultured unto an appropriate tissue culture medium. The pollen grains at this stage would be in the uninucleate microspore stage. In rice the late uninucleate stage is preferred. Callus formation starts within 2–6 weeks, depending on the species, genotype, and physiological state of the parent source. A high nitrogen content of the donor plant and exposure to low temperature at meiosis reduces albinos and enhances the chance of green plant regeneration. Pre-treatment (e.g., storing buds at 4–10 °C for 2–10 days) is needed in some species. This and other shock treatments promote embryogenic development. The culture medium is sometimes supplemented with plant extracts (e.g., coconut water, potato extract). To be useful for plant breeding, the haploid pollen plants are diplodized (by articifial doubling with 0.2% colchicines, or through somatic callus culture).

Applications

- **Development of new cultivars.** Through diplodization, haploids are used to generate instant homozygous true breeding lines. It takes only two seasons to obtain doubled haploid plants, versus about seven crop seasons using conventional procedures to attain near homozygous lines. The genetic effect of doubling is that doubled haploid lines exhibit variation due primarily to additive gene effects and additive × additive epistasis, enabling fixation to occur in only one cycle of selection. Heritability is high because dominance is eliminated. Consequently, only a small number of doubled haploid plants in the F_1 is needed, versus several thousands of F_2 for selecting desirable genotypes.
- **Selection of mutants.** Androgenic haploids have been used for selecting especially recessive mutants. In species such as tobacco, mutants resistant to methionine analogue (methionine sulfoxide) of the toxin produced by *Pseudomonas tabaci* have been selected.
- **Development of supermales in asparagus.** Haploids of *Asparagus officinalis* may be diplodized to produce homozygous males or females.

Limitations

- The full range of genetic segregation of interest to the plant breeder is observed because only a small fraction of androgenic grains develop into full sporophytes.
- High rates of albinos occur in cereal haploids (no agronomic value).
- Chromosomal aberrations often occur, resulting in plants with higher ploidy levels, requiring several cycles of screening to identify the haploids.
- Use of haploids for genetic studies is hampered by the high incidence of nuclear instability of haploid cells in culture.

8.17.2 Ovule/ovary culture

Gynogenesis, using ovules or ovaries has been achieved in species such as barley, wheat, rice, maize, tobacco, sugar beet, and onion. The method is less efficient than androgenesis because only one embryo sac exists per ovary as compared to thousands of microspores in each anther. Ovaries ranging in developmental stages from uninucleate to mature embryo sac stages are used. However, it is possible for callus and embryos to develop simultaneously from gametophytic and sporophytic cells, making it a challenge to distinguish haploids from

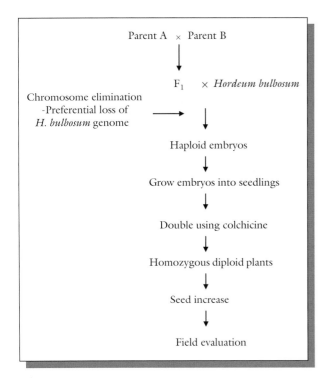

Parent A × Parent B

F$_1$ × *Hordeum bulbosum*

Chromosome elimination
-Preferential loss of
H. bulbosum genome

Haploid embryos

Grow embryos into seedlings

Double using colchicine

Homozygous diploid plants

Seed increase

Field evaluation

Figure 8.1 Generating haploids in barley by the bulbosom method.

those of somatic origin. Generally, gynogenesis is selected when androgenesis is problematic (as in sugar beet and onion).

8.17.3 Haploids from wide crosses

Certain specific crosses between cultivated and wild species are known to produce haploids. Well established systems include the interspecific crosses between *Hordeum vulgare* ($2n = 2x = 14$, *VV*) × *Hordeum bulbosum* ($2n = 2x = 14$, *BB*), commonly called the **bulbosum method**, and also in wheat × maize crosses. The bulbosum method is illustrated in Figure 8.1. The F$_1$ zygote has $2n = 2x = 14$ (*7V + 7B*). However, during the tissue culture of the embryo, the *bulbosum* chromosomes are eliminated, leaving a haploid ($2n = x = 7V$). This is then doubled by colchicines treatment to obtain $2n = 2x = 14$ *VV*.

8.17.4 Doubled haploids

Researchers exploit haploidy generally by doubling the chromosome number to create a cell with the double dose of each allele (homozygous).

Key features

Inbred lines are homozygous genotypes produced by repeated selfing with selection over several generations. The technique of **doubled haploids** may be used to produce complete homozygous diploid lines in just one year (versus more than four years in conventional breeding) by doubling the chromosome complement of haploid cells. Such doubling may be accomplished *in vivo* naturally or through crossing of appropriate parents, or *in vitro*, through the use of colchicine. The success of doubled haploids as a breeding technique depends on the availability of a reliable and efficient system for generating haploids and doubling them in the species.

Applications

Doubled haploids have been successfully used in breeding species in which efficient haploid generation and doubling systems have been developed. These include canola, barley, corn, and wheat. Additionally, doubled haploids are used to generate general genetic information that can be applied to facilitate the breeding process. Such information includes gene action and interaction, estimating the number of genetic variances, calculating combining abilities, and detection of gene linkages, pleiotropy, and chromosome locations. Haploids are used in mutation studies (recessive mutants are observable instantly) and in selecting against undesirable recessive alleles.

Procedure

The first step in using doubled haploids in breeding is identifying the source of haploids.

- **Natural sources.** Haploids originate in nature through the phenomenon of **parthenogenesis** (gamete formation without fertilization). The haploids may be maternal or paternal in origin. It is estimated that haploids occur in corn at the rate of 1 in 1000 diploids, 99% of which arise from parthenogenesis of maternal origin. Spontaneous doubling occurs in corn at the rate of 10% of haploids developed. The key is distinguishing between haploid and diploid plants. A marker system for this purpose was first developed by S.S. Chase based on seedling color, purple plants being encoded by the dominant gene (*P*) while normal green plants are recessive (*p*). A cross of

$F_1(pp) \times PP$ would yield $999Pp$ (purple diploids) and $1pp$ (green haploid). Another marker used is the purple aleurone color.

To use this marker system, the breeder should cross a heterozygous female to a male with marker genes. The seed from those with dominant endosperm marker of the male is saved and planted, discarding seedlings with the dominant male marker. Next, cytological evaluation of plants with the recessive female marker (by root tip squash) is conducted. The haploid plants are retained and grown in the greenhouse or field, and self-pollinated to produce diploids.

- **Artificial sources.** Haploid production through interspecific and intergeneric crosses is in use, one of the most well known being the barley system (previously discussed). After doubling the chromosome, the diploid plants are grown to maturity. Seeds are harvested for planting plant rows. Because diploids produced by this method are normally completely homozygous, there is no need for growing segregating generations as obtains in conventional programs.

- **Advantages and disadvantages.** The technique of doubled haploids has certain advantages and disadvantages, the key ones including:
  - Advantages
    - Complete homozygosity is attainable in a shorter period.
    - Duration of the breeding program can be reduced by several (2–3) generations.
    - It is easier and more efficient to select among homogeneous progeny (versus heterogeneous progeny in conventional breeding).
    - The cultivar released is homogeneous.
  - Disadvantages
    - The procedure requires special skills and equipment in some cases.
    - Additional technology for doubling may increase the cost of a breeding program.
    - Frequency of haploids generated is not predictable.
    - There is a lack of opportunity to observe line performance in early generations prior to homozygosity.

- **Genetic issues.** Unlike the conventional methods of inbreeding, it is possible to achieve completely homozygous individuals. Using an F_1 hybrid or a segregating population as female parent in the production of maternally derived haploids increases genetic diversity in the doubled haploid line. It is advantageous if the female also has agronomically

desirable traits. F_1 hybrids are suitable because their female gametes will be segregating.

8.18 *In vitro* selection

In vitro selection is essentially selection of desirable genotypes under controlled environments in the Petri dish. Reports on the use of this technology have dwindled over the years, after an explosion of reports shortly following the awareness of its potential as a source of biological variation. As a tool, it is applicable to both the sporophyte and the gametophyte.

8.18.1 Using whole plants or organs

Whole plants, seedlings or embryos may be the units of *in vitro* selection. As previously indicated, an appropriate tissue culture system is needed for *in vitro* selection. The method has been used for screening for resistance to diseases, such as *Fusarium culmorum* in wheat, and susceptibility to fungal spores in disease breeding. Tolerance (or resistance) to inorganic salts (e.g., salt tolerance in sugar beet) has been reported.

8.18.2 Using undifferentiated tissue

The capacity for regeneration from callus makes the material attractive for use in selection. Plants can be multiplied using a callus system as well. Sometimes, spontaneous variability arises in callus or suspension culture, some of which may be heritable, while some is epigenetic, disappearing when plants are regenerated or reproduced sexually. Tissue culture originating variability is called **somaclonal variation**.

8.18.3 Directed selection

Rather than allowing the variation to arise spontaneously, sometimes plant breeders apply selection pressure during the *in vitro* cultural process, to influence the variability that might arise.

- **Selection for disease resistance.** Various toxin metabolites have been included in tissue culture for use as the basis for selection, assuming that such metabolites have a role in pathogenesis. The culture filtrates from various fungi (*Fusarium, Helminthosporium maydis*) have been used to exert selection pressure for cells that are resistant to the pathogen.

The main constraint to the use of directed selection in plant breeding for disease resistance is the inability of the *in vitro* system to be used to select for hypersensitivity, a major strategy in disease resistance

- **Selection for herbicide tolerance.** Mutants with 10–100 times levels of resistance to herbicides (e.g., imidazilinone in sugar beet) have been successfully isolated, characterized, and incorporated into commercial cultivar development. Many of the recorded successes with *in vitro* selection have been with herbicide tolerance.
- **Selection for tolerance to abiotic stresses.** Selection for tolerance to salinity, metals (zinc, aluminum) and temperature (cold tolerance) has been attempted with varying degrees of success.
- **Single cell selection system.** Some researchers use single cell tissue culture systems (suspension culture,

protoplast culture) for *in vitro* selection. The advantages of this approach include a lack of chimerism and higher chances of isolation of true mutants, ability to more effectively apply microbial procedures of the large number of individual cells than can be screened in a small space. Selection for biotic stress resistance, herbicide tolerance (the author did this for chlorsulfuron), and aluminum tolerance, are among successful application of direct selection using single a cell selection system.

8.19 Germplasm preservation

Germplasm preservation in tissue culture was discussed in chapter 6. This method of germplasm storage is often used for vegetatively propagated species.

Industry highlights

Maize and Tripsacum: Experiments in intergeneric hybridization and the transfer of apomixis – an historical review

Bryan Kindiger

USDA-ARS Grazinglands Research Laboratory, El Reno, OK 73036, USA

Research in maize–*Tripsacum* hybridization is extensive and encompasses a period of more than 60 years of collective research. A vast amount of literature exists on various facets of this type of hybridization ranging from agronomy, plant disease, cytogenetics, breeding, and genetic analysis. As a consequence, no single article can cover all the research relevant to this topic. This report will not address all the various issues but will focus primarily on specific research and experiments which would perhaps be of value to a student interested in this topic. The interested student is encouraged to review the references and follow the additional references cited by the various authors to obtain more information on this topic.

Interspecific hybrids or hybrids generated between species are utilized by plant breeders to discover and transfer genes from one plant to another plant of a related species which cannot be found in the particular species of interest. One of the most interesting instances of interspecific hybridization is that between maize (*Zea mays* L.) and its most distant relative gamagrass (*Tripsacum* spp.). *Tripsacum* L. is a perennial, warm season bunch grass found throughout most of the subtropical and tropical regions of the Western Hemisphere (de Wet *et al.*, 1981, 1982) (Figure B8.1).

Approximately 16 species have been classified taxonomically in *Tripsacum*. The most common species is *Tripsacum dactyloides* (L.) L. that can be found growing in much of the USA, Mexico, Central America and South America (de Wet *et al.*, 1981, 1982). The most commonly studied maize–*Tripsacum* interspecific hybrids are those generated between diploid maize (2n = 2x = 20) and tetraploid *Tripsacum dactyloides* (2n = 4x = 72). Regardless of their complete difference in chromosome number, plant phenotype and environmental niche, hybrids are relatively easy to generate.

In 1939, Mangelsdorf and Reeves published their historical monograph *The Origin of Indian Corn and Its Relatives*, in which they discussed their research and views on the relationship of cultivated *Zea mays* to its distant cousins, teosinte (the closest relative of maize) and *Tripsacum* spp. (Mangelsdorf and Reeves, 1939). Though these early views regarding the origins of maize and its relationship to teosinte and *Tripsacum* are controversial and open to discussion and further investigation, the procedures for generating such interspecific hybridizations remains relatively unchanged (Mangelsdorf and Reeves, 1931).

Figure B8.1 A stand of *Tripsacum dactyloides* (eastern gamagrass) in Woodward County, OK, USA. The individual standing in the gamagrass is Dr Victor Sokolov, Institute of Cytology and Genetics, Novosibirsk, Russia. Figure courtesy of Bryan Kindiger.

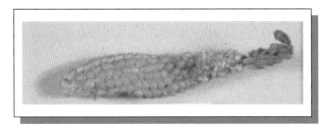

Figure B8.2 Hybrid seed set utilizing Ladyfinger popcorn as the maternal parent when pollinated by tetraploid *T. dactyloides*. Over 100 F₁ seeds can be readily obtained when an appropriate maize parent is utilized in the cross. Figure courtesy of Bryan Kindiger.

Generating maize ×tripsacum *hybrids*

Typically, the generation of a maize × *Tripsacum* hybrid is accomplished by covering the developing maize ear with a bag to prevent cross-contamination with corn pollen. Within a few days, ample silks are available on the ears and the silks are cut back to approximate 10–15 cm in length. *Tripsacum* pollen is gathered and dusted onto the maize ear. Typically, pollen from tetraploid *Tripsacum* ($2n = 4\times = 72$) provides superior seed set but, occasionally, a diploid *Tripsacum* genotype ($2n = 2\times = 36$) can be identified which also conditions superior seed development. When an appropriate maize genotype is utilized, embryo rescue techniques are not necessary. Experience has shown that a "Supergold popcorn" accession (PI222648) available from the USDA-ARS Plant Introduction Station, Ames, IA, gives abundant and viable F₁ seeds (Kindiger and Beckett, 1992) (Figure B8.2). Experimentation with other maize germplasm can provide similar if not superior results.

The F1 hybrids are completely pollen sterile and microsporogenesis is associated with a varying array of meiotic anomalies (Kindiger, 1993); they vary in seed fertility from completely sterile to highly seed fertile (Harlan and de Wet, 1977). To date, all seed fertile hybrids generated from tetraploid *Tripsacum dactyloides* resources exhibit some level of apomictic expression, which following backcrossing with maize is often lost. The most common or sexual pathway of genomic change in a series of maize–*Tripsacum* backcross hybrids has been clearly described by Harlan and de Wet (1977). Comparative genetics and other approaches that may result in the transfer of *Tripsacum* traits to maize, including apomixis, are described here.

Gene gransfer from **Tripsacum** to maize

Recent and past research strongly suggests that there is little homeology between the genomes of *Tripsacum* and maize. Maguire (1962) and Galinat (1973), each utilizing a set of recessive phenotypic maize markers, suggested that only maize chromosomes 2, 5, 8 and 9 have a potential for pairing and recombination and for gene introgression with *Tripsacum*. Additional research has confirmed conservation of loci specific to pistil development between maize and *Tripsacum* genomes (Kindiger *et al.*, 1995; Li *et al.*, 1997). Maguire (1957, 1960) successfully generated and identified a naturally occurring recombination event between an unknown *Tripsacum* chromosome and the short arm of maize chromosome 2. Studies using B-A translocation deletion lines suggested that the Mz9S region could pair and recombine with *Tripsacum* chromosome 5 (Kindiger and Beckett, 1990). Genomic *in situ* hybridization studies have also strongly suggested that only three regions of maize chromosomes have homeology with the *Tripsacum* genome: the subterminal regions of Mz2S, Mz6L, Mz8L (Poggio *et al.*, 1999). These regions correspond well with groups of conserved RFLP markers identified between maize and *Tripsacum* genomes (Blakey *et al.*, 1994; Leblanc *et al.*, 1995). As a consequence, few sites are available for *Tripsacum* introgression into the maize genome and, to date, only two instances are known where verifiable recombination/translocation events have occurred (Maguire, 1962; Kindiger *et al.*, 1996a).

Potential pathways for Tripsacum *introgression*

The 28–>38–>20 non-apomictic pathway

This pathway is the earliest known pathway of maize–*Tripsacum* hybridization first reported by Mangelsdorf and Reeves (1939) and repeated by several others (Harlan and de Wet, 1977). When crossing a diploid maize (2n = 2x = 20Mz) by a diploid *Tripsacum* (2n = 2x = 36Tr), the F₁ hybrid consists of 10Mz + 18Tr chromosomes. Backcrossing this hybrid by diploid maize typically results in the fertilization of an unreduced egg by the pollen source. This partially apomictic event (a 2n + n mating) results in what is called a B$_{III}$ derived hybrid (Bashaw and Hignight, 1990) and now possesses 20Mz + 18Tr chromosomes. This behavior is also commonly observed in apomictic tetraploid *Tripsacum dactyloides* (Kindiger and Dewald, 1997), which raises an interesting question regarding the potential of diploid *Tripsacum* to possess, but not utilize, the mechanisms of apomictic reproduction. In a subsequent backcross of this 38-chromosome individual with diploid maize, individuals possessing 20Mz + 1 through 17 Tr. chromosomes are generated. In some instances, the maize constitution can also be slightly aneuploid 20 + 1 or 2 maize chromosomes. At this point, the predisposition of these individuals is to rapidly lose most if not all of its *Tripsacum* chromosomes following additional backcrossing. The end result of continued backcrossing with maize is the recovery of maize, often completely lacking any level of *Tripsacum* genome introgression through homoeologous pairing and/or recombination. Though, in this pathway, *Tripsacum* introgression is rare, a method for enhancing the opportunity for introgression has been suggested but not pursued (Kindiger and Beckett, 1990).

The 28–>38 apomictic transfer pathway

This pathway has been described only once and has been little examined or discussed in the literature. Consequently, some detail regarding the results in this research will be presented. In 1958, Dr M. Borovsky, of the Institue of Agriculture, Kishinev, Moldova, performed a series of hybridizations between a diploid popcorn line identified as Risovaia 645 and a sexual diploid (2n = 2x = 36) *T. dactyloides* clone with the first maize–*Tripsacum* hybrids being generated in 1960 (Borovsky, 1966; Borovsky and Kovarsky, 1967). The F1 hybrids generated from the experiments possessed 28 chromosomes (10Mz + 18Tr). The F₁ plants were completely male sterile and were highly seed sterile. Backcrossing with diploid maize identified that some of the F₁ hybrids were approximately 1–1.5% seed fertile and resulted in the production of progeny possessing 28 chromosomes (10Mz + 18Tr) and 38 chromosomes (20Mz + 18Tr). When the F₁ was backcrossed to the *Tripsacum* parent, the fertile F₁ hybrids generated progeny with 28 chromosomes (10Mz + 18Tr) and 46-chromosomes (10Mz-18Tr + 18Tr). The complete set of backcrosses with maize and *Tripsacum* resulted in a ratio of approximately ten (28-chromosome plants) to one (38- or 46-chromosome plant). Phenotypic observations suggested that the 28-chromosome progeny were not different from their 28-chromosome parent while the 38- and 46-chromosome progeny were clearly different. In addition, some seed generated by the 28-chromosome F₁ hybrids were polyembryonic. Additional evaluations on the 28-chromosome F1 and its 28-chromosome progeny suggested that these F₁ plants and their progeny were apomictic. This early, non-replicated experiment is to date the only report where a 28-chromosome F₁ hybrid was maintained by apomixis. Polyembryony was noted and a diploid sexual *Tripsacum* was used to generate the interspecific hybrid.

The 46–>56–>38 non-apomictic pathway

This sexual or non-apomictic pathway, as discussed by Harlan and de Wet (1977), is believed to offer the greatest opportunity for *Tripsacum* introgression into maize and represents results of an early attempt to transfer apomixis to maize (Petrov *et al.*, 1979, 1984). In this pathway, a diploid maize (2n = 2x = 20Mz) is crossed with a tetraploid *Tripsacum* resource (2n = 4x = 72Tr). The resultant F₁ hybrid possesses 10Mz + 36Tr chromosomes. To date, published reports indicate all of these hybrids are pollen sterile and vary considerably in their levels of seed fertility. In this particular pathway, when the 46-chromosome F₁ is backcrossed to diploid maize, 56-chromosome individuals result following fertilization of an unreduced egg. As in the prior pathways, this result is generated by a 2n + n mating event. Offspring of the 56-chromosome individual following a second backcross to maize generally possess 38 chromosomes (20Mz + 18Tr) and resemble those discussed in the 28–>38–>20 pathway previously. The generation of progeny with 38 chromosomes is the result of meiosis in the developing megaspore. In this instance, the maize and *Tripsacum* complements pair with their homologous sets (Mz-Mz, Tr-Tr). Following a complete occurrence of meiosis I and II divisions, the result is a reduced egg having 10Mz + 18Tr chromosomes, which, when backcrossed by a diploid maize, results in progeny having 20Mz + 18Tr chromosomes. Almost exclusively, the 38-chromosome individuals no longer express any level or form of an apomictic reproductive mechanism and subsequent backcrossing to maize results in the recovery of individuals possessing 20Mz and varying number of *Tripsacum* chromosomes. Upon backcrossing, the 38-chromosome individuals behave in an identical manner to their 28-chromosome cousins represented in the 28–>38–>20 pathway. Generally, *Tripsacum* introgression by homoeologous pairing and recombination does not occur and genetic transfer of *Tripsacum* genes to maize is not accomplished.

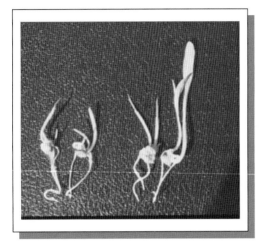

Figure B8.3 Polyembryony expression in germinating seed of an apomictic 46-chromosome F$_1$ maize-*Tripsacum* hybrid. Note that in the pairs one seedling is often larger and more vigorous than its sib. The larger sib of the pair is often the product of a 2n + n mating event. Figure courtesy of Bryan Kindiger.

The 46–>56–>38 apomictic transfer pathway

Though not addressed in Harlan and de Wet's 1977 research, this pathway is similar enough and worthwhile to discuss with regards toward its relevance to apomixis. First published by Petrov and colleagues as early as 1979, and replicated in similar style by others, a diploid or tetraploid maize line is pollinated by a tetraploid, apomictic *T. dactyloides* clone (Petrov *et al.*, 1979, 1984). If a diploid maize line is utilized, the resultant F$_1$ 46-chromosome hybrid possesses 10Mz and 36Tr chromosomes. Upon backcrossing with diploid maize, both apomictic 46-chromosome and 56-chromosome (20Mz + 36Tr) individuals can be obtained. The 46-chromosome offspring are products of apomixis. The 56-chromosome offspring are products of an unreduced egg being fertilized by the diploid maize pollen source, another 2n + n mating event. Often, these individuals exhibit polyembryony that results in the generation of "twins" being obtained from a single seed (Figure B8.3). In some instances, these polyembryonic events give rise to 46–46 pairs of twins (each apomictic clones of the other), 46–56 "twins", one arising from a unfertilized reduced egg, the other arising from a fertilized unreduced egg, and, in some instances, varying combinations of 46–46–46 or 46–46–56 triplets. Typically, as seedlings, the 56-chromosome individuals are more vigorous than their 46-chromosome sibs.

Backcrossing the 46-chromosome individuals by maize repeats the above cycle. Upon backcrossing the 56-chromsome individuals with maize, three types of progeny can be observed. Typically, progeny having 56 chromosomes are generated. However, in some instances, 2n+n matings occur, giving rise to individuals possessing 66 chromosomes (30Mz + 36Tr). Occasionally, a reduced egg will be generated and may or may not be fertilized by the available maize pollen. In rare instances of non-fertilization, a 28-chromosome individual is generated (10Mz + 18Tr). In instances whereby the maize pollen fertilizes the reduced egg, 38-chromosome individuals are obtained (20Mz + 18Tr). Generally, individuals possessing 38 chromosomes, rather than 28 chromosomes, are the most common product. What is unique about this pathway is that occasionally, the 38-chromosome individuals retain all the elements of apomixis which were present in the *Tripsacum* paternal parent and the F$_1$ and BC$_1$ individuals. The retention of apomixis to this 38-chromosome level has been well documented and repeated in several laboratories (Petrov *et al.*, 1979, 1984; Leblanc *et al.*, 1996; Kindiger and Sokolov, 1997). In addition, the occurrence of 2n + n matings, polyembryony and variation in apomixis expression is quite similar to that found in apomictic *Tripsacum* (Kindiger *et al.*, 1996a).

Following the generation and confirmation of apomictic 38-chromsome individuals (20Mz + 18Tr), it is apparently a difficult and uncommon occurrence to generate and maintain apomixis in backcross generations that have fewer *Tripsacum* chromosomes. Only one report has been published where apomictic individuals possessing only nine Tr chromosomes were obtained (Kindiger *et al.*, 1996a). Generally, by 2n+n mating events, the 38-chromosome individuals produce only apomictic 38-chromosome progeny and 48-chromsome progeny. Backcrossing the 48-chromsome individuals results in 48-chromosome apomictics and 58-chromosome apomictics. This accumulation of maize genomes continues until a point is achieved where the additional maize genomes eventually shift the individual from an apomictic reproductive mechanism to a traditional sexual mode of reproduction, whence apomixis is never again attained. This commonly occurs when five or six doses (50–60 maize chromosomes) are present. The result of a 78-chromosome individual (60Mz + 18Tr) losing apomixis is the return of meiosis and a highly seed sterile individual producing an array of highly maize-like aneuploids with a random set of *Tripsacum* chromosomes. Backcrossing these individuals, which are typically pollen sterile, generally results in the recovery of diploid maize lines with or without any Tr chromosomes. To date, the apomictic maize–*Tripsacum* line possessing 39 chromosomes (30Mz + 9Tr) represents the most advanced level of apomixis transfer to maize. An array of various ear types generated from a series of maize-*Tripsacum* hybrids is provided in Figure B8.4.

Figure B8.4 A series of maize-*Tripsacum* ear types. Left to right: dent corn; a apomictic 39-chromosome hybrid; an apomictic 38-chromosome hybrid (Yudin); an apomictic 56-chromosome hybrid; two apomictic 46-chromosome hybrids; and two tetraploid *Tripsacum dactyloides*. Figure courtesy of Bryan Kindiger.

Transfer of apomixis from Tripsacum to maize

Apomixis is an asexual mode of reproduction through the seed and is a genetically controlled process that bypasses female meiosis and fertilization to produce seed identical to the maternal parent. Apomixis is common in polyploid species where reproduction through classic meiotic reproductive events often result in high levels of seed abortion. Apomictic seed development is often viewed as an escape mechanism whereby ovule abortion in polyploid species can be reduced or eliminated through the omission of meiosis. Essentially, three forms of gametophytic apomixis are recognized: diplospory, apospory and adventitious embryony. In *Tripsacum*, the dominating form of apomixis is diplosporous pseudogamy of the *Antennaria* type. In this form of apomixis, the embryo sac originates from the megaspore mother cell either directly by mitosis or following an interrupted meiosis. In addition, an infrequent occurrence of a *Taraxacum* type of diplosporous pseudogamy has also been observed (Burson *et al.*, 1990).

Tripsacum has been suggested as a model system for the study of apomixis (Bantin *et al.*, 2001). As of this report, the prevailing wisdom suggests that apomixis (at least for *Tripsacum*) is controlled by no more than one or two genes, likely linked on a particular *Tripsacum* chromosome (Leblanc *et al.*, 1995; Grimanelli *et al.*, 1998). These results seem to be in agreement with molecular studies focused on understanding apomixis expression in other species (Noyes and Reiseberg, 2000; Albertini *et al.*, 2001).

Cytogenetic studies and GISH studies suggest this region may be Tr16L in the vicinity of the nucleolus organizing region which has homeology with the distal region of Mz6L (Kindiger *et al.*, 1996b; Poggio *et al.*, 1999). Still others suggest, from data generated from a *Tripsacum dactyloides* seed fertility study, that apomixis is a multigenic system that portends a difficult transfer to maize (Blakey *et al.*, 2001). An additional theory, though not necessarily directed toward maize–*Tripsacum* hybrids, suggests that asychronous meiosis in wide-cross hybrids may induce apomictic behavior (Carmen, 1997).

Regardless of the favorable light academics and researchers alike shed upon the prospects in this area of study, the research endeavor continues to be difficult, time consuming and expensive. The wide range of views regarding the inheritance of the trait suggests that there remains much work to be accomplished. In addition, various non-profit organizations suggest that the development of this technology for agriculture is a monopolistic tool for the agri-chemical industry. Several commercial–academic research alliances are indeed active and utilizing several molecular approaches to attempt the mapping and actual cloning of the apomixis loci in *Tripsacum* maize and/or other species. Consequently, research progress in this area has become difficult to monitor following the development of confidentiality agreements, plant patents, commercial-academic research funding collaborations, and so on.

However, some recent information and research progress remain in the public domain. Recent evaluation of several apomictic maize–*Tripsacum* hybrids has identified the occurrence of one line that does not possess an intact Tr16 chromosome or the Mz6L-Tr16L translocation (Figures B8.5a and B8.5b). RAPD markers previously known to be associated to apomixis (Kindiger *et al.*, 1996a) continue to be present in this germplasm. Cytological analysis of this particular chromosome suggests the chromosome carries the nucleolus organizing region and the Tr16L satellite. This small, isochromosome-appearing entity may indeed possess the loci conferring apomixis in this material.

Much has been written, scientific and otherwise, on the benefits of apomixis or the potential "sexual revolution" in cereal and crop species. The most obvious benefit of introducing apomixis into crops would be to allow the selection of a particular individual and propagate it indefinitely by seed. Theoretically, in an apomictic system, hybrids could be maintained indefinitely if the first division restitution (FDR) events discussed above did not occur. Most likely, apomixis will be first utilized to stabilize genetic combinations that otherwise could not occur naturally or are difficult or impossible to

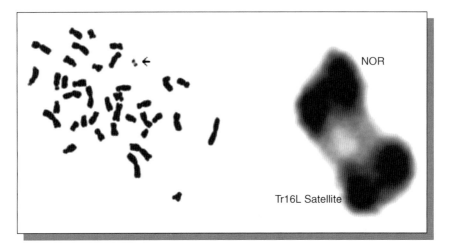

Figure B8.5 (a) (left) The satellite region of Tr16L (arrow) that confers apomixis in the V31 apomictic line. No normal or intact Tr16 is present in this line. (b) (right). An enlargement of the isochromosome with the nucleolus organizing region (NOR) and satellite regions identified. Figure courtesy of Bryan Kindiger.

maintain in nature. The facts are that even if a prototype apomictic system is generated, such as in "apomictic maize" US Patent No. 5,710,367, traditional breeding and gene transfer through backcrossing is questionable. Pollen sterility is, so far, the rule for all such hybrids and is likely caused by the presence of the same *Tripsacum* chromosome detailed in an earlier study (Maguire, 1957, 1960). In addition, a near obligate level of apomictic seed development does not provide an opportunity to transfer apomixis to other maize germplasm. Also, in this apomictic maize germplasm, seed set remains poor and uncharacterized problems associated with endosperm development persist. Essentially, in the case of the apomictic maize patent, a closed breeding system exists.

Pitfalls in the development of an apomictic maize

FDR in apomictic maize–Tripsacum hybrids

One unique attribute found only in the apomictic backcross hybrids, irrespective of their possessing a 38-chromosome (20Mz + 18Tr) or 39-chromosome (30Mz + 9Tr) constitution, is the maintenance of their genetic composition. Theoretically, apomictic individuals will reproduce a genetic copy of themselves through the seed they produce. However, studies focused on this behavior in maize–*Tripsacum* hybrids have proven this will not necessarily be the case.

First division restitution (FDR) can lead to chromosome doubling in intergeneric hybrids of various grasses, of which many are of polyploid or of complex polyploid origin (Xu and Joppa, 1995). Consequently, unreduced gametes bypass the reductional division and are generated only by a normal equational division. This behavior can occur during megasporogenesis or microsporogenesis. However, in the case of pollen sterile apomictic maize–*Tripsacum* hybrids, this behavior impacts genetic change by way of the megaspore. Studies utilizing both molecular and phenotypic evaluations have suggested that an FDR event can occur in apomictic *Tripsacum dactyloides* resulting in genome alterations in an apomictic genotype (Kindiger and Dewald, 1996). This behavior has been visually verified by the occasional discovery of major chromosome rearrangements involving Mz6L, Tr16L, Mz2S, and other unknown maize and *Tripsacum* chromosomes during routine cytogenetic investigations involving both apomictic 38- and 39-chromosome maize–*Tripsacum* hybrids (Kindiger, unpublished). This component of diplosporous apomixis in *Tripsacum* and maize–*Tripsacum* hybrids has not been well studied or addressed.

Evaluation of both apomictic 38-chromosome and 39-chromosome individuals that have been increased over several locations for a minimum of 15 years (Figure B8.6), has shown that a low level of genome reorganization does occur, sometimes resulting in genome loss. This behavior is well documented in some of the apomictic Petrov materials and can only be visualized following the passage of time. The generation of a Mz6L–Tr16L translocation is quite likely a product of this type of behavior (Kindiger and Dewald, 1996; Kindiger *et al.*, 1996a). In addition, a long term (15+ years) selection program in a 38-chromosome (20Mz + 18Tr) apomictic genotype has resulted in a pollen sterile, apomictic line that has a near perfect resemblance to maize in its plant and ear characteristics (Yudin and

Figure B8.6 A series of 39-chromosome maize-*Tripsacum* (30Mz + 9Tr) hybrids growing at the Japanese National Livestock and Grassland Research Institute, Nishinasuno, Japan. Figure courtesy of Bryan Kindiger.

Figure 8.7 (a) (left) A highly maize-like 38-chromosome apomictic maize–*Tripsacum* hybrid. This selection has none or few tillers and exhibits a distinct maize phenotype. (b) (right) A top and second ear taken from one of these highly maize-like apomictic individuals. Note the eight-rows on the ear is rarely found in other apomictic maize–*Tripsacum* hybrids. Figure courtesy of Bryan Kindiger.

Sokolov, 1989) (Figures B8.7a and B8.7b). Consequently, the occurrence of FDR events in apomictic maize–*Tripsacum* hybrid, should generate some concern regarding the development of apomictic off-types which, over time, would increase the genetic non-uniformity of a particular genotype.

Potential advantages of apomictic hybrid corn

Utilizing a traditional hybrid corn production methodology, two inbred lines are typically required to produce an F_1 hybrid. At present, apomictic hybrids would likely utilize one or both inbred lines that would carry the necessary genes and genetics to develop a true breeding, apomictic F_1 hybrid corn cultivar. Land, labor and storage space are also required to maintain these inbred lines. If true breeding, apomictic hybrids can be developed, the yearly seed increase of inbreds, the generation of hybrids, the necessary time allowed for such production, land, fertilizer and required field isolations necessary for producing a hybrid corn line could be omitted. With apomictic hybrid corn, seed generated from that crop would reproduce seed and individuals possessing the identical genetics of the parental hybrid. As such, the development of an apomictic seed crop from an apomictic hybrid would lead to a substantial savings in cost and time to commercial producers and hopefully a decrease seed price to farmers.

Under present agricultural patents and plant variety protection legislation, farmers are able to retain a limited amount of their produced seed for their own use. This could be particularly advantageous to poorer farmers in developing countries. In addition, farmers that live in remote regions where government or commercial seed suppliers are not available could essentially guarantee themselves a yearly supply of quality hybrid seed. These farmers would not be dependent on government or commercial seed suppliers once a suitable cultivar was introduced into their area and could provide farmers greater control regarding the production and use of their locally generated product. However, farmers utilizing such *"on the farm seed production rational"* need to be aware of seed quality issues as well as the enforceable and legal ramifications of selling and or distributing such seed protected by license, patent or Plant Variety Protection (PVP). Growers should also be aware that the lack of genetic diversity or uniformity that is embedded in their apomictic seed crop could eventually make that variety susceptible to a particular disease infestation. Retaining and sowing an apomictic seed crop over the long term could eventually result in major disease infestations and a dramatic reduction in crop yield.

At this time, it is difficult to visualize how the application of cytogenetics, molecular genetics, mapping and genetic engineering, coupled with traditional breeding procedures, will be readily integrated to incorporate apomixis into maize and develop a apomictic hybrid corn cultivar anytime in the near future. Some have given the development of apomictic hybrid corn an approximate twenty year horizon; others much longer. As has been discussed above, the research process to be utilized to develop apomictic corn or other species is time consuming, difficult and generates several legal, social, breeding and genetic issues that will need to be resolved. Though the transfer of apomixis holds much promise, it will be some time before a commercial apomictic maize hybrid will reach the market place.

References

Albertini, E., Barcaccia, G., Porceddu, A., Sorbolini, S., and Falcinelli, M. (2001). Mode of reproduction is detected by Parth1 and Sex1 SCAR markers in a wide range of facultative apomictic Kentucky bluegrass varieties. *Molecular Breeding*, **7**:293–300.

Bantin, J., Matzk, F., and Dresselhaus, T. (2001). *Tripsacum dactyloides* (Poaceae): a natural model system to study parthenogenesis. *Sexual Plant Reproduction*, **14**:219–226.

Bashaw, E.C., and Hignight, K.W. (1990). Gene transfer in apomictic buffelgrass through fertilization of an unreduced egg. *Crop Science*, **30**:571–575.

Blakey, C.A., Coe, E.H.Jr., and Dewald, C.L. (1994). Current status of the *Tripsacum* dactyloides (Eastern gamagrass) RFLP molecular genetic map. *Maize Genetics Coop. Newsletter*, **68**,University of Missouri, Columbia, MO, pp. 35–37.

Blakey, C.A., Goldman, S.L., and Dewald, C.L. (2001). Apomixis in *Tripsacum*: Comparative mapping of a multigene phenomenon. *Genome*, **44**:222–230.

Borovsky, M. (1966). Apomixis in intergeneric maize-*Tripsacum* hybrids in *Meeting on problems on Apomixis in plants*. Saratov State University, Saratov, Russia, pp. 8–9.

Borovsky, M., and Kovarsky, A.E. (1967). Intergenus maize-*Tripsacum* hybridizations. *Izvestia Akademii Nauk Moldovaski, SSR*, **11**:25–35.

Burson, B.L., Voigt, P.W., Sherman, R.A., and Dewald, C.L. (1990). Apomixis and sexuality in eastern gamagrass. *Crop Science*, **30**:86–89.

Carmen, J.G. (1997). Asychronous expression of duplicate genes in angiosperms may cause apomixis, bispory, tetraspory and polyembryony. *Biological Journal of the Linnean Society*, **61**:51–94.

de Wet, J.M.J., Timothy, D.H., Hilu, K.W., and Fletcher, G.B. (1981). Systematics of South American *Tripsacum* (Gramineae). *American Journal of Botany*, **68**:269–276.

de Wet, J.M.J., Harlan, J.R., and Brink, D.E. (1982). Systematics of *Tripsacum dactyloides* (Gramineae). *American Journal of Botany*, **69**:1251–1257.

Galinat, W.C. (1973). Intergenomic mapping of maize, teosinte and Tripsacum. *Evolution*, **27**:644–655.

Grimanelli, D., Leblanc, O., Espinosa, E., Perotti, E., Gonzalez de Leon, D., and Savidan, Y. (1998). Mapping diplosporous apomixis in tetraploid *Tripsacum*: one gene or several genes? *Heredity*, **80**:33–39.

Harlan, J.R., and de Wet, J.M.J. (1977). Pathways of genetic transfer from *Tripsacum* to *Zea mays*. *Proceedings of the National Academy of Sciences USA*, **74**:3494–3497.

Kindiger, B. (1993). Aberrant microspore development in hybrids of maize × *Tripsacum* dactyloides. *Genome*, **36**:987–997.

Kindiger, B., and Beckett, J.B. (1990). Cytological evidence supporting a procedure for directing and enhancing pairing between maize and *Tripsacum*. *Genome*, **33**:495–500.

Kindiger, B., and Beckett, J.B. (1992). Popcorn germplasm as a parental source for maize × *Tripsacum dactyloides* hybridization. *Maydica*, **37**:245–249.

Kindiger, B., and Dewald, C.L. (1996). A system for genetic change in apomictic eastern gamagrass. *Crop Science*, **36**:250–255.

Kindiger, B., and Dewald, C.L. (1997). The reproductive versatility of eastern gamagrass. *Crop Science*, **37**:1351–1360.

Kindiger, B., and Sokolov, V. (1997). Progress in the development of apomictic maize. *Trends in Agronomy*, **1**:75–94.

Kindiger, B., Blakey, C.A., and Dewald, C.L. (1995). Sex reversal in maize × *Tripsacum* hybrids: Allelic non-complementation of *ts2* and *gsf1*. *Maydica*, **40**:187–190.

Kindiger, B., Bai, D., and Sokolov, V. (1996a). Assignment of gene(s) conferring apomixis in *Tripsacum* to a chromosome arm: cytological and molecular evidence. *Genome*, **39**:1133–1141.

Kindiger, B., Sokolov, V., and Dewald, C.L. (1996b). A comparison of apomictic reproduction in eastern gamagrass (*Tripsacum dactyloides* (L.) and maize-*Tripsacum* hybrids. *Genetica*, **97**:103–110.

Leblanc, O., Griminelli, D., Gonzalez-de-Leon, D., and Savidan, Y. (1995). Detection of the apomictic mode of reproduction in maize-*Tripsacum* hybrids using maize RFLP markers. *Theoretical and Applied Genetics*, **90**:1198–1203.

Leblanc, O., Griminelli, D., Islan-Faridi, N., Berthaud, J., and Savidan, Y. (1996). Reproductive behavior in maize-*Tripsacum* polyhaploid plants: Implications for the transfer of apomixis into maize. *Journal of Heredity*, **87**:108–111.

Li, D., Blakey, C.A., Dewald, C.L., and Dellaporta, S.L. (1997). Evidence for a common sex determination mechanism for pistil abortion in maize and its wild relative *Tripsacum*. *Proceedings of the National Academy of Sciences USA*, **94**:4217–4222.

Maguire, M. (1957). Cytogenetic studies of a *Zea* hyperploid for a chromosome derived from *Tripsacum*. *Genetics*, **42**:474–486.

Maguire, M. (1960). A study of homology between a terminal portion of *Zea* chromosome 2 and a segment derived from *Tripsacum*. *Genetics*, **45**:195–209.

Maguire, M. (1962). Common loci in corn and *Tripsacum*. *Journal of Heredity*, **53**:87–88.

Mangelsdorf, P.C., and Reeves, R.G. (1931). Hybridization of maize and Tripsacum and Euchlaena. *Journal of Heredity*, **22**:329–343.

Mangelsdorf, P.C., and Reeves, R.G. (1939). The origin of Indian corn and its relatives. Texas Agric. Exp. Stn. Bull. No. 574.

Noyes, R.D., and Rieseberg, L.H. (2000). Two independent loci control agamospermy (apomixis) in the triploid flowering plant *Erigeron annuus*. *Genetics*, **155**:379–390.

Petrov, D.F., Belousova, N.I., and Fokina, E.S. (1979). Inheritance of apomixis and its elements in maize × *Tripsacum dactyloides* hybrids. *Genetika*, **15**:1827–1836.

Petrov, D.F., Belousova, N.I., Fokina, E.S., Laikova, L.I., Yatsenko, R.M., and Sorokina, T.P. (1984). Transfer of some elements of apomixis from *Tripsacum* to maize, in *Apomixis and its role in evolution and breeding* (ed. D.F. Petrov). Oxonian Press Ltd., New Delhi, India, pp. 9–73.

Poggio, L., Confalonieri, V., Comas, C., Cuadrado, A., Jouve, N., and Naranjo, C.A. (1999). Genomic in situ hybridization (GISH) of *Tripsacum dactyloides* and *Zea mays* ssp. mays with B chromosomes. *Genome*, **42**:687–691.

Xu, S.J., and Joppa, L.R. (1995). Mechanisms and inheritance of first division restitution in hybrids of wheat, rye, and *Aegilops squarrosa*. *Genome*, **38**:607–615.

Yudin, B.F., and Sokolov, V.A. (1989). Towards regular apomixis in maize achieved by experiment. *Genetic Manipulation in Plants*, **5**:36–40.

Key references and suggested reading

Jain, M., Chengalrayan, K., Gallo-Meacher, M., and Misley, P. (2005). Embryogenic callus induction and regeneration in a pentaploid hybrid bermudagrass. cv Tifton 85. *Crop Science*, **45**:1069–1072.

Scityavathi, V.V., Janhar, P.P., Elias, E.M., and Rao, M.B. (2004). Effects of growth regulators on *in vitro* plant regeneration in Durum wheat. *Crop Science*, **44**:1839–1846.

Tae-Seok, K.O., Nelson, R.L., and Korban, S.S. (2004). Screening multiple soybean cultivars (MG 00 to MG VIII) for embryogenesis following *Agrobacterium*-mediated transformation of immature cotyledons. *Crop Science*, **44**:1825–1831.

Mohammadi, S.A., Prasanna, B.M., and Singh, N.N. (2003). Sequential path model for determining interrelationship among grain yield and related characters in maize. *Crop Science*, **43**:1690–1697.

Trigiano, R.N., and Gray, D.J.,(eds) (1996). Plant tissue culture concepts and laboratory exercises. CRC Press, New York, NY.

Stefaniak, B. (1994). Somatic embryogenesis and plant regeneration of gladiolus. *Plant Cell Reports*, **13**:386–389.

Murashige, T., and Skoog, T. (1962). A revised medium for rapid growth and bioassays with tobacco tissue culture. *Physiologia Plantarum*, **15**:473–497.

Kamo, K. (1995). A cultivar comparison of plant regeneration from suspension cells, callus, and cormel slices of *Gladiolus. In Vitro Cellular and Developmental Biology*, **31**:113–115.

Kindinger, B., Bai, D., and Sokolov, V. (1996). Assignment of a gene(s) conferring apomixis in Tripsacum to a chromosome arm: cytological and molecular evidence. *Genome*, **39**:1133–1141.

Chaudhury, A.M., Ming, L., Miller, C., Craig, S., Dennis, E.S., and Peacock, W.J. (1997). Fertilization-independent seed development in Arabidopsis thaliana. *Proceeding of the National Academy of Sciences USA*, **94**:4223–4228.

Koltunow, A.M., Bicknell, R.A., and Chaudhury, A.-M. (1995). Apomixis: Molecular strategies for the generation of genetically identical seeds without fertilization. *Plant Physiology*, **108**:1345–1352.

Hanna, W.W., and Bashaw, E.C. 1987. Apomixis: Its identification and use in plant breeding. *Crop Science*, 27:1136–1139.

Internet resources

http://aggie-horticulture.tamu.edu/tisscult/microprop/microprop.html – Links to numerous aspects of plant micropropagation (accessed March 3, 2012).

http://www.sprrs.usda.gov/apomixis.htm – Comments from foremost scientists in field of apomixis (accessed March 3, 2012).

Outcomes assessment

Part A

Please answer the following questions true or false.

1 MS medium was developed by Morris and Stevenson.
2 Agar is used as a gelling agent in tissue culture.
3 A protoplast is all the cellular component of a cell plus the cell wall.
4 An auxin-cytokinin ratio in favor of auxin promotes rooting.
5 Propagation by cuttings is a form of clonal propagation.
6 Diplospory is the most common mechanism of apomixis in higher plants.
7 Facultative apomicts reproduce exclusively by apomixis.
8 Pathogenesis is equivalent to haploidy.

Part B

Please answer the following questions.

1 What is the part of the plant that is used to start tissue culture called?
2 After who is the MS tissue culture medium named?
3 What is a mass of undifferentiated cells such as meristematic cells called?
4 What is clonal propagation?
5 What is apomixis?
6 Distinguish between apospory and displospory as mechanisms of apomixis.
7 What are species that reproduce exclusive (or nearly so) by apomixis described as?
8 Give a specific advantage of clonal propagation.

Part C

Please write a brief essay on each of the following topics.

1 Why is it possible (at least theoretically) to raise a full plant or for that matter any organism from just one of its own cells?
2 All cells are not totipotent. Explain.
3 What it the importance of a callus phase in plant tissue culture research?
4 Discuss the rationale for the composition of a tissue culture medium.
5 Describe the *in vitro* production of hybrids.
6 Discuss practical applications of tissue culture in plant breeding.
7 Discuss the key breeding implications of clonal propagation.
8 Discuss the benefits of apomictic cultivars in crop production.
9 Apomixis can be a two-edged sword in plant breeding. Explain.
10 Discuss the occurrence of apomixis in nature.

Section 4

Germplasm for breeding

Chapter 9 Variation: types, origin and scale
Chapter 10 Plant domestication
Chapter 11 Plant genetic resources

Germplasm or plant variability is indispensable to crop improvement. Plant breeders assemble germplasm for their work from various sources. To facilitate their efforts, governments usually collect and maintain large numbers of germplasm in repositories called germplasm banks. Some of this variability originates directly from the wild and possesses many undesirable traits; some of the variability arises as germplasm has undergone modifications by humans to adapt them to modern crop production systems. This section is devoted to discussing the types, sources and maintenance of germplasm for crop improvement.

9

Variation: types, origin and scale

Purpose and expected outcomes

Biological variation is a fact of nature. No two plants are exactly alike. Plant breeders routinely deal with variability in one shape or form. It is indispensable to plant breeding, and hence breeders assemble or create it as a critical first step in a breeding program. Then, they have to discriminate among the variability, evaluate and compare superior genotypes, and increase and distribute the most desirable to consumers. After completing this chapter, the student should be able to:

1 Discuss the types of variation.
2 Discuss the origins of genetic variation.
3 Discuss the scale of genetic variation.
4 Distinguish between qualitative and quantitative variation.
5 Discuss the rules of plant classification.

9.1 Classifying plants

Plant taxonomy is the science of classifying and naming plants. Organisms are classified into five major groups (kingdoms) – Plantae, Animalia, Fungi, Protista, and Monera (Table 9.1). Plant breeders are most directly concerned about Kingdom Plantae, the plant kingdom. However, one of the major objectives of plant breeding is to breed for resistance of the host to diseases and economic destruction caused by organisms in the other four kingdoms that adversely impact plants. Plant breeding depends on plant variation or diversity for success. It is critical that the appropriate plant material is acquired for a breeding program. An international scientific body sets the rules for naming plants. Standardizing the naming of plants eliminates the confusion from the numerous culture-based names of plants. For example, corn in the United States is called maize in Europe, not to mention the thousands of other names worldwide.

The **binomial nomenclature** was developed by Carolus Linnaeus and entails assigning two names

---

Principles of Plant Genetics and Breeding, Second Edition. George Acquaah.
© 2012 John Wiley & Sons, Ltd. Published 2012 by John Wiley & Sons, Ltd.

Table 9.1 The five kingdoms of organisms as described by Whitaker.

1. **Monera** (have prokaryotic cells)
   Bacteria

2. **Protista** (have eukaryotic cells)
   Algae
   Slime molds
   Flagellate fungi
   Protozoa
   Sponges

3. **Fungi** (absorb food in solution)
   True fungi

4. **Plantae** (produce own food by the process of
   photosynthesis)
   Bryophytes
   Vascular plants

5. **Animalia** (ingest their food)
   Multicellular animals

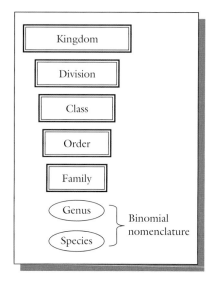

Figure 9.1 Taxonomic hierarchy of plants. Plant breeders routinely cross plants without problem within a species. Crosses between species are problematic and often impossible between genera and beyond.

based on the genus and species, the two bottom taxa in taxonomic hierarchy (Figure 9.1). It is important for the reader to understand that plant breeding by conventional tools alone is possible primarily at the species level. Crosses are possible within species and occasionally between species (but often problematic). However, plant breeding incorporating molecular tools allows gene transfer from any taxonomic level to another. It is important to emphasize that such a transfer is not routine and has its challenges.

The kingdom Plantae comprises **vascular plants** (plants that contain conducting vessels – xylem and phloem) and **non-vascular plants** (Table 9.2) Vascular plants may be seeded or seedless. Furthermore,

Table 9.2 Divisions in the kingdom Plantae.

	Division	**Common Name**
Bryophytes	Hepaticophyta	Liverworts
	Anthocerotophyta	Hornworts
	Bryophyta	Mosses
Vascular plants		
Seedless	Psilotophyta	Whisk ferns
	Lycophyta	Club mosses
	Sphenophyta	Horsetails
	Pterophyta	Ferns
Seeded	Pinophyta	Gymnosperms
	Subdivision: Cycadicae	Cycads
	Subdividion: Pinicae	
	Class: Ginkgoatae	*Ginko*
	Class: Pinatae	Conifers
	Subdivision: Gneticae	*Gnetum*
	Magnoliophyta	**Flowering plants**
	Class: Liliopsida	Monocots
	Class: Magnoliopsida	Dicots

Table 9.3 Important field crop families in the division Magnoliophyta (flowering plants).

Monocots

1 **Poaceae** (*grass family*)
In terms of numbers, the grass family is the largest of flowering plants. It is also the most widely distributed.
Examples of species: wheat, barley, oats, rice, corn, fescues, and bluegrass.

2 **Aracaceae** (*palm family*)
The palm family is tropical and subtropical in adaptation.
Examples of species: oil palm (*Elaeis guineensis*), coconut palm (*Cocos nucifera*).

3 **Amaryllidaceae** (*amaryllis family*)
Plants with tunicate bulbs characterize this family.
Examples of species: onion, garlic, and chives.

Dicots

1 **Brassicaceae** (*mustard family*)
The mustard family is noted for its pungent herbs.
Examples of species: cabbage, radish, cauliflower, turnip and broccoli.

2 **Fabaceae** (*legume family*)
The legume family is characterized by flowers that may be regular or irregular. The species in this family are an important source of protein for humans and livestock.
Examples of species: dry beans, mung bean, cowpea, pea, peanut, soybean, and clover.

3 **Solanaceae** (*night shade family*)
This family is noted for the poisonous alkaloids many of them produce (e.g. belladonna, nicotine, atropine, and solanine).
Examples of species: tobacco, potato, tomato, pepper, and eggplant.

4 **Euphorbiaceae** (*spurge family*)
Members of the spurge family produce milky latex and include a number of poisonous species.
Examples of species: cassava (*Manihot esculenta*) and castor bean.

5 **Asteraceae** (*sunflower family*)
The sunflower family has the second largest number of flowering plant species.
Example of species: sunflower and lettuce.

6 **Apiaceae** (*carrot family*)
Plants in this family usually produce flowers that are arranged in umbels.
Examples of species: carrot, parsley, and celery.

7 **Cucurbitaceae** (*pumpkin family*)
The pumpkin or gourd family is characterized by prostrate or climbing herbaceous vines with tendrils and large, fleshy fruits containing numerous seeds.
Examples of species: pumpkin, melon, watermelon, and cucumber.

seeded plants may be **gymnosperms** (have naked seed) or **angiosperms** (have seed borne in a fruit). Flowering plants may have seed with one cotyledon, called **monocots** (these include grasses such as wheat, barley and rice), or seed with two cotyledons, called **dicots** (these include legumes such as soybean, pea, and peanut) (Table 9.3). The strategies for breeding flowering species are different from those for non-flowering species. Flowering species can be genetically manipulated through the sexual process (sexually reproducing) by crossing, whereas non-flowering species (asexually reproducing) cannot. Furthermore, even within flowering plants, the method for breeding differs according to the mode of pollination – self-pollination or cross-pollination.

9.2 Rules of classification of plants

The science of plant taxonomy is coordinated by the **International Board of Plant Nomenclature**, which makes the rules. The Latin language is used in naming plants. Sometimes, the names given reflect specific plant attributes or use of the plant. For example, some specific epithets indicate color, e.g. *alba* (white), *variegata* (variegated), *rubrum* (red) and *aureum* (golden); others are *vulgaris* (common) *esculentus* (edible), *sativus* (cultivated), *tuberosum* (tuber bearing), and *officinalis* (medicinal). The ending of a name is often characteristic of the taxon. Class names often end in **opsida** (e.g. Magnoliopsida), orders in **ales** (e.g. Rosales), and families in **aceae** (e.g. Rosaceae). There are certain specific ways of writing the binomial name that are strictly adhered to in scientific communication. These rules are:

- It must be underlined or written in italics (being non-English).
- The genus name must start with an upper case letter; the species name always starts with a lower case letter. The term "species" is both singular and plural, and may be shortened to sp. or spp.
- Frequently, the scientist who first named the plant adds his or her initial to the binary name. The letter 'L' indicates that Linnaeus first named the plant. If revised later, the person responsible is identified after the 'L', for example, *Glycine max* L. Merr (for Merrill).
- The generic name may be abbreviated and can also stand alone. However, the specific epithet cannot stand alone. Valid examples are *Zea mays, Zea, Z. mays,* but not *mays.*
- The cultivar or variety name may be included in the binomial name. For example, *Solanum lycopersicum* cv. "Big Red", or *S. lycopersicum* "Big Red". The cultivar (cv) name, however, is not written in italics.

9.3 Operational classification systems

Crop plants may be classified for specific purposes, for example, according to seasonal growth, kinds of stem, growth form, and economic part or agronomic use.

(a) **Seasonal growth cycle.** Plants may be classified according to the duration of their lifecycle (i.e. from seed, to seedling, to flowering, to fruiting, to death, and back to seed). On this basis, crop plants may be classified as **annual, biennial, perennial**, or **monocarp**, as previously discussed in Chapter 4.

(b) **Stem type.** Certain plants have non-woody stems, existing primarily in vegetative form (e.g., onion, corn, or sugar beet) and are called **herbs** (or herbaceous plants). **Shrubs** are plants with multiple stems that arise from the ground level (e.g., dogwood, azalea, kalmia) while **trees** (e.g. apple, citrus, palms) have one main trunk or central axis.

(c) **Common stem growth form.** Certain plants can stand upright without artificial support; others cannot. Based on this characteristic, plants may be classified into groups. The common groups are.
- **Erect.** Erect plants can stand upright without physical support, growing at about a 90° angle to the ground. This feature is needed for mechanization of certain crops during production. Plant breeders develop erect (bush) forms of non-erect (pole) cultivars for this purpose. There are both pole and bush cultivars of crops such as bean (*Phaseolus vulgaris* L.) in cultivation.
- **Decumbent.** Plants with decumbent stem growth form, such as peanuts (*Arachis hypogea*), are extremely inclined with raised tips.
- **Creeping (or repent).** Plants in this category, such as strawberry (*Fragaria* spp.), have stems that grow horizontally on the ground.
- **Climbing.** Climbers are plants with modified vegetative parts (stems or leaves) that enable them to wrap around a nearby physical support, so they do not have to creep on the ground. Examples are yam (*Dioscorea* spp.) and ivy.
- **Despitose (bunch or tufted).** Grass species, such as buffalo grass, have a creeping form whereas others, such as tall fescue, have a bunch from and hence do not spread by horizontal growing stems.

(d) **Agronomic use.** Crop plants may be classified according to agronomic use as follows:
- **Cereals** – These are grasses such as wheat, barley, and oats that are grown for their edible seed.
- **Pulses** – Legumes grown for their edible seed (e.g. peas, beans).
- **Grains** – Crop plants grown for their edible dry seed (e.g. corn, soybean, cereals).
- **Small grains** – Grain crops with small seed (e.g. wheat, oats, barley).
- **Forage** – Plants grown for their vegetative matter that is harvested and used fresh or preserved as animal feed (e.g. alfalfa, red clover).
- **Roots** – Crops grown for their edible modified (swollen) roots (e.g. sweet potato, cassava).
- **Tubers** – Crops grown for their edible modified (swollen) stem (e.g. Irish potato, yam).
- **Oil crops** – Plants grown for their oil content (e.g. soybean, peanut, sunflower, oil palm).

- **Fiber crops** – Crop plant grown for use in fiber production (e.g. jute, flax, cotton).
- **Sugar crops** – Crops grown for use in making sugar (e.g. sugar cane, sugar beet).
- **Green manure crops** – Crop plant grown and plowed under the soil while still young and green, for the purpose of improving soil fertility (e.g. many leguminous species).
- **Cover crops** – Crops grown between regular cropping cycles, for the purpose of protecting the soil from erosion and other adverse weather factors (e.g. many annuals).
- **Hay** – Grasses or legume plants that are grown, harvested and cured for feeding animals (e.g. alfalfa, buffalo grass).

There are other operational classifications used by plant scientists.

(e) **Adaptation.** Plants may be classified on the basis of temperature adaptation as either cool season or warm season plants.

- **Cool season or temperate plants.** These plants, such as wheat, sugar beet, and tall fescue prefer a monthly temperature of between 15 and 18 °C (59–64°F) for growth and development.
- **Warm season or tropical plants.** These plants, such as corn, sorghum, and buffalo grass require warm temperatures of between 18 and 27 °C (64–80°F) during the growing season.

Whereas some of the operational classifications are applicable, horticultural plants have additional classification systems. These include the following:

- **Fruit type**
  1 **Temperate fruits** (e.g., apple, peach) versus **tropical fruits** (e.g., orange, coconut).
  2 **Fruit trees** – have fruits borne on trees (e.g., apple, pear).
  3 **Small fruits** – generally woody perennial dicots (e.g., strawberry, blackberry).
  4 **Bramble fruits** – non-tree fruits that need physical support (e.g., raspberry).
- **Flowering** (sunflower, pansy) versus **foliage** (non-flowering – coleus, sansevieria) plants.
- **Bedding plants** – annual plants grown in beds (e.g., zinnia, pansy, petunia).
- **Deciduous trees** (shed leaves seasonally) versus **evergreen plants** (no leaf shedding).

9.4 Types of variation among plants

As previously indicated, the phenotype (the observed trait) is the product of the genotype and the environment ($P = G + E$). The phenotype may be altered by altering G, E, or both. There are two fundamental sources of variation in phenotype: genotype and the environment – and hence two kinds of variation – **genetic** and **environmental**. Later in the book an additional source of variation, $G \times E$ (interaction of the genotype and the environment), will be introduced.

9.4.1 Environmental variation

When individuals from a clonal population (i.e., identical genotype) are grown in the field, the plants will exhibit differences in the expression of some traits because of non-uniform environment. The field is often heterogeneous with respect to plant growth factors – nutrients, moisture, light, and temperature. Some fields are more heterogeneous than others. Sometimes, non-growth factors may occur in the environment and impose different intensities of environmental stress on plants. For example, disease and pest agents may not uniformly infect plants in the field. Similarly, plants that occur in more favorable parts of the field or are impacted to a lesser degree by an adverse environmental factor would perform better than disadvantaged plants. That is, even clones may perform differently under different environments and inferior genotypes can outperform superior genotypes under uneven environmental conditions. If a breeder selects an inferior genotype by mistake, the progress of the breeding program will be slowed. Consequently, plant breeders use statistical tools and other selection aids to help in reducing the chance of advancing inferior genotypes, and thereby make rapid progress in the breeding program.

As previously noted, environmental variation is not heritable. However, it can impact heritable variation. Plant breeders want to be able to select a plant on the basis of its nature (genetics) not nurture (growth environment). To this end, evaluations of breeding material are conducted in a uniform environment as much as possible. Furthermore, the selection environment is often similar to the one in which the crop is commercially produced.

9.4.2 Genetic variability

Variability that can be attributed to genes that encode specific traits and can be transmitted from one generation to the next is described as **genetic** or **heritable variation**. Because genes are expressed in an environment, the degree of expression of a heritable trait is impacted by its environment, some more so than others (Figure 4.1). Heritable variability is indispensable to plant breeding. As previously noted, breeders seek to change the phenotype (trait) permanently and heritably by changing the genotype (genes) that encode it. Heritable variability is consistently expressed generation after generation. For example, a purple-flowered genotype will always produce purple flowers. However, a mutation can permanently alter an original expression. For example, a purple-flowered plant may be altered by mutation to become a white-flowered plant.

Genetic variation can be detected at the molecular as well as the gross morphological level. The availability of biotechnological tools (e.g., DNA markers) allows plant breeders to assess genetic diversity of their materials at the molecular level. Some genetic variation is manifested as visible variation in morphological traits (e.g., height, color, size), while compositional or chemical traits (e.g., protein content, sugar content of a plant part) require various tests or devices for evaluating them.

9.5 Origins of genetic variability

There are three ways in which genetic or heritable variability originates in nature – gene recombination, modifications in chromosome number, and mutations. The significant fact to note is that, rather than wait for them to occur naturally, plant breeders use a variety of techniques and methods to manipulate and make these three phenomena more and more targeted, as they generate genetic variation for their breeding programs. With advances in science and technology (e.g., gene transfer, somaclonal variation), new sources of genetic variability have become available to the plant breeder. Variability generated from these sources is, however, so far limited.

9.5.1 Genetic recombination

Genetic recombination applies only to sexually reproducing species and represents the primary source of variability for plant breeders in those species. As previously described, genetic recombination occurs via the cellular process of meiosis. This phenomenon is responsible for the creation of non-parental types in the progeny of a cross, through the physical exchange of parts of homologous chromosomes (by breakage-fusion). The cytological evidence of this event is the characteristic crossing (X-configuration or **chiasma**) of the adjacent homologous chromosome strands, as described in Chapter 3, allowing genes that were transmitted together (non-independent assortment) in the previous generation to become independent. Consequently, sexual reproduction brings about gene reshuffling and generation of new genetic combinations (recombinants). Unlike mutations that cause changes in genes themselves in order to generation variability, recombination generation variability by assembling new combinations of genes from different parents. In doing this, some gene associations are broken.

Consider a cross between two parents of contrasting genotypes $AAbb$ and $aaBB$. A cross between them will produce an F_1 of genotype $AaBb$. In the F_2 segregating population, and according to Mendel's law, the gametes (AB, Ab, aB, and ab) will combine to generate variability, some of which will be old (like the parents – parental), while others will be new (unlike the parents – recombinants) (Figure 9.2). The

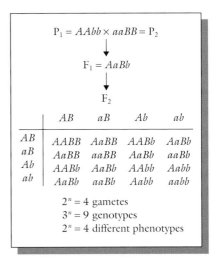

Figure 9.2 Genetic recombination results in the production of recombinants in the segregating population. This phenomenon is a primary source of variability in breeding flowering species. The larger the number of genes (n) the greater the amount of variability that can be generated from crossing.

larger the number of pairs of allelic genes by which the parents differ, the greater the new variability that will be generated. Representing the number of different allelic pairs by n, the number of gametes produced is 2^n, and the number of genotypes produced in the F_2 following random mating is 3^n with 2^n phenotypes (assuming complete dominance). In this example, two new homozygous genotypes (*aabb, AABB*) are obtained.

It should be pointed out that recombination only includes genes that are already present in the parents. Consequently, if there is no genetic linkage, the new gene recombination can be predicted. Where linkage is present, knowledge of the distance between gene loci on the chromosomes is needed for estimating their frequencies. As previously discussed in Chapter 3, additional variability for recombination may be observed where intra-allelic and inter-allelic interactions (epistasis) occur. This phenomenon results in new traits which were not found in the parents. Another source of genetic variability is the phenomenon of **gene transgression**, which causes some individuals in a segregating population from a cross to express the trait of interest outside the boundaries of the parents (e.g., taller than the taller parent, or shorter than the shorter parent). These new genotypes are called **transgressive segregants**. The discussion so far has assumed diploidy in the parents. However, in species of higher ploidy levels (e.g., tetraploid, hexaploid), it is not difficult to see how additional genetic variability could result where allelic interactions occur.

One of the tools of plant breeding is hybridization (crossing of divergent parents), whereby breeders selectively mate plants to allow their genomes to be reshuffled into new combinations to generate variability in which selection can be practiced. By carefully selecting the parents to be mated, the breeder has some control over the nature of the genetic variability to be generated. Breeding methods that include repeated hybridization (e.g., reciprocal selection, recurrent selection) offer more opportunities for recombination to occur.

The speed and efficiency with which a breeder can identify (by selecting among hybrids and their progeny) desirable combinations, is contingent upon the number of genes and linkage relationships that are involved. Because linkage is likely to exist, the plant breeder is more likely to make rapid progress with recombination by selecting plant genotypes with high chiasma frequency (albeit unconsciously). It follows then that the cultivar developed with the desired recombination would also have higher chiasma frequency than the parents used in the breeding program.

9.5.2 Ploidy modifications

New variability may arise naturally through modifications in chromosome number as a result of hybridization (between unidentical genotypes), or abnormalities in the nuclear division processes (spindle malfunction). Failure of the spindle mechanism, during karyokinesis or even prior to that, can lead to errors in chromosome numbers transmitted to cells, such as **polyploidy** (individuals with multiples of the basic set of chromosomes for the species in their cells) (Figure 9.3). Sometimes, instead of variations involving complete sets of chromosomes, plants may be produced with multiples of only certain chromosomes or deficiencies of others (called **aneuploidy**). Sometimes, plants are produced with half the number of chromosomes in the somatic cells (called haploids). Like genetic recombination, plant breeders are able to induce various kinds of chromosome modification

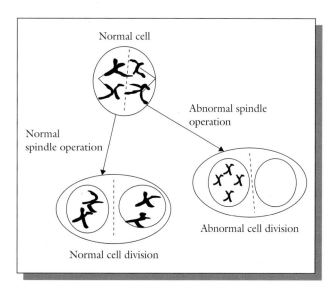

Figure 9.3 Failure of the genetic spindle mechanism may occur naturally or be artificially induced by plant breeders (using colchicine), resulting in cell division products that inherit abnormal chromosomes numbers. Plant breeders deliberately manipulate the ploidy of cells to create polyploids.

to generate variabiliy for breeding. The subject is discussed in detail in Chapter 24.

9.5.3 Mutation

Mutation is the ultimate source of biological variation. Mutations are important in biological evolution as sources of heritable variation. They arise spontaneously in nature as a result of errors in cellular processes such as DNA replication (or duplication) and by **chromosomal aberrations** (deletion, duplication, inversion, translocation). The molecular basis of mutation may be described by mechanisms such as modification of the structure of DNA or a component base of DNA, substitution of one base for a different base, deletion or addition of one base in one DNA strand, deletion or addition in one or more base pairs in both DNA strands, and inversion of a sequence of nucleotide base pairs within the DNA molecule. These mechanisms are discussed further in Chapter 23 on mutation breeding.

Mutations may also be induced by plant breeders using agents such as irradiation and chemicals. Many useful mutations have been found in nature or induced by plant breeders (e.g., dwarfs, nutritional quality genes). However, many mutations are deleterious to their carriers and are hence selected against in nature or by plant breeding. From the point of view of the breeder, mutations may be useful, deleterious, or neutral. Neutral mutations are neither advantageous nor disadvantageous to individuals in which they occur. They persist in the population in the heterozygous state as recessive alleles and become expressed only when in the homozygous state, following an event such as selfing.

9.5.4 Transposable elements

The phenomenon of **transposable elements** (genes with the capacity to relocate within the genome), creates new variability. **Transposable genetic elements** (transposable elements, transposons, or "jumping

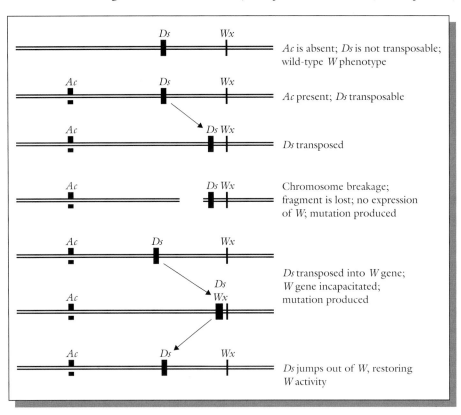

Figure 9.4 In the *Ac–Dc (activator–dissociation)* system of transposable elements in maize, the transposition of the *Ds* to *W* causes chromosome breakage, leading to the production of a mutant. In another scenario, the *Ds* is transposed into the *W*, causing a mutant to be produced.

genes") are known to be nearly universal in occurrence. These mobile genetic units relocate within the genome by the process called **transposition**. The presence of transposable elements indicates that genetic information is not fixed within the genome of an organism. Barbara McClintock, working with corn in the 1940s, was the first to detect transposable elements, which she initially identified as **controlling elements**. This discovery was about 20 years ahead of the discovery of transposable elements in prokaryotes. Controlling elements may be grouped into families. The members of each family may be divided into two classes: autonomous elements or non-autonomous elements. Autonomous elements have the ability to transpose whereas the non-autonomous elements are stable (but can transpose with the aid of an autonomous element through *trans*-activation).

McClintock studied two mutations: **dissociation** (*Ds*) and **activator** (*Ac*). The *Ds* element is located on chromosome 9. *Ac* is capable of autonomous movement, but *Ds* moves only in the presence of *Ac*. *Ds* has the effect of causing chromosome breakage at a point on the chromosome adjacent to its location (Figure 9.4). The *Ac* element has an open reading frame. The activities of corn transposable elements are developmentally regulated. That is, the transposable elements transpose and promote genetic rearrangements at only certain specific times and frequencies during plant development. Transposition involving the *Ac*–*Ds* system is observed in corn as spots of colored aleurone. A gene required for the synthesis of anthocyanin pigment is inactivated in some cells whereas other cells have normal genes, resulting in spots of pigment in the kernel (genetic mosaicism).

9.6 Biotechnology for creating genetic variability

9.6.1 Gene transfer

The **rDNA technology** is the state-of-the-art in gene transfer to generate genetic variability for plant breeding. With minor exceptions, the DNA is universal. Consequently, DNA from an animal may be transferred to a plant! The tools of biotechnology may be used to incorporate genes from distant sources into adapted cultivars. An increasing acreage of cotton, soybean, and maize are being sown to genetically modified (GM) cultivars, indicating the importance of this technology for creating variability for plant breeding. Economic gene transfers have been made from bacteria to plants to confer disease and herbicide resistance to plants. The most common GM products on the market are RoundupReady® cultivars (e.g., cotton, soybean) with herbicide tolerance, and *Bt* products (e.g., corn) with resistance to lepidopteran pests. The technique of **site-directed mutagenesis** allows scientists to introduce mutations into specified genes, primarily for the purpose of studying gene function, and not for generating variability for breeding *per se*. Other tissue culture based techniques include protoplast fusion, cybrid formation, and use of transposons. Chapter 25 is devoted to the application of biotechnology in plant breeding.

9.6.2 Somaclonal variation

In vitro culture of plants is supposed to produce clones (genetically identical derivatives) from the parent material. However, the tissue culture environment has been known to cause heritable variation called **somaclonal variation**. The causes cited for these changes include karyotypic changes, cryptic chromosomal rearrangements, somatic crossing over and sister chromatid exchange, transposable elements, and gene amplification. Some of these variations have been stable and fertile enough to be included in breeding programs.

9.7 Scale of variability

As previously indicated biological variation can be enormous and overwhelming to the user. Consequently, there is a need to classify it for effective and efficient use. Some variability can be readily categorized by counting and placing into distinct non-overlapping groups; this is said to be discrete or **qualitative variation**. Traits that exhibit this kind of variation are called **qualitative traits**. Other kinds of variability occur on a continuum and cannot be placed into discrete groups by counting. There are intermediates between the extreme expressions of such traits. They are best categorized by measuring or weighing and are described as exhibiting continuous or **quantitative variation**. Traits that exhibit this kind of variation are called **quantitative traits**.

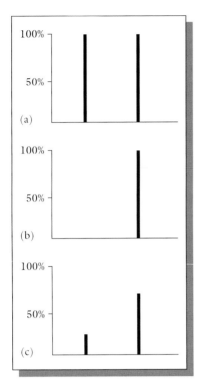

Figure 9.5 Qualitative variation produces discrete measurements that can be placed into distinct categories.

However, there are some plant traits that may be classified either way. Sometimes, for convenience, a quantitative trait may be classified as though it were qualitative. For example, an agronomic trait such as earliness of plant maturity is quantitative in nature. However, it is possible to categorize cultivars into maturity classes (e.g., in soybean, maturity classes range from 000 (very early) to VIII (very late)). Plant height can be treated in similar fashion, so too is seed coat color (that may be expressed as shades of a particular color).

9.7.1 Qualitative variation

Qualitative variation is easy to classify, study, and utilize in breeding. It is simply inherited (controlled by one or a few genes) and amenable to Mendelian analysis (Figure 9.5). Examples of qualitative traits include diseases, seed characteristics, and compositional traits. Because they are amenable to Mendelian analysis, the chi square statistical procedure may be used to determine the inheritance of qualitative genes. The success of gene transfer using molecular technology so far has involved the transfer of single genes (or a few at best), such as the *Bt* and *Ht* (herbicide tolerant) products.

Breeding qualitative traits

Breeding qualitative traits is relatively straight forward. They are readily identified and selected. Breeding recessive traits is a little different from breeding dominant traits (Figure 9.6). It is important to have a large segregating population, especially if several loci are segregating, to increase the chance of finding the

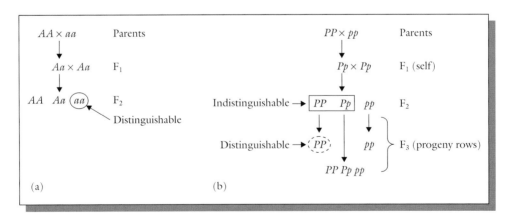

Figure 9.6 Breeding a qualitative trait conditioned by a recessive gene. The desired - aa genotype - can be observed and selected in the F_2. Breeding a qualitative trait conditioned by a dominant gene. The desired trait cannot be distinguished in the F_2, requiring another generation (progeny row) to distinguish between the dominant phenotypes.

desired homozygous recessive genotypes. For example, if two loci are segregating, a cross between $AA \times aa$ would produce 25% homozygous recessive individuals in the F_2 ($1AA$:$2Aa$:$1aa$). It is important to note that the desired genotype can be isolated from the F_2 without any further evaluation. In case of a dominant locus, (e.g., the cross $PP \times pp$), 25% of the F_2 will be homozygous recessive, whereas 75% would be of the heterozygous dominant phenotype (of which only 25% would be homozygous dominant). The breeder needs to advance the material one more generation to identify individuals that are homozygous dominant.

9.7.2 Quantitative variation

Most traits encountered in plant breeding are quantitatively inherited. A principal distinguishing feature of this variation is that the trait, whether controlled by few or many genes is influenced significantly by environmental variability, thus resulting in phenotypes that cannot be readily placed in distinct categories like the case for qualitative traits. Variation in quantitative trait expression is hence without natural discontinuities. Traits that exhibit continuous variations are also called **metric traits**. Any attempt to classify such traits into distinct groups is only arbitrary. For example,

height is a quantitative trait. If plants are grouped into tall versus short plants, one could find relatively tall plants in the short group and similarly short plants in the tall group (Figure 9.7).

Quantitative traits are conditioned by many to numerous genes (**polygenic inheritance**) with effects that are too small to be individually distinguished. They are sometimes called **minor genes**. Quantitative trait expression is very significantly modified by the variation in environmental factors to which plants in the population are subjected. Continuous variation is caused by environmental variation and genetic variation due to the simultaneous segregation of many genes affecting the trait. These effects convert the intrinsically discrete variation to a continuous one. Quantitative genetics is used to distinguish between the two factors that cause continuous variability to occur.

Breeding quantitative traits

Breeding quantitative traits is more challenging than breeding qualitative traits. A discussion of quantitative genetics will give the reader an appreciation for the nature of quantitative traits and a better understanding of their breeding. Quantitative genetics is discussed in Chapter 4.

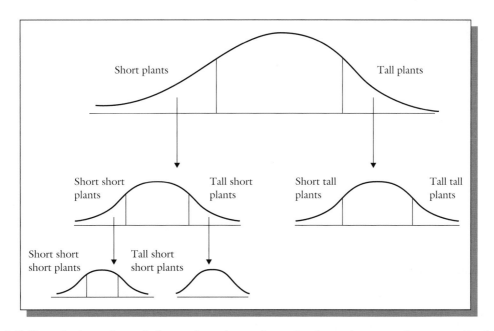

Figure 9.7 Quantitative traits are influenced to a larger degree by the environment than are qualitative traits.

Key references and suggested reading

Acquaah, G. (2004). *Horticulture: Principles and practices*, 3rd edn. Prentice Hall, Upper Saddle River, NJ.

Klug, W.S., and Cummings, M.R. (1997). *Concepts of genetics*, 5th edn. Prentice Hall, Upper Saddle River, NJ.

Falconer, D.S. (1981). *Introduction to quantitative genetics*, 2nd edn. Longman, London.

Outcomes assessment

Part A

Please answer the following questions true of false.

1 The bionomial nomenclature was discovered by Gregor Mendel.
2 Angiosperms have naked seed.
3 Environmental variation is heritable.
4 Qualitative traits exhibit continuous variation.

Part B

Please answer the following questions.

1 Define plant taxonomy. Why is it important to plant breeding?
2 . . . . is the international body responsible for coordinating plant taxonomy.
3 Phenotype = ... + environment.
4 What are transposable elements?

Part C

Please write a brief essay on each of the following topics.

1 Discuss genetic recombination as a source of variability for plant breeding.
2 Discuss the nature of qualitative variation.
3 Discuss the role of environmental variation in plant breeding.
4 Describe the role of biotechnology for creating variability for plant breeding.

10

Plant domestication

Purpose and expected outcomes

A long time ago, people lived as hunters and gatherers. Then, gradually, they began to produce the crops they ate on location and farming was born. Wild plants have certain traits that keep them protected and productive without human intervention. However, most of these traits are not desired for food or other use by humans. By spontaneous mutagenesis and hybridization, occasionally variant plants were found that lacked some of the undesired traits, such as bitter taste, thorns, or had particularly attractive features, such as larger fruits and more regular tuber shape. Consequently, with time, such variation was picked up by humans by selection of such variants, keeping and advancing them. After studying this chapter, the student should be able to:

1 Define domestication and discuss how it is similar to evolution.
2 Discuss the concept of the domestication syndrome.
3 Discuss the genetic basis of domestication.
4 Discuss the importance of the knowledge of crop origin to plant breeding.

10.1 The concept of evolution

Evolution is a population phenomenon. Populations, not individuals, evolve. Evolution is concerned with the effect of changes in the frequency of alleles within a gene pool of a population, such changes leading to changes in genetic diversity and the ability of the population to undergo evolutionary divergence. Simply stated, evolution is descent with modification. Proposed by Charles Darwin in 1859, there are certain key features of the concept or **theory of evolution**. Variation exists in the initial population of organisms, both plants and animals. As Darwin stated, variation is a feature of natural populations. More individuals are produced each generation than can be supported by or survive in the environment. Many individuals die before having participated in reproduction, or they produce less offspring than their peers. Part of these differences in mortality and reproduction are due to fortuitous events but mortality or low reproduction may also be the result of lower genetic adaptation of certain individuals to relevant environmental stresses that place them in the population at a disadvantage. The individuals with the best genetic fitness for the specific environment will survive and reproduce more successfully and become more competitive than other

---

Principles of Plant Genetics and Breeding, Second Edition. George Acquaah.
© 2012 John Wiley & Sons, Ltd. Published 2012 by John Wiley & Sons, Ltd.

individuals. The more competitive individuals will (on average) leave more offspring to participate in the next generation. Such a trend, in which the advantageous traits increase, will continue each generation, with the result that the population will be predominated by these favored individuals and is said to have evolved. The discriminating force, called **natural selection** by Darwin, is the final arbiter in deciding which individuals are advanced. When individuals in a resulting subpopulation become reproductively isolated from the original population or a sister population that developed in a different region or environment, new species will eventually form.

Patterns of such evolutionary changes have been unconsciously mimicked by plant breeders in the development of new cultivars. Plant breeders have selected against unfavorable plant types (bitter taste, small fruits, etc.) and favored others (sweet, nice color, regular shape, no seed shattering) to be represented in next plant generations. This way, natural selection (like for example frost tolerance) was complemented with artificial (man-guided) selection resulting in plant populations that better and better fitted the needs of man.

The process of evolution has parallels in plant breeding. Darwin's theory of evolution through natural selection can be summed up in three principles that are at the core of plant breeding. These are the principles of:

(i) **Variation**—variation in morphology, physiology, and behavior exist among individuals in a natural population.
(ii) **Heredity**—offspring resemble their parents more than they resemble unrelated individuals.
(iii) **Selection**—some individuals in a group are more capable of surviving and reproducing than others (i.e., more fit).

A key factor in evolution is time. The changes in evolution occur over *extremely long* periods.

Plant breeders depend on biological variation as a source of desired alleles. The first breeders were farmers that grew genetically heterogeneous plant populations that were harvested for their desired products like tubers, roots, berries or seeds. This was the beginning of domestication.

Plant breeding may be described as directed or targeted and accelerated evolution because the plant breeder, with a breeding objective in mind, deliberately and genetically manipulates plants (wild or domesticated) to achieve a stated goal, but in a *very short* time. Conceptually, breeding and evolution are the same, a key difference being the duration of the processes. Plant breeding has been described as evolution directed by humans. Compared to evolution, a plant breeding process is completed in the twinkle of an eye! Also, unlike evolution, plant breeders do not deal with closed populations. They introgress new variability from different genotypes of interest and, for practical and economic purposes, deal with limited population sizes.

10.2 What is domestication

Domestication is the process by which genetic changes (or shifts) in wild plants are brought about through a selection process imposed by humans. Because of the role of humans, the process results in characteristics that are beneficial to humans but some that would be disadvantageous for plants in their natural habitats. It is an evolutionary process in which selection (both natural (frost tolerance) and artificial (large pretty fruits)) operates to change plants genetically, morphologically, and physiologically. The results of domestication are plants that are adapted to supervised cultural conditions and which possess characteristics that are preferred by producers and consumers. In some ways, a domesticated plant may be likened to a tamed wild animal that has become a pet.

There are degrees of domestication. Species that become completely domesticated often are unable to survive when re-introduced into the wild. This is because the selection process that drives domestication strips plants of some natural adaptive features and mechanisms that are critical for survival in the wild but undesirable according to the needs of humans. Modern corn, for example, is stripped completely of its seed dispersal ability.

Like evolution, domestication is also a process of genetic change in which a population of plants can experience a shift in its genetic structure in the direction of selection imposed by the domesticator. New plant types are continually selected for in the domesticates as new demands are imposed, thereby gradually moving the selected individuals farther away (genetically, morphologically, and physiologically) from their wild progenitors. Both wild and domesticated populations are subject to evolution.

10.3 Evolution versus domestication

It was previously indicated that plant breeding and evolution are similar processes. The mechanisms that govern them are essentially the same, except that the creation of variation and the selection among the variability are conducted by humans. Another key difference is in the duration of the process. Compared to evolution, plant breeding occurs in the twinkle of an eye. Similarly, evolution and domestication have some things in common. They both are about change. The mechanisms of change (e.g., mutation, selection, hybridization, polyploidization) are similar for both processes. The major difference rests with the objective, evolution bringing about change for better reproduction in the natural environment (wild), whereas domestication changes wild species to make them adapted to better reproduction in cultural environments because of being favored by humans (agricultural production). Evolution changes plants to become dependent on nature while domestication changes plants to become dependent on humans (driven by the needs and preferences of humans). Some modern crops are incapable of establishing a new generation without the intervention of humans. Corn is a good example. The husks so tightly enclose the mature seeds – the cob, that natural dispersal is precluded. Only by human intervention can the seeds get free and be sown to produce the next year's crop.

10.4 Conscious selection versus unconscious selection

Diversity may arise by natural causes or be created or induced by humans. Evolution is driven by natural selection, while modern plant breeding is driven by artificial selection. Selection by humans is described by some as "conscious". However, Charles Darwin recognized that human selection could have an unconscious component. According Darwin, human-mediated selection may either be methodical or conscious (e.g., sweet fruits, beautiful flowers, large fruits), pursued with a goal to modify a breed according to a predetermined or preconceived objective, or unconscious. Unconscious selection may include traits such as fast and early germination of seeds (no dormancy), early maturation (late maturing plants are ploughed under before they mature), self-compatibility (single plants or trees only produce

fruits and seeds when self-compatible). The mere action of humans to change the cultural conditions of cultivated species has the lingering selective effect on the population even in the absence of a pre-conceived goal by the cultivator.

Regarding the domestication syndrome, it is very likely that most of the composite traits, excepting obviously attractive traits like color, taste and fruit size, resulted from unconscious selection. Some traits would have been difficult for early cultivators to observe or deliberately alter. For example, seed dormancy would have been unnoticed and problematic for early cultivators and hence selected against, unconsciously, without an expressed goal to plant only non-dormant propagules. Seeds with long dormancy were plainly not participating in that year's crop, and hence got lost. Unconscious selection was not limited to only visible phenotypes. As early domesticators propagated their plants under various cultural conditions, it is likely that these environments caused some physiological or developmental changes to occur in the plants that had adaptive value.

Men have also probably "domesticated" in some way the weeds in the crop fields. For example, weedy grasses in rice fields tend to mimic rice in color and shape. Grasses with high anthocyanin (red color) run much more risk to be removed by hand-weeding than grasses that have the same green color and same shape as young rice plants. Such mimicry also occurs in seed shape and size. Weeds that produce seeds that look like the seed crop will not be separated easily and will be sown in the next season with the crop. This poses a man-derived selection on weed populations, without the latter becoming useful crops. This may be compared to man not only domesticating animals to become pets but also to having "invited" animals like house mice to become dependent on man, and so, in a way, to become domesticated.

10.5 Patterns of plant domestication

Domestication has been conducted for over 10 000 years, and ever since agriculture was invented. Archeological and historical records provide some indications as to the period certain crops may have been domesticated, even though such data are not precise. Archeological records from arid regions are better preserved than those from the humid regions of the world.

However, in most cereal species, most experts believe that domestication occurred after the start of some form of cultivation. In wheat and barley, for example, tough rachis, which hampers natural seed dispersal and characterizes domesticates, would have been selected for during domestication and not necessarily before.

Most of the domesticated crops arose from wild plant species, pioneers that are adapted to disturbed and nutrient rich (dung) places, called ruderal environments. Such wild plants probably grew abundantly around settlements of hunter/gatherers and might have been exploited spontaneously by mankind. Selection of the most tasteful or attractive type to be sown in next seasons has resulted in the first crop species. Secondary crops are those that evolved from weeds that arose in cultivated fields of primary crops. For example, wild oats are a common weed (still today!) in wheat and barley fields. Such weedy species were harvested with the intended crop but gradually took over as the most important product of such a field. They made it to become a crop on their own merit. This is because it is difficult to use a single characteristic to differentiate between wild and cultivated species of these horticultural plants.

10.6 Centers of plant domestication

Centers of plant domestication are of interest to researchers from different disciplines, including botany, genetics, archaeology, anthropology, and plant breeding. Plant breeders are interested in centers of plant domestication as regions of genetic diversity, variability being critical to the success of crop improvement. De Candolle was the first to suggest, in 1886, that a crop plant originates from the area where its wild progenitor occurs. He considered archeological evidence to be the direct proof of the ancient existence of a crop species in a geographic area. Of course, the most important centers of domestication are also the regions where farming started.

Several scientists, notably N. Vavilov of Russia and J.R. Harlan of the United States, have provided the two most enduring views of plant domestication. Vavilov, on his plant explorations around the world in the 1920s and 1930s, noticed that extensive genetic variability within a crop species occurred in clusters within small geographic regions separated by geographic features such as mountains, rivers, and deserts. For example, whereas he found different forms of diploid, tetraploid, and hexaploid species of wheat in the Middle East and the Mediterranean region, he observed that only hexaploid cultivars were grown in Europe and Asia. Vavilov proposed the concept of **centers of diversity** to summarize his observations. He defined the **center of origin of a crop plant** as the geographic area(s) where it exhibits maximum diversity of botanical varieties occur. Additional evidence is that the presumed ancestral species occurs in that region in the wild. Vavilov identified eight major centers of diversity. In the 1960s many more scientists came with modifications and proposed other and additional areas, some of which were subdivided (subcenters). Centers of domestication always seem to be the presumed regions of domestication of several crop species. So, the near-eastern region is considered the center of domestication of barley, wheat, cabbage, pea en many more crops. Such centers of domestication are almost always also centers of diversity, but the reverse is not true. Some centers of diversity do not coincide with the center of domestication. China, for example, is a (secondary) region of diversity for waxy corn, which was domesticated in Central America.

Important centers of diversity are shown in Table 10.1.

Table 10.1 gives mostly the areas where crops have been domesticated. These are called "primary centers of diversity" and usually show the highest diversity. However, the region where crops were domesticated is not always unequivocal. For some crops (like papaya, faba bean) it is not sure which species is the wild ancestor. For others, like carrot, the wild ancestor has a huge geographical range: Europe, North Africa and central Asia. Some crops may have been domesticated simultaneously in several regions. Also, very early trading brought early domesticates to several regions. Such crops have several regions of diversity or disputed regions of domestication. For example, the melon, *Cucumis melo*, is presumably of African origin, but the Near East and India are (secondary) centers of diversity. Banana may have originated in the Malay peninsula, but East Africa is a secondary center of diversity. The primary center of corn is Mexico, but China is a secondary center of waxy types of corn. Barley and durum wheat are very diverse in Ethiopia, but the wild ancestors do not occur there. Therefore, it is unlikely to be a center of origin of these two crops, but rather it is a secondary

Table 10.1 Important centers of diversity (Zeven and de Wet, 1982).

Region of diversity	Crops having their primary center of diversity in region
1 Chinese–Japanese	rhubarb, soybean, Chinese cabbage, Japonica rice, tea, millet
2 Indochinese–Indonesian	ginger, rice, *Piper*-pepper, sugarcane, taro, banana, wax gourd, cardamom, various citrus crops
3 Australian	macadamia nut, *Eucalyptus*, some tobacco species
4 Hindustani	mango, tree cotton, pigeon pea, okra, black (piper) pepper, eggplant, rice, cucumber
5 Central Asian	onion, garlic, pistachio, spinach, walnut, carrot, almond
6 Near Eastern	alfalfa, chickpea, pea, lentils, flax, cabbage, rye, barley, wheat, leek
7 Mediterranean	celery, oats, beet, kale, rape seed, radish, olive, poppy, grape
8 African	coffee, grain sorghum, pearl millet, watermelon, melon, yam, castor bean, cowpea, oil palm, date palm
9 European–Siberian	lettuce, forage grasses, *Ribes* berries, hemp, hops, beet, Brussels witloof (chicory), turnip, black-mustard, peppermint, apple, pear, strawberry
10 South American	cashew, peanut, rubber, passion fruit, cacao, tomato, tobacco, potato, Egyptian cotton
11 Central American and Mexican	potato, sweet potato, pumpkin, zucchini, Opuntia, papaya, maize, avocado, phaseolus bean, runner bean, upland cotton, vanilla, Capsicum pepper
12 North American	sunflower, pecan, strawberry

center of diversity. In the case of several primary and secondary centers of diversity, each center usually shows it range of morphotypes of the crops.

Vavilov made other observations from his plant explorations. He found that the maximum amount of variability and the maximum concentration of dominant genes for crops occurred at the center and decreased toward the periphery of the cluster of diversity. Also, he discovered there were parallelisms (common features) in variability among related species and genera. For example, the various cotton species,

Gossypium hirsutum and *G. barbadense*, have similar pubescence, fiber color, type of branching, color of stem, and other features. Vavilov called this the **law of homologous series** in heritable variation (or parallel variation). In other words, species and genera that are genetically closely related are usually characterized by a similar series of heritable variations, such that it is possible to predict what parallel forms would occur in one species or genera from observing the series of forms in another related species. The breeding implication is that if a desirable gene is found in one species, it likely would occur in another related species. For example, in small grains, tough rachis, naked seed, awnless trait, black or white seed types are found in wheat, barley, and oats. Through comparative genomic studies, mapping of such domestication-related trait genes has revealed significant homology regarding the chromosomal location genes among species of the Poaceae family (specifically, rice, corn, sorghum, barley, and wheat); this condition is called synteny, the existence of highly conserved genetic regions of the chromosome.

10.7 Roll call of domesticated plants

It is estimated that 230 plant species have been domesticated belonging to 180 genera and 64 families. Some families, such as Gramineae (Poaceae), Leguminoseae (Fabaceae), Cruciferae, and Solanaceae, have yielded more domesticates than others. Further, culture plays a role in the types of crops that are domesticated. For example, the major world tuber and root crops, Irish potato, sweet potato, yam, cassava, and aroids, have similar cultural uses or purposes but represent distinct taxonomic groups.

10.8 Changes accompanying domestication

Selection exerted by humans on crop plants during the domestication process causes changes in the plants as they transit from wild species to domesticates. The assortments of morphological and physiological traits that are modified in the process and differentiate between the two types of plants were collectively coined the **domestication syndrome** by J.R. Harlan. Although the exact composition of the domestication syndrome traits depends on the particular species, certain basic characteristics are common

Table 10.2 Characteristics of domestication syndrome traits.

General effect	Specific traits altered
1 Increased seedling vigor (more plants germinating)	- loss of seed or tuber dormancy - large seeds
2 Modified reproductive system	- increased selfing - vegetatively reproducing plants - altered photoperiod sensitivity
3 Increased number of seeds harvested	- non-shattering - reduced number of branches (more fruits per branch)
4 Increased appeal to consumers	- attractive fruit/seed colors and patterns - enhanced flavor, texture, and taste of seeds/fruits/tubers (food parts) - reduced toxic principles (safer food) - larger fruits - reduced spikiness
5 Altered plant architecture and growth habit	- compact growth habit (determinacy, reduced plant size, dwarfism) - reduced branching

(Table 10.2). At the time Harlan described this suite of traits in his publication, they represented traits that were closely associated with increasing ease of harvest (e.g., no spontaneous seed dispersal (pods remain closed, tough rachis in cereals), larger seeds or fruits, less branching plant habit because of strong apical dominance, loss of seed dormancy, determinate growth habit, larger inflorescence size). It also includes other characters, like lack of spines and thorns, greater tendency for autogamy, lack of bitter tasting compounds, lack of fibers (like in the pods of common bean). (Since then, the domestication syndrome appears to have been expanded to include any gene that represents a difference between a crop and its wild relative.) Plant improvement efforts after domestication have also contributed to fixing traits that might not have been of interest to early farmers and, hence, were not selected for during domestication. An example is the genetic trait of formation of single seeds rather than clusters of seed in sugar beet.

The wild-type clusters required each cluster of seedlings to be thinned to one by hand. The domestication traits that are relevant affect characters that can be selected at seedling stage, plant growth and reproduction and harvest stage.

At the seedling stage, the goal of domestication is to get more seeds to germinate. This entails a loss of seed dormancy as well as increased seedling vigor. At the reproductive stage, the goal of domestication includes a capacity for vegetative reproduction and increased selfing rate. Plant traits relevant at the harvest or after harvest stage include elimination of seed dispersal (no shattering), uniform seed maturity, more compact plant architecture, and modification in photoperiod sensitivity. Modifications targeted at the consumer preferences include fruit size, color, taste, and reduction in toxic substances.

The genetic control of the traits comprising the domestication syndrome has been studied in many crops. Generally, these traits are controlled by one or few qualitative genes or quantitative genes with major phenotypic effects. For example, lack of fibers in common bean, rough rachis in barley, lack of spines on seeds of spinach are all monogenically inherited traits. Quantitative trait loci (QTLs) research has indicated that a few loci control traits such as flowering time, seed size and seed dispersal in maize, rice, and sorghum; and growth habit, photoperiod sensitivity and dormancy in common bean. Furthermore, linkage blocks of adaptation traits have been found in some species. A study by E.M.K Koinange and collaborators indicated that the domestication syndrome genes in common bean were primarily clustered in three genomic locations, one for growth habit and flowering time, a second for seed dispersal and dormancy, and a third for pod and seed size.

The domestication process essentially makes plants more dependent on humans for survival. Consequently, a difference between domesticates and their wild progenitors is the lack of traits that ensure survival in the wild. Such traits include dehiscence, dormancy, and thorns. Plants that dehisce their pods can invade new areas for competitive advantage. However, in modern cultivation, dehiscence or shattering is undesirable because seeds are lost to harvesting when it occurs. Some directions in the changes in plant domesticates have been dictated by the preferences of consumers. Wild tomato produces numerous tiny and hard fruits that are advantageous in the wild for survival. However, consumers prefer succulent

Industry highlights

The use of the wild potato species, solanum etuberosum, in developing virus and insect resistant potato varieties

Richard Novy

USDA-Agricultural Research Service, Aberdeen, ID 83210, USA

Historical background of the potato

Often referred to as the "Irish" potato, *Solanum tuberosum* subsp. tuberosum might more aptly be termed the "Inca" potato. The origin of the fourth most-widely grown cultivated crop in the world is thought to be the Central Andes region of South America, with the possibility of independent domestication in Chile as well. Potato was an important food crop for the Incas but it is unlikely that they were the civilization responsible for its domestication. Potato food remnants have been found in pre-ceramic archeological sites in South America that date to over 5000 years ago, indicating that potato truly is an ancient food crop.

The Spanish and English are thought to have brought this "New World" crop back to Europe in the late sixteenth century. Adapted to form tubers under the short day conditions near the equator (approximately 12 hours), potato did not successfully produce tubers in most northern latitudes prior to being killed by freezing temperatures in the fall. The exceptions were the milder climates of Spain, Italy, southern France, and Ireland, where potato was maintained as a botanical oddity in private and botanical gardens. Over the course of 150–200 years, potato clones (with the help of man) that formed tubers under the longer days of the northern latitudes were identified and propagated. This environmental adaptation allowed for the expansion and adoption of potato as a food crop in Europe, and eventually throughout the world.

Viruses of potato

Accompanying the introduction of the potato to Europe were pathogens that had co-evolved with the crop, the most notable being potato viruses X and Y (PVX and PVY) and potato leafroll virus (PLRV). These viruses are transmitted from an infected plant to the tubers it produces. When virus-infected tubers are cut and used to establish the potato crop the following growing season (asexual propagation), plants developing from the tuber seed are infected with the virus as well. Symptoms of virus infection in potato include stunting, chlorosis/necrosis of leaf tissue, and, in the case of PLRV (as indicated by the name), rolling or cupping of leaves. Total yield in a growing season can be reduced by as much as 80% if virus-infected seed is used. Transmission of PVY and PLRV from infected to healthy plants is mediated by aphids, most notably green peach aphid; PVX is transmitted to healthy plants via mechanical contact with an infected plant or with PVX-contaminated field equipment. The detrimental impact of viruses on potato was termed "degeneration" or "running-out" by early growers of potato who did not yet know of the existence of viruses.

Potato varieties with resistances to viruses can be effective in reducing crop losses. Cultivated potato is fortunate in having >200 wild *Solanum* relatives, many of which have been identified as virus resistant. In the United States, potato species collected in Mexico, Central and South America are maintained at the Potato Genebank in Sturgeon Bay, Wisconsin. This species collection has been systematically screened for resistance to the major pests and diseases of potato. Many species have been identified with high levels of resistance to PVX and PVY, with a lesser number identified as having desirable levels of PLRV resistance.

Ideally, from the standpoint of a potato breeder, it is desirable to work with species that have a high level of resistance to all three viruses. A search of the Potato Genebank collection consisting of 5634 introductions representing 168 species identified only one introduction (PI 245939) of the wild potato species, *S. etuberosum*, as having a high level of resistance to PVY, PVX, and PLRV. This accession also was identified as having resistance to green peach aphid – a primary insect vector of PVY and PLRV. The introgression of these multiple virus and insect vector resistances from *S. etuberosum* into cultivated potato is the focus of the remainder of this section.

Solanum etuberosum: its use in the genetic improvement of potato

Solanum etuberosum is a wild potato species endemic to Chile. Its natural habitat is among rocks on slopes with seepage or along streams. It is generally found in the open, or in the shade of trees and shrubs (Correll, 1962). It is notable for its large, deep purple flowers (Figure B10.1). The attractiveness and abundance of its flowers and its striking foliage led a taxonomist in 1835 to propose that it be grown as a hardy perennial for ornamental purposes in England (Correll, 1962). It also is notable among wild potato species in that it does not form tubers.

Figure B10.1 Flower and foliage of *S. etuberosum*. Considered a weed in Chile, this wild relative of potato has desirable genetic resistances to viruses and insects that plague cultivated potato. Figure courtesy of Richard Novy.

A diploid (2n = 2x = 24) species, *S. etuberosum* does not cross readily with either tetraploid (2n = 4x = 48) or dihaploid (2n = 2x = 24) forms of cultivated potato. Bridging species and ploidy manipulations were used by breeders in synthesizing *etuberosum-tuberosum* hybrids with limited success.

Another means of circumventing reproductive barriers is to bypass them completely by the use of somatic hybridization. This technique involves the isolation of potato cells from leaf tissue of parental clones, the enzymatic digestion of their cell walls to form protoplasts, and the fusion (using chemicals or electrical currents) of parental protoplasts. Fused protoplasts are then cultured on medium whereby they re-form a cell wall and are allowed to divide to form undifferentiated tissue called callus. Calli are then placed on culture media that promotes cell differentiation and the formation of plants. These hybrid plants can then be excised from the calli, induced to form roots, and grown as a normal plant in field or greenhouse environments.

Using somatic hybridization, Novy and Helgeson 1994a successfully generated hybrids between a *S. etuberosum* clone from virus resistant PI 245939 and a subsp. tuberosum dihaploid x *S. berthaultii* hybrid clone (2n = 2x = 24). The tri-species hybrids, based on cytological and molecular analyses, were at or near the expected 2n = 4x = 48. Somatic hybrids had very vigorous foliar growth in the field with limited tuberization (Figure B10.2); poor tuber type and yield was not unexpected in that half the genome of the somatic hybrids was from non-tuber-bearing *S. etuberosum*.

Backcrossing of somatic hybrids to potato cultivars was undertaken to improve tuberization and yield. Crosses using somatic hybrids as the male parent yielded few berries and no seeds. Stylar analyses showed that blockage of somatic hybrid pollen tuber growth generally occurred in the upper one-third of Gp. Tuberosum styles. Pollen tube blockage of cultivated potato was not observed in the styles of somatic hybrids – five hundred and three pollinations produced 99 berries containing 24 seeds. Five of the seed germinated to produce viable BC$_1$ progeny that were at or near the tetraploid level (48–49 chromosomes).

The five progeny obtained had much improved tuberization relative to the somatic hybrid parent, while still retaining 11–13 *S. etuberosum* chromosomes (Figure B10.2). One of the five progeny produced an average of six seeds per berry when crossed to cultivated potato. Viable BC$_2$ progeny were obtained from this seed. Tubers of BC$_2$, now looking like those of cultivated potato, are shown in Figure B10.2.

Virus and green peach aphid resistances of somatic hybrids and their progeny

Novy and Helgeson (1994b) analyzed the fusion parents, their somatic hybrids, and the sexual progeny of the somatic hybrids for resistance to PVY following their mechanical inoculation in the greenhouse over a two-year period. The *S. etuberosum* fusion parent was highly resistant to PVY infection whereas the *tuberosum-berthaultii* fusion parent was highly susceptible. Three somatic hybrids analyzed in this study did not show the high level of resistance found in the *S. etuberosum* parent; however, they were significantly more resistant than the cultivars Katahdin (moderate field resistance to PVY) and Atlantic (PVY susceptible). Five progeny of the somatic hybrids were analyzed also in this study. Three displayed PVY resistance comparable to the somatic hybrid parents, while the remaining two were more susceptible with absorbance means comparable to the potato varieties included in the study.

S. etuberosum also had been identified as having resistance to potato leafroll virus (PLRV) and green peach aphid – a primary vector of PLRV and PVY. Resistance to green peach aphid (*Myzus persicae*) can aid in decreasing the transmission of viruses by decreasing aphid population size and subsequent opportunities for virus transmission. However, green peach aphid resistance alone is not adequate to confer the necessary level of resistance needed by the industry. This is especially true in the case of PVY, which can be quickly transmitted by the stylar probings of many different aphid

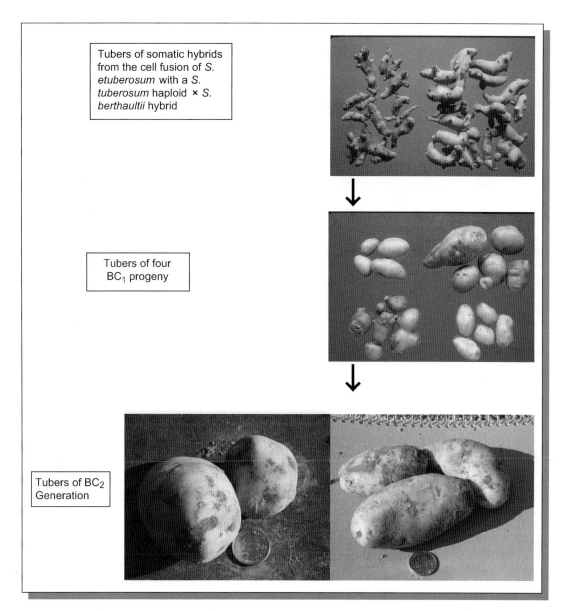

Tubers of somatic hybrids from the cell fusion of *S. etuberosum* with a *S. tuberosum* haploid × *S. berthaultii* hybrid

Tubers of four BC₁ progeny

Tubers of BC₂ Generation

Figure B10.2 The progression of tuber type in somatic hybrids of *S. etuberosum* and their backcross progeny in successive hybridizations with cultivated potato. Figure courtesy of Richard Novy.

species – species which may not include potato as a primary host and, therefore, will not be adversely impacted by host plant resistance.

A combination of green peach aphid and PVY/PLRV resistances is the most effective means to reduce virus infection and spread. Novy *et al.* (2002) evaluated five BC₂ progeny of the *S. etuberosum* somatic hybrids (the recurrent parents being potato varieties) for green peach aphid, PLRV, and PVY resistance. Virus resistances were evaluated in both open field and field cage trials; aphid resistance was evaluated in the field and greenhouse.

The authors identified resistance to green peach aphid in all *S. etuberosum*-derived BC₂ progeny. Resistance was characterized by reduced adult body size and fecundity. One BC₂ individual also exhibited reduced nymph survival. Prolonged development from nymph to adult also appeared to contribute to reduced aphid populations on the BC₂ relative to susceptible checks.

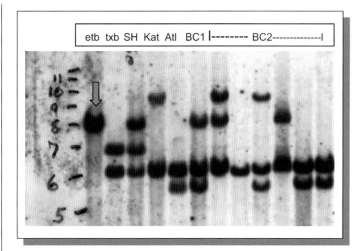

etb txb SH Kat Atl BC1 |-------- BC2-------------|

Figure B10.3 Segregation of a RFLP specific to *S. etuberosum*; RFLP probe used was TG65. The yellow arrow indicates RFLP unique to *S. etuberosum* that also is present in the somatic hybrid (SH), BC₁ and two of six BC₂. This molecular marker is not present in the *S. tuberosum* × *S. berthaultii* fusion parent (txb) or the potato cultivars Katahdin (Kat) or Atlantic (Atl) that were used as parents in the generation of the backcross progeny. The numbers on the side are approximations of RFLP fragment sizes. Figure courtesy of Richard Novy.

Analogous to observations in three of the BC_1, all BC_2 (derived from a PVY resistant BC_1) exhibited statistically significant reduced PVY infection relative to the PVY-susceptible potato variety, Russet Burbank. BC_2 were found to segregate for resistance to PLRV, with two of five displaying resistance to infection on the basis of field and cage evaluations. Progeny of a PLRV-resistant BC_2 clone also have shown high levels of resistance to PLRV in field evaluations in Idaho (author, unpublished data). *S. etuberosum*-derived PLRV resistance is highly heritable on the basis of its expression in third generation progeny of the somatic hybrids.

Backcross progeny are being analyzed with molecular markers to identify chromosomal regions from *S. etuberosum* associated with its observed PVY and PLRV resistances (Figure B10.3). Once identified, prospective regions will be further saturated with additional markers to identify those that are tightly linked to the virus resistances. Such markers can then be used for marker-assisted selection (MAS) in our breeding program. These correlated DNA markers can be used to assess whether an individual will express resistance to PLRV or PVY, thereby speeding the development of potato varieties with the virus resistances of *S. etuberosum*.

Wireworm resistance of backcross progeny derived from somatic hybrids

Wireworm, a larval stage of a beetle, is the most damaging soil-dwelling pest of potato. The larvae spend 3–4 years in the soil and their entry and feeding into potato tubers detrimentally impacts the marketability of the tuber for fresh consumption or processing.

Insecticide use has been the primary means of controlling wireworm damage. However, insecticides currently used for wireworm control may lose their registration in the future. Genetic resistance to wireworm could be an important component of an integrated pest management (IPM) program and it was decided to evaluate the progeny of *S. etuberosum* for wireworm resistance.

In collaboration with Dr Juan Alvarez, an entomologist with the University of Idaho, one BC_1 and four BC_2 clones derived from the *S. etuberosum* somatic hybrids were evaluated for wireworm resistance; comparisons of these clones were made relative to a susceptible cultivar, Russet Burbank. An additional treatment also was included in the evaluation; susceptible Russet Burbank was treated with Genesis, an insecticide commonly used for wireworm control. Four of the five backcross clones had a percentage of damaged tubers comparable or lower than that observed with the use of Genesis. Wireworm entry points or "holes" per tuber among the five clones were comparable to the numbers observed with the use of Genesis.

The resistance of two of the five clones was attributable to high levels of certain chemical compounds called glycoalkaloids naturally produced in the tuber. Total tuber glycoalkaloid concentrations need to be less than 20 mg/100 g tuber fresh weight for safe consumption by humans – these two highly resistant clones had levels ≥47 mg/100 g. However, the remaining three clones had acceptable total tuber glycoalkaloid levels of ≤13 mg/100 g; all three, relative to susceptible Russet Burbank, had reduced wireworm entry damage and two of the three had a reduced percentage of wireworm damaged tubers. These data indicate that high total tuber glycoalkaloid levels are not necessary for conferring wireworm resistance – an important finding if wireworm resistant potato cultivars with acceptable glycoalkaloid levels are to be developed.

References

Correll, D.S. (1962). *The potato and its wild relatives.* Texas Research Foundation. Renner, TX.

Novy, R.G., and Helgeson, J.P. (1994a). Somatic hybrids between *Solanum etuberosum* and diploid, tuber bearing *Solanum* clones. *Theor. Appl. Genet.,* **89**:775–782.

Novy, R.G., and Helgeson, J.P. (1994b). Resistance to potato virus Y in somatic hybrid between *Solanum etuberosum* and *S. tuberosum* x *S. berthaultii* hybrid. *Theor. Appl. Genet.,* **89**:783–786.

Novy, R.G., Nasruddin, A., Ragsdale, D.W., and Radcliffe, E.B. (2002). Genetic resistances to potato leafroll virus, potato virus Y, and green peach aphid in progeny of *Solanum etuberosum. Amer. J. Potato Res.,* **79**:9–18.

and juicy fruits. Consequently, domesticated tomato (whether small or large fruited) is juicy and succulent. Thorns protect against predators in the wild, but are a nuisance to modern use of plants. Hence, some varieties of ornamentals with pronounced thorns in the wild have been bred to be thornless of have less visible thorns (such as roses that are grown for cut flowers).

10.9 Genetic bottleneck

A notable characteristic of most domesticated plants is a reduction in the genetic diversity due to the so-called genetic bottleneck. This bottleneck occurred both during and after domestication. In the widely studied maize crop, diversity reduction between teosinte and modern corn is about 30%. Whereas molecular data (markers, sequence data) indicate a reduction, phenotypic data show an increase in diversity. Being neutral, molecular sequence variation may be reduced by genetic drift, while phenotypic data (domestication traits) are subject to selection.

Domestication has occurred only at one place, based on a limited sample of the ancestral species. This was a bottleneck for variation to pass on from the ancestral species to the crop. Therefore, the crop species contains only a limited amount of the original variation in DNA of the botanical species. Examples are tomato and lettuce. Typically variation studies in such crops at DNA level indicate limited polymorphism compared to the wild ancestral species. In other crops (notably potato and barley), polymorphism at DNA level is much higher than in lettuce and tomato. Probably those crops were domesticated at several places on the basis of many more ancestral plants. In potato much interspecific hybridization probably also took place during domestication. Another bottleneck may have occurred when crops

were introduced to other continents, like potato and tomato from America to Europe. Probably only a limited sample of the domesticate was introduced, representing only part of the variation available in the centre of domestication.

10.10 Tempo of domestication

Evolution is a gradual process. Domestication shares similarities with evolution, as previously discussed. The conventional view of domestication had for a long time being one of a rapid process. Domestication in the Near East had been thought to be a rapid process that followed the climatic transition between the Pleistocene and Holocene periods of evolution, with little pre-domestication, and a rapid rise of domesticated crops coupled with an explosive expansion of farmers out of the centers of crop origin. The domestication syndrome was thought to have arisen rather quickly because the conscious or artificial selection pressure was thought to have been stronger and more influential in modifying genetic patterns of diversity.

Recent advances in knowledge and technology, specifically archaeobotanical evidence and genome-wide multilocus studies, have challenged the rapid transition model. Similarly, mathematical models have suggested a rapid tempo for domestication. Nonetheless, the finding that domestication syndrome genes tend to occur in clusters has led some to propose that domestication could have been more rapid, at least in some species, than in species in which the genes for these beneficial traits were more dispersed. Further, the rapidity with which a given species can be domesticated is limited by the incidence of new, favorable phenotypes (mutation rates), the intensity of both deliberate (conscious) selection and unconscious selection, as well as the rate at which the linkages between unfavorable and favorable phenotypes are broken (recombination rate).

10.11 Genetic architecture and domestication

Genetic architecture was important in the domestication of crop plants. Earlier predictions had been made to the effect that domestication-related trait (domestication syndrome traits) were likely under the control of a few genetic loci with large effects. Recent mapping of quantitative trait loci has supported these predictions. Further, these loci occur in clusters within a genome. For example, Koinange and colleagues found that the domestication syndrome genes were concentrated in only three genomic locations, one controlling growth habit and flowering time, another controlling seed dispersal and dormancy, and the third controlling pod and seed size. There is also evidence that these gene clusters are more readily fixed, especially in outcrossing species, than when they are weakly linked. In other words, species with a cluster of favorable genes are more likely to be rapidly domesticated than those with dispersed genes. In one study, a single 16cM interval of chromosome contained loci for as many as six of the domestication syndrome traits, including shattering and plant height. An exception to the few loci notion is the sunflower plant, in which studies have consistently shown quantitative gene control (many small loci contributing small to moderate effects) of the key traits. Domestication traits are also highly heritable.

The long-held notion is that domestication-related traits are controlled by recessive, loss-of-function alleles. This appears to be the case for non-shattering in some cereals and fruit weight in some solanaceous crops. Recent studies indicate that many traits in the syndrome are conditioned by non-recessive genes. Similarly, recent research has shown that these traits do not condition loss-of-function, but rather change-of-function? On the other hand, reports indicate that loss-of-function genes are common in crop improvement. For example, flowering time, dwarf habit, and transition from two-rowed to six-rowed barley are conditioned by recessive loss-of-function mutations. Sequence information indicated that mutations in the regulatory regions rather than in the coding regions of the genes were responsible for many of these changes.

10.12 Models of domestication

Various mechanisms played a key-role in domestication of plant species. These include events such as mutation and hybridization, forces that change allele frequencies (e.g., selection, genetic drift), and phenomena that result in reproductive isolation (e.g., polyploidization and reproductive barriers). Selection (conscious or unconscious) plays a key role in all domestications.

- **Model 1 (bread wheat model).** Polyploidization is the most significant mechanism in the domestication process. Wild and domesticated genotypes differ in ploidy levels, which caused them to be reproductively isolated.
- **Model 2 (cotton model).** In this model, polyploidization occurred prior to domestication. The crop and its wild relatives are reproductively isolated from the diploid relatives of the genus. Further, the genetic effects of selection and inbreeding in both polyploids and diploids are different.
- **Model 3 (soybean model).** Drift is the most important factor in this model. Domestication occurred at the diploid level of the crop out of the gene pool of the wild ancestor. Crops domesticated under this model are primarily autogamous. It is possible that some introgression may have occurred in the process.
- **Model 4 (chili pepper model).** Selection, drift and hybridization are all important factors in this model. It is characterized by an abundant weedy gene pool. There is unrestricted gene exchange among crop, wild relatives, and weedy intermediates. A key difference between model 3 (soybean) and model 4 (chili pepper) is the possibility of spontaneous hybridization between wild and domesticated races.

10.13 Modern breeding is continuation of the domestication process

During the millennia of domestication, farmers depended on spontaneous mutation and hybridization, on trading material with far-away regions, allowing new hybridization and new variation to occur and to select superior types. Only in the past two centuries have agriculturists applied deliberate hybridization (including crosses with exotic plant introductions), selection schemes and the induction of mutations to more efficiently obtain superior crops: deliberate plant breeding came into being. Plant breeding may be described as directed or targeted and accelerated domestication because the plant breeder, with a breeding objective in mind, deliberately and genetically manipulates plants (wild or domesticated) to achieve a stated goal, but in a *very short* time. Conceptually, breeding and evolution are the same, a key difference being the duration of the processes. Compared to rates by which evolution and domestication progress, breeding speeds up developments much more.

Domestication entails genetics shifts in plants induced by humans. These shifts occurred during domestication. The level of human involvement in the orchestration of the genetic shifts depends on the level of information and technology available. In the era that some describe as pre-domestication, humans caused genetic shifts by selecting plants based on their desirable traits (general appeal or preference and functionality without understanding the genetic principles). When deliberate cultivation (farming) of crops began, the choice of crops to grow was a conscious selection, also based on appeal or preference and

functionality. By this time the choices had been narrowed down significantly.

Evidence of the deliberate manipulation of plants for improvement of desired traits has been found in archaeological records, suggesting that people of ancient times went beyond selecting among natural variability to creating new variability. There is no evidence that such an activity was conducted with an understanding of the hereditary basis of the traits. Scientific breeding of crops came about after the works of Koelreuter (hybridization) and Mendel (transmission genetics) laid the foundation for modern plant breeding. Breeders devised methods of selection (breeding or selection schemes) for improving traits according to their modes of reproduction. During this era, crops were modified based on information.

Up to this time, hereditary manipulation, though genetically based, was indirect in the sense that breeders did not directly manipulate the genetic material, DNA. With the discovery of technologies such as recombinant DNA, breeders have had a more sophisticated tool to pursue genetic modifications. The DNA at this stage was the direct target for manipulation.

Key references and suggested reading

Allaby, R.G., and Brown, T.A. (2003). AFLP and the origins of agriculture. *Genome*, **46**:448–453.

Allaby, R.G., Fuller, D.Q., and Brown, T.A. (2008). The genetic expectations of a protracted model of domestication of crops. *Proceeding of the National Academy of Sciences USA*, **105**(37): 13982–13986.

Burger, J.C., Chapman, M.A., and Burke, J.M. (2008). Molecular insights into the evolution of crop plants. *American Journal of Botany*, **95**:113–122.

Eyre-Walker, A., and Gaut, B. (1998). Investigation of the bottleneck leading to the domestication of maize. *Proceeding of the National Academy of Sciences USA*, **95**:4.

Fuller, D. (2007). Contrasting patterns in crop domestication and domestication rates: Recent archaeological insights from the Old World. *Annals of Botany*, **100**:903–924.

Gepts, P. (2004). Crop Domestication as a Long-term Selection Experiment, in *Plant Breeding Reviews* (ed. J. Janick), Volume **24**,Part 2. John Wiley & Sons, Inc., NJ, pp. 1–44.

Le Thierry D'Ennequin, M., Toupance, B., Godelle, B., and Gouyon, P.H. (1999). Plant domestication: A model for studying the selection of linkage. *Journal of Evolutionary Biology*, **12**:1138–1147.

Londo, J.P., Chiang, Y-C., Hung, K-H., Chiang, T-Y., and Schaal, B.A. (2006). Phylogeography of Asian wild rice, Oryza rufipogon, reveals multiple independent domestications of cultivated rice, Oryza sativa. *Proceeding of the National Academy of Sciences USA*, **103**:9578–9583.

Morrell, P.L., and Clegg, M.T. (2007). Evidence for a second domestication of barley (Hordeum vulgare) east of the Fertile Crescent. *Proceeding of the National Academy of Sciences USA*, **104**:3289–3294.

Raamsdonk, L.W.D. (1995). The cytological and genetical mechanisms of plant domestication exemplified by four crop models. *The Botanical Review*, **4**:367–399.

Ross-Ibarra, J., Morrell, P.L., and Gaut, B.S. (2007). Plant domestication, a unique opportunity to identify the genetic basis. *Proceeding of the National Academy of Sciences USA*, **104**(Suppl. 1): 8641–8648.

Spooner, D.M., McLean, K., Ramsay, G., Waugh, R., and Bryan, G. (2005). A single domestication for potato based on multilocus amplified fragment length polymorphism genotyping. *Proceeding of the National Academy of Sciences USA*, **102**:14694–14699.

Xiong, I., Liu, K., Dai, X., Xu, C., and Zhang, Q. (1999). Identification of genetic factors controlling domestication related traits of rice using an F2 population of a cross between Oryza sativa and O rufipogon. *Theoretical and Applied Genetics*, **98**:243–251.

Zeven, A.C., and de Wet, J.M.J. (1982). *Dictionary of Cultivated Plants and Their Regions of Diversity*. PUDOC, Wageningen.

Zhu, Q., Zheng, X., Luo, J., Gaut, B., and Song, G. (2007). Multilocus analysis of nucleotide variation of Oryza sativa and its wild relatives: Severe bottleneck during domestication of rice. *Molecular Biology and Evolution*, **24**:875–888.

Zohary, D. (1999). Monophyletic vs. polyphyletic origin of crops found in the Near East. *Genetic Resources and Crop Evolution*, **46**:133–142.

Zohary, D., and Hopf, M. (2000). *Domestication of Plants in the Old World*, 3rd edn. Oxford University Press, Oxford.

Outcomes assessment

Part A

Please answer the following questions true or false.

1 Evolution is a population phenomenon.
2 Land grant institutions conduct plant breeding in the private sector in the United States.
3 Traditional plant breeding tools are obsolete.
4 Plant breeding causes heritable changes in plants.

Part B

Please answer the following questions.

1 . . . . is the arbiter of evolution.
2 . . . . is the process by which wild plants are genetically changed through human selection.
3 Compare and contrast evolution and plant breeding.
4 What is plant domestication?
5 Give four specific examples of ways in which domesticated plants may differ from their wild progenitors.

Part C

Please write a short essay on the following topics.

1 Discuss the concept of breeder's eye.
2 Discuss the concept of the breeder as a decision maker.
3 Discuss the general steps in a plant breeding program.
4 Discuss the qualifications of a plant breeder.
5 Distinguish between public sector and private sector plant breeding.
6 Discuss the molecular and classical plant breeding approaches as complementary approaches in modern plant breeding.

11

Plant genetic resources

Purpose and expected outcomes

Human societies are constantly changing. Primitive societies change to become modern. Modern societies are more technologically advanced than their preceding counterparts. As societies evolve, their needs and preferences also change. The need for food, quantity and quality, as well as other plant products are constantly changing. The way food is produced has also changed over time. Some of these changes have been necessitated by the changes in the cultural environment as a result of humans interfering with the natural balance of nature. New pathogens have emerged in modern agriculture. Some cultivation practices adversely impact the environment. With the current environmental consciousness in society, there is a demand to reduce pesticide use. Consumer needs and preferences have changed over the years. All these changes require that plant breeders stay vigilant in developing new cultivars on a regular basis to respond to these changes. Plant breeders depend on variability (germplasm) to develop new cultivars. Some of it is obtained from local sources, while some comes from distant regions, each source with its unique advantages and disadvantages. It is important to note that, being a natural resource, germplasm is susceptible to erosion. To facilitate its use, germplasm is collected, characterized and properly stored and managed for use by plant breeders.

After completing this chapter, the student should be able to:

1 Discuss the importance of germplasm to plant breeding.
2 Describe the various sources of germplasm available to breeders, their advantages and disadvantages.
3 Define the term genetic vulnerability and discuss its implications to plant breeding.
4 Discuss the mechanisms for conservation of germplasm.
5 Discuss the international role in germplasm conservation.

11.1 Importance of germplasm to plant breeding

Germplasm is the lifeblood of plant breeding without which breeding is impossible to conduct. It is the genetic material that can be used to perpetuate a species or population. It not only has reproductive value, but through genetic manipulation (plant breeding), germplasm can be improved for better performance of the crop. Germplasm provides the materials (parents)

Principles of Plant Genetics and Breeding, Second Edition. George Acquaah.
© 2012 John Wiley & Sons, Ltd. Published 2012 by John Wiley & Sons, Ltd.

used to initiate a breeding program. Sometimes, all plant breeders do is to evaluate plant germplasm and make a selection from existing biological variation. Promising genotypes that are adapted to the production region are then released to producers. Other times, as discussed in Chapter 5, breeders generate new variability by using a variety of methods such as crossing parents, mutagenesis (inducing mutations) and, more recently, gene transfer. This base population is then subjected to appropriate selection methods, leading to the identification and further evaluation of promising genotypes for realease as cultivars. When breeders need to improve plants, they have to find a source of germplasm that would supply the genes needed to undertake the breeding project. To facilitate the use of germplasm, certain entities (germplasm banks) are charged with the responsibility of assembling, cataloguing, storing, and managing large numbers of germplasm. This strategy allows scientists ready and quick access to germplasm when they need it.

11.2 Centers of diversity in plant breeding

The subject of centers of diversity was first discussed in Chapter 10. Whereas the existence of centers of crop origin or domestication is not incontrovertible, the existence of natural reservoirs of plant genetic variability has been observed to occur in certain regions of the world. These centers are important to plant breeders because they represent pools of diversity, especially wild relatives of modern cultivars.

Plant breeding may be a victim of its own success. The consequence of selection by plant breeders in their programs is the steady erosion or reduction in genetic variability, especially in the highly improved crops. Modern plant breeding tends to focus on a small amount of variability for crop improvement. Researchers periodically conduct plant explorations (or collections) to those centers of diversity where wild plants grow in their natural habitats, to collect materials that frequently yield genes for addressing a wide variety of plant breeding problems, including disease resistance, drought resistance, and chemical composition augmentation.

11.3 Sources of germplasm for plant breeding

Germplasm may be classified into five major types – **advanced (elite) germplasm**, **improved germplasm**, **landraces**, **wild or weedy relatives**, and **genetic stocks**. The major sources of variability for plant breeders may also be categorized into three broad groups – **domesticated plants**, **undomesticated plants**, and **other species or genera**.

11.3.1 Domesticated plants

Domesticated plants are those plant materials that have been subjected to some form of human selection and are grown for food or other uses. There are various types of such material:

- **Commercial cultivars.** There are two forms of this material – **current cultivars** and **retired** or **obsolete cultivars**. These are products of formal plant breeding for specific objectives. It is expected that such genotypes would have superior gene combinations, be adapted to a growing area, and have a generally good performance. The obsolete cultivars were taken out of commercial production because they may have suffered a set-back (e.g., susceptible to disease) or higher performing cultivars were developed to replace them. If desirable parents are found in commercial cultivars, the breeder has a head start on breeding since most of the gene combinations would already be desirable and adapted to the production environment.
- **Breeding materials.** Ongoing or more established breeding programs maintain variability from previous projects. These intermediate breeding products are usually genetically narrow-based because they originate from a small number of genotypes or populations. For example, a breeder may release one genotype as a commercial cultivar after yield tests. Many of the genotypes that made it to the final stage or have unique traits will be retained as breeding materials to be considered in future projects. Similarly, genotypes with unique combinations may be retained.
- **Landraces.** Landraces are farmer-developed and maintained cultivars. They are developed over very long periods and have co-adapted gene complexes. They are adapted to the growing region and are often highly heterogeneous. Landraces are robust, having developed resistance to the environmental stresses in their areas of adaptation. They are adapted to unfavorable conditions and produce low but relatively stable performance. Landraces, hence, characterize subsistence agriculture. They may be used as starting material in mass selection or pure line breeding projects.

- **Plant introductions.** The plant breeder may import new, unadapted genotypes from outside the production region, usually from another country (called plant introductions). These new materials may be evaluated and adapted to new production regions as new cultivars, or used as parents for crossing in breeding projects.
- **Genetic stock.** This consists of products of specialized genetic manipulations by researchers (e.g., by using mutagenesis to generate various chromosomal and genomic mutants).

11.3.2 Undomesticated plants

When desired genes are not found in domesticated cultivars, plant breeders may seek them from wild populations. When wild plants are used in crosses, they may introduce wild traits that have an advantage for survival in the wild (e.g., hard seed coat, shattering, indeterminacy) but are undesirable in modern cultivation. These undesirable traits have been selected against through the process of domestication. Wild germplasms have been used as donors of several important disease and insect resistance genes and genes for adaptation to stressful environments. The cultivated tomato has benefited from such introgression by crossing with a variety of wild *Licopersicon* species. Other species such as potato, sunflower, and rice have benefited from wide crosses. In horticulture, various wild relatives of cultivated plants may be used as rootstock in grafting (e.g., citrus, grape) to allow cultivation of the plant in various adverse soil and climatic conditions.

11.3.3 Other species and genera

Gene transfer via crossing requires that the parents be cross-compatible or cross-fertile. As previously stated, crossing involving parents from within a species is usually successful and unproblematic. However, as the parents become more genetically divergent, crossing (wide crosses) is less successful, often requiring special techniques (e.g., embryo rescue) for intervening in the process in order to obtain a viable plant. Sometimes, related species may be crossed with little difficulty.

11.4 Concept of gene pools of cultivated crops

J.R. Harlan and J.M.J. de Wet proposed a categorization of gene pools of cultivated crops according to the feasibility of gene transfer or gene flow from those species to the crop species. Three categories were defined, primary, secondary, and tertiary gene pools:

(i) **Primary gene pool (GP1).** GP1 consists of biological species that can be intercrossed easily (interfertile) without any problems with fertility of the progeny. That is, there is no restriction to gene exchange between members of the group. This group may contain both cultivated and wild progenitors of the species.

(ii) **Secondary gene pool (GP2).** Members of this gene pool include both cultivated and wild relatives of the crop species. They are more distantly related and have crossability problems. Nonetheless, crossing produces hybrids and derivatives that are sufficiently fertile to allow gene flow. GP2 species can cross with those in GP1, with some fertility of the F1, but more difficulty with success.

(iii) **Tertiary gene pool (GP3).** GP3 involves the outer limits of potential genetic resources. Gene transfer by hybridization between GP1 and GP3 is very problematic, resulting in lethality, sterility, and other abnormalities. To exploit germplasm from distant relatives, tools such as embryo rescue and bridge crossing may be used to nurture an embryo from a wide cross to a full plant and to obtain fertile plants.

A classification of dry bean and rice is presented in Figure 11.1 for an illustration of this concept. In assembling germplasm for a plant breeding project, the general rule is to start by searching the domesticated germplasm collection first, before considering other sources, for reasons previously stated. However, there are times when the gene of interest occurs in undomesticated germplasm, or even outside the species. Gene transfer techniques enable breeders to transfer genes beyond the tertiary gene pool. Whereas all crop plants have a primary gene pool that includes the cultivated forms, all crops do not have wild forms in their GP1 (e.g., broad bean, cassava, and onions whose wild types are yet to be identified). Also, occasionally, the GP1 may contain taxa of other crop plants (e.g., almond belongs to the primary gene pool of peach). Most crop plants have a GP2, which consists primarily of species of the same genus. Some crop plants have no secondary gene pools (e.g., barley, soybean, onion, broad bean).

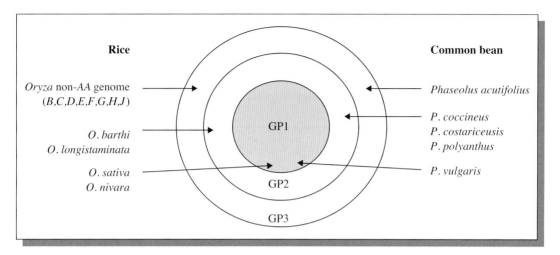

Figure 11.1 Crop gene pools. Harlan proposed the crop gene pools to guide the germplasm use by plant breeders. The number of species in each of the pools that plant breeders are using varies among crops. Harlan suggested that breeders first utilize the germplasm in the GP1 and proceed outwards.

11.5 Concept of genetic vulnerability

Genetic vulnerability is an important issue in modern plant breeding, brought about largely by the manner in which breeders go about developing new and improved cultivars for modern society.

11.5.1 What is genetic vulnerability?

Genetic vulnerability is a term used to indicate the genetic homogeneity and uniformity of a group of plants that predisposes it to susceptibility to a pest, pathogen, or environmental hazard of large-scale proportions. It is a complex problem that involves issues such as crop evolution, trends in breeding, trends in biological technology, decisions by crop producers, demands and preferences of consumers, and other factors. As a result of a combination of the above factors, a certain kind of crop cultivar (genotype) is developed for the agricultural production system. A case in point is the 1970 epidemic of southern leaf blight (*Helminthosporium maydis*) in the United States that devastated the corn industry. This genetic vulnerability in corn was attributed to uniformity in the genetic background in corn stemming from the widespread use of the T-cytoplasm in corn hybrid seed production.

Prior to the use of the T-cytoplasm, hybrid corn breeders had to detassel their non-pollen parent. The T-cytoplasm contained a gene in the mitochondrion (mtDNA) that conferred male sterility upon the plant, eliminating the need for detasseling. However, a male fertility restorer gene in parent B was needed to restore fertility to the F_1 (AxB) seed. Large acreages of hybrids developed this way were grown. Unfortunately, a new strain of *H. maydis* surfaced around 1970 with devastating consequences due to the large acreage of the crop.

Another classic case of crop vulnerability occurred in Europe. The late blight of potato was introduced from the Americas into Europe around 1845. It quickly spread in the area, causing major epidemics, in Ireland especially. The consequence of this disaster included starvation, disease, death, and the emigration of over five million Irish nationals to the United States.

Genetic uniformity and the consequence of vulnerability to a large extent are created, maybe inadvertently, in response to the preference of both consumers and producers for uniform products in some situations. Further, it appears the key factor is the gene shared in common rather than the genetic background in which the gene occurs. For example, a gene that may trigger a disease epidemic in one species may occur in another species. Consequently, when there is a disease outbreak, it is not confined but spreads to different species. *Phenotypically* dissimilar crops can share a trait that is simply inherited and that predisposes them to susceptibility to an adverse biotic or abiotic factor. A case in point is the chestnut blight (*Cryphonectria parasitica*) epidemic that occurred in the United States in which different species of the plant were affected.

It should be pointed out that genetic vulnerability is an issue in hindsight, since the disaster it causes cannot be predicted. For example, the use of the T-cytoplasm was a great strategy until the southern leaf blight appeared in the corn fields. Similarly, the *mlo* mildew resistance in barley against powdery mildew is beneficial to farmers in Europe, where about 70% of all spring barley incorporated that gene. However, should an unforeseen event similar to the southern leaf blight occur, would this once outstanding achievement be viewed in the negative light? The message is that breeders should be consider diversity when selecting parents for breeding programs so that crop vulnerability would be reduced.

Germplasm in which insufficient diversity is significant includes grapes, sweet potato, cucurbits, tropical fruits and nuts, cool-season food legumes, peach, cherry, walnuts, herbaceous and woody ornamentals.

11.5.2 Key factors in genetic vulnerability induced crop failure

The key factors that are responsible for the disastrous epidemics attributable to genetic vulnerability of crops are:

- The desire by growers and consumers for uniform and blemish-free products, which is often achieved by breeders by applying time and again the same (sets of) gene(s).
- The acreage devoted to the crop cultivars carrying such popular genes and method of field production.

Where the cultivars with the susceptible gene are widely distributed in production (i.e., most farmers use the same cultivars), the risk of disaster will be equally high but unpredictable! Further, where the threat is biotic, the mode of dispersal of the causal agent and the presence of a favorable environment for the pathogen to develop would increase the risk of disasters (e.g., wind mode of dispersal of spores or propagules will cause a rapid spread of the disease). In biotic disasters, the use of a single source of resistance to the pathogen is perhaps the single most important factor in vulnerability. However, the effect can be exacerbated by practices such as intensive and continuous monoculture using one cultivar or various cultivars carrying the same cytoplasm or resistance genes. Under such production practices, the pathogen only has to overcome one resistance gene, resulting in rapid disease advance and great damage to crop production.

11.6 What plant breeders can do to address crop vulnerability

The issue of the importance of genetic vulnerability is best left to plants to decide.

11.6.1 Reality check

First and foremost, plant breeders need to be convinced that genetic vulnerability is a real and present danger. Without this first step, efforts to address the issue are not likely to be taken seriously. A study by D.N. Duvick in 1984, albeit dated, posed the question "How serious is the problem of genetic vulnerability in your crop?" to plant breeders. The responses by breeders of selected crops (cotton, soybean, wheat, sorghum, maize, and others) indicated a wide range of perception of crop vulnerability, ranging from 0 to 25% thinking it was serious, and 25–60% thinking it was not a serious problem (at least at that time). Soybean and wheat breeders expressed the most concern about genetic vulnerability. Their fears are most certainly founded since, in soybean, it is estimated that only six cultivars constitute more than 50% of the crop acreage of North America. Similarly, more than 50% of the acreage of many crops in the United States is planted to less than 10 cultivars per crop.

An early assessment of genetic uniformity of major crops in the United States by the National Research Council showed that, in peanuts, 95% of the acreage was dominated by nine major cultivars. In corn, 71% of the acreage was sown to only six major varieties, while in the case of dry bean 60% of the acreage was cultivated to only two major cultivars. The message here is that several major crops in the United States are vulnerable to epidemics. However, there is reversal in the longtime trend toward increased diversification.

11.6.2 Use of wild germplasm

Many of the world's major crops are grown extensively outside the centers of origin where they

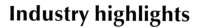

Industry highlights
Plant genetic resources for breeding

K. Hammer[1], F. Heuser[1], K. Khoshbakht[1], Y. Teklu[1], and M. Hammer-Spahillari[2]

[1]Institute of Crop Science, Agrobiodiversity Department, University Kassel, Steinstr. 19, D-37213 Witzenhausen, Germany

[2]Dr. Junghanns GmbH, D-06449 Aschersleben, Germany

Introduction

In recent years the maintenance of plant genetic resources (PGR) has attracted growing public and scientific interest as well as political support, since it is accepted that there is a close relationship between biological diversity and the health of the biosphere (Callow *et al.*, 1997). The resources are also necessary as a precondition for global food security (Rosegrant and Cline, 2003; Hazell, 2008).

PGR can be divided into two groups (Evenson amd Gollin, 2003): The first one consists of the cultivated plants themselves, their wild forms and related species. The value of this group for improving crop plants is well recognized and expressed in the investment made by collections, evaluations and conservation of these PGR in past and present times.

The second group is composed of PGR from other, non-cultivated plants like weeds and even in species outside the plant kingdom (Hammer *et al.*, 2003a). This group was neglected until modern methods of biotechnology appeared, which allowed the integration of "alien" genetic material into valuable plant species. While a great amount of the PGR of the first group is conserved in gene banks (*ex situ*), there is an increasing requirement of conserving or better preserving the non-cultivated PGR of the second group *in situ*, meaning to save biodiversity as a whole in the natural habitat. All kind of plants might dispose useful properties (in the future) to be of value for maintaining. Already Frankel (1974) pointed out that there is "no doubt primitive and wild gene pools will continue to serve as important sources of genes for resistance to parasites or for characteristics indicated by advances in science or technology or by changing demands of the consumer". Alongside raising the impact of PGR for providing food there also exist traditional issues like medicine, feed, fiber, clothing, shelter and energy.

Over a five-year period, 6.5% of all genetic research within the plant breeding and the seed industry that resulted in marketed innovation was concerned with germplasm from wild species and landraces, compared with only 2.2% of new germplasm originating from induced mutation (Swanson, 1996, Callow *et al.*, 1997).

Since the beginning of agriculture, natural diversity has declined due to agricultural domestication, breeding and distribution of crops (Becker, 2000). But in recent years crop species and varieties have also become threatened or even extinct (Khoshbakht and Hammer, 2010). In agriculture the widespread adoption of few varieties leads to a drastic decrease of landraces with their potential valuable genetic resources. Among the cultivated plants which represent less than 3% of the vascular plants only about 30 species feed the world (Hammer, 2003).

Conservation and monitoring of PGR

For conservation of PGR, monitoring and evaluation of plant material is required There may be a big difference between the phenotype and the genotype in a population. With improved biotechnology methods, like the assignment of molecular markers the gene level is of increasing interest.

Ex situ conservation

Ex situ conservation stands for all conservation methods in which the species or varieties are taken out of their natural habitat and are kept in surroundings made by humans. Large collections started with the activities of the Russian scientist N.I. Vavilov at the beginning of the last century. At this time already the employment of *ex situ* measures was necessary because of the rapidly increasing gene erosion of landraces (Hammer and Teklu, 2008) and other plants (Coats, 1969). Alongside, plant breeders contributed to the maintenance by collecting breeding material. This material was often kept in specific institutions, first called "gene banks" in the 1970s. They have been established for the collection (Guarino *et al.*, 1995), maintenance, study and supply of genetic resources of cultivated plants and related wild species. Gene banks maintain the plant material as seeds, *in vivo* (when the storage of seeds is difficult) or *in vitro* (mostly cryo-conservation). In contrast to the cultivation of plants in botanical gardens, the work in gene banks is more engaged in infraspecific variability. Unfortunately, a lot of the material stored in gene banks is not in good condition and urgently needs to be rejuvenated (Hammer, 2003).

Table B11.1 Conservation methods for different categories of diversity rated by their importance for specific group of diversity (Hammer, 2003).

Importance			Crops				Wild relatives				Weeds			
			1	2	3	4	1	2	3	4	1	2	3	4
Ex situ		Infraspecific diversity			X			X					X	
		Diversity of species		X				X					X	
		Diversity of ecosystems	X				X				X			
On farm	Developing countries	Infraspecific diversity				X			X					X
		Diversity of species				X			X					X
		Diversity of ecosystems				X			X					X
	Developed countries	Infraspecific diversity			X			X				X		
		Diversity of species			X			X					X	
		Diversity of ecosystems		X					X					X
In situ		Infraspecific diversity	X						X			X		
		Diversity of species	X						X			X		
		Diversity of ecosystems	X						X			X		

1 = no importance; 2 = low importance; 3 = important; 4 = very important

In situ conservation

As well as *ex situ* conservation there is also the attempt to save biodiversity, and therefore PGR, in ecosystems (*in situ*). This can occur in the natural habitat (especially wild relatives and forestry species) or in locations where the plants (land-races and weeds) have evolved (*on farm = in agro-ecosystems*). In contrast to *ex situ* conservation in gene banks, where only a section of the whole diversity is covered, the *in situ* approach is able to save larger parts of biological diversity (Hammer *et al.*, 2003b).

Table B11.1 summarizes the methods of conservation for the different categories of diversity and evaluates their relative importance. It differentiates between cultivated plants, wild growing resources and weeds.

Characterization and evaluation of plant genetic resources

The yield levels of many crops have reached a plateau due to the narrow genetic base of these crops. Although the results of some surveys (Brown, 1983; Chang, 1994) indicate that the genetic base of several important crops has begun to increase over the years, breeding programs of many important crops continue to include only a small part of genetic diversity available and the introduction of new and improved cultivars continues to replace indigenous varieties containing potentially useful germplasm. To widen the genetic base for further improvement, it is necessary to collect, characterize, evaluate and conserve plant biodiversity, particularly in local (Joshi and Witcombe, 2003; Veteläinen *et al.*, 2009), underutilized and neglected crop plants (Hammer *et al.*, 2001) and crop-wild relatives (Maxted and Guarino, 2006).

Morphological and agronomic characteristics are often used for basic characterization, because this information is of high interest to users of the genetic diversity. Such characterization requires considerable amounts of human labor, organizational skills, and elaborate systems for data documentation although it can be done by using simple techniques and can reach a high sample throughput. Quantitative agronomic traits can be used to measure the differences between individuals and populations with regard to genetically complex issues such as yield potential and stress tolerance. The diversity of a population, considering such complex issues, can be described by using its mean value and genetic variance in statistical terms. The traits detected are of great interest but are frequently subject to strong environmental influence, which makes their use as defining units for the measurement of genetic diversity problematical.

Molecular methods can be employed to characterize genetic resources and for the measurement of genetic variation. The major advantage of molecular methods for characterization is their direct investigation of the genotypic situation, which allows them to detect variation at the DNA level, thereby excluding all environmental influences. They can also be employed at very early growth stages. The advantages and disadvantages of some commonly used techniques for characterization of PGR have been summarized by Hammer (2003, p. 127). The future prospects for modern scientific advances are usually considered as high, for example for proteomics, transcriptomics, genomics, metabolomics and phylogenomics, but in spite of all good prospects many programs, especially in developing countries, are unable to apply them in plant breeding (FAO, 2010).

Germplasm enhancement

PGR are fundamental to improving agricultural productivity. These resources, fortunately stored in gene banks around the world, evolved an assortment of alleles needed for resistance and tolerance to the diseases, pests, and harsh environments found in their natural habitats. Though gene flow between crops and their wild relatives may be high (Anderson and de Vicente, 2010), only a small amount of this variability has been introgressed to crop species (Ortiz, 2002). Most cereal breeders do not make heavy use of germplasm of landraces and wild and weedy relatives existing in active collections. The valuable genetic resources are essentially "sitting on the shelf" in what has been dismissively termed "gene morgues" (Hoisington *et al.*, 1999). Germplasm enhancement may be one of the keys to maximizing utilization of this germplasm. It has become an important tool for the genetic improvement of breeding populations by gene introgression or incorporation of wild and landrace genetic resources into respective crop breeding pools. The terms "germplasm enhancement" or "pre-breeding" refer to the early component of sustainable plant breeding that deals with identifying a useful character, "capturing" its genetic diversity, and the transfer or introgression of these genes and gene combinations from non-adapted sources into breeding materials (Peloquin *et al.*, 1989).

The gene pools as defined by Harlan and de Wet (1971) have formed a valid scientific basis for the definition and utilization of plant genetic resources (Figure B11.1). More recently, however, plant transformation and genomics have led to a new quality, which has been defined by Gepts and Papa (2003) as a fourth gene pool, whereas Gladis and Hammer (2002) earlier concluded that information and genes from other species also belong to the third gene pool. The fourth gene pool should contain any synthetic strains with nucleic acid frequencies, DNA and RNA, that do *not* occur in nature.

The most widespread application of germplasm enhancement has been in resistance breeding with genetic resources of wild species. Backcross followed by selection has been the most common method for gene introgression from wild germplasm to breeding materials.

However, some problems still remain for genetic enhancement with wild species: linkage drag, sterility, small sample size of interspecific hybrid population, and restricted genetic recombination in the hybrid germplasm (Ortiz, 2002). Transgenesis allows us to bypass sexual incompatibility barriers altogether and introduce new genes into existing

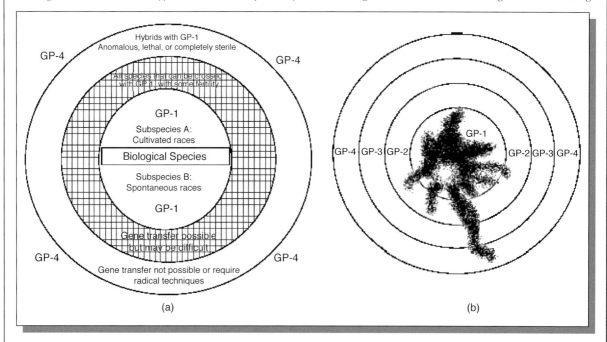

(a)

(b)

Figure B11.1 (a) The gene pool concept, established by Harlan and de Wet (1971), modified. (b) Example of an organismoid or a hypothetically designed crop with a genome composed of different gene pools and synthetic genes [for an explanation of this complicated matter, see Gladis and Hammer (2002)]. Figure courtesy of Karl Hammer.

cultivars. In recent years, transgenic plants have been incorporated as parents of hybrids in breeding programs in the United States for crops such as maize and oilseed rape. Molecular markers are being used to tag specific chromosome segments bearing the desired gene(s) to be transferred (or incorporated) into the breeding lines (or populations) (Gepts, 2002).

Examples of successful use of plant genetic resources

Over the last few decades, awareness of the rich diversity of exotic or wild germplasm has increased. This has led to a more intensive use of this germplasm in breeding (Kearsey, 1997) and thereby yields of many crops have increased dramatically.

GP-1: The biological species, including wild, weedy and cultivated races. GP-2: All species that can be crossed with GP 1, with some fertility in individuals of the F_1 generation; gene transfer is possible but may be difficult. GP-3: Hybrids with GP-1 do not occur in nature; they are anomalous, lethal, or completely sterile; gene transfer is not possible without applying radical techniques. Information from other genes refers to comparative genomic information on gene order and DNA sequence of homologous genes. GP-4: Any synthetic strains with nucleic acid frequencies (DNA or RNA) that do not occur in nature

The introgression of genes that reduce plant height and increase disease and viral resistance in wheat provided the foundation for the "Green Revolution" and demonstrated the tremendous impact that genetic resources can have on production (Hoisington *et al.*, 1999).

In Germany, plant genetic resource material stored in the Gatersleben gene bank has been successfully used for the development of improved varieties (Table B11.2).

Table B11.2 Varieties registered from 1973 to 1990 proved to have been developed with material from the gene bank Gatersleben (Hammer, 1991).

Crop	Number of varieties
Spring barley	30
Winter barley	3
Spring wheat	1
Winter wheat	12
Dry soup pea	2
Fodder pea	3
Lettuce	1
Vegetable pea	4
Total	*56*

Developing improved varieties using gene bank materials has a long history. For instance, for developing disease resistant material, the resistance must be located with great expenditure of time and effort from extensive collections. The experience in Gatersleben indicated that between the first discovery of the material and the launch of a new variety, roughly 20 years pass by, even if modern breeding methods are employed (Hammer, 2003). A positive correlation has been observed between the number of evaluated accessions in gene banks and the number of released varieties on the basis of evaluated material (Hammer, 1993).

The use of Turkish wheat to develop genetic resistance to diseases in western wheat crops was valued, in 1995, at US$50 million per year. Ethiopian barley has been used to protect Californian barley from dwarf yellow virus, saving damage estimated at $160 million per year. Mexican beans have been used to improve resistance to the Mexican bean weevil, which destroys as much as 25% of stored beans in Africa and 15% in South America (Perrings, 1998).

Conclusion

PGR are useful for present and future agri- and horticultural production. Particularly, they are needed for the genetic improvement of crop plants. Because of their usefulness and ongoing erosion in agro-ecosystems, it was necessary to establish large collections of PGR. The material in these collections has to be characterized and evaluated in order to introduce it into breeding programs. Pre-breeding and germplasm enhancement are necessary as first steps for the introduction of primitive material into modern varieties (FAO, 2010).

References

Anderson, M.S., and de Vincente, M.C. (2010). *Gene Flow between Crops and Their Wild Relatives.* The Johns Hopkins University Press, Baltimore, MD.

Becker, H.C. (2000). Einfluß der Pflanzenzüchtung auf die genetische Vielfalt. *Schriftenr. Vegetationskunde,* **32**:87–94.

Brown, W.L. (1983). Genetic diversity and genetic vulnerability – an appraisal. *Economic Botany,* **37**:4–12.

Callow J.A., Ford-Lloyd, B.V., and Newbury, H.J. (eds) (1997). *Biotechnology and Plant Genetic Resources – Conservation and Use.* Biotechnology in Agriculture Series No. 19, CAB International, Wallingford, UK.

Chang, T.T. (1994). The biodiversity crisis in Asia crop production and remedial measure, in *Biodiversity and Terrestrial Ecosystems* (eds C.I. Peng and C. H. Chou) Institute of Botany, Academia Sinica, Monograph Series 14. Taipei, pp. 25–41.

Coats, A. (1969). *The Quest for Plants: A History of the Horticultural Explorers*. Studio Vista, London.

Evenson, R.E., and Gollin, D. (eds) (2003). *Crop Varieties Improvement and its Effect on Productivity*. CAB International, Wallingford, UK.

FAO (2010). The Second Report on the State of the World's Plant Genetic Resources for Food and Agriculture. FAO, Rome.

Frankel, O.H. (1974). Genetic Conservation: Our evolutionary responsibility. *Genetics*, **78**:53–65.

Gepts, P. (2002). A comparison between crop domestication, classical plant breeding and genetic engineering. *Crop Science*, **42**:1780–1790.

Gepts, P., and Papa, R. (2003). Possible effects of (trans)gene flow from crops on the genetic diversity from landraces and wild relatives. *Environmental Biosafety Research*, **2**:89–103.

Gladis, T., and Hammer, K. (2002). The relevance of plant genetic resources in plant breeding. *FAL Agriculture Research*, Special Issue, **228**:3–13.

Guarino, L., Ramanatha-Rao, V.R., and Reid, R. (eds) (1995). *Collecting Plant Genetic Diversity: Technical Guidelines*. CAB International, Wallingford, UK.

Hammer, K. (1991). Die Nutzung des Materials der Gaterslebener Genbank für die Resistenzzüchtung – eine Übersicht. *Vortr. Pflanzenzüchtung*, **19**:197–206.

Hammer, K. (1993). The 50th anniversary of the Gatersleben genebank. *Plant Genetic Resources Newsletter*, **91/92**:1–8.

Hammer, K. (2003). Resolving the challenge posed by agrobiodiversity and plant genetic resources – an attempt. *Journal of Agriculture and Rural Development in the Tropics and Subtropics*, Supplement 76, 184 pp.

Hammer, K., Arrowsmith, N., and Gladis, Th. (2003a). Agrobiodiversity and plant genetic resources. *Naturwissenschaften*, **90**:241–250.

Hammer, K., Gladis, Th., and Diederichsen, A. (2003b). In situ and on-farm management of plant genetic resources. *European Journal of Agronomy*, **19**:509–517.

Hammer, K., Heller, J.A., and Engels, J. (2001). Monographs on underutilized and neglected crops. *Genetic Resources and Crop Evolution*, **48**:3–5.

Hammer, K., and Teklu, Y. (2008). Plant genetic resources: selected issues from genetic erosion to genetic engineering. *Journal of Agriculture and Rural Development in the Tropics and Subtropics*, **109**:15–50.

Harlan, J.R., and de Wet, J.M.J. (1971). Towards a rational classification of cultivated plants. *Taxon*, **20**:509–517.

Hazell, P.B.R. (2008). *An Assessment of the Impact of Agricultural Research in South Africa since the Green Revolution*. Consultative Group on International Agricultural Research (CGIAR), Science Council Secretariat, Rome, Italy.

Hoisington, D., Khairallah, M., Reeves, *et al.* (1999). Plant genetic resources: What can they contribute toward increased crop productivity? *Proceedings of the National Academy of Sciences*, **96**:5937–5943.

Joshi, K.D., and Witcombe, J.R. (2003). The impact of participatory plant breeding (PPB) on landrace diversity: A case study for high-altitiude rice in Nepal. *Euphytica*, **134**:117–125.

Kearsey, M.J. (1997). Genetic resources and plant breeding (identification, mapping and manipulation of simple and complex traits), in *Biotechnology and Plant Genetic Resources – Conservation and Use* (eds J.A. Callow, B.V. Ford-Lloyd, and H.N.Newberry). CAB International, Wallingford, UK, pp. 175–202.

Khoshbakht, K., and Hammer, K. (2010). *Threatened Crop Species Diversity*. Shahid Beheshti University Press, Tehran.

Maxted, N., and Guarino, L. (2006). Genetic erosion and genetic pollution of crop wild relatives, in *Genetic Erosion and Pollution Assessment Methodologies* (eds B.V. Ford-Lloyd et al.) IPGRI, Rome, Italy, pp. 35–46.

Ortiz, R. (2002). Germplasm enhancement to sustain genetic gains in crop improvement, in *Managing Plant Genetic Diversity* (eds J.M.M. Engels, V.R. Ramanatha, A.H.D. Brown, and M. Jackson). CABI Publishing, Wallingford, UK, pp. 275–290.

Peloquin, S.J., Yerk, G.L., Werner, J.E., and Darmo, E. (1989). Potato breeding with haploids and 2n gametes. *Genome*, **31**:1000–1004.

Perrings, C. (1998). The economics of biodiversity loss and agricultural development in low income countries. Paper presented at the American Association of Agricultural Economics International Conference on Agricultural Intensification, Economic Development, and the Environment, Salt Lake City, USA, 31 July 31 to August 1, pp. 52–75.

Rosegrant, M.W., and Cline, S.A. (2003). Global food security challenges and policies. *Science*, **302**:1917–1919.

Swanson, T. (1996). Global values of biological diversity: the public interest in the conservation of plant genetic resources for agriculture. *Plant Genetic Resources Newsletter*. **105**:1.

Veteläinen, M., Negri, V., and Maxted, N. (eds) (2009). *European landraces: on-farm conservation, management and use*. Bioversity Technical Bulletin No. 15, Biodiversity international, Rome, Italy.

co-evolved with pests and pathogens. Breeders should make deliberate efforts to expand the genetic base of their crops by exploiting genes from the wild progenitors of their species that are available in various germplasm repositories all over the world.

11.6.3 Paradigm shift

As D. Tanksley and S.R. McCouch of Cornell University point out, there is a need for a paradigm shift regarding the use of germplasm resources. Traditionally, breeders screen accessions from exotic germplasm banks on phenotypic basis for clearly defined and recognizable features of interest. Desirable genotypes are crossed with elite cultivars to introgress genes of interest. However, this approach is effective only for the utilization of simply inherited traits (conditioned by single, mostly dominant genes). The researchers proposed a shift from the old paradigm of looking for phenotypes to a new paradigm of looking for genes. To accomplish this, the modern techniques of genomics may be used to screen exotic germplasm by a gene-based approach. They proposed the use of molecular linkage maps and a new breeding technique called **advanced backcross QTL** introgression that allows the breeder to examine a subset of genes from the wild exotic plant in the genetic background of an elite cultivar.

11.6.4 Use of biotechnology to create new variability

The tools of modern biotechnology, such as recombinant DNA, cell fusion, somaclonal variation, and others, may be used to create new variability for use in plant breeding. Genetic engineering technologies may be used to transfer desirable genes across natural biological barriers.

11.6.5 Gene pyramiding

Plant breeders may broaden the diversity of resistance genes as well as introducing multiple genes from different sources into cultivars using the strategy of **gene pyramiding**, that is, introducing more than one resistance gene into one genotype. This approach will reduce the uniformity factor in crop vulnerability.

11.7 Wild (exotic) germplasm in plant breeding

Domestication has narrowed the genetic basis of many modern crop species. Breeders are returning to the reservoirs of genetic variations that were left behind by the process of domestication and thus exist in the wild, as well as in primitive germplasm such as landraces. Using wild germplasm in plant breeding has many challenges. It is time consuming and laborious; the unadapted germplasm has many undesirable genes; cross incompatibility between the wild and cultivated species often occurs; a successful cross often results in F_1 hybrid sterility and subsequent infertility of the segregating generations. Further, crosses exhibit reduced recombination between the wild and cultivated chromosomes as well as linkage drag (tight linkage of trait of interest in the wild genome to undesirable genes which are transferred to the cultivated genome).

Notwithstanding these and other challenges, plant breeders have used the method of introgression breeding to transfer agriculturally valuable traits from the wild into commercial cultivars. However, most of the successes have been achieved in the transfer of monogenic traits (e.g., disease resistance) leaving many other major traits like yield, quality, response of abiotic stress, and others relatively unattended to because of the complex genetics that underlie their inheritance (quantitative traits). These complex traits are controlled by quantitative trait loci (QTLs) which are being characterized through the use of modern molecular technologies.

The success and effectiveness of introgression of disease resistance genes into crop species from wild relatives varies by crop. Factors to consider include the amount of diversity within the crop species, ease of hybridization with wild relatives, and the complexity of the genetic control of the trait. Some crop breeders (e.g., tomato breeders) use genes from wild relatives more frequently than other breeders, such as sorghum breeders, who appear to find their needs in the domesticated crop species. The recurrent backcrossing approach previously mentioned has been used to improve over a dozen commercial cultivars of tomato. In tomato, wild relatives provided genes for enhancing the nutritional value (vitamin C and beta-carotene) and solids content, significantly boosting the commercial value of the crop. A specific example is the transfer of a gene, *B*, from the wild tomato,

Lycopersicon pennellii, which increased the level of provitamin A (beta-carotene) in the fruit by over 15-fold.

The impact of introgression of genes from the wild into adapted cultivars has been dramatic in some crops. For example, the resistance to the devastating late blight of potato was found in a wild species. Similarly, resistance to the root knot nematode in peanut was obtained from three wild species. A wild relative of rice, *Oryza nivara*, growing in the wild in Uttar Pradesh, India, was found to have one single gene for resistance to the grassy stunt virus, a disease that devastated the crop in South and South East Asia in the 1970s. In wheat, yield improvements associated with the transfer of a chromosomal segment carrying a rust resistance gene (*Lr19*) from *Agropyron elongatum* (tall wheat grass) has been reported. Similarly, a high-grain protein QTL from *Triticum dicoccoides*

(wild emer wheat) was introgressed into cultivated wheat to improve pasta flour quality.

In spite of these and other successes with introgression of wild genes into cultivated species, the use of this approach is anything but routine in plant breeding. Dani Zamir proposed the development of **exotic libraries** to facilitate and accelerate the exploitation of wild germplasm in breeding (Figure 11.2). His approach centers on developing a library of introgressed lines to be made available to breeders. At the present, should a breeder desire to re-screen the progeny of an interspecific cross that had been previously generated for a new trait, one would have to start from the parents to develop the required generation.

Even though modern technology has made it possible to develop a large number of informative genetic markers and high-density marker maps for

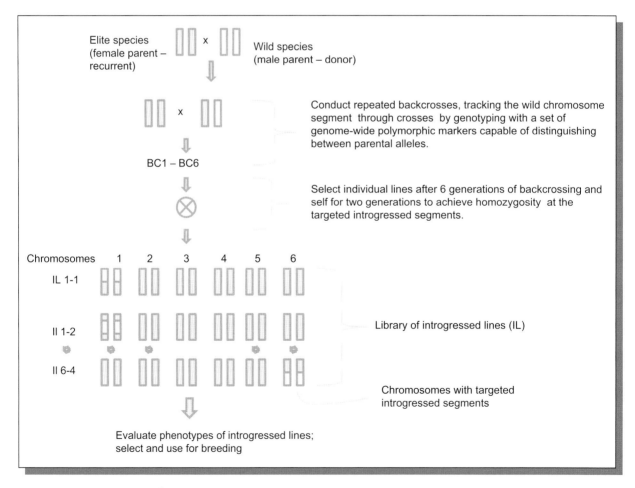

Figure 11.2 Generating and screening an exotic genetic library.

monogenic and QTLs that affect phenotypic variation, the problem still remains that in F_2, backcross, advanced backcross, and recombinant inbred populations resulting from divergent genomes, genetic analysis remains challenging because the presence of large proportion of wild genomes renders these populations partially sterile. Consequently, their use in identification of QTLs for improving yield and other important agronomic traits is problematic. Creating an exotic library would solve this problem.

An exotic library essentially consists of a set of lines, each carrying a single defined chromosome segment from a donor species in an otherwise uniform elite genetic background. It would take about 10 generations to create these lines, which would become immediately available for direct use in breeding. The lines would be excellent tools for detecting and mapping agronomic traits. Because the lines differ from the elite variety by only a single defined chromosomal segment, they would resemble the cultivated variety and have reduced sterility problems.

11.8 Plant genetic resources conservation

Plant breeders manipulate variability in various ways; for example, they assemble, recombine, select, and discard. The preferential use of certain elite genetic stock in breeding programs has narrowed the overall genetic base of modern cultivars. As already noted, pedigree analysis indicates that many cultivars of certain major crops of world importance have common ancestry, making the industry vulnerable to disasters (e.g., disease epidemic, climate changes). National and international efforts have been mobilized to conserve plant genetic resources.

11.8.1 Why conserve plant genetic resources?

There are several reasons why plant genetic resources must be conserved:

- Plant germplasm is exploited for food, fiber, feed, fuel, and medicines by agriculture, industry, and forestry.
- As a natural resource, germplasm is a depletable resource.
- Without genetic diversity, plant breeding cannot be conducted.

- Genetic diversity determines the boundaries of crop productivity and survival.
- To cater for future needs.

As society evolves, its needs will keep changing. Similarly, new environmental challenges might arise (e.g., new diseases, abiotic stresses, mechanical harvesting due to rise in labor costs) for which new variability might be needed for plant improvement.

When a genotype is unable to respond fully to the cultural environment, or able to resist unfavorable conditions thereof, crop productivity diminishes. The natural pools of plant genetic resources are under attack from the activities of modern society – urbanization, indiscriminate burning, clearing of virgin land for farming, to name a few. These and other activities erode genetic diversity in wild populations. Consequently, there is an urgent need to collect and maintain samples of natural variability. The actions of plant breeders also contribute to genetic erosion as previously indicated. High-yielding narrow genetic based cultivars are penetrating crop production systems all over the world, displacing the indigenous high-variability landrace cultivars. Some 20 000 plant species are suggested to be endangered.

11.8.2 Genetic erosion

Genetic erosion may be defined as the decline in genetic variation in cultivated or natural populations largely through the action of humans. Loss of genetic variation may be caused by natural factors, and by the actions of crop producers, plant breeders, and others in the society at large.

- **Natural factors.** Genetic diversity can be lost through natural disasters such as large-scale floods, wild fires, and severe and prolonged drought. These events are beyond the control of humans.
- **Action of crop producers.** Right from the beginnings of agriculture, farmers have engaged in activities that promote genetic erosion. These include clearing of virgin land in, especially, germplasm-rich tropical forests and the choice of planting material (narrow genetic base cultivars). Farmers, especially in developed economies, primarily grow improved seed that are genetically pure, having replaced most or all landraces with these superior cultivars. Also, monoculture, instead of mixed cropping, tends promotes the narrowing of genetic diversity as large

tracts of land are planted to uniform cultivars. Extending grazing lands into wild habitats by livestock farmers destroys wild species and wild germplasm resources.

- **Action of breeders.** Farmers plant what breeders develop, and breeders develop what farmers and consumers and industry demand. Some methods used for breeding (e.g., pure lines, single cross, multilines) promote uniformity and narrower genetic base. When breeders find superior germplasm, the tendency is to use it as much as possible in cultivar development. In soybean, as previously indicated, most of the modern cultivars in the United States can be traced back to about half a dozen parents. This practice causes severe reduction in genetic diversity.
- **Problems with germplasm conservation.** In spite of good efforts by curators of germplasm repositories to collect and conserve diversity, there are several ways in which diversity in their custody may be lost. The most obvious loss of diversity is attributed to human errors in the maintenance process (e.g., improper storage of materials leading to loss of germination, and hence loos off accessions and the variation they represent). Also, when germplasm is planted in the field, natural selection pressure may eliminate some unadapted genotypes. Hybridization, as well as genetic drift cause shifts in allele frequencies in small populations. They occur as consequences of periodic multiplication of the germplasm holdings by curators.
- **General public action.** As previously indicated, there is an increasing demand on land with increasing population. Such demands include settlement of new lands and the demand for alternative use of the land (e.g., for recreation, industry, roads) to meet the general needs of modern society. These actions tend to place wild germplasm in jeopardy. Such undertakings often entail clearing of virgin land where wild species occur.

11.8.3 Selected impact of germplasm acquisition

Impact on North American agriculture

Very few crops have their origin in North America. It goes without saying that North American agriculture owes its tremendous success to plant introductions, which brought major crops such as wheat, barley, soybean, rice, sugar cane, alfalfa, corn, potatoes, tobacco, and cotton to this part of the world. North America currently is the world's leading producer of many of these crops. Spectacular contributions by crop introductions to US agriculture include the following:

- Avocados – Introduced in 1898 from Mexico, this crop has created a viable industry in California.
- Rice – Varieties introduced from Japan in 1900 laid the foundation for the present rice industry in Louisiana and Texas.
- Spinach – A variety introduced from Manchuria in 1900 is credited with saving the Virginia spinach industry from blight disaster in 1920.
- Peach – Many US peach orchards are established by plants growing on root stalks obtained from collections in 1920.
- Oat – One of the world's most disease-resistant oat varieties was developed from germplasm imported from Israel in 1960s.

Other parts of the world

A few examples include dwarf wheats introduced into India, Pakistan, and Philippines as part of the Green Revolution, and introduction of soybeans and sunflower into India, have benefited the agriculture of these countries.

11.9 Nature of cultivated plant genetic resources

Currently, five kinds of cultivated plant materials are conserved by concerted worldwide efforts – **landraces (or folk, primitive varieties), obsolete varieties, commercial varieties (cultivars), plant breeder's lines,** and **genetic stocks**. Landraces are developed by indigenous farmers in various traditional agricultural systems or are products of nature. They are usually very variable in composition. Obsolete cultivars may be described as "ex-service" cultivars because they are no longer used for cultivation. Commercial cultivars are elite germplasm currently in use for crop production. These cultivars remain in production usually from 5–10 years before becoming obsolete and replaced. Breeder's lines may include parents that are inbred for hybrid breeding, genotypes from advanced yield tests that were not released as commercial cultivars, and unique mutants. Genetic stocks are genetically characterized lines of various species. These are advanced genetic materials developed by breeders, and are very useful and readily accessible to other breeders.

11.10 Approaches to germplasm conservation

There are two basic approaches to germplasm conservation – *in situ* and *ex situ*. These are best considered as complementary rather than independent systems.

11.10.1 *In situ* conservation

This is the preservation of variability in its natural habitat in its natural state (i.e., on site). It is most applicable to conserving wild plants and entails the use of legal measures to protect the ecosystem from encroachment by humans. These protected areas are called by various names (e.g., nature reserves, wildlife refuges, natural parks). Needless to say, there are various socioeconomic and political ramifications in such legal actions by the government. Environmentalists and commercial developers often clash on such restricted use or prohibited use of natural resources, imposed by the government. This approach to germplasm conservation is indiscriminatory with respect to species conserved (i.e., all species in the protected area are conserved).

11.10.2 *Ex situ* conservation

In contrast to *in situ* conservation, *ex situ* conservation entails planned conservation of targeted germplasm. Germplasm is conserved not in the natural places of origin but under supervision of professionals off site in locations called germplasm or gene banks. Plant materials may be in the form of seed or vegetative materials. The advantage of this approach is that small samples of the selected species are stored in a small space indoors or in a field outdoors, and under intensive management that facilitates their access to breeders. However, the approach is prone to some genetic erosion. The special care needed is expensive to provide. Other aspects of this approach are discussed later in this chapter.

11.11 Germplasm collection

Planned collections (germplasm explorations or expeditions) are conducted by experts to regions of crop origin. These trips are often multidisciplinary, comprising members with expertise in botany, ecology, pathology, population genetics, and plant breeding.

Familiarity with the species of interest and the culture of the regions to be explored are advantageous. Most of the materials collected are seeds, even though whole plants and vegetative parts (e.g., bulbs, tubers, cuttings, etc.) and even pollen may be collected. Because only a small amount of material is collected, sampling for representativeness of the population's natural variability is critical in the collection process, in order to obtain maximum possible amount of genetic diversity. For some species whose seed is prone to rapid loss of viability, or are bulky to transport (e.g., tropical fruits), *in vitro* techniques may be available to extract small samples from the parent source. Collectors should bear in mind that the value of the germplasm may not be immediately discernible. Materials should not be avoided for lack of obvious agronomically desirable properties. It takes time to discover the full potential of germplasm.

Seed materials vary in viability characteristics. These have to be taken into account during germplasm collection, transportation, and maintenance in repositories. Based on viability, seed may be classified into two main groups – **orthodox** and **recalcitrant** seed.

(i) **Orthodox seeds.** These are seeds that can prolong their viability under reduced moisture content and low temperature in storage. Examples include cereals, pulses, and oil seed like rape.

(ii) **Recalcitrant seeds.** Low temperature and decreased moisture content are intolerable to these seeds (e.g., coconut, coffee, cocoa). *In vitro* techniques might be beneficial to these species for long-term maintenance.

The conditions of storage of gemplasm differ depending on the mode of reproduction of thespecies.

- **Seed propagated species.** Seeds from seed propagated species are first dried to about 5% moisture content and then usually placed in hermetically sealed moisture-proof containers before storage. Seeds so stored can retain viability for 10–15 years.

- **Vegetatively propagated species.** These materials may be maintained as full plants for long periods in field gene banks, nature reserves, or botanical gardens. Alternatively, cuttings and other vegetative parts may be conserved for a short period under moderately low temperature and humidity. For long-term storage, *in vitro* technology is used.

11.12 Types of plant germplasm collections

Germplasm is stored in four categories – **base collections, backup collections, active collections, and breeders' or working collection**. These categorizations are only approximate since one group can fulfill multiple functions.

11.12.1 Base collections

These collections are not intended for distribution to researchers, but are maintained in long-term storage systems. This collection is the most comprehensive of the genetic variability of a species. Entries are maintained in the original form. Storage conditions are low humidity at freezing temperatures (-10 to $-20\,°C$) or cryogenic (-150 to $-195\,°C$), depending on the species. Materials may be stored for many decades under proper conditions.

11.12.2 Backup collections

The purpose of backup collections is to supplement the base selection. In case of a disaster at a center responsible for a base collection, a duplicate collection is available as insurance. In the United States, the National Seed Storage Laboratory at Fort Collins, Colorado, is a backup collections center for portions of the accessions of CIMMYT and IRRI (Chapter 30). On the international level, the Svalbard Global Seed Vault (the so-called doomsday seed bank) is a secure seed bank established in 2008 and located in an underground cavern in Norway, near the North Pole. The seed inventory represents duplicate samples of seed currently held in gene banks worldwide. This purpose of this unique collection is to provide an insurance against the loss of seed in the event of a large scale catastrophic regional or global crisis.

11.12.3 Active collections

Germplasm in base and backup collections is designed for long-term unperturbed storage. Active collections usually comprise the same materials as in base collections. However, the materials in active collections are available for distribution to plant breeders or other patrons upon request. They are stored at $0\,°C$ and about 8% moisture content, and remain viable for about 10–15 years. To meet this obligation, curators of active collections at germplasm banks must increase the germplasm to fill requests expeditiously. Because the accessions are more frequently increased through field multiplication, the genetic integrity of the accession may be jeopardized.

11.12.4 Working or breeders' collections

Breeders' collections are primarily composed of elite germplasm that is adapted. They also include enhanced breeding stocks with unique alleles for introgression into these adapted materials. In these times of genetic engineering, breeders' collections include products of recombinant DNA (rDNA) research that can be used as parents in breeding programs.

11.13 Managing plant genetic resources

The key activities of curators of germplasm banks include regeneration of accessions, characterization, evaluation, monitoring seed viability and genetic integrity during storage, and maintaining redundancy among collections and opening-up the collection to potential users. Germplasm banks receive new materials on a regular basis. These materials must be properly managed so as to encourage and facilitate their use by plant breeders and other researchers.

11.13.1 Regeneration

Germplasm needs to be periodically rejuvenated and multiplied. Regeneration of seed depends on the life cycle and breeding system of the species as well as cost of the activity. To keep costs to a minimum and to reduce loss of genetic integrity, it is best to keep regeneration and multiplication to a bare minimum. It is good strategy to make the first multiplication extensive so that ample original seed would be available for depositing in the base and duplicate or active collections. The methods of regeneration vary for self-pollinated, cross-pollinated, clonal (or vegetatively propagated) and apomictic species. A major threat to genetic integrity of accessions during regeneration is contamination (from outcrossing or accidental migration), which can change the genetic structure. Other factors include differential survival of alleles or genotypes within the accession, and random drift. Isolation of accessions during regeneration is critical to maintaining genetic integrity. This is achieved through proper spacing, caging, covering with bags, hand pollination, and other techniques. Regeneration of wild species is problematic because of high seed dormancy, seed shattering, high variability in flowering time, and low seed production. Some species have special environmental requirements (e.g., photoperiod, vernalization), and hence it is best to rejuvenate plants under conditions similar to those in the places of their origin, to prevent selection effect, which can eliminate certain alleles.

11.13.2 Characterization

Users of germplasm need some basic information about the plant materials to aid them in effectively using these resources. Curators of germplasm banks characterize their accessions, an activity that entails a systematic recording of particular traits of an accession. Traditionally, these data are limited to highly heritable morphological and agronomic traits. However, with the availability of molecular techniques, some germplasm banks have embarked upon molecular characterization of their holdings. For example, CIMMYT has used the simple sequence repeat (SSR) marker system for characterizing the maize germplasm in its holding. **Passport data** are included in germplasm characterization. These data include an accession number, scientific name, collection site (country, village), source (wild, market), geography of the location, and resistance to disease and insect pests. To facilitate data entry and retrieval, characterization includes the use of **descriptors**. These are specific pieces of information on plant or geographic factors that pertain to the plant collection. The International Plant Genetic Resources Institute (IPGRI) has prescribed guidelines for the categories of these descriptors. Descriptors have been standardized for some species. For example, for corn, the descriptors of kernel type are: floury, semi-floury, dent, semident, semi-flint, flint, pop, sweet, opaque2/QPM, tunicate, waxy. The person characterizing the corn kernel may indicate up to three kernel types.

11.13.3 Evaluation

Genetic diversity is not usable without proper evaluation. Preliminary evaluation consists of readily observable traits. Full evaluations are more involved and may include obtaining data on cytogenetics, evolution, physiology, and agronomy. More detailed evaluation is often done outside of the domain of the germplasm bank by various breeders and researchers using the specific plants. Traits such as disease resistance, productivity, and quality of product are important pieces of information for plant breeders. Without some basic information of the value of the accession, users will not be able to make proper requests and receive the most useful materials for their work. Curators that send the accessions often request their customers to report back in return information on the performance of the accessions.

11.13.4 Monitoring seed viability and genetic integrity

During storage, vigor tests should be conducted at appropriate intervals to ensure that seed viability remains high. During these tests, abnormal seedlings may indicate the presence of mutations or deterioration of viability.

11.13.5 Exchange

The ultimate goal of germplasm collection, rejuvenation, characterization, and evaluation is to make available and facilitate the use of germplasm. There are various computer-based genetic resource documentation systems worldwide, some of which are crop-specific. These systems allow breeders to rapidly search and request germplasm information. There are various laws regarding, especially, international exchange of germplasm. Apart from quarantine laws, various inspections and testing facilities are needed at the point of germplasm introduction.

11.14 Issue of redundancy and the concept of core subsets

Collections for major crops such as wheat and corn can be very large. Some of these accessions are bound to be duplicates. Because of the cost of germplasm maintenance, it is important for the process to be efficient and effective. Redundancy should be minimized in the collections. However, eliminating duplicates may be as expensive as maintaining them. The sources of duplication are numerous. Some names could be misspelled (Henni vs Hennie) or may be transcription errors from poor penmanship. Language (alphabet) difference could result in variation in spelling (e.g., K used in place of C). To facilitate the management of huge accessions, the concept of **core subsets** was proposed. A core subset comprises a sample of the base collection of a germplasm bank that represents the genetic diversity in the crop and its relatives, with minimum redundancy. The core would be well characterized and evaluated for ready access by users. However, some argue that maintaining a core subset might distract from maintaining the entire collection, leading possibly to loss of some accessions.

11.15 Germplasm storage technologies

Once collected, germplasm is maintained in the most appropriate form by the gene bank with storage responsibilities for the materials. Plant germplasm may be stored in the form of pollen, mostly seed, or plant tissue. Woody ornamental species may be maintained as living plants, as occurs in arboreta. Indoor maintenance is done under cold storage conditions, with temperatures ranging from -18 to $-196\,°C$.

11.15.1 Seed storage

Seeds are dried to the appropriate moisture content before placing in seed envelopes. These envelopes are then arranged in trays that are placed on shelves in the storage room. The storage room is maintained at $-18\,°C$, a temperature that will keep most seeds viable for up to 20 years or more. The curator of the laboratory and the staff periodically sample seeds of each accession to conduct a germination test. When germination falls below a certain predetermined level, the accession is regrown to obtain fresh seed.

11.15.2 Field growing

Accessions are regrown to obtain fresh seed or to increase existing supply (after filling orders by scientists and other clients). To keep the genetic purity, the accessions are grown in isolation, each plant covered with a cotton bag to keep foreign sources of pollen out and also to ensure self-pollination.

11.15.3 Cryopreservation

Cryopreservation or freeze preservation is the storage of materials at extremely low temperatures of between -150 and $-196\,°C$ in liquid nitrogen. Plant cells, tissue, or cuttings may be stored this way for a long time without losing regenerative capacity. Whereas seed may also be stored by this method, cryopreservation is reserved especially for vegetatively propagated species that need to be maintained as living plants. Shoot tip cultures are obtained from the material to be stored and protected by dipping in a cryoprotectant (e.g., a mixture of sugar and polyethylene glycol plus dimethylsulfoxide).

11.15.4 *In vitro* storage

Germplasm of vegetatively propagated crops is normally stored and distributed to users in vegetative forms such as tubers, corms, rhizomes, and cuttings. However, it is laborious and expensive to maintain plants in these forms. *In vitro* germplasm storage usually involves tissue culture. There are several types of tissue culture systems (suspension cells, callus, meristematic tissues). To use suspension cells and callus materials, there must be an established system of regeneration of full plants from these systems, something that is not available for all plant species yet. Consequently, meristem cultures are favored in *in vitro* storage because they are more stable. The tissue culture material may be stored by the method of slow growth (chemicals are applied to retard the culture temperature), or cryopreservation.

11.15.5 Molecular conservation

The advent of biotechnology has made it possible for researchers to sequence DNA of organisms. These sequences can be searched (see bioinformatics in Chapter 25) for genes at the molecular level. Specific genes can be isolated by cloning and used in developing transgenic products.

11.16 Using genetic resources

11.16.1 Perceptions and challenges

Breeders, to varying degrees, acknowledge the need to address the genetic vulnerability of their crops. Further, they acknowledge the presence of large amounts of genetic variation in wild crop relatives.

However, much of this variability is not useful to modern plant breeding. In using wild germplasm, there is a challenge to sort out and detect those that are useful to breeders. Modern cultivars have resulted from years of accumulation of favorable alleles that have been gradually assembled into adapted interacting multilocus combinations. Introgression of new genes may jeopardize these combinations through segregation and recombination. Hence, some breeders are less inclined to use unadapted germplasm. However, there are occasions on which the breeder has little choice but to take the risk of using unadapted germplasm (e.g., specific improvement of traits such as new races of disease, quality issues), because alleles for addressing these problems may be non-existent in the adapted materials. Plant breeders engaged in breeding of plant species that have little or no history of improvement are among the major users of active collections in germplasm banks. For such breeders, they may have no alternative but to evaluate primitive materials to identify those with promise for use as parents in breeding.

Plant breeders may use germplasm collections in one of two basic ways: (i) as sources of cultivars or (ii) as sources of specific genes. Breeder's collections contain alleles for specific traits that breeders can transfer into adapted genotypes using appropriate breeding methods. Accessions must be properly documented to facilitate the search by users. This means there should be accurate passport and descriptor information for all accessions. Unfortunately, this is not the case for many accessions.

The redundancy in germplasm banks is viewed by some breeders as unacceptable. A study showed that of the 250 000 accessions of barley at that time in repositories, only about 50 000 were unique, the rest being duplicates. Such discrepancy leads to false estimation of the true extent of diversity in the world collection. A large number of the accessions are also obsolete. Germplasm evaluation at the level of germplasm banks is very limited, making it more difficult for users to identify accessions with promise for breeding.

11.16.2 Concept of pre-breeding

Plant breeders usually make elite × elite crosses in a breeding program. This practice, coupled with the fact that modern crop production is restricted to the use of highly favored cultivars, has reduced crop genetic diversity and predisposed crop plants to disease and pest epidemics. To reverse this trend, plant breeders need to make deliberate efforts to diversify the gene pools of their crops to reduce genetic vulnerability. Furthermore, there are occasions on which breeders are compelled to look beyond the advanced germplasm pool to find desirable genes. The desired genes may reside in unadapted gene pools. Breeders are often reluctant to use such materials because the desired genes are often associated with undesirable effects (unadapted, unreproductive, yield-reducing factors). Hence, these exotic materials often cannot be used directly in cultivar development. Instead, the materials may be gradually introduced into the cultivar development program through crossing and selecting for intermediates with new traits, while maintaining a great amount of the adapted traits.

To use wild germplasm, the unadapted material is put through a preliminary breeding program to transfer the desirable genes into adapted genetic backgrounds. The process of the initial introgression of a trait from an undomesticated source (wild) or agronomically inferior source to a domesticated or adapted genotype is called **pre-breeding** or **germplasm enhancement**. The process varies in complexity and duration, depending on the source, the type of trait, and presence of reproductive barriers. It may be argued that pre-breeding is not an entirely new undertaking, considering the fact that all modern crops were domesticated through this process. The early farmer-selectors did what came naturally, discriminating among natural variation without deliberately hybridizing genotypes, and gradually moving them from wild to adapted domesticated domain.

The traditional techniques used are hybridization followed by backcrossing to the elite parent, or the use of cyclical population improvement techniques (recurrent selection). The issues associated with wide crossing are applicable (e.g., infertility, negative linkage drag, incompatibility), requiring techniques such as embryo rescue to be successful. The modern tools of molecular genetics and other biotechnological procedures are enabling radical gene transfer to be made into elite lines without linkage drag (e.g., transfer of genes from bacteria into plants; Chapters 14). This new approach to development of new breeding materials is more attractive and profitable to private investors (for-profit breeding programs). Such creations can be readily protected by patents for commercial exploitation. Further, these technologies are enabling plant breeders not only to develop new and improved highly productive cultivars but also to assign new roles to cultivars (e.g., plants can now be used as bioreactors for producing novel traits such as specialized oils, proteins, and pharmaceuticals).

The major uses of germplasm enhancement may be summarized as:

- Preventions of genetic uniformity and the resulting genetic vulnerability.
- Potential crop yield augmentation. History teaches us that some of the dramatic yield increases in major world food crops such as rice, wheat, and sorghum, were accomplished through introgression of unadapted genes.
- Introduction of new quality traits (e.g., starch, protein).
- Introduction of disease and insect resistance genes.
- Introduction of abiotic stress resistance genes (e.g., drought resistance)

Pre-breeding can be expensive to conduct and time consuming as well. With the exception of high value crops, most pre-breeding is conducted in the public sector. The Plant Variety Protection law (Chapter 28) does not provide adequate financial incentive for for-profit (commercial) breeders to invest resources in germplasm enhancement.

11.17 Plant explorations and introductions and their impact on agriculture

11.17.1 Plant explorations

Plant exploration is an international activity. Recent political developments are making germplasm collections less of an open access activity. Explorers must obtain permission to enter a country to collect plant material. Most of these germplasm rich places are located in developing countries, which frequently complain about not reaping adequate benefits from contributing germplasm to plant breeding. Consequently, these nations are increasingly prohibiting free access to their natural resources.

US historical perspectives

The USDA plant germplasm collection efforts began in 1898 under the leadership of David Fairchild. Fairchild collected pima cotton, pistachios, olives, walnuts, and other crop materials. Other notable personnel in the plant exploration efforts by the United States include S.A. Knapp, whose rice collection from Japan is credited with making the United States a rice exporting country. Tropical fruits collected by W. Popenoe from South and Central America, also created new industries in the United States. F.N. Meyer made outstanding collections between 1905 and 1918, mainly from Asia and Russia (e.g., alfalfa, apples, barley, melons, elms, dwarf cherry). One of his most notable collections was the soybean. Prior to his Chinese explorations, there were only eight varieties of soybean grown in the United States, all for forage. This picture changed between 1905 and 1908 when Meyer introduced 42 new soybean varieties into the United States, including seed and oil varieties, that helped to make the United States a world leader in soybean production. The current US system of plant inventory was established by Fairchild. The first accession, PI 1, was a cabbage accession from Moscow, Russia, collected in 1898. PI 600,000 is a pollinator sunflower with dwarf features, developed by ARS breeders.

Other efforts

Potato introduction to Europe and the introduction of maize and millet to Africa and Asia are examples of the impact of plant introductions on world food and agriculture. In fact, the Green Revolution depended on introductions of dwarf wheat and rice into India, Pakistan, and the Philippines.

11.17.2 Plant introductions

Plant introduction is the process of importing new plants or cultivars of well-established plants from the area of their adaptation to another area where their potential is evaluated for suitability for agricultural or horticultural use. Firstly, the germplasm to be introduced is processed through a plant quarantine station at the entry port, to ensure that no pest and diseases are introduced along with the desired material. Once this is accomplished, the material is released to the researcher for evaluation in the field for adaptation. The fundamental process of plant introductions as a plant breeding approach is acclimatization. Acclimatization is the process whereby the breeder exposes the plant to a gradual change in its environment (e.g., change in humidity, photoperiod, temperature, pH) over a short period of day or weeks. The purpose of this treatment is to allow the plant to maintain performance over a range of environmental condition. The inherent genetic variation in the introduced germplasm serves as the raw material for adaptation to the new environment, enabling the breeder to select superior performers to for cultivar development.

When the plant introduction is commercially usable and was introduced without any modification, it is called a **primary introduction**. However, more often than not, the breeder makes selections from the variable population, or uses the plant introduction as a parent in crosses. The products of such efforts are called **secondary introductions**. Some PIs may not be useful as cultivars in the new environment. However, they may be useful in breeding programs for specific genes they carry. Many diseases, plant stature, compositional traits, and genes for environmental stresses have been introduced by plant breeders.

As a plant breeding method, plant introductions have had a significant impact on world food and agriculture, one of the most spectacular stories being the transformation of US agriculture as previously indicated. One of the most successful agricultural nations in the world, US agriculture is built on plant introductions, since very few plants originated on that continent. The United States either leads the world or is among the top nations in the production of major world crops such as wheat, maize, rice, and soybean.

11.18 International conservation efforts

The reality of the matter of germplasm transactions is that a truly international cooperation is needed for success. No one country is self-sufficient in its germplasm needs. Most of the diversity resides in tropical and subtropical regions of the world where most developing nations occur. These germplasm-rich nations, unfortunately, lack the resources and the technology to make the most use of this diversity.

International cooperation and agreements are needed for the exploitation of these resources for the mutual benefit of donor and recipient countries.

Vavilov collected more than 250 000 plant accessions during the period of his plant collection expeditions. This collection currently resides at the All-Union Institute of Plant Industry in St. Petersburg. The Food and Agricultural Organization (FAO) of the United Nations (UN) is credited with the initial efforts to promote genetic conservation and assistance in establishing the International Board of Plant Genetic Resources (IBPGR) based in the FAO in Rome, Italy. Founded in 1974, the IBPGR is funded by donor countries, development banks, and foundations. It is a center in the Consultative Group of International Agriculture Research (CGIAR). The primary role of this board is to collect, preserve, evaluate, and assist with the exchange of plant genetic material for specific crops all over the world.

A major sponsor of these genetic conservation activities is the International Agricultural Research Centers (IARCs) strategically located throughout the tropics (Chapter 25). Gene banks at these centers focus on starchy crops that feed the world (wheat, corn, rice, potato, sorghum). These crops are often grown with high-tech cultivars that have narrow genetic base as a result of crop improvement.

There are other regional and country-based plant germplasm conservation programs. The EUCARPIA (The European Association for Plant Breeding Research), started in 1960, serves Europe and the Mediterranean region. Similarly, the Vegetable Gene Bank at the National Vegetable Research Station in the United Kingdom was established in 1981 to conserve vegetable genetic resources.

11.19 An example of a national germplasm conservation system

The US plant genetic conservation efforts are coordinated by the National Plant Germplasm System (NPGS). It is a cooperative effort by public (state and federal) and private organizations. The collections are stored at 20 different locations all over the United States, and coordinated by the Agricultural Research Service (ARS) located in Beltsville, Maryland. Over

400 000 accessions exist in the inventory of NPGS in the form of seed and vegetative material. As of July 1, 2008, there were 505 770 samples, 384 876 unique accessions, 1180 genera, 6968 species, 380 727 seed accessions, and 4177 vegetatively propagated accessions. The accessions in the NPGS are estimated to increase at the rate of 7000–15 000 new entries per year. The system has certain component units with specific functions as follows:

11.19.1 Plant introduction

Located in Beltsville, Maryland, the Plant Introduction Office is part of the Plant Genetics and Germplasm Institute of the United States Department of Agriculture/Agricultural Research Services (USDA/ARS). Each entry is given a plant introduction (PI) number, but this unit does not maintain any plant material collection. The responsibilities to maintain, evaluate, and release plant materials are assigned to four Regional Plant Introduction Stations (Western, North Central, North Eastern, and Southern) (Figure 11.3). The Plant Quarantine Facility of the USDA and the Animal and Plant Health Inspection Service

(APHIS) both operate from the Plant Introduction Station at Glenn Dale, Maryland. These regional centers were established in 1946 for several purposes: (a) to determine the germplasm needs within the region, (b) to assist with foreign explorations to fill regional needs, (c) to multiply, evaluate, and maintain new plant and seeds collections of crops adapted to the regions with minimal loss of genetic variability within the strains, and (d) to distribute the seed and plant accessions to plant scientists worldwide. These collections come from many countries. For example, in the National Small Grains Collection, the accessions for wheat, barley, and rice come from over 100 countries or regions.

11.19.2 Collections

The base collections of the United States are maintained at the National Seed Storage Laboratory at Fort Collins, Colorado. These collections are seldom regrown to avoid possible genetic changes. The laboratory provides long-term backup storage for the NPGS. In addition to seed, there are National Clonal Repositories for maintaining clonal

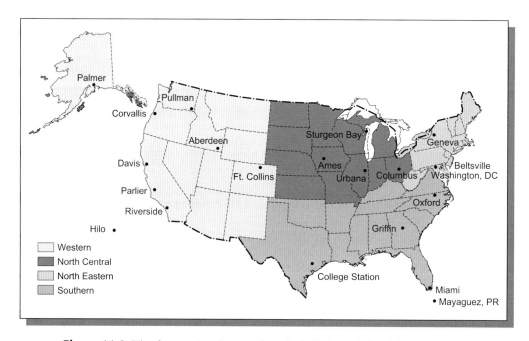

Figure 11.3 The four regional germplasm jurisdictions defined by the USDA.

Table 11.1 Germplasm holdings in germplasm banks in the United States.

Repository	Location	Germplasm	Holding
Barley Genetic Stock Center	Aberdeen, ID	Barley	3262
Clover Collection	Lexington, KY	Clover	246
Cotton Collection	College Station, TX	Cotton	9536
Database Management Unit	Beltsville, MD		
Desert Legume Program	Tuscon, AZ	Various	2585
Maize Genetics Stock Center	Urbana, IL	Maize	4710
Pea Genetic Stock Center	Pullman, WA	Pea	501
National Arctic Plant Genetic Resources Unit	Palmer, AK	Various	515
National Arid Land Plant Genetic Resources Unit	Parlier, CA	Various	1177
National Clonal Germplasm Repository	Corvallis, OR	Various	12 943
National Clonal Germplasm Repository	Riverside, CA	Citrus, dates	1167
National Clonal Germplasm Repository	Davis, CA	Tree fruit, nuts, grape	5397
National Germplasm Resources Laboratory	Beltsville, MD	Various	252
National Center for Genetic Resources Preservation	Fort Collins, CO	Various	23 007
National Small Grains Collection	Aberdeen, ID	Barley, others	126 883
National Temperate Forage Legume Genetic Resources Unit	Prosser, WA	Various	
North Central Regional Plant Introductions Stations	Ames, IA	Various	47 684
Northeast Regional Plant Introduction Station	Geneva, NY	Various	11 690
Ornamental Plant Germplasm Center	Columbus, OH	Various	2271
Pea Genetic Stock Collection	Pullman, WA	Pea	501
Pecan Breeding and Genetics	Brownwood and Somerville, TX	Pecan	881
Plant Genetic Resources Conservation Unit	Griffin, GA	Various	
Plant Genetic Resources Unit	Geneva, NY	Various	5243
Plant Germplasm Quarantine Office	Beltsville, MD	Various	4641
Rice Genetic Stock Center	Stuttgart, AR	Rice	19
Southern Regional Plant Introduction Station	Griffin, GA	Various	83 902
Soybean/Maize Germplasm, Pathology and Genetics Research Unit	Urbana, IL	Soybean, maize	20 601
Subtropical Horticulture Research Station	Miami, FL	Various	4779
Tobacco Collection	Oxford, NC	Tobacco	2106
Tomato Genetics Resource Center	Davis, CA	Tomato	3381
Tropical Agriculture Research Station	Mayaguez, Puerto	Rico Various	652
Tropical Plant Genetic Resource Management Unit	Hilo, HI	Various	692
United States Potato Genebank	Sturgeon Bay, WI	Potato	5648
Western Regional Plant Introduction Station	Pullman, WA	Various	72 190
Wheat Genetic Stocks Center	Aberdeen, ID	Wheat	334
Woody Landscape Plant Germplasm Repository (National Arboretum)	Washington, DC	Various	1904
Total germplasm holdings as of August 29, 2004			**460 799**

germplasm. These include Davis, California (for grapes, nuts, and stone fruits) and Miami, Florida (for subtropical and tropical fruits and sugarcane). The locations and crop mandates of 32 plant germplasm conservation programs in the NPGS are presented in Table 11.1. A summary of the germplasm holdings at each location as of August 2004 is provided in Table 11.2. Plant breeders have access to the accessions in these active collections.

Table 11.2 Germplasm holdings at the International Agricultural Research Centers.

International Center	Germplasm type	Holdings
International Rice Research Institute (IRRI)	Rice	80 617
International Center for the Improvement of Maize and Wheat (CIMMYT)	Wheat	95 113
	Maize	20 411
International Center for Tropical Agriculture (CIAT)	Forages	18 138
	Bean	31 718
International Institute of Tropical Agriculture (IITA)	Bambara groundnut	2029
	Cassava	2158
	Cowpea	15 001
	Soybean	1901
	Wild *Vigna*	1634
	Yam	2878
International Potato Center (CIP)	Potato	5057
	Sweet potato	6413
	Andean roots/tubers	1112
International Center for Agricultural Research in the Dry Areas (ICARDA)	Barley	24 218
	Chickpea	9116
	Faba bean	9074
	Forages	24 581
	Lentil	7827
	Wheat	30 270
International Crops Research Institute for the Semi-Arid Tropics (ICRISAT)	Chickpea	16 961
	Groundnut	14 357
	Pearl millet	21 250
	Minor millets	9050
	Pigeon pea	12 698
	Sorghum	35 780
West African Rice Development Association (WARDA)	Rice	14 917
International Center for Research in Agroforestry (ICRAF)	*Sesbania*	25
International Livestock Research Institute (ILRI)	Forages	11 537
International Plant Genetic Resources Institute (IPGRI)/International Network for the Improvement of Banana and Plantain (INIBAP)	Musa	931
Total		**532 508**

11.19.3 Information

The information on the accessions in the NPGS has been computerized to facilitate its dissemination. The system, Germplasm Resources Information Network, is located at the Beltsville USDA research center.

11.20 Who owns biodiversity?

Biodiversity has been defined as the sum of genetic differences existing in living organisms at the molecular, individual, population, and ecosystem levels. It is indispensable to the development of new and improved cultivars. The need to conserve and study biodiversity stems from the fact that biodiversity is being rapidly eroded by human activities, not the least of which are overpopulation and habitat destruction. The question of ownership of biodiversity became more important beginning in the 1970s, partly for political as well as economic reasons. Prior to that time, biodiversity was treated as common property to humankind (common heritage) and, hence, held in the public domain (not owned by an individual, group, or country). Further, it was supposed to be used in a manner such that its use by one person does not compete with the use by another, or exclude another person from using it. It was this principle of common heritage (open access and exchange) that caused the free diffusion of genetic resources from centers of plant domestication into new production areas by explorers and farmers. Similarly, the establishment and success of germplasm repositories (gene banks) at the international centers (IARCs) and in different countries owe their presence in large part to the common heritage principle, which in 1972 received legal status through the UNESCO Convention for the Protection of the World Cultural and Natural Heritage.

Notwithstanding legal status or assumed communal ownership, it must be pointed out that biodiversity resides within the borders of sovereign nations. Consequently, access to biodiversity is not without restrictions. For example, the governments of Peru and Bolivia restrict the export of planting materials of quinine (*Cinchona*) while the Ethiopia restricts *Coffea arabica* export. On the other hand, some of the international interventions include provisions for the management of biodiversity for the benefit of humankind, the exclusive use for peaceful purposes, and the ready access for scientific research. The common heritage principle is currently under assault, nations exercising their sovereignty to impose intellectual property rights on biodiversity occurring within their borders. Needless to say, this is controversial, bringing to the open the question of ownership of biodiversity.

The advent of recombinant DNA technology and other advances in science and technology have introduced a new variable into the biodiversity discussion. Genetic modification technologies have enabled novel genetic diversity (transgenic plants) to be developed to accomplish the purpose for which natural biodiversity is sought by plant breeders. Genes can now be routinely cloned and characterized; some genomes have been entirely sequenced. The molecular technologies offer scientists the capacity to obtain and store desired genes (DNA) in chemical form instead of obtaining and storing the entire organism (living form) from which the gene originated. The US Patent and Trademark Office recognizes genes and gene sequences as man-made novelties that are patentable. The question then is why cannot organisms be patented – if genes being natural entities can be patented, why not whole organisms? Patentability of whole life is discussed in Chapter 27.

The North–South global disparity, manifested in technology and natural resources, is at the heart of the open access to biodiversity debate. The North is technologically advanced, rich in other resources, but poor in biodiversity (gene-poor); while the South is less technologically advanced, poor, but rich in biodiversity (gene-rich). Consequently, the South lacks the ability to effectively exploit the wealth of biodiversity, while the North, with its technological superiority, is able, through the open access concept, to obtain and effectively exploit biodiversity in ways that some deem exploitative of the nations of origin of the materials. The challenge is finding equity in the biodiversity transaction between the North and South. In this "fight" for ownership of biodiversity, the hand of the South is significantly weakened by the fact that international centers (IARCs) that have been strategically located in centers of crop diversity already have most of the biodiversity in their banks, which are accessible to researchers worldwide. Further, the private companies that develop cultivars solely for profit seldom use wild germplasm, thus making any restrictions on exchange of biodiversity for crop improvement of major concern. When all is said and done, what the South seeks is a form of "profit sharing" resulting from commoditization of products that were developed, at least in part, from genetic resources obtained from these nations. Not surprisingly, the idea of compensatory payments is not warmly embraced by the multinational corporations that gain the most from commercializing products derived from exotic genetic resources. On the other hand, some developing countries are vigorously pursuing intellectual property rights to some of their major species that have high commercial value.

11.21 Understanding the genetic architecture of germplasm for crop improvement

To pursue crop improvement in an effective way, breeders need to assemble appropriate germplasm, identify the specific loci that influence the traits of interest, and further understand the genetic architecture and molecular mechanisms that govern these traits. It is important to know how these desirable trait-enhancing alleles behave in various genetic backgrounds and environments. There are established and emerging approaches used by breeders to identify the genetic components that control traits of interest to breeders. The subject is discussed in detail in various chapters in this book. This section only introduces the subject.

11.21.1 Established approaches for gene identification

Molecular genetic approaches (both forward and reverse genetics screens) and quantitative trait loci (QTLs) mapping of complex traits (a category in which most of the traits of interest to breeders fall) have been the major traditional methods that have been used for crop improvement.

- Molecular genetics approach

  Molecular approaches are becoming increasingly more accessible with the advent of new and less expensive technologies for genetic manipulation and analysis. Model crops provide large amounts of information that can be applied to other species. Access to genomic sequence information is becoming easier. The subject is discussed in detail in Chapters 18 and 19.

- Quantitative trait loci mapping

  QTL analysis is discussed in Chapter 20. QTL mapping uses statistical linkage analyses and quantitative traits of interest and genetic markers in gene identification. Unlike molecular genetics approach, the use of QTLs to identify genetic variation provides breeders a basis for crop improvement of through conventional breeding methods.

11.21.2 Emerging approaches for gene identification

Two of the newer approaches to gene identification in plant improvement are as:

(i) Association mapping

Also called linkage disequilibrium mapping, association mapping depends on correlation between a genetic marker and a phenotype among a set of diverse germplasm. The subject is discussed in detail in Chapter 20.

(ii) Selection screening

Selection screening depends on DNA sequence polymorphism while association mapping is trait-oriented. The underlying theory of selection screening is that the loci of interest (target loci) for selection shows a decrease in nucleotide diversity and increased linkage disequilibrium following strong selection. The approach identifies genes that show signatures of selection.

Key references and suggested reading

Astley, D. (1987). Genetic resource conservation. *Experimental Agriculture*, **23**:245–257.

Alston, J.M., and Venner, R.J. (1998). The effects of U.S. Plant Variety Protection Act on wheat germplasm improvement. CIMMYT Symposium on IPR and Agricultural Research Impact, March 5–7 El Batán, Mexico.

Andrews, L.B. (2002). Genes and patent policy: rethinking intellectual property rights. *Nature Reviews Genetics*, **3**:803–808.

Boisvert, V., and Caron, A. (2002). The convention on biological diversity: an institutionalist perspective of the debates. *Journal of Economic Issues*, **36**:151–166.

Brush, S.B. (2003). The demise of 'common heritage' and protection for traditional agricultural knowledge. Conference on Biodiversity, Biotechnology and the Protection of Traditional Knowledge, April 4–6 Washington University, St. Louis, MI.

Brush, S.B. (1995). In situ conservation of landraces in center of crop diversity. *Crop Science*, **35**:346–354.

Day, P.R. (ed.) (1991). *Managing global genetic resources: The US National Plant Germplasm System*. National Academic Press, Washington, DC.

Franco, J., Crossa, J., Ribant, J.M., Betran, J., Warburton, M.L., and Khairallah, M. (2001). A method of combining molecular markers and phenotypic attributes for classifying plant genotypes. *Theoretical and Applied Genetics*, **103**:944–952.

Gepts, P. (2004). Who owns biodiversity, and how should the owner be compensated? *Plant Physiology*, **134**:1295–1307.

Kloppenburg, J.J., and Kleinman, D.L. (1987). The plant germplasm controversy. *BioScience*, **37**:190–198.

Mohammadi, S.A., and Prasanna, B.M. (2003). Analysis of genetic diversity in crop plants – Salient statistical tools and considerations. *Crop Science*, **43**:1235–1248.

National Plant Germplasm System (2003). Germplasm Resources Information Network (GRIN) Database. Management Unit (DBMU), NPGS, USDA, Beltsville, MD.

Peterson, A.H., Boman, R.K., Brown, S.M., *et al.* (2005). Reducing the genetic vulnerability of cotton. *Crop Science*, **14**:1900–1901.

Stoskopf, N.C. (1993). *Plant breeding: theory and practice.* Westview Press, Boulder, CO.

Takeda, S., and Matsuoka, M. (2008). Genetic approaches to crop improvement: responding to environmental and population changes. *Nature Reviews*, **9**:444–457.

Warburton, M.L., Xianchun, X., Crossa, J., *et al.* (2002). Genetic characterization of CIMMYT inbred maize lines and open pollinated populations, using large scale fingerprinting methods. *Crop Science*, **42**:1832–1840.

Zamir, D. (2001). Improving plant breeding with exotic genetic libraries. *Nature Reviews Genetics*, **2**:983–989.

Internet resources

http://www.ars.usda.gov/main/site_main.htm?mod-ecode=12-75-15-00 – Website of National Germplasm Resources Lab (accessed March 5, 2012).

http://www.ciesin.org/docs/002-256a/002-256a.html – Paper on current status of biological diversity by E.O. Wilson of Harvard University (accessed March 5, 2012).

www.plantphysiol.org/cgi/doi/10.1104/pp.103.038885 – Who Owns Biodiversity, and How Should the Owners Be Compensated? by Paul Gepts (accessed March 5, 2012).

Outcomes assessment

Part A

Please answer the following questions true or false.

1 The US Nation Seed Storage Laboratory at Fort Collins maintains a base collection of germplasm.
2 Generally, the first source of germplasm considered by a plant breeder is undomesticated plants.
3 Without variability, it is impossible to embark upon a plant breeding project.
4 CIMMYT is responsible for maintaining wheat and corn germplasm.
5 Only seeds are stored at a germplasm bank.

Part B

Please answer the following questions.

1 Who is the Russian scientist who proposed the concept of centers of diversity?
2 What is a landrace?
3 Distinguish between a base collection of germplasm and an active collection of germplasm at a gene bank.
4 What is Vavilov's law of homologous series in a heritable variation?
5 Give specific sources of germplasm erosion.

Part C

Please write a brief essay on each of the following topics.

1 Discuss the importance of plant introductions to US agriculture.
2 Discuss the importance of domesticated germplasm to plant breeding.
3 Discuss the US germplasm conservation system.
4 Discuss the role of the CGIAR in germplasm conservation.
5 Discuss the need for germplasm conservation.
6 Discuss Vavilov's concept of the centers of diversity of crops.

Section 5
Breeding objectives

Chapter 12 Yield and morphological traits
Chapter 13 Quality traits
Chapter 14 Breeding for resistance to diseases and insect pests
Chapter 15 Breeding for resistance to abiotic stresses

A successful plant breeding program depends on clearly defined breeding objectives. Some objectives (e.g., yield) are broad and generally part of most breeding programs. All new cultivars should yield highly and resist major disease in the production area. However, there are objectives that are specialized and may be designed for specialized markets or consumers. This chapter focuses on some of the most common breeding objectives pursued by plant breeders.

12

Yield and morphological traits

Purpose and expected outcomes

Physiological processes are common to all plants. However, there are morphological and physiological differences among plants. Morphological and anatomical traits are products of physiological processes. Yield is the ultimate goal of plant breeding programs. It is the product of complex biochemical processes. Plant breeders rarely select solely on yield without regard to some morphological trait(s). After completing this chapter, the student should be able to:

1 Define yield.
2 Discuss the biological pathway to economic yield.
3 Discuss the concept of yield potential.
4 Discuss the concept of yield plateau.
5 Discuss the concept of yield stability.
6 Discuss the breeding of lodging resistance.
7 Discuss the breeding of shattering resistance.
8 Discuss the breeding of plant stature.
9 Discuss the breeding for early maturity.
10 Discuss the breeding of photoperiod response.

12.1 Physiological traits

Plant growth and development depend on a complex interaction of biochemical and physiological processes. Plant physiological processes are under genetic control as well as under the influence of the environment. The genotype of the plant determines the total complement of enzymes in the cell and, hence, is a determining factor in plant growth. As D.C. Rasmusson and B.G. Gengenbach stated, physiological gene functions determine the manner and extent of the genotypic contribution to the

Principles of Plant Genetics and Breeding, Second Edition. George Acquaah.
© 2012 John Wiley & Sons, Ltd. Published 2012 by John Wiley & Sons, Ltd.

phenotype of the plant. Physiological gene action also reflects gene differences that provide the basis for selection of desirable genotypes in plant breeding.

The major physiological processes are photosynthesis, respiration, translocation and transpiration. Crop yield and productivity depend on the proper functioning of these processes. These traits are quantitatively inherited. Physiological traits may be broadly defined to include the major physiological processes, yield and its components. It also includes plant environmental responses (to photoperiod and environmental stresses). Some of the specific physiological traits that have been bred by plant breeders with varying degrees of success are photosynthetic rate, leaf angle, leaf area, stomatal frequency, water utilization, photoperiod, harvest index, tolerance to environmental stresses (drought, cold, salt, heat) and mineral nutrition. Some of the significant achievements with breeding for physiological traits have resulted in the modification of plant architecture, specifically, short stature (semidwarf) in cereals (e.g., rice and wheat), with all the advantages of such a plant architecture. Photoperiod response is discussed in this chapter because of its association with maturity and plant stature.

12.2 What is yield?

Yield is a generic term used by crop producers to describe the amount of the part of a crop plant of interest that is harvested from a given area at the end of the cropping season or within a given period. The plant part of interest is that for which the crop producer grows the crop. It could be the leaves, fruits, stems, roots, or flowers, or any other morphological part. It could also be the chemical content of the plant, such as oil, sugar, or latex. In certain industrial crops such as cotton, the plant part of economic interest to the producer is the fiber, while for the producer of tea or tobacco the part of interest is the leaf. It should be added that a producer might harvest multiple parts of the plant (e.g. grain and leaf) for use or sale (i.e., multiple economic parts). Plant breeders seldom select solely on yield basis without some attention to other morphological features of the plants. Yield is the best measure of the integrated performance of a plant.

Biological yield may be measured by breeding for physiological and morphological traits. All crop production ultimately depends on photosynthesis (as well as other physiological processes, for example, respiration and translocation). Over the years, various researchers have attempted to improve biological yield by (a) increasing the photosynthetic capacity of the individual leaf, (b) improving light interception characteristics of plants, and (c) reducing wasteful respiration. In addition to increasing plant biomass, the goals of breeding for physiological and morphological traits include redistribution of assimilates to the economic products within the plant as well as alleviating or avoiding the effects of adverse environmental conditions.

The term **biomass** is used by scientists to describe the amount or mass of organic matter in a prescribed area at a given point in time. This measure of biological matter includes material formed above and below ground. Yield of liquid products (e.g., latex, syrup) are measured by quantifying the volume of the product harvested. Depending on the type of product and the purpose of producing it, harvesting may be undertaken at various stages of maturity for various product quality preferences as demanded by the targeted market. Plant breeders may breed certain crops for early harvesting (for the fresh market) and others for dry grain. The yields at various stages of harvesting will differ between premature and fully mature products. Sometimes, scientists eliminate the moisture factor by measuring the weight of the harvested product on dry matter basis after drying the product in an oven prior to being weighed.

12.3 Biological versus economic yield

Yield may be divided into two types – **biological** and **economic**.

12.3.1 Biological yield

Biological yield may be defined as the total dry matter produced per plant or per unit area (i.e., biomass). Researchers use this measurement of yield in agronomic, physiological, and plant breeding research to indicate dry matter accumulation by plants. All yield is firstly biological yield.

12.3.2 Economic yield

The **economic yield** represents the total weight per unit area of a specified plant product that is of marketable value or other use to the producer. The producer determines the product of economic value. A producer of corn for grain is interested in the grain; a producer of corn for silage is interested in the young, fresh stems and leaves. All yield is biological yield, but all biological yield is not necessarily economic yield. For example, the above ground parts of corn may be totally useful in one way or another (e.g., the grain for food or feed, and the remainder also for feed or crafts). The roots are of no practical or economic use. However, in certain root crops, such as sugar beet, the total plant is of economic value (root for sugar extraction and the leafy tops for livestock feed).

Yield depends on biomass and how it is partitioned. To increase yield, the breeder may breed for increased biomass and efficient partitioning of assimilates. The potential biomass of a crop is determined by factors including genotype, local environment (soil, weather), and the agronomic practices used to grow it. N.W Simmonds identified three strategies for enhancing biomass:

(i) **Seasonal adaptation.** The objective of this strategy is to optimally exploit the growing season by sowing early and harvesting late to maximize biomass accumulation. Of course, this will have to be done within reasonable agricultural limits, as dictated by weather and cropping sequence. Genotypes can be adapted to new growing conditions (e.g., cold tolerance to allow the farmer to plant earlier than normal).

(ii) **Tolerance of adverse environmental factors.** Because of the vagaries of the weather and the presence of other inconsistencies or variation in the production environment (climate, product management, etc.), biomass can be enhanced by breeding for tolerance to these factors. Such breeding efforts may be directed at developing tolerance to abiotic stresses (e.g., drought, heat, cold). This would allow the cultivar to produce acceptable yields in the face of moderate to severe adverse environmental conditions.

(iii) **Pest and disease resistance.** Diseases and pests can reduce biomass by killing plant tissue (or even an entire plant in extreme cases) and stunting or reducing the photosynthetic surface of the plant. Disease and pest resistance breeding will enhance the biomass potential of the crop.

Breeding to control pests is one of the major undertakings in plant breeding.

12.4 The ideotype concept

Plant breeders may be likened to plant structural and chemical engineers who manipulate the genetics of plants to create genotypes with new physical and biochemical attributes for high general worth. They manipulate plant morphology (shape, size, number of organs) to optimize the process of photosynthesis, which is responsible for creating the dry matter on which yield depends. Once created, dry matter is partitioned throughout the plant according to the capacity of meristems (growing points of the plant) to grow. **Partitioning** (pattern of carbon use) is influenced by both intrinsic (hormonal) and extrinsic (environmental) factors. Certain plant organs have the capacity to act as **sinks** (importers of substrates) while others are **sources** (exporters of substrates). However, an organ may be a source for one substrate at one point in time and then a sink at another time. For example, leaves are sinks for nutrients (e.g. nitrates) absorbed from the soil while they serve as sources for newly formed amino acids.

Plant genotypes differ in patterns for partitioning of dry matter. This means plant breeders can influence dry matter partitioning. Pole (indeterminate) cultivars of legumes differ in patterns of partitioning of dry matter from bush (determinate) cultivars. Similarly, in cereals crops, tall cultivars differ from dwarf cultivars in the pattern of dry matter partitioning. The concept of the **plant ideotype** was first proposed by C.M. Donald to describe a model of an ideal phenotype that represents optimum partitioning of dry matter according to the purpose for which the cultivars will be used. For example, dwarf (short statured) cultivars are designed to channel more dry matter into grain development whereas tall cultivars produce a lot of straw. Tall cultivars are preferred in cultures where straw is of economic value (for crafts, firewood). Consequently, ideotype development should target specific cultural conditions (e.g. monoculture, high density mechanized production, or production under high agronomic inputs). All breeders, consciously or unconsciously, have an ideotype in mind when they conduct selection within a segregating population.

Plant morphological and anatomical traits (e.g., plant height, leaf size) are relatively easy for the breeder to identify and quantify. They do not vary in the short term and also tend to be highly heritable. Consequently, these traits are most widely targeted for selection by breeders in these programs. The wheat ideotype defined by Donald comprised the following:

- A relatively short and strong stem.
- A single culm.
- Erect leaves (near vertical).
- A large ear.
- An erect ear.
- Simple awns.

It is not practical to specify every trait in modeling an ideotype. The degree to which a plant model is specified is left to the discretion of the breeder. A more accurate ideotype can be modeled if the breeder has adequate information about the physiological basis of these traits. The traits used to define the ideotype presumably are those that would contribute the most to crop economic yield under the range of environmental and management conditions that the crop would encounter during its life. As previously indicated, physiological processes are common to all plants, but there is no universal ideotype in plant breeding. This is because of the vast morphological and physiological diversity of crops range in their economic end products as well as the cultural environments.

The role of partitioning in determining crop yield depends on the species and the products of interest. In forages, the total aboveground vegetative material is harvested as the end product. The importance of partition in the economic yield in this instance is small. In other crops, the desired product is enhanced at the expense of the rest of the plant. N.W. Simmonds has identified three outcomes of competition among plant parts for assimilates and their implications in plant breeding:

(i) **Vegetative growth is sacrificed for reproductive growth.** In this outcome, the breeder reduces vegetative growth by reducing plant structure (breeding for dwarf cultivars or breeding determinate cultivars). Other strategies include the reduction in foliage, as in the okra leaf cotton phenotype. Dwarf cultivars have been developed for many major cereal crops of world importance (e.g., wheat, rice, sorghum, barley). Dwarf cultivars are environmentally responsive (i.e., respond to agronomic inputs – fertilizer, irrigation). The success of the Green Revolution was in part due to the use of dwarf cultivars of wheat and rice.

(ii) **Reproductive growth is sacrificed for vegetative growth.** In crops in which the desired part is vegetative, flowering and seed set are either reduced (e.g., yam, cassava, potato) in order to channel resources into the vegetative parts, or suppressed (e.g. sugar beet, carrot).

(iii) **One vegetative growth is sacrificed for another desired vegetative growth.** The objective is to allocate dry matter to the harvestable underground vegetative structures (e.g., potato).

12.5 Improving the efficiency of dry matter partitioning

Proposed by C.M. Donald, **harvest index** is the proportion of the plant that is of economic value. It is calculated as a ratio as follows:

Harvest index = (economic yield/biological yield)

For cereals, for example, the ratio will be grain yield to total plant weight (grain + straw). The theoretical value of harvest index ranges from 0.0 to 1.0 (the value may be converted to a percentage by multiplying by 100). The higher the value, the more efficient the plant is in directing assimilates to the part of the plant of economic value. Harvest index is hence also referred to as **coefficient of effectiveness**. The higher the harvest index, the more economically desirable the genotype, because it translocates more of the available assimilates into the economic parts of the plant.

Some researchers indicate that the dramatic increase in the grain yield of major world cereal crops is due mainly to increases in harvest index and to a lesser extent the biological yield. In maize, for example, harvest index changed from 24% in 1950 to 43% in 1970, increasing yield from 3 tons/ha to 8 tons/ha. Generally, tall cultivars have high biological yield and low harvest index, whereas semidwarf cultivars have high harvest index and high biological yield. On the other hand, full dwarf cultivars have low biological yield and low economic yield. The breeding question is how effectively can harvest index be selected to make it a breeding objective for increasing yield? One difficulty with selecting harvest index is that it is not a phenotypic trait that can be readily evaluated. It is calculated from data obtained from two separate weighings. Such data are problematic to obtain if experimental plants are harvested mechanically, as is the case in many large breeding programs.

The developmental pathway followed by the plant part or chemical component of economic value affects harvest index. In cereal crops (e.g., corn and wheat), the economic part, the grain, fills in a linear fashion up to a definite point, and then ceases. Harvest index in these crops depends on the relative duration of vegetative and reproductive phases of the plant life cycle. However, in crops such as sugar beets and Irish potato, the economic part follows a protracted developmental pathway. In these crops, harvest index depends more on genetics than environmental factors.

Harvest index can be decreased or increased by manipulating the cultural environment. For example, increasing plant density and drought or soil fertility (e.g., nitrogen application) are known to lower harvest index in corn. However, planting early-maturing cultivars under good management increased harvest index in rice in some studies. This happened because the plant was able to allocate assimilates to the seed sooner, thereby leading to reduced accumulation of reserves in the leaves.

Harvest index has also been increased in small grain cereals partly through decreasing plant stature (e.g., by using the *Rht* dwarfing gene in wheat or by selection technique).

12.6 Harvest index as a selection criterion for yield

In spite of the role of harvest in increasing crop yield, using this trait as a selection criterion for grain yield is problematic for the breeder. This is largely due to the effect of the environment and cultural conditions on harvest index as previously described. Sometimes, the breeder selects on the basis of single plants (e.g., in a space-planted segregating population in the early part of a breeding program) and on families at some point. Also, sometimes plants are evaluated in microplots and at other times in large field plots at commercial density. The challenge is for the breeder to predict yield ability from one plant arrangement (isolated plants) to another (field crops). A much more severe restriction to the practical use of harvest index as a selection criterion is the fact that it is tedious to measure than grain yield *per se*.

12.7 Selecting for yield *per se*

A plant breeder seeking to improve crop yield affects the trait through manipulating biomass or partitioning or both. Furthermore, because yield is a complex trait, an outcome or product of the interaction of numerous physiological processes, breeders seek effective and efficient ways of selecting superior genotypes in a breeding program. As also previously discussed, biomass and partition are tedious to estimate. The rationale of yield components as a basis for selecting for yield has not proven useful because of the occurrence of compensatory negative correlations (i.e., increase in one component produces a decrease in another). Similarly, certain physiological parameters (e.g., photosynthetic rate or net assimilation) that had been proposed as possible indicators of improved biomass have not materialized. Breeders have also resorted to a variety of statistical procedures to help the selection process to become more efficient and effective for yield. A notable application is the development of indices for selection (selection index). Other multivariate techniques, such as path coefficients, have been attempted with little success. Modern molecular technology is attempting to identify QTLs associated with complex traits. In view of the foregoing, it is of little wonder that many breeders select for yield *per se* in their breeding programs. As previously stated, it is the best measure of the integrated performance of a plant or crop.

12.8 Biological pathway to economic yield

Yield is a very complex trait. A good crop yield reflects a genotype with high yield potential, growing in a good environment. It reflects, also, proper growth and development, processes that are very complex in themselves. In an effort to manipulate crop yield, plant breeders attempt to construct the path by which the reproductive, developmental, and morphological features of plants in a crop stand contribute to yield of a specified product. The pathways to yield are collectively called **yield components**. In theory, the total yield can be increased by increasing one component while holding the others constant. By breaking down a complex trait into components, breeders hope to find selection criteria for improving it.

For grain yield, a model of yield components is as follows:

$$\text{Yield/unit area} = (\text{plants/unit area}) \times (\text{heads/plant}) \\ \times (\text{mean number of seeds/head}) \\ \times (\text{mean weight/seed})$$

Where the plant produces tillers, the model may be modified as follows:

$$\text{Yield/unit area} = (\text{plants/unit area}) \\ \times (\text{mean number of tillers} \\ \text{with ears/plant}) \\ \times (\text{mean number of grains/ear}) \\ \times (\text{mean grain weight})$$

These plant characteristics describe yield. They all depend on energy in a fixed pool that is furnished through photosynthesis. Plant breeders seek to influence yield by manipulating its components to positively affect photosynthesis.

It is important to mention that in interpreting the correlation between yield and its components, the breeder should evaluate the components in terms of relative importance. The seasonal sequence of environmental conditions that affect plant development should also be considered. Growing conditions may be ideal in the early growth and development of the crop, leading to good initiation of reproductive features. However, if there is an onset of drought, few pods may complete their development and be filled with seed, leading to low correlation between yield and the number of seeds per pod.

Yield components vary from one species to another in terms of optimum value relative to other components. Further, yield components affect each other to varying degrees. For example, if increasing plant density drastically reduces the number of pods per plant, the number of seeds per pod may only be moderately affected whereas seed size remains unchanged or only slightly affected. Whereas a balance among yield components has great adaptive advantage for the crop, the components are environmentally labile. High yield usually results from one component with extreme value. Furthermore, yield components are determined sequentially. As such, they tend to exhibit **yield compensation**, the phenomenon whereby deficiency or low value for the first component in the sequence of developmental events is made up by high values for the subsequent components. The net effect is that yield is maintained at a certain level. However, yield compensation is not a perfect phenomenon. For example, it may occur over a wide range of plant densities in certain species. In beans, reduction in pod number can be compensated for by an increase in seed number per pod and weight per seed.

12.9 The concept of yield potential

A given crop has an inherent optimum capacity to perform under a given environment. This capacity is called its **yield potential**. It can be measured through yield trials. It is the maximum attainable crop yield from a specific soil–water regime under ideal production conditions (experimental conditions whereby there is no limit on access to any needed production input). It is suggested that only 20–40% of this yield potential can be attained economically in actual production on farmers' fields. When a farmer has reached an economic yield potential for the crop, attempts can be made to increase field yields in a variety of ways. The farmer can use production resources more efficiently; agronomic innovations that are more responsive to local needs and conditions can be introduced to the farmer by extension agents; the government may also institute incentive policies (e.g., credit) for farmers. However, improved cultivars constitute perhaps the most effective approach. To do this, a breeder would have to assemble appropriate variability (genotypes with complementary genes contributing to yield potential) and hybridize them to generate transgressive segregates with superior yield. Biotechnology can be used to develop new cultivars to cope with the constraints to the rise in field yields (abiotic and biotic stresses). Molecular markers may be used to assist the breeding of especially complex traits while recombinant DNA technology may be used to incorporate desirable unique genes from unrelated sources.

Industry highlights
Bringing Roundup Ready® technology to wheat

Sally Metz

Monsanto Corporation, MO

Once a technology is proven successful in a crop, for example Roundup Ready® soybean (Padgette *et al.*, 1996), researchers can theoretically transfer the technology to other crops. Industry refers to this as product extension. Given the complexity of the crop and trait, researchers determine how much additional or less optimization is required to achieve commercial success in subsequent crops. This development process can take 5–8 years and involves many different aspects of science. Costs can approach $40 million (Context Network, 2004) over this period, requiring researchers to be strategic, focused and precise. The development timeline and process described below occurred between 1997 and 2004 (Figure B12.1).

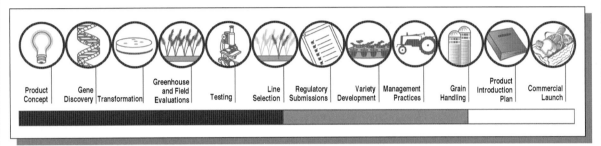

Product Concept | Gene Discovery | Transformation | Greenhouse and Field Evaluations | Testing | Line Selection | Regulatory Submissions | Variety Development | Management Practices | Grain Handling | Product Introduction Plan | Commercial Launch

Figure B12.1 The steps involved in the development and commercialization of Roundup Ready® wheat. Figure courtesy of Sally Metz.

Product concept

Understanding the needs of the customer, in this case the wheat growers, and then determining the technical criteria to meet those needs, are critical components in developing the product concept. In focus group sessions conducted in 2000, spring wheat growers identified numerous challenges with their current weed control options. These challenges included:

- Incomplete control of many tough weeds.
- Annual variability in product performance.
- Crop damage/setback caused by herbicides.
- The need for multiple products to broaden the spectrum of control.
- Significant price variability based on the quantity and type of weeds.

The product concept for Roundup Ready® wheat is complete, dependable, cost effective control of all weeds, often with only one herbicide application. Four years of field trials indicate that a single 26 oz/acre application of Roundup UltraMax® over the top of Roundup Ready® spring wheat provides 95–100% control of nearly all broadleaf and grassy weeds (Figure B12.2). Before a project is ever initiated, a market assessment is completed to understand the commercial potential for that trait/crop combination. If the results of that study are promising, then the project is initiated.

Figure B12.2 Product concept demonstration: control of broadleaf and grassy weeds. Figure courtesy of Sally Metz.

The development team

Because there are so many aspects to bringing a new product to commercialization, a wide diversity of expertise is needed amongst team members. Early-stage development teams usually combine expertise in molecular biology, transformation, genetics, plant breeding, agronomy and regulatory science. Bench scientists who have expertise in one of the team areas often lead early-stage teams. For instance, the early-stage Roundup Ready® wheat technical team was led by a molecular biologist. Mid-stage development teams evolve to incorporate additional expertise in regulatory affairs, seed production, industry affairs, public affairs, government affairs and commercial development. People who specialize in managing product development typically lead these teams. Late-stage development teams evolve again – phasing out the expertise in molecular biology, transformation and genetics, while adding expertise in marketing. Someone from the commercial arena usually leads these teams.

Transformation

To be commercially successful, a crop not only has to be transformed, but the process has to be amenable to making hundreds of transgenic plants. Monsanto developed a process to transform wheat, using the trait of interest – glyphosate tolerance – as the selectable marker. The transformation method also must result in transgenic plants that are "clean" events – meaning that the DNA insertion can be registered with many regulatory bodies around the world. It has been our experience that Agrobacterium transformation results in more useful events than does particle gun transformation (Hu *et al.*, 2003).

The gene and its expression in the plant

In Roundup Ready® crops, the principle gene that has been used is the CP4 EPSPS gene, which was isolated from Agrobacterium. This is the gene present in Roundup Ready® wheat. The expression of this gene allows the plant to continue to produce aromatic amino acids after the application of one of the Roundup® family of herbicides. The challenge is enabling the gene to be expressed in the right tissues at the right time. The promoters control expression. Some promoters, like the cauliflower mosaic virus enhanced 35S promoter (e35S), are expressed strongly in wheat leaves but at lower levels in specific tissues critical for reproduction. Other promoters, like rice actin, are expressed at higher levels in these critical reproductive tissues but at lower levels during plant regeneration and in the wheat leaves. In wheat, both the e35S and rice actin promoters were used to achieve plant selection and a commercial level of vegetative and reproductive tolerance to glyphosate.

Usually many different constructs, with different combinations of promoters, genes and stop regions are tested to find the combination that provides the desired phenotypic result.

First, tests for the function of the introduced gene – for example, in the case of Roundup Ready® wheat, glyphosate spray tests – are conducted to select transgenic plants that express the trait. Seed of each event must be increased in greenhouses or growth chambers to provide seed for field testing. Any "environmental release" (field trial) is conducted under the rules and regulations of the US Department of Agriculture – Animal and Plant Health Inspection Service (USDA-APHIS) as well as individual state departments of agriculture. Similarly, the Canadian Food Inspection Agency (CFIA) controls field testing in Canada. For wheat, these regulations include specifications regarding minimum isolation distances, volunteer monitoring requirements and crop rotations restrictions. Field trials are conducted over several locations and years to obtain enough performance data to select the commercial transgenic event (Figure B12.3) (Zhou *et al.*, 2003).

Figure B12.3 Field trials demonstrating gene efficacy. Figure courtesy of Sally Metz.

Selecting the commercial transgenic event

To facilitate the selection of the eventual commercial event, it is most useful for the team to set selection criteria before the data are analyzed. Selection criteria involves four areas: (i) molecular biology – single insertion, no vector backbone, intact insertion; (ii) genetics – trait inherited as a single dominant gene, expression maintained over generations; (iii) agronomics – yield, maturity, disease reactions (i.e. selection of plants that are not affected by the transformation/regeneration process); and (iv) trait performance – for herbicide tolerance, at least a 2× safety margin to the commercial rate and timing of application (Figure B12.4). In the case of Roundup Ready® wheat, the development team invoked an additional selection criterion, to select a transgenic event that was present on the A or B genome. Molecular breeding techniques were used to map the insertion of every potential commercial event. By selecting away from the D genome, research demonstrated there was a high likelihood that crossing between wheat and its only North American wild relative, jointed goatgrass (*Aegilops cylindrica*), would not result in introgression of the trait into the wild jointed goatgrass population.

Figure B12.4 Selecting the commercial event. Figure courtesy of Sally Metz.

Developing the trait

In parallel, the push is often underway in four primary areas: (i) developing the regulatory packages for the Environmental Protection Agency (EPA), Food and Drug Administration (FDA), USDA and other regulatory agencies such as those in Canada, Europe and Japan; (ii) developing commercial varieties which contain the new trait; (iii) developing agronomic and stewardship practices for the trait; and (iv) developing benefits data for growers, consumers and the internal marketing and sales people who will promote the product.

Questions are asked such as: When do you apply the herbicide? At what rate? Are split applications beneficial? How does rate and timing affect residue levels in grain or forage? How do you control volunteers? To what extent is outcrossing an issue? Are there any data published in the literature? Has end-use quality been affected? Can you optimize yields in the system? What about crop rotations? What is the most productive way to use this technology in a total cropping system? Has the feed efficiency of livestock been affected?

Monsanto conducted more than 48 formal regulatory studies, which demonstrated the substantial equivalence and safety of Roundup Ready® wheat compared to conventional wheat. These studies fell into two categories: (i) compositional analysis to ensure that Roundup Ready® wheat is substantially equivalent to standard wheat varieties in components important for human and animal nutrition, and that no unapproved food additives are present (Obert *et al.*, 2004), and (ii) safety studies to ensure that Roundup Ready® wheat contains no increase in allergens, toxicants or anti-nutrients when compared to standard wheat varieties (Goodman *et al.*, 2003).

Additional marketplace-support studies demonstrated feeding equivalence for hogs (Peterson *et al.*, 2004) and chickens (Kan and Hartnell, 2004), as well as weed control and yield optimization (Blackshaw and Harker, 2003). Between 2001 and 2004, Monsanto scientists or public scientists who were developing a basic database or evaluating the technology published more than 39 scientific presentations, posters and peer-reviewed journal articles.

Years of work and research demonstrates the successful transfer of the Roundup Ready® technology to wheat – meeting or exceeding all selection criteria and resulting in a product that provides 95–100% dependable, cost effective control of all weeds, often with only a single 26 oz/acre application of Roundup UltraMax®. In addition, Roundup Ready® wheat optimizes the yield potential by incurring less stress due to herbicide injury (Figure B12.5) and by decreasing the quantity of weeds capable of competing with the wheat crop for available water and nutrients. This has resulted in a consistent 5–15% yield increase over conventional wheat treated with conventional herbicides (Figure B12.6).

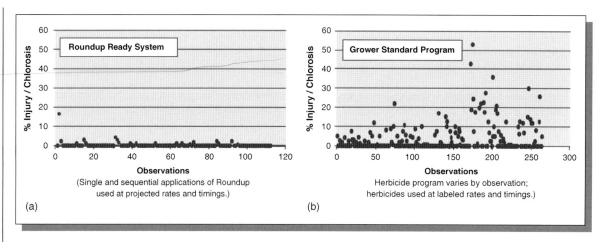

Figure B12.5 The crop safety of (a) the Roundup Ready® system versus (b) grower standard programs, showing that there is virtually no crop injury from the use of Roundup Ready® wheat. (Source: Monsanto and University 1999–2003 US field trials.)

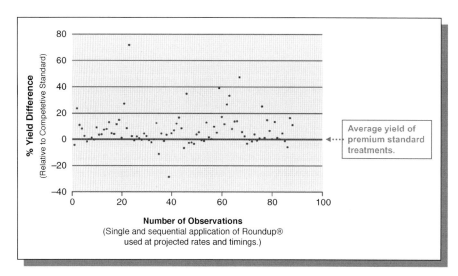

Figure B12.6 Results showing that Roundup Ready® wheat consistently yields better than wheat treated with the best commercial standard treatments. (Source: Monsanto and University 1999–2003 US field trials.)

References

Blackshaw, R.E., and Harker, K.N. (2003). Selective Weed Control with Glyphosate in Glyphosate-Resistant Spring Wheat (*Triticum aestivum*). *Weed Technology*, **16**(4): 885–892.

Context Network (2004). Biotech Traits Commercialized, 2004. CD ROM.

Goodman, R.E., Bardina, L., Nemeth, M.A., *et al.* (2003). Comparison of IgE binding to water-soluble proteins in genetically modified and traditionally bred varieties of hard red spring wheat. World Allergy Organization Congress, Vancouver, September 2003 Abstract. World Allergy Organization.

Hu, T., Metz, S., Chay, C., *et al.* (2003). Agrobacterium-mediated large-scale transformation of wheat (*Triticum aestivum* L.) *using glyphosate selection. Plant Cell Reports*, **21**:1010–1019.

Kan, C.A., and Hartnell, G.F. (2004). Evaluation of broiler performance when fed Roundup Ready® wheat (event MON 71800), control, and commercial wheat varieties. *Poultry Science*, **83**:1325–1334.

Obert, J., Ridley, W., Schneider, R., *et al.* (2004). The Composition of Grain and Forage from Glyphosate Tolerant Wheat, MON 71800, is Equivalent to that of Conventional Wheat (*Triticum aestivum* L.). *Journal of Agriculture and Food Chemistry*, **52**(5): 1375–1384.

Padgette, S.R., Re, D.B., Barry, G.F., *et al.* (1996). New weed control opportunities: Development of soybeans with a Roundup Ready gene, in *Herbicide-Resistant Crops* (ed. S.O. Duke). CRC Press, Boca Raton, FL, pp. 54–84.

Peterson, B.A., Hyun, Y., Stanisiewski, E.P., Hartnell, G.F., and Ellis, M. (2004). Performance of growing-finishing pigs fed diets containing Roundup Ready wheat (MON 71800), a non-transgenic genetically similar wheat, or conventional wheat varieties. *Journal of Animal Science*, **82**:571–580.

Zhou, H., Berg, J.D., Blank, S.E., *et al.* (2003). Field Efficacy Assessment. *Crop Science*, **43**:1072–1075.

12.10 The concept of yield plateau

It is important for food production to keep pace with population growth. Even though global total crop yields are continually rising, the rate of yield growth is slowing. This trend is termed **yield stagnation** or **yield plateau,** and has been observed in many of the crops that feed the world, especially cereals. For example, yield growth rates for wheat declined from 2.92% (between 1961 and 1979) to 1.78% (between 1980 and 1997). The decline in corn yield for the same periods was from 2.8 to 1.29%.

Several key factors may be responsible, at least in part, for yield stagnation. From the research perspective, many agricultural research programs do not focus on yield increases *per se* but rather on improving specific traits, such as drought tolerance, insect resistance, and disease resistance. A major factor is a shift away from cereal production to the production of more profitable crops with the decline in world cereal prices. There are no incentives for producers to pump production inputs into an enterprise that could raise field yield, only to take a loss at harvest time. Another factor is crop intensification, whereby multiple crops are being grown where previously only one was grown. This practice is suspected to cause rapid decline in soil fertility.

The genetic potential of many important crops remains untapped. For example, the global average yield of wheat is 2 metric tons/ha compared to the record yield of 14 metric tons/ha, and even a possible 21 metric tons/ha. Plant breeding is needed to increase crop yields by developing high yielding cultivars for various ecotypes.

12.11 Yield stability

Plant breeders are not only interested in developing high yielding cultivars. They are interested in developing cultivars with sustained or stable high performance over seasons and years (**yield stability**). One of the key decisions made by the plant breeder at the end of the breeding program is the genotype to release as a cultivar. This decision is arrived at after yield trials conducted over locations, seasons, and years, as applicable. When different genotypes exhibit differential responses to different sets of environmental conditions, a genotype x environment (G x E) interaction is said to occur. This subject is fully discussed elsewhere. Genotypes that are more responsive to the environment are less stable in performance, doing well in a good production environment and poorly under less optimal production conditions. On the contrary, less responsive genotypes generally perform well across varied environments.

Stability is rather difficult to determine or breed for in a breeding program. While it might be desirable to set as an objective to breed for either more or less environmental responsiveness, it is more practical and realistic to exploit whatever turns up during yield trails. Each kind of adaptation has its value. A cultivar that is responsive to the environment may be released for a narrowly-defined area of adaptation, whereas another that has a low G x E interaction is suitable for release for use over a wider region of production.

12.12 Lodging resistance

Lodging resistance may be defined as the leaning, bending, or breaking of the plant prior to harvesting.

12.12.1 Nature, types, and economic importance

There are two basic types of **lodging** that may be caused by biotic or abiotic factors. Lodging may originate at the root level (**root lodging**) or at the stem or stalk level (**stalk lodging**). Soil-borne disease and insect pests may destroy plant roots, causing the plant to lean over starting at the root level. Disease and insect attack can also cause the stem or stalk to lodge. For example, the European corn borer weakens the stalk and predisposes it to bending. Strong winds and other weather factors, such as hail, ice, or snow, are common causes of lodging in plants.

Certain plant characteristics make plants susceptible to lodging. These include tall plant height, thin stems, excessive vegetative growth, and succulence. In addition, the crop cultural environment may encourage lodging. For example, high nitrogen fertilization promotes luxuriant vegetative growth, and hence excessive amounts produce top-heavy and succulent plants that are also prone to disease and insect attack. Stems that have been attacked by pests are weak and lodge easily.

Breeding for lodging resistance is important because lodging results in yield reduction. Lodging prior to pod filling results in partial fruit or seed development. Lodging at maturity may also make pods or cobs inaccessible to mechanized harvesters (combines) and, hence, are left unharvested. Lodged plants are exposed to disease infestation. For example, in cereal crops such as wheat, contamination by mycotoxins produced by *Fusarium* species may increase.

12.12.2 Genetics and breeding

Breeding for lodging resistance is challenging, partly because it is a quantitative trait, and hence conditioned by many genes. Furthermore, its expression is significantly impacted by the environment. It is difficult to score for lodging resistance on a phenotypic basis in the field because the factors that cause lodging may occur at different stages in plant development, or never at all. Over the years, researchers have tried to find above-ground morphological traits that strongly correlate with lodging resistance in various species. Even though culm parameters (e.g., weight, diameter) have been reported to correlate significantly with lodging resistance in wheat, such an observation is not universal. Consequently, developing a reliable morphological index for field selection in a breeding program is difficult.

Nonetheless, resistance to lodging may be improved by targeting a combination of the following traits (depending on species) – short stature, stronger stalk, sturdy stem, thick rind, stiff culm, strong root system, resistance to stalk or stem diseases and insect

pests, and other traits. It should be pointed out that no single trait or group of traits has proven to be universally reliable as an index of lodging resistance.

To improve selection, QTLs for lodging resistance have been identified and used as molecular markers in some breeding programs using marker assisted selection. Furthermore, dwarfing genes have been used to breed short-statured cultivars in crops such as rice and wheat, resulting in increased lodging resistance. In corn, for example, recurrent selection has been used to improve lodging resistance. The success of the Green Revolution is attributed to the development of semidwarf cultivars that were high yielding and responsive to the environment without lodging.

12.13 Shattering resistance

Dry fruits that split open upon maturity to discharge their seeds are called **dehiscent fruits**. Whereas shattering is advantageous in the wild, it is undesirable in modern crop production.

12.13.1 Nature, types, and economic importance

Some fruits split along only one side (called a **follicle**), while others split along two sides (called a **legume**), or multiple sides (called a **capsule**). This natural mechanism of seed dispersal has adaptive value to plants in the wild. In crop production, the splitting of dry fruits to release their seeds prior to harvesting is called **grain shattering.** In serious cases, some cultivars can lose over 90% of their seed to shattering, if harvesting is delayed by just a few days. Whereas shattering is often identified with pod-bearing species (e.g., soybean, peas), it also occurs in cereal crops such as wheat and rice.

Shatter-sensitive cultivars are susceptible to high loses during harvesting. The physical contact of the harvesting equipment with the plant may be enough pressure to trigger shattering. However, most susceptible cultivars spontaneously shatter their seeds when the environmental condition is right (dry, sunny, and windy). Shattered seeds are not only lost but also become a nuisance when they germinate as volunteer plants in the next year's crop. Being weeds, the volunteer crops are controlled at additional production cost.

12.13.2 Breeding grain shattering resistance

Like lodging resistance, grain shattering resistance is a complex trait. There is a large variation in the degree of shattering in existing rice gene pools, ranging from extremely shattering to extremely hard to thresh. Furthermore, researchers have identified at least five genes that condition shattering in rice, including *sh1, sh2, sh4*.

12.14 Reduced plant height

Modern production of certain cereal crops is dominated by semidwarf or dwarf cultivars (e.g., rice, wheat, sorghum). These cultivars have advantages in mechanized agriculture and high input production systems.

12.14.1 Nature, types, and economic importance

Reduced plant height is associated with, or promotes, lodging resistance. Similarly, early maturity also reduces plant internode length. Producers desire crop cultivars with reduced plant stature because they are easier to harvest mechanically. They produce less straw after the economic product has been harvested. However, in certain cultures, the straw is used for crafts or firewood, and hence tall cultivars are preferred.

Reduced stature also increases the harvest index. Dwarf cultivars can be more closely spaced in the field for increased crop yield. These cultivars are also environmentally more responsive, responding to agronomic inputs, especially fertilizers, for increased productivity.

12.14.2 Genetics and germplasm resources

A variety of height-reducing genes has been discovered in various species. These genes reduce plant height when in the recessive form. In castor bean, the dwarf internode gene, *di*, reduces plant height by 25–50%. This reduction in height makes the plant sturdier and more resistant to lodging. It reduces the swaying of the pods in the wind, thereby reducing shattering. In wheat, the reduced-stature plants are called **semidwarfs**. The dwarfing genes in wheat are designated *Rht* (reduce height), of which about 20 have been discovered. These genes differ in their

effects on the plant. For example, *Rht8*, discovered in wheat cultivars from Yugoslavia, is widely used because of its less adverse effect on kernel density and weight. Furthermore, the wheat dwarfing genes increase grain yield by increasing plant tillering and number of seeds per plant. Sometimes, wheat breeders intensify the height reduction by including two different *Rht* genes in a cultivar. Such cultivars, called **double dwarfs**, are shorter and tiller more than single dwarfs. Using monosomic analysis, scientists have associated *Rht8* with chromosome 4A and *Rht2* with chromosome 4D.

Dwarfing genes have been used in rice cultivar development. One set of genes, designated *d*, has been known to reduce kernel size and grain yield along with reducing internode length, and hence is not used in commercial cultivar development. Instead, the *sd1* gene, a spontaneous mutant discovered in a Taiwanese cultivar, Dee-geo-woo-gen, has been used to develop successful cultivars of rice. The gene has been induced by mutagenesis and used in cultivar development. There are several major dwarfing genes in oats, of which, so far, the dominant *Dw6* and semidwarf *Dw7* are readily available for use in breeding. However, use of these genes to improve lodging resistance has been limited because their use results in reduced grain quality in many environments.

Sorghum breeding has also benefited significantly from the discovery and use of dwarfing genes. Four dwarfing genes – *dw1, dw2, dw3,* and *dw4* – have been discovered. These genes produce a type of dwarfism described as **brachytic dwarfism**, which reduces plant stature without significantly affecting leaf number, leaf size, plant maturity, and yield. The gene *dw3* is observed to be mutable (to *Dw3*) resulting in one tall mutant occurring out of about every 600–1200 plants in a field of dwarf plants. These rogues must be removed before harvesting to keep the seed as pure as possible. Furthermore, the reduction in plant stature is dependent upon the genotype with respect to the dwarfing loci. Cultivars may have one dwarfing gene (e.g., *Dw1, Dw2, Dw3, dw4*) two dwarfing genes (e.g., *Dw1, Dw2, dw3, dw4*), three dwarfing genes (e.g., *dw1, dw2, Dw3, dw4*), or four dwarfing genes (*dw1, dw2, dw3, dw4*). Most sorghum cultivars in the United States are three-dwarfs (have three dwarfing genes). They are superior in yield to four-dwarfs. It is also known that the manifestation of the dwarf trait is influenced by modifying genes such that cultivars with identical set of dwarfing genes vary in plant stature.

Plants that have dwarfing genes are unable to respond to gibberellin or gibberellic acid, the plant hormone that promotes stem elongation. Using transposon tagging, scientists have cloned a gene, *GAI*, in *Arabidospsis* that is responsible for dwarfism in that species. DNA sequence analysis of the gene followed by bioinformatics has revealed that other species have equivalent genes. This indicates the possibility of using rDNA technology to transfer the *GAI* gene into other species.

12.15 Breeding determinacy

Another way in which plant form is manipulated is through determinacy.

12.15.1 Nature, types, and economic importance

Plant growth form may be **indeterminate** or **determinate**. Some species, such as corn and wheat, are determinate in growth form, whereas soybeans have both determinate and indeterminate types. In indeterminate cultivars, new leaves continue to be initiated even after flowering has begun. Flowering occurs in axillary racemes. On the contrary, determinate cultivars or species do not initiate new leaves after flowering has begun. Also, flowering occurs in both axillary and terminal racemes.

Generally, determinate plant types tend to have stiffer and shorter stems (called bush types). In soybean, both determinate and indeterminate commercial cultivars are used in production. Determinate cultivars are used in areas with long growing seasons (i.e., southern latitudes), while indeterminate cultivars are grown in areas with short growing season (i.e., northern latitudes).

12.15.2 Genetics and germplasm resources

Determinacy in soybean is conditioned by a recessive gene, designated dt_1. This gene has been used to breed determinate cultivars for the northern growing regions. Semi-determinacy is conditioned by the Dt_2 allele, while the S allele conditions short plant internodes.

12.16 Photoperiod response

Plants exhibit environmentally determined developmental switches from the initiation of leaves (vegetative phase) to flowering (reproductive phase). The two developmental switches that plant breeders pay attention to are **photoperiod** and **vernalization**. The key environmental factors are temperature and day length. In some plants, flowering is not promoted by temperature and day length but occurs regardless of the conditions (called facultative plants), whereas in others, flowering will not occur without the appropriate temperature–day length combination (called obligate plants). Photoperiodism is a photomorphogenic responsive to day length (actually, plants track or measure the duration of dark period rather than duration of daylight). Three categories of responses are known:

(i) **Long-day (short-night) plants.** These plants require a light period longer than a certain critical length in order to flower. They will flower under continuous light. Cool season species (e.g., wheat, barley, alfalfa, sugar beet) are examples of long-day plants.

(ii) **Short-day (long-night) plants.** These plants will not flower under continuous light, requiring a photoperiod of less than a certain critical value in a 24-hour daily cycle. Examples include corn, rice, soybean, peanut, and sugarcane.

(iii) **Day-neutral (photoperiod insensitive).** Photoperiod insensitive plants will flower regardless of duration of day length. This trait is very desirable, enabling producers to grow the crops in a broad range of latitudes. Examples include tomato, cucumber, cotton, and sunflower.

Plant breeders need plants to flower at the appropriate time for hybridization in a breeding program and also to influence the cultivars they develop for different growing areas. Photoperiod influences the duration of the vegetative phase versus reproductive phase, and hence crop yield at different latitudes. Soybean, for example, has 13 recognized adaptation zones, ranging from 000 in the northern latitudes to IX in the southern latitudes. Day length increases as one goes north in summer in the northern hemisphere. Consequently, a cultivar developed for the southern latitudes may not be as productive in the northern latitudes where reproductive growth is not initiated until the fall (autumn) season when day length is short.

Vernalization is the process by which floral induction in some plants is promoted by exposing plants to chilling temperatures for a certain period. Plant breeders of crops such as wheat either sow in the fall so the plant goes through a natural vernalization in the winter, or they place trays of seedlings in a cold chamber for the same purpose, prior to transplanting.

12.16.1 Nature, types and economic importance

Photoperiod and temperature are two major environmental factors that influence crop adaptation through their effect on days to flowering. Photoperiod is also known to affect photosynthate partitioning in some species, such as peanuts, in which researchers found a reduction in the partitioning of dry matter to pods in certain genotypes under long photoperiods. Decreasing partitioning to grain favors partitioning to vegetative parts of the plant, resulting in increased leaf area and dry matter production. Crop cultivars that are developed for high altitudes should mature before the arrival of winter as well as be less sensitive to a long photoperiod so that seed yield would be high.

12.16.2 Genetics and germplasm

In soybean, a number of maturity genes that influence flowering under long day length have been reported. Of these, the E_3 locus, has the most significant effect on flowering under long day length. Plants with genotypes of e_3e_3 have insensitivity to fluorescent long day length. Genetic control of photoperiod insensitivity in wheat is variable among cultivars, including two genes with major effects, one dominant gene, or one recessive gene controlling the condition. In sorghum, four genes (Ma_1, Ma_2, Ma_3, Ma_4) affect plant maturity and response to long day lengths. As J.R. Quimby observed, the recessive alleles of these genes condition a degree of sensitivity to longer day lengths. The genotype did not affect time of initiation of flower primordial under short day (10 hours). However, under long days (14 hours), the genotype Ma_1 Ma_2 ma_3 Ma_4 in sorghum induces flowering in 35 days, while Ma_1 ma_2 Ma_3 Ma_4 flowers in 44 days. The genotype Ma_1 Ma_2 Ma_3 Ma_4 flowers in 70 days.

12.17 Early maturity

Crop maturity in general is affected by a variety of factors in the production environment, including photoperiod, temperature, altitude, moisture, soil fertility, and plant genotype. Early maturity could be used to address some environmental stresses in crop production, such as drought and temperature. Maturity impacts both crop yield and product quality. Sometimes, the producer desires the crop to grow to attain its maximum dry matter possible under the production conditions before harvesting is done. In some crops, premature harvesting produces the product quality for premium prices.

12.17.1 Nature, types, economic importance

There are two basic types of maturity – **physiological** and **harvest (market) maturity**. Physiological maturity is that stage at which the plant cannot benefit from additional production inputs (i.e., inputs such as fertilizer and irrigation will not translate into additional dry matter or gain in economic product) because it has attained its maximum dry matter. In certain crops, the product is harvested before physiological maturity to meet market demands. This stage of maturity is called harvest maturity. For example, green beans are harvested before physiological maturity to avoid the product becoming "stringy" or fibrous.

It is desirable for a producer to grow a cultivar that fully exploits the growing season for optimal productivity. However, under certain production conditions, it is advantageous for the cultivar to mature early (i.e., exploit only part of the growing season). Early maturity may allow a cultivar to escape environmental stresses (e.g., disease, insects, early frost, early fall rain storms, drought) that may occur later in the season. Also, early maturing cultivars are suitable for use in multiple cropping systems, allowing more than one crop to be grown in a production season. Early maturity has made it possible to extend the production of some crops to regions in higher altitudes and with shorter summers, as well as low rainfall.

Early maturity has its disadvantages. Because the plant only partially exploits the growing season, economic yield may be significantly reduced in species including corn, soybean, wheat, and rice. In cotton, earliness is negatively correlated with traits such as fiber length.

12.17.2 Genetics and germplasm

Research reports indicate earliness to be controlled by either dominant or recessive genes. Modifier genes of major genes have also been reported. The inheritance appears to differ among species. Researchers have conducted QTL analysis for improvement of earliness. Furthermore, a flowering-promoting factor (*fpf1* gene) has been cloned in *Arabidopsis*. This gene is being used to transform other species to create early maturity in those species. In sorghum, four gene loci, Ma_1, Ma_2, Ma_3, and Ma_4, influence time to maturity. These are the same genes that are implicated in photoperiod sensitivity. Tropical sorghum species are dominant at the Ma_1 locus, and do not flower during the long summer photoperiods in the United States.

Key references and suggested reading

Frey, K.J. (1959). Yield compensation of oats II: Effect of nitrogen fertilization. *Agronomy Journal*, **51**:605–608.

Lee, E.A., Ahmadzadeh, A., and Tollenaar, M. (2005). Quantitative genetic analysis of the physiological processes underlying maize grain yield. *Crop Science*, **45**:981–987.

Milach, S.C.K., and Federizzi, L.C. (2001). Dwarfing genes in plant improvement. *Advances in Agronomy*, **73**:35–65.

Simmonds, N.W. (1979). *Principles of crop improvement*. Longmans, New York.

Rasmusson, D.C., and Gengenbach, B.G. (1983). Breeding for physiological traits, in *Crop breeding*. American Society of Agronomy. Wisconsin, Madison.

Rosielle, A.A., and Hamblin, J. (1981). Theoretical aspects of selection for yield in stress and non- stress environments. *Crop Science*, **21**:943–946.

Worland, A.J. (1996). The influence of flowering time genes on environmental adaptability in European wheats. *Euphytica*, **89**:49–57.

Outcomes assessment

Part A

Please answer the following questions true or false.

1 All yield is biological, but all yield is not economic.
2 Sinks are exporters of substances.
3 Biomass is the yield per unit area of the grain.
4 The higher the harvest index, the more efficient the plant is in directing assimilates to the part of the plant of economic value.
5 Lodging may occur at either the root or stem levels.
6 Grain shattering occurs only in legumes.
7 Most sorghum cultivars in the US are 2-dwarfs.
8 The dwarfing gene in wheat is designated *Rht*.
9 A bush type cultivar is indeterminate in growth form.

Part B

Please answer the following questions.

1 What are the two types of lodging?
2 Give the genotype of 3-dwarf sorghum.
3 Distinguish between single dwarf and double dwarf sorghum cultivars.
4 What is the proportion of the plant that is of economic?

5 What is yield?
6 Distinguish between economic yield and biological yield.
7 What is yield potential?
8 Define harvest index.

Part C

Please write a brief essay on each of the following topics.

1 Discuss the genetics of dwarfing in sorghum.
2 Distinguish between determinate and indeterminate growth in plants.
3 Discuss the importance of lodging resistance in crop production.
4 Discuss the concept of plant ideotype.
5 Discuss the concept of yield components.
6 Discuss the importance of yield stability in cultivar development.
7 Discuss the concept of yield component composition.

13

Quality traits

Purpose and expected outcomes

The market value and utilization of a plant product is affected by a variety of factors. For example, for grain crops, the factors that affect grain quality and form the basis of crop quality improvement programs include market quality, milling quality, cooking and processing qualities, and nutritional quality. The specific breeding goals with respect to each of these factors differ among crop species and agricultural production regions. For example, for the same crop, the quality aspect of importance in developed economies may be very different and even opposite to that of developing economies. Some cultures prefer white, non-scented, or non-sticky rice, whereas other cultures prefer colored, scented, or sticky rice. After completing this chapter, the student should be able to:

1 Discuss the breeding for improved protein content.
2 Discuss the breeding for improved fatty acid content.
3 Discuss the breeding for seedlessness in fruits.
4 Discuss the breeding for delayed ripening.
5 Discuss the breeding for novel traits in plants using biotechnology.

13.1 Concept of quality

Quality means different things to different people. The terms used to describe quality vary from crop production to food consumption, and include terms for appearance, storage quality, processing quality, and nutritional worth or quality. Of these, most attention has been devoted to processing quality for the major crops, such as milling quality and baking quality of wheat, canning quality of beans, chipping or baking quality of potatoes, malting quality of barley, and fermenting quality of grapes. These crops and others are of such high economic value that special laboratories operated at the private, state, and national levels have been set up to research and develop standards for these specialized processes for the benefit of plant breeders and industry.

Plant breeders should be very familiar with the market quality standards for their crops. These standards are based on complex interaction of social, economic, and biological factors, and are highly crop specific.

Principles of Plant Genetics and Breeding, Second Edition. George Acquaah.
© 2012 John Wiley & Sons, Ltd. Published 2012 by John Wiley & Sons, Ltd.

Table 13.1 Essential amino acids that are low in selected major world food crops.

Crop	Deficient amino acid
Corn	Tryptophan
	Lysine
Wheat	Lysine
Rye	Tryptophan
	Lysine
Rice (polished)	Lysine
	Threonine
Millet	Lysine
Soybean	Methionine
	Cystine
	Valine
Lima bean	Methionine
	Cystine
Peanut	Lysine
	Methionine
	Cystine
	Threonine
Potato	Methionine
	Cystine

13.2 Nutritional quality of food crops

Plant parts used for food differ in nutritional quality. Different species and cultivars of the same species may differ significantly in total protein as well as the nutritional value of the protein. The amino acid profiles of cereal grains and legumes differ according to certain patterns. Cereals tend to be low in lysine while legumes tend to be deficient in tryptophan (Table 13.1).

Three of the crops that feed the world are cereals (corn, wheat, rice). Other important species are roots or tubers. Cereals and tubers are generally low in protein content. Rice averages about 8% protein, corn 10%, potato 2%, versus 38–42% in soybean, and 26% in peanut. Protein augmentation is a major breeding objective in many major world crops.

13.3 Brief history of breeding for improved nutritional quality of crops

Breeding for high protein content in crop plants is perhaps the highest priority in improving the nutritional quality of plants because about 70% of the protein supply of human consumption is of plant origin. Further, cereals are deficient in some essential amino acids and low in total protein. Maize was one of the first crops on which formal nutritional augmentation work was done. In 1896, C.G. Hopkins initiated a project to breed for high protein and oil content at the Illinois Agricultural Experimental Station. Work by T.B. Osborne in the early 1900s resulted in the fractionation and classification of proteins according to solubility properties. He and his colleague discovered **zein** (the prolamin or alcohol soluble fraction) as comprising the bulk of protein of maize endosperm. Later work in the mid 1900s by K.J. Frey demonstrated that breeding for protein augmentation primarily increased the zein content. There was a need to find a way to enhance the useful part of the protein. E.J. Mertz in 1964 discussed the nutritional effects of the *opaque-2* gene in maize. The mutant gene increased the lysine content, called **high lysine**. High lysine research has since been conducted in sorghum. Another cereal food of world importance is rice. However, it has significant nutritional problems, being low in protein as well as completely lacking vitamin A. Rice nutritional augmentation was initiated in 1966 at IRRI (the International Rice Research Institute) in the Philippines. The vitamin A deficiency is being addressed using genetic engineering.

13.4 Breeding for improved protein content

The key components of food that impact nutrition are carbohydrates, fats, proteins, minerals, water, vitamins, and fiber. The first three components provide calorific energy, while proteins, minerals and water play a role in the body tissue and structure. The roles of regulation and utilization are played by proteins, minerals, water, vitamins, and fiber. After satisfying calorific energy needs, proteins are the next most important nutritional component of a diet. Twenty-two amino acids are generally recognized in human nutrition, of which eight are essential for monogastric animals (Table 13.2). The utilization efficiency of the entire protein is diminished if the diet is deficient in any of the essential amino acids.

13.4.1 Breeding high lysine content grain

Breeders using conventional methods of ear-to-row selection were able to increase total protein content of corn kernel from 10.9 to 26.6%. Unfortunately, because the protein of corn is about 80% zein and

Table 13.2 Essential amino acids in animal/human nutrition.

Isoleucine	Alanine	Serine
Leucine	Arginine	Tryosine
Lysine	Cysteine	Asparagine
Methoinine	Glutamic acid	Glutamine
Phenylalanine	Glycine	Cystine
Threonine	Histidine	Hydroxyglutamic acid
Tryptophan	Proline	Norleucine
Valine		

hence nutritionally inadequate, the high increase in total protein was nutritionally unprofitable to non-ruminant animals. The zein fraction of the total protein is deficient in lysine and tryptophan. This deficiency was corrected in 1964 when researchers at Purdue University discovered mutant genes, called *opaque-2* and *floury-2*, which increased the lysine content of the kernel. The patterns of expression of the mutant genes differ slightly. The *opaque-2* gene has a recessive gene action, whereas the "*floury-2*" gene exhibits a dosage effect. The resulting corn is called **high lysine** corn and has a characteristic soft and starchy endosperm. Consequently, the softer endosperm predisposes high lysine kernels to breakage, cracking, and rot. Generally, high

lysine cultivars yield lower than their conventional counterparts. Cross-pollination with normal dent corn reverses the soft endosperm to normal dent endosperm. High lysine corn production must be done in isolated fields. The *opaque-2* recessive gene increased the lysine content of the kernel from about 0.26–0.30% to about 0.34–0.37%. High lysine has been transferred into sorghum.

13.4.2 Quality protein maize (QPM)

Quality protein maize(QPM) may be described as an extension of the improvement of high lysine maize. It is a high lysine product because it uses the *opaque-2* gene. However, it is unlike the traditional high lysine maize because it lacks all the undesirable attributes of high lysine products (i.e., low yields, chalky-looking grain, and susceptibility to diseases and insects pests). It looks like regular maize but has about twice the levels of lysine and tryptophan. QPM was developed by two researchers, K.V. Vasal and E. Villegas over three decades. They used conventional breeding methods to incorporate modifier genes to eliminate the undesirable effects of the lysine gene. The two scientists were rewarded with the World Food Prize in 2001 for their efforts.

Industry highlights
QPM: enhancing protein nutrition in sub-Saharan Africa

Twumasi Afriyie

International Maize and Wheat Improvement Center (CIMMYT), PO Box 5689, Addis Ababa, Ethiopia

Introduction

Maize is a major staple in sub-Saharan Africa and also constitutes an important source of food for children in particular. For example, children in Ghana grow well during the first six months of life but, thereafter, when breast milk ceases to be sufficient to sustain their rapid growth, malnutrition becomes normal. This nutritional trend is explained by the pervasive use of a thin gruel porridge made from maize or millet as the first weaning food fed to children. Few mothers supplement such cereal diets with other sources of protein, such as beans, fish or milk, due to ignorance about proper nutrition, high cost or lack of time. The cereals alone do not provide a balanced diet because they are low in lysine and tryptophan, essential amino acids, which cannot be synthesized by monogastric animals including humans (National Research Council, 1988). Normal maize, for example, has approximately 10% protein but the full amount is not utilizable by monogastric animals because the protein is low in lysine and tryptophan. When children are fed normal maize without any better-balanced protein supplement, they become malnourished and develop the protein deficiency disease called *Kwashiorkor*.

In 1963, Mertz and his coworkers at the University of Purdue discovered a recessive mutant maize gene, opaque-2, which resulted in grain protein with approximately twice the quantities of lysine and tryptophan, the two limiting amino acids in ordinary maize (Mertz *et al.*, 1964). There was an immediate upsurge of worldwide interest to develop nutritionally improved maize varieties. However, it was soon discovered that the gene conferring the improved nutritional quality also resulted in several undesirable agronomic traits, including low grain yield potential, unacceptable chalky grain type, high moisture at harvest, and high susceptibility to insects and disease attacks (National Research Council, 1988). When farmers rejected the early "high lysine" hybrids quickly released to them, the research in opaque-2 or high lysine maize waned markedly worldwide. However, unrelenting research continued for some 30 years at the International Maize and Wheat Improvement Center (CIMMYT), Mexico, resulting in the development of maize germplasm combining better protein quality with desirable grain yield potential and agronomic characteristics similar to normal maize (Bjarnason, 1990). This new source of nutritionally improved maize was the result of the accumulation of modifier genes in a rather complex breeding technique supported by a strong laboratory for analysis of protein quality (Vasal *et al.*, 1993). The new material had normal looking hard endosperm grain type and was designated Quality Protein Maize (QPM). At the time QPM germplasm became available, there were still doubts about the usefulness of QPM for farmer production and since researchers – especially those in developing countries – continued to show little interest in QPM, in 1991 CIMMYT therefore closed its research on QPM. It was at this time, however, that the Ghana Grains Development Project (GGDP) within the Crops Research Institute, Kumasi, Ghana, assigned one breeder full-time to initiate a QPM development project for Ghana. The Government of Ghana, the Canadian International Development Agency (CIDA), and Sasakawa Global 2000 (SG2000) provided research funding. The main objectives of QPM research in Ghana was to develop high and stable yielding QPM varieties with comparable performance to their normal counterparts, to demonstrate their nutritional advantages, and to promote their production, marketing and utilization.

Breeding approaches

The QPM germplasm used to initiate QPM breeding in Ghana was collected from CIMMYT, Mexico, in 1991. The germplasm included open-pollinated experimental varieties and early generation inbred lines from CIMMYT Populations 62 (white flint) and 63 (white dent). The maize streak virus disease was a major problem in Ghana at the onset of the QPM program. Therefore, the QPM germplasm was converted to streak virus disease resistance by backcrossing the susceptible materials to resistant sources obtained from the International Institute of Tropical Agriculture (IITA), Ibadan, Nigeria. This was followed by growing the lines under artificial streak pressure and self-pollinating resistant lines to produce S_1 lines. Half-sib (HS) and S_1 recurrent improvement schemes were employed to develop the first QPM variety in Ghana. In the HS procedure, about 300 families were grown in an isolation block. Three rows were detasseled before anthesis to constitute female rows and these were grown in alternation with one male row (not detasseled), which served as the pollen source. Equal numbers of seed from each family were composited to plant the male rows. At harvest, HS families with desired agronomic traits were selected. These families were planted again as ear-to-row and selfed to produce S_1 lines. Twenty seeds of selected S_1 lines were sent for protein quality analysis (basically measuring tryptophan levels in total protein) at CIMMYT, Mexico. At the same time, light tables were used to select modified grains in the S_1 lines. On the basis of visual selection of HS lines in the field, the result of the analysis of protein quality (tryptophan level) in the laboratory and grain modification under light tables, S_1 lines were selected and recombined in isolation to constitute the next cycle of improvement. A streak resistant QPM variety named Obatanpa (literally "Good Nursing Mother") was released in 1992. This was a white dent medium maturity (105 days) variety.

We also initiated a hybrid development program. We developed several inbred lines through inbreeding by self-pollination, and conducted early generation testing and topcross evaluation. Diallel cross evaluations were used to determine the combining abilities of advanced inbred lines. Several single and three-way hybrids were developed. Open-pollinated varieties and hybrids developed in the project were first tested at the research stations located in the major agro-ecological zones in Ghana. The Ghanaian QPM hybrids performed equal to or better than local check hybrids. Consequently, three of the hybrids were released in Ghana and one was released in South Africa for production in Southern African countries. QPM germplasm development was supported by biochemical analysis of the grain at CIMMYT, Mexico and Ghana.

QPM nutritional studies

Despite the progress obtained in QPM germplasm development in Ghana, widespread doubts persisted about the usefulness of QPM technology. We conducted several collaborative animal feeding studies on pigs, chickens and rats to ascertain the nutritional advantages of QPM when used as human or animal food. The collaborative institutions included the CRI, SG2000 and the Animal Science Department of the Kwame Nkrumah University of Science and Technology, and the Health and Nutrition Department of the Ministry of Health.

Feed ingredient for pigs

Fourteen starter pigs from two litters (8.4 kg average weights) were divided into two equal groups (each containing three females and four males) and were fed similar diets (*ad libitum*) for 16 weeks (Okai *et al.*, 1994). Group 1 diet contained

91% QPM (Obatanpa) and Group 2 diet contained 91% normal maize (NM) (Okomasa). The balance of both diets comprised equal quantities of mineral and vitamin supplement. Pigs fed a QPM diet averaged a higher weight gain (13.9 kg) than pigs fed NM (5.9 kg). The corresponding daily gains were 124.9 g and 52.7 g for the QPM and NM pigs, respectively, showing that QPM-fed pigs had a growth rate 2.30 times that of pigs on NM diet. The feed conversion efficiency of QPM pigs was also greater than NM pigs (Osei *et al.*, 1994).

Feed ingredient for broiler chickens

A series of three feeding trials was conducted with broiler chickens to assess the commercial viability of Obatanpa in poultry feed (Osei *et al.*, 1994). The initial studies focused on using QPM or NM as the sole source of protein and energy. Subsequent studies investigated the feasibility of reducing the levels of fish meal in commercial diets when Obatanpa was the source of maize. Results of the studies indicated that the use of Obatanpa allowed the level of fish meal (an expensive high protein ingredient in Ghana) to be reduced from 19.5 to 13.5% while still maintaining good performance. The use of Obatanpa in broiler diets resulted in significant economic advantage because of reduced use of fish meal – due mainly to the huge price disparity between QPM and fish meal.

Processed food from normal maize and QPM

Kenkey is a popular local food in Ghana. It is made from fermented maize meal. We studied the effect of processing and cooking on the nutritional quality of *Kenkey* made from normal maize or QPM (Ahenkora *et al.*, 1995). The Quality Protein Maize, Obatanpa, and the normal maize, Okomasa, were processed into *Kenkey*. Weaning rats were fed *ad libitum* on *Kenkey*-based diets, which served as the sole source of protein and amino acid for 28 days. Analysis of samples of the *Kenkey* revealed that processing and cooking raw grains into *Kenkey* reduced the lysine content by 13% and the tryptophan content by 22% (Ahenkora *et al.*, 1995). However, *Kenkey* from QPM contained 51% more lysine and 63% more tryptophan than *Kenkey* from normal maize. The individual average gain by rats fed on QPM *Kenkey* diet was 37.2 g compared with 16.2 g for normal-maize *Kenkey* – a 2.3-fold difference. Rats fed a QPM diet had better feed conversion ratio and higher protein efficiency ratio (PER) values than their counterparts fed a normal maize *Kenkey* diet.

Agricultural technology/nutrition impact study

A series of studies investigated the impact of QPM utilization on community-based agricultural technology interventions in the Ejura-Sekodumasi district, Ashanti Region, Ghana. The study was done through the Ministry of Health Nutrition Division with collaboration from other agricultural and health institutions in Ghana, particularly CRI, Ministry of Food and Agriculture (MOFA), and SG2000. The results showed that QPM enhanced growth relative to normal maize when fed to children.

Breeding challenges related to QPM

A series of experiments were conducted to dispel some of the doubts, myths and fallacies concerning QPM (Twumasi-Afriyie *et al.*, 1996).

- **QPM produces a lower grain yield than normal maize counterparts.** In Ghana, it was shown that QPM varieties could produce better than their normal counterparts.
- **Lysine and tryptophan were "heat labile" and would be destroyed during processing, thus QPM will lose its nutritional advantage during processing into local dishes.** We demonstrated that the nutritional advantage was maintained when QPM was processed into the most popular local dishes (Ahenkora *et al.* 1995).
- **QPM is conferred by a recessive gene and thus will lose its nutritional advantage in farmers' production plots, which are normally planted on small areas.** We conducted an experiment in which we surrounded a one-acre field of QPM with a yellow endosperm normal maize with the same maturity, and allowed the two to cross freely. Results from two years of data at several locations showed a maximum of 10% contamination by the normal maize. The contamination was most pronounced within the 12 m of the QPM field nearest the normal maize and was most serious at the southwestern sector of the field due to the prevailing wind. The nutritional quality of the bulked grain from the most contaminated lot was still not significantly different from the non-contaminated QPM (based on a rat feeding study).
- **QPM will not store well at the farm level.** From our study, when weevils were introduced into grains of normal or QPM grain, there was no difference in the extent of damage incurred. All samples were equally damaged in a short period. Moreover, it was detected that, in general, post-harvest handling was very poor in Ghana and that available improved technology, if followed, could enable farmers to store both NM and QPM with minimal problems.

- **Marketing will be difficult because QPM lacked visible identifiers that could facilitate sale at a higher price to offset additional cost of production.** In fact, there was no additional cost of production of Obatanpa as it produced higher yields than its normal counterparts under identical recommended practices. Agronomic performance *per se* became a driving force behind the adoption of this variety. Special marketing channels were developed for Obatanpa. Private purchasing agents began to market Obatanpa to satisfy the demands of commercial users such as food and feed processors and relief agencies. The private purchasing agents linked with producers and guaranteed the quality of Obatanpa to the users.

Current efforts in QPM development in Africa

CIMMYT re-commenced QPM development in 1997, partly due to the success achieved in Ghana. Current efforts in QPM development in sub-Saharan Africa involve several National Agricultural Research Institutes and increasing numbers of private seed companies. QPM development and deployment largely follows the Ghana model, involving multi-disciplinary and multi-institutional approaches in germplasm development, nutritional studies, variety releases, seed production and agricultural extension (Figure B13.1). To date, QPM varieties have been released in about 15 sub-Saharan African countries. Current efforts led by CIMMYT and IITA seek to incorporate QPM into elite and local cultivars through a process of conversion of normal maize germplasm to QPM. Conversion involves the use of backcrossing of normal maize to QPM using donor QPM populations or inbred lines. After one or two backcrosses, the plants are selfed or sib-pollinated; segregating opaque-2 phenotype grains, which appear partly opaque on light tables, are selected for further backcrossing. Two to three backcrosses are often enough to recover the recurrent parent genotype and phenotype in addition to the modified opaque-2 grain character. However, in some cases, additional cycles of improvement may be required to accumulate enough modifier genes to recover the normal maize endosperm phenotype. Marker Assisted Selection (MAS) has been employed to speed up the backcrossing. In this procedure, a DNA marker that is very closely linked to opaque-2 gene is used – as a replacement of the phenotypic test – to select progeny carrying the desired allele based on the analysis of leaf samples from young plants.

References

Ahenkora, K., Twumasi-Afriyie, S., Haag, W., and Dzah, B.D. (1995). Ghanaian *kenkey* from normal maize and quality protein maize: Comparative chemical composition and rat growth trials. *Cereals Communications*, **23**:299–304.

Bjarnason, M. (1990). CIMMYT's Quality Protein Maize program: present status and future strategies. International Maize and Wheat Improvement Center (CIMMYT), Mexico.

Mertz, E.T., Bates, L.S., and Nelson, O.E. (1964). Mutant genes that changes protein composition and increases lysine content of maize endosperm. *Science*, **145**:279–280.

National Research Council (1988). *Quality-Protein Maize.* Academy Press, Washington, DC.

Okai, D.B., Osei, S.A., Tua, A.K.,*et al.* (1994). The usefulness of Obatanpa, a quality protein maize variety in the feeding of pigs in Ghana. *Proc. Ghana Anim. Sci. Symp.*, **22**:37–43.

Osei, S.A., Atuahene, C.C., Donkoh, A.,*et al.* (1994). Further studies on the use of Quality protein maize as a feed ingredient for broiler chickens. *Proc. Ghana Anim. Sci. Symp.*, **22**:51–55.

Twumasi-Afriyie, S., Dzah, B.D., and Ahenkora, K. (1996). Why QPM moved in Ghana, in *Maize Productivity Gains through Research and Productivity Dissemination* (eds J.K. Ransom *et al.*). Proceedings of the Fifth Eastern and Southern Africa Maize Conference, Arusha, Tanzania, pp. 28–31.

Vasal, S.K., Srivisan, G., Pandey, S., Gonzalez, F., Crossa, J., and Beck, D.L. (1993). Heterosis and combining ability of CIMMYT's protein maize germplasm: Lowland tropical. *Crop Science*, **33**:46–51.

Figure B13.1 A farmer (left) and researcher (center) admire a bag of QPM maize being offered for sale at store in Ghana, West Africa. Figure courtesy of Twumasi Afriyie.

QPM has less of the indigestible prolamine-type amino acids that predominate the protein of normal maize. Instead, QPM cultivars have about 40% of the more digestible glutelins and a balanced leucine–isoleucine ratio for enhanced niacin production upon ingestion. Research also indicates that QPM has better food and feed efficiency ratings (grain food intake/grain weight gain) following feeding tests with animal (e.g., pigs and poultry). QPM cultivars have been released for production in over 20 developing countries since 1997.

13.5 Improving protein content by genetic engineering

Nutritional quality augmentation through addition of new quality traits, removing or reducing undesirable traits, or other manipulations, is an important goal in bioengineering of food crops. Crops that feed the world are primarily cereals, roots and tubers, and legumes. Unfortunately, they are nutritionally inadequate in providing certain amino acids required for proper growth and development of humans and monogastric animals. For example, cereals are generally deficient in lysine and threonine, whereas legumes are generally deficient in sulfur amino acids. In some species (e.g., rice) in which the amino acid balance is relatively appropriate, the overall protein quantities are low.

Molecular genetic approaches are being adopted for genetically engineering seed protein. They may be categorized as follows:

- Altering amino acid profile of the seed.
- Selective enhancement of expression of existing genes.
- Designing and producing biomolecules for nutritional quality.

13.5.1 The making of "Golden Rice"

"Golden Rice" is so called because it has been genetically engineered to produce **β-carotene** (responsible for the yellow color in certain plant parts like carrot roots) in its endosperm. This rice produces β-carotene or pro-vitamin A, the precursor of vitamin A, which does not occur in the endosperm of rice.

An estimated three billion people of the world depend on rice as staple food. Of this number, about 10% are at risk of vitamin A deficiency and the associated health problems that include blindness and deficiency of other micronutrients, such as iron and iodine. The effort to create Golden Rice was led by Dr. Ingo Potrykus, a professor of plant science at the Swiss Federal Institute of Technology. In 1990, Garry Toenniessen, the director of food security for the Rockefeller Foundation, recommended the use of the sophisticated tools of biotechnology to address the problem of lack of vitamin A in rice. Later, at a Rockefeller-sponsored meeting, Potrykus met Peter Beyer of the University of Freiburg in Germany, an expert on the β-carotene pathway in daffodils. In 1993, and with seed money of US$ 100 000 from the Rockefeller Foundation, the two embarked upon an ambitious project to create a transgenic plant in a manner unlike any before. After seven years, the duo announced to the world their outstanding achievement, the Golden Rice, at a cost of $2.6 million. The bill was partly footed by the Swiss government and the European Union.

The scientific feat accomplished in engineering β-carotene into rice is that it marks the first time a complete metabolic pathway has been engineered into an organism. Rice lacks the metabolic pathway to make β-carotene in its endosperm. Potrykus and Beyer had to engineer a metabolic pathway consisting of four enzymes into rice (Figure 13.1). Immature rice endosperm produces geranylgeranyl-diphosphate (GGPP), an early precursor of β-carotene. The first enzyme engineered was phytoene synthase, which converts GGPP to phytoene (a colorless product). Enzyme number 2, called phytoene desaturase, and enzyme number 3, called ζ-carotene desaturase, each catalyzes the introduction of two double bonds into the phytoene molecule to make lycopene (has red color). Enzyme number 4, called lycopene β-cyclase, converts lycopene into β-carotene. A unit of transgenic construct (called an expression cassette) was designed for each gene for each enzyme. These expression cassettes were linked in series or "stacked" in the final construct.

The source of genes for enzymes 1 and 4 was the daffodil, while genes for enzymes 2 and 3 were derived from the bacterium *Erwinia uredovora*. Three different gene constructs were created, the first and most complex combining enzyme 1 with an enzymes 2–3 combo, together with an antibiotic resistance marker gene that encodes hygromycin

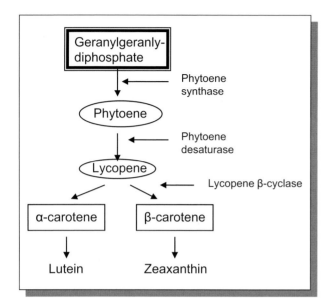

Figure 13.1 The key enzymes involved in the development the Golden Rice.

resistance, along with its promoter, CaMV 35S. The second gene construct was like the first, except that it lacked any antibiotic resistance marker gene. The third gene construct contained the expression cassette for enzyme 4, plus the antibiotic marker. By separating the genes for the enzymes and antibiotic resistance marker into two different constructs, the scientists reduced the chance of structural instability following transformation (the more cassettes that are stacked, the more unstable the construct).

These gene constructs were transformed into rice via *Agrobacterium* mediated gene transfer in two transformation experiments. In experiment 1, the scientists inoculated 800 immature rice embryos in tissue culture with *Agrobacterium* containing the first transgenic system. They isolated 50 transgenic plants following selection by hygromycin marker. In the second experiment, they used 500 immature rice embryos, inoculating them with a mixture of *Agrobacterium* (T-DNA) vectors carrying both the second and third constructs. This experiment yielded 60 transgenic plants. The second experiment was the one expected to yield the anticipated results of a golden endosperm. This was so because it had all the four enzymes required for the newly created metabolic pathway. However, the scientists also recovered transgenic plants with yellow endosperm from

experiment 1. Subsequent chemical analysis confirmed the presence of β-carotene, but no lycopene. This finding suggests that enzyme 4 may be present in rice endosperm naturally, or it could be induced by lycopene to turn lycopene into β-carotene. Analysis also showed the presence of lutein and zeaxanthin, both products derived from lycopene. None of the above was found in the control (non-engineered) plants.

13.5.2 Matters arising from the development of Golden Rice

The initial Golden Rice lines produced 1.6–2.0 micrograms of β-carotene per gram of grain. The recommended daily allowance (RDA) set by health agencies for children is 0.3 mg/day. Estimates of bioavailability of β-carotene have been put at less than 10% in some cases. The scientists intend to refine their invention to make it produce threefold to fivefold of its present level of β-carotene.

Upon an international intellectual property rights (IPR) audit commissioned by the Rockefeller Foundation through the International Service for the Acquisition of Agri-Biotech Applications, Potrykus and his team realized their invention utilized 70 IPRs (Intellectual Property Rights) and TPRs (technical property rights) owned by 32 different companies and universities. Because of the humanitarian goal of the project, the development team negotiated with owners of these patents to allow the use of their inventions under the "freedom to operate" clause. Further, because of public pressure and need for big business to tone down its profit-oriented public image, the key companies (e.g., Monsanto) offered free licenses for their IPRs involved with golden rice.

Chronologically, the idea of the Golden Rice project was first introduced at a conference in the Philippines in 1984. The first generation Golden Rice was produced in 1999 (patent WO2000/053768). In January of 2001, a new effort was launched in Philippines for a comprehensive set of tests to determine the efficiency, safety and usefulness of Golden Rice for people in the developing world. The joint effort includes the Philippines-based International Rice Research Institute (IRRI), Syngenta, and the Rockefeller Foundation. Further, IRRI has set up a Humanitarian Board to oversee this project and to ensure that the highest standards for testing, safety and support are achieved. The Board includes several public and private organizations, such as the World Bank, Cornell University, the Indo-Swiss Collaboration in Biotechnology, and the Rockefeller Foundation.

The second generation rice with significantly higher concentrations of the desired carotenoids was produced in 2004 by Syngenta (patent WO2004/085656). A Golden Rice 2 (GR2) variety, reported to have 23 times more β-carotene than the original variety of Golden Rice, was announced in 2005. Golden Rice has been bred with local rice varieties in the Philippines, Taiwan and the USA (at Louisiana State University in 2004). Tests to find out whether the carotenoids could be biologically absorbed were first conducted in 2009 in China. A trial Golden Rice crop was harvested in the Philippines in 2011. The Gates Foundation awarded IRRI another US$10 million for the Golden Rice project. Commercial cultivation is expected to begin in the Philippines in 2013, and 2017 in Bangladesh.

Golden Rice is far from becoming a widely grown rice cultivar. Apart from overcoming controversies surrounding GM cultivars, agronomic and other studies will have to be conducted to determine how well Golden Rice yields, its palatability and digestibility, and public acceptance.

13.6 Breeding improved oil quality

Oil quality improvement is a major breeding objective for major oil crops such as soybean and rape. Soybean oil accounts for about 22% of the world's total edible oil production.

13.6.1 Chemical composition of seed oil

By chemical composition, soybean oil consists primarily of palmitic (C16:0), stearic (C18:0), oleic (C18:1), linoleic (C18:2), and linolenic (C18:3) acids, of which palmitic acid is one of the two major components of saturated fatty acids. Consequently, the physical, chemical, and nutritional quality of soybean depends significantly on palmatic acid content. Breeders are interested in lowering the palmitic acid content of soybean oil, so as to lower its total saturated fatty acid content for higher desirability for human consumption. On the average, soybean seed oil contains about 120 g/kg of palmitic acid.

On the other hand, seed oil of modern oil seed rape contains about 60% oleate (C18:1), 20% linoleate (C18:2), 10% linolenate (C:18:3), and only 4% palmitate (C16:0) and 2% stearate (C18:0). The major breeding goal for oil seed rape quality improvement is to increase oleate content, which decreases the amounts of the polyunsaturated fatty acids, linoleate and linolenate. This would enhance its nutritional quality and increase its potential value as industrial oil.

Generally, food products with less than 7% total saturated fatty acids (C16:0 + C18:0 + C20:0 + C22:0) qualifies to be labeled "low in saturated fatty acids" in the United States.

13.6.2 Approaches to breeding oil quality

Conventional breeding approaches have been successfully applied to improve seed quality of various oil crops. Soybean lines with reduced palmitic acid content have been developed using traditional approaches of hybridization, recurrent selection, and chemical mutagenesis. Studies have shown that reduced palmitic acid is conditioned by at least two loci, without maternal effects. Some of these genes have been designated *fap1*, *fap2*, and *fap3*. Genes modifying the major palmitic acid loci have also been found to influence the trait by about 2–23 g/kg. Molecular markers, including RFLP, SSR and QTLs associated with palmitic acid reduction have been cloned and mapped.

Germplasm containing between about 75 and 90% of oleate have been developed in soybean, sunflower, and rape, among other species. Transgenic high oleate soybean has been developed. Soybean is also a major source of protein for animal feed. Reduced linolenate soybean is conditioned by three independent mutants – *fan1(A5)*, *fan 2*, and *fan 3*.

Research has shown that seed protein and oil content are negatively correlated. Consequently, developing high protein and high oil seed has limited success. Alternatively, breeders have devoted efforts to breeding cultivars with high protein and low oil, and those with high oil but low protein.

In sunflower, a breeding objective is to increase the stearic acid (C18:0) content of the seed oil. Mutagenesis was used to develop different lines with a higher C18:0 content of about 50 g/kg. The results of one genetic analysis indicated the inheritance of high stearic content to be under the control of one locus (*Es₁*, *es₁*) and partial dominance. Further studies indicated the presence of a second locus (*es₂*, *es₂*). Generally, fatty acid composition is controlled by the genotype of the developing embryo. Hence, selection for fatty acid may be conducted at the single seed level, by non-destructive techniques such as gas–liquid chromatography.

13.7 Breeding low phytate cultivar

Human and monogastric animals produce little amounts of the phytate enzyme needed to utilize phytate phosphorus. Soybean, for example, contains about 4.3 g/kg of phytate phosphorus and only 0.7 g/kg inorganic phosphorus. In humans, diets high in phytic acid decrease the absorption of essential minerals such as calcium, iron, and zinc. It would be desirable to remove phytate from cereals and oil seeds.

In soybean, low phytate mutants with about 1.9 g/kg phytate phosphorus and 3.1 g/kg inorganic phosphorus have been discovered. The trait is conditioned by recessive alleles designated *pha₁* and *pha₂* at two independent loci that exhibit duplicate dominant epistasis. Both of the alleles must be homozygous in order for low-phytate seed to be obtained. In wheat, low phytic acid mutants *lpa₁* and *lpa₂* have been identified.

13.8 Breeding end use quality

Breeders set breeding objectives to meet the needs of producers as well as consumers. The crop producer needs to focus on traits that facilitate crop production and increase crop yield (e.g., pest resistance, maturity, high yield, lodging resistance, drought resistance). Consumers are more concerned about nutritional quality traits (taste, protein content, appearance). Another group of consumer needs that is not nutritional but concerned with how they are used or stored is end use quality. Certain cultivars are bred for specific industrial quality traits (e.g., for processing, cooking).

13.8.1 Extended shelf life

Plant products that are harvested and used fresh (e.g., fruits, vegetables) are perishable and highly susceptible to spoilage soon after harvesting. In cases where production is far from marketing centers, the fresh produce has to be transported over long distances, and hence requires protection from bruising and rotting in transit. Fully vine- or plant-ripened fruits, even though desirable for their superior taste, are more susceptible to damage under such conditions than unriped fruits. Grocery stores need to display their produce for a period in good condition while waiting to make a sale. Extended shelf life is hence an important plant trait from the point of view of producers, wholesalers, and consumers.

Delayed ripening

Delayed ripening is desired in crops such as tomato and banana. Biotechnology has been successfully used to develop this quality in some crops. Certain fruits exhibit elevated respiration during ripening with concomitant evolution of high levels of ethylene. Called **climacteric fruits** (e.g., apples, bananas, tomatoes), the ripening process of these fruits involves a series of biochemical changes leading to fruit softening. Chlorophyll, starch, and cell wall are degraded. There is an accumulation of lycopene (red pigment in tomato), sugars, and various organic acids. Ripening is a complex process that includes fruit color change and softening. Ripening in tomato has received great attention because it is one of the most widely grown and eaten fruits in the world. Ethylene plays a key role in tomato ripening. When biosynthesis of ethylene is inhibited, fruits fail to ripen, indicating that ethylene regulates fruit ripening in tomato. The biosynthesis of ethylene is a two-step process in which s-adenosyl methionine is metabolized into aminocyclopropane-1 carboxylic acid, which in turn is converted to ethylene. Knowing the pathway of ethylene biosynthesis, scientists may manipulate the ripening process by either reducing the synthesis of ethylene or reducing the effects of ethylene (i.e., plant response).

In reducing ethylene biosynthesis, one successful strategy has been the cloning of a gene that hydrolyzes s-adenosyl methionine (SAM) called SAM hydrolase, from a bacterial virus, by Agritope of Oregon. After bioengineering the gene to include, among other factors, a promoter that initiates expression of the gene in mature green fruits, *Agrobacterium* mediated transformation was used to produce transgenic plants. The effect of the chimeric gene was to remove (divert) SAM from the metabolic pathway of ethylene biosynthesis. The approach adopted by researchers was to prevent the aminocy-clopropane-1-carboxylic acid (ACC) from being converted to ethylene. A gene for ACC synthase was isolated from a bacterium and used to create a chimeric gene as in the Agritope case.

The technology of antisense has been successfully used to develop commercial tomato that expresses the antisense RNA for ACC synthase and ACC oxidase. USDA scientists pioneered the ACC synthase work, while scientists from England in collaboration with Zeneca pioneered the ACC oxidase work. Because transgenic tomato with incapacitated ethylene biosynthetic pathway produced no ethylene, they failed to ripen on their own, unless exposed to artificial ethylene sources in ripening chambers. The technology needs to be perfected so that fruits can produce some minimum amount of ethylene for autocatalytic production for ripening over a protracted period.

The "FlavrSavr" tomato

Another application of antisense technology is in preventing an associated event in the ripening process, fruit softening, from occurring rapidly. Vine-ripened fruits are tastier than green-harvested and forced-ripened fruits. However, when fruits vine ripen before harvesting, they are prone to rotting during shipping or have a short shelf life in the store. It is desirable to have fruits ripen slowly. In this regard, the target for genetic engineering is the enzyme **polygalacturonase (PG)**. This enzyme accumulates as the fruit softens, along with cellulases that break cell wall cellulose and pectin methylesterase that together with PG breaks the pectic cross-linking molecules in the cell wall. Two pleiotropic mutants of tomato were isolated and studied. One mutant, never ripe (*Nr*), was observed to soften slowly and had reduced accumulation of PG, while the second mutant, ripening inhibitor (*rin*), had very little accumulation of PG throughout the ripening process. This and other research evidence strongly suggested a strong association between PG and fruit ripening. PG is biosynthesized in the plant and has three isoenzymes (PG1, PG2, PG3).

This technology was first successfully used by Calgene to produce the **FlavrSavr tomato**, the first bioengineered food crop, in 1985. The protocol was previously described. As previously noted, this pioneering effort by Calgene flopped for several reason, among which was the poor decision to market a product intended for tomato processing as a fresh market variety.

13.8.2 Breeding cooking and processing qualities

Cooking food changes its texture, color, taste (palatability), and digestibility, among other changes. The heat treatment applied during cooking breaks down some toxic compounds in food, where applicable. What is considered good cooking or processing quality depends on the product and the culture in which the product is used. As previously cited as an example, some cultures prefer sticky rice versus non-sticky rice for certain food preparations. Similarly, some potato cultivars are suitable for frying, others for baking, and yet others for cooking.

Similarly, canning or processing quality is an important breeding objective in crops that are grown for that purpose. It is desirable for canned produce to retain its texture and color to an appreciable degree. Some cultivars remain firm and of good color, whereas others crack or become mushy after canning.

Other products are crushed, ground or milled during processing. In corn for example, milling may be dry or wet. For dry milling white endosperm and semi-hard kernel is preferred, while wet milling (for starch, oil) requires softer kernels.

13.9 Breeding seedlessness

Fresh fruits without seeds are more convenient to eat, because there are no seeds to spit out. Common fresh fruits in which seedless cultivars exist include watermelon, grape, orange, and strawberry.

The conventional way of producing seedless fruits is the use of triploid hybrids. A tetraploid (4x) parent crossed with a diploid (2x) line is exemplified by seedless watermelon breeding (Figure 13.2). In the watermelon, 4x = 44 and 2x = 22. The tetraploid is always the female parent; the reciprocal cross (with the diploid as male parent) does not produce seed. The resulting triploid (3x = 33) is female sterile, and hence the fruit is seedless. Furthermore, because the triploid is also male sterile, producers of seedless watermelon must plant rows of diploid lines as pollinators for stimulation of fruit formation. In commercial production fields, growers usually plant a ratio of three triploid rows to one diploid row. It is important that the

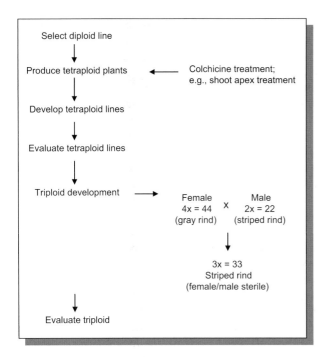

Figure 13.2 Generalized steps in the breeding of seedless melon.

rows be marked to enable harvesters pick fruits from only triploid lines.

Breeding of seedless watermelon will be described to illustrate the conventional production of seedless fruits. The process involves a number of steps.

13.9.1 Selection of diploid lines for developing tetraploids

Many producers use tetraploids with gray-colored rind and diploid with stripped rind. This way, it is easy to identify and discard selfed tetraploids (gray rinds) in the progeny.

13.9.2 Tetraploid induction

Chromosome doubling to produce tetraploids from the diploid line is accomplished by using colchicine. Other methods are available. Colchicine is applied to the shoot apex of diploid plant seedlings, just as the cotyledons first emerge from the soil. This causes the chromosomes at the shoot apex to double, resulting in the tetraploid seeding. The mutagen is applied at a concentration ranging between 1 and 2%, the lower concentration for small-seeded cultivars. Similarly, one or two drops of colchicine is applied, one drop to smaller seedlings and two drops to larger seedlings. The success rate of chromosome doubling is low (about 1%). Detection of tetraploids is by chromosome count or other method, such as a count of the number of chloroplasts in each side of the guard cell (diploids have 5–6 per side), whereas tetraploids have 10–14 per side. Morphologically, tetraploids tend to have thicker leaves, shorter stems, and slower growth than diploids.

13.9.3 Development of tetraploid lines

The putative tetraploids from the mutagenesis (T_1 plants) are evaluated in the next generation (T_2) to authenticate their polyploidy. Selected plants are selfed or sibbed repeatedly in isolation to stabilize the new genotypes and increase fertility for high seed yield to produce sufficient seed for evaluation of the tetraploid lines.

13.9.4 Evaluation of tetraploid lines

The tetraploid lines in the T_4 generation are evaluated especially for performance in a hybrid by making selected crosses to produce triploids. These triploids are evaluated to identify tetraploids with high potential as parents. Successful parents produce hybrids with high yield, good rind color (gray), and that lack empty seed coats.

13.9.5 Development and evaluation of triploid lines

Successful tetraploid parents are used in producing triploids by hand pollination (expensive) or insect pollination in isolation blocks. Hand pollination enables the breeder to exercise control over the pollination process to ensure high success rate of desirable crosses. Pistilate flowers from the female parent may be tagged. When developing triploids by insect pollination, the two parents should have different rind colors for easy identification of hybrids. As indicated previously, the tetraploid female usually has a gray rind whereas the diploid is stripped. The triploid produced by hybridization will be striped.

13.10 Breeding for industrial uses

Some crops have multiple uses – food, feed, and industrial. Corn, for example, may be milled for flour, extraction of oil, starch production, or sugar production. Breeders have developed special-purpose hybrids for these uses. For example, by inserting the endosperm gene, sugary (*su*), and another mutant gene, shrunken (*sh2*), in the same genotype, the resulting hybrid has increased sugar content (called supersweet or extra sweet corn). Similarly, waxy corn is developed for use in the production of adhesives, gums, and pudding, because of its high amylopectin content.

Edible oils that are high in polyunsaturated fatty acids are considered more healthy because they have the capacity to lower blood cholesterol. However, such oils are unstable for high-temperature cooking because they oxidize and break down readily at high temperatures. The hydrogenation process that is used to artificially stabilize oils converts the polyunsaturates back to monounsaturates. Oil plants, such as soybean and cotton, have genes that naturally convert oleic acid, a monounsaturated fatty acid, into polyunsaturated fatty acid. In cotton, about 25% of the seed oil comprises two fatty acids – palmitate and stearate. Conventional cotton seed oil contains primarily palmitate, which is thought to raise blood cholesterol more than stearate. Scientists at CSIRO, Australia, turned off the genes for polyunsaturates to produce cotton seed with high levels of oleic acid (monounsaturates). This high oleic oil is more stable for high-temperature cooking and can be used in place of hydrogenated oils. Further, the scientists modified the plants to produce more of stearate than palmitate, making the GM product healthier for human use.

Similarly, in soybean, other scientists used the technology of gene silencing to develop high oleic cultivars. Using biolistic techniques, a second copy of the *fad2* gene (fatty acid desaturase gene) that encodes the enzyme delta-12 desaturase that is involved in fatty acid synthesis was introduced into the genome of soybean. This event switched off the desaturase gene, causing the accumulation of oleic acid in the seed only (normal fatty acid biosynthesis occurs in other plant parts). Consequently, only small amounts of the polyunsaturated fatty acids, linoleic and linolenic, are produced in the seed. The high oleic soybean GM cultivars such as G94-1 contain about 80% more oleic acid than conventional seed, and higher than in olive oil and rapeseed oil.

13.11 Breeding plants for novel traits

An application of genetic engineering to breed novel traits is the use of organisms as bioreactors to produce pharmaceuticals. One of the earliest applications of this technology was the commercial production of human insulin in microbial systems. Similarly, certain pharmaceuticals are commercially produced in mammalian milk of sheep, goats, and rabbits. The application is being applied to plants to produce selected chemical compounds. Plant-made vaccines are currently under development for protection against cholera, diarrhea (Norwalk virus), and hepatitis B. The most common plants that are being used in plant-made pharmaceuticals are corn, tobacco, and rice. Other crops being investigated include alfalfa, potato, safflower, soybean, sugarcane, and tomato. To be usable, the plant should be readily amenable to genetic engineering and capable of producing high levels of protein. Further, there should be an efficient method for extracting the protein products from the plant tissues. Another example of plant-manufactured pharmaceutical is taxol, a secondary product derived from the Pacific Yew tree. This product has been found to be effective against certain cancers.

Key references and suggested reading

Hulke, B.S., Fehr, W.R., and Welke, G.A. (2004). Agronomic and seed characteristics of soybean with reduced phytate and palmate. *Crop Science*, **44**:2027–2031.

Helms, T.C., and Orf, J.H. (1998). Protein, oil, and yield in soybean lines selected for increased protein. *Crop Science*, **38**:707–711.

Mertz, E.T., Bates, L.S.E.T., and Nelson, O.E. (1964). Mutant genes that changes protein composition and increases lysine content of maize endosperm. *Science*, **145**:279–280.

National Research Council (1988). *Quality-Protein Maize.* Academy Press, Washington, DC.

Wang, T.L., Domoney, C., Hedley, C.L., Casey, R., and Grusak, M.A. (2003). Can we improve the nutritional quality of legume seeds? *Plant Physiology*, **131**:886–891.

Internet resources

http://www.ext.colostate.edu/PUBS/CROPS/00307.html – Brief overview of bio-pharming in plants (accessed March 6, 2012).

http://www.molecularfarming.com/molecular-farming-patents.html – Site to various products and companies engaged in pharming (accessed March 6, 2012).

Outcomes assessment

Part A

Please answer the following questions true or false.

1 Golden rice is rice bred for high protein content.
2 Triploid hybrids fruits are seedless.
3 The FlavrSavr tomato has high protein content.
4 Genes were obtained from the daffodil in the development of Golden Rice.
5 Cereals tend to be high in lysine.

Part B

Please answer the following questions.

1 Quality protein maize is bred by incorporating the . . . . gene.
2 The high lysine corn was producing using the . . . . gene.
3 Give a specific problem encountered in the development of Golden Rice.

Part C

Please write a brief essay on each of the following.

1 Describe the breeding of seedless watermelon.
2 Discuss the QPM.
3 Discuss the breeding of super-sweet corn.
4 Discuss high lysine corn.

14

Breeding for resistance to diseases and insect pests

Purpose and expected outcomes

Plants are plagued by a host of pathogens and pests that often must be controlled or managed in crop production. To control these organisms that "consume plants" effectively requires an understanding of their biology, epidemics, spread, and damage they cause. A variety of methods are used in pathogen and pest control, each with advantages and disadvantages. These methods are chemical, biological, cultural, legislative, and physical controls. A specific tactic in the method of biological pest control is the use of disease-resistant cultivars in crop production. Breeding for disease and insect resistance is one of the primary objectives in plant breeding programs. After completing this chapter the student should be able to:

1 Discuss the economic importance of plant pests.
2 Discuss the genetic basis of disease resistance.
3 Discuss resistance breeding approaches.
4 Discuss specific applications of biotechnology in plant breeding to control pathogens and pests.
5 Discuss disease epidemic and its breeding implication.

14.1 Selected definitions

Definitions of certain key terms will facilitate the understanding of this chapter. Additional terms will be defined in various sections of the chapter.

- **Pest.** An organism that is damaging to a crop. Broadly defined, a pest may include plant pathogenic fungi, bacteria, viruses, insects, and mammals.
- **Parasite.** An organism that feeds, grows, and is sheltered on or in a different organism while

Principles of Plant Genetics and Breeding, Second Edition. George Acquaah.
© 2012 John Wiley & Sons, Ltd. Published 2012 by John Wiley & Sons, Ltd.

contributing nothing to the survival or well-being of the host.

- **Pathogen.** An agent that causes disease. Usually, the term refers to living microorganisms (e.g., bacteria, viruses, fungi).
- **Disease.** Any condition caused by the presence of an invading organism or a toxic component that damages the host.
- **Pathogenecity.** The ability of an organism to enter a host and cause disease.
- **Virulence.** The degree of pathogenecity (the comparative ability to cause disease) is called virulence.
- **Infection.** The invasion by and multiplication of a pathogenic mircrooganism in a bodily part or tissue; may lead to tissue injury and disease.
- **Resistance.** A response to a cause (such as an attempted infection).
- **Immunity.** A genotype is said to be immune to a pathogen if it is completely or totally resistant to it, not showing any sign of infection.
- **Host resistance.** The ability of specific plant species to resist specific insects and pathogens because of a certain genetic architecture in the plant.
- **Non-host resistance.** The phenomenon of immunity against the majority of pathogens.

14.2 Groups of pathogens and pests targeted by plant breeders

Plant diseases and pests are caused by organisms that vary in nature and may be microscopic or readily visible (e.g., virus versus insect pests like beetles). These pathogens may be airborne or soil borne. Important groups of causal agents of diseases are microbes – **air-borne fungi, soil-borne fungi, oomycetes, bacteria, viruses, viroids, phytoplasmas, parasitic plants, nematodes,** and **insects**. Through an understanding of the biology, epidemics, spread, and damage caused by these organisms in each category, breeders have developed certain strategies and methods for breeding cultivars to resist certain types of biotic stress in plant production.

Plant breeders devote varying amounts of resources to breeding for resistance in these categories with varying degrees of success. Plant species vary in their susceptibility to diseases caused by pathogens or pests in each group. Some crop production pathogens and pests are conspicuously absent from the list because they are relatively unimportant (e.g., mites), relatively more easily controlled by application

of pesticides, or are not practical to breed against (e.g., birds).

Cereal crops tend to have significant air-borne fungal disease problems, while solanaceous species tend to experience viral attacks. Diseases that afflict crops of world importance and cause major economic losses and are readily transmitted across geographic boundaries receive funding from major donors. Breeding for resistance to fungi, especially air-borne fungi, is the most prominent resistance breeding activity. N.W. Simmonds suggested that the relative importance of the several groups of pathogens and pests of importance to plant breeders might be something like this: air-borne fungi > soil-borne fungi > viruses > bacteria = nematodes = insects.

Weed control is a major activity in crop production. However, crop tolerance to weeds is seldom, if ever, a breeding objective and hence will not be discussed in this chapter.

14.3 Biological and economic effects of plant pathogens and pests

Disease is basically a change from the state of metabolism necessary for normal development and functioning of an organism. An abnormal growth and development of the plant will cause a reduction in biological yield and product quality and, hence, invariably economic yield. Therefore, the need to control pathogens and pests in crop production is obvious. There are several basic ways by which diseases and pests adversely affect plant yield and general performance, and, eventually, reduce economic value. They include the following:

- **Complete plant death.** Certain pathogens and pests sooner or later will completely kill the afflicted plant. When this occurs early in crop production, gaps are left in the crop stand. Where such gaps cannot be compensated for by plants in the vicinity, additional nutrients may become available to the existing plants due to reduced competition. However, the reduced plant density normally results in reduced total crop biomass. Crop stand-reducing pathogens and pests include those that cause damping off diseases (the *Oomycete pythium*, for example), tosspot viruses, vascular wilts, and insects such as cutworms that cause a seedling to fall over and die.

- **Partial plant death.** Some diseases that afflict adult plants do not completely kill them. Rather, only certain parts of the plant (e.g., branches) are killed (e.g., as observed in fungal diebacks).
- **Stunted growth.** Viruses are known to reduce the metabolic performance of plants without killing them outright. Afflicted plants grow to only a fraction of their normal size and usually cause a severely reduced economic yield of plants.
- **Vegetative destruction.** Herbivores such as caterpillars, locusts, and rabbits eat plant tissue.
- **Direct product damage.** Pathogens and pests may directly affect the economic or harvested product (e.g., fruit, seed, or modified organs (tubers, bulbs)). Some pathogens and pests directly injure these products completely (e.g., by causing rotting of fruits) or reduce quality (e.g., by causing blemishes or holes in fruits).
- **Other effects.** Other effects that are less dramatic but nonetheless important include damage to plants caused by the sucking insects (e.g., aphids, thrips, mites) and by rust fungi and powdery mildews, which deplete plants of photosynthates and decrease photosynthetic area. Root nematodes also decrease root capacity of plants.

It is a principle in pathogen and pest control that a method of control (especially chemical method) is warranted only when economic loss is eminent. It does not make economic sense to spend $1000 in disease and pest control to save $100 of harvestable product. A breeder will only invest in breeding for resistant cultivars if the pathogen/pest organism is economically important and pesticides or biological control measures are expensive, outlawed or insufficiently effective.

14.4 Overview of the methods for control of plant pathogens and pests

Crop producers use as variety of pathogen and pest control methods that may be classified as follows:

- **Quarantine.** This method involves legislation that imposes crop isolation or inspection to prevent the pathogen or pest from making initial contact with the host plant. Quarantine laws help to prevent the introduction of new pathogens and pests into a production region. The laws are commonly enforced at the points of entry for people and goods into a

country or a specific region. Of course such measures are not relevant for pathogens and pests with a long range natural dispersal.
- **Phytosanitation and Cultural Practices.** In the event that a pathogen gains access into a production area, various methods may be used to remove it or reduce the inoculum to contain its spread. A method such as crop rotation reduces disease buildup in the field, while observance of sanitation (e.g., removing diseased plants and burning them) also reduces the spread of the pathogen or pest. The producer may also implement management practices that discourage the growth and spread of the disease agents (e.g., soil drainage, weeding, soil sterilization, and seed treatment).
- **Improvement of host resistance.** This is the method of most concern to plant breeders. It entails breeding to introduce genetic resistance to pathogens and pests into adapted cultivars. This is the primary subject of this chapter.
- **Application of pesticides.** Economic plants may be protected from pathogens and pests by using chemicals (pesticides). While this is widely used, the method is environmentally intrusive and expensive.
- **Biological control.** Biological control methods become increasingly more effective especially in protected cultivation (i.e., greenhouses); for example, the case of the predatory mite *Amblyseius swirskii* against white fly and thrips in cucumber.

14.5 Concepts of resistance in breeding

Breeding for yield and other morphological traits, as well as breeding for resistance to abiotic stress are conceptually different from breeding for resistance to biotic stresses. Breeding in the former scenarios entails the manipulation of a single genetic system – plants. Breeding for resistance to biotic stresses on the other hand involves the manipulation of two genetic systems – one for plants (host) and the other for the organism (pathogen or pest) – not independently, but with regard to the interaction between the two systems. The breeder needs to understand the interrelationships between plants and pathogens/pests that have persisted through co-evolution and co-existence.

There are degrees of resistance. In terms of disease, resistance is relative to a benchmark. A genotype is said to be **immune** to a pathogen if it is completely or totally resistant to it, not showing any sign of infection. However, in many cases the breeder will find in his or her germplasm a continuous variation in levels of resistance. A resistant cultivar has less disease than the standard cultivar – it could be a little or a lot less disease. True disease resistance has a genetic basis and is, hence, amenable to plant breeding methodologies. It is manifested in two basic forms – inhibition of infection or inhibition of subsequent growth of the pathogen, the former being the more common form of resistance. Resistance can be a qualitative or quantitative trait based on the underlying genetic mechanism.

Breeding for disease resistance is an integral part of any approach to crop disease and pest control. If yield and other desirable traits are maintained, resistance is an ideal method of breeding to control plant pathogens and pests. Development and use of resistant cultivars has several advantages. It is easy to deploy and has no adverse environmental consequences. Also it is convenient for farmers to use. However, when resistance is not satisfactory, farmers may need to supplement it with other disease and pest management practices (e.g., use of pesticides).

14.6 Concepts of pathogen and host

14.6.1 The pathogen

The **pathogen** is a living organism that is capable of inflicting a distinct disease in another organism (the host). There is significant confusion in the literature in the use of terminology pertaining to pathogens. The capacity of a pathogen to cause disease in a plant species is called its **pathogenicity**, while the extent of disease development it causes is its **aggressiveness. Virulence** is the ability of a pathogen to infect certain plant genotypes in spite of the presence of a gene for resistance. **Avirulence,** the converse of virulence, is the inability of a pathogen to infect a susceptible host. The pathogenicity and aggressiveness of a pathogen vary among pathogen types. The presence of a pathogen on a susceptible host is not enough to cause disease symptoms to occur. A third factor – a favorable environment – is needed; the trio (pathogen + susceptible host + favorable environment) are sometimes referred to as the disease triangle.

14.6.2 The host

The host (plant genotype) is the organism in which a pathogen or pest may produce an infection. If the pathogen or pest is successful in causing an infection the host is described as a susceptible host. A host may employ one of several mechanisms to cope with the threat of pathogens and pests.

14.7 Mechanisms of defense in plants against pathogens and pests

Plants exhibit a wide variety of mechanisms of defense against pathogens and pests that may be classified as **avoidance, resistance** and **tolerance**.

14.7.1 Avoidance

Avoidance is a mechanism that reduces the probability of contact between pathogens or pests and the plant. That is, it operates prior to the establishment of any intimate host–pest/pathogen contact. This should be distinguished from **escape,** the chance that a plant would not be infected. Avoidance is rare in pathogen attacks and, hence, is primarily applicable to attacks of animal pests and also to insect pests where entomologists describe it as **antixenosis**. The plant using this mechanism has characteristics (morphological or chemical) that make it unattractive, for example, to insects for feeding, oviposition, or shelter. In cabbage, a variant leaf color makes the plant unattractive to the cabbage aphid (*Brevicoryne brassicae*), while pubescence in cotton prevents oviposition in cotton boll weevil (*Heliothis zea*). Also, in cotton, a repulsive odor deters the boll weevil from feeding on the plant.

Avoidance may not be effective if there are no alternative hosts. Given the choice, insect pests may prefer one host to another. Where there is no choice (e.g., monocropping) a pest would attack the available host. Non-preference conditioned by allelochemicals is not effective in modern agriculture where monocropping and breakdown of resistance are common. However, morphological traits that interfere with feeding and breeding of insects are important breeding goals as a first line of defense against insect pests. Crop producers may help their crops to avoid pathogen and pest attack by timing the planting of the crop (early or late planting), so that the most susceptible or vulnerable growth stage does not coincide with the time when the insect pest or pathogen is most abundant.

14.7.2 Resistance

Resistance mechanisms have in common that they manifest after a host has been attacked by a pathogen or pest. The mechanism operates to curtail the invasion or to reduce the growth and/or development of the causal agent. The nature of the mechanism may be biochemical, physiological, anatomical or morphological. The equivalent term used to describe this mechanism as it relates to insect pests is **antibiosis**. Resistance is a relative term for the genetic-based capacity of a host to reduce the adverse effect of a pathogenic attack. It does not necessarily imply complete blocking/eliminating of infection. Some genotypes are more resistant than others, in the face of equal amount of initial inoculum. Further, host resistance mechanisms may be constitutive (called **passive resistance**). For example, the onion is resistant to the smudge disease (caused by *Colectotrichum circinaus*) because of the presence of various chemicals (e.g., catechol) in its outer scales. Resistance may also be induced or activated (called **active or induced resistance**) in response to an infection. Various resistance mechanisms exist.

Pre-existing defense mechanisms

These include morphological features that pose as barriers to the penetration of the pathogen into the plant (e.g., presence of lignin, cork layer, callose layers) or secondary metabolites (phenols, alkaloids, glycosides) that have antimicrobial properties.

Infection-induced defense mechanisms

Upon infection, the plant may quickly produce chemical products (e.g., peroxidases, hydrolases, phytoalexins, etc.) to combat the infection attempt. Also a common induced defense is the formation of papillae. These are cell wall appositions that are associated with plant cell penetration by pathogens. Infection-induced defense reactions manifest in various ways – **hypersensitive, overdevelopment, or underdevelopment**.

- **Hypersensitive reaction.** This mechanism acts to prevent pathogen establishment. An infection evokes a rapid localized reaction whereby the cells immediately surrounding the point of attack die so that the infection is contained. The pathogen eventually dies, leaving a necrotic spot (**necrosis**). This infection inhibition response is very common and is evoked in infections by pathogens causing leaf spots, blights, cankers, powdery mildews, downy mildews, late blight of potato, viral diseases, bacterial diseases. It is even found to occur against some insects, nematodes and parasitic plants.
- **Overdevelopment of tissue.** Meristematic activity may be stimulated, resulting in excessive abnormal tissue growth as exemplified by galls and also leaf curling. In some cases, a layer of suberized cells forms around the invaded tissue to curtail the spread of the pathogen.
- **Underdevelopment of tissue.** Afflicted plants become stunted in growth or organs become only partially developed. Viral infections produce this kind of effect.

14.7.3 Tolerance

Unlike avoidance and resistance mechanisms that operate to reduce the levels of infection by the pathogen or pest, tolerance (or endurance) operates to reduce the extent of damage inflicted. The afflicted host attempts to perform normally in spite of the biotic stress. A host may be highly susceptible (i.e., support a high population of pathogens or pests) and yet exhibit little reduction in economic yield (i.e., the plant host is tolerant of the pathogen or pest). Because tolerance is measured in terms of economic yield, it is not applicable to pathogens or insect pests that directly attack the economic part of the plant (e.g., grain, tuber or fruit). The mechanism may be applied to a situation in which a plant recovers quickly following a pest attack, such as grazing by a mammal. From the point of view of virologists, plants with little or no symptoms are described as tolerant. In a strict sense, plants showing hardly or no symptoms after having been exposed to a virus may be resistant (having interfered with the reproduction and spread of the virus in the plant) or tolerant (in case the virus reproduced and spread, but produced hardly any or no symptoms). In the latter scenario the plant is a symptomless carrier that may act as hidden source of inoculum to other crop genotypes nearby. To distinguish resistance and true tolerance to viral infection, the amount of virus (virus concentration in the plant) should be determined.

The exact mechanism(s) of tolerance is not understood. However, some attribute it to plant vigor, compensatory growth, changes in photosynthate partitioning, among other factors. Because it is difficult to determine the yield reduction per unit of infection, it is very difficult to determine the difference in tolerance between genotypes.

14.7.4 Host versus non-host disease resistance

Most plant pathogens are very narrowly specialized such that they can infect only a small number of host species (i.e., they have a **narrow host range**). Consequently, most plant species are completely resistant to nearly all pathogen species. This phenomenon of immunity against the majority of pathogens is called **non-host resistance** (NHR). NHR is, therefore, the most common form of disease resistance demonstrated by plants against most potential pathogens. It provides durable protection to plants against given species of pathogens or pests. The molecular basis of NHR is not well understood.

Host plant resistance (HR), on the other hand, entails the ability of specific plant species to resist specific insects and pathogens because of a certain genetic architecture in the plant. The host resistance is mediated by resistance (R) genes, which in turn define pathogen race- or cultivar-specific resistance. Unlike NHR, the effectiveness of HR is typically transient. One of the most important breeding goals is identifying and transferring genes that confer resistance to insects and pathogens into new cultivars.

14.7.5 Specificity of defense mechanisms

Almost all plant species are non-host for almost all pathogens, as previously noted. Most avoidance mechanisms are non-specific (i.e., are general or broad) in that plants are capable of avoiding groups of pathogens with similar ecological requirements and life style. For example, the closed flowering habit of some barley cultivars that excludes pathogens that invade through the open flower, such as *Ustilago nuda* (causes loose smut), also excludes *Claviceps purpurea* (causes ergot). However, some specificity may occur in the case of certain repellents. Resistance mechanism may be specific or non-specific. Resistance that is usually specific may not only be at the pathogen species level but even at the pathotype level. Race non-specific resistance is mostly highly pathogen

species specific. Plant breeders usually apply the race-specific and race-non-specific types of resistance, and much less avoidance or types of resistance that are based on high levels of constitutive defense like toxins.

14.7.6 Specificity of the pathogen

Another distinction between breeding for disease and pest resistance and breeding for other traits is that the former is influenced by genetic variability of the pathogen or pest, especially in diseases. When pathogen genotypes share a group of cultivars to which they are virulent, they are said to be belong to a **physiological race**. The equivalent term applied to viruses is **strain**, while it is **biotype** in insects and **pathotype** in nematodes. These terms have other usage in other contexts. Physiological races of pathogens occur in many pathogenic organisms, such as rusts, powdery mildew, and some insects. They may be identified by using **differential cultivars** (they contain known genes for disease reaction). Breeders use a series of differentials to determine what genes would be most effective to incorporate into a cultivar. The concept of differentials stems from the ability of a cultivar to differentiate between races of a pathogen on the basis of disease reaction. If a cultivar has resistance to one race but is susceptible to another, it has differential properties to identify two races of a pathogen and, hence, is called a **differential**. In the case of two categories of reaction (R = resistant, S = susceptible), n differentials may be used to differentiate 2^n races of a pathogen (where $n = 2$). In that case, four (2^2) races of a pathogen can be differentiated as follows:

		Races			
		R_1	R_2	R_3	R_4
Cultivars	C_1	−	−	+	+
	C_2	−	+	−	+

The ideal set of differential cultivars is one in which each cultivar carries a gene for resistance.

Physiological races are an abstraction, since they are not pure biotypes of an organism, but simply groups of genotypes that express the same reaction upon inoculation over a set of differential cultivars determined by experimentation. The differential cultivars are helpful to carry out virulence surveys. Such surveys sample a large number of isolates of the pathogen and test them one by one on the set of differential cultivars. The reaction patterns (compatible or incompatible) will indicate for each R-gene whether the isolate carries the avirulence Avr or the virulence avr allele. Together all data of all isolates indicate the frequency by which the virulence and avirulence alleles occur in the pathogen population.

14.7.7 Gene-for-gene reaction (genetics of specificity)

R.H. Biffen is credited with providing the first report on the genetics of resistance. Working on stripe rust (*Puccinia striiformis*) he reported that resistance to disease was controlled by a single Mendelian gene. Since then many reports have supported this finding that much host resistance to pathogens is genetically simply inherited. Dominance gene action is very common for resistance to pests and pathogens (especially, hypersensitive resistance). Monogenic recessive resistance is less common. However, it is known that resistance may be controlled by any number of genes whose effects may be large or small. Further, the genes may interact epistatically or additively. Commonly, resistance and avirulence genes usually operate on a gene-for-gene basis. Also, genes for resistance are often reported to be clustered in linkage groups.

Working on flax rust (caused by *Melampspora lini*), H.H. Flor discovered that the major genes for resistance in the host interacted specifically with major genes for avirulence in the fungus. He proposed the **gene-for-gene concept**, which states that for each gene conditioning resistance in the host there is a specific gene conditioning virulence in the pathogen. However, more recent studies demonstrated that the "virulence" allele is not "conditioning virulence", but the allele is a defect (by "loss-of-function" mutation) of the alternative, avirulence, allele. In other words, a resistance gene can act only if a locus in the pathogen carries a matching allele for avirulence but is ineffective if the pathogen carries the allele for virulence (usually, a recessive allele). The specific interaction occurs between dominant resistance alleles and the dominant avirulence alleles (Figure 14.1).

The original concept proposed in 1950 by Flor suggests that the virulence of a pathogen and the resistance of a host have a genetic basis. For each gene that confers resistance in the host, there is a corresponding gene in the pathogen that confers virulence

Host	Pathogen			
	AB	*Ab*	*Ab*	*ab*
rrss	+	+	+	+
RRss	-	+	-a	+
rrSS	-	-	+	+
RRSS	-	-	-	+

'a' – differential interaction

Figure 14.1 Demonstration of the concept of gene-for-gene-interaction in disease resistance.

to the pathogen, and vice versa. This host–pathogen/pest interaction is called the **gene-for-gene hypothesis**.

14.8 Types of genetic host resistance and their breeding approaches

Host resistance is generally categorized into two major kinds – **vertical** or **horizontal** – based on several associated criteria.

14.8.1 Vertical resistance

One of the major contributors to the elucidation of this type of resistance was Van der Plank. **Vertical resistance** is characterized by several features: it is based on a hypersensitivity reaction, it is race- or pathotype-specific in its effect, its inheritance is based on major genes resistance, and its effect in agricultural applications is usually not durable (non-durable resistance). Vertical resistance or race-specific resistance is oligogenic. Examples are the Sr-genes (for stem rust), Lr (for leaf rust), Pr (for powdery mildew) genes of wheat (cause hypersensitive reaction for resistance), and the Dm (for downy mildew) genes of lettuce. Major gene resistance is often complete, conferring immunity against the avirulent pathotypes. It delays the start of an epidemic and disappears when the pathotype virulent to the vertical resistance gene is prevalent. These genes control specific races or genotypes of pathogens and pests, and hence do not necessarily protect against new races of the pathogens and pests.

The pathotype specificity is the most striking feature of this type of resistance; plants are susceptible to some isolates of a pathogen species but resistant to other isolates. In the case of resistance, a necrotic spot typically appears due to the hypersensitive reaction. Cultivars equipped with this narrow-spectrum genetic protection are often vulnerable when a genetically variable population of a pathogen occurs in the environment. They may not succumb immediately to the new pathotypes of the pathogen but under certain cultural systems (e.g., monoculture without crop rotation, a situation in which the prevalent cultivar(s) carry one and the same *R*-gene) the populations of the new strains could build up high enough to cause economic injury to the crop. *R*-genes for vertical resistance occur in a wide range of crops against a wide diversity of pathogens and pests, such as rusts, viruses, mildews, bacteria and nematodes.

Vertical resistance is susceptible to **boom and bust** cycles, the cyclical failure of vertical resistance genes to deter newly evolved pathotypes of the pathogen. Boom and bust cycles arise when major genes for vertical resistance against a major economic race of a pathogen are used in cultivar development for a region, leading to widespread adoption of the resistant cultivars by most producers in the region (the **boom phase**). Selection pressure on the pathotypes of the pathogen present in the cultivars reduces the relative occurrence of the avirulent ones. However, the occasionally occurring virulent one against which the cultivars carry no major genes continues to increase until it becomes epidemic in the vast region of production of the crop (the **bust phase**).

Vertical resistance is more suitable for annuals than perennials. It is very effective against immobile or localized pathogens (e.g., soil-borne pathogen). However, it must be deployed strategically to be useful against mobile air-borne pathogens such as epidemic fungi of annual plants. This resistance is easy to select for because of the typically qualitative phenotypic contrast between resistant and susceptible plant genotypes. It is easy to breed because the major genes are easy to identify and transfer through simple crosses.

14.8.2 Horizontal resistance

Horizontal resistance is also known as **slow rusting** (of course only used for diseases caused by rusts), **partial resistance, field resistance, race-non-specific resistance, minor gene reactions,** or **polygenic inheritance**. This resistance is typically not complete

and it is effective against all genotypes of the pathogen species without cultivar x isolate interaction (hence its qualification to be race-non-specific). Following the initial establishment of the pathogen, the plant may resist its spread and reproduction such that the disease develops at a slower rate. Unlike with vertical resistance, this resistance cannot be broken down easily by a loss-of-function mutation. Therefore, evolution of new races followed by selection in favor of a novel more aggressive or virulent race probably does not occur. Indeed, generally this type of resistance is considered to be durably effective.

Horizontal resistance is mostly controlled by polygenes (i.e. several genes, each with a minor effect). Hence, the resistance is also called **minor gene resistance**. Polygenic resistance is widespread and can be identified by critically comparing the level of susceptibility of several "susceptible" plant accessions. A quantitative evaluation of a "susceptible" germplasm collection will in most plant–pathosystems reveal subtle to substantial differences in growth and reproduction rate of the pathogen. Polygenic resistance protects plants by slowing down the spread of disease and the development of epidemics in the field.

The inheritance of horizontal resistance is more complex than for vertical resistance. Examples include the partial resistance in potato to *Phytophthora infestans* and in maize to *Puccinia sorghi*. Any number of genes could be involved in horizontal resistance. It is a durable resistance because a pathogen has to change for a number of genes to overcome the defense mechanism in that one pathogen (non-specific). Horizontal resistance may arise when either the host genes do not operate in a gene-for-gene fashion with the pathogen genes (no differential interactions are possible) or when several to many host genes with small effects operate on a gene-for-gene basis with an equivalent number of genes in the pathogen population (differential effects are too small to be detectable and apparent to be horizontal resistance).

Breeding polygenic resistance is more challenging than monogenic traits. The many minor genes cannot be transferred through crossing in a predictable fashion. However, generally there is a high diversity of minor genes for resistance in any germplasm. Crossing of cultivars of the crop frequently results in transgressive segregation for levels of resistance, since the parents contain resistance genes at different loci. Segregants with higher levels of quantitative resistance can be identified and participate in further rounds of crossing and selection in order to increase levels of quantitative resistance even further. The heritability for polygenic resistance is high if careful quantitative observations are made and the inoculum is administered in a very homogeneous way. It is suitable for both annuals and perennials and applicable to all pathogens, but it is pathogen/pest species specific in the sense that horizontal resistance to one pathogen species is not effective to another pathogen species. It is laborious to develop breeding stocks with horizontal resistance. However, it is easy to improve on the very low level of horizontal resistance that normally underlies a failed vertical resistance. Such improvements may be accomplished by using recurrent selection methods.

In some pathosystems vertical resistance and horizontal resistance may be viewed as extremes of a continuum, if one assumes that gene-for-gene relationships are common in host-specific pathogen system. When few genes with large effects occur, differential interactions are readily identified and the result is vertical resistance.

14.8.3 Combining vertical and horizontal resistance

Vertical resistance is not preferred for improving perennials. Tree breeding is a much longer process than breeding annuals species. Errors are costly to correct. Should vertical resistance be overcome, it cannot be quickly replaced with a new genotype with new resistance genes. It is tempting to think that combining vertical resistance and horizontal resistance will provide the best of two worlds in the protection of plants. A complication is that the level of horizontal resistance cannot be assessed in the presence of effective vertical resistance.

Van der Plank suggested that in breeding programs that focus on race-specific vertical resistance the level of horizontal resistance tends to erode. He called this the **vertifolia effect** (after the potato cultivar "Vertifolia"). While the breeder focuses on vertical resistance, no evaluation and selection for horizontal resistance is possible, eventually leading to the loss of horizontal resistance. However, the vertifolia effect is not of universal occurrence. Some researchers have reported race-specific resistance in addition to high level polygenic resistance to leaf rust in barley. To reduce the incidence of the vertifolia effect, some suggest that breeders select and discard only the extremely susceptible plants in segregating populations, rather than selecting highly and fully resistant genotypes. Also, others suggest to first breed for a high level of horizontal resistance in a genotype then cross it with one that has high vertical resistance.

14.8.4 Durability and breakdown of resistance

An aspect of resistance of concern to plant breeders is the **durability**, a concept that was introduced earlier in this chapter. Resistance that is durable is one that remains effective in a widely grown cultivar for a long time in an environment that favors the pathogen. Horizontal resistance generally is more durable than vertical resistance. Adaptations of the disease agent to newly introduced resistant cultivars (i.e., **breakdown of resistance**) occurs frequently in pathogens but less so in insect pests. Durability of resistance is variable even within pathosystems. For example, in resistance to stripe rust in wheat durability ranged from one year to 18 years, depending on the cultivars and the R-genes involved.

In nature, the lifetime of an effective R gene depends on the evolutionary rate of the interacting Avr gene of the respective pathogen. Pathogen populations with high evolutionary potential typically have mixed reproduction modes with a high potential for gene flow, large effective population sizes and high mutation rates. These pathogens are more likely to overcome R genes.

14.9 Resistance breeding strategies

Disease breeding is a major objective for plant breeders all over the world. Breeders may use conventional or genetic modification strategies for breeding resistance. It is estimated that 95–98% of cultivars of small grains grown in the United States have at least one gene for disease resistance. Further, an estimated 75% of cultivated land in US crop production grows a crop with at least one disease resistance gene. A combination of traits rather than just one trait makes a cultivar desirable. Yield, quality and resistance to diseases are top considerations in breeding programs, the first trait being usually the most important breeding objective. Breeding for vertical resistance and horizontal resistance was discussed briefly in earlier sections of this chapter.

The first step in breeding for resistance to pathogens or insect pests is to assemble and maintain sources of resistance genes. The sources of resistance genes include commercial cultivars, landraces, wild progenitors, related species and genera, mutagenesis, and biotechnology. As indicated elsewhere in the book, obsolete and current commercial cultivars are most preferred because they have minimum undesirable traits (but lack novelty and hence the pathogen may break the same resistance in all recipient cultivars at once). Once a desirable source has been found, the backcross method of breeding is commonly used to transfer resistance genes into adapted cultivars. There should be an effective and efficient screening technique for disease resistance breeding. For cross-pollinated species and also in autogamous crops, recurrent selection is effective for increasing the level of resistance in a population of genetically heterogeneous population.

14.9.1 General considerations for breeding resistance to pathogens and pests

In spite of the advantages of host resistance as a plant breeding approach to controlling pests, the approach is not always the best choice, for several reasons.

- Breeding is an expensive and long duration undertaking that makes it justifiable (from at least the economic stand point) for major pests that impact crops that are widely produced or have significant benefit to society.
- Natural resistance is not available for all pests. Sometimes, the resistance is available in unadapted gene pools, requiring additional costs of pre-breeding.
- Breeding for resistance varies in ease and level of success from one pathogen or pest to another. Resistance to vascular pathogens, viruses, smuts, rusts, and mildews are relatively easier to breed than breeding against pathogens that cause rots (root rot, crown rot, storage rot) and ectoparasitic nematodes. This is partly due to the challenges of the methods used and the biology of the organisms. Similarly, it is relatively easier to have success with breeding resistance to aphids, green bugs, and hoppers, than to breed resistance to root-chewing, or grain-storage pests.
- Lack of durability of pest resistance is a key consideration in breeding for pest control because diseases and insect pests continue to change (evolve). New pathogenic races may arise, or the cultural environment may modify the resistance of the cultivar.

- The techniques of biotechnology may be effective in addressing some breeding problems more readily than traditional methods.
- After being satisfied that breeding for disease resistance is economical, the breeder should select the defense mechanism that would be most effect for the crop, taking into account the market demands. For example, horticultural products and produce for export usually require that they be free from blemish. In these cases, breeding for major gene resistance with complete expression is desirable. It is also easy to breed for this type of reaction. However, the breeders should note that generally this resistance is not durable.
- When a crop is grown for food or feed, breeding for mechanisms that increase the levels of chemical toxins in plant tissues is not suitable.
- In some crops routine application of fungicides and pesticides occurs. Some of those will protect against several pathogen or pest species. Introduction of resistance to one of the target pathogens species does not pay, it the farmers have to spray anyhow against the other pathogens.

14.9.2 Planned deployment of resistance genes

One particular feature of *R*-genes is that in most plant pathosystems there are many different *R*-genes available. In barley, for example there are at least 19 R-genes known to occur against barley leaf rust (*Puccinia hordei*) and at least 85 R-genes against barley powdery mildew (*Blumeria graminis*). The risk of deployment of such *R*-genes, that is their limited durability, may be reduced by following certain strategies in breeding and crop cultivation. Without a strategy, breeders might introduce just the resistance from a popular resistant barley cultivar. The result may be that all cultivars developed have the same resistance gene because it is the most readily available source to breeders. If that gene is broken-down by the pathogen, all cultivars will succumb simultaneously. Instead, it is recommended to have a planned release (consecutive release of different resistance genes) of resistance genes (rather than having several cultivars in production at one time with the same resistant gene). The virulence gene composition of the pathogen population should be monitored annually using a set of differential cultivars that carry different resistance genes. Once a current cultivar succumbs to a new race of a pathogen (i.e., a new race that is virulent on the resistance gene in use), breeders then release new cultivars that carry another effective gene. This way, plant breeding stays ahead of the pathogen. However, this will eventually use up all *R*-genes available in the crop germplasm.

For this (and other) approach(es) a thorough knowledge is required on which *R*-genes occur in the presently cultivated cultivars, and breeders need to collaborate and communicate on which *R*-genes may be deployed in their present selection programs. Also reliable data are required on the frequency of corresponding virulence alleles. In a competitive and commercial society such exchange of information is in conflict with secrecy because of commercial interests.

14.9.3 Gene pyramiding

The concept of transferring several *R*-genes simultaneously into one plant is called **gene pyramiding**. Because there are different races of pathogens, plant breeders may want to transfer a number of genes for conferring resistance to different races of a disease into a cultivar. Three major genes conditioning resistance to blast in rice, *Pi-1*, *Pi-2*, and *Pi-3*, were pyramided through pairwise crosses of the isogenic lines carrying the genes. Similarly, genes for resistance to Biotype L of the Hessian fly (*Mayetiola destructor*) of wheat were successfully pyramided into the crop. The rationale behind this approach is that the pathogen needs to overcome simultaneously a combination of *R*-genes, so it needs a double or triple loss-of-function mutation in the relevant *Avr* genes to become able to infect the cultivar carrying the stacked *R*-genes. This strategy is applicable to regions where plant breeding is centrally coordinated, and where the production region using the cultivar with the multiple resistance genes is isolated from other areas not using the system. Simultaneously releasing cultivars with single resistance genes along with the multiple gene cultivars would reduce the success of the latter approach.

14.9.4 Multilines

The rationale of a multiline is that a host population that is heterogeneous for resistance genes would provide a buffering system against destruction from diseases. Multiline breeding procedures are discussed in Chapter 16.

14.10 Challenges of breeding for pest resistance

Breeding for pest resistance differs fundamentally from breeding for other traits because the introduced resistance may cause a change in the evolving and variable pest or pathogen population. Further, the genes for resistance cannot be identified unless the plant containing the genes is interacting with the pathogen or insect pest in an environment in which plants are normally susceptible to disease. Plant breeders must develop a segregating population with adequate diversity to include the desired combination of genes of interest. In disease breeding, the challenge is identifying and selecting a desired genotype in a form in which it would genetically remain effective after release and many years of cultivation of the cultivar. The breeder must utilize reliable methods for detecting differences in resistance level among segregants. Whereas natural infection may be used, frequently artificial inoculation is more reliable.

A major problem in breeding for resistance to disease and insect pests is the fact that, over time, the crop cultural environment changes (e.g., different production methods and inputs) as well as pathogen and pest (through evolution). Breeders need to keep up with these changes by developing new cultivars with appropriate resistance genes, in order to ensure stability of crop production, by preventing the development of destructive epiphytotics and infestations, and to reduce the annual loss of products from pathogens and pests.

The breeder should guard against breeding highly resistant cultivars that have no economic worth. A good strategy is to breed for **middling resistance** with high yield. To this end breeding for polygenic horizontal resistance is the most desirable strategy since it accounts for most middling resistance. It should be pointed out that some single gene resistance effects do not confer immunity on the cultivar.

14.11 Role of wild germplasm in disease and pest resistance breeding

Wild relatives of cultivated crops have been a source of genes for solving a variety of plant breeding problems, especially, disease resistance (Chapter 11).

14.12 Screening techniques in disease and pest resistance breeding

One of the critical activities in breeding for resistance to diseases and insect pests is screening or testing for resistance. Various facilities, techniques and approaches are used, depending on the parasite and host characteristics.

14.12.1 Facilities

Disease, as previously indicated, depends on the interaction among three factors – pathogen, host, and environment. Whereas field screening has the advantage of representing the conditions under which the resistant cultivar would be grown, it has its limitations. The weather (or the environmental component of the disease triangle) is unpredictable, making it difficult to have uniformity and consistency of the parasite population. In field screening also other pathogens may appear that interfere with the assessment of infection by the target pathogen. In some years, the weather may not favor adequate pathogen population for an effective evaluation of plants. Also, the breeder depends on the naturally occurring pathotype or pathotype mixture that happens to arrive in the fields. Controlled environmental tests provide reliable, uniform and consistent evaluation of disease, but they have less field correspondence. Screening for resistance to mobile pests is challenging, requiring special provisions to confine the parasites.

14.12.2 Factors affecting expression of disease and insect resistance

Certain specific factors may complicate breeding for resistance. Such factors may be environmental or biological in nature.

- **Environmental factors** should, of course, be within a reasonable range to allow the development of the pathogen or pest species.
  - **Temperature.** Some resistance genes are not expressed when temperature is a few degrees centigrade too high or too low.
  - **Light.** Light intensity affects the levels of chemical compounds in plants that are relevant for defense to pests (e.g., glycoside in potato), which in turn may affect the level of resistance to a pathogen.
  - **Soil fertility.** Generally high soil fertility makes plants more succulent and more susceptible to infection. Other pathogens (generalist opportunists) however, are more successful on plants that suffer from malnutrition.
- **Biological factors**
  - **Age.** The response of a plant to a pathogen or insect pest may vary with age. Some diseases are more intense at the early stage in plant growth than others.
  - **Tissue type.** Some resistances are better expressed in, for example, tubers, others better in the leaf canopy.
  - **New pathotypes or biotypes.** As already explained, a breeder should always be aware that many cases of resistance are affective to some pathotype but not to others.
  - **Induced resistance and susceptibility.** Prior infection or infestation by a pest or pathogen (inducer) may induce a systemic resistance to subsequent infection by other pathogens or pests (challenger). There are also cases where prior infection by a pathogenic pest or pathogens induces local susceptibility to a subsequent challenging pathogen that normally cannot infect the plant genotype.

14.13 Applications of biotechnology in pest resistance breeding

One of the successful applications of agricultural biotechnology is in pest resistance breeding. The first disease resistance gene, *Pto* (binds with products of the pathogen to give resistance), was cloned in 1993 by Greg Martin and co-workers. Since then hundreds of *R*-genes have been cloned in many more crops to many more pathogen and pest species.

14.13.1 Engineering insect resistance

There are two basic approaches to genetic engineering of insect resistance in plants:

(i) The use of protein toxins of bacterial origin.
(ii) The use of insecticidal proteins of plant origin.

Protein toxins from Bacillus thuringiensis (Bt)

Bacillus thuringiensis (*Bt*) forms endotoxins that are crystalline proteins. It was first identified in 1911 when it was observed to kill the larvae of flour moth. It was registered as a biopesticide in the United States in 1961. *Bt* is very selective in action, that is, one strain of the bacterium kills only certain classes of insects. Formulations of sporulating bacteria are widely used as biopesticide spray for biological pest control in organic farming. There are several major varieties of the bacterial species that produce toxins against certain target pests: *B. thuringiensis* var *kurstaki* (for controlling lepidopteran pests of forests and agriculture), var. *Berliner* (wax moth), and var. *israelensis* (dipteran vectors of human disease). The most commercially important type of the crystalline proteinaceous inclusion bodies are called **δ-endotoxins**. To become toxic, these endotoxins, which are predominantly protoxins, need to be proteolytically activated in the midgut of the susceptible insect to become toxic to the insect. These endotoxins act by collapsing the cells of the lining of the gut regions.

Bt resistance development has been targeted especially at the European corn borer, which causes significant losses to corn in production over the world. Previous efforts developed resistance in crops like tobacco and cotton. The effort in corn was more challenging because it required the use of synthetic versions of the gene (rather than microbial *Bt per se*) to be created.

Two genes, *cryB1* and *cryB2* were isolated from *B. thuringiensis* subsp *kurstaki* HD1. These genes were cloned and sequenced. The genes differed in toxin specificities, *cryB1* gene product being toxic to both dipteran (*Aedes aegypti*) and lepidopteran (*Manduca sexta*) larvae, while *cryB2* affects only the latter. The *Bt* toxin is believed to be environmentally safe as an insecticide. In engineering *Bt* resistance in plants, scientists basically link the toxin to a constitutive (unregulated) promoter that will express the toxin systemically and continuously (i.e., in all tissues).

Transgenic plants expressing the δ-endotoxin gene have been developed. The first attempt to develop a transgenic plant for this toxin involved the fusion of the *Bt* endotoxin to a gene for kanamycin resistance to aid in selecting plants (conducted by a Belgian biotechnology company, Plant Genetic Systems in 1987). Later, Monsanto Company researchers expressed a truncated *Bt* gene in tomato directly by using the CaMV 35S promoter. Agracetus Company followed with the expression of the *Bt* endotoxin in tobacco with the CaMV 35S promoter linked to an

alfalfa mosaic virus (AMV) leader sequence. Since these initial attempts were made, modifications to the protocols have increased expression of the toxin in transgenic plants by several hundred fold. Transformation for expressing the chimeric *Bt* genes was *Agrobacterium* mediated, using the TR2' promoter. This promoter directs the expression of manopine synthase in plant cells transformed with the TR DNA of plasmid pTiA6.

The original *Bt* coding sequence has since been modified to achieve insecticidal efficacy. The complete genes failed to be fully expressed. Consequently, truncated (comprising the toxic parts) genes of *Bt* var *kurstaki* HD-1 (*cry1A*[b]) and HD-73 (*cry1A*[c]) were expressed in cotton against lepidopteran pests. In truncating the gene, the *N*-terminal half of the protein was kept intact. For improved expression, various promoters, fusion proteins, and leader sequences have been used. The toxin protein usually accounts for about 0.1% of the total protein of any tissue, but this concentration is all that is needed to confer resistance against the insect pest.

Genetically engineered *Bt* resistance for field application is not of just one kind. For example, Ciba Seed Company has developed three versions of synthetic *Bt* genes capable of selective expression in plants. One is expressed only in pollen, another in green tissue, and the third in other parts of the plant. This selectivity is desirable for several reasons. The European corn borer infestation is unpredictable from year to year. The life cycle of the insect impacts the specific control tactic used. The insect attack occurs in broods or generations. The *Bt* genes with specific switches (pollen and green tissue) produce the *Bt* endotoxin in the parts of the plants that are targets of attack at specific times (i.e., first and second broods). This way, the expression of the endotoxin in seed and other parts of the plant where protection is not critical is minimized. Monsanto's YieldGard corn produces *Bt* endotoxin throughout the plant, and protects against both first and second broods of the pest. The commercially available *Bt* corn cultivars were developed by different transformation events, each with a different promoter.

Bt cotton is another widely grown bioengineered crop. The pest resistance (to pink bollworm) conferred by the *Bt* gene has led to dramatic reduction in pesticide use, and consequently reduced adverse impact on the environment form agropesticides. As indicated previously, *Bt* sprays are widely used in organic farming for pest control. The advantage of genetically modified cotton compared to application of *Bt* spraying is that the spraying cannot reach the cotton ball worm inside the cotton balls. Also, the spraying may also affect some free-living non-target insects in the cotton field. Further, *Bt* sprays have short duration activity.

14.13.2 Engineering viral resistance

A virus is essentially nucleic acid encased in a protein coat. In some viruses that infect plants the nucleic acids are DNA, but most of them are RNA viruses. One of the most important plant virus in biotechnology is the Cauliflower Mosaic Virus (CaMV) from which the widely used 35S promoter was derived (CaMV 35S promoter). A key method of control of viral infections is through breeding of resistance cultivars. Also, plants can be protected against viral infection by a strategy that works like inoculation in animals. Plants may be protected against certain viral infections upon being infected with a mild strain of that virus. This strategy, called **cross-protection**, provides protection to the plant against future more severe infections.

Industry highlights
Breeding for durable resistance against an oomycete in lettuce

Marieke Jeuken

Wageningen UR Plant Breeding, Droevendaalsesteeg 1, 6708 PB Wageningen, The Netherlands

"I spend 80% of my time on downy mildew resistance breeding, and the rest of my breeding activities/goals I have to fit in 20% time!"

This quote from a lettuce breeder shows the urgent need to control disease in lettuce cultivation (and the pressure on breeders to come with new resistant cultivars).

Background

Lettuce is major fresh vegetable and its leaves are commonly found in salads. Downy mildew is a foliar disease in cultivated lettuce (*Lactuca sativa*) caused by the oomycete *Bremia lactucae*. It is the most destructive disease in lettuce cultivation with severe/significant loss of economic value of the crop.

Pesticides can be used to control the downy mildew but are not durably effective and not desirable because of health issues for consumers and environment issues.

Vertical resistance to *B. lactucae* is employed. The genetics for this host–pest interaction is based on the gene-for-gene interaction between resistance and avirulence genes (Crute and Johnson, 1976). Since the 1920s breeders in Europe and the USA have introduced more than 20 single dominant *R* genes to control *B. lactucae*. The appearance and identification of new races in Europe and the USA are monitored by joint initiatives of lettuce breeding companies and national institutes.

The known *R* genes are found in five clusters on the lettuce genome; the majority of the *R* genes reside in one cluster. This fact limits the possible *R* gene combinations. (Combination of two closely linked *R* genes from different sources will be more difficult as the *R* genes are in the repulsion phase. A recombination event between the genes in their progeny is needed to obtain them in the coupling phase.) One *R* gene, *Dm3*, has been cloned and is a NBS-LRR type gene (Shen *et al.*, 2002).

Unfortunately, this vertical resistance based on *R* genes is not durable, since the resistance is usually overcome by rapid adaptation of *B. lactucae* races 1–3 years after introduction of a new *R* gene. This low durability of *R* genes is partly explained by the high evolutionary potential of *B. lactucae* because of its mixed reproduction system, sexual and asexual, and its high gene flow by windblown spread (McDonald and Linde, 2002). At present more than 30 races/pathotypes of *B. lactucae* have been found to occur. An additional factor that may contribute to the low durability is the genetic vulnerability of cultivated lettuce to downy mildew/*B. lactucae*, which is attributed to uniformity in the *R* gene use and an almost threefold increase in lettuce acreage since 1980 (Still, 2007).

Aim

Other forms of resistance that might be more durable are desired by breeders. Durable is defined as "resistance that remains effective when deployed over extensive acreage and time, in an environment favorable for the disease" (Johnson, 1984).

One challenging way is the introgression of "non-host" resistance from wild species. Non-host resistance is the most common form of resistance in a given plant species to provide both complete and broad spectrum protection against potential pathogens (Niks, 1988). Compared with host resistance, the mechanisms underlying non-host resistances are not as well understood. Usually it is difficult to study the genetics of non-host resistance, as host and non-host plant species can either not be crossed or give aberrant or sterile progeny when crossed. One of the very few examples where a host and non-host species are sufficiently related to allow inheritance studies exists in the *Lactuca* germplasm. The wild lettuce *Lactuca saligna* is considered to be a non-host to *B. lactucae* because, as far as is known, all *L. saligna* accessions are completely resistant to all isolates of *B. lactucae*. Still it is sexually compatible with the cultivated host species *L. sativa*. If it was possible to identify the genes responsible for the non-host status and introgress them in cultivated lettuce, resistance to downy mildew might be durable. To investigate which genes determine host and non-host status a genetic dissection was initiated.

From the beginning, this research contained several uncertainties/risks: unknown resistance mechanism, populations suffering from potential inviabilities, no genetic map and markers available. To spread the risk and investments a pre-competitive research project by a university group was started; this was financed by joint breeding companies and governmental subsidies. Agreements on a possible embargo on publication, transfer of resulting material to third parties, confidentiality, and so on were put down in a contract.

Figure B14.1 Parental lettuce species used in research. Left: *Lactuca sativa*, susceptible cultivated parent; right: *Lactuca saligna*, resistant wild parent. Figure courtesy of Marieke Jeuken.

Methods/results

F₂ Strategy

L. sativa and *L. saligna* are highly self-pollinated species, they produce easily more than 10 000 seeds and their haploid genome consists of nine chromosomes (2n = 18). *L. saligna* belongs to the secondary gene pool of cultivated lettuce, because of some crossability problems. Many interspecific crosses were made between lettuce cultivars and *L. saligna* accessions. Four of such interspecific hybrids produced at least some F₂ (n < 163) (Figure B14.1).

To genetically dissect the non-host resistance, the cross with the largest F₂ population (n = 126, 23% was lethal) was phenotyped for disease resistance and genotyped by molecular markers. Amplified Fragment Length Polymorphism (AFLP) markers were chosen to start with as no prior sequence information was needed for amplification. A pilot experiment showed a very high polymorphism rate of 81% between *L. sativa* and *L. saligna*. Initially a genetic map was constructed of 500 randomly distributed AFLP markers (Jeuken *et al.*, 2001). Later, when new marker technologies and lettuce sequences became available, the map was extended to thousands of markers with marker types like microsatellites, gene derived markers and SNP markers.

Only few non-destructive disease tests with two *B. lactucae* races could be performed on the 126 F₂ plants because of the temporary character of the population.

Backcross inbred lines/infinite population strategy

To be able to perform more tests on different plant stages, with different pathogen pathotypes/isolates and under different conditions, such as outdoors (as proof of concept that *L saligna* nonhost resistance may effectively control cultivated lettuce in the field), an "immortal" population type was desired. The development of recombinant inbred lines (RILs) by single seed descent (SSD) seemed impossible because of the reduced germination and vigor (23% lethality), the reduced fertility (37% completely sterile) and the severely distorted segregations for several chromosome regions in the F₂ population.

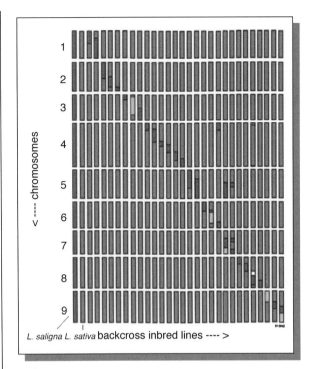

Figure B14.2 Genotype presentation of parental lines and set of backcross inbred lines (BILs); red= homozygous *L. saligna*, blue= homozygous *L. sativa*, yellow = heterozygous. Figure courtesy of Marieke Jeuken.

Figure B14.3 One of the field tests. Figure courtesy of Marieke Jeuken.

We followed another strategy of backcross inbred lines (BILs), in which single chromosome segments of *L. saligna* were introgressed into *L. sativa*. If successful this would lead to *L. sativa*-like lines, as in each line about 95% of the genome is from *L. sativa*. Lines were developed by four to five backcrosses and one to two generations of selfing. Marker assisted selection was used (mainly in the later generations) to select for the fewest chromosome segments of *L. saligna* (Figure B14.2). This required genome-wide analysis of each backcross line to distinguish the *L. sativa* background from the *L. saligna* foreground. The linkage map that was constructed on the basis of the F_2 was indispensible for this. In practice the whole process took three years, with 2–3 generations per year in the greenhouse, several crosses and many genotyping and selection (Figure B14.3).

The result was a set of 29 BILs, each with a single introgressed chromosomal segment from *L. saligna* in the *L. sativa* genetic backgrounds; this set of BILs represented 96% of the *L. saligna* genome (Jeuken and Lindhout, 2004). All lines were fertile and relatively similar to *L. sativa* except for a few lines that had a moderate aberrant phenotype (Figure B14.4).

Multiple disease tests were performed on the the set of BILs to evaluate the resistance at different plant stages, against *B. lactucae* races, and outdoors at various locations and seasons (Figure B14.5).

Disease test data for F_2 population were analyzed by QTL mapping analysis and the data for the set of BILs were analyzed for an effect of introgressed *L. saligna* chromosome segments on infection severity.

Figure B14.4 Left:Susceptible lettuce cultivar; right: resistant lettuce cultivar. Figure courtesy of Marieke Jeuken.

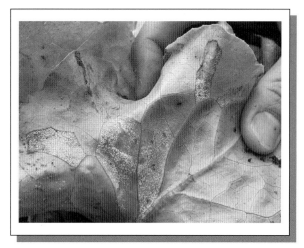

Figure B14.5 Symptom/sporulation of downy mildew on lettuce. Figure courtesy of Marieke Jeuken.

Conclusions

No dominant race-specific *R* genes were detected. Only quantitative effects for at least 14 chromosome regions were implicated to contribute to resistance to *B. lactucae* (Zhang, 2009a). A combination of three introgression segments from *L. saligna* in a *L. sativa* background gave full resistance to all *B. lactucae* isolates in field tests (Zhang, 2009b). The full resistance in *L. saligna* appeared, therefore, to be due to the combined effect of minor genes (polygenic inheritance).

Limitations

The *L. saligna* genome coverage in the set BILs was not (complete and not) always in a homozygous state. Plants that were homozygous for certain *L. saligna* segments were apparently lethal. Such chromosomal regions were only obtained in a heterozygous state. Hybrid incompatibilities by two or more interactive loci possibly explain this. One line even showed necrotic lesions and this case was recognized as hybrid necrosis (Jeuken *et al.*, 2009).

Quantitative resistance effects (QTLs) of specific BILs were still ascribed to large regions of 20–40 cM. The absence of an exact QTL position and the risk of linkage drag (i.e. "hitchhiking" of genes for undesirable traits with the gene for resistance) warranted the development of lines with smaller introgressions and further fine mapping.

By testing for resistance on lines with smaller introgression lengths or an increase in introgressions (by crossing BILs to combine their *L. saligna* derived introgression fragments) analysis revealed often complex interactions between more loci.

Future work

The university research group will focus on further fine mapping, cloning and studying the genes responsible for non-host resistance to unravel the genetics, the mechanism and the evolutionary processes behind the non-host status of *L. saligna*. The prospective availability of the whole genome sequence of lettuce will facilitate these processes. The breeders have to introduce the *L. saligna* derived resistance genes into their elite germplasm by marker assisted selection. Time will tell if the resistance is durable. If so, the breeders can afford to spend more time on the improvement of other traits.

References

Crute, I.R., and Johnson, A.G. (1976). The genetic relationship between races of *Bremiae lactucae* and cultivars of *Lactuca sativa*. *Annals of Applied Biology*, **83**:125–137.

Jeuken, M.J.W., and Lindhout, P. (2004). The development of lettuce backcross inbred lines (BILs) for exploitation of the *Lactuca saligna* (wild lettuce) germplasm. *Theoretical and Applied Genetics*, **109**:394–401.

Jeuken, M., van Wijk, R., Peleman, J., and Lindhout, P. (2001). An integrated interspecific AFLP map of lettuce (*Lactuca*) based on two *L. sativa* x *L. saligna* F$_2$ populations. *Theoretical and Applied Genetics*, **103**:638–647.

Jeuken, M.J.W., Zhang, N.W., McHale, L.K., *et al.* (2009). Rin4 Causes Hybrid Necrosis and Race-Specific Resistance in an Interspecific Lettuce Hybrid. *Plant Cell*, **21**:3368–3378.

Johnson, R. (1984). A Critical Analysis of Durable Resistance. *Annual Review of Phytopathology*, **22**(1):309–330.

McDonald, B.A., and Linde, C. (2002). The population genetics of plant pathogens and breeding strategies for durable resistance. *Euphytica*, **124**:163–180.

Niks, R.E. (1988). Nonhost plant species as donors for resistance to pathogens with narrow host range 2. Concepts and evidence on the genetic basis of nonhost resistance. *Euphytica*, **37**:89–99.

Still, D.W. (2007). Lettuce, in *Genome Mapping and Molecular Breeding in Plants, Volume 5, Vegetables* (ed. C. Kole). Springer, Berlin and Heidelberg, pp. 127–140.

Shen, K.A., Chin, D.B., Arroyo Garcia, R., *et al.* (2002). *Dm3* is one member of a large constitutively expressed family of nucleotide binding site-leucine-rich repeat encoding genes. *Molecular Plant Microbe Interactions*, **15**:251–261.

Zhang, N.W., Lindhout, P., Niks, R.E., and Jeuken, M.J.W. (2009a). Genetic dissection of *Lactuca saligna* nonhost resistance to downy mildew at various lettuce developmental stages. *Plant Pathology*, **58**:923–932.

Zhang, N.W., Pelgrom, K., Niks, R.E., Visser, R.G.F., and Jeuken, M.J.W. (2009b). Three combined quantitative trait loci from nonhost *Lactuca saligna* are sufficient to provide complete resistance of lettuce against *Bremia lactucae*. *Molecular Plant Microbe Interactions*, **22**:1160–1168.

Engineering transgenic plants with resistance to viral pathogens is accomplished by the method called **coat protein-mediated resistance**. First, the viral gene is reverse transcribed (being RNA) into DNA from which a double-stranded DNA is then produced. The product is cloned into a plasmid and sequenced to identify the genes in the viral genome. A chimeric gene is constructed to consist of the open reading frame for the coat protein to which a strong promoter is attached for high level of expression in the host. This gene construct is transferred into plants to produce transgenic plants.

Successes with this strategy have been reported in summer squash (the first product developed by this approach), and for resistance to papaya ring spot virus (a lethal disease of papaya), among others.

14.14 Epidemics and plant breeding

Modern agriculture is drastically different from the agriculture of early plant domesticators. Plant breeding, in adapting wild species for cultivation, often removes the natural means of crop protection that are needed for survival in the wild but which are undesirable in modern production. Starting with the early domesticators, mass production of a few desirable genotypes has become the norm of crop production. The monocultures of modern agriculture are extreme versions of mass production, in which genetic diversity is further restricted. It is not difficult to see how such breeding and production practices predispose crops to widespread disease and insect pest damage. With the occurrence of a devastating pest attack, breeders counter with the production of resistant cultivars, thus setting the stage for sequential cycles of pest resistance and pest susceptibility of crop plants.

The patterns of disease incidence vary from one class of pathogen to another. Soil-borne pathogens tend to be localized and persistent in the soil, season after season. Air-borne pathogens have a different biology and spread pattern, as well as the potential to become epidemic. An **epidemic** may be defined as an outbreak of disease characterized by an infection that starts from a low level and then progresses to a high one. Air-borne pathogens tend to have annual epidemic cycles. These cycles depend on seasonal weather patterns and the presence or absence of a susceptible host (crop). Soil-borne pathogens usually stabilize in a new location after several years and persist in the chronic phase. Each type of disease pathogen and the epidemic it causes is characterized by its own ecology and spatial and temporal dynamics. Some pathogens and pests are monocyclic (e.g., cyst nematodes, loose smut, *Fusarium oxysporum*) having only one infection cycle per crop season. Others (e.g., powdery and downy mildews, rusts, blight, aphids) are polycyclic, having several cycles of reproduction within the crop season, leading initially to exponential increase in infection. Plant fungal diseases are especially significantly impacted by weather conditions and climatic patterns. Epidemics in agricultural crop production impact the US agriculture by affecting the economic value, quantity, and quality of food and fiber produced.

J.E. Van der Plank provided the underlying scientific principles of disease epidemics (for polyclic infection). An epidemic occurs in phases. It starts with a small inoculum of the pathogen (that may be from overwintering spores, or freshly introduced into the area from outside). The initial rate of infection is proportional to the infection level. As time proceeds, the infection rate slows down as uninfected tissue decreases. Van der Plank expressed the initial infection rate by this mathematical relationship:

$$dI/dt = rI \text{ or } I = I_o exp^{rt}$$

where I is infection at time t, I_o is initial infection, and r is the multiplication rate of the pathogen. A general and more realistic expression is:

$$dI = rI(1 - I)$$
$$I = 1/[1 + (exp^{-rt}) \text{ or } \log_e (1/1 - I)$$
$$= rt + \log_e (1/C)$$

where C is a constant that depends upon I_o but is not equal to it.

In plant breeding, the breeder's goal in countering a disease epidemic is to reduce r (the multiplication rate of the pathogen) to a level such that, at the time of crop maturity, the final attack by the pathogen can inflict only a small and acceptable yield loss. It may be best if r = 0 (i.e., immunity), but this is not necessary. In practice, breeders should not aim for immunity but for reducing r below that of the current commercial cultivar that is susceptible to the pathogen.

Key references and suggested reading

Alam, S.N., and Cohen, M.B. (1998). Detection and analysis of QTLs for resistance to the brown planthopper, Nilaparvata lugens, in a doubled haploid rice population. *Theoretical and Applied Genetics*, **97**:1370–1379.

Bogdanove, A.J. (2002). Protein-protein interactions in pathogen recognition in plants. *Plant Molecular Biology*, **50**:981–989.

Flor, H.H. (1956). The complementary genic system in flax and flax rust. *Advances in Genetics*, **8**:29–54.

Flor, H.H. (1971). Current status of the gene-for-gene concept. *Annual Review of Phytopathology*, **9**:275–296.

Hill, C.B., Li, Y., and Hartman, G.L. (2004). Resistance to the soybean aphid in soybean germplasm. *Crop Science*, **44**:98–106.

Hooker, A.L. (1980). Breeding to control pests, In *Crop breeding*. American Society of Agronomy/Crop Science Society of America, Madision, WI.

Khush, G.S., and Brar, D.S. (1991). Genetics of resistance to insects in crop plants. *Advances in Agronomy*, **45**:223–274.

Niks, R.E., Ellis, P.R., and Parlevliet, J.E. (1993). Resistance to parasites, in *Plant breeding: principles and prospects* (eds M.D. Hayward, N.O. Bosemark, and I. Ramagosa), Chapman and Hall, UK.

Parlevliet, J.E. (1981). Disease resistance in plants and its consequences for plant breeding, in *Plant breeding II* (ed. K.J. Frey). Iowa State University Press, Ames, IA.

Powles, S.B., Preston, C., Bryan, I.B., and Jutsum, A.R. (1997). Herbicide resistance: Impact and management. *Advances in Agronomy*, **58**:57–93.

Simmonds, N.W. (1985). A plant breeder's perspective of durable resistance. *FAO Plant Protein Bulletin*, **33**:13–17.

Simmonds, N.W. (1991). Genetics of horizontal resistance to diseases of crops. *Biological Reviews*, **66**:189–241.

Simmonds, N.W. (1993). Introgression and incorporation strategies for the use of crop genetic resources. *Biological Reviews*, **68**:539–562.

Smith, G.M. (1989). *Plant resistance to insects: A fundamental approach*. John Wiley & Sons, Inc., New York.

Outcomes assessment

Part A

Please answer the following questions true or false.

1 Major gene resistance is durable.
2 Most of the viruses that affect plants are RNA viruses.
3 The genes for Bt products were obtained from bacteria.
4 Vertical resistance is conditioned by numerous minor genes.

Part B

Please answer the following questions.

1 . . . . is the concept of transferring several specific genes into one plant.
2 What is durable resistance.
3 Explain the strategy of cross-protection in viral resistance engineering.
4 Give two specific reasons for engineering herbicide resistance.
5 Describe the mechanism of hypersensitive reaction used by plants in response to the invasion of certain pathogens.

Part C

Please write a brief essay on each of the following topics.

1 Discuss the method of coat protein mediated resistance in plant breeding.
2 Discuss the concept of gene-for-gene hypothesis in disease genetics.
3 Discuss the mechanism plants use to respond to pathogen invasion.
4 Discuss the mechanisms commonly used by plants to respond to insect pest attacks.
5 Discuss the approach to breeding for vertical resistance.
6 Discuss the breeding of Bt resistance in plants.

15

Breeding for resistance to abiotic stresses

Purpose and expected outcomes

Climate is the variation in meteorological factors over a large area over many years. It determines plant adaptation to a growing region. Weather on the other hand is the environmental condition described by the short-term variations in meteorological factors within a local area. It determines crop development and productivity. Weather affects the choice of specific crop production activities (i.e., how the crop is produced in the field). Plant breeders usually develop breeding programs to produce cultivars for specific production regions. Crop production is subject to the vagaries of the weather. Unpredictable weather can drastically reduce crop yield. Crop varieties used in regions that are prone to adverse weather during production need to have the capacity to resist or tolerate environmental stresses to an extent that they produce acceptable economic yield. After completing this chapter, the student should be able to:

1 Discuss the types of environmental stresses that reduce plant performance in crop production.
2 Discuss the breeding for drought resistance.
3 Discuss the breeding for cold tolerance.
4 Discuss the breeding for salt tolerance.
5 Discuss the breeding for resistance to heat stress.
6 Discuss the breeding for tolerance to aluminum toxicity.
7 Discuss the breeding for tolerance to oxidative stress.
8 Discuss the breeding for resistance to water-logging.

15.1 Importance of breeding for resistance to abiotic stresses

Only about 30% of the earth is land. Of this, about 50% is not suitable for economic crop production mainly because of constraints of temperature, moisture and topography. Of the remaining portion of arable land, optimum production is further limited by a variety of environmental stresses, requiring mineral and moisture supplementation for economic crop

---

Principles of Plant Genetics and Breeding, Second Edition. George Acquaah.
© 2012 John Wiley & Sons, Ltd. Published 2012 by John Wiley & Sons, Ltd.

production. As the world population increases more food will have to be produced by increasing the productivity of existing farm lands as well as bringing new lands into production. This means marginal lands will have to be considered. Plants breeders will have to develop cultivars that are adapted to specific environmental stresses. It is estimated that abiotic environmental stresses are responsible for about 70% of yield reduction of crops in production.

In modern agricultural production, producers (especially those in developed countries) are able to alleviate moisture stress by providing supplemental irrigation. Also, fertilizers are used to alleviate nutritional stress while soil reaction (pH) can be modified by applying soil amendments such as lime or sulfur. Excessive soil moisture can be removed by installing drainage systems. However, in poor regions of the world, which incidentally are the places where food increases are needed the most, such technologies are not readily available or affordable. Further, fresh water is increasingly becoming scare and modern agriculture cannot rely on supplementing water to achieve genetic potential. The alternative is to breed cultivars that are able to resist these environmental stresses enough to produce acceptable crop yield.

Each species has natural limits of adaptability. There are tropical plants and temperate plants. Breeders are able, within limits, to adapt certain tropical plants to temperate production, and vice versa. To achieve this, plant breeders use various strategies and techniques (e.g., genetic modification, introgression of traits from adapted relatives) to develop new cultivars that are able to resist environmental stresses in their new environment. Sometimes, in modern crop production, breeders develop cultivars to resist certain environmental stresses to make better use of the production environment. For example, cold resistance enables producers to plant early in the season while the soil is still too cold for normal planting. This may be done to extend the growing season for higher productivity or some other reasons.

15.2 Resistance to abiotic stress and yield potential

Plant growth and development is the product of the interaction between the genotype (genetic potential) and the environment in which the plant grows. Any stress in the environment will adversely impact growth and development. Plants perform well in environments to which they are well adapted. Yield potential was previously defined as the highest yield attainable by a genotype growing in an environment to which it is adapted and in which there is no environmental stress (i.e., optimum growing conditions). Except, perhaps, under controlled environment culture, it is very difficult to find a production environment (especially on a large scale) in which some environmental stress of some sort does not occur. Stress resistance is an inherent part of all cultivar development programs. Prior to releasing a cultivar, genotypes of high potential are evaluated at different locations and over several years to determine the degree of adaptation.

High yield potential has been achieved in cereal crops through enhanced harvest index whereby dry matter has been redistributed into the grain (by reducing its distribution to vegetative parts – roots, stems, leaves). However, this pattern of dry matter distribution also reduces the ability of these genotypes with high yield potential to cope with environmental stresses. Studies on genotype × environment (G × E) interaction have shown crossover effects whereby, under conditions of severe stress, genotypes of high yield potential perform poorly. Further, as the abiotic constraints intensify (e.g., drought, low temperature, high temperature), a report by CIMMYT indicates that it becomes more difficult to improve genetic and agronomic yield of crops.

R.A. Fisher and R. Maurer partitioned stress effects on yield (Y) into parameters that measure sensitivity to stress (S) and the extent of the stress (D) and yield potential (Y_p):

$$Y = Y_p(1 - S \times D)$$

where $D = (1 - \overline{X}/\overline{X}_p)$ and $\overline{X}$ and $\overline{X}_p$ are the mean yield of all cultivars under stressed and optimal conditions, respectively. This relationship may be manipulated algebraically to obtain:

$$S = (1 - Y/Y_p)/D$$
$$= (Y_p - Y)(Y_p \times D)$$

D is a constant for a particular trial. Hence, S is a measure of the yield decrease due to the stress relative to the potential yield, with a low value of S being desirable.

The problem with using S as a measure of adaptation to stress is that there are cases where S has been positively correlated with Y_p (i.e., cultivars whose yields were affected little by the stress also had very low yield potential). In other words, cultivars with low S also may have low stress resistance (Y) and would not have been useful to the producer in the first place. The correlation between S and Y_p indicates that it may not be possible (or it would be challenging) to combine the desirable traits associated with a low S and high yield potential.

15.3 Types of abiotic environmental stresses

J. Lewitt observed that plants respond to stress by strain reactions, which take a form that may be classified as either plastic or elastic, and manifest themselves as $G \times E$ interaction. Plastic responses produce a permanent change in the phenotype, whereas elastic responses are flexible and permit the normal state of the plant to return. Plants respond physiologically to stress by changing this reaction norm. Plant growth and development depend on biochemical processes (e.g., photosynthesis, respiration) that, in turn, depend on factors in the environment in order to proceed optimally. As previously indicated, when conditions in the environment are less than optimal, the plant experiences stress, which adversely affects its growth and development, and ultimately its productivity and economic value. The common stresses that plants may be exposed to in agro-ecological systems include the following:

- **Drought.** This is the stress perceived by a plant as a consequence of water deficit.
- **Heat.** Heat stress occurs when temperatures are high enough to cause irreversible damage to plant function.
- **Cold.** Cold stress manifests itself when plants are exposed to low temperatures that cause physiological disruptions that may be irreversible.
- **Salinity.** Stress from salinity occurs when the dissolved salts accumulate in the soil solution to an extent that plant growth is inhibited.
- **Mineral toxicity.** Mineral toxicity occurs when an element in the soil solution is present at a concentration such that plants are physiologically impaired.

- **Oxidative.** Secondary stress (induced by other stresses) caused by oxygen free radicals (or activated oxygen) that are known to induce damage in plant cells.
- **Water-logging.** Excessive soil moisture as a result of prolonged rainfall can cause anoxic soil conditions, leading to roots suffering from lack of oxygen.
- **Mineral deficiency.** Inadequate amounts of essential soil minerals available for plant growth causes growth inhibition and crop injury.

Over the years, plant breeders have devoted attention and resources to addressing these environmental stresses to varying degrees and with varying success. The most widely studied stresses aree discussed in more detail in the next sections.

15.4 Tolerance to stress or resistance to stress?

The terms "**tolerance**" and "**resistance**" are used in the literature to describe the mechanisms by which a plant responds to stress. Often, they are used as though they were interchangeable. According to J. Lewitt, from a physiological standpoint, a plant's response to stress may be characterized as "**avoidance**" (i.e., the environmental factor is excluded from the plant tissue), or "tolerance" (i.e., the factor penetrates the tissue but the tissue survives). The term resistance, from a physiological standpoint, is mechanism-neutral (implying neither tolerance nor exclusion). When the term is applied to bacteria, the development of resistance to an antibiotic has evolutionary stages. Full antibiotic resistance is not necessarily conferred by an immediate change in the bacterial genome. It is preceded by tolerance. Because bacteria can survive in the presence of an antibiotic, they have the opportunity to develop resistance.

Researchers who use resistance to describe the plant response to a stress (cause) appear to view resistance as a generic term for describing a number of mechanisms of which tolerance is one. In breeding for response to a stress, the ultimate goal of the breeder is to transfer genes to the cultivar that would enable it to perform to a desirable degree in spite of the stress. In this chapter, the term that is most widely associated with a particular stress in the literature is used.

15.5 Screening for stress resistance

Because of the complexity of environmental stress as previously discussed, simple, practical and effective tests that can readily be used by breeders as selection aids are not widely available. A. Acevedo and E. Fereres summarized the criteria for the development and use of screening tests as follows:

- Genetic variation should occur in the germplasm pool for the trait under consideration.
- The heritability for the trait should be greater than the heritability for yield *per se.*
- The trait should be correlated with a yield-based stress resistance index.
- It is desirable for the trait to be causally related to yield.
- The screening test should be easy, rapid, and economical to apply.

Traits associated with resistance to abiotic stress for which genetic variation has been found in wheat include osmotic adjustment, proline accumulation, leaf area per plant, epicuticular wax content, organ pubescence, tolerance in translocation, root growth, and many more. However, few of these traits can be readily applied as selection aids by plant breeders.

15.6 Drought stress

Water is the most limiting factor in crop production. In tropical regions of the world, moisture extremes are prevalent. There is either too much of it, when the rain falls, or there is little or no rainfall. Drought is responsible for severe food shortages and famine in developing countries.

15.6.1 What is drought stress?

Drought occurs both above ground (**atmospheric drought**) and below ground (**soil drought**). The duration of drought is variable, sometimes lasting for a short time and without severe adverse physiological impact, whereas it might last throughout an entire growing season or even years, resulting in complete devastation of crops. The efforts of breeders are directed at short-duration drought that often is experienced when crop production is rain-fed. Under rainfed conditions, rainfall is often erratic in frequency, quantity, and distribution. To avoid disruption in

growth and development processes, and consequently in crop performance (yield), plants need to maintain a certain level of physiological activity during the adverse period.

The effect of drought varies among species and also depends on the stage of plant growth and development at which the moisture stress occurs. Drought at flowering may cause significant flower drop and low fruit set. Similarly, when drought occurs at fruiting, there will be fruit drop and/or partially filled or shriveled fruits. In the end, both quality and product yield will be decreased.

15.6.2 An overview of drought stress concepts

Scientists have developed crop simulation models that a researcher may use to estimate and quantify the impact of specific drought stress conditions on crop productivity. Models are available for cereals (e.g., maize, wheat, rice), grain legumes (e.g., soybean, dry bean, peanut), root crops (e.g., cassava, potato), and other crops (e.g., tomato, sugarcane).

As the demand for water exceeds supply, a plant develops what is called **plant water deficit**. The supply of water is determined by the amount of water held in the soil to the depth of the crop root system. The demand for water is determined by plant transpiration rate or crop evapotranspiration (both plant transpiration and soil evaporation). The rate of transpiration is influenced by solar radiation, ambient air temperature, relative humidity, and wind. These factors control transpiration at the single leaf level. The most important factor that controls transpiration at the whole plant or crop level is total leaf area.

It is difficult to sense and estimate plant water stress at whole crop levels because of the need to integrate an estimate based on the whole canopy. Various plant-based methods are used to obtain direct measurement of water status, stress status, and other physiological consequences of water deficit. These include leaf water potential, relative water content, and osmotic potential.

Soil moisture is measured in a variety of ways. The soil moisture content (volume) is measured by gravimetric methods (soil is weighed before and after drying to determine water content). The soil water status is measured in terms of potential and tension. The amount of water a given crop can extract from the soil at a given water potential and depth is called the extractable soil moisture. Most crops usually extract between 50 and 80% of the extractable soil moisture before crop transpiration is reduced and symptoms of water deficit occur.

The relative importance of the major mediators of cellular response to water deficit and their relative importance are not exactly known. Plant hormones (e.g., abscisic acid (ABA)), are believed to be a key factor in water stress response. Numerous water stress responsive genes have been identified. At the whole plant or crop level, water deficit effects manifest in various ways – phenology, phasic development, growth, carbon accumulation, assimilate partitioning, and reproductive processes. These manifestations are primarily responsible for the variations in crop yield attributable to drought stress. Water deficit causes reduced cell expansion, reduced plant water use, and reduced plant productivity. Reduced cell expansion also adversely impacts meristematic development of yield components (e.g., inflorescence or tiller initials in cereals), leading to potentially small reproductive organs, and hence reduced yield. Damage to the meristem is usually irreversible and cannot be corrected by irrigation. Water deficit also causes advanced or delayed flowering, depending on the species. This alteration in phasic development is known to be critical to maize yield under stress. Water deficit can cause reproductive failure whereby the pollen may desiccate, creating an effect similar to male sterility, and leading to reduced grain set. The duration of grain filling is reduced under stress. Root/shoot dry weight ratio increases as plants slip into water stress.

15.6.3 Managing drought stress

Irrigation is the primary means of addressing drought in crop production, provided this approach is economical. Crop production may be designed for irrigated or dryland (rain-fed) environments. In irrigated production, the common practice is to implement a supplemental irrigation regime, whereby irrigation is applied when rainfall is inadequate. Various soil and water conservation practices may be adopted to conserve soil moisture from one season to another. Practices such as leaving fallow and the use of ground cover are recommended practices for soil and water conservation.

15.7 Breeding drought resistance

Drought resistance is highly specific to crop region. Drought resistance in crops is a major breeding objective in dryland farming systems. In more advanced agricultural production systems, the goal of combating drought is to obtain economic plant production in spite of the stress; plant survival is not the goal. However, in less advanced agriculture economies, plant survival is critical to producing some yield. The yield, albeit small, may mean the difference between famine and livelihood.

15.7.1 Underlying principles

To formulate proper and effective breeding objectives in a drought breeding program, the breeder should understand the nature of the trait to be manipulated. Earlier plant breeding efforts were directed at developing a genotype that had water-saving capacity. However, this concept has proven to be inadequate in addressing modern crop development. Researchers (J.B. Passioura and C.T. de Wit) have produced a model to describe the relationship between crop yield and water use as follows:

$$Yield = T \times WUE \times HI$$

where T is the total seasonal crop transpiration, WUE is the crop water use efficiency, and HI is the crop harvest index. This relationship clearly indicates that the focus in breeding for drought resistance should be on water use (efficient water use), rather than water saving. In order for a crop to sustain yield, it should be able to use water under stress. Plants cannot live without water. Drought resistance is a finite trait. What is desirable in modern crop production is a plant with the ability to use water efficiently when this resource is limited by drought.

A successful drought breeding program depends on the formulation of an appropriate and relevant ideotype for a drought target environment. This task is very challenging, requiring the breeder to put together a conceptual description of detailed morphological, physiological, and developmental attributes of the ideal genotype.

15.7.2 Characterization of the drought environment

Generally, constraints to crop production manifest themselves as a combination of physiological stress factors (rather than one), even though one may predominate. For example, drought spells may be associated with low or high temperature, depending on the

Industry highlights
Discovering genes for drought adaptation in sorghum

Andrew Borrell[1], David Jordan[1], John Mullet[2], Patricia Klein[2], Robert Klein[3], Henry Nguyen[4], Darrell Rosenow[5], Graeme Hammer[6], and Bob Henzell[1]

[1] DPI&F, Hermitage Research Station, Warwick, QLD 4370, Australia
[2] Texas A&M University, Institute for Plant Genomics & Biotechnology, College Station, USA
[3] USDA-ARS, Southern Agricultural Research Station, College Station, USA
[4] University of Missouri, Plant Sciences Unit and National Center for Soybean Biotechnology, Columbia, MO 65211, USA (previously Texas Tech University, Lubbock, USA)
[5] Texas A&M Agricultural Research & Extension Center, Lubbock, TX 79403-9803, USA
[6] University of Queensland, School of Land and Food, QLD 4072 Australia

Preface

"The heat shimmered above the dry earth, engulfing all in its wake. The black clay soil was deeply cracked, and like a cosmic vacuum cleaner, the sun sucked the last hint of moisture from its depths. Yet this desolate landscape was not without life. A crop of sorghum stood defiantly, thrusting its red grain into the copper sky. It was the dry season in southern India and no rain had fallen for many weeks. The crop was almost ready to harvest, but the yield would be low due to the severe drought (Figure B15.1). I walked through my experimental plots at the International Crops Research Institute for the Semi-Arid Tropics (ICRISAT) on the outskirts of Hyderabad. A young Indian scientist accompanied me on this stroll through our 280 plots. We stopped briefly at each plot, noting if any leaves still remained alive. We were examining a population of sorghum lines that varied in the stay-green drought resistance trait.

Plants containing the stay-green trait maintain more green leaf and stem under drought compared with senescent (non-stay-green) plants, resulting in stronger stems and higher grain yield. In most plots we found all the leaves had died. It was not unusual to find whole plots laying on the ground, their stems greatly weakened by the drought. I will never forget what happened next. Looking up from my notebook, I was stunned to see a plot of sorghum with a number of large green leaves and strong green stems (Figure B15.2). Balancing on the end of these stems were large panicles yielding about three times as much grain as the other plots. How could this be? Our field-plan revealed that this particular line was B35, a stay-green line from Ethiopia that was first documented by Dr. Darrell Rosenow, a plant breeder at Texas A&M University. B35 is derived from a durra landrace, an ancient type of sorghum that was eaten in Egypt over 4000 years ago. I knelt in the dust to examine the soil, making sure this plot wasn't receiving any additional water. Like the other plots, deep cracks snaked through the soil indicating the severity of drought. Now it was time to check the other replicates. A critical component of experimental design is the existence of replicates to verify outcomes. Excitedly, we made our way over to the next replicate of B35. Before we even arrived I could see it standing out from its lifeless neighbors. Yes! Quickly we headed to the third and fourth replicates. The same again! I was amazed at the resilience of B35, and determined to find out what drought adaptation mechanisms contributed to its remarkable survival. For me, the adventure of cracking the stay-green phenomenon had just begun."

Andrew Borrell, 1996

Figure B15.1 Harvesting our sorghum experiment in India at the end of the dry season. B35 displayed remarkable resilience under extremely dry conditions in this experiment. Figure courtesy of Andrew Borrell.

Introduction

Producing more grain with less water is one of the greatest challenges facing crop scientists in the twenty-first century.

Figure B15.2 Photographs of senescent (left) and stay-green (right) sorghum taken on the same day just prior to harvesting our field experiment in southern India. The stay-green variety is B35, derived from a durra landrace in Ethiopia. Figure courtesy of Andrew Borrell.

Globally, the availability of fresh water per capita has declined 37% since 1970, as population growth and degradation of water supplies has surpassed the capacity to develop new sources (Downer, 2000). Governments all over the world are choosing carefully how they allocate water between agricultural, urban and industrial uses. In a contest between these three, agriculture is often the loser because water used for irrigation generally produces a smaller economic return than water diverted to industry (Dupont, 2000), with urban requirements being even more important for many governments. Yet in the face of diminishing water resources, the world is expected to consume twice as much food in the next 50 years as it has in the past 10 000 years. To meet this demand, world grain production will have to increase 40% by 2020 (Dupont, 2000).

This case study describes how a multidisciplinary team of Australian and United States scientists is collaborating to discover genes for drought adaptation in sorghum. The potential to utilize these genes in the world's other major cereals is also discussed. Sorghum is a repository of drought resistance mechanisms and has developed biochemical, physiological and morphological characteristics such as C_4 photosynthesis, deep roots and thick leaf wax that enable growth in hot and dry environments. Sorghum is the dietary staple of more than 500 million people in over 30 countries, making it the world's fifth most important crop for human consumption after rice, wheat, maize, and potatoes (Miller, 1996).

Multidisciplinary approach

In many areas of human endeavor, it is often the integration of fields of knowledge that proves to be the fertile ground for innovation. So it is with "gene discovery" in the world's most important cereal crops. The pursuit of drought-resistance genes in sorghum is a multidisciplinary effort involving plant breeders, crop physiologists, molecular biologists, biometricians, functional genomicists and simulation modelers. Scientists from Australia and the United States are collaborating in the search for genes (*Stg1*, *Stg2*, *Stg3* and *Stg4*) associated with the "stay-green" trait in grain sorghum. Keeping leaves alive for longer is a fundamental strategy for increasing crop production, particularly under water-limited conditions. During post-anthesis drought, genotypes possessing the stay-green trait maintain more photosynthetically active leaves than genotypes not possessing the trait. The broad objective of this research is to identify and understand the function of genes and gene networks that contribute to improved plant drought resistance under water-limited conditions.

Approaches to gene discovery

There are two general approaches to identify and isolate genes involved in drought resistance (Mullet *et al.*, 2001). First, genes that show relatively rapid changes in expression at the RNA level in response to water limitation are targeted. Second, sorghum genes involved in drought adaptation are identified and isolated using map-based gene discovery. The current stay-green project primarily utilizes map-based gene discovery undertaken by scientists at Texas A&M University, although micro-array analysis is being used simultaneously to assist in gene discovery.

Phenotyping, genotyping and physiological characterization

Phenotyping driving genotyping

Map-based cloning requires the accurate screening of phenotype and genotype of large segregating populations (Tanksley *et al.*, 1995), highlighting the need for collaboration between plant breeders, crop physiologists and molecular biologists.

Typically, plant breeders develop a range of populations for mapping (e.g., recombinant inbred lines), fine mapping (e.g., segregating populations with breakpoints across the loci of interest), and physiological dissection (e.g., near-isogenic lines). Such populations are systematically phenotyped and genotyped by crop physiologists and molecular biologists, respectively, resulting in the identification of regions of genomes (trait loci) that modulate the expression of traits such as stay-green.

Genotyping driving phenotyping

Following the mapping of drought resistance loci, efficient map-based cloning requires the availability of a high-resolution integrated genetic and physical map, large populations and careful phenotyping (Mullet et al., 2001). The construction of an integrated sorghum genome map is well underway. A genetic map with about 3000 points has been constructed with AFLPs, SSRs, and RFLPs using a Sorghum bicolor recombinant inbred population (Menz et al., 2002). In addition, physical maps of the sorghum genome are being constructed using BAC libraries that provide about 20× coverage of the sorghum genome (Klein et al., 2003) (Chapter 20). The resulting integrated genome maps are being aligned to the rice genome sequence (Klein et al., 2003).

Physiological characterization

The aim is to dissect complex traits such as stay-green into their functional components. The characteristics of such complex traits can be viewed as emergent consequences of the interactions between underlying determinants and the prevailing environmental conditions (Hammer, 1998). Integration of knowledge from gene to cropping system is also necessary. For example, stay-green can be viewed at a cell, leaf, whole plant, crop and system level. Understanding how gene networks respond to water deficits across these levels is critical to capturing traits like stay-green in plant breeding programs.

Stay-green can be viewed as a consequence of the balance between nitrogen demand by the grain and nitrogen supply during grain filling at a whole plant level (Figure B15.3). Matching nitrogen supply from age-related senescence and nitrogen uptake during grain filling with grain nitrogen demand found that the shortfall in nitrogen supply for grain filling was greater in the senescent than stay-green hybrids, resulting in more accelerated leaf senescence in the former (Borrell et al., 2001). Preliminary simulation modeling to assess the value of stay-green in a range of environments found that improved nitrogen dynamics alone could not explain the observed yield increases under drought; enhanced transpiration efficiency (TE) was also required (Chapman et al., 2003).

Identifying candidate genes

Initial mapping activities can generally map targeted loci to 1–5 cM regions of the sorghum genetic map (Klein et al., 2001). Analysis of large segregating populations (~1000 plants) is usually required to provide sufficient genetic resolution for efficient map-based cloning. Fine mapping can then reduce the target locus in euchromatic regions to less than 100 kbp, a size that can be readily sequenced using standard BAC-based shotgun sequencing approaches. Interestingly, ~100 kbp of sorghum DNA, on average, will encode ~10 sorghum genes. There are several ways to identify genes within the target genomic interval. First, if the targeted region is less than 500 kbp, shotgun sequencing of BAC DNAs spanning the region followed by BLASTX analysis can be used to identify sorghum genes that are related to other known protein coding genes. Second, the sorghum sequence can be compared to the sorghum EST database to identify the transcribed portions of the BAC sequence. Third, other genes encoded by the sorghum BAC sequence can be identified by aligning the sorghum sequence with orthologous rice or maize sequences. Finally, gene prediction programs such as FGENESH (http://www.softberry.com/) and riceGAAS (http://ricegaas.dna.affrc.go.jp/) can be used to identify regions of the sorghum sequence that may encode genes.

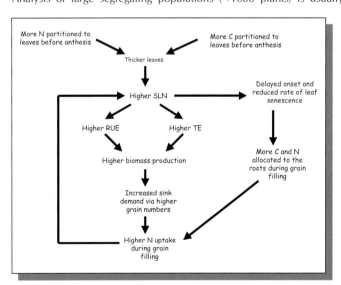

Figure B15.3 Nitrogen dynamics and the stay-green phenomenon in grain sorghum (Borrell et al., 2003). Figure courtesy of Andrew Borrell.

Proof of gene function

Identification of a candidate gene(s) is not the end point. Direct validation that a candidate gene causes variation in the trait under investigation is still required. Currently, this step is difficult in sorghum because sorghum gene transformation technology is time consuming, throughput is low, and not all genotypes are easily transformed (Mullet *et al.*, 2001). Other technologies such as RNAi are now under development in order to accelerate candidate gene validation (Holzberg *et al.*, 2002).

Assessing gene function across environments via simulation modeling

Crop modeling contributes to the genetic regulation of plant performance and improvement in a number of ways (Hammer *et al.*, 2002). For the purpose of this discussion, however, it is worth noting the role of simulation modeling as a means of both determining and assessing gene function *in silico*: the complexity arising from genes interacting with each other and their environments (both natural and managed) requires a mechanism to assimilate the many and varied combinations. Simulation modeling provides such a mechanism.

A blue future

As a result of this research, we will know the identity and function of the different genes involved in stay-green, enabling sorghum breeders to develop a new generation of crops adapted to our increasingly water-scarce world. Ultimately, we hope to transfer the stay-green genes to grain crops less well adapted to dry environments (i.e. rice, wheat and maize), or to identify similar drought resistant mechanisms in those crops. The global conservation of gene order among sorghum, rice and other grass species is clear from RFLP mapping information (Wilson *et al.*, 1999). Utilizing stay-green genes in rice could revolutionize rain-fed rice production.

The 1960s and 1970s were dominated by the "Green Revolution", when semi-dwarf varieties of wheat and rice were grown with high water and nitrogen inputs to produce record yields of grain. The early twenty-first century needs to be dominated by a "Blue Revolution" in which genetic, agronomic and management solutions are integrated to ensure food security in the face of global water shortages.

Acknowledgements

The Grains Research and Development Corporation and the Queensland Department of Primary Industries and Fisheries funded the Australian aspects of this research. Components of this research undertaken in the United States were funded by the National Science Foundation.

References

Borrell, A.K., Hammer, G.L., and Van Oosterom, E. (2001). Stay-green: A consequence of the balance between supply and demand for nitrogen during grain filling? *Annals of Applied Biology*, **138**:91–95.

Borrell, A.K., Van Oosteron, E., Hammer, G., Jordan, D., and Douglas, A. (2003). The physiology of "stay-green" in sorghum, in *Proceedings of the 11th Australian Agronomy Conference*. Australian Society of Agronomy, Gosford, Australia.

Chapman, S.C., Cooper, M., Podlich, D., and Hammer, G.L. (2003). Evaluating plant breeding strategies by simulating gene action and dryland environment effects. *Agronomy Journal*, **95**:99–113.

Downer, A. (2000). Sustenance and Security: Australia's multi-layered approach, in *Proceedings of a conference on Food, Water and War: Security in a World of Conflict*, The Crawford Fund for International Agricultural Research, Parliament House, Canberra, August 15, 2000. ACIAR Monograph No. 73, pp. 9–15.

Dupont, A. (2000). Food, Water and Security: What are the connections? in *Proceedings of a conference on Food, Water and War: Security in a World of Conflict*, The Crawford Fund for International Agricultural Research, Parliament House, Canberra, August 15, 2000. ACIAR Monograph No. 73, pp. 39–59.

Hammer, G.L. (1998). Crop modelling: Current status and opportunities to advance. *Acta Horticulturae*, **456**:27–36.

Hammer, G., Kropff, M.J., Sinclair, T.R., and Porter, J.R. (2002). Future contributions of crop modelling – from heuristics and supporting decision-making to understanding genetic regulation and aiding crop improvement. *European Journal of Agronomy*, **18**:15–31.

Holzberg, S., Brosio, P., Gross, C., and Pogue, G.P. (2002). Barley stripe mosaic virus-induced gene silencing in a monocot plant. *Plant Journal*, **30**:315–327.

Klein, P.E., Klein, R.R., Vrebalov, J., and Mullet, J.E. (2003). Sequence-based alignment of sorghum chromosome 3 and rice chromosome 1 reveals extensive conservation of gene order and one major chromosomal rearrangement. *Plant Journal*, **34**:605–621.

Klein, R.R., Klein, P.E., Chabra, A.K., *et al.* (2001). Molecular mapping of the *rf1* gene for pollen fertility restoration in sorghum (*Sorghum bicolor* L.). *Theoretical and Applied Genetics*, **102**:1206–1212.

Menz, M.A., Klein, R.R., Mullet, J.E., Obert, J.A., Unruh, N.C., and Klein, P.E. (2002). A high-density genetic map of *Sorghum bicolor* (L.) Moench based on 2926 AFLP®, RFLP and SSR markers. *Plant Molecular Biology*, **48**:483–499.

Miller, F. (1996). Sorghum, in *Lost Crops of Africa, Volume I, Grains* (eds National Academy of Sciences). National Academy Press, Washington, DC, pp. 127–143.

Mullet, J.E., Klein, R.R., and Klein, P.E. (2001). *Sorghum bicolor* – an important species for comparative grass genomics and a source of beneficial genes for agriculture. *Current Opinion in Plant Biology*, **5**:118–121.

Tanksley, S.D., Ganal, M.W., and Martin, G.B. (1995). Chromosome landing: a paradigm for map-based gene cloning in plants with large genomes. *Trends in Genetics*, **11**:63–68.

Wilson, A.W., Harrington, S.E., Woodman, W.L., Lee, M., Sorrells, M.E., and McCouch, S.R. (1999). Inferences on the genome structure of progenitor maize through comparative analysis of rice, maize and the domesticated panicoids. *Genetics*, **153**:453–473.

region. In irrigated production, drought is often associated with salinity stress. Further, the soil may introduce additional factors that limit crop production (e.g., unfavorable pH, toxic levels of elements such as aluminum, and deficiencies of other such micronutrients). To complicate the stress system, the predominant factors may be unpredictable in occurrence, intensity and duration. Two drought conditions may have significantly different composite properties. For example a drought event may occur with either a low vapor pressure deficit or high vapor pressure deficit.

Characterization of the drought environment is needed for several reasons, including the following:

- Drought resistance is highly location specific. This is because various factors (edaphic, biotic, agronomic) that are often location specific are involved.
- The drought-resistant genotypes to be developed need to be evaluated under a specific set of environmental conditions. Genotypes that are adapted to a given set of drought conditions may not be equally adapted to a different set of drought conditions.

It is clear then that breeders will have to characterize the target production region to know the most effective combination of adaptation features in genotypes best suited for successful production. Environmental characterization is a challenging undertaking. Consequently, breeders often limit this activity to only assembling qualitative meteorological data and soil description. These data may be obtained at the regional level or even as specific as farm level, and collected over seasons and years. Breeders engaged in breeding for resistance to drought should work closely with experts, including

agronomists, soil scientists, and meteorologists, or at least be familiar with techniques for computing soil moisture variations and rainfall probabilities. The modern technology of GIS is helpful in detailed characterization of environments.

15.7.3 Plant traits affecting drought response

Researchers have identified numerous stress response genes. However, these genes are not necessarily stress adaptive in terms of drought resistance. Nonetheless, both adaptive and non-adaptive genes are important in plant performance under drought stress. A good example of this fact is the role of genes that condition early flowering. Whereas these genes are expressed in any environment (i.e., not drought induced), early flowering may play a role in plant response to, and performance under, drought stress. Major plant traits that play a role in plant drought resistance response include phenology, development and size, root, plant surface, non-senescence, stem reserve utilization, photosynthetic systems, and water use efficiency.

Phenology

Phenology is the study of the times of recurring natural phenomena, especially as it relates to climate. Botanical phenology is the study of timing of vegetative activities, flowering, fruiting, and their relationship to environmental factors. For the plant breeder developing a cultivar with short growth duration (i.e., early flowering) may help the cultivar escape drought that may occur in the late season of crop production. However, it is known that longer growth duration is associated with high crop yield. Hence, using earliness for drought escape may reduce crop yield. Optimizing phenology is easier when there is a predictable environment. If the environment is highly unpredictable, using an early flowering indeterminate cultivar for crop production may be disadvantageous because stress would be unavoidable. Under such conditions, cultivars that are either determinate or indeterminate but have longer growth duration (late flowering) may have a chance to recover after a drought episode to resume growth. However, late maturing cultivars may face late season stress from disease and frost.

Research indicates a genetic linkage between growth duration and leaf number, and often leaf size. Consequently, early maturing genotypes have a small leaf area index and reduced evapotranspiration over most of the growth stages. Corn breeders make use of a phenological trait called **anthesis-to-silk interval** (ASI). A short ASI is desirable, whereas a longer ASI results in poor pollination. Crop plants differ in their phenological response to drought stress, some (e.g., wheat) advancing flowering, whereas others (e.g., rice) delay flowering.

Plant development and size

Water use by plants is significantly influenced by single plant leaf area or leaf area index (LAI). Genotypes with small size and reduced leaf area are generally conducive to low productivity, as they limit water use. These genotypes may resist drought but their growth rate and biomass accumulation are severely slowed down. Modern plant breeders tend to select for genotypes with moderate productivity and moderate size for use in dryland production.

Plant root characteristics

Root characteristics are critical to ideotype development for combating drought. The most important control of plant water status is at the root. The primary root characteristics of importance are root depth and root length density, of which root depth is more important to breeding drought resistance. In cereals where tillering occurs, deeper root extension occurs when fewer tillers are produced. Unfortunately, many drought prone regions of the world have shallow soils. Under such conditions, breeders may focus on other factors, such as shoot developmental characteristics, root hydraulics and osmotic adjustment. Also, development of lateral roots at very shallow soil depth

may be advantageous in capturing the small amounts of rainfall that may occur.

Plant surface

Plants interact with the environment through surface structures. The form and composition of such structures determine the nature of the interactions. Dehydration avoidance is positively correlated with yield under stress. The reflective properties of leaves and resistance to transpiration depend on plant surface structures. Stomata activity primarily determines resistance of plant leaves to transpiration but cuticular properties (e.g., wax load) also play a role. In sorghum, genotypes with lower epicuticular wax load (*bm*) had greater total leaf transpiration than genotypes with higher wax load (*Bm*). Another leaf surface feature with an implication in drought resistance is pubescence. In soybean, high leaf pubescence genotypes had higher water use efficiency, stemming from lower net radiation and transpiration with sustained photosynthesis. In wheat and barley, the role of leaf color in drought resistance has been noted. Yellow-leaf cultivars (with about one-third less leaf chlorophyll) tend to perform better than cultivars with normal green color, under drought stress.

Non-senescence

Environmental factors such as drought, heat and nitrogen deficiency are known to accelerate plant senescence. Certain genotypes, called "non-senescence", "delayed senescence", or "stay-green", have been identified to have the capacity to delay or slow senescence. These genotypes have high chlorophyll content and leaf reflectance. The expression of stay-green properties varies among species. In sorghum, the condition is better expressed when plants have been exposed to drought. Consequently, phenotypic selection for the trait may be more efficient under post-flowering drought stress.

Stem reservation utilization

Small grains and cereals store carbohydrates in their stems. The size of storage depends on favorable growing conditions before anthesis and the genotype. The potential stem storage is determined by stem length and stem weight density. It has been found that the presence of *Rht₁* and *Rht₂* dwarfing genes in wheat limits the potential reserve storage by about half, as a result of the reduction in stem length. This stem reduction may be partly responsible for the observed greater drought susceptibility of the dwarf high yielding wheat cultivars. The percentage of grain yield accounted for by stem reserves varies according to the environment and the genotype, and ranges between 90 and 100%. Stem reserve mobilization is a major source of carbon for grain filling under any stress. High reserve utilization, however, accelerates shoot senescence, a consequence of export of stored carbohydrates into the grain. This finding suggests that a breeder may not be as successful in selecting for both traits (delayed senescence and high reserve mobilization) as a strategy for developing a cultivar with high grain filling under stress.

15.7.4 Mechanisms of drought resistance

Plant species differ in the stages at which they are most susceptible to drought stress. Some species are most prone to stress damage during the early vegetative stage, while others are most susceptible during the pre- or post-anthesis stage, with others in between. Several general mechanisms may be identified by which plants resist drought:

Escape

Using early maturing cultivars may allow the crop to complete its lifecycle (or at least the critical growth stage) before the onset of drought later in the season. The plants use the optimal conditions at the beginning of the season to develop vigor.

Avoidance

Some plants avoid drought stress by decreasing water loss, for example by having cuticular wax or by having the capacity to extract soil moisture efficiently.

Tolerance

In species such as cereals, in which grain filling is found to depend on both actual photosynthesis during this stage as well as dry matter distribution from carbohydrates in pre-anthesis, terminal drought significantly reduces photosynthesis. This shifts the burden of grain filling to stored carbohydrate as the source of dry matter for the purpose. Consequently,

such species may be more tolerant of post-anthesis drought, being able to produce appreciable yield under the stress.

Recovery

Because drought varies in duration, some species are able to rebound (recover) after a brief drought episode. Traits that enhance recovery from drought include vegetative vigor, tillering, and long growth duration.

15.8 Approaches for breeding drought resistance

Plant breeders have two basic approaches for breeding for drought resistance – indirect breeding and direct breeding.

15.8.1 Indirect breeding

In this strategy, the breeder exposes genotypes to an environmental stress, even though they are not being directly evaluated for environmental stress. Indirect selection pressure is applied to these genotypes by conducting performance trials at locations where stress conditions exist. This approach is not advisable if the cultivars to be released are not intended for cultivation in the location where the evaluation was conducted. Under such atypical conditions, it is possible the cultivars might exhibit susceptibility to other stresses in their area of production.

15.8.2 Direct breeding

Direct selection for drought is best conducted under conditions where the stress factor occurs uniformly and predictably. Temperature and moisture are highly variable from one location to another, and hence difficult to predict. There are several methods used for direct breeding.

Field selection

Field selection is often problematic in drought resistance breeding. Water requirement is variable from year to year, and may be too severe in one year and cause loss of breeding materials. Further, drought in different seasons can occur at different growth stages.

Without special management for stress, inconsistencies in the field may result in inconsistent selection pressure from one cycle to the next. The ideal field selection site would be the desert, where irrigation may be used to dispense water at desired rates. Breeders may conduct multi-environment multi-year trials to assess G × E interactions and find genetic factors that enhance drought tolerance.

Selection under managed stress environments

Most drought breeding research is conducted under managed stress environments. A common facility for such research is the rainout shelter. This is essentially a mobile roof that protects selected plots from rain.

Selection based on yield per se

The concept of genotype × environment (G × E) is used to evaluate genotypes at the end of a breeding program. The desired genotype is one with minimal G × E interaction and stable yield in its target environment and across other environments. Selection for yield under stress is generally an inefficient approach. The generally low heritability of yield becomes even lower under stress. To get around this problem, some breeders resort to the tedious and expensive approach of screening very large populations. Some also use molecular markers to tag and select certain yield-related loci (QTLs) to aid in the selection process.

Selection based on developmental traits

Breeders tend to select for drought resistance on the basis of certain developmental traits, major ones including root size and development, and stem reserve utilization. Others are canopy development and leaf area. Root studies are tedious to perform. The most widely used technologies for detailed studies of roots are the **rhizotrons** and **lysimeters**. These pieces of equipment are available in various forms and provide various kinds of information. The **minirhizotron** even has a minute video camera included in the rig to record roots as they appear on the external surface of the tubes in which the plant is growing. Some researchers grow plants in soil-filled tubes (polyethylene tubes) and remove and wash away the soil, usually at flowering time, to measure root length. To select by using stem reserve utilization for grain filling, researchers completely inhibit the photosynthetic source (e.g., spraying plants with oxidizing chemicals such as potassium iodide) and measuring grain filling without photosynthesis, to compare with normal plants.

Selection based on assessment of plant water status and plant function

Methods used in this category of selection include assessing stress symptoms (e.g., leaf rolling, leaf desiccation, leaf tip burning). An infrared thermometer is also used to measure canopy temperature, whereas infrared photography is used to measure spectral reflection from leaves. Selection based on plant function includes the measurement of cell membrane stability, and chlorophyll fluorescence.

In breeding for drought resistance, breeders may select for early maturity for use in avoiding the stress. Also, genotypes with proven drought resistance may be used in hybridization programs to transfer the resistance genes into superior cultivars.

15.9 Cold stress

Plants experience cold stress in different ways. In some climatic zones, late frosts in springtime may harm flowering in fruit trees. Extreme low temperatures in California during wintertime may harm developing citrus fruits with considerable yield loss. Crops grown in regions with a relatively low average temperature tend to have relatively low yields.

15.9.1 An overview of cold stress concepts

Plants use a variety of adaptive mechanisms to combat environmental stress caused by cold temperature. Seed dormancy is a physiological condition that delays seed germination until the embryo has gone through an after-ripening period, during which certain biochemical and enzymatic processes occur for the seed to attain full maturity. In many temperate zone trees and shrubs, buds undergo a dormant stage, starting in late summer or early fall, ending only when the buds have been exposed to an extended period of cold or increasing day length in spring. Most winter annuals and biennials have a cold temperature requirement (vernalization) before they will flower. When plants are exposed to gradually decreasing temperatures below a certain threshold, they acclimatize (low-temperature acclamation) to the stress, a process called **cold hardening**.

In spite of various adaptations to cold, plants may be injured through exposure to cold temperatures in a variety of ways, depending on the temperature range. One type of injury, called **chilling injury**, occurs at exposures to temperature of between 20 and 0 °C. Some injuries are irreversible. Common chilling injuries include interruption of normal germination, flowering, and fruit development, which eventually adversely impact crop yield. Stored products may also suffer chilling injury. A more severe low-temperature injury is **freezing injury**, which occurs when temperatures drop below the freezing point of water. Sometimes, ice crystals form in the protoplasm of cells, resulting in cell death and possibly plant death.

Plants may be classified into three groups according to the tolerance to low temperatures. **Frost tender** plants are intolerant of ice in their tissues, and are hence sensitive to chilling injury. The plant (e.g., beans, corn, tomato) can be killed when temperatures fall just below 0 °C. **Frost resistant** plants can tolerate some ice in their cells and can survive cold temperatures of up to −40 °C. **Cold-hardy** plants are predominantly temperate woody species.

Most crops that originate in the tropics and subtropics are sensitive to chilling temperatures. However, some temperate fruits are also susceptible to chilling injury. The temperature at which chilling injury starts varies among species and depends on where they originate. Temperate fruits exhibit chilling injury starting at 0–4 °C, whereas the starting temperature is 8 °C for subtropical fruits and 12 °C for tropical fruits. Grains such as corn and rice suffer chilling injury at temperatures below 10 °C. When chilling temperatures occur at the seedling stage, susceptible crops suffer stand loss. Also, crop maturity is delayed while yield is reduced.

15.9.2 Genetic basis of low-temperature stress tolerance

The capacity of a genotype to tolerate low temperature has been extensively studied. It is agreed that low-temperature tolerance is a complex (quantitative) trait that may be influenced by several mechanisms. Reports indicate recessive, additive, partial dominance, and overdominance as the modes of action that occur in nature for cold stress. The diversity in the results is partly blamed on the way research is often conducted. Some workers use controlled-freeze tests while others use field tests. Furthermore, various reports indicate a role of cytoplasmic factors and non-additive gene effects, even though such effects are generally believed to be minor. Genes that condition varying levels of low-temperature tolerance occur within and among species. This genetic variability has been exploited to a degree in cultivar development within production regions.

A large amount of low-temperature tolerance research has been conducted in wheat. Low-temperature tolerance in cereals depends on a highly integrated system of structural, regulatory, and developmental genes. Several vernalization genes have been identified (e.g., vrn_1, vrn_4). The vrn_1 is homeo-allelic to the locus Sh_2 in barley and Sp_1 in rye. These two genes have been linked to genetic differences in low-temperature tolerance. Winter cereals also produce several proteins in response to low-temperature stress, for example, the dehydrin families of genes (dh_5, $Wcs120$).

15.10 Mechanisms of resistance to low temperature

Like drought resistance, certain physiological or morphological adaptations can make plants either avoid or tolerate stress due to low temperatures. Plants are described as cold-hardy (concept of cold-hardiness) when they have the capacity to withstand freezing temperatures. On the other hand, winter-hardy (winter-hardiness) species are able to avoid or tolerate a variety of weather-related effects associated with winter (e.g., freezing, heaving, desiccation, frost resistance, etc.).

The mechanisms of low-temperature resistance may be grouped into two.

(i) **Chilling resistance.** The factors that confer resistance to chilling are believed to operate at the cell membrane level where they influence membrane fluidity. Chilling resistant seeds are known to imbibe moisture slowly. The presence of phenols in the seed coat of legumes is implicated in conferring chilling resistance.

(ii) **Freezing resistance.** Several mechanisms are used by plants to resist freezing, including:
  - **Escape.** Like drought, cultural practices may be adopted by producers to prevent the vulnerable stage of growth coinciding with the presence of the stress factor.

- **Avoidance.** One of the injuries of low temperature results from the intracellular formation of ice following nucleation of ice in the tissue. Water may remain super cooled without forming ice crystals. Certain compounds that are capable of promoting ice nucleation are active at low temperatures. Bacteria such as *Pseudomonas syringae* are capable of producing ice nucleating proteins. The first field test of a bioengineered organism was the testing of "**ice minus**", a microbe genetically engineered to be incapable of producing the bacterial protein that causes ice nucleation. This was intended to be an approach to help frost-sensitive plants survive frost.
- **Tolerance.** Freezing tolerance occurs when a plant is able to withstand both intracellular and extracellular ice formation. A group of genes called the cold-response (COR) genes is believed to play a significant role in freezing tolerance. These genes are activated when the plant is exposed for a period to low but non-freezing temperatures (0–10 °C or 32–50 °F) conferring hardening on the plant. It is the hardening that makes plants survive freezing temperatures.

15.11 Selection for low-temperature tolerance

Field survival trials have proven to be inefficient for selecting genotypes with low-temperature tolerance. Because low-temperature stress brings about many changes in plants (including morphological, biochemical, and physiological changes), researchers are pursuing traits related to these changes in search of selection aids. Some factors that have shown promise in predicting low-temperature tolerance include plant erectness in winter (for cereals), tissue water content, and cell size. Unfortunately, these tests are not effective in discriminating among small differences that are of practical breeding importance.

Researchers commonly use controlled-freeze tests conducted in artificial environments to measure low-temperature tolerance. For example, the Field Survival Index developed by Fowler and Gusta is often used. The researchers found that the crown and leaf water content of field acclimated plants was a good indication of field survivability. Molecular marker technology is being pursued in the quest for QTLs associated with low-temperature tolerance. Other

biotechnology tools are being explored to help transfer low-temperature tolerance genes into cultivars. In spite of not being as efficient and as desirable as controlled environment tests, field testing remains a widely used screening approach in low-temperature tolerance breeding. When other selection approaches are used, field testing is what is used as a final measure of plant winter survival. Researchers can take various precautions to improve the efficiency of field tests.

15.12 Breeding for tolerance to low-temperature stress

Whereas the genetics of low-temperature tolerance has been studied to a reasonable degree, breeders have only had minimal success in applying research knowledge to practical breeding. Breeding superhardy cultivars remains a challenge. Several explanations for this lack of success have been proposed by D.B. Fowler and A.E. Limin. These are:

- Exploitable genetic variability for low-temperature tolerance has been largely exhausted within the existing gene pools of most species.
- A large number of genes with small effects and complex interaction is assumed to determine the phenotypic expansion of low-temperature tolerance, making selection difficult.
- Current methodologies for measuring low-temperature tolerance give poor resolution of small phenotypic differences.
- Measurements of low-temperature tolerance lack the precision for single plant analysis and many are destructive, making selection procedures complicated.
- Poor expression of low-temperature tolerance in alien genetic background has prevented the expansion of gene pools through interspecific and intergeneric transfers (e.g., the superior low-temperature tolerance of rye is suppressed in wheat background).

15.13 Salinity stress

Soil salinity constraints to crop production occur in an estimated 95% million hectares worldwide. **Salinity** is the accumulation of dissolved salts in the soil solution to a degree that it inhibits plant growth and development.

15.13.1 Overview of salinity stress concepts

Soils with salinity problems are described as **salt-affected**. When the salt concentration measured in terms of electrical conductivity (EC_e) is more than 4 dS/m and the pH is less than 8.5, the soil is called a **saline soil**. When the EC_e value is less than 4 dS/m and the pH is more than 8.5, the soil is a **sodic soil**. Sodic soils are high in sodium (Na^+) but low in other soluble salts. Semi-arid regions have saline/sodic soils, whereby salts accumulate in subsoils because of the low permeability of the subsoil.

Salinity may have natural origin (called **primary salinity**) as a result of weathering of parent materials that are rich in soluble salts. Human-aided salinity (called **secondary salinity**) occurs as a result of agricultural activities, especially, irrigation with impure (salt-rich) water. Salinity is often caused by a rising water table.

Plant growth is inhibited in salt-affected soils because the high salt concentration in the soil solution inhibits the process of water absorption by osmosis (osmotic stress). In addition, the concentration of Na^+ will increase intracellularly and interfere with essential biological processes, causing irreversible damage. When excessive amounts of salts enter the transpiration stream, plant cells may be injured. Plants that are tolerant to high soil salt concentration are called **halophytes**. Wheat is among the more salt-tolerant crops, while rice is one of the most salt-sensitive crops. Maize is moderately sensitive to salts in the soil solution (Table 15.1).

15.13.2 Breeding for salt tolerance

A common approach to breeding for salt tolerance starts with assembling and screening germplasm for salinity tolerance. The selected genotypes are used as parents to transfer the trait to desired cultivars, followed by selecting desirable recombinants from the segregating population. This approach has yielded some success in species such as rice, wheat, and lucerne. The challenge in breeding for salt tolerance is how to measure salinity tolerance. Screening is commonly based on growth of plants under salt stress. Two distinct mechanisms exist for salinity tolerance:

(i) Tolerance to the osmotic effect of the saline solution (osmotic effect makes it harder for plants to extract water from the soil).

(ii) Tolerance to the cytotoxic effect of Na^+ entering the cell. In addition, the salt-specific nature of the soil saline solution (high sodium concentration makes it difficult for the plant to exclude NaCl while taking up other ions).

Screening for salinity tolerance, just like other stress factors, is a long process and requires a large amount of space to screen progeny from crosses. Screening for specific traits is quicker and more effective (they are less influenced by the environment than growth rates). One of the most successful traits to incorporate in salt-tolerance breeding to combat salt-specific effects is the rate of Na^+ accumulation in leaves. This is measured as the increase in salt in a given leaf over a period. Selection for these ions was used in breeding rice and lucerne cultivars with high salt tolerance. Traits for osmotic effects are related to growth, for example, leaf elongation, root elongation, shoot biomass, and leaf area expansion, the latter two being the most effective indices. Molecular marker technology and genetic engineering techniques are being used in salt-tolerance breeding efforts. Salinity tolerance has been found in the wild species of crops such as tomato, pigeon pea, and common bean.

Table 15.1 Relative salt-tolerance in plants.

Tolerant	Moderately tolerant	Moderately sensitive	Sensitive
Barley (grain)	Barley (forage)	Alfalfa	Strawberry
Cotton	Sugar beet	Peanut	Bean
Bermudagrass	Wheat	Corn	Potato
	Oat	Rice (paddy)	Tomato
	Soybean	Sweet clover	Pineapple
	Sorghum	Sweet potato	Onion

15.14 Heat stress

Heat stress may be defined as the occurrence of temperatures high enough for sufficient time to cause irreversible damage to plant function or development. A heat resistant genotype is one that is more productive than most other genotypes in environments where heat stress occurs.

15.14.1 Overview of heat stress concepts

Heat stress occurs to varying degrees in different climatic zones. High temperatures can occur during the day or during the night. Also, temperature effects can be atmospheric or in the soil. Air temperature varies during the day and during the night. Annual crop species may be classified into two categories according to maximum threshold temperatures as either cool-season annuals or warm-season annuals (Table 15.2). Cool-season species are more sensitive to hot weather than warm-season species.

High night temperatures have detrimental effects on the reproductive function of plants. It has been shown that there is a distinct period during the 24-hour day cycle when pollen development is most sensitive to high night temperatures. In cowpea, plants that were exposed to high temperature during the last six hours of the night show significant decrease in pollen viability and pod set. Further, this damage was more pronounced in long days than short days. Other researchers also show that the stage of floral development most sensitive to high night temperature was between 7 and 9 days before anthesis.

Table 15.2 Examples of warm-season and cool-season crops.

Cool-season plants	Warm-season plants
Sugar beet	Okra
Cabbage	Eggplant (aubergine)
Apple	Corn
Wheat	Cotton
Barley	Sugarcane
Cauliflower	Peanut
	Sunflower
	Sorghum

Note: Some species have wide adaptation with varieties that are adapted to both cool and warm growing regions.

Excessive heat in the soil affects emergence of seedlings of both cool-season and warm-season crops, causing reduced crop stand.

15.14.2 Breeding for resistance to heat stress

Breeding for resistance to heat stress has not been as widely addressed as other environmental stresses that plants face in crop production. Some plant breeders use a direct measure of heat resistance approach to breeding whereby advanced lines are grown in a hot target production environment. Genotypes with greater yield than current cultivars are selected as superior. This breeding approach is more applicable to species that can be efficiently yield tested in small pots (e.g., wheat). Breeders may also use this approach in environments where heat is the only major stress. When other stresses occur, evaluation of heat damage is less conclusive (e.g., insect pest can cause damage to developing flower buds, similar to what would occur under heat stress).

An approach to breeding for heat resistance that is deemed by some to be more efficient is to select for specific traits that confer heat tolerance during reproductive development. To do this, genotypes with heat tolerance have to be discovered and the trait amenable to effective measurement. This would involve screening large numbers of accessions from germplasm collections. Then these genotypes may be crossed with desirable cultivars, if they lack the yield and other desired plant attributes.

The use of a controlled environment (hot greenhouse) has the advantage of providing a stable high nighttime temperature and stable air temperature from day to day and over a longer period. It is conducive to screening for reproductive stage heat tolerance. However, the facility can handle only a limited number of plants, compared to thousands of plants in a field evaluation. Selection aids (e.g., leaf-electrolyte-leakage) have been used by some researchers to identify genotypes with heat tolerance.

15.15 Mineral toxicity stress

Plants obtain most of their nutrient requirements from the soil, largely from the products of weathering of mineral rocks or the decomposition of organic matter. Uptake in improper amounts may lead to toxic consequences to plants.

Table 15.3 Summary of selected plant essential mineral nutrients and their roles.

Macronutrients	
Nitrogen (N)	Used in synthesis of amino acids and proteins; component of chlorophyll and enzymes.
Phosphorus (P)	Found in proteins and nucleic acids; critical in energy transfer process (ATP).
Potassium (K)	Catalyst for enzyme reactions; important in protein synthesis, translocation, and storage of starch.
Micronutrients	
Calcium (Ca)	Important in cell growth, cell division, and cell wall formation.
Magnesium (Mg)	Central atom of chlorophyll molecule; essential in formation of fats and sugars.
Sulfur (S)	An ingredient in vitamins and amino acids.
Boron (B)	Role in flowering, fruiting, cell division and water relations.
Iron (Fe)	Component of many enzymes; catalyst in synthesis of chlorophyll.
Molybdenum (Mo)	Role in protein synthesis and some enzymes.
Manganese (Mn)	Role in phosphorylation, activation of enzymes, and carbohydrate metabolism.
Zinc (Zn)	Role in enzyme activation.
Copper (Cu)	Catalyst for respiration and carbohydrate and protein metabolism.

15.15.1 Soil nutrient elements

Metals occur naturally in soils; some are beneficial and essential for plant growth and development, while others are toxic. About 16–20 elements have been identified as essential to plant nutrition. These may be broadly classified into two groups, based on the amounts taken up by plants, as **major (macro) nutrient elements** (these are required in large amounts) and **minor (micro) nutrient elements** (required in very small amounts) (Table 15.3). Each element has an optimal pH at which it is most available in the soil for plant uptake. However, at extreme conditions of soil reaction, excessive amounts of some elements become available. Some micronutrients are required in only trace amounts; their presence in large quantities in the soil solution may be toxic to plants. Some of the known toxicities of metallic elements occur at low pH (high acidity) and include iron and aluminum toxicities.

15.15.2 Aluminum toxicity

Aluminum (Al) is one of the most abundant elements in the earth's crust. One of the important metal toxicities of economic importance to crop production is aluminum toxicity that occurs when the aluminum concentration is greater than 2–3 ppm. At low pH, Al^{3+} ions predominate in the soil. Aluminum is not a plant essential nutrient. At a pH of five or less, aluminum inhibits plant growth by interfering with cell division in root tips and lateral roots, increasing cell wall rigidity, reducing DNA replication, decreasing respiration, and other effects. In some cases, excess aluminum induces iron deficiency in some crops (e.g., rice, sorghum, wheat). A visual symptom of aluminum toxicity is the so called **root pruning**, whereby root growth is severely inhibited. Stunting of roots leads to chronic drought and nutrient stress in afflicted plants.

15.15.3 Breeding aluminum tolerance

Aluminum-tolerant genotypes have been identified. Based on the patterns of aluminum accumulation in the plant tissue, three groups of aluminum-tolerant plants may be identified: (i) those with apparent exclusion mechanisms allowing lower accumulation of aluminum in their roots than aluminum-sensitive plants (e.g., wheat, barley, soybean); (ii) those with less aluminum in shoot but more in roots (e.g., wheat, barley, potato); and (iii) those with high aluminum accumulation in the shoot (e.g., pine trees). Research in wheat suggests the possibility of more than one aluminum-tolerance gene and more than one aluminum-tolerance mechanism. In one study, two QTLs associated with aluminum tolerance were identified in the F_2 population of diploid alfalfa and confirmed in the backcross population. Breeding for aluminum tolerance helps to expand crop productivity to acidic soils.

15.16 Mineral deficiency stress

15.16.1 Concepts associated with mineral deficiency

Mineral deficiencies or toxicities are widespread. A CIAT (International Center for Tropical Agriculture in Peru) report estimates that about 60% of the soils

Table 15.4 Common deficiency symptoms of selected plant essential nutrient elements.

Nitrogen (N)	Chlorosis or yellowing of leaves; stunted growth.
Phosphorus (P)	Dark green leaves; purpling of plant parts.
Potassium (K)	Marginal necrosis; weak stem and lodging; leaf curling.
Calcium (Ca)	Terminal bud growth ceases or is defective.
Magnesium (Mg)	Chlorosis of older leaves.
Sulfur (S)	Chlorosis; weak stems.
Boron (B)	Death of terminal bud producing growth called witches' broom.
Iron (Fe)	Interveinal chlorosis of young leaves.
Molybdenum (Mo).	Whip tail growth in coniferous species.
Manganese (Mn)	Interveinal chlorosis.
Zinc (Zn)	Mottled leaves.
Copper (Cu)	Stunting; interveinal chlorosis.

in the common bean production regions of the world have some soil mineral problem. Soils that are high in calcareous minerals tend to have high amounts of basic elements (e.g., Ca, Mg, K, Na) that tend to raise soil pH. A high soil pH in turn causes mineral deficiency problems (e.g., Fe, Zn, Fe, P). Common mineral deficiency symptoms are summarized in (Table 15.4). Zinc deficiency in common bean has been reported in production areas such as southern Idaho and Michigan.

15.16.2 Breeding efforts

Cultivars vary in their sensitivity to zinc deficiency. Sensitive cultivars take up and store less zinc in various plant parts and the seed than resistant cultivars. Researchers in common bean identified a zinc deficiency resistant cultivar, "Matterhorn", and subsequently determined that a single dominant gene, *Znd*, conditioned resistance to soil zinc deficiency.

Improving nitrogen use efficiency (NUE) is an important breeding objective, being a vital component of yield and end-use quality. The price of nitrogen fertilizer is increasing, so are the associated environmental concerns. Classical and molecular techniques are being used in this search for genotypes with improved NUE (e.g., genetically engineering plants that overexpress alanine aminotransferase). QTLs involved with NUE are being sought by researchers.

15.17 Oxidative stress

15.17.1 Concepts associated with oxidative stress

Oxygen free radicals (or activated oxygen) have been implicated in a variety of environmental stresses in plants. They are involved in many degenerative conditions in eukaryotic cells (e.g., peroxidation of lipids, cross-linking and inactivation of protein, and mutation in the DNA). However, the biosynthesis of some complex organic molecules, detoxification of xenobiotic chemicals, polymerization of cell wall constituents, and defense against pathogens are examples of essential cellular activities that depend on oxygen free radicals. Hence, the issue is not preventing their formation but how to control and manage the potential reactions of activated oxygen. The plant has a system of complex scavenging of activated oxygen that is highly conserved among plants.

Numerous sites of oxygen activation occur in the plant cell. These sites are highly controlled and coupled to prevent the release of intermediate products. It is presumed that such a control or coupling breaks down when a plant is under stress, resulting in leaking of activated oxygen. Injuries to the plant occur when the production of activated oxygen exceeds the plant's capacity to detoxify it. Symptoms of oxidative stress include loss of osmotic responsiveness, wilting, and necrosis.

There are two forms of activated oxygen that are produced via distinctly different mechanisms. Most biological systems produce activated oxygen via reduction of oxygen to form superoxide, hydrogen peroxide, and hydroxyl radicals. In photosynthetic plants, the activated oxygen form is also produced by photosynthesis.

15.17.2 Applications and breeding efforts

Several herbicides are designed to function by the involvement of activated oxygen. These herbicides promote the accumulation of metabolic intermediates and the energy used to create singlet oxygen that kills the plant. These herbicides are described as photobleaching herbicides (e.g., p-nitrodiphenyl ethers). Other herbicides that depend on light and chlorophyll are paraquat and diquat (both bipyridylium herbicides). So far, few plants have been selected for tolerance to oxygen free radicals.

15.18 Flood stress (water logging)

Whereas some plants are adapted to water-logged conditions (e.g., flooded rice culture), most plants need well-drained soils to grow properly.

15.18.1 Concepts associated with water-logging stress

In soybean, stress due to water-logging can reduce crop yield by 17–43% when it occurs at the vegetative stage and about 50–56% if the stress occurs at the reproductive stage. Floods are often caused by excessive rainfall due to a prolonged seasonal rainfall, or excessive rainfall after a long period of drought. The excessive amount of water quickly creates anoxic (oxygen deficient) soil conditions causing flood-sensitive plants to suffer anoxia or hypoxia. Fermentation occurs in plant roots under such conditions. Photosynthetic capacity of plants is significantly inhibited. Flood-tolerant species have certain adaptive mechanisms, such as the formation of aerenchyma and adventitious roots. Some studies indicate that root tissue survival under hypoxia depends on the fermentation rate and sufficient sugar supply to maintain cell energy and membrane function.

15.18.2 Breeding efforts

Tolerance to water-logging appears to be quantitatively inherited. QTLs for tolerance to water-logging have been reported in rice and soybean. Researchers at the International Rice Research Institute isolated the SUB1 locus that promotes tolerance to flooding. In conjunction with the Central Rice Research Institute in India, a flood-resistant paddy rice variety "Swarna Sub1" that can tolerate submergence due to flash flood for 14 days has been developed.

Key references and suggested reading

Cinchy, K.A., Foster, S., Grafton, K.F., and Hosfield, G.L. (2005). Inheritance of seed zinc accumulation in navy bean. *Crop Science*, **45**:864–870.

Chinnusamy, Y., Jagendorf, A., and Zhn, J-K. (2005). Understanding and improving salt tolerance in plants. *Crop Science*, **45**:437–448.

Fischer, R.A., and Maurer, R. (1978). Drought resistance in spring wheat cultivars I. Grain yield response. *Australian Journal of Agricultural Research*, **29**:897–912.

Ibrahim, A.M.H., and Quick, J.S. (2001). Heritability of heat tolerance in winter and spring wheat. *Crop Science*, **41**:1401–1405.

Lewitt, J. (1972). *Responses of plants to environmental stresses*. Academic Press, New York.

Lee, E.A., Staebler, M.A., and Tollenaar, M. (2002). Genetic variation in physiological discriminators of cold tolerance – early autotrophic phase of maize development. *Crop Science*, **42**:1919–1929.

McKersie, B.D., and Leshem, Y.Y. (1994). *Stress and coping in cultivated plants*. Kluwer Academic Publishers, Dordrecht, The Netherlands.

Voigt, P.W., and Stanley, T.E. (2004). Selection of aluminum and acid-soil resistance in white clover. *Crop Science*, **44**:38–48.

Yordanov, I., Velikova, V., and Tsonev, T. (2000). Plant responses to drought, acclimation, and stress tolerance. *Photosynthetica*, **38**:171–186.

Outcomes assessment

Part A

Please answer the following questions true or false.

1. Sodic soils have high accumulation of potassium.
2. Harvest maturity is determined by crop producers.
3. Early maturity is recessive to late maturity.
4. Most irrigated farms in the world have salinity problems.

Part B

Please answer the following questions.

1. Distinguish between physiological maturity and marker maturity.
2. Distinguish between weather and climate.
3. Distinguish between drought tolerance and drought avoidance.
4. Briefly explain the role of plant growth regulators in drought tolerance.
5. Give four examples of crop plants that are susceptible to salts in the soil.
6. What is drought resistance?

Part C

Please write a brief essay on each of the following topics.

1. Discuss the advantages and disadvantages of early maturity.
2. Discuss the breeding of early maturity in crop plants.
3. Discuss the rationale for breeding drought resistance.
4. Discuss the mechanisms used by plant to avoid the effects of drought.

Section 6

Selection methods

Chapter 16 Breeding self-pollinated species
Chapter 17 Breeding cross-pollinated species
Chapter 18 Breeding hybrid cultivars
Chapter 19 Breeding clonally propagated species

_____

Plant breeders depend on variability for success in their breeding programs. Once assembled or created, breeders used selection strategies or methods to discriminate among the variability to identify those with the desired genotypes that can be developed into cultivars. Selection strategies used depend on the modes of reproduction of the species being genetically improved. This section of the book is devoted to discussing the various methods of selection commonly used in plant improvement.

_____

16

Breeding self-pollinated species

Purpose and expected outcomes

As previously discussed, self-pollinated species have a genetic structure that has implication in the choice of methods for their improvement. They are naturally inbred and hence inbreeding to fix genes is one of the goals of a breeding program for self-pollinated species in which variability is generated by crossing. However, crossing does not precede some breeding methods for self-pollinated species. The purpose of this chapter is to discuss specific methods of selection for improving self-pollinated species. After studying this chapter, the student should be able to discuss the characteristics, application, genetics, advantages, and disadvantages of the following methods of selection:

1 Mass selection.
2 Pure line selection.
3 Pedigree selection.
4 Bulk population.
5 Single seed descent.

The student should also be able to discuss:

6 The technique/method of backcrossing.
7 The method of multiline breeding.
8 The method of breeding composites.
9 The method of recurrent selection.

16.1 Types of cultivars

There are six basic types of cultivars that plant breeders develop. These cultivars derive from four basic populations used in plant breeding – **inbred pure lines, open-pollinated populations, hybrids,** and **clones**. Plant breeders use a variety of methods and techniques to develop these cultivars.

Principles of Plant Genetics and Breeding, Second Edition. George Acquaah.
© 2012 John Wiley & Sons, Ltd. Published 2012 by John Wiley & Sons, Ltd.

16.1.1 Pure-line cultivars

Pure-line cultivars are developed for species that are highly self-pollinated. These cultivars are homogeneous and homozygous in genetic structure, a condition attained through a series of self-pollination. These cultivars are often used as parents in the production of other kinds of cultivars. Pure-line cultivars have a narrow genetic base. They are desired in regions where uniformity of a product has a high premium. It should be pointed out, though, that genetic uniformity occurs in other types of cultivars besides pure lines, for example hybrids and vegetatively propagated cultivars.

16.1.2 Open-pollinated cultivars

Contrary to pure-lines, **open-pollinated cultivars** are developed for species that are naturally cross-pollinated. The cultivars are genetically heterogeneous and heterozygous. Two basic types of open-pollinated cultivars are developed. One type is developed by improving the general population by **recurrent** (or repeated) **selection** or bulking and increasing material from selected superior inbred lines. The other type, called a **synthetic cultivar**, is derived from planned matings involving selected genotypes. Open pollinated cultivars have a broad genetic base. Another important type of cultivar developed for open-pollinated species is the **hybrid cultivar**.

16.1.3 Hybrid cultivars

Hybrid cultivars are produced by crossing inbred lines that have been evaluated for their ability to produce hybrids with superior vigor over and above those of the parents used in the cross. Hybrid production exploits the phenomenon of hybrid vigor (or heterosis) to produce superior yields. Heterosis is usually less in crosses involving self-pollinated species than those involving cross-pollinated species. Hybrid cultivars are homogeneous but highly heterozygous. Pollination is highly controlled and restricted in hybrid breeding to only the designated pollen source. In the past, physical human intervention was required to enforce this strict pollination requirement, making hybrid seed expensive. However, with time, various techniques have been developed to capitalize on natural reproductive control systems (e.g., male sterility) to facilitate hybrid production. Hybrid production is more widespread in cross-pollinated species (e.g., corn, sorghum), because the natural reproductive mechanisms (e.g., cross fertilization, cytoplasmic male sterility) are more readily economically exploitable than in self-pollinated species.

16.1.4 Clonal cultivars

Seeds are used to produce most commercial crop plants. However, a significant number of species are propagated by using plant parts other than seed (vegetative parts such as stems and roots). By using vegetative parts, the cultivar produced consists of plants with identical genotypes and is homogeneous. However, the cultivar is genetically highly heterozygous. Some plant species are sexually reproducing but are propagated clonally (vegetatively) by choice. Such species are improved through hybridization, so that when hybrid vigor exists it can be fixed (i.e., the vigor is retained from one generation to another) and then the improved cultivar propagated asexually. In seed propagated hybrids, hybrid vigor is highest in the F_1, but is reduced by 50% in each subsequent generation. In other words, whereas clonally propagated hybrid cultivars may be harvested and used for planting the next season's crop without adverse effects, producers of sexually reproducing species using hybrid seed must obtain a new supply of seed, as previously indicated.

16.1.5 Apomictic cultivars

Apomixis is the phenomenon of production of seed without the benefit of the union of sperm and egg cells (i.e., without fertilization). The seed harvested are thus genetically identical to the mother plant (apomixis is asexual reproduction via seed). Hence, apomictic cultivars have the same benefits of clonally propagated ones, as previously discussed. In addition, they have the convenience of vegetative propagation through seed (versus propagation through cuttings or vegetative plant parts). Apomixis is common in perennial forage grasses.

16.1.6 Multilines

Multilines are developed for self-pollinating species. These cultivars consist of a mixture of specially developed genotypes called **isolines** (or **near isogenic lines**) because they differ only in a single gene (or a defined set of genes). Isolines are developed primarily for disease control, even though these cultivars, potentially, could be developed to address other environmental stresses. Isolines are developed by using the techniques of backcrossing in which the F_1 is repeatedly crossed to one of the parents (recurrent parent) that lacked the gene of interest (e.g., disease resistance).

16.2 Genetic structure of cultivars and its implications

The products of plant breeding that are released to farmers for use in production vary in genetic structure and, consequently, the phenotypic uniformity of the product. Furthermore, the nature of the product has implications in how it is maintained by the producers regarding the next season's planting.

16.2.1 Homozygous and homogeneous cultivars

A cultivar may be genetically homozygous and, hence, produce a homogeneous phenotype or product. Self-pollinated species are naturally inbred and tend to be homozygous. Breeding strategies in these species are geared toward producing cultivars that are homozygous. The products of economic importance are uniform. Furthermore, the farmer may save seed from the current season's crop (where legal and applicable) for planting the next season's crop, without loss of cultivar performance, regarding yield and product quality. This attribute is especially desirable to producers in many developing countries where the general tradition is to save seed from the current season for planting the next season. However, in developed economies with well-established commercial seed production systems, intellectual property rights prohibit the re-use of commercial seed for planting the next season's crop, thus requiring seasonal purchase of seed by the farmer from seed companies.

16.2.2 Heterozygous and homogeneous cultivars

The method of breeding of certain crops leaves the cultivar genetically heterozygous yet phenotypically homogeneous. One such method is hybrid cultivar production, a method widely used for production of, especially, outcrossing species such as corn. The heterozygous genetic structure stems from the fact that a hybrid cultivar is the F_1 product of a cross of highly inbred (repeatedly selfed; homozygous) parents. Crossing such pure lines produces highly heterozygous F_1 plants. Because the F_1 is the final product released as a cultivar, all plants are uniformly heterozygous, and hence homogeneous in appearance. However, the seed harvested from the F_1 cultivar is F_2 seed, consequently producing maximum heterozygosity and heterogeneity upon planting. The implication for the farmer is that the current season's seed cannot be saved for planting the next season's crop for obvious reasons. The farmer who grows hybrid cultivars must purchase fresh seed from the seed company for planting each season. Whereas this works well in developed economies, hybrids generally do not fit well into the farming systems of developing countries where farmers save seed from the current season for planting the next season's crop. Nonetheless, the use of hybrid seed is gradually infiltrating crop production in developing countries.

16.2.3 Heterozygous and heterogeneous cultivars

Other approaches of breeding produce heterozygous and homogeneous (relatively) cultivars, for example, synthetic and composite breeding. These approaches allow the farmer to save seed for planting. Composite cultivars are suited to production in developing countries, while synthetic cultivars are common in forage production all over the world.

16.2.4 Homozygous and heterogeneous cultivars

An example of such a breeding product is the mixed landrace types that are developed by producers. The component genotypes are homozygous but there is such a large amount of diverse genotypes included that the overall cultivar is not uniform.

16.2.5 Clonal cultivar

Clones, by definition, produce offspring that are not only identical to each other but also the parent. Clones may be very heterozygous but whatever advantage heterozygosity confers is locked in for as long as propagation is clonally conducted. The offspring of a clonal population is homogeneous. Once the genotype has been manipulated and altered in a desirable way, for example through sexual means (since some species are flowering but are not propagated through seed but vegetatively) the changes are fixed for as long as clones are used for propagation. Flowering species such as cassava and sugarcane may be genetically improved through sex-based methods, and thereafter commercially clonally propagated.

16.3 Types of self-pollinated cultivars

In terms of genetic structure, there are two types of self-pollinated cultivars:

(i) Those derived from a single plant.
(ii) Those derived from a mixture of plants.

Single plant selection may or may not be preceded by a planned cross but often it is the case. Cultivars derived from single plants are homozygous and homogeneous. However, cultivars derived from plant mixtures may appear homogeneous but, because the individual plants have different genotypes, and because some outcrossing (albeit small) occurs in most selfing species, heterozygosity would arise later in the population. The methods of breeding self-pollinated species may be divided into two broad groups – those preceded by hybridization and those not proceeded by hybridization.

16.4 Common plant breeding notations

Plant breeders use shorthand to facilitate the documentation of their breeding programs. Some symbols are standard genetic notations, while others were developed by breeders. Unfortunately, there is no one comprehensive and universal system in use, making it necessary, especially with the breeding symbols, for the breeder to always provide some definitions to describe the specific steps in a breeding method employed in the breeding program.

16.4.1 Symbols for basic crosses

F

The symbol F (for **filial**) denotes the progeny of a cross between two parents. The subscript (x) represents the specific generation (F_x). If the parents are homozygous, the F_1 generation will be homogeneous.

Crossing of two F_1 plants (or selfing an F_1) yields an F_2 plants ($F_1 \times F_1 = F_2$). Planting seed from the F_2 plants will yield an F_2 population, the most diverse generation following a cross, in which plant breeders often begin selection. Selfing F_2 plants produce F_3 plants, and so on. It should be noted that the seed is one generation ahead of the plant, that is, an F_2 plant bears F_3 seed.

The symbol $\otimes$ is the notation for selfing.

S

The S notation is also used with numeric subscripts. In one usage $S_o = F_1$; another system indicates $S_o = F_2$.

16.4.2 Symbols for inbred lines

Inbred lines are described by two systems. System I describes an inbred line based on the generation of plants that is being currently grown. System II describes both the generation of the plant from which the line originated as well as the generation of plants being currently grown. The following examples are used to distinguish between the two systems.

- Example 1: The base population is F_2. The breeder selects an F_2 plant from the population and plants the F_3 seeds in the next season.
    System I: The planted seed produces an F_3 line.
    System II: The planted seed produces F_2 derived line in F_3 or $F_{2:3}$ line.
  If seed from the F_3 plants is harvested and bulked, and the breeder samples the F_4 seed in the next season, the symbolism will be as follows:
    System I: The planted seed produces an F_4 line.
    System II: The planted seed produces an F_2 derived line in F_4 or $F_{2:4}$ line
- Example 2: The breeder harvests a single F_4 and plants F_5 seed in a row.
    System I: The planted row produces an F_5 line.
    System II: The planted row constitutes F_4 derived line in F_5 or $F_{4:5}$ line.
  Similarly the S notation may be treated likewise. Taking example 1 for example:
    System I: S_1 line.
    System II: S_o derived line in S_1 or $So_{:1}$ line.

16.4.3 Notation for pedigrees

Knowing the **pedigree** or ancestry of a cultivar would enable the plant breeder to retrace the steps in a breeding program to reconstitute a cultivar. Plant breeders follow a short-hand system of notations to write plant pedigrees. Some pedigrees are simple, others are complex. Some of the common notations are as follows:

A '/' indicates a cross; a figure between slashes, /2/, indicates the sequence or order of crossing. A/2/is equivalent to//and indicates the second cross. Similarly, /is the first cross, and///the third cross. A backcross is indicated by *; *3 indicates genotype backcrossed three times to another genotype. The following two examples are used to illustrate the concept.

- **Pedigree 1.** MSU48-10/3/Pontiac/Laker/2/ MS-64
  Interpretation:
  1. The first cross was Pontiac (as female) × Laker (as male).
  2. The second cross was [Pontiac/Laker (as female)] × MS-64 (as male)
  3. The third cross was MSU48-10 (as female) × [Pontiac/Laker//MS-64 (as male)]
- **Pedigree 2.** MK2-57*3/SV-2
  Equivalent formula: MK2-57/3/MK2-57/2/ MK2-57/SV-2.
  Interpretation: The genotype MK2-57 was backcrossed three times to genotype SV-2

16.5 Mass selection

Mass selection is an example of selection from a biologically variable population in which differences are genetic in origin. The Danish biologist, W. Johansen, is credited with developing the basis for mass selection in 1903. Mass selection is often described as the oldest method of breeding self-pollinated plant species. However, this by no means makes the procedure outdated. As an ancient art, farmers saved seed from desirable plants for planting the next season's crop, a practice that is still common in the agriculture of many developing countries. This method of selection is applicable to both self- and cross-pollinated species, provided there is genetic variation.

16.5.1 Key features

The purpose of mass selection is population improvement by increasing the gene frequencies of desirable genes. Selection is based on plant phenotype and one generation per cycle is needed. Mass selection is imposed once or multiple times (recurrent mass selection). The improvement is limited to the genetic variability that existed in the original populations (i.e., new variability is not generated during the breeding process). The goal in cultivar development by mass selection is to improve the average performance of the base population.

16.5.2 Application

As a modern method of plant breeding, mass selection has several applications:

- It may be used to maintain the purity of an existing cultivar that has become contaminated, or is segregating. The off-types are simply rogued out of the population and the rest of the material bulked. Existing cultivars become contaminated over the years by natural processes (e.g., outcrossing, mutation) or by human error (e.g., inadvertent seed mixture during harvesting or processing stages of crop production).
- It can also be used to develop a cultivar from a base population created by hybridization, using the procedure described in Section 16.5.3.
- It may be used to preserve the identity of an established cultivar or soon to be released new cultivar. The breeder selects several hundred (200–300) plants (or heads) and plants them in individual rows for comparison. Rows showing significant phenotypic differences from the other rows are discarded, while the remainder is bulked as breeder seed. Prior to bulking, sample plants or heads are taken from each row and kept for future use in reproducing the original cultivar.
- When a new crop is introduced into a new production region, the breeder may adapt it to the new region by selecting for key factors needed for successful production (e.g., maturity). This, hence, becomes a way of improving the new cultivar for the new production region.
- Mass selection can be used to breed horizontal (durable) disease resistance into a cultivar. The breeder applies low densities of disease inoculum (to stimulate moderate disease development) so that

quantitative (minor gene effects) genetic effects (instead of major gene effects) can be assessed. This way, the cultivar is race-non-specific and moderately tolerant of disease. Further, crop yield is stable and the disease resistance is durable.

- Some breeders use mass selection as part of their breeding program to rogue out undesirable plants, thereby reducing the materials advanced and saving time and reducing cost of breeding.

16.5.3 Procedure

Overview

The general procedure in mass selection is to rogue out off-types or plants with undesirable traits. This is called by some researchers **negative mass selection**. The specific strategies for retaining representative individuals for the population vary according to species, traits of interest, or creativity of the breeder to find ways to facilitate the breeding program. Whereas rouging out and bulking appears to be the basic strategy of mass selection, some breeders may rather select and advance a large number of plants that are desirable and uniform for the trait(s) of interest (**positive mass selection**). Where applicable, single pods from each plant may be picked and bulked for planting. For cereal species, the heads may be picked and bulked.

Steps

The breeder plants the heterogeneous population in the field, looks for off-types to remove and discard (Figure 16.1). In this way the original genetic structure is retained as much as possible. A mechanical device (e.g., using a sieve to determine which size of grain would be advanced) may be used, or selection may be purely on visual basis according to the

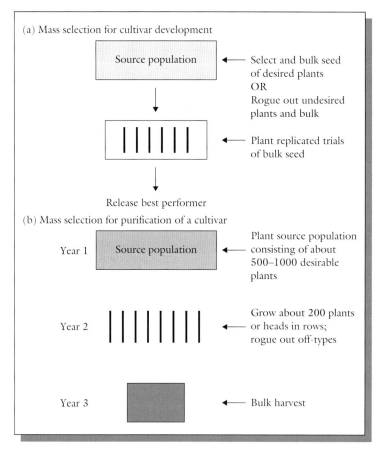

Figure 16.1 Generalized steps in breeding by mass selection: (a) for cultivar development and (b) for purification of an existing cultivar.

breeder's visual evaluation. Further, selection may be based on targeted traits (direct selection) or indirectly by selecting a trait correlated with the trait to be improved.

- **Year 1.** If the objective is to purify an established cultivar, seed of selected plants may be progeny-rowed to confirm the purity of the selected plants prior to bulking. This would make a cycle of mass selection have a two-year duration instead of one year. The original cultivar needs to be planted alongside for comparison.
- **Year 2.** Evaluate composite seed in replicated trial, using original cultivar as check. This test may be conducted at different locations and over several years. The seed is bulk-harvested.

16.5.4 Genetic issues

Contamination from outcrossing may produce heterozygotes in the population. Unfortunately, where the dominance effect is involved in the expression of the trait, heterozygotes are indistinguishable from homozygous dominant individuals. Including heterozygotes in a naturally selfing population will provide material for future segregations to produce new off-types. Mass selection is most effective if the expression of the trait of interest is conditioned by additive gene action.

Mass selection may be conducted in self-pollinated populations as well as cross-pollinated populations, but with different genetic consequences. In self-pollinated populations, the persistence of inbreeding will alter population gene frequencies by reducing heterozygosity from one generation to the next. However, in cross-pollinated populations, gene frequencies are expected to remain unchanged unless the selection of plants was biased enough to change the frequency of alleles that control the trait of interest.

Mass selection is based on plant phenotype. Consequently, it is most effective if the trait of interest has high heritability. Also, cultivars developed by mass selection tend to be phenotypically uniform for qualitative (simply inherited) traits that are readily selectable in a breeding program. This uniformity notwithstanding, the cultivar could retain significant variability for quantitative traits. It is helpful if the selection environment is uniform. This will ensure that genetically superior plants are distinguishable from mediocre plants. When

selecting for disease resistance, the method is more effective if the pathogen is uniformly present throughout the field without "hot spots".

Some studies have shown correlated response to selection in secondary traits as a result of mass selection. Such a response may be attributed to linkage or pleiotropy.

16.5.5 Advantages and disadvantages

There are both major advantages and disadvantages of mass selection for improving self-pollinated species.

Advantages

- It is rapid, simple, and straightforward. Large populations can be handled and one generation per cycle can be used.
- It is inexpensive to conduct.
- The cultivar is phenotypically fairly uniform even though it is a mixture of pure lines.

Disadvantages

- To be most effective, the traits of interest should have high heritability.
- Because selection is based on phenotypic values, optimal selection is achieved if it is conducted in a uniform environment.
- Phenotypic uniformity is less than in cultivars produced by pure line selection.
- With dominance, heterozygotes are indistinguishable from homozygous dominant genotypes. Without progeny testing, the selected heterozygotes will segregate in the next generation.

16.5.6 Modification

Mass selection may be direct or indirect. Indirect selection will have high success if two traits result from pleiotropy or if the selected trait is a component of the trait targeted for improvement. For example, researchers improved the seed protein or oil by selecting on the basis of density separation of the seed.

16.6 Pure-line selection

The theory of the pure line was developed in 1903 by the Danish botanist Johannsen. Studying seed weight

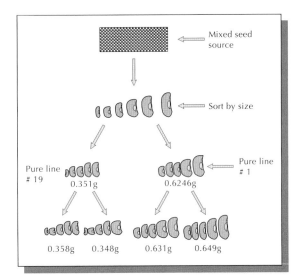

Figure 16.2 The development of the pure line theory by Johannsen.

of beans, he demonstrated that a mixed population of self-pollinated species could be sorted out into genetically pure lines. However, these lines were subsequently non-responsive to selection within each of them (Figure 16.2). Selection is a passive process, since it eliminates variation but does not create it. The pure-line theory may be summarized as follows:

- Lines that are genetically different may be successfully isolated from within a population of mixed genetic types.
- Any variation that occurs within a pure line is not heritable but due to environmental factors only. Consequently, as Johansen's bean study showed, further selection within the line is not effective.

Lines are important to many breeding efforts. They are used as cultivars or as parents in hybrid production (inbred lines). Also, lines are used in the development of genetic stock (containing specific genes such as disease resistance, nutritional quality) and synthetic and multiline cultivars.

16.6.1 Key features

A **pure line** suggests that a cultivar has identical alleles at all loci. Even though plant breeders may make this assumption, it is one that is not practical to achieve in a breeding program. What plant breeders call pure-line cultivars are best aptly called "near" pure-line cultivars because, as researchers such as K.J. Frey observed, high mutation rates occur in such genotypes. Line cultivars have a very narrow genetic base and tend to be uniform in traits of interest (e.g., height, maturity). In case of proprietary dispute, lines are easy to unequivocally identify.

16.6.2 Application

Pure-line breeding is desirable for developing cultivars for certain uses:

- Cultivars for mechanized production that must meet a certain specification for uniform operation by farm machines (e.g., uniform maturity, uniform height for uniform location of economic part).
- Cultivars developed for a discriminating market that puts a premium on eye-appeal (e.g., uniform shape, size).
- Cultivars for the processing market (e.g., with demand for certain canning qualities, texture).
- Advancing "sports" that appear in a population (e.g., a mutant flower for ornamental use).
- Improving newly domesticated crops that have some variability.
- The pure-line selection method is also an integral part of other breeding method,s such as the pedigree selection and bulk population selection.

16.6.3 Procedure

Overview

The **pure-line** selection in breeding entails repeated cycles of selfing following the initial selection from a mixture of homozygous lines. Natural populations of self-pollinated species consist of mixtures of homozygous lines with transient heterozygosity originating from mutations and outcrossing.

Steps

- **Year 1.** The first step is to obtain a variable base population (e.g., introductions, segregating populations from crosses, land race) and space plant it in the first year, select, and harvest desirable individuals (Figure 16.3).

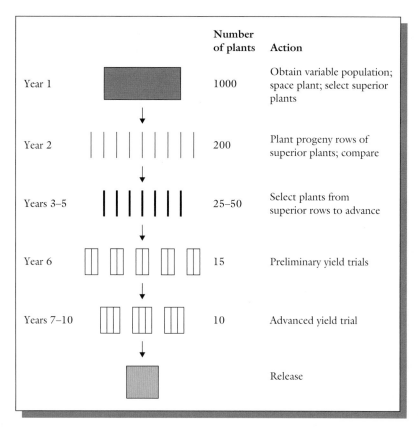

Figure 16.3 Generalized steps in breeding by pure-line selection.

- **Year 2.** Grow progeny rows of selected plants. Rogue out any variants. Harvest selected progenies individually. These are experimental strains.
- **Year 3–6.** Conduct preliminary yield trials of the experimental strains including appropriate check cultivars.
- **Year 7–10.** Conduct advanced yield trials at multi-locations. Release highest yielding line as new cultivar.

16.6.4 Genetic issues

Pure-line breeding produces cultivars with a narrow genetic base and, hence, that are less likely to produce stable yields over a wider range of environments. Such cultivars are more prone to being wiped out by pathogenic outbreaks. Because outcrossing occurs to some extent within most self-pollinated cultivars, coupled with the possibility of spontaneous mutation variants may arise in commercial cultivars over time. It is tempting to select from established cultivars to develop new lines, an action that some view as

unacceptable and unprofessional practice. As previously discussed, pure-line cultivars depend primarily on phenotypic plasticity for production response and stability across environments.

16.6.5 Advantages and disadvantages

There are both major advantages and disadvantages of the application of the pure-line method for improving self-pollinated species.

Advantages

- It is a rapid breeding method.
- The method is inexpensive to conduct. The base population can be a landrace. The population size selected is variable and can be small or large, depending on the objective.
- The cultivar developed by this method has great "eye appeal" (because of the high uniformity of, e.g., harvesting time, height, etc.).

- It is applicable to improving traits of low heritability, because selection is based on progeny performance.
- Mass selection may include some inferior pure lines. In pure line selection, only the best pure line is selected for maximum genetic advance.

Disadvantages

- The purity of the cultivar may be altered through admixture, natural crossing with other cultivars, and mutations. Such off-type plants should be rogued out to maintain cultivar purity.
- The cultivar has a narrow genetic base and, hence, is susceptible to devastation from adverse environmental factors because of uniform response.
- A new genotype is not created. Rather, improvement is limited to the isolation of the most desirable or best genotype from a mixed population.

- The method promotes genetic erosion because most superior pure lines are identified and multiplied to the exclusion of other genetic variants.
- Progeny rows takes up more resources (time, space, funds).

16.7 Pedigree selection

Pedigree selection is a widely used method of breeding self-pollinated species (and even cross-pollinated species such as corn and other crops produced as hybrids). A key difference between pedigree selection and mass selection or pure-line selection is that hybridization is used to generate variability (for the base population), unlike the other methods in which production of genetic variation is not a feature. The method was first described by H.H. Lowe in 1927.

Industry highlights

Utilizing a dihaploid-gamete selection strategy for tall fescue development

Bryan Kindiger

USDA-ARS, Grazinglands Research Laboratory, El Reno, OK 73036, USA

Introduction

Gamete selection as originally defined by Stadler (1944) is based on the principle that selection exerted at the gametophytic level can increase desirable allelic frequencies detectable at the sporophytic level. If superior gametes can be recognized with certainty through a selection cycle, then such a system would be theoretically more efficient than one based on zygotic selection (Richey, 1947). In practice, gamete selection ordinarily involves two steps: (i) selection on the basis of outcross performance testing of individual plants of a variety or population; and (ii) a similar controlled selection for outstanding individuals exhibiting desirable agronomic attributes. Following the identification of superior genotypes, such individuals would undergo continued selfing, followed by phenotypic selection, to generate a homozygous line fixed for the desired agronomic characteristics. In instances where haploids can be generated through microspore culture, followed by genome doubling, or some other method to induce homozygosity, homozygous or dihaploid lines will result.

Dihaploid (DH) selection methods can result in more rapid and efficient gains than other forms of selection (Niroula and Bimb, 2009). Briefly, when utilizing a DH approach, in a diploid organism, only two types of genotypes occur for a pair of alleles, A and a, with the frequency of $\frac{1}{2}$ AA and $\frac{1}{2}$ aa. When utilizing a selfing or backcross approach, three genotypes occur with the frequency of $\frac{1}{4}$ AA, $\frac{1}{2}$ Aa, $\frac{1}{4}$ aa. Therefore, if AA represents a desirable genotype, the probability of obtaining the desirable AA genotype is higher when utilizing a DH approach. Also, if 'n' advantageous loci are segregating in a population, the probability of obtaining that desirable genotype is (1/2)n when employing a DH approach, and (1/4)n when utilizing selfing or backcrossing methods. As a consequence, the efficiency of a DH selection approach will be higher when the number of genes governing a trait are quantitative in their inheritance and expression (Kotch *et al.*, 1992). When applied to a polyploid species such as tall fescue, the gain in efficiency is more profound.

Stadler (1944) is among the earliest who reviewed the potential application of gamete selection in corn breeding. More recently, the importance of a gamete form of selection, utilizing early generation selection, has demonstrated successful simultaneous selection of multiple traits, including effective selection for quantitative trait loci governing characteristics such as seed yield, maturity, and tolerance to disease (Singh, 1994; Ravikumar and Patil, 2004). In these instances, gamete selection has emerged as a superior and efficient form of selection and the feasibility of this approach appeared to be a promising method for facilitating the incorporation of desirable alleles within a shortened breeding period (Singh, 1994). Through the utilization of a paternal monoploid or dihaploid generation process, gamete selection is a proven and efficient method of selection for several species (Stadler 1944; Fehr, 1984; Snape et al., 1986; Schon et al., 1990; Lu, et al., 1996; Rotarenco and Chalyk, 2000). Though DH breeding methods have been suggested and occasionally employed as a means to develop new festulolium lines within the *Festuca-Lolium* complex of polyploid grasses (Guo and Yamada, 2004; Guo et al., 2005), there have been few commercially successful cultivar releases. In addition, the time consuming and genotypic specific method of microspore culture were employed to generate the new festulolium's (Humphreys et al., 2003).

Previous research utilizing homozygous tall fescue derivatives, generated through a traditional program of selfing, were used to study the inheritance of palatability. These studies suggested that selection within homozygous lines can represent a useful methodology for the improvement of tall fescue (Henson and Buckner, 1957; Buckner and Fergus, 1960). As a consequence, when a gamete selection approach can be applied in a breeding program, the generation of DH tall fescue lines should result in the development of efficient breeding advances and superior germplasm (Bouchez and Gallais, 2000). Additional research also indicates that genetic mapping studies focused on the elucidation of a variety of complex genetic traits can be more readily accomplished through the use of homozygous lines (Riera-Lizarazu et al., 2010).

Sporophytically expressed traits are transmitted as genetic information through the gametes (sperm or egg nuclei) and contain half the information that is contained in the sporophytic tissue. In some instances, genetic overlap, a situation where gametophytic genes can regulate both gamete and sporophytic traits can occur. In the presented gamete selection approach, a slight modification of the methodology presented by Stadler and his contemporaries is utilized. Employing the method of gamete selection considered in this text, traits transmitted by the gamete are under the control of genes that are expressed specifically in the sporophyte. A single gamete from the tall fescue parent fertilizes the egg of an inducer line (IL) which generates an F_1 hybrid. Within the F_1, genome loss and a parthenogenic response give rise to DH recoveries. Essentially, this protocol constitutes a gamete-based dihaploid technology that greatly enhances the production of complete homozygous dihaploid lines in a single generation that increases the precision and efficacy of selection in tall fescue breeding and genetic studies. This approach for generating DH is more efficient for producing recovered dihaploid lines than typical microspore culture approaches. It is also simple, less genotype dependent, less labor intensive and less time consuming than other breeding approaches. The discussed approach represents a dihaploid inducement technology that is similar to the wheat × maize DH inducer technology (Niroula and Bimb, 2009). In this approach, embryo culture or rescue is not necessary and selection is based on the F_1 hybrid, prior to DH generation. In the described method, selection applied upon the F_1 sporophyte and the tall fescue genome that it possesses constitutes a form of gamete selection. DH individuals derived from these events are generally recovered ryegrass, tall fescue or varied intermediate genotypes of ryegrass–fescue (e.g. festuloliums). A few amphidiploid selections are also generated.

Tall fescue (*Lolium arundinaceum* (Schreb.) S.J. Darbyshire) ($2n = 6x = 42$) represents the predominant perennial cool-season grass forage in the USA. Its wide adaptation, excellent spring, summer and fall production, deep root system, tolerance to heat and persistence over summer conditions make this a highly desirable species for hay, pasture and turf. Tall fescue tolerates short-term flooding, moderate drought and heavy livestock grazing and machinery traffic. It responds well to fertilizer but can maintain itself under limited fertility conditions and is adapted to moderately acid and wet soils (Jennings et al., 2008). Gamete selection, as applied here, will be more efficient and effective in developing superior tall fescue cultivars than traditional recurrent selection techniques.

Methods

The USDA-ARS has recently released two ryegrass (*Lolium perenne* L. subsp. *multiflorum* (Lam.) Husnot (syn. *Lolium multiflorum* Lam.) ($2n = 2x = 14$) genetic stocks, identified as IL1 and IL2. Each is characterized by a genome loss phenomenon following hybridization with tall fescue (*Lolium arundinaceum* (Schreb.)) S.J. Darbyshire (syn. *Festuca arundinacea* Schreb.) ($2n = 6x = 42$), then followed by a low level of parthenogenic development. The application of the gamete selection breeding approach using these lines is relatively straightforward and efficient. The IL1 and IL2 genetic stocks exhibit few advantageous agronomic characteristics and are essentially notable only for their ability to induce chromosome or genome loss following hybridization. Both lines are free of the fungal endophytes *Epichloë* sp. or *Neotyphodium* sp. (Carroll, 1988; Moon et al., 1994; Pederson and Sleper, 1988).

Pollinations are generated by hand or in field or greenhouse isolations utilizing the IL genetic stock as the maternal parent. Following germination of the hybrid seed, numerous ryegrass-tall fescue F_1 hybrids are generated, with each F_1 hybrid being derived from a single tall fescue pollen grain sperm nucleus fertilizing the IL egg. Hybrids are generally

IL (2n = 2x = 14) × Tall Fescue (2n = 6x = 42) → F1 (7R + 21TF)

Types of Individuals Recovered from the F$_1$:

- Recovered dihaploid ryegrass with 14 chromosomes
- Recovered dihaploid tall fescue with 42 chromosomes
- Recovered amphidiploids (possessing 14 ryegrass + 42 tall fescue chromosomes)
- Recovered 28 chromosomes (the original F$_1$ plants, 7R + 21 TF chromosomes)
- Recovered 35 chromosomes individuals (possess balanced but uncharacterized doses of the ryegrass and tall fescue genomes)
- Aneuploids (individuals with infinite combinations of unbalanced chromosome numbers. Generally die as seedlings)
- Possible festulolium with varying degrees of ryegrass/tall fescue introgression

Figure B16.1 Critique of various outcomes from IL × tall fescue F$_1$ hybrids. Figure courtesy of Bryan Kindiger.

sterile; however, a low incidence of fertility can occasionally occur (Buckner, 1960; Buckner *et al.*, 1961). Due to chromosome or genome loss, F$_1$ hybrids will occasionally generate ryegrass or tall fescue sectors and exhibit chimeras. Possible outcomes of an IL × tall fescue hybridization are illustrated in Figure B16.1. Each type can be identified through phenotypic evaluations, chromosome counts or molecular marker analysis. The DH ryegrass and DH tall fescue recoveries represent the most abundant types (>90%).

The generated F$_1$ (IL × tall fescue) hybrid seed can be sown in low density space planting nurseries or grown in the greenhouse where various induced or natural selection pressures can be applied to the hybrid individuals to identify superior genotypes (Figure B16.2). Multiple locations are desirable to apply an appropriate level of selection on a plant weakness, such as stress or rust tolerance, in the region where the germplasm is to be eventually released. If disease tolerance is a selection criterion, then the hybrids should be grown in a region where the specific disease under study is prevalent. Once exceptional hybrid individuals are identified, the F$_1$s can be transferred to the greenhouse to exclude any chance of cross-pollination with any tall fescue pollen in the field. If an abundance of tall fescue pollen is not a problem in the nursery, then the hybrids can remain in the nursery for a future harvest of each selected plant inflorescence.

Figure B16.2 F$_1$ evaluations for rust susceptibility at the Barenbrug Research Center, Albany, OR, USA. Figure courtesy of Bryan Kindiger.

Hybrid individuals that do not possess a viable or tolerant genotypic contribution from the paternal tall fescue parent are culled from the nursery. If multiple years of selection are to be performed, seed heads are not to be retained until the final year of selection. The surviving hybrids are allowed to flower; then inflorescences are gathered at maturity, broken up by hand or machine, and sown to trays. A light cleaning is applied to remove stems. The cleaned seed heads are then placed in germination trays for identification and selection of either recovered ryegrass or tall fescue seedlings. Typically, following two weeks of germination, a few seedlings will appear and are allowed to grow to appropriate size for transplanting. The germinating seedlings will generally be ryegrass recoveries possessing a chromosome number of 2n = 2x = 14, tall fescue recoveries possessing a chromosome number of 2n = 6x = 42, or various tall fescue DH recoveries with ryegrass introgression or ryegrass recoveries with tall fescue introgression. Occasional amphidiploids (2n = 8x = 56) possessing full ryegrass (2n = 2x = 14) and full tall fescue (2n = 6x = 42) complements are generated. Since the seedlings obtained from the sterile F$_1$ hybrids are generated through parthenogenic development following spontaneous chromosome doubling, each recovery will

Figure B16.3 Uniformity growout trials of offspring generated by selfing DH lines at the Barenbrug Research Center. Figure courtesy of Bryan Kindiger.

possess a full genome contribution of either the ryegrass or tall fescue parent. Genome loss followed by spontaneous doubling and subsequent parthenogenic development represent the important and essential contributions of the IL1 and IL2 lines. All recovered lines derived from this process will have all genes, alleles or quantitative loci conferring a trait fixed in the DH recovery (Kindiger and Singh, 2011; Kindiger 2012). Each recovered individual will be free of the *Lolium* sp. fungal endophyte since the IL1 and IL2 lines do not possess an endophyte. Recovered tall fescue DH lines can then be evaluated under additional selection schemes and, eventually, inter-crossed to perhaps generate a synthetic possessing various advantageous attributes, or be utilized as breeding lines for the development of cultivars. Seed increase of superior DH lines can be accomplished by selfing. Since tall fescue is generally an out-crossing species, varying levels of self-incompatibility are observed; however, appropriate selection for recovered DH lines having low or no levels of self-incompatibility can be identified and seed generated (Figure B16.3).

Though IL1 and IL2 can be utilized to generate DH lines, the frequency of generation is low, less than 1%. However, the ability to generate IL × tall fescue hybrids is rapid and inflorescences on the F_1 hybrids are abundant. When numerous F_1 inflorescences are harvested from the F_1, the recovery of DH lines is quite efficient. Depending on the quality and number of inflorescences, it is not unusual to obtain 1–8 seedlings from each IL × tall fescue hybrid. When multiple hybrids are screened and placed in a commercial DH production situation, hundreds of DH recoveries may be obtained each season. What the approach may lack in DH generation frequency is compensated for by its efficiency in recovering DH lines. The actual mechanism of genome loss and subsequent parthenogenic behavior is unknown. Assumptions based on the occasional generation of chimeral sectors in F_1 individuals (Kindiger, 2012) indicate some form of mitotic nuclear fusion events (Segui-Simarro and Nuez, 2008).

Figure B16.4 Offspring generated by selfing DH recovery R4P14. Seedlings phenotypes such as size, leaf angle, color, and so on are an indicator of uniformity. Figure courtesy of Bryan Kindiger.

There are two methods utilized to verify that the recovered tall fescue materials are DH homozygous recoveries. The first step is to self-pollinate the particular DH individual for seed increase and perform nursery or greenhouse grow-outs (Figures B16.3 and B16.4). If the offspring from the selfing do not segregate for size, inflorescence morphology, maturity or other obvious phenotypic characteristics, they are likely DHs. A second approach is to utilize molecular markers known to exhibit a banding pattern that is consistent with disomic inheritance and codominant expression. Since tall fescue possesses a polyploid genome constitution (Alderson and Sharp, 1994) the species will possess a considerable level of genome duplication. As a consequence, it is important to utilize markers that are known to amplify only a single or specific locus within the tall fescue genome. When such markers are utilized on offspring generated by selfing a presumed DH line, only a single amplification product will be formed at that locus, since that locus would be homozygous (Figure B16.5). Such

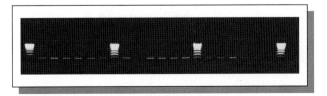

Figure B16.5 Replicated EST-SSR marker NF24 evaluation of six offspring generated by selfing DH recovery R4P15. Size markers are in lanes 1, 7, 14 and 21. Figure courtesy of Bryan Kindiger.

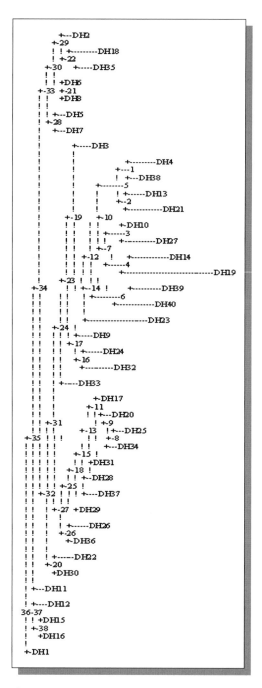

Figure B16.6 A dendogram illustrating the genetic variability or genetic distance of 40 DH tall fescue lines. DH lines with greater distances between branches or within branches represent a visual indicator of genetic distance. Figure courtesy of Bryan Kindiger.

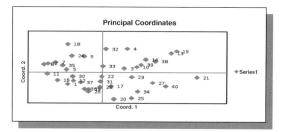

Figure B16.7 Principal coordinate analysis of forty DH lines evaluated across 40 EST-SSR markers. Figure courtesy of Bryan Kindiger.

markers will be most useful since segregation will be predictable and these markers will be easy to score (Stift, *et al.*, 2008). Once DH lines are confirmed, they can be utilized in the development of new wide- or narrow-based cultivars or even hybrids. Selection of superior DH lines exhibiting superior trait qualities can be followed by the performance of a genetic distance analysis (GDA) through the use of molecular markers. DNA markers provide a powerful tool for genetic evaluation and marker-assisted breeding of crops and, especially, for cultivar identification. The greater the GDA, the greater the potential for advantageous genome interactions and an advantageous heterotic response. Recent studies have suggested that advances can be made in yield performance of forages, including tall fescue, through the use of molecular markers (Brummer and Casler, 2009; Amini *et al.*, 2011). A preliminary analysis of forty DH lines is provided in a GDA dendogram tree and subjected to a principal component analys based on 40 EST-SSR markers (Figures B16.6 and B16.7). The observed differences across the DH lines can be utilized to combine agronomically superior DH lines to maximize the potential for heterosis. The squared-euclidean distance (horizontal axis) between objects was used as the genetic distance measure and the dendogram tree clustering was performed following the method of Ward (1963).

Conclusions

Dihaploid breeding methods provide an alternative approach to the traditional backcross, selfing or recurrent selection breeding programs since they can greatly simplify gene or trait inheritance ratios and intralocus effects (Maluszynski *et al.*, 2003). The generation of haploid or dihaploid lines through a gamete selection approach represents an under-utilized but important breeding approach for plant breeding and genetic analysis research (Dunwell, 2010). The materials and procedures described apply directly to the breeding and selection of superior tall fescue germplasm. However, they could be expanded to ryegrass or additional fescue species. The sampling of hundreds of thousands of gametes via the pollen grains, each segregating for a myriad of genotypes from a single tall fescue individual or population, requires less labor input and represents a low cost, rapid selection strategy that can be implemented across a diversity of environments. More importantly, the breeder applying this technique to generate tall fescue DH recoveries will only rely on the potential genotypic contributions of the paternal parent. The simultaneous selection of multiple traits or complementary traits can be applied quite effectively and the prescribed approach does not require any prior genetic information regarding the inheritance or expression of the quantitative trait. The breeder can essentially "skim the cream" from these germplasm resources when applying this method to improved or previously selected populations or cultivars. The retention and fixation of the qualitative or quantitative traits of interest are easily maintained by selfing the DH line. It is anticipated that, when applied correctly, this approach will be effective for selection of both qualitative and quantitative traits such as maturity, drought tolerance, disease resistance, grazing persistence, forage yield and superior forage quality.

References

Alderson, J., and Sharp, W.C. (1994). *Grass varieties in the United States*. Agricultural Handbook 170. Soil Conservation Service, USDA, Washington, DC.

Amini, F., Mirlohi, A., Majidi, M.M., Far, S.S., and Killiker, R. (2011). Improved polycross breeding of tall fescue through marker-based parental selection. *Plant Breeding*, **130**:701–707.

Bouchez, A., and Gallais, A. (2000). Efficiency of the use of doubled-haploids in recurrent selection for combining ability. *Crop Science*, **40**(1):23–29.

Brummer, E.C., and Casler, M.D. (2009). Improving selection for forage, turf and biomass crops using molecular markers, in *Molecular Breeding of Forage and Turf, The proceedings of the 5th International Symposium on the Molecular Breeding of Forage and Turf* (eds T. Yamada, and G. Spangenberg) Springer, New York, NY, pp. 193–209.

Buckner, R.C. (1960). Cross compatibility of annual and perennial ryegrass with tall fescue. *Agronomy Journal*, **52**:409–410.

Buckner, R.C., and Fergus, E.N. (1960). Improvement of tall fescue for palatability by selection within inbred lines. *Agronomy Journal*, **52**:173–176.

Buckner, R.C., Hill, H.D., and Burrus, P.B. (1961). Some characteristics of perennial and annual ryegrass × tall fescue hybrids and of the amphidiploid progenies of annual ryegrass × tall fescue. *Crop Science*, **1**:75–80.

Carroll, G.C. (1988). Fungal endophytes in stems and leaves: From latent pathogen to mutualistic symbiont. *Ecology*, **69**:2–9.

Dunwell, J.W. (2010). Haploids in flowering plants: Origins and exploitation. *Plant Biotechnology Journal*, **8**:377–424.

Fehr, W. (1984). Homozygous lines from double haploids, in *Principles of cultivar development*, Vol. **1**. Macmillan Publishing Company, New York, pp. 337–358.

Guo, Y., and Yamada, T. (2004). High frequency androgenic embryogenesis and genome size variation in pollen-derived Festulolium, in *Development of a novel grass with environmental stress tolerance and high forage quality through intergeneric hybridization between Lolium and Festuca* (eds T. Yamada, and T. Takamizo) Natl. Agric. Bio-oriented Res. Org. (NARO), Ibaraki, Japan, pp. 49–54.

Guo, Y., Mizukami, Y., and Yamada, T. (2005). Genetic characterization of androgenic progeny derived from *Lolium perenne* × *Festuca pratensis* cultivars. *New Phytologist*, **166**:455–464.

Henson, L., and Buckner, R.C. (1957). Resistance to *Helminthosporium dictyoides* in inbred lines of *Festuca arundinacea*. *Phytopathology*, **47**:523.

Humphreys, M.W., Canter, P.J., and Thomas, H.M. (2003). Advances in introgression technologies for precision breeding within the *Lolium-Festuca* complex. *Annals of Applied Biology*, **143**:1–10.

Jennings, J.A., Beck, P., and West, C. (2008). Tall fescue for forage. University of Arkansas Extension Bulletin, Dept. of Agricultural and Natural Resources. FSA 2133-PD-1-08RV, University of Arkansas, Fayetteville, AR.

Kindiger, B., and Singh, D. (2011). Registration of Annual Ryegrass Genetic Stock IL2. *Journal of Plant Registrations*, **5**:254–256.

Kindiger, B. (2012). Notification of the Release of Annual Ryegrass Genetic Stock IL1. *Journal of Plant Registrations*, **6**:117–120.

Kotch, G.P., Ortiz, R., Peloquin, S.J. (1992). Genetic analysis by use of potato haploid populations. *Genome*, **35**:103–108.

Lu, C., Shen, L., Tan, Z., *et al.* (1996). Comparative mapping of QTL for agronomic traits in rice across environments using a doubled haploid population. *TAG*, **93**:1211–1217.

Maluszynski, M. Kasha, K. J., Forster, B.P., and Szarejko, I. (2003). *Doubled haploid production in crop plants: A manual.* Kluwer Academic Publications, Dordrecht, The Netherlands.

Moon, C.D., Tapper, B.A., and Scott, B. (1994). Identification of Epichloë Endophytes in planta by a microsatellite-based PCR fingerprinting assay with automated analysis. *Applied and Environmental Microbiology*, **65**:1268–1279.

Niroula, R.K., and Bimb. (2009). Overview of Wheat × Maize System of Crosses for Dihaploid Induction in Wheat. *World Applied Sciences Journal*, **7**:1037–1045.

Pedersen, J.F., and Sleper, D.A. (1988). Considerations in breeding endophyte-free tall fescue forage cultivars. *Journal of Production Agriculture*, **1**:127–132.

Ravikumar, R.L., and Patil, B.S. (2004). Effect of gamete selection on segregation of wilt susceptibility-linked DNA marker in chickpea. *Cur. Sci.*, **86**:642–643.

Richey, F.D. (1947). Corn breeding: Gamete selection, the Oenothera method and related miscellany. *Journal of the American Society of Agronomy*, **39**:403–411.

Riera-Lizarazu, O., Peterson, C.J., and Leonard, J.M. (2010). Registration of the OS9XQ36 mapping population of wheat (*Triticum aestivum* L.). *Journal of Plant Registrations*, **4**:98–102.

Rotarenco, V.A., and Chalyk, S.T. (2000). Selection at the level of haploid sporophyte and its influence on the traits of diploid plants in maize. *Genetika*, **32**:479–485.

Schon, C., Sanchez, M., Blake, T., and Hayes, P.M. (1990). Segregation of Mendelian markers in doubled haploid and F2 progeny of barley cross. *Hereditas*, **113**:69–72.

Seguí-Simarro, J.M., and Nuez, F. (2008). Pathways to doubled haploidy: chromosome doubling during androgenesis. *Cytogenetic and Genome Research*, **120**:358–369.

Singh, S.P. (1994). Gamete selection for simultaneous improvement of multiple traits in common bean. *Crop Science*, **34**:352–355.

Snape, J.W., Simpson, E., and Parker, B.B. (1986). Criteria for the selection and use of dihaploid systems in cereal breeding programmes, in *Genetic manipulation in plant breeding* (eds W. Horn, C.J. Jensen, W. Odenbach, and O. Schieder) Walter de Gruyter, Berlin, pp. 217–229.

Stadler, L.J. (1944). Gamete selection in corn breeding. *Journal of the American Society of Agronomy*, **36**:988–989.

Stift, M., Berenos, C., Kuperus, P., and van Tienderen, P. (2008). Segregation models for disomic, tetrasomic and intermediate inheritance in tetraploids: A general procedure applied to Rorippa (Yellow Cress) microsatellite data. *Genetics*, **179**:2113–2123.

Ward, J.H. (1963). Hierarchical grouping to optimize an objective function. *Journal of the American Statistical Association*, **58**:236–244.

16.7.1 Key features

Pedigree selection is a breeding method in which the breeder keeps records of the ancestry of the cultivar. The base population, of necessity, is established by crossing selected parents, followed by handling an actively segregating population. Documentation of the pedigree enables breeders to trace parent-progeny back to an individual F_2 plant from any subsequent generation. To be successful, the breeder should be able to distinguish between desirable and undesirable plants on the basis of a single plant phenotype in a segregating population. It is a method of continuous individual selection after hybridization. Once selected, plants are reselected in each subsequent generation. This process is continued until a desirable level of homozygosity is attained. At that stage, plants appear phenotypically homogeneous.

The breeder should develop an effective, easy to maintain system of record keeping. The most basic form is based on numbering of plants as they are selected, and developing an extension to indicate subsequent selections. For example, if five crosses are made and 750 plants are selected in the F_2 (or list the first selection generation), a family could be designated 5-175 (meaning it was derived from plant 175 selected from cross number five). If selection is subsequently made from this family, it can be named, for example, 5-175-10. Some breeders include letters to indicate the parental sources or the kind of crop (e.g., NP-5-175-10) or some other useful information. The key is to keep it simple, manageable, and informative.

16.7.2 Application

Pedigree selection is applicable to breeding species that allow individual plants to be observed, described, and harvested separately. It has been used to breed species including peanuts, tobacco, tomato, and some cereals, especially where readily identifiable qualitative traits are targeted for improvement.

16.7.3 General guides to selection following a cross

The success of breeding methods preceded by hybridization rest primarily on the parents used to initiate the breeding program. Each generation has genetic characteristic and is handled differently in a breeding program.

F_1 generations

Unless in hybrid seed programs in which the F_1 is the commercial product, the purpose of the F_1 is to grow sufficient F_2 population for selection. To achieve this, F_1 seed is usually space-planted for maximum seed production. It is critical also to be able to authenticate hybridity and identity and remove seeds from self-pollination. Whenever possible plant breeders use genetic markers in crossing programs.

F_2 generation

Selection in the plant breeding program often starts in the F_2, the generation with the maximum genetic variation. The rate of segregation is higher if the parents differ by a larger number of genes. Generally, a large F_2 population is planted (2000–5000). Of the genotypes in the F_2 50% are heterozygous and, hence, selection intensity should be moderate (about 10%) in order to select plants that would likely include those with the desired gene combinations. The actual number of plants selected depends on the trait (its heritability) and resources. Traits with high heritability are more effectively selected, requiring lower numbers than for traits with low heritability. The F_2 is also usually space-planted to allow individual plants to be evaluated for selection. In pedigree selection, each F_2 selected plant is documented.

F_3 generation

Seed from individual plants are progeny-rowed. This allows homozygous and heterozygous genotypes to be distinguished. The homozygosity in the F_3 is 50% less than in the F_2. The heterozygotes will segregate in the rows. The F_3 generation is the beginning of line formation. It is helpful to include check cultivars in the planting to help in selecting superior plants.

F_4 generation

F_3 plants are grown in plant-to-row fashion as in the F_3 generation. The progenies become more homogeneous (homozygosity is 87.5%). Lines are formed in the F_4. Consequently, selection in the F_4 should focus more on progenies rather than individuals plants.

F_5 generation

Lines selected in the F_4 are grown in preliminary yield trials (PYT). F_5 plants are 93.8% homozygous. These are replicated trials with at least two replications (depending on the amount of seed available). The seeding rate is the commercial rate (or as close as possible), receiving all the customary cultural inputs. Evaluation of quality traits and disease resistance can be included. The PYT should include check cultivars. Best performing lines are selected for advancing to the next stage in the breeding program.

F_6 generation

The superior lines from F_5 are further evaluated in competitive yield trials or advanced yield trials (AYT), including a check.

F_7 and subsequent generations

Superior lines from F_6 are evaluated in AYT for several years, at different locations, and in different seasons as desirable. Eventually, after F_8, the most outstanding entry is released as commercial cultivar.

16.7.4 Procedure

Overview

The key steps in the pedigree selection procedure are:

1 Establish a base population by making a cross of selected parents.
2 Space-plant progenies of selected plants.
3 Keep accurate records of selection from one generation to the next.

Steps

- **Year 1.** Identify desirable homozygous parents and make about 20–200 crosses (Figure 16.4).
- **Year 2.** Grow 50–100 F_1 plants including parents for comparison to authenticate its hybridity.
- **Year 3.** Grow about 1000–2000 F_2 plants. Space-plant to allow individual plants to be examined and documented. Include check cultivars for comparison. Desirable plants are selected and harvested separately, keeping records of their identities. In some cases, it may be advantageous not to space plant F_2s to encourage competition among plants.
- **Year 4.** Seed from superior plants are progeny-rowed in F_3 to F_5 generations, making sure to space-plant in the rows for easy record keeping. Selection at this stage is both within and between

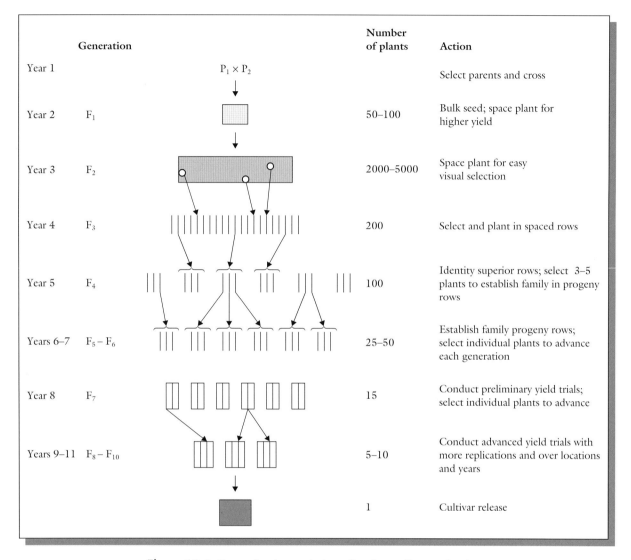

Figure 16.4 Generalized steps in breeding by pedigree selection.

rows by first identifying superior rows and selecting 3–5 plants from each progeny to plant the next generation.

- **Year 5.** By the end of the F_4 generation, there would be between 25 and 50 rows with records of the plant and row. Grow progeny of each selected F_3.
- **Year 6.** Family rows are planted in the F_6 to produce experimental lines for preliminary yield trials in the F_7. The benchmark or check variety is a locally adapted cultivar. Several checks may be included in the trial.
- **Year 7.** Advanced yield trials over locations, regions, and years are conducted in the F_8 to F_{10}

generations, advancing only superior experimental material to the next generation. Ultimately, the goal is to identify one or two lines that are superior to the check cultivars for release as a new cultivar. Consequently, evaluations at the advanced stages of the trial should include superior expression of traits that are deemed to be of agronomic importance for successful production of the particular crop (e.g., lodging resistance, shattering resistance, disease resistance). If a superior line is identified for release, it is put through the customary cultivar release process (i.e., seed increase and certification).

Comments

- Growing parents, making a cross, and growing F_1 plants may take 1–2 years, depending on the facilities available for growing multiple experiments in a year (e.g., greenhouse) and the growing period of the crop.
- The number of plants selected in the F_2 depends on resources available (labor, space, time), and may be even 10 000 plants.
- F_3 family rows should contain large enough number of plants (25–30) to permit the true family features to be evident so the most desirable plant(s) can be selected. Families that are distinctly inferior should be discarded, while more than one plant may be selected from exceptional families. However, generally, the number of plants advanced does not exceed the number of F_3 families.
- From F_3–F_5, selection is conducted between and within rows, identifying superior rows and selecting 3–5 of the best plants in each family. By F_5, only about 25–50 families are retained.
- By the F_5, plant density may reflect the commercial seeding rate. Further, the plants from this generation that are going forth would be significantly homozygous to warrant conducting preliminary and later advanced yield trials.

16.7.5 Genetic issues

Detailed records are kept from one generation to the next regarding parentage and other characteristics of plants. The method allows the breeder to create genetic variability during the process. Consequently, the breeder can influence the genetic variation available by the choice of parents. The method is more conducive for breeding qualitative disease resistance, than for quantitative resistance. The product (cultivar) is genetically relatively narrow-based but not as extremely so as in pure-line selection. The records help the breeder to advance only progeny lines with plants that exhibit genes for desired traits occur.

16.7.6 Advantages and disadvantages

The pedigree method of breeding has advantages and disadvantages.

Advantages

- Record keeping provides a catalog of genetic information of the cultivar unavailable from other methods.
- Selection is based not only on phenotype but also on genotype (progeny row) making it an effective method for selecting superior lines from among segregating.
- Using the records, the breeder is able to advance only the progeny lines in which plants that carry the genes for the target traits occur.
- A high degree of genetic purity is produced in the cultivar, an advantage where such property is desirable (e.g., certification of products for certain markets).

Disadvantages

- Record keeping is slow, tedious, time consuming, and expensive. It places pressure on resources (e.g., land for space-planting for easy observation). Seeding and harvesting are tedious operations. However, modern research plot equipment for planting and harvesting are versatile and sophisticated to allow complex operations and record taking to be conducted, making pedigree selection easier to implement and hence widely used. Large plant populations can now be handled without much difficulty.
- The method is not suitable for species in which individual plants are difficult to isolate and characterize.
- Pedigree selection is a long procedure, requiring about 10–12 years or more to complete, if only one growing season is possible.
- The method is more suited for qualitative than for quantitative disease resistance breeding. It is not effective for accumulating the number of minor genes needed to provide horizontal resistance.
- Selecting in the F_2 (early generation testing) on the basis of quantitative traits such as yield may not be effective. It is more efficient to select among F_3 lines planted in rows than selecting based on individual plants in the F_2.

16.7.7 Modifications

The pedigree selection method is a continuous selection of individuals after hybridization. A discontinuous method has been proposed but is not considered practical enough for wide adoption. The breeder may modify the pedigree method to suit specific objectives and resources. Some specific ways are as:

- The numbers of plants to select at each step may be modified according to the species, the breeding objective, and the genetics of the traits of interest, as well as the experience of the breeder with the crop, and resources available for the project.

- The details of records kept are at the discretion of the breeder.
- Off-season planting (e.g., winter nurseries), use of the greenhouse, and multiple plantings a year (where possible) are ways of speeding up the breeding process.
- Early generation selection for yield in pedigree selection is not effective. This is a major objection to the procedure. Consequently, several modifications have been introduced by breeders to delay selection till later generations (e.g., F_5). Mass selection or bulk selection is practiced in the early generation.

16.8 Bulk population breeding

Bulk population breeding is a strategy of crop improvement in which natural selection effect is solicited more directly in the early generations of the procedure by delaying stringent artificial selection until later generations. The Swede, H. Nilsson-Ehle, developed the procedure. H.V. Harlan and colleagues provided additional theoretical foundation for this method through their work in barley breeding in 1940s. As proposed by Harlan and colleagues, the bulk method entails yield testing of the F_2 bulk progenies from crosses and discarding whole crosses based on yield performance. In other words, the primary objective is to stratify crosses for selection of parents based on yield values. The current application of the bulk method has a different objective.

16.8.1 Key features

The rationale for delaying artificial selection is to allow natural selection pressure (e.g., abiotic factors such as drought, cold) to eliminate or reduce the productivity of less fit genotypes in the population. Just like the pedigree method, the bulk method also applies pure line theory to segregating populations to develop pure line cultivars. Genetic recombination in the heterozygous state cannot be used in self-pollinated species because self-pollination progressively increases homozygosity. By F_6 the homozygosity is about 98.9%. The strategy in plant breeding is to delay selection until at the high level of homozygosity.

16.8.2 Applications

It is a procedure used primarily for breeding self-pollinated species, but can be adapted to produce inbred populations for cross-pollinated species. It is most suitable for breeding species that are normally closely spaced in production (e.g., small grains – wheat, barley). It is used for field beans and soybeans. However, it is not suitable for improving fruit crops and many vegetables in which competitive ability is not desirable.

16.8.3 Procedure

Overview

After making a cross, several hundreds to several thousands of F_2 selections are planted at a predetermined (usually conventional rate) close spacing. The whole plot is bulk harvested. A sample of seed is used to plant another field block for the next selection, subjecting it to natural selection pressure through the next 2–3 generations. In the F_5, the plants are space-planted to allow individual plant evaluation for effective selection. Preliminary yield trials may start in the F_7 followed by advanced yield trials, leading to cultivar release.

Steps

- **Year 1.** Identify desirable parents (cultivars, single crosses, etc.) and make sufficient number of crosses between them (Figure 16.5).
- **Year 2.** Following a cross between appropriate parents, about 50–100 F_1 plants are planted and harvested as a bulk, after rouging out selfs.
- **Year 3.** The seeds from the second year are used to plant a bulk plot of about 2000–3000 F_2 plants. The F_2 is bulk harvested.
- **Year 4–6.** A sample of the F_2 seed is planted in bulk plots, repeating the steps for year 2 and year 3 until about F_4 or when a desired level of homozygosity has been attained in the population. Space plant about 3000–5000 F_5 plants and select about 10% (300–500) superior plants for planting F_6 progeny rows.
- **Year 7.** Select and harvest about 10% (30–50) progeny rows that exhibit genes for desired traits for planting preliminary yield trails in the F_7.
- **Year 8 and later.** Conduct advanced yield trials from F_8 through F_{10} at multiple locations and

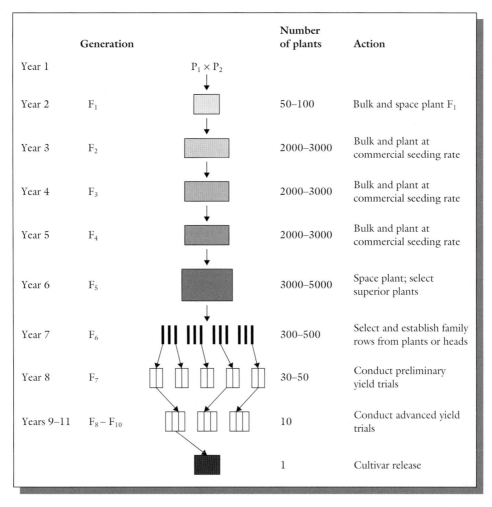

Figure 16.5 Generalized steps in breeding by bulk selection.

regions, including adapted cultivars as checks. After identifying a superior line, it is put through the customary cultivar release process.

Comments

- Space planting of the F_1 will increase the yield of F_2 seed.
- The breeder may screen the bulk population under different natural environments in a rotation (e.g., soil condition – salinity, acidity; disease resistance; temperature – winter kill, etc.). There may be an increase in broad adaptation of the cultivar. However, care should be exercised to avoid the evaluation of plants under a condition that could eliminate genotypes that are of value at different sets of environmental conditions.

- Screening for photoperiodic response is desirable and advantageous in the early stages to eliminate genotypes that are incapable of reproducing under the environmental conditions.
- Natural selection may be aided by artificial selection. Aggressive and highly competitive but undesirable genotypes may be physically rogued out of the population to avoid increasing the frequency of undesirable genes, or to help select benign traits such as seed color or fiber length of cotton. Aiding natural selection also accelerates the breeding program.
- The degree of selection pressure applied, its consistency, duration, and the heritability of traits, are all factors that impact the rate at which unadapted segregates are eliminated from the bulk population.

16.8.4 Genetic issue

Applying the theories of population genetics (Chapter 3), repeated self-pollination and fertilization will result in three key outcomes:

  (i) In advanced generations, the plants will be homozygous at nearly all loci.

  (ii) The mean population performance will be improved as a result of natural selection.

  (iii) Genotypes with good agricultural fitness will be retained in the population.

Bulk selection promotes intergenotypic competition. By allowing natural selection to operate on early generations, the gene frequencies in the population at each generation will depend upon:

- The genetic potential of a genotype for productivity.
- The competitive ability of the genotype.
- The effect of the environment on the expression of a genotype.
- The proportions and kinds of genotypes advanced to the next generation (i.e., sampling). The effects of these factors may change from one generation to the next. More importantly, it is possible that desirable genotypes may be out-competed by more aggressive undesirable genotypes. For example, tall plants may smother short desirable plants. It is not possible to predict which F_2 plant's progeny will be represented in the next generation, nor predict the genetic variability for each character in any generation.

The role of natural selection in bulk breeding is not incontrovertible. It is presumed to play a role in genetic shifts in favor of good competitive types, largely due to high fecundity of competitive types. Such an impact is not hard to accept when traits that confer advantage through resistance to biotic and abiotic stresses are considered. For example, if the bulk population were subjected to various environments (e.g., salinity, cold temperature, water logging, drought, photoperiod), the fecundity may be drastically low for ill-adapted genotypes. These are factors that affect adaptation of plants. Some traits are more neutral in competition (e.g., disease resistance). If two genotypes are in competition, their survival depends on the number of seed produced by each genotype as well as the number of seeds produced by their progeny.

Using the natural relationship developed by W. Allard for illustration, the survival of an inferior genotype may be calculated as:

$$A_n = a \times S^{n-1}$$

where A_n is the proportion of inferior genotypes, n is the generation, a is the initial proportion of the inferior genotype, and S is the selection index. Given two genotypes, A (superior) and B (inferior), in equal proportions in a mixture (50% A: 50% B), and of survival capacities A = 1, B = 0.9, the proportion of the inferior genotype in F_5 would be:

$$
\begin{aligned}
A_5 &= (0.5) \times (0.9)^{5-1} \\
&= 0.3645 \ (\text{or } 36.45\%)
\end{aligned}
$$

This means the inferior genotype would decrease from 50 to 36.45% by F_5. Conversely, the proportion of the superior genotype would increase to 63.55%.

As previously indicated, the bulk selection method promotes intergenotypic competition, it is important to point out that the outcome is not always desirable because a more aggressive inferior genotype may out-compete a superior (desirable) but poor competitor. In a classic study by C.A. Suneson, an equal mixture (25%) of four barley cultivars was followed. After more than five generations, the cultivar Atlas was represented by 88.1%, Club Mariot by 11%, Hero by 1%, while Vaughn was completely eliminated. However, in pure stands, Vaughn out-yielded Atlas. It may also be said that if the genotypes whose frequency in the population increased over generations are the ones of agronomic value (i.e., desired by the breeder), then the competition in bulking is advantageous to plant breeding. The effect of natural selection in bulk population can be positive or negative, and varies according to the traits of interest, the environment under which the population is growing, and the degree of intergenotypic competition (spacing among plants). If there is no competition between plants, genotype frequencies would not be changed significantly. Also, the role of natural selection in genetic shifts would be less important when the duration of the period is less (6–10 generations) as is the case in bulk breeding. This is so because natural selection acts on the heterozygotes in the early generations. However, the goal of bulk breeding is to develop pure lines. By the time this is released, the breeding program would have ended, giving natural selection on time to act on the pure lines.

16.8.5 Advantages and disadvantages

There are key advantages and disadvantages of the bulk breeding method.

Advantages

- It is simple and convenient to conduct.
- It is less labor intensive and less expensive in early generations.
- Natural selection may increase frequency of desirable genotypes by the end of the bulking period.
- It is compatible with mass selection in self-pollinated species.
- Bulk breeding allows large amounts of segregating materials to be handled. Consequently, the breeder can make and evaluate more crosses.
- The cultivar developed would be adapted to the environment, having been derived from material that had gone through years of natural selection.
- Single plant selections are made when plants are more homozygous, making it more effective to evaluate and compare plant performance.

Disadvantages

- Superior genotypes may be lost to natural selection, while undesirable ones are promoted during the early generations.
- It is not suited to species that are widely spaced in normal production.
- Genetic characteristics of the populations are difficult to ascertain from one generation to the next.
- Genotypes are not equally represented in each generation because all plants in one generation are not advanced to the next generation. Improper sampling may lead to genetic drift.
- Selecting in off-season nurseries and the greenhouse may favor genotypes that are undesirable in the production region where the breeding is conducted, and hence is not a recommended practice.
- The procedure is lengthy, but cannot take advantage of off-season planting.

16.8.6 Modifications

Modifications of the classic bulk breeding method include the following:

- The breeder may impose artificial selection sooner (F_3 or F_4) to shift the population toward an agriculturally more desirable type.

- Rouging may be conducted to remove undesirable genotypes prior to bulking.
- The breeder may select the appropriate environment to favor desired genotypes in the population. For example, selecting under disease pressure would eliminate susceptible individuals from the population.
- Preliminary yield trials may be started even while the lines are segregating in the F_3 or F_4.
- The **single seed descent** method may be used at each generation to reduce the chance of genetic drift. Each generation, a single seed is harvested from each plant to grow the next bulk population. The dense planting makes this approach problematic in locating individual plants.
- The **composite cross bulk population** breeding, also called the **evolutionary method of breeding**, was developed by C.A. Suneson and entails systematically crossing a large number of cultivars. First, the pars of parents are crossed, then pairs of F_1s are crossed. This continues until a single hybrid stock containing all parents is produced. The method has potential for crop improvement but it takes a very long time to complete.

16.9 Single seed descent

The method of **single seed descent** was born out of a need to speed up the breeding program by rapidly inbreeding a population prior to beginning individual plant selection and evaluation, while reducing a loss of genotypes during the segregating generations. The concept was first proposed by C.H. Goulden in 1941 when he attained the F_6 generation in two years by reducing the number of generations grown from a plant to one or two, while conducting multiple plantings per year, using the greenhouse and the off season. H.W. Johnson and R.L. Bernard described the procedure of harvesting a single seed per plant for soybean in 1962. However, it was C.A. Brim who, in 1966, provided a formal description of the procedure of single seed descent, calling it a **modified pedigree method.**

16.9.1 Key features

The method allows the breeder to advance the maximum number of F_2 plants through the F_5 generation. This is achieved by advancing one randomly selected seed per plant through the early segregating stages. The focus on the early stages of the procedure is on attaining homozygosity as rapidly as possible, without selection. Discriminating among plants starts after attainment of homozygosity.

16.9.2 Applications

Growing plants in the greenhouse under artificial conditions tends to reduce flower size and increase cleistogamy. Consequently, single seed descent is best for self-pollinated species. It is effective for breeding small grains as well as legumes, especially those that can tolerate close planting and still produce at least one seed per plant. Species that can be forced to mature rapidly are suitable for breeding by this method. It is widely used in soybean breeding to advance the early generation.

16.9.3 Procedures

Overview

A large F_1 population is generated to ensure adequate recombination among parental chromosomes. A single seed per plant is advanced in each subsequent generation until the desired level of inbreeding is attained. Selection is usually not practiced until F_5 or F_6. Then, each plant is used to establish a family to help breeders in selection and to increase seed for subsequent yield trails.

Steps

- **Year 1.** Crossing is used to create the base population. Cross selected parents to generate adequate number of F_1 for the production of a large F_2 population.
- **Year 2.** About 50–100 F_1 plants are grown in a greenhouse ground bench or in pots. They may also be grown in the field. Harvest identical F_1 crosses and bulk.
- **Year 3.** About, 2000–3000 F_2 plants are grown. At maturity, a single seed per plant is harvested and bulked for planting F_3. Subsequently, F_2 is spaced enough to allow each plant to produce only a few seeds.
- **Year 4–6.** Single pods per plant are harvested to plant the F_4. The F_5 is space-planted in the field, harvesting seed from only superior plants to grow progeny rows in the F_6 generation.
- **Year 7.** Superior rows are harvested to grow preliminary yield trails in the F_7.

- **Year 8 and later.** Yield trials are conducted in the F_8 to F_{10} generations. The most superior line is increased in the F_{11} and F_{12} as a new cultivar.

Comments

- If the sample is too small, superior genetic combinations may be lost because only one seed from each plant is used.
- It may be advantageous to use progeny rows prior to yield testing to produce sufficient seed as well as to help in selecting superior families.
- The breeder may choose to impose some artificial selection pressure by excluding undesirable plants from contributing to the subsequent generations (in the early generations). This is effective for qualitative traits.
- Record keeping is minimal and so are other activities such as harvesting, especially in early generations.

16.9.4 Genetic issues

Each individual in the final population is descendent from a different F_2 plant. Each of these plants undergoes a decrease in heterozygosity at a rapid rate, each generation. Barring the inability of a seed to germinate or a plant to set seed, the effect of natural selection is practically non-existent in the single seed descent procedure. Only one seed per plant is advanced, regardless of the number produced. That is, a plant producing one seed is equally represented in the next generation as one producing 1000 seeds. Selection is conducted on homozygous plants rather than segregating material. An efficient early generation testing is needed to avoid genetic drift of desirable alleles. Single seed descent is similar to bulk selection in that the F_6/F_7 comprises a large number of homozygous lines, prior to selection among progenies. A wide genetic diversity is carried on to relatively advanced generations (F_6/F_7).

16.9.5 Advantages and disadvantages

Single seed descent has certain advantages and disadvantages.

Advantages

- It is an easy and rapid way to attain homozygosity (2–3 generations per year).

- Small spaces are required in early generations (e.g., can be conducted in a greenhouse) to grow the selections.
- Natural selection has no effect (hence it can't impose adverse impact).
- The duration of the breeding program can be reduced by several years by using single seed descent.
- Every plant originates from a different F_2 plant, resulting in greater genetic diversity in each generation.
- It is suited to environments that do not represent those in which the ultimate cultivar will be commercially produced (no natural selection imposed).

Disadvantages

- Natural selection has no effect (hence no benefit from its possible positive impact).
- Plants are selected based on individual phenotype not progeny performance.
- Inability of seed to germinate or plant to set seed may prohibit every F_2 plant from being represented in the subsequent population.
- The number of plants in the F_2 is equal to the number of plants in the F_4. Selecting a single seed per plant runs the risks of losing desirable genes. The assumption is that the single seed represents the genetic base of each F_2. This may not be true.

16.9.6 Modifications

The procedure described so far is the classic single seed descent breeding method. There are two main modifications of this basic procedure. The multiple seed procedure (or **modified single seed descent**) entails selecting 2–4 seeds per plant, bulking and splitting the bulk into two, one for planting the next generation, and the other half held as reserve. Because some soybean breeders simply harvest one multi-seeded pod per plant, the procedure is also referred to by some as **bulk-pod method**.

Another modification is the **single hill method** in which progeny from individual plants are maintained as separate lines during the early generations by planting a few seeds in a hill. Seeds are harvested from the hill and planted in another hill the next generation. A plant is harvested from each line when homozygosity is attained.

16.10 Backcross breeding

This application in plants was first proposed by H.V. Harlan and M.N. Pope in 1922. In principle, **backcross breeding** does not improve the genotype of the product, except for the substituted gene(s).

16.10.1 Key features

The rationale of backcross breeding is to replace a specific undesirable gene with a desirable alternative, while preserving all other qualities (adaptation, productivity, etc.) of an adapted cultivar (or breeding line). Instead of inbreeding the F_1 as normally done, it is repeatedly crossed with the desirable parent to retrieve (by "modified inbreeding") the desirable genotype. The adapted and highly desirable parent is called the **recurrent parent** in the crossing program, while the source of the desirable gene missing in the adapted parent is called the **donor parent**. Even though the chief role of the donor parent is to supply the missing gene, it should not be significantly deficient in other desirable traits. An inferior recurrent parent will still be inferior after the gene transfer.

16.10.2 Application

The backcross method of breeding is best suited to improving established cultivars that are later found to be deficient in one or two specific traits. It is most effective and easy to conduct when the missing trait is qualitatively (simply) inherited, dominant, and produces a phenotype that is readily observed in a hybrid plant. Quantitative traits are more difficult to breed by this method. The procedure for transferring a recessive trait is similar to that for dominant traits, but entails an additional step.

Backcrossing is used to transfer entire sets of chromosomes in the foreign cytoplasm to create a cytoplasmic male sterile (CMS) genotype that is used to facilitate hybrid production in species including corn, onion, and wheat. This is accomplished by crossing the donor (of the chromosomes) as male until all donor chromosomes are recovered in the cytoplasm of the recurrent parent.

Backcrossing is also used for the introgression of genes via wide crosses. However, such programs are often lengthy because wild plant species possess significant amounts of undesirable traits. Backcross breeding can also be used to develop **isogenic lines** (genotypes that differ only in alleles at specific a locus) for traits (e.g., disease resistance, plant height) in which phenotypes contrast. The method is effective for breeding when the expression of a trait depends mainly on one pair of genes, the heterozygote is readily identified, and the species is self-fertilizing. Backcrossing is applicable in the development of multilines.

16.10.3 Procedure

Overview

To initiate a backcross breeding program, the breeder crosses the recurrent parent with the donor parent. The F_1 is grown and crossed with the recurrent parent again. The second step is repeated for as long as it takes to recover the characteristics of the recurrent parent. This may vary from two to five cycles (or more in some cases) depending on how easy the expression of the transferred gene is to observe, how much of the recurrent parental genotype the breeder wants to recover, and the overall acceptability of the donor parent. A selection pressure is imposed after each backcross to identify and discard the homozygous recessive individuals. Where the desired trait is recessive, it will be necessary to conduct a progeny test to determine the genotype of a backcross progeny before continuing with the next cross.

Steps

Dominant gene transfer

- **Year 1.** Select donor (RR) and recurrent parent (rr) and make 10–20 crosses. Harvest F_1 seed (Figure 16.6).

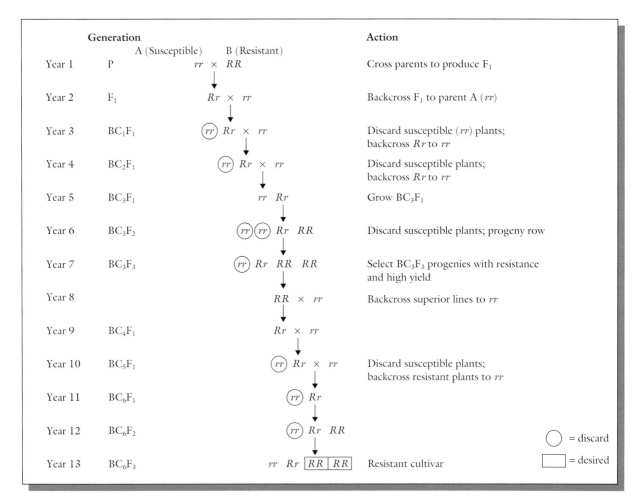

Figure 16.6 Generalized steps in breeding a dominant trait by the backcross method.

- **Year 2.** Grow F_1 plants and cross (backcross) with recurrent parent to obtain first backcross (BC_1).
- **Year 3–7.** Grow the appropriate BC (BC_1–BC_5) and backcross to the recurrent parent as female. Each time, select about 30–50 heterozygous parents (BCs) that most resemble the recurrent parent to be used in the next backcross. The recessive genotypes are discarded after each backcross. The breeder should use any appropriate screening techniques to identify the heterozygotes (and discard the homozygous recessive). For disease resistance breeding, artificial epiphytotic conditions are created. After six backcrosses, the BC_5 would very closely resemble the recurrent parent and express the donor trait. As generations advance, most plants would be increasingly more like the adapted cultivar.
- **Year 8.** Grow BC_5F_1 plants to be selfed. Select several hundred (300–400) desirable plants and harvest individually.
- **Year 9.** Grow BC_5F_2 progeny rows. Identify and select about 100 desirable non-segregating progenies and bulk.
- **Year 10.** Conduct yield tests of backcross with recurrent cultivar to determine equivalence before releasing.

Comments

The steps for transferring a dominant gene are straight forward. Following the first cross between the parents, phenotypic selection is adequate for selecting plants that exhibit the target trait. Recessive genotypes are discarded. The recurrent parent traits are not selected at this stage. The next cross is between the selected F_1 and the recurrent parent. This step is repeated for several cycles (BCn). After satisfactory recovery of the recurrent parent, the selected plant ($BCnF_1$) will be homozygous for other alleles but heterozygous for the desired traits. The last backcross is followed by selfing to stabilize the desired gene in the homozygous state. All homozygous ($BCnF_2$) recessive segregates are discarded.

Recessive gene transfer

- **Year 1–2.** Same as for dominant gene transfer. The donor parent has the recessive desirable gene (Figure 16.7).
- **Year 3.** Grow BC_1F_1 plants and self, harvest, and bulk the BC_1F_2 seed. In disease resistance breeding, all BC_1s will be susceptible.

- **Year 4.** Grow BC_1F_2 plants and screen for desirable plants. Backcross 10–20 plants to recurrent parent to have BC_2F_2 seed.
- **Year 5.** Grow BC_2 plants. Select 10–20 plants that resemble the recurrent parent and cross with recurrent parent.
- **Year 6.** Grow BC_3, harvest and bulk BC_3F_2 seed.
- **Year 7.** Grow BC_3F_2 plants, screen and select desirable plants. Backcross 10–20 plants with the recurrent parent.
- **Year 8.** Grow BC_4, harvest and bulk BC_4F_2 seed
- **Year 9.** Grow BC_4F_2 plants, screen and select desirable plants. Backcross 10–20 plants with the recurrent parent.
- **Year 10.** Grow BC_5, harvest and bulk BC_5F_2 seed.
- **Year 11.** Grow BC_5F_2 plants, screen and backcross.
- **Year 12.** Grow BC_6, harvest and bulk BC_6F_2.
- **Year 13.** Grow BC_6F_2, grow and screen; select 400–500 plants and harvest separately for growing progeny rows.
- **Year 14.** Grow progenies of selected plants, screen and select about 100–200 uniform progenies; harvest and bulk seed.
- **Year 15/16.** Follow procedure as done in breeding for dominant gene.

The key difference between the transfer of dominant and recessive alleles is that, in the latter case, phenotypic identification is not possible after a cross. Each cross needs to be followed by selfing so that the progeny with the homozygous recessive genotype can be identified and backcrossed to the recurrent parent.

Comments

- Backcrossing does not have to be conducted in the environment in which the recurrent parent is adapted because all that is needed is to be able to identify and select the target trait.
- Extensive advanced testing is not necessary in a backcross because the new cultivar already resembles the adapted cultivar, except for the newly incorporated trait.
- It is possible to transfer two or more genes by simultaneous selection among the progeny. This undertaking requires a larger population that would be necessary if two genes are transferred independently.
- Introgression of genes from weedy, adapted, exotic, or wild germplasm is possible by backcrossing. However, such transfers are often longer than the typical transfers, because of the time needed to

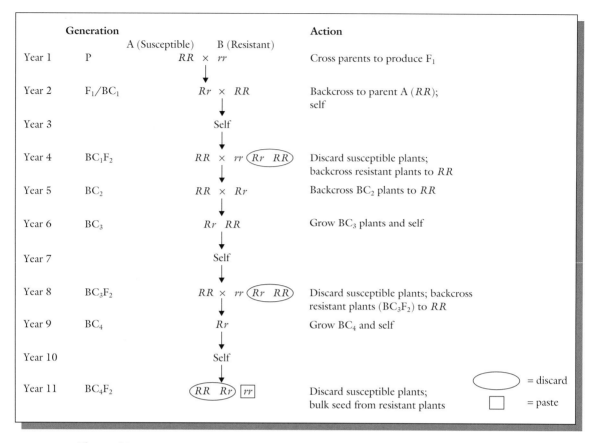

Figure 16.7 Generalized steps in breeding a recessive trait by the backcross method.

remove the undesirable agronomic traits brought in by these distantly related sources.

16.10.4 Genetic issues

With each backcross, the progeny becomes more like the recurrent parent. In theory, the BC_4 genotype will be 93.75% identical to the recurrent parent. The mathematical relationship for the recovery of the recurrent parent is presented by W. Allard is:

$$1 - \left(\tfrac{1}{2}\right)^{m-1}$$

where m is the number of generations of selfing or backcrosses. In another way, the proportion of the donor genes is reduced by 50% following each generation of backcrossing. This is obtained by the relationship $\left(\tfrac{1}{2}\right)^{m+1}$, where m is the number of crosses and backcross to the parent. For example, in the BC_4,

the value is $\left(\tfrac{1}{2}\right)^5 = 3.125\%$. To obtain the percentage of homozygotes for alleles of recurrent parents in any generation, the mathematical relationship is:

$$[(2^{m-1})/2^m]^n$$

where n is the number of genes.

Because of cytoplasmic inheritance, it is sometimes critical which of the two parents is used as female. For example, to use CMS in breeding, the male fertile inbred lines with normal cytoplasm and non-restorer genes (B lines) are converted to sterile cytoplasm (A lines) to be used as male sterile female lines in a cross.

The resulting cultivar from a backcross breeding program could differ from the starting cultivar beyond the transferred gene(s) because of **linkage drag** from the association of undesirable traits with the genes from the donor. Backcrossing is more effective in breaking linkages over selfing, especially where heritability is low for the undesirable trait.

A certain number of individuals is needed for a chance to recover the desired genes in a backcross program. This number increases as the number of genes controlling the donor trait increases. Furthermore, for multiple gene traits, it will be necessary to grow backcross progeny through F_2 or later generations to obtain the desired genotypes for advancing the program. When the trait is governed by a dominant gene, it is easy to identify plants carrying the desired gene. However, when the desired trait is conditioned by a recessive gene, an additional step is needed after each backcross to produce an F_2 generation in order to identify the recessive trait. The genetic advance in backcross breeding depends on several factors:

- **Heritability of the trait.** As previously indicated, traits that are conditioned by major genes and have high heritability are easier to transfer by backcross.
- **Sustainable intensity of trait expression.** Progress with selection will be steadier where the expression of the trait of interest remains at a high intensity throughout the program (i.e., no modifier gene action).
- **Availability of selection aids.** Ability to identify and select desirable genotypes after the backcross is critical to the success of the procedure. Depending on the trait, special selection techniques may be needed. For disease resistance breeding, artificial disease epiphytotic may be necessary. Molecular markers may be helpful in selection reducing the number of backcrosses needed for the program.
- **Number of backcrosses of marker.** The genetic distance between the parents is important to the progress made in backcrossing. If both are closely related cultivars, fewer backcrosses would be needed than if the gene transfer is from a wild genotype to an adapted one.

16.10.5 Advantages and disadvantages

There are major advantages and limitations of backcross breeding.

Advantages

- The method reduces the amount of field testing needed, as the new cultivar will be adapted to the same area as the original cultivar (especially true when both parents are adapted).
- Backcross breeding is repeatable. If the same parents are used, the same backcrossed cultivar can be recovered.
- It is a conservative method, not permitting new recombination to occur.
- It is useful for introgressing specific genes from wide crosses.
- It is applicable to breeding both self-pollinated and cross-pollinated species.

Disadvantages

- Backcrossing is not effective for transferring quantitative traits. The trait should be highly heritable and readily identifiable in each generation. However, the application of molecular markers is helping to change the application of backcrossing to improving quantitative traits (Chapter 22).
- The presence of undesirable linkages may prevent the cultivar being improved from attaining the performance of the original recurrent parent.
- Recessive traits are more time consuming to transfer.

16.10.6 Modifications

When transferring a recessive gene (rr) the BC will segregate for both homozygous dominant and heterozygous genotypes (e.g., RR and Rr). To identify the appropriate genotype to advance, it will be necessary to self the BC to distinguish the two segregants for the Rr. Alternatively, both segregants may be used in the next cross, followed by selfing. The BC progenies from the plants that produce homozygous (rr) segregates are heterozygous and are kept while the others are discarded. This is actually not a modification *per se*, since it is the way to transfer a recessive allele.

If a breeding program is designed to transfer genes for multiple traits, it will be more efficient to conduct separate backcross programs for each trait. The backcross-derived lines are then used as parents in a cross to develop one line that contains the multiple traits.

16.11 Special backcross procedures

16.11.1 Congruency backcross

Congruency backcross technique is a modification of the standard backcross procedure whereby multiple

Table 16.1 The concept of congruency backcross.

Cross	Hybrid type	Genetic constitution		
		A	:	B
A × B	F_1	50		50
F1 × A	BC_1	75		25
BC1 × B	CBC_2	37.5		62.5
CBC2 × A	CBC_3	68.8		31.3
CBC3 × B	CBC_4	34.4		65.6
CBC4 × A	CBC_5	67.2		32.8

backcrosses, alternating between the two parents in the cross (instead of restricted to the recurrent parent), are used. The technique has been used to overcome the interspecific hybridization barrier of hybrid sterility, genotypic incompatibility, and embryo abortion that occurs in simple interspecific crosses. The crosses and their genetic contribution are demonstrated in Table 16.1.

16.11.2 Advanced backcross QTL

The advanced backcross QTL developed by S.D. Tanksley and J.C. Nelson allows breeders to combine backcross with mapping to transfer genes for QTL from unadapted germplasm into an adapted cultivar. This method was developed for the simultaneous discovery and transfer of desirable QTLs from unadapted germplasm into elite lines. It has been briefly discussed previously.

16.12 Multiline breeding and cultivar blends

N.F. Jensen is credited with first using this breeding method in oat breeding in 1952 to achieve a more lasting form of disease resistance. Multilines are generally more expensive to produce than developing a synthetic cultivar, because each component line must be developed by a separate backcross.

16.12.1 Key features

The key feature of a multiline cultivar is disease protection. Technically, a **multiline** or **blend** is a planned seed mixture of cultivars or lines (multiple pure lines) such that each component constitutes at least 5% of the whole mixture. The pure lines are phenotypically

uniform for morphological and other traits of agronomic importance (e.g., height, maturity, photoperiod), in addition to the genetic resistance for a specific disease. The component lines are grown separately, followed by composting in a predetermined ratio. Even though the term multiline is often used interchangeably with blend, sometimes the former is limited to mixtures involving **isolines** or near **isogenic lines** (lines that are genetically identical except for the alleles at one locus). The purpose of mixing different genotypes is to increase heterogeneity in the cultivars of self-pollinated species. This strategy would decrease the risk of total crop loss from the infection of one race of the pathogen, or some other biotic or abiotic factor. The component genotypes are designed to respond to different versions or degrees of an environmental stress factor (e.g., different races of a pathogen).

16.12.2 Applications

One of the earliest applications of multilines was for breeding "variable cultivars" to reduce the risk of loss to pests that have multiple races and whose incidence is erratic from season to season. Planting a heterogeneous mixture can physically impede the spread of disease in the field as resistant and susceptible genotypes intermingle.

Mixtures may be composited to provide stable performance in the face of variable environment. Mixtures and blends are common in the turf-grass industry. Prescribing plants for conditions that are not clear-cut is challenging. Using mixtures or blends will increase the chance that at least one of the component genotypes would match the environment.

In backcross breeding, the deficiency in a high yielding and most desirable cultivar is remedied by gene substitution from a donor. Similarly, the deficiency of an adapted and desirable cultivar may be overcome by mixing it with another cultivar that may not be as productive but has the trait that is missing in the desirable cultivar. Even though this strategy will result in lower yield per unit area in favorable conditions, the yield will be higher than it would be under adverse conditions if only a pure adapted cultivar was planted.

Multilines composited for disease resistance are most effective against air-borne pathogens with physiological races that are explosive in reproduction. An application of blends and mixtures that is not directly related to plant breeding is marketing. Provided a label "variety not stated" is attached to the seed bag, blends of two or more cultivars can be sold under various brand names, even if they have identical composition.

16.12.3 Procedure

The backcross is the breeding method for developing multilines. The agronomically superior line is the recurrent parent, while the source of disease resistance constitutes the donor parent. To develop multilines by isolines, the first step is to derive a series of backcross-derived isolines or near-isogenic lines (since true isolines are illusive because of linkage between genes of interest and other genes influencing other traits). A method for developing multilines is illustrated in Figure 16.8. The results of the procedure

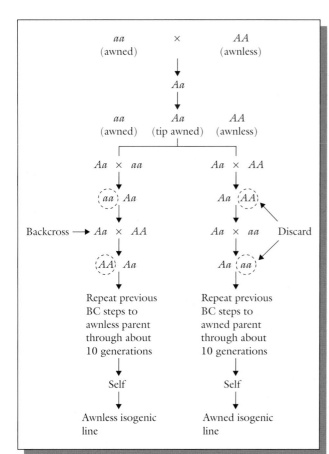

Figure 16.8 Generalized steps in breeding multiline cultivars.

are two cultivars that contrast only in a specific feature. For disease resistance, each isoline should contribute resistance to a different physiological race (or group of races) of the disease.

The component lines of multilines are screened for disease resistance at multilocations. The breeder then selects resistant lines that are phenotypically uniform for selected traits of importance to the crop cultivar. The selected components are also evaluated for performance (yield ability), quality, and competing ability. Mixtures are composited annually based on disease patterns. It is suggested that at least 60% of the mixture comprise isolines resistant to the prevalent disease races at the time. The proportion of the component lines are determined by taking into account the seed analysis (germination percentage, viability).

16.12.4 Genetic issues

Multiline cultivars consist of one genetic background but different genes for the trait of interest. A multiline is hence spatially differentiated, plant-to-plant. When planted, the cultivar creates a mosaic of genotypes in the field to provide a buffering against the rapid development of disease.

Two basic mechanisms are used by multiline cultivars to control disease – stabilization of the patterns of virulence genes and population resistance (Chapter 13). By stabilizing the patterns of virulence genes in the pathogen, it is supposed that genes for resistance would retain their value in protecting the cultivar for an extended period. The concept of population resistance is the delay in the buildup of the pathogen in the multiline cultivar.

Spore trapping has also been proposed to explain disease buildup in the population of a multiline by reducing the effective inoculation load in each generation. Following the primary inoculation (the initial spores to infect the field), spores that land on resistant genotypes will not germinate. Similarly, progeny spores from susceptible genotypes landing on resistant genotypes will not germinate. The sum effect of these events is a reduction in the inoculum load in each generation.

16.12.5 Advantages and disadvantages

Multilines have certain key advantages and disadvantages.

Advantages

- A multiline would provide protection to a broad spectrum of races of a disease-producing pathogen.
- The cultivar is phenotypically uniform.
- Multilines provide greater yield stability.
- A multiline can be readily modified (reconstituted) by replacing a component line that becomes susceptible to the pathogen, with new disease resistant line.

Disadvantages

- It takes a long time to develop all the isolines to be used in a multiline, making it laborious and expensive to produce.
- Multilines are most effective in areas where there is a specialized disease pathogen that causes frequent severe damage to plants.
- Maintaining the isolines is labor intensive.

16.12.6 Modifications

Cultivars can be created with different genetic backgrounds (instead of one genetic background). When different genetic lines (e.g., two or more cultivars) are combined, the mixture is a **composite** called a **variety blend**. Blends are less uniform in appearance than a pure line cultivar. They provide buffering effect against genotype × environment interactions.

16.13 Composites

As previously stated, a **composite cultivar**, like a multiline, is a mixture of different genotypes. The difference between the two lies primarily in the genetic distance between the components of the mixture. Whereas a multiline is constituted of closely related lines (isolines), a composite may consist of inbred lines, all types of hybrids, populations, and other less similar genotypes. However, the components are selected to have common characters, such as similar growth period, degrees of resistance to lodging or to a pathogenic agent. This consideration is critical to having uniformity in the cultivar.

A composite cultivar should be distinguished from a composite cross that is used to generate multiple-parent crosses by successively crossing parents (i.e., single cross, double cross, etc.) until the final parent contains all parents. Composites may serve as a continuous source of new entries for a breeding nursery. Any number of entries may be included in a composite, provided selection is judiciously made after evaluation. New entries may be added at any time. Technically, a composite may derive from a single diverse variety, a progeny from a single cross, or even several hundreds of entries. However, a good number of entries is between 10 and 20. The breeder's objectives determine the kind of entries used for breeding a composite. Using elite and similar genotypes would make the composite more uniform, robust (at least initially), but less genetically diverse. The reverse would be true if diverse entries are included. As a population improvement product, the yield of a composite can be improved by advancing it through several cycles of selection.

In species such as sorghum, which are predominantly self-pollinated, recessive male sterility gene that is stable across environments may be incorporated into the composite (e.g., the ms_3 in sorghum) by crossing each entry to the source of the sterility gene prior to mixing. The F_1 (fertile) is first selfed and then backcrossed to the male sterile segregates. The recurrent parents are then mixed to create the composite.

16.14 Recurrent selection

Recurrent selection is a cyclical improvement technique aimed at gradually concentrating desirable alleles in a population. It is one of the oldest techniques of plant breeding. The name was coined by F.H. Hull in 1945. It was first developed for improving cross-pollinated species (maize) and has been a major breeding method for this group of plants. Hence, detailed discussion of this method of breeding is deferred to Chapter 17. It is increasingly becoming a method of improving self-pollinated species. It has the advantage of providing additional opportunities for genetic recombination through repeated intermating after the first cross, something not available with pedigree selection. It is effective for improving quantitative traits.

16.14.1 Comments

- Recurrent selection requires extensive crossing, which is a challenge in autogamous species. To overcome this problem, male sterility system may be incorporated into the breeding program. With male sterility, natural crossing by wind and/or insects will eliminate the need for hand pollination.
- Adequate seed may be obtained by crossing under controlled environment (greenhouse) where the crossing period can be extended.

16.14.2 Advantages and disadvantages

There several advantages and disadvantages of the application of recurrent selection to breeding autogamous species.

Advantages

- Opportunities to break linkage blocks exist because of repeated intercrossing exists.
- It is applicable to both autogamous grasses (monocots) and legumes (dicots).

Disadvantages

- Extensive crossing is required, something that is a challenge in autogamous species. Male sterility system may be used to facilitate this process.
- Sufficient seed may not be available after intercrossing. This also may be resolved by including male sterility in the breeding program.
- More intermatings may prolong the duration of the breeding program.
- There is also the possibility of breaking desirable linkages.

Key references and suggested reading

Agrawal, R.L. (1998). *Fundamentals of plant breeding and hybrid seed production*. Science Publishers, Inc., Enfield, NH.

Chahal, G.S., and Gosal, S.S. (2000). *Principles and procedures of plant breeding: Biotechnological and conventional approaches*. CRC Press, New York.

Degago, Y., and Caviness, C.E. (1987). Seed yield of soybean bulk populations grown for 10–18 years in two environments. *Crop Science*, 27:207–210.

Eaton, D.L., Busch, R.H., and Youngs, V.L. (1986). Introgression of unadapted germplasm into adapted spring wheat. *Crop Science*, 26: 473–478.

Fehr, W.R. (1987). *Principles of cultivar development*. McMillan, New York.

Frey, K.J. (1983). Plant population management and breeding, in *Cultivar breeding* (ed. D.R. Wood). American Society of Agronomy, Madison, WI.

Hamblin, J. (1977). Plant breeding interpretations on the effects of bulk breeding on four populations of beans (*Phaseolus vulgaris* L.) *Euphytica*, 25:157–168.

Jensen, N.F. (1978). Composite breeding methods and the DSM system in cereals. *Crop Science*, 18:622–626.

Poehlman, J.M., and Slepper, D.H. (1995). *Breeding field crops*. Iowa State University Press, Ames, IA.

Tigchelaat, E.C., and Casali, V.W.D. (1976). Single seed descent: Applications and merits in breeding self-pollinated crops. *Acta Horticulturae*, 63:85–90.

Wilcox, J.R., and Cavins, I.F. (1995). Backcrossing high seed protein to a soybean cultivar. *Crop Science*, 35:1036–1041.

Outcomes assessment

Part A

Please answer the following questions true or false

1 The adapted and highly desirable parent in a backcross is the donor parent.
2 With each backcross, the progeny becomes more like donor parent.

3 Isogenic lines differ in alleles at a specified locus.

4 A composite may consist of hybrid cultivars.

5 The donor parent is used only once in a cross in a backcross program.

6 Single seed descent is the oldest plant breeding method.

7 Record keeping is critical part of pedigree selection method of breeding.

8 Nilsson-Ehle developed the mass selection procedure.

9 Bulk population breeding is suited to breeding plants that are closely spaced in commercial planting.

10 Mass is most effective is the trait of interest has high heritability.

Part B

Please answer the following questions.

1 . . . . is the adapted parent in a backcross.

2 Give a specific advantage of multiline cultivars.

3 Give a specific disadvantage of backcross breeding method.

4 . . . . developed the pure-line method of plant breeding.

5 Discuss a specific genetic issue involved with mass selection.

6 Give a specific disadvantage of pedigree method of breeding.

Part C

Please write a brief essay on each of the following topics.

1 Discuss the key features of a backcross breeding.

2 Discuss the application of multiline breeding.

3 Distinguish between composite breeding and multiline breeding.

4 Discuss the advantages of bulk breeding.

5 Discuss the advantages of single seed descent method of selection.

6 Compare and contrast the mass selection and pure line selection methods of breeding.

7 Describe the steps involved in mass selection method of breeding.

17

Breeding cross-pollinated species

Purpose and expected outcomes

As previously noted, breeding cross-pollinated species tends to focus on population improvement rather than the improvement of individual plants, as is the focus in breeding self-pollinated species. In addition to methods such as mass selection that are applicable to both self-pollinated and cross-pollinated species, there are specific methods that are suited to population improvement. Some methods are used less frequently in breeding. Further, certain methods are more effective and readily applied for breeding certain species than others. After studying this chapter, the student should be able to:

1 Discuss the method of mass selection in cross-pollinated species.
2 Discuss the concept of recurrent selection.
3 Describe the methods of half-sib and full-sib selection.
4 Discuss the method of S_1 and S_2 selection.
5 Discuss the development of synthetic cultivars
6 Discuss the application of backcross technique in cross-pollinated species.

17.1 The concept of population improvement

As stated in the introduction, the methods of selection for improving self-pollinated species tend to focus on improving individual plants. In contrast, the methods of improving cross-fertilized species tend to focus on improving a population of plants. A population is a large group of interbreeding individuals. The application of the principles and concepts of population genetics are made to effect changes in the genetic structure of a population of plants. Overall, breeders seek to change the gene frequency such that desirable genotypes predominate in the population. Also, in the process of changing gene frequencies, new genotypes (that did not exist in the initial population) will arise. It is important for breeders to maintain genetic variability in these populations, so that further

---

Principles of Plant Genetics and Breeding, Second Edition. George Acquaah.
© 2012 John Wiley & Sons, Ltd. Published 2012 by John Wiley & Sons, Ltd.

improvements of the population may be achieved in the future.

To improve the population, breeders generally assemble germplasm, evaluate, self-selected plants, cross the progenies of the selected selfed plants in all possible combinations, bulk, and develop inbred lines from the populations. In cross-pollinated species, a cyclical selection approach, called recurrent selection, is often used for intermating. The cyclical selection was developed to increase the frequency of favorable genes for quantitative traits. Various methods of recurrent selection are used for producing progenies for evaluation, as will be discussed here.

The procedures for population improvement may be classified in several ways, such as according to the unit of selection – either individual plants or family of plants. Also, the method may be grouped according to the populations undergoing selection as either **intrapopulation** or **interpopulation**.

Intrapopulation improvement

Selection is practiced within a specific population for its improvement for specific purposes. Intrapopulation improvement is suitable for improving populations for:

- Which the end product will be a population or synthetic cultivar.
- Developing elite pure lines for hybrid production.
- Developing mixed genotype cultivars (in self-fertilized species).

Interpopulation improvement

Methods of interpopulation improvement entail selection on the basis of the performance of a cross between two populations. This approach is suitable for use when the final product will be a hybrid cultivar. Interpopulation heterosis is exploited.

17.2 Concept of recurrent selection

The concept of **recurrent selection** was introduced in Chapter 4 as a cyclical and systematic technique in which desirable individuals are selected from a population and mated to form a new population; the cycle is then repeated. The purpose of a recurrent selection in a plant breeding program is to improve the performance of a population with respect to one or more traits of interest, such that the new population is superior to the original population in mean performance and in the performance of the best individuals within it.

The source material may be random mating populations, synthetic cultivars, single cross, or double cross. The improved population may be released as new cultivar or used as a breeding material (parent) in other breeding programs. The most desirable outcome of recurrent selection is that the improved population is produced without reduction in genetic variability. In this way, the population will respond to future improvement.

The success of a recurrent selection program rests on the genetic nature of the base population. Several key factors should be considered in the development of the base population. First, the parents should have high performance regarding the traits of interest and should not be closely related. This will increase the chance of maximizing genetic diversity in the population. It is also recommended to include as many parents as possible in the initial crossing to increase genetic diversity. Crossing provides opportunity for recombination of genes to increase genetic diversity of the population. More rounds of mating will increase the opportunity for recombination but it increases the duration of the breeding program. The breeder should decide on the number of generations of intermating that is appropriate for a breeding program.

17.2.1 Key features

A recurrent selection cycle consists of three main phases:-

(i) Individual families are created for evaluation. Parents are crossed in all possible combinations.
(ii) The plants or families are evaluated and a new set of parents selected.
(iii) The selected parents are intermated to produce the population for the next cycle of selection.

This pattern or cycle is repeated several times (3–5). The first (original) cycle is labeled C_0, and is called the base population. The subsequent cycles are named consecutively as C_1, C_2 . . . . C_n (Figure 17.1). It is possible, in theory, to assemble all the favorable genes in a population in a single generation if plant

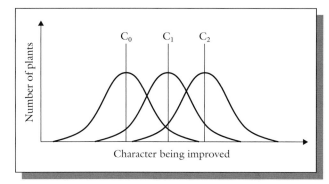

Figure 17.1 The concept of recurrent selection.

breeders could handle a population of infinite size. However, in practice, as J.K. Frey pointed out, the technique of recurrent selection is applied to breeding with the hope that desirable genes will be gradually accumulated until there is a reasonable probability of obtaining the ultimate genotype in a finite sample.

17.2.2 Applications

Recurrent selection may be used to establish a broad genetic base in a breeding program. Because of multiple opportunities for intermating, the breeder may add new germplasm during the procedure when the genetic base of the population rapidly narrows after selection cycles. Research has indicated that recurrent selection is superior to classical breeding when linkage disequilibrium exists. In fact, the procedure is even more effective when epistatic interactions enhance the selective advantage of new recombinants. Recurrent selection is applied to legumes (e.g., peanut, soybean) as well as cereals (e.g., barley, oats).

17.3 Genetic basis of recurrent selection

Various recurrent selection schemes are available. They exploit additive, partial dominance to dominance and overdominance types of gene action. However, without the use of testers (as in simple recurrent selection) the scheme is effective for only traits of high heritability. Hence, only additive gene action is exploited in the selection for the trait. Where testers are used, selection for general combining ability (GCA) and specific combining ability (SCA) are applicable, permitting the exploitation of other gene effects. Recurrent for GCA is more effective than other schemes when additive gene effects are more important. Recurrent selection for SCA is more effective than other selection schemes when overdominance gene effects are more important. Reciprocal recurrent selection is more effective than others when both additive and overdominance gene effects are more important. All three schemes are equally effective when additive with partial to complete dominance effects prevail. The expected genetic advance may be obtained by the following general formula:

$$\Delta G = [CiV_A]/y\sigma_p$$

where ΔG is the expected genetic advance per cycle, C is the measure of parental control ($C = \frac{1}{2}$ if selection is based on one parent, and $C = 1$ when both parents are involved), i is the selection intensity, V_A is the additive genetic variance among the units of selection, y is the number of years per cycle, and σ_p is the phenotypic standard deviation among units of selection. Increasing the selection pressure (intensity) will increase gain in selection provided the population advanced is not reduced to a size where genetic drift and loss of genetic variance can occur. Other ways of enhancing genetic advance per cycle include selection for both male and female parents, maximizing available additive genetic variance, and management of environmental variance among selection units. The formulas for various schemes are presented at the appropriate times in this textbook.

The role of parental control in genetic gain can be manipulated by the breeder through the exercise of control over parents in a breeding program. When the breeder controls the genetic contributions of both parents to the selection population, the genetic gain can be twice as much as when only one of the parents is under control. Both parents may be controlled in one of several ways – (a) selfing of selected individuals (instead of being pollinated by selected and unselected males), (b) select before pollination and recombination among selected plants only, and (c) recombination occurs among selected clones.

17.4 Types of recurrent selection

There are four basic recurrent selection schemes, based on how plants with the desired traits are identified:

(i) **Simple recurrent selection.** This is similar to mass selection with one or two years per cycle. The procedure does not involve the use of a tester. Selection is based on phenotypic scores. This procedure is also called phenotypic recurrent selection.

(ii) **Recurrent selection for general combining ability.** This is a half-sib progeny test procedure in which a wide genetic based genotype (e.g., a cultivar) is used as a tester. The test cross performance is evaluated in replicated trials prior to selection.

(iii) **Recurrent selection for specific combining ability.** This scheme uses an inbred line (narrow genetic base) for a tester. The test cross performance is evaluated in replicated trails before selection.

(iv) **Reciprocal recurrent selection.** This scheme is capable of exploiting both general and specific combining ability. It entails two heterozygous populations, each serving as a tester for the other. Two genetically different populations are altered to improve their crossbred mean. To achieve this, individual plants from two populations (A and B) are selfed and also crossed with plants from the reciprocal female tester population (B and A, respectively).

17.5 Intrapopulation improvement methods

Common intrapopulation improvement methods in use include mass selection, ear-to-row selection, and recurrent selection. Intrapopulation methods may be based on single plants as unit of selection (e.g., as in mass selection), or family selection (e.g., as in various recurrent selection methods).

17.5.1 Individual plant selection methods

Mass selection

Mass selection for line development (Chapter 16) is different from mass selection for population improvement. Mass selection for population improvement aims at improving the general population performance by selecting and bulking superior genotypes that already exist in the population.

Key features

The selection units are individual plants. Selection is solely on phenotypic performance. Seed from selected plants (pollinated by the population at large) are bulked to start the next generation. No crosses are made, but progeny test is conducted. The process is repeated until a desirable level of improvement is observed.

Genetic issues

The effectiveness of the method depends on the heritability of the trait since selection is solely on the phenotype. It is also most effective where additive gene action operates. Effectiveness of mass selection also depends on the number of gene involved in the control of the trait of interest. The more additive genes that are involved, the greater the efficiency of mass selection. The expected genetic advance through mass selection is given by the following (for one sex – female):

$$\Delta G_m = [(1/2)i\sigma_A^2]\sigma_p$$
$$= [1/2i\sigma_A^2]/[\sigma_A^2 + \sigma_D^2 + \sigma_{A\times E}^2 + \sigma_{D\times E}^2 + \sigma_e^2 + \sigma_{me}^2]$$

where σ_p is the phenotypic standard deviation in the population, σ_A^2 is the additive variance, σ_D^2 is the dominance variance, and the other factors are interaction variances. ΔG_m doubles with both sexes. This large denominator makes mass selection inefficient for low heritability traits. Selection is limited to only the female parents since there is no control over pollination.

Procedure

- **Year 1.** Plant the source population (local variety, synthetic variety, bulk population, etc.). Rogue out undesirable plant before flowering, and then select several hundreds of plants based on phenotype. Harvest and bulk.
- **Year 2.** Repeat year 1. Grow selected bulk in a preliminary yield trial, including a check. The check is the unselected population (original), if the goal of the mass selection is to improve the population.
- **Year 3.** Repeat year 2 for as long as progress is made.
- **Year 4.** Conduct advanced yield trial.

The mass selection may be longer, depending on progress being made.

Advantages

These are highlighted in Chapter 16.

Disadvantages

- Using phenotypic selection makes selection of superior plant often difficult.
- Lack of pollen control means both desirable and undesirable pollen will be involved in pollination of the selected plants.
- If selection intensity is high (small population size advanced) the possibility of inbreeding depression is increased, as well as the probability of losing individuals with desirable combinations.

Modifications

- **Stratified or grid system.** Proposed by C.O. Gardener, the field is divided into small grids (or subplots) with little environmental variance. An equal number of superior plants is selected from each grid for harvesting and bulking.
- **Honeycomb design.** Proposed by A. Fasoulas, the planting pattern is triangular rather than the conventional rectangular pattern. Each single plant is at the center of a regular hexagon, with other six equidistant plants, and is compared to the other six equidistant plants. There are modifications that are sometimes complex to apply and have variable effects on selection response.

17.5.2 Family selection methods

Family selection methods are characterized by three general steps:

(i) Creation of a family structure.
(ii) Evaluation of families and selection of superior ones by progeny testing.
(iii) Recombination of selected families or plants within families to create a new base population for the next cycle of selection.

Generally, the duration of each step is one generation, but variations exist.

1 Half-sib family selection methods

The basic feature of this group of methods is that half-sib families are created for evaluation and recombination, both steps occurring in one generation. The populations are created by random pollination of selected female plants in generation 1. The seed from generation 1 families are evaluated in replicated trials and in different environments for selection. There are different kinds of half-sib family selection methods, the simplest one being ear-to-row selection.

This method is applicable to cross-pollinated species (Figure 17.2).

Steps

- **Season 1.** Grow source population (heterozygous population) and select desirable plants (S_0) based on phenotype according to the traits of interest. Harvest plants individually. Keep remnant seed of each plant.
- **Season 2.** Grow replicated half-sib progenies ($S_0 \times$ tester) from selected individuals in one environment (yield trial). Select best progenies and bulk to create progenies of the next cycle. The bulk is grown in isolation (crossing block) and random mated.
- **Season 3.** The seed is harvested and used to grown the next cycle.

Alternatively, the breeder may bulk the remnant seed of S_0 plants whose progeny have been selected and used that to initiate the next cycle.

Genetic issues

The expected genetic gain from half-sib selection is given by:

$$\Delta G_{HS} = [\tfrac{1}{4} \, i\sigma_A]/\sigma_{PHS}$$

where σ_{PHS} is the standard deviation of the phenotypic variance among half-sibs. Other components are as before. The tester is the parental population, and hence selection or control is over only one sex. The genetic gain is hence reduced by half (the available additive genetic variance is reduced by half because of the control over the female parent). Genetic gain can be doubled by selfing each parent to obtain S_1, then crossing to obtain half-sibs.

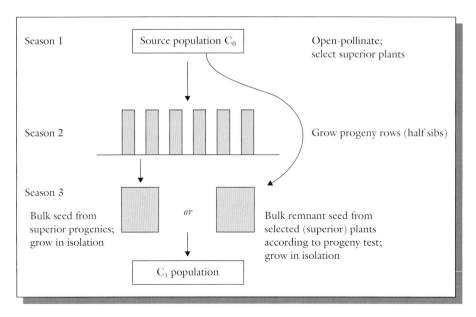

Figure 17.2 Generalized steps in breeding by ear-to-row selection.

Application

Half-sib selection is widely used for breeding perennial forage grasses and legumes. A polycross mating system is used to generate the half-sib families from selected vegetatively maintained clones. The families are evaluated in replicated rows for 2–3 years. Selecting traits of high heritability (e.g., oil and protein content of maize) is effective.

Modification

The basic or traditional ear-to-row selection method did not show much gain over mass selection. An improvement was proposed by J.H. Lohnquist in which creation of family structure, evaluation and recombination are conducted in one generation. The half-sib families are evaluated in replicated trials in many environments. The approach was to better manage the environmental and G × E interactions.

Steps

- **Season 1.** Select desirable plants from the source population. Harvest these open-pollinated (half-sibs) individually.
- **Season 2.** Grow progeny rows of selected plants at multiple locations and evaluate for yield performance. Plant female rows within seed from individual half-sib families alternating with male rows (pollinators) planted with bulked seed from the entire population. Select desirable plants (based on average performance over locations) from each progeny separately. Bulk the seed to start next cycle.

Genetic issues

The genetic gain has two components – among ear-rows across environments (inter-family selection) and within families (intra-family selection). The total genetic gain is given by:

$$\Delta G_{mHS} = [1/8 i\sigma_A]/[\sigma_{PHS}] + [3/8 i\sigma_A]/[\sigma_{wc}]$$

where σ_{wc} is the square root of the plant-to-plant within plot variance. Others components are as before.

Application

This method of breeding has been applied to the improvement of perennial species as previously indicated for the traditional ear-to-row selection. It has been used in maize for yield gains of between 1.8 and 6.3% per cycle.

2 Full-sib family selection

Full sibs are generated from biparental crosses using parents from the base population. The families are evaluated in a replicated trial to identify and select superior full-sib families, which are then recombined to initiate the next cycle.

Steps

Cycle 0

- **Season 1.** Select random pairs of plants from the base population and intermate, pollinating one with the other (reciprocal pollination). Make between 100 and 200 biparental crosses. Save the remnant seed of each full-sib cross (Figure 17.3).
- **Season 2.** Evaluate full-sib progenies in multilocation replicated trails. Select promising half-sibs (20–30).
- **Season 3.** Recombine selected full-sib.

Cycle 1 Same as cycle C_0

Genetic issues

The genetic gain per cycle is given by:

$$\Delta G_{FS} = [i\sigma_A^2]/2\sigma_{FS}$$

where σ_{FS} is the phenotypic standard deviation of the full-sib families.

Application

Full-sib family selection has been used for maize improvement. Selection response per cycle of about 3.3% has been recorded in maize.

3 Selfed (S_1 or S_2) families selection

An S_1 is a selfed plant from the base population. The key features are generations S_1 or S_2 families, evaluating them in replicated multi-environment trials, followed by recombination of remnant seed from selected families (Figure 17.4).

Steps

- **Season 1.** Self pollinate about 300 selected S_0 plants. Harvest the selfed seed and keep the remnant seed of each S_1.
- **Season 2.** Evaluate S_1 progeny rows to identify superior progenies.
- **Season 3.** Random mate selected S_1 progenies to form C_1 cycle population.

Genetic issues

The main reason for this scheme is to increase the magnitude of additive genetic variance. In theory the

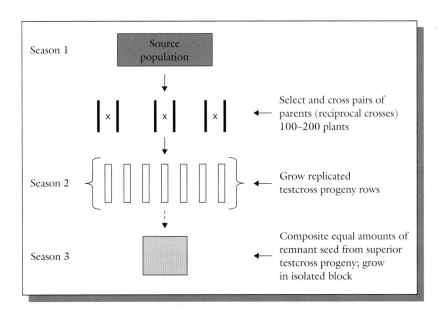

Figure 17.3 Generalized steps in breeding by full-sib method.

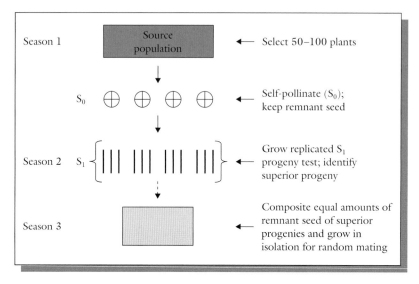

Figure 17.4 Generalized steps in breeding based on S_1/S_2 progeny performance.

genetic gain is given by:

$$\Delta G_{S1} = [i\sigma^2_{A1}]/\sigma_{PS1}$$

where σ^2_{A1} is the additive genetic variance among S_1, and σ_{PSA} is the phenotypic standard deviation among S_1 families. The additive genetic variation among S_2 is two times that of S_1. The S_1 and S_2, theoretically, have the highest expected genetic gain per cycle for intrapopulation improvement. However, various reports have indicated that, in practice, full-sib and test cross selection have produced greater genetic gain for both populations *per se* and the population crosses.

Application

The S_1 appears to be best suited for self-pollinated species (e.g., wheat, soybean). It has been used in maize breeding. One cycle is completed in three seasons in S_1 and four seasons in S_2. Genetic gain per cycle 3.3% has been recorded.

17.5.3 Family selection based on test cross

The key feature of this approach to selection is that it is designed to improve both the population *per se* as well as its combining ability. The choice of the tester is most critical to the success of the schemes. Using a tester to aid in selection increases the duration of a cycle by one year (i.e., three-year cycle instead of two years as in the phenotypic selection). The choice of a tester is critical to the success of a recurrent selection breeding program. The commonly used testers may be classified into two: (i) a narrow genetic base tester (e.g., an inbred line), and (ii) a broad-genetic base tester (e.g., open pollinated cultivars, synthetic cultivar or double-cross hybrid). Broad-base testers are used for testing GCA in the population under improvement, whereas narrow genetic base testers are used to evaluate SCA and possibly GCA.

Generally, plants are selected from the source population and selfed in year 1. Prior to intermating, the selected plants are crossed as females to a tester in year 2. Intermating of selected plants occur in year 3.

1 Half-sib selection with progeny test

Half-sib or **half-sib family selection** is so-called because only one parent in the cross is known. In 1899, C.G. Hopkins first used this procedure to alter the chemical composition of corn by growing progeny rows from corn ears picked from desirable plants. Superior rows were harvested and increased as a new cultivar. The method as applied to corn is called **ear-to-row breeding**.

Key features

There are various half-sib progeny tests, such as, topcross progeny test, open-pollinated progeny test, and polycross progeny test. A half-sib is a plant (or family

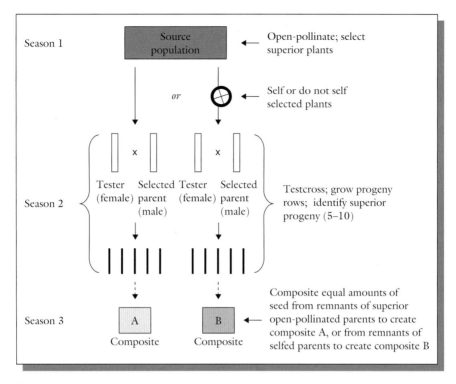

Figure 17.5 Generalized steps in breeding by half-sib selection with progeny test.

of plants) with a common parent or pollen source. Individuals in a half-sib selection are evaluated based on their half-sib progeny. Unlike mass selection in which individuals are selected solely on phenotypic basis, the half-sibs are selected based on the performance of their progenies. The specific identity of the pollen sources is not known.

Applications

Recurrent half-sib breeding has been used to improve agronomic traits as well as seed composition traits in corn. It is suited for improving traits with high heritability and species that can produce sufficient seed per plant to grow a yield trial. Species with self-incompatibility (no self-fertilization) or some other constraint of sexual biology (e.g., male sterile) are also suited to this method of breeding.

Procedure

A typical cycle of half-sib selection entails three activities – crossing the plants to be evaluated to a common tester, evaluating the half-sib progeny from each plant, and intercrossing the selected individuals to form a new population. In the second season, each separate seed pack is used to plant a progeny row in an isolated area (Figure 17.5). The remnant seed is saved as backup. In season 3, select 5–10 superior progenies, harvest and composite seed; alternatively, do the same with the remnant seed. Grow the composites in an isolation block for open pollination. Harvest seed as a new open pollinated cultivar or for use to start a new population.

Genetic issues

Like mass selection, half-sib selection is based on maternal plant selection without pollen control. Consequently, heritability estimates are reduced by 50%. Half-sib selection is hence less effective for changing traits with low heritability.

Modifications

The modifications or alternative procedures for recurrent half-sib selection differ by (a) testers used, (b) selection of one or both parents, and (c) seed used for intercrossing . Some procedures use a population as tester and half-sib seed for intercrossing. Selfed seed or clones may be used for intercrossing.

Where genetic male sterility is incorporated in the scheme, male sterile plants may be tagged in the field at the time of flowering. After open pollination, each head is harvested and threshed separately. Seed from one head (family) is evaluated as one entry in a yield trial, saving remnant seed superior entries are identified and the remnant seed bulked and planted for recombination to occur. Male sterile plants are tagged again and harvested individually to form the next cycle of evaluation.

Advantages and disadvantages

There are major advantages and disadvantages of half-sib selection.

- **Advantages**
  - Rapid to conduct.
  - Progeny testing increases success of selection especially quantitative gene action occurs or heritability is low.
- **Disadvantages**
  - Trait of interest should have high heritability for success.
  - Not readily applicable to species that cannot produce enough seed per plant to conduct a yield trial.
  - Lack of pollen control reduces heritability by half.

2 Half-sib selection with test cross

Another way of evaluating genotypes to be composited is by conducting a test cross.

Key features

This variation of half-sib selection allows the breeder to more precisely evaluate the genotype of the selected plant by choosing the most suitable test cross parent. The half-sib lines to be composited are selected based on a test cross evaluation not progeny performance. The tester may be an inbred, in which case all the progeny lines will have a common parental gamete.

Application

Like half-sib selection with progeny test, this procedure is applicable to cross-pollinated species in which sufficient seed can be produced by crossing to grow a replicated testcross progeny trial. However, in procedures in which self-pollination is required, the method cannot be applied to species with self-incompatibility.

Procedure

In season 1, the breeder selects 50–100 plants from the source population. A tester parent is pollinated with pollen from each of the selected plants (Figure 17.6). The crossed seed from the tester as well as the open-pollinated selected plants are harvested separately. In season 2, the test cross progenies are grown in replicated plots. In season 3, an equal amount of open-pollinated seed from 5–10 superior plants is composited and grown in isolation for open-pollination to occur.

Advantages and disadvantages

There are major advantages and disadvantages of half-sib selection with test cross.

- **Advantages**
  - Control over the testcross parents permits a more precise evaluation of the genotype of the selected plant than would be obtained by open-pollination as in the half-sib selection with progeny test.
  - It is rapid to conduct.
- **Disadvantages**
  - This method of breeding is applicable to species that can produce sufficient seed by crossing, for replicated testcross progeny trials.
  - When self-pollination is used, the method is applicable to species without self-incompatibility issues.

Modification

Pollen from each selected plant may be used to pollinate a tester plant and self-pollinate the selected plant. Also, in season 3, equal quantities of selfed seed may be composited and planted in isolation.

3 Interpopulation improvement methods

The purpose of this group of recurrent selection schemes is to improve the performance of a cross between two populations. To achieve this, interpopulation heterosis is exploited. The procedures are

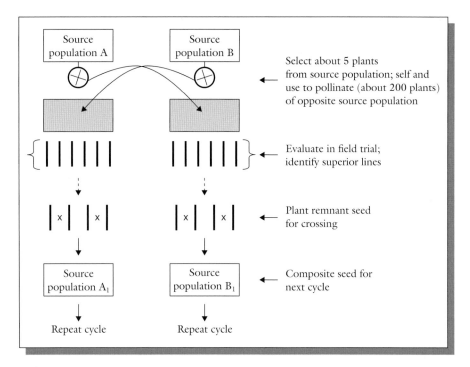

Figure 17.6 Generalized steps in breeding by half-sib selection with a test cross.

appropriate when the breeder's goal is hybrid production (i.e., the final product or cultivar is a hybrid). Developed by P.E. Comstock and his colleagues, the procedures allow the breeder to improve two genetically different populations for GCA and SCA, thereby improving their crossbred mean.

1 Half-sib reciprocal recurrent selection

Key features

The half-sib recurrent selection scheme involves the making of S_1 plant test crosses and evaluating them to identify and select superior progenies.

Procedure

Cycle 0

- **Season 1.** Select and self-pollinate about 200–300 S_1 plants in each of two populations, A and B.
- **Season 2.** Grow 200–300 of the selected S_1 progenies and produce half-sibs of population A by crossing a number of plants with B as females, and vice versa. Self-pollinate the S_1 plants used in making the test crosses. Save S_1 seed.

- **Season 3.** Evaluate about 100 half-sibs in replicated trials. Select about 20 promising test cross families. This is done for both populations, A and B.
- **Season 4.** Randomly mate the plants selected from S_1 families within A and B to obtain new seed to initiate Cycle 1.

Cycle 1   Repeat cycle 0

Genetic issues

This scheme makes use of additive, dominance, and overdominance gene action. It is effective for selecting favorable epistatic gene combinations in the population. The change in the crossbred mean may be calculated as follows:

$$\Delta G_{(A \times B)} = [i\sigma^2_{A(HSA)}]/4\sigma_{P(HSA)} + [i'\sigma^2_{A(HSA)}]/4\sigma_{P(HSB)}$$

where i and i' are the selection intensities in populations A and B respectively; $\sigma^2_{A\ (HSA)}$ and $\sigma^2_{A\ B\ (HSB)}$ are the additive variances for populations A and B, respectively. Similarly, $\sigma_{p(HSA)}$ and $\sigma_{P(HSB)}$ are the phenotypic standard deviations among half-sibs.

2 Full-sib reciprocal recurrent selection

Key features

Developed by Hallauer and Eberhart as modifications of the method by Comstock and colleagues, the full-sib method requires at least one of the populations to be prolific. The recombination units are half-sibs (instead of S_1 families). Developed for maize, full-sib families are produced by pairing plants from two populations, A and B. The top ear of a plant from population A is crossed with a plant from populations B. The lower ear is selfed to be saved as remnant seed. The same is done for the reciprocal plant from population B, if they have two ears, otherwise, they are selfed.

Procedure

Cycle 0

- **Season 1.** Plant population A as females (detassel) in an isolated block and population B as males in field 1. Plant population B as females and population A as males in field 2. The upper ears in each field are open pollinated, while the lower ears are protected and pollinated manually. The result is that the upper ear is an interpopulation half-sib family while the lower ear is an intrapopulation half-sib family.
- **Season 2.** Evaluate 100–200 A × B and B × A half-sibs in replicated trials. Select best half-sibs from both sets of crosses.
- **Season 3.** Plant the remnant seed of lower ears selfed by hand pollination that corresponds to the best A × B half-sibs in ear-to-row as females (detassel). The males are the bulk remnant half-sib seed from population B corresponding to the best B × A crosses. They are randomly mated. The open pollinated seed in populations A and B are harvested to initiate the next cycle.

Advantages

- As compared to the half-sib method, one half of the families are evaluated in each cycle because the evaluation of each full-sib reflects the worth of two parental plants, one from each population.
- Superior $S_0 \times S_0$ crosses may be advanced in further generations and evaluated as $S1 \times S_1$, $S_2 \times S_2$, ..., $S_n \times S_n$ to allow the breeder to simultaneously develop hybrids while improving the populations.

Genetic issues

Another advantage of this method is that additive genetic variance of full-sib families is twice that of the half-sib families. The expected genetic gain is given by:

$$\Delta G = [i\sigma_A^2]/[2\sigma_{PFS}]$$

where σ_{PFS} is the phenotypic standard deviation of the full-sib families.

Application

The scheme has been used in crops such as maize and sunflower with reported genetic gains of the magnitudes of 2.17% for population *per se* and 4.90% for the population hybrid.

17.6 Optimizing gain from selection in population improvement

The goal of the breeder is to make systematic progress in the mean expression of the trait of interest from one cycle to the next. Achieving progressive gains in yield depends on several factors.

- **Genetic variance.** As previously indicated, additive genetic variance is critical to increase in gains per cycle. Additive genetic variance can be increased through increasing diversity in the entries used in population improvement.
- **Selection intensity.** The rate of gain with selection is increased when selection intensity is increased. The number of individuals selected for recombination in each cycle should be limited to the best performers.
- **Generations per cycle.** Breeder's choice of the breeding system to use in a breeding project is influenced by how rapidly each cycle of selection can be completed. When possible, using 2–3 generations per year can increase yield gains. Multiple generations per year is achieved by using off-season nurseries (winter nurseries), or planting in the dry season using irrigation.
- **Field plot technique.** Breeders select in the field, often handling large numbers of plants. Heterozygosity in the field should be managed by using proper experimental designs to reduce random variation. Whenever possible, uniform fields should be selected for field evaluations. The cultural conditions

should be optimized as much as possible (proper fertilization, irrigation, disease and pest control, weed control, etc.). This practice will reduce variation between replications. Other factors to consider are plot sizes, number of plants per plot, number of replications per trial, and number of locations. Implemented properly, these factors reduce random variations that complicate experimental results.

17.7 Development of synthetic cultivars

17.7.1 Synthetic cultivar versus germplasm composites

There are two basic types of open-pollinated populations of crops – those produced by population improvement, and synthetics. As previously discussed, population improvement methods can be categorized into two – those that depend on purely phenotypic selection (mass selection) and those that involve selection with progeny testing. A **synthetic cultivar** may be defined as an advanced generation of cross-fertilized (random mating in all combinations) seed mixture of parents that may be strains, clones, or hybrids. The parents are selected based on GCA. The primary distinction between these basic types of populations mentioned in this section is that population improvement cultivars can be propagated indefinitely as such. However, a synthetic cultivar is propagated for only a limited number of generations and then must be reconstituted from the parental stock. A synthetic population differs from a natural population by consisting of breeder-selected parental stocks.

Germplasm composites is a broad term used to refer to the mixing together of breeding materials on the basis of some agronomic trait (e.g., yield potential, maturity, disease resistance), followed by random mating. There are many ways to put a composite together. Germplasm composites are by nature genetically broad based and very complex. They can be used as for commercial cultivation over a broad range of agroecological environments. However, they can also be used as reservoirs of useful genes for use in breeding programs.

17.7.2 Desirable features of a synthetic cultivar

K.J. Frey summarized three major desirable features of synthetic cultivars as:

(i) Yield reduction in advanced generations is less than with a single or double cross. For example, in maize an estimated 15–30% reduction occurs between F_1 and F_2, as compared to only a reduction of 5–15% from *syn-1* to *syn-2*. This slow rate of reduction in yield makes it unnecessary for producers to obtain new seed of the cultivar for planting in each season.

(ii) A synthetic cultivar may become better adapted to the local production environment over time, as it is produced in successive generations in the region.

(iii) A synthetic cultivar is genetically heterogeneous, a population structure that makes it perform stably over changing environmental conditions. Further, because of this heterogeneity, both natural and artificial selection can modify the genotypic structure of synthetic cultivars. That is, a breeder may achieve gain in performance by practicing selection in *syn-2* and subsequent generations.

17.7.3 Application

The synthetic method of breeding is suitable for improving cross-fertilized crops. It is widely used to breed forage species. Successful synthetic cultivars have been bred for corn, sugar beets, and other species. The suitability of forage species for this method of breeding stems from several biological factors. Forages have perfect flowers, making it difficult to produce hybrid seed for commercial use. The use of male sterility may facilitate controlled cross-pollination, which is difficult to achieve in most forage species. To test individual plants for use in producing the commercial seed, it is essential to obtain sufficient seed from these plants. The amount of seed obtained from single plants of these species is often inadequate for a progeny test. Furthermore, forage species often exhibit self-incompatibility, a condition that inhibits the production of selfed seed. Synthetic cultivars are also used as gene pools in breeding progeny. Synthetic cultivars are advantageous in agricultural production systems where farmers routinely save seed for planting. One of the well-known and widely used synthetic is the Iowa stiff-stalk synthetic (BSSS) of maize.

17.7.4 Key features

There are three primary steps in the development of a synthetic:

(i) Assembly of parents.
(ii) Assessment of GCA.
(iii) Random mating to produce synthetic cultivars.

The parents used in synthetics may be clones (e.g., in forage species) or inbred lines (e.g., corn, sugar beet). Whereas forages can be increased indefinitely by clonal propagation, inbred lines are needed to perpetuate the genotypes used in hybrid production. The parental materials are reproducible and may be substituted with new genotypes as they become available, for some improvement in the synthetic cultivar. The parents are selected after progeny testing or general combining ability analysis using a test cross or topcross, but most frequently a polycross, for evaluation.

1 Test for GCA

Testers

- **Polycross. A polycross test** is generally preferred because it is simple and convenient to conduct and also, by nature, provides an efficient estimate of GCA, a desired attribute in synthetic production. Furthermore, it allows an adequate amount of seed to be obtained for more comprehensive testing using commercial standards. It provides a greater insurance to cultivars against genetic shifts that could arise during seed increase. However, any significant amount of selfing (especially if unequal among the component parents) or non-random cross-pollination could result in bias. The component clones may vary in self-fertility and other biological characteristics that impact fertilization. To minimize such deviations from a perfect polycross, the Latin square design (Chapter 4) may be used to establish the polycross nursery. In theory, the polycross allows each clone in the nursery to be pollinated by about the same pollen sources as a result of random pollination from all the entries in the same plot.
- **Topcross.** Selected clones are grown in alternative rows with an open-pollinated cultivar as tester. The test cross seed includes both selfs and intercrosses among the clones being evaluated.
- **Diallel cross.** A diallel cross entails achieving all possible single crosses involving all the parents. This is laborious to conduct. It requires that each parent

be grown in isolation. It provides information on both GCA and SCA.

Procedure

A procedure for crops in which selections are clonally propagable is as follows:

- **Year 1: The source nursery.** The source population consists of clones. The source nursery is established by planting several thousands (5000–10 000) of plants assembled from many sources to provide a broad genetic base of the clonal lines for selection. The germplasm in the nursery is screened and evaluated to identify superior individuals according to the breeding objectives.
- **Year 2: Clonal lines.** The breeder first selects 100 to 200 superior plants on phenotypic basis to multiply clonally to produce clonal lines. A clonal line nursery is established, each line consisting of about 20–25 plants derived from the same parental line. The breeder may impose various biotic and abiotic selective pressure (e.g., drought, specific disease epidemic, severe clipping) to aid in identifying about 25–50 most desirable clones.
- **Year 3: Polycross nursery.** The selected clonal lines are planted in a polycross nursery to generate seed for progeny testing. Ideally, the layout of the polycross in the field should allow each clone to be pollinated by a random sample of pollen from all the other entries. A method of layout to achieve this objective is a square plot (e.g., 12×12) in which every clone occurs once in every row. Covering with a fine mesh tent or separating the plots by adequate distance isolates each square plot. The mesh is removed once the pollination period is over. A large number of replications (10 or more) of the single randomized clones is suggested for achieving a highly mixed pollination. Seed from each clone is harvested separately. The polycross test is valid if the layout ensures random interpollination. Alternative methods of evaluating clones for quantitatively inherited traits are available. Self-fertilization may be used but it often yields a little amount of seed. A diallel cross is cumbersome to conduct, especially for large entries. A topcross evaluates SCA. The polycross is used because it evaluates GCA.
- **Year 4: Polycross progeny test.** Seed is harvested from the replicated clones and bulked for planting progeny rows for performance evaluation. The progeny test evaluates yield and other traits,

according to the breeding objective. The top performing 5–10 clones are selected for inclusion in the synthetic cultivar.

- **Year 5: *Syn-0* generation.** The selected clones are vegetatively propagated and randomly transplanted into an isolated field for cross-fertilization to produce *syn-0* seed. Leguminous species may be isolated in an insect-proof cage and cross-fertilized by using insects.
- **Year 6: *Syn-1* generation.** The syn-0 seed is increased by planting in isolation. Equal amount of seed is obtained from each parent and mixed to ensure random mating in the field. Bulk seed is harvested from seed increased in syn-1 generation that may be released as a commercial cultivar provided sufficient seed is produced.
- **Year 7: Subsequent *syn* generations.** Frequently, the *syn-1* seed is not sufficient to release to farmers. Consequently, a more practical synthetic breeding scheme is to produce *syn-2* generation by open-pollinated increase of seed from *syn-1*. The syn-2 seed may be likened to a breeder seed. It is further increased to produce *syn-3* (foundation seed) and *syn-4* (certified seed). Commercial seed classes are discussed in detail in Chapter 27. The pattern of loss in vigor, progressively with advancement of generations from *syn-1, syn-2*, and *syn-n*, is similar to what occurs when hybrids are progressively selfed from $F_1, F_2 \ldots F_n$ generations. It is important to maintain the original clones so that the synthetic can be reconstituted as needed. The steps described are only generalized and can be adapted and modified according to the species and the objectives of the breeder.

17.7.5 Genetic issues

The highest yield performance is obtained in the *syn-1* generation, hybrid vigor declining with subsequent generations. It is generally estimated that a synthetic forage cultivar of cross-fertilized diploid or polyploid species will experience a maximum yield decline of 10–12% from *syn-1* to *syn-2* generation, as previously stated. The yield decline is less in subsequent generations. Sewall Wright proposed a formula to predict the F_2 yield of a group of inbred lines as follows:

$$F_2 = F_1 - [(F_1 - P)]/n$$

where F_2 is the expected performance of the F_2, F_1 is *the* mean F_1 hybrid performance from combinations of inbred lines, P is the average performance of inbred

lines, and n is the number of inbred lines. That is, one can increase F_2 yield by increasing the average F_1 yield, increasing yield of parental lines, or increasing the number of lines used to create the synthetic. This formula assumes that the species has diploid reproduction and that the parents are inbred. Hence, even though shown to be accurate for maize, it is not applicable to polyploid species and those that are obligate outcrossers.

The formula may also be written as follows:

$$syn\text{-}2 = syn\text{-}1 - [syn\text{-}1 - syn\text{-}0]/n$$

Studies involving inbred lines and diploid species have indicated that as the number of parental lines increase the F_1 performance is increased. Parental lines with high combining ability will have high F_1 performance. In practice, it is a difficult task to find a large number of parents with very high combining ability. Furthermore, predicting yield performance of synthetic cultivars of cross-fertilized diploid and polyploid forage species is more complicated than is described by their relationship in the equation. Given a set of n inbred lines, the total number of synthetics, N, of size ranging from 2 to n is given by:

$$N = 2n - n - 1$$

As inbred lines increase, the number of possible synthetics increases rapidly, making it impractical to synthesize and evaluate all the possible synthetic cultivars.

The theoretical optimum number of parents to include in a synthetic is believed to be about 4–6. However, many breeders favoring yield stability over yield ability tend to use large numbers of parents ranging from about 10–100 or more. Large numbers are especially advantageous when selecting for traits with low heritability.

Synthetics of autotetraploid species (e.g., alfalfa) are known to experience severe and widespread decline in vigor between *syn-1*, and *syn-2*, which has been partly attributed to a reduction in triallelic and tetrallelic loci. Higher numbers of tetrallelic loci have been shown to be associated with higher agronomic performance (e.g., forage yield, seed yield, height) of alfalfa. The number of selfed generations is limited to one. Selfed seed from selected S_0 plants are intermated to produce the synthetic population. The rationale is that S_0 plants with high combining ability would contain many favorable genes and gene combinations. Selecting specific individuals from the segregating population to self could jeopardize these desirable combinations.

Additive gene action is considered more important than dominance genotypic variance for optimum performance of synthetic cultivars. In autotetraploids in which intralocus and allelic interactions occur, high performing synthetic cultivars should include parents that have a high capacity to transfer their desirable performance to their offspring. Such high additive gene action coupled with a high capacity for intralocus or allelic interactions will likely result in higher-performing synthetics.

Synthetic cultivars exploit the benefits of both heterozygosity and heterosis. J.W. Dudley demonstrated that yield was a function of heterozygosis by observing that, in alfalfa crosses, the F_1 yields reduced as generations advanced. Further, he observed that allele distribution among parents used in a cross impacted heterozygosity. For example, a cross of duplex × nulliplex ($A^2 \times A^0$) always had higher degree of heterozygosity than say a cross of simplex × simplex ($A^1 \times A^1$) regardless of the clonal generation used to make the cross.

Natural selection changes the genotypic composition of synthetics. The effect can have significant consequences when the cultivar is developed in one environment and used for production in a distinctly different environment. There can be noticeable shifts in physiological adaptation (e.g., winter hardiness) as well as morphological traits (e.g., plant height). For example, growing alfalfa seed in the western states (e.g., California) for use in the Midwest, the cultivars may lose some degree of winter hardiness, a trait desired in the production region of the Midwest. A way to reduce this adverse impact is to grow seed crop in the west using foundation seed from the Midwest.

17.7.6 Factors affecting performance of synthetic cultivars

Three factors are key in determining the performance of a synthetic cultivar.

(i) **Number of parental lines used.** Synthetic cultivars are maintained by open pollination. Consequently, the F_2 (*syn-2*) yield should be high to make it a successful cultivar. The Hardy–Weinberg equilibrium is reached in *syn-2* for each individual locus and, hence, should remain unchanged in subsequent generations. It follows

then that the F_3 (*syn-3*) should produce as well as *syn-2*. Some researchers have even shown that F_3 and F_4 generations yielded as much or slightly better than F_2 generations, provided the number of lines included in the synthetic cultivar is not small. With $n = 2$, the reduction in performance will be equal to 50% of the heterosis. Consequently, n has to be increased to an optimum number without sacrificing high GCA. When n is small, the yields of *syn-1* are high, but so are the decline in *syn-2* yields. On the other hand increasing n decreases *syn-1* yields and *syn-2* decline. A balance needs to be struck between the two effects. Some researchers estimate the optimum number of lines to include in a synthetic to be five or six.

(ii) **Mean performance of the parental lines (in syn-0).** The lines used in synthetics should have high performance. A high value of parental lines reduces the reduction in performance of *syn-2* over *syn-1*. Preferably, the parents should be non-inbreds or have minimum inbreeding (e.g., S_0 or S_1).

(iii) **Mean syn-1.** In theory, the highest value of *syn-1* is produced by a single cross. However, alone it will suffer from a higher reduction in performance. It is important for the mean F_1 (*syn-1*) yield of all the component F_1 crosses to be high enough such that the *syn-2* yield would remain high in spite of some decline.

17.7.7 Advantages and limitations of development of synthetics

Advantages

- The method is relatively easy to implement.
- It can be used to produce variability for hybrid breeding programs.
- Advanced genotypes of synthetics show little yield reduction from *syn-1*, making it possible for farmers to save and use seed from the current season to plant the next season.

Disadvantages

- Because inadequate seed is often produced in *syn-1*, the method fails to exploit to the maximum the effects of heterosis, as is the case in conventional F_1 hybrid breeding. The method of synthetics is thus a compromise to the conventional means of exploiting heterosis.

- Natural selection changes the genotypic composition of synthetics, which may be undesirable.

17.8 Backcross breeding

The key concern in the application of the backcross technique to cross-pollinated species is the issue of inbreeding. Selfing cross-pollinated species leads to inbreeding depression (rapid loss of vigor). The use of a recurrent parent in a backcross with cross-pollinated species is tantamount to inbreeding. To minimize the loss of vigor, large populations should be used to enable the breeder sample and maintain the diversity of the cultivar and to insure against the harmful effects of inbreeding.

Just like self-pollinated species, it is relatively straight forward to improve a qualitative trait conditioned by a single dominant gene. The breeder simply selects and advances individuals expressing the trait. Where a recessive gene is being transferred, each backcross should be followed by one round of intercrossing to identify the recessive phenotype.

Improving inbred lines (to be used as parents in crosses) is equivalent to improving self-pollinated species. The key to success is for the breeder to maintain a broad gene base by using adequate number (2–3) of backcrosses and a large segregating population.

Key References and Suggested Reading

Agrawal, R.L. (1998). *Fundamentals of plant breeding and hybrid seed production.* Science Publishers, Inc., Enfield, NH.

Avey, D.P., Ohm, H.W., Patterson, F.L., and Nyquist, W.E. (1982). Three cycles of simple recurrent selection for early heading in winter wheat. *Crop Science*, **15**: 908–912.

Burton, J.W., and Brim, C.A. (1981). Recurrent selection in soybeans III: Selection for increased percent oil in seeds. *Crop Science*, **21**: 31–34.

Chahal, G.S., and Gosal, S.S. (2000). *Principles and procedures of plant breeding: Biotechnological and conventional approaches.* CRC Press, New York.

Dudley, J.W. (1964). genetic evaluation of methods of utilizing heterozygosis and dominance in autotetraploids. *Crop Science*, 4:410–413.

Dudley, J.W. (ed.) (1974). *Seventy generations of selection for oil and protein in maize.* CSSA, Madison, WI.

Dudley, J.W., and Lambert, R.J. (1992). Ninety generations of selection for oil and protein in maize. *Maydica*, **37**: 81–87.

Fehr, W.R. (1987). Principles of cultivar development. McMillan, New York.

Feng, L., Burton, J.W., Carter, T.E.Jr., and Pantalone, V.R. (2004). Recurrent half-sib selection with testcross evaluation for increased oil content in soybean. *Crop Science*, **44**: 63–69.

Hallauer, A.R. (1985). Compendium of recurrent selection methods and their applications. *Critical Reviews in Plant Sciences*, **3**: 1–33.

Moreno-Gonzalez, J., and Cubero, J.I. (1993). Selection Strategies and choice of the breeding method, in *Plant breeding: Principles and prospects* (eds M.D. Hayward, N. O. Bosemark, and I. Romagosa). Chapman and Hall, London, pp. 281–313.

Narro, Pandey, L.S., Crossa, J., DeLeon, C., and Salazar, F. (2003). Using line x tester interaction for the formation of Yellow Maize synthetics tolerant to acid soils. *Crop Science*, **43**: 1718–1728.

Poehlman, J.M., and Slepper, D.H. (1995). *Breeding field crops.* Iowa State University Press, Ames, IA.

Sumarno, T., and Fehr, W.R. (1982). Response to recurrent selection for yield in soybean. *Crop Science*, **22**: 295–299.

Vogel, K.P., and Pedersen, J.F. (1993). Breeding systems for cross-pollinated perennial grasses. *Plant Breeding Reviews*, **11**: 251–174.

Wilcox, J.R., and Cavins, I.F. (1995). Backcrossing high seed protein to a soybean cultivar. *Crop Science*, **35**: 1036–1041.

Outcomes assessment

Part A

Please answer the following questions true or false.

1 The synthetic method of cultivar breeding is suitable for cross-pollinated species.
2 The highest yield performance is obtained in the syn-1 generation.
3 A polycross evaluates SCA.
4 The theoretical optimum of parents to include in a synthetic is about 10–15.
5 Additive gene action is considered more important that dominance genotypic variance for optimum performance of synthetic cultivars.
6 The success of a recurrent selection program depends on the genetic nature of the base population.
7 The ear-to-row breeding in corn is equivalent to full-sib selection.
8 Combining ability is important in half-sib selection method of breeding.
9 Half-sib selection is based on maternal plant selection.
10 In full-sib selection, both parents in a cross are known?

Part B

Please answer the following questions.

1 Why is it more practical to release syn-2 rather than syn-1 to farmers?
2 Give a specific disadvantage of synthetic breeding.
3 Why is the Latin square design a preferred layout of a polycross nursery?
4 Define a synthetic cultivar.
5 Mass selection as applied to cross-pollinated species is also called.
6 In recurrent selection the base population is designated cycle.
7 Give a specific disadvantage of mass selection as applied to cross-pollinated species.
8 What is the technique of gridding in mass selection, and what is its effect on the selection process in breeding?

Part C

Please write a brief essay on each of the following questions.

1 Discuss the key features of a polycross nursery.
2 State and discuss the implications of Sewal Wright's proposed formula for predicting F_2 yield of a group of inbred lines.
3 Discuss the application of synthetic breeding.
4 Compare and contrast the methods of half-sib selection with progeny test and half-sib selection with a testcross.
5 Discuss the issue of heritability in recurrent selection.
6 Describe the method half-sib selection with a testcross.
7 Discuss the concept of recurrent selection.

18

Breeding hybrid cultivars

Purpose and expected outcomes

The methods of breeding discussed so far that are preceded by crossing go beyond the F_1. As previously indicated, the F_2 is the most variable population following a cross. Consequently, selection often starts in this population. In contrast, hybrid breeding ends with the F_1. The purpose of this chapter is to discuss the rationale of a hybrid and the genetics underlying the development of this type of cultivar. After studying this chapter, the student should be able to:

1 Discuss the historical background of hybrid seed development.
2 Discuss the concept of hybrid vigor and its role in hybrid development.
3 Discuss the genetic basis of hybrid seed development.
4 Discuss the steps in the procedure of hybrid breeding.

18.1 What is a hybrid cultivar?

A **hybrid cultivar**, by definition, is the F_1 offspring of a planned cross between inbred lines, cultivars, clones, or populations. Depending on the breeding approach, the hybrid may comprise two or more parents. A critical requirement of hybrid production is that the parents are not identical. As will be discussed next, it is this divergence that gives hybrids their superior performance. The outstanding yields of certain modern crops, notably corn, owe their success to the exploitation of the phenomenon of heterosis (hybrid vigor), which is high when parents are divergent. Much of what we know about hybrid breeding came from the discoveries and experiences of scientists engaged in corn hybrid cultivar development. However, commercial hybrids are now available for many crops, including self-pollinating species.

18.2 Brief historical perspective

One of the earliest records on hybridization dates back to 1716 when American Cotton Mather observed the effects of cross fertilization in maize, attributing the multicolored kernels to wind-borne inter-mixture of different colored cultivars. However, it was the German T.G. Koelreuter who conducted the first systematic studies on plant hybridization in 1766. Even though previous observations had been

---

Principles of Plant Genetics and Breeding, Second Edition. George Acquaah.
© 2012 John Wiley & Sons, Ltd. Published 2012 by John Wiley & Sons, Ltd.

made to the effect that offspring of crosses tended to exhibit superior performance over the parents, it was G.H. Shull who, in 1909, first made clear scientific-based proposals for exploiting heterosis to produce uniform and high yielding cultivars. Unfortunately, the idea was at that time impractical and potentially expensive to commercially exploit. In 1918, D.F. Jones proposed a more practical and cost-effective approach to producing hybrid cultivars by the method of the double-cross. Double-cross hybrids produced significantly more economic yield than the single-cross hybrids originally proposed by Shull. Single-cross hybrid seed was then produced on weak and unproductive inbred parents, whereas double-cross seed was produced on vigorous and productive single-cross plants. The corn production industry was transformed by hybrids, starting in the 1930s.

Other notable advances in the breeding of hybrids were made by researchers, including M.I. Jenkins in 1934 who devised a method (topcross performance) to evaluate the effectiveness of parents in a cross (i.e., combining ability). Through this screening process, breeders were able to select a few lines that were good combiners (productive in a cross) for use in a hybrid breeding.

The next significant impact on hybrid production also came in the area of techniques of crossing. Because corn is outcrossed and monoecious, it is necessary to emasculate one of the parents (i.e., make one female) as part of the breeding process. In the early years of corn hybrid breeding, emasculation was accomplished by the labor-intensive method of mechanical detasseling (removal of the tassel). The discovery and application of **cytoplasmic male sterility (CMS)** to corn hybrid programs eliminated the need for emasculation by the late 1960s. Unfortunately, the success of CMS was derailed when the **Texas cytoplasm (T-cytoplasm),** which was discovered in 1938 and was at that time the dominant form of male sterility used in corn breeding, succumbed to the southern leaf blight epidemic of 1970 and devastated the corn industry. It should be mentioned that mechanized detasselers (rather than CMS) are used by some major seed companies in hybrid seed production of corn today.

Realizing that the limited number of inbred lines used in hybrid programs did not embody the complete genetic potential of the source population, and the need to develop new inbred lines, scientists embarked on cyclical recombination (by recurrent selection) to generate new variability and to improve parental lines. Breeders were able to develop outstanding inbred lines to make single-cross hybrids economical enough to replace double-cross hybrids by the 1970s. By this time, corn hybrid production programs had developed a set of standard practices consisting of the following, as observed by N.W. Simmonds:

- Maintenance and improvement of source population by open pollinated methods (recurrent selection).
- Isolation of new inbreds and improvement of old ones (by backcrossing).
- Successive improvement of single-cross hybrids by parental improvement.
- CMS-based seed production.

The application of hybrid methodology in breeding has socioeconomic implications. The commercial seed industry has rights to its inventions that generate royalties. More importantly, because heterosis is maximized in the F_1, farmers are generally prohibited from saving seed from the current season's crop to plant the next year's crop. They must purchase seed from the seed suppliers each season. Unfortunately, poor producers in developing countries cannot afford annual seed purchase. Consequently, local and international (e.g., international agricultural research centers such as CIMMYT in Mexico) efforts continue to be largely devoted to producing propagable improved open-pollinated cultivars for developing countries.

The idea of commercializing hybrid seed production is traced to Henry A. Wallace, an Iowa farmer, who studied self-pollination and selfing of corn in 1913. His industry led to the founding of Pioneer Hi-Bred Corn Company in Iowa, in 1925. In 1933, only about 0.1% of US corn production was devoted to hybrid seed. Today, hybrid seed is planted on almost all corn fields. Hybrids are also gradually being embraced by developing countries (Chapter 26).

18.3 The concepts of hybrid vigor and inbreeding depression

As previously stated, the hybrid production industry thrives on the phenomenon or **hybrid vigor.**

Industry highlights

Pioneer Hi-Bred, a DuPont business – Bringing seed value to the grower

Jerry Harrington

Pioneer Hi-Bred, a DuPont business, Des Moines, IA, USA

Pioneer Hi-Bred, a DuPont business, is the world's largest seed company with annual sales of over US$5 billion. With headquarters in Des Moines, IA, Pioneer sells seed to growers in over 90 countries. The company has over 10 600 employees with over 78 seed conditioning plants and 110 plant research centers throughout the world.

Founded in 1926, Pioneer was the first company to develop, produce and market hybrid seed corn. Before the advent of hybrid corn, farmers saved grain from one year's crop to use as seed for the next with annual yields averaging between 20 and 40 bushels an acre. With the new hybrids, yields improved dramatically, and corn was larger, stronger and better able to stand up to the elements, resulting today in yields averaging 140 bushels/acre in the United States and commonly 200 bushels/acre throughout the Corn Belt.

By combining seed research with programs to show the value of hybrid seed corn to growers, Pioneer played a significant role in ushering in the modern age of farming.

The company looks quite a bit different than it did over 85 years ago. Pioneer has expanded beyond corn and now develops and markets seed for the soybean, sorghum, sunflower, canola, alfalfa, rice, millet and soft red winter wheat markets. Rapid advances in biotechnology and genetics have dramatically changed research tools and procedures and the very products offered to growers.

In this box, we will concentrate only on corn/maize hybrid development, production and sales.

Research

Pioneer Crop Genetics Research & Development develops hybrids of corn, sorghum, sunflower and canola, and varieties of soybean, alfalfa, wheat, and canola for worldwide markets. Hybrids and varieties are developed at primary research locations and tested at thousands of other locations in order to be sure products are adapted to a wide range of growing environments.

Like all successful businesses, Pioneer R&D has specific goals outlined for the marketplace. The goals of the maize research team are to:

- Develop hybrids with greater than 5% yield performance advantage.
- Reduce crop losses, grower input costs and risk through biotechnology that provides insect, disease and herbicide resistance within maize.
- Create more value and new uses for maize by improving the quality of the grain and forage produced.
- Use available, appropriate technologies that result in improved products for customers.

Customers actually start the hybrid development process. Corn growers, processors, livestock producers and commodity grain users, along with sales and marketing staffs, identify specific traits they want in a hybrid. Then plant genetic researchers draw upon the universal gene pool, proprietary germplasm and genetic technologies to develop new hybrids.

Laboratory and field researchers work together to develop products, and scientists apply the latest crop production technology all the way along the product development cycle. While scientists in the laboratory use technologies to test genes and proteins, scientists in the field evaluate germplasm combinations in numerous environments. Scientists developing Pioneer products lead the seed industry in the development and application of genomic tools. The information gathered through the use of genomic tools, when used in conjunction with other technologies, helps researchers understand gene functions. This is critical because the information helps scientists to better understand which genes determine important traits, gain valuable knowledge of how the genes work together and get insight into genes that control complex traits such as drought resistance and maturity.

Researchers at Pioneer have already discovered some genes that impact important traits such as disease and insect resistance, drought tolerance and grain traits, and are searching for more. Genomic tools also allow researchers to look across species for traits that are important for corn. Scientists use this information, in addition to an extensive library of elite genetics, to develop better products.

A series of additional tools within maize has led to the introduction of transgenic traits that provide resistance to damaging insects and low-cost herbicides. These can lower production costs, enhance grower efficiency and increase yields. Before introducing seed products with these traits onto the marketplace, Pioneer must make sure that these traits are registered within the country where the products are sold and follow all legal and regulatory guidelines within each market. This also includes following product stewardship guidelines within product development and educating its grower customers on these rules, so that the technology is preserved long into the future.

Another application of biotechnology tools is the gene mapping technique. This provides information for direct genetic selection of the gene combinations in breeding lines. Using this information, researchers make more accurate decisions about which lines to use in developing new hybrids.

Researchers at Pioneer spend four to seven years evaluating products in the laboratory and in a wide variety of growing conditions before new hybrids are released for sale to farmers.

To some extent, plant breeding is a numbers game: the more genetic combinations that are developed and tested, the greater the odds of developing improved products faster. Every year, Pioneer maize researchers around the world evaluate about 130 000 new experimental hybrids. These hybrids enter a four- to five-generation testing cycle. One might think of these experimental products as "applicants" to college. The top 10% – 13 000 from the first season of testing – make up the "freshman college class".

During each of the next four generations, the hybrids are tested at more locations and in a range of soil types, stresses and climate conditions. At every stage of testing, researchers look for high, stable yield; standability; tolerance to stresses and other agronomic characteristics. Only hybrids that meet Pioneer standards are advanced to the "sophomore" class. Typically, there are about 6000 sophomores each season, less than half of the freshman class. Each year, approximately 200 Pioneer experimental hybrids make it to the "junior" level; fewer than 100 hybrids make it to the "senior" level. And, finally, from about 130 000 original candidates, only about 15–20 hybrids "graduate" to commercial status.

By the time a Pioneer[R] brand hybrid is offered for sale, it has been tested at more than 150 locations, and in more than 200 customers' fields. This rigorous testing system helps Pioneer researchers develop leading edge, new genetics with a total package of traits that add value for customers.

Supply management

When the decision is made to sell a new hybrid, production of that seed is undertaken by Pioneer's Supply Management division. The mission of this group is to reliably provide the highest quality seed for Pioneer's customers.

The process begins with a small number of seeds of the new product being handed off from Research to Supply Management. From these seeds, parent inbred lines are produced and multiplied by the division's parent seed operations, located in the United States mainland, in Hawaii and at locations around the globe. As the anticipated demand for the hybrid is identified, commercial volumes of seed are grown, conditioned, and stored to fill customer needs.

The volume of seed for a given product hybrid can number into the hundreds of thousands of units (80 000 seeds constitute a unit). However, regardless of volumes, Pioneer maintains high standards of quality – for germination rates, genetic purity and physical purity – and regularly monitors seed quality throughout production for all products.

In North America, there are over 20 commercial corn production facilities, allowing Pioneer to grow and condition products in a number of different environments to spread risk, while producing in the most cost-effective manner. In addition to a summer production cycle in North America, Pioneer manages a "winter" production cycle, using its facilities in South America, to accelerate the production of the latest products, that is, those advanced to commercialization in the fall of one year for sale the next spring. This winter production option also allows for recovery of seed supplies in case of a reduced summer production of a given hybrid.

The Supply Management team makes use of state-of-the-art technology and science throughout production. Research agronomists provide the latest scientific information regarding optimum agronomic conditions to grow and harvest seed crops. Pioneer's sophisticated physical facilities are designed by its own engineers to ensure seed quality is maintained through harvesting, conditioning, and shipping. Despite the millions of units annually conditioned, the equipment is designed to carefully handle every seed with minimal physical damage. A wide range of technologies – global positioning systems, bar codes, automated sampling, and sophisticated inventory management systems – assure that Pioneer produces, ships and tracks seed efficiently and accurately around the world.

The annual production cycle begins with seed fields being contracted from high quality growers around a production location. Pioneer production agronomists work closely with these growers to ensure each field meets Pioneer's standards and is managed to optimum production. For example, to attain the very highest seed purity, seed fields are preferred to be on rotated ground – that is, the field had a crop other than corn during the prior growing season. Attention is also given to how well the seed crop can be isolated from other corn crops and corn pollen. The majority of Pioneer's seed fields are irrigated, which assures adequate moisture is available during critical stages of the crop's development.

Production agronomists help ensure that each field is planted with the correct ratio of individual rows of male or female inbred parent seed and that the seed is planted at the right soil depth and population densities. Field planters are thoroughly inspected before being filled to make sure there is no foreign grain present. Agronomists and contract seed growers work with neighboring farmers, whose fields adjoin Pioneer's seed ground, to guarantee that Pioneer's strict isolation requirements can be achieved.

As the plants emerge and grow, fields are regularly checked for rogues or off-type plants to ensure that only the intended corn plants are present. If any others are found, they are destroyed. All this is done to make sure that customers receive the superior genetics they expect.

Most modern hybrids are single cross, meaning that the pollen comes from one male inbred and fertilizes – or crosses – with a different female inbred. To assure that this cross – and only this cross – occurs, the female tassel is removed prior to shedding pollen so that it does not pollinate itself. Automated mechanical cutters and wheel puller type detasselers are used to handle some of this work initially. Pioneer also employs thousands of teenagers and adults to complete female detasseling by hand. This work is finished only when Pioneer inspectors certify that at least 99.5% of the female tassels in a field have been pulled. Fields not meeting this standard are abandoned as far as seed production is concerned.

Harvest is an especially busy time of year at the production locations and generally begins when the seed corn moisture in the field reaches its physiological maturity level of 35– 40%, this being much higher than the 15–20% moisture level for the harvest of grain. Pioneer research has shown that harvesting seed corn at this higher moisture level and drying it gradually in a controlled environment results in improved yield and quality. However, a freeze at this level of moisture could result in reduced germination levels. At this time of year, production locations run multiple shifts and longer workweeks to accommodate getting the crop in before the weather can adversely impact the crop.

Throughout its growth, the crop has been tracked with sophisticated systems. This careful supervision continues during conditioning. These systems will continue to keep the local production team informed of a wide range of product aspects – origin, amount, location, quality, conditioning activities underway, and so on.

When it's ready, the crop is mechanically picked on the cob with the husk on – much like sweet corn – rather than combined and shelled in the field as grain is harvested. "Husk-on" harvesting ensures gentler handling and helps protect the relatively soft corn seed during this harvest and transportation stage.

A given hybrid, and that hybrid alone, is brought to the unloading area at the production location where the ears are gently unloaded. Between hybrids, the area is thoroughly cleaned to assure there is no mixing of products.

Next is the husking and sorting stage, where the husk is removed and a visual inspection is made of every ear. Any that fail are discarded. Hybrid seed is then dried slowly at low temperatures to just under 13% moisture.

The dried corn moves by conveyer to the sheller where seed kernels are carefully removed from the cob while avoiding damage to the living seed. Semi-finished seed is then moved to special bulk storage bins that are equipped with electronic sensors to monitor the seed temperature and which can quickly engage fans to keep the seed cool during storage.

From here, seed is gently moved by bucket elevators and belt conveyors to seed sizing. This stage physically categorizes seed by width and thickness, so that packaged seed will be uniform. Some farmers request uniform seed size to maintain proper plant populations.

All seed is treated with a fungicide to protect it from soil-borne fungi. Depending on customer requirements, other insecticide seed treatments may be subsequently applied to protect seed from insects once it's out in farmers' fields.

As the seed moves through the production location, quality tests are run to evaluate the seed's status. Seed lots meeting minimum quality standards proceed, while seed lots failing the standard are held aside for further evaluation or discard. The overriding concern is maintaining purity and assuring seed quality.

The final stage of production is packaging. Seed can be packaged into bags of 80 000 kernels, packaged into larger containers, such as PROBOX (holding from 25–50 units) or handled in an unpackaged manner. Regardless of package type, however, every one contains a tag that gives the grower important information about the hybrid type, size, special traits, the date of germination tests, the origin of the seed, and other information.

Seed is then shipped to Pioneer sales agents around North America and other markets throughout the globe. Computer technology provides the tools needed to track the increasingly complex and vast inventory of seed from Pioneer, allowing the company to ship seed when and where it's needed.

At the end of the customers' planting season, unused seed is returned to Pioneer's Supply Management warehouses and stored in large coolers during the hot summer months to better assure that seed germination is preserved. Quality tests continue to be run on each seed lot even while in storage so that the company understands the seed vigor of all of its products. Seed lots that deteriorate beyond acceptable limits during this extended storage period are discarded. The company regularly checks its product supply against anticipated customer demand and lays out plans for growing new crops if needed to assure supplies meet what customer needs.

Sales

The Pioneer sales organization is among the most well trained and equipped in the industry. In North America, Pioneer distributes its seed products through a network of more than 2500 independent sales representatives and several thousand retail dealer outlets in the South and West. Many of the Pioneer sales representatives are qualified as Certified Crop Advisors (CCAs), requiring extensive study of plant agronomy, written exams and follow-up courses.

When the Pioneer sales organization was created in the late-1920s and 1930s, many of the salesmen were farmers who saw the value of the hybrid corn, offering credible testimonials to their neighbors. This was key to getting farmers to accept the new concept of buying hybrid seed corn, rather than using grain produced in their fields. Initially, most of these salespeople were full-time farmers. Because the seed market has become increasingly complex, the modern Pioneer sales organization has evolved into a system of professional salespeople who are dedicated to more full-time selling, with extensive training programs and a demand for computer technology tools and skills. This sales force of independent agents can connect to a Pioneer intranet site that provides them a convenient way to record and store customer information, as well to access product and agronomic information.

In North America alone, Pioneer offers about 300 different types of corn hybrids, with the complexity increasing each year. While each region sells only hybrids developed for that area, the sales force still needs to become familiar with a vast array of crop production knowledge to help each customer decide which set of hybrids works best on his or her farm.

In addition to the wide offerings of diverse genetics, corn hybrids are differentiated by agronomic characteristics, insect-resistant and herbicide-resistant traits, seed size and insecticide seed treatments. In order for growers to get the specific hybrids they want, they are urged to order from Pioneer earlier and earlier each year. Often, the decision making and seed ordering time for growers is at the end of the previous year's harvest. Farmers will evaluate next year's decisions at field days and through yield comparisons on their own farms and those throughout their area.

The independent sales agents in North America are supported by a staff of regional Pioneer agronomists who provide the sales force and customers with agronomic and production information to help get the greatest value from each bag of seed. The Pioneer agronomists, in turn, are supported by Pioneer research information, as well as information generated by Pioneer Agronomy and Nutritional Sciences, which conducts ongoing research on production and animal nutrition issues to support growers.

The Pioneer sales representatives and dealers are also supported by a staff of full-time regional, area and district sales personnel. The sales effort is enhanced by national and regional advertising and a communications effort that positions Pioneer value with its customers and provides information to growers to help make their seed purchase decisions. This includes regional seed catalogs, a national magazine and a Web site offered only to Pioneer customers which provides valuable crop production information for growers.

All these efforts – within Pioneer research, supply management and sales – have one goal in mind. It is to provide the greatest value to Pioneer customers year in and year out.

18.3.1 Hybrid vigor

Hybrid vigor may be defined as the increase in size, vigor, fertility, and overall productivity of a hybrid plant over the mid-parent value (average performance of the two parents). It is calculated as the difference between the crossbred and inbred means:

$$\text{Hybrid vigor} = \{[F_1 - (P_1 + P_2)/2]/[(P_1 + P_2)/2]\}$$

The estimate is usually calculated as a percentage (i.e., $\times 100$).

The synonymous term, **heterosis**, was coined by G.H. Shull. It should be pointed out immediately that, as it stands, heterosis is of no value to the breeder (and hence farmer), if a hybrid will only exceed the mid-parent in performance. Such advantageous hybrid vigor is observed more frequently when breeders cross parents that are genetically diverse. The *practical* definition of heterosis is hybrid vigor that greatly exceeds the better or higher parent in a cross. Heterosis occurs when two inbred lines of outbred species are crossed, as much as when crosses are made between pure lines of inbreeders.

Heterosis, though widespread in the plant kingdom, is not uniformly manifested in all species and for all traits. It is manifested at a higher intensity in traits that have fitness value, and also more frequently among cross-pollinated species than self-pollinated species. All breeding methods that are preceded by crossing make use of heterosis to some extent. However, it is only in hybrid cultivar breeding and the breeding of clones in which the breeder has opportunity to exploit the phenomenon to full advantage.

Hybrids dramatically increase yields of non-hybrid cultivars. By the early 1930s (before extensive use of hybrids), maize yield in the United States averaged 1250 kg/ha. By the early 1970s (following the adoption of hybrids), maize yields had quadrupled to 4850 kg/ha. The contribution of hybrids (genotype) to this increase was estimated at about 60% (the remainder being attributed to production practices.

18.3.2 Inbreeding depression

Heterosis is opposite to **inbreeding depression** (reduction in fitness as a direct result of inbreeding). In theory, the heterosis observed on crossing is expected to be equal to the depression upon inbreeding, considering a large number of crosses between lines derived from a single base population. In practice, plant breeders are interested in heterosis expressed by specific crosses between selected parents, or between populations that have no known common origin.

Reduction in fitness is usually manifested as a reduction in vigor, fertility, and productivity. The effect of inbreeding is more severe in the early generations (5–8). Just like heterosis, inbreeding depression is not uniformly manifested in plants. Plants including onions, sunflower, cucurbits, and rye are more tolerant of inbreeding with minimal consequences of inbreeding depression. On the other hand plants such as alfalfa and carrot are highly intolerant of inbreeding.

18.4 Genetic basis of heterosis

Two schools of thought have been advanced to explain the genetic basis for why fitness lost on inbreeding tends to be restored upon crossing. The two most commonly known are the **dominance theory**, first proposed by C.G. Davenport in 1908 and later by I.M. Lerner, and the **overdominance theory**, first proposed by Shull in 1908 and later by K. Mather and J.L. Jinks. A third theory, the mechanism of epistasis (non-allelic gene interactions), has also been proposed. A viable theory should account for both inbreeding depression in cross-pollinated species upon selfing and increased vigor in F_1, upon hybridization.

18.4.1 Dominance theory

The dominance theory assumes that vigor in plants is conditioned by dominant alleles, recessive alleles being deleterious or neutral in effect. It follows then that a genotype with more dominant alleles will be more vigorous than one with few dominant alleles. Consequently, crossing two parents with complementary dominant alleles will concentrate more favorable alleles in the hybrid than either parent. The dominance theory is the more favored of the two theories by most scientists, even though neither is completely satisfactory. In practice, linkage and a large number of genes prevent the breeder from developing inbred lines that contain all homozygous dominant alleles. If too many deleterious alleles are present it makes it difficult to inbreed to recover sufficient loci with homozygous dominant alleles. Inbreeding depression occurs upon selfing because the deleterious recessive alleles that are protected in the heterozygous condition (heterozygous advantage) become homozygous and are expressed. It should be pointed out that highly productive inbred lines have continued to be produced for hybrid production, the reason why single-cross hybrids have returned to dominance in corn hybrid production.

To illustrate this theory, assume a quantitative trait is conditioned by four loci. Assume that each dominant genotype contributes two units to the phenotype, while a recessive genotype contributes one unit. A cross between two inbred parents could produce the following outcome:

$$
\begin{array}{ccc}
P_1 & \times & P_2 \\
(AAbbCCdd) & & (aaBBccDD)
\end{array}
$$

Phenotypic value $2 + 1 + 2 + 1 = 6 \qquad 1 + 2 + 1 + 2 = 6$

$$\downarrow$$

$$
\begin{array}{c}
F_1 \\
(AaBbCcDd) \\
2 + 2 + 2 + 2 = 8
\end{array}
$$

With dominance, each locus will contribute two units to the phenotype. The result is that the F_1 would be more productive than either parent.

D.L. Falconer developed a mathematical expression for the relationship between the parents in a cross that leads to heterosis as follows:

$$HF_1 = \sum dy^2$$

where HF_1 is the the deviation of the hybrid from the mid-parent value, d is the the degree of dominance, and y is the the difference in gene frequency in the parents of the cross. From the expression, maximum mid-parent heterosis (HF_1) will occur when the values of the two factors (d, y) are each unity. That is, the populations to be crossed are fixed for opposite alleles (y = 1.0) and there is complete dominance (d = 1.0).

18.4.2 Overdominance theory

The phenomenon of the heterozygote being superior to the homozygote is called **overdominance** (i.e., heterozygosity *per se* is assumed to be responsible for heterosis). The overdominance theory assumes that the alleles of a gene (e.g., *A, a*) are contrasting but each has a different favorable effect in the plant. Consequently, a heterozygous locus would have greater positive effect than a homozygous locus and, by extrapolation, a genotype with more heterozygous loci would be more vigorous than one with less heterozygotes.

To illustrate this phenomenon, consider a quantitative trait conditioned by four loci. Assume that recessive, heterozygote, and homozygote dominants contribute 1, 2, and 1½ units to the phenotypic value, respectively:

$$P_1 \quad\quad \times \quad\quad P_2$$
$$\text{(aabbCCDD)} \quad\quad\quad \text{(AABBccdd)}$$

Phenotypic value   1 + 1 + 1 ½ + 1 ½ = 5   1 ½ + 1 ½ + 1 + 1 = 5

$$\downarrow$$
$$F_1$$
$$\text{(AaBbCcDd)}$$
$$2 + 2 + 2 + 2 = 8$$

Heterozygosity *per se* is most superior of the three genotypes.

18.5 Biometrics of heterosis

Heterosis may be defined in two basic ways:

(i) **Better-parent heterosis.** This is calculated as the degree by which the F_1 mean exceeds the better parent in the cross.

(ii) **Mid-parent heterosis.** Previously defined as the superiority of the F_1 over the mean of the parents.

For breeding purposes, the breeder is most interested to know whether heterosis can be manipulated for crop improvement. To do this, the breeder needs to understand the types of gene action involved in the phenomenon as it operates in the breeding population of interest. As Falconer indicated, in order for heterosis to manifest for the breeder to exploit, some level of dominance gene action must be present, in addition to the presence of relative difference in gene frequency in the two parents.

Given two populations (A, B), in Hardy–Weinberg equilibrium, with genotypic values and frequencies for one locus with two alleles *p* and *q* for population A, and *r* and *s* for population B as follows:

Genotypes	Gene frequency		Genotypic values	Number of A_1 alleles
	Population A	Population B		
A_1A_1	p^2	r^2	$+a$	2
A_1A_2	$2pq$	$2rs$	d	1
A_2A_2	q^2	s^2	$-a$	0

After a cross ($A \times B$) between the populations in Hardy–Weinberg equilibrium and genotypic values and frequencies in the cross as follows:

Genotypes	Frequencies	Genotypic values
A_1A_1	pr	$+a$
A_1A_2	$ps + qr$	d
A_2A_2	qs	$-a$

where *p* and *r* are the frequencies of allele A_1 and q and s are frequencies of alleles A_2 in the two populations. Also, $q = 1 - p$ and $s = 1 - r$. The mean values of the populations are P_A and P_B.

$$\begin{aligned} P_A &= [(p - q)a] + 2pqd \\ &= [(2p - 1)a] + [2(p - q^2)d] \\ P_B &= (r - s)a + 2rsd \\ &= (2r - 1) + [2(r - r^2)d] \\ F &= [(pr - qs)a] + [(ps + qr)d] \\ &= [(p + r - 1)a] + [(p + r - 2pr)d] \end{aligned}$$

Calculating heterosis as a deviation from the mid-parent values is as follows:

$$H_{MP} = F_1 - (P_1 + P_2)/2$$
$$= [(p + r - 1)a + (p + r - 2pr)d]$$
$$- \tfrac{1}{2}[(2p - 1)a + 2(p - q^2)d$$
$$+ (2r - 1)a + 2(r - 1)a + 2(r - r^2)d]$$
$$= (p - r)^2 d$$

From the foregoing, if d = 0 (no dominance), heterosis = 0 (i.e., F = MP, the mean of mid-parents). On the other hand, if in population A p = 0 or 1 and by the same token in population B r = 0 or 1 for the same locus, depending on whether the allele is in homozygous recessive or dominant state, there will be a heterotic response. In the first generation, the heterotic response will be due to the loci where p = 1 and r = 0, or vice versa. Consequently, heterosis manifested will depend on the number of loci that have contrasting loci as well as the level of dominance at each locus. The highest heterosis will occur when one allele is fixed in one population and the other allele in the other. If gene action is completely additive, the average response would be equal to the mid-parent, and hence heterosis will be zero. On the other hand, if there is dominance and/or epistasis, heterosis will manifest.

Plant breeders develop cultivars that are homozygous (according to the nature of method of reproduction). When there is complete or partial dominance, the best genotypes to develop are homozygotes or heterozygous, where there could be opportunities to discover transgressive segregates. On the other hand, when non-allelic interaction is significant, the best genotype to breed would be a heterozygote.

Some recent views on heterosis have been published. Some maize researchers have provided evidence to the effect that the genetic basis of heterosis is partial dominance to complete dominance. A number of research data supporting overdominance suggest that it resulted from pseudo-overdominance arising from dominant alleles in repulsion phase linkage. Yet, still, some workers in maize research have suggested epistasis between linked loci to explain the observance of heterosis.

18.6 Concept of heterotic relationship

Genetic diversity in the germplasm used in a breeding program affects the potential genetic gain that can be achieved through selection. The most costly and time consuming phase in a hybrid program is the identification of parental lines that would produce superior hybrids when crossed. Hybrid production exploits the phenomenon of heterosis, as already indicated. Genetic distance between parents plays a role in heterosis.

In general, heterosis is considered an expression of the genetic divergence among cultivars. When heterosis or some of its components are significant for all traits, it may be concluded that there is genetic divergence among the parental cultivars. Information on the genetic diversity and distance among the breeding lines, and the correlation between genetic distance and hybrid performance, are important for determining breeding strategies, classifying the parental lines, defining heterotic groups, and predicting future hybrid performance.

18.6.1 Definition

A **heterotic group** may be defined as a group of related or unrelated genotypes from the same or different populations, which display similar combining ability when crossed with genotypes from other germplasm groups. A **heterotic pattern,** on the other hand, is a specific pair of heterotic groups, which may be populations or lines, which express in their crosses high heterosis and, consequently, high hybrid performance. Knowledge of the heterotic groups and patterns is helpful in plant breeding. It helps breeders to utilize their germplasm in a more efficient and consistent manner through exploitation of complementary lines for maximizing the outcomes of a hybrid breeding program. Breeders may use heterotic group information for cataloging diversity and directing the introgression of traits and creation of new heterotic groups.

The concept of heterotic groups was first developed by maize researchers who observed that inbred lines selected out of certain populations tended to produce superior performing hybrids when hybridized with inbreds from other groups. The existence of heterotic groups has been attributed to the possibility that populations of divergent backgrounds might have unique allelic diversity that could have originated from founder effects, genetic drift, or accumulation of unique diversity by mutation or selection. Interallelic interaction (overdominance) or repulsion phase linkage among loci showing dominance (psuedo-overdominance) could explain the observance of significantly greater heterosis following a cross between genetically divergent populations. Experimental evidence supports the concept of heterotic patterns. Such research has demonstrated that intergroup hybrids significantly out-yielded intragroup hybrids. In maize, one study showed that intergroup hybrids between Reid Yellow Dent x Lancaster Sure Crop out-yielded intragroup hybrids by 21%.

D. Melchinger and R.R. Gumber noted that heterotic groups are the backbone of successful hybrid breeding, and hence a decision about them should be made at the beginning of a hybrid crop improvement program. They further commented that once established and improved over a number of selection cycles, it is extremely difficult to develop new and competitive heterotic groups. This is because, at an advanced stage, the gap in performance between improved breeding materials and unimproved source materials is often too large. However, the chance to develop new heterotic groups could be enhanced with a change in breeding objectives. Once developed, heterotic groups should be broadened continuously by introgressing unique germplasm in order to sustain medium- and long-term gains from selection.

18.6.2 Methods for developing heterotic groups

A number of procedures may be used by breeders to establish heterotic groups and patterns. These include pedigree analysis, geographic isolation inference, measurement of heterosis, and combining ability analysis. Some have used diallel analysis to obtain preliminary information on heterotic patterns. The procedure is recommended for use with small populations. The technology of molecular markers may be used to refine existing groups and patterns or for expediting the establishment of new ones, through the determination of genetic distances.

To establish a heterotic group and pattern, breeders make crosses between or within populations. Intergroup hybrids have been shown to be superior over intragroup hybrids in establishing heterotic relationships. In practice, most of the primary heterotic groups were not developed systematically but rather by relating the observed heterosis and hybrid performance with the origin of parents included in the crosses. One of the earliest contributions to

knowledge in the areas of developing heterotic patterns was made in 1922. Comparing heterosis for yield in a large number of intervarietal crosses of maize, it was discovered that hybrids between varieties of different endosperm types produced a higher performance than among varieties with the same endosperm type. This discovery, by F.D. Richey, suggested that crosses between geographically or genetically distant parents expressed higher performance and, hence, increased heterosis. This information led to the development of the most widely used heterotic pattern in the US Corn Belt – the Reid Yellow Dent x Lancaster Sure Crop.

18.6.3 Heterotic groups and patterns in crops

Heterotic patterns have been studied in various species. For certain crops, breeders have defined standard patterns that guide in the production of hybrids. As previously indicated in maize, for example, a widely used scheme for hybrid development in temperate maize is the Reid × Lancaster heterotic pattern. These heterotic populations were discovered from pedigree and geographic analysis of inbred lines used in the Corn Belt of the United States. In Europe, a common pattern for maize is the European flint × Corn Belt Dent, identified based on endosperm types. In France, $F_2 \times F_6$ heterotic pattern derived from the same open pollinated cultivars was reported. Other patterns include ETO-composite × Tuxpeno and Suwan 1 × Tuxpeno in tropical regions. Alternate heterotic patterns continue to be sought.

In rice, some research suggests two heterotic groups within O. indica, one including strains from S.E. China and another containing strains from S.E. Asia. In rye, the two most widely used germplasm groups are the Petkus and Carsten, while in faba bean three major germplasm pools are available, namely, Minor, Major, and Mediterranean.

Even though various approaches are used for the identification of heterotic patterns, they generally follow certain principles. The first step is to assemble a large number of germplasm sources and then make parent populations of crosses from among which the highest performing hybrids are selected as potential heterotic groups and patterns. If established heterotic patterns already exist, the performance of the putative patterns with the established ones is compared. Where the germplasm accession is too large to permit the practical use of a diallel cross, the germplasm may first be grouped based on genetic similarity. For these groups, representatives are selected for evaluation in a diallel cross.

According to Melchinger, the choice of a heterotic group or pattern in a breeding program should be based on the following criteria:

- High mean performance and genetic variance in the hybrid population.
- High *per se* performance and good adaptation of parent population to the target region.
- Low inbreeding of inbreds.

18.6.4 Estimation of heterotic effects

Consider a cross between two inbred lines, A and B, with population means of X_{P1} and X_{P2}, respectively. The phenotypic variability of the F_1 is generally less than the variability of either parent (Figure 18.1). This indicates that the heterozygotes are less subject to environmental influences than the homozygotes. The heterotic effect resulting from the crossing is roughly estimated as

$$H_{F1} = \overline{X}_{F1} - \tfrac{1}{2}(\overline{X}_{P1} + \overline{X}_{P2})$$

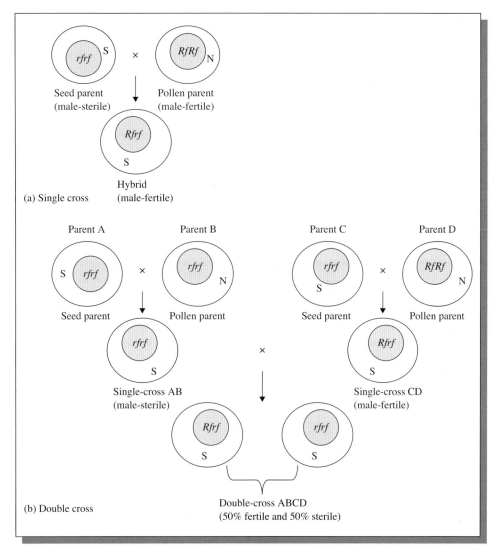

Figure 18.1 Breeding by CMS. (a) single cross and (b) double cross. N, normal cytoplasm; S, sterile cytoplasm. Parent A = A-line; parent B = B-line, and parent D = R-line.

This equation indicates the average excess in vigor exhibited by F_1 hybrids over the midpoint (mid-parent) between the means of the inbred parents. K.R. Lamkey and J.W. Edward coined the term **panmictic midparent heterosis** to describe the deviation in performance between a population cross and its two parent populations in Hardy–Weinberg equilibrium.

18.7 Types of hybrids

As previously discussed, the commercial applications of hybrid breeding started with a cross of two inbred lines (a single cross – $A \times B$) and later shifted to the more economic double cross, $([A \times B] \times [C \times D])$ and then back to a single cross. Other parent combinations in hybrid development have been proposed, including the three-way cross $([A \times B] \times C)$ and modified versions of the single cross, in which closely related crosses showed that the single cross was superior in performance to the other two in terms of average yield. However, it was noted also that the genotype × environment interaction (hybrid × environment) mean sum of squares (from the ANOVA table in Chapter 26) for the single cross was more than twice that for the double crosses, the mean sum of squares for the three-way cross being intermediate. This indicated that the single crosses were more sensitive or responsive to environmental conditions than the other crosses. Whereas high average yield is important to the producer, consistency in performance across years and locations (i.e., yield stability) is also important. As R.W Allard and A.D. Bradshaw explained, there are two basic ways in which stability may be achieved in the field. Double and three-way crosses have a more genetically divergent population for achieving buffering. However, a population of single cross genotypes that are less divergent can also achieve stability on the basis of individual buffering, whereby individuals in the population are adapted to a wide range of environments.

Today, commercial hybrids are predominantly single crosses. Breeders continue to develop superior inbred lines. The key to using these materials in hybrid breeding is identifying pairs of inbreds with outstanding combining ability.

18.8 Germplasm procurement and development for hybrid production

As previously indicated, the breeder needs to obtain germplasm from the appropriate heterotic groups, where available. It is critical that the source population has the genes needed in the hybrid. Plant breeders in ongoing breeding programs often have breeding lines in storage or in nurseries from which potential parents could be selected for future programs. These materials should be evaluated for performance capabilities and, especially, for traits of interest in the proposed breeding program. Germplasm may be introduced from germplasm banks and other sources. Such material should also be evaluated as done with local materials.

18.8.1 Development and maintenance of inbred lines

An **inbred line** is a breeding material that is homozygous. It is developed and maintained by repeated selfing of selected plants. In principle, developing inbred lines from cross-pollinated species is not different from developing pure lines in self-pollinated species. About 5–7 generations of selfing and pedigree selection are required for developing an inbred line. As previously indicated, inbreeders tolerate inbreeding, whereas outbreeders experience varying degrees of inbreeding depression. Consequently, the extent of inbreeding in developing inbred lines varies with the species. Species such as alfalfa and red clover that are more intolerant of inbreeding may be selfed only a few times. Alternatively, sib mating may be used to maintain some level of heterozygosity in these sensitive species.

Hybrid breeding, as previously stated, exploits the phenomenon of heterosis. Heterosis will be highest when one allele is fixed in one parent to be used in a cross and the other allele fixed in the other parent.

Inbred lines of inbreeding species

Inbred lines in self-pollinated species have been discussed previously. They are relatively easy to maintain. The breeder should be familiar with the material to be able to spot off-types that may arise from admixtures or outcrossing in the field. Off-types should be rogued out and discarded, unless they are interesting and warrant additional observation and evaluation.

Physical mixtures occur at harvesting (e.g., due to equipment not cleaned properly before switching to another line), threshing, processing and handling, storage, and at planting. When maintaining certain lines, especially those developed from wild species, it may be necessary to be more vigilant and harvest promptly, or bag the inflorescence before complete maturity occurs to avoid losing seed to shattering.

Inbred lines of cross-pollinated species

Because of the mode of reproduction, breeding lines from cross-pollinated species are more challenging to develop and maintain. Inbred lines may be developed from heterozygous materials obtained from a natural population, or from F_2 selected genotypes. Depending on the breeding procedure, parents for hybrid production may be developed in the conventional fashion, or non-conventional fashion.

Conventional or normal inbreds. Normal inbreds are developed by repeatedly self-pollinating selected plants, from $S_0–S_n$ (for materials drawn from natural populations) or from F_1 to F_n (for materials obtained from crossing), the latter being akin to the pedigree breeding method previously described for self-pollinated species. The S_1 or F_2 populations are heterogeneous, as are results of segregation of traits. Superior plants are selected and progeny-rowed to expose inferior genotypes. Superior individuals are selected for the next cycle of selfing. By S_3, the plants in the progeny would be fairly uniform. After about 6–8 generations of selfing, the negative effects of inbreeding ceases. The next step then is to compare different lines. The value of n, the number of generations of self-pollination, varies from about 5–8. The goal is to attain a level of homozygosity at which the inbred lines are uniform in characteristics and will remain so under continued selfing, with no further loss of vigor. At this stage, the inbred line may be maintained by self-pollination.

Inbred lines should be evaluated for performance and other general agronomic qualities (e.g., drought resistance, lodging resistance, disease resistance), especially those that are basic to the specific crop industry and the production region. This way, the final lines developed would have high desirability and productive potential. These materials are maintained by conventional selfing or sibbing procedures.

Non-conventional inbred lines. To facilitate hybrid production, cytoplasmic male sterility may be incorporated into lines to eliminate the need for mechanical emasculation. Three different inbred lines are required to implement a CMS breeding project. Different kinds of parent materials need to be developed and maintained when making use of a cytoplasmic-genetic male sterility system in breeding. Two kinds of female parents are needed (Figure 18.1) – an **A-line** (male sterile, sterile cytoplasm (s), non-restorer genes (*rfrf*) in the nucleus), and a **B-line** (male fertile, fertile cytoplasm (N), with non-restorer genes (*rfrf*) in the nucleus). The A-line is the seed-producing parent. To develop an A-line, cross a B-line as male to a male sterile female with sterile cytoplasm and fertility restorer genes, followed by repeated backcross (5–7) to the B-line. The A-line and B-line are hence **isogenic** (genetically different at only a specific locus).

To maintain the A-line, the breeder crosses sterile A-line (*S rf rf*) with the B-line (*N rf rf*) to obtain all male sterile offspring. The B-lines (maintainer lines) are maintained by selfing, sibbing or by open pollination in isolated crossing blocks. Another inbred parent, the fertility restorer parent, is required in a CMS breeding program. This parent (called the **R-line**) has the genotype *Rf Rf* and is the pollen parent in the hybrid. It is produced by first selecting a desirable line to be converted into an *R*-line. It is crossed to the restorer gene donor parent using backcrossing, with the R-line as the recurrent parent. The process can be simplified if the donor has sterile cytoplasm (*S rf rf*) and is used as the female parent in the cross. This strategy will eliminate the need of a test cross or selfing after each backcross to distinguish between fertile and sterile individuals (*S Rr rf* x *N rf rf*, *S Rf rf* (fertile), or *S rf rf* (sterile), which are phenotypically distinguishable). Once produced, the R-line is maintained like the B-line.

It should be mentioned that inbred lines of self-pollinated species may also be developed to incorporate a male sterility system for hybrid production.

Genetically modified inbreds. With the advent of genetic engineering, special lines with specific transgenes are developed for use in producing transgenic hybrids. The details of the techniques are discussed in Chapter 25. Inbred lines are transformed with these special genes to develop transgenic breeding stocks.

18.8.2 Storage of seed

It is critical that germplasm be stored such that its viability is retained over the duration of storage. Seed germplasm should be stored at low seed moisture content in an environment in which the humidity, temperature, and oxygen content are low. To this end, seed to be stored is usually dried to about 10.0–12.5% moisture content, and stored at a temperature of less than 21 °C. The specific requirements differ among species. The Harrington's rule of thumb suggests that seed viability is retained for a longer time if the sum of the storage temperature (°F) and relative humidity (%) is less than 100. Relative humidity is more important in the storage of soybean. A sum of 60 is desired for long-term storage of corn. The rate of decline in seed viability in storage also varies among species. Storage in a household freezer may suffice for certain species, especially, small-seeded legumes (alfalfa, clover). The oxygen level in the storage environment may be reduced by introducing gasses such as carbon dioxide, nitrogen, or argon. The seed may also be stored in a vacuum.

18.9 Selection of parents (inbred lines)

The choice of parents to be used in a cross is the most critical step in a plant breeding program for the development of hybrids. The choice of parents depends on the specific objectives of the breeding program and what germplasm is available. Once the inbred lines have been developed, the breeder has the task of identifying a few lines with potential for use as parents in hybrid production. The number of inbred lines that would emerge from a random mating population in which a number of loci are segregating is given by $2n$. Hence, for $n = 10$, there will be 1024 inbreds. First, the large pool of inbreds needs to be significantly reduced by phenotypic selection to identify a small number of high performing inbreds. This is effective for traits of high heritability. The next step is to subject the promising lines to a more rigorous test of their performance in crosses (combining ability test; Chapter 4). Combining ability tests entails crossing each inbred with all other inbreds to be evaluated. Suppose 50 inbreds were selected, the cross combinations required in a combining ability test is given by $n(n-1) = 50(50-1) = 2450$ crosses! To handle this large number, the practice is to use a common tester. As previously indicated, the breeder should select parents from different heterotic groups (interheterotic cross) rather than within the same group. A GCA test should be conducted first, to be followed by a specific combining ability (SCA) test to identify specific pairs of inbreds with exceptional performance in crosses. This sequence of activities is of practical and strategic importance in quickly reducing the large number of inbreds to a manageable size by the time of the more involved evaluations. Certain inbreds have high general combining ability, being able to produce high performing hybrids with a series of other inbreds. On the other hand, certain inbreds are able to "nick" with only a few in the set of inbreds tested. The key decision in combining ability testing is the type of tester to use. A tester can have broad genetic base (e.g., open-pollinated cultivars) or a narrow genetic base (e.g., elite inbreds, related inbred lines).

Where a hybrid breeding program already exists, breeders may want to develop one or two new inbreds to replace those in the program that have been shown to have weaknesses. To replace an inbred in an established single cross, for example, the opposite inbred should be used as tester. Substitute inbred lines may be developed by backcross procedures (so the inbred is least genetically reorganized), or by isolating new inbreds from the same genetic source. New inbreds may also be developed from completely new sources.

18.10 Field establishment

Once a breeder has identified superior inbreds, these lines are used as parents for producing hybrid seed. Considerations for maximizing hybrid seed production in the field include the following:

- **Field preparation.** The field should be properly prepared to obtain a seedbed suitable for the seed size. The field should be free from weeds (use pre- and post-emergence weed control as appropriate). Competition from weeds will adversely affect crop establishment.

- **Planting time.** It is important that the planting be timed such that the seed will germinate promptly for good establishment. Also, the time of pollination should coincide with good weather. In fact, the whole operation, from planting to harvesting, should occur within the growing season, making maximum use of the growing condition for optimum seed yield. The breeder may use **heat units** to calculate the best time for planting the parents (Box 18.1).
- **Synchronization of flowering.** Because a hybrid depends on two different genotypes, the breeder should synchronize the flowering of these inbreds, so that both male and female plants would be ready at that same time for effective pollination. A technique that is often used is **staggered planting**, whereby the inbred lines are planted at different times to insure that pollen will be available for both early and later flowering females.
- **Field layout.** Female and male inbred parents should be arranged in the field such that pollen distribution is effective and efficient. Effective layout patterns vary among species. Male and female lines may be planted in alternating rows, or a certain number of males will alternate with a certain number of females according to pollen producing ability of the pollinator.

Box 18.1 Calculating heat units

Heat units and degree-days

Scientists use the correlation between temperature and plant growth to predict crop harvesting times and also to determine adaptability to a given area. To do this, plant growth is measured in **heat units** (the number of degrees Farenheit the mean daily temperature exceeds a base minimum growth temperature):

Heat units = ((Maximum temperature + Minimum temperature)/2 − Threshold temperature

The base temperature varies with the species (e.g. 40 °F for small grains, 50 °F for corn and soybean, and 60 °F for cotton). A modification of the heat units called **growing degree-days** (GDD) is used where, for the particular species, limits are placed on the daily minimum and maximum temperature (obtained from weather records). For example, if the critical maximum temperature is 86 °F and the minimum is 50 °F, the GDD for that particular day is:

$$GDD = (50 + 86)/2 - 50 = 18$$

The GDD are summed over the growing season to give the total needed for commercial production of these crops.

A generalized equation for estimating accumulated temperature above a threshold is:

$$GDD = T_{mean} - T_{base}$$

The application of this technology is in determining the staggering of planting dates of male and female parents in a hybrid production program in the field to ensure high cross-pollination. GDD is also used for predicting vegetative growth and other plant growth processes. Certain horticultural crops, such as temperate fruits, need to accumulate a certain critical amount of **chilling days** without which fruiting would not occur. Estimation of chilling dates is the opposite of GDD. Field crops such as spring wheat are sown in fall so that they would receive a cold treatment (**vernalization**) during the winter. This treatment triggers the reproductive phase of plant development in spring.

Some species are frost tolerant while others are frost sensitive. Frost damage is most critical when flower buds start to open. The grower should pay attention to and use frost forecast information provided by the USDA. Temperature is also an important requirement for germination. Some species (e.g., oat) can germinate in cool soils (3 °C or 37 °F) while others (e.g., sorghum) prefer a warm temperature (5 °C or 58 °F) for germination.

Temperature is a key factor in the Hopkins Bioclimatic Law. This law states that phenological events (e.g., bloom time) show a lag of four days for each degree of latitude, five degrees of longitude, and 400 feet of altitude, northward, eastward, and upwards in spring and early summer. In autumn, they are westward or downward and four days earlier.

• **Plant density.** Hybrid seed is harvested from only the female parent (i.e., the space occupied by the male is not available for seed production). It is important to maximize plant populations in seed fields. The female plants may be increased relative to the pollen source.

18.11 Maintenance

Once established, the crop should be properly maintained for optimal growth and development. The field care should include weed control, irrigation, proper fertilization, and disease and insect control. To enhance the development of the female line, and reduce the chance of contamination at harvesting, the male line may be removed after pollination is completed. This "thinning" of the field provides additional growth resources for the female line, as a result of reduced competition.

Where pollen control was implemented by mechanical means (as opposed to the use of male sterility system), workers need to walk through the field several times during hybrid seed production to remove any pollen source that may have survived mechanical emasculation. Further, it is good practice to walk through the field to rogue out any off-types prior to pollination.

18.12 Harvesting and processing

The seed should be harvested at the proper maturity and moisture content. A physiological maturity in corn, for example, the kernel moisture is about 30–40%. Safe harvesting is done at a moisture content of 20% or less. The timing of this phase in hybrid seed production is critical because the seed is intended for use as planting material and must be of the highest possible germination capacity. Mechanical damage, physiological immaturity, and improper seed moisture adversely impact seed quality and reduce the germination capacity. Further, improper seed moisture may predispose the seed to rapid deterioration in storage.

The processing needed for the seed varies with the crop. In corn for example, workers first clean the ears to remove diseased and discolored ears before shelling. All seed must be cleaned to remove weed seeds and debris as much as possible. It is required that the

Figure 18.2 Using a mechanical detasseler to emasculate corn. (Courtesy of Pioneer Hi-Bred Seed Company.)

producer of the seed attach a label providing specific information including the seed analysis results.

18.13 Hybrid seed production of maize

Commercial hybrid seed production in maize is used as an example because hybrid production in maize is one of the earliest and most successful exploitation of heterosis. Both CMS and mechanical detasseling are used in commercial seed production. In the United States the single cross is used in maize hybrid breeding. The female plant is male sterile (A-line). It is maintained by crossing with the B-line in isolation. The A-line is grown in alternating rows with the pollinator (R-line) in a ratio of 1 : 2, 2 : 3, or 2 : 4. Some seed companies mechanically detassel their maize instead of using a CMS system (Figure 18.2).

18.14 Hybrids in horticulture

A review by J. Janick indicates that hybrid seed is significantly used in horticultural production. A wide variety of mating systems are used in hybrid seed products of these species. These include hand emasculation (e.g., in sweet pepper, tomato, eggplant), CMS (e.g., in sugar beet, carrot, onion), self-incompatibility (e.g., in cauliflower, broccoli), and monoecy (e.g., in muskmelon, cucumber). The importance of hybrids is variable among species. The approximate percentage of hybrid seed in use in commercial production of selected plants are: carrot – 90% of fresh market and 40–60% of canning and freezing cultivars; broccoli – 100%; cauliflower – 40%; sugar beet – 70%; spinach – 90%; muskmelon – 80–100%; sweet corn – 99%; tomato – 100% of fresh market; and onion – 65%.

In the ornamental industry a similar picture prevails. F_1 hybrid seed is used in begonia (100% by emasculation), impatiens (100% by CMS), petunia (100% by CMS), seed geranium (100% by genetic male sterility), carnation (80% by genetic male sterility), and dianthus (70% by genetic male sterility).

18.15 Exploiting hybrid vigor in asexually reproducing species

18.15.1 Plants with vegetative propagations

Asexual (vegetative) reproduction is the propagation of plants using propagules other than seed. Many horticultural plants are vegetatively propagated. The economic parts of many important world food crops are non-seed, such as tubers (e.g., potato), stems (e.g., sugar cane), and roots (e.g., cassava). Heterosis can be effectively exploited in species that have the capacity to bear seed and yet be propagated vegetatively. In such species, the plant breeder only needs to create one superior genotype. There is no need for progeny testing. The hybrid vigor and other traits assembled in the F_1 can be maintained indefinitely, as long as the genotype is propagated asexually, thereafter.

In horticulture, the superior genotype may be propagated by using techniques such as micropropagation, grafting, budding, sectioning, and cutting (Chapter 8). Successful hybrids have been developed in species such as sugarcane (most commercial cultivars), turfgrasses (e.g., "Tifway"), and forage crops.

18.15.2 Apomixis

Apomixis is the vegetative propagation through the seed. The seed in this instance is genetically identical to the female plant. Using the apomictic propagule is similar to reproducing the plant by other vegetative means, as previously described, only more

convenient. Hybrid vigor is fixed and expressed indefinitely through vegetative propagation. The genetics and mechanisms of apomixis are described in detail under breeding clonally propagated species (Chapter 8).

18.15.3 Monoecy and dioecy

The reproductive biology of monoecy and dioecy has been described previously. Monoecious species bear male and female flowers (imperfect) on the same plants but on different parts of the plant. Environmental conditions (e.g., photoperiod and temperature) can influence sex expression by making one plant more female or more male. In cucumber, short day and low night temperature promotes femaleness, while application of gibberellic acid promotes maleness. Breeders may manipulate the environment to produce hybrid seed.

18.16 Prerequisites for successful commercial hybrid seed production

As briefly reviewed, commercial hybrid seed production is undertaken for a wide variety of species, including field crops and horticultural species. Some species are more suited to commercial hybrid seed production than others. Generally, the following are needed for a successful commercial hybrid seed production venture.

- **High heterosis.** Just as plant breeding cannot be conducted without variability, hybrid seed production is not meaningful without heterosis. The F_1 should exhibit superior performance over both parents. The degree of heterosis is higher in some species (e.g., corn) than others (e.g., wheat).
- **Pollen control and fertility restoration system.** An efficient, effective, reliable and economic system should exist for pollen control to exclude unwanted pollen from a cross. Some species have natural pollination control mechanisms (S_1, male sterility) or reproductive behaviors (monoecy, dioecy) that can be exploited to facilitate the crossing program. A sterility system should have a fertility restoration system to restore fertility to the commercial seed. In the absence of natural pollen control mechanisms, mechanical or hand emasculation should be feasible on a large scale.

- **High F$_1$ yield.** The F$_1$ seed is the commercial product. Species such as corn that bear a large number of seed per F$_1$ plant are more suited to hybrid seed production than species that produce small amounts of seed on an F$_1$ plant (e.g., wheat).
- **Economic seed production.** Hybrid seed production is more expensive overall than conventional seed production. The cost of seed production may be significantly higher when hand emasculation is the method used in the crossing process. In this latter scenario, hybrid seed production would be economic only in high-priced crops (e.g., tomato), or where labor is cheap (e.g., cotton production in India).

Key references and suggested reading

Burton, G.W. (1983). Utilization of hybrid vigor, in *Crop breeding* (ed. Wood D. R.). American Society of Agronomy and Crop Science Society of American. Madison:WI.

Crow, J.F. (1998). 90 years ago: The beginning of hybrid maize. *Genetics*, **148**:923–928.

Hallauer, A.R., and Miranda J.B. (1988). *Quantitative genetics in maize breeding*. Iowa State University Press, Ames, IA.

Janick, J. (1996). Hybrids in horticultural crops. CSSA/ASHS Workshop on heterosis, Indianapolis, November 3, 1996. Department of Horticulture, Purdue University, West Lafayette, IN.

Melchinger, A.E., and Gumber R.K. (2000). Overview of heterosis and heterotic groups in agronomic crops, in *Concepts and breeding of heterosis in crop plants* (eds Lamkey, K.R., and Staub J.E.). Crop Science Society of America, Madison, WI, pp. 29–44.

Norskog, C. (1995). *Hybrid seed corn enterprises. A brief history*. Curtis Norskog, 2901 15th St. S.W, Willmar, MN.

Sinobas, J., and Monteagudo I. (1996). Heterotic patterns among US Corn Belt and Spanish maize populations. *Maydica*, **41**:143–148.

Riday, H., and Brummer E.C. (2002). Heterosis of agronomic traits in alfalfa. *Crop Science*, **42**:1081–1087.

Outcomes assessment

Part A

Please answer the following questions true or false.

1 Hybrid vigor is highest in a cross between two identical parents.
2 CMS may be used in hybrid breeding to eliminate emasculation.
3 The A inbred line is male sterile.
4 G.H. Shull proposed the dominance theory of heterosis.
5 A hybrid cultivar is the F$_1$ offspring of a cross between inbred lines.

Part B

Please answer the following questions.

1 Define a hybrid cultivar.
2 What is hybrid vigor, and what is its importance in hybrid breeding.
3 What is an inbred line?
4 What is a heterotic group?
5 Explain the dominance of single cross hybrids in modern corn hybrid production.

Part C

Please write a brief essay on each of the following topics.

1 Discuss the dominance theory of heterosis.
2 Discuss the importance of synchronization of flowering in hybrid breeding.
3 Discuss inbred lines and their use in hybrid breeding.
4 Discuss the contributions of G.H. Shull and D.F. Jones in hybrid breeding.

19

Breeding clonally propagated species

Purpose and expected outcomes

The modes of reproduction in plants were discussed in detail in Chapter 5. Further, the need for variability for plant improvement to be undertaken has been discussed in various preceding chapters. The most common source of variation is through recombination that comes via the meiotic process in flowering species. Because obligate clonally propagated species may lack the meiotic mechanism, breeding such species has to rely on other means to create variability. After completing this chapter, the student should be able to discuss:

1 The categories of clonally propagated species.
2 Genetic issues in breeding clonally propagated species.
3 Mutation breeding of clonal species.
4 Breeding apomictic species.

19.1 Clones, inbred lines and pure lines

As previously discussed, plants may be naturally sexually or asexually propagated. Further, sexually propagated species may be self-fertilized or cross-fertilized. These natural modes of reproduction have implications in the genetic structure and constitution of plants and breeding implications, as already discussed. Plant breeders are able to manipulate the natural reproductive systems of species to develop plants that have atypical genetic constitution. The terms **pure line**, **inbred line**, and **clone** are applied to materials developed by plant breeders to connote sameness of genetic constitution in some fashion. However, there are some significant distinctions among them.

- **Pure lines**. These genotypes are developed as cultivars of self-pollinated crops for direct use by farmers. As products of repeated selfing of single plants, pure lines are homogeneous and homozygous and can be naturally maintained by selfing.
- **Inbred lines**. These are genotypes that are developed to be used as parents in the production of hybrid cultivars and synthetic cultivars in the

Principles of Plant Genetics and Breeding, Second Edition. George Acquaah.
© 2012 John Wiley & Sons, Ltd. Published 2012 by John Wiley & Sons, Ltd.

breeding of cross-pollinated species. They are not meant for direct release for use by farmers. They are homogenous and homozygous, just like pure lines. However, unlike pure lines, they need to be artificially maintained because they are produced by forced selfing (not natural selfing) of naturally cross-pollinated species.

- **Clones**. Clones are identical copies of a genotype. Together, they are phenotypically homogeneous. However, individually, they are highly heterozygous. Asexually or clonally propagated plants produce genetically identical progeny.

19.2 Categories of clonally propagated species for breeding purposes

Asexually propagated species were previously grouped into two according to economic use as those cultivated for vegetative products and those cultivated for their fruits. For breeding purposes, vegetatively propagated crops may be grouped into four based on flowering behavior.

(i) **Those with normal flowering and seed set**. Species in this category produce normal flowers and are capable of sexual reproduction (to varying extents) without artificial intervention (e.g., sugar cane). However, in crop production the preference is to propagate them sexually. Such species enjoy the advantages of both sexual and asexual reproduction. Hybridization is used to generate recombinants (through meiosis) and introduce new genes into the adapted cultivar, while vegetative propagation is used to maintain indefinitely the advantages of the heterozygosity arising from hybridization.

(ii) **Those with normal flowers but poor seed set**. Some plant species produce normal looking flowers that have poor seed set, or set seed only under certain conditions but not under others. These restrictions on reproduction make it unattractive to use seed as a means of propagation. However, the opportunity for hybridization may be exploited to transfer genes into adapted cultivars.

(iii) **Produce seed by apomixis**. The phenomenon of apomixis results in the production of seed without fertilization, as was first discussed in Chapter 4. Over 100 species of perennial grasses have this reproductive mechanism.

(iv) **Non-flowering species**. These species may be described as "obligate asexually propagated species" because they have no other choice. Without flowers (or with sterile flowers) these species can only be improved by asexual means. Genetic diversity is not obtained via recombination but by other sources (e.g., mutation).

19.3 Breeding implications of clonal propagation

There are certain characteristics of clonal propagation that have breeding implications.

- Clonal species with viable seed and high pollen fertility can be improved by crosses.
- Unlike crossing in sexually reproducing species, which often requires additional steps to fix the genetic variability in a genotype for release as a cultivar (except for hybrid cultivars), clonal cultivars can be released immediately following a cross, provided a desirable genotype combination has been achieved.
- When improving species whose economic parts are vegetative products, it is not important for the product of crossing to be fertile.
- Because of the capacity to multiply from vegetative material (either through methods such as cuttings or micropropagation), the breeder only needs to obtain a single desirable plant to be used as stock.
- Heterosis (hybrid vigor), if it occurs, is fixed in the hybrid product. That is, unlike hybrid cultivars in seed-producing species, there is no need to reconstitute the hybrid. Once bred, heterozygosity is maintained indefinitely.
- It is more difficult to obtain large quantities of planting material from clones in the short time.
- Plant species that are vegetatively parthenocarpic (e.g., banana) cannot be improved by the method of crossing parents.
- Species such as mango and citrus produce polyembryonic seeds. This reproductive irregularity complicates breeding because clones of the parent are mixed with hybrid progeny.
- Many clonal species are perennial outcrossers and intolerant of inbreeding. These are highly heterozygous.
- Unlike sexual crop breeding in which the genotype of the cultivar is determined at the end of the breeding process (because it changes with inbreeding),

the genotype of a clone is fixed and determined at the outset.

- Both general combining ability (GCA) and specific combining ability (SCA), that is performance in crosses, can be fully exploited with appropriate breeding approaches.

19.4 Genetic issues in clonal breeding

Major genetic issues that breeders confront in breeding clonal species include the following:

- **Genetic make-up.** All the progeny from an individual propagated asexually are genetically identical (clones) and uniform. Clones are products of mitosis. Any variation occurring among them is environmental in origin.
- **Heterozygosity and heterosis.** Many species that are asexually propagated are highly heterozygous. They are highly heterotic. Consequently, they are susceptible to inbreeding depression. For those species that can be hybridized without problems, any advantage of asexual propagation is that heterosis, where it occurs, is fixed for as long as the cultivar is propagated asexually.
- **Ploidy.** Many known species that are asexually propagated are interspecific hybrids or have high ploidy.
- **Chimerism.** Clones are stable over many generations of multiplication. The only source of natural variation, albeit rare, is somatic mutation in the bud. Plant breeders may generate variability by the method of mutagenesis. Whether natural or artificial, somatic mutations are characterized by tissue mosaicism, a phenomenon called chimerism.

A chimera or chimeric change occurs when an individual consists of two or more genetically different types of cells. Though heritable changes, these mosaics can only be maintained by vegetative propagation (not transferable to progenies by sexual means).

There are four basic types of chimeras.

1 **Sectorial.** This chimera is observed in a growing shoot as two different tissues located side-by-side. The effect of this modification is that the stem develops with two distinct tissues on each half.
2 **Periclinal.** This type of chimera consists of two thin layers of different genetic makeup, one over the other.

3 **Mericlinal.** When an outer layer of different genetic tissue does no completely extend over the layer below, the chimera is mericlinal.
4 **Graft chimeras.** Unlike the first three chimeras that have genetic origin, a graft chimera is a non-heritable mixture of tissues that may occur after grafting is made.

Whereas chimeras are undesirable in crop plants, they may be successfully exploited in horticulture.

19.5 Breeding approaches used in clonal crops

Several breeding approaches are used in the breeding of asexually propagated species.

19.5.1 Planned introduction

Just like seed, vegetative material (whole plants or parts) may be introduced into a new production area for evaluation and adaptation to the new area. Seedlings or cuttings may be introduced. However, the technology of tissue culture allows a large variety of small samples to be introduced in sterile condition. These disease-free samples are easier to process through plant quarantine.

19.5.2 Clonal selection

Clonal selection has two primary goals – to maintain disease-free and genetically pure clones, and the development of new cultivars.

Purifying an infected cultivar

Clonal cultivars may become infected by pathogens, some of which may be systemic (e.g., viruses). Two general approaches may be used to purify a cultivar to restore it to its disease-free original genetic purity.

1 **Screening for disease-free material.** Plant materials may be visually inspected for the presence of pathogens. However, because some pathogens may be latent, a variety of serological and histological techniques are used to detect the presence of specific pathogens. Called **indexing**, these techniques can detect latent viruses (**viral indexing**) as well as other pathogens. A negative test may not always be proof of the absence of pathogens. It could be that

the particular assay is not effective. The clean clonal material is then used as starting material for multiplication for propagation.

2 **Elimination of pathogen**. A positive test from indexing indicates the presence of a pathogen. Should this be the only source of planting material, the breeder has no choice but to eliminate the pathogen from the plant tissue by one of several methods.

(a) **Tissue culture**. Even when the pathogen is systemic, it is known that tissue from the terminal growing points is often pathogen-free. Tissue from these points may be ascetically removed and cultured under tissue culture conditions to produce disease-free plantlets. Through micropropagation, numerous disease-free plants can be obtained.

(b) **Heat treatment**. This may be short- or long-duration. Short-duration heat treatment is administered to the plant material for about 30 minutes to 4 hours at 43–57 °C. This could be in the form of hot air treatment or by soaking the material in hot water. This works well for fungal, bacterial, and nematode infection. For viruses, a longer treatment of about several weeks (2–4 weeks) is used. Potted plants are held at 37 °C in a controlled environment for the duration of the treatment. Cuttings from the treated plants may be used as scions in grafts, or rooted into a seedling.

(c) **Chemical treatment**. This surface sterilization treatment is suitable for elimination of pathogens that are external to the plant material (e.g., in tubers).

(d) **Use of apomictic seed**. Viral infections are generally not transmitted through seed in cultivars that are capable of apomixis (e.g., citrus).

Clonal selection for cultivar development

This procedure is effective if variability exists in the natural clonal population.

- **Year 1**. Assemble clonal population. Plant and expose to diseases of interest. Select resistant clones with other superior traits and harvest individually.
- **Year 2**. Grow progenies of selected clones and evaluate as in year 1. Select superior clones.
- **Year 3**. Conduct preliminary yield trials. Select superior clones.
- **Year 4–6**. Conduct advanced yield trials at multilocations for cultivar release.

19.5.3 Hybridization with clonal selection

This procedure is applicable to species that are capable of producing seed in appreciable quantities. Because heterosis can be fixed in clonal populations, the breeder may conduct combining ability analysis to determine the best combiners to be used in hybridization.

- **Year 1**. Cross selected parents. Harvest F_1 seed.
- **Year 2**. Plant and evaluate F_1s. Select vigorous and healthy plants.
- **Year 3**. Space plant clonal progeny rows of selected plants. Select about 100–200 superior plant progenies.
- **Year 4**. Conduct preliminary yield trials.
- **Year 5–7**. Conduct advanced yield trials for cultivar release.

Other techniques that are applicable include backcrossing to transfer specific traits and wide crossing. The challenges with backcrossing are several. As previously indicated, clonal species are very heterozygous and prone to inbreeding depression. Backcrossing to one parent (the recurrent parent) provides opportunity for homozygosity and consequently inbreeding depression. To prevent this, breeders may cross the backcross to another clone instead of the recurrent parent, followed selection to identify superior plants. The process is repeated as needed.

19.5.4 Mutation breeding

The subject is discussed in detail in Chapter 23. Inducing variability via mutagenesis is challenging for two key reasons. Being rare events, a large population of M_1V_2 is needed to have a good chance of observing desired mutants. Obtaining a large number of vegetative propagules is difficult. Also, mutations occur in individual cells. Without the benefit of meiosis, the mutated clonal material develops chimeras. Using adventitious buds as starting material reduces the chance of chimeras. A mutation in the epidermal cell (usually there is one) would result in an adventitious shoot that originated from a single mutant cell. This technique is not universally applicable.

19.6 Advantages and limitations of clonal propagation

There several advantages and limitations of breeding clonally propagated species.

Advantages

- Sterility is not a factor in clonal propagation because seed is not involved.
- Because clonal plants are homogeneous, the commercial product is uniform.
- Micropropagation can be used to rapidly multiply planting material.
- Heterozygosity and heterosis are fixed in clonal populations.

Disadvantages

- Clonal propagules are often bulky to handle (e.g., stems, bulbs).
- Clones are susceptible to devastation by an epidemic. Because all plants in the clonal population are identical, they are susceptible to the same strain of pathogen.
- Clonal propagules are difficult to store for a long time because they are generally fresh and succulent materials.

19.7 Breeding apomictic cultivars

Apomixis, the phenomenon of seed development without fertilization was discussed in Chapter 8, including its occurrence in nature, mechanisms, and benefits to the farmer and breeder.

Genetic control of apomixis has been demonstrated, implicating a few genes. Efforts using modern molecular genetic tools continue to be made to isolate those apomictic genes. Apomixis can be a two-edge sword – it can hinder breeding progress or it can be an effective breeding tool. To improve an apomict, there should be suitable materials, that is, sexual or partially sexual plants for use as female plants for crossing. Generally, an obligate apomict cannot be used as a female parent in a hybridization program. However, most apomictic plants produce adequate amounts of viable pollen to be usable as males in crossing. This leaves the identification of a suitable female the first critical step in an apomictic improvement program.

Once suitable parents have been selected, crossing can be conducted as for regular plants. A sexual female plant may be crossed with an apomictic male to produce F_1 hybrids, some of which will be obligate and true breeding apomicts, while others will be asexual

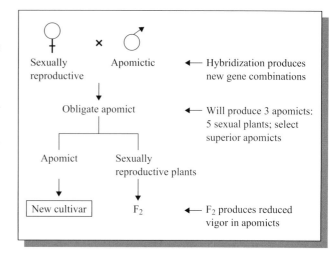

Figure 19.1 Steps in a method for breeding apomictic species.

hybrids that will segregate in the F_2 (Figure 19.1). Because apomicts are generally heterozygous, selfing of sexual clones will yield variability in which the breeder can practice selection. The use of markers and precautions in emasculation will help in distinguishing hybrids from other heterozygous plants. It should be pointed out that the breeder should aim for identifying superior genotypes in the F_1 where heterosis is maximum, rather than in later sexual generations.

As previously indicated, facultative apomicts are more challenging to improve, partly because the breeder cannot control variation in their progenies (produce both sexual and apomictic plants). Furthermore, the stability of the reproductive process is influenced by environmental factors (especially photoperiod). Photoperiod has been observed to significantly affect the relative frequency of sexual versus apomictic embryo sacs in ovules of certain species.

19.8 *In vitro* selection

In vitro selection is essentially selection of desirable genotypes under controlled environments in the Petri dish. This technique is not widely used in modern breeding. Reports on the use of this technology have dwindled over the years, after an explosion of reports shortly following the awareness of its potential as a source of biological variation. As a tool, it is applicable to both the sporophyte and the gametophyte.

19.8.1 Using whole plants or organs

Whole plants, seedlings or embryos may be the units of *in vitro* selection. As previously indicated, an appropriate tissue culture system is needed for *in vitro* selection. The method has been used for screening for resistance to diseases such as *Fusarium culmorum* in wheat and susceptibility to fungal spores in disease breeding. Tolerance (or resistance) to inorganic salts (e.g., salt tolerance in sugar beet) has been reported.

19.8.2 Using undifferentiated tissue

The capacity for regeneration from callus makes the material attractive for use in selection. Plants can be multiplied using a callus system as well. Sometimes, spontaneous variability arises in callus or suspension culture, some of which may be heritable, while some is epigenetic, disappearing when plants are regenerated or reproduced sexually. Tissue culture originating variability is called **somaclonal variation**.

19.8.3 Directed selection

Rather than allowing the variation to arise spontaneously, sometimes plant breeders apply selection pressure during the *in vitro* cultural process, to influence the variability that might arise.

Selection for disease resistance

Various toxin metabolites have been included in tissue culture for use as the basis for selection, assuming that such metabolites have a role in pathogenesis. The culture filtrates from various fungi (*Fusarium, Helminthosporium maydis*) have been used to exert selection pressure for cells that are resistant to the pathogen. The main constraint to the use of directed selection in plant breeding for disease resistance is the inability of the *in vitro* system to be used to select for hypersensitivity, a major strategy in disease resistance.

Selection for herbicide tolerance

Mutants with about 10–100 times the levels of resistance to herbicides (e.g., imidazilinone in sugar beet) have been successfully isolated, characterized, and incorporated into commercial cultivar development. Many of the recorded successes with *in vitro* selection have been with herbicide tolerance.

Selection for tolerance to abiotic stresses

Selection for tolerance to salinity, metals (Zn, Al) and temperature (cold tolerance) has been attempted with varying degrees of success.

Single cell selection system

Some researchers use single cell tissue culture systems (suspension culture, protoplast culture) for *in vitro* selection. The advantages of this approach include a lack of chimerism, higher chances of isolation of true mutants, and the ability to more effectively apply microbial procedures to the large number of individual cells than can be screened in a small space. Selection for biotic stress resistance, herbicide tolerance (the author did this for chlorsulfuron), and aluminum tolerance, are among successful application of direct selection using single a cell selection system.

Key references and suggested reading

Chaudhury, A.M., Ming, L., Miller, C., Craig, S., Dennis, E.S., and Peacock, W.J. (1997). Fertilization-independent seed development in Arabidopsis thaliana. *Proceedings of the National Academy of Sciences USA*, **94**:4223–4228.

Hanna, W.W., and Bashaw, E.C. (1987). Apomixis: Its identification and use in plant breeding. *Crop Science*, **27**:1136–1139.

Jain, M., Chengalrayan, K., Gallo-Meacher, M., and Misley, P. (2005). Embryogenic callus induction and regeneration in a pentaploid hybrid bermudagrass. cv Tifton 85. *Crop Science*, **45**:1069–1072.

Kamo, K. (1995). A cultivar comparison of plant regeneration from suspension cells, callus, and cormel slices of *Gladiolus. In Vitro Cellular and Developmental Biology*, **31**:113–115.

Kindinger, B., Bai, D., and Sokolov, V. (1996). Assignment of a gene(s) conferring apomixis in Tripsacum to a chromosome arm: cytological and molecular evidence. *Genome*, **39**:1133–1141.

Koltunow, A.M., Bicknell, R.A., and Chaudhury, A.-M. (1995). Apomixis: Molecular strategies for the generation of genetically identical seeds without fertilization. *Plant Physiology*, **108**:1345–1352.

Mohammadi, S.A., Prasanna, B.M., and Singh, N.N. (2003). Sequential path model for determining interrelationship among grain yield and related characters in maize. *Crop Science*, **43**:1690–1697.

Murashige, T., and Skoog, T. (1962). A revised medium for rapid growth and bioassays with tobacco tissue culture. *Physiologia Plantarum*, **15**:473–497.

Scityavathi, V.V., Janhar, P.P., Elias, E.M., and Rao, M.B. (2004). Effects of growth regulators on *in vitro* plant regeneration in Durum wheat. *Crop Science*, **44**:1839–1846.

Stefaniak, B. (1994). Somatic embryogenesis and plant regeneration of gladiolus. *Plant Cell Reports*, **13**:386–389.

Tae-Seok, K.O., Nelson, R.L., and Korban, S.S. (2004). Screening multiple soybean cultivars (MG 00 to MG VIII) for embryogenesis following *Agrobacterium*-mediated transformation of immature cotyledons. *Crop Science*, **44**:1825–1831.

Trigiano, R.N., and Gray, D.J. (eds) (1996). *Plant tissue culture concepts and laboratory exercises*. CRC Press, New York, NY.

Internet resources

http://aggie-horticulture.tamu.edu/tisscult/microprop/microprop.html – links to numerous aspects of plant micropropagation (accessed March 10, 2012).

http://www.sprrs.usda.gov/apomixis.htm – Comments from foremost scientists in field of apomixis (accessed March 10, 2012).

http://www.blogontheweb.com/tissue_culture – Excellent discussion on tissue culture (accessed March 10, 2012).

Outcomes assessment

Part A

Please answer the following questions true or false.

1 Propagation by cuttings is a form of clonal propagation.
2 Diplospory is the most common mechanism of apomixis in higher plants.
3 Facultative apomicts reproduce exclusively by apomixis.
4 Pathogenesis is equivalent to haploidy.

Part B

Please answer the following questions.

1 What is clonal propagation?
2 What is apomixis?
3 Distinguish between apospory and displospory as mechanisms of apomixis.
4 Species that reproduce exclusive (or nearly so) by apomixis are described as . . . .
5 Give a specific advantage of clonal propagation.

Part C

Please write a brief essay on each of the following topics.

1 Discuss the key breeding implications of clonal propagation.
2 Discuss the benefits of apomictic cultivars in crop production.
3 Apomixis can be a two-edged sword in plant breeding. Explain.
4 Discuss the occurrence of apomixis in nature.

Section 7

Molecular breeding

Plant breeding is a numbers game. Classical breeding procedures often entail the production of large numbers of genotypes from which breeders attempt to identify a few with the most desirable combinations of genes. Modern DNA modification technologies allow breeders to generate new variability that was previously impossible to obtain. Similarly, molecular marker technologies allow them to more effectively and efficiently discriminate among variability to identify the most desirable individuals to advance in a breeding program. DNA marker technology is not only one of the more readily accessible technologies in modern breeding but also it is less controversial in the ongoing biotechnology safety debate.

20

Molecular markers

Purpose and expected outcomes

*Plant breeders use **genetic markers** (or simply markers) to study genomic organization, locate genes of interest, and facilitate the plant breeding process. After studying this chapter, the student should be able to:*

1 Discuss the concept of markers.
2 Define genetic markers and discuss their importance in plant breeding.
3 Discuss the types of markers.
4 Compare and contrast different kinds of molecular markers.

20.1 The concept of genetic markers

Genetic markers are simply landmarks on chromosomes that serve as reference points to the location of other genes of interest when a genetic map is constructed. Breeders are interested in knowing the association (linkage) of markers to genes controlling traits they are trying to manipulate. The rationale of markers is that an easy-to-observe trait (marker) is tightly linked (located in close proximity on the chromosome) to a more difficult-to-observe and desirable trait (the target of the breeding programs or study). Selection (discriminating among variability) is the method used to advance the breeding population. Hence, in a breeding program, breeders select for the trait of interest by indirectly selecting for the marker (that is readily assayed or detected or observed). When a marker is observed or detected, it signals that the trait of interest is present (by association). Markers do not affect the phenotype of the trait *per se*.

Genetic markers can be detected at both the morphological level and the molecular or cellular level, which is the basis for classification of markers into two general categories as **morphological markers** and **molecular markers**. Morphological markers are manifested as adult phenotypes, the products of the interaction of genes and the environment (e.g., seed shape, flower color, and growth habit). On the other hand, molecular markers are detected at the subcellular level and can be assayed before the adult stage in the lifecycle of the organism. Molecular markers are assayed by chemical procedures of necessity. Morphological markers, while still useful in plant breeding, are less so because they are limited in number and also influenced by the environment. Further, the phenotypes of some morphological markers depend upon

Principles of Plant Genetics and Breeding, Second Edition. George Acquaah.
© 2012 John Wiley & Sons, Ltd. Published 2012 by John Wiley & Sons, Ltd.

the developmental stage of the plant at which they are assayed. The bulk of this chapter is devoted to discussing various aspects of molecular markers.

20.2 Use of genetic markers in plant breeding

Heritable (genetic) variation is central to the success of plant breeding. One of the goals of breeders is to first understand the nature of genetic variation in the plant species of interest, so it can be properly utilized for plant improvement. Molecular markers are used for several purposes in plant breeding.

- **Gaining a better understanding of breeding materials and breeding system.** The success of a breeding program depends to a large extent on the materials used to initiate it. Molecular markers can be used to characterize germplasm, develop linkage maps, and identify heterotic patterns. An understanding of the breeding material will allow breeders to select the appropriate parents to use in crosses. Usually, breeders select genetically divergent parents for crossing. Molecular characterization will help to select parents that are complementary at the genetic level. Molecular markers can be especially useful in identifying markers that co-segregate with quantitative trait loci (QTLs) to facilitate the breeding of polygenic traits.
- **Rapid introgression of simply inherited traits.** Introgression of genes into another genetic background involves several rounds of tedious backcrosses. When the source of desirable genes is a wild species, the issue of linkage drag becomes more important because the dragged genes are often undesirable, requiring additional backcrosses to accomplish breeding objectives. Using markers and QTL analysis, the genome regions of the wild genotype containing the genes encoding the desirable gene can be identified more precisely, thereby reducing the fragment that needs to be introgressed, and consequently reducing linkage drag.
- **Early generation testing.** Unlike phenotypic markers that manifest often in the adult stage, molecular markers can be assayed at an early stage in the development of the plant. Breeding for compositional traits, such as high lysine and high tryptophan genes in maize, can be advanced with early detection and selection of desirable segregants.

- **Unconventional problem solving.** The use of molecular markers can bring about novel ways of solving traditional problems, or solving problems traditional breeding could not handle. When linkage drag is recessive and tightly linked, numerous rounds of backcrosses may never detect and remove it. Disease resistance is often a recessive trait. When the source of the gene is a wild germplasm, linkage drag could be difficult to remove by traditional backcross procedure. Marker analysis can help to solve the problem, as was done by J.P.A. Jansen when he introgressed resistance to the aphid *Nasonovia ribisnigi* from a wild lettuce *Lactuca virosa* by repeated backcross. The result of the breeding was a lettuce plant of highly undesirable quality. The recessive linkage drag was removed by using DNA markers flanking the introgression to pre-select for individuals that were recombinant in the vicinity of the gene.

  The life span of new cultivars can be extended through the technique of **gene pyramiding** (i.e., transferring multiple disease resistance genes into one genotype) for breeding disease resistant cultivars. Marker assisted backcross can be used to achieve this rapidly, especially for genes with indistinguishable phenotypes.
- **Plant cultivar identification.** Molecular markers are effective in cultivar identification for protecting proprietary rights as well as authenticating plant cultivars.

20.3 Concept of polymorphism and the origin of molecular markers

Polymorphism is the term used to describe the frequent and simultaneous occurrence within a single interbreeding population of more than one discontinuous genotype or variation. They are observable when a single gene with a distinct and recognizable effect occurs in a population at a rate that is too frequent to be attributable to chance or mutation alone. Consequently, a genetic variant should appear in at least 1% of the population to be declared polymorphic. Polymorphisms are very common in natural populations. In terms of markers, polymorphism may be defined as the simultaneous occurrence of more than one allele or genetic marker at the same locus, with the least frequent allele or marker occurring at a rate that cannot be attributed to mutation alone. Polymorphisms are elucidated via

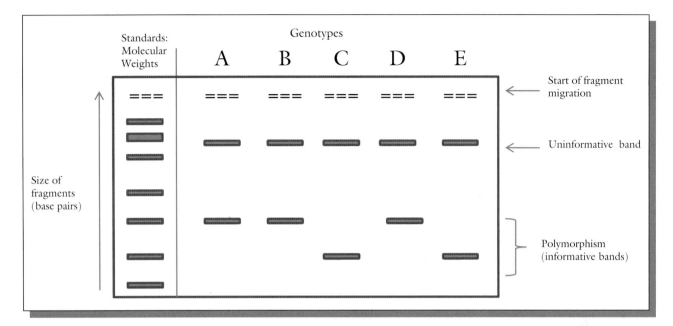

Figure 20.1 Eletrophoretic mobility patterns reveal differences among genotypes. Some bands show polymorphism and can be used as the basis for distinguishing among genotypes.

electrophoresis. Electrophoretic mobility patterns reveal the different forms of the gene or DNA that are present (Figure 20.1). In the case of enzymes, the electrophoretic variants are called **allozymes**.

Polymorphism in the genome can arise from a variety of DNA mutations, including substitution events (point mutations), chromosomal re-arrangement events (insertions or deletions), and errors in the replication of tandemly repeated DNA. The most common type of polymorphism derives from variation at a single base pair and is called a single nucleotide polymorphism (SNP). Depending upon the mutational event, the polymorphism may be due to variation in length of the DNA fragment (fragment length polymorphism). Because molecular markers are usually located in non-coding regions of DNA, they are selectively neutral. Further, they are limitless and unaffected by the environment or the developmental stage of the plant part used for assaying.

20.4 Brief history of molecular markers

Chemical assay for **isozymes** (multiple forms of enzymes) or **allozymes** (allelic variation of enzymes) marked the beginning of the practical application of molecular markers. The **enzyme-based marker** assays detect variations in protein products (products of translation) not variation in DNA *per se* (encodes the proteins). Their use is limited by the insufficient number of assays available (only about several dozen protocols exist) and their uneven distribution on the genetic map. **DNA-based markers** have the advantage of directly surveying DNA variation *per se*, rather than the electrophoretic mobility of proteins. Further, this technology allows researchers to quantify the number of mutation between alleles.

In the 1960s, Arber, Smith, and Nathans discovered and isolated DNA-cleaving enzymes from bacteria. Called restriction endonucleases, these "molecular scissors" had the capability of cleaving DNA molecules at characteristic sites into pieces, called restriction fragments. The pattern of restriction gave rise to variations in lengths that were the result of single nucleotide substitutions in the recognition sequence of the restriction enzyme, the basis of a new marker called the **restriction fragment length polymorphisms (RFLPs)**. A significant advantage of this marker system is that it allowed researchers to assay genetic variation that occurred in the non-coding parts of the DNA sequence (unexpressed or silent) as well as in coding regions. With this capability, the number of markers that could theoretically be generated was infinite, except that the methodology of RFLP is technically challenging because it required a suitable hybridization probe to be available for the assay to detect an existing polymorphism. Consequently, by the mid-1990s, RFLP had been marginalized as a marker system in favor of the more versatile **minisatellite markers**.

Minisatellites consist of tandem repeats of nucleotides that yield polymorphisms on the basis of the length of repeats. Just like their predecessor RFLPs, minisatellite polymorphisms are generated by subjecting the DNA to restriction enzyme digest followed by a hybridization step. The technique yields an extremely large number of polymorphisms. However, associated with this prolific yield of detectable variation is a complex electrophoretic banding pattern that makes it impossible to assign alleles to a specific locus. Single locus minisatellites were eventually developed. The technique of DNA fingerprinting was boosted by the copious amounts of polymorphism generated by minisatellites. However, the non-random distribution of minisatellites on the genome also reduced its versatility as a tool for genetic analysis, especially, mapping and association studies, gradually becoming unimportant as a marker system by the late 1990s.

Perhaps the workhorse in marker technology is electrophoresis, which is the basis for detecting genetic variants. However, in 1965, a game-changing technology, the **polymerase chain reaction (PCR)**, revolutionized the development and application of molecular markers. Prior to this, the requirement of cloning or isolating large amounts of pure DNA posed severe technical challenges and limitations to molecular maker technology. With the advent of PCR, researchers were able to gain access to the entire genome as a source of genetic variation and, thereby, freely amplify any region for investigation without the previous restrictions. This gave birth to a new set of molecular markers, the first of which were **microsatellites**, which, like minisatellites, consisted of tandem repeats of nucleotides, only shorter sequences. Other PCR-based marker systems are RAPDs, AFLPs, and ISSRs.

As biological knowledge abounds, researchers continue to advance molecular marker technology by modifying existing methods for assaying genetic variation or developing additional types which are essentially variations on existing themes. The trend is to increase throughput and make new molecular marker technologies more informative, easy to use, and less expensive.

20.5 Classification of molecular markers

Molecular markers are classified in various ways, including on genetic basis and operational basis. On the basis of genetic characteristics, molecular markers may be grouped into two general categories:

(i) **Single locus, multi-allelic, codominant markers.** Examples are RFLPs and microsatellites (SSR). Microsatellites are capable of detecting higher levels of polymorphisms than RFLPs.

(ii) **Multilocus, single allelic, dominant markers.** Examples are AFLP (amplified fragment length polymorphism) and RAPD (random amplified polymorphic DNA).

As discussed previously, genetic variation in the genome may be assayed at the DNA level or the protein level. Further, a variety of mutations in the genome gives rise to the variations that cause polymorphisms in the population. Figure 20.2 shows the difference between the results of electrophoresis involving a codominant marker and a dominant marker. Whereas a codominant marker can discriminate between homozygote and heterozygote genotypes, a dominant marker cannot. These mutations cause variations in the DNA sequence, or cause certain sequences to be repeated. In view of the foregoing, molecular markers may be classified into three classes according to the nature of genome variation or the methods used for assaying variation:

(i) Variation in proteins (allozymes).
(ii) Variation in DNA sequence (DNA sequence polymorphism, e.g., RFLPs).
(iii) Variation in DNA repeats (e.g., microsatellites).

Various methods are used to detect molecular markers. In this regard, four distinct conceptual classes of molecular markers (marker systems) may be identified:

(i) Enzyme-based markers.
(ii) Hybridization-based markers.
(iii) PCR-based markers.
(iv) DNA-sequence based.

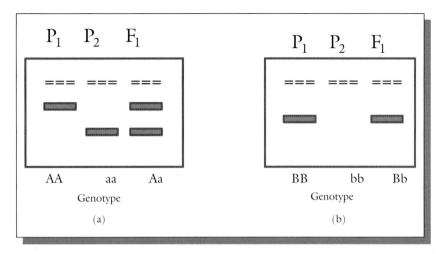

Figure 20.2 Markers may be (a) codominant or (b) dominant, the former showing a heterozygote with both bands.

20.6 Enzyme-based markers

Enzymes are macromolecular compounds that catalyze biochemical reactions. Most enzymes are proteins. Enzymes catalyze specific chemical reactions. Scientists have developed methods that allow the coupling of certain chemical reactions to the biochemical processes for colorimetric detection and location of specific enzymes. **Isozymes** are multiple forms of an enzyme that differ from each other by the substrate they act on, maximum activity, or regulatory properties. The term refers to enzyme polymorphisms that result from different loci. The term, **allozyme**, is reserved for allelic enzymes.

Proteins can exist at one of four levels of structural complexity, of which the primary structure is the simplest. The most complex form, the quaternary structure, is attained through the folding and aggregation of polypeptide units. When an enzyme comprises of one polypeptide chain, it is called a **monomer**. An enzyme comprising aggregates of polypeptide chains is called a **multimer** (or polymer). If an allozyme is multimeric, both homomers and heteromers will be produced in heterozygous individuals.

The enzyme-based marker system has certain limitations, the major ones being the paucity of isozymes in plants and the tendency to be limited to certain chromosomes (not evenly distributed in the genome). Also, isozymes are sensitive to tissue type and age. However, the technology is inexpensive and relatively easy to apply. Some of the earlier successful applications were made in tomato (e.g., tagging of the *Aps-1* locus for acid phosphatase, and the exploitation of its linkage to nematode resistance). In spite of advances in molecular marker technology, protein markers are used for certain purposes, such as simple way of authentication of hybridity in hybrid seed development.

20.7 Hybridization-based markers

DNA-based marker systems may be described as hybridization-based makers when the DNA profiles are visualized by hybridizing fragments to labeled probes.

20.7.1 Restriction fragment length polymorphisms (RFLPs)

The **restriction fragment length polymorphisms (RFLPs)** marker technology is the first generation of DNA markers and one of the best for plant genome mapping. These variations are codominantly inherited. The markers became possible as a result of the discovery of restriction endonucleases, bacterial enzymes with characteristic restriction or recognition sites (cleave DNA only at sections where a unique and recognizable nucleotide sequence occurs, e.g., AATT). As indicated previously, mutation events (e.g., insertion, deletion) cause natural variations to occur in the genome. These variations may cause alterations (abolish) in the recognition sites for restriction enzymes. As a consequence, when homologous chromosomes are subjected to restriction enzyme digest, different restriction products varying in length are produced, which in turn generate variable electrophoretic mobility patterns (hence called restriction fragment length polymorphisms).

RFLPs are randomly distributed throughout the genome of an organism and may occur in both exons and introns. DNA profiles or fingerprints produced are specific to the combination of restriction enzyme and probe (used to detect the polymorphism by Southern blotting). Probes may be derived from a random genomic DNA library, cDNA library, or minisatellites from other organisms.

There are different types of RFLP polymorphisms, the simplest being the 2-allele system involving the presence or absence of a recognition site for a single restriction enzyme. Screening reveals three different types of banding patterns: a large band (homozygous), two smaller bands (restriction site occurs on both homologs), and all three bands (heterozygote). It is assumed that a single base pair change within the recognition site will result in a chromosome that would either have the restriction site or not. In another allele system, one band corresponds to one allele. This system is also easy to score. One variable band corresponds to a homozygote. An individual inherits only two of the variant types of fragment sizes. Tomato was one of the first plant species to be characterized by RFLPs. The disadvantage of this marker system is that it is expensive and has low throughput.

One of the advantages of RFLPs is that the sequence used as a probe need not be known. All that a researcher needs is a genomic clone that can be used to detect the polymorphism. Very few RFLPs have been sequenced to know what sequence variation is responsible for the polymorphism. In the absence of sequence information, interpreting complex RFLP allelic systems may be problematic. A key disadvantage of this marker is the large amount of DNA required for restriction digest and Southern blotting. Also, a RFLP marker system is laborious and time consuming to implement. Polymorphism is not only low but this marker is not effective for detecting single base changes.

RFLP markers may be converted to various PCR-based markers.

20.7.2 Minisatellites (SSRs)

A repetitive DNA sequence is common in the eukaryotic genome, varying between 30 and 90% of the genome according to the species. A minisatellite is a section of DNA that consists of variant repeats of bases, the repeat length varying between 10 and 60 bp, and sometimes over 100 bp. The variant repeats are heterogeneous and tandemly intermingled. The repeats tend to be GC-rich with a strong strand asymmetry. Further, minisatellites comprise a consensus or core sequence (e.g., GGGCAGGANG″ where N could be any base). Minisatellites arise through tandem duplications and tandem deletions of variants. Even though sometimes found in tandem array, they tend to be more commonly scattered throughout the genome of the species. Minisatellites are also called **variable number tandem repeats (VNTRs)**. They yield a high level of polymorphism and are, hence, useful as markers in linkage analysis and population studies. Locus-specific probes are required to detect these highly polyallelic fragment length variation.

To use as an assay of genomic variation, researchers hybridize genomic DNA with a minisatellite core sequence to generate what has been described by some as a "bar-code-like" hybridization pattern. The application opened the way to DNA fingerprinting championed by Allec Jeffreys in the 1980s. However, the complex bar-code-like patterns preclude the application of microsatellites in some standard genetic analysis.

20.8 PCR-based markers

Polymerase chain reaction (PCR) is a technology discovered in 1986 for directly amplifying a specific short segment of DNA without the use of a cloning method. This eliminates the tedious process of repeated cloning to obtain ample quantities of DNA for a study. An attractive feature of a PCR-based marker system is that only a minute amount of DNA is needed for a project. Also, it has a higher throughput than RFLP. Because of the sensitivity of PCR technology to contamination, it is common to observe a variety of bands that are not associated with the target genome but are artifacts of the PCR condition. Consequently, certain bands may not be reproducible.

Industry highlights

Molecular marker survey of genetic diversity in the Genus Garcinia

George N. Ude[1], Brian M. Irish[2], and George Acquaah[1]

[1] Natural Science Department, Bowie State University, Bowie, MD 20715, USA

[2] USDA-ARS, Tropical Agriculture Research Station (TARS) Mayaguez, 2200 Pedro Albizu Campos Ave. Suite. 201, Mayaguez, Puerto Rico 00680-5470

Introduction

The Genomics Laboratory at Bowie State University is an NSF-supported facility. Research activities in this laboratory include genetic diversity analysis of tropical plant germplasm (Genera – *Garcinia* and *Annona*) using molecular techniques, for example, Random Amplified Polymorphic DNA (RAPD) and Amplified Fragment Length Polymorphisms (AFLP). Also, researchers conduct molecular marker assisted selection (MAS) for disease resistance, molecular mapping, and DNA sequencing. Some of the work is done in collaboration with researchers at the USDA-ARS Tropical Agricultural Research Center at Mayaguez, Puerto Rico.

Garcinia is a large genus with about 400 species. The species are found in regions of West Africa and South East Asia and are mostly cultivated for their edible fruits (Figure B20.1). *Garcinia mangostana* (mangosteen) produces the most treasured fruit in the Guttiferae family, with a total world production estimated to be about 150 000 tons/annum. Despite the usefulness of this crop and the dependency of many local economies on the commercial export of the fruits, very little is known about genetic variation within and among the species in the genus. The objective of this study was to use the RAPD molecular marker technique to study genetic variation within 43 accessions derived from 14 species in the genus *Garcinia*.

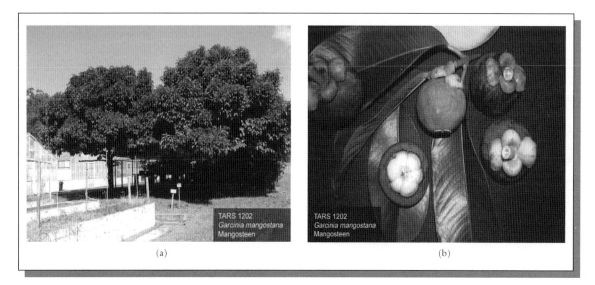

Figure B20.1 *Garcinia mangostana* tree (a) and fruits (b). Figure courtesy of George Ude.

Study 1: RAPD marker

Materials and method

DNA samples extracted from leaves derived from 14 species (Table B20.1) of the genus *Garcinia*, from the USDA-ARS TARS collection, were subjected to PCR using nine Operon primers. Agarose gel electrophoresis was used to separate

Table B20.1 *Garcinia* accession studied.

Sample number	Scientific name	TARS number
1	*Clusia Major*	16300
2	*Mammea americana*	3305
3	*G. benthamii*	6813
4	*G. benthamii*	6813
6	*G. cochinensis*	1769
7	*G. cochinensis*	1769
8	*G. cochinensis*	1769
9	*G. cochinensis*	1769
10	*G. cornea*	6815
11	*G. cymosa*	15219
12	*G. cymosa*	15219
13	*G. dulcis*	16312
14	*G. dulcis*	16312
15	*G. dulcis*	16312
16	*G. dulcis*	16312
17	*G. livingstonei*	1910
18	*G. livingstonei*	1910
19	*G. madruño*	18000
20	*G. madruño*	18000
21	*G. portoricensis*	18011
22	*G. subelliptica*	4560
23	*G. subelliptica*	4560
24	*G. subelliptica*	4560
25	*G.venulosa*	1611
26	*G. xanthochymus*	4823
27	*G. mangostana*	1202
28	*G. mangostana*	1202
29	*G. mangostana*	1202
30	*G. mangostana*	1202
32	*G. mangostana*	1202
33	*G. mangostana*	1202
34	*G. mangostana*	1790
35	*G. mangostana*	1202
36	*G. mangostana*	1202
37	*G. mangostana*	1202
38	*G. mangostana*	—
39	*G. mangostana*	—
40	*G. mangostana*	—
41	*G. mangostana*	—
42	*G. mangostana*	—
43	*G. mangostana*	—
44	*G. mangostana*	—
45	*G. mangostana*	—

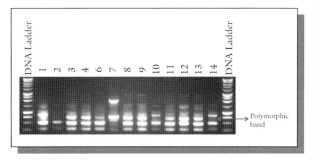

Figure B20.2 RAPD – accessions from the genus *Garcinia*. Figure courtesy of George Ude.

the bands and only polymorphic bands were scored (Figure B20.2). UPGMA-SAHN cluster analysis was done using NTSYS-pc (version 2.2) to produce the dendrograms and scatter plots that indicate phylogenetic relationships.

Results and discussion

The nine Operon primers produced 73 scorable polymorphic DNA bands. Average polymorphic information content (PIC) per primer was 0.35 (range: 0.26–0.43). Clustering and principal coordinate analysis using all 73 polymorphic bands showed three groups (Figures B20.3 and B20.4). All accessions of *G. mangostana* clustered closely together in group 2 but still showed variation despite apomixis. Generally, accessions with similar morphological characteristics (leaf, flower and fruit traits) grouped together.

The accessions of *G. mangostana* clustered closely within the similarity index range of 93–100%. However, our data resolved three subgroups among *G. mangostana* accessions, despite apomictic reproduction. *Garcinia benthamii* and *G. portoricensis* clustered with accessions of *G. mangostana* in Group 2, indicating that they may be closely related and could potentially serve as sources of rootstocks for grafting or for useful genes for the agricultural improvement of *G. mangostana*. Group 3 contains *G. cornea, G. intermedia, G. livingstonei,* and *G. madruno*. *Garcinia cymosa* was genetically far removed from all accessions studied and that agrees with its unique morphological features, amongst these a very columnar growth and pendulant branches, that distinguish it from the other species evaluated. The clustering of *Mammea americana* with other accessions in Group 2 raises identity questions that calls for more research. Overall, the genetic diversity and relationship information obtained from this research will enhance the future conservation and improvement efforts of species of the genus *Garcinia*.

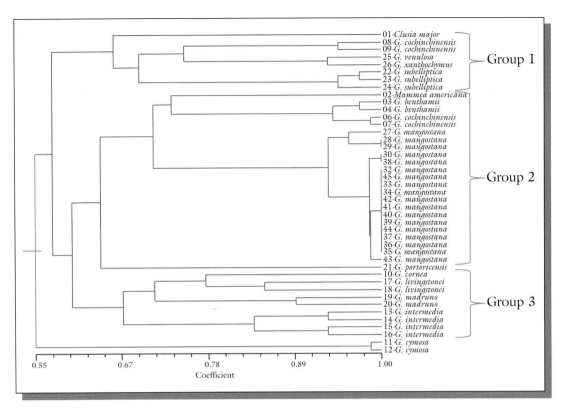

Figure B20.3 Dendrogram showing the grouping/genetic relationships of *Garcinia* species using the RAPD technique. Figure courtesy of George Ude.

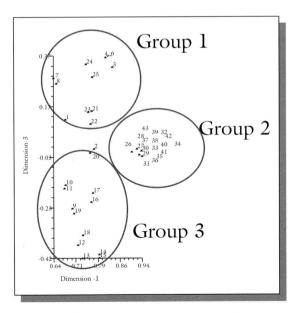

Figure B20.4 Scatter plot showing three main clusters of all studied accession of the genus *Garcinia spp.* with RAPD. Figure courtesy of George Ude.

Study 2: AFLP marker

Materials and methods

Nine primers were used in this study. A total of 773 polymorphic bands were generated (Figure B20.5) and used for clustering analysis.

Results and discussion

The results show five clusters that could be compressed into two groups (Figures B20.6 and B20.7). All the accessions of *Garcina mangostana* were clustered very closely while *Mammea americana* was distantly removed from the members of the genus garcinia. *Garcinia cymosa* showed closer relationship to *G. cornea, G. intermedia, G. livingstonei, G. madruno* and *G. subelliptica* than it did to the *G. mangostana* and *G. benthamii* accessions, which clustered in a different group.

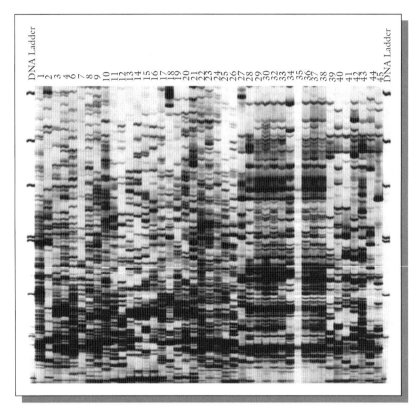

Figure B20.5 AFLP – accessions from the genus *Garcinia*. Figure courtesy of George Ude.

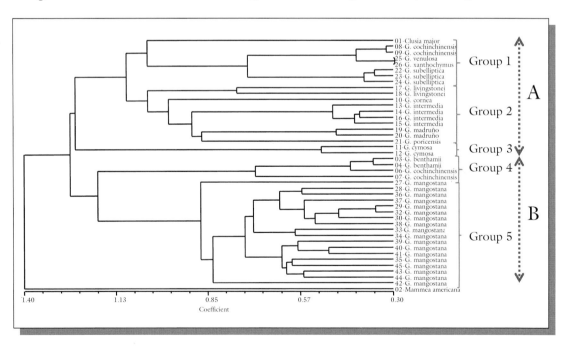

Figure B20.6 Dendrogram showing the grouping/genetic relationships of *Garcinia* species using 773 AFLP polymorphic bands. Figure courtesy of George Ude.

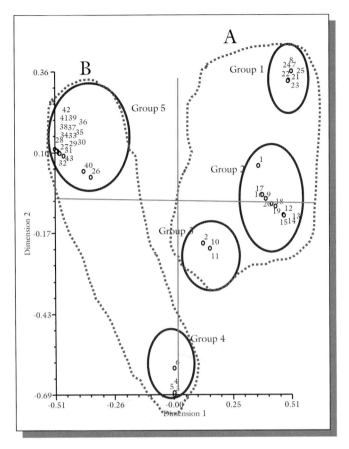

Figure B20.7 Scatter plot showing three main clusters of all studied accession of the genus *Garcinia spp.* with AFLPs. Figure courtesy of George Ude.

Conclusion

RAPD and AFLP molecular markers were effective for surveying genetic diversity in *Garcinia* germplasm. Whereas both markers types were effective in distinguishing among diversity in the germplasm accessions, AFLP appeared more sensitive, identifying subgroups in addition to the two general clusters.

Procedures for this marker system require nucleotide information for primer design for PCR and sophisticated electrophoresis systems and computer software for accurate separation and scoring of bands.

20.8.1 Microsatellites

Like minisatellites, **microsatellites** (or **simple sequence repeats (SSRs), sequence-tagged microsatellite sites (STMS), simple sequence repeat polymorphisms (SSRP))** are repetitive DNA sequences, only of shorter motifs. These VNTRs are about 2–5 bp long each. The repeats may be di-, tri- or tetranucleotides (e.g., GT, GAC, GACA; or generally $(CA)_n$ repeat where n varies among alleles) in the nuclear genome. Mononucleotide repeats (e.g., AA) occur in the chloroplast genome. The copy number of these repeats varies among individuals, ranging between 9 and 100, and is a source of polymorphism in plants. Microsatellites were the first most successful and widely exploited PCR-based markers. In addition to being highly polymorphic, SSRs occur frequently and randomly distributed throughout eukaryotic genomes. Their high variability is due to the fact that they have a higher mutation rate as compared to the neutral parts of the DNA. Microsatellites commonly have many alleles at each locus. This property allows genotypes within pedigrees to be fully informative, such that the progenitor of a specific allele can usually be identified. Microsatellites are very popular for recombination mapping and population genetic studies. They also provide the researcher with clues about which alleles are more closely related.

Because the DNA sequences that flank microsatellite regions are usually conserved, primers specific for these regions are designed for use in the PCR reaction. These primers are based on the unique sequences that flank these regions. Microsatellites developed for one species may be applied to another species, the success being higher when the genetic distance between them is low. Like minisatellites, the mutation pattern is often too complex to make this marker system readily amenable to population genetic studies. The high susceptibility to mutation (by replication slippage) can interrupt microsatellites, reduce polymorphism, and cause amplification errors leading to PCR artifacts during assays. Other events of concern include null alleles that may arise when microsatellites fail to amplify in the assay. Null alleles complicate the interpretation of allele frequencies.

20.8.2 Inter simple sequence repeats (ISSR)

As the name implies, **inter simple sequence repeats (ISSRs)** are genome regions that occur between microsatellites. Primers used for assay are either anchored at the 5′ or 3′ end of a repeat region and subsequently extended to amplify the region between two neighboring microsatellites. This is the opposite of the SSR that uses primers in the flanking region. Sequence diversity from ISSR is lower than from SSR.

An ISSR may be conserved or unconserved. This property limits its application in distinguishing among individuals. However, it is useful for other studies, including genetic diversity, phylogeny, genome mapping and evolutionary biology. It is simple and quick to use and combines the advantages of SSRs, AFLP and RAPD markers. However, its major limitation as a marker system is its inability

to distinguish heterozygotes as loci (i.e., the markers are dominant).

20.8.3 Random amplified polymorphic DNA (RAPD)

Random amplified polymorphic DNA (RAPD) is a PCR-based marker system in which the total genomic DNA is amplified using a single short (about 10 bps) random primer (Figure 20.3). It differs from traditional PCR analysis in that it does not require specific knowledge of the DNA sequence of the target organism (arbitrary sequence). The primer will or will not amplify a segment of DNA, depending on whether the positions are complementary to the primer's sequence. Consequently, if a mutation has occurred in the template DNA at the site that used to be complementary to the primer, no amplification will occur to produce a PCR product, resulting in a different electrophoretic mobility pattern for the amplified DNA segments. Success of an RAPD assay depends on the selection of the right sequence for the primer.

RAPD markers are mostly dominant markers (impossible to distinguish between DNA amplified from a heterozygous locus or homozygous locus). Further, the results of an assay is laboratory dependent, influenced by the concentration of the template DNA and PCR parameters and cycling conditions (i.e., they are very difficult to reproduce). This method yields high levels of polymorphism and is simple and quick to conduct. When using RAPD markers, using only the reproducible major bands for identification may minimize its shortcomings. Further, parental genomes may be included where available to help determine bands of genetic origin.

Locus-specific, codominant markers may be developed from RAPDs. First, the polymorphic RAPD marker band is isolated from the gel following

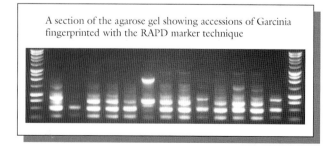

A section of the agarose gel showing accessions of Garcinia fingerprinted with the RAPD marker technique

Figure 20.3 A typical RAPD gel product.

electrophoresis. It is then amplified in the PCR reaction, cloned and sequenced.

20.8.4 DNA amplification fingerprinting (DAF)

DNA amplification fingerprinting (DAF) is a variation of the RAPD methodology. It produces more variation than RAPD because it uses very short (5–8 bases) random primers. Because of the great capacity for producing polymorphisms, DAF is best used where plants are genetically closely related (e.g., used to distinguish among GM cultivars that differ only in transgenes). It is less effective for distinguishing among species of plants at a higher taxonomic level where genetic variation is already pronounced. The procedure can be made more effective and efficient by digesting the template DNA with restriction enzymes prior to conducting the PCR technique and optimizing the PCR environment for reproducible results.

20.8.5 Sequence characterized amplified regions (SCAR) and sequence tagged sites (STS)

Sequence characterized amplified region (SCAR) and **sequence tagged site (STS)** markers are derived from PCR-based markers by sequencing the ends of fragments to develop longer primers (about 22–24 bp). They have higher reproducibility than RAPDs. SCAR markers are obtained by sequencing the ends of RAPD fragments, whereas STS markers are obtained by sequencing the ends of RFLP markers. SCARs are usually dominant markers.

20.8.6 Amplified fragment length polymorphism (AFLP)

Amplified fragment length polymorphism (AFLP) is simply RFLPs visualized by selective PCR amplification of DNA restriction fragments (Figure 20.4). The technique uses primers that are 17–21 nucleotides in length and are capable of annealing perfectly to their target sequences (the adapter and restriction sites) as well as a small number of nucleotides adjacent to the restriction sites. This property of AFLP technology makes it very reliable, robust, and immune to small variations in PCR amplification parameters (e.g., thermal cyclers, template concentration). The technique also produces a high marker density. A typical AFLP fingerprint (the restriction fragment patterns

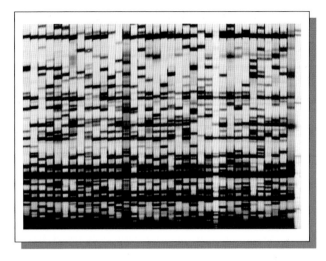

Figure 20.4 A typical AFLP gel product.

generated by the technique) contains 50–100 amplified fragments, of which up to 80% may serve as genetic markers. Another advantage of the technology is that it does not require sequence information or probe collections prior to generating the fingerprints. This is particularly useful when DNA markers are scarce. The markers generated are unique DNA fragments (usually exhibit Mendelian inheritance) and mostly mono-allelic (the corresponding allele is not detected). AFLPs are very useful for detecting polymorphism between closely related genotypes.

Some of the applications of AFLP markers include biodiversity studies, analysis of germplasm collections, genotyping of individuals, identification of closely-linked DNA markers, construction of genetic DNA marker maps, construction of physical maps, gene mapping, and transcript profiling.

20.8.7 Single nucleotide polymorphisms (SNPs)

A single nucleotide polymorphism (SNP) is a single base pair site in the genome that is different from one individual to another. SNPs are the most abundant and widely distributed molecular markers in the genome. Their occurrence and distribution vary among species, estimated at one SNP per 60–120 bp in corn, one SNP per 20 bp in some regions of wheat, and one SNP per 1000 bp in humans. SNPs are mostly biallelic, a property that makes them less informative than multiallelic markers like RFLP and microsatellites. However, this deficiency is overcome by their relative abundance, which makes it possible to use more loci in studies. SNPs are more prevalent in non-coding regions of the genome. SNPs are often linked to genes, making them very attractive subjects to study by scientists interested in locating, for example, disease genes. Sometimes, SNPs have no detectable phenotypic effect. However, in other cases, SNPs are responsible for dramatic changes.

A variety of methods are used to identify SNPs in the genome. A simple method is the screening of ESTs (expressed sequence tags) for polymorphic sites, but the use of whole genome shotgun sequencing is a more comprehensive approach. Prior to use, SNPs must be validated. This is achieved by DNA sequencing or *in silico* protocols. PCR-based validation methods include ARMS PCR, AS-PCR, and PCR-CTPP.

SNPs are less mutable compared to other markers like microsatellites. They are evolutionarily stable and suitable for studying complex traits, evolutionary genomics, population structure analysis, identification of crop cultivars, construction of ultra high density genetic maps, linkage disequilibrium (LD) mapping, association studies, and map-based gene isolation. They are readily amenable to automation. The shortcomings of SNP technology include ascertainment bias (isolation strategy has bias for certain parameters such allele frequency and linkage disequilibrium), reduced information content (from the biallelic nature), and that they are costly and time intensive. If the SNP occurs at a highly mutable site, the biallelic nature assumption may not hold true.

20.8.8 PCR-based markers from RFLPs

Just like mini-and microsatellites that may be converted into PCR-based markers (e.g., ISSR, STMS), RFLP markers may also be converted into PCR-based markers to overcome some of the limitations of the former. Some of these conversions are as follows: (1) *Sequence-tagged sites (STS)* – these are about 200–500 bp long and generally codominant markers. Broadly defined, STS include other markers such as microsatellites, SCARs, CAPs, and ISSRs. (2) *Expressed sequence tags (EST)* – these markers are generated by partial sequencing of random cDNA clones. (3) *Allele-specific associated primers (ASAP)* – these markers are similar to SCARs. (4) *Single strand conformation polymorphism (SSCP)*.

20.9 DNA sequence-based markers

DNA sequencing does not qualify as a marker system by the conventional definition. However, it can be applied to achieve the goal of a marker, only more precisely, for it produces the most complete information for the target region analyzed. This property can expand the information obtainable from species with low natural polymorphism. Sequence information is of the highest possible resolution and free of biases. Once not readily accessible because of cost and time intensiveness, modern automation for high throughput has made DNA sequencing affordable.

20.10 Comparison of selected molecular markers

The advantages and disadvantages of some of the commonly used molecular markers are summarized in Table 20.1. As new technologies are developed and knowledge abounds in biological sciences, new markers and applications will certainly emerge.

20.11 Desirable properties of a molecular marker system

The applications of molecular markers in plant breeding and biology are diverse and increasing. Based upon the concept of markers discussed previously, certain properties are desirable in order for a marker to be useful. The key ones include:

- High degree of polymorphism.
- High frequency of occurrence in the genome (abundant).
- Random distribution in the genome.
- Selectively neutral.
- Codominant inheritance.
- Low mutation rate.
- Low cost to use.
- Easy and quick to isolate (readily assayable).
- Highly reproducible.
- Low ascertainment bias.
- Amenable to automation (high throughput application).
- Existence of extensive data repositories.
- Ease of cross-study comparisons.
- Easy information management.

Table 20.1 A comparison of different molecular markers.

Marker	Genetic control	Advantages	Disadvantages
Allozyme	Codominant	– Low cost of assay – Universal protocols – Direct survey of DNA variation *per se*	– Low number of markers – Requires fresh/frozen samples – Some loci show unstable protein – Uneven distribution on genetic map
RFLPs	Codominant	– Easy to score – Sequence used as probe need not be known – Transferable across populations	– Requires large amounts of DNA – Expensive – Low throughput – Tedious – Low polymorphism
RAPD	Dominant	– Inexpensive – Small amount of DNA needed – Quick and easy to use – Large number of bands	– Low reproducibility – Difficulty to automate – Low transferability – Difficult to analyze
Microsatellites	Codominant	– Easy to isolate – Large number of alleles – Transferable across populations – Can be converted into single locus markers	– Primer preparation is time consuming – Difficult to automate – High mutation rate
AFLP	Dominant	– Reliable – High levels of polymorphism – Multiple loci	– Requires large amount of DNA – Methodology is complex
SNPs	Codominant	– Abundant – Low mutation rate – Easy to automate – Cross-study comparison easy	– Time consuming – Expensive – Low information content of a single SNP – Ascertainment bias

No one marker system will satisfy all these desirable qualities. The choice of a marker system depends on the specific application and resources available to the researchers.

20.12 Readying markers for marker assisted selection

One of the most important applications of genetic markers is to facilitate plant breeding by making the selection process more effective, efficient and speedy. Called marker assisted selection (MAS), its success depends on the identification of reliable marker systems. Two of the general ways in which markers are readied for application are validation and conversion.

20.12.1 Marker validation

Markers are identified from specific experimental populations with specific genetic backgrounds. The purpose of marker validation is to ascertain the reliability and efficacy of a putative marker in detecting the target phenotype or trait in different genetics and populations independent of the one in which it was discovered. Some markers are polymorphic only in certain genetic backgrounds (what I loosely call lack of "marker penetrance"). If a marker is not stable and reliable in predicting a phenotype it is supposed to be associated with, it is of no use as a tool in breeding program to aid in selection. Without validation, false association between markers and the trait of interest can arise, with adverse consequence on a breeding program. Population stratification may be a source of false associations, depending on the nature of the parental genotypes that comprised the discovery population. If the discovery population comprised genetically distant germplasm, markers may not be useful in all populations.

Prior to use, markers should be validated by testing for their presence and polymorphism in a wide range of cultivars and genotypes, as well as different environments, especially if these marker associations involve complex or quantitative traits (QTLs).

20.12.2 Marker conversion

All the different types of markers discussed in this chapter vary in different ways, including cost of development and use, technical challenges in operation, time required, genetic nature (e.g., dominant or codominant), reliability, reproducibility, robustness, interpretation of results, and amenability to automation. To capitalize on the good aspects of some markers, while minimizing or eliminating their weaknesses, researchers have converted some markers from one type to another. AFLP technology is useful for a wide variety of applications (e.g., genetic diversity analysis, local marker saturation, construction of genetic maps and QTL mapping) but is less useful for single locus assays (e.g., allele frequency studies, marker-assisted selection, or map-based cloning). Many AFLP markers are redundant and, consequently, too expensive and laborious for large-scale single locus screening applications. Consequently, AFLP markers have been converted into high throughput, single locus, PCR-based markers such as CAPS (cleaved amplified polymorphic site) and SCAR (sequence characterized amplified region). Similarly, AFLP markers have been converted into polymorphic STS (sequence tagged site) markers for high-throughput applications. Other known marker conversions are from RAPD markers to the more stable STS markers. These conversions differ in their technical difficulty to achieve.

20.13 DNA sequencing

DNA sequencing is not a genetic marker in the narrow sense of the definition. However, DNA sequence analysis is a powerful tool in plant breeding and plant research. It provides the highest resolution possible of all markers and is not biased. It is amenable to cross-study comparisons. Data for various sequences are already available in repositories. A critical consideration in its application in research is cost, which is still relatively high compared to other marker systems.

Two basic approaches are used in DNA sequencing – whole genome shotgun and clone-by-clone. The clone-by-clone approach simplifies a large genome into small pieces of known order and delimits uncertainties to about 100 kb. After these pieces are sequenced, they have to be ordered and assembled at a high cost. On the other hand, whole genome strategy samples the entire genome such that each nucleotide in the genome is covered by about 6–8 (redundant sampling) paired end sequences. The clone-by-clone approach is prone to not sampling highly repetitive regions of the genome (e.g., the centromere or near it). Often, researchers skip such problematic regions when using this approach. Even though relatively replete of genes, the ability to characterize these repetitive regions is of value to researchers. However, the clone-by-clone approach, which allows the assembly of alleles one at a time, excludes the possibility of heterozygosity. It offers higher contiguity, but the shotgun approach can use various techniques to reduce the incidence of gaps.

There are general challenges to sequencing crop genomes. The first full plant genome sequence obtained was that of *Arabidopsis thaliana*, a plant with a tiny (125 Mb haploid) genome. However, many crops of economic interest have large genomes that complicate sequencing. Another factor is the presence of large amounts of repetitive DNA. Non-redundant sequences in plants range from about 13% in onion (*Allium cepa*) to about 77% in tomato (*Lycopersicon esculentum*). Many plants of economic importance are polyploids (the entire genome is duplicated), a condition that complicates sequence assembly.

Two rice (*Oryza*) subspecies have been sequenced using both approaches. Crop genera in which sequence projects have been completed or are ongoing includes *Populus, Medicago, Sorghum, Lycopersicon, Zea, Brassica*, and *Solanum*.

Key references and suggested reading

Collard, B.C.Y., Jahurfer, M.Z.Z., Brouwer, J.B., and Pang, E.C.K. , (2005). An introduction to markers, quantitative trait loci (QTL) mapping, and marker-assisted selection for crop improvement: The basic concepts. *Euphytica*, **142**:169–196.

Ellegren, H., (2004). Microsatellites: Simple sequences with complex evolution. *Nature Reviews Genetics*, **5**:435–445.

Schlotterer, C., (2004). The evolution of molecular markers – just a matter of fashion? *Nature Reviews Genetics*, **5**:63–69.

Outcomes assessment

Part A

Please answer the following questions true or false.

1 Isozymes markers are DNA markers.
2 AFLP is a PCR-based marker.
3 SCAR markers are derived from RFLP markers.
4 RFLP is a multilocus marker.

Part B

Please answer the following questions.

1 Please give the full name for each of the following acronyms:
RFLP. . . . .
AFLP. . . . .
SCAR . . . .
SNPs . . . .
DAF . . . .
2 Give two specific disadvantages of isozyme markers.
3 Give two specific applications of molecular markers in plant breeding.
4 What is marker assisted selection?

Part C

Please write a brief essay on each of the following questions.

1 Give the rationale of markers.
2 Discuss the use of markers in plant cultivar identification.
3 Discuss the advantages of the SNP technology.
4 Distinguish between microsatellite markers and minisatellite markers.

21

Mapping of genes

Purpose and expected outcomes

The concept of a map, as alluded to previously, is the pictorial or visual depiction of an area with symbols that represent elements distributed in the area, their relative positions and relationship between them.

After studying this chapter, the student should be able to:

1 Discuss the importance of genetic maps in plant breeding.
2 Discuss the types of genetic maps.
3 Discuss the types of mapping populations.
4 Discuss the mapping of quantitative trait loci.

21.1 Why map genes?

The primary purpose of a traditional geographic map is to facilitate navigation and access to features and places or elements of interest in a geographic area. In genetics, the genome represents such an area, the chromosomes being the key landmarks, while individual genes represent the places or elements of interest. To facilitate plant breeding, researchers benefit from knowing the genetic landscape of the genome as it pertains to the relative position of genes along a chromosome. Because most traits of interest to plant breeders (in fact most biological traits) are genetically complex, mapping or analyzing these complex traits (called quantitative trait loci (QTL)) helps in understanding their genetic architecture and, thereby, facilitates their manipulation by breeders and other

researchers. The ultimate goal of gene mapping is to clone the mapped genes.

21.2 Types of gene maps

There are two conceptual ways of generating a map of genes, genetic mapping and physical mapping.

21.2.1 Genetic mapping

In this approach, the classical genetic tool of pedigree analysis is used to determine the relationship between individuals in the progeny of a cross. The underlying principle (Mendelian law) is that genes and markers segregate via chromosome recombination during meiosis, and the pattern of segregation can be

Principles of Plant Genetics and Breeding, Second Edition. George Acquaah.
© 2012 John Wiley & Sons, Ltd. Published 2012 by John Wiley & Sons, Ltd.

analyzed to reveal the strength of association between genes. Called **linkage analysis**, the frequency at which two gene loci become separated during chromosomal recombination (by crossing-over) is used to calculate distances between the loci on the chromosome. The map produced by this process is called a **linkage map** or **genetic map**, and the process **linkage mapping** or **genetic mapping.**

21.2.2 Physical mapping

There are three basic types of physical map – chromosomal or cytogenetic, radiation hybrid, and sequence maps. These types vary in their degree of resolution (ability to measure the separation between elements that are close together). Of these three, sequence maps have the highest resolution, being able to show genetic markers as well as the sequence between them (measured in base pairs). Whereas linkage mapping attempts to assign gene positions through breeding experiments or pedigree analyses which are not precise, physical mapping by sequencing provides absolute positions of mapped genes. It is achieved via various methods, including the use of STS (e.g., ESTs and SSLPs) for physical mapping. Radiation hybrid mapping relies on artificially induced (by radiation) breaks to determine the distances between markers (linkage mapping depends on natural recombination).

It is recommended that genetic and physical maps be integrated for higher accuracy.

21.3 Principles of linkage mapping

Genetic linkage occurs when loci or alleles for a gene are inherited together. The genes on a chromosome tend to be transmitted together and are hence said to be genetically linked. However, whether or not genes on the same chromosome will stay together through meiosis depends on how closely or tightly linked they are. A genetic linkage map shows the positions of known genes or genetic markers relative to one another. These positions are estimated by calculating the recombination frequencies between marker genes or loci (the likelihood of a chromosomal crossover occurring between these genes). The greater the recombination frequency, the farther apart the genes are on the chromosome. A genetic linkage map is constructed by calculating recombination frequencies

between genes on chromosomes. It should be pointed out that a linkage map does not correspond to a fixed length of chromosome for a variety of reasons, including the non-randomness of crossover location and also the number of intervening genes being considered in the calculations. The reader may refer to Supplementary Chapter 2 for details on how to calculate recombination frequencies. Basically, researchers calculate the relative distance between two genes from the offspring of the species that show two linked traits. The higher the percentage of the offspring that show no linkage, the farther apart the two genes are on the chromosome.

Genetic linkage map construction entails several key steps:

   (i) Creation of a mapping population.
 (ii) Identification of polymorphism.
(iii) Calculation of pair-wise recombination frequencies from the mapping population.
 (iv) Establishment of linkage groups and estimation of map distances.
  (v) Determination of map order of the genes.

21.4 Mapping population

The first and most important activity to conduct in a gene mapping undertaking is to create an appropriate genetic population that will provide the fundamental information for map construction. A population created solely for gene mapping is called a **mapping population.** It is traditionally created from controlled crosses (researchers select parents and cross them according to a selected mating design). Consequently, the success of map construction depends on the judicious selection of parents. A key consideration in this exercise is the source of parents, specifically, whether they are adapted or exotic. Whereas crosses among parents that are genetically distant would create segregating populations with large amounts of polymorphism, the high level of cytogenetic and genetic irregularities that accompany such crosses may reduce the linkage distances observed. Further, wide cross mapping is less useful unless it is collinear with maps generated from adapted crosses.

Another important characteristic of parents used to generate an effective mapping population is that they possess adequate variation, both at the DNA and phenotype levels, for the trait(s) of interest. The more pronounced the difference is in the phenotypic variation, the more likely that this variation will have a genetic basis. This variation should include as many qualitative traits of economic value to which markers are associated as possible.

21.4.1 Types of mapping populations

A mapping population of necessity should be a segregating population (Figure 21.1). To optimize segregation, inbred lines (highly homozygous) are preferred as parents. These are readily obtained for mapping involving self-pollinating species. Generating highly inbred lines in cross-pollinating species is a challenge because of their general intolerance to inbreeding. Consequently, a cross between a heterozygous parent and a haploid or homozygous parent may be used. Another challenge in mapping of cross-pollinating species is the prevalence of polyploidy, which complicates the process. The number of individuals from the segregating population required for mapping varies from about 50–250 plants. Larger numbers are needed to

achieve maps of high resolution. Further, if the objective is to map quantitative trait loci (QTL), the mapping population must be phenotypically evaluated or characterized before use.

There are a number of mapping populations used in linkage mapping, each with advantages and disadvantages. These include:

- **F_2 population.** This population is simple to generate and requires a short breeding time. It is produced by selfing or sib-mating F_1 hybrids. In characterizing the population, the simple Mendelian ratios expected are 3:1 for dominant markers and 1:2:1 for codominant markers. This population is ideal for preliminary mapping. However, F_2 mapping populations have serious limitations that reduce their use for fine mapping. Recombination is limited in these populations because they are produced from only one cycle of meiosis. It is impossible to exactly duplicate this kind of population or increase the quantity of seed. Most maps are generated for QTL studies, something for which this mapping population is ineffective. A major

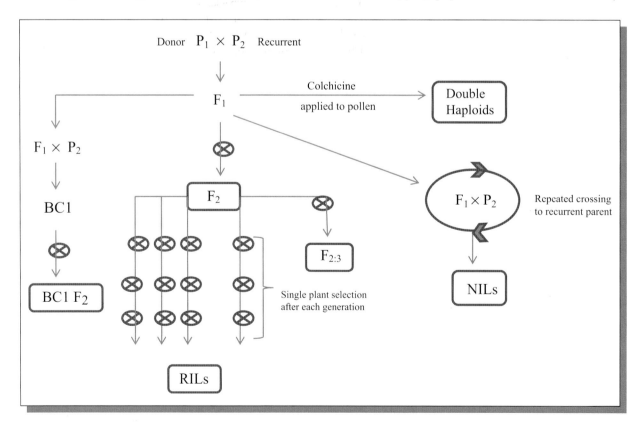

Figure 21.1 Mapping populations can be created using a variety of methods.

drawback of F_2 mapping populations is that they are not "immortal" or "eternal" ("everlasting"), as they are colloquially described. This means that, in molecular mapping, the source of tissue for either DNA or protein isolation may be exhausted before the study is over, compelling the researcher to map in a different population all together.

- **F_2 derived F_3 ($F_{2:3}$) population.** This breeding population benefits from an additional meiotic cycle by selfing F_2 plants. It is useful for mapping recessive genes and QTLs. This population is still not immortal.

- **Backcross population.** A **backcross** is produced by crossing the F_1 to one of the parents, which is quick to perform. When characterizing backcross populations with molecular markers, the segregation patterns differ with respect to dominant and recessive alleles. Dominant markers will segregate in a ratio of 1:0, while codominant markers would segregate in a ratio of 1:1, when the backcross has the dominant parent. In the case of the recessive parent, the segregation pattern would be 1:1. Apart from having only two electrophoretic mobility patterns to score, another advantage of a backcross mapping population is that the number of individuals with the recessive trait is greater than it occurs in an F_2 population. Drawbacks of this population include lack of immortality. Each individual is unique and, hence, only a limited amount of tissues is available from each plant. Further, data cannot be correlated across research laboratories when a universal mapping strategy is being used.

- **Recombinant inbred lines (RILs).** **Recombinant inbred lines (RILs)** are produced by repeated selfing of selections from the F_2 population until these lines are homozygous and can be propagated without segregation. The breeding technique commonly used for this is single seed descent, even though alternative procedures may be used. The repeated selfing increases the amount of recombination observed, making this population useful for detecting tightly linked markers. This population is immortal and allows replicated trials to be conducted over years and locations. This makes RILs highly suited for mapping QTLs. The genetic segregation ratio is 1:1 for both dominant and codominant markers. Large quantities of seed can be produced for each line. Further, each line is fixed for many recombination events. The drawbacks of this mapping population include the length of time required to attain homozygosity. Further, it is not suited to species that are intolerant of inbreeding. Also, for mapping

analyses, RILs can be limited because of masking effects of major QTLs and epistatic interaction of multiple QTLs.

- **Doubled haploids.** **Doubled haploids** are created through chromosome doubling of pollen grains in another culture. This process creates instant homozygous lines. Pollen from F_1 plants is frequently used. Information from recombination is limited because of the fact that only one cycle of meiosis is involved in producing the parent plant. However, all F_1 parents are genotypically identical and have the same linkage phase. Consequently, doubled-haploid individuals are completely informative with maximum linkage equilibrium. This population is also permanent and can be maintained indefinitely without genotypic change. It allows replicated trials to be conducted over years and locations. The comprehensive homozygosity of doubled haploids gives them an advantage over RILs in mapping of genes. Further, the expected genetic segregation ratio for both dominant and codominant markers is 1:1. The population offers eternal resources materials for mapping both qualitative traits and QTLs. One of the drawbacks of this population for mapping include the fact that only recombination from the male source (pollen) is represented in the population. Also, an effective *in vitro* culture system should be available for success, adding to the skill level required to use this population.

- **Near-isogenic lines (NILs).** **Near isogenic lines (NILs)** are genetic materials that are identical except at one or a few loci. Essentially, pairs of lines differ only in the genomic region of interest, all other genes that affect the trait of interest being the same in both lines. Any difference between lines can, therefore, be ascribed to a single gene as if the locus were Mendelian. NILs that differ at QTL are useful for detailed mapping and characterization of individual loci. NILs are an important step in the cloning of QTL. NILs may be produced by repeated backcrossing of the F_1 to recurrent parents or simply by repeated selfing of the F_1. When, for example, the backcross method is used, the NILs are similar to the recurrent parent except for the gene of interest. They are immortal experimental populations that are useful for genetic and physiological studies and QTL analyses. The genetic segregation ratio of markers is 1:1. A drawback of these populations is the time and effort required for their development. Also, linkage drag may complicate their development. NILs are useful directly for molecular tagging of the gene of interest but not for linkage mapping.

21.5 Identification of polymorphic markers

Adequate polymorphism is critical to the success of linkage mapping. Polymorphism is identified by the use of markers. Consequently, the next step in linkage mapping is to identify polymorphic markers (i.e., markers that can reveal difference between parents). The choice of suitable markers for mapping depends on the species being mapped and the availability of characterized markers. Once identified, these markers are screened (marker genotyping) across the entire mapping population (including parents, F_1). To accomplish this, DNA is extracted from each of the plants selected (50–250). The expected segregation in different mapping populations has been discussed previously. Information on each plant is coded properly for the ensuing linkage analysis.

21.6 Linkage analysis of markers

Manual linkage analysis is feasible if only a few markers are being studied. Modern linkage map construction is a computerized operation, popular mapping packages include Linkage1, GMendel, Joinmap, Mapmaker and MapManager, the last two being available free of charge via the Internet. These computer software packages use the coded information from the segregating population to determine recombination frequencies. The basic calculation is a ratio (odds ratio) of linkage versus no linkage, expressed as the log of the ratio (logarithm of odds or LOD). A LOD value or score measures the likelihood of linkage between two markers, a score of more than three usually being the cut-off minimum for mapping. A LOD of three indicates a 1000:1 odds in favor of genetic linkage (linkage between the two markers is a thousand times more likely than no linkage). The researcher may vary the stringency of mapping by, for example, lowering the LOD score to detect a greater level of linkage.

21.7 Rendering linkage maps

Genes are located on chromosomes. Genes in a chromosome are transmitted together as a group because they are linked (linkage groups). Consequently, there are as many linkage groups as there are homologous pairs of chromosomes. After calculating the LOD scores, the next steps in linkage mapping are to establish linkage groups and then estimate map distances within each group. This allows the genes or loci to be ordered to show their relative positions on a chromosome. The distance between a pair of markers on a map is measured in terms of the chance of recombination occurring between them (recombination frequency). However, because the recombination frequency and the frequency of crossing-over are not linearly related, the recombination frequencies are converted to map units or centiMorgans (cM) using mapping functions. Commonly used mapping functions are the Kosambi (assumes interference between crossover events) and the Haldane (assumes no interference between crossover events). Map distances of less than 10 cM are equal to the recombination frequency but this relationship is not true for distances exceeding 10 cM.

An example of a linkage map is shown in Figure 21.2. The distance between markers is not directly related to the physical distance of DNA between them but depends on the size of the genome. The frequency of recombination is not equal along chromosomes but there are hot spots (high recombination frequency) and cold spots (low recombination frequency). Consequently, the relationship between genetic and physical distances also varies along a chromosome. Similarly, the markers used in linkage analysis are not evenly distributed over the chromosome but display patterns whereby they cluster in some regions while being absent in others. The consequence of this is that it is difficult to obtain an equal number of linkage groups and chromosomes in linkage mapping.

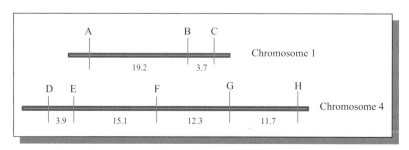

Figure 21.2 A hypothetical linkage map.

21.8 Mapping quantitative trait loci (QTL)

Many plant traits of interest to plant breeders are complex with complex inheritance. Most, but not all, complex traits are conditioned by more than one locus or gene and are said to be quantitative or polygenic traits. The phenotype of the quantitative trait is the result of small contributions from many individual genes rather than the effect of one or a few genes, as is the case in qualitative traits. The expression of complex traits can be influenced by non-genetic factors. Because of the large number of genes perceived to condition complex traits, it is customary to refer to these clusters of genes as **quantitative trait loci (QTL)**. QTLs may be viewed as hypotheses that specific chromosomal regions contain genes that significantly contribute to the expression of a complex trait. Consequently, their locations are likely temporary as their nature is subject to reinterpretation and revision. Similarly, reference to QTL in such terms as phenotype or trait is semantically imprecise.

Industry highlights

The use of haplotype information in QTL analysis

Herman J. van Eck

Laboratory for Plant Breeding, Wageningen University, Droevendaalsesteeg 1, 6708PB Wageningen, The Netherlands

Introduction

When students are introduced to genetics and plant breeding, most attention is paid to Mendelian or single gene inheritance of qualitative traits. With such discrete characters many basic concepts can be illustrated, including independent assortment of genes, gene action (dominant or recessive alleles), mutation, and gene interactions, such as epistasis or complementary genes. Qualitative traits are also used to explain the concept of genetic linkage and the methods to construct eukaryote chromosome maps based on recombination frequencies.

Without diminishing anything of the importance of these concepts, most of the plant breeding activities focus on traits with a quantitative inheritance. The earliest attempts to grasp the underlying concepts (Johannsen, 1909; Nilsson-Ehle, 1909) offer merely a description of the phenotypic variation in statistical terms. The term polygenic inheritance was first coined in a 1913, published German textbook on *Vererbungslehre* by Plate. In an elaborate and influential paper on quantitative genetic variation (Mather, 1941), the term polygenetic inheritance was adopted to describe the inheritance of quantitative traits. Since the publication of this paper it has generally been accepted that quantitative traits are controlled by an indefinitely large number of genes, many of which have approximately equal effects, and the effects of which are influenced by the quality of the environment (the infinitesimal model). For a long time these traits could not be studied using the common techniques of genetics. Statisticians offered some help by distinguishing between variance components, that is, genetic and environmental variance. In spite of the agricultural importance of quantitative traits, little progress was made for decades. Many contemporary textbooks on genetics still offer an explanation for quantitative trait inheritance in terms of polygenes, although it is only one extreme of the spectrum.

Resolution into quantitative trait loci (QTLs)

A breakthrough was achieved by Paterson *et al.* (1988), who demonstrated in their highly cited *Nature* paper a method to resolve quantitative traits into Mendelian factors by using a complete linkage map comprised of DNA markers. The black box of quantitative traits was now illuminated and the underlying genetic loci were reclaimed into the Mendelian domain. Indeed, genes involved in quantitative traits were cloned soon after. A gene contributing to tomato fruit weight (*fw2.2*) was among the first examples (Frary *et al.*, 2000). Even the DNA sequence variant causing the quantitative differences in trait values (Quantitative Trait Nucleotide, QTN) has been identified. A gene increasing sugar yield of tomatoes appeared to encode an invertase and recombinant analysis allowed the identification of the QTN causing an amino acid change that influenced enzyme function (Fridman *et al.*, 2004). In almost 25 years Paterson *et al.* 1988 has been cited more than 500 times and the term quantitative trait locus (QTL) has appeared in more than 50 000 articles. In plants several dozens of genes have been cloned that explain QTL involved in yield, quality and disease resistance.

What are the most important lessons learned from these QTL studies?

(i) **Number of loci.** Quantitative traits differ considerably in the number of loci involved. This can range from truly polygenic to oligogenic, but also monogenic quantitative traits exist. Tuber shape in potato may range from round to oval to long, but a single locus on chromosome 10 explaining 75% of the trait variance suggests a monogenic inheritance of a quantitative trait (Van Eck *et al.*, 1994). Similar situations are true for potato tuber flesh color and plant maturity. In fact, there is a thin line between the distinction of qualitative and quantitative traits. If a trait cannot be classified in two distinct phenotypes (e.g. red and white flowers), because there seems to be a continuous range, and if the trait is measured on a metric scale, even then a sharp "breeders' eye" might classify the different shades. Discrete classes underlying seemingly continuous trait values may be recognized by a bimodal distribution of the trait values. However, usually it is the method of data collection (classification or metric observations) and not the genetic architecture that decides if a trait is monogenic and qualitative or polygenic and quantitative. Also, quantitative or field resistance of potato to late blight has been explained by a monogenic *R*-gene (Tan *et al.*, 2008).

(ii) **Effect of loci.** There is no reason to assume that QTLs should have a more or less equal contribution. On the contrary, many studies have demonstrated a mixture of a few major effect QTLs, each explaining a considerable proportion of the genetic variance, along with a larger number of minor QTLs. Many minor QTLs are just about the threshold of significance and require confirmation by independent studies and/or independent germplasm.

(iii) **Effect of alleles.** In the early days of genetics, Sirks 1929 offered an alternative explanation for continuous variation in trait values. Rather than polygenes, he proposed a series of multiple alleles at a single locus, where alleles could differ in their effect on the phenotype. Intra-locus allele interactions (e.g. complete-, under-, and overdominance) have to be considered as well. The number of different alleles in the gene pool is crop specific, depending on mode of reproduction and bottlenecks during domestication and breeding history.

(iv) **Environmental noise.** There are large contrasts between quantitative traits with respect to the amount of environmental variance (σ_e^2) relative to the genetic variance (σ_g^2). When the total variance ($\sigma_{total}^2 = \sigma_g^2 + \sigma_e^2$) is decomposed into a genetic and environmental component, the heritability (h^2) of a trait can be estimated using the formula $h^2 = \sigma_g^2/(\sigma_g^2 + \sigma_e^2)$. Some traits have a high heritability, suggesting strong environmental stability in trait realization. But often a heritability is not an inherent feature of the trait but the result of how trait observations were recorded in an experiment with little experimental noise (like homogeneous field plots and homogeneously administered inoculum in disease tests) and in sufficient replications. Greenhouse experiments usually result in higher heritabilities than field studies unless very homogeneous soils and irrigation are available.

Obviously, these issues on the number and effect of loci and alleles and the experimental noise are interrelated. When the heritability is low, of course the QTLs will not explain much of the phenotypic variance. If the number of QTLs is high (polygenetic) it is not expected that each locus can explain a large proportion of the phenotypic variance. In the example on tuber shape (van Eck *et al.*, 1994) the heritability was high ($h^2 = 80\%$), but also largely explained by a single locus. The ability to classify tubers into distinct shape categories was hampered by the number of different alleles (up to four in non-inbred diploids; up to eight in non-inbred tetraploids), which segregate in mapping populations from non-inbred parents. In many studies, however, there is a large gap between the trait heritability and the genetic variance collectively explained by the QTLs. This gap is referred to as the "missing heritability" (Brachi *et al.*, 2011).

In summary, the term polygenic inheritance is considered as a misrepresentation of quantitative inheritance, because it includes the assumption of the involvement of many underlying genes. The term quantitative inheritance for metric traits does not include this assumption but it does imply that observations could not be classified, not even with a skilled breeders' eye.

The next step: resolution of QTL into trait alleles

For a biologist it is a challenge to understand the genes underlying quantitative traits. Such knowledge may elucidate that certain genes involved in carbohydrate metabolism explain the difference between, for example, potato cultivars suitable or unsuitable for the production of French fries or crisps (chips). Likewise, genes in anthocyanin biosynthesis and regulation may underlie QTLs for intensity of flower color, or genes in cell wall metabolism may explain fiber properties in cotton or paper quality of trees.

However, for a breeder it is not the locus that matters. It is the allele that matters! In my opinion the ability to recognize the potentially many different alleles in breeders germplasm has received insufficient attention. Those plant breeders who aim to use marker assisted selection are not selecting loci but alleles. The locus will always be there but alleles with a negative contribution to a trait value can be substituted by an allele with a positive contribution. Marker assisted breeding is about allele selection.

The experimental design determines number of different quantitative trait alleles (QTA) that can be analyzed

The identification of QTLs in crop species can be pursued with different experimental approaches; for example, (i) bi-parental mapping populations of full-sibs or (ii) unrelated germplasm in genome wide association studies (GWAS). In GWAS experiments the association between alleles of traits and molecular markers is not based on genetic linkage after one cycle of meiotic recombination but on the amount of linkage disequilibrium that remains after immeasurable numbers of meiotic recombination events in the gene pool since crop domestication. In addition, crops can also differ in their mode of reproduction (inbreeders or (obligatory) outbreeders) and ploidy level. Table B21.1 lists the combinations of properties or limitations of two experimental designs for the identification of alleles that matter at QTLs. It will be obvious that these approaches will have different statistical power to establish associations between trait and marker alleles. Other experimental designs have been proposed which lie in between a completely structured (full-sibs) and unstructured population of individuals, such as groups of half-sibs and multiparent advanced generation intercross (MAGIC) lines.

The binary information of molecular genetic methods

Several types of molecular marker system and their utility for the identification of alleles at QTLs have been reviewed (Collard *et al.*, 2005). Marker systems generate binary data, resulting from the usually two character states of nucleotide variation at the DNA level. The absence or presence of restriction sites, nucleotide substitutions, insertions or deletions, or primer annealing polymorphisms separates alleles in two classes. The microsatellite or SSR markers are an exception, because the technology to detect them captures data from multiple alleles. However, SSR markers and other gel-based marker systems are being increasingly replaced by massively parallel methods to genotype single nucleotide polymorphisms (SNP). Popular SNP genotyping platforms have been reviewed in Appleby *et al.* (2009) but in the near future the decreasing costs for sequencing may replace SNP assays with genotyping-by-sequencing approaches. Thus, molecular marker variants are binary variants, with SSR an exception.

In our laboratory, the genetic composition of potato alleles was studied using a re-sequencing approach. This resulted in more insight into DNA variation of alleles underlying trait variation and the consequences for marker development. Figure B21.1 shows in vertical columns the SNP positions for ten alleles of a functional gene. The polymorphic positions

Table B21.1 Some remarks on the properties and limitations of two experimental designs for QTL identification.

	Bi-parental mapping populations	**Genome wide association studies (GWAS)**
diploid inbreeders	Linkage map construction and QTL analysis is straightforward. Quantitative Trait Alleles (QTA) can be identified only if parents carry two different alleles	Unstructured populations can be scanned with markers at known map positions. Associations between markers and QTA require haplotype specific markers. Allele frequencies are variable and affect the statistical power to detect QTL.
diploid outbreeders	Linkage mapping and QTL analysis is more complicated, but can be performed with available standard software. Identification of the QTA-that-matters requires markers or marker combinations that offer full classification of the four (or less) different alleles in the four (or less) offspring classes; e.g. AB × CD → AC, AD, BC, BD	Multiple QTL alleles and marker alleles severely complicate association analysis. Codominant marker techniques will allow the identification of allele dosage in three classes (homozygous, heterozygous and absent)
polyploid outbreeders	The large number of alleles severely complicates linkage mapping and QTL analysis, requiring dedicated software. Linkage phase information is no longer obvious. Full genetic classification is hardly feasible. QTA will go unnoticed.	Multiallelism severely complicates association analysis. Identification of the genotype of a marker with two alleles in five classes (AAAA, AAAa, AAaa, Aaaa, aaaa) requires quantification of the allele dosage. Multiple allele combinations lead to a plethora of possible genotypes (e.g. AABC, ABCE, BDEF, etc.)

6	14	17	69	89	93	96	99	105	110	138	142	143	144	153	156	171	222	226	231	232	279	304	318	321	369	396	434	435	510	513	516	525	559	561	564	591	599	615	678	679	681	691	692	702	723		QTA
T	C	T	T	T	T	G	C	A	C	A	G	T	C	A	C	G	C	A	A	T	T	T	G	T	G	C	A	T	T	T	T	C	T	T	T	A	A	T	A	A	T	A	T	G	T		-
A	C	C	C	C	G	A	C	A	T	A	A	C	C	G	C	A	C	A	A	T	G	T	G	T	G	T	A	C	T	T	T	C	T	G	T	A	C	C	A	G	T	A	C	G	T		+
A	C	C	C	T	T	A	G	G	C	T	A	C	T	A	T	G	G	G	A	T	G	T	G	T	G	C	A	C	T	T	T	C	T	T	T	T	C	C	A	A	T	A	C	G	T		+
A	C	C	C	T	T	A	G	G	C	A	A	C	T	A	T	G	G	G	A	T	G	T	G	T	G	C	A	C	T	T	T	C	T	T	T	A	C	C	A	A	T	A	C	G	T		+
T	C	T	T	T	T	G	C	A	C	A	A	C	C	A	C	G	C	A	A	T	G	T	G	T	G	C	T	C	T	T	C	C	T	T	T	A	A	C	A	A	T	G	G	G	G		+
A	C	C	C	T	T	A	G	A	C	A	A	C	C	A	C	G	G	G	A	T	G	T	G	T	T	C	A	C	T	T	T	T	T	T	T	A	C	C	A	G	A	A	C	G	T		–
A	T	C	C	T	T	A	G	G	C	A	A	C	T	A	C	G	G	G	A	T	G	T	T	T	T	C	A	C	T	T	T	T	T	T	T	A	C	C	A	T	A	A	C	A	T		–
A	C	C	C	T	T	A	G	G	C	A	A	C	T	A	C	G	G	G	A	T	G	T	G	T	G	C	A	C	T	T	T	C	T	T	T	A	C	C	A	A	T	A	C	G	T		+
A	C	C	C	T	T	A	C	A	C	A	A	T	C	A	C	G	G	G	A	T	G	T	G	T	G	C	A	C	C	C	T	C	T	T	T	A	C	C	A	A	T	A	C	G	T		–
A	C	C	C	T	T	A	G	G	C	A	A	C	T	A	T	G	G	A	T	A	G	G	T	A	T	C	A	C	T	T	T	C	C	T	C	A	C	C	G	A	A	G	G	G	T		–
	↑			↑	↑				↑	↑	↑			↑		↑			↑	↑	↑	↑		↑		↑	↑	↑	↑	↑	↑		↑	↑	↑	↑		↑	↑	↑				↑	↑		

Figure B21.1 DNA sequence alignment of 10 alleles only showing the SNP positions in the first 858 bp from the start codon. The boxed and colored nucleotides represent haplotype specific SNPs (hs-SNPs) that uniquely diagnose the presence of that haplotype. The shaded nucleotides represent SNPs that lump two or more of the many alleles in potato germplasm in a group. In the last column, the plus and minus sign indicates whether the Quantitative Trait Allele (QTA) has a positive or negative contribution to the agronomical value of the quantitative trait. Figure courtesy of Herman J.van Eck.

are indicated at base pair coordinates from the start codon. A total of 46 SNPs was observed across 723 bp across 10 alleles. A haplotype is defined as the joint observation across a set of marker loci. In this context, each SNP locus has two alleles but this gene has 10 haplotypes or alleles. The figure was obtained from the DNA sequence alignment of Sanger EST reads obtained from various alleles within and across potato cultivars; invariable nucleotide positions were removed. The SNPs highlighted with arrows and boxes are indicative for a unique haplotype and are referred to as haplotype specific SNPs (hs-SNPs). The other SNPs, indicated without arrows, are not unique to one haplotype, and thus represent an SNP marker that will lump a variable number of alleles. A complete statistical model of the genetic effects of individual alleles and the allele interactions on trait values requires a comprehensive set of these hs-SNPs. In the example below, the presence of the first haplotype is best monitored with A142G, G279T, C435T, C615T or C692T, whereas all other SNPs are unable to distinguish this haplotype unambiguously. From this figure it is clear that the binary classification offered by molecular markers is insufficient to account for the variation between haplotypes.

Why is marker development and validation not always successful?

Suppose that two homozygous individuals with the first and second haplotype are used as parents of a mapping population. A molecular marker exploiting the first T/A SNP on position 6bp will be adequate to map the QTL and identify that the T-allele has a negative and the A-allele has a positive contribution to the trait value. If then this marker is validated in a wider range of germplasm, the QTL is probably not confirmed. The fifth haplotype also has a T-allele at the SNP at 6bp position, but with a positive contribution. Figure B21.1 shows that, considering all the SNP markers detected, there are some haplotypes (the fourth, sixth, and eighth) that cannot be identified unambiguously, because none of the SNPs are diagnostic for these alleles. Only with a combination of several SNP loci will it be feasible to characterize all individual alleles. A simplified example with three haplotypes and two SNP loci is shown in Figure B21.2. Haplotype analysis is sufficiently powerful to allow the conclusion that

repeated observations of the same haplotype are indicative for "identity-by-descent" of that haplotype. In potato we could confirm identity-by-descent using the potato pedigree database (http://www.plantbreeding.wur.nl/potatopedigree/), and indeed related germplasm carried the tenth allele/haplotype, which was derived from the wild *Solanum demissum*. The higher number of SNPs in this *S. demissum* haplotype relative to the other reflects the larger genetic distance from *S. tuberosum*.

Figure B21.1 also illustrates the reason of failing SNP assays due to flanking SNPs affecting the annealing of DNA primers. The position information of each SNP illustrates that multiple SNPs may so localize within the approximately 20 bp distance occupied by a PCR primer (e.g. 510bp, 513bp, 516bp). These SNPs do not necessarily belong to the same haplotype. Lack of primer annealing on one or more haplotypes will introduce null alleles upon PCR amplification. When a G/T SNP assay is hit by annealing polymorphisms caused by flanking SNPs, the diploid individuals will not only display GG, GT or TT, but also G0, T0 and 00 observations, depending on the underlying haplotypes. At the tetraploid level, SNP analysis requires the ability to cluster data in five groups – GGGG, GGGT, GGTT, GTTT, TTTT – with signal intensities of 0, $\frac{1}{4}$, $\frac{1}{2}$, $\frac{3}{4}$ and 1, according to the signal capture method of the SNP assay. Individuals with null alleles will display, for example, GGT0 and GTT0 with signal intensities of 1/3 and 2/3, effectively blurring the discriminatory power of the SNP assay.

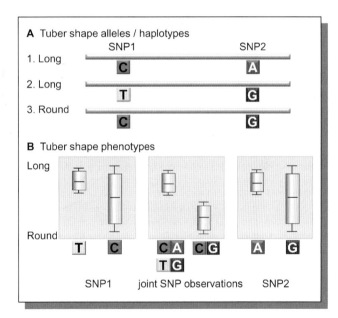

Figure B21.2 Simplified case where two SNPs are not haplotype specific SNPs (hs-SNPs) but lump alleles. Single SNP–trait association studies will have limited statistical power as compared with haplotype–trait associations. Haplotypes can be identified with hs-SNPs (if available), or with combinations of lump-SNPs. The figure also illustrates the issue of "missing heritability", because the joint SNP observations will offer more explained variance than either SNP1 or SNP2 alone can explain. Figure courtesy of Herman J.van Eck.

From this example it will be clear that markers developed on the basis of parental clones are not necessarily suitable for the entire gene pool, or may not explain the phenotypic variation in genome-wide association studies. Efforts to extrapolate marker–trait linkage from experimental bi-parental mapping populations to breeders' germplasm may fail often. In crops with narrow gene pools it is perhaps not a serious issue, but in highly heterogeneous germplasm it is discouraging.

Multiallelism or "Allelic heterogeneity" may also explain some of the "missing heritability" (Bergelson and Roux, 2010; Brachi *et al.*, 2011). In potato large numbers of SNPs allow the reconstruction of haplotypes based on moderate read lengths. In many other organisms with SNPs at larger distances, the reconstruction of haplotypes must be achieved with linkage disequilibrium estimates. In polyploid outbreeders the development of haplotype specific molecular markers will be very complicated.

Conclusion

Molecular markers opened the black box of quantitative inheritance and QTLs have been identified. The QTL is in itself also a black box. The identification of haplotypes based on hs-SNPs or multiple lump-SNPs will allow the identification of the many possible quantitative trait alleles (QTA) and estimation of the contribution of each allele on the agronomic properties of the crop.

References

Appleby, N., Edwards, D., and Batley, J. (2009). New technologies for ultra-high throughput genotyping in plants. *Methods in Molecular Biology*, **513**:19–39.

Bergelson, J., and Roux, F. (2010). Identifying the genetic basis of complex traits in Arabidopsis thaliana. *Nature Review Genetics*, **11**:867–879.

Brachi, B., Morris, G.P., and Borevitz, J.O. (2011). Genome-wide association studies in plants: the missing heritability is in the field. *Genome Biology*, **12**(10):232.

Collard, B.C.Y., Jahufer, M.Z.Z., Brouwer, J.B., and Pang, E.C.K. (2005). An introduction to markers, quantitative trait loci (QTL) mapping and marker-assisted selection for crop improvement: The basic concepts. *Euphytica*, **142**(1–2):169–196.

Frary, A., Nesbitt, T.C., Frary, A.,C *et al.* (2000). fw2.2: a quantitative trait locus key to the evolution of tomato fruit size. *Science*, **289**:85–88.

Fridman, E., Carrari, F., Liu, Y.-S., Fernie, A.R., and Zamir, D. (2004). Zooming in on a quantitative trait for tomato yield using interspecific introgressions. *Science*, **305**(5691):1786–1789.

Johannsen, W. (1909). *Elemente der exakten Erblichkeitslehre*. Fischer, Jena, Germany.

Mather, K. (1941). Variation and selection of polygenic characters. *Journal of Genetics*, **41**(2–3):159–193.

Nilsson-Ehle, H. (1909). Kreuzunguntersuchungen an Hafer und Weizen, Lund, Sweden.

Paterson, A.H., Lander, E.S., Hewitt, J.D., Peterson, S., Lincoln, S.E., and Tanksley, S.D. (1988). Resolution of quantitative traits into Mendelian factors by using a complete linkage map of restriction fragment length polymorphisms. *Nature*, **335**(6192):721–726.

Sirks, M.J. (1929). Multiple allelomorphs versus multiple factors, in *Proceedings of the International Congress of Plant Science*, Vol. 1, Ithaca, New York, pp. 803–814.

Tan, M.Y.A., Hutten, R.C.B., Celis, C.,C *et al.* (2008). The *RPi-mcd1* locus from *Solanum microdontum* involved in resistance to *Phytophthora infestans*, causing a delay in infection, maps on potato chromosome 4 in a cluster of NBS-LRR genes. *Molecular Plant-Microbe Interactions*, **21**(7):909–918.

van Eck, H.J., Jacobs, J.M.E., Stam, P., Ton, J., Stiekema, W.J., and Jacobsen, E. (1994). Multiple alleles for tuber shape in diploid potato detected by qualitative and quantitative genetic analysis using RFLPs. *Genetics*, **137**(1):303–309.

Plant breeders routinely select for quantitative traits of economic value directly, a method that is ineffective because these traits are influenced by the environment. A method of indirect selection that is not influenced by the environment is desired. This calls for the identification of polymorphic markers that are linked to putative QTLs. The general method of identifying QTLs is by comparing the linkage of these markers and phenotypic measurements. Mapping of QTLs has become more accessible because of advances in molecular genetics and statistical techniques. It has been reported that most of the successes with mapping the genes of the domestication syndrome have come about thought QTL mapping, including the efforts by Steve Tanksley and colleagues who localized six QTLs in tomato.

QTL mapping has shortcomings. It is challenging to develop mapping populations for perennials, inbreeding species, and vegetatively propagated crops. The results of QTL analysis often depend on the environment. Further, there is what is called the Beavis effect, associated with the limited power of statistical procedures to accurately estimate the number and size of QTLs. The statistical challenges are of more concern when the traits are conditioned by a large number of QTLs of smaller phenotypic effect and where epistasis is involved.

21.8.1 Principles of QTL mapping

QTL mapping basically entails finding an association between a genetic marker and a measurable phenotype (Figure 21.3). Researchers work from the phenotype to the genotype, using statistical techniques to localize chromosomal regions that might contain genes contributing to the phenotypic variation in a quantitative trait of interest in a population. Conceptually, if all the large-seeded plants from among 1000 plants with different seed sizes all have a particular allele of a genetic marker, the researcher can safely infer that there is a high likelihood that a QTL for large seed size is associated with the marker in that particular population.

A mapping population is created by crossing two parental strains that differ in alleles that affect the phenotypic trait of interest (measurable trait). Commonly used populations for mapping QTLs are F_{2-3} families and RILs, the former having an advantage over the latter because it is possible to measure additive and dominance effects at specific loci. Even though RILs cannot measure additive gene action, they are immortal and can allow researchers to conduct experiments over different locations and years. Polymorphic markers are used to genetically partition the population into groups to ascertain whether significant differences exist between them on the basis of the phenotype measured (Figure 21.4). A significant difference between mean trait values is interpreted to mean a linkage between the marker locus and a QTL conditioning the trait of interest. This is so because if a marker is closely linked to a QTL, the chance of recombination between them is low, resulting in them being inherited or transmitted together to cause their mean to be significantly different from that of the group without a marker. In the case of weak or no association of a marker to a QTL, the two will segregate independently resulting in no significant difference between the means of the diverse genotype groups.

Generally, mapping QTLs does not accurately position genes underlying polygenic traits on the genome.

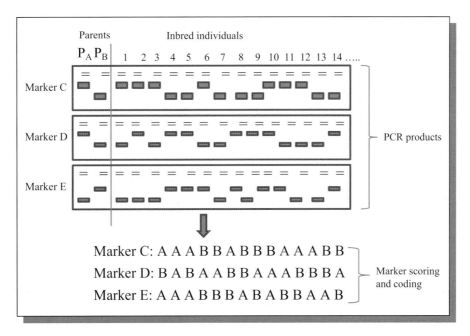

Figure 21.3 Constructing a linkage map using molecular markers. Once markers are scored, the results are fed into a mapping program to generate a linkage map.

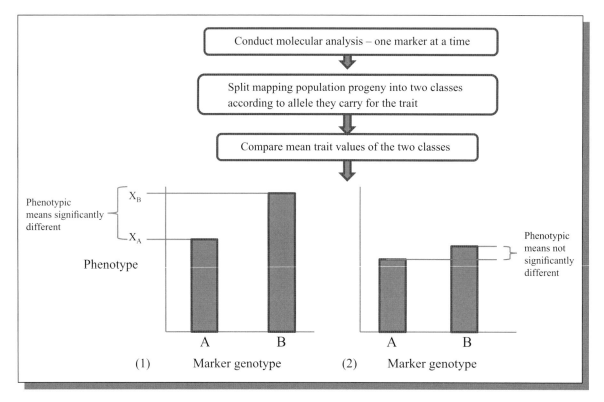

Figure 21.4 Using molecular markers to map QTL. The mapping population is placed into two groups (A and B) based on the genotype of the marker. The difference between the two group means is tested for significance, statistical significance indicating that the marker is linked to a QTL.

Also, large numbers of individuals and genotypes per individuals are needed to detect and localize QTL in a single mapping undertaking. Consequently, QTL mapping is an iterative procedure in which the first phase (primary or coarse mapping) determines the general locations of QTLs. The second phase focuses on developing high-resolution maps of the regions containing the putative QTLs by sampling more individuals to obtain the required recombinations and identify molecular markers in the region of interest. High-resolution mapping frequently shows that single QTLs fractionate into multiple and closely linked QTL, which often have different effects.

21.8.2 Steps in QTL mapping

There are several general requirements for QTL mapping (Figure 21.5), including that there is a:

- Quantitative trait of interest.
- Genetically variable population (mapping population).
- Complete map of molecular markers (polymorphic markers).
- Well-defined method for phenotyping.
- Appropriate set of statistical methods (to localize chromosomal regions that might contain QTLs).

With reference to mapping population, there are two common approaches to QTL mapping:

(i) **Linkage mapping.** Involves mapping in families or the segregating progeny of crosses between genetically divergent strains. Because the procedure relies on crosses between two strains, only a small fraction of the genetic diversity in the population is captured. The procedure can detect chromosomal regions containing one or more QTLs that affect the trait of interest but it only imprecisely localizes the QTLs.

(ii) **Association mapping.** Involves unrelated individuals from the same population. The procedure captures a wider genetic diversity than

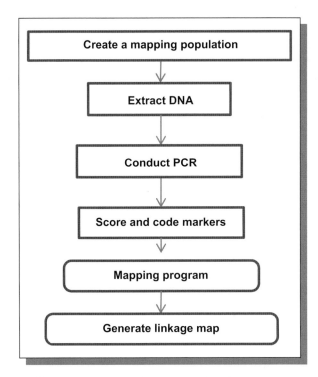

Figure 21.5 General steps in molecular mapping.

linkage mapping and uses fewer individuals for the analysis. However, a key drawback is the reduced power to detect QTLs when marker alleles are not common, requiring the expense of high-resolution genotyping (using a large number of markers) to increase the precision of QTL localization. Further, this procedure is susceptible to false positives stemming from the mixing of populations with different allele frequencies and values of the quantitative trait of interest.

The mapping process itself entails two general activities – detection and localization. The power to detect QTL depends on their effects and allele frequencies. Effects refer to the average difference in the phenotype of the trait between marker allele genotypes (*a*), scaled by the phenotypic standard deviation of the trait within marker genotype classes (*s*). The number of individuals needed for QTL mapping increases as *a/s* decreases and as allele frequencies depart from 0.5. The principal practical limitation to localizing QTLs appears to be the availability of markers. The accuracy and precision of locating QTLs depend to a large extent on the density of the linkage map created, the

higher the density, the higher the precision of locating putative QTLs. On the other hand, the higher the map density, the higher the likelihood is of detecting false positive QTLs. Nonetheless, a denser map, whereby QTL are mapped to a relatively small chromosomal region or regions would greatly facilitate the isolation of specific genes (by methods like positional cloning).

21.8.3 Methods of QTL detection and localization

There are three commonly used detection methods – single marker analysis, simple interval mapping, and composite interval mapping. The MapQTL 4.0 software package may be used for mapping QTL.

Single marker analysis

Single marker analysis (single point analysis) is the simplest of the QTL detection methods associated with single markers. It requires a linkage map and can be performed with basic statistical software packages (ANOVA, t-test, linear regression). The ANOVA procedure is sometimes called the marker regression analysis. The most commonly used technique is the simple regression analysis in which the phenotype is correlated with each marker genotype (Figure 21.6). The coefficient of determination (R^2) from the marker explains the phenotypic variation arising from the QTL being linked to it. This method underestimates the effect of the QTL because the farther away it is from the marker, the weaker its effect will be and the less likely it would be detected. The procedure is not applicable when marker information is missing (must discard individuals whose genotypes are missing the marker). A larger number of segregating markers with wide genomic coverage may improve the effectiveness of this method in QTL detection.

Simple interval mapping

Making use of linkage maps, the **simple interval mapping (SIM)** method uses a pair or two pairs of linked flanking markers at a time to detect a QTL located within a chromosomal interval between the markers. Like ANOVA, it assumes the presence of a single QTL. Consequently, each location in the genome is posited, one at a time, as the location of the putative QTL. A large number of markers are scored, followed by an assessment of the probability that an interval between two markers is associated

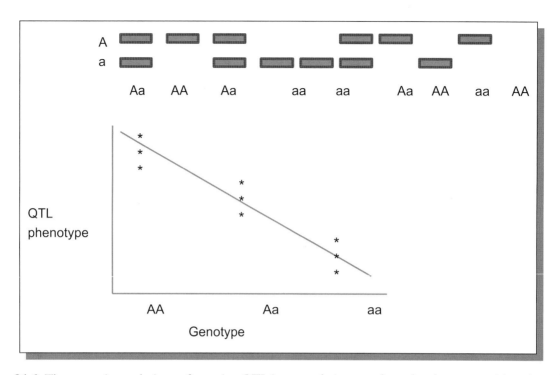

Figure 21.6 The regression technique of mapping QTL in a population correlates the phenotype with each marker genotype. In this example, a single marker, "A" is scored. A significant relationship between marker and the phenotype indicates that the trait is probably linked to the marker. The method does not localize the chromosomal segment that contains the QTL and may underestimate the effect of the QTL.

with a QTL that affects the trait of interest. Using the maximum likelihood test of hypothesis, the results are plotted as a likelihood-ratio test statistic versus the chromosomal map position, measured in recombinational units (Morgans) (Figure 21.7). LOD scores are used to measure the strength of evidence for the presence of a putative. The statistically significant threshold is indicated as the dotted line. Permutation tests are widely used to determine the significance thresholds. Peaks above this line indicate significance, and the chromosomal location that corresponds to the highest significant likelihood ratio is the estimate of the location of a QTL. Where other QTL are linked to the interval of interest, false peaks (false detection of QTL) may occur.

Composite interval mapping

The **composite interval mapping (CIM)** technique combines SIM with multiple regression statistical analysis in an attempt to improve the accuracy of QTL detection by controlling the effects of other

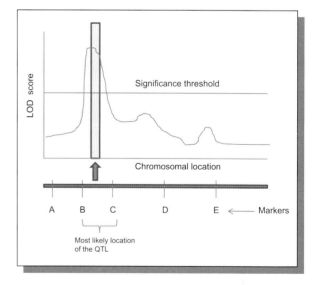

Figure 21.7 A typical output from interval mapping showing the test statistic on the y-axis and the markers comprising the linkage group on the x-axis. The distinct and pronounced peak between flanking markers B and C indicates the most likely position of the QTL.

markers on the traits of interest. The ANOVA technique, though very simple to use, is deficient when the markers are widely spaced and also when missing data occur. CIM, though complicated and computationally intensive, can accommodate missing genotype data.

Multiple interval mapping

Interval mapping assumes the presence of only one QTL. Even though it is possible to use interval mapping techniques to detect multiple QTLs, especially when they are located on different chromosomes, there are statistical methods that model multiple QTLs simultaneously. These methods have the advantage of controlling for the presence of a QTL, and thereby reducing the residual variation to obtain a higher power for detecting additional QTLs. Also, these methods allow for a better separation of linked QTLs and the identification of interaction between QTLs (epistasis). The technique of **multiple interval mapping** is simply an extension of interval mapping to multiple QTLs, just as multiple regression analysis extends ANOVA.

Presenting interval mapping results

Information from interval mapping procedures may be presented in tabular form or line drawings (Figure 21.8).

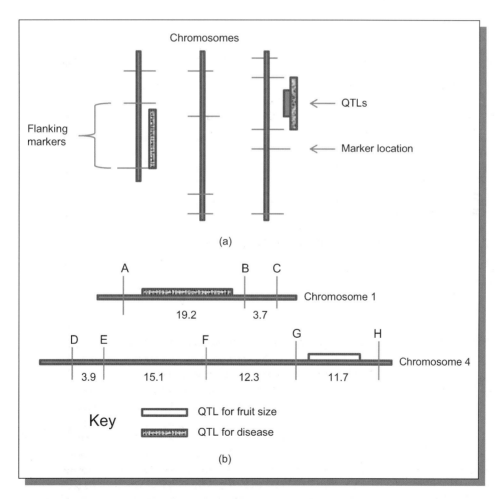

Figure 21.8 Interval mapping may be presented typically in on one of two ways, tabular or line drawing. (a) and (b) show line drawing QTL for fruit size and disease resistance are located on chromosomes 1 and 4, respectively.

21.9 High-resolution QTL mapping

As indicated previously, it is challenging to precisely locate a QTL on the chromosome, necessitating QTL mapping to be conducted iteratively, a primary map followed by fine mapping. The accuracy of QTL mapping is influenced by such factors as population size and type, level of replication of phenotypic data, genotyping errors, and environmental effects. The preliminary map (also called a "framework" or "scaffold" linkage map) is designed to be comprehensive in coverage of all the chromosomes in the genome, albeit in less detail. One of the most important applications of genetic markers is to facilitate plant breeding by making the selection process more effective, efficient and speedy. Called marker assisted selection (MAS), its success depends on the identification of reliable markers systems.

It is customary to conduct QTL confirmation and validation subsequent to generating the preliminary map to ensure that a QTL is not an artifact of experimental procedure or false positive but that it is stable and effective or reliable in predicting the trait of interest in different genetic backgrounds. The goal of high-resolution mapping is to establish the level of polymorphism of most tightly-linked markers to within less than 1 cM (no more than 5 cM) of the trait of interest. This requires the use of larger population sizes and includes additional markers to saturate the preliminary map. The use of high-throughput marker techniques is needed for processing large numbers of markers. Further, the procedure of bulk segregant analysis is helpful in this regard. As stated previously, the ultimate goal of genetic mapping is gene isolation. With fine mapping of QTLs, the goal is to saturate the map to such high density that observed recombination approaches the resolution of single genes, so that the causal genetic change for a QTL can be determined. When QTLs can and are molecularly isolated, breeders can develop markers at the potential resolution of the nucleotide, thereby increasing the specificity and precision by which favorable genetic effects can be estimated and manipulated in a breeding program to increase genetic gain.

21.10 Bulk segregant analysis (BSA)

A genetic map has applications in plant breeding and genetics. However, researchers may not always desire to develop a comprehensive molecular map but rather have interest in a few markers that are closely linked to a specific trait of interest. To identify these markers, a methodology called **bulk segregant analysis (BSA)** is commonly used. This methodology of rapid gene mapping is suitable for monogenic qualitative traits. Essentially, the researcher creates a bulk sample of DNA by pooling DNA from individuals with similar phenotypes. If the desire is to find molecular markers associated with a disease-resistant locus, the researcher would create two bulk DNA samples, one containing DNA from disease-resistant lines and the other containing DNA from disease-susceptible plants. The premise is that two bulked DNA pools of segregants will contain random samples of all the loci or alleles in the genome, except for those that are in the region of the gene that is the basis for bulking the DNA. Consequently, two pools differing for one trait of interest would differ only at the locus harboring that trait, and any difference in marker pattern would be linked to the locus which was the basis for bulking of the DNA.

21.11 The value of multiple populations in mapping

What has been called the "global genetic architecture" refers to all of the loci, their effects and potential interactions that condition the natural variation for a trait within a species. In an effort to further our understanding of the genetic and molecular basis of complex traits, scientists may employ whole genome scans to identify QTL and to obtain an overview of the genetic architecture of a trait (genetic underpinnings of phenotypic variation – number of loci, type and magnitude of their effects, interactions between genes, and gene × environment interactions). Analysis of QTL is commonly done in one natural or experimental mapping population. Genetic architecture is largely population-specific. Further, most species exhibit some degree of population genetic structure. Consequently, an analysis based on one population is likely to capture only a fraction of the global genetic architecture of the species presented by the progenitors of the population. More comprehensive information is likely to emerge with the use of multiple populations.

21.12 Comparative genome mapping

The theory of evolution suggests common ancestry of all living things, with branches off the evolutionary tree as accumulated mutations causing variation and subsequent divergence into distinct species. Consequently, is it expected that different species would share certain genes in common. With the advent of throughput genome sequencing capabilities, many organisms of varying levels of genetic complexity have either been completely or partially sequenced. In addition to the human genome, there are a number of organisms, designated as model organisms, whose complete genome sequences are known. This includes *E. coli*, yeast, zebra fish, *Arabidopsis*, mouse, rice and corn. The essence of comparative genomics is to compare the finished reference sequence of one species with the genomes of other organisms, so regions of similarity and difference may be identified. Specific benefits of comparative genetics (comparative maps) include:

- **Evolutionary studies.** Comparative genomics is a powerful tool for studying evolutionary changes (genome rearrangements) among organisms by helping to identify genes that have been conserved among species through time, as well as what gives individual organisms their uniqueness. Such studies could also assist in pinpointing the signals that control gene function.
- **Development of linkage maps.** Using anchor markers developed from the mapping of several species, researchers are able to construct molecular linkage maps of new species.
- **Gene cloning.** Map-based cloning or positional cloning is tedious even in species with small genomes (e.g., rice), let alone in species with large genomes (e.g., wheat, barley). Comparative genomics can help in cloning genes in complex species by using information from species with smaller genomes. This strategy could be especially help in cloning genes in the grass family to which many crop species are associated.

Examples of comparative genome maps available include one for Solanaceae (tomato–potato–pepper–eggplant–jimsonweed–tobacco), tomato–potato, and wheat–rye–barley.

21.13 Synteny

Gene order in chromosomes is conserved over wide evolutionary distances. In some comparative studies, scientists discovered that large segments of chromosomes, or even entire chromosomes in some cases, had the same order of genes. However, the spacing between the mapped genes was not always proportional. The term colinearity is used to refer to the conservation of the gene order within a chromosomal segment between different species. The term **synteny** is technically used to refer to the presence of two or more loci on the same chromosome that may or may not be linked. The modern definition of the term has been broadened to include the homoeology (homoeologous chromosomes are located in different species or in different genomes in polyploid species and originate from a common ancestral chromosome) of originally completely homologous chromosomes. Whole genome comparative maps have been developed for many species (including many in the Fabaceae – soybean, mugbean, etc.) but are most advanced in the Gramineae family (Poaceae).

Researchers may use such correlated arrangements (synteny maps) among taxa to provide a framework for inference of common ancestry of genes (i.e., whether two genomes are evolutionarily closely related). These maps may also be used to understand how genomes have changed and diverged over time. Further, information from model organism may be used to study other species that are not well understood. Research has shown that about 75% of genes in dicot species and about 40% in monocot species occur in regions that have synteny or colinearity with *Arabidopsis*. Other applications of colinearity to plant breeding include the prediction of genes controlling a specific function in crop species and facilitating genome mapping (markers from a well studied and mapped species to one less studied).

The preservation of synteny in large portions of a chromosome is sometimes referred to as **macrosynteny**, to be distinguished from **microsynteny,** the preservation of synteny for only a few genes at a time. Such larger blocks of conserved gene order occur beyond what would be expected under a random breakage model of chromosome evolution. Some researchers have attempted to clone a gene in one plant species based on the detailed and sequence information (microsynteny) in a homoeologous region of another genus. Microsynteny-based comparative mapping, combined with CAPS analysis of recombinant plants, has been used to narrow mapping regions to develop high-density maps to aid in positional cloning and marker assisted selection in some plants.

Synteny mapping is computationally intensive. Special software has been developed to assist with this process. This includes GridMap (for representing similarities and differences between pairs of objects, such as genomes and sequences, in a grid form) which is suitable for graphical analysis of comparative mapping data. The pairwise comparative map (PCM) software allows a researcher to display two maps of any type and draw lines between the homologous loci on each map.

21.14 Linkage disequilibrium and haplotypes

Linkage disequilibrium (LD), or gametic phase disequilibrium, is said to occur when there is non-random association of alleles at two or more loci in the genome. Some combinations of alleles (or genetic markers) occur more or less frequently in a population than would be expected if there were random formation (accounted for by chance) of **haplotypes** (combination of alleles at multiple loci that are transmitted together on the same chromosome) from alleles based on their frequencies. When a particular allele at one locus is found together on the same chromosome with a specific allele at a second locus more often than expected if the loci were segregating independently in a population, the loci are in disequilibrium. Another way of stating this fact is that the observed frequencies of haplotypes is not in agreement with the haplotype frequencies predicted by multiplying the frequency of individual genetic markers in each haplotype. The occurrence of LD is an indication that the two marker alleles are physically close on the DNA strand. Haplotypes are often associated with traits of interest that have complex genetic origins.

Genetic linkage (GL) and LD are related concepts but differ in several significant ways. In terms of origin, GL is a product of a more recent event (describes the association of two or more loci on a chromosome with limited recombination between them) while LD is evolutionary or ancestral in origin (Figure 21.9).

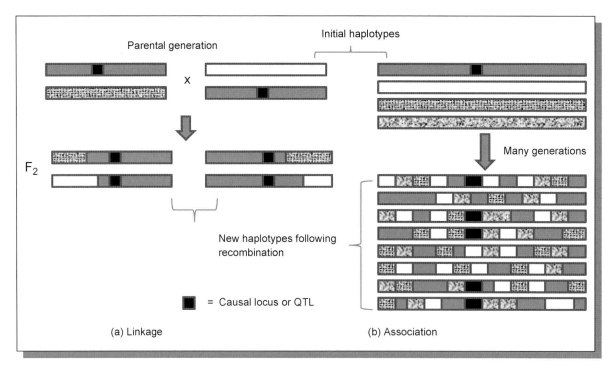

Figure 21.9 Both linkage (a) and association (b) take into account recombination that has the effect of shuffling the initial haplotypes to uncouple all but the most tightly linked markers from the causal locus. Because only the tightly linked markers will predict the organismal phenotype, the causal locus can be localized with precision. The difference between the genetic events is that linkage utilized recent recombinational events while association is based on historical events (many generations of recombination).

GL is an association related to a locus and measures the cosegregation occurring within a pedigree, while LD emphasizes an allele at a locus and measures the cosegregation within a large population. LD is detected for markers that are located much closer (0.01–0.02 cM) to the target gene than GL (1–5 cM).

The level of LD depends on factors such as mutation, recombination rate, genetic linkage, selection, genetic drift, non-random mating, and population structure, the first two being the most influential. Designated as D, LD may be mathematically represented for two adjacent loci, A and B, with two alleles (A, a and B, b) at each locus and observed frequency of the haplotype consisting of alleles A and B (P_{AB}), as follows:

$$D = P_{AB} - P_A \times P_B$$

where P_A and P_B are respective gene frequencies and P_{AB} the haplotype frequency.

21.15 Linkage disequilibrium mapping (association mapping)

LD mapping, also known as **association mapping**, has some advantages in overcoming some of the limitations of QTL mapping. Its goal is to identify genome-wide variations that have associations with phenotypic variation. One of the key advantages of LD mapping is that it does not require crosses and production of large numbers of progeny; it can rely on population samples. This property makes it useful for studying long-lived perennial species (e.g., banana, palm) expeditiously. Using a population sample also has the advantage of having materials that may contain more meiotic information (from the ancestral or evolutionary history of the sample) than the mapping populations commonly used in QTL mapping. This could result in the phenotype of interest being associated with a significantly smaller chromosomal segment than would be the case in a QTL population and, thereby, potentially lead to a higher mapping resolution.

To be successful, LD mapping depends on a well thought out experimental design. Large sample sizes are needed to reduce background noise. Further, the effects of individual mutations may have to be assumed to be additive. The procedure requires measures of variability in markers representing most of the genome, as well as tests of phenotype–genotype associations for each marker. It also entails an analysis of the type of association that attempts to pinpoint the causative genetic mutation(s) that condition the phenotype. The challenge with this second activity is associated with marker (especially single nucleotide polymorphism (SNP)) density in mapping. LD mapping is more powerful when the causative mutation is genotyped. Where this is not the case, the researcher may identify association using markers that are in LD with the causative mutation. It must be noted that the extent of LD can vary significantly among plant species and population samples, as well as among genomic regions within plants.

It is a challenge to test for epistatic interaction between the multitude of markers. Researchers may facilitate this activity by testing for association between phenotypes and haplotypes instead of individual markers, assuming haplotypes can be inferred experimentally as is the case in self-pollinating species (e.g., barley, rice), thereby avoiding avoid the use of computational inference. Additional challenges in LD mapping arise when phenotypes of interest vary by geographic origin (e.g., flowering time, photoperiod sensitivity). Where such is the situation, it is possible for false associations to be observed as a result of the genotype associating with the geographic region instead of with a phenotype. Such problems may arise with species that are derived from wild populations with extensive geographic structure, including barley, rice and soybean. In spite of these challenges, LD mapping has been successfully used, for example, to link phenotypic variation in malting quality in barley to haplotype variation at the B-amylase2 gene (involved in starch hydrolysis). Differences in the coding region of this enzyme affect thermostability of the enzyme. Using SNP genotyping, researchers showed that cultivars with high malting quality and the high-thermostability enzyme share a common haplotype.

Association mapping may be placed into two broad categories according to the focus and scale of the study:

(i) **Candidate-gene association mapping.** This strategy relates polymorphisms in selected candidate genes that have known roles in the control of phenotypic variation for specific traits. This approach is often pursued when researchers have interest in a specific trait or suite of traits.

Further information regarding the location and function of genes involved in either biochemical or regulatory pathways that lead to final trait variation often is required (annotated genome sequences from several model species are available as well as a wide array of genomic technology – sequencing, genotyping, gene expression profiling, comparative genomics, bioinformatics, linkage analysis, mutagenesis, biochemistry).

(ii) **Genome-wide association mapping.** Also called genome scan, this strategy surveys genetic variation in the whole genome to identify signals of association for various complex traits. Often undertaken by a consortium of researchers, this approach is comprehensive and entails the testing of a large number of molecular markers (hundreds of thousands) distributed across the genome for association.

Association mapping is being used by some plant breeders to identify genes responsible for quantitative variation of complex traits that have agronomic and evolutionary importance. This has become possible because of the recent advances in genomic technology, reduced costs, the availability of robust statistical methods, as well as the need to exploit natural diversity. As the strategy is better understood (to identify quantitative trait loci by examining the marker-trait associations that can be attributed to the strength of linkage disequilibrium between markers and functional polymorphisms across a set of diverse germplasm) more researchers will adopt it in plant improvement undertakings. Whereas conventional linkage analysis is widely used in mapping research (e.g., single marker analysis, interval mapping, multiple interval mapping, and Bayesian interval mapping, which use experimental populations derived from a bi-parental cross to obtain information about traits that tend to be specific to the same or genetically related populations), association mapping has the advantage of allowing the researcher to search for functional variation in a much broader germplasm context (information is applicable to a wider germplasm base). The strategy is promising in plant breeding because it is relatively easy to produce a large number of progenies from controlled crosses, and also to conduct replicated trials using immortal individuals (e.g., inbreds and recombinant inbred lines).

The need to survey natural diversity is great because it has implications on future trait improvement and germplasm security. Older strategies for this kind of undertaking include advanced backcross QTL and introgression libraries that have been successfully used to mine alleles from exotic germplasm for use in improving crop productivity, adaptation, quality, and nutritional value of crops. Association mapping may be considered complementary to these strategies, with the added advantage of the larger scale of the survey of natural diversity. Also, this strategy can leverage emerging genomic technologies with sequencing, re-sequencing (hundreds of SNPs can be identified by re-sequencing a core set of diverse lines), and genotyping as the intermediate steps to the ultimate goal of linking functional polymorphisms to complex variation.

Association mapping is a long-term commitment. Consequently, it is important to consider all genetic aspects of the species and the associated germplasm available (e.g., ploidy level of the species, available molecular markers, status of linkage analysis for the targeted traits) prior to commencing the project. It should be pointed out that genetic diversity, extent of genome-wide linkage disequilibrium, and the relatedness within the population are factors that affect the mapping resolution, marker density, statistical methodology to be used, and mapping power. Materials that can be used for association mapping may be obtained from germplasm banks, synthetic populations as well as elite germplasm. Many populations readily accessible to the researcher would be samples with both population structure and familial relationships (because of local adaptation, selection, and breeding history). However, other categories of germplasm that can be used fall into one of the following additional categories: (a) ideal sample with subtle population structure and familial relatedness, (b) multifamily sample, (c) sample with population structure, and (d) sample with severe population structure and familial relationships.

21.15.1 Breeding applications of association mapping

Most of the traits of interest to plant breeders are complex in genetic underpinnings. Breeding of complex traits that are significantly influenced by population structure, such as flowering time, disease resistance, plant architectural traits, and seed quality will benefit from association mapping. Using pedigree-based germplasm in association mapping will facilitate marker assisted selection by identifying superior alleles that have been assembled by breeding practices. In addition to using assembled germplasm, researchers can apply association mapping data (phenotypic, genotypic, and pedigree) generated from plant breeding using computer-aided approaches. Further, association mapping has the capacity to identify superior alleles (superior allele mapping) that were missed by breeding practices and introgress these alleles into elite genotypes for breeding purposes.

References and suggested reading

Broman, K.W. (2001). Review of statistical methods for QTL mapping in experimental crosses. *Lab Animal*, **30** (7):44–52.

Collard, B.C.Y., Jahurfer, M.Z.Z., Brouwer, J.B., and Pang, E.C.K. (2005). An introduction to markers, quantitative trait loci (QTL) mapping, and marker-assisted selection for crop improvement: The basic concepts. *Euphytica*, **142**:169–196.

Mackay, T.F.C., Stone, E.A., and Ayroles, J.F. (2009). The genetics of quantitative traits: Challenges and Prospects. *Nature Reviews Genetics*, **10**:565–577.

Ross-Ibarra, J., Morrell, P.L., and Gaut, S. (2007). Plant domestication, a unique opportunity to identify the genetic basis of adaptation. *Proceedings of the National Academy of Sciences USA*, **104**:8641–8648.

Symonds, V.V., Godoy, A.V., Alconada, T.,C *et al.* (2005). Mapping quantitative trait loci in multiple populations of Arabidopsis thaliana identifies natural allelic variation for trichome density. *Genetics*, **169**(3): 1649–1658.

Outcomes assessment

Part A

Please answer the following questions true or false.

1 A population created solely for gene mapping is called a clonal population.
2 F_2 mapping populations are "immortal".
3 A linkage map is the same as genetic map.

Part B

Please answer the following questions.

1 Give an example of a mapping population.
2 What is an "immortal population"?
3 Define synteny.

Part C

Please write a brief essay on each of the following questions.

1 Distinguish between a genetic map and a physical map.
2 Describe how association mapping is applied in plant breeding.
3 Describe how linkage mapping is conducted.
4 Describe the steps in QTL mapping.
5 What is the advantage of using multiple populations in mapping?
6 Compare and contrast backcross and F_2 populations for mapping.

22

Marker assisted selection

Purpose and expected outcomes

In developing new cultivars, plant breeders practice selection in segregating populations on the basis of phenotype. As previously stated, most biological traits are genetically complex (QTL) and their expression highly influenced by the environment. Selection efficiency and effectiveness, and therefore breeding progress, could be adversely impacted if the trait of interest in the breeding program has low heritability. Selecting on the basis of markers linked to traits of interest could greatly facilitate the breeding of quantitative traits, except that QTL mapping remains challenging to allow its widespread application at this time. Nonetheless, molecular markers are used to a considerable extent in plant breeding. **Marker assisted selection (MAS)** *is the application of molecular markers in plant breeding.*
After studying this chapter, the student should be able to:

1 Explain the advantages of MAS over conventional selection in breeding.
2 Describe the general steps in developing markers for a MAS QTL.
3 Describe marker assisted backcross breeding.
4 Describe marker assisted gene pyramiding.
5 Describe how backcross inbred lines (BILs) are used for introgressing wild genes.
6 Discuss the limitations of MAS.

22.1 The concept of molecular breeding

Molecular breeding is the application of biotechnological tools in crop improvement. As a proven method of plant breeding, molecular breeding is best considered not as a single breeding approach like bulk breeding or backcross breeding, but rather as a collection of tools and efforts for improving trait phenotypes via direct manipulation of the genotype at the DNA level. The tools include genomic analysis, functional genomics, proteomics, and metabolic profiling. One of the more specific molecular breeding approaches is the use of molecular markers, especially DNA markers, for crop improvement. This approach

Principles of Plant Genetics and Breeding, Second Edition. George Acquaah.
© 2012 John Wiley & Sons, Ltd. Published 2012 by John Wiley & Sons, Ltd.

is called **marker assisted selection (MAS)**. One of the earliest documented applications of MAS in plant breeding was the breeding for resistance to soybean cyst nematode (*Heterodera glycines* Ichinohe). Since then, there has been an explosion in applications of this breeding approach. Large scale MAS programs have been undertaken by CIMMYT in Mexico for wheat improvement; similar applications for this crop have been undertaken in Australia. Similarly, MAS for cultivar development in corn and many other crops have been reported.

22.2 Choosing molecular markers for MAS

Molecular markers are discussed in detail in Chapter 20. Also, over a dozen desirable properties of molecular markers that make them useful for applications in biological research are listed. For application in MAS, the key requirements of molecular markers include:

- **High degree of polymorphism.** The marker selected should be effective in distinguishing between the genetic variability present in the breeding population. Marker validation should be conducted in the appropriate population.
- **High reliability.** Beyond being highly polymorphic, it is critical that the marker be reliable in predicting the trait or phenotype of interest. Reliable markers are those that are tightly linked to the target trait (less than 5 cM desired).
- **Ease of use.** Large numbers of plants are frequently assayed in selection during crop breeding. Consequently, detection methods that are simple and quick to use, amenable to automation, easy to score and interpret, and with reproducible results, are most desired. For MAS application in QTL breeding programs, markers are seldom useful without additional steps of QTL confirmation and validation. Sometimes, fine mapping is required.
- **Cost effective.** Detection methods should be affordable and cost effective. Marker systems that can use crude protein and small amounts of DNA are advantageous in cutting down on the pre-application preparation time.

22.3 Advantages of MAS over conventional breeding protocols

Molecular markers are applicable to plant breeding from program initiation (selection of parents for the creation of the breeding population), to discrimination among individuals in a segregating population, and cultivar identification. The key areas in which MAS is advantageous over conventional breeding include:

- **Distinguishing between heterozygote and homozygote genotypes.** Discriminating between genotypes in conventional breeding is on the basis of phenotypic (visual) selection. Visual selection is seldom effective in distinguishing between homozygous and heterozygous genotypes. Ability to make this distinction is required at some steps in some breeding programs.
- **Early generation discrimination.** In conventional breeding, plant breeders frequently advance large numbers of plants until advanced stages in order to effectively identify and select the desired genotypes. This extends the length of the breeding programs. MAS allows breeders to eliminate undesirable genotypes early in the breeding program by screening plants in the seedling stage.
- **Convenient screening.** Some breeding objectives, like breeding for resistance to root-borne diseases, are labor intensive to achieve because the plants need to be dug out of the soil for evaluation. Markers linked to such traits would make it easier for breeders to make quicker progress in their breeding program and reduce screening cost.
- **Reduced space for screening.** Because of the effect of the environment on the phenotype of especially complex traits, conventional breeding routinely requires selection to be done on segregating plants grown in groups (families, rows, plots). The early generation of MAS may be conducted on seedlings in a small space in the greenhouse. Further, MAS can be conducted on single plants versus selecting a number of plants from groups of plants grown.
- **Reduced breeding time.** Early generation testing and ease of screening during selection may hasten the duration of the breeding program.

In sum, MAS is simpler to conduct than phenotypic screening, can be carried out at the seedling stage, and allows single plants to be selected with high reliability, making for higher efficiency and accelerated line development in plant breeding. In some cases, using DNA markers in breeding may be more cost effective than screening for target trait directly.

22.4 MAS schemes

Like some new technologies, there are plant breeders who believe MAS is a revolutionary tool in the toolkit of breeders and that is superior to conventional breeding, while others think it has limited utility. Because quantitative traits dominate the traits of interest to plant breeders, and because QTL mapping is still evolving, MAS is currently considered by some as of a limited application in breeding. Further, MAS is commonly used not as a stand-alone breeding procedure, but rather as part of conventional breeding. To this end, there are certain activities in conventional breeding that can benefit from the application of molecular marker technology.

The goal of the breeder is to make progress in the trait of interest with each cycle of selection. Called expected **genetic gain** or **genetic advance**, this is the amount of increase in performance that is achieved through a breeding program after each cycle of selection. The response in one generation may be mathematically expressed as

$$\overline{X}o - \overline{X}p = R = ih^2\sigma \ (or \ \Delta G = ih^2\sigma)$$

where $\overline{X}o$ is the mean phenotype of the offspring of selected parents, $\overline{X}p$ is the mean phenotype of the whole parental generation, R is the the advance in one generation of selection, h^2 is the heritability (probability that a trait phenotype will be transmitted to the offspring), σ is the phenotypic standard deviation of the parental population, i is the intensity of selection (proportion of the population selected as parents for the next generation), and ΔG is the genetic gain or genetic advance.

Another expression of this concept is $\Delta G = ih^2\sigma/L$, where L represents the length of time needed to complete one cycle of selection. Usually, L = 1 if one year is the duration of a cycle. This equation has been suggested by some to be one of the fundamental equations of plant breeding that must be understood by all breeders, hence called the **breeder's equation**. Consequently, it is appropriate and expedient to examine the equation to see how MAS may be used to improve genetic advance or genetic gain by changing the effects of the components of the equations. It is clear from the equation that the expected genetic gain can be increased by increasing the degree of phenotypic variation in the breeding population, the heritability of the trait, selection intensity, or by decreasing the duration of a cycle of selection.

The application of MAS will be discussed in two general ways:-

(i) How MAS can be used to improve genetic gain by addressing the components of the breeders equation.
(ii) How MAS can be used to improve genetic gain by using the common breeding methods or schemes – backcrossing, pedigree selection, recurrent selection, and hybrid production.

The general steps in developing markers for QTL MAS are described in Figure 22.1. Factors such as sampling bias may cause QTL positions and effects to be inaccurate. Consequently, it is recommended to validate QTL (test to see if QTL is effective in different genetic backgrounds).

22.4.1 Assessment of genetic diversity and selection of parents for crossing

The success of a breeding program hinges on the initial population used or the parents selected to create such a population. Molecular markers can be used to characterize genetic diversity to assist the breeder in the judicious selection of parents used to create the original breeding population. They can provide a common measure of assessing divergence in germplasm. The genetic base of the breeding population can be broadened by crossing genetically diverse parents. Markers can be used to guide the introgression of novel genes from genetically distant sources (e.g., from wild species, interspecific crosses, and other exotic materials).

Molecular markers have been used to assist in the discovery of heterotic patterns in order to guide the selection of parents for use in a hybrid breeding program (e.g., in maize). Marker analysis (e.g., by Roger's distance) may be used to ascertain genetic dissimilarity among lines of different heterotic groups to enable the breeder predict the performance of hybrids to be developed from different intergroup crosses.

22.4.2 Increasing favorable gene action

From the breeder's equation, heritability, especially in the narrow sense (calculated as the proportion of

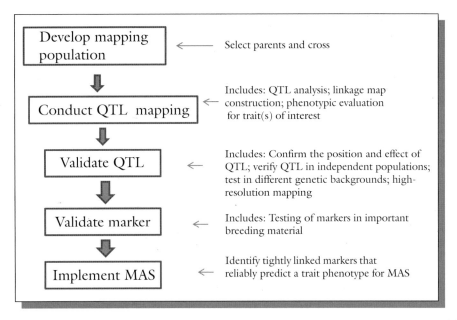

Figure 22.1 General steps in the use of QTL for marker assisted selection (MAS).

genetic variation due to additive genetic effects), is a key determinant of the size of genetic gain with selection. Because additive genetic effects are predictably transferred to progeny, they are also referred to as the breeding value of the individual as a parent in a cross. Breeders can use molecular markers to characterize gene action and breeding values by conducting progeny testing (test the value for selective breeding of an individuals' genotype by examining the progeny produced from self-pollination or cross-pollination). With the proliferation of molecular markers and technologies for high throughput genetic analysis, high density genetic maps are more readily available for more species, thus expanding our understanding of the genetic architecture of traits of interest to breeders. A better understanding of favorable gene action (additive gene action) through the use of molecular markers can assist breeders in better manipulating heritability for greater genetic gain in a breeding program. For example, breeders can add or delete specific alleles that contribute to breeding value.

Because most of the traits of interest to breeders are quantitative in nature, and because quantitative traits are highly influenced by the environment and hence have low to moderate heritability, it is generally believed that MAS is more beneficial to breeding of quantitative traits. QTL analysis was previously discussed (Chapter 21). High density QTL mapping and validation were noted as key to the effective use of QTL in breeding. For example, QTL information can be used to improve genetic gains by breaking up undesirable genetic linkages or epistasis among loci with antagonistic effects on a trait that tend to be a drag on genetic gain. Further, the more simplified the genetic architecture of quantitative traits can be, the easier it will be for breeders to manipulate them to their advantage. In Chapter 21, it was stated that when QTL are molecularly isolated, breeders can develop markers at the potential resolution of the nucleotide, thereby increasing the specificity and precision by which favorable genetic effects can be estimated and manipulated in a breeding program to increase genetic gain.

MAS for complex traits is still challenging to conduct, partly because of the difficulty of effective detection, estimation, and utility of QTL and their effects. The problem is more significant with more complex traits (such as grain yield) that are controlled by many genes under the influence of epistasis (gene-by-gene interaction) and gene-by-environment (G × E) interaction effects. On the contrary, most researchers engaged in evaluations of mapping and MAS tend to assume that QTLs act independently (i.e., no interaction with other genes and/or environment).

To overcome this problem, Podlich and colleagues proposed a new approach to MAS from the conventional one, which assumes that desirable QTL alleles, once identified, will remain relevant throughout many cycles of selection during plant breeding. In the conventional procedure, researchers tend to estimate QTL effects at the beginning of the project and continue to apply the estimates to new germplasm created during the breeding process (i.e., mapping start only). The assumption of fixed QTL values is appropriate if the traits are controlled by additive genes. This will allow MAS to be conducted by independently assembling or stacking desirable alleles. This assumption is not applicable to situations in which context dependencies (changes in genetic background) occur. On such occasions, the value of QTL alleles can change depending on the genetic structure of the current set of germplasm in the breeding program. In other words, QTL values will change over the cycles of selection, as the background effects change. These progressive changes in genetic structure may make the initial combinations of alleles no longer the best target or significant to increasing the trait performance in future breeding cycles. Podlich and colleagues proposed the mapping-as-you-go approach to MAS of complex traits. QTL effects are cyclically, re-estimated each time a new set of germplasm is created. This ensures that the basis of MAS remains relevant to the current set of germplasm.

22.4.3 Increasing selection efficiency

Conventional breeding is a lengthy process, requiring over 10 years in some cases to develop a new cultivar. Part of the reason for the long duration of breeding programs is due to the fact that most of the traits of interest are quantitative in genetic structure, as noted several times previously, and therefore under significant environmental influence. Selecting on a phenotypic basis to identify individuals with the highest breeding value is complicated by the effects of the environment and the G × E (genotype × environment) interaction, not to mention operator error. Other practical challenges to hastening the breeding process include that it is essential in some cases, such as disease breeding, to evaluate genotypes in different environments to subject the materials to varying disease pressure, or the need to wait to an advanced developmental stage in order for the desired phenotype to be optimally expressed for effective selection.

Another reality to note is that breeders commonly consider more than one trait in the selection process, a practice that could reduce the genetic gain from selection. To minimize some of these extraneous effects on the accurate measurement of phenotypes, breeders resort to a variety of strategies, including the use of large populations and samples, and the evaluation of genotypes in multiple environments (multiple locations and multiple years). MAS may be employed in conjunction with conventional breeding to reduce the L factor in the breeder's equation. Where markers are tightly linked to QTL, a breeder may reduce the number of phenotypic cycles by substituting genotypic for phenotypic selection during some cycles. This way, off-season nurseries may be used to reduce the number of selection cycles.

Other ways of improving the efficiency of conventional breeding with MAS include early generation testing to eliminate inferior genotypes, thereby reducing the number of genotypes to be evaluated in the advanced stages of the breeding program. MAS is uniquely suited to efficiently addressing some breeding situations that are challenging to conventional breeding. For example, some traits may be expensive to evaluate phenotypically, or may require special environmental conditions for the phenotype to be properly expressed for evaluation. In these cases, genotypic evaluation using molecular markers could be the answer.

22.4.4 Marker assisted backcrossing

The use of molecular markers can reduce the number of backcrosses by three to four generations. To achieve this involves three levels of selection. First, molecular markers may be used to screen for the target trait, especially when the trait is conditioned by a recessive allele or is laborious to screen phenotypically. At the next level, the breeder may use markers to select backcross progeny containing the target gene and the tightly linked flanking markers so as to reduce linkage drag. Finally, markers can be used to select backcross progeny that have previously been selected for the target trait and possess the background markers (Figure 22.2). Using markers this way, the breeder selects against the donor genome, thereby possibly accelerating the recovery of the recurrent parent genome.

The method of backcrossing is described in detail in Chapter 16. Conceptually, backcrossing is conducted

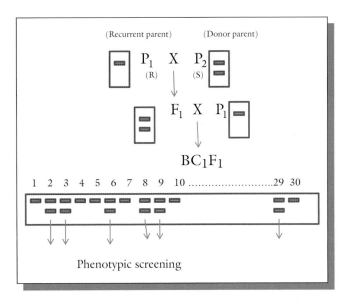

Figure 22.2 Using backcrossing in marker assisted selection.

to transfer a major gene (or a few genes) into an adapted (elite, commercial) cultivar, such that the only change in the elite cultivar is due to the transferred gene(s). Sometimes, the transfer involves a dominant gene, while in other cases it involves a recessive gene; the protocol is slightly different for these two scenarios. MAS may be used to aid backcrossing when recessive genes are involved or when the multiple genes being transferred can epistatically mask each other's effects (e.g., when pyramiding multiple disease resistance genes). MAS is also useful when the experimental environment is not ideal for the expression of the target trait (e.g., disease breeding) or when phenotypic assays are laborious and expensive (e.g., quality traits).

In addition to these more general applications, research indicates that transferring of genes from the wild may be complicated by the phenomenon of linkage drag. However, markers may be used to select for the rare progeny that contain little of the undesirable wild DNA as possible. Also, in situations where markers are not linked to target traits, MAS may be used to accumulate more of the desired recurrent parent's genetic background. A more contemporary use of MAS is the transfer of transgenes into elite or commercial cultivars.

The breeder may calculate the probability of success in carrying out a backcross breeding program by calculating the probability (p) of losing the target allele to recombination when selection is conducted on a linked marker locus as $p = 1 - (1 - r)t$, where r is the recombination frequency between the marker and the target trait and t is the number of generations of backcrossing. Whereas it is preferred to have a tight linkage and an as low as possible recombination frequency, such an event is not easy to discover. Alternatively, breeders may select on the basis of two markers flanking the target trait, even if these markers are not tightly linked to the target, to accomplish the goal of successfully conducting a marker assisted backcross.

Selection in a backcross progeny is for a heterozygous progeny. Therefore, codominant markers are more effective for marker assisted backcrossing. For example, if a dominant marker is used for selection, it would be informative if the dominant allele, conferring band presence, is linked to the donor parent allele. On the other hand, if the recessive allele, conferring band absence, is linked to the donor parent allele, the marker would be uninformative because all backcross progeny will either be heterozygous or homozygous. It would require progeny testing of each individual in each backcross generation to detect those segregating for the recessive marker allele. Such an undertaking would double the number of generations required for marker assisted selection, thereby losing the advantage of increased efficiency.

22.4.5 Backcross inbred lines (BILs) for introgression of wild genes

Exploiting the rich source of genetic diversity in exotic germplasm is met with a host of challenges, not the least of which are those that are genetic in origin. Wild × domesticated crosses have problems such as F_1 hybrid sterility, infertility of the segregating generations, linkage drag, and suppressed recombination between chromosomes of the two species. Tanksley and Nelson developed a procedure, **advanced backcross breeding**, for the simultaneous discovery and transfer of desirable QTL from unadapted germplasm into elite lines. Basically, this procedure postpones QTL mapping until the BC_2 or BC_3, applying negative selection during these generations to reduce the occurrence of undesirable alleles from the donor (unadapted genotype). The advantage of this strategy is that BC_2/BC_3 provides adequate statistical power for QTL identification, while at the same time being sufficiently similar to the recurrent parent to allow selection for QTL-NIL (near isogenic lines) in a short time (1–2 years). The QTL discovered can be verified and the NILs used directly as improved cultivars or as parents for hybrid breeding.

Backcross inbred lines (BILs) are populations of plants derived from the repeated backcrossing of a recombinant line with the wild type, using phenotypic or molecular marker selection techniques to generate introgression lines. This is conceptually the same as the advanced backcross strategy. BILs are immortal lines and, hence, can be used in replicated experimentation at multiple locations and years. Being near-isogenic, BILs have high genetic and morphological similarity with the recurrent parent to allow precise estimates of traits. Genes that are not detected in an F_2 analysis are more likely to be picked up in BIL studies. Single gene detection is more favored in this strategy. BIL studies tend to reveal QTL that are not involved in interactions, thus making the introgression of the wild trait into commercial cultivars straightforward. However, they are laborious and costly to generate, taking six generations to develop BILs to cover the entire genome (F_1, BC_1, BC_2, BC_3, BC_4 and BC_4S_1). They are superior to an F_2 population for introgression of quantitative traits, except when a single dominant gene conditions the trait or when the trait is expressed through epistatic interactions of a few genes.

22.4.6 Marker assisted "forward selection"

In spite of the advantages discussed, MAS for backcrossing represents a very limited application of molecular markers in plant improvement, because the breeding method is extremely conservative, improving the current cultivar only one or a few genes at a time. On the other hand, "forward selection" programs allow the breeder to recombine alleles throughout the genome to produce new allele combinations. To be cost effective and efficient, molecular markers used in forward breeding programs are tightly linked to a few loci that have large effects on traits that are challenging or expensive to phenotype. More importantly, the linkage between specific markers and target trait loci are stable or consistently diagnostic for target alleles across different populations, thus eliminating the need to re-establish such associations in every population. This is important because the parental stocks with these linkage phases are used in new combinations from time to time to create new breeding experimental populations for selection (i.e., the same set of markers are effective for use in future crosses).

22.4.7 Linkage disequilibrium

Another key requirement for the success of MAS in forward crossing or breeding is the presence of genome-wide **linkage disequilibrium** between markers and the target trait loci. Linkage disequilibrium (LD) is said to occur when alleles occur together more often than it can be accounted for by chance (Chapter 21).

22.4.8 Marker assisted gene pyramiding

Gene pyramiding is the concept of transferring several genes of a kind into a single genotype. It is most commonly practiced for disease breeding where, for example, the genes for more than two races of a pathogen may be systematically transferred into one genotype (Figure 22.3). The purpose of this strategy in breeding is to develop cultivars with durable resistance (horizontal resistance, slow rust, broad spectrum resistance). Achieving this goal via

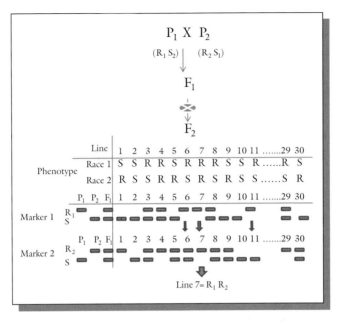

Figure 22.3 Using markers in gene pyramiding.

conventional breeding methods is challenging, because of the difficulty in identifying plants with more than one trait in phenotypic selection. The breeder would have to evaluate individual plants for all the traits tested, something that can be challenging in some populations (e.g., F_2) as well as traits that are assessed by destructive bioassays. Gene pyramiding is difficult or even impossible in early generation of plant development. MAS is suited to addressing this problem because it uses a non-destructive assay and more than one specific gene can be tested using the same DNA sample without phenotyping. When the conventional method is used for gene pyramiding, breeders need to conduct progeny testing to determine which plants have multiple disease genes, since the genes often have the same phenotype. However, this step is unnecessary when MAS is employed because each gene is linked to a specific marker, making it easy to determine the number of desired genes a plant may have. Further, with MAS, it is possible to pyramid QTL along with qualitative genes.

22.4.9 Marker assisted early generation selection

Plant breeding is said to be numbers game. Traditionally, large numbers of segregating populations dominate the early stages of a breeding program. With selection, the numbers are steadily reduced until the ideal genotype is found and multiplied. Consequently, it is obvious that a breeding program would benefit from an ability to reduce the numbers of plants in the early generation (F_2 or F_3). Markers may be used to eliminated and discard as many plants as possible that lack desired gene combinations early in the program, thereby making latter stages of the breeding program more efficient and less expensive in terms of labor and space (Figure 22.4). When the linkage between the marker and the target QTL is weak, MAS has the greatest efficiency when applied in the early generation. This is because the weak linkage between the two units will make recombination between them more likely. On the other hand, applying MAS in the early generation would require one to work with a large population (high cost of genotyping).

22.4.10 Mas and seed purity/cultivar identity

Physical mixing of seed is common occurrence when handling large numbers of genotypes in a breeding program. In hybrid breeding programs where genetic purity is critical for the right genotypes to be used for maximum exploitation of heterosis, some breeders

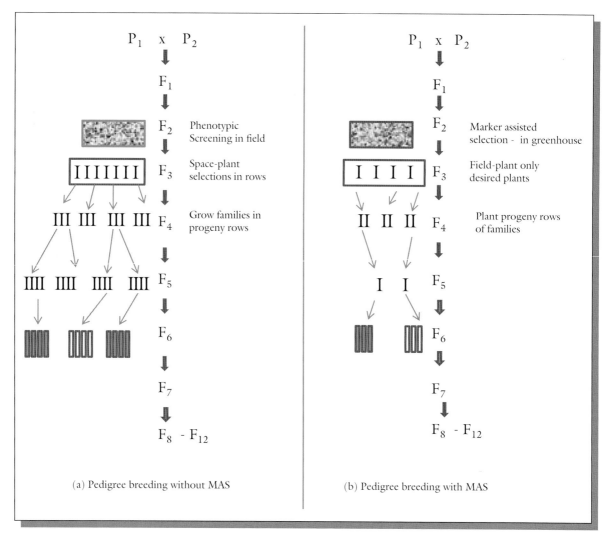

Figure 22.4 Pedigree breeding (a) without and (b) with marker assisted selection. With marker assisted selection the breeder handles a lot less material.

precede their programs with genetic purity confirmation using molecular markers (e.g., SSR and STS markers in some rice breeding programs) in lieu of field confirmation on morphological basis.

22.5 Limitations of MAS

Whereas MAS has had distinct successes in plant breeding, there are situations in which significant limitations to its wide deployment exist. An area in which significant barriers exist is in MAS for

polygenetic trait improvement. The key reason for the persistence of this barrier is the challenge of accurately localizing QTL on maps. Because QTL lack discrete phenotypic effects, it is not possible to map them as if they were qualitative or discrete Mendelian loci. Even though sophisticated statistical tools are employed, the possibility always exists that the prescribed maximum likelihood position may not be the precise location of the QTL. An attempt to improve the size of QTL confidence intervals by saturating QTL maps with markers to enhance their resolution is not effective beyond an inter-marker spacing of 10 cM. The strategy of selecting for marker loci covering large genomic segments to ensure that the target QTL alleles are retained in the selected progeny may inadvertently maintain undesirable linkages between QTL that affect one or more target traits. A case in point is a study in which a presumed QTL contained not only the one targeted QTL but additional ones in the region, tightly linked in repulsion phase. A conventional strategy using a phenotypic selection method would have provided opportunities for uncovering a rare recombinant that had desirable alleles at both linked QTL.

One of the chief challenges in breeding quantitative traits is the strong influence of the environment on their expressions, and consequently their low heritability. It has been shown by research that MAS is most effective when breeding values are predicted by an index of QTL genotypic values, as inferred from linked marker genotypes and QTL effect estimates, and phenotypic values. When heritability increases, phenotypic information becomes a more effective estimator of genotypic values and, consequently, diminishes the importance of marker scores in the selection index. Many research reports indicate that MAS has no advantage over conventional phenotypic selection when heritability is 50% or greater. This is because when heritability is high, gain from phenotypic selection approaches the maximum possible given the genetic variance. This leaves little room for additional improvement with marker assisted selection. To increase heritability, breeders boost the effectiveness of phenotypic selection in estimating the breeding values of quantitative traits by evaluating the experimental materials at multiple locations and over years, activities that are time and cost intensive. For this reason, some suggest that MAS could be considered as a more cost effective alternative. However, without an effective map, a QTL-based MAS is not effective. In sum, if the phenotypic data available are poor indicators of the genotypic values, it is challenging to produce a QTL map that will be adequate for MAS to proceed. On the other hand, when the phenotypic data are good, MAS for genetic improvement of quantitative traits is not warranted.

A major consideration in the implementation of MAS is cost. An initial cost of equipment (new research infrastructure) is needed to support the marker system selected for use. However, it may be cheaper to outsource such activities to commercial laboratories. Some marker technologies may attract intellectual property rights and patent fees. In some cases, a trained person (new intellectual capacity in the form of technician, postdoctoral fellow, etc.) versed in marker technology may need to be hired for the MAS breeding program. Some studies show that the cost/benefit ratio of MAS is dependent upon factors including the inheritance of the trait, the cost of field and greenhouse evaluation, and the method of phenotypic evaluation.

New marker systems that are more efficient, easier to use, and amenable to automation are being developed. The SNP (single nucleotide polymorphism) based genetic markers have great potential for increasing scale and efficiency enough to reduce the unit cost of MAS because they can readily be automated and also include simultaneous analysis of multiple loci.

22.6 MAS and developing countries

Breeders in developing countries face the problem of focusing in crops that often lack international significance. These regional crops (orphan crops) usually receive limited funding support for breeding, especially from for-profit multinational companies. Marker assisted selection (MAS) is one of the relatively accessible biotechnologies for plant breeding in developing countries. Nonetheless, cost remains an issue for many countries. To effectively implement MAS, breeders in developing countries should prioritize their traits of interest and seek cost effective genotyping methods. Some suggest that the number of DNA makers for orphan crop breeding could be increased via extensive data mining of genomics databases, as has been achieved in pearl millet with the development of single-strand conformational polymorphism (SSCP)–SNP markers.

22.7 Enhancing the potential of MAS in breeding

Some of the suggested approaches to achieving higher success with MAS include:

- Well planned and executed QTL mapping studies, including validation of results before application in MAS.

- An integrated approach to breeding whereby conventional strategies, QTL mapping/validation and MAS are used in a complementary fashion.
- Developing cost effective and efficient marker systems by optimizing the methods entailed in MAS, such as DNA extraction and genotyping of markers.
- Using markers to reduce linkage drag through recombinant selection.

- Screening of multiple traits per unit of DNA for rapid identification of gene functions in the major cereal crops.
- Continued use of MAS for breeding programs that involve high-priority traits that are difficult or expensive to measure.

Key references and suggested reading

Bertrand, C. Collard, Y. and Mackill, D.J. (2008). Marker-assisted selection: an approach for precision plant breeding in the twenty-first century. *Philosophical transactions of the Royal Society of London. Series B, Biological Sciences*, **363**:557–572.

Dreher, K. Khairallah, M. Ribaut, J. and Morris, M. (2003). Money matters (I): costs of field and laboratory procedures associated with conventional and marker-assisted maize breeding at CIMMYT. *Molecular Breeding*, **11**:221–234.

Goodman, M.M. (2004). Plant breeding requirements for applied molecular biology. *Crop Science*, **44**:1913–1914.

Hittalmani, S. Parco, A. Mew, T.V. Zeigler, R.S. Huang, N. (2000). Fine mapping and DNA marker-assisted pyramiding of the three major genes for blast resistance in rice. *Theoretical and Applied Genetics*, **100**:1121–1128.

Holland, J.B. (2004). Implementation of molecular markers for quantitative traits in breeding programs - challenges and opportunities, in *New directions for a diverse planet*. Proceedings of the 4th International Crop Science Congress, Sep 26 –Oct 1, 2004, Brisbane, Australia. Published on CDROM (www.cropscience.org.au).

Moose, S.P. and Mumm, R.H. (2008). Molecular plant breeding as the foundation for 21st century crop improvement. *Plant Physiology*, **147**:969–977.

Podlich, D.W. Winlker, C.R. and Cooper, M. (2004). Mapping as you go: an effective approach to marker assisted selection of complex traits. *Crop Science*, **44**:1560–1571.

Young, N.D. (1999). A cautiously optimistic vision for marker-assisted breeding. *Molecular Breeding*, **5**:505–510.

Outcomes assessment

Part A

Please answer the following questions true or false.

1 MAS is more effective for improving quantitative traits than qualitative traits.
2 QTLs lack distinct phenotypes.

Part B

Please answer the following questions.

1 Give three specific factors to consider in selecting makers for MAS.
2 Give a specific application in plant breeding which MAS is more effective than conventional breeding procedure.
3 What is the breeder's equation?
4 Give a specific application backcross inbred lines (BILs) in plant breeding.

Part C

Please write a brief essay on each of the following questions.

1 Describe how MAS is applied in gene pyramiding.
2 Discuss the limitation of MAS in breeding.
3 Describe how MAS is applied in backcross plant breeding.
4 What is linkage disequilibrium and what is its importance to MAS?

23

Mutagenesis in plant breeding

Purpose and expected outcomes

*It was previously pointed out that mutation is the ultimate source of variation. Without adequate variation, plant breeding is impossible. To start a breeding program, the breeder must find the appropriate genotype (containing the desired genes) from existing variation, or create the variation if it is not found in nature. **Mutagenesis** is the process by which new alleles are created. The purpose of this chapter is to discuss mutagenesis as both a technique and a breeding method. The newly created mutants may be used as parents in future breeding programs, in which case mutagenesis is a breeding technique as a source of variation. However, an induced mutant can be systematically processed through conventional breeding steps to be released as a cultivar, hence making it a breeding method (mutation breeding). Mutations arise spontaneously in nature and are pivotal in natural evolution. After completing this chapter, the student should be able to:*

1 Define mutation and mutagenesis.
2 Discuss mutagenic agents.
3 Discuss the steps in a mutation breeding program.
4 Discuss the limitations of mutation breeding.

23.1 Brief historical perspectives

The discovery of the mutagentic effects of X-rays on the fruit fly (*Drosophila*) by H. Muller in the 1920s paved the way for researchers to experiment with its effects on various organisms. In 1928, H. Stubbe demonstrated the use of mutagenesis in producing mutants in tomatoes, soybeans, and other crops. The first commercial mutant was produced in tobacco in 1934. Reports by B. Sigurbjornsson and A. Micke indicated 77 cultivars that were developed via mutagenesis prior to 1995. In 1995, the number was 484. This number has since been significantly exceeded. They include food crops (e.g., corn, wheat, pea), ornamentals (e.g., chrysanthemum, poinsettia, dahlia), and fruit trees (e.g., citrus, apple, peach).

Principles of Plant Genetics and Breeding, Second Edition. George Acquaah.
© 2012 John Wiley & Sons, Ltd. Published 2012 by John Wiley & Sons, Ltd.

Traits modified include agronomic ones, such as plant maturity, winter hardiness, lodging resistance, and product quality (e.g., protein and lysine content), and numerous ornamental ones.

The role of mutations in plant breeding over the years has gone from skepticism, as was demonstrated by the reaction of L.J. Stadler, who is said to have advised his students against using mutation breeding for commercial crop improvement, to the over-optimism by protagonists who saw it as a revolutionary plant breeding method. Currently, induced mutations are used more often in a supplementary role as a source of new alleles. However, it is still important in breeding vegetatively propagated species, including field crops, ornamentals, and fruit and forest species. It is especially useful in ornamental plant breeding where novelty is often advantageous and can become commercially significant. Furthermore, with the advent of genetic engineering and its radical tools, which allow targeted genetic alteration (versus the random genetic alteration produced by conventional mutagenesis), it appears that breeders are gravitating towards this truly revolutionary technology for creating new variability. Nonetheless, every now and then, some breeders find good reason to use a technique or technology that has been marginalized by advances in science and technology.

In conventional breeding of sexual plants, genetic variability is derived from recombination. Parents must not be identical, or else there would be no segregation in the F_2. Even when parents are dissimilar, they often have similar "housekeeping genes", which are common to both parents. Whereas segregation will not occur for these common genes, mutagenesis can create variability by altering them.

23.2 Types of mutations

In terms of origin, mutations may be **spontaneous** (natural) or **induced** (artificial, with the aid of agents). Spontaneous mutations arise at the very low rate of about 10^{-5} or 10^{-6} per generation for most loci in most organisms. This translates to 1 in 100 000 or 1 in 1 000 000 gametes that may carry a newly mutated allele at any locus. They are caused by mistakes in molecular processes associated with the replication of DNA, recombination, and nuclear division. However, because mutagenic agents are common in the general environment, induced mutations,

as a result of these agents (natural radiations) are hard to distinguish from spontaneously induced mutations due to cellular processes.

Mutations may also be classified according to the type of structural change produced as:

- **Ploidy variation.** involving changes in chromosome number (grain or loss in complete sets of chromosomes or parts of a set).
- **Chromosome structure variation.** involving changes in chromosome structure (e.g., duplications of segments, translocation of segments).
- **Gene mutation.** changes in nucleotide constitution of DNA (by deletion or substitution).

Mutation may occur in the nuclear DNA or chromosomes, or in extra-nuclear (cytoplasmic) genetic systems. A good example of the practical application of mutations in plant breeding is cytoplasmic – the genetic male sterility gene, which occurs in chloroplasts.

Mutations that convert the **wild type** (the common phenotype) to the mutant form (the rare phenotype) are called **forward mutations**, while those that change a mutant phenotype to wild phenotype are called **reverse mutations**. Forward mutations are more common than reverse mutations. Recessive mutations are the most common types of mutations. However, recessive alleles in a diploid are expressed only when they are in the homozygous state. Consequently, an organism may accumulate a **genetic load** (the sum of deleterious genes that are carried, mostly hidden, in the genome and may be transmitted to descendants) without any consequence because of heterozygous advantage. As previously discussed, outcrossing species are susceptible to inbreeding depression (loss of vigor) because of the opportunities for expression of deleterious recessive alleles.

23.2.1 Induced mutations versus spontaneous mutations

Spontaneous mutations produce novel alleles for the evolutionary process. Natural mutations have the benefit of being processed through the evolutionary process, whereby viable mutants become recombined with existing forms and become adapted under the guidance of natural selection. Mutagenesis can be used to create new alleles that can be incorporated into existing cultivars through recombination following hybridization and under the guidance of artificial selection. Modern crop production systems are capable of providing supplemental care to enable a mutant that would not have survived natural selection to become productive. As previously discussed, a significant number of commercial cultivars originated from mutation breeding techniques. Furthermore, the rate of spontaneous mutation is low (10^{-6} per locus). Artificial mutagenesis aims at increasing mutation rates for desired traits.

23.2.2 Cell type: Gametic mutations versus somatic mutations

Mutations may originate in the gametic or somatic cells. **Gametic mutations** are heritable from one generation to the next and expressed in the entire plant. However, mutations in a somatic tissue will affect only that portion of the plant resulting, in a condition called **chimera**. In species that produce tillers, it is possible to have a tiller originate from a chimeric tissue, while others are normal. A chimera consists of two genetically distinct tissues and may produce two distinct flowers on the same plant. However, the dual color is impossible to reproduce by either sexual or asexual propagation. Commercial use of chimera is not attractive because the vegetative propagules must, of necessity, comprise both kinds of tissues in order to reproduce the maternal features. It is also well known that cells at the G_2 (gap phase) and M (mitosis) stages are more sensitive to radiation than those at the G_1 and S (synthesis) stages.

23.2.3 Gene action: Dominant versus recessive mutations

In terms of gene action, a mutation may be a:

- **Recessive mutation.** change of a dominant allele to a recessive allele ($A \rightarrow a$).
- **Dominant mutation.** change of a recessive allele to a dominant allele ($a \rightarrow A$).

Open pollinated species may accumulate a large amount of recessive mutant alleles without any adverse effects. However, upon selfing, the recessive alleles become homozygous and are expressed, leading to the phenomenon of inbreeding depression. Using recessive genes in breeding takes a longer time because it requires an additional step of selfing in

order to identify and select the desired recombinants. On the other hand, dominant mutations manifest in the current generation, needing no additional regeneration to be observable.

23.2.4 Structural changes at the chromosomal level

Three types of structural changes in the chromosome can occur as a result of mutation.

1. Gene mutation

(a) Kind Gene mutations entail a change in the nucleotide constitution of the DNA sequence, adding or deleting nucleotides.

- **Transitions and transversions.** DNA consists of four bases – A, T, C, and G – that pair in a specific pattern, G–C and A–T. The structure of the DNA may be modified in two ways – **transition** or **transversion** of bases (Figure 23.1). Mutation by transition entails the conversion of one purine base to another purine (or a pyrimidine to another pyrimidine). During replication, the second purine (different purine), having altered base pairing properties, guides an incorrect base into position. Consequently, one normal base pair is converted to another pair that is genetically incorrect. An agent of mutation (mutagen) such as nitrous acid has been known to cause deamination of adinine to hypoxanthin, cytosine to uracil, and guanine to xanthine, the net effect being a replacement of A–T with G–C in the DNA structure. A transversion involves the substitution of a purine by a pyrimidine and vice versa.
- **Tautomeric shifts.** It is known that each of the bases of DNA can exist in rare states as a result of the redistribution of electrons and protons in the molecule, events called **tautomeric shifts**. When

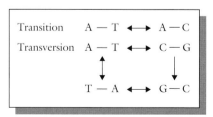

Figure 23.1 Mutations may occur by transition of transversion.

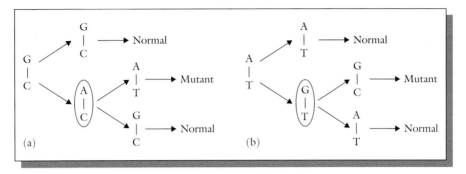

Figure 23.2 Mutations may be caused by tautomeric shifts: (a) shift involving cytosine; (b) shift involving thymine.

this occurs, the base sometimes is unable to hydrogen bond with its complementary base. Instead, some of these altered bases succeed in bonding with the wrong bases, resulting in mutations when, during replication, one purine (or pyrimidine) is substituted for the other (Figure 23.2).

- **Effect of base analogues.** Certain analogues of the naturally occurring bases in the DNA molecule have been shown to have mutagenic effects. For example, the natural base thymine (T), a 5-methyluracil, has a structural analog, 5-bromouracil (5-BU). The two bases are so similar that 5-BU can substitute for T during replication, leading to base pair transition (Figure 23.3).
- **Single base deletions or additions.** A variety of alkylating agents (e.g., sulfur and nitrogen mustards) can act on the DNA molecule, reacting mainly with guanine (G) to alkylate and remove it from the DNA chain. The missing spot may be occupied by any of the four bases to create mutations, usually by transition. Acridine is also known

to express its mutagenic effect through addition of deletion of bases.

(b) Effect at the protein level Four different effects are known to occur as a result of nucleotide substitution.

- **Silent mutation.** Because the genetic code is degenerate (one amino acid can be coded by more than one triplet), a change from $ACG \rightarrow CGG$ has no effect as both triplets code for arginine.
- **Neutral mutation.** This kind of mutation involves an altered triplet code that codes for a different but chemically equivalent amino acid. For example, CAC may change to CGC, altering histidine to a chemically equivalent amino acid, arginine. The change causes a change in the primary structure of the protein (amino acid sequence) but the form of the resultant protein may be unchanged.
- **Missense mutation.** Unlike neutral mutations, a missense mutation results when an altered triplet

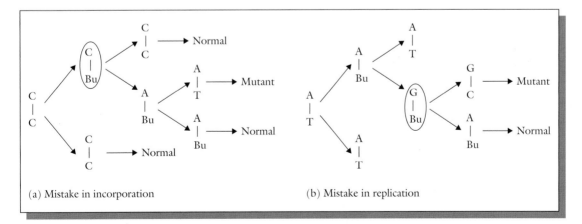

(a) Mistake in incorporation   (b) Mistake in replication

Figure 23.3 Mutations may be caused by transition resulting from the substitution of base analogues for a natural base: (a) mistake in incorporation; (b) mistake in replication.

codes for a different amino acid. Not all missense mutations lead to appreciable protein changes. For example, an amino acid may be replaced by another of very similar chemical properties, thereby altering the structure without impacting its normal function (a neutral mutation). Also, the amino acid substitution could occur in a region of the protein which does not significantly affect the protein secondary structure or function. On the other hand, some missense mutations have devastating consequences. For example, in hemoglobin of humans, a change of *GAG (Glu)* to *GTG (Val)* results in serious consequences (sickle cell anemia).

- **Nonsense mutation.** A nonsense mutation causes an existing amino acid to be changed to a stop codon (e.g., *TAA, TAG)*, resulting in premature termination of protein synthesis.

(c) Frame shift mutation Insertion–deletion mutations (indels) may cause significant changes in the amino acid composition of a protein, and hence in its function. For example, GAG–CCG–CAA–CTT–C (corresponding to *Glu–Pro–Glu–Leu*) may be altered by deletion of G that shifts the reading frame to the right by one nucleotide to produce AGC–CGC–AAC–TTC (corresponding to *Ser–Arg–Asi–Phe*). Very simply: CAT–CAT–CAT–CAT + A (at the beginning) ------→ ACA–TCA–TCA, and so on.

2. Genomic mutation

Errors in cell division resulting from disorders in the spindle mechanism may result in improper distribution of chromosomes to daughter cells. Such errors may cause some cell division products to inherit more or less of the normal chromosome number. These errors, called **chromosomal mutations**, are of two main kinds: **euploidy** (cells inherit additional complete set of the basic chromosome set – n) and **aneuploidy** (certain chromosomes are missing from the basic set or added to the set in some cell division products).

3. Structural chromosomal changes (aberrations)

Changes in chromosome structure begin with a physical break that may be caused by ionizing radiation (e.g., X-rays). After a break, several events may occur:

- The ends of the segment may be disunited.
- The break may be repaired to restore the chromosome to its original form (restitution).
- Or, one or both ends of a break may join to the ends produced by a different break event (non-restitutional union). These events may result in one of four types of rearrangement – **deletions, duplications, inversions,** or **translocations**. The resulting consequences are variable.

23.3 Mutagenic agents

Agents of artificial mutations are called **mutagens**. They may be grouped into two broad categories – **physical mutagens** and **chemical mutagens**. The specific agents vary in ease of use, safety issues, and effectiveness in inducing certain genetic alterations, suitable tissue, and cost, among other factors.

23.3.1 Physical mutagens

The principal physical mutagens are ionizing radiations (Table 23.1). X-rays were first to be used to induce mutations. Since then, various subatomic particles (neutrons, protons, beta particles, alpha particles)

Table 23.1 Examples of commonly used physical mutagens.

Mutagen	Characteristics
X-rays	Electromagnetic radiation; penetrates tissues from a few millimeters to many centimeters.
Gamma rays	Electromagnetic radiation produced by radioisotopes and nuclear reactors; very penetrating into tissues; sources are ^{60}Co and ^{137}Ce.
Neutrons	A variety exists (fast, slow, thermal); produced in nuclear reactors; uncharged particles and penetrates tissues to many centimeters. Source ^{235}U.
Beta particles	Produced in particle accelerators or from radioisotopes; are electrons; ionize; shallowly penetrating; sources include ^{32}P and ^{14}C.
Alpha particles	Derived from radioisotopes; a helium nucleus capable of heavy ionization; very shallowly penetrating.
Protons	Produced in nuclear reactors and accelerators; derived from hydrogen nucleus; penetrates tissues up to several centimeters.

have been generated using nuclear reactors. Gamma radiation from radioactive cobalt (^{60}Co) is widely used. It is very penetrating and dangerous. Neutrons are hazardous and less penetrating, but they are known to seriously damage chromosomes. They are best used for materials such as dry seed, whereas the gentler gamma radiation is suited also for irradiating whole plants and delicate materials such as pollen grains. The relative biological effectiveness (RBE) of fast neutrons is higher than for gamma rays and X-rays. The breeder is interested in identifying and using treatments with high RBE for maximizing the number of mutants produced. Treatments with high RBE have high ionization density. Modifying the treatment environment (e.g., oxygen, moisture content of tissue) can increase the effectiveness of the treatment.

Ionizing radiations cause mutations by breaking chemical bonds in the DNA molecule, deleting a nucleotide or substituting it with a new one. The radiation should be applied at the proper dose, a factor that depends on radiation intensity and duration of exposure. The dosage of radiation is commonly measured in Roentgen (r or R) units. The exposure may be **chronic** (continuous low dose administered for a long period) or **acute** (high dose over a short period). The quality of mutation (proportion of useful mutations) is not necessarily positively correlated with dose rate. A high dose does not necessarily yield the best results. A key limitation of use of physical mutagens is the source. Special equipment or facilities are required for X-rays and nuclear-based radiations.

About 70% of the world's successful cultivars derived from mutagenesis were products of gamma-ray radiation.

23.3.2 Chemical mutagens

Chemical mutagens are generally milder in their effect on plant material (Table 23.2). They can be applied without complicated equipment or facilities. The ratio of mutational to undesirable modifications is generally higher for chemical mutagens than for physical mutagens. However, practical success with chemical mutagens lags behind achievements with physical mutagens. Usually, the material is soaked in a solution of the mutagen to induce mutations. Chemical mutagens are generally carcinogenic and must be used with great caution. One of the most effective chemical mutagenic groups is that of alkylating agents (they react with the DNA by alkylating the phosphate

Table 23.2 Examples of commonly used chemical mutagens.

Mutagen group	Examples
Base analogs	5-bromo-uracil, 5-bromo-deoxyuridine
Related compounds	Maleic hydrazide, 8-ethoxy caffeine
Antibiotics	Actinomycin D, mitomycin C, streptonigrin
Alkylating agents	
– Sulfur mustards	Ethyl-2-chloroethyl sulfide
– Nitrogen mustards	2-chloroethyl-dimethyl amine
– Epoxides	Ethylene oxide
– Ethyleneimines	Ethyleneimine
– Sulfonates, etc.	Ethyl methanesulfonate (EMS)
– Diazoalanes	Diazomethane
– Nitroso compounds	N-ethyl-N-nitroso urea
Azide	Sodium azide
Hydroxylamine	Hydroxylamine
Nitrous acid	Nitrous acid
Acridines	Acridine orange

groups as well as the purines and pyrimidines). Another group is the base analogues (they are closely related to the DNA bases and can be wrongly incorporated during replication); examples are 5-bromo-uracil and maleic hydrazide. Other chemical mutagens commonly used are ethyl methane sulfonate (EMS) and diethylsulfonate (DES). The half-life (time taken for degradation of half the initial amount of alkylating agent) for EMS in water is about 93 hours at 20 °C but only 10 hours at 37 °C. Consequently, chemical mutagens are best freshly prepared for each occasion. Plant breeders who undertake wide crosses often need to manipulate the chromosome number to have fertility. To do this, the most common chemical mutagen used is colchicine. Chemical mutagens produce more gene mutations and less chromosome damage.

23.4 Types of tissues used for mutagenesis

Whole plants and plant parts may be used for mutagenesis, depending on the mutagen to be used, the species, and the research objective, among other factors. Seeds are the most commonly used material for mutagenesis involving physical or chemical mutagens. Whole plants are suited to chronic exposure to physical mutagens. A major drawback to seeds and whole plants is that chimeras are commonly produced. Pollen grains can be used to avoid chimeras. The resulting plants from pollen mutagenesis are heterozygous for any genetic change. However, sufficient quantities of pollen are difficult to obtain; also, they do not stay viable for long.

The plant breeder may also use cuttings and cell or tissue culture explants for mutagenesis. In tissue culture, cells or tissue are treated with the appropriate mutagens and regenerated into new whole plants.

23.5 Factors affecting the success of mutagenesis

Mutations are random events, even when scientists induce them. The plant breeder using the technique can increase the success rate by observing the following:

- **Clear objective.** A program established to select one specific trait is more focused and easier to conduct with a higher chance of success than a program designed to select more than one trait.
- **Efficient screening method.** Mutation breeding programs examine large segregating populations to increase the chance of finding the typically rare desirable mutational events. An efficient method of screening should be developed for a mutation breeding program.
- **Proper choice of mutation and method of treatment.** Mutagens, as previously discussed, vary in various properties including source, ease of use, penetration of tissue, and safety. Some are suitable for soft tissues, whereas others are suited to hard tissue.
- **Dose and rate.** The breeder should decide on the appropriate and effective **dose rate** (dose and duration of application). The proper dose rate is determined by experimentation for each species and genotype. Plant materials differ in sensitivity to mutagenic treatment. It is difficult to find the precise dose (intensity), but careful experimentation can identify an optimum dose rate. Mutagenic treatments invariably kill some cells. Overdose kills too many cells and often produces crippled plants, whereas underdose tends to produce too few mutants. Not only are dose–response relationships rarely known, but they are influenced by the experimental conditions. It is recommended that three dose levels be used in a project – the optimal rate, plus one dose above and below this rate.

- **Proper experimental conditions.** It is known that the oxygen level in the plant material affects the amount of damage caused by the mutagen, the higher the oxygen level, the greater the injury to the material tends to be. The change in the effect of a mutagen with oxygen supply is called the **oxygen enhancement ratio**. In studies in which enhanced mutagen frequency is desired, the experimental conditions may be supplemented with oxygen (e.g., bubbling oxygen through the mutagen solution in the case of chemical mutagenesis). Where such enhancement is undesirable, an oxygen-free environment should be used. The effect of oxygen on mutagenesis is dependent upon the moisture content of the tissue. The higher the tissue moisture, the lower the tissue oxygen supply. Mutagens vary in the importance of moisture in their effectiveness in inducing mutation, X-rays being more affected than gamma rays, for example. Dry seeds are better to use when enhanced mutation frequency due to oxygen status of the research environment is not desired. In chemical mutagenesis, temperature has an effect on the half-life of the mutagen, high temperature accelerating the reaction.

Another experimental factor affecting the success of mutagenesis is the pH of the environment in chemical mutagenesis. For example, EMS is most effective at pH 7.0, whereas sodium azide is most effective at pH 3.0. Sometimes, dry seeds need to be pre-soaked to prepare the cells to initiate metabolic activity. It is important to mention that it is easier to modify the experimental condition when seeds are being used for the mutation program than when other materials are used.

23.6 Mutation breeding of seed-bearing plants

23.6.1 Objectives

The breeder should have clear objectives regarding the trait desired to be induced. Induced mutations are equivalent to natural mutations and hence, once observed, the normal procedure for plant breeding is applicable. It should be borne in mind that mutagenic treatments have both primary and secondary effects. In addition to the specific desired alteration (primary effect), mutagens tend to alter the background genotype (secondary effect). Mutagenesis may cause significant variation in quantitative traits.

23.6.2 Genotypes and source of material

Both self- pollinated and cross-pollinated species can be improved with induced mutations. Seeds from self-pollinated species are homozygous and homogeneous, making it easier to identify mutants in the field. Outcrossed species are heterozygous and heterogeneous, making it more challenging to identify mutants. Dormant seeds are easier to handle than vegetative material.

Mutation breeding is often used to correct a specific deficiency in an adapted and high yielding genotype. This otherwise desirable genotype may be susceptible to a disease or may need modification in plant architecture. A mutant is easier to spot if the parent (source genotype) is genetically pure (as opposed to being a mixture). Also, mutants are easier to identify if the mutant trait is distinctly different from the parental trait (e.g., it is easier to spot a true dwarf mutant if a tall parent is used for the project). Some suggest using an F_1 in some instances because it contains two genomes, which may increase gene recombination and thereby produce a greater amount of diversity in gene mutations.

23.6.3 Treatment

The goal in seed treatment is to treat enough seed to eventually produce a large enough segregating population in the second generation (M_2). This means the number of treated seed should generate sufficient first generation plants (M_1) that have sufficient fertility to produce the size of M_2 population needed. The appropriate number is determined through preliminary experimentation. One of the side effects of mutagen treatment is sterility, which should be factored into the decision process to determine the number of seeds to treat. M_2 population may be 20 000–50 000 plants.

The seed is multicellular. Hence, a mutation in a single cell will give rise to a chimeric plant. Thus, an M_1 plant is subject to both **diplontic** (competition between mutated and normal tissue during the vegetative phase of development) and **haplontic** (competition between mutated and normal pollen during fertilization) selection in order to be included in the tissues forming seed on the plants. Treating seed with physical agents is done by placing the seed in an appropriate container, positioning it at a proper distance from the source of radiation, and exposing it at a predetermined dose rate. Gamma irradiation doses as low as 0.5 krad (e.g., in corn) and as high as 25 krad (in dry seed of wheat) have been used, whereas the dose rate for X-rays ranges between 10 and 25 krad. When using chemicals, the proper concentration of mutagen is prepared at the desirable pH and temperature. Seeds are soaked in the mutagenic solution for the appropriate duration. EMS concentrations that have been used range from less than 1% (e.g., in tomato) to about 4% (e.g., in wheat).

The common physiological injuries caused by the mutagenic treatment are reduction in seedling height (most frequently used identification of injury in the M_1), cytological effects (chromosomal aberrations), and sterility (evaluated by counting the number of inflorescence per plant, etc.).

23.6.4 Field planting and evaluation

Treated seed (M_1) may be planted in a small plot to produce M_2 seed that is harvested for planting an M_2 spaced population. It is advisable to plant plots of the untreated genotype (M_o) used in the project for comparison, to aid in readily identifying mutants. Dominant mutants are identifiable in the M_1 generation; recessive mutants are observable after selfing to produce the M_2 generation.

Furthermore, in species that produce tillers, it is possible that only one of several tillers produced would carry an induced mutation. To produce the mutation throughout the plant, gametes (pollen grain) should be treated. Using seed also frequently produces chimeras, since mutations are induced in single cells that divide and differentiate into parts of the plant. Consequently, the stem may be a mutant tissue while the leaf is normal.

It should be pointed out that M_1 and F_1 generation plants could be genetically different. F_1 plants produced from inbred lines are genetically identical. However, M_1 plants in a population may have different mutations. Hence, it is logical and proper to handle an M_1 population as a segregating population such as an F_2.

To use mutagenesis as a breeding method for producing new cultivars, various cultivar selection strategies may be used, including:

- **Bulk selection.** First, the breeder grows M_1 plants, harvests, and bulks all the seed. A sample of seed is planted in the next season, harvested and the seed bulked. Alternatively, individual M_2 plants may be harvested and bulked for progeny testing in the next season. Seed from progeny rows showing the desired mutant phenotype are identified and harvested. If rows are segregating, the M_3 plants may be harvested and advanced individually. The breeder conducts replicated tests, evaluating on the basis of the desired mutant and other desirable agronomic traits. A weakness in the bulk method is that mutant plants often have low productivity. Consequently, by planting only a sample of the M_2, it is likely to exclude the desired mutant.

- **Single seed descent.** In this method, one or a few M_2 seeds are selected from each plant and bulked for planting M_2 plants. Desirable M_2 plants are harvested and progeny-rowed; alternatively, seeds from desirable plants may be harvested and bulked. This method also has the potential of excluding desirable mutants, even though it allows a larger number of M_1 plants to be sampled.
- **Pedigree method.** Each M_1 plant is harvested separately. M_2 progeny is grown from M_1 plants. Desirable M_2 plants are harvested and progeny rowed. Desirable rows are harvested and bulked separately. The $M_{2:4}$ lines are evaluated in replicated tests.

23.7 Mutation breeding of clonally propagated species

Species that reproduce vegetatively or by apomixis may also be improved through mutation breeding. Vegetatively propagated species tend to be highly heterozygous. Consequently, selfing is accompanied by inbreeding depression. Mutation breeding offers a method of crop improvement whereby the genetic structure is largely unperturbed. Physical mutagens are more frequently used in such species than chemical mutagens. The plant parts targeted for mutagen treatment are those that can produce a bud from which a plant can develop. These parts include modified parts (e.g., tubers, rhizomes, shoot apex, cuttings, bulbs). The exposure to the mutagen must occur as early in the development of the bud and target the meristematic cell. A mutation in the meristematic cell is critical to avoiding chimeras, which can delay selection because the breeder may not be able to identify the mutant without propagating all the material additional times.

Chimeras are desirable in the breeding of certain ornamental species, such as African violet (*Saintpaulia ionantha*). Induced mutations in clones are, of necessity, dominant mutations, unless the starting material was heterozygous and, hence, could yield recessive mutants. Furthermore, clonally induced mutants are primarily chimerical and start as **sectorial** mutants, later becoming **periclinal**. The growing point commonly has two layers (some have three), the outer layer generating the epidermis as well as some leaf mesophyll, while the inner layer generates the remaining parts of the plant (Figure 23.4). The innermost layer (L-I) has one cell per layer, while

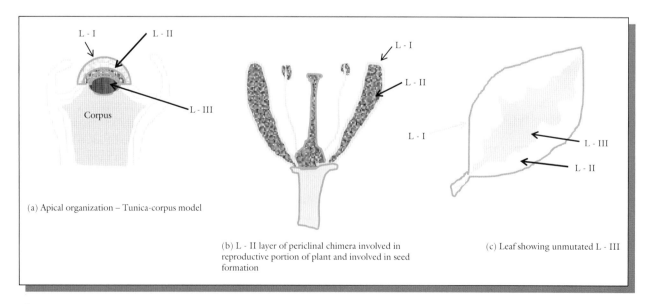

(a) Apical organization – Tunica-corpus model

(b) L - II layer of periclinal chimera involved in reproductive portion of plant and involved in seed formation

(c) Leaf showing unmutated L - III

Figure 23.4 The tunica corpus model of the growing point of a plant, showing three layers, L-I, L-II and L-III, each giving rise to a specific anatomical part of the plant. The effect of a mutation depends on the layer in which it occurs.

L-III has multiple layers. Each layer of the tunica corpus gives rise to specific anatomical regions of the plant body:

L-I   This is the first or outer layer of the tunica. It contributes to the epidermis in all organs. In dicots, this layer is colorless, and hence an L-I mutation is invisible. In monocots, it gives rise to a strip along the leaf margin.

L-II   This is the second layer of the tunica. It contributes to the outer and inner cortex of stem and root, as well as the mesophyll in the outer region of the perimeter of the leaf. It gives rise to the mesophyll of the petals and sepals as well as the internal tissues of the stamens and pistil, hence the pollen and seeds. A mutation in this layer is what gets transmitted to the seed.

L-III This layer contributes to the corpus and is not involved in any part of the flower, fruit or seed.

Mutations, being one-celled events, result in part of whole layers exhibiting chimeras. A periclinal chimera is one that involves the entire layer (whole layer is mutated), whereas a **mericlinal** chimera involves only part of one layer. Periclinal mutations are stable. The number of initial cells in the layers in the growing tip determines the sectorial patterns. Sectorial patterns are mostly unstable, eventually producing periclinal structures. Because of this behavior, the breeder needs to grow several clonal generations successively and select for the desired mutant as well as vegetative stability. Vegetative stability is attained when uniformly periclinal lines of subclones have been isolated. In vegetative system, the treated plant meristem is called the M_1V_1, the next vegetative plant being M_1V_2, and so on.

23.8 Mutations from tissue culture systems

Tissue culture systems may be used for more efficient mutation breeding. For example, somatic embryogenesis technique, coupled with *in vitro* mutagenesis and pollen mutagenesis, is being used in cocoa to induce virus-resistant cocoa trees in Ghana.

23.9 Using induced mutants

A plant may be selected for a specific mutant trait. However, it does not mean that this plant contains only one genetic change. It is recommended to backcross this mutant to its original parents to reselect a pure version of the mutant. Mutants may be directly used in cultivar development. Like the backcross method, an adapted line can be improved through mutation breeding to correct a short coming. In sexually reproducing species, the target mutant traits can be fixed and isolated in the M_2 or M_3 generation (compared to F_6 or F_7 generations in conventional breeding). The mutant lines may be used in genetic studies as markers, among other uses.

23.10 Limitations of mutagenesis as a plant breeding technique

23.10.1 Associated side effects

Mutations induced by mutagens are very diverse in nature. Invariably, the mutagen kills some cells outright while surviving plants display a wide range of deformities. Even those plants with the desirable mutations always inherit some undesirable side effects (akin to linkage drag, or is it a "mutation drag"?). Therefore, it is often necessary to transfer a new mutant into a stable genetic background, free of some of the associated undesirable side effects such as sterility. To do that, the new mutant must be successively propagated for several generations and sometimes even crossed with other genotypes.

23.10.2 Large segregating population needed

Another weakness is the need to produce and sort out a large segregating population in order to have a good chance of finding a desirable mutant. Most mutations are deleterious or undesirable. Mutation breeding may be likened to finding a needle in a haystack. To sort through all the garbage, the plant breeder should have an easy method of screening the enormous variation. Morphological changes (e.g., shape, color) are easy to screen. However, subtle changes require more definitive tests to evaluate and hence are more expensive to undertake.

23.10.3 Recessivity of mutants

Most mutations are recessive and, hence, are observed only when the homozygous genotype occurs. This condition is readily satisfied in species that are naturally inbreeding. The situation practically excludes species that are polyploid, propagated clonally, or have a fertility regulating mechanism (e.g., self-incompatibility) from being amenable to mutation breeding (except when dominant mutations are targeted).

23.10.4 Limited pre-existing genome

The researcher cannot change what does not exist. Consequently, mutations can only be induced in existing genes.

23.10.5 Mutations are generally random events

A modern biotechnology technique, site-directed mutagenesis, has been used to induce targeted mutations. However, conventional mutagenesis remains unpredictable, not being able to be directed to specific genes.

23.11 Selected significant successes of mutation breeding

Mutation breeding has been used to improve a number of important crops (Table 23.3). The first commercial cultivar derived from mutagenesis was "Chlorina", a tobacco cultivar. It was developed with X-ray radiation in 1930. Since then, several hundreds of commercial cultivars or ornamentals, field crops, fruit crops, and other plant kinds have been released for production. These include barley cultivars "Pallas" and "Mari" that were developed by Swedish scientists. These genotypes and other cultivars developed through hybridizations involving the two mutants, significantly impacted barley production in Denmark and Sweden. In the United States, "Pemrad" and "Luther" cultivars of barley were significant in the production of the crop in the 1970s. Dwarf cultivars of cereals have been developed by mutagenesis. Classic examples are "Norin 10" of the Green Revolution fame and "Calrose 76" rice, which strongly impacted California rice production.

23.12 Molecular techniques for enhancing efficiency of induced mutagenesis

Conventional mutation breeding requires the rapid generation of large mutant populations of desirable genetic backgrounds (homozygous for the mutation events and without chimeras) within which selection of desirable mutants is pursued. The need to plant large field trials or conduct laboratory assays on the large numbers of plants in order to discover mutational events is laborious and expensive. Molecular techniques are able to directly query target genes for changes (mutations), thus reducing or eliminating the need for fields trials involving large populations. The polymerase chain reaction (PCR) technique may be used to amplify target regions of the genome by using enzymatic mismatch cleavages.

23.12.1 Reverse genetics

Classical genetics is an approach to understanding heredity by finding out the gene(s) responsible for a particular trait of phenotype in an organism. This is also called forward genetics. Reverse genetics, as the name implies, is an approach to understanding

Table 23.3 Selected general areas of achievement in mutation breeding.

1.	Disease resistance – e.g., *Verticilium* wilt resistance in perpermint, Victorial blight resistance in barley, downy mildew resistance in pearl millet.
2.	Modification of plant structure – e.g., bush habit in dry bean, dwarf mutants in wheat and other cereals.
3.	Nutritional quality augmentation – e.g., opaque and floury endosperm mutants in maize.
4.	Chemical composition alteration – e.g., low-euricic acid mutants of rape seed.
5.	Male sterility – for use in hybrid breeding in various crops.
6.	Horticultural variants – development of various floral mutants.
7.	Breeding of asexually propagated species – numerous species and traits.
8.	Development of genetic stock – various lines for breeding and research.
9.	Development of earliness in many species.

heredity whereby scientists start with a gene or an unknown genetic sequence and attempt to find the phenotype(s) associated with it. A common approach to reverse genetics is to alter or disrupt the DNA sequence or gene and look for the effects of such an action. Several techniques are used for disrupting the DNA element; the common ones include:

- **Site-directed mutagenesis.** Scientists can change either the regulatory regions in the promoter or a gene, or make minor changes in the codon in the open reading frame. These changes comprise of deletions and point mutations. Such changes may result in null allele, in which the gene is no longer functional. This deletion mutational event is referred to as **gene knockout**. Rather than delete genes right away, scientists may devise a technique to include "conditional alleles" that have normal functions until they are activated. These are called **gene knock-in** (e.g., *Cre* or *Flp* recombinases that are activated by chemical, heat shock, etc.).
- **Gene silencing.** In contrast to gene knockout, which permanently inactivates a gene, genes may be temporarily inactivated (**gene knockdown**). This is achieved by the technique of gene silencing involving double stranded RNA (called **RNA interference or RNAi**) or Morpholino oligosaccharides. RNAi interferes with gene expression without actually mutating the DNA of interest. It relies on cellular components (Dicer proteins, RISC complex) for efficacy. The Morpholino antisense oligosaccharides bind and block access to the target mRNA without any helper molecules or accelerating the degradation of mRNA.
- **Interference using transgenes.** Interference of gene expressions may be accomplished through the use of trangenes to overexpress a normal gene of interest or overexpress mutant forms of a gene that is known to interfere with the normal gene function.

23.12.2 Targeted induced local lesions in genomes (TILLING)

Modern advances in genome research have resulted in the availability of whole genome sequences for many model organisms. Coupled with advances in genomics technologies and availability of cost effect and high throughput DNA genotyping and chemical mutagenesis protocols, researchers can now identify mutations in nearly every gene of interest. Mutational analysis has been a major tool for researchers in understanding the basic mechanisms of disease, development, cell biology and metabolism. Earlier efforts employed the forward genetics approach that involved systematic screening for mutations that produced visible phenotypes as a result of defective developmental processes. Though effective, this approach has severe limitations. Many mutations can remain undetected in phenotypic screening because of sheer genetic complexity. Further, practical challenges occur because it is not feasible to screen the enormous number of genomes that will be needed in order for rare mutations or phenotypes to be identified.

To overcome some of the limitations of the forward genetics approach, researchers resort to the generation or targeted discovery of a mutation within a gene of known sequence, an approach called reverse genetics. The key limitation of this approach is the fact that it tends to be organism specific. Well established reverse genetic systems include those for *Drosophila*, *Xenopus*, *Caenorhabditis elegans*, and zebrafish. A new approach to mutational analysis developed by Henikoff and colleagues, dubbed **TILLING (targeted induced local lesions in genomes)**, and first applied to *Arabidopsis thaliana*, provides a more generalized approach to identifying mutations in genes that are only known by their sequence. An advantage of this approach is its ability to generate an allelic series of mutations that is not biased by phenotypic selection.

TILLING starts with first assembling chemically induced mutation-carrying gametes or living individuals (usually thousands, but this depends on the size of the gene and frequency of induced changes) (Figure 23.5). Chemical mutagens that are best suited to this approach are those that generate point mutations (e.g., EMS and ENU). Different mutagens may be used to diversify an allelic series. A sample of DNA is taken from each individual for high throughput analysis of heterozygosity. Genotyping involves gene-specific PCR with labeled primers, melting and re-annealing, followed by treatment with Cel 1 endonuclease and analysis on automated sequencing machines (verification of identified mutants). The differential double-end labeling technique employed allows mutants to be detected on complementary strands. Upon identifying a mutant in a pool, genotyping is repeated for the DNA samples in that pool to identify the individual mutants.

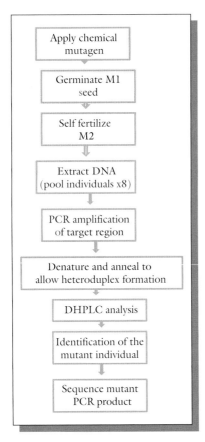

Figure 23.5 A process for TILLING. DNA collected from individuals is pooled for genotyping (involving gene-specific PCR with labeled primers, melting and re-annealing, then treatment with Cel 1 endonuclease, and analysis on automated sequencing machines). Chemicals useful for TILLING are those that generate point mutations (e.g., alkylating agents like EMS).

23.12.3 T-DNA Mutagenesis

The advances in genome sequence technology have significantly facilitated the efforts of researchers in structural genomics, making partial and even complete genome sequences of model organisms available. The discipline of bioinformatics provides tools for *in silico* analysis and annotation of genome sequences through comparison of a new DNA sequence with sequences in existing databases. After ascertaining some indication of gene function from such comparative genetic analysis, experimental confirmation is usually required for definitive functional assignment.

Determining the phenotype caused by the mutation of a specific gene has long been used a as tool by geneticists for discovering gene function. **Insertional mutagenesis** is a genetic tool for rapid identification of a tagged gene responsible for a specific phenotype. Inserting a piece of DNA into a functioning gene deactivates it (knocks it out, creating the so-called knockout mutants). The transposon and *Agrobacterium*-mediated DNA integration are the most widely used biological mutagens in plants. The DNA element is allowed to integrate at random locations in the genome of the organism of interest. The resultant mutants are screened for unusual phenotypes. Once found, a mutant phenotype is assumed to be have been caused by the insertion by way of inactivating the gene relating to that phenotype. Because the sequence of the insertion is known, researchers can identify the gene by either the method of whole genome sequencing to search for the sequence or by PCR to specifically amplify the mutated gene. The sequence of the genomic DNA flanking the inserted element can be readily cloned and sequenced, leading to the identification of mutated gene. The method of insertional mutagenesis facilitates the identification of the culprit gene because of a molecular tag the site of the mutagenic lesion.

An advantage of T-DNA as an insertional mutagen versus transposons is that it does not transpose following insertion into the genome and is chemically and physically stable through several generations. The fact that a marker is left at the site of the mutagenic lesion (T-DNA tagging) is also advantageous. However, like all gene disruption techniques, some limitations occur. This technology yields a relatively low frequency of mutations as compared to mutagenesis via conventional mutagens. Also, it is difficult to identify the functions of genes required in early embryogenesis or gametophytic development, as well as being difficult to identify redundant genes. To overcome some of these limitations, scientists use modified insertional elements, such as activation trap, enhancer trap, and gene or promoter trap, as mutagens. These are complementary to the loss-of-function studies, because they may generate new types of gain-of-function mutants. The promoter and enhancer traps have the advantage of providing a means of identifying genes and characterizing their expression patterns *in vivo* throughout the plant life.

23.13 Horticultural applications of mutagenesis

One of the areas in which induced genetic change is of significant use is in horticulture. Applications for crop plants are applicable in this area. However, oddities (also called sports) that may not have implication in crop productivity can be of significant interest in horticulture for aesthetic reasons. Many flowers have been developed via mutagenesis.

Mutagenesis is of importance in citrus breeding where facultative apomixis, self-incompatibility, cross-incompatibility, long juvenile period, and high heterozygosis are some of the key challenges breeders must address. The occurrence of polyembryony (multiple embryos) in oranges and grapefruits makes it difficult if not impossible for breeders to produce enough zygotic offspring for selection of superior genotypes in conventional breeding. Consequently, most of the commercially successful cultivars of these citrus species are derived from natural or induced mutagenesis. Examples include the Star Ruby grapefruit and Fairchild mandarin. One of the uses of mutagenesis in citrus fruit breeding is the induction of seedlessness.

Flower color modification is a popular objective of mutagenesis in horticulture. Chronic radiation is known to yield a higher mutation rate and a wider spectrum of colors than acute radiation. Further, plants regenerated from petals and buds are more susceptible to color mutation than those derived from leaves. Combining chronic irradiation with floral organ culture is useful for induction of mutants both flowering and non-flowering species.

23.14 General effects of mutagenesis

The exposure of plant materials to mutagens results in three key effects – point mutations, chromosomal aberrations, or physiological damage.

23.14.1 Disadvantages of induced mutagenesis for breeding vegetatively propagated species

The application of mutagenesis to breeding of vegetatively propagated species has certain drawbacks including (some of these are applicable to seed-producing species):

- **Absence of "meiotic sieves".** The overall effect of mutagens depends on various factors, including the mode of reproduction of the plant. Breeding of seed-producing crops could involve the meiotic cycle, which then impacts how putative mutants are advanced through subsequent generations. For example, lethal mutations are eliminated (sieved out) at the M1 generation, leaving only non-lethal and point mutations to advance. However, in vegetatively propagated species, the consequence of the absence of meiosis is that some deleterious mutations can be advanced and accumulate in advanced generations.
- **Concurrent fixation of deleterious alleles.** Genetic effects such as pleiotropy and linkage drag can be mitigated through conventional techniques like hybridization and backcrossing in seed-producing species. However, these techniques are not applicable to vegetatively propagated species, resulting in the accumulation of undesirable genic effects in the mutant, which could jeopardize its success.
- **Presence of chimeras.** The issue of chimeras is problematic in mutagenesis when the tissue used for the induction is multicellular, as is the case in vegetatively propagated plant mutagenesis. The consequence is that the mutant population is not uniform for the mutation event, requiring additional efforts to identify desirable mutants.
- **Complications of systemic pathogens.** Pathogens that invade plant tissues can persist and be systemically transmitted from one generation to the next in mutation breeding of vegetatively propagated species. Such a transmission can occur through seed. More importantly, as the pathogens accumulate over generations, the possible consequence could be masking of the true effects of the mutation event.
- **Bulkiness of starting material.** Seeds are usually more convenient than the usually bulky vegetative propagules that are used in mutation breeding of vegetatively propagated species. This requires more space, time and other resources during breeding.
- **Large mutant populations.** Induced mutations are a function of statistical probability. Consequently, breeders must generate and handle large populations in order to increase their chance of detecting desirable mutations. The bulkiness of vegetative materials used increases the cost and tedium of the operations.

23.14.2 Overcoming drawbacks

Some of the drawbacks can be mitigated by using cellular and tissue biological techniques. Single cells produced in tissue culture (suspension culture) can be used as starting material for induction of mutations. Whole mutant plants may be generated from mutant cells via somatic embryogenesis. In species in which this technique is not feasible, *in vitro* multiplication techniques may be used to rapidly multiply the meristematic tissue of plants and used for induction of mutations. Repeated regeneration helps to dissociate chimeras. This latter approach was used in developing a successful commercial cultivar of banana (*Musa* sp) in Malaysia.

23.15 Key successes of induced mutagenesis

Perhaps the most significant accomplishment of mutation breeding is the development of the isolation of semi-dwarf characteristics in wheat by researchers at CIMMYT (International Center for Maize and Wheat) in Mexico. The semi-dwarf was found to be conditioned by the mutant *sd1* gene, which caused a deficiency in the growth hormone, gibberellic acid. The deployment of wheat varieties with this gene led to the Green Revolution. Mutations played a significant role in crop speciation and domestication. Abolishment of shattering in cultivated peas and cereals such as wheat and barley, as well as removal of bitter principle in edible almonds, parthenocarpy that confers seedlessness in grapes and other fruits (e.g., banana) are all mutational events. A 2008 FAO/IAEA report indicated that of a total entry of 2797 mutant varieties, only 310 were vegetatively propagated plant varieties. Commercially viable varieties of Mandarin orange, apple, sweet cherry, banana, pear, cherry, grape and cassava, are among the crops in cultivation. In ornamental breeding, a large number of chlorophyll variegation leaf mutants have been developed, as well as a large number of flower colors.

Key references and suggested reading

Anonymous (1991). Plant Mutation Breeding for Crop Improvement. Proceeding of the FAO/IAEA Symposium, Vienna. IAEA, Vienna, Austria.

Broertjes, C., and Van Harten, A.M. (1988). *Applied Mutation Breeding for Vegetatively Propagated Crops.* Elsevier, Amsterdam, The Netherlands.

Micke, A. (1992). 50 years induced mutations for improving disease resistance of crop plants. *Mutation Breeding Newsletter,* 39:2–4.

Peloquin, S.J. (1982). Meiotic mutants in potato breeding. *Stadler symp,* 14:99–109.

Pozniak, C.J., and Huel, P.J. (2004). Genetic analysis of imidazoline resistance in mutation-derived lines of common wheat. *Crop Science,* 44:23–30.

Rasmusson, D.C., and Phillips, R.L. (1997). Plant breeding progress and genetic diversity from de novo variation and elevated epistasis. *Crop Science,* 37, 303–310.

Stemple, D.L. (2004). TILLING- a high-throughput harvest for functional genomes. *Nature Reviews Genetics,* 5 (2):145–150.

Internet resources

http://users.rcn.com/jkimball.ma.ultranet/Biology-Pages/M/Mutations.html – Detailed discuss of mutations in nature (accessed March 12, 2012).

http://www.plantmutations.com/– Excellent discussion of mutagenesis in plant breeding (accessed March 12, 2012).

Outcomes assessment

Part A

Please answer the following question true or false.

1 Gamma rays are chemical mutagens.
2 A forward mutation converts a wild type to a mutant phenotype.
3 An acute application of a mutagen occurs over a short period.
4 EMS is a chemical mutagen.
5 Most mutations are recessive.

Part B

Please answer the following questions.

1 Mutations may be spontaneous or . . . .
2 Mutagens may be chemical or . . . .
3 Applications of mutagens may be acute or . . . .
4 Mutations may be forward or . . . .
5 Give three examples of common mutagens.

Part C

Please write a brief essay on each of the following topics.

1 Discuss mutations according to the structural change produced.
2 Discuss the factors that impact the success of mutagenesis.
3 Discuss the importance of mutation in plant breeding.

24

Polyploidy in plant breeding

Purpose and expected outcomes

Hybridization, as previously discussed, is a means of reorganizing genes from the parents involved in the cross in a new genetic matrix. Whereas the contents of the chromosomes may change because of the phenomenon of genetic recombination, normal hybridization does not alter the chromosome number of the species. However, certain natural processes can result in altered chromosome numbers. Similarly, the breeder may develop new variability by altering the number of chromosomes in the species through various processes. Furthermore, a number of the major crop species contain altered chromosome numbers. After studying this chapter, the student should be able to:

1 Define the term polyploidy.
2 Discuss the variations in chromosome number in plants.
3 Discuss the effects of polyploidy on plants.
4 Discuss the importance of autoploidy in crop production.
5 Discuss the genetics of autoploidy.
6 Discuss the implications of autoploidy in plant breeding.
7 Discuss the occurrence of alloploidy in nature.
8 Discuss the genetics and breeding of alloploidy in plant breeding.
9 Discuss the applications of aneuploidy.

24.1 Terminology

Ploidy refers to the number of copies of the entire chromosome set in a cell of an individual. The complete chromosome set is characteristic of, or basic to, a species. A set of chromosomes (the genome) is designated by "x". Furthermore, the basic set is called the **monoploid** set. The **haploid number** (*n*) is the number of chromosomes that occurs in gametes. This represents half the chromosome number in somatic cells, which is designated 2*n*. A diploid species such as corn, has *n* = 10 and

---

Principles of Plant Genetics and Breeding, Second Edition. George Acquaah.
© 2012 John Wiley & Sons, Ltd. Published 2012 by John Wiley & Sons, Ltd.

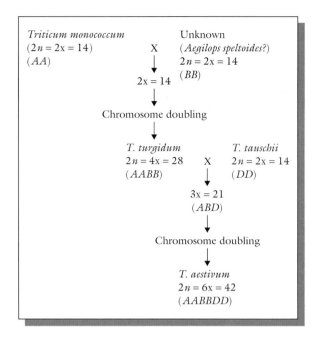

Figure 24.1 The proposed origin of common wheat *Triticum aestivum*.

$2n = 20$. Also a diploid species has $2n = 2x$ in its somatic cells and $n = x$ in its gametes. A species with a higher ploidy, for example autotetraploid (four basic sets of chromosomes) has somatic cells with $2n = 4x$ and gametes with $n = 2x$. For corn, for example, $2n = 2x = 20$, while for wheat, a hexaploid with 42 chromosomes and a basic set of 7, $2n = 6x = 42$. Sometimes species that have more than two genomes comprise sets from different origins. To distinguish the source, each genome is designated by a different letter. For example, wheat has chromosome sets from three different origins and, hence, has a genetic designation (**genomic formula**) of AABBDD (Figure 24.1). To indicate the number of haploids derived from individuals of

different ploidy levels for a single genome, a prefix is added to the term "haploid'" to denote the number of sets (x) of the basic genome present. For example a **monohaploid** ($n = 1x$) is derived from a diploid, while a **dihaploid** ($n = 2x$) is from a tetraploid, and so on.

In some species of higher plants, a pattern of ploidy emerges whereby the gametic (haploid) and somatic (diploid) chromosome numbers increase in an arithmetic progression, as illustrated by oats and wheat (Table 24.1). The set of species displaying this pattern constitute a **polyploid series**.

24.2 Variations in chromosome number

In nature, there exist two types of variation in chromosome number. In one type, called **euploidy**, the individuals contain multiples of the complete set of chromosomes that is characteristic of the species (the basic number, x). In another, called **aneuploidy**, individuals contain incomplete sets of chromosomes that may be equivalent to the euploid number plus or minus one or more specific chromosomes (Table 24.2). The state of having multiples of the basic set in the somatic cell in excess of the diploid number is called **polyploidy**, and the individuals with such cells, **polyploids**. Polyploids are euploids. When euploids comprise multiples of the genome (i.e., duplicates of the genome from the same species) they are called **autoploids** and the condition **autoploidy** (or **autopolyploidy**). However, when a combination of genomes from different species are involved, the term **alloploid** or **allopolyploid** (and, similarly, **alloploidy** or **allopolyploidy**) is used. Alternatively, the term **amphiploid** or **amphidiploid** (and, similarly, **amphiploidy** or **amphidiploidy**) is also used to describe polyploids with different genomes. It should be pointed out that autoploidy and alloploidy are

Table 24.1 Polyploid series of selected species.

	Oat (*Avena* species)	Wheat (*Triticum* species)
Diploid ($2n = 2x = 14$)	*A. brevis* (short oat)	*T. monococcum* (einkorn)
	A. strgoga (sand oat)	*T. tauschii* (wild oat)
Tetraploid ($2n = 4x = 28$)	*A. barbata* (slender oat)	*T. timopheevii* (wild)
	A. abyssinica (Abyssinia oat)	*T. turgidum* (emmer)
Hexaploid ($2n = 6x = 42$)	*A. sativa* (common oat)	*T. aestivum* (common bread wheat)
	A. byzantina (red oat)	

Table 24.2 Classification of polyploidy.

Ploidy		Genome		Description
Diploidy		*AA*	*BB*	Contains two of a basic chromosome set
Euploidy				
(a)	Autoploidy	*AAA*	*BBBB*	Multiples of a basic set (n) of one specific genome.
(b)	Alloploidy	*AAB*	*AABB*	Multiples of the basic number but of different genomes
(c)	Segmental Alloploidy	*AA'B*	*AB'B'*	Multiples of the basic number but the genomes have similar parts
Aneuploidy		**AA**		2n +/− 1,2, ... k

extreme forms of polyploidy. Intermediates occur between them on a continuum of genomic relationships. C.L. Stebbins called the intermediates **segmental alloploids**. Polyploids are named such that the prefix to the standard suffix (ploid) refers to the basic chromosome set (Table 24.3). For example "triploid" refers to a cell with three genomes (3x) while "hexaploid" refers to a cell with six genomes (6x).

24.3 General effects of polyploidy of plants

In terms of general morphology, an autoploid would resemble the original parent whereas an alloploid would tend to exhibit a phenotype that is intermediate between its parental species. Autoploidy increases cell size, especially in meristematic tissues. Autoploids usually have thicker, broader, and shorter leaves. Other plant organs may increase in size compared to their corresponding parts in diploids, an effect called **gigas** features. The gigas luxuriance contributes more to moisture content of the plant parts than to biomass. The plants tend to be determinate in growth.

Table 24.3 Naming of polyploids.

	Genome formula	General name	Specific name
n	*A*	haploid (monoploid)	
2n	*AA*	diploid	
3n	*AAA*	triploid	autotriploid
	AAB	triploid	allotriploid
4n	*AAAA*	tetraploid	autotetraploid
	AABB	tetraploid	allotetraploid
6n	*AAAAAA*	hexaploid	autohexaploid
	AABBDD	hexaploid	allohexaploid

The growth rate of polyploids is less than that of diploids. This may be due to their lower auxin content than their diploid counterparts, as found for tetraploids. Polyploids tend to flower later and over a longer period. In grasses, autoploidy tends to reduce branching or tillering.

Polyploidy also affects the chemical composition of plant parts. For example, artificial polyploidy increased the synthesis of artemisinin (an antimalarial sesquiterpene) produced by diploid *Artemisia annua*, L) twofold to five fold in induced tetraploids. Similarly, the vitamin C content of vegetables and fruits has been known to increase following chromosome doubling. The nicotine content of tetraploid tobacco is about 18–% higher than in diploid species. Autoploids, generally, have fertility problems and have poor pollen production. In some cases, reduction in fertility as compared to their diploid counterparts may be as high as 80–95%. This reduction in fertility is attributed to genetic imbalance following chromosome doubling that leads to disharmonies in development (e.g., abnormal pollen sac, failure of fertilization). Some changes in ecological requirements such as photoperiod and heat requirements have been reported in some species following chromosome doubling.

24.4 Origin of polyploids

The strategies employed in the breeding of polyploids are determined primarily on their origin. J.R. Harlan and J.M. de Wet concluded from an extensive review of the literature that nearly all polyploids arise by the path of unreduced gametes. They pointed out that the most common factor leading to polyploidy is the fusion of 2*n* and *n* gametes to form a triploid, followed by either backcrossing or selfing to produce a tetraploid. Further, they observed that the occurrence of unreduced gametes is variable and pervasive in the plant kingdom.

The unreduced ($2n$) gametes arise by one of two mechanisms – **first division restitution(FDR)** or **second division restitution(SDR)** – during meiosis (Figure 24.2). Each mechanism has a different genetic consequence. In the FDR, the $2n$ gametes result from parallel spindle formation after the normal first division of meiosis. The cleavage furrows occur across the plane of the parallel spindles, producing dyads and $2 \times 2n$ pollen. The genetic consequence of the mechanism is that most of the heterozygosity of the diploid hybrid is conserved in the $2n$ gametes. In the SDR mechanism, the first meiotic division if followed by cytokinesis, but the second division is absent. This results in a dyad with $2 \times 2n$ gametes. However, in terms of genetic consequence, SDR results in significantly reduced heterozygosity in the $2n$ gametes. Researchers such as T. Bingham have proposed the fusion of two FDR $2n$ gametes to harness the heterosis that results in breeding potato. This heterosis can be fixed; the elite lines produced will then be clonally propagated. Seed-propagated species (e.g., alfalfa) cannot benefit from this strategy.

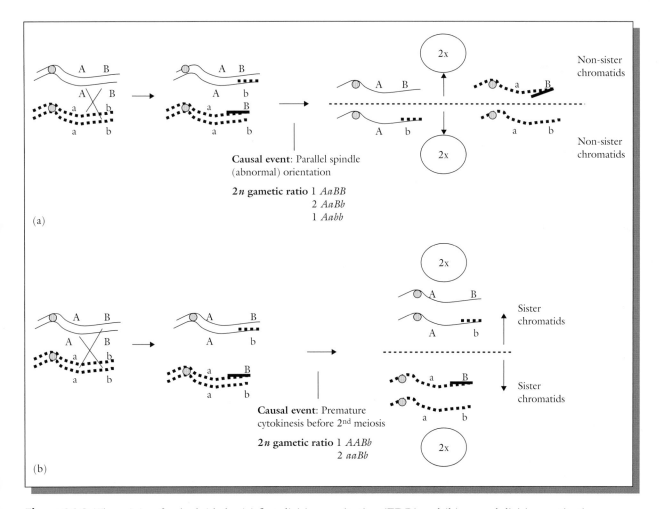

Figure 24.2 The origin of polyploidy by (a) first division restitution (FDR) and (b) second division restitution (SDR). Dyads occur in telophase II. FDR is caused by the presence of parallel or fused spindles, while SDR is caused by the presence of a cell plate before anaphase II. 2n pollen tends to be bigger in size than 1n pollen.

24.5 Autoploidy

As previously defined, autoploids comprise duplicates of the same genome. Autoploids are useful in making alloploids and wide crosses.

24.5.1 Natural autoploids of commercial importance

Autoploidy is not known to have profoundly impacted the evolution of species. Having increased sets of chromosomes does not necessarily increase performance. Autoploids of commercial value include banana, a triploid, which is seedless (diploid bananas have hard seeds and not desirable in production for food). Other important autoploids are tetraploid crops such as alfalfa, peanut, potato, and coffee. Spontaneous autoploids are very important in the horticultural industry where the gigas feature has produced superior varieties of flowering ornamentals of narcissus, tulips, hyacinths, gladiolus, and dahlia among others. Autoploid red clovers and ryegrasses with lusher and larger leaves, taking advantage of the gigas feature of polyploidy, have been bred for commercial use as palatable and digestible livestock forage. It should be mentioned that there is no overwhelming evidence to suggest that autotetraploids are productively superior to their diploid counterparts.

24.5.2 Cytology of autoploids

Autoploids contain more than two homologous chromosomes. Consequently, instead of forming bivalents during meiosis as in diploids, there are also multivalents (Figure 24.3). For example, autoploids have mostly trivalents but some bivalents and univalents are also present. Tetraploids have quadrivalents or bivalents as well as some trivalents and univalents. These meiotic abnormalities are implicated in sterility to some extent, more so in triploids. The microspores and megaspores with x or 2x genomes are usually viable.

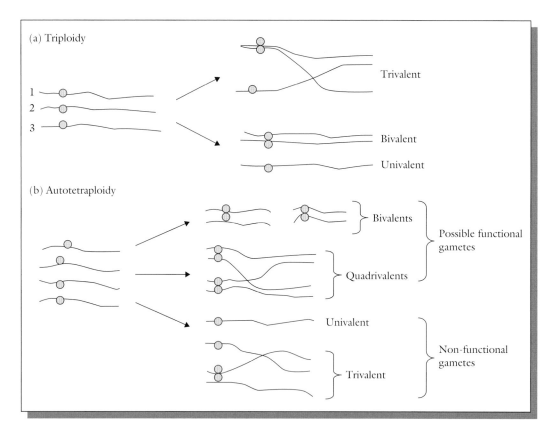

Figure 24.3 Cytology of polyploids: (a) triploidy and (b) autotetraploidy. Bivalents and quadrivalents usually produce functional gametes while univalents and trivalents produce sterile gametes.

The amount and nature of chromosome pairing directly impacts the breeding behavior of autoploids. Autoploids are induced artificially by chromosome doubling using colchicine. Doubling a hybrid between two diploid cultivars would produce a tetraploid in which there may be a tendency for the doubled set of chromosomes from one parent to pair independently of the doubled set of chromosomes of the other parent. This propensity is called **preferential** or **selective pairing**, a phenomenon with genetic consequences. If preferential pairing is complete, there would be no new genetic recombination, and hence the progeny would look like the doubled F_1. Furthermore, the bivalent pairing would contribute to sterility originating from meiotic disorders, while preserving heterosis indefinitely, should there be any produced by the original cross. The concept of preferential pairing is applied in the modern breeding of polyploids whereby alloploids are stabilized and made reliable as diploids, a process called **diploidization**.

24.5.3 Genetics of autoploids

The ploidy level may also be defined as the number of different alleles that an individual can possess for a single locus on a chromosome. A diploid can have two alleles per locus, whereas an autotetraploid can have four different alleles. The genetics of autoploids is complicated by multi-allelism and multivalent association of chromosomes during meiosis. Consider the segregation of alleles of a single locus (A, a). In a diploid species, there would be three possible genotypes AA, Aa, and aa. However, in an autotetraploid there would be five genotypes ranging from multiplex (aaaa) to quadriplex ($AAAA$) (Table 24.4). The proportion of dominant (A) to recessive (a) genes is

different in two of the five genotypes ($AAAA$ and $Aaaa$) in autotetraploids from what obtains in diploids. The number of phenotypes observed depends on dominance relationship of A and a. If allele A is completely dominant to allele a, there would be only two phenotypes. If dominance is incomplete or the effect of allele A is cumulative, there could be up to five phenotypes. Upon selfing, a dominant phenotype in a diploid (AA, Aa) would produce a progeny that is all dominant, or segregate in 3 : 1 ratio. Selfing each of the five categories would produce much different outcomes in autotetraploids, assuming random chromosome segregation (Table 24.5).

An autoploid individual can have up to four alleles (*abcd*) per locus. Five different genotype categories are similarly possible, except that there may be only four multiplex genotypes (*aaaa, bbbb, cccc, dddd*), only one tetragenic genotype (*abcd*), but numerous

Table 24.5 Genetic frequencies following chromosome segregation of autotetraploid.

Genotype	Gametic frequency		
	AA	Aa	aa
AAAA	1	0	0
AAAa	1/2	1/2	0
AAaa	1/6	4/6	1/6
Aaaa	0	1/2	1/2
Aaaa	0	0	1

Note Chromatid segregation occurs less frequently than chromosomes segregation and produces alternative types of segregation. For example the simplex (*Aaaa*) can produce gametes that are homozygous (*AA*) by the process called double reduction.

Assuming complete dominance and chromosome segregation the following phenotypic ratios are observed. Certain segregation ratios are sometimes indicative of the nature of autotetraploid inheritance.

Cross	Progeny (dominant : recessive)
AAAA × AAAA	1 : 0
AAAa × AAAa	1 : 0
AAaa × AAaa	35 : 1
AAaa × Aaaa	11 : 1
AAaa × aaaa	5 : 1
Aaaa × Aaaa	3 : 1
Aaaa × aaaa	1 : 1
Aaaa × aaaa	0 : 1

Table 24.4 Genetics of autoploids.

Diploid	Polyploidy		Name
Cross			
$Aa \times Aa$	$AAaa \times AAaa$		
Products			
1/4 AA	1/36	$AAAA$	Quadruplex
2/4 Aa	8/36	$AAAa$	Triplex
1/4 aa	18/36	$AAaa$	Duplex
	8/36	$Aaaa$	Simplex
	1/36	$aaaa$	Nulliplex

Table 24.6 Multiple allelism in autotetraploids.

Tetrasomic condition

$a_1a_1a_1a_1$	All alleles are identical; monoallelic; balanced.
$a_1a_1a_1a_2$	Two different alleles; diallelic; unbalanced.
$a_1a_1a_2a_2$	Two different alleles; diallelic; balanced.
$a_1a_1a_2a_3$	Three different alleles; triallelic.
$a_1a_2a_3a_4$	Four different alleles; tetraallelic.

The number of possible interactions are (a) first order (e.g., a_1a_2, a_1a_3), (b) second order (e.g., $a_1a_2a_3$, $a_1a_3a_4$), and (c) third order interaction ($a_1a_2a_3a_4$). This depends on the tetrasomic condition.

Tetrasomic condition	1st	2nd	3rd	Total
$a_1a_2a_3a_4$	6	4	1	11
$a_1a_1a_2a_3$	3	1	0	4
$a_1a_1a_2a_2$	1	0	0	1
$a_1a_1a_1a_2$	1	0	0	1
$a_1a_1a_1a_1$	0	0	0	0

combinations for the intermediate (Table 24.6). The possible gametic array is shown for each genotype. Interallelic and intrallelic interactions may occur for as many as four alleles per locus in an autotetraploid. The degree to which intrallelic interaction occurs determines the expression of heterosis and inbreeding depression in an autotetraploid. Because four identical alleles are required to achieve homozygosity in an autotetraploid compared to only two in a diploid, homozygosity is achieved at a less rapid rate in autotetraploids (Figure 24.4).

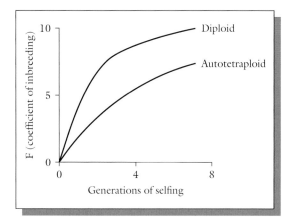

Figure 24.4 The effect of ploidy on the inbreeding as demonstrated by diploids and autotetraploid.

Another aspect of autoploid genetics with plant breeding implication is the difficulty of distinguishing between a triplex and a quadriplex on the basis of a progeny test (assuming random chromosome segregation). Both genotypes (*AAAA* and *AAAa*) will breed true for the dominant allele. To identify a triplex plant, the breeder would have to advance the progeny one more generation to identify the duplex plants of the S_1. Achieving genetic purity in autotetraploid stocks is difficult, not only because it is challenging to identify triplex plants, but also because deleterious genes may persist in an autotetraploid, manifesting themselves only rarely in the homozygous. The breeder would need an additional two generations in order to identify the homozygous dominant genotype unequivocally.

24.5.4 Induction of autoploids

Plant breeders were initially attracted to induce polyploidy primarily because of the gigas effects, which increased cell size (but also reduced fertility). This pro and con of the gigas effects make the induction of autoploids more suited to crops whose economic part is vegetative. The primary technique for inducing autoploids is the use of colchicine ($C_{22}H_{25}O_6$), an alkaloid from the autumn crocus (*Colchicum autumnale*). This chemical compound works by disrupting the spindle mechanism in mitosis, thereby preventing the migration of duplicate chromosomes to opposite poles at anaphase. Consequently, the nucleus is reconstituted with twice the normal number of chromosomes, without any nuclear or cell division.

Ryegrass is one of the species that has been successfully improved by the induction of autoploidy. Rye (*Secale cereale*) is perhaps the only grain producing crop for which synthetic autoploids have been developed. Meristematic tissue is most susceptible to colchicine treatment. Hence, a germinating seed, a young seedling, or a developing bud are the commonly used plant material for autoploid induction. The chemical may be applied in aqueous solution or through various media (e.g., agar lanolin paste). Seeds may be soaked in aqueous colchicine at a concentration of 0.05–0.4% for 30 minutes to three hours. Buds are treated differently, for example, by intermittently exposing the selected plant material for 2–6 days at concentrations of 0.2–0.5%. The breeder should determine the best treatment condition by experimentation. The material treated should be thoroughly washed after application to remove excess chemicals.

24.6 Breeding autoploids

In developing and using autoploids in plant breeding, certain general guidelines may be observed.

- Generally, species tend to have an optimum chromosome number (optimum ploidy number) at which they perform best. Because chromosome doubling instantly and drastically increases chromosome number, selecting parents with low chromosome number for autoploid breeding will reduce the risk of meiotic complications that are often associated with large chromosome numbers. This will increase the chance of obtaining fertile autoploids.
- Autoploids tend to have gigas features and also a high rate of infertility. Consequently, autoploidy is more useful for breeding species in which the economic product is not seed or grain (e.g., forage crops, vegetables, ornamental flowers).
- Producing autoploids from cross-fertilizing species promotes gene recombination among the polyploids for a better chance of obtaining a balanced genotype.

D.R. Dewey summarized the properties of a species suited for induction of polyploidy as:

- The species has low chromosome number.
- The economic part of the plant is the vegetative material (e.g., forage grasses).
- The plant is cross-pollinated (allogamous).
- The plant is perennial in habit.
- The plant has the ability to reproduce vegetatively.

24.6.1 Autotetraploids and autotriploids

Tetraploid rye ($2n = 4x = 28$) has about 2% more protein and superior baking qualities than diploid cultures. However, it also has about 20% higher incidence of sterility per spike, resulting in lower grain yield than diploids. Autotriploids of commercial importance include sugar beet ($2n = 2x = 18$; $3x = 27$, $2n = 4x = 36$). Triploidy is associated with the genetic consequences of sterility because of the odd chromosome number that results in irregular meiosis. The sterility favors species that are grown for vegetative commercial parts (e.g., grasses) and ornamentals and fruits (seedless). In sugar beet, triploid

cultivars of monogerm types have significantly impacted the sugar industry.

Triploid hybrids are produced by crossing diploids with tetraploids. Breeders use three kinds of genotypes. The diploid is male sterile (female, cms) while the tetraploid is the pollinator. The third component is a male sterility maintainer (a diploid, N). The tetraploid is derived from a diploid by colchicines treatment of the seed (soak in 0.2% for 15 hours at 30 °C). Seedless watermelon ($3x = 33$) is also produced by crossing diploid ($2n = 2x = 22$) with tetraploid ($2n = 4x = 44$).

24.7 Natural alloploids

A number of economically important crops are alloploids. These include food crops (e.g., wheat and oat), industrial crops (e.g., tobacco, cotton, and sugarcane), and fruits crops (e.g., strawberry and blueberry). These crops, by definition, contain a combination of different genomes. Researchers over the years have attempted to elucidate the ancesteral origin of some alloploids. One of the most widely known successes was the work of Nagaharu U, the Japanese scientist who described the genomic relationships among naturally occurring mustard (*Brassica*) species (Figure 24.5). Dubbed the **triangle of U**, it describes the origins of three *Brassica* species by alloploidy. The diploid species involved are turnip or rape (*B. campestris*, $n = 10$), cabbage or kale (*B. oleracea*,

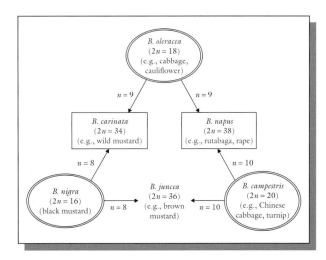

Figure 24.5 The triangle of U showing the origins of various alloploids in *Brassica*.

$n = 9$), and black mustard (*B. nigra*, $n = 8$). For example, rutabaga (*B. napus*) has $2n = 38$, being a natural amphiploid of *B. olercea* and *B. campestris*.

In cereal crops, wheat is a widely studied alloploid that comprises genomes from three species. Cultivated common wheat (*Triticum aestivum*) is a hexaploid with 21 pairs of chromosomes and designated *AABBDD*. The *AA* genome comes from Einkorn (*T. monococcum*). Tetraploid wheats have the genomic formula *AABB*. Emmer wheat (*T. dicoccum*) crossed naturally with *Aegilops squarrosa* (*DD*) to form the common wheat.

24.7.1 Genetics of alloploids

As previously indicated, alloploids arise from the combination and subsequent doubling of different genomes, a cytological event called **alloploidy**. The genomes that are combined differ in degrees of homology, some being close enough to pair with each other, whereas others are too divergent to pair. Sometimes, only segments of the chromosomes of the component genomes are different, a condition that is called **segmental alloploidy**. Some of the chromosomes of one genome may share a function in common with some chromosomes in a different genome. Chromosomes from two genomes are said to be **homeologous** when they are similar but not **homologous** (identical).

Most alloploids have evolved certain genetic systems that ensure that pairing occurs between chromosomes of the same genome. A classic example occurs in wheat ($2n = 6x = 42$) in which a gene on chromosome 5B, designated *Ph* enforces this diploid-like pairing within genomes of the alloploid. When this gene is absent, pairing between homoeologous chromosomes as well as corresponding chromosomes of the three genomes occurs, resulting in the formation of multivalents at meiosis I.

Alloploids exhibit a variety of meiotic features. Sometimes chromosomes pair as bivalents, and thereby produce disomic ratios. Where the component genomes have genes in common, duplicate factor ratios will emerge from meiosis, an event that sometimes is an indication of alloploid origin of the species. Whereas significant duplications of genetic material have been found in wheat, the genomes of upland cotton have little duplication. Tetrasomic ratios are expected for some loci where multivalent associations are found in allotetraploids.

24.7.2 Breeding alloploids

Alloploids may be induced by crossing two species with different genomes, followed by chromosome doubling of the hybrid. Compared to autoploids, inducing alloploids is not commonly done by plant breeders. If successful, the newly induced amphiploid instantly becomes a new species (unable to cross to either parent). It also becomes reproductively isolated from its parents. Success of induced alloploids is enhanced by the proper choice of parents. Using parents with low ploidy levels increases the chance of high fertility and seed set in the amphiploid especially. Commercially successful induced alloploids are few. The most noted success with induced alloploidy is the commercially grown amphiploid, triticale (*X Triticosecale*), derived from a cross between wheat (*Triticum*) and rye (*Secale*) (Figure 24.6). The objective of developing triticale was to obtain a product that combines the qualities of wheat with the hardiness of rye. In lieu of doubling the F_1 to produce the desired synthetic product, wheat x rye cross may be undertaken. The

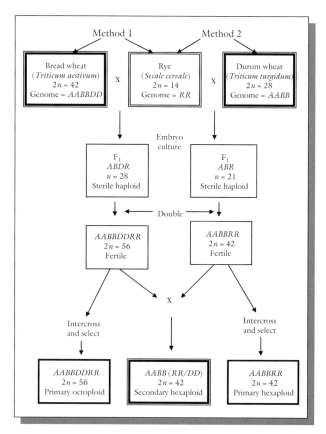

Figure 24.6 Steps in the development of triticale.

Table 24.7 Aneuploid nomenclature.

Chromosome number	Term	Nature of chromosomal change
2n	diploid	normal
Aneuploidy		
2n − 1	monosomy	1 copy of a pair of chromosome missing
2n + 1	trisomy	3 copies of one chromosome (i.e., an extra copy)
2n + 2	tetrasomy	4 copies of one chromosome (i.e., two extra copies)
2n + 3	pentasomy	5 copies of one chromosome (i.e., three extra copies)

The individual with the condition, e.g., trisomy, is called a trisomic.

F_1 plant possesses 28 chromosomes and exhibits intermediate traits that favor rye (hairy neck, spike length). All F_1s are sterile because of the formation of univalents and irregular gametogenesis. F_1s are backcrossed to wheat to produce progenies containing 42 chromosomes (seven from rye and the rest from wheat). The wheat chromosomes form bivalents at meiosis, while the rye chromosomes form univalents. The bivalent wheat chromosomes are irregularly arranged. Fertilization of an ovule with $21 + 7$ chromosomes by pollen with the same genomic constitution will contain the full complement of chromosomes for wheat and rye (56 chromosomes). This product would be the synthetic alloploid called **triticale**. Hexaploid triticale ($AABBRR$, $2n. = .6x. = .42$) is superior agronomically to octoploid triticale ($AABBDDRR$, $2n. = .8x. = .56$), but it requires embryo culturing to obtain F_1s between durum wheat and rye.

All amphiploid breeding is a long-term project because it takes several cycles of crossing and selection to obtain a genotype with acceptable yield and product quality. Common undesirable features encountered in triticale breeding include low fertility, shriveled seeds, and weak straw. Even though tetraploid ($2n = 2x = 28$), hexaploid ($2n = 6x = 42$) and octoploid ($2n = 8x = 56$) forms of triticale have been developed, the hexaploid forms have more desirable agronomic traits, and hence are preferred. Alloploids have been used to study the genetic origins of species. Sometimes, amphiploidy is used by breeders as bridge crosses in wide crosses.

24.8 Aneuploidy

Whereas polyploidy entails a change in ploidy number, **aneuploidy** involves a gain or a loss of one or a few chromosomes that make up the ploidy of the species (i.e., one of a few chromosomes less or more than the complete euploid complement of chromosomes). Just like polyploidy, aneuploidy has its own nomenclature (Table 24.7).

24.8.1 Cytogenetics of autoploids

The diploid complement of chromosomes is designated $2n$. A **nullisomic**, for example, is an individual with a missing pair of chromosomes ($2n − 2$), while a **tetrasomic** has gained a pair of chromosomes ($2n + 2$). Similarly, a **monosomic** has lost one chromosome from a homologous pair ($2n − 1$), while a **trisomic** has gained an extra chromosome ($2n + 1$).

Aneuploidy commonly arises as a result of irregular meiotic mechanisms, such as **non-disjunction** (failure of homologous chromosomes to separate), leading to unequal distribution of chromosomes to opposite poles (Figure 24.7). Consequently, gametes resulting from such aberrant meiosis may have a loss or gain of chromosomes. Furthermore, chromosome additions often cause chromosome imbalance and reduced plant vigor.

24.8.2 Applications of aneuploidy

Aneuploidy is used in various genetic analyses.

Chromosome additions

Chromosome addition lines are developed by backcrossing the synthetic alloploid (F_1) as seed parent to a cultivated species as pollen parent. This strategy is preferred because male gametogenesis is more readily perturbed by chromosomal or genic disharmonies than is the case in the female gametophyte. For example, E.R. Sears transferred the resistance of leaf rust of *Aegilops umbellulata* to *Triticum aestivum* (bread

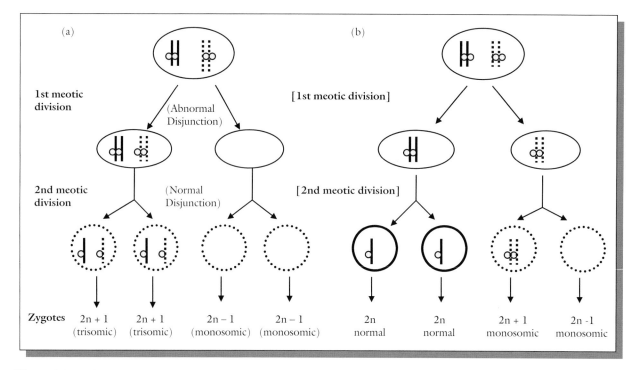

Figure 24.7 The origin of anueploidy. Abnormal disjunction may occur at the first meiotic division (a) or at the second meiotic division (b) producing gametes with a gain or loss in chromosomes.

wheat) via bridge crossing with *T. dicocoides* as follows:

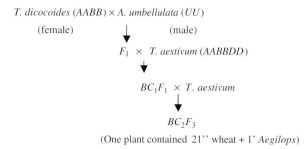

T. dicocoides (*AABB*) × *A. umbellulata* (*UU*)

(female)   (male)

F_1 × *T. aestivum* (*AABBDD*)

BC_1F_1 × *T. aestivum*

BC_2F_3

(One plant contained 21'' wheat + 1' *Aegilops*)

However, it had draws backs (sterile pollen, brittle spike axis, etc.). Subjecting these chromosome addition lines to irradiation successfully translocated the segment of the *Aegilops'* chromosome with the desired resistance genes to chromosome 6*B* of wheat, effectively removing the negative effects. The new genotype has been used in breeding as a source of resistance to leaf and stem rust.

Trisomics are important in genetic analysis. There are several types of trisomics. The term **primary trisomic** is used to refer to a case in which the euploid complement is increased by one complete chromosome. A **secondary trisomic** is one in which the extra chromosome has identical arms (i.e., one particular arm occurs as a quadruplicate). Such a chromosome is called an **iso-chromosome**. Sometimes, the extra chromosome added is derived from parts of different chromosomes. Such additives arise from chromosome breakage-fusion events.

Primary trisomics may be used to assign genes to chromosomes. Theoretically, there are as many possible trisomics, as there are chromosome pairs. Scientists can generate trisomic stocks for a species. To assign a gene, the mutant (e.g., *a*) that is homozygous for the allele of interest is crossed to all the trisomic tester stocks. Assuming that all stocks are homozygous for the wild type and assuming normal meiotic segregation, two F_1 types will be produced. Those produced from union of normal gametes (*n*) will segregate with the normal diploid ratio 3*A*:1*aa*. However, where a trisomic plant (*n* + 1 gamete) is involved, an aberrant ratio would result. Because a trisomic stock is unique, the gene of interest would be located on the chromosome that the trisomic stock represents. These results assume random segregation of the three chromosomes of the trisomic and equal viability of pollen regardless of genetic constitution. In reality there is a preponderance of n gametes and reduced function of $n + 1$ gametes. The consequence of reduced functionality is that, sooner or later, a trisomic will revert to a diploid, unless the scientist makes special efforts to maintain it. Trisomics have been applied in creative ways in plant breeding, including their use in hybrid seed production in barley, using a balanced tertiary trisomic that carries a recessive male sterility gene. The addition of chromosomes from other species (called **alien addition lines**) has been explored in interspecies crosses such as wheat x rye. Chromosome addition lines may be unstable enough to be developed as cultivars.

Chromosome deletions

Unlike chromosome addition, in which gene duplication occurs (hence an implied duplication in function), **chromosome deletion** leads to a loss of function. The consequence of a deletion depends on the functional roles of the genes of the chromosome that is lost. Invariably, surviving plants have less vigor and more sterility problems. However, in polyploids, the presence of homeologous (in alloploids) or homologous (in autoploids) chromosomes may make up for the missing functions.

Monosomics ($2n - 1$) may be used just like trisomics to assign genes to chromosomes in a polyploid species. This requires the development of monosomics for all the existing chromosome pairs in the species, as was done by E.R. Sears for the Chinese Spring cultivar of wheat. Nullisomics ($2n - 2$) may also be used in this fashion, but with less success because of severe reduction in vigor and fertility.

Chromosome substitution

Whereas alien chromosome addition entails adding an alien chromosome to the genome of an existing genotype, chromosome substitution entails replacing or substituting a chromosome of the recipient species with an alien chromosome. Intervarietal (between varieties of the same species) and interspecific chromosome substitutions are more important in plant breeding than addition of chromosomes. One of the well known substitutions involves chromosome $1B$ of wheat and chromosome $1R$ of rye. The resultant wheat cultivar provided resistance to disease (leaf rust, stripe rust, powdery mildew). To use this technique, there

have to be monosomic lines for the species (lines are available for wheat, cotton, tobacco, oats).

The backcross breeding procedure may be used to substitute one chromosome for another in monosomics or nullisomics. Such chromosome substitution may be done within the species and involves other species (i.e., alien substitution). Researchers such as Sears have used the technique to assign numerous genes to chromosomes. However, the technique is challenging and requires a great amount of cytological analysis.

Supernumerary chromosomes

Also called **accessory** or **B-chromosomes, supernumerary chromosomes** are natural additions of varying numbers of small chromosomes to the normal genome. They have been found in all major taxonomic groups of organisms. These chromosomes are often predominantly heterochromatic and unstable in behavior. Although largely considered as genetically inert, studies in some species have indicated that the B-chromosomes increase the recombination frequency of **A-chromosomes** (the normal set of chromosomes) in species in which they occur. It is possible to use certain breeding techniques to increase their number. In some species, such rye, fertility is reduced by the presence of one or two supernumerary chromosomes. However corn plants can accumulate at least 10 such materials before an adverse effect on fertility is noticeable.

24.9 General importance of polyploidy in plant improvement

Polyploidy played a significant role in plant evolution and domestication. Induced polyploids received significant attention when mitotic inhibitors (oryzalin, trifluralin, amiprophos-methyl, and N_2O gas, colchicine) were discovered. Unfortunately, the induced polyploids seldom outperformed their diploid progenitors. In terms of generating genetic variability for breeding, somatic doubling does not produce genetic material but only produces additional copies of existing chromosomes. Abnormalities noted by researchers to be associated induced polyploidy include erratic fruit bearing, brittle wood, watery fruit, stunting anatomical imbalances (resulting from the gigas feature), somatic instability, and extreme genetic redundancy (resulting in chimeric tissues in high order ploidy).

Using polyploidy for cultivar development as the ultimate goal may not be a worthy undertaking. On the other hand, induced polyploids may be used in other ways in a breeding program, including:

- **Enhancing heterozygosity.** Induced polyploids may used as germplasm in a breeding program to increase heterozygosity.
- **Overcoming interploid blocks.** Barriers to hybridization resulting from differences in ploidy levels commonly have their origin in endosperm imbalance. Instead of the normal maternal:paternal genomic makeup of the endosperm $(2:1)$, meiotic errors may cause a deviation from this ratio, resulting in underdevelopment or abortion of the embryo. Genomic manipulation such as chromosome doubling may be used to make the parents cross-compatible.
- **Development of sterile cultivars.** Irregularities in the meiotic apparatus often end in the production of sterile plants. In ornamental horticulture, sterile plants either do not bear flowers or bear sterile flowers that do not produce seed. These seedless plants remain longer in the landscape and also have no potential to become an invasive species. In fruit production, sterility results in seedlessness of fruits. Triploidy, as previously discussed, is a method of inducing seedless fruits. Triploids may also be developed from the nutritive tissue of the embryo in most angiosperms. This tissue is triploid (produced from the fusion of three haploid nuclei). Autotetraploids can also produce sterility from meiotic complications.
- **Enhancing pest resistance.** Some research has shown that increasing the chromosome number (or gene dose) may increase the expression of certain secondary metabolites and chemicals that promote pest resistance. Autotetraploid rye is more disease tolerant than its diploid counterparts. Some evidence exists to suggest that polyploidy may enhance stress tolerance in some species.
- **Restoring fertility.** Polyploidy can also be used to restore fertility in sterile plants by doubling the chromosome number. Restoring fertility allows the plant to be used as germplasm in a breeding program, or even to be released as a cultivar.
- **Producing large fruits.** In species where large fruit size is desirable, the gigas effect of polyploidy is advantageous. Triploidy in apples is known to result in increased fruit size without loss of quality and appearance. Unfortunately, tetraploid apples have larger fruits than diploid apples, but have poor fruit quality.
- **Increasing vigor.** Polyploidy flowers are known to be larger and last longer than their diploid counterparts before withering.

24.10 Inducing polyploids

The most commonly used meiotic inhibitor in the induction of polyploids is colchicine. It is applied to growing regions of plants (meristems). Young seedlings (or their epical meristems) may be soaked (or sprayed) in an appropriate concentration of the chemical for an appropriate amount of time (could be a few hours or even days). Axillary or subaxillary buds may also be used. Verification of the success of the treatment may be conducted in several ways, the easiest being visual inspection. Such an inspection produces only tentative results. Early physical evidence of polyploidy includes thicker and broader leaves, larger flowers and fruits, distorted growth, and shortened internodes. Putative polyploids need to be authenticated by more reliable, albeit more time consuming, methods, such as examination of the pollen (larger) and the chloroplast count (more chloroplasts per guard cell), flow cytometry (measures DNA content) or chromosome count (the ultimate and definitive evidence of polyploidy). For chromosome count, root tips and anthers are popular materials to use. It should be noted that the induction of polyploidy is not uniform throughout the treated material. The convention is to recognize three histogenic layers (L-1, L-2, L-3), which may be altered differently. Guard cells reflect the L-1 layer, while the cortical layer and reproductive tissues (e.g., anthers) reflect the L-2 and L-3 layers, respectively.

24.11 Use of 2n gametes for introgression breeding

When gametes have the somatic chromosome number, $2n$, they are called **unreduced gametes**. This condition occurs widely among angiosperms and some researchers believe this to be origin of polyploidy plant species. Whereas artificially induced allopolyploids have fixed heterozygosity, sexual polyploids lack this characteristic and, hence, recombination occurs between the alien parental genomes. Also, because of recombination, introgression can be achieved. Successful introgression breeding has been achieved in species such as *Alstroemeria*, *Lilium*, *Medicago*, *Solanum*, and *Primula*.

$2n$ gametes may be detected in plants through approaches such as pollen size examination, flow cytometry, and progeny analysis. Intergenomic recombination is critical for introgression. GISH (genomic *in situ* hybridization) is one of the most effective techniques used to detect chromosomal recombination. *In situ* is used to locate the chromosomal location of a specific DNA or RNA probe (labeled with a fluorescent probe). The 2n gamete introgression technique has been used successfully in potato, *Brassica, Musa, Allium cepa, Lilium longiflorum* and others.

24.12 Haploids in breeding

Haploids contain half the chromosome number of somatic cells. Anthers contain immature microspores or pollen grains with the haploid (n) chromosome number. If successfully cultured (anther culture), the plantlets resulting will have a haploid genotype. Haploid plantlets may arise directly from embryos or indirectly via callus, as previously discussed. To have maximum genetic variability in the plantlets, breeders usually use anthers from F_1 or F_2 plants. Usually, the haploid plant is not the goal of anther culture. Rather, the plantlets are diplodized (to produce diploid plants) by using colchicine for chromosome doubling. This strategy yields a highly inbred line that is homozygous at all loci after just one generation.

Methods used for breeding self-pollinated species generally aim to maintain their characteristic narrow genetic base through repeated selfing over several generations for homozygosity. The idea of using haploids to produce instant homozygotes by artificial doubling has received attention.

Haploids may be produced by one of several methods:

- Anther culture to induce androgenesis.
- Ovary culture to induce gynogenesis.
- Embryo rescue from wide crosses.

24.12.1 Anther culture

Flower buds are picked from healthy plants. After surface sterilization, the anthers are excised from the buds and cultured unto an appropriate tissue culture medium. The pollen grains at this stage would be in the uninucleate microspore stage. In rice the late uninucleate stage is preferred. Callus formation starts within 2–6 weeks, depending on the species, genotype, and physiological state of the parent source.

High nitrogen content of the donor plant and exposure to low temperature at meiosis reduces albinos and enhances the chance of green plant regeneration. Pre-treatment (e.g., storing buds at 4–10 °C for 2–10 days) is needed in some species. This and other shock treatments promote embryogenic development. The culture medium is sometimes supplemented with plant extracts (e.g., coconut water, potato extract). To be useful for plant breeding, the haploid pollen plants are diplodized (by articifial doubling with 0.2% colchicines, or through somatic callus culture).

Applications

(a) **Development of new cultivars.** Through diplodization, haploids are used to generate instant homozygous true breeding lines. It takes only two seasons to obtain doubled haploid plants, versus about seven crop seasons using conventional procedures to attain near homozygous lines. The genetic effect of doubling is that doubled haploid lines exhibit variation due primarily to additive gene effects and additive x additive epistasis, enabling fixation to occur in only one cycle of selection. Heritability is high because dominance is eliminated. Consequently, only a small number of doubled haploid plants in the F_1 is needed, versus several thousands of F_2 for selecting desirable genotypes.

(b) **Selection of mutants.** Androgenic haploids have been used for selecting especially recessive mutants. In species such as tobacco, mutants resistant to methionine analogue (methionine sulfoxide) of the toxin produced by *Pseudomonas tabaci* have been selected.

(c) **Development of supermales in asparagus.** Haploids of *Asparagus officinalis* may be diplodized to produce homozygous males or females.

Limitations

(a) The full range of genetic segregation of interest to the plant breeder is observed because only a small fraction of androgenic grains develop into full sporophytes.

(b) High rates of albinos occur in cereal haploids (no agronomic value).

(c) Chromosomal aberrations often occur, resulting in plants with higher ploidy levels, requiring several cycles of screening to identify the haploids.

(d) Use of haploids for genetic studies is hampered by the high incidence of nuclear instability of haploid cells in culture.

24.12.2 Ovule/ovary culture

Gynogenesis using ovules or ovaries has been achieved in species such as barley, wheat, rice, maize, tobacco, sugar beet, and onion. The method is less efficient than androgenesis because only one embryo sac exists per ovary as compared to thousands of microspores in each anther. Ovaries ranging in developmental stages from uninucleate to mature embryo sac stages are used. However, it is possible for callus and embryos to develop simultaneously from gametophytic and sporophytic cells, making it a challenge to distinguish haploids from those of somatic origin. Generally, gynogenesis is selected when androgenesis is problematic (as in sugar beet and onion).

24.12.3 Haploids from wide crosses

Certain specific crosses between cultivated and wild species are known to produce haploids. Well established systems include the interspecific crosses between *Hordeum vulgare* ($2n = 2x = 14$, *VV*) x *Hordeum bulbosum* ($2n = 2x = 14$, *BB*), commonly called the **bulbosum method**, and also in wheat x maize crosses. The F_1 zygote has $2n = 2x = 14$ (*7V + 7B*). However, during the tissue culture of the embryo, the *bulbosum* chromosomes are eliminated, leaving a haploid ($2n = x = 7V$). This is then doubled by colchicines treatment to obtain $2n = 2x = 14$ *VV*.

24.12.4 Haploid breeding versus conventional methods

The potential of haploid breeding to hasten crop improvement is attractive. However, the method is yet to become a mainstream breeding approach. Haploids cannot be obtained in the high enough frequency that is needed for selection. Further, haploids will express recessive deleterious traits. The method may not be cost effective, overall. Nonetheless, haploids have been used in breeding of many crops, including asparagus, barley, citrus, corn, grape, cucumber, pepper, peanut, wheat and potato.

Haploid breeding applied to species with polysomic inheritance (e.g., potato) is different from that applied to species like barley and rice. Polyhaploids obtained from polyploids have polysomic inheritance and may be homozygous of heterozygous. In an auto-tetraploid like potato, dihaploids may be *AA, Aa,* or *aa*, depending on the genotype of the tetraploid, which could be heterozygous (*AAAa, AAaa, Aaaa*). Consequently, doubling of haploids will not always result in homozygosity, unless dihaploids are used again for the production of monoploids.

24.13 Doubled haploids

Researchers exploit haploidy generally by doubling the chromosome number to create a cell with the double dose of each allele (homozygous).

24.13.1 Key features

Inbred lines are homozygous genotypes produced by repeated selfing with selection over several generations. The technique of **doubled haploids** may be used to produce complete homozygous diploid lines in just one year (versus more than four years in conventional breeding) by doubling the chromosome complement of haploid cells. Such doubling may be accomplished *in vivo* naturally, or through crossing of appropriate parents, or *in vitro* through the use of colchicine. The success of doubled haploids as a breeding technique depends on the availability of a reliable and efficient system for generating haploids and doubling them in the species.

24.13.2 Applications

Doubled haploids have been successfully used in breeding species in which efficient haploid generation and doubling systems have been developed. These include canola, barley, corn, and wheat. Additionally, doubled haploids are used to generate general genetic information that can be applied to facilitate the breeding process. Such information includes gene action and interaction, estimating the number of genetic variances, calculating combining abilities, and detection of gene linkages, pleiotropy, and chromosome locations. Haploids are used in mutation studies (recessive mutants are observable instantly) and in selecting against undesirable recessive alleles.

24.13.3 Procedure

The first step in using doubled haploids in breeding is identifying the source of haploids.

Natural sources

Haploids originate in nature through the phenomenon of **parthenogenesis** (gamete formation without fertilization). The haploids may be maternal or paternal in origin. It is estimated that haploids occur in corn at the rate of 1 in 1000 diploids, 99% of which arise from parthenogenesis of maternal origin. Spontaneous doubling occurs in corn at the rate of 10% of haploids developed. The key is distinguishing between haploid and diploid plants. A marker system for this purpose was first developed by S.S. Chase based on seedling color, purple plants being encoded by the dominant gene (P) while normal green plants are recessive (p). A cross of $F_1(pp)$ x PP would yield $999Pp$ (purple diploids) and $1pp$ (green haploid). Another marker used is the purple aleurone color.

To use this marker system, the breeder should cross a heterozygous female to a male with marker genes. The seed from those with dominant endosperm marker of the male is saved and planted, discarding seedlings with the dominant male marker. Next, cytological evaluation of plants with the recessive female marker (by root tip squash) is conducted. The haploid plants are retained and grown in the greenhouse or field, and self-pollinated to produce diploids.

Artificial sources

Haploid production through interspecific and intergeneric crosses is in use, one of the well known being the barley system (previously discussed). After doubling the chromosome, the diploid plants are grown to maturity. Seeds are harvested for planting plant rows. Because diploids produced by this method are normally completely homozygous, there is no need for growing segregating generations as obtains in conventional programs.

Advantages and disadvantages

The technique of doubled haploids has certain advantages and disadvantages, the key one including:

Advantages
- Complete homozygosity is attainable in a shorter period.
- Duration of the breeding program can be reduced by several (2–3) generations.
- It is easier and more efficient to select among homogeneous progeny (versus heterogeneous progeny in conventional breeding).
- The cultivar released is homogeneous.

Disadvantages
- The procedure requires special skills and equipment in some cases.
- Additional technology for doubling may increase the cost of a breeding program.
- Frequency of haploids generated is not predictable.
- There is a lack of opportunity to observe line performance in early generations prior to homozygosity.

Genetic issues

Unlike the conventional methods of inbreeding, it is possible to achieve completely homozygous individuals. Using an F_1 hybrid or a segregating population as female parent in the production of maternally derived haploids increases genetic diversity in the doubled haploid line. It is advantageous if the female also has agronomically desirable traits. F_1 hybrids are suitable because their female gametes will be segregating.

Key references and suggested reading

Borojevic, S. (1990). *Principles and methods of plant breeding*. Elsevier, New York.

Bingham, E.T. (1980). Maximizing heterozygosity in autotetraploids, in *Polyploidy: Biological relevance* (ed. Lewis W.H.). Plenum Press. New York, pp. 471–489.

Chase, S.S. (1964). Analytic breeding of amphipolyploid plant varieties. *Crop Science*, **4**:334–337.

Dewey, D.R. (1980). Some applications and misapplication of induced polyploidy to plant breeding, in *Polyploidy: Biological relevance* (ed. W.H. Lewis). Plenum Press. New York, pp. 445–470.

Haynes, K.G. (1992). Some aspects of inbreeding in derived tetraploids of potatoes. *Journal of Heredity*, **83**:67–70.

Hermsen, J.G. Th. (1984). Nature, evolution, and breeding of polyploids. *Iowa State Journal of Research*, **58**:411–420.

Hill, R.R. (1971). Selection in autotetraploids. *Theoretical and Applied Genetics*, **41**:181–186.

Poehlman, J.M., and Sleper D. A. (1995). *Breeding field crops*, 5th edn. Iowa State University Press. Ames, IA.

Sterett, S.B., Henninger, M.R., Yencho, G.C., Lu, W., Vinyard, B.T., and Haynes, K.G. (2003). Stability of internal heat necrosis and specific gravity in tetraploid x diploid potatoes. *Crop Science*, **43**:790–796.

Thomas, H. (1993). Chromosome manipulation and polyploidy, in *Plant breeding: Principles and prospects* (eds. Hayward M.D., Bosemark N.O. and Ramagosa I.). Chapman and Hall, London.

Sears, E.R. (1954). The aneuploids of common wheat. *Missouri Agric. Exp. Stn. Res. Bull.*, **572**:1–58.

Singh, A.K., Moss, J.P., and Smartt, J. (1990). Ploidy manipulations for interspecific gene transfer. *Advances in Agronomy*, **43**:199–240.

Internet resources

http://users.rcn.com/jkimball.ma.ultranet/BiologyPages/P/Polyploidy.html – A good discussion on polyploidy (accessed March 15, 2012).

http://wheat.pw.usda.gov/ggpages/BarleyNewsletter/42/oral04.html – Application of doubled haploids in barley breeding (accessed March 15, 2012).

Outcomes assessment

Part A

Please answer the following questions true or false.

1 A genomic formula of $2n - 1$ refers to a trisomic.
2 The regular set of chromosomes of a species is called A chromosomes.
3 An individual in which the euploid complement of chromosomes is increased by one complete set of chromosomes is called a secondary trisomic.
4 The triangle of U describes genomic relationships among naturally occurring species of wheat.
5 Triticale is an euploid.
6 Colchicine is used for reducing the number of chromosomes in a cell.
7 Aneuploids have a duplicate of the entire chromosome set.
8 The genotype $AAAA$ represents a duplex tetraploid.
9 The genotype $AADDEE$ represents an alloploid.
10 A hexaploid consists of six genomes.

Part B

Please answer the following questions.

1 Describe the triangle of U.
2 Distinguish between homologous and homeologous chromosomes.
3 Distinguish between a primary trisomic and a secondary trisomic.
4 Discuss a common mechanism of aneuploidy.
5 Distinguish between an aneuploid and euploid.
6 . . . . is polyploidy with chromosomes from different genomes.
7 Write the genetic formula for a triplex genotype.
8 What is a segmental alloploid?

Part C

Please write a short essay on each of the following topics.

1 Discuss the effect of polyploidy on plants.
2 Discuss, with an example, a polyploidy series.
3 Discuss the artificial induction of polyploids.
4 Discuss the importance of doubled haploids to plant breeding.

25

Molecular genetic modifications and genome-wide genetics

Purpose and expected outcomes

Heritable variation, as previously noted, can arise through natural phenomena or can be artificially induced via mutagenesis. Another way of creating artificial variation is to directly modify the genetic material. Traditional crossing and mutagenesis have limitations as to how to reorganize genes in a new genetic matrix to create new cultivars. Conventional hybridization is limited in its capacity to transfer genes from one parent to another because of biological barriers to crossing. Mutagenesis has the weakness of not being precise or sufficiently targeted. A variety of modern tools is available for molecular genetic modifications, the most radical, perhaps, being recombinant DNA (rDNA) technology, which theoretically allows scientists to transfer genes from one organism to any other, circumventing the sexual process. For example, a gene from a bacterium can be transferred into the corn genome. Consequently, rDNA technology allows scientists to treat all living things as theoretically belonging to one giant breeding gene pool. Since the discovery of this radical technology, along with new biological knowledge, the genome of plants can now be manipulated in dramatic ways. The discussion in this chapter is meant to be an overview of the array of technologies of molecular genetic manipulations and how they are applied in plant breeding.

After studying this chapter, the student should be able to discuss:

1 The fundamental differences between conventional breeding and genetic engineering.
2 The concept of molecular breeding.
3 Principles of breeding genetically engineered plants.
4 Genetically engineering of selected traits.

Principles of Plant Genetics and Breeding, Second Edition. George Acquaah.
© 2012 John Wiley & Sons, Ltd. Published 2012 by John Wiley & Sons, Ltd.

25.1 What is biotechnology?

Etymologically, **biotechnology** is the study of tools from living things. In its current usage, the term is defined either broadly or narrowly. It may be defined broadly as the use of techniques based on living systems to make products or improve other species. This would include the use of microbes to make products via fermentation, an age-old practice. In a narrower definition, biotechnology refers to the genetic manipulation of organisms for specific purposes. The term **genetic engineering** is sometimes used to describe this practice. Some argue that classical plant breeding is genetic engineering, since the genetics (DNA) of plants are manipulated by breeders, albeit *indirectly*. Consequently, a much narrower definition of genetic engineering is used to describe the manipulation of organisms at the molecular level, *directly* involving the DNA. However, it is the revolutionary technology of **recombinant DNA (rDNA),** which enables researchers to transfer genes from any organism to another that some accept as genetic engineering. The term **molecular breeding** is used to describe the use of a variety of tools for manipulating the DNA of plants (which may or may not involve rDNA) to improve them for specific purposes.

25.2 Antisense technology

Watson and Crick showed the DNA structure to be a double helix, comprising two complementary strands. Protein formation entails the processes of DNA transcription and translation. Although the two strands are exact mirror images (complementary), only one of them contains the information for making proteins. This strand is called the **antisense strand** (transcribed strand; template DNA), the other the **sense strand** (Figure 25.1). The product of DNA transcription, the mRNA transcript, has the same sequence as the sense sequence (except that T bases in DNA are substituted with U bases in RNA). In order for protein to be formed from a gene, it must first be successfully transcribed and then translated. These and other requirements offer opportunities for scientists to manipulate the process of protein formation or gene expression.

The basic principle of antisense technology is to prevent protein production from a targeted gene. Research is ongoing to elucidate the precise

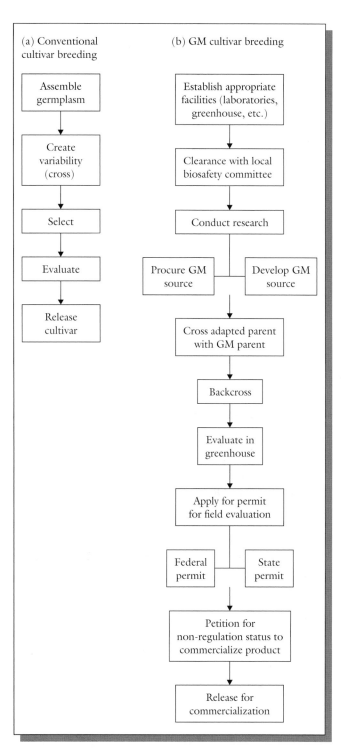

Figure 25.1 A comparison of the general steps involved in breeding cultivars by (a) the conventional method and (b) by the use of genetic engineering technology. The specific steps will vary among breeders and situations.

mechanism by which such an event may occur. In the interim, several mechanisms have been proposed, including blocking of RNA splicing, triplex formation, prevention of the transport of mRNA transcript into the cytoplasm for translation, increasing RNA degradation, and blocking of the initiation of translation. The rationale of the application of antisense technology is that the undesirable consequences of a gene can be eliminated or reduced by modifying the process of gene expression. This application was first demonstrated in plants when the technology was successfully used to alter the levels of various degradative enzymes. Antisense technology, hence, provides scientists a tool for targeted modulation of gene expression. This tool is useful for investigation the function of any protein in the cell.

The strategy of antisense technology is to design and synthesize an oligonucleotide that is complementary to the antisense DNA strand, to be delivered into the cellular matrix to interfere with the translation process by binding onto the mRNA transcript. The mechanism by which this first step (the successful penetration of the oligonucleotide into the targeted cell) in the application of the technology occurs is not clearly understood. Further, once in the cell, the oligonucleotides encounter cellular nucleases that can degrade these molecules and, thereby, reduce their effectiveness. Some of the challenges in antisense technology include the development of appropriate synthetic oligonucleotides as well as effective and efficient delivery systems. Most importantly, scientists are most concerned about the specificity of the action of antisense oligonucleotides (i.e., to engage and bind to only the target gene or RNA without side effects).

Some of the delivery issues are being mitigated with the modification of oligonucleotides by altering phosphodiester bond by replacing an oxygen atom with sulfur atom to create a class of oligonucleotides called phosphorothionates. These oligonucleotides are more stable (resistant to degradation by intracellular endonucleases and exonucleases) than their predecessors, the methylphosphonates. However, they have limitations of their own that reduce their effectiveness as antisense effector molecules. New classes of effector molecules continue to be developed, including the phosphorodiamidate morpholine oligomers. On the basis of mechanism of action, two general classes of antisense oligomers may be distinguished: (i) the

RNase H-dependent oligonucleotides, which induce degradation of mRNA and (ii) the steric-blocker oligonucleotides, which physically inhibit the progression of splicing or the translation mechanism. The former class is generally about 80–95% effective in down-regulation of protein and mRNA expression. Most of the research reports to date focus on this class. Further, most of the antisense oligomers used have targeted the translation initiation codon, even though this target is not always the most idea.

With the discovery of **small interfering RNAs (siRNAs)**, scientists now have a new tool in the antisense toolkit (RNAi, RNAi technology). The siRNAs are 21–23mer double-stranded RNA molecules that can silence gene expression. They occur in very low concentration in the cell and are very highly specific in terms of target-binding.

25.3 Plant genomics

The basic set of chromosomes of an organism is called its **genome.** Traditional geneticists generally investigate single genes, one at a time, as snapshots. Studying genetics whereby the totality of the genes in an individual is considered together would be advantageous. Genes seldom, if ever, work independently in higher organisms.

25.3.1 What is genomics?

Genomics is the approach of investigating the totality of genes in an individual as a dynamic system, not as snapshots, but rather over time and to determine how these genes interact and impact biological pathways and the general physiology of an organism, from a global perspective (i.e., the "big picture").

Genomics may be broadly categorized into two – structural and **functional**, each with its set of tools and functions. However, as the field advances, new terms, categories, and subcategories continue to emerge in the literature.

25.3.2 Structural (classical) genomics

Structural genomics, at the very basic, is concerned with activities at the initial phase of genome analysis – **mapping** (the construction of high resolution genetic, physical, and transcript maps of an organism). The initial phase of genome analysis is the establishment of a physical map of the genome. The ultimate physical map of an organism that can be achieved is the complete sequence of its total DNA (**genomic sequence**). Genomic sequencing projects yield linear amino acid sequences. The most visible genome sequencing project in recent times is the Human Genome Project. Genome sequencing projects have been initiated or completed for some major food crops, such as rice, maize, and wheat.

Genome mapping entails determining the order of genes (or other genetic markers) and the spacing between them on each chromosome. Two categories of mapping strategies are employed in genome mapping: (i) genetic linkage mapping and (ii) physical mapping.

25.3.3 Genome sequencing

The purpose of **DNA sequencing** is to determine the order (sequence) of bases in a DNA molecule. Classical mapping approaches involve generating and mapping mutants, thus limiting characterization to those genes for which mutants have been isolated. rDNA techniques provide a direct approach to genetic analysis, whereby a genomic library is created, from which overlapping clones are then assembled to construct genetic and physical maps for the entire genome. Ultimately, the entire genome is sequenced such that all genes are identified by both their location in the chromosome as well as their nucleotide sequence.

Allan Maxam and Walter Gilbert invented the **chemical degradation method** of DNA sequencing. Another method, the Sanger **enzymatic (dideoxy) method (chain terminating)**, named after its inventor, is more commonly used. However, in these days of more advanced technology, there are high throughput methods of DNA sequencing, making it more convenient for most researchers to contract sequencing jobs to outside companies. **Capillary array electrophoresis** (CAE) systems coupled with high sensitivity detection provided by energy-transfer labeling reagents is now the standard for high throughput DNA sequencing facilities.

Genome sequencing is involved, and hence often undertaken on collaborative basis (i.e., consortium), over several years. Modern supercomputers and powerful algorithms are making genome sequence more efficient and through put.

25.3.4 Comparative genomics

Biotechnology makes use of certain organisms as **models** for comparative research. A fundamental premise of rDNA is that DNA is universal and hence transfer of DNA is possible, in theory, across biological barriers. Eukaryotes are genetically complex organisms and not directly amenable to certain biotechnological procedures. Consequently, scientists pick plants and animals that are easy to manipulate, and invest in a thorough genetic analysis of their genomes. The information derived from such studies is extrapolated to more complex organisms. *Arabidopsis thaliana* was the first plant to be completely sequenced. This flowering plant has a short lifecycle of just about six weeks from germination to seed maturity. It produces seed profusely. Because of its small size, arabidopsis can be cultured in trays or pots in limited space. Two of the most important higher plants used for food that are being developed as model plants are rice (*Oriza sativa*) and maize (*Zea mays*). There are several subspecies of rice, the most widely eaten being the *indica*, followed by *japonica*. Through both public and private sector efforts, the genome sequences of the two subspecies have been developed. Genomes of all cereals (wheat, barley, rice, corn, etc.) are structurally similar (see Synteny later in this chapter).

25.3.5 Genomic colinearity and its application in plant breeding

Genomic colinearity is a term used to refer to the conservation (through evolutionary history) of gene content, order, and orientation between chromosomes of different species, or between non-homologous chromosomes within a single species. The term is sometimes used as synonymous with synteny, the phenomenon of conservation of gene order within related genomes. Synteny has been observed in the Brasicaseae using molecular techniques. An important discovery of synteny is that between dicots and monocots. The sequence information for *Arabidopsis thaliana* (a brasica species) and rice has revealed a conserved gene order across this vast evolutionary distance. It is estimated that about 75% of the genes in dicot species and about 40% in monocots occur in regions that have colinearity with *Arabidopsis* that could be exploited by breeders.

Synteny has also been discovered in Fabaceae involving soybean, peanuts, mungbean, lentil, common bean, common pea, and alfalfa. Some conservation of the genomic region in this family with *Arabidopsis* has been established. Synteny in the Poaceae is one of the clearest to be documented. Using the rice genome at the inner circle, researchers have graphically aligned the genomes of several cereals (sorghum, maize, wheat) by colinearity into concentric circles such that collinear regions in different species lie along any radius. Rice has become a model species for comparative genomics studies of species with large genomes (e.g., wheat, sugarcane) that are difficult to study by traditional genetics. Another advantage is that the so-called "orphan crops" (e.g., millet), which would otherwise not receive significant attention like the major cereals, can now benefit from the information obtained from rice.

The applications of colinearity to plant breeding include:

- Prediction of the location of genes controlling a particular function in crop species.
- Facilitate genome mapping by transferring markers from a well-mapped genome to a less-studied one.
- Mapping will facilitate marker assisted selection or map-based cloning.
- After cloning a locus of agronomic importance, colinearity may provide opportunities for accumulating alleles of that locus from other distantly related species. This collection of genes may be a source for genetic engineering applications for crop improvement.

25.3.6 Functional genomics

Once DNA sequences have been obtained, the next important task is to understand their function. Structural genomics focuses on sequencing of the genome; **functional genomics** focuses on gene function. The genome is essentially a set of instructions for making proteins of various kinds. Because most genes are expressed as proteins, one of the common ways of understanding gene function is by tracking protein expression by cells (called **proteomics**). Genes may provide instructions for making specific proteins. However, in the process of carrying out these instructions, additional proteins can be produced, as has already been noted. Linking gene to function is a complex undertaking. Understanding the genome structure alone is insufficient; it is critical to identify

the proteins they encode. Many techniques have been developed for deciphering gene function.

25.4 Bioinformatics in plant breeding

Bioinformatics may be defined as a knowledge-based theoretical discipline that attempts to make predictions about biological function using data from DNA sequence analysis. It is an application of information science to biology. It uses supercomputers and sophisticated software to search and analyze databases accumulated from genome sequencing projects and other similar efforts. Bioinformatics allows scientists to make predictions based on previous experiences with biological reality. In one application, the biological information bank is searched to find sequences with known function that resemble the unknown sequence, and thereby predict the function of the unknown sequence. Databases are critical to bioinformatics, and hence repositories exist in various parts of the world for genetic sequence data.

25.4.1 Types of bioinformatics ddatabases

Information used in bioinformatics research may be grouped into two categories:

(i) **Primary databases.** These databases consist of original biological data such as raw DNA sequences and protein structure information from crystallography.

(ii) **Secondary databases.** These databases contain original data that have been processed for value added to suit certain specific applications.

To be useful, a good database should have two critical parts: (1) the **original sequence** and (ii) an **annotation description** of the biological context of the data. It is critical that each entry be accompanied by a detailed and complete annotation, without which a bioinformatics search becomes an exercise in futility, since it would be difficult to assign valid meaning to any relationships discovered. Some databases include taxonomic information, such as the structural and biochemical characteristics of organisms.

The bulk of data in repositories consist of primary data. Three major entities are collaboratively responsible for maintaining gene sequence databases. These entities are the European Molecular Biology Lab

Industry highlights

Bioinformatics for sequence and genomic data

Hugh B. Nicholas, Jr., David W. Deerfield II, and Alexander J. Ropelewski

Pittsburgh Supercomputing Center, 4400 Fifth Avenue, Pittsburgh, PA 15213, USA

Introduction

Bioinformatics is the application of informatics techniques to biological data. These techniques include acquiring, annotating, analyzing, and archiving the biological data, using concepts from biology, computer science and mathematics. Bioinformatics has its roots in the work of people like Margaret Dayhoff (collecting known sequences (Dayhoff *et al.*, 1965) and a mathematical model of protein evolution (Dayhoff *et al.*, 1978)) and David Sankoff (sequence alignment and statistical tests for homology) in molecular phylogenetics.

Bioinformatics analyses aim to discover precise, testable hypotheses to supplement or redirect biology experiments. The results of these bioinformatics derived experiments then influence the next round of computer-based studies. This interaction between experiment and computation speeds scientific progress.

The first large collections of biological data were protein sequences, followed by nucleic acid sequences, and were collected by individual laboratories and stored on computer punch cards, along with annotations, such as species, biochemical function and physiological role, and functional and structural domains. As the amount of sequence data increased, the need to share this data between multiple laboratories grew, and the collection and curation were taken over by specialized groups, such as NBRF/PIR, GenBank, EMBL and Swiss-Prot, where curation included standardizing the format and ancillary data. With increased size also came the need for tools to search, analyze, and annotate these databases.

Pairwise alignment and database searching

Molecular biologists commonly isolate and sequence molecules based on their association with particular biological phenomena, such as disease resistance in plants. Typically, the biochemical function of the newly determined sequence is not known and one compares the newly determined sequence to all known sequences whose biochemical functions are known to generate a testable hypothesis about its function. Thus, an early bioinformatics tool was to search a database of annotated sequences with a newly determined sequence to find all similar sequences. Sequences similar enough were inferred to be homologous, that is, to have descended from a common evolutionary ancestor. The inference of homology generates the hypothesis that the two molecules carry out the same biochemical function and, perhaps, the same physiological role. A complete discussion of database searching is given elsewhere (Nicholas *et al.*, 2000).

The power of a successful database search is demonstrated by comparing the histories of Cystic Fibrosis (CF) and type I Neurofibromatosis (NF-1) research. Both disease genes were isolated in 1988. The CF gene was identified as a chloride ion transport protein which led to the development of a number of therapies, with many now in final clinical trials. In 1988 the database search with the NF-1 gene failed and no homologues were found. It was not until 1998 that NF-1 was identified as a growth suppressor, which has rapidly lead to improved diagnosis and many potential therapies for which clinical trials are just beginning. Thus, the successful identification of CF gene as chloride ion transporter accelerated this research area by a decade compared to the time required to discover the biochemical function of the NF-1 gene through biological experiments.

Multiple sequence alignment

A database search results in discovering many similar sequences from which one would like to create a multiple sequence alignment that simultaneously shows the relationship among its homologous residues in the other sequences. This alignment is a map of the evolution of the protein family. The multiple sequence alignment is a rich source of hypotheses to guide experimental work, since the alignment contains patterns of conservation and variation of residues among the sequences, which provides insights into functional and structural positions for either the family of proteins or the genes encoding them. Such inferences are strongest if the alignment contains sequences from widely diverse species.

Multiple sequence alignment implies that the residues in each column of the alignment are all evolutionarily related to each other. Thus, accuracy is most commonly considered improved by maximizing the observed degree of conservation in the alignment as a whole (Nicholas *et al.*, 2002).

Analysis

Two evolutionary concepts underlie the inferences from analyses of multiple sequence alignments. The first is that the mutation of one sequence residue to another is a random event in nature and that all residues are more or less equally subject to mutation. The second is that the conservation of specific residues is maintained through evolutionary selection; that is, mutations that adversely affect the ability of a molecule to carry out its biochemical activity or physiological role will be eliminated by reducing the ability of the organism to live. Details of the evolutionary history between different pairs of sequences can lead to different inferences about the properties of the sequences.

In the evolutionary history of some pairs of sequences, orthologs or orthologous sequences, the sequences have only speciation events in their evolutionary history. Other pairs of sequences, paralogs or paralogous sequences, have one or more gene duplication events in their common evolutionary history as well as having speciation events. In general, paralogs will carry out the same biochemistry on different substrates or with different cofactors; while orthologs will carry out the same biochemistry on the same substrates and often serve the identical physiological role in the same pathways in different organisms. Homologs include both orthologs and paralogs. Since paralogs carry out the same basic biochemistry, for example reduce an aldehyde, the residues responsible for this activity are conserved. But the residues responsible for the physiological role (e.g., substrate recognition, which specific aldehyde to reduce or lipid to bind) will be under different evolutionary pressures and will often diverge. A complete analysis of the multiple sequence alignment includes identifying residues responsible for the common, shared properties of the entire family and the residues that discriminate between paralogous groups (Figure B25.1).

```
Bpi_Bovin : VTCSSCSSHINSVHVHISKSK-VGWLIQLFHKKIESALRNKMNSQVCEKVTNSVSSKLQPYFQTLP
Bpi_Human : ITCSSCSSDIADVEVDMSGD--SGWLLNLFHNQIESKFQKVLESRICEMIQKSVSSDLQPYLQTLP
Lbp_Human : GYCLSCSSDIQNVELDIEGD--LEELLNLLQSQIDARLREVLESKICRQIEEAVTAHLQPYLQTLP
Lbp_Rabit : VTTSSCSSRIRDLELHVSGN--VGWLLNLFHNQIESKLQKVLESKICEMIQKSVTSDLQPYLQTLP
Lbp_Rat   : VTASGCSNSFHKLLLHLQGEREPGWIKQLFTNFISFTLKLVLKGQICKEI-NVISNIMADFVQTRA
Cetp_Human: TDAPDCYLSFHKLLLHLQGEREPGWIKQLFTNFISFTLKLVLKGQICKEI-NIISNIMADFVQTRA
Cetp_Macfa: TDAPDCYLAFHKLLLHLQGEREPGWLKQLFTNFISFTLKLILKRQVCNEI-NTISNIMADFVQTRA
Cetp_Rabit: TNAPDCYLAFHKLLLHLQGEREPGWLKQLFTNFISFTLKLILKRQVCNEI-NTISNIMADFVQTRA
Pltp_Human: VSNVSCQASVSRMHAAFGGT--FKKVYDFLSTFITSGMRFLLNQQICPVLYHAGTVLLNSLLDTVP
Pltp_Mouse: VSNVSCEASVSKMNMAFGGT--FRRMYNFFSTFITSGMRFLLNQQICPVLYHAGTVLLNSLLDTVP
Lplc3_Rat : LILKRCNT----LLGHISLT--SGLLPTPIFGLVEQTLCKVLPGLLCPVV-DSVLSVVNELLGATL
Lplc4_Rat : LVIERCDT----LLGGIKVKLLRGLLPNLVDNLVNRVLANVLPDLLCPIV-DVVLGLVNDQLGLVD
Consensus        C                       i        l   C             t
```

Figure B25.1 A section of a multiple sequence alignment of 12 sequences from the BPI/LBP/LPI superfamily of lipid binding protein sequences is shown. The 12 sequences belong to six different paralogous families, each of which binds a different lipid and performs a different physiological role, generally a transport function. Each sequence is identified by a label that identifies the paralogous family (they have a gene duplication event separating the families) before the underscore character in the protein name and the species after the underscore. The protein names are taken from the Swiss-Prot database. The annotation associated with each sequence in the Swiss-Prot database describes the lipid bound and the physiological role of the protein. The sequences within each paralogous family are orthologous (related to each other only by speciation events). The dash sequence character indicates an alignment column, where either the amino acids replaced by dashes have been lost through a deletion event or amino acids shown by the single letter codes have been inserted into the sequence during the course of evolution. Figure courtesy of Hugh B. Nicholas, Jr.

The section marked with a black background and white letters is a moderately well conserved motif. Such conservation often marks regions important to the structure or function of the protein. Amino acids highlighted with black letters on a light grey background are identical within the highlighted family of orthologous sequences and have quite different chemical or physical properties to those in the same column in other families. They may mark positions that are responsible for the differences among the paralogous families.

The results from the analysis of multiple sequence alignments is illustrated in results obtained in the decades long quest to understand how each of the twenty aminoacyl transfer RNA synthetase (aaRS) enzymes map to each of the sixty transfer RNA (tRNA) to its correct cognate amino acid. The problem, as outlined by Sir Francis Crick in 1957, is where is the information that allows an aaRS to recognize only the tRNAs encoding the correct amino acid? Experimental groups employed a number of creative and innovative approaches towards answering this question (Söll and Schimmel, 1974).

McClain, Nicholas and co-workers applied computational techniques, beginning with creating an accurate multiple sequence alignment of the tRNAs. The tRNA multiple sequence alignment was divided into twenty subsets, each based on their amino acid acceptor activity (McClain and Nicholas, 1987). First, these authors identified the sequence positions in the on 66 tRNA sequences from *E. coli*, its bacteriophages and near relatives that were conserved. Second, these authors examined the remaining positions and asked the question: what position or combination of positions discriminate one subset from the others? Using a number of statistical techniques (e.g., multivariate analysis, group theory), they were able to develop a model for how tRNA sequences allowed the aaRS enzymes to identify the correct tRNA molecules and reject the incorrect (McClain, 1995). Experimental verification (started prior to the computational work in one case) of these results were published subsequently (McClain and Foss, 1988; Hou and Schimmel, 1988).

Additional use of the multiple sequence alignment

In addition to the above analysis of a well-crafted multiple sequence alignment, there are two additional areas that can use the information contained within a multiple sequence alignment. The first is the creation of a phylogenetic tree for evolutionary studies, while the second is to allow more sensitive database searches using a representation that incorporates the pattern of substitution seen in the multiple sequence alignment to allow the researcher to find more highly diverged homologous sequences.

Pattern identification

Pattern identification has been developed to identify small, unique sections of the several unaligned sequences. Often, these contiguous regions of conserved residues, called motifs, are important for molecular interactions, such as regulatory regions or binding sites. Thus, motifs are often essential for the correct functioning of the molecule.

The classic example of pattern identification is to collect DNA sequences from the region just upstream (on the 5′ side) of the coding region of a gene and examine these for a conserved pattern of nucleotides (Sadler *et al.*, 1983) involved in regulating the transcription of the genes. What distinguishes this problem from global multiple sequence alignment is that outside of the conserved patterns there is no expectation that the sequence is conserved and thus alignable.

Modern pattern identification programs (Bailey and Elkan, 1994) make use of a modern statistical processes designed to deal with the fact that we do not know where the patterns are located (Expectation Maximization), and a sophisticated sampling routine (Stochastic Sampling) that reduces the number of combinations must be tried.

Other techniques

As the biomedical sciences have expanded their repertoire of research methods and the kinds of data that can be collected, the field of Bioinformatics has created techniques for dealing with these new kinds of data. The advent of the complete genome sequences for many organisms has been accompanied by software to allow the manipulation, annotation, analysis and comparison of these large sequences. Complex mathematical models of genes try to find and identify all of the genes in each genome (Rogic *et al.*, 2001).

Techniques have been developed to measure the change in expression for cDNA (microarrays) or amount of proteins in cells over time or between mutants and wild types. It is not unusual for a research group to monitor thousands of molecules simultaneously, looking for either increases or decreases in the relative levels between the standard and the state under investigation. These large-scale experiments are being analyzed with a number of statistical techniques (Wetzel *et al.*, 2000), such as analysis of variance, which produces a statistical model of the changes observed (Kerr and Churchill, 2001). Other researchers are using multivariate statistical techniques to identify which molecules vary their presence in a coordinated manner in response to changing conditions. Interestingly, a number of these techniques were first developed many years ago to study the factors influencing crop growth.

Conclusion

Ultimately, though, the field of bioinformatics does have some general themes that should continue to run throughout it in the future. First, the bioinformaticist tool chest in not complete – the tool chest of tomorrow will have only minimal relationship to today's set of tools, with better and more sensitive tools continuing to be developed. Second, the numbers and types of databases of experimental data will continue expanding at an alarming rate. The majority of the databases will be developed to describe one type of experimental data, like the sequence data or microarray data, with only minimal references or consistencies (vocabulary) to the other databases. Third is that diverse data must be integrated across ranges of scale, both temporal and spatial. For example, a single point mutation in a mouse might cause kidney deformations that result in blood chemistry being incorrect. Thus, you have a single point mutation that causes effects at the cellular and organ levels. Biological scientists must learn the techniques necessary to manage and make use of the new data resources that their research is creating.

References

Bailey, T.L., and Elkan, C. (1994). *Fitting a mixture model by expectation maximization to discover motifs in biopolymers*, in *ISMB-94, Proceedings Second International Conference on Intelligent Systems for Molecular Biology* (eds. R. Altman, D. Brutlag, P. Karp, R. Lathrop, and D. Searls), AAAI Press, Menlo Park, CA, pp. 28–36.

Crick, F.H.C. (1957). The structure of nucleic acids and their role in protein sysnthesis, in *Biochemical Society Symposium*, No. 14 (ed. E.M. Crook). Cambridge University Press, Cambridge, pp. 25–26.

Dayhoff, M.O., Eck, R.V., Chang, M.A., and Sochard, M.R. (1965). *Atlas of Protein Sequence and Structure*, Vol. 1 National Biomedical Research Foundation, Silver Spring, MD.

Dayhoff, M.O., Schwartz, R.M., and Orcutt, B.C. (1978). A model of evolutionary change in proteins, in *Atlas of Protein Sequences and Structure* (ed. M. O. Dayhoff), Vol. 5 National Biomedical Research Foundation, Washington, pp. 345–352.

Hou, Y.M., and Schimmel, P. (1988). A simple structural feature is a major determinant of the identity of a transfer RNA. *Nature*, **333**:140–145.

Kerr, K.M., and Churchill, G.A. (2001). Statistical design and the Analysis of Gene Expression Microaarays. *Genetical Research*, **77**:123–128.

McClain, W.H. (1995). The tRNA Identity Problem: Past, Present, and Future, in *tRNA: Structure, Biosynthesis and Function* (eds D. Söll, and U.L. RajBhandary). ASM Press, Washington, DC, pp. 335–347.

McClain, W.H., and Nicholas, H.B. Jr. (1987). Differences between transfer RNA molecules. *Journal of Molecular Biology*, **194**:635–642.

McClain, W.H., and Foss, K. (1988). Changing the identity of a tRNA by introducing a G-U wobble pair near the 3′ Acceptor end. *Science*, **240**:793–796.

Nicholas, H.B., Jr., Deerfield, D.W., II, and Ropelewski, A.J. (2000). Strategies for Searching Sequence Databases. *Biotechniques*, **28**:1174–1191.

Nicholas, H.B., Jr., Ropelewski, A.J., Deerfield, D.W., II (2002). Strategies for Multiple Sequence Alignment. *Biotechniques*, **32**:592–603.

Rogic, A., Mackworth, A.K., and Oullette, F.B. (2001). Evaluation of gene-finding programs on mammalian sequences. *Genome Research*, **11**:817–832.

Sadler, J.R., Waterman, M.S., and Smith, T.F. (1983). Regulatory pattern identification in nucleic acid sequences. *Nucleic Acids Research*, **11**:2221–2231.

Söll, D., and Schimmel, P.R. (1974). Aminoacyl tRNA Synthetases. *The Enzymes*, **X**:489–538.

Wetzel, A.W., Gilbertson, J., Zheng, L., *et al.* (2000). Three-dimensional reconstruction for genomic analysis of prostate cancer. *Proceedings of the SPIE*, **3905**:204–212.

(EMBL) of Cambridge, UK, the GeneBank of the National Center for Biotechnology Information (NCBI) that is affiliated with the National Institute of Health, USA, and the DNA Databank of Japan.

Databases for both protein sequences and structure are being maintained. The Department of Medical Biochemistry (University of Geneva) and the European Bioinformatics Institute collaboratively maintain properly annotated translations of sequences in the EMBL databases. This is called the SWISSPROT. TREMBL is another protein database consisting solely of protein-coding regions of the EMBL database (called the translated EMBL or TREMBL). The NCBI also maintains a database of the translations of the GeneBank. Another kind of protein database consisting of experimentally derived 3-D structures of proteins is kept at the protein databank where these structures are determined by X-ray diffraction and nuclear magnetic resonance.

25.5 Breeding genetically modified (GM) cultivars

Steps in the breeding of conventional cultivars are summarized in Figure 25.1. The breeder starts by assembling germplasm to create the base population. Selection is practiced in the segregating population to identify and advance the more desirable genotypes, which are evaluated to identify and release the most promising one as a commercial cultivar. A breeding program can be implemented without oversight by any entity, except at the stage of certification, when certifying agencies inspect the product for conformity with set standards.

On the other hand, breeding genetically modified (GM) cultivars is a highly regulated activity, from the inception to the completion of the project. Before biotechnology development and application activities can proceed in a country, there has to be a national biosafety policy in place (Chapter 28). Furthermore, the institution at which the breeding program is to be conducted must also have its own Institutional Research Policy guidelines, prescribing procedures for conducting research. Some of the guidelines pertain to the use of hazardous materials. There are specific guidelines for recombinant DNA (rDNA) research (some variation may occur from one institution to another).

25.5.1 Clearance

The first step in a genetic engineering project is to submit the proposal to the local Institutional Review Board and other specialized committees. (e.g., Institutional Biosafety Committee). These bodies review the project materials and methods for scientific quality and compliance with established safety and ethical guidelines. There has to be some provisions for containment and/or disposal of hazardous materials. Depending on the nature of the project, a biosafety level is assigned by the committee according to federal guidelines.

The biosafety levels and their restrictions are:

BL1-P. Basic containment level. Restricted access to greenhouse; insect, weed, rodent control mechanisms required; screens recommended.

BL2-P. For agents of moderate potential hazard. BL1-P requirements; concrete floor; screens restricting movement of small insects but not pollen; autoclave to sterilize transgenic materials before removal.

BL3-P. For agents of serious potential hazard. BL2-P requirements; plus collection and sterilization of liquid run-off; sealed windows; ventilation filters; security fence; protective clothing.

BL4-P. For work with extremely hazardous agents; including certain exotic plant pathogens. Similar to BL3-P but more stringent.

25.5.2 Conduct research

Large seed companies, such as Monsanto and Pioneer, develop their own GM breeding stock. They transfer the gene of interest into the appropriate genetic background. Other breeding programs may procure the cloned gene or genetic stock at a fee, for use in their projects. Developers of transgenic breeding lines must evaluate them initially for:

- Activity of the introduced gene (transgene).
- Stable inheritance of the gene.
- Unintended effects of the growth and development of the plant.

25.5.3 Hybridize (cross)

Once an appropriate GM breeding line has been obtained, it is crossed with a conventional adapted cultivar to transfer the transgene into the desirable cultivar.

25.5.4 Backcross

The first cross is followed by cycles of repeated backcrosses to the adapted cultivar to recover the traits of this parent as much as possible. All these activities are conducted under restricted environment (greenhouse).

25.5.5 Evaluation

The breeder evaluates the success of the project in the greenhouse. This may include evaluating the proper expression of the transgene, yield and quality and overall performance of the new product.

25.5.6 Field testing

When the breeder is satisfied that the breeding project is successful, the next step is to apply for permission to field test the cultivar. An application to transport or field test a transgenic plant is submitted to the Animal and Plant Health Inspection Service (APHIS). Under the Federal Plant Pest Act, APHIS must determine if a transgenic plant variety has the potential to become a pest (Chapter 28).

Before permission is granted for field testing, some of the basic criteria to be satisfied by the breeder include presenting the pieces of evidence to show:

- Stable integration in the plant chromosome of the transgene.
- Non-pathogenic to animals or humans.

- Unlikely to be toxic to other non-target organisms.
- Low risk of creating new plant viruses.

Field testing, if permission is granted, is conducted at multiple locations and over several years. The breeder should conduct the test in a manner that will not permit contamination of the environment or food supply system. At the end of the evaluation, a comprehensive report is submitted to APHIS, including data on the gene construct, effects on plant biology, effects on ecosystem, and spread of the gene to others species. Depending on the product and its intended use, other federal agencies (EPA, FDA) may be involved in the process of field evaluation (Chapter 28). In addition to satisfying federal agency requirements, some states may have their own regulations that must be satisfied for field testing of GM varieties.

25.5.7 Commercialization

Once the field evaluation is successfully completed, the breeder may apply for non-regulatory status to commercialize the cultivar. If successful, the cultivar may be released as a commercial cultivar to producers. Of course, the breeder may seek appropriate protection of the invention through patent application or other appropriate measures.

25.6 Vectors

A vector is a DNA molecule that is used to transfer foreign genetic material into another cell. There are four major types of vectors – plasmid, viral and other bacteriophages, cosmid, and artificial chromosomes, of which the first two are most common.

25.6.1 Plasmid vectors

Plasmids are double-stranded, self-replicating extrachromosomal molecules with antibiotic selectable markers. They are small molecules that can handle small inserts of about 10 kb in size. They replicate prolifically but also easily lose larger inserts (approaching 20 kb).

25.6.2 Bacteriophage λ-derived vectors

These are viruses that infect bacteria. They have a cloning capacity of about 23 kb. They are known for

their low tendency to lose inserts. Viral vectors carry modified viral DNA or RNA, and even though attenuated (rendered noninfectious) they contain viral promoters for translating the target gene in the host cell. They are often designed to leave a footprint (distinct genetic markers) after the incorporation of the target gene into the host genome (e.g., retroviruses leave a characteristic and detectable retroviral integration pattern to indicate successful transfection). The lack of infectious sequences requires that viral vectors be provided assistance (packaging) for large-scale transfection.

25.6.3 Cosmids

Cosmids are plasmid vectors with *cos*-sequence (for packing) from bacteriophage. They are capable of carrying about 42 kb of inserts. They rarely lose inserts and also replicate intensely.

25.6.4 Yeast artificial chromosomes (YAC)

These are useful for cloning large fragments in the megabase range. While they may be used to even clone whole chromosomes, they are notorious for generating chimeras (recombinant molecules in which non-contiguous donor fragments are joined together), which are laborious to separate from the desired recombinants.

25.6.5 PI-derived artificial chromosome (PAC) and bacterial artificial chromosome (BAC)

PAC vectors can handle about 350 kb while BACs can handle about 150 kb inserts. BAC vectors form fewer chimeras than YACs.

Engineered vectors vary in architecture, function, and capacity for foreign gene, among other factors. They differ in the size of fragments they can stably incorporate, the procedure for screening the insert DNA or target DNA, and the number of recombinant copies they can produce per cell. However, they have three basic component features in common – an origin of replication, a multicloning site (restriction enzyme site), and a selectable marker. In addition to the transgene insert and the backbone, modern vectors include the following features:

Promoter. critical to all vectors and used for transcription of the transgenes they carry.

Genetic markers. this is used for confirming that the vector has been successfully integrated into the host genomic DNA.

Antibiotic resistance. vectors may contain antibiotic resistance open reading frames for identification via antibiotic selection, the cells that take up the vector.

Epitope. also called an antigenic determinant, an epitope is part of a macromolecule that is recognized (bound to) by the immune system (antibodies, B or T cells). It may consist of sugars, lipids or amino acids.

β-galactosidase. this enzyme digests galactose and its sequence is located on either side of the transgene insertion site. In the event of unsuccessful insertion or full ligation into the vector, cells containing "empty vectors" will generate this enzyme to digest galactose. However, if successful (positively transformed), the vector with the transgene will lack the sequence for this enzyme (in active), and hence be unable to digest (hydrolyze) the sugar which is introduced as X-gal (colorless) into the selective agar plate, and would be identified by a color dye for this sugar (transformant colonies are blue).

Targeting sequence. this feature may be included in an expression vector for encoding a sequence in the finished protein that directs the expressed protein to a specific organelle in the cell.

25.7 Categories of vectors by functions

The types of vectors previously described can be designed to perform different functions.

25.7.1 Cloning vectors

Cloning vectors (or simply **vectors**) are replicating units into which isolated fragments of DNA may be integrated for maintenance. For convenience and increased efficiency in transgene insertion (transfection; transduction if viral vector), the original restriction sites of most cloning vectors are replaced by a synthetic multiple cloning site (contains many restriction sites). Other additional features that may be engineered into vectors include vir genes (for plant transformation), integrase sites (for chromosomal insertion), and *lacZ*a fragment (for a complementation).

The choice of cloning vector to use for a particular project depends on factors including the insert size, copy number, cloning sites, selectable marker, and incompatibility. The size of the insert to be cloned appears to be the key factor that determines the choice of a cloning vector. A desirable vector must be a relatively small molecule for convenience of manipulation and should also be capable of prolific replication to enable the target DNA to be sufficiently amplified in the living cell. Cloning vectors may be used for specialized function such as transcription and expression.

25.7.2 Transcription vectors

Transcription is a necessary component of all vectors. Stable expression of an insert depends on stable transcription (which depends on the promoters in the vector). Transcription vectors are designed to only be transcribed (replicated or amplified) but not translated (expressed). They are relatively simple in their construction. Plasmid transcription vectors lack the required sequences that encode polyadenylation and termination sequences in translated mRNAs, thereby making it impossible to make protein expression from these vectors.

25.7.3 Expression vectors

Expression vectors (expression constructs) are specifically designed for the controlled expression of the DNA insert (or transgene) in the target cell or organism. Once inside the host cell, the protein that is encoded by the insert is produced by the regular cellular transcription and translation machinery. An efficient expression vector will produce large amounts of stable mRNA. They generally include regulatory sequences that serve as enhancer and promoter regions for the efficient transcription of the insert gene on the vector. The promoter sequence is required to drive the expression of the insert. It is desirable to insert the transgene into a site that is under the control of a specific promoter. Commonly used promoters include the T7 promoters, lac promoters, and cauliflower mosaic virus's 35s promoter for plant viruses. In addition to a strong promoter, these vectors have a strong termination codon, the insertion of a transcription sequence, and a portable translation sequence (PTIS). They require sequences that create a polyadenylation tail at the end of the transcribed pre-mRNA to stabilize mRNA production (protects is from exonucleases and ensures transcriptional and translational termination). Optimal expression vectors also include minimal UTR length (UTRs contain specific characteristics that may impede transcription or translation; hence for optimal expression, none at all or the shorted UTRs are encoded). The capacity to encode a Kozak sequence in the mRNA (assembles the ribosome for translation of the mRNA) is required. These latter set of requirements apply to eukaryotes not prokaryotes. Expression vectors are used in techniques such as site-directed mutagenesis.

25.7.4 Intragenic vectors

Intergenic vector transfers have dominated the cloning of DNA fragments ever since the advent of recombinant DNA technology. This vector system is based on DNA from prokaryotic origin and has been controversial since its first use because alien DNA (from bacteria or viruses) is transferred to plants during the transformation process. More recently, vector systems based on fragments identified within the specific genome of the crop species (intragenic) have been constructed for plant transformation research. These vectors are called intragenic vectors. They are constructed by assembling functional equivalents of vector components in plant genomes. Consequently, the target DNA transfer is accomplished without the inclusion of alien DNA (any DNA concomitantly transferred originates from the plant genome (not from prokaryotes). Though the standard protocols of molecular biology are used, the products of transformation using intragenic vectors are non-transgenic

The preferred method of gene transfer is via *Agrobacterium* mediation. To construct an intragenic vector from plant-derived sources for this mode of gene transfer requires plant-derived:

- T-DNA region with two T-DNA border-like sequences in the proper orientation and also suitable restriction sites for cloning the target gene.
- Origins of replication to allow the multiplication of the vector in bacteria.
- Selection system to select for the presence of the vector in bacteria.

If transfer is to be accomplished via direct DNA uptake, only plant-derived origins of replication and selectable marker system are required for multiplication in E. coli.

Intragenic vectors have distinct advantages over intergenic vectors. Because alien DNA is not incorporated in the transformation process, these vectors are suited for highly targeted genetic improvement of plants. Gene transfer into an adapted (elite) cultivar can be accomplished in one step without linkage drag (inclusion of unwanted neighboring genes) that occurs in traditional breeding (e.g., backcrossing). New gene configurations can be designed and constructed whereby promoters are switched to regulate gene expression (gene silencing) and also the possibility of inserting an additional locus to effect gene pyramiding. The intragenic vector approach could be especially useful for breeding of clonally propagated and cross-pollinated species in which inclusion of a new gene by traditional breeding is a challenge. This technique only introduces minor rearrangements of the endogenous genome, something that can occur naturally or via mutagenesis.

The question that now has to be asked is whether or not the use of intragenic vectors for gene transfer would more acceptable to those who oppose transgenic products on ethical grounds (even if a transgene is transferred, the amount of alien DNA is reduced). Further, would there be the need to exempt products of intragenic vectors from the restrictions imposed on GM products? A case for exemption has been made for products resulting from cisgenesis (see later in this chapter) because the genetic modification derives from natural gene(s) from a sexually compatible plant. However, the gene transfer process still uses standard transfer vectors that contain alien DNA. It is clear that combining cisgenesis and intragenic vector approaches will be most desirable in achieving genetic engineering without inclusion of alien DNA. There are other efforts to eliminate foreign DNA from gene transfers, including the recombinase-mediated auto-excision of transgenes during pollen development, by a tightly regulated microspore-specific promoter to eliminate either selectable marker genes or all of the transgenes.

25.8 Cisgenesis

The transfer of genes across very wide taxonomic boundaries (e.g., from bacteria to plants), called **transgenesis**, remains a controversial issue in society. **Cisgenesis** is the term coined by Schouten and colleagues to represent the transfer of genes within the traditional gene pools available to plant breeders. That is, breeders engineer plants using genes derived from the plant itself or from a close sexually compatible relative. The product of transgenesis is a **transgenic plant**; the product of cisgenesis is called a **cisgenic plant**. Similarly, the genetic construct transferred via transgenesis is called a **transgene,** while that from cisgenesis is called a **cisgene**. Similar terminologies for cisgenesis or cisgenic transfers include intragenesis, "intragenic", "all-native DNA", and "P-DNA". However, the originators of the concept of cisgenesis would rather refer to intragenesis and cisgenes as "sisters" rather than identical twins.

Originally, Schouten and colleagues defined a cisgenic plant as a "crop plant that has been genetically modified with one or more genes (containing introns and flanking regions such as native promoter and terminator regions in a sense orientation) isolated from a crossable donor plant". Since then, some authors have presented modifications to this definition that have caused some confusion. This confusion may be clarified by distinguishing between cisgenesis and **intragenesis**. In cisgenesis, the gene has its native promoter, introns and terminator (i.e., it is a complete DNA copy of a natural gene including its promoter and terminator). Contrarily, intragenesis has no requirements regarding introns or terminators. In fact, the promoters and coding sequences may be newly reconstituted from the genetic elements that are drawn from the sexual compatibility group.

There is debate about whether or not a cisgene can introduce important unanticipated phenotypic changes that do not occur in the wild type parents or conventionally bred counterparts. Some argue that cisgenic plants should not be classified like a transgenic plant and subjected to the same regulation as the latter because the former is close to traditional breeding.

25.9 Zinc finger proteins

A zinc finger protein or motif is a sequence of small protein structures (fingers) that contains one or more zinc ions, which are crucial for structural stability, and has the capacity to bind to a specific DNA sequence (notably transcription factors – proteins that bind to DNA and control the transcription of information to RNA). A significant difference between zinc fingers and oligonucleotides, which bind to single-stranded DNA (e.g., helix-turn-helix and leucine zipper) at the Watson–Crick interface (they make use of the twofold symmetry of the double helix), is that the former bind to the major groove of the double-stranded DNA (dsDNA), as they recognize DNA base pairs from the side (they can be linked linearly in tandem to recognize nucleic acid sequences of varying lengths). This unique mode of binding enables sequence-specific interaction to occur without the need to pry the DNA strands apart. Their modular property allows researchers to design them to bind to specific sequences. In addition to being a novel protein fold, zinc fingers provide a novel principle of DNA (or RNA) recognition.

The classical structure of zinc fingers is the *Cys2-His2* structure. The finger is a self-contained domain that is stabilized by a zinc ion that is ligated to a pair of cysteines and a pair of histidines, and by an inner hydrophobic core. The most common zinc fingers consist of an alpha helix and beta sheet). This is actually a novel protein fold. Structurally, there are three general types of zinc fingers:

(i) **C_2H_2 zinc finger.** characterized by the sequence $CX_{2-4}C$..............$HX_{2-4}H$ (where C = cysteine, H = histidine, and X = any amino acid).

(ii) **C_4 zinc finger.** characterized by the consensus sequence of $CX_2CX_{13}CX_2CX_{14-15}CX_5CX_9CX_2C$.

(iii) **C_6 zinc finger.** with the consensus sequence of $CX_2CX_6CX_{5-6}CX_2CX_6C$.

These naturally occurring proteins (abundant in eukaryotes) can be purposefully re-engineered to target DNA sequences of interest and encourage a cell to repair genetic mutations (as alternative gene therapy). Synthetic versions can be engineered (zinc finger nuclease or ZFN – has an engineered zinc finger binding domain fused to a restriction endonuclease, or DNA-cleaving enzyme). For example, a zinc finger can be engineered to cut at the site where a disease-causing gene is located, and then the cell's own DNA repair machinery can take over and repair the damage, based on the template engineered by the researcher, thereby correcting the disease problem. Alternatively, the proteins can be engineered to down regulate the overexpression of a gene that contributes to a disease process.

Zinc fingers are a useful tool for promoting site-specific recombination of DNA. They can be engineered to be used as artificial transcription factors. They can be used to precisely modify engineered report genes in plants. In spite of the advances in modern agricultural biotechnology, targeted genome modification remains intractable. Plant trait engineering is laborious, time consuming and unpredictable. Conventional random mutagenesis and transgenesis are largely inefficient gene targeting tools, limiting the success of researchers at dissecting plant function and engineering crop plants. High frequency ZFN-stimulated gene targeting at endogenous plant genes has been achieved in some species.

Zinc fingers afford research a creative approach to DNA editing. This partly because a restriction nuclease that is capable of cutting dsDNA (can be engineered to induce double-stranded breaks in specific DNA sequences) can be inactivated by splitting into domains and regain its activity by simply putting the domains back together. The domains are complementary and are linked to the zinc fingers that bind sequences that flank the site of interest. Where these sequences occur together, the two zinc fingers bind to each other such that the nuclease domains are together, restoring functionality while cutting at a predetermined site without modifying the genome. This capability gives zinc fingers high reliability and specificity (precision), whether they are used to inactivate or edit a selected genetic component.

25.10 Engineering pest resistance

Crop domestication and subsequent plant breeding efforts have stripped plants of their natural protection that served them well in the wild but reduced their usefulness to humans in modern times. To increase productivity, crops must be artificially protected in modern production, hence the use of pesticides. Pesticide use has environmental consequences. The creation of plants that are able to act as pesticides would eliminate the need for toxic chemicals that have adverse side effects. Genetic engineering procedures have been used to accomplish this goal.

25.10.1 Engineering insect pest resistance (*Bt*)

The most significant and widespread applications of recombinant DNA technology in practical plant breeding to date are the development of *Bt* cultivars and herbicide tolerance in plants. The soil-borne

bacterium, *Bacillus thuringiensis*, is the source of the gene used in the development *Bt* products. The gene encodes the inactive form of a protein, *Bt* toxin, this is toxic to various herbivorous insects when ingested and converted to it toxic form (delta endotoxin) in the gut of the insect. Over 100 different variation of the Bt toxin have been identified, as well as a variety of associated target insect specificity. The toxins classified as *Cry1a* group target Lepidoptera or butterfly group, while the toxins in *Cry3* group target beetles. Scientists have cloned the bacterial genes, which are then transferred into plants to provide resistance to target pest, thereby eliminating the need for pesticides. The major crops that have received such treatment include corn, cotton, and potato.

Researchers are pursuing additional naturally occurring insecticidal compounds as alternatives to the Bt technology. These included chitinase, lectins, alpha-amylase inhibitors, cystatin, and proteinase inhibitors.

25.10.2 Engineering herbicide resistance

GMOs engineered for herbicide resistance are among the major applications of biotechnology in plant food biotechnology.

Why engineer herbicide resistant crops?

A successful herbicide should destroy weeds only, leaving the economic plant unharmed. Broad-spectrum herbicides (non-selective) are attractive but their use in crop production can be problematic, especially in the production of broadleaf crops such as soybean and cotton. There is a general lack of herbicides that will discriminate between dicot weeds and crop plants. Pre-plant applications may be practical to implement. However, once the crop is established and too tall for the safe use of machinery, chemical pest management becomes impractical. Grass crops (e.g., wheat, corn) may tolerate broadleaf herbicides better that the reverse situation. Consequently, when cereal crops and broadleaf crops are grown in rotation or adjacent fields, the broadleaf plants are prone to damage from residual herbicides in the soil, or drift from herbicides applied to grasses. When a crop field is infested by weed species that are closely related to the crop (e.g., red rice in rice crop or nightshade in potato crop), herbicides lack sensitivity enough to distinguish between the plants.

To address these problems, one of two approaches may be pursued: (i) the development of new selective post emergent herbicides or (ii) genetic development of herbicide resistance in crops to existing broad-spectrum herbicides. The latter strategy would be advantageous to the agrochemical industry (increased market) and farmers (safer alternative of pesticides that are already in use). New herbicides are expensive to develop and take time.

Modes of action and herbicide resistance mechanisms

Most herbicides are designed to kill target plants by interrupting a metabolic stage in photosynthesis. Because all higher plants photosynthesize, most herbicides will kill both weeds and desirable plants. Plants resist phytotoxic compounds via one of several mechanisms:

- The plant or cell does not take up toxic molecules because of external barriers such as cuticles.
- Toxic molecules are taken but sequestered in a subcellular compartment away from the target (e.g., protein) compounds the toxin was designed to attack.
- The plant or cell detoxifies the toxic compound by enzymatic processes into harmless compounds.
- The plant or cell equipped with resistance genes against the toxin may produce a modified target compound that is insensitive to the herbicide.
- The plant or cell overproduces the target compound for the phytotoxin in large amounts such that it would take a high concentration of the herbicide to overcome it.

25.10.3 Concerns with the deployment of GM cultivars

Environmental impact

One of the complaints launched against the deployment of GM crops was that using plants as "pesticides" could have ecological consequence by being harmful to non-target organisms. This concern has not been significantly substantiated, at least in the case of *Bt* products. The tendency is to increase the dose rate of pesticides as pests develop resistance to them. However, this is counterproductive in that such an action only serves to intensify the selection pressure for resistant pests.

Pest resistance

Repeated and widespread use eventually results in resistance of pests to pesticides. The high rates of pesticide use create a situation for intense selective pressure that increases resistance to pesticides. Also, whereas older pesticides were designed to attack multiple sites in their target organism, modern pesticides are designed to be more specific in action (often one biochemical pathway), thereby increasing the chance for development of resistance.

Resistance of insects to the defense mechanisms of plant is well-documented. Resistance to a number of commonly used insecticides has been reported in hundreds of insects and mites. There has been the concern that, sooner or later, resistance of insects to *Bt* cultivars would surface. Laboratory resistance to *Bt* has been demonstrated for some major pests, such as the tobacco budworm, Colorado potato beetle, and the diamondback moth. Some reports have indicated field resistance for the diamondback moth to *Bt*. Also, resistant populations of the bollworm (*Helicoverpa zea*) have been reported in some fields in Mississippi and Arkansas.

Researchers are investigating how to extend the *Bt* durability in transgenic cultivars. Current approaches include the engineering of cultivars with very high levels of insecticidal crystal proteins. This way, only insects that have a high-level resistance gene can survive after feeding on these new cultivars. Another approach is to search for new insecticidal genes for developing new transgenic plants that can express multiple insecticidal genes that target different sites in the insect. Insects that can overcome this strategy are those with multiple resistance genes.

Resistance to herbicides is also growing, the confirmed cases approaching 300. The concern for growers and researchers is that some pests are resistance to multiple pesticides, while some are resistant to all the pesticides that are legally approved for their control!

25.11 Is there a future in breeding GM cultivars?

In spite of the protestation from certain quarters of society against GM cultivars, the acreage to these high-tech cultivars increased 60 times in the last decade in the United States, which is one of the world's major producers of GM crops. It is estimated that about 70% of all food sold in the United States contains GM ingredients. In 2005, the GM acreage in the United States was about 50 million hectares, representing about 55% or the global crop acreage. By 2006, 100 million acres were planted to GM cultivars worldwide. For there to be a future in the continuing breeding of GM crops, there has to be a sustained interest in their adoption by growers and acceptance by consumers. Studies available in the United States indicate that growers of GM cultivars do not realize higher profits than counterparts using conventional seed. In herbicide tolerant soybean, for example, researchers concluded that using GM seed resulted in lower weed management and herbicide costs but this was offset by the higher cost of the seed and slightly lower yield. This notwithstanding, growers were attracted by convenience of using GM seed. A USDA report covering 1996–2003 indicated that the adoption of GM cultivars by growers actually was associated with an increase in the use of pesticides.

Key references and suggested reading

Conner, A.J. and Jacobs, J.M.E. (2006). GM Plants without Foreign DNA: Implications from New Approaches in Vector Development, in *Proceedings of the 9th International Symposium on the Biosafety of Genetically Modified Organisms: Biosafety Research and Environmental Risk Assessment* (ed. A. roberts), Jeju Island, Korea, pp. 197–203.

Ahloowalia, B.S., Maluszynski, M., and Nichterlein, K. (2004). Global impact of mutation derived varieties. *Euphytica*, **135**:87–204.

Conner, A.J., Glare, T.R., and Nap, J.P. (2003). The release of genetically modified crops into the environment: II. Overview of ecological risk assessment. *Plant Journal*, **33**:19–46.

Townsend, J.A., Wright, D.A., Winfrey, R.J., et al. (2009). High-frequency modification of plant genes using engineered zinc-finger nucleases. *Nature*, **459**:442–445.

Mlynárová, L., Conner, A.J., and Nap, J.P. (2006). Directed microspore-specific recombination of transgenic alleles to prevent pollen-mediated transmission of transgenes. *Plant Biotechnology Journal*, **4**:445–452.

Nuffield Council on Bioethics (1999). *Genetically modified crops: the ethical and social issues.* Nuffield Council on Bioethics, London.

Nielsen, K.M. (2003). Transgenic organisms: time for a conceptual change. *Nature Biotechnology*, **21**:227–228.

Nap, J.P., Metz, P.L.J., Escaler, M., and Conner, A.J. (2003). The release of genetically modified crops into the environment: I. Overview of current status and regulations. Plant Journal, **33**:1–18.

Porteus, M.H. (2009). Plant biotechnology: Zinc fingers on target. *Nature*, **459**:337–338.

Rommens, C.M. (2004). All-native DNA transformation: a new approach to plant genetic engineering. *Trends Plant Science*, **9**:457–464.

Rommens, C.M., Humara, J.M., Ye, J.S. *et al.* (2004). Crop improvement through modification of the plant's own DNA. *Plant Physiology*, **135**:421–431.

Schouten, H.J., Krens, F.A., and Jacobsen, E. (2006a). Do cisgenic plants warrant less stringent oversight? *Nature Biotechnology*, **24**:753.

Schouten, H.J., Krens, F.A., and Jacobsen, E. (2006b). Cisgenic plants are similar to traditionally bred plants. *EMBO Reports*, 7:750–753.

Outcomes assessment

Part A

Please answer the following questions true or false.

1 CaMV 35S is a constitutive promoter.
2 A transgene cannot function without a promoter.
3 Restriction enzymes have a common recognition site.

Part B

Please answer the following questions.

1 What is cisgenesis?
2 What is a transgenic organism?
3 What is a transgene?
4 Describe how microprojectile bombardment is used in biotechnology.
5 Give two examples of scorable markers used in biotechnology.
6 What is a promoter?
7 Discuss the types of promoters used in genetic engineering.
8 What is genomics?
9 What is bioinformatics and what is its role in biotechnology?

Part C

Please write a brief essay on each of the following questions.

1 Discuss how breeding of GM cultivars differs from breeding conventional cultivars.
2 Describe gene transfer by *Agrobacterium* mediation.
3 Give the general characteristics of restriction endonucleases.
4 Discuss plasmid vectors used for rDNA research.
5 Describe the method of gene transfer by biolistics.
6 Describe how the antisense technology works in plant breeding.
7 What are zinc finger proteins and how are they used in plant breeding?

Section 8

Marketing and societal issues in breeding

Chapter 26 Performance evaluation for crop cultivar release
Chapter 27 Seed certification and commercial seed release
Chapter 28 Regulatory and legal issues
Chapter 29 Value-driven concepts and social concerns
Chapter 30 International plant breeding efforts

Plant breeders use scientific principles and methods to design and create new and improved cultivars that meet the needs of various consumers. To protect their inventions from unauthorized use, breeders may seek a variety of legal protections. Further, before new cultivars reach farmers, the breeder's product must go through steps of seed multiplication, regulation, and release. Methods used by breeders may be controversial and, hence, embroiled in social debate. Also, the laws that govern the activities of plant breeders may vary from one nation to another. In this section, the student will be introduced, among other things, to social and legal issues associated with plant breeding.

26

Performance evaluation for crop cultivar release

Purpose and expected outcomes

The ultimate goal of the plant breeder is to be able to identify a superior genotype that can be released as a new cultivar to farmers for commercial production. To arrive at this goal, many experimental genotypes of high genetic potential are evaluated for performance capabilities under various environmental conditions, over several seasons and years, and at different locations. Detailed records are compiled and analyzed to help in the decision process. The breeding materials are evaluated by using appropriate statistical tools that entail design of the trials, collection of data, analysis and interpretation of results. This calls for an understanding of how genotypes interact with the environment and field plot technique. After studying this chapter, the student should be able to:

1 Discuss the concept and role of genotype × environment interaction in plant breeding.
2 Discuss field plot technique in plant breeding.
3 Discuss the process of crop registration.

26.1 Purpose of performance trials

As indicated in all the breeding schemes discussed in this book, the breeder conducts performance or field trials of the advanced generation of the materials developed in a breeding program, primarily to identify a genotype to be released as cultivar to producers. In a sense, these tests or field trials are designed to forecast the performance of the genotype to be released as a cultivar. These trials are usually called **yield trials** because yield is usually the most important trait in a breeding program. Breeding programs are frequently undertaken to address "secondary" traits (disease resistance, early maturity, high seed protein, etc.). If these traits are successfully transferred but the genotype has poor yield, it would not be released as a cultivar. The field trial also enables the breeder to collect data about the characteristics of the potential cultivar, for other uses (e.g., registration of the cultivar). It is the primary source of information for the breeder to use in the decision making process of cultivar release.

Principles of Plant Genetics and Breeding, Second Edition. George Acquaah.
© 2012 John Wiley & Sons, Ltd. Published 2012 by John Wiley & Sons, Ltd.

The breeder is not only concerned about the level of yield of the cultivar but also the stability of yield. Consequently, stability analysis is often part of the performance evaluation of genotypes prior to cultivar release. The concepts of adaptation and yield stability analysis are further discussed later in this chapter.

26.2 Kinds of field trials

There are two basic kinds of field trials conducted in plant breeding – **breeder's trial** and **official trial**.

26.2.1 Breeder's trials

These trials are conducted for the primary purpose of evaluating the performance of the final set of genotypes (advanced generations in a breeding program) to allow the breeder to make a decision as to which genotype to release as a cultivar. Some breeders conduct these trials in two stages. The first stage, called the **preliminary yield trial (PYT)**, starts at an earlier generation (e.g., F_6, depending on the objectives and method of breeding) and consists of a larger number of entries (genotypes). Further, these entries may be planted in fewer rows per plot (e.g., two rows without borders) and fewer replications (2–3) than would be used in the final trial, the **advanced yield trial (AYT)**. Superior genotypes are identified for more detailed evaluations. The PYT is designed to be a quick evaluation of the breeding efforts. The PYT consists of a fewer number of promising genotypes (10–20), depending on resources. It is conducted for several years at different locations, using more replications and plots with more rows and with borders rows. It is also subjected to more detailed statistical analysis.

Breeder's trials vary in scope, according to the crop, its distribution and importance, and resources available to the breeder. Some breeders (especially in the public sector) limit their evaluations to within the state or mandate region. Commercial breeders may conduct regional, national, and even international trials through established networks. Public breeders may have wide networks for trials (e.g., INTSOY – the International Soybean program in the United States). In terms of management, breeder's trails may also be conducted in one of two ways – **research managed** or **farmer managed**.

Research managed

This is a trial conducted at a research station or experimental farm under the supervision of researchers. These are usually replicated, full scale, and self-contained trials designed to collect data that can be published in a scientific publication. Extensive data are often collected.

Farmer managed

The trials are conducted on farmers' fields (see participatory breeding in Chapter 30). Often, it is a scaled down version of what is conducted at a research station (fewer replications, fewer plots, etc.), so that the farmer may not be overburdened. The breeder may use some creative analysis to obtain valid data from these trials. For example, different farmers in the same location may be considered as a block in a randomized complete block design.

26.2.2 Official trial

After a genotype has been identified as a potential cultivar, the breeder may seek legal protection by applying for protection under the Plant Variety Protection (PVP) Act (Chapter 28) and/or registration of the cultivar with an official seed agency. This trial is more detailed than the performance trial for yield and provides information needed to establish legal identity for the cultivar, showing its distinctness from existing ones. Data must be collected to also indicate its uniformity and stability (i.e., the genotype breeds true from year to year) (cultivar registration, Chapter 27).

26.3 Designing field trials

One of the phases of plant breeding in which statistical analysis is used extensively is the design and conduct of performance evaluations. The key considerations in the design of a field trial are:

- **Number of genotypes to evaluate.** As previously indicated, PYTs have more entries than AYTs. Whereas research-managed trials have the full complement of genotypes, farmer-managed trials may be reduced to a small number.
- **Where to conduct the trials (locations).** Breeders usually conduct trials at multiple locations. These

locations, ideally, should be representative of the target region for which the cultivars are to be released. In practice, test locations are seldom randomly selected. Breeders are limited to sites where they have collaborators (e.g., institutes, research stations, universities) or farmers who have interest in participating in the project. Where possible, the breeder should endeavor to test at both research stations (where optimal selection environment can be obtained) as well as sites that reflect the major cropping areas and farming practices. Even when trials are conducted at research-managed sites, efforts should be made to replicate the actual production conditions in the farmer's field (e.g., crop management practices). Research institutions often strategically locate a few research farms in target regions that represent the climatic and soil conditions of the area. The total number of sites is variable (about 5–10), but it depends on the extent of variability in the target region. Areas of major production should have more sites than those with less production of the crop.

- **What statistical design to use for field layout.** Randomized complete block designs are commonly used in breeder's trials. Research-managed trials may adopt more sophisticated designs, but farmer-managed trials should be as simple as possible. The former should have more replications than the latter, as previously stated.
- **What data to collect.** Researchers at experimental stations may use equipment and machinery designed for research (e.g., plot planter and combines). They have the time and the expertise to collect a wide variety of data in addition to yield. Farmer-managed trials should be designed to permit farmers to utilize existing equipment and machinery already at their farms. Also, data collection, if it is to be done by the farmer, should be made easy with as few as possible collected. Some breeders sometimes request a farmer to make available the land for the trial. Planting, management and data collection are done by the researcher.
- **Number of seasons/years to conduct the trial.** For effective evaluation of genotype × location (G × E) interaction, at least two years of testing (more for repeatability) are needed for annual crops.
- **How to analyze and interpret the analysis to draw valid conclusions.** The breeder may use more efficient designs based on incomplete blocks. There may be unbalanced data (e.g., missing plots). Analyzing data over seasons, years, and locations is complex. The breeder should have the software for these

and other analyses (e.g., stability analysis) and be familiar with statistical methods (or at least have assistance to correctly analyze and interpret the results).

Before describing the steps involved in conducting field trials, some key concepts that are critical to the design and analysis of such experiments are discussed here.

26.4 The role of the environment in field trials

The terms site and location are used interchangeably to indicate spatial variation. The term environment is used to represent the conditions under which plants grow and includes locations, years, and management practice adopted. A location/year constitutes one environment. The nature and effect of the environment has implications in the design and conduct of field trials. Test environments may be artificial (e.g., different levels of fertilizer) or natural (e.g., seasons, location) or both.

26.4.1 Types of environmental variables

The environmental variables that plant breeders face during genotype evaluation may be divided into two general categories – predictable and unpredictable factors.

Predictable factors

Predictable environmental factors are those which occur in a systematic fashion or can be controlled and manipulated by the breeder. These include natural variables such as soil type (e.g., clay, sandy, organic, etc.) that are immutable over a short period and breeder-imposed variations (e.g., planting dates, intra- and inter-row spacing, rates of fertilizer or irrigation application). Breeders may design studies to evaluate each of the imposed factors separately or several simultaneously. Variations in the soil are managed through the way plots are oriented, their shape, and sizes, as well as how the breeder-imposed variations (called **treatments**) are allocated to plots in the field.

Unpredictable factors

Unpredictable factors of the environment are those which vary erratically over short or long periods. The local weather (the short-term meteorological characteristics of a place) is more fickle and relatively unpredictable than climate (the long term patterns in meteorological characteristics of a region). As previously indicated, climate is the basis of crop adaptation; weather is the basis for crop production. Key unpredictable environmental factors of interest in genotype evaluation include rainfall, temperature, and relative humidity. To evaluate the effect of these factors, breeders test their materials at different locations (genotype × location) or in different years (genotype × years) or a combination of these factors (genotype × location × years).

26.4.2 Scale

Another way in which environments are categorized is according to scale.

Microenvironment

This is the immediate environment often pertaining to the organism (plant). It includes soil and meteorological factors (e.g., light, moisture, temperature), and biotic factors in a limited space, intimately associated with the organism.

Macroenvironment

This refers to the abiotic and biotic factors on a larger scale (location, region) at a particular time.

26.5 Genotype × environment (G × E) interaction

Genotype × environment (G × E) interaction is said to occur when two or more genotypes are compared across different environments and their relative performances (responses to the environment) found to differ. That is, one cultivar may have the highest performance in one environment but perform poorly in others. Another way of stating this is that, over different environments, the relative performance of genotypes is inconsistent. G × E is a differential genotypic expression across multiple environments. The effect of this interaction is that the association between phenotype and genotype is reduced. This raises the important issue of adaptation, because a breeder's selection in one environment of superior performers may not hold true in another environment. By measuring the G × E interaction, the breeder will be better equipped to determine the best breeding strategy to use to develop the genotype that is most adapted to the target region.

26.5.1 Classification of G × E interaction

The type of G × E interaction influences the nature of the cultivar the breeder eventually releases for the production region. The environment, as previously described, can be complex; so can the genotype of the plant. Consequently, the biological basis of G × E interaction is complex by nature. Environmental factors are constantly changing. The interaction between the genotype (cultivar) and the environment is ongoing. As the number of environments (n) and number of genotypes (m) increase, the number of possible G × E interactions is given by $mn!/m!n!$ Of this, there is theoretically only one genotype that is the best performer under all environments, odds that make a search for it futile. Allard and Bradshaw classified this interaction into three common patterns using two genotypes (A, B) and two environments (E_1, E_2) for a graphical illustration of the concept of G × E interaction. In statistical terms, a G × E interaction has occurred when the difference in performance between the two genotypes is inconsistent over the environment:

$$A_1 - B_1 \neq A_2 - B_2 \ (\text{or } A_1 - B_1 - (A_2 - B_2) \neq 0$$

A G × E interaction exists when:

$$A_1 - B_1 - A_2 + B_2 \neq 0$$

Three basic types of G × E interaction – **no interaction, non-crossover interaction** (quantitative interaction), and **crossover interaction** (qualitative interaction) – are recognized. A numeric example can be used to distinguish these classes of interactions (Table 26.1). A graphical illustration may also be used to demonstrate the nature of these interactions (Figure 26.1). Consider two genotypes, A and B, in a field trial analysis.

Table 26.1 Demonstration of G × E interaction.

(a) No interaction

	Environment 1	Environment 2	Difference
Genotype A	10	14	+4
Genotype B	16	20	+4
Difference	+6	+6	

(b) Non-crossover interaction

	Environment 1	Environment 2	Difference
Genotype A	10	14	+4
Genotype B	16	24	+8
Difference	+6	+10	

(c) Crossover interaction

	Environment 1	Environment 2	Difference
Genotype A	16	14	−2
Genotype B	10	20	+10
Difference	−6	+6	

1 **No G × E Interaction:** A no G × E interaction occurs when one genotype (e.g., A) consistently performs better than the other genotype (B) by about the same amount across all the environments included in the test.

2 **A non-crossover G × E interaction:** A non-crossover G × E interaction is said to occur when a genotype (A) consistently outperforms genotype B, across the entire test environment. However, the differential performance is not the same across the environment. That is whereas there is no change in rank, genotype A may exceed genotype B by 20 units in one environment and 60 units another.

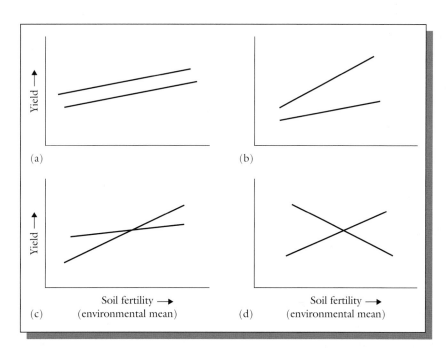

Figure 26.1 Graphical presentation of: (a) genotype × environment (G × E) interaction, (b) heterogeneity, (c) crossover interactions, and (d) combined interactions.

3 A crossover G × E interaction: This is the most important G × E interaction to plant breeders. A crossover G × E interaction occurs when a genotype (A) is more productive in one environment, but a different genotype (B) is more productive in another environment. The basic test for crossover interaction (also called qualitative interaction) is to compare the performance of two genotypes in two environments and to determine if the difference in performance is significantly less than zero in one environment and significantly greater than zero in the other.

4 Combined G × E interaction: The first three interactions previously described are the ones commonly discussed. If the one of the factors considered on the axes increases for one genotype and reduces for the other genotype, there is combined G × E interaction.

The axes in the graph may be for any relevant factor of interest to the breeder. For example, the x-axis may be rainfall, while the vertical axis (y) may be grain yield. In spite of the complexity of the environment, sometimes one factor may predominate to characterize the environment (or may be imposed by design). It should be pointed out these four graphs are only a selected unique few of the numerous patterns that may occur in reality. The breeder is most interested in repeatable G × E interactions.

26.5.2 Measurement of G × E interaction

Interactions occur at various biological levels, such as genotypic, QTL, and phenotypic levels, the first two requiring genetic analysis. G × E interaction at the phenotypic level requires observations at the plant or crop level. The G × E interactions can also be partitioned into linear trends (e.g., G × location, G × year, G × time, etc.). Statistical methods are used to assess G × E interactions. Consequently, the proper field plot design and analysis are required for an effective assessment of the interactions. These methods include both parametric and nonparametric procedures – partitioning of variance, regression analysis, nonparametric methods, and multivariate techniques.

Analysis of variance (ANOVA)

To ascertain the presence of G × E interaction, breeders conduct a network of comparative trials, as previously described, in which prospective cultivars are compared with standard cultivars at multiple locations or agroecological regions. The premise for such trials, according to Mather and others is expressed by a linear equation:

$$X = \mu + g + e + ge$$

where X is the yield of some other quantitative traits, μ is the the mean value of the population (trial), g is the the value of the genotype (cultivar), e is the the value of the environmental effect, and ge is the the genotype × environment interaction.

Different models of ANOVA are used for partitioning variance. The genotypes are never chosen at random since they are deliberately selected by the breeder as prospective cultivars. Similarly, locations are often not randomly chosen, as previously discussed. However, they may be considered random if there are many of them spread over a large region. The genotypes are evaluated over several years. The effects are random since the environment is not controlled. Also, where the genotypes are a random sample from a large population, their effects are random.

For a two-factor mixed model (fixed genotypes + random environment) the ANOVA table is:

Source	df	MS	EMS
Reps (R)	ly(r − 1)		
Years (Y)	y − 1		
Location (L)	l − 1		
Y × L	(y − 1)(l − 1)		
Genotypes (G)	g − 1	M_1	$\sigma^2_e + r\sigma^2_{gly} + rl\sigma^2_{gy} + ry\sigma^2_{gl} + rly\sigma^2_g$
G × Y	(g − 1)(y − 1)	M_2	$\sigma^2_e + r\sigma^2_{gly} + ry\sigma^2_{gl} + rly\sigma^2_g$
G × L	(g − 1)(l − 1)	M_3	$\sigma^2_e + r\sigma^2_{gly} + ry\sigma^2_{gl}$
G × Y × L	(g − 1)(y − 1)(l − 1)	M_4	$\sigma^2_e + r\sigma^2_{gly}$
Error (G × R)	(g − 1)(r − 1)ly	M_5	σ^2_e

Variances are calculated as follows:

$$\sigma^2_c = M_5$$
$$\sigma^2_{gly} = (M_4 - M_5)/r$$
$$\sigma^2_{gy} = (M_2 - M_4)/rl$$
$$\sigma^2_{gl} = (M_3 - M_4)/ry$$
$$\sigma^2_g = (M_1 - M_2 - M_3 + M_4)/rly$$

Breeding implications of ANOVA results

Lack of significant $G \times E$ interactions for genotype × location or location × years indicates that they breeder may be able to select a superior genotype for release for use throughout a specified production region, following genotype evaluations in just one year. Crossover $G \times E$ interactions are those that require careful interpretation by breeders. In this instance decisions are based on the practical significance of the results of the analysis. Breeders need to include in their decision making process factors such as the magnitude of the change in rank. Consequently, it is not uncommon for different conclusions to be drawn by different breeders examining the same results. General interpretations of $G \times E$ interactions resulting from unpredictable causes are:

- If significant genotype × location effects are observed and the rankings fluctuate by wide margins, the results indicate that the breeder should consider establishing separate breeding programs for the different locations (i.e., to develop different cultivars for different locations). However, before making a decision, it is wise to examine the data to see what specific factors are responsible for the variation. If stable factors such as soil are the source of variation, separate breeding efforts may be warranted.
- A significant genotype × year interaction is similar in effect to genotype × location. However, because the breeder cannot develop programs for different years, a good decision would be to conduct tests over several years and select the genotype with superior average performance over the years for release. Because conducting one trial per year for more years will prolong the breeding program, the breeder may include more locations and decrease the number of years.
- The breeding implications for a complex interaction like genotype × years × location is for the breeder to select genotypes with superior average performance across locations and over years for release as new cultivars for the production region.

Farmers will benefit from growing more than one cultivar each cropping season. This strategy will reduce the effects of the fluctuations attributed to genotype × year interaction.

- The magnitude of $G \times E$ interaction is influenced by the genetic structure of the genotype. Genotypes with less heterogeneity (e.g., pure lines, single cross hybrids, clones) or heterozygosity (e.g., pure lines) generally interact more with the environment than open-pollinated genotypes or mixtures because of lower amounts of adaptive genes.
- Also, it is widely known that only $G \times L$ interaction (rather than all the kinds of $G \times E$ interactions) is useful for depicting adaptation patterns. This is because it is the only interaction that can be exploited by selecting for specific adaptation or by growing specifically adapted genotypes. For example, a significant $G \times L$ interaction cannot be exploited because the climatic conditions that generate year-to-year environmental variation are not known in advance. Consequently, the analysis of multiple environments yield trials should focus primarily on $G \times L$ interactions, the other interaction being considered in terms of yield stability.

26.6 Stability analysis

An ANOVA identifies statically significant and specific interactions without telling the breeder anything about them. If $G \times E$ interaction is significant and the environmental variations are unpredictable the breeder needs to know which genotypes in the trials are stable. A **stability analysis** is used to answer this question. **Field stability** may be defined as the ability of a genotype to perform consistently, whether at a high or low level, across a wide range of environments. Stability may be static or dynamic. **Static stability** is analogous to the biological concept of homeostasis (i.e., a stable genotype tend to maintain a constant yield across environments). **Dynamic stability** is when a stable genotype has a yield response in each environment such that it is always parallel to the mean response of the genotypes evaluated in the trials (i.e., $G \times E = 0$). Static stability is believed to be more useful than dynamic stability in a wide range of situations, and especially in developing countries. The genotype produces a better response in unfavorable environments or years. Whenever the $G \times E$ interaction variation is wide, breeding for high yield stability is justifiable.

As previously stated, ANOVA only detects the existence of G × E effects. Breeders will benefit from additional information that indicates the stability of genotype performance under different environmental conditions. The stability of cultivar performance across environments depends on the genotype of individual plants and the genetic relationship among them. Generally, heterozygous individuals (e.g., F_1 hybrids) are more stable in their performance than their homozygous inbred parents.

A variety of methods has been proposed for genotype stability analysis. Examples are the regression analysis and the method of means.

(a) **Regression analysis.** This method of simple linear regression (also called **joint regression analysis**) stability analysis was developed by K.W Finlay and G.N. Wilkinson (1963) and later by S.A. Eberhart and W.A. Russell (1966). It is preceded by an ANOVA to assess the significance of G × E. The breeder proceeds to the next step of regression analysis only if the G × E interaction is significant.

Statistically, the observed performance (Y_{ij}) of the ith genotype (i = 1, . . . ,5) in the jth environment (j = 1, ...,e), may be expressed as:

$$Y_{ij} = \mu + g_i + e_j + ge_{ij} + \varepsilon_{ij}$$

where μ is the the grand mean over all genotypes and environments, g_i is the additive contribution of the ith genotype (calculated as the deviation from μ of the mean of the ith genotype averaged over all environments, e is the additive environmental contribution of the jth environment (calculated as the deviation μ of the mean of the jth environment averaged over all genotypes); ge_{ij} is the G × E interaction of the ith genotype in the jth environment, and ε_{ij} is the error term attached to the i_{th} genotype in the jth environment. The regression coefficient for a specific genotype is obtained by regressing its observed Y_{ij} value against the corresponding mean of the jth environment.

If the yield was conducted over a wide range of environments, and hence a wide range of yields obtained, it is reasonable to assume that individual trial means sufficiently summarize the effects of the environments. The mean performance of each genotype over all the test environments constitutes the **environmental index** (Table 26.2). In effect, this method of analysis produces a scale of environmental quality. The results for each genotype are plotted, trial by trial, against trial means, to obtain a regression line. According to Eberhart and Russell, an average performing genotype will have a regression coefficient of 1.0 and deviations from regression of 0.0, since it will tend to agree with the means. However, the genotypes that were responsible for the G × E interactions detected in the ANOVA, will have slopes that are unequal to unity. Furthermore, a genotype that is unresponsive to environments (i.e., a stable performer) will have a low slope (b < 1) while a genotype that is responsive to the environments (i.e., an unstable performer), will have a steep slope (b > 1).

The regression analysis technique has certain limitations, both physiological and statistical, which can result in inaccurate interpretation and recommendations for cultivar release. For example, the use of mean yield in each environment as an environmental index flouts the statistical requirement of independent variable. Also, when regression lines do not adequately represent data, the point of intersection of these lines has little biological meaning, unless considered with the associated statistical error, something often ignored by many breeders. Limitations notwithstanding, the regression technique is simple and has biological significance. Complex interactions are reduced to linear responses.

(b) **Plot of means versus coefficient of variation.** Proposed by Francis and Kannenburg in 1978, this method entails calculating, for each variety, the overall mean and the coefficient (CV) of variations across the environments. A plot of means versus CVs yields a scattergram that can be divided into four sections by transecting the average CV and the grand mean yield (Figure 26.2). The most desirable genotype will be found in Group I (high yield low CV) while the least desirable (low yield, high CV) will occur in Group IV.

26.7 Non-parametric methods

Multivariate procedures used to analyze G × E interactions include clustering, PCA, factor analysis. These procedures perform uniformly across environments. A recent addition to these techniques is the additive mean effects multiplicative interaction (AMMI). These procedures are discussed in Supplementary Chapter 2.

Table 26.2 Stability analysis.

(a) ANOVA

Source	df	SS	MS	F	Probability
Environment (E)	4	27103.87	6775.96	104.03	0.0001
Reps (R) × E	15	22580.65	1505.37	23.11	0.0001
Genotype (G)	9	4595.65	510.65	7.84	0.0001
E × G	36	6068.63	168.57	2.59	0.0001
Pooled Error	135	8792.85	65.13		

(b) Regression analysis

Genotype	Mean	Regression coefficient
N1	105.2	0.906
N2	103.3	0.759
N3	108.3	1.741
N4	99.9	0.972
N5	101.9	0.559
N6	108.5	1.141
N7	106.7	0.926
N8	103.7	0.999
N9	94.8	0.968
N10	112.9	1.028

Genotypes N2 and N5 are stable performers; N3 and N6 are responsive to the environment and unstable in performance; N8 and N10 are average performers.

(c) Environmental index

N1	93.47
N2	96.3
N3	112.4
N4	123.6
N5	96.7

26.8 Adaptation

According to P.M.A Tigerstedt, in the context of evolutionary biology, **adaptation** is a process, **adaptiveness** is the level of adaptation of a plant material to an environment, while **adaptability** is the ability to show adaptedness in a wide range of environments. However, according to Cooper and Byth, in a plant breeding context, adaptation and adaptedness relate to a condition rather than a process. They indicate the ability of a plant material to be high yielding with respect to a specific environment, one to which it is adapted. One of the purposes of a G × E analysis is to help breeders decide whether to breed for narrow or wide adaptation or adaptability. Breeding for narrow adaptability, the breeder's goal is to release a cultivar for a specific part of the target region (with a unique set of conditions), whereas breeding for wide adaptability, the

breeder focuses on releasing a cultivar with high performance across all environments.

As previously stated, only G × L interaction is useful for determining genotype adaptation patterns in an ANOVA analysis. Further, only repeatable G × L interactions are of practical importance. However, if the G × L variance is small in magnitude compared to other sources of variance, even though significant, it minimizes the specific advantage for breeding for narrow adaptation.

26.9 Field plot technique in plant breeding

The subject of **field plot technique** deals with decision making processes and other activities conducted by the plant breeder to fairly and effectively evaluate genotypes from a breeding program, to select cultivars for release. Field trials are designed to evaluate

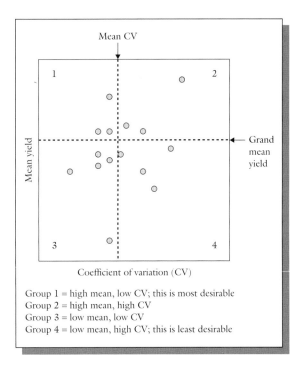

Figure 26.2 G × E interaction based on coefficient of variation (CV).

genotypes in a natural environment that is similar to that which farmers operate in. To evaluate the genetic potential of genotypes, they must be grown in an identical environment, so that any observed differences can be attributed to the genotype not chance variations. Unfortunately, outside of a controlled environment condition such as a greenhouse, it is impossible to find a truly homogeneous environment in the field. One of the goals of field plot technique is to evaluate genotypes error free (or in practice, with minimum error). It is important, therefore, to identify the source of error in a field study.

To collect data in the proper manner for analysis and decision making, the breeder follows a set of statistically-based rules for laying out an experiment in the field, called the **experimental design**. A decision not only allows the breeder to estimate random error in the field but also the choice of a good design can minimize this error.

26.9.1 Sources of experimental error

In an experiment a researcher may deliberately impose conditions or factors that would generate variance in outcomes, such a factor is called a **treatment**. The

unit to which a treatment is applied is the **experimental unit**. In the field, there are plots (of rows of plants). In the greenhouse, a unit could be a pot. For example, applying different levels of fertilizer, planting at different spacing, or evaluating different genotypes (as in breeding), as the research objective might be, are treatments that elicit variable response from plants. However, in addition to the expected variation from a treatment, there is always some variation that is unintended and cannot be accounted for; this is known in research as **experimental error**. By definition, this is the variation among plots treated alike. Major sources of error are those due to soil heterogeneity, operator's inability to conduct the experiment uniformly as intended and interplant competition.

Soil (site) variability

Soils are naturally heterogeneous, some more so than others. Natural variation may originate from differences in soil minerals, soil moisture, organic matter, or topography. The tops of slopes are drier than the bottom parts; nutrients wash down and accumulate at the bottom; depressions may drain poorly. In addition to these natural sources of soil variation, humans create additional variation through how they manage the soil in production. Before selecting and using a site for genotype evaluation, the breeder should know about previous use of the land (history of use). The parcel of land of interest may have been differentially managed (e.g., different plant species were grown, or different tillage practices were imposed). The breeder should use all available data (e.g., yield records, management records) and visual observations to identify general patterns of variations at the proposed site. Experimental design techniques and other tactics can be used to minimize the effects of soil heterogeneity in field trials.

As previously, indicated, breeders usually have selected sites at which they conduct their yield trials (e.g., experimental stations of universities). Are locations then a fixed variable in ANOVA? Some argue that locations are random effects, since the breeder has no control over the meteorological factors that occur at locations. Most breeders also consider years of testing random effects. It is important to have at least two replications of each treatment (genotypes) in each trial for estimating error variance.

Industry highlights

An example of participatory plant breeding: barley at ICARDA

S. Ceccarelli and S. Grando

The International Center for Agricultural Research in the Dry Areas (ICARDA), PO Box 5466 Aleppo, Syria

Decentralized selection, defined as selection in the target environment, has been used by ICARDA's barley breeding program to avoid the risk of discarding useful lines because of their relatively poor performance at the research stations. Decentralized selection is a powerful methodology to fit crops to the physical environment. However, crop breeding based on decentralized selection can still miss its objectives if its products do not fit the farmers' specific needs and uses.

Participation of farmers in the initial stages of breeding, when the genetic variability created by the breeders is untapped, will fully exploit the potential gains from breeding for specific adaptation through decentralized selection by adding farmer's perception of their own needs and farmers' knowledge of the crop. Therefore, farmers' participation has been the ultimate conceptual consequence of a positive interpretation of genotype × environment interactions, that is, of breeding for specific adaptation (Ceccarelli, 2009).

At ICARDA, the gradual change from centralized non-participatory to decentralized-participatory barley breeding was implemented in Syria between 1997 and 2003 in three steps, and the model and concepts developed during these developments were gradually applied in Tunisia, Morocco, Eritrea, Ethiopia, Yemen, Jordan, Iran, and Egypt.

Step 1: selection phase

The first step was an exploratory step with the main objectives of building human relationships, understanding farmers' preferences, measuring farmers' selection efficiency, developing scoring methodology, and enhancing farmers' skills. The exploratory work included the selection of farmers and sites, and the establishment of one common experiment for all participants. The experiment, described by Ceccarelli *et al.* [2, 3], included 208 plots and was grown in two research stations and nine villages. All possible combinations of selection were conducted, namely centralized non-participatory (breeders on station), centralized-participatory (farmers on station), decentralized non-participatory (breeders on farm) and decentralized-participatory (farmers on farm). The results indicated that (i) farmers can handle large number of entries, can take a number of observations during the cropping season, and develop their own scoring methods; (ii) farmers select for specific adaptation; (iii) for some broad attributes, selection is mostly driven by environment; (iv) there is more diversity among farmers' selections in their own fields than among farmers' selections on research stations, and among breeder's selections, irrespective of where the selection was conducted; (v) the selection criteria used by the farmers are nearly the same as those used by the breeder; and (vi) in their own fields, farmers are slightly more efficient than the breeder in identifying the highest yielding entries – the breeder is more efficient than the farmer in selecting in the research station located in a high rainfall area but less efficient than the farmers in research stations located in a low rainfall area. Therefore, the first step indicated that there is much to gain, and nothing to lose, in implementing a decentralized participatory plant breeding program.

Step 2: methodology

The second step was mostly about methodologies. It consisted of the implementation of the breeding plan, the choice and testing of experimental designs and statistical analysis, the refinement of farmers' selection methodology (Ceccarelli and Grando, 2007), and, eventually, initiating village-based seed production activities.

From a breeding point of view, the major features of the second phase were (i) a different role for the two research stations, one of which was not used while the second, located in an area with more reliable rainfall, was used for seed multiplication; (ii) the increase in the number of farmers involved in the project; and (iii) the initiation of village-based seed production. The details of the second phase, such as number of lines, plot size, type of germplasm, selection criteria, and seed production issues, were discussed in meetings with farmers in each village. The host farmers and a number of neighbors attended these meetings that were organized by the host farmers. In the case of the type of germplasm, the farmers generally expressed preferences for the seed color (black or white) and the row type. In one village farmers wanted to test the breeding lines in two different rotations (barley–barley and vetch–barley), and in another village in deep and shallow soil. The model of plant breeding we use in Syria and in a number of other countries is a bulk-pedigree system, in which the crosses are done on station, where we also grow the F_1 and the F_2, while in the farmers' fields we

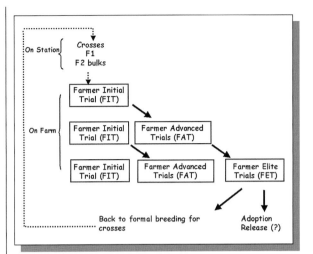

Figure B26.1 The scheme of decentralized participatory barley breeding implemented in Syria. Figure courtesy of S. Ceccarelli.

yield test the bulks over a period of three years (Figure B26.1) starting from the F_3 bulks in trials called Farmers Initial Trials (FIT), which are unreplicated trials with 165 entries, 5 common checks and 30 systematic check plots (with one or two check cultivars). This allows the evaluation of 165 new breeding materials every year.

In parallel, we conduct on station pure line selection by collecting heads within the F_3 bulks selected by the farmers. The F_4 head rows are promoted to the F_5 screening nursery only if farmers select the corresponding F_4 bulks. The process is repeated in the F_5 and the resulting families, after one generation of increase, return as F_7 in the yield testing phase. Therefore, when the model is fully implemented, the breeding material that is yield tested includes new bulks as well as pure lines extracted from the best bulks of the previous cycle.

The breeding materials selected from the FIT are yield tested for a second year in the Farmer Advanced Trials (FAT); these are grown by between four and eight farmers in each village. Within a village the FAT contain the same entries but the type and the number of entries and checks vary in each village. The number of FAT in each village depends on how many farmers are willing to grow this type of trial. Each farmer decides the rotation, the soil type, the amount and the time of application of fertilizer. Therefore, the FAT are planted in different conditions and managements. During selection farmers exchange information about the agronomic management of the trials and rely on this information before deciding which lines to select. Therefore, one of the advantages of the program is that the lines start to be characterized for their responses to environmental or agronomic factors at an early stage of the selection process.

The entries selected from the FAT are yield tested for a third year in the Farmer Elite Trials (FET), which are grown by between four and eight farmers in each village. These entries are also used on station as parents in the crossing program. The three types of trials are planted by scientists using plot drills and are entirely managed by the farmers.

During selection some farmers are assisted by a researcher (Figure B26.2). Some farmers do selection at various stages but the majority do selection when the crop is close to full maturity, and using a scoring method from 0 = discarded to 4 = the most desirable, farmers expressed their opinion on each individual entry.

In each trial, the scientists record plant height, spike length, grain yield, total biomass and straw yield, harvest index, and 1000 kernel weight. On station scientist record days to heading and days to maturity. The data are subjected to different types of analysis, such as the spatial analysis of un-replicated or replicated trials (Singh *et al.*, 2003). The environmental standardized Best Lineal Unbiased Predictors (BLUPs) obtained from the spatial analysis are used to analyze Genotype × Environment Interactions with the GGEbiplot software (Yan *et al.*, 2000).

One farmers' concern was the seed multiplication of the selected lines. Farmers requested a full control of this operation to avoid mechanical mixture. To address this concern, we established, in four of the eight villages, small seed units consisting of a seed cleaner, which also treat the seed with fungicides against seed-born diseases. The unit has a limited capacity (about 400 kg/h) but allows farmers a full control of the seed quality of their selections in the various stages of the breeding program. This is the first step towards the creation of village-based seed production activities.

Figure B26.2 Farmers in Eritrea doing visual selection assisted by researchers. Figure courtesy of S. Ceccarelli.

Step 3: project extension

We soon recognized that the work described above would not be able to reach a large number of villages and farmers and, hence, have an impact at national level. Therefore, in the third phase the emphasis was on institutionalization and scaling-up.

The first step in this direction was the organization of a Workshop, with the participation of the farmers, researchers (including heads of research stations of Agricultural Offices and Research Policy makers), the Seed Organization, the Extension Service and the Minister of Agriculture. The discussions covered the relationships between participatory plant breeding (PPB), seed production, and variety release. The mechanism agreed upon for scaling-up PPB was a gradual transfer of responsibilities from ICARDA scientists to the General Commission for Scientific and Agricultural Research (GCSAR) scientists and the staff of the Extension Service in a way that at the end of the process each province would implement all the various PPB activities within its boundaries, with the overall coordination shared between ICARDA and the Ministry of Agriculture. Therefore, one component of the initial steps of scaling-up was a training program for the researchers and extension staff on all the aspects of PPB.

As a result the PPB program was extended from five to seven provinces and from 11 to 25 villages (Figure B26.3) with between 15 and 30 farmers per village Such a large network of farmers will facilitate the access of non-participating farmers to the products of PPB, and to their large scale adoption. For this to be possible, village-based seed production will play a key role.

One of the examples of the success that the PPB project is having is offered by the variety Zanbaka, which about 10 years ago went through the conventional system and was rejected from being released. When it entered the PPB program it began to be slowly adopted, until the drought in year 2000 forced the farmers to use all the available seed to feed their sheep. We then distributed 5 t of seed, which was planted on about 50 ha. Within two years the variety has reached 3500 ha in an area receiving 150–250 mm rainfall and where conventional breeding never had any impact. Similar initial successes have been observed in Egypt, where new barley varieties out-yielding the local by between 30% and over 100% are multiplied in four villages, in Yemen, where two varieties of barley and two of lentil have been adopted by farmers, in Eritrea and in Ethiopia.

As more and more of the breeding programs are becoming private, a further development of participatory plant breeding is evolutionary plant breeding, which can be largely handled by the farmers themselves. Evolutionary plant breeding (Suneson, 1956; Ceccarelli *et al.*, 2010) consists in deploying in farmers' fields largely variable heterogeneous populations which, by continuously evolving over time, represent a permanent source of novel diversity adapted to the conditions where the populations evolve as well as to climate changes. This is a dynamic, inexpensive and extremely powerful strategy to specifically adapt crops to climate changes, and in general to all possible agronomic environments.

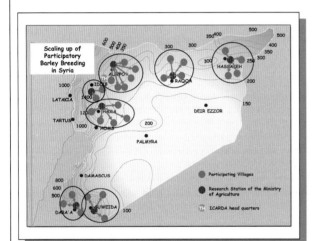

Figure B26.3 The participatory barley breeding in Syria has been extended to seven provinces and 25 villages. Figure courtesy of S. Ceccarelli.

References

Ceccarelli, S. (2009). Main stages of a plant breeding programme, in *Plant Breeding and Farmer Participation* (eds S. Ceccarelli, E.P. Guimaraes, and E. Weltzien). FAO, Rome, Italy, pp. 63–74.

Ceccarelli, S., Grando, S., Tutwiler, R., *et al.* (2000). A Methodological Study on Participatory Barley Breeding. I. Selection Phase. *Euphytica*, **111**:91–104.

Ceccarelli, S., Grando, S., Singh, M., *et al.* (2003). A Methodological Study on Participatory Barley Breeding. II. Response to Selection. *Euphytica*, **133**:185–200.

Ceccarelli, S., and Grando, S. (2007). Decentralized-Participatory Plant Breeding: An Example of Demand Driven Research. *Euphytica*, **155**:349–360.

Ceccarelli, S., Grando, S., Maatougui, M., *et al.* (2010). Plant breeding and climate changes. *Journal of Agricultural Science*, **148**:1–11.

Singh, M., Malhotra, R. S., Ceccarelli, S., Sarker, A., Grando, S., and Erskine, W. (2003). Spatial variability models to improve dryland field trials. *Experimental Agriculture*, **39**:1–10.

Suneson, C.A. (1956). An evolutionary plant breeding method. *Agronomy Journal*, **48**:188–191.

Yan, W., Hunt, L.A., Sheng, Q., and Szlavnics, Z. (2000). Cultivar Evaluation and Mega-Environment Investigation Based on the GGE Biplot. *Crop Science*, **40**:597–605.

Further reading

Ceccarelli, S., Bailey, E., Grando, S., and Tutwiler, R. (1997). Decentralized, Participatory Plant Breeding: a Link Between Formal Plant Breeding and Small Farmers, in *New Frontiers in Participatory Research and Gender Analysis.* Proceedings of an International Seminar on a Participatory Research and Gender Analysis for Technology Development, Cali, Colombia, pp. 65–74.

Ceccarelli, S., Grando, A., Amri, F.A., *et al.* (2001). Decentralized and participatory plant breeding for marginal environments, in *Broadening the Genetic Base of Crop Production* (eds H.D. Cooper, C. Spillane, and T. Hodgink). CABI, New York (USA)/FAO, Rome (Italy)/IPRI, Rome (Italy), pp. 115–136.

Ceccarelli, S., Grando, A., and Capettini, F. (2001). La partecipation de los agricultores en el mejoramiento de cebada en al ICARDAS, in *Memorias de la Conferencia Internacional sobre: Futuras Estrategias para Implementar Mejoramiento Participativo en los Cultivos de las Zonas Altas en la Region Andina* (ed. D. Daniel), September 23–27, Quito, Ecuador, pp. 25–54.

Ceccarelli, S., Grando, S., Bailey, E., *et al.* (2001). Farmer Participation in Barley Breeding in Syria, Morocco and Tunisia. *Euphytica*, **122**:521–536.

Ceccarelli, S., and Grando, S. (2002). Plant breeding with farmers requires testing the assumptions of conventional plant breeding: Lessons from the ICARDA barley program, in *Farmers, scientists and plant breeding: Integrating Knowledge and Practice* (eds D.A. Cleveland and D. Soleri). CABI Publishing International, Wallingford, UK, pp. 297–332.

Ceccarelli, S., Guimaraes, E.P., and Weltzien, E. (eds.) (2009). *Plant Breeding and Farmer Participation.* FAO, Rome, Italy.

Ceccarelli, S., Galié, A., Mustafa, Y., and Grando, S. (2012). Syria: Participatory barley breeding—farmers' input becomes everyone's gain, in *The Custodians of biodiversity: sharing access and benefits to genetic resources* (eds M. Ruiz and R. Vernooy). Earthscan, Abingdon, UK, pp. 53–66.

Soleri, D., Cleveland, D.A., Smith, S.E., *et al.* (2002). Understanding farmers' knowledge as the basis for collaboration with plant breeders: methodological development and examples from ongoing research in Mexico, Syria, Cuba and Nepal, in *Farmers, scientists and plant breeding: Integrating Knowledge and Practice* (eds D.A. Cleveland and D. Soleri). CABI Publishing International, Wallingford, UK, pp. 19–60.

Tactics for reducing experimental error

To correctly and effectively evaluate the desired variation, the researcher should eliminate or, more realistically, minimize extraneous variation. Some errors come from natural soil variability whereas others are human in origin. The plant breeder may use certain field plot techniques to minimize experimental errors.

1 **Use of border rows:** Different genotypes differ in various plant characteristics to varying extents (e.g., growth rate, size, height, nutrient, and moisture uptake). When planted next to each other, interplot completion may cause the performance of one genotype to be influenced by another in the adjacent plot. In early generation or preliminary trials, which usually include large numbers of genotypes, breeders often use fewer rows (e.g., two rows) in planting a plot to save resources. In advanced yield trails, four row plots are customarily used. Data are collected on the middle rows only, because they are protected from **border effects**. Some breeders minimize border effects by increasing interplot spacing or using a common genotype for planting the border of all plots in a test in which row plots are few (one or two rows). To reduce interplot competition, the materials may be grouped according to competitive abilities

2 **Proper choice of plot size and shape:** Several factors affect the optimum plot size to use in field evaluation of genotypes (breeding objectives, stage of breeding, resources available, equipment). Evaluation of an F_2 segregating population is often based

on individual plant performance (i.e., individual plants are essentially plots). Consequently, the plants are adequately spaced to allow the breeder enough room to examine each plant.

Some breeders use what are called **microplots**, planting hills or short rows of test plants. This tactic is used in early stages of genotype evaluation as a quick and inexpensive way of eliminating inferior genotypes.

Row plots are commonly used by breeders for genotype evaluation. The size and shape depends on the plant species, the land available, and the method of harvesting. Generally, row plots are rectangular in shape.

3 **Adequate number of replication:** Replication or repetition of treatment in a test is critical to statistical analysis, providing a means of estimating statistical error. The number of replications used usually varies between two and four; fewer replications may be used in early evaluation of genotypes while advanced yield tests usually have four replications. The number of replications depends on the accuracy desired in the analysis, resources available (land, seed, labor), and the statistical design used. Replications increase the precision of the experiment and help the plant breeder to more effectively evaluate the genotypes to identify superior ones.

4 **Minimizing operator errors:** Where a machine or equipment is to be used, it should be properly serviced (cleaned, calibrated). The plots should be planted uniformly and managed uniformly (i.e., uniform spacing, fertilizing, irrigating). Border rows should be planted uniformly. Each plot is not enclosed in border rows. The plants at the end of rows have a competitive advantage over those in the inner part of the rows. Mechanized harvesting usually starts at the first plants of the middle row and proceeds to the last plants. This may introduce yield inflation and may require adjustment of plot yields.

26.9.2 Principles of experimental design

There are many experimental designs used to allocate treatments to experimental units. However, they are all based on three basic principles – replication, randomization, and local control.

Replication

The principle of replication is critical for estimating experimental error, as previously indicated. It is also used for reducing the magnitude of error in an experiment.

Randomization

This is the principle of assigning treatments to experimental units such that each unit has an equal opportunity of receiving each treatment. This action eliminates bias in the estimation of treatment effects and makes the experimental error independent of treatment effects, a requirement for valid test of significance of effects. Systematic arrangement allocates treatments to experimental units according to a predetermined pattern.

Local control

Sometimes, researchers find it more efficient to impose restrictions on randomization to further minimize experimental error. This is appropriate when there is a gradient in an environmental factor (e.g., fertility, moisture). Fertility is different at the top of a slope than at the bottom as previously indicated. Rather than ignoring this obvious variation, a technique called **blocking** may be used to divide the field into distinct areas, maximizing the variation between blocks and increasing the homogeneity within blocks. Statistical analysis is then used to extract interblock variation, thereby reducing the total error in the experiment.

26.10 Field plot designs

Plant breeders use experimental designs to arrange genotypes in a trial to minimize experimental error. The designs vary according to purpose of the evaluation, the nature of the genotypes (e.g., segregating or non-segregating), number of genotypes, stage of a breeding program (e.g., preliminary or advanced yield trials), and resources.

26.10.1 Evaluating single plants

No design arrangement

Breeders using certain methods may select among segregating plants, starting in the F_2 generation, on single plant basis. Generally plants are spaced in a completely random arrangement.

Advantages
- Inexpensive and easy to conduct.
- Large number of genotypes can be evaluated at one time.

Limitations
- If a large land area is involved, the chance of soil heterogeneity effect increases. Inferior plants growing in fertile soil may outperform superior plants growing in less fertile spots in the field.
- It is suitable for evaluating plants on the basis of traits with high heritability but less effective for evaluating traits with low heritability.

Modifications. It is helpful to plant rows of standard cultivars in adjacent plots for comparison, to aid in efficient and effective selection of superior genotypes.

Use of experimental designs

Grid Design First proposed by C.O. Gardener, the grid design entails subdividing the land into smaller blocks. The rationale is that smaller blocks are likely to be more homogeneous than larger blocks. Plants are selected based on comparison among plants within each block only.

- **Advantages**
  - Reduces effects of soil heterogeneity.
  - No precise arrangements needed, hence conventional plot planters can be used.
  - Suitable also for selecting on the basis of plant traits with low heritability.
  - Easier and more effective to compare plants within a small group than in a large group.
  - Selection intensity can be varied by selecting more than one plant per block.

Honeycomb (hexagonal) design Proposed by A. Fasoulas, a key feature of this design is a planting arrangement in which each plant is equidistant from others in a hexagonal pattern. Furthermore, the spacing is determined to remove interplant competition. Plants are selected if only they are superior to all in their hexagonal units. The selection intensity can be varied by widening the hexagon. When a plant in the immediate hexagon out-yields its surrounding six plants, it is selected by 14.3% selection intensity. If it out-yields the 18 plants within the second concentric hexagon, it is selected by 5.3% selection intensity, and so on. When using this method in breeding, the unit

of selection at all stages in the breeding program are single plants, not plots.

- **Disadvantages**
  - Conventional equipment is not usable for planting.
  - It is more complex to conduct.

26.10.2 Evaluating multiple plants

Unreplicated tests

When conducted, unreplicated tests often entail planting single rows of genotypes with check cultivars strategically located for easy comparison. A breeder may use this design to evaluate a large number of genotypes quickly, to eliminate inferior ones prior to more comprehensive field trials.

- **Advantages**
  - It saves space.
  - It is less expensive to conduct.
  - Large number of genotypes can be quickly evaluated.
- **Disadvantages**
  - Susceptible to the effects of field heterogeneity. There is no other plot of the genotype in the test for confirming performance. Poor soil may mask the genotypic potential of a superior genotype by causing it to perform poorly.
  - Experimental error estimation is problematic.

Replicated tests

Replication in field plot technique entails the representation of a particular entry (genotype) multiple times (usually 2–4) in a test. With multiples of each entry, the important and critical design consideration is how to arrange duplicates of genotypes in the field. The statistical concept of randomization requires treatment allocation to be by chance, such that each genotype has an equal chance of being allocated to each available plot. Even though randomization may not be imposed on certain occasions for practical reasons, plant breeders normally use randomization in advanced trials.

Different types of experimental design are used to conduct replicated trials. Designs impose varying degrees of restriction on randomization. A major consideration in plant breeding research is the number of entries to include in an evaluation. As stated elsewhere, plant breeding is a numbers game. The numbers are larger in the early part of the program. Three categories of experimental designs are used in plant breeding.

(i) **Complete block designs.** These designs are suited to evaluating a small number of entries. Each block contains at least one complete set of entries (genotypes). That is, the number of replications and the number of blocks are the same.

(ii) **Incomplete block designs.** These designs are suited for evaluating a very large number of entries. Under such conditions, complete blocking is impractical because of the large numbers. Instead, each block contains only part of the complete set of entries being evaluated in the study. Hence, the number of replications and the number of blocks are not the same.

(iii) **Partially balanced designs.** These designs are generally complex to use. Some pairs of treatments occur in the same block an equal number of times and hence comparisons among treatments are not equally precise.

Complete block designs

(a) **Completely randomized design (CRD).** This design assumes that the entire experimental area is homogeneous, hence there is no need for local control.
  - **Advantages**
    ○ It is the simplest of the designs to use and analyze.
    ○ It yields the highest number of degrees of freedom for error, making the error of the relatively lowest magnitude compared to other designs.
    ○ Missing data, when they occur, do not complicate the analysis.
  - **Disadvantages**
    ○ It is not conducive for field studies for obvious reasons of soil heterogeneity.

A typical ANOVA table for RCD is:

Source	df	SS	MS	F
Treatment	$k-1$	SS_{Tr}	$SS_{Tr}/k-1 = (M_{Tr})$	MS_{Tr}/M_E
Error	$N-k$	SS_E	$SS_E/(N-k) = (M_E)$	
Total	$N-1$	SS_T		

(b) **Randomized complete block design (RCBD).** In this design, there are as many blocks as there are replicates. Each replicate of a genotype is represented in each block. This design is suitable for a test involving a small number of entries. It is the most widely used experimental design in plant breeding.
  - **Advantages**
    ○ It is flexible, being applicable to small as well as moderately large entries.
    ○ It is relatively easy to conduct. The field layout and statistical analysis are relatively straight forward.
    ○ Unbiased error can be effectively estimated.
  - **Disadvantages**
    ○ It is applicable to evaluations in a homogeneous environment.
    ○ It is not suitable for a large number of entries

A typical ANOVA for RCBD is:

Source	df	SS	MS	F
Between treatment	$k-1$	SS_{Tr}	$SS_{Tr}/k-1 = (M_{Tr})$	MS_{Tr}/M_E
Between blocks	$b-1$	SS_B	$SS_B/b-1 = (M_B)$	M_B/M_E
Error	$(k-1)(b-1)$	SS_E	$SS_E/(k-1)(b-1) = (M_E)$	
Total	$bk-1$	SS_T		

(c) **Latin square.** In Latin square design, blocking is used to control environmental variation in two directions. It is best used when field variation occurs in two directions, perpendicular to each other. A layout is a follows:

		Column blocks			
		I	II	III	IV
Row blocks	1	A	B	C	D
	2	B	A	D	C
	3	C	D	A	B
	4	D	C	B	A

Treatment $= 4 + 2$ blocking factors $= (2 \times 2 \times 4 = 16$ units); however, for a factorial design it will be $4 \times 4 \times 4 = 64$ units.
  - **Advantages.** It reduces the number of subjects, as previously indicated (from 64 to 16 subjects), thereby reducing the cost of the investigation.

- **Disadvantage.** It is best used when the entries are less than 10.

Incomplete block designs

(a) **Lattice design.** It is a preferred design for evaluating a large number of genotypes.
  - **Advantages**
    o Allows unbiased error to be estimated to determine the origin (genetic or environmental) of observed variation among genotypes.
    o Provides more effective comparison among genotypes.
  - **Disadvantage**
    o Randomization and statistical analysis for design can be challenging (without the use of a computer).

26.11 Materials, equipment, and machinery for field evaluation of genotypes

Plant breeding, even on a small scale, entails the evaluation of large numbers of genotypes. Modern technology has enabled certain aspects of the field testing process to be mechanized. Some of the materials, equipment, and machinery commonly used in plant breeding field tests are as follow.

26.11.1 Materials

(i) **Field plan.** A notebook or field record book is prepared to show the layout of plots according to the experimental design. The layout also shows the treatment assignments to plots.

(ii) **Labels/stakes.** Computer software such as MSTAT will allow the plant breeder to select a design, allocate treatment to plots, print labels for seed envelopes, print a field layout, and print record books for the study, among other options. Stakes (wooden, plastic) are prepared to identify the plots.

(iii) **Seed envelopes.** The computer labels are pasted on paper envelopes. Some breeding programs use cloth bags, in which case labels are pasted on card labels and fastened to the sack by means of wires. Breeders may use permanent ink pens to prepare their labels.

(iv) **Seed treatment.** Treatment of the seed with fungicides may be needed. A bucket (or an appropriate container) may be needed for the process.

(v) **Record books.** Data collection is facilitated by having record books printed according to the field plan and traits to be scored.

(vi) **Statistical package for computers.** Data collected from the field is analyzed on the computer using appropriate software.

26.11.2 Equipment

(i) **Seed Counter.** Based on the plot length and spacing, seed packets are prepared containing the appropriate number of seeds for the appropriate number of rows.

(ii) **Seed trays/boxes.** To facilitate planting, the seed envelopes are prearranged in order (according to the field layout) and set in trays. The trays are appropriately numbered.

(iii) **Plot planters.** Mechanized planting is necessary for a large breeding program. Customized planters may be purchased for specific crops. Often, plot planters, designed for various types of crop (e.g., small grains, bean) may be obtained and adjusted appropriately for the grain size and spacing desired. In the absence of tractor-based planting, other smaller motorized planters are available for various crops.

(iv) **Harvesting.** Plot combines are available for harvesting, threshing, and bagging of seeds from small plots. The preparation of seed packets (labels, bags, etc.) is required for harvesting.

(v) **Computer/data loggers.** Some breeders may be able to computerize their field data collection to enter data directly into the computer. Where this is not possible, data from field record books are entered into the computer at a later date.

26.12 Crop registration

After the formal release of the cultivar, it may be **registered**. In the United States, this voluntary activity is coordinated by the Crop Science Society of America (CSSA).

26.12.1 Objectives of crop registration

According to the CSSA, crop registration is designed to inform the scientific community of the attributes and availability of the new genetic material, and to provide readily accessible cultivar names or designations for a given crop. Further, crop registration helps to prevent duplication of cultivar names. Complete guidelines for crop registration may be obtained from the CSSA. Excerpts of these guidelines are discussed in this chapter.

26.12.2 What can be registered?

Over 50 crops and groups of crops may be registered. Other categories include grasses, legumes, oilseeds. Subcommittees have been established to review the registration manuscripts for various crops. Hybrids may not be registered. Eligible materials may be cultivars, parental lines, elite germplasm, genetic stocks, and mapping populations. The cultivar to be registered must have demonstrated its utility and provide a new variant characteristic (e.g., disease or insect resistance, tolerance to stress). For breeding lines (parental lines), the breeder must include information on sources of cytoplasm and restoration information. Germplasm needs not be commercially valuable in its present form, but must possess a demonstrable merit (e.g., unique trait, exotic background) that has potential for commercial utility when used in a breeding program. Genetic stocks are primarily used for basic genetic studies. They should be useful and unique. Key mapping populations should have high intrinsic value and utility (e.g., may be used to establish species-representative or landmark molecular maps).

26.12.3 Registration procedure

First, the material to be registered should have been released by the breeder or the organization. A seed sample must be deposited in the National Seed Storage Laboratory (NSSL) prior to or at the time of submission of the manuscript. A registration packet is obtained, completed and submitted to the appropriate subcommittee responsible for the crop. The

packet includes a manuscript to be prepared by the breeder and a signed copy of the NSSL Storage Application Form.

The manuscript should include the following information (see CSSA guidelines for details):

1 Name or identification assigned at time of release.
2 Scientific name (complete).
3 Experimental number or designation during development.
4 Names of agencies, organizations, or institutions involved in the development and evaluation; names of those officially releasing the plant materials.
5 Brief description of the material (including distinguishing features from like types); breeding procedures; pedigree;, comparative performance data (if applicable).
6 Probable regions of adaptation, generations of seed increase, and area of production for cultivars.
7 Agency or institution with responsibility for maintenance of the basic stock of these materials.
8 Any limitation on availability of the materials

Mapping populations have additional specific requirements.

The manuscript is required to be prepared using the ASA Style Manual and following specific instructions regarding order of topics, spacing, font and other instructions as customary for publishing in scientific journals.

26.13 Variety protection

In addition to registration, a breeder may seek legal protection of the cultivar in one or several ways, as discussed in detail in Chapter 28. A common protection, the Plant Variety Protection, or the Plant Breeder's Rights, is a *sui generi* (of its kind) legal protection, as discussed.

Key references and suggested reading

Annicchiarico, P. (2002). Genotype × environment interaction. Challenges and opportunities for plant breeding and cultivar recommendation. Plant Production and Protection paper 174, FAO, Rome, Italy.

Ceccarelli, S. (1994). Specific adaptation and breeding for marginal conditions. *Euphytica*, 77:205–219.

Cooper, M., and Byth, D.E. (1996). Understanding plant adaptation to achieve systematic applied crop improvement – A fundamental challenge, in *Plant adaptation and crop improvement* (eds M. Cooper and G.L. Hammer). CABI, Wallingford, UK, pp. 2–23.

Eberhart, S.A., and Russell, W.A. (1966). Stability parameters for comparing varieties. *Crop Science*, **6**:36–40.

Finlay, W., and Wilkinson, G.N. (1963). The analysis of adaptation in a plant breeding programme. *Australian Journal of Agricultural Research*, **14**:742–754.

Koemel, J.E. Jr., Guenzi, A.C., Carver, B.F., Payton, M.E., Morgan, G.H., and Smith, E.L. (2004). Hybrid and pure line hard winter wheat yield and stability. *Crop Science*, **44**:107–113.

Tigerstedt, P.M.A. (1994). Adaptation, variation, and selection in marginal areas. *Euphytica*, **77**:171–174.

Tollenaar, M., and Lee, E.A. (2002). Yield potential, yield stability and stress tolerance in maize. *Field Crops Research*, **75**:161–169.

Yan, W., and Tinker, N.A. (2005). An integrated biplot analysis system for displaying, interpreting, and exploring genotype × environment interaction. *Crop Science*, **45**:1004–1016.

Yan, W., and Hunt, L.A. (2001). Interpretation of genotype × environment interaction for winter wheat yield in Ontario. *Crop Science*, **41**:19–25.

Outcomes assessment

Part A

Please answer these questions true or false.

1 A breeder is required to register his or her cultivar upon release.
2 A non-crossover G × E interaction indicates a lack of differential performance.
3 Hybrids may not be registered.

Part B

Please answer the following questions.

1 Give the organization responsible for crop registration in the United States.
2 Give the purpose of a field trial in plant breeding.
3 Give the kinds of cross that can be registered.
4 Distinguish between the breeder's trial and the official trial.

Part C

Please write a brief essay on each of the following topics.

1 Discuss the G × E interactions in plant breeding.
2 Discuss the importance of stability analysis in plant breeding.
3 Discuss the crop registration process.
4 Discuss how plant breeders minimize experimental error in field trials.

27

Seed certification and commercial seed release

Purpose and expected outcomes

The ultimate goal of a plant breeder is to make available the new cultivar that has been developed to crop producers to be multiplied for use by consumers. This tail end of the plant breeding process constitutes a huge industry with its customs and regulations. After the breeder has conducted yield trials and selected the genotype for release as a commercial cultivar, there are certain customary steps that are followed by the breeder and other appointed certifying agencies to release the cultivar into the public domain for access by consumers. Farmers can purchase seed that has been certified for use by the appropriate authorities. Consequently, plant breeders should be familiar with this process. After studying this chapter, the student should be able to discuss:

1 Discuss the concept of seed certification.
2 Discuss the official agencies and their roles in the seed certification process.
3 Discuss the commercial classes of seed.
4 Discuss the certification process.
5 Discuss the role of improved seed in agriculture.

27.1 The role of improved seed in agriculture

Seed is critical to the success of the agricultural enterprise. In traditional agriculture, seed is usually an integral part of the farmer's operation. The technology used in such production systems is essentially developed *in situ* – seed is obtained from the crop produced on the farm; additional fertility is obtained from livestock. In modern agriculture, however, seed development and production is a separate enterprise. Farmers are supplied seed by seed companies. Agrochemicals are widely used to provide the needed cultural environment for the high yield potential of the improved cultivars to be fully expressed for optimal crop productivity.

Principles of Plant Genetics and Breeding, Second Edition. George Acquaah.
© 2012 John Wiley & Sons, Ltd. Published 2012 by John Wiley & Sons, Ltd.

27.1.1 Yield gains

The major food crops of the world (wheat, corn, sorghum, rice, and soybean) and industrial crops (e.g., cotton) are produced by seed. In these major crops, dramatic increases in yield have been observed over the past 70 years due to a combination of the role of plant breeding, mechanization, and improved management practices (especially, pest control, fertilizers). In the United States, most significant increases have been recorded in corn, in which average yield per acre rose from 30 bushels in 1930 to about 70 bushels in 1970 and then 140 bushels by middle 1990s. The trends in other major crops are similar, albeit not as dramatic. Soybean yield rose fourfold between 1930 and 1998, while wheat expanded by a modest 2.5% during the period.

27.1.2 Seed market

The seed market is very lucrative worldwide. The Unites States leads the world in the seed market size, with about $15.7 billion in 1997, representing about 20% share of the world market. In the same year, China and Japan used $3.0 and $2.5 billion of improved seed, respectively. In the Unites States, total seed expenditure rose from $500 million in 1960 to over $6.7 billion in 1997. Soybean and wheat led the total amount of seed used in 1997 with 2.06 million and 3.08 million tons, respectively. Seed use is a factor of acreage planted, seeding rate per acre, cropping practice, and variation in agro-ecological factors. In 1997, seed use in the Unites States for the major field crops totaled 6.5 million tons.

The Unites States is a net exporter of seed, attaining an export growth from $305 million in 1982 to $698 million in 1996. Major importers of US seed are Mexico, Canada, Italy, Japan, France, The Netherlands, and Argentina, accounting for about 72% of the total US export in 1996. On the other hand, US seed import grew from $87 million in 1985 to $314 million in 1996.

27.1.3 Regulations in the seed industry

Plant breeders are particularly at risk of having their creation or invention illegally used by competitors. This is because their inventions (cultivars) or the knowledge that led to them, are readily transported, imitated, or reproduced with little difficulty and at low cost. Once released for sale and planted, it is easy for competitors to have access to the seed. Consequently, there are laws (intellectual property rights) in place to protect seed developers (Chapter 28).

Apart from regulations that protect the plant breeder, there are regulations that protect the consumer. The two key avenues for consumer protection are varietal registration and seed certification, as discussed in this chapter. Also, there are regulations that protect the environment, especially, when the invention derives from genetic engineering.

27.2 Role of the private sector in the seed industry

27.2.1 Early history

The first seed company in North America was established by David Landreth in 1784. He published a seed (vegetables) catalog in 1799. In 1876, W. Atlee Burpee established vegetable and flower seed companies. US crop producers practiced saving seed from the previous crop and sharing seed with neighbors for planting in the early 1900s. Even though there were commercial seed producers, commercial seed was not well patronized by farmers until seed certification programs that provided quality assurance to farmers were introduced. These early entrepreneurs were small-scale family owned enterprises focusing on multiplying improved cultivars for the public breeding system.

The single most influential crop in the growth of the seed industry was corn. Open-pollinated varieties of corn dominated production till the end of the nineteenth century. The development of superior yielding hybrid corn marked the turning point in the shift toward commercial seed by the 1930s. An estimated 150 companies were in operation in the early 1930s, some of them devoting resources to plant breeding in addition to replication of seed in the system. By 1995, an estimated 95% of corn acreage was planted to commercial hybrid seed. Hybrid corn was the primary business of the seed industry by 1944. The domination of the seed industry by the private sector began in the late nineteenth century. Research and development was intensified, resulting in the continuous development of higher performing hybrids.

27.2.2 The growth of the seed industry

The Plant Variety Protection Act of 1970 and other amendments gave the impetus for growth in the seed industry. Heretofore, the private industry mostly had no proprietary rights over their inventions (except for the hybrid industry). This situation provided no incentive for commercial seed companies to develop new cultivars. The landscape of the seed industry has experienced repeated alterations over the past 30 years, as mergers and acquisitions eliminated many smaller companies from the scene and large companies became dominant. A significant fact to note is that many of these acquisitions were made by pharmaceutical and petrochemical companies, many of which were multinational (e.g., Ciba-Geigy, Sandoz, Upjohn, Royal Dutch Shell). However, by the end of the 1980s, many of these chemical and industrial manufacturing companies exited the seed market. By 1989, the industry leaders were Pioneer HiBred, Sandoz, Asgrow and Limagrain.

In an attempt for the exiting companies to increase their market shares for high profits, mergers and acquisitions continued in the 1990s. From this shuffle,

Monsanto, Norvartis, and AgrEvo gained significant grounds, structuring their businesses around the "life science" image (i.e., they developed chemicals, seeds, foods and food ingredients, and pharmaceuticals based on the application of the principles and practices of biotechnology and genetics). In terms of US market shares, the leaders (total sales) in 1997 for the major seed crops were Pioneer HiBred ($1178 million), Monsanto ($541 million), Novartis ($262 million), Dow Agroscience/Mycogen ($136 million), Golden Harvest ($93 million), AgrEvo/Cargill ($93 million), and Delta and Pine Land ($79 million). The total market share for Pioneer was 33.6%.

The shuffling continues, the most recent being the acquisition of Novartis by Monsanto in 2005, and prior to that (in 1999) Syngenta (formerly Novartis/AstraZeneca), Aventis (formerly Hoechst and Rhone Poulenc – later acquired by Bayer), Dupont (incorporated Pioneer), and Monsanto/-Pharmacia merged. An example of the evolution of multinational companies engaged in the seed industry is presented in Figure 27.1, as put together by J. Fernandez-Cornejo of the USDA.

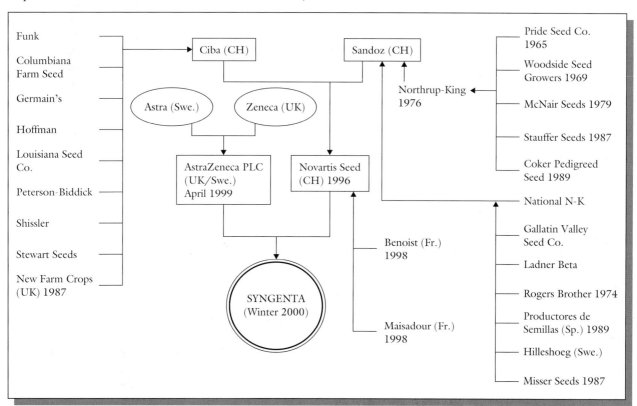

Figure 27.1 The evolution of the Syngenta AG seed company (after Fernandez-Cornejo, 2004).

27.3 General steps of operation of the seed industry

The specific details vary among crop species. However, four basic elements, as identified by Fernandez-Cornejo, occur in the seed industry from development to marketing of the seed:

- **Seed development.** This stage in the seed industry entails research and development of technologies and the application of science to develop new seed. This is the plant breeding stage where improved seed is developed. It is very capital intensive and occurs in both the private and public sectors, as previously noted.
- **Seed production.** Once a cultivar with potential has been developed, certain steps are followed to increase the seed to make it available to farmers. These steps are described in the next section.
- **Seed conditioning.** Seed conditioning is the process of readying the seed for market, whereby the certified seed is properly dried, cleaned, sorted, treated (where applicable), and packaged for sale. Tests for quality standards are conducted (discussed later in this chapter).
- **Seed marketing and distribution.** Seeds are marketed at different levels, including direct marketing by the seed companies, or through licensed distributors. Local distribution may involve farmer-dealers, farmer's association, sales persons of the seed companies, private wholesalers, and retailers.

27.4 The cultivar release process

This is the ultimate goal of a plant breeding program – to release a cultivar with high yield and superior performance to existing cultivars. After analyzing the data from the field test, taking into account the G × E interactions as well as the stability analysis, and considering local issues (e.g., market needs), the breeder will elect to declare one genotype a **cultivar** for release to the agricultural system for use by producers. As previously discussed, this cultivar may be released for specific adaptation or broad adaptation. One of the decisions during cultivar release is naming the new cultivar. Whereas there are no rules for naming cultivars, it is often done thoughtfully and strategically. Some breeders name cultivars in honor of people or special things of meaning to the breeder or the organization. Sometimes, the name may reflect a special attribute (e.g., physical appeal, color, size, nutritional quality, etc.) of the cultivar that producers can readily identify with or it may be a town or locality in the production region. In the case of small grains in the United States, a cultivar name must be approved by the National Small Grains authority.

The specific steps of the cultivar release process vary among countries and even among crops in the same country. In addition to natural seed organizations or committees, there may be regional, state, and crop-specific bodies that oversee the process. The breeder (or organization) is required to submit certain specific data to the appropriate review board for review and recommendation for release. Generally, the committees will evaluate the data to ascertain the genetic distinctness, uniformity, stability of performance, and general agricultural merit.

27.5 Multiplication of pedigree seed

Pedigree seed is seed of a named cultivar that is produced under supervision of a certifying agency for compliance with genetic purity and identity of the original source. After developing, naming, and registering (and securing other legal rights), the breeder makes the cultivar available for commercial multiplication for sale to the public. In introducing a new cultivar into the food production system, the obligation to maintain it in a marketable form so that it can be released at intervals, as needed, in its authentic form for commercial multiplication, is the responsibility of the breeder.

27.5.1 Classes of seed in a certification system

This is a limited generation system for maintenance of pedigree seed. To maintain the original genetic identity and purity, multiplication of a newly released cultivar is limited to four generations or classes of seed, the main difference among them being the quantity of material generated. These classes are **breeder seed, foundation seed, registered seed**, and **certified seed**. These categories essentially represent multiplication classes in the ancestry of certified seed. The Organization for Economic Cooperation and Development (OECD) terminology for these classes are **pre-basic, basic, first generation certified** and **second or successive-generation certified seed**, respectively.

Industry highlights

Public release and registration of "Prolina" soybean

Joe W. Burton

USDA Plant Science Building, 3127 Ligon Street, Raleigh, NC 27607, USA

Release and registration procedures for public soybean varieties will vary somewhat depending on the institution (s) that a soybean breeder is associated with. "Prolina" was developed in a USDA-ARS program in cooperation with the North Carolina Agricultural Research Service (NCARS). Thus, approval from both institutions was required to release the "Prolina" soybean to the public for farm production. The procedures for this are outlined here.

1. **Obtain enough data to demonstrate that the new soybean line is different from other available varieties** and can potentially contribute to soybean production in some geographical area (usually a state or growing region).
2. **Prepare a case justifying release** that includes data summaries with appropriate test statistics comparing the new line with current varieties.
3. **Present this case to the Breeders Release Board** of the College of Agriculture and Life Sciences at North Carolina State University. The Board evaluated the case for "Prolina" and approved its release.
4. **Prepare a release notice** which will be distributed to other soybean breeders (mainly in the USA). The release notice for "Prolina" was prepared showing that it was a release of the USDA-ARS administrator who has oversight for release of germplasm. That person reviewed and approved the notice and added their signature to the notice. (Occasionally, the administrator will require some revision of the notice prior to signing it). The release notice was then sent to the Director of NCARS for his signature.
5. **Distribute the approved and signed notice** to all US soybean breeders. The release notice for "Prolina" was sent directly to all public soybean breeders. It was also sent to the secretary of the Commercial Soybean Breeders organization, who distributed it to the other publicity organizations of USDA-ARS and NCARS, which then prepared and distributed press releases.
6. **Write a registration article** for the journal *Crop Science*. The registration manuscript for "Prolina" was prepared after its release following journal guidelines for registration articles. It was submitted to the editor who handles the review and acceptance of soybean germplasm registrations. It was peer reviewed and accepted contingent on some revisions. After revision, the journal accepted the article, gave "Prolina" a registration number, and published it in *Crop Science* (Figure 27.1B).
7. **Submit seed samples** to the USDA-ARS soybean germplasm collection in Urbana, Illinois, and to the National Seed Storage Laboratory at Fort Collins, Colorado. Seed samples of "Prolina" were sent to both these institutions, as requested.

The Crop Science Society of America then issued a registration certificate for "Prolina" (Figure 27.2B).

Breeder seed

Breeder seed (or vegetative propagating material) is in the direct control of the plant breeder (or the organization) responsible for developing the cultivar. The breeder assumes responsibility for preserving the genetic constitution of the cultivar being the one uniquely familiar with the true genetic identity of the product. This class of seed is expected to have the highest level of genetic purity of any class of seed. The quality on hand is variable but may range from a few pounds to several sacks. It is used to establish the next class of seed (the foundation seed). The breeder seed plot (or especially self-pollinating species) comprises progeny rows of seed from individual plants (called **nucleus seed**) from the previous planting. Upon inspection, all deviant rows (possibly from natural outcrossing) are discarded. Seed from rows that are true-to-type are bulked to form the breeder seed. The nucleus seed may be maintained as stock seed.

Producers and consumers do not have access to the breeder seed, partly because of its small quantity, but mainly because of laws governing the marketing of seed. The breeder's seed is to satisfy the official certification process for purity, quality, health, and uniformity so that the producer has access to high quality seed.

Foundation seed

Foundation seed (or basic seed) is the immediate or first generation increase of the breeder seed. The genetic purity should be very close to the breeder seed. It is used to produce certified seed directly or through registered seed. It is produced under supervision (e.g., of breeder or developer of the cultivar or their representative).

Registered seed

Registered seed is produced directly from breeder seed or from foundation seed. It is the source of certified seed. It is often grown by farmers under contract with a seed company. The material is subject to inspection both in the field and as harvested seed. Farmers may commercially plant this seed.

Certified seed

Certified seed may be produced from foundation, registered, or certified seed. It is grown in isolation under prescribed conditions for the crop such that it meets the genetic identity and purity of the cultivar required for approval by certifying agency upon both field and seed inspection. This class of seed is what is usually available for planting by producers.

27.5.2 Maintaining genetic identity of the breeder seed

Properties of the breeder seed

The breeder must increase the original seed to have enough to store and supply to growers in the commercial seed multiplication chain. This seed (breeder seed) must be of high purity, quality, health, and uniformity.

- **Purity.** The seed must be free from contamination from other cultivars of the same species, as well as other foreign materials (e.g., weed seeds, pollen). The breeder seed is the most authentic and genetically pure seed. To obtain this high level of purity, the seed must be produced in a location distant from all sources of possible contamination and harvested and processed with care (using clean equipment).
- **Quality.** High quality seed is one that is fully mature and produced under optimal environment so that all the traits of the cultivar are optimally expressed. It should be harvested at optimal moisture content.
- **Health.** The seed should be free from diseases and damage from insects. This is difficult to achieve, especially if the disease is seed-borne.
- **Uniformity.** Cultivar uniformity depends on the kind of cultivar that was bred. Where clonal cultivars, hybrids, and inbred lines can be uniform to a great extent, synthetics and other open-pollinated cultivars are less uniform.

Causes of loss of genetic purity of seed

Genetic deterioration of the breeder seed may arise from several sources:

- **Mechanical admixture.** Failure to clean planting, harvesting and processing machinery and equipment used in seed production may result in physical transfer of seeds of one cultivar to another. When the contaminated cultivar is planted, more of the offending seed will be produced to further pollute the genetic purity of the cultivar of interest (unless the offending genotype has distinct morphological features that would enable the breeder to readily rogue them how, should that be desired, before harvesting).
- **Natural outcrossing.** Natural outcrossing by wind or insects may occur where the new cultivar is planted close to another cultivar. As previously indicated, the breeder seed should be planted in isolation to avoid or decrease the chance of natural outcrossing with other cultivars in the vicinity.
- **Mutations.** Spontaneous mutations may arise during multiplication of seed.
- **Growing in area of non-adaptation.** Growing the cultivar in an area to which it is not adapted could induce developmental variations.

Prevention of loss of genetic-purity

- Plant in isolation: planting the breeder seed in isolation will reduce the opportunities for outcrossing.
- Rogue out off-types: all atypical material should be rogued out of the field before harvesting.
- Plant only in areas of crop adaptation to reduce incidence of developmental changes.
- Enforce quality control measures (e.g., sanitation, cleaning of equipment).

27.6 Concept of seed certification

Seed certification is a legal mechanism established to ensure that pedigree seed produced by a plant breeder reaches the public in its highest quality, original genetic identity, and highest genetic purity. A certified seed is required to be identified by a label that attests to the quality of the product by guaranteeing its genetic constitution, level of germination, and the results of seed analysis. Certification is conducted by certifying agencies according to prescribed guidelines for the crop and the Association of Official Seed Certifying Agencies (AOSCA) (most certifying agencies are members of this agency). It includes field inspection and seed inspection of crops grown under prescribed conditions. Following a request by the AOSCA, the Federal Seed Act of 1970 was enacted to set the minimum standards for all certified seeds produced in the United States.

Usually, seed certification agencies in the United States are organized as state crop improvement associations that operate in collaboration with state agricultural extension stations. The operation often includes agricultural extension services and the state department of agriculture. The expenses of their operation are covered partly by the inspection and certification fees they charge growers.

The specific guidelines vary among crop species and regions of production. For example, the kinds and proportion of weed seed in commercial seed that is tolerable varies from one region to another. Certification services are available for field crops, turf grasses, vegetables, fruits, vegetatively propagated species, woody plants, and forbs.

27.7 The seed certification process

There are several key events in the seed certification process that includes paper and fieldwork. The specifics may vary among countries.

27.7.1 Application for certification

The developer (breeder, agency, company) wishing to certify a new cultivar must first apply and receive approval from the appropriate authority in the state. In the United States, it is the responsibility of each state to establish the specific protocol for producing each class of seed, as well as the standards for genetic purity. These local standards, however, should not be inferior to those set by AOSCA.

To apply for certification, the applicant is required to supply information including: history or origin of the cultivar, documentation of evaluations conducted in comparison with other cultivars, a detailed description of how the classes of seed (breeder, foundation, registered, certified) will be maintained, the number of generations that can be grown from breeder seed (or number of harvests from one generation in case of perennial crops), source of seed for planning and cropping history of the land.

27.7.2 Source of seed

The grower must use a seed class such as foundation or registered seed of an approved cultivar for initiating the certification process. The seed requirement is that the starting seed must produce in the next generation a class of seed that can be verified by a crop certificate (e.g., foundation seed to produce registered seed or registered seed to produce certified seed).

27.7.3 Site selection (land)

A key requirement is that the land should not be a source of contamination from volunteer plants or noxious weeds. To this end, the land must not have grown the crop in the recent (five years) past (unless the previous crop was the same cultivar and of properly certified class). Both primary and secondary noxious weeds (especially those whose seeds are hard to separate from the crop seeds) are intolerable. The site should be adequately isolated to exclude contamination. This is especially critical for open-pollinated species.

27.7.4 Management in the field

Once planted, the field must be kept free of weeds. Off-types must be rogued out, preferably before they flower.

27.7.5 Field inspection

While growing in the field the crop is subject to inspection by authorized personnel from seed certifying agencies. The inspector looks for the presence of noxious weeds, seed-borne diseases, and any factors that might affect the purity of the cultivar.

27.7.6 Harvesting and processing

The equipment for harvesting must be thoroughly cleaned to avoid mechanical mixtures from occurring. The harvested seed is cleaned, conditioned, and bagged. Representative samples are drawn from each lot by officials from the seed certifying agency. Once determined to meet or exceed the minimum standards set by the association (e.g., Crop Improvement Association), the seed is declared officially certified, and receives the official tag. The tag indicates the results of seed analysis (percentage germination, inert matter, weed seed, etc.).

27.8 Seed testing

The information mandated on a seed tag that accompanies a certified seed is obtained from laboratory evaluations collectively called **seed testing**. The Federal Seed Act of 1939 mandated the use of seed tags or labeling. Seed testing provides the information to meet legal standards, determines seed quality, and establishes the rate of sowing for a given stand of seedlings. The USDA is the source of the procedures for seed testing in the United States. The **Association of Official Seed Analysts** also publishes additional seed testing procedures for flowers, trees, and shrub species. At the international level, the **International Seed Testing Association** publishes international rules for seed testing.

A seed testing lab conducts tests in five primary categories – germination (viability), purity, vigor, seed health, and noxious weed seed contamination. The first step in seed testing is **seed sampling**. The rules of seed testing provide guidelines for the proper sample size to be submitted for seed testing pertaining to a particular species. Seed testing or **seed analysis** consists of the following tests:

- **Seed germination test (viability).** Seed viability is determined by conducting the standard germination test to determine the germination percentage (the percentage of normal seedlings produced by the pure seed). The common methods of germination tests are the **rolled towel** (seeds rolled in moist paper towel) and **Petri dish test** (seeds placed on absorbent paper in dishes). The rolled towel is kept at 20 °C for 16 hours, followed by 30 °C for 8 hours. The germinated seedlings are counted.

Seed viability may also be determined by a biochemical test, the **tetrazolium test**. Seeds are soaked in 2,3,5-triphenyltetrazolium chloride solution. Living tissue changes color to red while non-living tissue remains uncolored.

- **Purity test.** Seed purity is determined at two levels – genetic and physical. The sample seed should reflect the physical features of the cultivar and be distinguished from other seeds and weed seeds. A genetic purity test requires a chemical test that may be as relatively simple as isozyme analysis, or as sophisticated as DNA profiling or fingerprinting (Chapter 20).
- **Vigor testing.** Seed vigor determines the capacity of seed to emerge rapidly and uniformly and develop into normal seedlings under a range of conditions. Common tests of vigor include **accelerated aging** (seeds are subjected to high temperature of between 40 and 45 °C, and high humidity, before conducting a germination test), a cold test (seeds in rolled towel or containers are held at 10 °C for 7 days before moving to a 25 °C environment), and **electrical conductivity** of seed leakage.
- **Seed health.** The seed sample is examined for the presence of pathogens by visual inspection, seed incubation, or biochemical testing.
- **Noxious weed seed.** This is seed from a species that sooner or later becomes so aggressive and difficult to control. The weed is officially prohibited from being reintroduced into a production region. A noxious seed in one state may not be classified as such in another state.

27.9 Tagging commercial seed

Once tested, the seed is ready to be made available to consumers, but not until it is properly identified with a tag or label. A tag on the seed bag identifies each class of seed. Customarily, a **white tag** identifies breeder or foundation seed, while a **purple tag** is used to identify registered seed. A certified seed receives a **blue tag**. On some occasions, a **green tag** is used to identify a cultivar that the developer opts not to have certified but is nonetheless subjected to the standards of certified seed. Such a cultivar may be simultaneously released by different companies under different brand names (called **branded cultivars**) for marketing purposes. An example of a tag and the information it displays is presented in Figure 27.2 and Table 27.1.

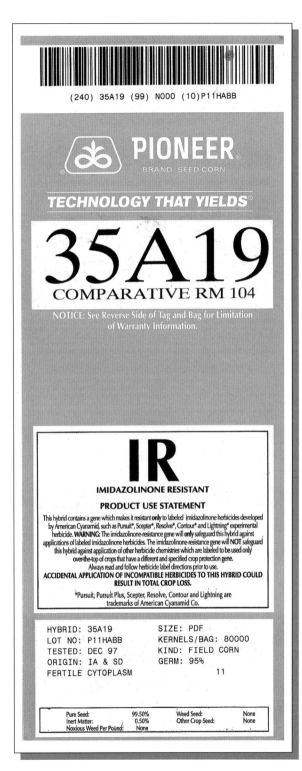

Figure 27.2 A sample seed tag (courtesy of Pioneer Seed Company).

Table 27.1 Information on a seed tag.

Item	What it means
Kind and variety	States variety name; term "mixture" is used to refer to seed with more than one component.
Lot identification	Identifies a specific amount of seed of uniform quality associated with a specific seed test.
Pure seed	Indicates percentage purity as it relates to the kind and variety of crop species indicated.
Other crop seed	Percentage by weight of other crop seed as contaminants.
Weed seed	Percentage by weight of weed seed present.
Inert material	Percentage by weight of foreign material such as chaff, stones, cracked seed, soil, etc.
Prohibited noxious weeds	Weed seeds prohibited in the variety to be sold (e.g., field bindweed, Canada thistle).
Restricted noxious weeds	Weed seeds that may be present up to an allowable limit (90 seeds per pound of pure seed), e.g., quackgrass, dodder, and hedge bindweed.
Germination	Percentage of seed that germinates to produce normal seedlings during a standard seed analysis.
Date of test	Year and month in which seed test was conducted.
Origin	Source of seed – where seed was produced.
Name and address	Name of seed company or seller of the seed.

27.10 International role in seed certification

International involvement in seed certification is critical because of international trade of seeds. In North America, the first official international effort was the International Crop Improvement Association (ICIA) founded in 1919. Its objectives were:

(a) To assist members in promoting the production, identification, distribution, and use of certified seed and other propagating materials of superior crop varieties.

(b) To establish minimum standards of seed production, storage, and handling.

(c) To assist in standardization of certification requirements and procedures to the end that all certified seed will be as good or better than an accepted, minimum quality.

(d) To inform the public as to the value of certified seed and encourage its wide-scale use through approved educational media.

(e) To develop cooperation with all individuals, groups, and organizations directly or indirectly interested in the improvement of crops.

The ICIA gave birth to the Association of Official Seed Certifying Agencies (AOSCA) to coordinate, standardize, and establish minimum standards for genetic purity and to identify minimum standards for seed quality for all classes of pedigreed seed.

At the international level, the Organization for Economic Cooperation and Development (OECD) was established in 1966 to facilitate international movement of certified seed. The OECD has developed five schemes for (i) oil seed and forage seed, (ii) cereal seed, (iii) vegetable seed, (iv) sugar beet and fodder, and (v) corn. Each participating country is responsible for enforcing and supervising the certification process to conform to international standards and to issue appropriate official documentation to accompany commercial seed.

There are various national and regional seed associations around the world. In Canada, there is the Canadian Seed Trade Association, while the United States has the American Seed Trade Association. Within the United States, there are several regional groups, such as the Western Seedmen's Association and the Southern Seedmen's Association, and so on.

27.11 Production of conventional seed

Conventional seed production is often done in regions with a drier climate. Drier growing conditions reduce the incidence of various diseases, and hence result in higher quality seeds. Furthermore, if the growing season is long, growers can use proper irrigation management to maximize crop yield. In the United States, companies that produce vegetable seeds, flower seeds, and seeds of forage legumes and grass crops are concentrated in the western regions.

The key considerations in commercial seed production are:

- **Site (location) selection.** The location should be such that the seed can be produced in isolation to reduce contamination from natural outcrossing. Even though drier climates are desired for seed production, it is best to produce seed in regions of crop adaptation to reduce genetic shifts. The site should be well drained and of good fertility. Cropping history is critical, since a piece of land that previously carried a different cultivar of the crop may produce volunteer plants in the field that could be a source of admixture.

- **Field preparation.** Apart from developing a good seed bed, the key in field preparation is to control weeds to avoid contamination of the harvested commercial seed with seed that reduces the quality of the product.

- **Management.** This includes additional weed control, disease and pest control, fertilization, and irrigation as needed for optimal seed yield and healthy, disease-free products.

- **Harvesting.** The seed should be harvested at the proper stage of maturity for the species and the right moisture content, without being in jeopardy of lodging and shattering, where applicable. It is best to harvest at a moisture content that would require no post-harvest drying. Sorghum and wheat can be safely stored at 13% or less seed moisture, whereas sunflower is best stored at 9.5–13% seed moisture. Where artificial drying is needed, seed may be dried at 40 °C for 6–8 hours.

- **Drying and storage.** It is best for seed to dry in the field to the desired moisture content. However, when needed, seed should be dried after harvesting, to the level prescribed for the species.

- **Conditioning.** Conditioning is done to remove inert material and weed seeds. Also, seed is sampled and analyzed (seed testing) to provide the data needed for producing a seed tag to accompany the product. Foundation fields require isolation to avoid contamination. Off-types should be rogued before pollination, as previously indicated. Harvesting should be done carefully to avoid the mixing of seed of different cultivars. Harvesting equipment should be thoroughly cleaned to remove any seed from the previous harvesting operation.

Seed conditioning entails cleaning, sorting (by size), seed treatment with pesticides, and packaging (bagging). Seed is cleaned to remove inert material, weed seeds, broken seed, and other undesirable materials (plant debris). The grain is sorted into size classes where applicable. Fungicides may be used to dress seeds to protect them from soil-borne pathogens. Clean, treated seed are packaged for the market. When needed, seed is stored in cold dry place (e.g., RH of 50% and temperature of 10 °C).

- **Packaging.** Clean seed is bagged and tagged for marketing.

Isolation is a key provision in the field production of crop seed to prevent mechanical admixture of seed, especially when several cultivars of the same species or related species are being simultaneously produced. In the production of seed from self-pollinated species, cultivars only need to be separated by about 3–6 meters of uncropped or mowed strips. On the other hand, spacing between different cultivars of cross-pollinated species is usually in excess of 200 meters, reaching over 1000 meters in certain cases. Techniques used to reduce mixing from cross-pollination include the use of windbreaks between cultivars, and laying plots such that prevailing wind effect is minimized. Because insects tend to first visit flowers at the edge of the field, crop fields should be square in shape for insect-pollinated species. Furthermore, the border plants may be discarded at harvesting.

27.12 Production of hybrid seed

Hybrid seed production was discussed in detail in Chapter 18. Most commercial hybrids are single crosses (A × B). The success of commercial hybrid seed production is the availability of adequate foundation seed. Foundation seed is derived from crossing inbred lines. CMS may be incorporated to eliminate the need for emasculation. In corn, artificial emasculation by detasseling may be used. It is critical that adequate pollen be available for maximum seed set. Pollen shed varies from one growing environment to another. Corn producers often use a planting pattern consisting of a ratio such as $1:4$ of parent row:seed parent rows, or a ratio of $1:2:1:4$ of pollinator to seed producing rows. The female:male ratio for sorghum ranges from $3:1$ to $6:1$, while sunflower producers use ratios of $2:1$ to $7:1$ to optimize seed set. Producing hybrids of insect-pollinated species may require the aid of artificial colonies of bees for effective pollination. Seed-set may also be optimized by synchronizing the flowering of the parents in a cross. It is important for the parent lines to be genetically pure to reduce the need for rouging, which can be expensive if the production field is large.

Key references and suggested reading

Fernandez-Cornejo, J. (2004). The seed industry in the US agriculture. Agriculture Information Bulletin No. 786, USDA/ERS, Washington, DC.

Boswell, V.R. (1961). The importance of seeds, in *Seeds: The yearbook of agriculture, 1961*. USDA, Washington, DC.

Duvick, D.N. (1977). Genetic rate of gain in hybrid maize yield during the past 40 years. *Maydica*, **22**:187–196.

Outcomes assessment

Part A

Please answer the following questions true or false.

1 Farmers plant registered seed.
2 Seed certification in the United States is conducted by the AOSCA.
3 The breeder seed has the highest genetic purity of any class of seed.

Part B

Please answer the following questions.

1. What does the acronym AOSCA stand for?
2. Give the class of commercial seed that is released to farmers.
3. A white seed tag identifies a certified seed.
4. Give two specific pieces of information on a seed tag.

Part C

Please write a brief essay on each of the following questions.

1. Discuss the commercial classes of pedigree seed.
2. Discuss the commercial production of hybrid seed.
3. Discuss the international role in seed certification.
4. Discuss the role of multinational corporations in the seed industry.

28

Regulatory and legal issues

Purpose and expected outcomes

Intellectual property rights are necessary for a company that owns an invention to preserve its competitive edge over its competitors. These rights also protect a valuable resource that the property owner can later license to a third party for profit. Development and application of biotechnology raises ethical questions, some of which are serious enough to generate significant opposition from the consuming public. Breeders need to be aware of the local and international issues pertaining to public acceptance and rights and regulations affecting the biotechnology industry. The purpose of this chapter is to discuss intellectual property and the ethical issues associated with the breeding of plants, especially as they pertain to the use of biotechnology. After studying this chapter, the student should be able to:

1 Discuss the concept and importance of intellectual property in plant breeding.
2 Explain what a patent is, discuss the types, and what can be patented.
3 Discuss patent infringement.
4 Discuss how ethics impacts the development and application of plant biotechnology.
5 Discuss the agencies and their specific roles in the regulation of biotechnology industry.
6 Discuss international biotechnology regulation issues.
7 Discuss the public perception of biotechnology and its implication to breeding.

28.1 The concept of intellectual property

Intellectual property consists of principles that a society observes to ensure that an inventor is protected from unfair use of his or her invention by others. To achieve this, a variety of legal provisions are made to protect against improper use of other's original ideas and creations. The most common of such provisions are **copyrights, confidential information, breeders' rights, trademarks** and **patents**. The concept of intellectual property takes into account that fact that innovation has a price tag, and hence the inventor must be adequately compensated. However, it also acknowledges that the invention should not be a secret

Principles of Plant Genetics and Breeding, Second Edition. George Acquaah.
© 2012 John Wiley & Sons, Ltd. Published 2012 by John Wiley & Sons, Ltd.

but rather should be made available to the general public (society) so that others may improve upon it.

- **Copyrights©.** Copyright protection is commonly sought by people or entities for the protection of such things as aesthetic creations, music, painting, works of literature, and computer software. It is of little use to plant breeders except for the protection of published research results or inventions such as computer-based tools.

- **Confidential information.** This protection is limited in its scope and pertains to an organization. Its success depends on the extent people can be trusted to keep a secret. The term "confidential information" is used to apply to the variety of strategies used by companies to protect their unpatented inventions, in the hope that they will not be leaked into the public domain. The common strategies include **trade secrets** and **proprietary information**. An invention may not be patented because it is unpatentable, or perhaps the inventor may choose not to, or for financial reasons. Keeping trade secrets is cheaper than applying for a patent. To ensure that employees who know the privileged information do not divulge it for any purpose, some companies enforce a policy that all employees sign a confidentiality agreement.

- **Breeders' rights.** This is one of the common protections sought by plant breeders. It was specifically designed for the benefit of plant and animal breeders to protect them from the unlawful use of their creation by other scientists, competitors, or producers. Exclusive rights to the sale and multiplication of the reproductive material may be obtained, provided the cultivar is new and not previously marketed, different from all other varieties, uniform (all plants in the cultivar are the same) and stable (remains the same from one generation to the another).

- **Trademarks**^{TM}. Pharmaceutical companies use trademark protection extensively. Trademarks are important to all businesses.

- **Patents.** Plant breeders protect their inventions by seeking patents. The right to patent an invention is one of the most widely applicable distinct rights provided by intellectual property.

28.2 Patents

28.2.1 Definition

A **patent** may be defined as an exclusive right granted for an invention of a product or process that provides a novel way of doing something or offers a new technical solution to a problem. The owner of a patent is the **patentee**. A patent provides protection for the invention to the patentee for a predetermined period. A patent may be described as a "negative right" because it confers upon the patentee the right to exclude others from commercially exploiting the invention without the owner's authorization. The protection stipulates that, without permission, no one should make, use, sell, offer for sale, or import the invention. The duration of such exclusive rights is 20 years, after which the invention is released into the public domain. It should be pointed out that while a patent prohibits unauthorized use by others, it does not authorize limitless use by the patentee. A patent is not a "positive right" because it does not empower or obligate the patentee to do something that he or she would otherwise be prohibited from doing. In other words, in making, using, or selling the invention, the patentee must operate within the limits of existing laws of society. Further, it goes to emphasize the fact that an inventor is under no obligation to patent an invention in order to exploit it commercially. However, without a patent for protection of the invention, a third party can commercially exploit the invention without authorization and no legal consequence.

28.2.2 The importance of patents to society

Patents are so pervasive in society that it is nearly impossible to find anything created by humans that is not directly or indirectly associated with a patent. As the US Constitution indicates (US. Const., Article 1, sec. 8, cl. 8) the **patent law** serves to "promote the progress of science and the useful arts". In exchange for exclusive rights to the invention, the patentee is obligated to disclose the invention and provide information that expands the existing technical knowledge base. In other words, whereas others cannot imitate the invention during the life of the patent, they can utilize the divulged information toward further innovations, advancing science and arts to enhance the quality of life of humans. Patents are not only prestigious to own, but they have potential financial value. They provide effective incentives for creativity and innovation, recognizing the achievement of the inventor and, in cases where the invention can be commercialized, financial reward. Because companies invest large sums of money in R&D (research and development) of new products, patents ensure that they can enjoy a monopoly for a period in which they would hopefully recoup their investments, at least. Patents are hence strong incentives for continued R&D. It goes without saying that for-profit breeding companies invest resources in improving crop species with high market potential. Hence, crops of low market potential but of high social value (e.g., important to poor developing countries) are often ignored (the so-called "orphaned crops").

28.2.3 What can be patented?

Patent law specifies what can be patented and the conditions under which a patent can be granted. One of the key functions of a patent is to define the scope of the protection. Ideas and suggestions no matter how brilliant and creative, cannot be patented. Similarly, mixtures of ingredients (e.g., medicines) are also excluded, unless such mixtures produce synergistic effects or some unique and unexpected advantage. Defining the scope of protection is straightforward in certain cases (e.g., a simple device) and very complex in others (e.g., biotechnology inventions). There are five basic classes of patentable inventions:

(i) **Compositions of matter.** A new chemical entity produced from the combination of two or more compounds (common in pharmaceutical and agrochemical research).

(ii) **Processes or procedures.** A series of steps that are followed to synthesize a new compound or a new method of making a product.

(iii) **Articles of manufacture.** Nearly every man-made object.

(iv) **Machines.** Any mechanical or electrical apparatus or device.

(v) **Improvements.** on any of the previous four categories.

28.2.4 Types of patents

Patents may be classified into three basic types – **utility, design**, and **plant**. The duration of each of these patents is 20 years.

Utility patent

A utility patent is one of the most common and also most difficult patents to obtain. The applicant is required to submit a comprehensive description of how to make and use the invention, including detailed drawings, where appropriate, among other requirements. The scope of protection includes the functional characteristics of machines, electronic devices, manufacturing processes, chemical compounds, composition of medical treatment, manufactured articles.

Design patent

A design patent may be sought for the protection of the shape of ornamental or artistic features of an article (e.g., unique shape of a bottle, the grill of an automobile).

Plant patent

The plant patent protects the invention of discovery of distinct and new plant variety via asexual reproductive methods.

28.2.5 National and international patents

Patent laws vary among nations of the world. National patents pertain to the applicant's country of origin. However, applicants may also seek worldwide protection for an invention (e.g., from the European Patent Convention (EPC)). National patents pertain to the applicant's country of origin or operation. The EPC allows patent rights to be obtained in one or more European countries that are parties to the EPC by making a single European patent application. If successful, the applicant must then use the patents in the countries that the applicant designated. **International patents** may be obtained by filing a single patent and designating countries in which protection is desired. Within a specified time limit, this single patent will be copied to the designated countries where they will be processed as national applications. Countries vary in the diligence with which they honor and enforce patent laws. Patent violations cost companies huge losses each year. Sometimes, it takes political pressure to encourage the enforcement of international patents.

28.2.6 Scope of protection of a patent

The applicant is responsible for defining the scope of protection desired. However, there must be ample evidence to justify the scope of protection being sought. The **scope of protection** may be defined narrowly or broadly, each scenario with its consequences. Sometimes, in an attempt to prevent the competition from copying the invention, an applicant may make claims to encompass the embodiments of the invention that were not existent at the time of the invention. Such a practice favors the inventor to the disadvantage of the competition. An example of a broad scope of protection is one sought by Monsanto to protect its highly successful herbicide, Roundup®.

28.2.7 Criteria for patentability

- **Conception.** The applicant should be able to paint a mental picture of the invention that is detailed enough to allow a person knowledgeable in the subject to which the invention relates to make and use the invention.
- **Reduction to practice.** The inventor should make or construct the invention and test it to demonstrate its usefulness.
- **Utility.** An invention must be useful (not merely aesthetic) to the user. That is, the invention must take some practical form (applicability).
- **Novelty.** The invention must not be a copy or repetition of an existing one. Among other things, it should not have been known, published, or used publicly anywhere previously. In the United Kingdom, an invention must not have been in the public domain anywhere in the world prior to application of a patent. Some countries (e.g., the United States) allow a grace period during which an invention that has already been introduced to the public, under certain conditions, could still be patented.
- **Obviousness.** This is a very difficult criterion to satisfy. The invention or product should neither be expected nor obvious. A person knowledgeable in the subject matter (skilled in the art) should not be able to readily figure out how to piece together components parts to make the product (Section 103 of US patent law).

28.2.8 Applying for a patent

In the United States, patent applications may be submitted to the Patent and Trademarks Office, US Department of Commerce, Washington, DC. Copyright applications may be submitted to the US Copyright Office, Library of Congress, Washington, DC. In Europe, applications may be submitted to the European Patent Office. The services of patent attorneys may be engaged in preparing and filing an application. An applicant must define the problem or objective addressed by the invention very clearly and in detail. There are certain general steps involved in applying for a patent:

- **Filing fee.** An applicant for a patent is required to pay a fee for the processing of the application.
- **Search and examination.** The patent examiner will conduct a "prior art" search to ascertain the novelty and non-obviousness of the invention. The claims defining the scope and monopoly being sought are rigorously examined. The examiner may accept or reject the application based upon the search and examination results. The applicant has the right to argue against an adverse judgment, or to amend the claims to the satisfaction of the examiner.
- **Publication.** Successful applications will be published along with the claims of the applicant.
- **Maintenance fees.** Most countries require a successful applicant to pay a periodic maintenance fee to guarantee or prevent the patent from lapsing.

28.2.9 Exploiting intellectual property

Apart from exploiting an intellectual property personally, the owner of the patent has the right to sell, license, or give it away.

Assignment

A patent or patent application of an invention may be sold or assigned to another party just like a piece of property. The method of payment for use or acquisition of the property is entirely up to the owner of the property (e.g., one-time payment or periodic royalties).

License

Rather than outright sale, the intellectual property may be licensed to another party according specified terms and conditions. The license agreement usually guarantees the licensor royalties from the licensee for use of the invention. License agreements have limitations to the extent of exploitation of the invention permitted, and may limit the application of the invention to certain uses. In this era of genetic engineering, a wide variety of genes have been cloned. Also, various biological tools (e.g., vectors, promoters) have been developed. Some of those undertakings were expensive and tedious. Rather than re-invent the wheel, it might be easier to pay a fee to use the technology.

Freedom of use

Freedom of use considerations may hinder a patent from being exploited without infringing upon existing patents. Such restrictions on free exploitation of an inventor's patent may derive from the scope of the patent. Scopes that are too narrow are susceptible to such infringements. An example of such a situation involving freedom of use may arise when an inventor patents a process that requires another patented compound in order to exploit the invention.

28.3 Patents in plant breeding and biotechnology: unique issues and challenges

28.3.1 Patenting organisms

Patenting in the biological sciences or non-empirical sciences is very challenging, especially when life is involved (e.g., organisms). Whereas it is not easy to duplicate and proliferate a mechanical invention, it is easy to reproduce an organism (just like a computer software and copyrighted music). Consequently, the early application of patent laws to biology tended to be strongly and broadly interpreted in favor of inventors.

Generally, it is easier to satisfy the patent requirement when inventions are the results of empirical discovery, as pertains in the pharmaceuticals and agrochemical industries. It is easy to lay claim to discovery of a compound with potential for use as an active ingredient in pesticides, antibiotics, and other therapeutics. Development of truly new and unexpected phenomenon is not common. Progress is made incrementally. It is often a challenge to satisfy the traditional criteria stipulated by the patent office – obvious phenomenon, specific utility, and teaching others how to make and use the invention. Another limitation is that ideas and properties of nature are not patentable.

28.3.2 Patenting the hereditary material

The turning point in patenting genes and other biological resources occurred in 1980, when the US Supreme Court decision in Diamond v. Chankrabarty was a patent granted for an oil-dissolving microbe. The technologies of genetic engineering and genomics have resulted in the discovery of millions of genes and fragments of genes (expressed sequence tags or ESTs) that have been submitted for patenting. However, not all players are satisfied with the scope of protection provided by the patent laws. A microscopic view will allow nearly anything novel to be patentable, while opening up the doors for competitors to easily circumvent the narrow claims. Some scientists are opposed to the granting of broad patents to what they describe as the early stages of the biotechnology game. Some of the genes submitted for patents have not been characterized, neither have the applicants determined their functions and specific uses. The concern is that large-scale and wholesale patenting of genes by biotechnology researchers or companies who have no clue about the functions of these genes is tantamount to staking a claim to all future discoveries associated with those genes (the so-called "**reach-through patents**").

This concern is a genuine one. For, until not long ago, genomics companies had a field day staking claims to the genome landmark (the "genome run"). But with the focus now on understanding gene function, the proteomics companies now have their chance to do likewise. This is stirring up new controversies in the patenting of biotechnology inventions.

Another issue with biotechnology patents is **"patent stacking"**, a situation in which a single gene is patented by different scientists. This situation is not favorable to product development because users are deterred by the possibility that they would have to pay multiple royalties to all owners of the patent. Further, because patent applications are secret, it is possible for an R&D team in a different company to be working on development of a product only to be surprised at a later date by the fact that a patent (called a **"submarine patent"**) has already been granted.

28.3.3 Patenting proteins

A patent on a specific DNA sequence and the protein it produces may not cover some biologically important variant. It is estimated that the top genomics companies have collectively filed over 25 000 DNA-based patents. The business rationale to their strategy includes the potential to receive royalties from third parties that use any of them. But this may not be as simple as it sounds, unless one gene makes one mRNA, which in turn makes one protein, something that is not true anymore (alternative splicing, Supplementary Chapter 1).

It is most likely and perhaps inevitable that some protein discovery projects will turn out proteins that correlate better with disease than those for which patent claims are already in existence. In such cases, litigation seems the likely recourse. However, it is also likely that potential litigants may opt for the less costly route of **cross-licensing**, whereby each party can cross-license another's patents.

28.3.4 Patenting products of nature

Patent laws protect the public by enforcing "the product of nature" requirement in patent applications. The public is free to use things found in nature. That is, if for example a compound occurs naturally but it is also produced commercially by a company via a biotechnology method, the genetically engineered product is technically identical to the natural product. However, in the case of Scripp v. Genentech, a US court ruled that a genetically engineered factor VIIIc infringed a claim to VIIIc obtained by purification of a natural product. This indicates that a previously isolated natural product had first claim to patent rights over a later invention by genetic engineering. If a company seeks to apply for a patent for an invention to produce a rare naturally occurring compound in pure form, the argument will have to be made for the technique used for extraction, purification or synthesis, not for the material *per se*.

28.3.5 Moral issues in patenting

Biotechnology also faces a moral dilemma in patent issues. Specifically, is it moral to patent any form of life? Further, if the discovery has medical value, should it be patented? Then is the issue of the poor. Is it moral to demand that the poor pay royalties they can ill afford for using patented products for survival purposes? A debated issue is the Plant Breeder's Rights. Should breeders be permitted to incorporate seed sterilizing technology (e.g., the so called "terminator technology") in their products to prevent

farmers from using seed from harvested proprietary material for planting the field the following season? For balance, is it fair to expect a company to invest huge amounts of resources in an invention and not recoup its investments? There are no easy answers to these questions.

28.3.6 International issues in patenting

Courts in Europe and the United States as well as other parts of the world differ in their positions on patent issues. Patent laws and how they are enforced may differ among nations. For example, the European Directive on the Legal Protection of Biotechnological Inventions passed in 1998 declares that a mere discovery of the sequence or partial sequence of a gene does not constitute a patentable invention. Genes are not patentable while they are in the body (*in situ*). However, genes isolated from the organism or artificial copies of the genes produced by some technical process may be patentable, provided the novelty, inventive step, and utility are clearly demonstrated. The US laws have been tightened to include a clause to the effect that the utility of the invention must be "specific, substantial, and credible" (i.e., readily apparent, and well-established utility).

In addressing the issue of morality, the European Directive also specifically excludes certain inventions from patentability. These include processes for reproductive cloning of human beings, processes for modifying the germ line genetic identity of human beings, and uses of human embryos for industrial or commercial purposes. Essentially, if the publication or exploitation of an invention would generally be considered immoral or contrary to public order, it cannot be patented.

28.4 Protecting plant varieties

Laws protecting plant varieties vary among countries. For example, in the Netherlands there is strict separation between the public sector breeding efforts, engaged in by the state institutions, and private sector efforts. By mandate, the public sector is restricted to breeding research and the release of unfinished products. The release of commercial cultivars is restricted to only private sector companies, which are self-supporting. This provision ensures fair competition between the two entities.

28.4.1 International efforts

Intellectual property rights issues impact plant breeders, researchers, producers, and seed companies as well as consumers. A formal specific government protection for plant varieties in the United States was first implemented in 1930. The **Plant Patent Act** was (and still is) limited to clonally (vegetatively) propagated plants. This was essentially a provision made within an existing patent law (i.e., *not sui generis* or a system of its own kind). The first plant breeders' rights legislation in the world was developed in the Netherlands in 1941. The **International Convention for the Protection of New Varieties of Plants (l'Union pour la Protection des Obstentions Vegetables – UPOV)** was established in 1961. Membership of UPOV is currently 67 nations. Each of these ratifying nations has the obligation to produce independent national laws, consistent with general guidelines of UPOV, for their sovereign use.

According to the UPOV guidelines, a plant breeder may receive protection for a new variety by meeting the following requirements:

(a) **The "DUS" criteria.** Distinct, Uniform, and Stable.
(b) **Other criteria.** Novelty, Variety denomination (name), and Rightful person (breeder's name).

To satisfy DUS, it is necessary to understand the UPOV definition for a "plant variety." A plant variety is a plant grouping within a single botanical taxon of the lowest rank, which grouping, irrespective of whether the conditions for the grant of a plant breeders' right (PBR) are fully met, can be:

- defined by the expression of the characteristics resulting from a given genotype or combination of genotypes,
- distinguished from any other plant grouping by the expression of at least one of the said characteristics, and
- considered as a unit with regard to its suitability for being propagated unchanged.

It is informative to note that by satisfying the definition of "plant variety" the breeder essentially satisfies the requirements for DUS. The details of the UPOV are highlighted briefly here.

Distinctness

The candidate variety must be clearly distinguishable from any other variety whose existence is a matter of common knowledge at the time of filing the application for a PBR (e.g., variety has been officially registered or the material is available on the market). This requirement is the most difficult to meet. To be distinct, the new variety must clearly differ from known ones in at least one characteristic (e.g., potato flower color or skin color).

Uniformity

A PBR is granted to only one particular plant variety, the plant grouping representing that variety must not be a mixture of several varieties. Natural variation that is expected from the particular mode of reproduction for the species is integrated into the concept of uniformity. In self-pollinated crops, uniformity implies the selection of a pure line; in cross-pollinated crops a higher degree of variation is accepted; in vegetatively propagated crops, uniformity means vegetative propagation does not affect the genotype; while in hybrids the variety is uniform as a result of the pure inbred lines.

Stability

The genome rearrangement as a result of the manipulations from the breeding method used to create the variety must have become stable (the relevant characteristics of the variety remain unchanged after repeated propagation). Achieving stability is easier for self-pollinating species and more problematic for cross-pollinating species.

Novelty

Novelty requires that the variety must not have been sold in that country for a minimum period prior to the application for the protection. The minimum time is one year, but a longer period is required for trees and vines.

Scope of plant breeders' rights

Upon granting, authorization is required from the breeder in order for any of the following activities involving the protected variety (propagating material) to be conducted:

- Reproduction of the propagating material or the conditioning for that purpose.
- Offering for sale, selling or marketing of the propagating material.
- Exporting and importing of the propagating material.
- Stocking of the propagating material for any of these purposes.

Breeders' exemption

PBR does not extend to the use of proprietary material for the purpose of breeding or research and developing other varieties. However, the outcomes of such activities are protected provided the offspring meet the DUS criteria.

Essentially derived variety

According to the UPOV, a variety is deemed essentially derived (EDV) from another variety when it is predominantly derived from the initial variety, or from a variety that is itself predominantly derived from the initial variety, while retaining the expression of the essential characteristics that result from the genotype or combination of genotypes of the initial variety. This protection is granted to deter variety pirating whereby unscrupulous breeders only "tweak" proprietary varieties and present them as novelty. For example, a proprietary variety of potato may be manipulated to produce a different skin or flower color, while retaining all the other qualities of the original variety.

Farmers' privilege

This concept is tolerated to some extent especially in developing countries where national laws may have some flexibility to allow farmers to use proprietary material within limits, provided the breeder's rights and interests are not severely infringed or jeopardized. Prior to 1991 when the UPOV amended its principles to protect against the exercise of this privilege, farmers could save seed from proprietary material for planting the subsequent season's crop, rather than purchasing fresh seed from the seed company each season.

Territorial limitations on plant protection

It should be pointed out that the juridical ways of protecting novel cultivars operate under territorial limitations. That is, patents and plant variety protection (PVP) certificates are only valid for the jurisdiction in which they were awarded. Consequently, US patents are only enforceable in the United States. For global trade purposes, it is desirable to broaden the geographic scope of protection of an invention. However, for this to be realized, counties around the world should not only have national legislation for protection (i.e., IPR), but such legislation must be enforced. Developing nations lag behind developed nations in the development and implementation of plant protection legislation. An effort made by the World Trade Organization in 1995, called the Trade-Related Aspects of Intellectual Property Rights (TRIPs) agreement, was geared toward encouraging developing countries to have proper legislation in place that would open up global markets for free trade.

Related legislation to TRIPs, the CBD (Convention on Biological Diversity) was enforced in 1993, with the goal to promote conservation of biological diversity, sustainable use of components of biological diversity, fair and equitable sharing of benefits arising out of the use of these natural resources, as well as the appropriate transfer of relevant technology. Over 180 countries (with the notable exception of the United States) are signatories to this framework agreement. A difference between the CBD and TRIPs is that the former addresses sovereignty over individual organisms (plants and animals) as tangible goods, while the latter is concerned with access to biological diversity through internationally standardized IPR rules. The CBD also makes provision for quid pro quo in the process of acquisition of germplasm, so that a country acquiring the biological diversity is required (not obligated) to transfer some technology to the country of origin of the germplasm. Another provision that favors the country of origin of the germplasm is the so-called traditional knowledge (TK) status (acknowledging the role of indigenous people in conserving and sustaining natural biodiversity). In so doing, TK is accorded some form of IPR, just like what scientific knowledge traditionally receives. The CBD is periodically revisited to develop appropriate protocols to guide its implementation, the most well known of such summits being the

Cartegena Protocol, which addresses the handling, transfer and release of transgenic organism.

Yet another international effort to facilitate the fair use of biodiversity is the International Treaty on Plant Genetic Resources for Food and Agriculture (ITPGRFA), signed in 2001 by 140 countries. The treaty, which was approved by the FAO Conference in November 2001, entered into force on June 29, 2004. The stated goal of this multilateral agreement is to provide governments or authorized individuals access to a list of over 100 selected genetic resources that are protected by IPR in exchange for an equitable contribution to an international fund (Global Crop Diversity Trust) from the returns accruing from the commercialization of an invention (cultivar) that was based on any of the listed materials. Funds would then be used to support germplasm conservation and capacity building programs. Notable crops included on the list are rice, maize, wheat, potato, banana (*Musa* spp.), and common bean. Notable crops that are absent from the list are soybean, sugarcane, groundnut, tropical forages, tomato, grape, cocoa, coffee, and industrial crops such as oil palm and rubber. They are excluded by countries that are rich in such biodiversity because independent bilateral agreements for their sale are potentially more profitable than participation in the international treaty.

With Japan and Australia, the United States is the only country awarding utility patents for plant cultivars. In other countries, such as the European Union, plant cultivars can only receive IP protection under the PVP legislation. The European Patent Convention explicitly prohibits patents on plant and also animal varieties.

28.4.2 US efforts

Protection of all new plant varieties was first adopted in the United States in 1970 under the **Plant Variety Protection** (PVP) Act of 1970. This act was amended in 1994, when the United States implemented the UPOV Act of 1991. As previously indicated, the use of utility patents for plant variety protection was made possible following the US Supreme Court ruling in favor of Chankrabaty in 1980, declaring that "anything under the sun that is made by man" may be patented. This ruling originally pertained to microorganisms. However, the US Patent and Trademark Office extended patent protection to plant varieties in 1985. It should be pointed out that the goal of PVP is not to protect the variety *per se*, but rather to protect the breeder of the variety. Consequently, plant breeders' right previously discussed is a more apt description of the purpose of PVP.

A plant breeder wishing to patent a plant variety in the United States currently has three options:

(i) **Plant patents.** This is limited to vegetatively propagated varieties.
(ii) **Plant Variety Protection.** This is applicable to all sexually or asexually propagated varieties, pure lines, and hybrids produced from pure lines.
(iii) **Utility patent.** This is applicable to all plant varieties (including pure lines and hybrids produced from pure lines).

Plant patents are honored in the United States only but members who are signatories to the International Convention for the Protection of New Varieties of Plants, established in 1961, operate similar laws. Utility patents are honored mainly in the United States, Japan, and Australia. Most plant breeders protect their inventions under either the Plant Variety Protection (PVP) or Utility Patents (UP). A major difference between the two protection systems is that PVP protected varieties may be used as breeding material by a researcher without requiring the permission of the rights holder. Should the breeder develop a product that is distinct from the parental germplasm, it may qualify for protection in its own rights. On the other hand, the claims in a UP may be so designed to exclude breeding use without express permission from the owner of the property. Breeding companies use UP to protect their germplasm base. This is because breeding progress usually proceeds by serial improvement on the current optimal materials. Whereas this is advantageous to commercial companies, the research exemption it denies to the wider scientific community is viewed by some as undesirable. However, patents would allow the company to recoup its investments in developing a breeding material that can be utilized by its competitor. For example, developing a breeding material by pre-breeding (germplasm enhancement; Chapter 11) can be risky, expensive, and of a long duration. The developer needs to protect his or her invention.

28.5 The concept of substantial equivalence in regulation of biotechnology

A major challenge facing regulatory agencies the world over is in deciding what constitutes a meaningful difference between a conventional crop cultivar and its genetically modified derivative. The concept of **substantial equivalence** originates from the general position taken by regulators to the effect that conventional cultivars and their GM derivatives are so similar that they can be considered "substantially equivalent". This concept, apparently, has its origins in conventional plant breeding in which the mixing of the genomes of plants through hybridization may create new recombinants that are equivalent. However, critics are quick to point out that this is not exactly what pertains with genetic engineering. In conventional breeding, the genes being reshuffled between domesticated cultivars have had the benefit of years of evolution during which undesirable traits have been selected out of most of the major crops, though the use of a wild relative for breeding purposes could introduce genes potentially detrimental to human health. In contrast, the genes involved in GM crop production have been introduced into unfamiliar genetic backgrounds.

The traditional method of evaluating new cultivars from breeding programs is to compare them with existing cultivars they would be replacing, if successful. To this end, evaluations include gross phenotype, general performance, quality of product, and chemical analysis (especially if the goal of the breeding program was to improve plant chemical content). It is expected (or at least hoped) that the new cultivar will not be identical to the existing cultivar, because of the investment of time and effort in breeding. Nonetheless, it is not expected that the new cultivar would produce any adverse effects for users of the product. On the other hand, there are still genes detrimental to human health in wild relatives used for breeding and, in principle, mutagenic methods used to generate diversity in breeding programs could also cause changes that impact food safety or quality.

In 1995, the World Health Organization released a report in which the concept of substantial equivalence was endorsed and promoted as the basis for safety assessment decisions involving genetically modified organisms and products. Since then, the concept has attracted both supporters and opponents. Some feel that, as a decision threshold, the concept is vague, ambiguous and lacks specificity, setting the standard of evaluation of GM product as low as possible. On the other hand, supporters believe that what is intended is for regulators to have a conceptual framework, not a scientific formulation, which does not limit what kinds and amounts of tests regulators may impose on new foods. The concept of substantial equivalence has since been revisited by the FAO and WHO and amended. It appears that opposition to the use of this concept as a regulatory tool would be minimized if a product is declared substantially equivalent after rigorous scientific analysis has been conducted to establish the GM product poses no more health or environmental risk than its conventional counterpart. Of course, the cost of such an evaluation may prohibit the use of biotechnology varieties unless there are great benefits from their use.

28.6 The issue of "novel traits"

Sometimes, conventional plant breeding may introduce a "novel trait" into the breeding program through wide crosses or mutagenesis. This notwithstanding, the new cultivar produced is still considered substantially equivalent to other cultivars of the same crop. On the other hand, even though the presence of a transgene in a GM cultivar is considered an incorporation of a "novel trait", the GM product and the conventional product differ. The novel traits incorporated in the major commercially produced GM crops are derived from non-plant origins (mainly from microorganisms). Secondly, only a single gene separates a GM product from its derivative. In conventional breeding, the desired genes are transferred along with numerous other unintended genes.

The question then is whether the more precise gene transfer of genetic engineering means that a GM crop and the traditional counterpart would differ only in the transgene and its products. If this were so, a simple linear model would be adequate to predict the phenotype of a genetically modified organism (GMO). Unfortunately, because of the role of the environment in gene expression, and the complex interactions that occur in a biological system, linear models are seldom adequate in predicting complex biological systems. Further, it is known that single mutations often produce pleiotropic effects (collateral changes) in the organism. Similarly, the collateral effects of a transgene have been demonstrated in the transgenic salmon carrying the transgene coding for human growth hormone, in which researchers found a range of phenotypes. Also, it is important to mention that the altered phenotypes may appear at particular growth times in the growth cycle of the organism or in response to specific environmental conditions. Further, these phenotypic changes induced by the transgene may only be minor alterations.

In view of the forgoing, it appears that the best way to assess any adverse effect of a transgene is to directly test for harmful outcomes. In food biotechnology, some believe that this should include testing for both short- and long-term human toxicity and allergenicity, among other things. There should also be an assessment of environmental impact over time and across relevant sites. Then, a final assessment should be made regarding the extent to which the transgenic cultivar deviates from the parental genotype. It is important that the analysis indicates whether such deviations, if any, are biologically significant. Otherwise, the GM cultivar would be substantially equivalent to the existing cultivars and would not need prior approval for introduction into the food chain.

However, as the transgene is derived from a source that would not necessarily be consumed with the crop or be present in the environment where the crop is grown, there is a novel exposure to the transgene product. This could provide justification for an increased assessment of human health and environmental effects of the transgenic crop compared to a conventionally bred crop.

28.7 The concept of the precautionary principle

The **precautionary principle** is an approach to handling uncertainty in the assessment and management of risk. This principle recommends that uncertainty, when it exits, be handled in favor of certain values (health and environment) over others. In other words, when our best predictions turn out to be in error it is better to err on the side of safety. Another way of putting it is that, all things being equal, it is better to have forgone important benefits of a technology by wrongly predicting its risks to health or the environment than to have experienced the harmful consequences by wrongly failing to predict them. In statistical terms, if an error in scientific prediction should occur, it is better to commit a Type I error of

declaring a false positive (that is erroneously predict an adverse effect where there is none), than a Type II error (erroneously predict no such effect when there actually is one). However, it is the custom of science that it is more serious a flaw in analysis to commit a Type I error (make a premature claim, e.g., reject the null hypothesis that GM crop poses no significantly greater risk than its conventional counterpart) without adequate scientific evidence.

In view of the forgoing, it not difficult to see why the precautionary principle has both proponents and opponents. Proponents see it as a proactive and anticipatory strategy for protecting the public, environment and animals from potential harm that is hard to predict by even the best science available. On the other hand, opponents view the precautionary principle as unscientific, a tool that promotes unfounded fear in the public and mitigates against research and development of new technologies.

This principle emerged in the 1970s and is currently invoked in numerous international laws, treaties and protocols (e.g., the Cartegena Protocol on Biosafety of 2000). It is more cautiously intepreted in Europe than the United States. There are certain common criticisms of the precautionary principle. Some feel it is ambiguous and lacks uniform interpretation. Also, it marginalizes the role of scientists in that, whenever it is invoked, it usually tends to relax the standards of proof normally required by the scientific community. Others see the precautionary principle as a veiled form of trade protectionism. Specifically, nations may invoke this principle to circumvent the science-based decisions established in trade agreements and enforced by the World Trade Organization. Such rules generally require that a nation provide reliable scientific evidence to support its decision (e.g., to ban importation of a product). For example, the decision by the European markets to ban American and Canadian beef treated with rBST (growth hormone) is considered to be colored by protectionism.

28.8 Regulation and the issue of public trust

The public needs to trust those who develop and implement regulations that govern the development and application of technologies. It is widely accepted that even the most minimal risks may be unacceptable if levels of public trust in those who manage these risks are low or eroding. In Europe, the general public apprehension about the risks of GM foods is blamed to a large extent on the loss of public trust in scientists and regulatory bodies resulting from the BSE crisis in the United Kingdom.

It is claimed that the assessment of biotechnology risks is a science-based activity. Consequently, it is important that the process be above reproach. The science should be high quality and the conduct of the assessment be independent and objective. There should be no conflict of interest in the regulatory process. Any association between producers and regulators is bound to cast doubt on the integrity of the process.

One factor in boosting public confidence is transparency of the regulatory process. During the application process, the applicant is required to submit certain data to the regulatory agency, though it can be claimed that parts of the submitted data are Confidential Business Information (CBI) and not be viewed by the public. Different regulatory agencies (in the US government and other governments) have somewhat different restraints on what sorts of information can be claimed as CBI. The question then is how much of the information should be divulged to the public and how much should remain proprietary information? Because the regulatory process is claimed to be science-based, and because the custom of scientific enquiry is to be open and completely transparent, and, further, because the decision of the regulatory authority is based on the scientific evidence, any attempt to withhold information involved in decision-making process may cast doubt on the integrity of the process.

28.9 Biosafety regulation at the international level

Because biotechnology products are accepted to varying extents in various countries, and because international trade involves crops that are targets for biotech, it is imperative that trading nations develop a consensus for biosafety regulation. An international delegation convened to draft global regulatory guidelines, called the **Cartagena Protocol on Biosafety**. An outgrowth of the Convention on Biological Diversity (which was adopted at the 1992 Earth Summit in Rio de Janeiro), the Biosafety Protocol, which entered into force in September 2004, is designed to provide guidelines for signatories of the Protocol and their trading partners on the transfer, handling, and use of what are described as living modified organisms (LMOs) that have the potential to impact the conservation and sustainable use of biodiversity. The Cartagena Protocol on Biosafety has been interpreted by some to mean that LMOs intended for food, feed or processing, must be identified as LMOs.

Basically, an exporter of a product will be under obligation to provide the importer with information about the LMO regarding risk assessment and obtain consent prior to shipment. Critics of the Biosafety Protocol say that its implementation will adversely impact international trade by imposing severe trade barriers on a wide variety of biotechnology products (bulk grain, processed food, drugs, etc.). The cost of goods will increase as shippers will have to segregate products, thereby increasing handling costs. Scientific development progress will also be impacted, as scientists are compelled to pay more attention to special interest groups.

Biosafety regulation stringency is variable from one nation to another. In Japan, the Ministry of Agriculture, Forestry and Fisheries is responsible for assessing environmental and feed safety, while the Ministry of Health and Welfare is responsible for food safety assessment. Basically, a product is subject to scrutiny if it was developed by rDNA technology. In Canada, the basis of assessment is the safety of the novel traits that have been incorporated, regardless of the technology used to produce the product. Gaining access to the European Union market is a complicated task. However, once approved, the product becomes legal in all the member countries of the European Union. The product manufacturer or importer must submit a notification to the competent authority of the member state of the European Union where the product is intended to be marketed. In China, the State Science and Technology Commission has the responsibility of developing a regulatory system for GMOs. Regulations in developing countries are generally lacking. The Biosafety Protocol might be beneficial in this regard to assist the less industrialized economies in gaining market access to developed economies. There is no denying that a unified regulatory system of GMOs would facilitate international trade involving these products. Unfortunately, a consensus that will be fully acceptable to all nations will be difficult to achieve in the near future.

28.10 Labeling of biotechnology products

Some propose that consumers should have the right of "informed choice" about exposure to risks of GM products. This push for labeling is partly because of the perception of lack of transparency from regulatory agencies and the absence of balanced risk/benefit analyses. Because the first generation products of biotechnology benefited the food producing industry directly, as previously indicated, consumers tend to view GM crops are geared towards enriching large corporations. Some consumer advocates would like to see all biotechnology foods labeled as such. The argument against labeling advanced by the biotechnology industry is viewed as an attempt to conceal information from the public. Opponents do not see a need for labeling, since the Food and Drug Administration (FDA) has ruled that there is no inherent health risk in the use of biotechnology to develop new food products. The food industry opposes mandatory labeling because of the concern that such labeling could be interpreted to be "warning labels", implying that biotechnology foods are less safe or nutritious than their conventional counterparts.

The FDA requires a food product (including biotechnology foods) to be labeled if the following apply:

- It contains a protein known to pose allergenic risk (e.g., milk, eggs, peanuts, tree nuts). Consequently, any genetic engineering involving gene transfer from any of these organisms must be labeled.
- Its nutrient content as a result of the genetic manipulation is significantly different from what occurs in a normal product. For example, if high protein is engineered into cereal or root crop, the product must be labeled.

Opponents argue that labeling all biotechnologically produced foods would increase the cost of products as a result of the added cost of product segregation for the purpose of the so-called **identity preservation** of GE and non-GE products. To avoid contamination, biotechnology and conventional products must be kept apart at all phases of production, storage, processing, and distribution at additional cost. This would impact bulk or commodity products like grains (corn, wheat, and soybean). However, specialty and high value fruits and vegetables are already identity preserved for premium prices.

Labeling of all products might be helpful to those who practice certain lifestyles or religious beliefs that impose strict dietary observances. A plant with an animal gene may not be acceptable to a strict vegetarian. However, studies have shown that both the kosher (Jewish) and halal (Muslim) communities have mechanisms in place to determine which products are acceptable to their adherents. Leadership of both religious groups has ruled that simple gene additions that lead to one or a few components in a species are acceptable for their religious practices. However, the Muslim community has not resolved the issue regarding acceptability of gene transfer from swine into species, should that happen. Both Jewish and Muslim communities accept the use of bioengineered chymosin (rennin) in the cheese production.

Many countries have some form of labeling regulations or guidelines, which could be mandatory or voluntary. The primary forum for the discussion of food labeling at the international level is the Codex Alimentarius Commission. Mandatory labeling has been implemented in the European Union and is being implemented in Japan. In Europe, all products containing GMOs must be labeled as such. Even where mixtures of conventional and GMOs are concerned, a label must be provided to indicate that GMOs may be present. The United States and Canada require GM food products that could have health or safety effects (possible allergens or changes in nutritional content from acceptable levels) to be labeled.

In North America, labeling is generally thought to be necessary only when there is some feature of the product itself that needs to be brought to consumers' attention (e.g., health risk or nutritional issue). The process by which the product is produced (e.g., by genetic modification) is considered inconsequential. This is described as product-based (as opposed to process-based) regulation. An exception to this approach in the United States and Canada is the requirement that food subjected to the processes of irradiation be labeled. In the United States, the FDA and the courts generally consider reference be made to a "material fact" about the product that is pertinent to nutritional value or safety. This affirms the concept of substantial equivalence in which a new food product that is substantially equivalent to existing products is exempt from labeling.

28.11 Economic impact of labeling and regulations

The economic impact of food regulations and labeling on trade depends on the products involved, the cost of labeling, and sometimes how consumers use such information. The cost of labeling will depend on the stringency imposed. That is, whether "zero tolerance" or "minimum tolerance" of GM product is the goal. Implementing the former standard would require expensive safeguards to be implemented to avoid cross-contamination. Harvesting, processing, shipping, and other product handling would require modifications.

Government approval can have severe adverse consequences on trade. For example, sale of United States corn in European Union countries was dealt a devastating blow in 1999 because certain GM corn varieties were not approved for sale in the European Union. This action caused US corn export to the European Union to drop from $190 million in 1997 to $35 million in 1998, and then to a low of $6 million in 1999. Consumer response to labeling has impact on product demand. Sometimes, products intended for use as feed may not require labeling.

28.12 Legal risks that accompany adoption of GM crops

As biotechnology products enter the food and agriculture production chain, their adoption is accompanied by a variety of legal risks. Some of these legal issues are briefly discussed next.

28.12.1 Tort liability versus regulatory approval

Biotechnology research and application are highly regulated by the federal government as previously discussed. These regulations include the conditions under which a certain GM crop or GM organism or products will be approved for safe use by humans and for the safety of the environment. Regulations also affect how the products are produced, marketed, and used.

However, before or after regulatory approval, a GM plant or organism could cause damage to property, persons, markets, environment, or social structure. **Tort** is a civil legal action whereby the claimant alleges injury or wrong, arising independent of contact, to the person or property of the claimant. Both producers and users of biotechnology products are subject to these kinds of liability that may exist for these kinds of possible damage, as discussed by D.L. Kershen.

28.12.2 Damage to property

Trespass

The most likely grounds on which property damage may be alleged are:

(i) Pollen flow from a GM crop to a non-GM crop.
(ii) Mechanical mixing of seeds or plant parts during harvesting, storage, or transportation.

Producers of specialty crops (e.g., organic crop producers) may allege this damage and bring a common law cause of action based on **trespass** (the physical invasion of the possessory interests of the property (land) of the person claiming damages). Physical spread of GM pollen to a neighbor's field is enough for this infringement to occur. Because pollen flow is a natural process that occurs freely, it is important that in such a law suit the plaintiff establishes that the physical invasion caused damage.

Strict liability

Damage to property may also be alleged and a tort claim in **strict liability** brought by a plaintiff. This liability is a case of liability without fault despite the exercise of utmost care and can be considered if it is shown that the activity of growing a GM crop is abnormally dangerous. The plaintiff would have to show that (i) a high degree of risk of some harm to the person, land, or chattels of others exist, (ii) there is a likelihood that the harm that results will be great, (iii) there is an inability to eliminate the risk by the exercise of reasonable care, (iv) the extent to which the activity is not a matter of common usage, (v) there is inappropriateness of the activity to the place where it is conducted, and (vi) the extent to which its value to the community is outweighed by its dangerous attributes.

Negligence

Negligence is a fault-based claim that alleges that property damage resulted because a neighbor growing a GM crop failed to take adequate precautions. In this case, both the farmer and the agbiotech company that developed the GM cultivar are liable. It is for this reason that refuge rows are required by growers of GM crops.

Private nuisance

The owner of a property has a right to the private use and enjoyment of land free from non-trespassory invasion by another. The plaintiff must prove that nearby fields of GM crops have unreasonably interfered with the use and enjoyment of one's own land.

28.12.3 Damage to person

A **damage-to-person claim** may be brought by a plaintiff alleging damage arising from the toxicity of a GM crop or its product, allergic response to these crops or their food products. The claim may also be based on alleged damage from long-term exposure to GM crops or that their food products cause ill-effects to health. A case in point is the Starlink® corn in human food, involving Aventis Corporation. Pharmaceutical GM crops have the highest likelihood to present such a risk.

28.12.4 Damage to economic interest (market)

Some producers have brought claims to the effect that the introduction of GM crops increased their production costs and equipment cost, while decreasing the prices of their agricultural products. In other words, the presence per se of GM crops in the agricultural sector has affected market access and market prices for their non-GM crops in a general way. Of relevance in such a case is the level or type of approval given by the regulating authority for the GM crop (i.e., limited use, all use, limited to USA, etc.). A case in point, again, is the Starlink® GM cultivar that was approved for limited use (animal feed), but found its way into the food chain. Such a case is relatively easy to prosecute with zero tolerance as the regulatory standard.

28.13 Overview of the regulation of the biotechnology industry in the United States

The rationale for regulation of biotechnology is to protect consumers from product risk and to promote and retain their confidence in biotechnology products, as well as promote trade. The agencies with regulatory oversight in biotechnology in the United

Table 28.1 US agencies and biotechnology regulatory oversight.

Agency	Products regulated
US Department of Agriculture	Plant pests, plants, veterinary biologics
Environmental Protection Agency	Microbial/plant pesticides, new uses of existing pesticides, novel microorganisms
Food and Drug Administration	Food, feed, food additives, veterinary drugs, human drugs, medical devices

Source: USDA

States are the Department of Agriculture (USDA), the Food and Drug Administration (FDA), and the Environmental Protection Agency (EPA). The products they regulate are summarized in Table 28.1. Manufacturers and developers of biotechnology products are required to meet certain minimum product standards stipulated in state and federal marketing statutes. These include state seed certification laws, the Federal Food and Drug Cosmetic Act (FFDCA), the Federal Insecticide, Fungicide, and Rodenticide Act (FIFRA), the Toxic Substances Control Act (TSCA), and the Plant Protection and the National Environmental Policy Act (NEPA). Depending upon the product, an agency may review it for its safety to grow, safety to eat or safety to the environment. The Animal and Plant Health Inspection Services (APHIS) conducts the USDA biotechnology evaluation. The EPA ensures the safety of pesticides and enforces the FIFRA (to regulate the distribution, sale, use and testing of plants and microbes producing pesticidal substances) and the FFDCA (to set tolerance limits or exemptions from tolerances for pesticides used in food, and feed production). The FDA is part of the Department of Health and Human Services, and enforces the FDA's regulations (to regulate foods and feeds derived from new plant varieties).

28.13.1 USDA-APHIS

APHIS is authorized to regulate the confined release, in particular the interstate movement, importation, and field testing, of organisms and products altered or produced through biotechnology processes that are plant pests or are suspected of being so, or have incorporated genetic components from organisms that are plant pests. An individual or an entity seeking to conduct any of the activities mentioned above must apply for and receive one of the permits from APHIS before proceeding.

- **Permit for movement and importation.** This requires the applicant to disclose the nature of the organism, its origin, and its intended use.
- **Permit for release into the environment.** APHIS oversees field-testing of biotechnology products. The applicant is required to provide information on the plant (including new genes and new gene products), its origin, the purpose of the test, experimental design, and precautions to be taken to prevent escape of pollen, plants, or plant parts from the experimental site.

For low risk or familiar organisms or classes of modification, the developer may apply for an expedited permit through the notification process. Furthermore, in order to be allowed to release a previously regulated organism or product in an unconfined manner and commercially sell the organism, the developer must petition APHIS for non-regulated status. When evaluating petitions for non-regulated status, APHIS considers the potential risks to agriculture due to release of the organism, under the Plant Protection Act, and to the environment, under NEPA and the Threatened and Endangered Species Act.

28.13.2 FDA

The decision to subject all biotechnology products to the same standards of regulation of traditional products was made by the FDA in 1997. In the Federal Register, vol. 57, the FDA directs that companies or researchers whose products meet one of the following criteria should submit them for testing:

- **Unexpected effects.** that is, the product produces unexpected genetic effects.
- **Known toxicants.** the product has higher than normal levels of toxicants than in other edible varieties of the same species.
- **Nutrient level.** the product has altered levels of essential nutrients.
- **New substances.** the chemical composition of the product is significantly different from existing normal products.
- **Allergenicity.** the product contains proteins that have allergenic properties.

- **Antibiotic resistance selectable marker.** the product is produced by a biotechnology process that utilizes genetic markers that could adversely impact current clinically useful antibiotics.
- **Plants developed to make specialty non-food substance.** the plants are engineered to produce pharmaceuticals or industrial compounds.
- **Issues specific to animal feed.** the product's chemical composition regarding nutrient and toxins is significantly different from levels in similar products used for feed.

In addition to these federal regulatory activities, individual states are at liberty to develop and implement additional regulations. Exempted from pre-market approval are products that are classified as **GRAS (generally recognized as safe)**. Such food products may have been engineered to express proteins. However, a GRAS substance is excluded from the definition of a food additive. FDA encourages developers of engineered foods and feeds to consult with them throughout the development process to ensure that the product does not need to be regulated by the FDA.

28.13.3 EPA

The EPA regulates pesticides through a registration process. Its definition of a pesticide is any substance or mixture of substances intended for preventing, destroying, repelling, or mitigating pests. A new category of pesticides is the **Plant Incorporated Protectants (PIPs)**. These are substances that have been genetically engineered into plants so that the plants become pests by producing the substance (pesticides) in its tissues (e.g., *Bt* crops). Although plants engineered to be herbicide resistant are not classified as PIPs, and thus are not regulated by the EPA, the use of herbicide-tolerant plants is nonetheless subject to EPA regulation. The authority for such regulation is provided under the Federal Insecticide, Fungicide, and Rodenticide Act (FIFRA) and the Federal Food and Drug Cosmetic Act (FFDCA). If a plant producing a plant pesticide is intended to be used for food, the EPA must establish a "safe level" of the pesticide residue allowed. The EPA defines a safe level as a reasonable certainty that no harm will result from aggregate exposure to the pesticide chemical residue, including all anticipated dietary exposures and all other exposures for which there is reliable information. If the EPA determines that no harm will result from the aggregate exposure to the pesticide residue,

Table 28.2 Some viral coat proteins that have EPA exemption.

Papaya ringspot virus coat protein.
Potato leaf roll virus (PLRV) replicase protein as produced in plants.
Potato virus Y coat protein.
Watermelon mosaic virus (WMV2) coat protein in squash.
Zucchini yellow mosaic (ZYMV) coat protein.
WMV2 and ZYMV in ASGROW ZWO.

it can exempt the pesticide from needing a tolerance level. The EPA has categorically exempted the nucleic acids that encode PIPs, and has also exempted genetically engineered products that utilize viral coat proteins to protect plants against viral infections from the requirement of a tolerance prior to being used. EPA has also exempted all Bt products currently registered for use as pesticides with the EPA. Some of the products are listed in Table 28.2.

28.13.4 The concept of biopiracy

Biopiracy is the term used to describe the illegal acquisition of biodiversity and the associated traditional knowledge and indigenous technologies (sometimes dubbed "take-and-run"). Even though this activity may be engaged in by a variety of people or groups, large corporations, which often have the capacity to commercialize products from such activities, are often singled out as the chief culprits. The activities are deemed illegal because perpetrators circumvent existing international conventions and their associated national laws. More importantly, the pirated materials become protected through the application of IPR, essentially requiring the owners to request or pay to use their own materials. Further, because of the perceived or realized profitability of such activities, the countries of origin of the biodiversity are seen to be economically disadvantaged or even deliberately robbed or taken advantage of. Whereas two wrongs do not make a right, some point to the fact that biopiracy is comparable to the intellectual piracy whereby developing countries are noted for infringing upon patents, copyrights, and other intellectual property owned by developed countries.

Another term, **bioprospecting**, is used to describe activities whereby individuals or groups pursue the exploration for biodiversity and relevant traditional knowledge that have potential for commercial exploitation. Some see such undertakings as only a precursor to biopiracy. Simply stated, bioprospecting identifies valuable biological resources without taking; biopiracy goes the extra step of taking the materials without appropriate permission or compensation. If biopiracy is illegal, then corporate entities that usually profit from it have accomplices in individuals or entities (local government, non-governmental agencies, researchers from universities and other agencies that pursue germplasm explorations, etc.) that have ready access to these valuable biological resources. These "middlemen" receive various kinds of compensation for their efforts (e.g., research funding, conservation support, up-front fees). However, if a good argument can be made for bioprospecting, it is that the process can lead to the discovery novel compounds or medicinal or other values that are unknown to indigenous populations.

Illegal acquisition of biodiversity can be deterred through enforcement of international treaties and national laws and punishing offenders. As previously discussed, the Convention on Biological Diversity is one such international treaty designed to recognize the sovereignty of member nations over indigenous biodiversity and to require them to commit to conservation practices, sustainable development and use of the resources, as well as equitable sharing of the benefits from their exploitation (called the Access and Benefits Sharing Agreements – ABAs). An example of patent cancellation as a result of a violation of an international treaty occurred in 1995. The patent for "Use of Turmeric in Wound Healing" was revoked after the Indian Council of Scientific and Industrial Research provided evidence to establish the fact that such a medicinal use of turmeric had been known in India for many generations.

Biopiracy and bioprospecting have been costly to developing countries. Specific cases that pertain to Mexico involve the yellow bean and pozol, a traditional drink. A Colorado-based seed company, POD-NERS, purchased yellow bean seeds in Sonoral, Mexico, and secured a patent (US #5984079) two years later in 1996 for bean seed. Dubbed Enola, the Colorado company sued two Mexican food producers (Productos Verde Valle and Tutuli Produce) as well US importers of the bean for patent infringement. This action had adverse financial consequences on the operation of the Mexican companies and the local farmers (export sales declined by 90% following the filing of the lawsuit). The patent was challenged in

2005 by CIAT – International Center for Tropical Agriculture. The US Patent Office ruled in favor of the farmers. The patent was finally revoked in 2009, after an appeal was unsuccessful. In another case, the Dutch corporation, Quest International, in conjunction with the University of Minnesota, jointly secured a patent (US #5919695) for "pozol", a drink made from fermented corn. This drink is attributed to the ancient Mayan culture and touted for its medicinal values. The defense of the patentees is that their claim for patentability is not for the drink *per se* but rather for the active ingredient (the microbe used in the fermentation process).

28.14 The impact of IPRs on plant breeding

Modern plant breeding is a science, art and business. It is a long way from farmers selecting and saving seed from good-looking plants or fruits for planting the subsequent year's crop. Development of modern cultivars can take over a decade and cost millions of dollars. The granting of intellectual property right (IPR) to plant breeders for their inventions was designed to permit them to recoup their investment in the cultivar development efforts. Further, it is hoped that such a legal protection would stimulate research and development. However, in most cases, research indicates that trade secrets and rapid innovation are stronger stimulants of research and development than patents. The exception to this reality is in areas where inventions are easy to copy (e.g., pharmaceuticals and living organisms). Patents are aggressively pursued in these two areas.

A clear impact of IPR on modern plant breeding is the increased involvement of the private sector. The private sector dominates in the breeding of crops with high economic returns. Similarly, about 75% of utility patents for plant biotechnologies originate in the private sector (since the Diamond v. Chakrabarty decision of 1980). Commercial companies are increasingly procuring utility patents for parents used in F_1 hybrid seed production (especially for GM products). The provision in the PVP legislation that allows breeders access to proprietary material for research and breeding purposes has greatly facilitated plant breeding by allowing breeders to build upon an invention (cultivar) rather than "re-inventing the wheel" as some put it. The development of transgenic crops can involve breeders in a complex web of IPRs.

For example, it was determined that the development of pro-vitamin A-rich rice, dubbed "golden rice", involved 70 IPs or property rights owned by 32 companies and universities! To proceed, the developers sought and were granted a license under the freedom-to-operate (FTO) provision of the IPR for humanitarian use (the rice was developed to benefit the poor in developing countries). What appears to be consequence of IPRs is that, even though researchers in the public sector have been major contributors to innovation through basic research in biotechnology, they are not exempt from the challenges associated with access to protected biotechnology tools.

On the other side of the equation, developed countries hold a disproportionate share of the capabilities in pharmaceutical sciences and genetic crop improvement (including plant breeding and genetic engineering), two of the major uses in developed countries of biodiversity from developing countries. For example, the 10 major pharmaceutical companies originate in the United States, United Kingdom, Switzerland, Sweden, France, and Germany. The 10 major seed companies come from the United States, Switzerland, United Kingdom, France, Mexico, Japan, and Germany (Fulton and Giannakas, 2001). With the exception of one company (Seminis, Mexico), they deal mostly with field crops, such as maize, soybean, and cotton (*Gossypium hirsutum*). Many of them have significant biotechnology investments, which have resulted in the release of transgenic cultivars that occupy a large portion of the cultivated area in the United States. Transgenic soybean, cotton, and maize were planted on 81%, 73%, and 40%, respectively, of their respective areas in 2003 (US Department of Agriculture Economic Research Service, 2003). Worldwide, the United States grew 66% of the total area grown with transgenic crops, followed by Argentina with 23%. All other countries had a participation of less than 10% (James, 2003).

The combination of molecular technology and the capability of protecting molecular inventions by IPR has led to significant activities in the private sector in the area of genetic engineering of crop plants. While large chemical companies did have the financial wherewithal to engage in genetic engineering research, they have had to complete their IPR portfolio by taking over biotechnology companies (often start-ups). They also had to acquire capabilities in classical plant breeding in order to develop cultivars as vehicles to deliver the results of their genetic engineering technology, such as herbicide tolerance or insect resistance. They also needed the necessary seed marketing channels. The last two objectives were achieved by buying smaller seed companies, which had neither the financial wherewithal nor technological track record to survive in this new environment. This has led to a situation in which only five major firms now sell genetically improved seeds: Monsanto, DuPont/Pioneer, Aventis, Syngenta, and Dow. These same companies account for about a quarter of total seed sales (Fulton and Giannakas, 2001; Falcon and Fowler, 2002; Pingali and Traxler, 2002). For example, in 1998, Monsanto and Pioneer HiBred controlled 15% and 39% of the US maize seed market, respectively. For soybean seed, these companies controlled around 24% and 17%, respectively, of the

market. For US cotton, Delta & Pine Land and Stoneville had 71% and 16%, respectively, of the seed market (Kalaitzandonakes and Hayenga, 2000). In 1999, 61% of the cotton area in the United States was planted to a small number of closely related cultivars in which transgenes had been introduced, such as Deltapine 90 and DES56 (US Department of Agriculture Agricultural Marketing Service, 1999). Thus, while the availability of IPR has allowed the private sector to step into the crop improvement area beyond hybridization (maize and vegetables), there has been an apparent price to pay in the reduction of the number of cultivars grown by farmers. This reduction would be of major concern if it were extended to centers of crop origins, where a wealth of genetic diversity among landraces is still grown by farmers (Gepts and Papa, 2003).

Key references and suggested reading

Acquaah, G. (2004). *Understanding biotechnology: An integrated and cyber-based approach*. Prentice Hall, Upper Saddle River, NJ.

Andrews, L.B. (2002). Genes and patent policy: rethinking intellectual property rights. *Nature Reviews Genetics*, 3:803–808.

Arnold, B.E., and Ogielska-Zei, E. (2002). Patenting genes and genetic research tools: good or bad for innovation? *Annual Review of Genomics and Human Genetics*, 3:415–432.

Atkinson, R.C., Beachy, R.N., Conway, G., *et al.* (2003). Public sector collaboration for agricultural IP management. *Science*, 301:174–175.

Baumgardt, B.R. and Martin, M.A. (eds) (1991). *Agricultural biotechnology: issues and choices*. Purdue University Agricultural Experiment Station, West Lafayette, IN.

Borlaug, N.E. (2000). Ending world hunger. The promise of biotechnology and the threat of antiscience zealotry. *Plant Physiology*, 124:487–490.

Brill, W. (1985). Safety concerns and genetic engineering in agriculture. *Science*, 227:381–384.

Burgiel, S., and Cosbey, A. (2000). *The Cartagena Protocol on Biosafety: An Analysis on Results*. IISD (International Institute for Sustainable Development), Manitoba, Canada.

Crespi, R.S. (1988). *A basic guide to patenting in biotechnology*. Cambridge University Press, Cambridge.

Falcon, W.P., and Fowler, C. (2002). Carving up the commons: emergence of a new international regime for germplasm development and transfer. *Food Policy*, 27:197–222.

Fulton, M., and Giannakas, K. (2001). Agricultural biotechnology and industry structure. *AgBioForum*, 4:137–151.

Gaythwaite, D.M. Intellectual Property and Technical Know-how (1999). in *Biotechnology. The Science and Business* (eds V. Moses and R.E. Cape). Harwood Academic Publishers, London, UK, pp. 67–88.

Gepts, P., and Papa, R. (2003). Possible effects of (trans) gene flow from crops on the genetic diversity from landraces and wild relatives. *Environmental Biosafety Research*, 2:89–103.

James C. (2003). Global Review of Commercialized Transgenic Crops: 2002 Feature: Bt Maize (Executive Summary). International Service for the Acquisition of Agri-Biotech Applications. http://www.isaaa.org/Press_release/Briefs29-2003/Briefs29-2003%20ESummary.pdf (accessed December 22, 2003).

Kalaitzandonakes, N., and Hayenga, M. (2000). Structural change in the biotechnology and seed industrial complex: theory and evidence, in *Transitions in Agbiotech: Economics of Strategy and Policy* (ed. W. Lesser). Food Marketing Policy Center, University of Connecticut, Storrs, CT

Kershen, D.L. (2004). Legal liability issues in agricultural biotechnology. *Crop Science*, 44:456–463.

Lesser, W., and Mutschler M.A. (2004). Balancing investment incentives social benefits when protecting plant varieties implementing initial variety systems. *Crop Science*, 44:1113–1120.

National Research Council (1997). *Intellectual property rights and plant biotechnology*. National Academy Press, Washington, DC.

Pingali, P.L., and Traxler, G. (2002). Changing locus of agricultural research: Will the poor benefit from biotechnology and privatization trends? *Food Policy*, 27:223–228.

Sechley, K.A., and Schroeder, H. (2002). Intellectual property protection of plant biotechnology innovations. *Trends in Biotechnology*, 20:456–461.

Sanders, P.R., Thomas, C.L., Groth, M.E., Astwood, J.M., and Fuchs, R.L. (1998). Safety Assessment of Insect-Protected corn, in *Biotechnology and Safety Assessment* (ed. J.A. Thomas), 2nd edn. Taylor and Francis, Philadelphia, PA.

The Royal Society of Canada, (2001). Expert Panel on the Future of Food Biotechnology Report. The Royal Society of Canada, Ottawa, Canada.

UPOV (1996). *Model law on the protection of new varieties of plants*. UPOV, Geneva.

US Department of Agriculture, Economic Research Service (2003). Adoption of genetically engineered crops in the US. USDA-ERS [http://www.ers.usda.gov/Data/BiotechCrops/ (accessed March 30, 2012)].

US Department of Agriculture, Agricultural Marketing Service (1999). Cotton Varieties Planted: 1999 Crop. Agricultural Marketing Service, Memphis, TN

Internet resources

http://www.wipo.int/patentscope/en/patents/– General information about patents (accessed March 18, 2012).

http://www.codexalimentarius.net/– Codex Alimentarius Commission (accessed March 19, 2012).

http://www.wto.org/english/tratop_e/trips_e/trips_e.htm – The TRIPs agreement (accessed March 18, 2012).

Two additional international treaties are relevant to this discussion. The Convention on B Parties has led to the development of "decisions" and "protocols" (e.g. http://www.biodiv.org/convention/cops.asp; accessed March 18, 2012). One of these protocols is the so-called Cartagena protocol, which deals with the handling, transfer, and release of transgenic organisms (http://www.biodiv.org/biosafety/(accessed March 18, 2012); Burgiel and Cosbey, 2000) the other is Biological Diversity (CBD; http://www.biodiv.org/; accessed March 18, 2012).

Outcomes assessment

Part A

Please answer the following questions true or false.

1 When a single gene is patented by more than one company the patent is called a reach-through patent.
2 Mixtures of ingredients can be patented.
3 A plant patent is awarded for 30 years.
4 A patent is described as a "negative right".
5 The Biosafety Protocol was drafted in 2002.

Part B

Please answer the following questions.

1 What is a patent?
2 What it the importance of intellectual property right in biotechnology?
3 All inventions are not patentable. Explain.
4 What is a utility patent?
5 A patent is not a "positive right". Explain.
6 What agencies oversee the regulation of biotechnology in the United States?
7 How is regulation of biotechnology in the United States different from what obtains in the European Union?
8 What is bioprospecting?

Part C

Please write a brief essay on each of the following topics.

1 Discuss the concept of "freedom of use" in patent law.
2 List and discuss the basic steps in applying for a patent.
3 What is the impact of IPRs on plant breeding?

29

Value-driven concepts and social concerns

Purpose and expected outcomes

In a highly heterogeneous global community comprising of a dizzying array of cultures, reaching a consensus on any issue is often short of miraculous. Societies operate different value systems. In democratic societies, debates of issues are influenced by many factors, including religion, politics, socioeconomics, and personal preferences. The subjects that generate most debate pertain to life and living. Is life sacred and untouchable or can it be manipulated or taken at will? Life depends on the environment. Who is responsible for a healthy environment? Some eat plants only; others eat both plants and animals as food sources. It is not difficult to see how plant breeding, the science and art of manipulating plants (living things), can generate controversy. This chapter is devoted to social issues of an ethical and value-driven nature that are associated with the improvement of plants.

After studying this chapter the student should be able to:

1 Discuss the impact of ethics and morality on the plant breeding industry.
2 Discuss selected controversial issues in society regarding the development and deployment of modern plant breeding tools and products.

29.1 Concepts of ethical, moral, and values

The following brief introduction to ethics, morals, and values is designed to help the student become better equipped to participate in debates that center around these issues in relation to genetic manipulation of plants. **Ethics** is the science of morals in human conduct (i.e., a study of moral principles). **Morals** are concerned with the accepted rules and standards of human behavior in a society. They involve the concept of right or wrong, goodness or badness of human character or behavior. **Value** is basically the worth attached to something. In other words, ethics is evaluative of the decisions people make and the actions they take as they are presented with dilemmas. **Morality** depends on values in order to determine the goodness or badness of an action. In a pluralistic society, there are differences in the sense

Principles of Plant Genetics and Breeding, Second Edition. George Acquaah.
© 2012 John Wiley & Sons, Ltd. Published 2012 by John Wiley & Sons, Ltd.

of values (i.e., relativism). Consequently, there is a variety of moral theories that do not necessarily constitute truth. Further, law, religion, and custom should be distinguished from morality. In law, lawmakers define what is right or wrong. Those who break the law are subject to punishment prescribed by the legislature. In religion, right or wrong is based on revelation or scriptural authority. Whatever choice that is made has eternal consequences. In the case of custom, tradition determines what is acceptable or not, and society expresses approval or disapproval of an action.

29.2 Evolution of social debates on science-based issues

Higher education is all about knowledge, its discovery or creation, application, and dissemination for the betterment of society. Society, however, is not an innocent bystander in the development and deployment of science and technology. When appropriate, people interject their opinions in the process. Social debates of this nature tend to follow a certain pattern. Scientists are their own toughest critics. The first level of discussion occurs among researchers who debate the technicalities and soundness of the innovation or the issue. As the new scientific innovation gains public attention, interest groups and individuals start to voice their opinions for or against it. If the issue being debated impinges on traditional hot-button issues, deep divisions begin to form, as people take sides and dig in their heels for the long haul. This is the second stage of the evolution of a public debate. As the debate continues and people become more passionate about their positions, a third phase may arise whereby deep divisions may cause undue polarization of the public. Politicians may capitalize on the state of affairs for political advantage. Other activist groups may also take sides. In this chapter, public debate of biotechnology is discussed.

29.3 Ethics in plant breeding

Manipulating plants through conventional plant breeding efforts have generally proceeded without fanfare and objections from the consumers. Significant protestations from the consuming public started when plant breeders added biotechnology to their tool bag.

29.3.1 The biotechnology debate

Public perceptions about biotechnology products are rooted in the perceived risks that these products pose to social and personal values. Public reaction to biotechnology often is influenced by activism and special interest propaganda. The biotechnology debate is rooted in three fundamental disagreements:

(i) **Scientific disagreements.** Society is concerned about the potential risk that the development and application of biotechnology poses to humans, animals and environmental health. These issues involve empirical questions and are usually resolved by scientific methods. However, they are not exclusively resolvable by the scientific method of enquiry. Sometimes, value judgment is critical in their resolution. For example, the way to handle uncertainties in scientific data and the definition of the levels of risk deemed acceptable are both value judgments.

(ii) **Political disagreements.** Political disagreements are generally about the social and economic impacts of biotechnology based on the various political viewpoints. Political disagreements usually intensify during an election season and may swing in favor of one side, depending on the dominant political ideology of the day.

(iii) **Religious, ethical and philosophical disagreements.** These disagreements are often faith-based and include issues about morality and whether scientists are playing God, or whether the biotechnology products are natural. Faith-based disagreements are difficult to resolve in a pluralistic society. Sometimes, such issues incite extremist acts or vigilantism (e.g., destroying field tests or laboratory projects).

29.3.2 Plant biotechnology: ethical and value issues

The issues raised in this section are only a sample of public opinion intended to generate discussion, not to support the author's point of view. The issue of ethics, morals, and values in society is more important in a society in which religion is important to a significant portion of the population. Where there is a belief in God the Creator, a major source of discord in society regarding biotechnology is the notion that scientists are playing God when they fail to respect human limitations. God, humanity and nature are linked, God being the creator of both. Some people see nature as God's creation for the benefit of humans who therefore can use plants, animals, and the eco-system for their purposes, as they deem necessary. Others see nature as a sacred creation that must be respected and not tempered with. Does this respect mean that humans cannot manipulate nature? What cannot be denied is that the Creator has endowed humans with considerable creative genius. The obvious question then is whether exercising creativity through plant breeding and biotechnology is within the scope of this endowment or whether it is tantamount to an infringement on divine prerogative? For those who see nature as a gift to humans for their use, recombining genetic materials may be justified as just another way of using natural resources.

to be able to correctly evaluative the choices, decisions and acts as they pertain to genetic manipulation, there is the need for certain basic sets of information to be available. One set pertains to the values we attribute to things and acts we perform, the other set being value-free. Scientists, traditionally, generate value-free information. However, both kinds of information (tested empirically and experientially) and their impact need to be accumulated for use in making choices and decisions about biotechnology.

The ethical issues and the passion with which they are debated in the public arena vary among applications. Manipulation of the food chain seems to attract more attention than clinical applications (e.g., xenografts). For example, heart valves from pigs have been used in humans without fanfare. However, GM grains have encountered considerable public opposition from certain quarters. In general, the ethical issues of concern to the public are the impacts of biotechnology on human health and safety, environmental impacts, intrusions into the natural order, invasion of privacy, issues of rights and justice, economics, and others. It is important that both benefits and risks of biotechnology be considered in making ethical decisions about the discipline.

The problem is that, at the moment, we are limited in our knowledge about the full benefits and risks of biotechnology. Consequently, we are in danger of either underestimating or overestimating the potential of biotechnology for good or evil. Further, public reaction may be rooted in undue fear or hope stemming from misunderstanding, misinformation, or lack of information about various aspects of genetic manipulation of plants.

New technologies often tend to tip the scales in favor of those with resources to acquire it. They are most likely to be adopted if they increase profitability to producers while lowering the cost to consumers. There is also the issue of the developing countries. Many of the germplasm resources used in plant and animal improvement are derived from these regions of the world. The debate over patenting biological material is often linked to this fact.

29.4 Risk analysis of biotechnology

Risk analysis of biotechnology is complicated by the fact the activity is unique for the crop species, the genetic modification, and the production environment. A useful and fair analysis of the impact of biotechnology would be obtained if risk analysis of a biotech product were done in comparison with competing products or technologies. Examples for fair analysis would be to compare chemical pesticides with *Bt* products; use of glyphosate herbicide with glyphosate resistant crops, compared to the use of herbicide atrazine or other weed management methods; or planting GM crops with high productivity compared to clearing new land to plant conventional lower productivity cultivars. In conducting risk assessment, it is important that the process enhances consumer confidence and trust, without which marketing GM products is bound to be problematic. In part, public perceptions and attitudes about biotechnology are shaped by concerns about the risks and safety (acceptability of risk) of genetically engineered foods and other products. These biotechnology products are perceived as posing risks to a variety of social and personal values.

An expert panel on the future of food biotechnology commissioned by the Canadian Food Inspection Agency and Environment Canada categorized the values that are perceived by the public as being placed at risk by biotechnology into three categories:

(i) **Potential risks to the health of human beings, animals, and natural environment.** The risks to human health and the environment are at the top of the list of public concerns about the impact of biotechnology on society.

(ii) **Potential risks to social, political, and economic relationships and values.** Commonly, the public is concerned about the monopoly of certain industries (e.g., seed) by multinational corporations to the detriment of small producers and the risk of increased dependency of developing economies on these monopolies. It is the opinion of many experts that the level of risk acceptable by the public depends on the overriding benefits to be achieved (risk–cost benefit).

(iii) **Potential risks to fundamental philosophical, religious or metaphysical values held by different individuals and groups.** This category addresses the issue the public takes with the *process* of biotechnology rather than the product or *impacts*. The concern is the risk of playing God by implementing processes that are unnatural to alter nature.

The extent to which the public is willing to be exposed to unknown or uncertain risks, and how much risk is acceptable, is influenced by social, economic, and philosophical factors. People will be more willing, for example, to accept a higher risk level if they are strongly convinced about the benefits of adopting biotechnology products or, on the other hand, the adverse consequences of not adopting biotechnology products.

29.5 Genetic use restriction systems

Plant breeders may protect their inventions (cultivars) by seeking patents or plant variety protection. However, the legal provisions are effective in protecting proprietary material from abuse only if they are enforced or the farmers are trusted to abide by the legal restrictions. Researchers have been working on protection systems that are self-regulatory, needing no policing for enforcement. The first of such systems was unveiled in 1998. Developed jointly by the USDA and the Delta and Pine Land Company, the **Technology Protection System (TPS)** was awarded a patent in 1998. The nature of the patent allows each party to act independently from the other. The original genetic molecular switch was inserted into tobacco and cotton. Delta and Pine Land is the world leader in cotton seed production. Soon after the announcement, the technology was greeted by negative attacks from activists and other sources. The Rural Advancement Foundation International

(RAFI) (now the ETC Group) described it in near derogatory terms as the **"Terminator Technology"**, a term that appears to have stuck. To avoid this unscientific term, a new term was proposed and introduced in 1999, the **Genetic Use Restriction System (GURT)**. The term is broadly used to describe the use of exogenous substances as inducers to control the expression of a plant's genetic traits (e.g., trait for sterility, color, ripening, and cold tolerance). The restriction of a specific trait in a plant is called the T-GURT (also derided by activist as the "Traitor Technology"); the V-GURT refers to the use of genetic engineering of plants to produce sterile seeds (i.e., the Terminator Technology).

29.5.1 How the TPS works

The TPS may be deployed in three basic steps:

(i) The terminator genes are spliced into the genome of the target crop.
(ii) The seed company initiates the terminator process prior to selling the seed to farmers, by treating the seed with a substance (an inducer).
(iii) Farmers plant and harvest the seed in the usual way; however, the seed is sterile and will not germinate upon replanting.

Seed sterilization by TPS may be accomplished by one of three scientific approaches. Generally, all approaches use known gene mechanisms to control the expression of genes of known functions. Three genes with on/off switches are strategically engineered into the plant to interact in a predetermined sequence, the last gene becoming activated very late in seed development. At this stage, the gene is turned on by an inducer, causing a toxin to be produced that kills the embryo. The three proposed approaches to accomplishing seed sterility are (Figure 29.1):

1 **Transfection of target plant cells with three different transgenes.** Three different but functionally related transcriptional units are used – a repressor gene, a recombinase gene, and toxin gene. The repressor gene codes for a protein called Ribosome Inhibitory Protein (RIP). A DNA fragment that is a binding site to the repressor gene is located between the promoter and the recombinase gene. In the absence of and exogenous inducer, the repressor binds to the binding site to prevent the

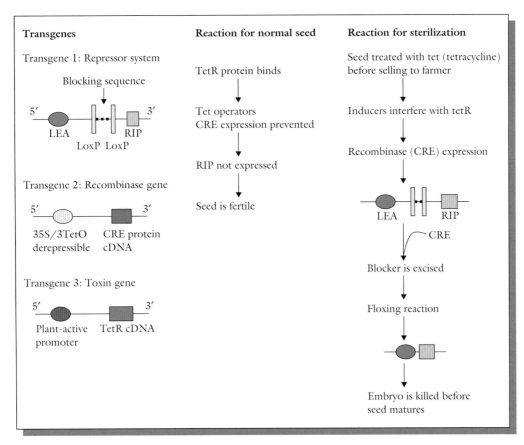

Figure 29.1 A diagrammatic presentation of how the technology protection system (TPS) works, as described by Marvin Oliver and colleagues. LEA (Late Embryogenesis Abundant) promoter is derived from cotton. It is active only during late stages of embryogenesis. RIP (Recombinase Inhibitory Protein) cDNA is derived from soapwort. It inhibits cellular translation, resulting in cell death. It expression is inhibited by the blocking sequence. LoxP sites are derived from bacteriophage P₁. It comprises direct repeats recognized by CRE (recombinase gene) protein mediates the removal of the blocker sequence. Column 1 shows three alternative genetic systems, while column 2 and 3 show the application of only one genetic system – the repressor system – in normal seed and in inducing seed sterility.

plant from producing the recombinase protein. The toxin gene is controlled by a late promoter. A blocker sequence is located between this promoter and the toxin gene to interfere with the ability of the promoter to turn the gene on. An inducer is needed to release the recombinase enzyme that has the capacity to snip out the blocker gene to allow the late promoter to turn on the toxin gene.

2 **Creation of a sterile hybrid.** In this approach, two fertile transgenic plants (A × B) containing two different sets of transgenes are developed. Plant A has a transcriptional unit made of the LEA promoter, LoxP sequences, blocking sequence, and RIP coding sequence. Plant B consists of a promoter that is active during germination and a CRE coding sequence. When the seed from a cross of A × B is planted, the floxing reaction occurs. Because the LEA promoter is inactive after late embryogenesis, the expression of the RIP is restricted to the seeds of the resulting mature plants from the hybridization.

3 **The use of an inducible promoter.** This strategy is similar to using the hybridization process. However, in this case, the germination-specific promoter is replaced by a promoter that is controlled directly by an exogenous substance.

29.5.2 The current status of GURTs

Since the first GURT patent was awarded jointly to the USDA and Delta and Pine Land, various entities, including universities and private corporations, have pursued the development of a variety of technologies for seed sterilization. These include Syngenta (with at least eight GURT patents), Dupont, Monsanto, BASF, Iowa State University, and Cornell University. TPS technology has so far not been commercially exploited. However, it appears various companies are working towards this objective. The Convention on Biological Diversity continues to discuss the issue. Like every technology, there are those who see the promise of the TSP and those who describe it in the most unflattering terms. Some of the stated potential advantages are:

- The TSP would be an incentive for further research and development of value added cultivars.
- It could possibly reduce the unintended gene flow from transgenic cultivars to conventional cultivars.
- It could reduce the incidence of volunteer weeds.

Distracters counter that:

- The only reason for developing and deploying the technology is to maximize the profits of the seed companies.
- Poor farmers cannot afford the seed; further, they cannot save seed to plant if they wanted to.
- The protection provided lasts longer than any other similar protection system already in place.

29.6 Public perceptions and fears about biotechnology

Stakeholders generally acknowledge that no technology is without potential for adverse impacts. Another truth is that the perception of technologies varies between scientists who develop them and the public who use them. It is tempting to contrast public perception, which is largely subjective, with scientific knowledge, which is largely objective. Further, public perception is viewed by some as basically illogical, irrational, and based on emotions rather than scientific facts. There is the temptation to condescend and disregard public opinion as uninformed. However, scientists also have perceptions of technologies they develop that are not necessarily rooted in facts,

because knowledge available may be scanty or inconclusive. There is the tendency, therefore, to either overestimate or underestimate the benefits and risks of technology.

Because of the imperfections that plague both camps, there are some misconceptions about biotechnology in the general community that hinder the development and application of the technology. A few of these are introduced for the sole purpose of initiating discussion on subject.

29.6.1 The techniques of biotechnology are alien, unnatural and too radical

It was made clear in Chapter 1 that the definition of biotechnology can be broad or narrow. In the broad sense, organisms have been used to make products for thousands of years (yeast in bakery products and bacteria in fermented products). In the narrow sense, biotechnology allows genes to be transferred unrestricted among living things, in effect disregarding natural genetic barriers. Whereas such interspecies gene exchange is not the norm in nature, there are examples in nature of various degrees of such horizontal gene transfers, especially on evolutionary time scales. Cross-pollinated species propagate through gene mixing, normally within the species, though. A more dramatic natural mixing occurred in wheat. Common wheat (*Triticum aestivum*) is an allopolyploid (hexaploid) consisting of three genomes of three different species. Certain microbes have the capacity to transfer some of their genetic material into the hosts they infect, even though the outcome may be undesirable for the hosts.

In biotechnology, the gene transfer system of choice in plants is *Agrobacterium* mediated transfer. This bacterium naturally transfers a portion of its genome into the plant it infects, a bacterium to plant gene transfer that forms galls in the plants. In biotechnology, scientists capitalize on this natural process to transfer genes of choice, only the result is not disease because the bacterium is disarmed prior to use.

The normal direction of genetic information transfer is from DNA to RNA to protein (the so-called central dogma of molecular biology). However, there are certain viruses that have RNA as genetic material. In retroviral infections, the single-stranded RNA is reverse transcribed to single-stranded DNA and then doubled to double-stranded DNA. This is incorporated into the host genome. Scientists can go a step further and replace the disease causing components of the retroviral genome with the desired genes for incorporation.

Mutations or heritable genetic changes occur naturally as spontaneous events. Such natural gene alterations produce variability for evolutionary processes to occur. Instead of haphazard and random events, scientists are able to induce genetic changes that are desired.

It is obvious from these selected examples that science merely imitates nature after studying to understand it. Rather than random events, scientists attempt to nudge nature purposefully to the advantage of humans. One may argue that just because nature does it does not mean humans should do the same. But then another may argue why not?

29.6.2 Genetic engineering is an exact science

It is true that specific genes can be identified, isolated, and characterized. However, the current gene transfer systems leave much to be desired. Once the DNA is delivered into the cell, scientists are not able to direct or predict where it will be inserted in the genome. Consequently, scientists cannot predict precisely the outcome of a transformation event. Where the gene inserts itself in the genome has a bearing on its expression. Even though this appears to be an uncontrollable event that could lead to unexpected phenotypes, scientists screen the products of transformation to identify the individual(s) in which the transgene apparently has been properly inserted and is functioning as desired. It should be pointed out that, compared to traditional breeding in which transfer of a desirable gene is usually accompanied by the transfer of numerous others, genetic engineering is relatively very precise.

29.6.3 Pesticide resistance in the agro-ecosystem as a result of the use of biotech crops is unavoidable

There are major crops in production with engineered targeted resistance to pests and herbicides. Pesticide resistance concerns (or "collateral resistance") center around four main aspects: (i) creating weeds out of cultivated cultivars, (ii) creating "super weeds" from existing weeds, (iii) creating resistant pests, and (iv) creating antibiotic resistance in harmful microbes.

The problems vary in the probability of occurring. The fact is that pests always manage to eventually adapt to any pest management strategy that is implemented repeatedly over a long period. This is especially true when the organism has a short life cycle (e.g., bacteria and many insects). The fear of creating a weed out of cultivated crops engineered to be herbicide tolerant stems from the fact that most modern herbicides are broad spectrum in action (i.e., kill many plant species). Bioengineered herbicide resistant crops are consequently resistant to broad-spectrum herbicides. If, for example Roundup Ready® soybean follows Roundup Ready® corn, volunteer corn plants will resist Roundup® and be a weed problem. Whereas crop rotations are desirable, it should not involve crops engineered with identical herbicide resistance. Further, using the same herbicide repeatedly for a long time is not a recommended agronomic practice, anyhow. Should one decided not to heed this advice, there are herbicides besides glyphosate that can control Roundup Ready® corn or soybean "weeds".

On the issue of biotechnology contributing to the development of "super weeds", the potential exists for gene escape from cultivated species engineered for herbicide tolerance to interbreed with wild relatives, thereby creating more competitive and difficult to control weeds (the so-called **super weeds**). While the movement of transgenes into wild relatives of transgenic crops is possible, this would occur only if the cultivated species of crops are grown where their weedy relatives with which they can interbreed also occur. This is not the case for the major crops that are transgenic for herbicide resistance in the United States (e.g., corn, soybean, potato). However, it is the case for squash and canola. It should be pointed out that the development of resistance to herbicides by weeds and other plants do occur, following prolonged exposure to certain herbicides. Further, irrespective of the herbicide or weed management tactic used, resistance to the chemical over the long haul is inevitable. This is why new herbicides will continue to be needed. There is no evidence to suggest that development of resistance is more problematic with the use of transgenic crops than direct use of herbicides. The issue of insects developing resistance to transgenic crops is similar to the weeds. Pests routinely overcome management tactics used against them.

To reduce the rate of insect resistance development to the Plant Incorporated Protectants, producers of transgenic crops require customers to grow a refuge of non-biotech and untreated crops around the Bt crops. The idea is that the refuge maintains a insect population susceptible to Bt that could interbreed with any pests on the biotech product that may develop resistance and prevent the resistance from becoming established in the population.

Certain biotechnology techniques utilize antibiotic resistance markers in developing transgenic crops. Consequently, the products contain these genes that can possibly be transferred to microbes in the environment. The fact is that the antibiotic resistance markers used in crop development do not provide resistance to most of the antibiotics used in the clinical setting.

29.6.4 Biotechnology products are unnatural and unsafe

Nature, and therefore anything natural, is perceived by some as superior to anything artificial. Modern foods that have been processed and modified in all and sundry ways are making us sick, they perceive; hence, the booming organic food and health food markets. Herbal medicine is being actively promoted in western societies. The public is concerned about biotechnology inadvertently introducing undesirable and unnatural chemicals into the food chain. The fact is that the public has embraced artificial components in food and medicine for a long time. Western therapy is almost exclusively dependent upon synthetic pharmaceuticals. Food additives and coloring are used routinely in both home and industrial food preparations. Instead of adding these materials to food during preparation, biotechnology seeks to make plants and animals produce nutrition-augmenting materials via natural processes to be included in the plant and animal tissues. For example, instead of post-harvest vitamin-enriching of the product (e.g., rice) for added value, the plant is engineered with the capacity to produce pro-vitamin A (Golden Rice).

29.7 Some concerns of plant breeders

Apart from these social concerns, two of the key concerns of interest to plant breeders include:

(i) That genetic engineering technology produces transgenic plants that are often useful as breeding

materials. The transgenic trait must be transferred into adapted cultivars (via backcrossing) to develop useful cultivars. Hence, breeding by genetic engineering, can be a lengthy process and expensive.

(ii) The transfer of proprietary genes into public germplasm has the potential to restrict the free access and distribution of germplasm among researchers.

29.8 GM foods and the issue of food allergy

Food allergy or food sensitivity is a common medical condition in society. It is described as an adverse immunologic reaction resulting from the ingestion, inhalation, or contact with a food or food additive. The specific factor that triggers the allergic reaction is the **allergen**. Food allergy should be distinguished from food intolerance (e.g., lactose intolerance). The most widely studied mechanism of food allergy is that mediated by **immunoglobulin E (IgE)**. Allergic reactions are produced immediately when IgE, an antibody, is exposed to an allergen (usually a protein substance). Such an exposure causes immune cells in the body (mast cells and basophils) to release different kinds of toxic mediators (e.g., histamine and leukotrienes), which then trigger an allergic reaction. The reaction may manifest in a variety of ways, ranging from minor itches to anaphylactic shock and death. There are some allergic reactions to foods that are not mediated by IgE (e.g., celiac disease or gluten-sensitive enteropathy).

The only current treatment for food allergy is avoidance. An estimated 34% of emergency room visits in the United States is for the treatment of anaphylaxis related to food allergy. Because of the rise in allergic disorders, the public is concerned about potential risk posed by bioengineered foods and food products. The most common food associated with food allergies (and accounting for over 90% of reported food allergies worldwide) are peanuts, tree nuts (e.g., almond, brazil nut, cashew, macadamia, hazelnut, pecan, pistachio, walnut), cow's milk, fish, shellfish (crustaceans and mollusks), soy, wheat, and sesame seed.

It is known that allergenic proteins can be transferred by genetic engineering from one organism to another. Such a transfer was confirmed in the case of the brazil nut 2S albumin storage protein that was transferred by rDNA technology to soybean to increase its methionine contents (legumes like soybean are low or deficient in the essential amino acid methionine). Tests for allergenic potential to humans by radioallergosorbent test (RAST) and skin prick test showed that the Brazil nut allergen had been transferred. It should be mentioned that commercial development of this food product was summarily discontinued. Further, there are no validated reports of allergic reactions to any of the currently marketed GM foods as a result of transgenic protein.

Genetic engineering of foods may have potential collateral changes (or pleiotropic effects) resulting from the transgene having an effect simultaneously on more than one characteristic of the host. In principle this may include an alteration of the intrinsic allergenicity of the protein by, for example, glycosylation, or alteration of the amount of allergenic protein produced. It is hence important to consider the potential changes in endogenous host allergens subsequent to gene transfer. Therefore, many of these physiochemical characteristics are assessed before registration of PIP products by the EPA and during the consultation process with the FDA. In the past, the EPA has denied approval for human consumption a PIP that demonstrated some allergenic characteristics.

29.9 The concept of organic plant breeding

The concept of organic breeding is relatively new and less developed than the concept of participatory plant breeding. **Organic agriculture** is simply agricultural production without the use of synthetic inputs (e.g., pesticides, fertilizers, herbicides). To achieve this, organic producers apply agro-ecological principles that promote the self-regulating capacity of the agro-ecological system (i.e., self-regulation of the soil, plant and animals).

29.9.1 The issue of "naturalness"

Organic production is also described as the "natural" way of production (alluding to the absence of synthetic inputs). Some researchers insist that naturalness of organic agriculture should not be limited to the absence of synthetic inputs or adherence to ecological principles but also to an acknowledgement of

integrity and wholeness in the production system. The concept of integrity implies a belonging to a specific natural entity (just like the concept of species) that can freely interbreed but which is genetically separated from others belonging to another natural group.

To be organic, this naturalness must be accorded proper consideration in how plants are propagated, cultivated, or genetically manipulated to be in conformity with the ideals of true organic agriculture. Further, plants are ascribed an *intrinsic value* that indicates that they are ethically relevant (in accordance with the attitude society has towards nature).

29.9.2 Need for organic plant breeding

Organic plant breeding is a relatively new concept of crop improvement. Currently, organic crop producers depend on seed and other propagules that are developed by conventional breeding procedures. If the integrity of an organic production system is to be maintained, the planting material used to initiate production should have organic origin.

There are several reasons why organic breeding is needed to service the organic production industry:

- Genetically modified (GM) crop varieties are not allowed in organic crop production. However, the trends in the production of some crops are to use genetically modified seed or other materials. Efforts should be made to develop non-GM varieties for organic production.
- Crop cultivars suited to organic production are different from those suited to conventional production. Successful cultivars should be adapted to specific soil and fertility conditions, be disease and insect pest resistant, and be competitive against weeds. Organic producers rely on natural fertility of the soil to a large extent, hence the need for cultivars that optimally interact with existing conditions. Crop cultivars with architecture and structure that reduce disease incidence are desirable.
- There is a need to preserve the integrity of plants. Conventional plant breeding methods sometimes violates natural barriers (genetic engineering, wide crosses) and consequently the integrity of plants.

29.10 Principles of organic plant breeding

The integrity approach to organic agriculture does not imply that breeders cannot manipulate plants but rather that the tools and techniques used in breeding, propagation, and cultivation of plants should not violate this integrity. Integrity of plants pertains to characters such as their nature, wholeness, species-specific characteristics, and their being in balance with species-specific environment.

Four levels of plant integrity have been proposed:

(i) **Integrity of life. Integrity of life** is defined as the state of wholeness or completeness of a living organism that allows it to perform all of its functions in a more or less autonomous fashion. Consequently, crop cultural practices that introduce synthetic chemicals may interfere with this self-regulating capacity of the plant, and hence incompatible with organic farming.

(ii) **Plant specific integrity. Plant specific integrity** is the state of wholeness or completeness of a plant that allows it to perform all of its plant-specific function. Plants and animals differ in specific ways at the cellular, whole organism and functional levels. Growing plants in artificial environments (tissue culture, hydroponics) infringes on the plant's ability to perform its natural functions (natural interaction with the soil). Using techniques that reduce the natural reproductive capacity of plants is unacceptable practice in organic breeding. For example, using cytoplasmic male sterility (cms) without fertility restorer genes will cause the progeny from cms hybrids to be sterile.

(iii) **Genotypic integrity. Genotypic integrity** is defined as the state of wholesomeness or completeness of the species-specific genome. Plant breeding depends on variability for success. Genotypic integrity is not violated as long as the variation is natural in origin. However, genetic engineering technology, which allows the transfer of genes across natural barriers, breaches this integrity principle.

(iv) **Phenotypic integrity. Phenotypic integrity** is defined as the state of wholesomeness or completeness of an individual plant, including its health. This principle is violated when plants are developed (or cultivated) in a manner that makes them unable to maintain themselves or complete their lifecycle in an organic production system without chemical protection. Chemical mutagenesis as a means of breeding violates this principle simply because chemicals are used in the process.

29.11 Acceptable organic plant breeding techniques

In terms of creating variability, techniques that do not violate the integrity of plants include crossing of cultivars, hybrid development with fertile F_1, testcrosses, backcrosses, and bridge crossing. However, techniques at the cell level (e.g., embryo cultures, somatic variation, and ovary culture), and the DNA level (e.g., genetic engineering, protoplast fusion) are not acceptable.

In terms of methods of selection, mass selection, pedigree selection, and even DNA diagnostics and marker-assisted selection are considered compatible with plant integrity. The diagnostic tools are acceptable because they do not cause genetic modification of plants.

29.12 Making agricultural biotechnology more acceptable to society

One of the enduring public misgivings about biotechnology is the perception that the technology violates natural laws. Consequently, researchers are constantly finding ways of allaying the fears of the public by developing new technologies for improving plants that are acceptable to the public. Whereas genetic engineering is generally viewed by the public as an unacceptable breeding technology, classical breeding methods are unquestioned as safe. However, there have been instances where traditional breeding has produced toxic products. As a case in point, two commercial varieties of potato were withdrawn from the market because they contained unsafe levels of solanine. Consequently, some scientists argue that cultivars produced by traditional means should be subjected to the same safety testing as genetically modified plants.

To address the concerns of the impact of selection markers on health and the environment, scientists are working on protocols for marker-free transgenic plants. The strategies being adopted include the avoidance of resistance marker genes by either using safer markers, using no markers at all, or a system in which the marker gene is eliminated (self-destruct) after the selection process. To eliminate the marker gene after selection, various strategies are currently in use (e.g., site-specific recombination, homologous recombination, transposition and cotransformation). Instead of marker genes, some efforts are directed at promoting regeneration after the transformation has occurred.

On another front, scientists are working on achieving gene transfer within biological boundaries without going through the sexual process. As previously stated, transgenesis is the genetic modification of a recipient plant with one or more genes from any non-plant organism, or from a donor plant that is sexually incompatible with the recipient plant. A new approach, called cisgenesis, is the genetic modification of a recipient plant with a natural gene from a cross-able – sexually compatible – plant. Cisgenic plants contain cisgenes but not transgenes.

All biotechnology, as previously indicated, is not genetic engineering. One of the commonly used approaches to facilitating crop improvement is via marker assisted selection (MAS). Molecular markers that are linked to desirable traits (e.g., disease resistance) are used to facilitate the selection process during breeding.

Another consumer friendly biotechnology is tissue culture. The technique allows scientists to rapidly multiply plant material without going through the sexual process, using any part of the plant, even cells, as materials of origin. Disease-free materials can be produced from infected tissue by using this technology.

29.13 The hallo effect of GM crops in the field

Research has shown that planting a genetically modified corn field inadvertently reduced the presence of pests in neighboring farms that did not plant GM corn. This so-called "halo effect" arises because the protection offered by the GM corn variety is so effective that it depresses the general population of insects in both the GM field and neighboring fields. The female moths of the corn borer cannot distinguish between Bt corn and non-Bt corn when they deposit their offspring. So, assuming that insects deposit their offspring evenly on corn in an area that is, say 75% sown to Bt corn, then 75% would be in Bt territory and would be killed. If this pattern is repeated over several years, the corn borer population would decline in the area, benefiting both GM and conventional corn seed growers. Further, even though the

conventional seed growers comprised a smaller percentage of the total corn crop in the area, the benefit realized from low insect population by the latter group was estimated to be disproportionately greater (about two-thirds the benefit) than for GM growers, because they did not have to bear the cost of GM seed.

29.14 The rise of minor pests in GM fields

Whenever a primary pest is targeted, other species are likely to rise in its place. The concept of biological pest control is that organisms generally have natural enemies, thereby maintaining a natural balance in the populations of organisms in an area. This balance is disturbed when an event causes a specific organism to be favored above the others. Such is the case when GM seed is deployed in cultivation. For example, in China, the use of GM cotton in production has steadily increased over the years. The effect is that the boll-worm (*Helicoverpa armigera*) population has been decimated. The consequence of this is that previously minor pests, the mirid bugs, have become a major pest in cotton farms. More significantly, researchers indicate that pesticide use is steadily increasing as this new economic pest has to be controlled. They attribute the rise of the mirid bugs to the fact that the deployment of GM varieties was accompanied by the use of narrow-spectrum insecticides that were ineffective against the minor pests of cotton. This trend threatens to erode and eliminate the benefits associated with the use of GM cotton varieties in cultivation. Further, it enforces the recommendation by some for GM variety deployment to be part of an integrated pest-management system to retain long-term benefits.

29.15 Who owns biodiversity?

Can life be patented? Can life be privatized? Privatizing life refers to the ownership of life forms and traditional knowledge, commonly through IPRs (international property rights), whereby individuals or corporations may claim legal ownership of biological resources and applicable processes. Arguments for and against ownership of life forms abound. Those who argue for privatization of life cite the huge investments they make in bioprospecting (searching for useful biodiversity) and developing processes for enhancing organisms or using them for making products. These activities are undertaken by individual researchers or corporations, who seek legal protection through patents of their innovations or discoveries.

On the other side of the debate, some claim that privatization leads to monopolistic control of natural resources upon which human survival depends (food, water, and health care). Ownership of biodiversity threatens food security because farmers are denied access to traditional agricultural resources for production (companies collect traditional seeds and sell back to famers value-added and chemically-dependent derivatives). Farmers will be compelled to pay royalties to use traditional medicines, seed (seed cannot be saved to plant the next season's crop), and livestock, thereby limiting their ability to diversify and advance their industry. Most of the useful biodiversity (an estimated 90%) resides in developing countries while the capacity (an estimated 90% of patents on life or life processes) to add value and transform raw biodiversity into derivatives resides in the develop countries. The cost of acquisition, maintenance and protection of patents, makes it financially unattainable for developing countries.

As biodiversity becomes commoditized, the developed countries with the capacity to pursue IP protection are reaching out to claim ownership of specific crop germplasm originating in developing countries, sometimes without proper authorization or compensation, action sometimes referred to by some as biopiracy. Some controversial patents involving foreign genetic resources have been awarded (and challenged) for some species such as yellow and popping beans (*P. vulgaris*), the neem tree (*Azadirachta indica*) oil, maca (*Lepidium meyenii*), and basmati rice (*O. sativa*). The current trend is to assign national sovereignty for biodiversity to individual countries, through events like the Convention on Biological Diversity and the work of the Commission on Genetic Resources for Food and Agriculture. There are international guidelines in place, emanating from the Convention on Biodiversity, which pertain to access and benefits sharing, called the Bonn Guidelines.

Furthermore, countries or groups of countries (e.g. The Philippines and Andean Pact countries) have or are in the process of adopting specific legislation governing bioprospecting activities. In these legislations, indigenous or local people must be given information about the ultimate use and purpose of the biological resource and they must give consent (prior informed consent). In a more general sense, international guidelines have now been developed by the CBD Conference of the Parties No. 6 at The Hague in 2002, which pertain to access and benefits sharing (the so-called Bonn Guidelines on Access to Genetic Resources and Fair and Equitable Sharing of the Benefits Arising Out of Their Utilization).

Key references and suggested reading

Ceccarelli, S. (1994). Specific adaptation and breeding for marginal conditions. *Euphytica*, 77:205–219.

Ceccarelli, S., and Grando, S. (1991). Environment of selection and type of germplasm in barley breeding for stress conditions. *Euphytica*, 57:207–219.

Heaf, D., and Wirt, J. (eds) (2001). *Intrinsic value and integrity of plants in the context of genetic engineering.* Proceedings of the Ifgene Workshop, May 9–11, 2001. International Forum for Genetic Engineering (Ifgene), Dornach, Switzerland, pp. 32–35.

Lammerts van Bueren, E.T., Struik, P.C., Tiemens-Hulscher, M., and Jacobsen, E. (2003). Concepts of intrinsic value and integrity of plants in organic plant breeding and propagation. *Crop Science*, **43**:1922–1929.

Lammerts van Bueren, E.T., Tiemens-Hulscher, M., Jongerden, J., van Mansvelt, J.D., den Nijs, A.P.M., and Ruivenkamp, G.T.P. (1999). *Sustainable organic plant breeding.* Louis Bolk Institute, Driebergen, The Netherlands.

Lu, Y., Wu, K., Jiang, Y., *et al.* (2010). Mirid Bug Outbreaks in Multiple Crops Correlated with Wide-Scale Adoption of Bt Cotton in China. *Science*, (advance online publication) doi: 10.1126/science.1187881.

Verhoog, H. (1992). The concept of intrinsic value and transgenic animals. *Journal of Agricultural & Environmental Ethics*, **5**(2):147–160.

Wang, S., Just, D., and Pinstrup-Anderson, P. (2008). Bt-cotton and secondary pests. *International Journal of Biotechnology*, **10**:113–120.

Internet resources

http://geography.berkeley.edu/ProjectsResources/BRP/ BRP.html – Directory of guidance documents relating to biodiversity and cultural knowledge research and prospecting (accessed March 19, 2012).

http://www.biodiv.org/programmes/socio-eco/benefit/ bonn.asp# – Bonn Guidelines (accessed March 19, 2012).

Outcomes assessment

Part A

Please answer the following questions true or false.

1 Ethics is the science of morals in society.
2 Biotechnology is an exact science.

Part B

1 What does the acronym GURT stand for?
2 Distinguish between ethics, morals and values.
3 In terms of ethics, what is your opinion about the development and application of biotechnology?
4 Give a disadvantage of the deployment of GURT technology to the producer.

Part C

Please write a brief essay on each of the following topics.

1 Describe how the TPS technology works.
2 Discuss the so-called "halo effect" associated with the use of GM cultivars in production.
3 Discuss ownership of biodiversity.
4 Describe how scientists are attempting to make biotechnology more acceptable to society.
5 Explain how food allergens may be a problem with GM breeding.
6 How would you respond to a charge that application of biotechnology could produce "supper weeds".
7 Discuss the risk associated with the development and deployment of GM cultivars.
8 Describe how the use of GM pest-resistant cultivars could result in the rise of minor pests in production.

30

International plant breeding efforts

Purpose and expected outcomes

The purpose of this chapter is to discuss the plant breeding efforts at international institutes and their impact on world food supply. Most of their efforts are directed at developing countries. Modern plant breeding is significantly responsible for the tremendous success of the agriculture of developed economies. It takes tremendous amounts of resources – human and financial – to undertake modern plant breeding research for developing new and improved cultivars for producers. The research infrastructure of most developing countries and the political support available limit the effectiveness of local scientists in addressing crop improvement needs. Because of the lack of economic markets in developing countries, the multinational corporations that dominate the commercial seed market in developed countries find it unattractive to invest in the improvement of crops that are of importance primarily to developing countries. Consequently, plant breeding efforts in developing countries depend on philanthropic organizations and the international agricultural centers they support for significant support of their local breeding programs.

After studying this chapter, the student should be able to:

1 List all the International Agricultural Research Centers (IARCs) and indicate their mandate crops.
2 Discuss the contributions of the International Agricultural Research Centers to world crop improvement.
3 Discuss the role of the International Agricultural Research Centers in germplasm collection and maintenance.
4 Discuss plant breeding efforts by national programs in developing countries.
5 Discuss the importance of orphan crops and the efforts being made to improve them.
6 Discuss the Green Revolution and its importance.

30.1 International Crop Research Centers

The frequent food deficits in developing countries often prompt the international community to intervene in the local food and agricultural production systems of these parts of the world. International involvement in the agriculture of developing countries led to a concerted effort to boost international

---

Principles of Plant Genetics and Breeding, Second Edition. George Acquaah.
© 2012 John Wiley & Sons, Ltd. Published 2012 by John Wiley & Sons, Ltd.

Table 30.1 The 16 centers supported by the Consultative Group on International Agricultural Research (CGIAR).

International Center	Founded	Location	Key mandate crop/activity
International Rice Research Institute (**IRRI**)	1960	Los Banos, Philippines	Rice
Centro Internacional de Mejoramiento de Maíz y Trigo (**CIMMYT**) (*International Center for the Improvement of Maize and Wheat*)	1966	El Batan, Mexico	Wheat, maize
International Institute of Tropical Agriculture (**IITA**)	1967	Ibadan, Nigeria	Maize, cassava, cowpea, soybean, yam
Centro Internacional de Agricultura Tropical (**CIAT**) (*International Center for Tropical Agriculture*)	1967	Cali, Columbia	Cassava, beans, tropical forages, rice
Centro Internacional de la Papa (**CIP**) (*International Potato Center*)	1970	Lima, Peru	Potato, sweet potato, Andean root and tuber crops
Africa Rice Center (**ARC**) (formerly **WARDA**) *West African Rice Development Association* (*WARDA*)	1970	Bouake, Ivory Coast	Rice improvement for Africa
International Crops Research Institute for the Semi-Arid Tropics (**ICRISAT**)	1972	Patancheru, India	Groundnut, pearl millet, sorghum, pigeonpea
International Plant Genetic Resources Institute (**IPGRI**) (formerly **IBPGR**) *International Board of Plant Genetic Resources (IBPGR)*	1974	Rome, Italy	Genetic resources, promote biodiversity
International Center for Agricultural Research in the Dry Areas (**ICARDA**)	1975	Aleppo, Syria	Barley, lentils, faba bean, wheat
International Food Policy Institute (**IFPRI**)	1975	Washington, DC, USA	Food policy issues
International Service for National Agricultural Research (**ISNAR**)	1979	The Hague, Netherlands	Institutional development, policy, dissemination of information
International Water Management Institute (**IWMI**) (formerly **IIMI**) *International Irrigation Management Institute (IIMI)*	1984	Colombo, Sri Lanka	Irrigation in developing countries
Center for International Forestry Research (**CIFOR**)	1993	Bogor, Indonesia	Forestry issues in the tropics
International Livestock Research Institute (**ILRI**) (formerly **ILCA, ILRAD**) *International Livestock Center for Africa (ILCA) International Laboratory for Research on Animal Diseases (ILRAD)*	1994	Nairobi, Kenya	Livestock production issues in Africa
World Fish Center (**WFC**) (formerly **ICLARM**) (*Int. Center for Living Aquatic Resources Management (ICLARM)*)	2002	Penang, Malaysia	Fisheries and other aquatic resources
World Agroforestry Center (**WAC**) (formerly **ICRAF**) (*International Center for Research in Agroforestry (ICRAF)*)	2002	Nairobi, Kenya	Improved agroforestry systems

agricultural research, especially in the tropical regions of the world, where the need is most urgent.

The initial efforts by the Ford Foundation and the Rockefeller Foundation led to the establishment of four international agricultural research centers (the acronyms are explained in Table 30.1):

1. CIAT in Columbia – focusing on general tropical agriculture.
2. CIMMYT in Mexico – focusing on tropical maize and wheat.
3. IITA in Nigeria – focusing on tropical agriculture.
4. IRRI in Philippines – focusing on rice.

One of the most dramatic impacts on tropical agriculture, dubbed the Green Revolution, was associated with two of these centers, CIMMYT and IRRI. As discussed later in this chapter, the Green Revolution was responsible for increased yields in wheat and rice through the breeding of high yielding and environmentally responsive cultivars of these major world food grains. This outstanding success prompted a discussion in the world community to extend the impact of the international agricultural research centers beyond Asia, which was the major beneficiary of the earlier efforts. Led by the World Bank and supported by the Food and Agricultural Organization (FAO), United Nations Development Program (UNDP), the Consultative Group on International Agricultural Research (CGIAR) was formed in 1971. The nucleus of 18 member nations has since increased to 62 (24 developing countries, 22 industrialized countries, 12 international/regional organizations, and 4 foundations). Similarly, the CGIAR centers have increased from the founding four to the current 16 (Table 30.1). Each of these centers is autonomous, with its own charter, international board of trustees, and staff. In 2001, the centers teamed up with an organization, the Future Harvest, to build support for international research. The IARCs then became known as **Future Harvest Centers**.

30.2 The CGIAR centers and their mission

The CGIAR centers are internationally recognized and respected for their leadership role in advancing agricultural research and crop productivity in developing countries. Unlike for-profit multinational companies, the CGIAR centers undertake the development of crops that may not be profitable to the private sector, but nonetheless are important to alleviating hunger in poor countries, the so-called **orphan crops**. A simple definition for orphan crops is crops that have regional food and economic importance but receive little or no attention from researchers because of lack of global trade importance. Other terminologies applied to these crops include "underutilized species", "neglected crops", "minor crops", or "niche crops". These crops tend to be of importance to developing countries and include tef (*Eragrostis tef*), finger millet, yam, sweet potato, bambara groundnuts, cowpea, plantain, and cassava. In sub-Saharan Africa, for example, sorghum and pearl millet are more important in addressing food security than rice and wheat! Further, cassava is ranked third in importance as a calorie source in Africa. The Global Facilitation Unit for Underutilized Crops described orphan or underutilized crops as:

- Representing an enormous wealth of agrobiodiversity and having great potential for contributing to improved incomes, food security and nutrition, and for combating the "hidden hunger" caused by micronutrient (vitamin and mineral) deficiencies.
- Are strongly linked to the cultural heritage of their places of origin.
- Are mainly local and traditional crops (with their ecotypes and landraces) or wild species whose distribution, biology, cultivation and uses are poorly documented.
- Tend to be adapted to specific agroecological niches and marginal land.
- Have weak or no formal seed supply systems.
- Are recognized to have traditional uses in localized areas.
- Are collected from the wild or produced in traditional production systems with little or no external inputs.
- Receive little attention from research, extension services, farmers, policy and decision makers, donors, technology providers and consumers.
- May be highly nutritious and/or have medicinal properties or other multiple uses.

30.2.1 Structural organization and mission

The current composition of member nations gives the CGIAR a strong north–south identity, with 22 countries from the south and 25 from the north (industrialized). Its original center-focused scientific research has shifted to a strategy that is partnership oriented. The CGIAR has established partnership committees with non-governmental organizations (NGO) and the private sector. Funding for its major research activities is in excess of $300 million annually. In 2008, its expenditure totaled $542 million. The centers attract highly qualified researchers from around the globe and include scientists from the region of its immediate influence.

The mission of the CGIAR has been modified over the years to reflect its priority focus on food security and poverty eradication. The current mission statement reads as follows:

" To achieve sustainable food security and reduce poverty in developing countries through scientific research and research-related activities in the fields of agriculture, forestry, fisheries, policy and environment."

30.2.2 Location and mandate of the CGIAR centers

The 16 CGIAR centers are located primarily in the tropical regions of the world. Four of these centers are located in Africa, with five in Asia. Each of these centers has mandate plants or animals on which it focuses its research efforts. These mandated areas are briefly summarized in Table 30.2. Some of these centers are located in regions of centers of diversity of their mandate crops.

30.2.3 Research emphasis

CGIAR research addresses issues involving plants, animals, soil, water, and policy as they impact

Table 30.2 The focus crops for the CGIAR research.

Cereals	Food legumes
Rice	Chickpea
Wheat	Cowpea
Maize (corn)	Beans
Barley	Lentil
Sorghum	Pigeon pea
Millet	Soybean
Roots, tubers, banana, and plantain	**Oil crops**
Cassava	Coconut
Potato	Groundnut (peanut)
Sweet potato	
Yam	
Banana and plantain	

productivity and the management of these natural resources for the benefit of developing countries.

The five areas of research focus are:

(i) **Sustainable production.** ensuring sustainable management and conservation of water, land and forests; focusing on crops, livestock, fisheries, forests and natural resources.

(ii) **Enhancing national capacities.** through joint research, policy support, training, and knowledge sharing; promoting opportunities for economic development through agricultural diversification and high value commodities and products.

(iii) **Germplasm improvement.** focusing on mandate or priority crops, livestock, trees and fish; reducing poverty and hunger by producing more and better food through genetic improvement.

(iv) **Germplasm collection.** sustaining agricultural biodiversity both in the field and in genebanks; currently maintains the world's largest seed collections in 11 genebanks.

(v) **Policy.** improving policies and facilitating institutional innovation; fostering research on policies that have a major impact on agriculture, food, health, spread of new technologies, and the management and conservation of natural resources.

30.2.4 The Alliance

A center-driven coalition of 15 International Research Centers – the Alliance – was created in 2006 to enhance collective action among the Centers and between the Centers and their partners. A key goal of the coalition is higher efficiency through programmatic collective actions that synergistically increase the overall impact on poverty alleviation and the environment. The Centers, their partners and stakeholders benefit from this coalition, which allows more efficient access to the expertise of the Centers for more effective program outputs and outcomes.

30.2.5 Mandate crops

The CGIAR scientists conduct research on cereals, roots, tubers, banana, plantain, food legumes, and oil crops. Other research areas are forestry, agroforestry, fisheries, livestock, and water management. A brief overview of research in selected crops is shown here.

Wheat

Research in wheat is concentrated at CIMMYT and ICARDA. CIMMYT is considered the world center for the breeding of bread wheat, durum wheat, and triticale. Over 50 million acres of the world's wheat are grown to cultivars developed at these CGIAR research centers. New wheat growing areas have been established in West Africa and North Africa. Cultivars have been developed with traits such as dwarf stature, disease resistance, efficient water and nutrient use, and tolerance to environmental stress.

Maize

Research in maize is conducted at CIMMYT and IITA. Major traits improved in maize are drought resistance, yield, protein quality, and resistance to maize streak and downy mildew.

Rice

Rice improvement is conducted at IRRI, WARDA, and CIAT. Research focus includes improving yield potential, developing hybrids for the tropics, and pest resistance.

Barley

Barley research is conducted mainly at ICARDA. Over 100 barley cultivars have been developed for use in over 30 countries.

Sorghum

ICRISAT is the world's center for grain sorghum research. The research objectives include developing early maturing cultivars, and disease resistance (e.g., sorghum midge).

Soybean

Soybean research is conducted at IITA. Breeding objectives include improved capacity to fix nitrogen without inoculation, high yields, and resistance to shattering.

Potato

Potato research is conducted at CIP where breeding objectives include disease resistance (*Phytophthora infestans*).

Industry highlights

Plant breeding research at ICRISAT

P.M. Gaur, K.B. Saxena, S.N. Nigam, B.V.S. Reddy, K.N. Rai, C.L.L. Gowda, and H.D. Upadhyaya

International Crops Research Institute for the Semi-Arid Tropics, Patancheru 502 324, AP, India

Introduction

The International Crops Research Institute for the Semi-Arid Tropics (ICRISAT) has a global mandate for the improvement of chickpea (*Cicer arietinum* L.), pigeonpea (*Cajanus cajan* (L.) Millsp.), groundnut (*Arachis hypogaea* L.), sorghum (*Sorghum bicolor* (L.) Moench), and pearl millet (*Pennisetum glaucum* (L.) R. Br.) (Figure B30.1). These crops are grown in about 100 million hectares globally, predominantly under rain-fed conditions by resource poor farmers of the semi-arid tropics (SAT).

ICRISAT has assembled over 104 000 accessions of these crops (17 258 chickpea, 13 632 pigeonpea, 15 419 groundnut, 36 774 sorghum, 21 594 pearl millet) through donations by various genebanks and national programs and joint explorations. These valuable genetic resources preserved in ICRISAT's genebank at Patancheru, India, have contributed significantly in strengthening breeding programs of ICRISAT and National Agricultural Research Systems (NARS) globally. Close to 1.2 million samples of these crops have so far been distributed to the NARS and ICRISAT scientists.

The crop improvement activities are conducted at ICRISAT's locations in India and Africa, and jointly with many national program scientists globally, wherever the mandate crops are cultivated. In African regions, development of varieties in all five crops continues to be the primary objective, while in Asia (specifically for India), the present emphasis is towards development of varieties in chickpea, groundnut and pigeonpea, and hybrids in sorghum, pearl millet, and pigeonpea. Towards this goal, ICRISAT develops segregating materials, populations, advanced breeding lines and

Figure B30.1 The five mandate crops of ICRISAT. Figure courtesy of P.M. Guar.

hybrid parents and supplies these to scientists in NARS, non-government organizations (NGOs) and the private sectors for evaluation and selection at their locations and utilization in their breeding programs. Based on performance in local, regional or national trials, varieties/hybrids are released or notified by the various national programs according to their own protocols and procedures.

Breeding objectives

The crop breeding priorities and strategies at ICRISAT have been dynamic, guided by the changing scenario of agriculture and development of new technologies; the priorities and strategies are reviewed and revised periodically based on the feedback from NARS scientists, extension personnel, farmers, consumers, and industry. Improved yield potential (mostly for grain but more recently also for fodder in sorghum, pearl millet and pigeonpea) is the common and foremost important breeding objective in all crops. The other major objectives include genetic improvement of resistance/tolerance to diseases (Fusarium wilt, Ascochyta blight, Botrytis gray mold in chickpea; wilt and sterility mosaic in pigeonpea; rust, early and late leaf spots, rosette and bud necrosis in groundnut; grain mold, anthracnose, and charcoal rot in sorghum; and downy mildew in pearl millet), insect pests (*Helicoverpa* pod borer in chickpea; *Helicoverpa* and *Maruca* pod borers in pigeonpea; shoot fly, stem borer, midge and head bug in sorghum) and abiotic stresses (drought and cold in chickpea, drought in groundnut; drought, salinity and acidity in sorghum); adaptation (early maturity in chickpea, pigeonpea and groundnut) and quality of grain (reduction in aflatoxin contamination in groundnut) and fodder (in sorghum, pearl millet and pigeonpea).

Breeding methods and techniques

Conventional breeding methods

Chickpea   A combination of pedigree and bulk method is generally used for selection after hybridization in this highly self-pollinated legume. The early segregating generations (F_2 and F_3) are invariably grown in a fusarium wilt-sick nursery and surviving plants (resistant to fusarium wilt) harvested as bulk. The selection of single plants starts from F_4 or later generations. Progeny evaluation is carried out in F_5 to F_7 generations. High yielding and nearly uniform progenies are bulked for replicated yield tests. The backcross method is used only occasionally to incorporate one or a few traits from a germplasm line, sometimes a wild species, to a well-adapted variety. Rapid generation advancement in greenhouse following a single seed descent (SSD) method is generally used for development of recombinant inbred lines (RILs).

Pigeonpea This is a partially out-crossed crop (out-crossing up to 30%) and the breeding methods generally recommended for self-pollinated crops are being used. Recurrent out-crossing and selfing within landraces has resulted in pigeonpea being heterozygous as well as heterogeneous for important agronomic traits. Such landraces contain a tremendous amount of genetic variability, which has been utilized very effectively to select/breed high yielding pure line varieties. Besides this, hybridization and pedigree selection is widely used. The limited natural out-crossing has been successfully exploited for increasing yield and stability through development of commercial hybrids using genetic male sterility. Currently, cytoplasmic male sterility is being used to develop and commercialize hybrids.

Groundnut Being a completely self-pollinated crop, the pedigree method is the most commonly used breeding method. This allows breeders to practice selection for highly heritable traits, such as plant type, pod and seed size and shape, and testa color, in early segregating generations. Selection for quantitative traits, such as yield and seed composition, are made in later generations. SSD and recurrent selection have been used very sparingly. Only limited use of backcrossing has been made, particularly in situations where one of the parents is a primitive land race or a compatible wild species.

Sorghum A trait-based pedigree breeding approach is being used in which the families are used as the selection units for resistance response, and individuals within the resistant families as selection units for grain yield. Also a simple mass selection-based recurrent method is being used to improve male sterile (ms_3 and ms_7 genes-based) populations to develop trait-based gene pools. Simultaneous test crossing and backcrossing the selected maintainer plants along with the selection for resistance trait and grain yield in the trait-based breeding programs are carried out for the purpose of improving male sterile lines for a specific resistance trait and high grain yield through heterosis.

Pearl millet Being a highly cross-pollinated crop (>85% outcrossing), pearl millet provides an opportunity for exploitation of heterosis. Various forms of recurrent selection have been used to develop open-pollinated varieties (OPVs). The availability of several alternative cytoplasmic-nuclear male sterility (CMS) systems and their restorers has enabled large-scale commercial exploitation of single cross hybrids in India. Pedigree breeding has been used in populations developed by recurrent selection, albeit on limited scale, to develop hybrid parents. Various forms of pedigree breeding have been extensively used in populations derived from hybridization between lines to develop hybrid parents. Backcross breeding has been extensively used in developing partially converted dwarf versions of several composites. Of course, backcross remains the only option to develop male sterile line (A-line) counterparts of maintainer lines (B-lines).

Marker assisted breeding

Marker assisted selection (MAS) is being considered as a potential method to hasten and improve precision and effectiveness of crop improvement. ICRISAT has established a high-throughput applied genomics laboratory and identified molecular markers for several important traits, such as stay green trait and resistance to shoot fly and *Striga* in sorghum, and downy mildew resistance and terminal drought tolerance in pearl millet. Research is underway to identify markers for root mass and resistance to ascochyta blight, botrytis gray mold, and Helicoverpa pod borer in chickpea, as well as Fusarium wilt resistance and fertility restorer genes in pigeonpea. MAS has been successfully practiced for some traits. For instance, marker assisted backcross breeding was used to incorporate resistance to downy mildew in pearl millet single cross hybrid HHB 67. Marker assisted breeding for terminal drought tolerance in pearl millet is in progress.

Transgenics

Transgenics have been developed in pigeonpea and chickpea with resistance to *Helicoverpa* pod borer by using *Bt Cry1A (b)* gene derived from the bacterium *Bacillus thuringiensis* and soybean trypsin inhibitor (*SbTI*) gene. The molecular characterization and insect bioassays are in progress. Efforts are also being made to develop transgenics in chickpea for tolerance to abiotic stresses, such as drought and low temperatures. Transgenics have been developed in groundnut for several genes, such as those encoding for viral coat protein of Indian peanut clump virus (IPCV) and groundnut rosette assistor virus (GRAV), replicase of IPCV, *Bt Cry1A (b)*, and chitinase from rice. In cereals, transgenics have been developed for resistance to stem borer in sorghum and are currently under greenhouse testing.

Major accomplishments

Chickpea

- Development of short-duration varieties (85–100 days at Patancheru, 17.4° N) that can escape terminal drought and provide wider adaptability to the crop, for example, ICCV 2 and KAK 2 in *kabuli* type and ICC 37 and JG 11 in *desi* type.
- Development of super-early *desi* chickpea lines, ICCV 96029 and ICCV 96030, which mature in 75–80 days at Patancheru. These lines are extensively being used in crossing programs as source of earliness by NARS in many countries.
- Identified high root biomass as an important trait for drought avoidance in terminal drought conditions. Lines with greater degree of drought tolerance (e.g. ICCVs 98901 to 98907) were developed by combining large root traits of ICC 4958 with few pinnules trait of ICCV 5680.
- Most chickpea cultivars are susceptible to chilling temperature at flowering. A number of cold tolerant lines (e.g. ICCV 88502, 88503, 88506, 88510, and 88516) have been developed which are able to set pod at low temperature.
- Several varieties with high and stable resistance to Fusarium wilt, the most important root disease of chickpea, have been developed (e.g., ICCV 10, ICCC 37, JG 11).
- Breeding lines with moderate to high level of resistance to the important foliar diseases, Ascochyta blight (e.g., ICCVs 04512, 04514, 04516) and Botrytis grey mold (e.g., ICCL 87322 and ICCV 88510), have been developed.
- Sources of high level of resistance are not available for pod borer (*Helicoverpa armigera* Hubner), which is the most important pest of chickpea worldwide. Several breeding lines/cultivars have been developed with some level of resistance, for example, ICCV 7, ICCV 10, ICCL 86102, and ICCL 86103. Further, efforts are being made to combine different mechanisms of resistance identified in the cultivated and wild germplasm.

Pigeonpea

- Extra-early and early maturing (90–120 days at Patancheru, 17.4° N) photo-insensitive varieties/lines have been developed that made cultivation of pigeonpea possible in a range of environments. Extra-early lines (e.g., ICPL 88039) allow farmers to take two crops (pigeonpea and wheat) in a year.
- Fusarium wilt (FW) and sterility mosaic (SM) are major pigeonpea diseases. A number of varieties with high resistance to FW and SM have been developed and some of these have combined resistance to both diseases (e.g., ICPL 87119).
- *Helicoverpa* and *Maruca* pod borers are the major insect pests. Sources of moderate resistance have been identified and a moderately resistant variety (Abhaya) has been released.
- The high protein trait was successfully transferred from wild species *C. scarabaeoides, C. sericeus,* and *C. albicans* to the cultivated species without sacrificing grain yield or seed size (e.g., HPL 40.)
- Commercial hybrids were initially developed using genetic male sterility (GMS) systems, for example, ICPH 8. These provided, on average, 30–35% more yield and greater stability in yield than the pure line cultivars. Recently, three stable cytoplasmic male sterility (CMS) systems have been developed and corresponding fertility restores have been identified to overcome the problems of hybrid seed production associated with GMS systems. Efforts are being made to develop commercial hybrids using CMS systems.

Groundnut

- Several drought tolerant and high yielding varieties have been developed, which perform well under rain-fed conditions, for eaxample, ICGS 5, ICGS 44, ICGS 76, and ICG (FDRS) 10 for mid-season drought and ICGS 11 and ICGS 37 for end-of-season drought in India and ICGV 86021 for end-of-season drought in Indonesia.
- Varieties with high levels of resistance to rust and moderate levels of resistance to late leaf spot have been developed and released by national programs to farmers in Asia (ICG (FDRS) 10 and ICGV 86590 in India) and Africa (ICG (FDRS) 4 in Mali). Resistance to early leaf spot has been introgressed from wild *Arachis* species.
- Varieties with field resistance to peanut bud necrosis disease (PBND), a widespread disease in Asia, have been developed (e.g., ICGS 11, ICGS 44, and ICGS 37). These varieties are generally resistant/tolerant to thrips, the vector of the disease. Some genotypes also show tolerance to the virus (ICGV 86031 and ICGV 86029).
- Groundnut rosette disease (GRD) is endemic to Africa and its nearby islands. A short-duration GRD resistant variety has been developed (ICGV-IS 96894) and released as Samnut 23 in Nigeria.
- The contamination of groundnut by aflatoxins is a serious problem in most groundnut producing countries. Genetic variation has been identified for pre-harvest seed infection, *in vitro* seed colonization, and aflatoxin

production. These resistances have been transferred to superior agronomic backgrounds, for example ICGV 88145 and 89104.

- Short-duration varieties are required where the growing season is short, crop suffers end-of-season drought, early frost occurs, and in multiple cropping system. Using the thermal time concept, several short-duration and high yielding cultivars have been developed, for example ICGV 86143 as BSR 1 in India, ICGV 86015 as Jayanti in Nepal, BARD 92 in Pakistan, HL 25 in Vietnam, ICGV 93382 as Sinpadetha 7 in Myanmar, ICGV 86072 as BARI Groundnut 5 in Bangladesh, and ICGV 93437 as Nyanda in Zimbabwe. It has also been possible to combine early maturity with high yield potential and tolerance to rust, late leaf spots and low temperature in ICGV 92267.

- Groundnut genotypes belonging to Spanish types have non-dormant seed and rain prior to harvest in such genotypes can cause seeds to sprout in the ground, resulting in loss of yield and poor quality of produce. Fresh seed dormancy trait has been successfully introduced in Spanish types from Virginia types, and several short-duration Spanish cultivars with fresh seed dormancy of 2–3 weeks have been developed, for example ICGVs 86155, 86156, 86158, 87378, 87921, and 93470.

Sorghum

- Several high yielding varieties have been developed and released in several countries in Africa and Asia for rain-fed drought-prone areas. Some varieties are popular in many countries, for example ICSV 112 in Zimbabwe, Kenya, Swaziland, Malawi, and Mozambique, ICSV 111 in Cameroon and Chad and Nigeria. Some varieties have been bred for dual purpose (grain and stover), for example ICSV 112 and ICSV745. A variety NTJ 2 is highly popular for its "roti" making quality.

- Several ICRISAT-bred improved hybrid parents have been extensively used by both public and private sector research organizations to develop and market hybrids in Asia. More than 30 hybrids based on ICRISAT-bred parents have been released in India and China. Notable among them are JKSH 22 in India, and Lio Za 4, Longsi 1, Jinza 12, and Gilaza 80 in China.

- *Striga*, an abnoxious obligate parasitic weed, is one of the most important biotic yield constraint in Africa, although less important in Asia. Several *Striga* resistant varieties (e.g., Framida in Burkina Faso and Ghana; SAR 1 in India) and seed parents (e.g., ICSA 579, ICSA 583, ICSA 584, ICSA 588, and ICSA 592) have been developed.

- Grain mold is an important disease of sorghum in Asia and Africa. Many grain mold resistant varieties have been developed. Among them PVK 801, besides being grain mold resistant, is dual-purpose variety with a good quality stover.

- Shoot fly, stem borer and midge are the major insect pests of sorghum. Midge resistant white grain varieties in tan color background (e.g., ICSV 745 and PM 13654) have been released in Australia. Several grain mold resistant (e.g., ICSA 300, ICSA 369, ICSA 400, ICSA 403 and ICSA 404) and shoot fly tolerant (e.g., ICSA 419 and ICSA 435 for rainy season and ICSA 445 and ICSA 452 for post rainy season) cytoplasmic-nuclear male sterility-based seed parents have been developed.

Pearl millet

- About 60 open-pollinated varieties (OPVs), 40 in Asia and 20 in African regions, have been developed primarily for grain yield and downy mildew resistance. The most popular OPVs include WC-C75 and ICTP 8203.

- Several hybrids and hybrid parents have been developed with resistance to downy mildew, ergot and smut. ICMH 451, the first ICRISAT-bred hybrid developed in 1986, covered an area of over one million hectare by mid 1990s. Later, ICRISAT developed and disseminated a wide range of breeding materials and over 90 male sterile lines for use by NARS and the private sector in hybrid development. Of the 70 pearl millet hybrids released in India, about 60 are based on ICRISAT-bred A-lines or on A-lines developed by public and the private sector from ICRSAT-bred germplasm.

- Alternative and more stable CMS sources were identified and characterized. The A_1 CMS system was not associated with susceptibility to downy mildew, ergot and smut. Using the stable A_4 CMS system, it was also shown that it is possible to quickly develop a male sterile population for use in breeding inter-population hybrids.

- Topcross, three-way and inter-population hybrids forms were identified to have numerous advantages over single cross hybrids, in terms of seed production economy and reduced vulnerability of downy mildew, ergot and smut diseases. Topcross hybrids were suggested to be the most efficient route to combine high yield potential of improved seed parents and adaptation of landrace-derived populations.

30.2.6 Selected accomplishments

The impact of the CGIARs on the agriculture of developing countries includes the following:

- **Food production and human nutrition.** Researchers have developed over 300 cultivars of wheat and rice and more than 200 cultivars of maize for farmers. Some of these cultivars, such as quality protein maize (QPM), have augmented nutritional profiles of major food crops, enhancing the health of consumers.
- **Natural resources conservation.** The CGIAR-developed technologies have been adopted by farmers resulting in reduced environmental degradation and conservation of water, soil, and biodiversity.
- **Reduction in pesticide use.** This has occurred through the development of pest-resistant cultivars and improved farming practices for use in developing countries.
- **Germplasm conservation.** The CGIAR maintains over 6000 accessions of germplasm, representing over 3000 species (including crops, forage, and pasture species).
- **Capacity building.** The CGIAR has trained over 75 000 scientists and technical staff from developing countries.

30.2.7 Education, training and technical assistance

The CGIAR centers provide opportunities for training of human resources in plant breeding and other areas for especially developing countries. This is achieved in a variety of ways, including short courses, research fellowships, internships and collaborations with local scientists. Local researchers participate in germplasm evaluations of materials from these centers. Capacity development for biotechnology and plant breeding is needed most developing countries.

30.3 Brief overview of plant breeding in developed countries

The purpose of this very brief overview is to provide a basis for contrasting plant breeding in developed countries with that in developing countries. Plant breeding in developed economies is conducted in both the public and private sectors. In the United States, the land grant university system ensures the training of professional plant breeders, whereas its researchers actively engage in plant breeding research, resulting in development of new technologies and development of new plant cultivars. Public sector agricultural research is well funded by the government at both state and national levels, and primarily has a not-for-profit philosophy. Researchers use both conventional and modern technologies in their research.

Private sector research in plant breeding is significant in developed economies. It is dominated by multinational corporations and is primarily for profit. Examples of such entities are Monsanto, Novartis, and Du Pont. These entities focus on high value crops (e.g., corn, wheat, rice, and soybean) that are grown widely over the world. Unlike the public sector, patent rights protect the inventions of private corporations. Even though patents exist in the public sector as well, access to such protected materials is often much easier than access to those in the private sector. The issue of intellectual property is a major one in plant breeding in developed countries, with consequences for crop improvement in developing countries.

30.4 Brief overview of plant breeding in developing counties

Whereas agricultural research in developed countries is generally well organized and well funded, formal agricultural research in most developing countries is limited and under-funded. Typically, plant breeding research in these countries is conducted primarily in the public domain at national agricultural research stations and the local universities. Further, the emphasis of breeding is on improving a few of the major food and cash crops of local importance. Many farmers use land races to grow many of the staple crops. Improved cultivars are obtained through plant introductions and the limited efforts of local research stations.

In terms of human capital, developing countries often depend on developed countries for the training of high level (graduate) researchers (e.g., plant breeders). Plant breeders in developing countries depend largely on conventional breeding technologies, since the exorbitant cost of some of the modern technologies (e.g., genetic engineering) prohibits their adoption. Cultivars are developed so that the farmer can maintain them, that is, seed can be saved for planting the next season's crop from the current season's harvest. Commercial seed companies are non-existent in most developing countries. It should be pointed out that the relatively more technologically advanced third world countries have very well funded plant breeding programs, which have produced outstanding results. Such countries include India, China, Brazil, and South Africa.

30.5 Plant breeding efforts in sub-Saharan Africa

One of the regions of the world that frequently experiences food deficits and famines is sub-Saharan Africa. This region also has some of the most heterogeneous agroecological conditions, coupled with some of the most unstable political systems. Research indicates that some progress has been made in the development of research infrastructure, including the development of human capital and research capabilities. There were about 2000 full-time equivalent (FTE) researchers in 1961 and about 9000 FTE researchers in 1991, over 90% of this human capital being Africans. Unfortunately, research expenditure over the same period declined dramatically. In 1991, agricultural research spending averaged about 0.73% of GDP, with donor funding for agricultural research accounting for about 43% of total expenditure. It should be pointed out that there is marked variability in these statistics. For example, whereas Nigeria received only 6% of its agricultural research funding from donors, countries such as Senegal and Zambia received more than 60% of their funds from donor sources. An estimated 40% of the budget of the International Agricultural Research Centers is disbursed to efforts in Africa.

The most important crops produced in developing countries by acreage are maize, sorghum/millet, and root and tuber crops. The increase in production of major crops from 1971–1997 averaged more than 2.0%. Yield increase from crop improvement accounted for 70% of the increase in wheat production. Compared to Asia and Latin America, African research emphasis is on crop management against the complex agroecological conditions that prevail, rather than the improvement of maize. Input from the International Agricultural Research Centers (IARCs) (through CIMMYT and IITA) is very strong for maize. Cultivar releases include open-pollinated varieties (OPV) and hybrids, the latter kind being more common in East and South Africa, while OPVs and local cultivars dominate production in West and Central Africa. Between 1981 and 1990, cultivars of maize used in the regions included about 62% OPVs and 38% hybrids. The yield increase of hybrids over local varieties averages about 40%. Yield gain from OPVs is about 14–25% over local varieties.

Sorghum and millet are the second and third most important cereal crops, respectively, in Africa. They have the dubious title of "poor man's crop". Crop improvement efforts have been significantly supported by ICRISAT and INTSORMIL-CRSP. Other success stories include hybrid sorghum in Sudan, semidwarf rice for irrigated production in West Africa and disease resistant potatoes in East and Central Africa.

On the whole, the National Agricultural Research Stations, with support from the IARCs, have made some progress in getting improved cultivars into agricultural production. However, a 1991 report by the Special Programme for African Agricultural Research (SPAAR) suggests that for agricultural research to remain a catalyst for modernizing African agriculture, important issues to be considered include the size of the National Agricultural Research Stations, commodity research programs, relative emphasis on testing versus breeding of cultivars, allocation of resources to different research activities in the various geographic regions, low salaries and, consequently, high turnover among local scientists.

Whereas the role and impact of the CGIARs in developing nations is ongoing and continues to expand, there have been some significant new and recent efforts to improve plant breeding efforts in Africa. In 2000, the African Center for Crop Improvement (ACCI) was created at the University of Kwazulu-Natal, South Africa, in collaboration with Cornell University, with the purpose of training plant breeders for Africa. About 50 breeders have been trained for about 13 countries. The effort was supported by the Bill and Melinda Gates Foundation, the Rockefeller Foundation, and later by AGRA. A parallel program the West Africa Center for Crop Improvement (WACCI), was established at the University of Ghana, Legon, in 2009.

30.6 Biotechnology efforts in developing countries

Whenever the subject comes up, the role bio-technology can play in the humid tropics is often identified to be alleviating hunger. Then, there is also the ongoing debate about whether or not developing economies and donor nations and agencies should exploit biotechnology in addressing the food security of developing nations.

30.6.1 Overview of world food issues

Because of the expected population expansion and increasing land erosion, food security in developing countries is a major concern to the international community. Whereas population growth is leveling off in developed countries, most of the estimated five billion additional people on Earth by 2030 will inhabit the poor regions of the humid tropics. It should be pointed out immediately that it is an oversimplification to equate hunger alleviation with food security. Associated issues such as effective and efficient distribution networks, effective management of production resources (land, water) and government pricing policy critically impact the success of any food security undertaking. Further, food security in these economically disadvantaged areas is intertwined inextricably with disease and environmental degradation. Poor soils and poor production management result in low crop yield, malnutrition and a variety of health issues. Some observe that the medical problems of Africa are inseparable from that of lack of food. Needless to say that tackling third world food security is a challenging proposition that requires careful planning and integrated approach.

Promotion of agricultural biotechnology in developing economies should be accompanied by a promotion of improved agricultural practices. This way the ecological limits of population growth can be expanded by utilizing existing farmland more productively and also improving crop harvest. It has been pointed out by some experts that current agricultural biotechnologies do not increase the productivity *per se* of plants. Instead, they lower pre- and post-harvest losses by up to 25%. In terms of strategy, it is suggested that in view of the problems with food distribution and local environments, agriculture in the humid tropics must be indigenous and very productive. Consequently, the infusion of foreign technologies must proceed cautiously. Further, the technology of gene transfer must be developed *in situ*, at least in some of the tropical developing countries, to ensure that it responds to local conditions. This strategy will also ensure that the technology is more readily acceptable to the local government, the local scientific community, plant breeders and also the local population.

Biotechnology is very capital and knowledge intensive. Such commitments are woefully inadequate in many developing economies. However, because the private sector plays a dominant role in setting biotechnology research and development targets, and because economic returns on investment is critical to investors, crops that benefit developing countries receive little attention because they are of little commercial interest. Consequently, the talk of improving tropical crops falls to other agencies (e.g., the United Nations) with little pressure to make profit. The need to infuse biotechnology into the agriculture of developing countries is further underscored by the fact that most of these countries have agricultural-based economies. Improving agriculture is hence a major avenue for improving such economies. The productivity potential of major food crops of the developing world is far from being attained. Doubling the current level of productivity of staple food crops will make significant impact on the food security of these nations.

30.6.2 Barriers to commercializing biotechnology in developing countries

Developing countries are as diverse as they are numerous. The common differences among them are based on political, socioeconomic and geographic factors. The existing technology capacity for biotechnology in these nations ranges from nil to adequate. The levels of resource (human, capital) endowment also vary widely, and so do domestic conditions regarding politics, government policies, scientific knowledge base, and macroeconomics. It is inaccurate, therefore, to lump all developing nations into one category. Rather, two general groups may be identified: (i) countries in economic transition and (ii) those in economic stagnation (with nothing or very limited existing infrastructural capacities to exploit biotechnology). Most of the countries in the lower tier are located in Africa. Even so, some countries in Africa have begun to put in place mechanisms for commercially exploiting biotechnology. These include Kenya, Zimbabwe, Nigeria, and South Africa. In the Caribbean region, Cuba has implemented significant biotechnology programs. There are also the newly industrialized nations in Asia (e.g., China, India) and Latin America (Brazil, Mexico).

There are several major barriers to commercializing biotechnology in developing countries:

- Lack of appropriate technology.
- Limited infrastructure for exploiting biotechnology.
- Intellectual property rights.
- Biosafety issues.
- Lack of market mechanisms.
- The biotech debate being waged in the potential donor countries.
- Local and regional politics.
- Poverty and disparities.

It might appear that appropriate technology would be a major barrier to attempts at applying biotechnology to benefit the needy in the developing world. This is so because most of the research and product development occur in developed countries and are targeted to solving problems in their regions. Existing technologies may be adapted for use in developing countries, while some new and unique technologies may have to be developed *in situ* in these nations to be effective. The issue of intellectual property rights is also a potential key barrier. Companies in developed countries own most of the patents for the technologies that would be deployed in poor regions. The commercial companies would have to be adequately compensated, in most cases, to have access to their inventions. These two factors notwithstanding, some experts believe that the primary barrier to successful exploitation of biotechnology in developing countries is the lack of market mechanisms that normally constitute the driving force behind the research and development process. In terms of agriculture, one of the clearly accessible markets is the seed market, especially those for cash crops. Major seed companies in the United States (e.g., Monsanto) and Europe (e.g., Sandoz) have interest in accessing this market. If profitable markets exist for biotechnology, companies in industrialized nations with resources will be enticed to invest in the third world oriented projects. However, if the objective of biotechnology exploitation in developing countries is to benefit the poor and

needy, then other avenues beside business ventures need to be sought. Developing countries also need to implement biosafety guidelines in order to conform to international regulations for conducting biotechnology research. As the biotechnology debate goes on in developed countries, attempts by developing countries to advance their biotechnology efforts become needlessly entangled in the debate and adversely affected. Some opponents of biotechnology tend to think that multinational corporations are only profit oriented and look for opportunities to exploit developing countries.

Apart from barriers that may originate outside the developing world, local and regional politics in developing countries pose a significant barrier to the adoption of biotechnology. Local governments are responsible for developing or implementing biosafety regulations, honoring intellectual property rights, supporting local research and development efforts, accepting biotechnology as a viable tool for helping local agriculture, and putting in place the environment for overseas partnerships to be successful. The issue of poverty is important in the adoption of any technology. Most of the agricultural production in developing countries is undertaken by the rural poor. The concern always is how they can afford new technologies. The other critical concerns are about the distribution of benefits or the impacts of technology. A criticism of the Green Revolution is that it marginalized the poorer producers, while bringing most of the economic benefits to the already richer producers.

30.6.3 The role of international initiatives in biotechnology

With proper caution and good planning, biotechnology can be successfully implemented in developing countries to improve agricultural production. It is important that any effort be approached from the angle of partnerships and collaboration. Overseas partnerships should include the public and private sectors, as well as the international entities. More importantly, every partnerships should involve the developing countries directly and be implemented in the social context. Including the developing countries would make the technology more readily acceptable and facilitate its adoption. It would also make the developing countries feel they are not being taking advantage of, or being forced to accept what they do not want.

Because of the prohibitive costs of participation in the exploitation of biotechnology by many developing countries, a variety of international initiatives exist for supporting countries to plan or implement research and extension programs in biotechnology. Currently, most of these efforts are directed towards food-based biotechnology and involve bilateral initiatives by governments from the developed world, private foundations, tripartite arrangements, and efforts by the UN system.

30.7 New approach to international– national collaborative breeding

The IARCs devote resources to assisting national agricultural programs to increase agricultural production by developing improved cultivars for farmers. Generally, the international–national collaboration has been, as some have observed, "top down", whereby the international programs develop new genotypes that are then evaluated by national programs. From testing these finished or near-finished products, the local breeders recommend superior ones for release as cultivars for local production. This trend of plant "introduction" has the tendency to displace locally adapted cultivars.

In view of the preceding discussion, it is being proposed that early generation breeding material be evaluated by local breeders in developing countries, as illustrated by Cecarelli in Chapter 26. This approach will increase the chance of exploiting positive $G \times E$ interactions. It will also make breeding programs in developing countries more self-reliant. The needs of small producers will be more effectively met with the release of cultivars that suit their specific production package.

30.7.1 Concept of centralized breeding

The traditional approach to plant breeding is for researchers to independently initiate and conduct a plant breeding program at a specific research station or institute, without input from farmers (**centralized plant breeding**). Then, they evaluate advanced generations of genotypes over selected locations, after which they release a genotype as a new cultivar. At the very best, farmers are occasionally invited to observe finished products during field days. Even on such occasions, their input is limited to choosing among different finished products. Some breeders believe that engaging farmers at some point in the actual breeding process is advantageous. There is a disconnect between the site of selection (site of breeding) and target environment (where the product will be used). Consequently, as S. Cicarelli pointed out, the selection efficiency decreases as the selection environment becomes increasingly different from the target environment. In developed countries, crop producers often have facilities to duplicate the favorable selection environments that occur at research stations. Unfortunately, in developing countries, breeders address problems of poor farmers who operate in unfavorable conditions, from their national research stations (and from the IARCs). As previously discussed (Chapter 23), plant breeders interpret $G \times E$ interaction in their cultivar release decision-making process. Some researchers, such as Cicarelli, contest that most farmers are interested in avoiding or reducing temporal variability, while most breeders focus on avoiding geographic variability. To achieve temporal variability, breeders should develop heterogeneous cultivars rather than uniform ones. This will give yield stability over time.

30.7.2 Decentralized-participatory plant breeding

Decentralized-participatory plant breeding is the concept of actively involving farmers in the plant breeding process, rather than delivering a pre-packaged seed product to farmers. Farmers are involved in the selection process in the early segregating populations so that the final products are adapted to the target environments in which they will be used for production. This approach to breeding should be distinguished from a multilocation performance evaluation of breeding lines that is part of most conventional cultivar development programs. Also, participation of the farmer should be emphasized in order to make decentralized breeding accomplish its objective of taking advantage of the farmer's knowledge of the crop and the production environment.

The concept of participatory plant breeding is not novel, since farmers, from the time of the invention of agriculture, have selected among existing variability to advance genotypes with appealing characteristics. In fact, farmers in developing countries continue this practice by selecting and saving seed from the most appealing plants in the current season's crop for planting the next season's crop. However, the modern application of decentralized-participatory plant breeding is attributed to N.W. Simmonds who, in 1984, used the term **"decentralized selection"** to refer to the selection process with emphasis on selecting for specific adaptation to specific environments, rather than evaluating the mean performance of genotypes across different environments.

The scientific basis of this approach to breeding rests on $G \times E$ and its interpretation. In Chapter 26 the concept of $G \times E$ and its importance in plant breeding was discussed in detail. It was indicated that the type of $G \times E$ interaction determines the type of cultivar a plant breeder releases for use by farmers. Where evaluations reveal a crossover $G \times E$ interaction (i.e., the rank in genotypes changes in different environments), the breeder is confronted with a more complex decision regarding the kind of cultivar to release. Some argue that the traditional action by most plant breeders is to release genotypes with the highest average performance (yield), discarding the best or worst performers at the extremes of the scale. This habit has been described as "negative interpretation of $G \times E$ interactions" and is motivated by a desire to release a cultivar with wide adaptation for seed production. This approach to plant breeding suits the crop production practices of developed countries where production conditions can be readily modified or supplemented to become conducive to optimal plant performance. However, in developing economies, most farmers produce crops under marginal conditions. Consequently, it is important to consider genotypes that perform best under favorable conditions as well as those that perform well under less favorable conditions. In other words, selection should be for specific adaptation, both favorable and less favorable. Such a selection approach is described by some as "positive interpretation of $G \times E$ interaction".

30.8 Conventional plant breeding versus decentralized-participatory plant breeding

There are certain key features, advantages and disadvantages of decentralized-participatory plant breeding.

30.8.1 Key features of conventional breeding

The key features of conventional plant breeding may be summarized as:

- Breeders formulate breeding objectives and initiate cultivar development at their research facility.
- Promising genotypes in advanced stages of cultivar development are evaluated at selected sites by breeders.
- Superior genotypes that are uniform and have wide adaptation are released through formal channels.
- Farmers may visit on-farm trials on field days at research stations, but are not actively involved in the breeding process.
- Breeders continue to develop superior cultivars to replace older cultivars.

Advantages

The key advantages of conventional plant breeding include:

- The process is generally simplified, not having the added management challenges of supervising farmers.
- Usually, only one genotype is released as a cultivar.

Disadvantages

- Genotypes that may perform well under marginal conditions are discarded during the early part of the selection process.
- The cultivar released is adapted to wider regions, rather than a specific region.
- Adoption of the new cultivar by farmers is not guaranteed.
- New cultivars are needed every so often.

30.8.2 Key features of decentralized-participatory breeding

The key features of decentralized-participatory plant breeding include:

- Breeders formulate breeding objectives and initiate the breeding process at the research station.
- Early segregating populations are evaluated in the target environments in which the products would be used for crop production.

- Farmers are actually involved in the breeding process, contributing their intimate knowledge of the local environment and the crop.
- Cultivars are released based upon specific adaptation to the growing environment.
- Focus is on adaptation over time (i.e., stability).

Advantages

- Cultivars released are readily adopted by farmers and are well adapted to the production environment.
- Variability that may have been discarded in the early stages of conventional breeding may be adapted to specific farmers' fields.
- The approach favors producers in developing countries where the practice of low input agriculture is prevalent, and farmers use highly adapted land races.
- Farmers are able to select for traits they actually need that may not be appealing to plant breeders.

Disadvantages

- Including farmers in the decision-making process may be time consuming.
- Several cultivars are released simultaneously.
- Replacing cultivars with new ones is not a straight-forward operation.

30.9 The Green Revolution

Producing enough food to feed the world's ever increasing population has been a lingering concern of modern societies. Perhaps the most notable essay on food and population dynamics was written by Thomas Malthus in 1798. In this essay, "Essay on the Principles of Population", he identified the geometric role of natural population increase in outrunning subsistence food supplies. He observed that unchecked by environmental or social constraints it appears that human populations doubled every 25 years, regardless of the initial population size. Because population increase, according to this observation, was geometric, whereas food supply at best was arithmetic, there was implicit in this theory pessimism about the possibility of feeding ever-growing populations. Fortunately, mitigating factors, such as technological advances, advances in agricultural production, changes in socioeconomics, and political thinking of modern society, have enabled this dire prophesy to remain unfulfilled.

Unfortunately, the technological advances in the twentieth century primarily benefited the industrialized countries, leaving widespread hunger and malnutrition to persist in most developing countries. Many of these nations depend on food aid from industrial countries for survival. In 1967, a report by the USPresident's Science Advisory Committee came to the grim conclusion that "the scale, severity and duration of the world food problem are so great that a massive, long-range, innovative effort unprecedented in human history will be required to master it". The Rockefeller and Ford Foundations, acting on this challenge, proceeded to establish the first international agricultural system to help transfer the agricultural technologies of the developed countries to the developing countries. The humble beginnings led to a dramatic impact on food production in the third world, especially Asia, which would be dubbed the **Green Revolution**, a term coined in 1968 by the USAID Administrator, William S. Gaud.

The Green Revolution started in 1943 when the Mexican government and the Rockefeller Foundation cosponsored a project, the Mexican Agricultural Program, to increase food production in Mexico. The first target crop was wheat and the goal was to increase wheat production by a large margin. Using an interdisciplinary approach, the scientific team headed by Norman Borlaug, a wheat breeder at the Rockefeller Foundation, started to assemble genetic resources (germplasm) of wheat from all over the world (East Africa, Middle East, South Asia, Western Hemisphere). The key genotypes used by Norman Borlaug in his breeding program were the Japanese "Norin" dwarf genotypes supplied by Burton Bayles of the USDA and a segregating (F_2) population of "Norin 10" crossed with "Brevor", a Pacific Northwest wheat, supplied by Orville Vogel of the USDA. These introductions were crossed with indigenous (Mexican) wheat that had adaptability (to temperature, photoperiod) to the region and were disease resistant but were low yielding and prone to lodging. The team was able to develop lodging-resistant cultivars through introgression of dwarf genes from semidwarf cultivars from North America. This breakthrough occurred in 1953. Further crossing and selection resulted in the release of the first Mexican semidwarf cultivars, "Penjamo 62" and

"Pitic 62". Together with other cultivars, these two hybrids dramatically transformed wheat yields in Mexico, eventually making Mexico a major wheat exporting country. The successful wheat cultivars were introduced into Pakistan, India, and Turkey in 1966, with similar results of outstanding performance. During the period, wheat production increased from 300 000 tons/year to 2.6 million tons/year; yields per unit area increased from 750 kg/hectare to 3200 kg/hectare.

The Mexican model (interdisciplinary approach, international team effort) for agricultural transformation was duplicated in rice in the Philippines in 1960. This occurred at the International Rice Research Institute (IRRI). The goal of the IRRI team was to increase productivity of rice in the field. Rice germplasm was assembled. Scientists determined that, like wheat, a dwarf cultivar that was resistant to lodging, amenable to high density crop stand, responsive to fertilization, and highly efficient in partitioning of photosynthates or dry matter to the grain was the cultivar to breed.

In 1966, IRRI released a number of dwarf rice cultivars to farmers in the Philippines. The most success was realized with IR8, which was early maturing (120 days), thus allowing double cropping in certain regions. The key to the high yield of the IR series was their responsiveness to heavy fertilization. The short, stiff stalk of the improved dwarf cultivar resisted lodging under heavy fertilization. Unimproved indigenous genotypes experienced severe lodging under heavy fertilization, resulting in drastic reduction in grain yield.

Similarly, cereal production in Asia doubled between 1970 and 1995, as the population increased by 60%. Unfortunately, the benefits of the Green Revolution barely reached sub-Saharan Africa, a region of the world with perennial severe food shortages, partly because of the lack of appropriate infrastructure and limited resources. Dr. Norman Borlaug received the 1970 Nobel Prize for Peace for his efforts at curbing global hunger.

Three specific strategies were employed in the green revolution: plant improvement, complementary agronomic package, and favorable return on investment in technology.

   (i) **Plant improvement.** The green revolution centered on the breeding of high yielding, disease-resistant and environmentally responsive (adapted, responsive to fertilizer, irrigation, etc.) cultivars.

  (ii) **Complementary agronomic package.** Improved cultivars are as good as their environment. To realize the full potential of the newly created genotype, a certain production package was developed to complement the improved genotype. This agronomic package included tillage, fertilization, irrigation, and pest control.

 (iii) **Return on investment in technology.** Favorable ratio between the cost of fertilizer and other inputs and the price the farmer received for using this product was an incentive for farmers to adopt the production package.

Not unexpectedly, the Green Revolution has been the subject of some intensive discussion to assess its sociological impacts and identify its shortcomings. Incomes of farm families were raised, leading to an increase in demand for goods and services. The rural economy was energized. Food prices dropped. Poverty declined as agricultural growth increased. However, critics charge that the increase in income was inequitable, arguing that the technology package was not scale neutral (i.e., owners of larger farms were the primary adopters because of their access to production inputs – i.e., capital, seed, irrigation, fertilizers). Furthermore, the Green Revolution did not escape the accusations often leveled at high yield agriculture – environmental degradation from improper or excessive use of agrochemicals. Recent studies have shown that many of these charges are overstated.

30.10 The Green Revolution and the impact of international breeding efforts

The Special Project on Impact Assessment (SPIA) of the CGIAR's Technical Advisory Committee conducted a comprehensive study spanning the period 1960–2000. The purpose of the study was to evaluate the impact of international breeding efforts on crop production in developing countries. International and national programs work in complimentary fashion in developing modern high yielding crop cultivars for use in production systems in developing countries. The SPIA study assessed the direct and indirect international contributions to varietal improvement, direct contributions being cultivars developed at the IARCs and then released by nationals programs without further crossing, while indirect contributions constituted cultivars that were developed by the various National Agricultural Research Stations (NARS) using germplasm developed at IARCs. Summarizing the SPIA study, Evenson and Gollin found that 35% of modern cultivars released and adopted were based on crosses made at the IARCs while 15% of NARS-crossed modern cultivars had an IARC-crossed parent. Further, 7% of the NARS-crossed cultivars had an IARC-crossed ancestor. Cultivars with IARC ancestry were more widely cultivated by farmers than other planting materials. For the 11 crops studied, over 8000 cultivars were released by more than 400 public breeding programs and seed boards in over 100 countries. Most of the early efforts focused on rice and wheat. Crops adapted to the semiarid and dryland conditions (sorghum, millet, barley) as well as pulses and root crops (especially cassava) were late comers, because of lack of prior existing breeding programs in developed countries to be capitalized upon.

Evenson and Gollin also note that introduction of modern varieties into production areas in did not translate into automatic adoption by farmers. In fact, it was only after breeders embarked on location-specific breeding efforts that adoption rates improved. They observed that the successful approach was for breeders to first develop a productive plant type (plant architecture), for example the high yielding semidwarf architecture), to be used as a platform for local adaptation, to which location-relevant traits (biotic and abiotic resistance) were systematically added. Another significant observation made by these researchers was

that, contrary to what is widely believed, the impact of the Green Revolution increased in the 1980s and 1990s more than when the original cultivars were introduced in the 1960s and 1970s. They suggested that the Green Revolution is best viewed in terms not of a one-time boost in production but rather as long-term trend in increase in productivity. This is because the original event provided a foundation for newer and more productive generations of cultivars to be developed. They further proposed two phases for the event, an "early Green Revolution" (1961–1980; mainly in Asia and Latin America) and a "late Green Revolution"' (1981–2000). Further analysis also revealed that production gains in the late revolution were attributable more to genetic factors (breeding efforts; higher yielding cultivars) than in the early phase. Although high input agriculture intensified in the late phase, the higher yielding capacity of the later generation cultivars allowed higher crop production gains to be achieved with only a small increased in new crop land.

The most notable success to date of international breeding efforts on crop production in developing countries is what became known as the Green Revolution. Counting on previous experiences in breeding for developed countries, the breeders at IRRI and CIMMYT developed stiff-strawed semidwarf cultivars by incorporating dwarfing genes. These environmentally responsive cultivars, were able to take advantage of high production inputs to significantly boost production. These cultivars were released to farmers in the late 1960s and early 1970s, first in Asia and Latin America, and later to other parts of the world.

Key references and suggested reading

Ceccarelli, S., Grando, S., Tutwiler, *et al.* (2000). A Methodological Study on Participatory Barley Breeding. I. Selection Phase. *Euphytica*, **111**:91–104.

Evenson, R.E., and Gollin, D. (eds)(2003). *Crop Variety Improvement and Its Effect on Productivity: The Impact of International Agricultural Research*. CAB International, Wallingford, UK.

Evenson, R.E., and Gollin, D. (2003). Assessing the Impact of the Green Revolution, 1960 to 2000, *Science*, **300**: 758–762.

Outcomes assessment

Part A

Please answer the following questions true or false.

1 CIAT is located in Mexico.
2 IRRI has a mandate for corn research.
3 There are 26 CGIAR centers.
4 The World Bank is a funding member of CGIAR.

Part B

Please answer the following questions.

1 Give the full name of the following acronyms:
 CIMMYT ....
 IITA ....
 CIAT ....
2 Give a specific research emphasis of the CGIAR.
3 Give three of the mandate crops of the CGIAR.
4 WARDA is located in ....

Part C

Please write a brief essay on each of the following topics.

1 Please discuss the funding support for the CGIAR.
2 Discuss the major selected accomplishments of the CGIAR.
3 Discuss the structural organization and mission of the CGIAR.

Section 9
Breeding selected crops

Chapter 31 Breeding wheat
Chapter 32 Breeding corn
Chapter 33 Breeding rice
Chapter 34 Breeding sorghum
Chapter 35 Breeding soybean
Chapter 36 Breeding peanut
Chapter 37 Breeding potato
Chapter 38 Breeding cotton
Chapter 39 Breeding tomato
Chapter 40 Breeding cucumber
Chapter 41 Breeding roses

The purpose of this section is to discuss the breeding of selected major world crops, including their economic importance, origin and history, genetics and cytogenetics, general botany, reproductive biology, and common breeding methods used. A unique feature of this section is that professionals engaged in the breeding of each of these selected crops have provided an overview of their breeding programs. The presentations are different in style and content, according to the breeder's preference. However, each presentation provides the student an opportunity to see how the principles and concepts of genetics and plant breeding are applied in conducting an actual breeding program.

Purpose and expected outcomes

The purpose of this section is to discuss the breeding of crops. After studying each of thesechapters, the student should be able to discuss:

1 The economic importance of the crop.
2 The origin and adaptation of the crop.
3 The genetics and cytogenetics of the crop that impact its breeding.
4 The germplasm resources for breeding the crop.
5 The general botany and reproductive biology of the crop.
6 The common breeding methods used in breeding crop.

31

Breeding wheat

Taxonomy

Family	Poaceae
Genus	*Triticum* L.
Species	*Triticum aestivum* L.

31.1 Economic importance

Wheat is the most important cereal grain crop in the world. It is the principal cereal grain crop used for food consumption in the United States and most parts of the world. In the United States, it usually ranks fourth after corn, hay, and soybeans in that order of importance. Wheat is grown commercially in nearly every state in the USA, with a concentration of production in the Great Plains, an area spanning the states from Texas to Montana. USDA production trends indicate that in 1866, wheat was harvested from an area of 15.4 million acres, yielding an average of 11 bushels/acre. By 1950, production occurred on 61.6 million acres, with an average yield of 16.5 bushels/acre. In 1990, the acreage was 69.2 million acres, yield average being 39.5 bushels/acre. The Central and Southern Plains (Texas, Oklahoma, Kansas, among others) produce more than the Northern Plains (e.g., Montana, North Dakota) the two regions accounting for two-thirds of the US wheat production and about 80% of wheat acreage. Kansas leads all states in wheat production. Hard white wheat is also grown in the Central Plains and Northern Plains.

World production of wheat in 2001 was 583.9 million metric tons, occurring on 219.5 million acres. World wheat consumption in that period was 590.6 million tons. Developing countries (excluding those in Eastern Europe or the former Soviet Union) account for nearly 50% of the world's wheat production, the leading producers being China, India, Turkey, Pakistan, and Argentina. The success of wheat production in these countries is credited to the impact of the Green Revolution that occurred in the 1960s and 1970s. In 2000, China produced 111.9 million metric tons while India produced 26.5 metric tons. Latin America and Asia (excluding China and India) each produce about 20 million metric tons a year. Wheat is produced in Europe, including the United Kingdom, Denmark, The Netherlands, Belgium, Switzerland, and West Germany.

31.2 Origin and history

The origin of wheat is believed to be southwestern Asia. A cross between wild Emmer wheat (*Triticum dicoccoides*) × *Aegilops squarrosa*, a grass, produced a spelt-like plant. This suggests that the common or bread wheat (*T. aestivum*) is descended from a cross between spelt and the progenitor of Persian wheat (*T. persicum*). Persian wheat occurs in the wild in the Russian Caucasus. The Persian wheat probably is descended from the wheat of the Neolithic Swiss lake dwellers, which in turn might have originated from a

Principles of Plant Genetics and Breeding, Second Edition. George Acquaah.
© 2012 John Wiley & Sons, Ltd. Published 2012 by John Wiley & Sons, Ltd.

cross between einkorn and a grass, *Agropyron triticum*. Archeological findings indicate that Emmer wheat was cultivated before 7000 BC. Similarly, wheat was cultivated in Europe in prehistoric times. In the United States, wheat was first cultivated along the Atlantic Coast in early seventeenth century, moving westwards as the country was settled.

31.3 Adaptation

Wheat is best adapted to cool temperate climates where rainfall is not excessive (15–24 inches/annum). Based on season of production, there are two types of wheat – winter wheat and spring wheat.

31.3.1 Winter wheat

Winter wheat is sown in the fall (autumn) so that it can have some growth before the onset of cold weather in winter. Growth ceases and the plants remain dormant through winter, resuming growth in spring for harvesting in summer. About two-thirds of US wheat is winter wheat. Winter wheat can survive cold temperatures as low as −40 °F if protected by snow.

31.3.2 Spring wheat

Spring wheat is less tolerant of low temperatures and is damaged by even a light frost of 28–30 °F. Spring wheat is planted in early spring and harvested in July–August.

Wheat is a long day plant. Short days of high temperatures stimulate tillering and leaf formation but delay flowering of wheat plants. Early maturing cultivars are available for production under any photoperiod conditions. However, the quality (nutritional uses such as baking) of wheat is influenced by the production environment. For example, growing hard wheats in soft wheat regions results in grains that are starchy or "yellow berry" (soft and starchy).

31.4 History of breeding in the United States

Wheat is one of a few food crops (the others being corn and rice) that have been associated with the Nobel Peace Prize. The 1970 Nobel Peace Prize awarded to Norman Borlaug, the father of the Green

Revolution, recognized his contribution to agricultural productivity through the introduction of superior genotypes of wheat. These superior varieties were high yielding, shorter (semidwarf wheat) more lodging resistant, and responsive to high levels of fertilizer. A significant contributor to this effort was Orville Vogel, a USDA wheat breeder stationed at Washington State University. Under his leadership, the first successful commercial semidwarf wheat variety in the Western Hemisphere was released to farmers in 1961. This variety, Gaines, was a soft white wheat and yielded in excess of 100 bushels/acre under both dry-land and irrigated production. In spite of its agronomic qualities, Gaines had milling quality problems. In response to the demands of the milling industry, a new selection with more desirable milling qualities, called Nugaines, was released in 1965. The role of wheat in the Green Revolution was discussed in Chapter 1.

31.5 Commercial wheat classes

Wheat breeders specialize in one of the special market classes of wheat. There is a genetic basis for this classification. Wheat may be classified into seven groups based on the time of year of planting and kernel characteristics (hardness, color, shape). However, for commercial production, the varieties may be narrowed down to six basic classes: hard red winter, soft red winter, hard red spring, hard white, soft white, and durum wheat. The hard red wheat accounts for about 40% of total US wheat production and is the dominant class in US wheat export.

31.5.1 Hard red winter wheat

This is grown mainly in the Great Plains (Kansas, Oklahoma, Nebraska, Texas, Colorado). It is also grown in the former Soviet Union, Argentina, and Danube Valley of Europe. It is used for bread flour.

31.5.2 Hard red spring wheat

This class of wheat is grown in regions with severe winters in the North Central states (North Dakota, Montana, South Dakota, Minnesota). It is also produced in Canada, Russia, and Poland. It is the standard wheat for bread flour.

31.5.3 Soft red winter wheat

This class of wheat is grown predominantly in the eastern United States (Ohio, Missouri, Indiana, Illinois, Pennsylvania). It is also grown in Western Europe. Soft red winter wheats are used mainly for pastry, cake, biscuit, and household flour. For bread making, it needs to be blended with hard red wheat flour.

31.5.4 White wheat

White wheat (hard or soft) is produced in the four western states (North Dakota, South Dakota, Nebraska, Minnesota) and in the northeastern states (Washington, Oregon, Michigan, California, New York). Some of this is club wheat. It is produced also in northern, eastern, and southern Europe, Australia, South Africa, South America, and Asia.

31.5.5 Durum wheat

Durum wheat is grown mainly in North Dakota, South Dakota, and Minnesota. Other smaller production states are California, Arizona, Oregon, and Texas. Elsewhere, it is grown in North Africa, Southern Europe, and the former Soviet Union. Durum wheat is used in making semolina, which is used for producing products such as macaroni and spaghetti.

31.6 Germplasm resources

Plant breeders have access to over 400 000 accessions in natural and international germplasm banks. These banks include the USDA National Seed Storage Lab at Fort Collins, Colorado, the CIMMYT, in Mexico, and the N.I. Vavilov All-Union Institute of Plant Industry, St Petersburg, Russia. Over 40 000 accessions are held at Aberdeen, Idaho, as a working collection and parts of the United States National Small Grains Collection.

31.7 Cytogenetics

The species of *Triticum* are grouped into three ploidy classes: Diploid ($2n = 2x = 14$), tetraploid ($2n = 2x = 28$), and hexaploid ($2n = 6x = 42$). The cytoplasmic male sterility (cms) gene used in modern wheat breeding is derived from *T. timopheevii*, a wild tetraploid variety. Three genomes (*A*, *B*, *D*) comprise the polyploid series of wheat. The A genome comes from *T. monococcum*, while the *D* comes from *Aegilops squarrosa* (or *T. tauschii*). The origin of the B genome is debatable. The genomic formula of the ploidy classes are *AA* or *BB* for diploids and *AABB* for the tetraploid or Emmer wheat. Common wheat (*T. aestivum*) is an allohexaploid of genomic formula *AABBDD*. In hexaploid wheat, the 21 chromosomes are divided into seven homeologous groups (partially homologous chromosomes) identified group numbers from 1 to 7. The three chromosomes within the *ABD* homeologous group usually share some loci in common for a specific trait. An example of this is the two genes for rust resistance that occur on chromosome 2*A*, three genes on 2*B*, and three genes on 2*D*.

Tetraploid and hexaploid wheat reproduce naturally as diploids ($2n = 28$ or $2n = 42$). This reproductive mechanism is made possible by the presence of a gene on chromosome 5*B*, *Ph1*, which enables diploid pairing to occur. The *Ph1* gene causes truly homologous paring within the same genome. When absent, paring between one chromosome and a homeologous chromosome from another genome is possible.

The homeology that exists in its three component genomes allows the species to tolerate a range of aneuploidy. *T. aestivum* exhibits vigor and morphology similar to disomic wheat. Among other applications, aneuploidy has been used to locate genes that confer agronomically important traits (e.g., the *mlo* locus for resistance to powdery mildew). Classical wheat genetics was advanced through the work of E. R. Sears of the University of Missouri. He developed a compatible set of the possible 21 monosomics ($2n - 1$) of wheat and sets of related aneuploid forms in the hexaploid wheat cultivar, "Chinese Spring".

Introgression of alien genes is problematic because of the lack of ability for crossing between hexaploid and diploid species, as well as the numerous problems that manifest at various stages in the ontogeny of the hybrid. Genes for crossing (*kr1kr1, kr2kr2*) located on chromosomes 5*B*, 5*A*, and 5*D*, respectively) have been identified in Chinese Spring wheat, which facilitates wheat × rye cross. Some breeders also use genetic bridges and chromosome number doubling to overcome problems with ploidy differences. Alien autotetraploids of *Agropyron cristatum*, *Psathyrostachys juncea*, especially, have been used to overcome hexaploid × diploid alien species crossability barriers. Generally, in practice, the parent with the higher ploidy is used as the female in crosses. However, successes with the reverse have also been recorded. Widening the genetic base of *T. aestivum* through intergeneric crosses often involves complex wheat and alien chromosome combinations. Research has shown that alien genes must be epistatic to those of wheat or interact with them to produce the desired effect. Modifications of the expression of disease and pest resistance genes usually occur when they are introduced into a new genetic background. Nonetheless, successes with spontaneous translocations have been reported in triticale/wheat crosses. One of the notable induced translocations was conducted by Sears and involved chromosome 6B and an *Ae. umbelluta* chromosome, resulting in leaf rust resistance in the release cultivar, "Transfer".

Fertile wheat × alien amphiploids can result from chromosome doubling, the most successful so far being triticale (wheat × rye). Other wheat x alien amphiploids are less successful, being of poor fertility and often exhibiting undesirable alien traits. The technique of alien chromosome additions has been used in an attempt to reduce the undesirable effects introduced by the wild species.

31.8 Genetics

As previously noted, dwarfing genes occur in wheat and have been used in breeding to develop cultivars with short stature (semidwarf wheat) (see Green Revolution in Chapter 1). Early work in Japan produced dwarfing genes. Designated *Rht*, over 20 dwarfing genes have been identified, the most commonly used in wheat breeding including Rht_1, Rht_2, and Rht_8. The first two, called the **Norin 10 dwarfing genes**, also belong to a group of dwarfing genes called gibberellic acid-insensitive dwarfing genes. Cultivars with these genes fail to respond to the application of gibberllic acid. Rht_3 and Rht_{10} genes confer extreme dwarfism on plants, the latter more so in its effect. Practical application to commercial breeding is yet to materialize. Rht_4 and Rht_8 plus others are called the gibberellic acid sensitive dwarfing genes. Monosomic analysis was used to locate the Rht_1 gene and Rht_2 gene on chromosomes *4A* and *4D*, respectively. Chromosome substitution can be used to transfer these genes in a breeding program. The dwarfing genes increase grain yield by increasing tillering and number of seeds per plant.

Other genes of interest in wheat breeding include awnedness, pubescence, grain color, and glume color. The awnedness trait is inhibited by three dominant alleles at three independent loci. *Hd* conditions hooded awn, while B_1 and B_2 condition awnless or tipped awned phenotype. A genotype of hdb_1b_2 produces a bearded or fully awned phenotype. Pubescence in the glume and other parts of the plant is conditioned by a variety of dominant alleles, for example, *Hg* producing hairy glume, while *Hp* conditions hairy peduncle. Red grain color is conditioned by three independent dominant alleles acting in additive fashion ($R_1R_2R_3$), while white grain occurs when the genotype is $r_1r_2r_3$. Consequently, when all three alleles occur in one genotype, the seed color is very dark red.

Anthocyanin pigmentation occurs in various parts of the plant. For example, red auricles are conditioned by a single dominant allele, *Ra*. The red color of glumes is controlled by two dominant alleles, Rg_1 and Rg_2, while photoinsensitivity is controlled by alleles at three independent loci, designated ppd_1, ppd_2, and ppd_3.

31.9 General botany

Wheat (*Tricticum* spp) is an annual plant. It has a spikelet inflorescence. A floret is composed of a lemma, palea, and a caryopsis or grain that has a deep furrow and a hairy tip or brush. The floret may be **awned** or **awnless**. Awned varieties are common in regions of low rainfall and warm temperatures. The presence of awns also tends to influence transpiration rate, accelerating the drying of ripe grain. Consequently, the tips of awnless spikes tend to be blasted in hot dry weather. The grain may also be amber, red, purple, or creamy white in color.

Under normal high density production conditions, a wheat plant may produce 2–3 tillers. However, when amply spaced on fertile soils, a plant may produce 30–100 tillers. The spike (head) of a plant may contain 14–17 spikelets, each spike containing about 25–30 grains. Large spikes may contain between 50 and 75 grains. The grain size varies within the spikelet, the largest being the second grain from the bottom and decreasing in size progressively towards the tip of the spike.

Wheat is predominantly self-pollinated. Anthers assume a pendant position soon after the flower opens. Blooming occurs at temperatures between 13 and 25 °F starting with the spikelet around the middle of spike and proceeding upwards and downwards. The wheat kernel or berry is a caryopsis varying between 3 and 10 mm in length and 3 and 5 mm in width. It has a multilayered pericarp that is removed along with the testa, nucellus and aleurone layers during milling. The endosperm makes up about 85% of a well-developed kernel. Below the aleurone layer occurs a complex protein called **gluten** that has cohesive properties. It is responsible for the ability of wheat flour to hold together, stretch, and retain gas as fermented dough rises. This property is available to the flour of only one other species, the rye flour.

Wheat is classified based on three primary characteristics – agronomic, kernel color, and endosperm quality. There are two seed coat colors – **red** or **white**. Red is conditioned by three dominant genes, the true whites comprising of recessive alleles of all three genes. Most wheat varieties in the United States are red. Kernel hardness is classified into two – **hard** or **soft**. Upon milling, hard wheat yields coarse flour. White wheats, lacking in this starch–protein complex, produce a higher yield of fine flour upon milling. Hard wheat is used for bread making because its gluten protein is cohesive and elastic.

31.10 Reproductive biology

31.10.1 Floral biology

Wheat has a determinate, composite spike inflorescence. Each spike bears 10–30 spikelets, which are borne singly at nodes on alternate sides of a zig-zag rachis. A spike may be awnless, awnleted, or awned. A spikelet consists of 1–5 flowers (or florets) attached alternatively to opposite sides of the rachilla (central axis). Except in some club wheats, only two or three kernels mature, because one or more of the upper florets are usually sterile. A spikelet is subtended by a pair of empty bracts and glumes.

A floret consists of a lemma and palea, which enclose these stamens and a pistil, plus two lodicules that regulate the opening of the flowers and anthers. Wheat flowers bloom under temperatures of 13–25 °F. The flowering is usually diurnal, the highest peak occurring in the morning, and a lower peak in the afternoon. Blooming begins in the spikelets located above the middle of the spike and proceeds both upward and downward. It takes about 2–3 days for a wheat spike to complete blooming after the appearance of the first anthers. The flowering period may last from 14–21 days.

31.10.2 Pollination

Wheat is predominantly self-pollinated with about 1–4% natural cross-pollination. Pollen shed usually starts inside the floret, but about 80% of anther dehiscence occurs outside the floret. The primary and secondary florets produce larger and more viable pollen grains than other florets. Wheat pollen remains viable for up to about 30 minutes after shedding. Once pollinated, the pollen tube growth starts within 15–60 minutes. Even though the stigma remains receptive for up to 13 days, it is most receptive within three days of anthesis. Xenia may occur when plants with the blue aleurone trait are used as males in a cross.

31.11 Common breeding methods

A sample of some of the steps used at CIMMYT are summarized below as examples.

F_1 Make simple crosses. Evaluate on the basis of disease resistance, agronomic traits, and hybrid vigor. Bulk and harvest seed for F_2.

F_2 Space plant 2000–3000 F_2 under optimal conditions (high fertility, moisture). Select plants based on disease resistance, lodging, tillering, maturity, and so on.

F_3 Grow progeny rows in 2-m long three-row plots at dense spacing. Select desirable plots and then select and bulk best heads in each plot. Selection environment is variable (irrigated, rainfed, acid soil, etc.)

F_4 Grow selected dense planted rows and treat as in F_3.

F_5 Space plant 100 plants per plot of selected F_4 families under favorable conditions. Evaluate on the basis of disease resistance, desirable agronomic traits, and spike fertility.

F_6 Grow selected plants individually as F_6 plots of three rows, 2 m long. Select and bulk agronomically superior lines for yield test under various conditions (irrigated, rainfed, hot climate, and acid soil, etc.).

Various approaches are adopted in wheat improvement. Some cultivars are developed though introduction of genotypes and adapting them to new production environments. Evaluation of germplasm is also a way of identifying genotypes for use as parents in future breeding.

Modern wheat breeding depends primarily on hybridization to create variability for selection. Being a self-pollinating species, pure line selection is often used in wheat improvement. As needed, backcrossing may be used to introgress desirable genes into existing commercial cultivars. Many traits in wheat are influenced by several genes rather than one or two, because wheat is a polyploid species. Consequently, it is common for breeders not to observe unexpected phenotypes in the F_1. Hybridization may also bring together the three independent complementary genetic systems that condition lethals, partial lethals, or reduced productivity in the F_1. Specific undesirable traits resulting from hybridization include hybrid chlorosis and grass-clump dwarfness. Hybrid necrosis is conditioned by a complementary two-gene, multi-allelic system, whereas hybrid chlorosis is controlled by a complementary two-gene system. Genotypes with these undesirable genes are known. Breeders can reduce the incidence of these F_1 defects by carefully selecting parents for hybridization. The incidence of grass-clump dwarfness may be reduced by growing plants under high temperature and also using gibberellic acid.

Hybrid wheat for commercial production has not been a practical breeding approach because of lack of sufficient heterosis upon crossing and low seed set. Other practical problems include the complexity of fertility restoration in wheat hybridization. Several major and minor genes are involved.

Industry highlights

Bringing genomics to the wheat fields

K.A. Garland-Campbell[1], J. Dubcovsky[2], J.A. Anderson[3], P.S. Baenziger[4], G. Brown-Guedira[5], X. Chen[1], E. Elias[6], A. Fritz[7], B.S. Gill[8], K.S. Gill[9], S. Haley[10], K.K. Kidwell[9], S.F. Kianian[6], N. Lapitan[10], H. Ohm[11], D. Santra[9], M. Sorrells[12], M. Soria[2], E. Souza[13], L. Talbert[14]

[1] USDA-ARS Wheat Genetics, Quality, Physiology & Disease Research Unit, Washington State University, 209 Johnson Hall, Pullman, WA 99164-6420, USA; [2] University of California at Davis, Department of Agronomy and Range Science, 281 Hunt Hall, Davis, CA 95616-8515, USA; [3] University of Minnesota, Twin Cities, Department of Agronomy and Plant Genetics, 411 Borlaug Hall, St Paul, MN 55108-6026, USA; [4] University of Nebraska-Lincoln, Department of Agronomy & Horticulture, 330 K, Lincoln, NE 68583-0915, USA; [5] USDA-ARS Plant Science Research Unit, Dept. of Crop Science, North Carolina State University, 840 Main Campus Drive, Box 7258, Raleigh, NC 27606, USA; [6] North Dakota State University, Department of Plant Sciences, 470G Loftsgard Hall, Fargo, ND 58105, USA; [7] Kansas State University, Department of Agronomy, 4012 Throckmorton Hall, Manhattan, KS 66506, USA; [8] Kansas State University, Wheat Genetics Resource Center, Department of Plant Pathology, 4024 Throckmorton Hall, Manhattan, KS 66506, USA; [9] Washington State University, Department of Crop and Soil Sciences, Johnson 277, PO BOX 646420, Pullman, WA 99164-6420, USA; [10] Colorado State University, Department of Soil and Crop Sciences, C101 Plant Sciences, Fort Collins, CO 80526, USA; [11] Purdue University, Department of Agronomy, 1150 Lilly Hall, West Lafayette, IN 47907-1150, USA; [12] Cornell University, Department of Plant Breeding, 252 Emerson Hall, Ithaca, NY 14853-1902, USA; [13] University of Idaho, Aberdeen Research and Extension Center, 1693 South 2700 West, Aberdeen, ID 83210, USA; [14] Montana State University, Bozeman, Department of Plant Sciences and Plant Pathology, 406 Leon Johnson Hall, Bozeman, MT 59717-3150, USA

Wheat (*Triticum aestivum* L. em Thell) is well adapted to diverse climatic conditions around the world, and is grown in all regions of the United States. Wheat genetic improvement was the foundation of the Green Revolution and recent progress in the area of wheat genomics has been referred to as the beginning of a new Green Revolution. While wheat improvement through plant breeding still requires a great deal of phenotypic, or trait assessment, new knowledge about the wheat genome enables plant breeders to target their breeding efforts with more precision than ever before.

Triticum aestivum (breadwheat) is hexaploid while *Triticum durum* L. (durum or macaroni wheat) is tetraploid. Hexaploid wheat has three genomes made up of seven chromosomes each labeled 1 to 7. The genomes are named A, B and D and closely related homoeologous chromosomes are named with the genome name, i.e. 1A, 1B, and 1D. Durum wheat has the A and B genomes. Genes are frequently found in multiples, as gene orthologues located on each genome, that are closely related in their structure and function. Orthologous sets of wheat genes are named with the gene name, the genome name and the orthologous set designation. Long and short arms of chromosomes are designated with L and S, respectively. For example, the genes for reduced height which were fundamental to the performance of wheat cultivars developed during the green revolution are *Rht-B1b*, and *Rht1-D1b* and are located on wheat chromosomes 4B and 4D.

Two growth habits are agriculturally important in wheat production, fall-seeded winter wheat and spring-seeded spring wheat. In warm climates, spring wheat is also seeded in the fall. Winter wheat requires a period of cold temperatures, or vernalization, to induce flowering. Wheat is classified in world markets according to its growth habit, grain texture (soft or hard), grain color (red or white) and gluten properties (strong or weak). Those characteristics result in specific end-use properties of each class of wheat, which has led to targeted breeding efforts (i.e. bread vs. confectionary uses).

Marker assisted selection (MAS) is a method of rapidly incorporating valuable traits into new cultivars. Molecular markers, or DNA tags, that have been shown to be linked to traits of interest are particularly useful for incorporating genes that are highly affected by the environment and genes that are resistant to diseases and pests, as well to accumulate multiple genes for resistance to specific diseases and pests within the same cultivar, a process called gene pyramiding. One of the first wheat cultivars to be developed using MAS was the soft winter wheat cultivar "Madsen", released in 1986 by the USDA-ARS and Washington State University. Madsen was developed using the isozyme marker from the endopeptidase protein, EpD1b, to incorporate a gene for resistance to eyespot (*Tapesia yallunde*) (Allan *et al.*, 1989). Since 1990, detailed molecular maps of wheat have been constructed that include more than 3000 molecular markers and several important traits have been associated with DNA markers. Additional markers can be developed from the 8000 expressed sequence tags (ESTs) that have been mapped in wheat. (Maps and references are available online at Graingenes: A database for Triticeae and Avena; USDA-ARS, 2005).

If selected genes or chromosome segments are incorporated from donor parents into adapted wheat breeding lines for six backcross (BC) generations using markers to select for the targeted gene from the BC$_2$ generation, more than 99% recovery of the adapted recurrent parent is expected (Figure B31.1) (Hospital *et al.*, 1992). Seven BC plants per generation are adequate to have a probability higher than 0.99 of recovering a single BC plant with the desired genotype. Linkage drag, which results when undesirable chromosome regions from the donor parent are carried along during backcrossing, can be reduced by advancing more than 10 plants per cross and selecting for those with higher percentages of the adapted recurrent parent genome.

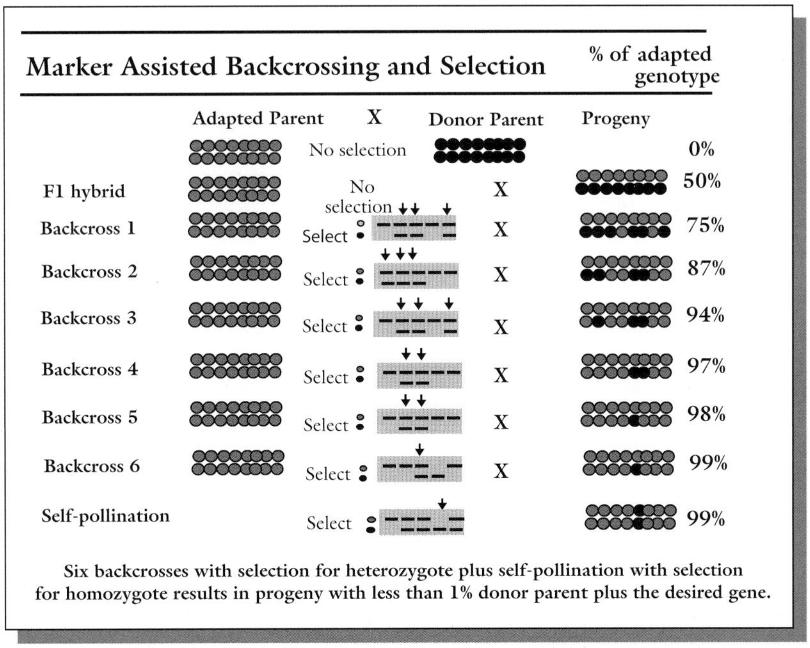

Six backcrosses with selection for heterozygote plus self-pollination with selection for homozygote results in progeny with less than 1% donor parent plus the desired gene.

Figure B31.1 Six backcrosses with marker assisted backcrossing and selection for heterozygote plus self-pollination with selection for homozygote results in progeny with less that 1% donor parent plus the desired gene. Figure courtesy of K.A. Garland-Campbell.

Current use of MAS in US wheat breeding

Several traits have been incorporated into advanced wheat breeding lines using MAS. We will focus on two of those examples that improve end-use quality and disease resistance. Additional details are available online (Dubcovsky and Soria, 2005).

End-use quality improvements

Grain protein content is one of the major factors affecting bread making and pasta quality. In spite of the importance of this character, progress in breeding for high grain protein content (HGPC) has been slow and difficult for two reasons: (i) most variation in protein content is due to environmental rather than genetic effects and (ii) there is a strong negative relationship between grain protein content and grain yield, so that cultivars selected for high grain protein content tend to have low grain yields. A promising source of high grain protein content was detected in a survey of the wild tetraploid wheat relative, *Triticum dicoccoides*. It was crossed into the durum cultivar "Langdon" and the responsible gene was mapped on the short arm of chromosome 6B (6BS) (Joppa *et al.*, 1997). This segment accounted for 66% of the variation in grain protein content observed in a cross between durum wheat and *T. dicoccoides*. The same chromosome segment was transferred to hexaploid wheat by R. Frohberg. The chromosome segment carrying the *HPGC* gene for high seed protein content from *T. dicoccoides* can be efficiently manipulated in tetraploid and hexaploid wheat with microsatellite markers (also known as simple sequence repeats or SSRs). The more useful markers include *Xgwm193* and a cleaved amplified polymorphic (CAP) marker for the NOR locus (Khan *et al.*, 2000). Results from field trials have been mixed, but generally indicate an increase in protein dependent upon genetic background and environment (Kidwell, personal communication).

Wheat grain end-use properties are affected markedly by endosperm texture. Hard wheat requires more grinding energy to reduce endosperm into flour and during this milling process a considerable number of starch granules become physically damaged. Soft wheats, by contrast, produce flours with smaller particles and lower levels of damaged starch. Damaged starch is of value in yeast-leavened products because, in addition to absorbing water, it acts as a substrate for alpha-amylase and creates a favorable environment for yeast growth. In contrast, chemically- leavened soft wheat products have better texture if they are made from flour with small particle size and low water retention capacity. Therefore, hard wheat lines have been selected for high damaged starch and higher hardness values and soft wheat flour for the opposite parameters. Differences in endosperm texture are associated with the complimentary action of proteins puroindoline A (pinA) and puroindoline B (pinB), which are coded by genes located on the distal part of chromosome arm 5DS. Most hard wheats possess a glycine to serine mutation in puroindoline B (allele *pinB-D1b*) or they are devoid of puroindoline A (allele *pinA-D1b*) (Giroux and Morris, 1998). Cultivars carrying these different mutations differ in their grain hardness and in their milling and baking characteristics. Hard red spring wheats with the *pinB-D1b* allele have improved flour yield, milling quality and loaf volume relative to sister lines with the *pinA-D1b* allele. Allelic variants at the *Ha* locus are available to modulate grain texture in soft wheats. The replacement of the distal part of 5AS by the distal 5A^{m}S segment from *T. monococcum* (active *pinB-A^{m}1a* and *pinA-A^{m}1a*) results in a reduction of the hardness of soft cultivars (Tranquilli and Dubcovsky, unpublished). These different alleles have been manipulated to confer particular levels of hardness.

Host plant resistance to disease and insect pests

Genetic resistance is the primary method of choice for controlling diseases in wheat and has been proven repeatedly as an effective and environmentally sound method to control serious yield-reducing pathogens. The use of disease resistant cultivars reduces the use of pesticides, and thus contributes to a reduction in environmental contamination (Anderson, 2000). Losses to all pests and diseases currently average 10–20% annually; therefore potential savings to growers, not counting the elimination of costs of applying pesticides, are in the hundreds of millions of dollars (10% of the US wheat production has been worth about $500 million annually since 2000).

Genetic resistance has frequently resulted in selection pressure on the pathogen population, which then mutates to overcome the resistance. Two strategies, either a pyramid or multiline approach, have been used to increase the durability of resistance genes. Gene pyramiding is the combining of two or more resistance genes. Resistance is more durable because additional mutations in the pathogen population are needed to overcome the resistance in the host plant. Multiline cultivars are made up of a series of closely related genotypes, each carrying different sources of resistance to a pest. The genetic diversity of the multiline results in balancing selection against the pest that enhances the durability of the resistance genes deployed in the multiline cultivar. Both strategies have been difficult to accomplish without MAS because once an effective resistance gene is present in a breeding line, it is difficult to screen for the incorporation of additional resistance genes using the plant phenotype. Few multilines have been released, in practice, because of the difficulty in introgressing so many different sources of resistance into a several visually and agronomically similar genotypes. MAS can be used effectively to combine different resistance genes into elite lines while maintaining pre-existing, effective resistance genes and to introgress several resistance sources into a recurrent parent.

On a global basis, the three rusts – leaf rust (*Puccinia recondita*), stem rust (*Puccina graminis*) and yellow or stripe rust (*Puccinia striiformis*) – are among the most damaging diseases of wheat and other small grain crops. Besides reducing yield, the rust diseases also seriously affect the milling and baking qualities of wheat flour. Multi-million dollar yield losses have been attributed to leaf and stripe rust every year since 2000. Stripe rust along caused a $360 million dollar loss to the wheat crop in 2004 (updates are available online (Long, 2005). Since 2000, new races of leaf and stripe rust that are virulent in many of the previously resistant cultivars have been identified in the United States. Fortunately, new resistance genes have been identified in wheat cultivars and wild relatives of wheat have been introgressed into hexaploid wheat. Using MAS, six new leaf rust resistance genes and five stripe rust resistance genes have been introgressed into 58 lines and cultivars from eight different market classes. Leaf rust resistance genes include *Lr21*, *Lr39* and *Lr40* from *T. tauschii*, *Lr37* from *T. ventricosum*, *Lr47* from *T. speltoides*, and *LrArm* from *T. timopheevi* subsp. *armeniacum*. Four major stripe rust resistance genes *Yr5*, *Yr8*, *Yr15*, and *Yr17* have been combined with the high temperature adult plant resistance gene identified in "Stephens". This resistance has been durable for over 25 years in the PNW. A quantitative trait locus (QTL) for this resistance has recently been identified, probably located on chromosome 6BS (Campbell, unpublished data). A new stripe rust resistance gene, Yr36, has also been identified on chromosome 6BS, closely linked to the HGPC gene. Both genes can be simultaneously selected using polymerase chain reaction (PCR) markers *Xuhw89* and *Xgwm193* (Dubcovsky, personal communication). PCR-specific markers are available for *Lr2*, *Lr47* and the linked group of resistance genes *Lr37-Yr17-Sr38* (Helguera *et al.*, 2003). Microsatellite marker *Xgwm210* is linked to *Lr39* and *Xgwm382* to *LrA^m*. Microsatellite markers *Xgwm18* and *Xgwm264* flank stripe rust resistance gene *Yr15* (Chen *et al.*, 2003).

Fusarium head blight (FHB) is an important disease in common and durum wheat producing areas of the United States and Canada. An epidemic of FHB from 1993 to 1997 resulted in devastating economic losses to the wheat industry of the region, with 1993 estimates alone surpassing one billion dollars. *Fusarium* head blight causes both severe yield reduction and decreases grain quality. In addition, infected grain may contain harmful levels of mycotoxins that prevent its use for human consumption or feed. Control of FHB has been difficult due to the ubiquitous nature and wide host range of the pathogen and dependence of the disease upon unpredictable climatic conditions. In some parts of the United States, fungicides have been used to reduce losses but this practice adds to grower costs, poses significant environmental risks, and is not always effective. Available resistance to FHB in wheat is quantitatively expressed, with a continuous distribution among progeny. Two major QTLs have been identified on chromosome 3BS from "Sumai 3" and on chromosome 3A from *T. dicoccoides*. Microsatellite markers, *Xgwm533* and *Xgwm493*, bracketing both QTL regions, have been used to introgress these genes into 28 durum and common wheat cultivars. The selected QTL region from the Sumai 3 chromosome arm 3BS explains up to 40% of the phenotypic variation in FHB resistance in one cross. Selection for the QTL region from chromosome arm 3AS in durum wheat has been done using SSR markers *Xgwm2* and *Xgwm674*. The QTL region on 3AL explains more than 37% of the phenotypic variation in a durum cross. Both FHB QTL regions are robust and are expressed in well-adapted genetic backgrounds (Liu and Anderson, 2003).

Practical use of MAS in forward breeding programs

MAS selection programs must be integrated at all times into existing breeding programs. Recurrent parents are selected from high yielding, elite germplasm. Intermediate products of MAS are returned to the breeding programs for evaluation and crossing purposes. After each generation of backcrossing, selected heterozygous plants are self-pollinated and the $BC_{1-3}F_2$ seeds planted as additional segregating populations. A strict backcrossing strategy is not expected to increase yield, except for the reduction of yield losses due to pathogens. Therefore, this backcrossing strategy should be used only as a complement of active "forward breeding" programs.

Completion of the trait introgression to the BC_6 generation projects not only provides superior cultivars for immediate use by growers but also provides a unique set off near-isogenic experimental materials to rigorously address scientific questions in wheat breeding and wheat genetics. One such question regards the potential costs of pest resistance genes in the absence of pest infestation. Several experiments have shown that some disease resistance genes in plants come at a cost in fitness, or yield potential. In general, however, measurement of cost has been difficult due to the lack of defined genetic populations. The development of isogenic lines carrying different genes affecting quality in the same genetic background will facilitate studies of epistatic interactions. A cross between two isogenic lines in the same recurrent parent generates a mapping population segregating only for the targeted genes in an isogenic background. Current public awareness of biotechnology has been shaped by the Genetically Modified Organism (GMO) debate. This is unfortunate because Genetically Modified Organisms are only one aspect of biotechnology. MAS is a valuable biotechnological tool for selection and re-assembly of genes that already exist in the *Triticeae*.

References

Allan, R.E., Petersen, C.J. Jr., Rubenthaler, G.L., Line, R.F., and Roberts, D.E. (1989). Registration of 'Madsen' wheat. *Crop Science*, **29**:1575–1576.

Anderson, J.A. (2000). Marker-Assisted selection of disease resistance genes in wheat, in *Marker-Assisted selection of disease resistance genes in wheat* (eds M.M. Kohli, and M. Francis). La Estanzuela, Uruguay, pp. 71–84.

Chen X., Soria, M.A., Yan, G., Sun, J., and Dubcovsky J. (2003). Development of user-friendly PCR markers for wheat stripe rust resistance gene *Yr5*. *Crop Science*, **43**:2058–2064.

Dubcovsky, J., and Soria, M.A. (2005). MASWheat: Bringing Genomics to the Wheatfields. Online: http://maswheat.ucdavis.edu/Index.htm (accessed March 21, 2012).

Giroux, M.J., and Morris, C.F. (1998). Wheat grain hardness results from highly conserved mutations in the friabilin components puroindoline a and b. *Proceedings of the National Academy of Sciences USA*, **95**:6262–6266.

Helguera M., Khan, I.A., Kolmer, J., Lijavetzky, D., Zhong-qi, L., and Dubcovsky, J. (2003). PCR assays for the *Lr37-Yr17-Sr38* cluster of rust resistance genes and their use to develop isogenic hard red spring wheat lines. *Crop Science*, **43**:1839–1847.

Hospital, F., Chevalet, C., and Mulsant, P. (1992). Using markers in gene introgression breeding programs. *Genetics*, **132**:1199–1212.

Joppa, L.R., Du, C., Hart, G.E., and Hareland, G.A. (1997). Mapping a QTL for grain protein in tetraploid wheat (*Triticum turgidum* L.) using a population of recombinant inbred chromosome lines. Crop Science, **37**:1586–1589.

Khan, I.A., Procunier, J.D., Humphreys, D.G., *et al.* (2000). Development of PCR based markers for a high grain protein content gene from *Triticum turgidum* ssp. *dicoccoides* transferred to bread wheat. *Crop Science*, **40**:518–524.

Liu, S., and Anderson, J.A. (2003). Targeted molecular mapping of a major wheat QTL for *Fusarium* head blight resistant using wheat ESTs and synteny with rice. *Genome*, **46**:817–823.

Long, D. (2005). Estimated Small Grain Losses Due to Rust in 2005. USDA-ARS Cereal Disease Laboratory, St. Paul, MN. [Online: http://www.ars.usda.gov/SP2UserFiles/ad_hoc/36400500Smallgrainlossesduetorust/2005loss/05rustloss.pdf (accessed March 30, 2012)].

USDA-ARS (2005). Graingenes: A database for Triticeae and Avena. Online: http://wheat.pw.usda.gov/GG2/index.shtml (accessed March 21, 2012).

31.12 Establishing a breeding nursery

31.12.1 Field nursery

Layout

A wheat breeding nursery may contain various materials, such as parental genotypes with desired traits, elite lines, special genetic stock, and sterility sources. It is advantageous to locate the crossing block close to the F$_1$ hybrid nursery to facilitate backcrossing, topcrossing, and hybrid–parent comparisons. Wide between-row and within-row spacing facilitates crossing and promotes good plant growth and development.

Planting

Researchers often raise their seedlings at the plant spacing of 30–60cm between rows and 15cm within rows for good tillering. Planting dates should be selected such that flowering is synchronized for crossing. The growth environment may also be modified to synchronize flowering. Some breeders of, especially, winter wheat, clip the plants (not below the spike primordial to avoid killing it).

31.12.2 Greenhouse nursery

Breeders may use a greenhouse to help synchronize flowering. Greenhouses enable plant breeders to manipulate temperature, photoperiod, light and other plant growth environmental conditions to synchronize flowering. Increasing the photoperiod tends to reduce days to heading. A greenhouse nursery also enables the plant breeder to move pots around to facilitate crossing.

31.13 Artificial pollination for hybridization

31.13.1 Materials and equipment

Emasculation is needed in wheat crossing. The equipment and materials used include scissors, curved and straight type forceps, soda straw, squeeze-type bottles, paper clips, tags, and glassine bags (5 × 17 cm).

31.13.2 Emasculation

Preparation of the female for hybridization by emasculation is done 1–3 days before the normal anthesis. The anthers at this stage should be light green but not yellow or cream-colored. The spikes to be emasculated should be carefully selected. Further, the stigmas should be clearly visible extending to about one quarter of the length of the floret. One to three of the upper and basal spikelets are usually non-functional and are removed with the scissors. In awned varieties, the awns are clipped. Only the primary and secondary florets are retained emasculated. To emasculate, gently but firmly grasp the spike between the thumb and index finger of one hand. Then, lightly press the thumb below the tertiary floret and pull the center florets downward and outward with the forceps. The three anthers in each floret are removed by carefully inserting a pair of forceps between the lemma and palea and spreading them. Care should be exercised to avoid damaging the stigma. Cover the emasculated spike with a bag.

Florets may also be emasculated by cutting off about one-third of the florets with the scissors to expose the anthers, which are then removed with the forceps. This procedure predisposes the florets to rapid desiccation and is favored under cool and humid environments. Another procedure entails cutting the primary and secondary florets at about 5–7 days before anthesis. The remaining parts of the anthers eventually degenerate without forming viable pollen. This procedure is quick but requires greater skill and experience for high success.

31.13.3 Pollination

Emasculated flowers should be pollinated within 2–4 days for best results. Also, a mature and receptive stigma is one with fully developed feathery features, whereas recently extruded anthers provide good pollen. Forceps are used to remove anthers from the florets. The glassine bag is removed from the emasculated spike so the pollen can be gently brushed on one or several stigmas. A soft brush may also be used to transfer pollen to stigmas. CIMMYT in Mexico developed a pollination technique that is suitable when pollen is present in copious amounts. First, the upper one-third of each primary and secondary floret of the male spike is removed to allow the mature anthers to freely extrude to release the pollen. The top of the bag covering the female spike is cut to allow the male spike to be inverted into the bag, parallel to the female spike. The male spike is then vigorously rotated by twirling the peduncle between the thumb and forefinger. The bag is released with the male spike in place, for an additional day for enhanced pollination.

In another method, the approach method, the female spike is positioned lower than the male spike and the two covered with a glassine bag. Such maneuvering is easier in the greenhouse. It can be used in the field if the female and male plants are planted in close proximity or the male spike is detached. To prolong the supply of pollen, culms of the detached spikes can be placed in water to keep them alive.

31.14 Natural pollination

As previously indicated, about 1–4% outcrossing occurs in wheat, depending on the variety. To facilitate large-scale hybridization and eliminate the tedium of emasculation, repeated backcrossing is used to convert breeding lines into female cytoplasmic male sterile or A lines. This scheme enables breeders to undertake composite cross breeding, production of synthetics, and the hybridizations needed for recurrent selection. However, the stability of the male fertility restoration gene is difficult to maintain, being conditioned by several genes and affected by the genetic background in which the genes occurs.

31.15 Seed development

A pollinated spike is properly targeted for identification. Successful pollination and fertilization can be verified after 3–5 days of pollination. Kernel development is detectable during the first week following pollination. At normal maturity, the tagged spikes are harvested and threshed.

31.16 Common breeding objectives

The success of modern wheat cultivars is largely due to high yield potential, wide agroecological adaptation and high responses to agronomic inputs (fertilizers, irrigation). Yield components of wheat are spikes × number of grains/spike × weight of grain (or number of grains/unit area × weight of grain in that unit). Breeders should determine what balance of these components to include in a cultivar for an agroecological niche. The right balance is determined taking into account the photoperiod, heat units, and the moisture and fertility status of the target area. For example, in one study, a top-yielding wheat cultivar, Seri 82, produced 3778 kg/ha in 98 days (maturity) at 22°12′N but 8544 kg/ha in 140 days at 30°53′N. It is not possible to simultaneously select for all the yield components because of the presence of negative intercorrelations among them.

31.16.1 Yield potential

Breeding for high yield potential in wheat can be accomplished by hybridizing high yielding genotypes and selecting transgressive segregants from the progeny with desired traits. However, it is the discovery and use of dwarfing genes that dramatically increased yield potential in wheat. Short-statured cultivars have high tillering capacity and also increased grain yield per spike. Manipulation of harvest index by incorporating semidwarf genes has resulted in high lodging resistance, high biomass, and high harvest index and, consequently, a high rate of partitioning of assimilates into the grain for higher grain yield.

31.16.2 Yield stability

Some breeders have used the concept of shuttle breeding (selecting F_2 segregating populations in one location and the F_3 in another, etc.) to develop cultivars with wide adaptation and high yield potential (e.g., cultivars such as Siete Cerros, and Pavon 76 developed at CIMMYT). The capacity to sustain high yield potential over a broad range of growing environments is desirable in a cultivar. Breeders conduct G × E evaluations of genotypes in yield trials to identify those with yield stability.

31.16.3 Agromorphological traits

As previously discussed, short stature and lodging resistance are important breeding objectives in wheat breeding. Semidwarf cultivars are lodging resistant. Selection for short stature often impacts other plant characteristics. For example, semidwarfs tend to be photoperiod insensitive and have reduced seed size and protein content.

31.16.4 Adaptation

Winter hardiness

Winter hardy wheat cultivars are needed in places where plants are likely to be exposed to unseasonable low temperatures. Regions with high rainfall tend to hold moisture in the soil for a longer time. Under wind chill temperatures and alternate freezing and thawing, wheat plants grown in such soils are prone to heaving. The red soft winter wheat types are more resistant to heaving injury than the hard red wheats. Selecting for winter hardiness should be done under natural conditions.

Drought resistance

Germplasm of Crimean origin has drought resistance and narrow leaves. Similarly, durum wheats have drought resistance and are adapted to the drier production regions of North Africa and the Middle East. At CIMMYT, breeders have used shuttle breeding approach in drought breeding. Generations F_2, F_5 and F_6 are tested under optimal conditions, while F_3 and F_4 are evaluated at reduced fertility and moisture conditions. The assumption is that input efficiency and input responsiveness can be incorporated into one genotype. Traits of interest under optimal conditions are disease resistance, good tillering capacity, head development, leaf retention, and grain plumpness. Under low input, breeders select for delayed leaf senescence, tiller viability, grain plumpness, reduced spike sterility, relative high yield, and relatively higher pest resistance.

Aluminum tolerance

Tolerance to aluminum is needed in cultivars grown for production regions where the soils are acidic. Breeders may select aluminum tolerance under artificial conditions in the laboratory. In fact, breeding effort in Brazil produced genotypes with high aluminum tolerance. However, yield of those genotypes were poor. Breeders at CIMMYT have improved upon the yield potential of the aluminum-tolerant genotypes.

31.16.5 Disease resistance

The strategy of using resistance genes singly in wheat breeding is not effective, and hence not widely practiced any longer. Rather, a combination of multiple hypersensitive resistance genes is preferred. Rust resistance genes (Lr_9, Lr_{19}, Lr_{24}) and stem resistance genes (Sr_{24}, Sr_{26}, Sr_{31}) once deployed singly are now used in combination to promote stability to stem rust in North America and Australia. However, this strategy is effective only when breeding is centrally coordinated in the country and when the production area is isolated from other areas where this system is not operational. Some failures of the system have occurred in leaf rust fungus in Mexico.

Rusts

Diseases of importance of wheat production include stem rust (caused by *Puccinia graminis* Pers.), leaf rust (by *P. recondite* Rob.), and stripe rust (*P. striformis* West). Stem rust is particularly devastating to wheat production. Researchers have identified numerous physiological races of rust species. Over 30 resistance genes have been discovered for *P. graminis* and *P. recondite*, and more than 16 genes for *P. striformis*.

Smuts

Wheat is attacked by a variety of smut diseases, the major ones being loose smut (caused by *Ustilago tritici*), common bunt (by *Tilletia tritici*), and dwarf bunt (by *Tilletia controversa*). Yield can be drastically reduced by smut. Loose smut can destroy the entire spike, whereas other smuts cause the grain to be replaced with smut spores. Genes of resistance have been identified for loose and common smut diseases.

Powdery mildew

Genes encoding resistance to powdery mildew (caused by *Erysiphe graminis*) have been identified.

Other diseases

Other diseases of interest in wheat breeding include viral infections, blights, and rots.

31.16.6 Insect resistance

Hessian fly

The Hessian fly (*Mayetiola destructor*) is known to cause a wide range of effects to wheat plants, including dwarfing of plants, reduced tillering, straw breakage, and increased susceptibility to winter injury. Some genotypes have tolerance to pest, being able to compensate for reduce tillering by producing additional tillers. Several biotypes of the fly have been identified. Resistance genes have been identified. Sources of resistance include genotypes such as "Ribeiro", "Marquillo", and "Kawvale".

Greenbug

About eight biotypes of the greenbug (*Schizaphis graminum*) have been identified. Genes for many of these biotypes have been identified.

31.16.7 End use quality

Wheat is used in bakery products and semolina. Various breeding programs focus on various market types. The hard wheats are used for bread, whereas the soft wheats are used for confectionary products and cookies. Durum wheats are used for pasta products.

Key references and suggested reading

Acquaah, G. (2005). *Principles of crop production: Theory, techniques, and technology.* Prentice Hall, Upper Saddle River, NJ.

Hall, M.D., and van Sanford, D.A. (2003). Diallel analysis of Fusarium head blight resistance in soft red winter wheat. *Crop Science*, **43**:1663–1670.

Mujeeb-Kazi, A.A. (2006). *Genetic Resources, Chromosome Engineering, and Crop Improvement Cereals*, Volume 2 (eds R.J. Singh, and P.P. Jauhar). CRC Press, Boca Raton, FL, pp. 61–97.

Outcomes assessment

Part A

Please answer the following questions true or false

1 Wheat is predominantly a self-pollinated species.
2 Wheat has spikelet inflorescence.
3 Cultivated wheat is a tetraploid.
4 Gluten is a complex starch found in wheat.

Part B

Please answer the following questions.

1 What is vernalization?
2 Give the two groups of wheat on the basis of adaptation.
3 What specific kind of polyploidy is cultivated wheat?
4 Give three of the top major wheat producing states in the United States.
5 Give the scientific name of cultivated wheat.

Part C

Please write a brief essay on each of the following topics.

1 Discuss three major breeding objectives in wheat breeding.
2 Discuss important diseases of wheat.
3 Discuss the market classes of wheat.
4 Briefly discuss the emasculation of wheat flowers for crossing.

32

Breeding corn

Principles of Plant Genetics and Breeding, Second Edition. George Acquaah.
© 2012 John Wiley & Sons, Ltd. Published 2012 by John Wiley & Sons, Ltd.

Taxonomy

Family	Poaceae
Genus	*Zea*
Species	*Zea mays* L.

32.1 Economic importance

Corn or maize is the single most important crop in the United States. It is grown on more than 20% of the cropland. Most of the production occurs in the region of the United States called the Corn Belt, where six states, Iowa, Illinois, Nebraska, Minnesota, Indiana, and Ohio account for about 80% of the national production. Corn is grown in every state except Alaska. Iowa leads the nation with 22% of the total production. Corn is the fourth most important crop in the world, behind wheat, rice and potato in total production. More than 327 million acres of corn are planted each year, worldwide. World yields average about 42 bushels/acre.

There are six "Corn Belts" in the world – the US Corn Belt, Danube Basin (South West Germany), Po Valley (North Italy), Plains of North China, North East Argentina, and South East Brazil. On the world scene, the United States, China, Brazil, Mexico, France, and Argentina together account for 75% of the world's corn production, the United States accounting for about 40% of this total. Other producers include Romania and South Africa. In 1866, the USDA reported 30 million acres of corn was harvested at an average yield of about 24 bushels/acre. In the early 1990s, about 100 million acres were grown, with an average harvested yield of 28 bushel/acre. However, in 2000, the USDA reported 67 million acres of harvested corn with an average yield of 118 bushels/acre. The dramatic increase in yield over the period is attributed to the adoption of hybrid seed and the use of fertilizers. Corn has the highest value of production of any crop in the United States, averaging 8 billion bushels worth $20 billion/year.

32.2 Origin and general history

Corn is arguably the most completely domesticated of field crops. Modern corn is incapable of existing as a wild plant. No wild form of corn is known. Its origin is probably Mexico, or Central America. It was produced as early as 6000 BC in Tehuacan, Mexico, by the Mayan and Aztec Indian civilizations. It was taken north by the native Americans. Corn was dispersed to the Old World in the sixteenth and seventeenth centuries. The explorers introduced Indian corn to Europe and Africa. Modern varieties have larger cobs and greater number and weight of kernels per year compared to the original Indian corn. The modern cultivated plant is believed to have been obtained through the process of mutation, coupled with natural selection, and mass selection by the American Indians. It is proposed that corn's progenitor may be a domesticated version of teosinte, a wild grass that occurs in Mexico and Guatemala.

32.3 Adaptation

Corn has a wide geographic adaptation. It is grown from as far north as 58°N latitude to 35–45°N latitude. It is grown at below sea level to 13 000 feet. Corn is adapted to warm temperatures.

32.4 History of corn breeding in the United States

Corn is believed to have been domesticated more than 7000 years ago. Deliberate attempts at improving corn began with the development of open-pollinated varieties that are still common in developing countries. Even though the first corn hybrid in the United States was developed in the early 1920s, most producers continued to use open-pollinated cultivars until the 1940s. Producers in the short season northern regions preferred flint (Northern flints) corn varieties. However, early production acreage was dominated by the Corn Belt Dent variety. However, growers in the mid-maturity production areas preferred the Southern Dent open-pollinated corn races for their superior high yield potential. Natural crossing between the Northern Flint and Southern Dent varieties produced "hybrids" (intervarietal hybrids) with superior performance. Farmers were able to select and develop numerous open pollinated varieties, one of the most successful being the Minnesota 13, which was later widely used in early hybrid variety development.

The pioneering work of G.H. Shull of the Carnegie Institution of Washington started the move toward the development and spread of hybrid corn. His landmark publication, *A pure line method of corn breeding*, laid the foundation for corn hybrid breeding. Later, E.M. East, while providing collaborative evidence for Shull's work, at the same time discouraged the commercial development and use of hybrids. This was because the first hybrids were inferior and, further, the single cross hybrid produced seed on an inbred line, making it an expensive undertaking. The first commercial hybrid, the Copper Cross, was a single cross variety. This first commercial variety was produced in Iowa under contract by H.A. Wallace in 1923. Wallace founded the "Hi-Bred Corn Company", which later became the Pioneer Hi-Bred Company. Commercial hybrid production became feasible when, in 1918, D.F. Jones proposed the use of the double cross hybrid, which was a product of two single crosses. This meant that a hybrid seed was produced on a relatively high yielding single cross hybrid female plant. Double cross hybrids dominated corn production until the 1960s. The dominance of hybrid varieties was due to their superior characteristics, especially greater uniformity, higher yield, tolerance to biotic and abiotic stresses, and amenability to mechanization.

Plant breeders later (after the 1960s) developed superior inbred lines with high yield potential and standability. Single cross hybrids became more productive and more uniform, eventually replacing the double crosses. Initially, corn breeding was primarily conducted in the public arena. However, private companies began developing superior lines in the 1960s for hybrid breeding, eventually dominating the commercial seed corn market. Corn yield has characteristically increased from about 64 bushels/acre with double cross in 1959 to about 129 bushels/acre using single crosses.

It has become clear to breeders that most of the corn hybrids in North America originate from a few inbred lines developed in the public arena, severely narrowing the genetic base of corn and making them vulnerable. There is an effort by breeders in both private and public sectors to enhance the genetic base of corn through the Germplasm enhancement of corn. In the latter parts of the twentieth century, the use of biotechnology became a major feature of corn breeding, resulting in transgenic products, such as, Bt-corn. Nutritional augmentation breeding has also produced products such as high lysine corn and quality protein corn (QPM).

32.5 Types of corn

Corn belongs to the family Poaceae, other members of this family being *Zea, Euchlaena, Tripsacum, Coix, Chionachne, Polytoca, Scherachne,* and *Triobachne.* Of these, the closest relatives to corn are *Euchlaena* (teosinte) and *Trypsacum*. Intergeneric crosses between *Zea* and *Euchlaena, Tripsacum, Saccharum,* and *Coix* have been successfully attempted, the products usually being sterile. *Zea × Euchlaena* (corn ×

teosinte) crosses are, however, often successful with fertile offspring.

Corn may be grouped into seven types on the basis of endosperm and glume characteristics –as dent, flint, flour, pop, sweet, waxy, and pod corns. Of these, five are commercially produced (dent, flint, flour, sweet, and waxy corns).

- **Dent corn** (*Z. mays indentata*). Dent corn is the most widely cultivated type in the United States. It is characterized by a depression (dent) in the crown caused by the rapid drying and shrinkage of the soft starch at the crown. Of the multiple colors available, the yellow or white kernels dominate the commercial production.
- **Flint corn** (*Z. mays indurate*). Flint corn is predominantly comprised of corneous or hard starch that encloses the soft starch on the center. The kernels are smooth, hard, and usually rounded at the top. This type of corn is grown widely in Europe, Asia, Central America, and South America. It is less widely grown in the United States.
- **Flour corn** (*Z. mays amylacea*). As the name implies, flour corn consists almost entirely of soft starch, making the kernels soft. It has the shape of dent corn but shrinks uniformly upon drying. It is grown in the drier sections of the United States, mainly by American Indians, and also in the Andean region of the Central and South America. Different kernel colors exist, the most common being white, blue, and variegated.
- **Pop corn** (*Z. mays everta*). Popcorn is an extreme form of flint corn. It has a very hard corneous endosperm with only a small portion of soft starch. The kernels are characteristically small and may either be pointed or have a rounded tip. Different colors exist, most corneous varieties being yellow or white. The kernel pops upon heating as a result of the unique quality of the endosperm that makes it resist the steam pressure generated, until it reaches explosive proportions.
- **Sweet corn** (*Z. mays saccharata*). This corn is characterized by a translucent and wrinkled appearance upon drying and a sweet taste when immature. The standard sweet corn is a mutant of the dent corn with a mutation at the *sugary* (*sy*) locus. This mutation causes the endosperm to accumulate about two times more sugar than the field corn. New mutants have been developed – *sugary enhanced* (*se*) and *shrunken-2* (*sh2*) or supersweet corn. Some sweet corn varieties are unable to convert sugar to starch. Sweet corn is grown as a winter crop in the southern United States, especially Florida.
- **Waxy corn.** Waxy corn has a uniformly dull appearance. Instead of amylose, the starch of waxy corn consists of amylopectin, the result of *waxy* (*wx*) mutation. The ordinary corn consists of about 78% amylopectin (a high molecular weight branched chain) and 22% amylose (a low molecular weight, straight chain).
- **Pod corn** (*Z. mays tunicata*). Pod corn has primitive features, each kernel being enclosed in a pod or husk, before the entire ear is enclosed in husks like other corns. Pod corn versions of the other types of corn (i.e., flint podcorn, dent podcorn, etc.) exist.

Similarly, corn that is indigenous to the United States (excluding sweet and pop corns) may be classified up to nine or ten races. Of these, the most important are the Corn Belt Dents, the Southern dents, and the Northern dents.

32.6 Germplasm resources

Over 13 000 accessions of corn are kept in storage at CIMMYT in Mexico, with duplicates of these accessions held in the United States (at the US National Seed Storage Laboratory, Fort Collins, Colorado), Columbia, and Peru. Many heterotic populations have been identified in corn, the most well known and exploited including Reid × Lancaster, Lancaster × Stiff Stalk. Others are the European flint × dents, Tuxpeno × ETO, and Pantap × Suwan 1.

32.7 Cytogenetics

Corn (*Zea mays* L.) is a diploid ($2n = 20$) and a monocot of the family Poaceae (Gramineae), or grass family. The genus has four species: *Zea mays* (cultivated corn and teosinte), *Z. diploperennis* Iltis et. al. (diploperennial teosinte), *Z. luxurians*, and *Z. perennis* (perennial teosinte). Of these four species, only *Zea mays* is widely grown commercially in the United States. The closest generic relative of *Zea* is *Tripsacum*, which has seven species, three of which are known to occur in the United States. Teosinte occurs in the wild in Mexico and Guatemala. Cultivated corn has 10 pairs of chromosomes ($n = 10$). However, plants with 1–8 sets of chromosomes have been developed for various purposes.

In addition to the autosomes or normal or standard chromosomes (A-chromosomes), corn has supernumerary elements such as the B-chromosomes (also called supernumerary chromosomes). The role of the B-chromosomes in the cell varies from being practically a nuisance to having some definite function, depending on the organism. However, when the B-chromosomes number 10–15 or more, certain abnormalities may occur (e.g., reduced fertility, decreased vigor, aborted pollen, defective kernels). Seed is rarely produced with the B-chromosomes in excess of 25. Corn B-chromosomes are among those widely studied in plants. They are suspected of influencing the frequency of crossing over, among other roles. Thousands of translocation events have been described in corn. They are used for locating genes on chromosomes.

Monoploids (haploids) may arise spontaneously by pathogenesis (unfertilized egg develops into a plant). Occasionally, paternal haploids develop by androgenesis. Haploid generation via these systems has a low frequency of occurring. The average frequency of this event in corn is estimated at one per thousand kernels. These lines may be used to develop homozygous diploid inbred lines for hybrid production. They can also be used to convert inbred lines with male fertility to male sterile cytoplasm. Tetraploid corn was shown to have gigas features (e.g., regarding leaves, tassels, ears, kernel size) but with reduced fertility. Tetraploid yellow corn produced also about 40% higher carotenoid pigment content than the diploid parent. Barbara McClintock conducted extensive cytological work on corn. She developed a complete primary trisomic series, using triploids, of which only trisomy 5, 7 and 8 can be distinguished and characterized phenotypically. Primary trisomics have been used to assign genes to specific chromosomes.

Intergeneric crosses between cultivated corn and related genera teosinte (*Euchlaena* spp.) and gammagrass (*Tripsacum* spp.) have been accomplished. The success is more common with annual teosinte (*Zea mexicana*) in which the annual strains (Chalco, Durango, Florida varieties) readily cross with cultivated corn. *Tripsacum* may be diploid ($2n = 36$) or tetraploid ($2n = 72$). It is a potential source of resistance to many diseases and insect pests of corn. The F_1 is backcrossed to corn to remove all the *Tripsacum* chromosomes, leaving a plant that exhibits pure corn

phenotype. However, yield is reduced, as is agronomic suitability, making this introgression, overall, less attractive at the present time.

32.8 Genetics

Corn is one of the plants that has been genetically widely studied. Hundreds of mutations have been identified in corn that impact traits such as plant height, endosperm characteristics, plant colors, insect resistance, disease resistance, stalk strength, and many other traits. Some of the significant genetic effects are as follows:

- **Xenia.** Xenia is the immediate effect of pollen on the developing kernel. It may be observed when two varieties differing in a single visible endosperm trait are crossed. Xenia occurs when the trait difference is conditioned by a dominant gene present in the pollen. However, when dominance is incomplete, xenia would occur when either variety is the pollen parent. Xenia is important because endosperm characteristics distinguish some of the major corn groups. For example, starchy endosperm is dominant over sugary (sweet) and waxy. A cross of starchy × sugary exhibits xenia. Similarly, a cross of shrunken × non-shrunken endosperm, waxy × non-waxy endosperm, purple × colorless aleurone, and yellow × white (colorless) endosperm all exhibit xenia.

  Whereas xenia may result from simple dominance gene action, the effect is different in some instances. In the cross of flinty × floury endosperm, the F_1 is flinty (*FFf*). However, the reciprocal cross of floury × flinty endosperm produces an F_1 with floury endosperm (*ffF*), indicating the ineffectiveness of the dominant allele (*F*) to overcome the double recessive (*rr*) floury genes. Similarly, xenia in aleurone color depends on the combined action of five dominant genes (designated *A1*, *A2*, *C*, *R*, and *Pr*).

- **Chlorophyll varieties.** Numerous leaf color abnormalities that affect the corn plant in both seedling and mature stages have been identified. Chlorophyll-deficient mutations cause a variety of leaf colors, such as albino, virescent, and luteus (yellow). Mature plants exhibiting golden, green-striped and other leaf patterns are known.

- **Transposable elements.** Genomes are relatively static. However, they evolve, albeit slowly, by either acquiring new sequences or rearranging existing sequences. Genomes acquire new sequences either by mutation of existing sequences or through introduction (e.g., by vectors, hybridization). Rearrangements occur by certain processes, chiefly genetic recombination and transposable genetic elements (Chapter 23).

The ancient allotetraploid origins and the presence of large numbers of transposable elements makes the corn genome complex. However, it is these very features that make the corn plant very suitable for functional genomics studies. The number and variety of transposable elements facilitate insertional mutagenesis projects. Further, its allotetraploid-based gene redundancy allows scientists to characterize mutants that may be lethal in a diploid species.

- **Male sterility.** Male sterility genes are among the most important mutations in corn from the standpoint of breeding. Designated ms_1, ms_2, ...ms_n, over 20 of these genes are known. Sterility controlled by nuclear genes is also known in corn, even though it is not of practical application in hybrid corn production. In corn, two main sources of male sterility, controlled by the cytoplasm known as the T (Texas) and S systems, are known. Though deemed superior to the S system, the T system, once most widely used, fell out of favor when overuse predisposed corn production in the United States to corn leaf blight in 1920, leading to devastation of the industry. CMS-containing stock is used as seed parents to eliminate the cost of detasseling in commercial hybrid seed production. In addition to sterility genes, fertility restoration (RF_n) genes are used in corn. Some seed corn companies use mechanical detasseling for hybrid seed production.
- **Plant height.** Studies have been conducted on the *brachytic 2 (br2)* gene. This gene drastically reduces plant height. However, its practical exploitation in breeding corn has been limited because of the undesirable effects associated with its use (e.g., reduced broad leaves, delayed maturity thick stem).

RFLP probes capable of detecting more than 500 polymorphic loci have been developed in corn. Corn RFLP map has been generated.

32.9 General botany

Corn is a monoecious annual and one of the largest of the cereals, capable of reaching 15 feet in height. The male flowers (staminate) occur in the terminal panicle or tassel at the top of the stalk, while the female inflorescence (pistillate) is borne in the axils of leaves as clusters, called a cob, at a joint of the stalk. Long silks (long styles) hang from the husk of each cob. These pollen tubes are the longest known in the plant kingdom. As pollen receptors, each silk must be individually pollinated in order to produce a fruit or kernel. A fertilized cob (also called an ear) may contain eight or more rows of kernels. Furthermore, a stalk may bear 1–3 cobs.

Corn has a variety of morphological features. Some early maturing types, maturing in 50 days, may attain a height of two feet and produce 8–9 leaves, while tall late maturing types (330 days) may attain a culm or stalk height of 20 feet and bear 42–44 leaves. The hybrid corn varieties grown in the northern United States attain a height of 3–8 feet, bear 9–18 leaves, and mature within 90–120 days. The Central Corn Belt hybrids varieties range between 8–10 feet in height, bear 18–21 leaves and mature in 130–150 days. The varieties used on the Gulf coast and South Atlantic regions are much taller (10–12 feet), produce more leaves (22–27), and tiller profusely, maturing late (170–190 days).

Corn has both seminal and adventitious roots. The seminal roots may number 3–5 and grow downward at the time of seed germination. The crown or coronal roots arise from the nodes of the stem, about 1–2 inches below the soil surface, and may number between 15–20 times as many as the seminal roots. The aerial roots (buttress, prop, or brace roots) arise at nodes on the stem above ground.

The number of rows of grain is variable among varieties, ranging between 8 and 28. Each row may contain between 20 and 70 kernels. Most of the corn varieties grown in the United States contain 14, 10, 12, or 18 rows of kernels, and average about 500 kernels per ear.

32.10 Reproductive biology

32.10.1 Floral morphology

Each spikelet consists of two flowers. Each flower has three anthers that are pushed out of the spikelet as the filament elongates at anthesis. The exsertion of the anthers is followed by the opening of these structures to shed the pollen. Complete pollen shed may occur in just a few minutes or over a longer period. A tassel may shed all of its pollen in one day or over a period of about one week. The pollen shed pattern is dependent on the genotype and environmental factors, such as temperature, humidity, and air movement. Corn produces pollen grains profusely. A normal plant tassel produces an estimated 25 000 pollen grains per each kernel on the ear of corn.

32.10.2 Pollination

Pollen

Corn pollen is primarily dispersed by wind. Consequently, corn is about 95% cross-pollinated, most of the effective pollination of an ear originating from sources in the immediate vicinity of the ear. Pollen dispersal is favored by warm temperature and low humidity. Under such conditions, pollen dispersal may start at about three hours after sunrise in the Corn Belt of the United States and last for 1–3 hours. High temperature, about 35 °C (95 °F), may kill the pollen grain. Further, once dispersed into the atmosphere, the pollen grain may lose viability due to rapid desiccation within a few minutes.

Pistillate flower

The pistillate flowers are borne on a cob that arises from a husk formed at the sixth or seventh node on the stem below the tassel. The female spikelet occurs in pairs of one fertile ovary and one sterile ovary. This is the reason for the even number of rows of kernels on a corn ear. Occasionally, both ovaries in a pair become fertilized, producing kernels, resulting in crowding and irregular rows of kernels on the ear. A silk grows from the tip of each ovary until it emerges at the tip of the ear husk. Silks at the basal part of the ear usually emerge first. A silk is structurally a stigma and style, and hence is receptive along its entire length. Temperature, soil moisture, and soil fertility affect the rate of silk emergence. Adverse weather, such as severe drought, may delay or cause complete cessation of silk emergence.

Receptivity of the stigma

Corn is generally a protandrous plant (the male spikelets usually mature before the female spikelets. For the same plant, pollen shed usually precedes silk emergence by about 1–3 days. Silks are receptive soon after emergence and remain receptive for up to about 10 days. For optimum results, emerged silk should be pollinated within 3–5 days after first silk emergence. Fertilization usually occurs within 12–24 hours of pollination. High temperatures or low humidities adversely affect stigma receptivity.

32.11 Genetic consequences of reproductive biology

(a) **Heterozygosity.** Being predominantly open-pollinated, natural populations of corn are highly heterozygous and genetically variable. In theory, each kernel on the ear could be produced from the fertilization of an ovule by a different pollen parent. A field of naturally pollinated corn in effect comprises a population of hybrids.
(b) **Xenia.** As previously discussed, xenia occurs in corn.

32.12 Common breeding approaches

Three general approaches are used in the world for corn improvement for cultivar development – germplasm introduction, population improvement, and hybrid breeding. These approaches are used to develop (i) open-pollinated cultivars, (ii) population improvement cultivars, and (iii) hybrid cultivars. Both intrapopulation and interpopulation improvements are used in corn breeding.

Mass selection has been used by plant breeders to modify corn characteristics such as height, maturity, ear characteristics, and grain yield. The method was used to develop cultivars for new production areas in the Corn Belt. The modified mass selection was used to achieve a 2.9% per cycle average yield gain. Mass selection in corn breeding in the United States is now practically non-existent. However, the method continues to be used in developing countries by farmers and plant breeders to produce open-pollinated cultivars.

Other intrapopulation breeding methods used in corn breeding include ear-to-row (half-sib family), modified ear-to-row, half-sib family, full-sib family, modified full-sib, and selfed family, with varying degrees of success. Genetic gains averaging 3.8% per cycle were recorded using modified ear-to-row, while a selfed family selection average of 6.4% per cycle for grain yield was recorded. Various interpopulation methods (recurrent selection – both half-sib and full-sib) have been used.

32.12.1 Development of inbred lines in hybrid corn production

Sources of inbreds for corn hybrid breeding have changed over the years. Before 1930, corn inbreds were isolated from open-pollinated varieties. Later, breeders used inbreds derived from single-crosses, modified single crosses and three-way crosses. Sometimes, backcrossing is used to enhance any inbred line in a specific way by introducing a gene to correct a deficiency. However, these inbreds were less effective for improving quantitative traits. To overcome this weakness, breeders resorted to developing inbred lines from recurrent selection populations that have

been improved for specific quantitative traits (e.g., grain yield, stalk quality, disease resistance). Superior inbred lines are also produced by crossing other inbred lines with superior complementary traits and select from the progeny.

Inbreds are developed by artificially controlled pollination. The F_1 plants of a cross are pollinated by hand, followed by pedigree selection through 5–7 generations, or until a desirable level of uniformity in appearance and performance is achieved. The inbred lines are evaluated for combining ability. Single cross hybrid corn is most commonly used in commercial production. Consequently, it is critical that the inbred lines on which the crop will be produced be high yielding. Selection at each generation should also be based on superior performance regarding disease and insect resistance, grain quality, lodging resistance and other traits that support the breeding objectives.

Industry highlights
Hybrid breeding in maize

F.J. Betrán

Texas A&M University, College Station, TX 77843, USA

Maize is a cross-pollinated species that shows high heterosis (i.e., superior performance of crosses relative to their parents) for grain yield (Figure B32.1). This high expression of heterosis is exploited in maize hybrids and constitutes the foundation of the maize seed industry. Maize hybrids were first developed in the United States in the mid-1930s and by the early 1960s practically all the maize area in the United States was planted to hybrids (Duvick and Cassman, 1999). Improved productivity and selection gain with the use of hybrids has stimulated increased investment in hybrid development, resulting in impressive genetic progress (Figure B32.2). Shull (1909) outlined the pure line method in maize breeding, suggesting the use of self-fertilization to develop homozygous lines that would be of use in hybrid production. This combination of inbreeding and hybridization constitutes the basis of maize improvement. The general process to develop maize hybrids starts with the creation of a source segregating breeding population that it is used to develop inbred lines through inbreeding and selection (Figure B32.3). Selected inbreds are then evaluated in hybrid

Figure B32.1 Heterosis in maize: A × B hybrid ears in the middle and corresponding parental inbreds A and B on either side. Figure courtesy of F.J. Betran.

Figure B32.2 Relative grain yield of an open-pollinated population and a single cross hybrid developed with selected inbreds from the same population. Figure courtesy of F.J. Betran.

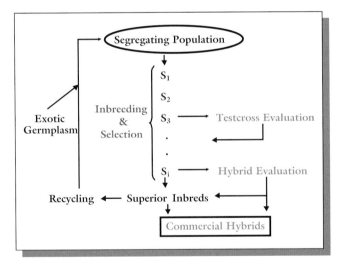

Figure B32.3 General scheme for the development of maize inbreds and hybrids. Figure courtesy of F.J. Betran.

combinations across locations to select superior hybrids and to estimate their combining abilities. The following it is a brief description of the main components of this process from the starting breeding population to commercial hybrids.

Source breeding populations

Different type of segregating populations can be used as the source in line development: open-pollinated cultivars (OPC), synthetic cultivars, single crosses, backcrosses, double crosses, related line crosses, and exotic germplasm. Overall, major emphasis goes to the use of breeding populations created by hybridization of complementary inbreds and the selection of progenies possessing the desirable traits from both parents (Hallauer, 1990). Selection within F_2 and backcross populations using pedigree breeding is the most important breeding method of developing maize inbreds. Breeding programs that emphasize pedigree selection within populations developed from elite inbred lines are therefore cyclical creating second, third, fourth, and so on generation recycled improved inbreds (Figure B32.3). The incorporation and introgression of exotic germplasm brings new desirable alleles and genetic diversity to this recycling of elite lines (Goodman *et al.*, 2000). A backcross or multiple backcross to the best parental inbred is used commonly to increase the probability of maintaining favorable combinations of alleles (Troyer, 2001). Maize breeders use multitrait, multistage, and multi-environment selection methods (Betrán *et al.*, 2003). Multi-environment and multiyear inbred general combining ability values with validated performance in hybrids are considered heavily when choosing parents to start breeding populations. The hybrid testing program provides information about the best inbreds to initiate breeding projects.

Development of parental inbreds

Inbreeding and inbreeding depression

Hybrid development requires development of parental inbred lines (Figure B32.3). The inbred parents used to produce the hybrids are developed through a process of inbreeding and selection. The consequence of inbreeding is an increase in homozygosity, which leads to homogeneous expression of traits and to inbreeding depression (i.e., loss of vigor and productivity). Self-pollination is the most common and fastest system of inbreeding. As inbreeding reduces the genetic variation within families and increases the genetic variation among families, the efficiency of selection among lines increases while it decreases within lines. The level of inbreeding depression depends on the trait. Traits that show high inbreeding depression also show high heterosis (e.g., grain yield). Vigor, plant size, grain yield components, and grain yield are reduced while time to flowering and incidence of barrenness increase with inbreeding (Hallauer and Miranda, 1988). The improvement of inbred lines over recent decades has reduced the inbreeding depression considerably in temperate maize. Therefore, source germplasm oriented to develop hybrids has been selected for low inbreeding depression (particularly in the female parent because high yielding line reduces the cost of the hybrid seed) in addition to high combining ability.

Breeding methods to develop inbreds

Development of inbred parents can follow different breeding methods, such as pedigree breeding, backcross, bulk, single seed descent (SSD), double haploids, and so on. *Pedigree Breeding* is the most widely used breeding system to develop maize inbreds. Typically, specific crosses are made between inbred lines and self-pollination is then applied to the F_1 and subsequent generations to develop inbred lines that are superior to either parent (transgressive segregants) through genetic segregation and recombination. Selection is applied among progeny rows and among plants within S_1 families. It is common to have replicated nurseries for the S_1 families exposed to different disease, insect, or abiotic stresses. This process of selfing and selection is repeated in successive generations ($S_2, S_3, S_4, S_5, \ldots . S_n$) until homozygous elite inbreds are developed. Effective phenotypic selection and greater selection intensity can be applied in initial inbreeding stages for traits with high heritability, such as pest resistance, maturity, morphological traits, and so on. The *Backcross Breeding* method is used widely in maize breeding to transfer one or a few traits/genes from the donor parent to the recurrent and most desirable parent. With the advent of genetically modified organisms, major emphasis is devoted to accelerate backcrosses to transfer the transgenes to elite inbreds. The use of DNA molecular markers has facilitated both the speed and accurate recovery of the recurrent parent, and the reduction of linkage drag. The *Bulk method*, where the seed for each selfing generation are harvested in bulk, and s*ingle seed descent*, where one or a few seeds from each genotype are advanced each generation until approximate fixation is reached, are also used because of their simplicity and low space requirements. *Double haploids* derived from maternal (e.g., *stock6*) or paternal (e.g., *indeterminate gametophyte ig*) gametes have been used to derive homozygous inbred lines instantaneously (Birchler, 1994). However, this method of developing inbred lines has not been used extensively because of the absence of any possibility for phenotypic evaluation and selection generates an unselected large sample of inbreds that needs to be evaluated for combining ability.

Breeding methodologies

Improvements in technologies such as off-season nurseries, managed environments for screening against biotic and abiotic stresses, adoption of experimental equipment (combines, planters, computers, etc.), and applications of molecular tools and biological research have increased the accuracy and efficiency of inbred development. Experimental screening techniques have been developed to increase heritability, such as artificial insect infestation, disease inoculation, and environments managed for higher plant densities or specific abiotic stress factors.

Development of hybrids

Heterotic groups

Heterosis is related to the level of heterozygosity. If two inbreds are crossed, heterosis is a function of the dominance in those loci with different alleles in the inbreds. Therefore, the identification and development of heterotic groups of elite inbreds having different alleles at loci regulating productivity can contribute to hybrid performance. A heterotic group is germplasm that when crossed to germplasm from another heterotic group, tends to exhibit a higher degree of heterosis than when crossed to a member of its own group (Lee, 1995). Heterotic patterns, which are composed of two reciprocal heterotic groups, were empirically established through testing and choice of lines to start breeding populations. Performance of lines in hybrids has been the main criteria of classification into heterotic groups: A × B hybrids were superior to A × A or B × B hybrids. Heterotic patterns for temperate maize are well established (e.g., Reid Yellow Dent × Lancaster Sure Crop in the United States and European flints × US dent in Europe). After establishment, heterotic patterns have been enhanced and optimized through selection and recombination. Multiple heterotic patterns have been developed as a result of intensive elite line recycling and specific emphasis across breeding programs. Testcross performance with representative testers has been used to group large number of inbreds to known heterotic groups. Recently, DNA molecular markers have being effective for assigning inbreds to heterotic groups (Melchinger, 1999). The enhancement of heterotic response is improved by subsequent cycles of inbred line development. Increasing degrees of heterosis are observed after several cycles of hybrid selection, due to increasing divergence of allele frequencies and selection of complementary alleles in the heterotic groups. In line recycling and in the development of source breeding populations, crosses among elite lines from the same heterotic group are preferred. Heterotic response is heritable and inbreds have heterotic reactions similar to their parents (Troyer, 2001).

Correlation between inbred and hybrids, and hybrid prediction

The number of potential single crosses to evaluate increases substantially with the number of parental inbreds. The possibility of using inbred line information, as indicative of hybrid performance, is desirable to reduce the number of hybrid evaluations. The correlation between parental inbreds and hybrids depends on the trait. In general, the correlation is relatively high for some additively inherited traits (e.g., plant morphology, ear traits, maturity, quality characters, etc.) but is relatively low for grain yield. The correlation for grain yield has been consistently positive and significant but not high enough to predict hybrid performance. The correlations between parental genetic diversity estimated with molecular markers, pedigree, or phenotypic traits and hybrid performance also have been too low to have predictive value (Melchinger, 1999). Methods based on linear mixed models have been adapted to maize to predict performance of inbreds in untested environments or hybrid combinations (Bernardo, 1999). Although these recent approaches facilitate hybrid selection, hybrid testing is required ultimately to identify the inbreds with the best breeding values.

Hybrid testing and screening

Hybrid testing in several environments representative of the target area is executed in several testing stages. A good example of testing stages within a commercial breeding program has been outlined by Smith *et al.* (1999):

Stage 1. Testcross performance of experimental lines in few locations (e.g., five).
Stage 2. Hybrid evaluation of selected lines in more hybrid combinations and locations (e.g., 20).
Stage 3. Hybrid evaluation in about 50 locations on research plots in several hybrid combinations.
Stage 4. Evaluation of best precommercial hybrids in about 75 research plot locations and about 200–500 on farm locations.
Stage 5. Hybrid performance verification in about 75 research plot locations and 300–1500 on farm strip plot tests.

Efforts are allocated in preliminary tests to evaluate as many hybrids as possible in a few locations with intensive selection, leaving relative few hybrids to proceed to the more advance pre-commercial stages. As the numbers of lines to be tested at various stages of inbreeding increase over time, their evaluation in all possible hybrid combinations is not feasible. Therefore, testcrossing with appropriate testers has been adopted extensively to evaluate the relative combining ability of experimental inbred lines. The most common tester used is an elite inbred line from the opposite heterotic group. Testers with low frequency for favorable alleles (e.g., susceptible to diseases) are also used. The level of inbreeding when testcross evaluation is conducted varies among breeders and depends on the traits under consideration and the effectiveness of visual selection. Two basic systems are used: late testing and early testing. In late testing, hybrid evaluation is delayed until advanced generations of selfing (e.g., S_5 or S_6) assuming that selection for additively inherited traits and seed yield during inbred line development will assist in reducing the number of lines for testcross evaluation. In early testing, evaluation of hybrid performance is conducted in early generations of inbreeding (S_1 to S_3). Approximately 60% of the maize breeders in the United States evaluate new lines in testcrosses in S_3 and S_4 (Bauman, 1981). Characterization and selection of inbreds is a sequential testing, with some lines discarded either by early testing or by performance *per se* in early generations, and others discarded later by general and specific combining ability in hybrid combinations.

Evaluation of hybrids in the late stages emphasizes wide-area evaluation with multiple environments. This extensive evaluation permits the selection of the hybrids for adaptation and stability. Managed environments may include cultural practices, such as different densities, planting dates, drought stress, fertilization levels, tillage, crop rotation, and so on. Environments are representative of the most common farmer field conditions. More weight is applied in decision making to those environments more correlated with the target environment. Traits that are emphasized during hybrid screening depend on the target environments. In the United States desired traits include high grain yield, low moisture content at harvest, good standability, ear retention, stability over years, drought tolerance, stay-green, disease and insect resistance, early vigor, and proper maturity (Troyer, 2001). The hybrid seed industry has concentrated on extensive and efficient testing, improve mechanization, reliable data collection and analysis, and high selection intensities (Coors, 1999). After all the processes described, only a very small percentage of the experimental hybrids tested make it all the way to commercial hybrids.

References

Bauman, L.F. (1981). Review of methods used by breeders to develop superior corn inbreds. *Proc. Annu. Corn Sorghum Ind. Res. Conf.*, **36**:199–208.

Bernardo, R. (1999). Prediction of single-cross performance, in *The genetics and exploitation of heterosis in crops* (eds J. G. Coors, and S. Pandey) ASA, Wisconsin, pp. 269–276.

Betrán, F.J., Bänziger, M., and Menz, M. (2003). Corn Breeding, in *Corn: origin, history, technology, and production* (eds W. Smith, F.J. Betrán, and E. Runge) John Wiley & Sons, Inc., New York, Chapter 2.3.

Birchler, J.A. (1994). Practical aspects of haploid production, in *The Maize Handbook* (eds M. Freeling, and V. Walbot) Springer-Verlag, New York, pp. 386–388.

Coors, J.G. (1999). Selection methodologies and heterosis, in *The genetics and exploitation of heterosis in crops* (eds J.G. Coors, and S. Pandey) ASA, Wisconsin, pp. 225–246.

Duvick, D.N., and Cassman, K.G. (1999). Post-green revolution trends in yield potential of temperate maize in the north-central United States. *Crop Science*, **39**:1622–1630.

Goodman, M.M., Moreno, J., Castillo, F., Holley, R.N., and Carson, M.L. (2000). Using tropical maize germplasm for temperate breeding. *Maydica*, **45**:221–234.

Hallauer, A.R. (1990). Methods used in developing maize inbreds. *Maydica*, **35**:1–16.

Hallauer, A.R., and Miranda, J.B. (1988). *Quantitative genetics in maize breeding*, 2nd edn. Iowa State University Press, Ames, IA.

Lee, M. (1995). DNA markers and plant breeding programs. *Advances in Agronomy*, **55**:265–344.

Melchinger, A.E. (1999). Genetic diversity and heterosis, in *The genetics and exploitation of heterosis in crops* (eds J.G. Coors, and S. Pandey) ASA, Wisconsin, pp. 99–118.

Shull, G.H. (1909). A pure line method of corn breeding. *Rep. Am. Breeders Assoc.*, **5**:51–59.

Smith, O.S., Hoard, K., Shaw, F., and Shaw, R. (1999). Prediction of single-cross performance, in *The genetics and exploitation of heterosis in crops* (eds J.G. Coors, and S. Pandey) ASA, Wisconsin, pp. 277–286.

Troyer, A.F. (2001). Temperate corn – background, behavior, and breeding, in *Specialty corn* (ed. A.R. Hallauer) CRC Press, New York, pp. 393–466.

32.13 Establishing a breeding nursery

A proper layout of the breeding nursery facilitates a breeding program. A comprehensive corn breeding program will have nurseries consisting of selfing nurseries for the maintenance of inbred lines, increasing inbred lines, and inbred line selection and breeding. There will also be crossing blocks (for single, three-way, double crosses, and topcrosses) and recurrent nurseries.

The plants should be planted in rows with alleys for easy movement of the breeder. Parents for crossing should be arranged in paired rows. Where several crosses have one row, the common parent may be used as the male parent and planted adjacent to a block of the female parent.

32.14 Other nurseries

As previously indicated, corn breeders use field nurseries exclusively for breeding. Greenhouses and growth chambers may be used for experimental purposes but they are not amenable to the practical breeding of corn. Breeding in cooler climates may use winter nurseries for additional breeding work.

32.15 Special environment

No special environmental conditions are needed if the breeding nursery is established in an area adapted for corn production. Temperature, soil moisture, and humidity are critical for proper flowering and pollination of corn. Temperatures above 30 °C and low humidity adversely impact pollen viability. Similarly, drought causes delay in silk emergence relative to tassel development. It is important to have a source of supplemental irrigation to provide moisture at critical times to avoid tassel firing and male flower abortion.

32.16 Artificial pollination for hybridization

32.16.1 Materials and equipment

Crossing corn requires the use of simple equipment, including a knife and stapler. Materials include tassel bags, ear shoot bags, paper clips, a pencil, tags, carrying case or apron with pockets. The bags should be water-repellent.

32.16.2 Preparing the female flower

Corn is a monoecious plant. The female flower should be covered and protected from unwanted pollen. The ear shoot should be covered with an ear bag before the silks emerge from the husk tips. Once covered, artificial pollination should be made within about three days, otherwise the silks should be trimmed to prevent them from growing out of the bag and becoming contaminated. Further, trimming the silks to within 2 cm of the husk tip is usually done the day before pollination.

32.16.3 Pollination

Where controlled pollination is desired, the tassel is also covered with a tassel bag and held securely in place by stapling or using paper clips at least one day before pollination. This would prevent contamination from unwanted sources and also save desirable pollen that would otherwise have been lost to the wind. Copious amounts of pollen are available on the second or third day following pollen dehiscence.

Care should be taken to eliminate contamination, even though corn produces large amounts of pollen. When selfing, the pollen may be taken directly from the tassel and deposited on the silk. In cross-pollination, the pollen may be collected in a bag by shaking the tassel and dusting the pollen on the ears. A pollen gun may also be used for multiple pollinations.

The optimum time for pollination depends on temperature and humidity but is usually about three hours after sunrise. The ear bag should be replaced immediately after pollination to avoid contamination. However, under some conditions, covering the ear shoot for a long time may cause ear tip rot. Pollination information, including parents involved in the cross and date of pollination, is written on the envelopes.

32.17 Natural pollination for hybridization

Controlled artificial pollination (and sometimes natural pollination) is used to develop commercial hybrid cultivars. Once developed, hybrid seed (or synthetic seed) production is conducted in the field using the wind as the agent of pollination. To maintain the genetic purity of the new cultivar, and to meet certification standards for corn, seed production must be conducted in isolation from other corn cultivars to keep the contamination below the mandated 1%. It is mandated that an isolation block for seed increase be located no less than 200 meters from corn of a different color or texture.

32.18 Common breeding objectives

The breeding objectives presented next are not in order of importance.

32.18.1 Grain yield

Grain yield of corn is a major breeding objective for this crop, which is among the major crops that feed the world. It is estimated that corn yield in the United States increased about 340% between 1940 and 1990. This increase is attributed to the development of cultivars with high yield potential. Over the period indicated, producers switched from open-pollinated cultivars to higher yielding cultivars. Yield is a complex trait and depends on the genes that are associated with basic physiological processes, plant structure, and morphology. Yield components of corn include number of ears, kernel rows and kernels per row, kernel test weight, and shelling percentage.

Days to maturity, standability and resistance to environmental stress (biotic and abiotic) affect the grain yield of corn. Features of modern corn hybrid cultivars compared to older cultivars were summarized by W.A. Russell as follows: longer grain filling period (i.e., early flowering and delayed senescence), rapid grain filling, increased sink size (i.e., more kernels per unit area, larger kernels, and induced bareness), higher harvest index and shelling percentage, short plant and tassel, upright leaves, shorter anthesis to silking interval, better standability, and better tolerance to abiotic stress. Further, modern day cultivars are more efficient at exploiting high soil nitrogen and also are amenable to cultivation under high plant density.

Another approach to yield breeding corn is to select for prolificacy (plant has more than one ear). Multiple ears per plant is known to improve standability under adverse conditions.

32.18.2 Yield stability

It is desirable for cultivars to have stable yield in the production region.

32.18.3 Agromorphological traits

Lodging resistance

To accomplish lodging resistance or good standability in corn, the specific focus of corn breeders is improved stalk quality. Lodging in corn may be at the root level or the stalk level. Rind thickness and crushing strength are indicators used to select stalk strength. Tall genotypes will see ears too high on the stem and become prone to lodging. Stalk breakage is caused by boring insects, such as the European corn borer (*Ostrinia nubilalis*). One of the most successful applications of genetic engineering in plant breeding is the development and deployment of *Bt* corn varieties. These GM varieties are resistant to attack by the borer. As previously indicated, the use of the *brachytic 2* gene in breeding is limited.

Resistance to ear dropping

When harvesting is delayed, mature corn may break off and drop to the ground before harvesting. Sometimes, the impact of the combine harvester may cause such ear drop to occur, leading to significant field losses. Disease may promote ear dropping. Genotypes with large, heavy ears are more susceptible to ear dropping. Long ear shanks are more susceptible to corn borer attack and are structurally weak and more prone to ear dropping. Plant breeders hence select for shorter and stronger shanks. However, it is desirable for the ear to bend downward when mature to accelerate drying.

Husk covering

The ear of corn is covered by a husk that has a protective role (against pests and weather) and also promotes rapid drying. Breeders select for long husk or complete covering or the tip.

Dry-down

The husk and grain must dry before harvesting of corn. The more rapid the drying process, the better, as it allows for early harvesting of the crop.

32.18.4 Adaptation

Early maturity

Maturity in corn is a qualitative trait. Corn is frost-insensitive. Hence, early maturity is an important objective in temperate climates. Early maturing cultivars are usable in areas of shorter growing season and higher altitudes. Maturity traits used in breeding corn include days to silk, tassel and brown husk. The black layer is an indicator of physiological maturity.

Drought resistance

Corn is adapted to both temperate and tropical climates. However, excessive temperature and drought at flowering promotes reduced seed set and bareness. Drought resistance traits in corn include small male tassel, small leaf area, prolificacy, leaf elongation, heat tolerance, high abscisic acid, and low temperature.

Cold tolerance

Cold tolerance is important for germination under unseasonably cold soil temperature, and also for quick dry down in cold weather.

32.18.5 Disease resistance

Corn is plagued by over 100 diseases and over 100 insect pests worldwide, some of which have global importance.

Seed rots and seedling blights

Corn is plagued by many pathogens that cause rots and blights, the common ones include *Diplodia maydis*, *Fusarium moniliforme*, and *Penicillium* spp.

Root, stalk and ear rots

Agents of common root rots include *Fusarium spp.* and *Pythium* spp. Common stalk rots include *Diplodia* rots and *Fusarium* rots. Corn is also attacked by ear rots caused by *Diplodia* and *Aspegillus*.

Leaf blight or spots

Leaf blights or spots of economic importance include bacterial wilt (caused by *Bacterium stewarti*),

Helminthosporium leaf diseases like *H. maydis* (causes southern corn leaf blight), *H. carbonum*, and *H. turcicum* (causes northern corn leaf blight). Common rusts include common corn rust (caused by *Puccinia sorghi*) and southern corn rust (caused by *Puccinia polysora*). Important viral diseases of corn include maize dwarf mosaic virus (MDMV) and maize chloric dwarf virus (MCDV).

Smuts

Important smuts include common smut by the fungus *Ustilago maydis* or *U. zeae*, and head smut by *Sphacelotheca reiliana*.

32.18.6 Insect resistance

Most of the research emphasis has been on breeding resistance to the European corn borer. A chemical found in corn with the acronym DIMBOA is reported to impart resistance to this pest to plants. Insect pests of economic importance include:

Soil-inhabiting insects

Important soil-inhabiting insect pests of corn include corn rootworms (*Diabratica longicornis*), black cutworms (*Agrostis ypsilon*), wireworms such as *Melanotus cribulosus*, and the sugarcane beetle (*Euetheola rugiceps*).

Leaf, stalk, and ear insects

An important pest of corn is the corn earworm (*Heliothis armigera* or *H. zea*), the European corn borer (*Pyrausta nubilalis*).

Stored corn grain insects

Major storage pests of corn include the rice or black weevil (*Sitophilus oryzae*) and the Angoumois grain moth (*Sitotroga cerealella*).

32.18.7 Product quality

One of the major achievements in cereal grain protein improvement is the breeding of quality protein maize (QPM), a high protein corn bred by incorporating the *opaque-2* gene. Also, high oil content is a goal of some breeding programs. High oil corn is of high industrial value. Selection in the long-duration Illinois High Oil Strain experiment has increased oil content from the initial 4.7% to over 21% over several decades. Other end-use quality breeding goals include improvement of milling quality and enhancing the sugar content of sweet corn (by incorporating the *shrunken-2* (sh_2) gene in breeding).

Key references and suggested reading

Acquaah, G. (2005). *Principles of crop production. Theory, techniques and technology.* Prentice Hall, Upper Saddle River, NJ.

Khehra, A.S., Dhillon, B.S., and Multani, D.S. (1997). Corn, in *Genetics, cytogenetics, and breeding of crop plants. Vol. 2 Cereal and commercial crops* (eds P.N. Bhal, P.M. Salimath, and A.K. Mandal) Science Publishers, Inc., Enfield, NH.

Russell, W.A. (1991). Genetic improvement of maize yields. *Advances in Agronomy,* **46**:245–298.

Outcomes assessment

Part A

Please answer the following questions true or false.

1 Corn is a dioecious plant.
2 E.M. East founded the Hi-Bred Corn Company of Iowa.
3 Corn belongs to the family Poaceae.
4 B-chromosomes occur in corn.

Part B

Please answer the following questions.

1 What is xenia?
2 Give three of the states in the Corn Belt of the United States.
3 Discuss the types of corn.
4 What does QPM stand for?

Part C

Please write a brief essay on each of the following topics.

1 Discuss major objectives in breeding pest resistance in corn.
2 Discuss the major agromorphological traits of importance in corn breeding.
3 Discuss the main methods of breeding corn.

33

Breeding rice

Taxonomy

Family	Poaceae
Genus	*Oryza*
Species	*Oryza sativa* L.

33.1 Economic importance

Rice accounts for about 20% of the world's total grain production, second only to wheat. It is the primary staple for more than 50% of the world's population. An estimated 90% of production and consumption of rice occurs in Asia where the per capita consumption is about 100 pounds, compared to a world average of about 60 pounds and a US average of about 20 pounds. World consumption in 1993/1994 was 358.5 million metric tons and has increased steadily to an estimated 408.8 million metric tons in 2002/2003. Similarly, the acreage devoted to rice increased from 145.2 hectares to 155.1 hectares in the period 1990/2000, with a slight drop thereafter. Production trends mirror the area cultivated. Rice production in the United States is dominated by six states – Arkansas, California, Louisiana, Texas, Mississippi, and Missouri, together accounting for about 99% of the total US production. Other minor producers in the United States include Florida, Tennessee, Illinois, South Carolina, and Kentucky. Arkansas leads the production in the United States, with about 100 cwt (conversion units are given in Appendix 1) of rice produced in 2001 on about 1.61 million acres.

The world's major producers of rice are China, India, Indonesia, and Bangladesh, which together account for more than 70% of the world's total production. China, the world's leading producer, regularly accounts for about 36% of the world's total production. The expansion in rice production is attributed largely to the impact of the Green Revolution that was implemented in the 1960s and 1970s in Asia and other parts of the tropical world. Other major producers in Asia include Vietnam, Thailand, Japan, and Burma, together accounting for about 12% of the total world production. Outside of Asia, rice is produced in substantial amounts in countries including the United States, Italy, Spain, Egypt, Australia, and countries in the West African region.

Thailand leads the world in rice exports, accounting for about 25% of the total world exports (about 4 million metric tons). The United States is the second largest exporter of rice, exporting about 2.7 million metric tons. Latin America is the largest market for US rice export. Other major exporters of rice include Vietnam, Pakistan, China, Australia, Burma, Italy, India, and Uruguay.

33.2 Origin and history

The origin of rice is not exactly known. Wild species of rice occur across South and East Asia, including India and South China. Rice was domesticated in the fifth millennium BC. The cultivated species of rice

Principles of Plant Genetics and Breeding, Second Edition. George Acquaah.
© 2012 John Wiley & Sons, Ltd. Published 2012 by John Wiley & Sons, Ltd.

Oryza sativa is believed to have derived from annual progenitors found in a wide area extending from the Gangetic Plains, through Burma, Northern Thailand, Northern Vietnam, and Southern China. Rice trade occurred between Egypt, India, and China. The Moors are credited with bringing rice to Spain, from where it was introduced to Italy (in the fifteenth century) and subsequently to Central America. The African cultigen (*O. glaberrima*) originated in the Niger River delta of Mali in West Africa.

A British sea captain brought rice to the United States from Madagascar. The first commercial planting occurred in South Carolina in about 1685. Early rice production concentrated in the South East United States, where South Carolina, Georgia, Louisiana, North Carolina, Mississippi, Alabama, Kentucky, and Tennessee were major producers in the early 1800s. However, in the early nineteenth century, shifts in rice production started in the United States, reaching its highest in the late 1900s. Production became more modernized and spread to Arkansas and the Mississippi River Delta, where the flatter lands permitted larger scale, high-tech production (mechanized, irrigated, etc.). In 1838, South Carolina produced about 75% of the total US rice crop. In 1903, Louisiana and Texas produced 99% of the US crop. However, by 1990, Arkansas, Louisiana and Mississippi accounted for more than two-thirds of US rice production, with Texas and California accounting for most of the balance. Rice production in the United States is currently concentrated in the Arkansas Grand Prairie, the Mississippi River Delta, Southwestern Louisiana, the coast prairie of Texas, and the Sacramento Valley of California.

33.3 Adaptation

Rice may be described as a semi-acquatic plant. It is adapted to very wide agroecological zones, ranging from dry to submerged root growing conditions. Four general ecosystems can be identified for commercial rice production around the world, based on elevation, rainfall pattern, depth of flooding, and drainage.

33.3.1 Rain-fed lowland rice ecosystem

This ecosystem is common in densely populated rural regions of the world, where producers face severe economic challenges in addition to burgeoning population. Rice production under these conditions account for about 25% of the world's harvested rice area and 17% of the world's production. Producers prepare the land by puddling the soil or dyking fields to hold water for a variable duration of flooding, according to the rainfall. The soils alternate between flooded to dry conditions during the growing season. Rice is direct seeded or transplanted into the field.

33.3.2 Upland rice ecosystem

Upland rice production occurs on well-drained, level to steeply sloping farmlands. These soils are frequently moisture deficient. Upland production occurs in regions of the world where slash-and-burn agriculture is common. The removal of vegetation from these slopes predisposes the soil to physical deterioration and nutrient depletion. Crop yields are generally low. Upland rice production constitutes about 13% of the world's harvested area and only 4% of the total rice produced. Rice is direct seeded to non-flooded soils.

33.3.3 Flood-prone rice ecosystem

Rice production in certain areas occurs on flooded soils throughout the growing season, until harvest time. Rice is direct seeded or transplanted into flooded fields (50–300 cm deep) during the rainy season. Flooded rice production occurs widely in the South and Southeast Asia, and some parts of West Africa and Latin America. Problems of salinity and toxicity from various ions are common in this ecosystem. Crop yields are unpredictable and generally low.

33.3.4 Irrigated rice ecosystem

The key feature of this system is that moisture is controlled in both dry and wet seasons. Various methods are used to provide and regulate soil moisture. About 55% of the world's harvested area and 75% of total production occurs in irrigated ecosystem. Production involves the use of modern technology with high production inputs (e.g., fertilizers). Consequently, yields are high, reaching about 5 tons/ha in the wet season to about 10 tons/ha in the dry season.

33.3.5 Other classification

T. Chang of the IRRI reported a classification of rice according to topohydrological, edaphic, cultural, and seasonal criteria. According to eco-goegraphic differentiation, rice may be classified into three races of *O. sativa* as **Indica, Javanica,** and **Sinica** (or **Japonica**). The Japonica race has upland and lowland cultivars, whereas the Indica has cultivars that span dryland to deep water and floating cultivars.

33.4 Commercial classes

Rice varieties are primarily classified according to the length of the grain – short (5.5 mm), medium (6.6 mm), and long (7–8 mm). The shorter-grained varieties are also called Japonica types, and have short stiff, lodging-resistant stalks, making them more responsive to heavy fertilization. The longer-grained varieties, called Indica types, have taller, weaker stems that lodge under heavy fertilization. The United States produces mainly Indica rice (about 65% of the annual production) primarily in Arkansas, Mississippi, Louisiana, and Texas, while California produces mostly medium and short grained rice.

Rice may also be classified in terms of maturity – early maturing (about 120–129 days), mid-season (about 130–139 days), or late maturing (about 140 days or more). Rice may be scented (aromatic) or unscented, the two most common scented types being basmati and jasmine. Basmati rice has a distinctive odor, doubles in grain size upon cooking, and is non-sticky (grains remain separate). It is cultivated mainly in the Punjab area of Central Pakistan and Northern India. Jasmine rice is grown mainly in Thailand and is preferred more by the Asian community in the United States. Cooked jasmine rice is soft, moist, and sticky. The stickiness derives from the types of starch in the grain. The endosperm starch of rice may be glutinous or commonly non-glutinous (non-sticky). The glutinous property is conferred by amylopectin type of starch.

33.5 Germplasm resources

Over 85 000 rice accessions are maintained at the IRRI in the Philippines, representing the largest rice germplasm collection in the world. The West African Rice Development Association (WARDA) in Ivory Coast maintains large numbers of accessions, especially, *O. glaberrima* genotypes. The USDA maintains about 16 000 accession of rice, including a large number of breeding lines. Genetic testers, developed mostly by Japanese researchers, are available at the Rice Genetics Cooperative.

33.6 Cytogenetics

Twenty species of the genus *Oryza* are known, with a basic chromosome number of 12. The genus has six genome groups – *A, B, C, D, E,* and *F*. The cultivated species *O. sativa* ($2n = 2x = 24$) has the genome formula AA, whereas *O. glaberrima* ($2n = 2x = 24$) is designated $A^g A^g$ because it does not pair well with *O. sativa*. There are two major species of cultivated rice – *Oryza sativa* and *O. glaberrima* – the latter being native to Africa and cultivated in West Africa and Central Africa. Other wild species are *O. alta* ($2n = 2x = 48$; *CCDD*), *O. australiensis* ($2n = 2x = 24$; EE), *O. longistaminata* ($2n = 2x = 48$; *CCDD*), and *O. brachyantha* ($2n = 2x = 24$; *FF*), and *O. punctata* ($2n = 2x = 48$; *BBCC*).

The 22 species are divided into diploids ($2n = 2x = 24$), for example, *australiensis, barthii, glaberrima,* and *sativa*, and tetraploids ($2n = 2x = 48$), for example, *alta, puncata,* and *grandiglamis*. Rice chromosomes have a tendency to form secondary associations during meiosis of normal diploids. Asynaptic plants occur in nature as well as haploids, triploids and tetraploids. They may be induced artificially. Primary trisomics have also been developed by various researchers for rice. Also, chromosome maps have been developed by different workers.

Crossing between plants from different ecogeographic races, or even within a race, may exhibit some hybrid sterility, inviability or weakness. In practice, plant breeders are able to overcome these sterility problems through rigorous selection to obtain fertile lines. Wild germplasm has proved a useful genetic resource for rice breeding. The wild abortive CMS cytoplasm that is commercially used was obtained from *O. sativa* f *spontanea*.

33.7 Genetics

Rice genetics is quite extensive. Color development in rice is complex. Anthocynin pigmentation is controlled by a complementary genetic system, the *C-A-P* gene system. *C* is the basic gene for chromogen production, while *A* controls the conversion of chromogen to anthocynin,. *P*, which is variable, determines the site of pigment expression (e.g., *Pg, Pm, Ps, Px,* etc., according to the site). The *CAP* gene system is affected by modifying genes, and occasionally an inhibitor gene in some species. Also, multiple alleles and various dominance levels are associated with *C, A,* and *Pl.* Color development is also affected by light intensity, growth stage, and fading and leaching. The color of the different layers of seed coat may be affected by different genes or sets of genes. A smooth hull is desirable for mechanical harvesting and processing. Pubescence on the surface of the blades and hull is controlled by a dominant gene, glabrous (*gl*).

Semidwarfism is desirable in rice and is controlled by a pair of recessive alleles, *sd1.* However, studies indicate that semidwarfism is a complex, quantitative trait. One three-gene unit with cumulative effect controls awning in rice, fully awned being $An_1 An_2 An_3$ (or $An_1 An_2 an_3$), while awnless is conditioned by *an1an2an3.* Photoperiod sensitivity is controlled by one or two genes (Se_1, Se_2), while one or two dominant genes control shattering. Genes for host resistance to many diseases have been identified, including bacterial blight (*Xa-1, Xa-2,* etc.), blast ($Pl_1, Pl_2,$ and others), yellow dwarf virus (*Ydv*), and brown spot (*He* or he_1).

One of the important mutations in rice breeding is the *sd*, the recessive gene that conditions semidwarf stature. It was discovered in a Taiwanese cultivar, "Dee-geo-woo-gen". Hybrid rice using CMS entails crossing a CMS line (*rfrf-S*) with a restorer (*RfRf-F*) to obtain an F_1 (*Rfrf-S*). The wild abortive or the Chinsurah Bone II are the preferred sources of CMS for rice hybrids.

33.8 General botany

An annual grass, rice has erect culms that may reach six feet in some varieties. It produces about five tillers. Rice inflorescence is a loose terminal panicle consisting of spikelets that are self-pollinated. The rice grain is enclosed by the lemma and palea (constitutes the hull) that may be straw yellow, red, brown, or black in color. Depending on the variety, the lemmas may be fully awned, partly awned, tip-awned, or awnless. Hulled kernels vary in length from 3.5–8 mm, 1.7–3 mm in breadth, and 1.3–2.3 mm in thickness. Furthermore, the kernels may be hard, semihard, or soft textured. The color of unmilled rice kernel is variable and may be white, brown, amber, red, or purple; lighter colors (white, light brown) may be preferred in the United States.

33.9 Reproductive biology

Rice has a loose terminal panicle with branches that arise singly or in whorls. It is predominantly self-pollinated with less than 1% outcrossing. A panicle may contain 75–150 spikelets, or even several hundred in some varieties. Rice has perfect flowers that are borne in single-flowered spikelets. The flower consists of two lodicules, six stamens (instead of three as in most cereals), two plumose stigmas on two styles, surrounded by floral bracts. The floral bracts (lemma and palea) may be straw, yellowish-gold, red, brown, or purplish in color. Further, the lemma may be fully awned, partly awned, tip-awned, or awnless.

Rice is predominantly self-pollinated. The time of anthesis is significantly influenced by environmental factors, especially temperature and, to a lesser extent, by genotype. The entire panicle completes blooming in 4–7 days. Peak anther dehiscence occurs between mid to late morning in tropical regions and at about noon in temperate regions.

33.10 Common breeding methods

Being a self-pollinated species, rice can be improved by any of the methods of breeding for self-pollinated species. Field tests may be conducted to evaluate introductions from various parts of the world to identify adapted varieties for commercial production. International breeding programs and non-profit organizations may engage in breeding genotypes for use in developing countries with little resources for embarking on elaborate breeding programs.

Hybridization is the principal procedure used to generate a segregating population for a breeding program with rice. Depending on the breeding objective and the underlying genetics of the trait of interest, breeders may use any of the breeding methods for self-pollinated species in rice breeding.

Commercial hybrid rice seed production was started in the People's Republic of China in the

1970s, where hybrid varieties occupy more than 50% of the 75 million acres of rice cultivated annually. The key factor in the success of the commercial hybrid program was the discovery of a source of cytoplasmic male sterility, called the "wild-abortive" in a wild rice plant (*O. sativa f spontanea*). Fertility restoring genes were discovered in indica cultivars. Hybrid rice production is practical in China largely because of the low cost of labor. To enhance flowering, cross-pollination and seed-set, seed producers implement practices such as cutting back the flag leaves of female plants to enhance pollination.

33.11 Establishing a breeding nursery

33.11.1 Field establishment

Rice is a warm season species and is successful under a mean temperature of about 20 °C or higher. High humidity encourages diseases and, hence, is undesirable. Heavy-textured soil with impervious subsoil for holding moisture is desirable for rice. The tolerable pH ranges between 4.5 and 7.5. Rice is also a short day plant.

Rice seeding is accomplished by using ground equipment (broadcast seeders or grain drills). Rice is drilled to about 1–2 inches in a good seedbed. Ground equipment may be used to apply fertilizer at the time of seeding. The rate of nitrogen application may vary between 30 and 100 lb/acre. High levels may cause lodging in some varieties. Where soils are deficient, application of moderate amounts of phosphorus and potassium may be beneficial.

A photoperiod of 10–12 hours is preferred for rice growth. Optimum temperature for growing rice is about 27 °C but the optimum temperature for flowering is dependent on photoperiod.

33.11.2 Greenhouse and growth chamber

Crossing indoors under a controlled environment in a greenhouse or growth chamber is especially convenient for breeders working in the monsoon tropics or flooded paddies. Under these conditions, the parents may be grown in the field and dug up and placed in pots for crossing in the greenhouse. Working indoors allows the breeder to optimize photoperiod, temperature, and light level for optimal growth and flowering.

33.11.3 Artificial pollination for hybridization

Materials and equipment

Various methods are used by rice breeders for emasculation and pollination. The equipment and materials differ for the method used, the common ones being scissors, fine forceps, glassine bags, pot labels, paper clips, wax pencil, tags, hot water bath, vacuum emasculator.

Preparation of the female flower

The plant is ready for the preparation of the female for pollination when 50–60% of the panicle has emerged from the boot. Emasculation of individual flowers is done before anthesis and after they emerge from the boot. Emasculation in the tropics is best done after mid-afternoon, when anthesis has ceased for the day. In the temperate regions, preparation of the female can be done in the morning or late afternoon. Hot water emasculation is done by soaking the panicle for five minutes in water maintained at 43 °C. This treatment must be followed by pollination within 30–60 minutes after emasculation. The hot water may be carried in a Thermos bottle. A simpler and more efficient emasculation technique is to clip the spikelets and remove the anthers with a pair of forceps or by a vacuum unit.

Where plants have to be transplanted from the field into pots, the relocation must be completed at least six hours before emasculation, to allow plants to recover from any transplanting shock. Once a panicle has been identified for emasculation, it is separated from the others nearby to facilitate the emasculation process. The flag leaf is carefully removed. Florets at the top of the panicle that may have already self-pollinated and the young flowers at the bottom are cut with a pair of scissors. Next, about one third to one half of each floret is cut off at a slant to expose the anthers. If cut too low, the stigma may be damaged. If the anthers are removed with forceps, the cut may be made across the anthers. The anthers may be extracted by using a vacuum pump or by forceps.

Pollination

The stigma remains receptive for about 4–5 days, and hence pollination should be made sooner rather than later. Pollen remains viable for a shorter period (a few minutes to about half a day). Pollination should be

Industry highlights
Breeding rice

Anna Myers McClung

USDA-ARS, Rice Research Unit, 1509 Aggie Dr., Beaumont, TX 77713

Introduction

The USDA-ARS rice breeding program located at the Texas A&M Research and Extension Center at Beaumont, TX, has been in operation since the early 1930s. The objective of this program has been to develop superior performing rice cultivars that are adapted to the southern rice growing region of the United States, which includes Texas, Louisiana, Arkansas, Mississippi, and Missouri. Some 80% of this region is dedicated to production of long grain cultivars while 20% produces medium grain cultivars, with a small percentage dedicated to specialty rice cultivars for niche markets. In the United States, the conventional market classes of rice are categorized according to grain dimensions and cooking quality, the latter being primarily determined by amylose content and starch gelatinization temperature (Webb, 1985).

Development of new rice cultivars must include selection for agronomic traits, resistance to disease and insect pests, and grain quality traits (McClung, 2002). In addition, cultivars that are produced along the Gulf Coast of Texas and Louisiana, where the growing season is relatively long, are also evaluated for second crop potential, which is called the ratoon crop. After the main crop is harvested, a second crop develops from the stubble of the first crop. About 60 days after the main crop harvest, the ratoon crop is cut, producing up to 50% of the first crop's yield.

Most public rice breeding programs use standard pedigree, bulk, and backcross breeding approaches. However, mutation breeding and recurrent selection methods are also used to a limited extent. Many of these programs are now using molecular markers that are associated with traits of economic importance to expedite the breeding process. In addition, there are private breeding programs that are developing rice hybrids and others that are using transgenic technology (McClung, 2004).

The development of "Saber" rice (McClung *et al.*, 2004) is an example of a recent breeding project conducted by the Beaumont rice variety development program. In 1989, a cross was made between "Gulfmont" and an experimental selection, RU8703196. At the time, Gulfmont was a new release from the breeding program that was characterized as an early maturing, semidwarf, long grain cultivar having excellent main and ratoon crop yield as well as good milling quality. It is rated moderately resistant to blast disease (caused by *Pyricularia grisea*) and very susceptible to sheath blight disease (caused by *Rhizoctonia solani*) (Bollich *et al.*, 1990), the two most common yield limiting diseases in the southern rice growing region. RU8703196 is a long grain germplasm source that was released as an improved source of resistance to blast and sheath blight diseases (Marchetti *et al.*, 1995). The F_1 of Gulfmont/RU8703196 was then crossed with TeQing (PI 536047). This is a medium grain cultivar from China that is characterized as having high yield potential, intermediate height, relatively late maturity, and excellent resistance to blast and sheath blight diseases when grown in the southern United States. The objective of the cross was to maintain the plant stature and grain quality of Gulfmont while improving its yield and disease resistance.

Timeline for development of saber rice

1989
Make cross B8910 (Gulfmont/RU8703196// TeQing) in the greenhouse and produce 27 F_1 seed (Figure B33.1). Plant each F_1 seed in separate containers in greenhouse. Harvest F_2 seed from each F_1 plant separately.

Plant the bulk of each F_1 derived F_2 population in a fall-planted nursery in Puerto Rico. Three of the F_1 derived populations appear to be selfs and are discarded. Strip bulk F_3 seed from F_2 plants at harvest.

1990
Plant the bulk of the F_1 derived F_3 populations in winter planted nursery in Puerto Rico. Strip bulk F_4 seed from F_3 plants.
Plant the bulk of the F_1 derived F_4 populations in a summer nursery in Beaumont, TX. Strip bulk F_5 seed from F_4 plants.

1991
Plant the bulk of the F_5 population in the summer nursery. Select panicles from a total of 215 plants for the B8910 project based upon field observations of plant height, days to heading, good tillering capacity, and good appearing grain dimensions.

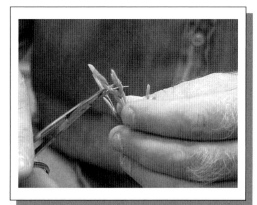

Figure B33.1 Rice panicle being prepared for emasculation prior to making a controlled cross. Figure courtesy of Anna Myers McClung.

Figure B33.2 Plant height is measured in yield plots after flowering has occurred. Figure courtesy of Anna Myers McClung.

1992

Plant 215 F_6 panicle rows in the summer nursery. Select 63 rows based upon field observations of plant height, days to heading, good tillering capacity, and good appearing grain dimensions. Harvest five panicles per row.

1993

Project not planted due to field space limitations.

1994

Plant five F_7 panicle rows from each of 63 F_6 derived families in the fall-planted nursery in Puerto Rico. Select 29 rows based upon field observations of plant height, maturity relative to check varieties, good tillering capacity, and good appearing grain dimensions. In each selected row harvest five panicles and bulk harvest the row.

1995

F_8 panicle rows are planted in the summer breeding nursery for each of the 29 families. Ten F_9 panicles are selected from one of the five rows for each of the 29 families.

In addition, the bulk harvested F_8 seed from Puerto Rico from each of the 29 families is planted in an unreplicated yield trial with some 600 other entries during the summer. In this trial, data is collected on height (Figure B33.2), heading, harvest maturity, yield, milling yield, and resistance to blast disease using inoculated disease screening nursery (Figure B33.3). Fourteen of the 29 yield plots from the B8910 project are selected to undergo further evaluation during the winter for grain cooking quality traits (i.e. amylose content and alkali spreading value) (Figure B33.4) and grain chalkiness (Webb, 1985). Using these data, five lines are selected for testing in the following year.

1996

The ten F_9 panicle rows from each of the five families selected based upon the yield trial are planted in the summer breeding nursery. One row is selected from each of the five families and ten panicles are picked.

Using the bulked yield plot seed from 1995, one of the five families is entered into the Uniform Rice Regional Nursery (URRN) as entry RU9603178. The URRN is an extensive replicated yield trial that includes a total of 200 entries from four southern rice breeding programs and is planted in Texas, Arkansas, Louisiana, and Mississippi. This study has two sections: Advanced entries, which are replicated four times, and Preliminary entries, which are replicated twice. RU9603178 is entered into the Preliminary trial section based upon limited availability of seed. In addition to the standard yield, agronomic, milling, and cooking quality traits that are evaluated, several states screen all 200 entries in inoculated nurseries for resistance to blast and sheath blight diseases as well as the physiological disorder, straighthead. Two of the states evaluate all entries for ratoon crop yield and for reaction to nine individual races of blast using controlled greenhouse conditions. All of the entries from the Beaumont program that are in the URRN trial also undergo additional yield testing at 2–3 other Texas locations that are representative of the state's rice growing area. The other four entries from the B8910 cross are planted only in these Texas replicated yield trials.

Results from the URRN analyzed during the winter demonstrate that RU9603178 has very good main and ratoon crop yield, good resistance to all races of blast disease (including IB49 rating of 1 on a scale of 0 to 9), good tolerance to sheath blight (rating of 4 on a scale of 0 to 9) and straighthead, very clear, non-chalky grain, cooking quality of typical United States long grains, and excellent milling quality. It is advanced for testing in 1997 along with one of the other four entries (BPRE 18) that had been tested in the Texas state-wide trials.

Figure B33.3 Rice cultivars are evaluated for their reaction to a mixture of several races of *Pyricularia grisea* disease screening nurseries. Figure courtesy of Anna Myers McClung.

Figure B33.4 Milled rice flour is used to determine grain amylose content. Figure courtesy of Anna Myers McClung.

1997

Ten F_9 derived F_{10} panicle rows of RU9603178 and of BPRE 18 are planted in the summer nursery and 20 panicles are selected from one row of the two entries.

Bulked F_{10} seed from the 1996 URRN yield trial in Beaumont of entry RU9603178 is used to plant in the Advanced section of the 1997 URRN as entry 23. The other B8910 project line (BPRE 18) is entered into the Advanced URRN as entry 32 using seed harvested from one of the 1996 Texas replicated trials. The Advanced entries of the URRN are tested using four replications at the four state locations and the Texas entries of the URRN are also evaluated in the multilocation replicated trials within Texas.

Results from these trials indicate that RU9603178 continues to demonstrate competitive yield potential, excellent milling quality, and has moderate tolerance (rating of 6) to the sheath blight pathogen. The other B8910 line, entry 32, does not perform as well as RU9603178 and is dropped from further testing. Moreover, RU9603178 demonstrates resistance to all races of the blast fungus. Based upon this spectrum of resistance and the pedigree of RU960178, this suggests that the line may possess a novel major resistance gene to *P. grisea*. Previous research had demonstrated success in developing genetic markers associated with major genes which control rice cooking quality (Ayres *et al.*, 1997). This spawned further research to identify the novel resistance gene and develop closely linked DNA markers.

During the fall of 1997, twenty F_{11} panicles from the summer breeding nursery selection of RU9603178 are planted in the Puerto Rico nursery. 287 F_{12} panicles are harvested from the 20 panicle rows in Puerto Rico.

1998

The 287 F_{12} panicles are planted in an isolation block during the summer at Beaumont (Headrow 1). Thirty rows are selected and 20 panicles per row are harvested. Twenty-two rows are removed that are earlier, later, or taller than the others and then the remaining rows are bulk harvested. In addition, detailed phenotypic data are collected from the Headrow 1 field as an objective description of cultivar, which is required for application to the state department of agriculture certified seed program.

RU9603178 is evaluated in the 1998 URRN and Texas state-wide trials using the bulked harvested F_{11} seed from the 1997 yield trial. Results demonstrate that RU9603178 has main crop and ratoon crop yield competitive with other cultivars in its maturity group, some lodging susceptibility at two locations, excellent milling yield, and good tolerance to sheath blight disease (rating of 2). Its reaction to eight races of blast shows excellent resistance; however, it appears to be segregating for resistance to race IB49 (rating of 2 and 4).

1999

In an effort to further clarify the level of resistance to the IB49 race of blast, F_{13} seed from the 265 bulk harvested 1998 Headrow 1 is evaluated for reaction to race IB49 during the 1999 spring greenhouse. Of some 300 seedlings that are screened, 37% are highly resistant (rating of 1–2), 61% are moderately resistant (rating of 3–4), and 2% are susceptible (rating of 5–6). This suggests that further selection in RU9601378 may allow for improvement in resistance to this pathotype.

During the summer, RU9603178 is evaluated in the same manner as before in the 1999 URRN and Texas state-wide trials using the bulked harvested F13 seed from the 265 rows in Headrow 1. Agronomic and milling quality results from this trial are consistent with the previous year. Its reaction to eight of the blast races shows excellent resistance; however, it is rated as susceptible (4) to IB49.

Headrow 2 is planted in the summer using 596 F_{13} panicle rows derived from the 30 row selections from 1998. The information collected for the objective description of the variety in the 1998 headrow is verified in the Headrow 2 field. Each of the 596 rows is harvested separately for verification of reaction to blast race IB49 during the fall greenhouse. Over 99% of the families are rated as highly resistant (rating of 1–2) and the remaining families are rated as moderately resistant (rating of 3). This indicates that the 30 families that had been selected in the 1998 Headrow 1, and which constitute Headrow 2, are highly resistant to this pathotype.

2000

Approximately 20 F_{14} families that had been rated as resistant (i.e., 1) and 20 families that had been rated as moderately resistant (i.e., 3) in the fall 1999 greenhouse are re-evaluated in the spring 2000 greenhouse with IB49. All plants in the 40 families are confirmed to be resistant or moderately resistant and no susceptible plants are identified. This verifies the resistant-moderate resistant rating to IB49 infection of Headrow 2 of RU9603178 and indicates that resistance is not due to escapes nor is there heterogeneity for susceptibility. However, as a precaution, families rated as 3 are eliminated and only families that had an IB49 rating of 1 are provided to the foundation seed program. The objective description of the variety is provided to the state inspector to evaluate the foundation seed field (Figure B33.5).

Families that had a resistance rating of 2 to IB49 are bulked and used for further field testing during 2000. The same battery of field trials, disease screenings, and quality tests are conducted as before. DNA markers are developed that are associated with the *Pi-b* blast resistance gene which is found in TeQing and explains the multirace resistance to blast that is observed in RU9603178 (Figure B33.6) (Fjellstrom et al., 2004).

By the end of the 2000 field season, RU9603178 has been evaluated at over 40 environments and it is decided to proceed with public release of the variety. The name "Saber" is selected in honor of the Texas A&M University Corps of Cadets. The complete set of data collected over the previous five field seasons is summarized to justify to the state seed board the release of Saber. In February 2001, the seed board accepts Saber into the state seed certification program; this allows the foundation seed that was produced in 2000 to be sold as certified (F_{15}) seed for the 2001 planting season.

Figure B33.5 A foundation seed field of Saber rice. Figure courtesy of Anna Myers McClung.

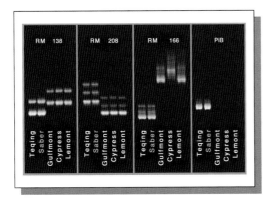

Figure B33.6 DNA markers on chromosome 2 that are associated with the *Pi-b* blast resistance gene are present in Saber and its parent, TeQing (Fjellstrom *et al.*, 2004). Figure courtesy of Anna Myers McClung.

References

Ayres, N.M., McClung, A.M., Larkin, P.D., Bligh, H.F.J., Jones, C.A., and Park, W.D. (1997). Microsatellites and a single-nucleotide polymorphism differentiate apparent amylose classes in an extended pedigree of U. S. rice germplasm. *Theoretical and Applied Genetics*, **94**:773–781.

Bollich, C.N., Webb, B.D., Marchetti, M.A., and Scott, J.E. (1990). Registration of Gulfmont rice. *Crop Science*, **30**: 1159.

Fjellstrom, R.G., Bormans, C.A., Marchetti, M.A., Shank, A.R., Park, W.D., and McClung, A.M. (2004). Development of DNA markers suitable for marker assisted selection of three *Pi* genes conferring resistance to multiple *Pyricularia grisea* pathotypes. *Crop Science*, **44**:1790–1798.

Marchetti, M.A., Bollich, C.N., McClung, A.M., Scott, J.E., and Webb, B.D. (1995). Registration of RU87030196 disease resistant rice germplasm. *Crop Science*, **35**:600.

McClung, A.M. (2002). Techniques for development of new cultivars, in *Rice: Origin, History, Technology, and Production* (eds C.W. Smith and R.H. Dilday). John Wiley and Sons, Inc., NJ, pp. 177–202.

McClung, A.M. (2004). The Rice Plant: Growth, Development and Genetic Improvement, in *Rice Chemistry and Technology*, 3rd edn (ed. E.T. Champagne). American Association of Cereal Chemists, Inc., St. Paul, MN, pp. 25–48.

McClung, A.M., Fjellstrom, R.G., Bergman, C.J., Bormans, C.A., Park, W.D., and Marchetti, M.A. (2004). Registration of "Saber" rice. *Crop Science*, **44**:693–694.

Webb, B.D. (1985). Criteria of rice quality in the United States, in *Rice Chemistry and Technology* (ed. B.O. Juliano). American Association of Cereal Chemists, Inc., St. Paul, MN, pp. 403–442.

conducted during the period of peak anther dehiscence. Pollen is gathered just before anther dehiscence. The male panicles are cut and the flag leaf removed. These panicles are watched closely for anther extrusion and used thereafter for pollination. The bag is taken from the female and the pollen shaken over it. The bag is replaced and clipped securely against the stem. Other techniques are used for pollination.

Natural cross-pollination

Rice is highly self-pollinated. Where commercial F_1 hybrid seed production is undertaken, natural cross-pollination is often inadequate. Consequently, natural pollination is supplemented with hand pollination. The flag leaf may be cut and the leaf sheaths that enclose the panicle torn to help the release of pollen. Also, a rope or pole may be dragged across the field at the level of the panicles each day during the flowering period to aid pollen dehiscence.

Seed development and harvesting

The success rate in artificial pollination of rice is about 50% or higher, depending on the technique used. Ovule swelling starts 3–4 days after pollination. The developing F_1 seed lacks a complete covering because the glumes were cut during the preparation of the female for pollination.

Rice does not mature uniformly in the head. Grain harvest moisture is critical to yield and produce quality. The recommended grain moisture content is between 23 and 28%. At this stage the grains in the top portion of the head would be ripe but those in the lower portion would be in the hard-dough stage. Harvesting at this stage will include some immature grain but delaying harvesting increases the chances of shattering and checking of grains in susceptible varieties.

33.12 Common breeding objectives

33.12.1 Grain yield

Grain yield in rice depends on yield potential under the most favorable conditions, yield stability (across seasons), and crop productivity. The components of yield are panicle number per unit area, the number of filled grains per panicle, and grain weight. Some breeders use spikelet number per unit area and grain weight. The dramatic yield increases observed in the tropics in the 1960s were as a result of the development and use of semidwarf (sd1 gene) cultivars that were environmentally responsive.

33.12.2 Grain quality

Grain quality traits of interest vary from one region to another. They include grain size and shape, color of kernel, aroma, stickiness, and protein content.

33.12.3 Disease resistance

Host resistance to many major diseases and insect pests have been identified in rice. Many of these are under oligogenic control and, hence, susceptible to changes in the pest growing a resistant a resistant cultivar for a long period,. The strategy of durable resistance is favored by many breeders. Resistance to insect-transmitted viral infections is complex to breed because the insect resistance aspect (e.g., non-preference of the vector) can mask the disease resistance component.

33.12.4 Resistance to environmental stresses

Breeding resistance and tolerance to various environmental stresses are important in rice breeding. Drought and flooding frequently alternate with each other. Research indicates that, regarding moisture stress, the escape and avoidance mechanisms are important in dryland culture, while tolerance and recovery mechanisms are important under rain-fed wetland culture. If should be pointed out that plant reaction to environmental stresses are site and growth stage specific.

Key references and suggested reading

Chang, T. (1997). Rice, in *Genetics, cytogenetics and breeding of crop plants*, Vol. **2**(eds P.N. Bahl, P.M. Salimath, and A.K. Mandel). Science Publishers, Inc., Enfield, NH, pp. 1–76.

Coffman, W.R., and Herrera, R.M. (1980). Rice, in *Hybridization of crop plants*. American Society of Agronomy, Madison, WI, pp. 511–522.

Outcomes assessment

Part A

Please answer the following questions true or false.

1 Commercial classification of rice is based primarily on grain length.
2 The basic chromosome number of the genus Oryza is 12.
3 Rice is predominantly cross-pollinated.
4 Most of the world's rice production occurs in irrigated ecosystems.

Part B

Please answer the following questions.

1 Give the three ecogeographic classes of rice.
2 List four major quality breeding objectives in rice breeding.
3 Describe the distinguishing features of the japonica class of rice.

Part C

Please write a brief essay on each of the following topics.

1 Discuss the commercial classes of cultivated rice.
2 Discuss the common methods of breeding rice.
3 Discuss the genetics of grain color in rice.
4 Discuss the importance of the "wild abortive" gene in rice breeding.

34

Breeding sorghum

Taxonomy

Family   Poaceae
Genus   *Sorghum* Moench
Species   *Sorghum bicolor* (L.) Moench

34.1 Economic importance

Sorghum was the world's fourth most important crop in 1995, accounting for 4% of the total cereal production and a total harvest of 53 million metric tons. Sorghum and millet are two of the major world food crops that originate from Africa. Even though sorghum has become important in the agricultural production of developed countries, it is still primarily a developing country crop, with 90% of the world's acreage found in Africa and Asia.

In the United States, 470 525 000 bushels were produced in 2000 on 9 195 000 acres. The leading producer was Kansas, with 3 500 000 acres and production of 188 800 000 bushels, followed by Texas with 3 000 000 acres and production of 143 350 000 bushels. Other important producers were Nebraska, South Dakota, Colorado, Oklahoma, Missouri, Louisiana, New Mexico, and Arizona. The varieties grown in the United States were primarily (78%) for grain. However, sorgho (sweet sorghum) is grown for forage, silage, and syrup (stalks crushed). The important sorgho states are Alabama, Mississippi, Georgia, Tennessee, and Iowa. Dual purpose cultivars (grain and forage) are produced in states such as Texas, Kansas, Nebraska, and Oklahoma.

On the world scene, sorghum is produced principally in Africa, Asia, the Americas, and Australia, Important producers in Africa include Nigeria, Sudan, Burkina Faso, Cameroon, Chad, Mali, and Rwanda. About 74% of sorghum produced in Africa is used for food. Most developing countries use sorghum for food, unlike the United States where sorghum is mainly grown for feed. Even though most of the acreage for sorghum occurs in developing countries, the United States leads the world in production, followed by India, Nigeria, China, Mexico, and Sudan.

34.2 Origin

Sorghum originated in north-eastern Africa (Ethiopia, Sudan, East Africa) where the greatest diversity, both wild and cultivated species, occurs. It was domesticated in Ethiopia and parts of Congo between 5000 and7000 AD with secondary centers of origin in India, Sudan, and Nigeria. It moved into East Africa from Ethiopia around 200 AD or earlier. It was distributed along trade and shipping routes throughout Africa and through the Middle East to India at least 3000 years ago. Sorghum was taken to India from eastern Africa during the first millennium BC. It arrived in China along the silk route. It was introduced into the Americas as guinea corn from

Principles of Plant Genetics and Breeding, Second Edition. George Acquaah.
© 2012 John Wiley & Sons, Ltd. Published 2012 by John Wiley & Sons, Ltd.

West Africa through the slave trade in about the middle of the nineteenth century.

34.3 History of breeding in the United States

Concerned about the narrow genetic base of germplasm available to US breeders, J.C. Stephens and J.R. Quimby of the US Department of Agriculture embarked upon a project to convert the tall, late maturing, tropical accessions in the World Collection to short, day-neutral genotypes for use in temperate regions, starting with 1500 accessions. A large part of this material is also maintained in long-term storage at the National Seed Storage Laboratory, Fort Collins, Colorado.

34.4 Genetic resources

In the United States, sorghum germplasm is maintained at the Regional Plant Introduction Station, Experiment, Georgia, with duplicate samples stored at the National Seed Storage Laboratory, Fort Collins, Colorado. A more comprehensive collection with accessions in excess of 250 000 is maintained at ICRISAT in India.

34.5 Cytogenetics

Sorghum comprises three known ploidy levels: x = 5, x = 10, and x = 20. Important sorghum species are: *S. bicolor* (Linn.) Moench ($2n = 2x = 20$), S. *propinquun (Kunth)* Hitche ($2n = 2x = 20$), and S. *halepense (Linn.)* Pers. ($2n = 4x = 40$). Of these three, the most important to crop production is S. *bicolor*. Grassy species such as S. *arundinuceuin, S. verticillijlorum*, and S. *aethiopicum*, have the same chromosome number ($2n = 2x = 20$) and can be crossed with S. *bicolor*. Sorghum has also been successfully crossed with sugarcane and corn.

Cultivated races are readily crossable with each other to produce fertile hybrids. In sorghum production, natural interspecific crossing between S. *bicolor* and S. *halepense* is a common source of off-types in the field. Two types of products arise from this type of cross - sterile plants with 30 chromosomes and fertile plants with 40 chromosomes, the former often persisting in the field as perennial weeds.

Cytogenetic male sterility and fertility restoration systems have been discovered in sorghum and are used in hybrid sorghum production. The CMS in sorghum genotypes was developed by backcrossing chromosomes of kafir into the cytoplasm of milo. Similarly, genetic male sterility *(ms)* has been discovered in sorghum male sterile plants *(Msms)*.

34.6 Genetics

Midrib color is controlled by a single dominant gene *(D)*, whereas resistance of hydrogen cyanide (HCN) appears to be controlled by more than one factor. Researchers have determined that grain color in sorghum is influenced by pericarp color, pericarp thickness, and presence of testa, color of testa, endosperm color, glume color, and plant color. Each of these characteristics is determined by one or a few genes. For example, two genes, R and Y, determine whether the pericarp would be red *(RY)*, colorless or white *(Ry, ry)* or lemon yellow *(rY)*. The kernel starch is conditioned by a dominant allele (*Wx); wxwx* conditions a waxy endosperm. Similarly, sugary endosperm is controlled by a single locus, *susu*.

Genetic male sterility genes occur in sorghum, the most commonly used being the *ms3*. It has stable expression over different environments. CMS also occurs in sorghum, conditioned by an interaction between two major genes, *mscl* and *msc2*.

Important genes that have impacted sorghum breeding are those affecting maturity and plant height. Sorghum is a short-day plant. Maturity is influenced by photoperiod and temperature. Genes that influence sorghum maturity are designated Ma_1, Ma_2, Ma_3, and Ma_4; tropical cultivars being dominant at the Ma_1, locus. A genotype of $Ma_1Ma_2Ma_3Ma_4$ takes about 90.5 days to anthesis, while ma_1ma_2-ma_3Ma_4 takes 55.3 days to anthesis. Similarly, four independent recessive genes, dw_1, dw_2, dw_3, and dw_4, reduce internode length of the sorghum stalk without affecting time of blooming and leaf size. A genotype, $Dw_1Dw_2Dw_3dw_4$, has a flag leaf at 127 cm, while $dw_1dw_2dwv_3dw_4$ has a flag leaf located at 43 cm above ground on the stalk. Sorghum cultivars containing dwarf genes are identified by the number of specific dwarf genes they contain as 1-dwarf (contains one dwarf gene, e.g., dw_2), 2-dwarf (contains two dwarf genes, e.g., dw_2dw_3), and so on. United States commercial sorghum hybrids are 3-dwarf.

34.7 General botany

Sorghum (*Sorghum bicolor* (L.) Moench) is known by common names such as milo, kafir, and guinea corn. The annual sorghums have $n = 10$ and include the grain sorghum, sorgo, broomcorn, and sudangrass. *Sorghum halepense* (johnsongrass) is a perennial sorghum with $2n = 20$. The sorghum plant has culms that may stand 2–15 feet tall, depending on the type and variety. It may produce two or more tillers. The stalk is solid. The center of the stem can be dry or juicy, insipid or sweet to taste. A dry-stalk variety has leaves with a white or yellow midrib, while a juicy-stalk variety has a dull, green midrib because of the presence of the juice instead of air spaces in the pithy tissues.

The number of leaves on the plant varies between 7 and 24 depending on the variety. The sorghum inflorescence is a panicle that may be loose or dense. It is usually erect but may curve to form a "gooseneck". The panicle has a central rachis, with short or long primary, secondary, and mature tertiary branches, which bear groups of spikelets. The length and closeness of the panicle branches determine the shape of the panicle, which varies from densely packed conical or oval spreading and lax. Sorghum is predominantly self-pollinated.

A fully developed panicle may contain 2000 grains, each one usually partly covered by glumes. The grain is rounded and bluntly pointed from 4–8 mm in diameter and of varying size, shape, and color according to the variety. Pigments occur in the pericarp, testa or both. Cultivars with a pigmented pericarp have yellow or red color. When the pericarp is white and a testa is present, the seed color may be buff or bluish white. When a colored pericarp and a testa are present, the seeds tend to have dark brown or reddish brown color.

34.8 Sorghum races

Five major races of sorghum are recognized - durra, kafir, guinea, bicolor, and caudatum. They differ in panicle morphology, grain size, and yield potential, among other characteristics. Durra sorghums developed primarily in Ethiopia and the Horn of Africa, from where they spread to Nigeria and the Savanna region of West Africa. Kafir types originated in eastern and southern Africa. Guinea sorghums are grown mainly in West and Central Africa, while bicolor types are the least important to African production, occurring in East Africa. Caudatum varieties originated in Kenya or Ethiopia.

34.9 Grain sorghum groups

Most of the grain sorghums in cultivation are hybrids, derived from kafir-milo crosses. The major commercial groups of grain sorghum are kafir, hegari, milo, feterita, durra, shallu, and kaoliang.

- **Kafir.** These have a thick juicy stalk, relatively large, flat, dark green leaves and awnless cylindrical heads. The seed color may be white, pink or red.
- **Hegari.** These types have a more nearly oval head, more abundant leaves than kafir, and sweeter juice, and hence are more desirable for forage.
- **Milo.** This group has a less juicy stalk, curly light green leaves, and smaller leaves and stalks. The head is short, compact and oval with large yellow or white seeds. The plant tillers more than kafir and is more drought tolerant.
- **Feterita.** This group has few leaves, relatively dry stalks, and an oval compact head, with very large chalk-white seeds.
- **Durra.** This group has dry stalks, flat seeds, and very pubescent glumes. The panicles are erect but may be compact or loose. The varieties are chiefly grown in North Africa, India, and the Near East.
- **Shallu.** This group is characterized by tall, slender, dry stalks, a loose head, and pearly white seeds. The varieties are late maturing.
- **Kaoliang.** The varieties in this group have dry, stiff, slender stalks, an open bushy panicle, and small brown or white seeds. They are grown exclusively in China, Korea, Japan, and southeastern Siberia.

34.10 Reproductive biology

Sorghum inflorescence is a panicle ranging from 3–20 inches in length and 0.5–8 inches in width. Morphologically, the panicle (or head) ranges from compact to open. The spikelets are borne in pairs on branches arranged in whorls. One spikelet is sessile, bisexual and fertile, whereas the other is pedicelled and male sterile. The sessile spikelet contains two florets, one perfect and fertile, the other sterile. The pedicelled spikelet is sterile. The fertile floret has a membranous lemma, a palea, two lodicules, three stamens, and an ovary with two long styles with plumose stigmas.

34.11 Pollination

Blooming of sorghum starts within three days of the emergence of the panicle from the boot. Blooming starts at or near the panicle apex and proceeds downward, the process lasting 4–7 days. Blooming is hastened by short day length and higher temperature. Optimal flowering occurs at temperatures of 21–35 °C. Depending on the environment, the stigma may remain receptive for 5–16 days after anthesis, if a flower is unpollinated. Anthesis usually occurs in the morning. The anthers dehisce as they are exserted or soon thereafter, usually becoming pendant. The pollen is most viable within the first 30 minutes; viability is negligible after four hours.

Industry highlights
Sorghum breeding

William Rooney

Texas A&M University, College Station, TX 77843, USA

Overview

Sorghum (*Sorghum bicolor* L. Moench) is one of the more important cereal grain crops in the world. In 2001, sorghum was produced on approximately 50 million hectares with an average yield of 1280 kg/ha worldwide (FAO, 2001). It is commonly grown in semi-arid tropical, subtropical, and temperate regions of the world. The crop is used for many different purposes. The grain is used as a food grain, feed grain, and for industrial purposes. In many production systems, the vegetation is used as forage. The location of production often defines the ultimate end use and the specific types of sorghum that will be grown.

Table B34.1 Genotypes and corresponding phenotypes for A-, B- and R-lines in the A1 cytoplasmic-genetic male sterility system in sorghum.

Line	Cytoplasm[a]	Genotype[b]	Phenotype
A-line	[A]	*rf rf*	Male sterile
B-line	[N]	*rf rf*	Male fertile
R-line	[A] or [N]	*RF RF*	Male fertile
Hybrid	[A]	*RF rf*	Male fertile

[a]Cytoplasm types: [A] = sterility inducing cytoplasm type, [N] normal cytoplasm.

[b]Genotype: *RF* = dominant allele for fertility restoration, *rf* = recessive allele for fertility restoration.

Sorghum is grown as a hybrid crop in the United States. Because it is a predominately self-pollinated crop, hybrid seed production requires the use of cytoplasmic-genetic male sterility (CMS). Without this system, hybrid seed production would not be economically feasible. The CMS system is based on male sterility-inducing cytoplasm that is complemented by alleles in the nuclear genome that either restore fertility or maintain sterility. In the CMS system, lines that have [A] cytoplasm must have a dominant allele present in the nuclear genome to restore male fertility (Table B34.1). If the line lacks the dominant allele for fertility restoration, the plant will be male sterile.

Hybrid seed production requires maintenance of the A, B, and R lines (Figure B34.1). Seed of a male sterile A-line is increased by pollination using the complementary B-line. The sole purpose of the B-line (also known as a maintainer) is to perpetuate or maintain the A-line. The A-line and B-line are genetically identical except that the A-line has a sterility-inducing cytoplasm while the B-line has normal fertile cytoplasm. Thus, A-line plants that are male sterile can be pollinated with pollen from B-line plants to regenerate seed of the A-line. To produce hybrid seed, the male sterile A-line is pollinated with pollen from the male fertile R-line plants. The R-line (also known as a restorer line) is genetically very different from the A-line and it carries the dominant fertility restoration alleles needed to restore fertility in the progeny of the A-line. The seed that is produced on the A-line from this pollination is the seed that is planted by the producer for commercial grain production.

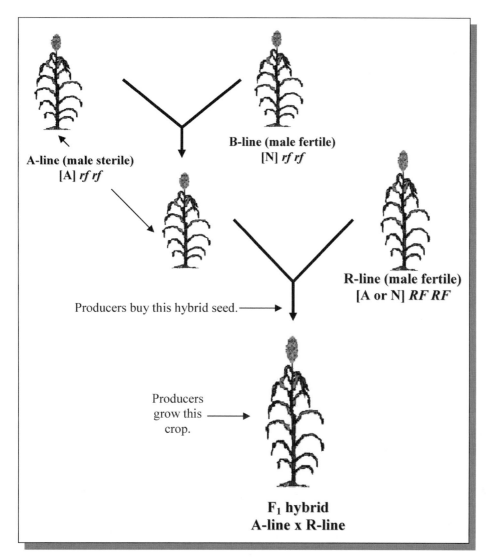

Figure B34.1 Schematic of the sorghum hybrid seed production process utilizing cytoplasmic-genetic male sterility. The genetics for each line are described in Table B34.1. The A-line parent is increased utilizing pollen from the B-line. The F₁ hybrid is produced by pollinating the A-line with an R-line pollinator. Both the B-line and R-line are maintained through self-pollination. ([A], sterility inducing cytoplasm; [N], normal cytoplasm.) Figure courtesy of William Rooney.

Private and public research

Because sorghum is grown as a hybrid, the hybrids that producers grow are produced and sold by private industry. Private industry also maintains a limited number of breeding programs for the production of new A/B and R lines for new hybrids. This work is supplemented and enhanced by researchers in public breeding programs, such as those at Kansas State University, Texas A&M University, and the USDA. These public programs do not produce or sell hybrids, but they develop parental lines and germplasms that are used by private industry in commercial hybrid production. In addition, public research programs in sorghum conduct research in long-term projects, such as the introgression and development of new germplasm, that may provide useful traits in the future.

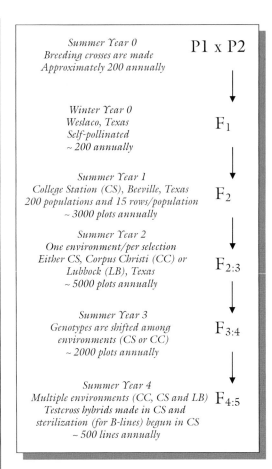

Summer Year 0
Breeding crosses are made
Approximately 200 annually

P1 x P2

Winter Year 0
Weslaco, Texas
Self-pollinated
~ 200 annually

F_1

Summer Year 1
College Station (CS), Beeville, Texas
200 populations and 15 rows/population
~ 3000 plots annually

F_2

Summer Year 2
One environment/per selection
Either CS, Corpus Christi (CC) or
Lubbock (LB), Texas
~ 5000 plots annually

$F_{2:3}$

Summer Year 3
Genotypes are shifted among
environments (CS or CC)
~ 2000 plots annually

$F_{3:4}$

Summer Year 4
Multiple environments (CC, CS and LB)
Testcross hybrids made in CS and
sterilization (for B-lines) begun in CS
~ 500 lines annually

$F_{4:5}$

Figure B34.2 Pedigree breeding scheme used by the TAES sorghum breeding program at College Station, Texas. This scheme is used for the development of new B- and R-lines and germplasm. Initial crosses are made using either plastic bag crosses or hand emasculations. Open-pollinated selections are made in each generation until the F_5 where the plot is self-pollinated and used to make testcross hybrids. At the F_5 generation, new B-lines enter sterilization and testcrossing while new R-lines are evaluated in testcrosses. Figure courtesy of William Rooney.

The Texas Agricultural Experiment Station (TAES) sorghum breeding program located at Texas A&M University in College Station, Texas, is part of a multiproject, multilocation sorghum improvement program supported by the TAES. Sorghum breeders work in conjunction with plant pathologists, entomologists, and grain quality and molecular geneticists to create an effective and important research and application oriented team. In terms of the sorghum breeding program at College Station, the breeding program has several objectives: (i) to develop and release germplasm and parental lines with improved adaptability, yield, quality, and stress resistances; (ii) to conduct research that increases understanding and knowledge of sorghum breeding and genetics; and (iii) to train undergraduate and graduate students in plant breeding.

Methodology of the TAES sorghum breeding program at college station

For improved hybrids, new and improved parental lines must be developed. First, genetic variability must be developed through the selection and hybridization of parent material. This is a crucial step in the process. Usually elite germplasm is crossed to other material (elite lines, germplasm or genetic stocks) to correct a perceived deficiency in the elite material. For example, if an otherwise good A/B pair is susceptible to lodging, it will be hybridized with several different sources of lodging resistance with the goal of producing a new A/B pair with improved lodging resistance. In our program, the A/B program is managed separately from the R-line program to maintain heterosis between the two groups and keep the fertility restoration and maintenance genetics separate. Based on the considerations listed above, specific crosses are made using the methodology described by Rooney (2004). These F_1 progeny are self-pollinated to produce an F_2 population.

Once F_2 populations are created, our program utilizes a pedigree breeding approach for the development of inbred lines (Figure B34.2). From the F_2 generation until the F_5 generation (in which uniform lines are selected), the progeny rows are grown and panicles in rows are visually selected on the basis of agronomic desirability, pest resistance, and abiotic stress tolerance. F_5 lines that are phenotypically uniform are testcrossed to measure their general and specific combining ability and their suitability as parent lines in hybrid combination.

The appropriate time for the selection of specific traits is dependent on the heritability of the trait and the environments in which the selection occurs. In our program, traits with higher heritability (maturity, height, grain color, etc.) are selected in the early generations while traits with lower heritability are selected in a more advanced generation (yield, drought tolerance, disease and insect resistance). These more complexly inherited traits must also be screened in multiple environments, because these traits may not be expressed in any given environment. Evaluation in multiple environments is crucial to the development of widely adapted sorghum genotypes. In our program, we use three basic regions for inbred selection: South Texas, Central Texas and the Texas High Plains (Figure B34.3). These regions are each unique and force different selection pressures on the material grown therein. For example, our South Texas nurseries are rain-fed and subject to drought stress and consistent disease pressure. In addition, this region is good for selecting genotypes that perform well in subtropical growing environments. In our High Plains nurseries, the environment is typical of temperate production regions and different traits are of importance. The evaluation of material at these locations allows us to select widely adapted germplasm.

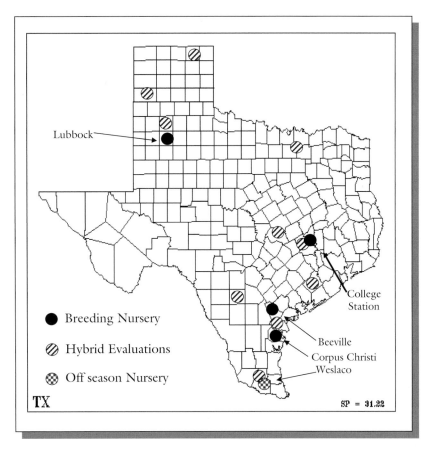

Figure B34.3 Map of Texas indicating locations used by the TAES sorghum breeding program for breeding, selection, and evaluation. Figure courtesy of William Rooney.

Many traits are important in our breeding program (both inbred and hybrids). Since grain sorghum producers are paid based on grain yield, it remains the most important trait. Almost every breeding program measures the yield potential in hybrids as the correlation between inbred yield and hybrid yield are rather poor (Rooney, 2004). In addition to breeding for yield, any factor that reduces yield becomes important as well. Therefore, breeding for drought tolerance, disease and insect resistance have priority as well. Diseases of importance in US sorghum include stalk lodging, grain mold, anthracnose, downy mildew, head smut, sooty stripe and leaf blights. Insects of economic importance include sorghum midge, greenbug, and chinch bug. Finally, grain quality in sorghum has become more important. Until recently, sorghum in the United States was used exclusively as feed grain, but the development of food quality sorghum hybrids has resulted in an increase in the sale and use of sorghum in food products. A major goal of our program is to produce sorghum lines that will make hybrids with grain quality suitable for both feed and food use.

Prior to testcrossing, the new B-lines must be male-sterilized. B-lines are sterilized using a backcross program in which a standard A-line with a similar pedigree to the new B-line is used as the source of the male sterile cytoplasm (Figure B34.4). The new B-line is then used as a recurrent parent to produce an A-line that is genetically identical to the B-line (except that it is male sterile). In each generation of backcrossing, plants and progenies that are fully male sterile and are the most similar to the B-line are selected. The sterilization process usually requires a minimum of five backcrosses and most sorghum breeding programs utilize winter nurseries to reduce the amount of time required for sterilization.

Our program begins testcrossing in the F_5 for R-lines and in the BC_3 of sterilization for A/B lines. New R-lines are testcrossed to A-line testers and new A-lines are testcrossed to R-line testers to determine each new line's general

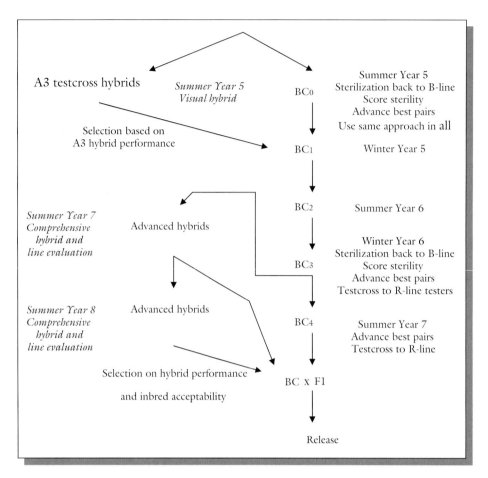

Figure B34.4 The advanced testing and sterilization scheme used by the TAES sorghum breeding program at College Station, Texas. From the F_5 generation (Figure B34.2), testcross hybrids and BC_0 sterilization lines are grown for evaluation and continued backcrossing. If testcross performance is acceptable, sterilization is continued through backcrossing until the A-line is identical in phenotype to the B-line. As the A-line becomes available additional hybrid evaluation is performed to confirm heterosis and line acceptability. The R-line testcross evaluation process is similar but eliminates the requirement of sterilization. Figure courtesy of William Rooney.

combining ability and the fertility of the hybrids. Lines that produce high yielding hybrids with appropriate agronomic parameters are advanced for additional testing. These lines are hybridized to several potential parental lines of the opposite group to identify those hybrids with good general and specific combining ability. These hybrids will be tested in multiple locations and those lines that produce hybrids with high yield, good stability, good agronomic characteristics (height, maturity, etc.), and acceptable abiotic and biotic stress tolerances will be released. These releases are then tested by private industry to determine if they will be used in hybrid releases for producers (Rooney, 2003a, 2003b).

In addition to hybrid performance, private companies must consider the parental line's performance with regard to seed production. Seed producers must be able to consistently coordinate flowering of the A-line and the R-line. If the two lines have a poor "nick", the A-line will have extremely low seed set and, consequently, poor yields of hybrid seed. The pollinator line should start shedding pollen prior to the emergence of stigmas in the A-line and the R-line must continue to shed pollen throughout the flowering of the A-line. In addition, the pollen shed from the R-line should be consistent and relatively unaffected by normal environmental conditions. Obviously, the R-line must consistently restore fertility to the hybrid and the A-line parent must produce seed yields high enough to justify seed production costs.

In addition to breeding for grain sorghums, there are directed breeding programs in sorghum to improve the crop for other uses. These include the improvement of sorghum for forages, such as silage, grazing and hay. Other programs focus on the use of sorghum for sweeteners, in which the crop is used much like sugarcane. All of these uses have resulted in the development to numerous genotypes suitable for the production environments and purposes of the crop. The challenge is for sorghum breeders to continue future improvements given the relatively small number of researchers actively engaged in this field of research.

References

FAO (2001). *State of food security in the world.* Food and Agricultural Organization, Rome, Italy.
Rooney, W.L. (2003a). Registration of Tx2912 to Tx2920 Sorghum Germplasm. *Crop Science,* **43**:442–443.
Rooney, W.L. (2003b). Registration of Tx2921 to Tx2928 Sorghum Germplasm. *Crop Science,* **43**:443–444.
Rooney, W.L. (2004). An Overview of Sorghum Breeding Methodology. *Advances in Agronomy,* **83**:37–109.

34.12 Common breeding methods

Early sorghum improvement depended on introductions. Introduction of short-day germplasm from the tropical regions into temperate regions with a longer photoperiod was unsuitable for commercial production because it usually matured too late or was too tall. Using backcrossing, USDA and Texas A&M University researchers embarked on the sorghum conversion program to convert the tropical varieties to temperate climate-adapted cultivars by substituting two recessive alleles for the dominant height alleles in the tropical varieties, as well as the recessive *mal* maturity alleles for the *Mal* dominant counterparts.

Pure line selections have been made in many countries in Africa and India. Pedigree selection is commonly used following hybridization. The population improvement methods discussed in Chapters 1–30 are applicable to sorghum breeding. To develop a random mating population the breeder starts by selecting 20–40 parents. The next step is to incorporate a male sterility gene by crossing each parent individually to a male sterile stock. The F_2 segregates for male sterility. The selected plants are backcrossed one or two times if the male sterile stock lacks good agronomic qualities. An equal amount of F_2 seed from all crosses is bulked and grown in isolation for random mating.

Sorghum is also bred using population improvement methods for enhancing quantitative traits. The pure line method of breeding has been used in sorghum improvement. With the discovery of CMS and fertility restoring genes in the 1950s, hybrid sorghum breeding became practical and profitable, using the A-B-R line breeding systems as in corn.

The sorghum conversion program (adapting tropical sorghum to temperate climate) has been a significant part of the success of sorghum breeding in the United States. A conversion program starts with a cross of tropical x temperate lines at a tropical station. The resulting F_1 is grown at a temperate experimental station to obtain F_2 seed. The F_2 is planted at a tropical station. Selections are made and backcrossed (five cycles of backcrosses) to adapt the tropical line to temperate conditions.

34.13 Establishing a breeding nursery

The specific layout of a nursery depends on the task to be performed and materials to be handled. A section of the nursery may be allocated to procedures such as selfing, crossing, population breeding, observations, forage breeding, and tropical conversion. To reduce walking, parents to be crossed are planted near each other. Plant density should be similar to what farmers use in production. Spacing varies from 10–100 cm between rows and 5–30 cm within rows.

34.14 Artificial pollination

34.14.1 Equipment

The equipment used includes pollinating bags (6 × 12–40 cm), stapler, knife, marking pencils, scissors, hot water container, clips string, apron.

34.14.2 Preparing the female

Sorghum breeders control pollen using one of four general methods - male sterility, hot water emasculation, hand emasculation, and control of anther dehiscence. The panicle of the male sterile plant is bagged just before anthesis. Using male sterility enables the breeder to undertake large-scale hybrid seed production. Hot water emasculation is also used, but usually when male sterility is not available in the parents, and when complete pollen control (i. e., presence of a few selfs) is not critical. To emasculate in the field, a suitable panicle (just beginning to flower) is selected and all opened spikelets removed. The panicle is enclosed in a rubber or plastic sleeve, tied tightly around the peduncle but open at the top. Water at 47–48 °C is poured into the sleeve and held for 10 minutes. In the greenhouse, the particle may be emasculated by directly immersing the panicle in hot water by inverting the potted plant. The panicle is left to dry before covering it with a pollination bag.

Breeders may use hand emasculation when only a small quantity of seed is needed and complete pollen control is desired. This method of emasculation requires skill to succeed. Hand pollination in the field is undertaken in the afternoon when contamination from other plants is least. Anthers may be removed using forceps, scissors or other pointed instruments. The bag is placed over the emasculated panicle.

Anther dehiscence can be delayed by high humidity. In regions of high humidity, placing polythene or paper over a particle can delay anther dehiscence by about 30 minutess the next morning. However, under hot conditions, heat might build up under the bag and injure the flowers.

34.14.3 Pollination

Pollen collection is best done in the morning between 7 and 12 a.m. The center of the particle yields the most pollen. It may be necessary to cover the male plant the day before pollination to keep out contaminants. When it is time to pollinate, the panicle is tilted so that the pollen can be shaken into the bag. Male sterile plants and those emasculated by hot water are ready for pollination about 5–8 days after bagging, when flowering is complete and the stigmas are extruded. The bag containing the pollen is inverted over the female panicle and shaken to pollinate. The operator may also remove a branch of the panicle and brush pollen onto the stigma. The pollinated panicle is covered with the bag. Pollination success varies with the method of pollination, operator skill,

environment, age of the stigma, and amount of pollen, among other factors.

34.15 Natural pollination

Cytoplasmic genetic male sterility is used for large-scale hybridization in an isolation block where random mating populations are grown. A fertility restorer pollinator is planted with several male sterile females.

34.16 Seed development

Successful pollination is evident by the seventh day following pollination. The grains are physiologically mature when the black layer forms at the base of the seed. This stage occurs between 35 and 45 days after pollination.

34.17 Harvesting

Grain sorghum is a perennial plant. Consequently, the plant remains green and alive, even after the grain is matured, until killed by tillage or freezing temperatures. The grain dries slowly. At the hard dough stage, the grain contains about 18–20% moisture. For effective combining, the grain moisture content should be about 13% or less. Waiting for the grain to dry in the field delays harvesting and increases the risk of damage by weather events and birds. Furthermore, delayed harvesting also delays the rotation of sorghum with a winter crop (i.e., wheat). Desiccants (e.g., diquat, 28% nitrogen urea-ammonium nitrate) are applied as a pre-harvest treatment or grain sorghum. Roundup Ultra® may be used as a defoliate when the grain is for feed.

34.18 Common breeding objectives

34.18.1 Grain yield

Grain yield is one of the principal objectives of sorghum breeding. Grain yield improvement in sorghum has made significant strides over the years. The success mainly stems from access to additional germplasm from tropical accessions through the Sorghum Conversion Program, which augmented existing genetic variability, and the development of hybrid cultivars. Also, disease and insect resistance protect grains in the field, reducing harvest losses.

34.18.2 Yield stability

Sorghum is more adapted to marginal production environments than cereal crops such as corn and rice. Breeders are interested in stable performance over variable environmental conditions in the production region.

34.18.3 Agromorphological traits

Lodging resistance

Tall cultivars are prone to lodging. Breeders use dwarfing genes to attain short stature. Sources of stiff stalk include the kafir and hegari genotypes. Another way lodging resistance is enhanced is through the development of resistance to stalk and root diseases.

Short stature

As previously discussed, breeders use the four recessive dwarfing genes – dw_1, dw_2, dw_3, and dw_4 - to develop short-statured cultivars. Most US cultivars are 3-dwarfs (have three recessive dwarfing genes). It should be mentioned that taller cultivars are preferred for production of forage and silage, and also in dry production systems. Short stature is desirable for mechanized harvesting. In addition to short stature, crosses between milo and kafir produce recombinants with erect head and stout stalk.

34.18.4 Adaptation

Early maturity

Early maturing cultivars are advantageous in low rain-fall regions, allowing the crop to escape damage from drought. These cultivars also allow an expansion of sorghum production to regions of high altitude and short growing season. Genes for early maturity tends to reduce plant stature.

Photoperiod insensitivity

Photoperiod insensitivity adapts the crop to regions with a shorter growing season.

Drought resistance

Sorghum is more resistant to heat and drought than corn. In breeding for drought and heat resistance, breeders select for extensive root systems that promote more extensive exploration of the soil for moisture, as well as for plant features that reduce moisture loss from the leaves (e.g., fewer stomata).

Tolerance of aluminum

High levels of soil aluminum reduce root development and predispose plants to drought injury. Genotypes with aluminum tolerances have been identified for breeding.

34.18.5 Disease resistance

Sorghum diseases of economic importance include:

- **Rots.** Sorghum is known to be attacked by both root and stalk rots, the common ones including *Fusarium* root rot (caused by *Fusarium moniliforme)* and stalk rot, charcoal rot (by *Macrophomina phoseolina*), and *Periconia* root and crown rot.
- **Blights.** An important blight is the northern leaf blight (caused by *Exserohilum turcicum*), a disease that is common in humid production regions. Sources of resistance include grain and sweet sorghum types.
- **Smuts.** An important smut disease is the head smut (caused by *Sporisorium redianum).*
- **Rust.** The *Pucciniapurpurea* is responsible for sorghum rust, a diseasethat is not economically very important in the United States.
- **Viral diseases.** One of the most economically important viral infections of sorghum is the maize dwarf mosaic (MDM), a disease that is spread by the greenbug and various species of aphids.

34.18.6 Insect resistance

Major insect pests of sorghum include:

- **Greenbug.** The greenbug *(Schizaphis grulninuln)* is a sucking insect pest with biologic races. Breeders, hence, have to continue to breed new cultivars as new races evolve.
- **Sorghum midge.** The sorghum midge *(Contarinia sorghicola)* is of economic importance worldwide. Resistant germplasrn have been identified among accessions from Ethiopia and Brazil. Resistance is

quantitatively inherited. Consequently, breeders should use resistant A and R lines in developing resistant hybrids.

- **Stalk borers.** The European stalk borer *(Ostrinia nubilis)* and the southwestern corn borer *(Diatraea grandiosella)* are the most economically important.
- **Shoot fly.** The shoot fly *(Antherigona soccata) is* of economic importance, primarily in the tropics.

34.18.7 Product quality

Sorghum is produced for grain for food, especially in developing countries. For food purposes, endosperm qualities are important in breeding objectives. Sorghum endosperm may be floury or corneous, the latter trait important in dry milling. The pericarp, may by pigmented or without pigment. To improve the nutritional quality of sorghum for feed, one objective is to reduce the tannin content.

Sorghum is also grown for industrial uses, including flour, syrup, and malt production. For feed, an objective is to reduce the cynogenetic glucoside content of the plant and improve its fodder, forage and silage qualities.

Key references and suggested reading

Acquaah, G. (2005). *Principles of crop production. Theory, techniques and technology.* Prentice Hall, Upper Saddle River, NJ.

Outcomes assessment

Part A

Please answer the following questions true of false.

1 Sorghum is a self-pollinated species.
2 Sorghum originated in Africa.
3 Sorghum bicolor is the most important species for crop production.
4 Most US sorghum cultivars are two-dwarfs.

Part B

Please answer the following questions.

1 Give the five major races of sorghum.
2 Give four examples of the grain sorghum groups.
3 Give two of the top sorghum producing states in the United States.

Part C

Please write a brief essay on each of the following topics.

1 Discuss the breeding of plant stature in sorghum.
2 Discuss the important diseases and insect pests of sorghum.
3 Discuss the pollination process in sorghum breeding.

35

Breeding soybean

Taxonomy

Family   Fabaceae
Genus   *Glycine* Willd.
Species   *Glycine max* (L) Merr.

35.1 Economic importance

Soybean (*Glycine max* L. Merrill) it the major world oil seed. It is also a major source of meal used for livestock feed. It consists of about 35–40% protein and less than 20% oil. The United States is the world's leading producer of soybean, accounting for nearly 50% of the world's total production and about 40% of the cultivated acreage. The average yield of US soybean was only about 11 bushels/acre in 1924 but increased to 25.4 bushels/acre in 1966. In 1994, the production reached a new high of 41.9 bushels/acre. In that year, producers in Iowa recorded a yield of 50.5 bushels/acre. Similarly, the acreage of soybean in 1924 was 1.8 million acres, 18.9 million acres in 1954, and 63.4 million acres in 1996. It rose to 72.7 million acres in 2000.

Soybean production in the United States occurs primarily in the North Central states, which overlaps the Corn Belt states. The major producing states are Iowa, Illinois, Minnesota, Indiana, Ohio, Missouri, and Nebraska; these states together accounted for 72% of the US total production in 2000. Iowa and Illinois produce more than 30% of the US crop.

Soybean is important in the Southern and Southeastern states of Arkansas, Mississippi, North Carolina, Kentucky, Tennessee, Louisiana, and Alabama; these states together produce about 10% of the US total crop. Other producers are South Dakota, Kansas, Missouri, Wisconsin, and North Dakota.

On the world scene, the United States has dominated production since the 1950s, growing more than 75% of the world crop by the 1970s. By 2000, the United States, though still the world's leader, had a share of 45%, Argentina and Brazil taking up about 15% and 21%, respectively. China ranks fourth in the world, with about 12% of total world production. Other producers in the world include Japan, Indonesia, and the former Soviet Union.

35.2 History and origin

The soybean is considered among the oldest cultivated crops. The first record of the crop is contained in a 2838 BC Chinese book in which Emperor Cheng-Nung described the plant. Soybean was a "Wu Ku", one of the sacred five grains (the others being rice, wheat, barley, and millet) considered essential for the existence of the Chinese civilization. Cultivated soybean is believed to have derived from a wild progenitor, *Glycine ussuriensis*, which occurs in eastern Asia (Korea, Taiwan, Japan, the Yangtze Valley of Central China, the northeastern provinces of

Principles of Plant Genetics and Breeding, Second Edition. George Acquaah.
© 2012 John Wiley & Sons, Ltd. Published 2012 by John Wiley & Sons, Ltd.

China, and the adjacent areas of Russia). The plant was first domesticated in the eastern half of north China in the eleventh century BC. It was introduced into Korea from this region and then into Japan between 200 BC and 300 AD.

Soybean was known to be grown in Europe in the seventeenth century. Its first introduction into the United States is traced to a Samuel Bowen, an employee of East India company, a seaman, who brought it to Savannah, Georgia, from China via England. Between 1804 and 1890, numerous soybean introductions were made into the United States from China, India, Manchuria, Korea, Taiwan, and Japan. In 1852, J.J. Jackson is reported to have first planted soybean as an ornamental plant in Davenport, Iowa. Most of the production of soybean in the United States prior to the 1920s occurred in the southern United States, mostly for hay, and then spread to the Corn Belt after about 1924.

Soybean is a subtropical plant. However, it is grown over a wide range of ecological zones, ranging from the tropics to 52 °N. Its climatic requirements are similar to those of corn.

35.3 History of breeding

Soybean was introduced into the United States in about the late 1700s. Initially, the crop was grown primarily as a forage species. The early introductions had weedy characteristics. They also shattered profusely, making them unsuitable for mechanized production. Producing soybean for seed started in the 1920s. Formal soybean research was initiated as a joint activity between the USDA research wing and the Agricultural Experiment Stations of land grant universities in the Midwest and southern states. In 1936, a cooperative state and federal program was established at the University of Illinois, Champaign/Urbana, to conduct soybean research. This program is now known as the US Regional Soybean Laboratory.

Seed companies became involved in soybean breeding as a result of the United States Plant Variety Act of 1970. With this protection against unauthorized use of proprietary material, seed companies could make money developing and selling improved seed. Modern cultivars were developed for seed, shifting the emphasis away from soybean as a forage crop to soybean as an oilseed crop.

35.4 Genetic resources

The USDA maintains about 15 000 accessions of *G. max* and a smaller number of other *Glycine* species at Urbana, Illinois, and Stoneville, Mississippi. Soybean accessions are also held by the Applied Genetics of Korea Atomic Energy Research Institute in Seoul, Korea, the National Institute of Agricultural Sciences at Hiratsuka, Japan, and the Asian Vegetable Research and Development Center, Taiwan.

35.5 Cytogenetics

The genus *Glycine* has two subgenera – *Soja* and *Gycine*. The subgenus *Soja* consists of two species: *G. max* ($2n = 2x = 40$), the cultivated species, and *G. soja* (L.) Sieb or *G. ussuriensis* ($2n = 2x = 40$), a wild species. These two species are cross-fertile. There are 15 wild species of soybean for which *G. tabacinia* and *G. tomentella* have polyploidy forms (including $2n = 4x = 80$).

35.6 Genetics

Several hundreds of genes for qualitative traits have been identified for soybean. Among them are four recessive genes for genetic male sterility, designated ms_1, ms_2, ms_3, ms_4. Linkage groups have been identified for 13 of the 20 soybean chromosomes. Soybean has pigmentation on various parts of the plant. Black and brown seed coat and hilum color are conditioned by two gene pairs, *Tr* and *Rr*. The expression of these genes is modified in some condition by brow seed coat (*TrO*), which is dominant over reddish-brown coat (*Tro*), and green seed coat (*G*), which is dominant over yellow seed coat (*g*).

Soybean may have either brown (tawny) or grey pubescence, the trait being conditioned by a single gene and brown being dominant over grey. Stem termination is controlled by two genes, Dt_1 and Dt_2, Dt_1 conferring intermediate stem while Dt_2 conditions semideterminacy, with dt_1 conditioning determinancy. Dt_1 and Dt_2 are dominant to dt_1 and dt_2, respectively; dt_1 is epistatic to Dt_2 and dt_2. Pod color at maturity is controlled by two genes; L_1L_2 and L_1l_2 produce black pods, while l_1L_2 produces brown color. Tan pods are conditioned by l_1l_2. Six independent recessive genes df_1 to df_6 are known to condition dwarfism in soybean, while five major and independent genes, E_1–E_5, condition flowering and maturity, with E_1 and E_2 delaying flowering and maturity.

35.7 General botany

Soybean (*Glycine max* L. Merrill) is an annual summer legume. Cultivated soybean is usually erect with a well-defined main stem and branches, and numerous leaves. Both determinate and indeterminate cultivars are used in production. The leaves and stems are usually pubescent. The flowers are either purple or white, and are borne in axillary racemes on peduncles at the nodes. The plant produces a large number of flowers, but only about two-thirds to three quarters of them produce pods. The pods are also pubescent. They range in color from light straw to black, containing 1–4 seeds (occasionally five). The seeds are usually unicolored and may be straw-yellow, greenish-yellow, green, brown, or black. Bicolored seeds exist, such as yellow with a saddle of black or brown. The hilum is also colored in various patterns – yellow, buff, brown, black. Soybean is primarily self-pollinated. In the proper soil environment, soybean is infected by the bacterium *Rhizobium*, resulting in roundish nodules on the roots, in which the nitrogen fixing bacteria live.

35.8 Cultivars

Soybean may be produced for forage, the varieties for this purpose generally being small-seeded, finer-stemmed, and very leafy. There are also cultivars for edible, dry or green-shelled beans. These cultivars usually have straw-yellow or olive-yellow seed and a yellow, brown, or black hilum. Soybean grown for grain is grouped into 13 maturity classes, ranging from 000 to X. The 000 group consists of the earliest maturing cultivars while the X consists of the latest maturing cultivars. Groups 000–IV are considered indeterminate, while groups V–X are determinate cultivars. Further, early maturing cultivars (000–IV) are adapted to the northern production regions, while those with a high maturity class designation are adapted to southern production regions. The 000 cultivars are adapted to the short summer growing seasons of northwest United States and Canada. Cultivars in groups II and III are best adapted to the

Midwest growing area. Maturity groups VIII and higher are grown in the southern or Coastal Plain counties.

35.9 Reproductive biology

35.9.1 Floral morphology

Soybean flowers are borne in the axil of a branch in clusters of 3–15. Soybean has the floral features of the family Papillionadae, comprising a large standard petal, two wing petals and two keel petals. The five petals enclose a pistil surrounded by 10 stamens, nine of which are fused into a tube, the one nearest the standard being free. The style curves towards the standard petal bearing a club-shaped stigma.

35.9.2 Pollination

The soybean flower is cleistogamous; self-pollination occurs shortly before or after the flower opens. Pollen shed depends mainly on temperature. Under warmer conditions (30 °C or higher) pollen shed may occur before 7–9 a.m. Anther dehiscence under moderate temperature starts later in the morning and continues for much of the day. Cool temperature reduces pollen shed and often causes flowers to prematurely self-pollinate before the breeder can emasculate stem.

35.10 Common breeding methods

As previously indicated, plant introductions played a significant role in early soybean breeding programs. Selections were made from introductions to develop commercial cultivars. Modern soybean breeders use a wide variety of methods in their programs, including backcrossing to transfer qualitative traits and single seed descent to accelerate a breeding program. Pedigree selection is also used because the plants can be well-spaced for individual observations.

Hybridization is most commonly used in soybean breeding programs for gene transfer. Both qualitative and quantitative traits have been improved in breeding programs using hybridization to create new variability for selection. Recurrent selection is possible if male sterility is incorporated into the breeding program.

Industry Highlights

Estimating inheritance factors and developing cultivars for tolerance to charcoal rot in soybean

James R. Smith

USDA-ARS, Crop Genetics and Production Research Unit, Stoneville, MS 38776, USA

Figure B35.1 Soybean plants infected with *M. phaseolina* that have wilted and died prematurely. Note the dead leaves and petioles attached to the plants. Courtesy of A. Mengistu.

Figure B35.2 Microsclerotia on the lower interior (vascular, cortical, and pith tissues) of a split soybean stem that was inoculated with *M. phaseolina* at the time of planting and harvested at physiological maturity. Figure courtesy of James R. Smith.

The following is a summary of the preliminary data for the purpose of understanding the inheritance of soybean [*Glycine max* (L.) Merr.] tolerance to charcoal rot and for developing improved varieties with tolerance to it. The summary demonstrates the kind of thinking and planning that needs to take place to successfully achieve the above objectives.

Charcoal rot is a disease of soybean caused by the fungus *Macrophomina phaseolina* (Tassi) Goidanich and occurs throughout the world; it can cause severe yield losses to soybean. Charcoal rot symptoms appear during midsummer under conditions of high temperature (28–35 °C) and low soil moisture availability. Diseased plants may wilt and prematurely die, with dead leaflets and petioles remaining attached to the plant (Figure B35.1). The most diagnostic symptom of charcoal rot is the black speckled (charcoal-like) appearance of microsclerotia on the exterior and interior (vascular, cortical, and pith tissues) of the lower stem of effected plants (Figure B35.2). Disease severity can be affected by heat, drought, and the presence of other pathogens. Attempts to control charcoal rot in soybean have included crop rotation, application of fungicides, biological controls, and adjustments to plant population, planting date, and irrigation protocols. Because these strategies have largely failed, researchers have become increasingly interested in the potential of varietal tolerance as a means of reducing yield losses. However, little progress has been made in developing charcoal rot-tolerant varieties, and the inheritance of varietal tolerance to charcoal rot is unknown.

To understand the inheritance of tolerance to charcoal rot and utilize it in breeding programs, it is necessary to generate segregating families (populations) of plants. Therefore, highly tolerant and highly susceptible soybean lines were chosen as parents, controlled pollinations between them were successfully made, and the appropriate segregating populations were generated.

Various options are available with segregating populations for estimating the importance of genetic factors in tolerance to charcoal rot. One way is by estimating heritability. Heritability can be expressed as the ratio of genetic variation over the total variation (genetic and non-genetic) influencing the trait. Traits with high heritabilities can be easily selected by plant breeders, whereas those with low heritabilities are not easily selected. High heritability, in the narrow sense (Φ_A^2/Φ_P^2), means that a high proportion of the total observed phenotypic variability (Φ_P^2) is accounted for by what is called additive genetic variance (Φ_A^2).

Additive genetic variance is the variance due to the average effects of alleles (Bernardo, 2002, p. 91). It measures the variation in the effects that are transmitted from one generation to the next. It does not imply that the alleles act in a purely additive manner. Rather, segregating loci with dominance, partial dominance, and overdominance gene actions, as well as additive gene action, can all contribute to Φ_A^2 (Bernardo, 2002, p. 92). Epistatic gene action can also contribute to Φ_A^2, as well as to Φ_D^2 (dominance genetic variance) and Φ_I^2 (epistatic genetic variance) (Bernardo, 2002, p. 96).

Φ_D^2 is the variance of dominance deviations and indicates dominance gene action, which describes the genotypic interaction of alleles within a locus. Φ_D^2 is not considered useful for making progress from selection because intralocus interactions are not passed on to progeny. Rather, meiosis determines that only one allele from each diploid intralocus interaction will be passed on to progeny through gametes. Hence, an individual's dominance genotypic interactions are not passed on to its progeny and that is why Φ_D^2 is not considered useful for making progress from selection. When $\Phi_D^2 = 0$, dominance gene action is absent and the intralocus variance is comprised solely of Φ_A^2, indicating purely additive gene action within the locus but not necessarily between loci (Bernardo, 2002, p. 92). Some components of Φ_I^2, such as Φ_{AA}^2 (additive by additive epistatic genetic variance), are useful in selection (Hanson, 1963, p. 133), because such interlocus interactions can be passed on to progeny. A broad sense estimate of heritability (Φ_G^2/Φ_P^2) includes the variance components of all types of gene action (Φ_G^2), some of which (i.e. dominance) are not passed on from parent to offspring. Exceptions to the above, where dominance gene action can be useful in selection, are when selecting among asexually-propagated clones or among single cross hybrids (Bernardo, 2002, p. 109).

If tolerance were controlled by one or a few major genes, with only minimal environmental influence, then heritability would be expected to be high and selection for charcoal rot tolerance effective in the F_2 generation among single plants or in the F_3 generation among $F_{2:3}$ progenies. If tolerance were controlled by multiple genes but the environment had only a minor influence, then heritability might still be high and selection in early generations still effective. However, if heritability for tolerance to charcoal rot were low, then selection might only be effective among advanced generation breeding lines grown in replicated experiments.

Creating segregating populations of multiple generations (F_2 and backcrosses to each parent) from two parents can serve multiple purposes in inheritance studies. Warner (1952) proposed utilizing the above generations (P1, P2, F_1, F_2, BC1P1, BC1P2) to estimate a narrow-sense heritability (Φ_A^2/Φ_P^2), but suggested the need to test at least several hundred F_2, BC1P1, and BC1P2 individuals in order to reduce potential sampling error. Reinert and Eason (2000) provide an example of estimating a narrow-sense heritability in a self-pollinated species using the above generations. Bernardo (2002, pp. 110, 146) listed three disadvantages of estimating Φ_A^2 and Φ_D^2 using the above generations: (i) as single F_2 and BC1 plants cannot be replicated, the lack of replication across environments causes these estimates of genetic variance to be confounded with the variance for genotype by environment; (ii) any linkage disequilibrium in the non-random-mated F_2 and BC1 populations will cause the relationship between genotypic values at two loci to be confounded with Φ_A^2 and Φ_D^2; and (iii) individual plant measurements of quantitative traits are prone to large non-genetic effects.

Another option for utilizing the above generations could be to determine the number of genes and their modes of action for simply inherited (one or two genes) traits. If all generations (P1, P2, F_1, F_2, BC1P1, and BC1P2) are assayed together for tolerance to charcoal rot, segregation ratios of the F_2, BC1P1, and BC1P2, along with the assay values of the P1, P2, and F_1 individuals, can determine if inheritance is simple and if dominance gene action is evident. The segregating F_2 generation can provide one estimate for an inheritance model, while the two backcross generations can provide a second and confirming estimate. Velez et al. (1998) and Singh and Westermann (2002) provide examples in dry bean (*Phaseolus vulgaris* L.) of utilizing the above generations in this way to determine the qualitative inheritance of resistance to bean disorders. However, when inheritance is affected by many genes and influenced greatly by environment, the above estimates may be inadequate.

An additional purpose for creating and utilizing the above generations could be to conduct an analysis of generation means. Generation mean analysis provides information on the relative importance of additive and dominance effects in populations created from two inbreds. It involves measuring the means of different generations (P1, P2, F_1, F_2, BC1P1, and BC1P2, etc.) derived from two inbreds and interpreting the means in terms of the different genetic effects (Bernardo, 2002, p. 144). Because the actual means of single loci are unobservable, generation means estimate the pooled genic effects across loci. Generation mean analysis is most useful when the two parents differ greatly in favorable alleles; that is, when one parent has most, if not all, of the favorable alleles and the other parent has few, if any, favorable alleles. The pooled estimates of effects are summed across all loci for which P1 and P2 differ. Generation mean analysis has been commonly used to study disease resistance, where one parent has high resistance and the other has high susceptibility (Bernardo, 2002, p. 145). Useful examples of generation mean analysis have been provided by Reinert and Eason (2000) in snap bean (*P. vulgaris* L.), Mansur et al. (1993) in soybean, and Campbell and White (1995) and Campbell et al. (1997) in maize (*Zea mays* L.).

Generation mean analysis can also be used to estimate effects due to epistasis, environment, genotype by environment interactions, and linkage (Mather and Jinks, 1971, pp. 83–119). However, the experiments can become much more complex and are unnecessary if the simpler additive-dominance model accounts for the variability present.

An important consideration for valid estimates of generation means is that there be sufficient sampling of segregating generations (Hallauer and Miranda, 1981, p. 109). Bernardo (2002, p. 138) recommended that sampled breeding populations have a minimum of 50–100 progenies.

There are several advantages of generation mean analysis. (i) It is relatively simple and statistically reliable (Mather and Jinks, 1971, p. 126). Sampling errors are inherently smaller when working with means than with variances for estimating inheritance. Smaller experiments can therefore be used to obtain the same level of precision (Hallauer and Miranda, 1981, p. 111; Campbell *et al.*, 1997). (ii) The estimation and interpretation of non-allelic interactions (epistasis) is more progressive for generation mean analysis than for variance estimates because mean effects are less confounded with one another and because the kinds of experiments required for analysis of means are smaller and easier to carry out than are those for variances (Mather and Jinks, 1971, p. 126). (iii) Populations evaluated in generation mean analysis can be used in applied breeding programs (Campbell *et al.*, 1997). (iv) It is equally applicable to both self- and cross-pollinated species (Hallauer and Miranda, 1981, p. 111).

However, generation mean analysis has several weaknesses. (i) It has limited value for quantitative traits whose parents have comparable mean performance (Bernardo, 2002, p. 146). (ii) As the information derived from the analysis is relevant to only a specific pair of parents, it has little application to other populations (Hallauer and Miranda, 1981, p. 111). (iii) Negative effects at some loci can cancel out positive effects at other loci, causing true genic effects to be underestimated. Generation mean analysis does not reveal opposing effects (Bernardo, 2002, p. 145). For example, a mean dominance effect of zero due to cancellation of opposing effects does not mean that there are not dominance effects. But it does mean that there are no evident dominance effects and that a determination of degree of dominance using variance components may be necessary. (iv) Generation mean analysis does not provide estimates of heritability, which are essential for estimating predicted gain from selection (Hallauer and Miranda, 1981, p. 111). (v) Finally, if epistatic effects are present, additive and dominance effects can be biased by the epistatic effects and by linkage disequilibrium (Hallauer and Miranda, 1981, p. 110).

The use of generation mean analysis should be considered as complementary, rather than as alternative, to variance component analyses (Mather and Jinks, 1971, p. 126). However, if one estimates additive and dominance genic effects using generation mean analysis, and also estimates Φ_A^2 and Φ_D^2, there may be little relation in the magnitude of the two sets of estimates (Hallauer and Miranda, 1981, p. 110). This might be expected because generation means estimate the sum of the genic effects, whereas variances are the squares of the genic effects. Further, the magnitude of Φ_A^2, Φ_D^2, and Φ_I^2 may be poor indicators of the underlying gene actions for quantitative traits (Bernardo, 2002, p. 144). For example, Moll *et al.* (1963) found that generation mean analysis detected epistatic effects that were not evident from estimates of Φ_A^2, Φ_D^2, and Φ_E^2 (environmental variance), but noted that variance components may detect genic variation not detected by generation mean analysis due to cancellation of mean effects.

Because of the weaknesses of single plant data and the complexities of multigene quantitative traits, it may be advisable to generate F_3 progenies and selfed progenies of both the BC1P1 and BC1P2 generations (Bernardo 2002, pp. 175–177). These generations are useful for estimating genetic variances, provided that they are developed without selection. The F_3 progenies can be grown in replicated trials, which can provide an estimate of environmental effects and genotype by environment interactions. Hamblin and White (2000) and Walker and White (2001) provide examples in maize of using the ANOVA of F_3 progeny means to estimate Φ_A^2, heritability, and predicted gain from selection.

Hallauer and Miranda (1981, p. 91) recommended the use of F_3 families for estimating Φ_A^2 in maize. Bernardo (2002, pp. 179–181) noted that an increase in either the number of environments or replications reduces the variance of an F_3 family mean and, in turn, increases heritability. Hence, selection for quantitative traits among F_3 families can be effective if each family is grown in extensive performance tests (Bernardo, 2002, p. 181).

A further advantage of developing F_3 progenies is that the segregation ratios within each F_3 progeny can be used to confirm applicable F_2 qualitative inheritance models. Thompson *et al.* (1997) used F_3 progeny ratios to help accurately categorize each F_2 plant genotype.

An additional genetic relationship that could be used to estimate heritability and predicted gain from selection from the above generations is that of parent–offspring relationship between F_2 individuals and $F_{2:3}$ progeny means (Hallauer and Miranda, 1981, p. 110). Using least squares regression of F_3 offspring means onto individual F_2 parental values, the slope (b) is equal to $(\Phi_A^2 + 1/2\Phi_D^2)/\Phi_P^2$ and can be interpreted as the change in breeding value per change in phenotypic value (Bernardo, 2002, p. 110). As were all the estimates of heritability previously discussed, this parent–offspring estimate is referenced to an F_2 population assumed to be in Hardy–Weinberg equilibrium and considered to be non-inbred (Bernardo, 2002, p. 34). This estimate of heritability may be biased upward by the presence of some potential amount of Φ_D^2 and any epistatic variance component involving Φ_D^2 (Φ_{AD}^2, etc.).

An advantage of using parent–offspring regression to estimate heritability is that it is straightforward (Hanson, 1963, p. 129). However, F_2 individuals and derived F_3 progenies are grown in separate environments (years), with no estimate of environmental variance and no estimate of genotype by environment interaction. Depending on the variability of the trait measured, biases can be significant (Hanson, 1963, p. 129).

In summary, the development and utilization of multiple generations (P1, P2, F_1, F_2, F_3, BC1P1, BC1P2, BC1P1S1, and BC1P2S1, etc.) can provide flexibility and opportunities for multiple estimates of inheritance factors, both qualitative and quantitative. Such populations are also complimentary for developing breeding populations.

In conjunction with the above early generation studies, F_2 plants were grown for the purpose of producing recombinant inbred lines (RILs) by single seed descent. The use of RILs in genetic studies requires that the finished inbred lines be an unselected sample of the F_2 population; ideally, an inbred line for each F_2 plant grown (Hallauer and Miranda, 1981, p. 89). If inbreeding depression is a problem, resulting in the death of some plants before becoming inbred, a random non-selected RIL population may be difficult to obtain. However, this is usually not the case with self-pollinated crops, such as soybean. The net result of using RILs to estimate inheritance, as opposed to an F_2 population, is that the Φ_A^2 among RILs is twice as large as the Φ_A^2 among F_3 families or among F_2 individuals. Because each RIL is theoretically completely inbred, Φ_D^2 is zero (there are no intralocus genotypic interactions), which also means that there are no unusable Φ_{AD}^2 interactions. Hence, selection among RILs is expected to be more effective than selection among $F_{2:3}$ families (Bernardo, 2002, p. 180).

However, Hallauer and Miranda (1981, p. 91) noted that the use of RILs has two serious handicaps. First, as already noted, RILs require the development of a set of unselected inbred lines that are representative of genotypes of a reference population (the F_2); this can be difficult if, during the selfing process, inbreeding produces weak plants that die and if high disease pressure kills plants. Second, the time required to develop RILs is much greater than that for developing and testing F_3 progeny rows. Hence, it is wise to make early generation estimates of predicted gain from selection so as to determine if early generation selection can be profitable. Early generation selection can be highly desirable, if possible, enabling a greater allocation of resources to the most promising families (Bernardo, 2002, pp. 119, 180, and 181).

However, where heritabilities are low (due to environment, genotype x environment interactions, Φ_D^2, etc.) RILs can provide useful estimates of genetic parameters and be complimentary to the breeding program. Replicated field trials across multiple locations and years among RILs will likely result in better estimates of heritability, better estimates of genetic gain from selection, and in the potential development of improved lines.

The development and utilization of RILs can facilitate the construction of a genetic map based on molecular marker linkages in the RIL population. If sufficient markers are located at enough strategic points, markers can be detected which are linked to genetic factors controlling the expression of quantitative traits. This was an important consideration in the development of RILs segregating for tolerance to charcoal rot and will allow for the development and release of charcoal rot-tolerant varieties, along with the release of molecular markers tightly linked to genetic factors affecting tolerance. Breeders can then create populations and test them in their own unique environments, while selecting for charcoal rot tolerance using molecular markers.

The most cost effective way to reduce losses from plant diseases and stresses is through the use of cultivars with tolerance to the appropriate stresses. It is anticipated that this will be the case with charcoal rot of soybean. However, at this time, insufficient information is available to determine how tolerance to charcoal rot is inherited: as a single gene, multiple genes with high heritability, or multiple genes with low heritability. But, because of proper planning and execution, sufficient quantities of the appropriate segregating populations have been developed to effectively determine the inheritance of tolerance to charcoal rot and to maximize the available genetic potential for developing improved soybean varieties with tolerance to charcoal rot.

References

Bernardo, R. (2002). *Breeding for quantitative traits in plants*. Stemma Press, Woodbury, MN.

Campbell, K.W., and White, D.G. (1995). Inheritance of resistance to Aspergillus ear rot and aflatoxin in corn genotypes. *Phytopath*, **85**:886–896.

Campbell, K.W., Hamblin, A.M., and White, D.G. (1997). Inheritance of resistance to aflatoxin production in the cross between corn inbreds B73 and LB31. *Phytopath*, **87**:1144–1147.

Hallauer, A.R., and Miranda, J.B. (1981). *Quantitative genetics in maize breeding*. Iowa State University Press, Ames, IA.

Hamblin, A.M., and White, D.G. (2000). Inheritance of resistance to Aspergillus ear rot and aflatoxin production in corn from Tex6. *Phytopath*, **90**:292–296.

Hanson, W.D. (1963). Heritability, in *Statistical genetics and plant breeding* (eds W.D. Hanson and H.F. Robinson). National Academy of Sciences–National Research Council. Washington, DC, pp. 125–140.

Mansur, L.M., Carriquiry, A.L., Rao-Arelli, A.P. (1993). Generation mean analysis of resistance to race 3 of soybean cyst nematode. *Crop Science*, **33**:1249–1253.

Mather, K., and Jinks, J. (1971). *Biometrical genetics, the study of continuous variation*, 2nd edn. Cornell University Press, Ithaca, NY.

Moll, R.H., Thompson, D.L., and Harvey, P.H. (1963). A quantitative genetic study of the inheritance of resistance to brown spot (*Physoderma maydis*) of corn. *Crop Science*, **3**:389–391.

Reinert, R.A., and Eason, G. (2000). Genetic control of O_3 sensitivity in a cross between two cultivars of snap bean. *Journal of the American Society for Horticultural Science*, **125**(2): 222–227.

Singh, S.P., and Westermann, D.T. (2002). A single dominant gene controlling resistance to soil zinc deficiency in common bean. *Crop Science*, **42**:1071–1074.

Thompson, J.A., Bernard, R.L., and Nelson, R.L. (1997). A third allele at the soybean dt1 locus. *Crop Science*, **37**:757–762.

Walker, R.D., and White, D.G. (2001). Inheritance of resistance to Aspergillus ear rot and aflatoxin production in corn from CI2. *Plant Disease*, **85**:322–327.

Warner, J.N. (1952). A method for estimating heritability. *Agronomy Journal*, **44**:427–430.

Velez, J.J., Bassett, M.J., Beaver, J.S., and Molina, A. (1998). Inheritance of resistance to bean golden mosaic virus in common bean. *Journal of the American Society for Horticultural Science*, **123**(4): 628–631.

35.11 Establishing a breeding nursery

Field nursery

To facilitate crossing in the field, the two parents to be crossed are planted in opposite rows with wide spacing (65–100 cm) between them for the operator to move freely. Having the two parents to be crossed planted side by side facilitates the crossing process and reduces the chance of crossing wrong parents.

Greenhouse nursery

Soybeans can be readily hybridized in the greenhouse. Potted plants can be moved around and positioned at levels that are convenient for the crossing operation.

35.12 Artificial hybridization

Equipment and materials

Equipment and materials used in soybean hybridization include fine tip forceps, tags, bottle with alcohol, petri dish, pencil, head-band mounted magnifiers.

Emasculation

Soybean is emasculated in the bud stage. The suitable buds are those ready to open the following day. At this stage the corolla is visible at the tip of the bud. To emasculate, the flower is grasped between the thumb and index finger, keeping the hand steady to avoid snapping the pedicel. The calyx is removed with a pair of forceps by grasping the sepal and gently pulling up with the forceps while at the same time gently wiggling them. It is important to remove all unemasculated flowers from the leaf axil. The leaf axil is tagged. If more than two cultivars are involved in the crossing program, the forceps should be sterilized between emasculations by dipping the tip in a bottle of alcohol.

Artificial pollinations

Pollination immediately follows emasculation. Recently opened flowers have fresh pollen. Flowers from the pollen source are picked. The stamen column and the style are together removed with forceps and used as a brush to dust pollen unto the stigma of the female flower. When a flower yields copious amounts of pollen, more than one stigma may be pollinated with one male stamenal column.

Under humid conditions, as often occurs in the southern United States, anther dehiscence occurs in the morning when the female parts are less mature and more difficult to emasculate and also the flowers may be damp from the morning dew. Sometimes, male flowers are collected in petri dishes and stored in desiccators until the afternoon. Pollen remains viable for about two days when stored in a cool dry place.

Pollinated flowers are tagged. The success rate of pollination depends on the skill of the operator and could be zero to about 40%. Failure of a cross may be due to factors including immature female, injury to the stigma, injury to the pedicel, high temperature, and inadequate pollen.

35.13 Natural hybridization

Natural pollination is facilitated by the use of genetic male sterility. Breeders using a recurrent selection method may benefit from this pollination method.

35.14 Seed development

Signs of successful cross are visible within seven days of pollination. Sometimes, new pods may develop in the axis where the artificially pollinated flowers occur. These should be removed. A successful cross can be identified by the presence of the calyx scar resulting from the emasculation process.

35.15 Harvesting

Mature soybean should be harvested on time to avoid field loses. The crop is ready to harvest when the seeds are at the hard dough stage. The moisture content of the seed should be 12–14%. Drier seed (less than 12%) increases the incidence of seed coat cracking, splitting of the seed, and shattering of the seed. The crop may be harvested at high seed moisture content (17–18%) provided post-harvest drying is available.

35.16 Common breeding objectives

35.16.1 Grain yield

Soybean grain yield is a major breeding objective. Progress has been made, but not at the rate achieved in cereal crops such as corn. The major yield components are numbers of nodes per plant, number of pods per node, number of seeds per pod, and seed size. Adapting soybean to new production environments is followed by improvement in yield potential for cultivars in those new regions.

35.16.2 Agromorphological traits

Lodging resistance

Lodging resistance is critical to mechanized production. An attempt to decrease plant height to improve lodging resistance by using the dt_1 gene for determinate growth, produced genotypes that were shorter but lower in plant yield.

Shattering resistance

Shattering resistance is important for production to allow the crops to attain full maturity before harvesting. Sometimes, inclement weather may force harvesting to be delayed after the crop is mature. The mechanical harvesting process involves physical contact with moving parts of the combine, which may promote shattering.

35.16.3 Adaptation

Maturity

Thirteen maturity groups have been identified for soybean. Late maturity in soybean has been found to be conditioned by three dominant genes, E_1, E_2, E_3. A gene, E_4, has been found to condition sensitivity to a long photoperiod.

Drought and other environmental stresses

Drought resistance is important in areas where production is rain-fed. Nutrient stress and heat stress may be important in areas where the production environment is not ideal.

Herbicide resistance

Modern technology of genetic engineering has enabled herbicide tolerant cultivars to be developed for commercial production (e.g., Roundup Ready®).

35.16.4 Disease resistance

Soybean is plagued by numerous diseases, the major ones include those shown here.

Bacterial blight

Caused by *Pseudomonas syringae*, this pathogen occurs worldwide. Resistance to the disease has been incorporated into various breeding programs.

Rots

Phytophthora root rot is the most important of the rots affecting soybean. Caused by *Phytophthora megasperma*, this problem affects all stages of development. Resistance is conditioned by six dominant genes – Rps_1–Rps_6.

Viral disease

The main viral disease of soybean is the soybean mosaic virus (SMV). It is spread by aphids. Genes for resistance, Rsv_1 and Rsv_2, have been discovered for SMV.

Nematodes

The major species of nematodes of importance to soybean producers are the cyst nematode (*Heterodera glycines*) and root knot nematode (*Meloidogyne* spp.). Three recessive genes, rhg_1, rhg_2, and rhg_3, and one dominant gene, Rhg_4, have been identified and incorporated into breeding programs to develop resistant cultivars.

35.16.5 Insect resistance

Major insect pests of soybean include the southern green stink bug and beanflies that are common in Asia and Africa.

35.16.6 Seed compositional traits and quality

Oil quality

Soybean is the world's leading source of vegetable oil, accounting for more than 75% of the market share.

Breeding objectives include increase in oil content as well as improvement in oleic acid and reduced linolenic acid for high oil quality.

Seed protein

Soybean is also the leading source of protein meal. Because seed oil and seed protein are negatively correlated, breeding has tended to focus on developing cultivars with high protein and low oil content.

References and Suggested Reading

Chen, Y., and Nelson, R.L. (2004). Genetic variation and relationship among cultivated, wild, and semiwild soybean. *Crop Science*, **44**:316–325.

Carlson, J.B., and Lersten, N.R. (1987). Reproductive morphology, in *Soybeans: Improvement, Production, and Uses*, 2nd edn (ed. J.R. Wilcox). American Society of Agronomy. Madison, Wisconsin, pp. 95–134.

Kisha, T., Sneller, C.H., and Diers, B.W. (1997). Relationship between genetic distance among parents and genetic variation in populations of soybean. *Crop Science*, **33**:1313–1325.

Specht, J.E., Hume, D.J., and Kumudini, S.V. (1999). Soybean yield potential – A genetic and physiological perspectives. *Crop Science*, **39**:1560–1570.

Internet resource

http://www.notrans.iastate.edu/research.html#breeding – Breeding low linolenic acid (accessed March 23, 2012).

Outcomes assessment

Part A

Please answer the following questions true or false.

1 Soybean is a subtropical plant.
2 Soybean maturity group VI is produced in the northern parts of the United States.
3 Soybean is long-day plant.

Part B

Please answer the following questions.

1 Give the scientific name of soybean.
2 Cultivated soybean is tetraploid.
3 Describe the reproductive biology of soybean.

Part C

Please write a brief essay on each of the following topics.

1 Discuss emasculation for crossing in soybean.
2 Discuss the breeding of agromorphological traits in soybean.
3 Discuss the genetic resources available for soybean breeding.
4 Discuss the history of soybean breeding in the United States.

36

Breeding peanut

Taxonomy

Family	Fabaceae
Genus	*Arachis* L.
Species	*Arachis hypogea* L.

36.1 Economic importance

Peanuts are an important legume crop in the warm climates of the world. The United States produces about 10% of the world's peanut crop on about 3% of the total world peanut acreage. This disproportionate share of the world's production is attributable to the high average yield per acre in the United States (2800–3000 lb/acre) compared to the world's average yield (800–1000 lb/acre). Nine states in the United States account for 99% of the US peanut crop. These states are Georgia, Texas, Alabama, North Carolina, Florida, Oklahoma, Virginia, South Carolina, and New Mexico. Georgia alone produces 39% of the US total production. In 2001, the US harvested acreage was 1 411 900 acres, a total production of 4 276 704 000 lb and an average yield of 3029 lb/acre. There are three main peanuts regions in the United States – Georgia–Florida–Alabama (southeast), Texas–Oklahoma–New Mexico (Southwest region), and the Virginia–South Carolina–North Carolina region, the southeast region accounting for about 55% of all US production.

On the world scene, peanuts are produced in Asia, Africa, Australia, and the Americas. India and China together account for more than 50% of the world's total production. Other substantial peanut producing nations include Senegal, Sudan, Brazil, Argentina, South Africa, Malawi, and Nigeria.

36.2 Origin and history

The peanut is native to the Western Hemisphere. It probably originated in South America, the center of origin most likely being Brazil, where about 15 wild species are found. The Spanish explorers are credited with its spread throughout the New World. They introduced it to Europe, from where traders spread it to Asia and Africa. Peanuts reached North America via the slave trade. Commercial production of peanuts in the United States began in about 1876. The demand for the crop increased after the Civil War, transforming it from a regional (southern) food to a national food. Production came to the Cotton Belt after 1900. The expansion of the peanut industry was driven by advances in technology that resulted in the development of equipment and machinery for planting, harvesting, and processing the crop.

36.3 Market types

There are four basic market types – Runner, Virginia, Spanish, and Valencia.

Principles of Plant Genetics and Breeding, Second Edition. George Acquaah.
© 2012 John Wiley & Sons, Ltd. Published 2012 by John Wiley & Sons, Ltd.

36.3.1 Runner

Runners have become the predominant peanut type in the United States following the introduction of the cultivar, the Florunner, which was responsible for the dramatic yield increase of the crop in the United States. They have uniform size and are grown mainly in Georgia, Alabama, Florida, Texas, and Oklahoma. About 54% of the crop is used for making peanut butter. Runners mature between 130 and 150 days, depending on the cultivar. The seeds are medium sized (900–1000 seeds/lb).

36.3.2 Virginia

Virginia cultivar have dark-green foliage and large pods. They have the largest seeds of all the types (about 500 seeds/lb) and have a russet testa. The pods usually have two seeds (occasionally 3–4). They are grown mainly in Virginia and North Carolina. They mature in about 135–140 days and may have runner or bunch types. Large seeds are sold as snack peanuts.

36.3.3 Spanish

The Spanish group of peanuts comprises bunch types with erect, light-green foliage. The pods rarely contain more than two seeds that are short with tan testa. The seeds are small sized (1000–1400 seeds/lb). They have a higher oil content compared with other types. They are grown mainly in Oklahoma and Texas and are used mainly for making peanut candies, and also snack nuts and peanut butter. They mature earlier than the runner types (about 140 days).

36.3.4 Valencia

The Valencia types typically bear many pods with 3–4 seeds and bright-red testa color. They are erect and sparsely branching with dark green foliage. They are very sweet peanuts and are usually roasted and sold as in-the-shell or boiled peanut. Valencias are grown mainly in New Mexico.

36.4 Genetic resources

Over 10 000 peanut accessions are held in germplasm banks in various countries. In the United States, the Southern Regional Plant Introduction Station at Experiment, Georgia, maintains about 4000 accessions. The International Crops Research Institute for Semi-Arid Tropics (ICRISAT) also maintains thousands of accessions.

36.5 Cytogenetics

The cultivated peanut belongs to section *Arachis*, which consists of three series – annuals, perennial, and amphidiploids. The series amphidiploide has two tetraploids – *A. hypogeae*, and *A. monticola*. Some intersectional crosses yield fertile hybrids. Others produce meiotic problems leading to embyo abortions. The genus, *Arachis*, is classified into seven sections. The section *Arachis* includes four annual diploids ($2n = 20$), several (at least five) perennial diploids, and two annual tetraploids ($2n = 40$). The cultivated peanut is a tetraploid with two subspecies and four interfertile varieties:

Subspecies

(i)	*A. hypogea*	- variety *hypogea* (US market types Virginia, Runner)
		- variety *hirsuta*
(ii)	*A. fastigiata*	- variety *fastigiata* (US market type Valencia)
		- variety *vulgaris* (US market type Spanish)

36.6 General botany

The peanut (*Arachis hypogaea*), also called the groundnut and earthnut, among other names, is technically a pea not a nut. It is an unusual plant in the sense that it flowers aboveground, but fruits belowground. The cultivated plant is an annual with a central upright stem that may stand up to about 18 inches tall. It bears numerous branches that vary from prostrate to nearly erect. It has pinnately compound leaves. The cultivars in cultivation may be grouped into two, based on the arrangement of the nuts at the base of the stem. In **bunch types**, the nuts are closely clustered about the base, while in **runner types** the nuts are scattered along prostrate branches that radiate from the base of the plant to the top. Peanuts have a strong taproot system. Most roots are

Industry highlights

Breeding peanut (Arachis hypogaea L.) and root-knot nematode resistance

Charles Simpson

Texas A & M University, College Station, TX 77843, USA

The program

The Texas peanut breeding program originated in 1939 with the principle objective being to improve yield stability of the crop. Later, objectives included development of disease (pest) resistance and, more recently, objectives to improve the edible quality of the product were implemented. Pedigree selection with and without single seed descent has been used extensively in the program but the mainstay of the disease resistance transfer has been accomplished via the backcross technique. Diseases and nematodes have been recognized as major constraints to production since the very early stages of the program. After discovery of moderate resistance in some plant introduction material in the late 1950s, a project was initiated to develop leaf spot resistance. To enhance the available gene pool, a concentrated effort began in 1970 to introduce leaf spot resistance genes from the wild species. In 1976 an international effort was initiated to collect, preserve, evaluate, and utilize wild *Arachis* from the area of origin of the genus in South America. Through this effort, more than 1500 new wild *Arachis* accessions were added to the germplasm collection, including more than 60 new species and additional representatives of the 22 previously described species. We also introduced more than 3900 land races of *A. hypogaea*.

Introgression program

Leaf spot

Our introgression program began in earnest in 1972. Abdou *et al.* (1974) had identified near immunity to two leaf spot pathogens [*Cercosporidium personatum* (Beck and Curtis) Deighton and *Cercospora arachidicola* Hori] in *A. cardenasii* Krapov. and W.C. Gregory and *A. chacoensis* nom. nud. (now classified as *A. diogoi* (Krapovickas and Gregory, 1994)), respectively. Most of the wild *Arachis* are $2n = 20$ but the cultigen is $2n = 4x = 40$. We had used both *A. cardenasii* and *A. diogoi* species individually in attempts to cross with *A. hypogaea* and form fertile hexaploids from interspecific triploid progenies, followed by crossing of the 6x progeny back to 4x individuals, then further backcrossing of progenies to 4x to accomplish chromosome segregation and to eventually develop tetraploid progenies that were cross-compatible with *A. hypogaea* and possessed the leaf spot resistance(s) (Simpson, 1990). With both species, widespread sterility occurred after one or two generations of backcrossing to the *A. hypogaea* recurrent parent; in several lines sterility prevented the second backcross. In 1973 we attempted to combine the two resistances into one diploid hybrid and use that hybrid to attempt the hexaploid route. A successful hybridization of *A. diogoi* × *A. cardenasii* was achieved only once in 3500 pollinations, resulting in a partially fertile hybrid with 50% pollen stain. However, that was the end of that pathway's success. Major sterility problems ensued as we doubled chromosome numbers of the triploid and attempted crosses/backcrosses to *A. hypogaea* (Simpson, 1990).

The next attempt consisted of doubling the chromosome number of the *A. diogoi* × *A. cardenasii* hybrid, then crossing with *A. hypogaea*. Even though this was a tetraploid × tetraploid, the process met with equal, if not greater sterility problems than previous attempts.

In this time-frame, Smartt *et al.* (1978a, 1978b) published their theory that the cultivated peanut was comprised of two genomes, made up of an A genome represented by *A. cardenasii* and *A. diogoi*, and a B genome represented by *A. batizocoi*. Based on this hypothesis, we crossed the *A. diogoi* × A. *cardenasii* hybrid on to *A. batizocoi* and obtained a sterile (pollen stain = 0.01%) hybrid. We doubled the chromosomes of this plant to yield a highly fertile complex amphiploid (tetraploid) hybrid with normal meiosis and a pollen stain >90%. From this point we initiated our crossing/backcrossing effort to establish fertile progenies with leaf spot resistance (the pathway is shown in Figure B36.1). Progress was slow primarily because we had not established a reliable laboratory technique for screening large numbers of plants; therefore, each cycle required more than 18 months when selections were made based on leaf spot resistance. However, we carried two sections of the program, with the hope that selection based on fertility and agronomic traits other than resistance would yield fertile plants with leaf spot resistance at an earlier date. Without selection for resistance, the program moved rapidly, with a cycle being made in 10 months or less. The initial complex amphiploid from this program was later released as the germplasm line TxAG-6 (Simpson *et al.*, 1993).

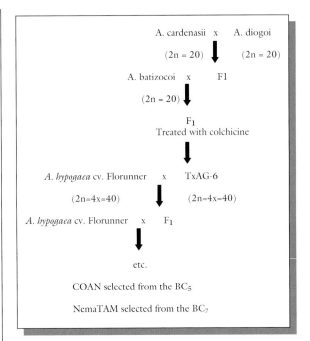

A. cardenasii x A. diogoi

(2n = 20) ↓ (2n = 20)

A. batizocoi x F1

(2n = 20) ↓

F$_1$
Treated with colchicine

↓

A. hypogaea cv. Florunner x TxAG-6

(2n=4x=40) ↓ (2n=4x=40)

A. hypogaea cv. Florunner x F$_1$

↓

etc.

COAN selected from the BC$_5$

NemaTAM selected from the BC$_7$

Figure B36.1 Tetraploid pathway for introgression of root-knot nematode resistance into *Arachis hypogaea*. Figure courtesy of Charles Simpson.

Nematodes

Root-knot nematodes (*Meloidogyne arenaria* (Neal) Chitwood) became a serious pest of peanut in the major growing area of Central Texas (as well as many other places in the USA and the world) by the early 1970s. However, no useful level of resistance had been identified in the cultigen, so we began testing for resistance in the wild species. Banks (1969) had demonstrated some resistance to *M. hapla* in the *Rhizomatosae*, but since the rhizomatous taxa would not cross with the cultivated peanut or its sectional relatives (Gregory and Gregory, 1979) these genes were unavailable. In 1989 numerous accessions of wild peanut were identified with resistance (near immunity) to the root-knot nematode (Nelson *et al.*, 1989). Fortunately one of those accessions was *A. cardenasii*, the primary A genome parent we had been working with in the leaf spot resistance program. Our next effort was to test the advanced backcross generation materials from the leaf spot program in hopes that the gene(s) for nematode resistance might have been retained. Unfortunately, the only materials that proved to be resistant to the root-knot nematode were TxAG-6 and the F$_1$ from the first backcross to *A. hypogaea*. Even so, the availability of a high level of resistance to root-knot in genotypes that were readily cross-compatible with the cultigen was a tremendous benefit to the effort to develop a nematode resistant cultivar (Simpson, 1990).

We then initiated an intense backcross breeding program with a selection from the BC$_1$F$_1$ materials, subsequently released as the germplasm line TxAG-7 (Simpson *et al.*, 1993), to transfer the gene(s) to the most desirable cultivated peanut available, which at that time was Florunner. Florunner, a revolutionary peanut because of its large increase in yield, was released from University of Florida in 1969 (Norden *et al.*, 1969) and by 1986 had become the most successful peanut variety ever developed, worldwide.

Backcross program

Our backcross effort (Simpson, 1990) consisted of: making crosses in the fall of the year at the Texas A&M University Research and Extension Center in Stephenville; harvesting and then planting the F$_1$ seeds as soon as they were mature enough to germinate (usually mid-December); harvesting seeds (F$_2$ or BC$_n$F$_1$) from these plants in late April; and sending the seeds to the Plant Nematology laboratory in College Station for resistance screening. Seeds were planted in 10 cm pots and the seedlings inoculated with 10 000 root-knot nematode eggs per pot. Plants were harvested eight weeks after inoculation and highly resistant plants were identified based on low levels of nematode reproduction (Starr *et al.*, 1990, 1995). Stem cuttings from resistant individuals were then sent back to Stephenville to be used as male parents in the next backcross generation that were initiated in September and October. In each cycle, the primary selection criterion was the nematode resistance, especially in the earlier generations. After the BC$_3$ generation we also began making selections based on the plant characters of pod shape and seed per pod. We continued this process through seven backcrosses. The pathway is shown in Figure B36.1 (Simpson, 1990).

At about the third backcross generation we determined that at least two resistance genes were present in the early generation materials but at this stage it was not certain that more than one dominant gene controlled the trait (Starr *et al.*, 1990) in our more advanced generation lines. At this time we also initiated a collaboration with A.H. Paterson and M.D. Burow to develop molecular markers linked to the resistance loci (Burow *et al.*, 1996).

Releases

After the fifth backcross, we also initiated seed increases of lines that appeared to have most of the plant, pod, and seed characteristics of Florunner and nematode resistance. Field tests for yield potential began in 1995 and from these tests we identified five lines with reasonable yield and excellent resistance. Following a second year of testing we made a winter increase of numerous sister seed rows in Puerto Rico from one line; from this material we released the world's first root-knot resistant peanut cultivar, COAN, in 1999 (Simpson and Starr, 2001). This release was rushed through the process and had one major flaw; the plant size was too small, limiting the seed production of the plants. Even though the yields under heavy nematode pressure had yields that were 150–200% better than susceptible cultivars, the total yields were still too low to be profitable for grower acceptance.

During this time the molecular marker work was progressing and preliminary tests with several restriction fragment length polymorphism (RFLP) loci indicated that we would be able to use this technology for more efficient screening for resistance. We would even be able to identify plants homozygous for the one major dominant gene that appeared to control resistance (Burow *et al.*, 1996).

After the BC_7 was made, selection and seed increase resulted in lines that performed much better than COAN in yield tests and still contained the resistance gene. After one year of testing we decided to do a Puerto Rico winter increase of three of the best lines in the hope of being able to select a line for release as a cultivar with greater yield potential.

In the second testing year we space planted 300 individual seeds from the winter nursery of the three lines in the field at Stephenville. We collected tissue for DNA analyses from the 900 plants and at harvest we had molecular data to identify the homozygous resistant plants which had the dominant (RR) gene (Church *et al.*, 2000). These data were utilized when we were evaluating the individual plants for plant, pod, and seed characters and selecting individual plants for a breeder seed increase. We selected numerous desirable plants that were homozygous resistant from each line. The seeds from these were planted as plant rows in a Puerto Rico winter nursery to gain another generation.

After a third year of yield testing and extensive other evaluations, the cultivar NemaTAM was released in 2001 (Simpson *et al.*, 2003). NemaTAM has about 30% greater yield potential than COAN and the same high level of root-knot resistance. Two major benefits of these resistant cultivars are that (i) the resistance will eliminate the need for use of nematicides, even at very high nematode population densities, and (ii) the inhibition of nematode reproduction due to the resistance results in lower nematode population densities such that a susceptible crop plant in rotation with the resistant cultivar will be subjected to less nematode disease pressure (Starr *et al.*, 2002).

The resistance in COAN and NemaTAM to *M. arenaria* is controlled by a single dominant gene (Burow *et al.*, 1996; Choi *et al.*, 1999; Church *et al.*, 2000), but evidence for additional genes in the same species used to form TxAG-6 indicates that we have the opportunity to pyramid genes for more stable resistance (Burow *et al.*, 1996; Garcia *et al.*, 1996; Choi *et al.*, 1999).

Further testing indicates that the resistance in TxAG-6 is conditioned by at least two genes, one dominant and one recessive (Church, 2002) We also discovered that COAN and NemaTAM are resistant to *M. javanica*, which is also parasitic on peanut and especially prevalent in India and northern Africa. This resistance to *M. javanica* has been confirmed in an independent study (Timper *et al.*, 2003). At present we can only assume the resistance is conditioned by the same gene; we have not tested this hypothesis.

The future

The program continues to try to identify, characterize, and locate flanking molecular markers for the second resistance gene so we can pyramid the genes. It would be desirable to move from the current RFLP marker assisted selection system to one based on polymerase chain reaction (PCR), which would increase the efficiency of the system. We are also continuing our efforts to identify genes and markers resistance to *M. hapla* as well as *M. javanica*. However, now our major efforts are to combine the nematode resistance gene(s) with other characters to develop cultivars with multiple traits, including high O/L (ratio of oleic free fatty acid to linoleic free fatty acid), tomato spotted wilt virus resistance, sclerotinia resistance (*Sclerotinia minor* Jagger) and leaf spot resistance.

References

Adbou, Y.A-M., Gregory, W.C., and Cooper, W.E. (1974). Sources and nature of resistance to *Cercospora arachidicola* Hori and *Cercosporidium personatum* (Beck and Curtis) Deighton in Arachis species. *Peanut Science*, **1**:6–11.

Banks, D.J. (1969). Breeding for northern root-knot nematode, *Meloidogyne hapla*, resistance in peanuts. *Journal of American Peanut Research and Education Association*, **1**:23–28.

Burow, M.D., Simpson, C.E., Paterson, A.H., and Starr, J.L. (1996). Identification of peanut (*Arachis hypogaea* L.) RAPD markers diagnostic of root-knot nematode (*Meloidogyne arenaria* (Neal) Chitwood) resistance. *Molecular Breeding*, **2**:369–379.

Choi, K., Burow, M.D., Church, G., *et al.* (1999). Genetics and mechanism of resistance to *Meloidogyne arenaria* in peanut germplasm. *Journal of Nematology*, **31**:283–290.

Church, G.T. (2002). Resistance to *Meloidogyne arenaria* in peanut: gene identification and molecular markers. PhD Dissertation, Texas A&M University, Department of Plant Pathology and Microbiology, College Station, TX.

Church, G.T., Simpson, C.E., Burow, M.D., Paterson, A.H., and Starr, J.L. (2000). Use of RFLP markers for identification of individuals homozygous for resistance to *Meloidogyne arenaria* in peanut. *Nematology*, **2**:575–580.

Garcia, G.M., Stalker, H.T., Shroeder, E., and Kochert, G. (1996). Identification of RAPD, SCAR, and RFLP markers tightly linked to nematode resistance genes introgressed from *Arachis cardenasii* into *Arachis hypogaea*. *Genome*, **39**: 836–845.

Gregory, M.P. and Gregory, W.C. (1979). Exotic germ plasm of *Arachis* L. interspecific hybrids. *Journal of Heredity*, **70**:185–193.

Krapovickas, A. and Gregory, W.C. (1994). Taxonomia del genero *Arachis* (*Leguminosae*). *Bonplandia*, **8**(1–4):1–186.

Nelson, S.C., Simpson, C.E., and Starr, J.L. (1989). Resistance to *Meloidogyne arenaria* in *Arachis* spp. germplasm. *Journal of Applied Nematology* (Suppl. Annals of Applied Nematology), **21**:654–660.

Norden, A.J., Lipscomb, R.W., and Carver, W.A. (1969). Registration of Florunner peanuts. *Crop Science*, **9**:850.

Simpson, C.E. (1990). Pathways for introgression of pest resistance into *Arachis hypogaea* L. *Peanut Science*, **18**:22–26.

Simpson, C.E. and Starr, J.L. (2001). Registration of 'COAN' Peanut. *Crop Science*, **41**:918.

Simpson, C.E., Starr, J.L., Nelson, S.C., Woodard, K.E., and Smith, O.D. (1993). Registration of TxAG-6 and TxAG-7 Peanut Germplasm. *Crop Science*, **33**:1418.

Simpson, C.E., Starr, J.L., Church, G.T., Burow, M.D., and Paterson, A.H. (2003). Registration of 'NemaTAM' peanut. *Crop Science*, **43**:1561.

Smartt, J., Gregory, W.C., and Gregory, M.P. (1978a). The Genomes of *Arachis hypogaea* I. Cytogenetic studies of putative genome donors. *Euphytica*, **27**:665–675.

Smartt, J., Gregory, W.C., and Gregory, M.P. (1978b). The Genomes of *Arachis hypogaea* II. The implications in interspecific breeding. *Euphytica*, **27**:677–680.

Starr, J.L., Schuster, G.L., and Simpson, C.E. (1990). Characterization of the resistance to *Meloidogyne arenaria* in an interspecific Arachis spp. hybrid. *Peanut Science*, **17**:106–108.

Starr, J.L., Simpson, C.E., and Lee, T.A. Jr. (1995). Resistance to *Meloidogyne arenaria* in advanced generation breeding lines of peanut. *Peanut Science*, **22**:59–61.

Starr, J.L., Morgan, E. and Simpson, C.E. (2002). Management of the peanut root-knot nematode, *Meloidogyne arenaria*, with host resistance. [Online] *Plant Health Progress*. doi: 10.1094/PHP-2002-1121-01-HM.

Timper, P., Holbrook, C.C., and Anderson, W.F. (2003). Reproduction of *Meloidogyne* spp. on resistant peanut genotypes from three breeding programs. *Journal of Nematology*, **35**:417–421.

nodulated for biological nitrogen fixation. The branching pattern in peanut may be alternate or sequential. Virginia types have alternate branching, a trait that is generally believed to be dominant over the sequential branching of Spanish and Valencia peanuts. Aerial podding is conditioned by dominance or partial dominance.

The flowers arise in the leaf axils above the ground. They are self-pollinated. Upon fertilization, the ovary begins to enlarge while the section behind it, called the **peg** or **gynophore,** elongates to push the ovary into the soil for fruit development. The fruit is an indehiscent pod that may contain 1–6 (but usually 1–3) seeds. The pods form predominantly underground. However, mutants with aerial podding have been developed. Consequently, it is critical that the pegs reach the soil. The seed has a thin papery testa that varies in color – brick red, russet, light tan, purple, white, black or multicolored. Some cultivars

exhibit seed dormancy. Well-developed nuts have a shelling percentage of 70–80%.

36.7 Reproductive biology

36.7.1 Floral morphology

The peanut inflorescence occurs in clusters of three or more flowers in the axil of the cataphyll of foliage leaves. Inflorescence may occur on either the main stem or lateral branches. Spanish and Valencia cultivars bear their inflorescence on the main stem, whereas the inflorescence of Virginia cultivars occurs on lateral branches. A flower is subtended by a bract and occurs on a minute branch of the inflorescence, which arises in the axil of a second bract. The calyx and corolla are borne at the top of the hypanthium, which surrounds the staminal column. The calyx has five lobes, whereas the staminal column is usually composed of ten filaments, eight of which are normally anther bearing. About 50–75% of the bottom parts of the filaments are fused.

36.7.2 Pollination

Normally, only one of the flowers in an inflorescence matures to anthesis. Anthesis generally occurs before flower opening. Bud opening occurs at the beginning of the light period. The stigma remains receptive for about 12–24 hours after flower opening. It takes only about 5–6 hours for an opened flower to stay open.

36.8 Common breeding methods

The common approaches to breeding peanuts include the use of plant introductions. Introductions of peanut germplasm can be a beginning point for crop improvement. Selections from introductions may provide a parental material for breeding. Selection may also be used to isolate pure lines from hybrid populations.

A widely used method of breeding peanut is hybridization of superior parents to create opportunities for transgressive segregation to occur. Pedigree selection may be used to advance generations. However, the use of single seed descent is more rapid when used in conjunction with winter nurseries. Backcross breeding may be used to incorporate specific genes of interest.

36.9 Establishing a breeding nursery

Peanut crossing is often done in the greenhouse using potted plants. However, wild species are more successfully hybridized in the field than in the greenhouse. The success of hybridization, whether under field or greenhouse conditions, depends on proper humidity. Drought causes low success. Breeders may emasculate the flowers in the evening and pollinate the next morning.

36.10 Artificial pollination

36.10.1 Materials and equipment

Equipment use includes forceps, sharp knife, scalpel, razor blade, magnifier (2–3x) attached to a head band, camel hair brush, bottle of alcohol.

36.10.2 Preparation of female plant

Flowers near the main stem are preferred for emasculation. Further, one flower in each inflorescence is selected for emasculation. The flower is emasculated in the bud stage. To emasculate, the bud is grasped between the thumb and index finger of the hand. Next, the petal in front of the keel and the sepal on the side of the standard are folded down. The standard petal is opened with the forceps and the wing petals pulled out and down. The standard is held back with the thumb and index finger while the operator pulls the keel free of the stigma and anthers. Alternatively, the keel and wing petals may be removed altogether. The anther and the stamens are removed to complete the emasculation. If an emasculated flower is not be immediately pollinated, the hypanthium is tagged (e.g., with a small thread).

36.10.3 Pollination

Pollen collected between 5 and 7 a.m. is the most viable. Flowers are artificially pollinated in the morning to early afternoon. The keel may be detached and used as a brush to directly deposit the pollen unto the stigma. Some operators transfer the pollen with a camel hair brush or the tip of the forceps. The environment may be humidified, but care should be taken not to dislodge the pollen. Pollen may be collected and stored under desiccated condition in a cool place (6 °C) for up to about a week without losing viability.

Fertilization occurs between 12 and 16 hours after pollination, after which the ovary elongates as the intercalary meristem at its base grows. The peg eventually penetrates the soil where it develops into a mature peanut fruit. An identification wire may be tied to the peg from the emasculated flower before it penetrates the soil.

36.10.4 Seed development and harvesting

Seed is mature for harvesting after about 55–65 days, depending on the environment and cultivar. Peanuts are ready to be harvested when 65–70% is mature. Early harvesting results in shriveled kernels. At proper maturity, the kernels display the distinct texture and color of the variety, the inside of the shell is beginning to color and show dark veins. Harvesting it is easier in sandy soils and occurs with less loss of pods with small-podded than large-podded varieties.

36.11 Common breeding objectives

Some of the major objectives in peanut breeding are:

- **Yield potential and stability.** Breeders are interested high crop yield *per se* and also stable yield with adaptation to various agroecological zones. High shelling percentage is important. Early maturity is desired in some production areas.
- **Disease resistance.** Important diseases of peanut include foliar ones such as leaf spot (caused by *Cercospora* spp). Stem and peg rots (caused by *Sclerotium rolfsii*) and charcoal rot (by *Rizoctonia* spp.) are economic diseases in some production area.
- **Insect resistance.** Important insect pests include leafhoppers, corn earworm, cutworms, and tobacco thrips.
- **Product quality.** Peanuts consist of 40–48% oil and 25–30% protein, making the crop a major source of vegetable oil and plant protein. Breeding high oil content is an important objective.

Key references and suggested reading

Boerma, H.R. and Hussey, R.S. (1992). Breeding plants for resistance to nematodes. *Journal of Nematology*, **24**: 242–252.

Hyman, B.C. (1990). Molecular diagnosis of *Meloidogyne* species. *Journal of Nematology*, **22**:24–30.

Isleib, G. and Kochert, G. (1994). Introgression from *A. cardenasii* germplasm lines. *Crop Science*, **33**:1418.

Outcomes assessment

Part A

Please answer the following questions true or false.

1 The cultivated peanut is a diploid.
2 Peanut is self-pollinated.
3 Virginia variety of peanuts is characterized by small pods.

Part B

Please answer the following questions.

1 Give the four basic market types of peanuts.
2 Give the scientific name of peanuts.
3 Distinguish between a runner and bunch type of peanuts.

Part C

Please write a brief essay on each of the following topics.

1 Discuss the reproductive biology of peanuts.
2 Discuss the common breeding methods for peanuts.
3 Discuss the emasculation of peanut flower.

37

Breeding potato

Taxonomy

Family	Solanaceae
Genus	*Solanum* L.
Species	*Solanum tuberosum* L

37.1 Economic importance

Potato is among the top five crops that feed the world, the others being wheat, corn, sorghum, and rice. In 1998, potatoes ranked as the fourth most important food crop in the United States. The total harvest in 1998 was 1.388 million acres and a total yield of 475.8 million cwt. About 35% of the production was processed into frozen products (primarily fries). Per capita consumption of potatoes in the United States in 1999 was 144.7 lb. The top five producing states in 1998 were Idaho, Washington, Oregon, Wisconsin, and North Dakota, with Idaho leading all production with about 450 million cwt, followed by Washington.

On the world scene, 293 million tons of potatoes were produced on 18 million hectares worldwide in 1998. With the break-up of the former Soviet Union, China becomes the world's leading producer of potatoes. Developing countries accounted for 36% of total production in 1998. In Asia, the key producers include China, India, Indonesia, and Nepal; in Africa, Egypt, South Africa, Algeria, and Morocco account for 80% of the region's total production. Latin American countries that produce a substantial amount of potatoes include Ecuador, Peru, Brazil, and Mexico. In Europe, the leading countries include the former Soviet Union, Poland, West Germany, and France.

37.2 Origin and history

Potato originates in the Andes mountains of Peru and Bolivia where the plant has been cultivated for over 2400 years. The Aymara Indians developed numerous varieties on the Titicaca Plateau, some 10 000 feet above sea level. The first written account of potato was made in 1553 by the Spanish Conquistador Pedro Cieza de Leon in his journal, *Chronicle of Peru*. The Spanish introduced the potato to Europe between 1565 and 1580. It was taken from England to Bermuda and later to Virginia, USA, in 1621. It was also introduced into Germany in the 1620s where it became a part of the Prussian diet by the time of the Seven Year War (1756–1763). Antoine Parmentier, a prisoner of war in Prussia, introduced the crop to France after the war. Potato was introduced into North America when Irish immigrants (hence the name "Irish potato") brought it to Londonderry, New Hampshire, where large-scale production occurred in 1719.

By the 1840s, the devastation by the fungus *Phytophthora infestans* (causes late blight) and heavy rains brought untold hunger and starvation to Ireland. This sparked a mass immigration of about two million Irish, mostly to North America. In America, Luther

Principles of Plant Genetics and Breeding, Second Edition. George Acquaah.
© 2012 John Wiley & Sons, Ltd. Published 2012 by John Wiley & Sons, Ltd.

Burbank was first to undertake improvement of the plant, subsequently releasing the "Burbank" potato to West Coast states in the late 1800s. A mutation of the Burbank potato was discovered in Colorado that was disease resistant. It had reticulated skin and became known as the Burbank Russet and is grown on most farms in Idaho.

37.3 Adaptation

Potato is a cool-season crop. The optimum temperature for shoot growth and development is 22 °C. In the early stages, soil temperature of about 24 °C is ideal. However, in later stages, a cooler temperature of about 18 °C is desired for good tuberization. Tuber formation and development is slowed when soil temperature rises above 20 °C and ceases at 29 °C. Above this temperature, the effect of respiration exceeds the rate of accumulation of assimilates from photosynthesis.

Potatoes are sensitive to hard frost that may occur in fall, winter, or early spring. Tubers will freeze at about −2 °C and loose quality upon thawing. Tuberization is best under short photoperiod, cool temperature, and low nitrogen, while vegetative shoot production is favored by long days, high temperature, low light intensity, and high amounts of nitrogen. However, tuberization can occur at 12 °C night air temperatures. Tuber yield decreases by 4% for each 0.6 °C above optimal temperature. The best potato production occurs in regions with daily growing season temperatures averaging between 15.5 °C and 18 °C. High soil temperatures result in knobby and malformed tubers. Potato flowers and sets seed best when long days and cool temperatures prevail. Consequently, potatoes set seed when grown in the northern states but not the southern states.

37.4 Genetic resources

Potato diversity is maintained in a number of germplasm banks in various parts of the world. The major repositories include the International Potato Center (CIP) in Peru (has the best collection of cultivated potatoes, especially, andigena, phureja, and "bitter potatoes"); the Commonwealth Potato Collection at Pentlandfield, Scotland, and the German–Dutch Potato Collection at Braunschweig, Germany. In the United States, the IR-1 project at Sturgeon Bay, Wisconsin, is important to US breeders.

37.5 Cytogenetics

The genus *Solanum* contains about 2000 species, of which only about 150 bear tubers. The cultivated potato, *Solanum tuberosum*, is a tetraploid ($2n = 4x = 48$). Five cytological groups of potato have been identified, with somatic numbers of 24, 36, 48, 60, and 72. About 70% of tuber-bearing potatoes are diploids, while 5% and 8% are tetraploids and hexaploids, respectively. Most of the diploids are self-incompatible, producing seed only when fertilized by pollen containing a different *S* allele.

A cultivated diploid that is used in South American production is the *S. phureja*. It is used in bridge crossing and other genetic studies. Triploid potato is sterile; a few are cultivated. Producing triploid potato by crossing 2n × 4n is seldom successful because of the so-called "triploid block". Hexaploids are self-fertile. A widely used hexaploid is *S. demissum*. It is the sources of the major *R* gene that confers resistance to late blight.

37.6 Genetics

Potato genetics is complex because of its autotetraploid origin. There can be four different alleles at a locus. Intralocus interactions (heterozygosity) and interlocus interaction (epistasis) occur and can be exploited by using the appropriate breeding procedure.

37.7 General botany

Potato (*Solanum tuberosum*) is an annual plant with short (1–2 feet), erect, and branched stems. Its compound leaves can be 1–2 feet long with a terminal leaflet. The flowers are borne in compound, terminal cymes with long peduncles. The flower color may be white, rose, lilac, or purple. The plant bears fruits (berries) called **potato balls**. The underground commercial part is a modified stem (or tuber) that is borne at the end of a stolon. The "eyes" on the tuber are actually rudimentary leaf scars favored by lateral branches. Each eye contains at least three buds

Industry highlights
The breeding of potato

John E. Bradshaw

James Hutton Institute, Invergowrie, Dundee DD2 5DA, UK

Evolution of the modern potato crop

Potatoes were domesticated between 7000 and 10 000 years ago in the Andes of South America (Bradshaw and Bonierbale, 2010), their wild species progenitors having been the subject of much discussion. Spooner *et al.* (2005) have provided molecular taxonomic evidence for a single domestication in the highlands of southern Peru from the northern group of members of the *S. brevicaule* complex of diploid species, such as *S. canasense, S.multidissectum* and *S. bukasovii*. The result was diploid *S. stenotomum*, also referred to as a form of *S. tuberosum* (Group Stenotomum), from which other cultivated species were derived; including diploid *S. phureja* (or Group Phureja), tetraploid *S. tuberosum* subsp. *andigena* (or Group Andigena) and tetraploid *S. tuberosum* subsp. *tuberosum* (or Group Tuberosum). Andigena potatoes became the most widely grown form in South America. Tuberosum potatoes were selected from Andigena types for tuber production in long days in coastal Chile and are referred to as Chilean Tuberosum. Phureja potatoes were selected from Stenotomum for lack of tuber dormancy, so that up to three crops per year could be grown in the lower, warmer, eastern valleys of the Andes.

Potatoes were introduced from South America into the Canary Islands around 1562 and from there to mainland Europe in the 1570s (Hawkes and Francisco-Ortega, 1993). It has often been assumed that these early introductions were Andigena types that evolved into Tuberosum types as the growing of potatoes spread north-eastward across Europe. However, it now seems safest to assume that the early introductions came from both the Andes and coastal Chile, as analysis of DNA from 49 herbarium specimens confirmed the presence in Europe of Andean potatoes from around 1700 and Chilean potatoes from 1811 (Ames and Spooner, 2008). Starting in the seventeenth century, potatoes were taken from Europe and cultivated in many other parts of the world. Today potatoes are grown on 18.5 million hectares of land in 149 countries from latitudes 65 °N to 50 °S and at altitudes from sea level to 4000 m (Hijmans, 2001). The potato is now the third most important food crop after wheat and rice,with annual production of over 300 million metric tons fresh weight of tubers (http://faostat.fao.org). As well as being a staple food, the potato is grown as a vegetable for table use, is processed into French fries and crisps (chips), and is used for dried products and starch production.

Potato breeding and the need for new cultivars

Potato breeding

The reproductive biology of potato is ideal for creating and maintaining variation. Potatoes flower and set true seed in berries following natural pollination by insects, particularly bumble bees. Outcrossing is enforced in cultivated (and most wild) diploid species by a gametophytic self-incompatibility system. Whilst this system does not operate in *S. tuberosum* because it is a tetraploid, 20% natural cross-pollination was estimated to occur in an artificially constructed Andigena population (Glendinning, 1976). This sexual reproduction creates an abundance of diversity by recombining the variants of genes which arose by mutation, and potatoes are highly heterozygous individuals which display inbreeding depression on selfing. The genetically unique seedlings which grow from true seeds produce tubers which can be replanted as seed tubers, and hence distinct clones can either be established and maintained by asexual vegetative reproduction or discarded. Domestication involved selection of less bitter and, hence, less toxic tubers and Andean farmers certainly retained a much wider variety of tuber shapes and skin and flesh colours than seen in wild species (Bradshaw and Bonierbale, 2010).

Modern potato breeding began in the nineteenth century when, as early as 1807, Knight in England (Knight, 1807) was advocating the use of artificial hybridization in the breeding of new cultivars. It flourished in Britain and elsewhere in Europe and North America during the second half of the nineteenth century when many new cultivars were produced by farmers, hobby breeders and seedsmen. A single Chilean Tuberosum cultivar, Rough Purple Chili, was introduced into the USA in 1851 (Goodrich, 1863); North America's most popular potato cultivar, Russet Burbank, was derived from it by three generations of open pollination with selection and released in 1914 (Ortiz, 2001). The descendents of Rough Purple Chili were widely employed as female parents in crosses with European Tuberosum at the end of the nineteenth century. Modern potato breeding started later in India and China but with rapid expansion since 1948 and 1978, respectively; furthermore, these countries are now two of the leading potato producers in the world. By 2009, the World

Catalogue of Potato Varieties (Pieterse and Hils, 2009) was able to list more than 4500 cultivars from 102 countries. This is a remarkable achievement since the genetic base of this potato breeding must be considered narrow, despite some introgressions of disease resistance genes from wild and cultivated relatives of Tuberosum potatoes and some base broadening with Andigena and Phureja/Stenotomum (Bradshaw and Bonierbale, 2010).

Need for new cultivars

Despite the large number of cultivars currently available there is a continuing need for new ones. At least two contrasting scenarios can be seen. In the European Union the potato industry is trying to increase potato usage in an economically and environmentally sustainable way. New cultivars must give more yield of saleable product at less cost of production. They must have inbuilt resistances to pests and diseases, and increased water and mineral use efficiency that allow reduced use of pesticides and fungicides and better use of water and fertilizers. Furthermore, they must have the quality traits demanded by processors and supermarkets. Finally, new cultivars must help meet consumer demands for convenience foods, improved nutritional and health benefits, improved flavor and novel products. In contrast, in Asia, Africa, and Latin America there is a need for increased and stable potato production to meet increased demand for food from human population growth. New cultivars must deliver higher yields under low inputs, disease and pest attacks, and environmental stresses such as drought, heat, cold and salinity, with water stress affecting potato production in most areas of the world. If possible, new cultivars should have improved nutritional and health properties, such as higher levels of micronutrients (biofortification). However, the greatest need is to raise fresh weight yields from a world average of 17 t/ha to European and North American levels where over 40 t/ha is achieved. Finally, the most important consequence of climate change is likely to be the need for cultivars that make better use of water, and either avoid drought (faster tuber bulking) or are tolerant of drought (Bradshaw and Bonierbale, 2010).

Breeding finished cultivars

Parents and crossing

Potato breeding worldwide has traditionally involved making crosses between pairs of parents with complementary features and this is still the main route to new cultivars. The aim is to generate genetic variation on which to practice phenotypic selection over a number of vegetative generations, for clones with as many desirable characteristics as possible for release as new cultivars. The choice of parents is all important as breeding can never simply be a numbers game. Crossing the 4500 cultivars in the world catalogue in all possible combinations would generate over ten million progenies and raising 500 seedlings of each would give a staggering total of over five billion for evaluation, an impossible task. In contrast, a phenotypic assessment of 4500 cultivars is feasible, and so is a genotypic assessment of diversity with molecular markers. Hence, breeders can now think in terms of capturing allelic diversity in a smaller core set of parents and of using association genetics to choose parents genotypically as well as phenotypically. They can also use genetic distance based on molecular markers to complement coancestry/pedigree analysis (Sun et al., 2003) in order to avoid closely related parents, and hence inbreeding depression, and to ensure genetic variation for continued progress.

As genetic knowledge accumulates, it will be possible to choose parents for use in pair crosses so that one or both parents have the desired major genes and alleles of large effect at quantitative trait loci (QTLs). Major genes have been mapped for flesh, skin, and flower color, for tuber shape and eye depth, and for resistances to late blight, nematodes, viruses (PVY, PVA, PVX, PVM, PVS and PLTV), and wart. QTLs of large effect have been mapped for total glycoalkaloid content, maturity and resistances to late blight, Verticillium wilt, cyst nematodes. and PLRV. In contrast, many economically important traits are still best viewed as complex polygenic traits, despite a number of QTLs being found, and these include tuber dormancy, dry matter and starch content, fry color, resistance to *Pectobacterium*, tuberization, and yield. For these traits, breeders will still have to rely primarily on phenotypic data and use knowledge of offspring–mid-parent regressions to determine crossing strategy. A statistically significant regression is evidence of heritable variation and the slope of the regression line is a measure of heritability. With a highly heritable trait like fry color the mid-parent value is a good predictor of the mean performance of the offspring and a few carefully chosen crosses can be made (Bradshaw, 2007). In contrast, with only a moderately heritable trait such as yield, offspring mean is less predictable and more crosses need to be made to ensure that they include the best possible.

Clonal generations

Potato breeding at the Scottish Crop Research Institute (SCRI) from 1965 to 1985 was typical of most relatively large programs, both then and now (Mackay, 2005; Bradshaw, 2009). The target duration of the breeding scheme was 12 years: year 1 – crosses (200–300); year 2 – visual assessment of seedlings (100 000) in a glasshouse; year 3 – visual assessment of

Table B37.1 James Hutton Institute (formerly SCRI) strategy for breeding finished cultivars.

Year	Strategy
0	**Decide objectives and evaluate potential parents** Midparent values used to predict mean performance of crosses for quantitative traits + other genetic information
	GLASSHOUSE
1	Choose parents and make 200 crosses
2	Seedling progeny tests on 200 progenies × 2 replicates × 25 seedlings (late blight, potato cyst nematodes, visual assessment of tubers) Select best 40 progenies (after first cycle will also have midparent values for fry color and other traits from year 5 of previous cycle)
	SEED SITE
3	Tuber progeny tests on 40 progenies × 2 replicates as 2,000 spaced plants (visual assessment of tubers and fry color). Select 500 spaced plants at harvest (four tubers of each plant) Sow more seed of 10 best progenies in the glasshouse to provide a further 10 × 250 = 2,500 (four tubers of each) clones for year 4 Select clones for use as parents in next cycle of crosses at random from those (500) advancing to year 4
4	3,000 unreplicated four-plant plots (including parents of next cycle of crosses)

Assessment for yield and quality and special disease tests

	Seed site (number of plants)	Clones (number)	Ware site (number of plots and plants per plot)
5	6	1,000	2 × 5; include parents used in crosses
6	20	360	2 × 10
7	100	120	Two harvests or sites of 2 × 10
8	300	40	Ware trials at a number of sites
9	700	20	Ware trials at a number of sites
10	2,000	2	Official trials
11	2,000	1	Official trials

Multiplication and commercialization from virus-free stock

single spaced plants (50 000) at seed site; year 4 – visual assessment of unreplicated small plots (4000) at seed site and limited post-harvest assessment of quality and disease resistance; years 5–7 – ware (yield) trials (1000, 500, and 200 clones) at breeding station, seed production at seed site and disease and quality testing; years 8–10 – multisite trials (60, 10, and 5 clones) in Britain and overseas and larger-scale seed production at seed site; years 11 and 12 – statutory National List (NL) trials (one or more clones), registration for Plant Breeders' Rights and multiplication from disease-free stock; year 13 – new cultivar(s) added to the National List, 12 years after crossing. Further information on all of these stages has been reviewed elsewhere (Bradshaw, 2007).

Research in the 1980s found that intense early generation visual selection for most quantitative traits was ineffective, particularly between seedlings in a glasshouse and spaced plants at a seed site. The solution to this problem developed and implemented at SCRI from 1985 was the use of progeny tests to discard whole progenies (= full-sib families) before starting conventional within-progeny selection at the unreplicated small-plot stage. Seedling progeny tests were developed for disease and pest resistance, and tuber progeny tests for quality traits. More true seed of the best progenies would be sown to increase the number of clones on which to practice selection in seeking new cultivars. In programs targeted to particular end users, duration was reduced by using progeny tests to discard whole progenies, by starting replicated trials earlier at more than one site and by using micropropagation to multiply promising clones for more extensive testing (Mackay, 2005). Once promising clones had been identified after both the first and second year of ware trials, they were used as parents in the next round of crossing to keep the momentum of the program going. This avoids delaying progress while waiting to obtain more information on potential parents over further clonal generations. In effect, combined between and within full-sib family selection is operated on a five or six year cycle, but it could be done on a three year cycle with limited within-family selection. The latter is desirable when one wishes to combine genes from more than two parents as quickly as possible at the start of a new program. Finally, such a recurrent selection program can accommodate new breeding objectives and germplasm whilst continuing to maintain progress (Bradshaw *et al.*, 2003, 2009). Table B37.1 shows how these principles are currently implemented in commercially funded breeding.

Genetic knowledge and molecular marker assisted selection

As knowledge increases about the number and chromosomal locations of genes affecting economically important traits, breeders should be able to design better breeding programs. They will be able to select parents that complement one another genotypically as well as phenotypically. In the offspring, they will be able to determine the seedling population size required for certainty of finding the genotype with the desired combination of genes (alleles) and hence traits. If this population size is too large to handle in practice, then the required number of cycles of crossing and selection can be determined. A big impact on the efficiency and rate of progress would be the identification of superior clones genotypically as seedlings in the glasshouse and the use of modern methods of rapid multiplication to progress them to commercialization. This will require molecular marker assisted selection or preferably direct recognition of the desired allele at a genetic locus (Bradshaw and Bonierbale, 2010). The discovery and tracking of such alleles should be easier now that the potato genome has been sequenced (http://www.potatogenome.net).

Widening the genetic base for future potato breeding

In future, for many traits, greater use can be expected of the world collection of 3527 potato cultivars native to Latin America which is maintained by the International Potato Centre (CIP) in Peru (Huaman *et al.*, 1997). In addition, further improvements in resistance to abiotic and biotic stresses should come from a greater use of wild species, given the wide range of habitats in which they have evolved. The Inter-genebank Potato Database (IPD) contains 7112 different accessions of 188 taxa (species, subspecies, varieties and forms) out of the 247 tuber-bearing wild potato taxa recognized by Hawkes (Huaman *et al.*, 2000) and data are available for more than 33 000 evaluations covering 55 traits. The species form a polyploid series from diploid ($2n = 2x = 24$) to hexaploid ($2n = 6x = 72$). By manipulation of ploidy, with due regard to Endosperm Balance Number, virtually any potato species can be utilized for the introgression of desirable genes into *S. tuberosum* (Ortiz, 2001; Jansky, 2006). In the past, it took up to five backcross generations and 30 years to transfer a major dominant resistance gene from a wild species into a successful cultivar but today molecular marker assisted introgression offers the possibility of faster progress. As potatoes are heterozygous outbreeders, use of the same recurrent parent during introgression would result in a self of the recurrent parent, and hence inbreeding depression. This can be avoided by using different Tuberosum parents for each backcross but would result in an entirely new cultivar, which may or may not be the desired outcome. The only way to introduce a gene into a known cultivar is by the transgenic route. Hence, the molecular cloning of natural resistance genes and their transfer by *Agrobacterium*-mediated transformation into well-adapted but susceptible cultivars is being pursued in a number of laboratories worldwide. Given the timescale of conventional breeding, the genetic improvement of popular potato cultivars such as Russet Burbank by transformation is an attractive proposition, despite public concerns about GM (genetically modified) crops in some countries. Transformation also offers the possibility of improvements through gene silencing and through the introduction of novel traits into potato, such as those mentioned by Bradshaw and Bonierbale (2010).

References

Ames, M., and Spooner, D.A. (2008). DNA from herbarium specimens settles a controversy about origins of the European potato. *American Journal of Botany*, **95**:252–257.

Bradshaw, J.E. (2007). Breeding potato as a major staple crop, in *Breeding major food staples* (eds M.S. Kang and P.M. Priyadarshan). Blackwell Scientific Publishing, Oxford, pp. 277–332.

Bradshaw, J.E. (2009). Potato breeding at the Scottish Plant Breeding Station and the Scottish Crop Research Institute: 1920–2008. *Potato Research*, **52**:141–172.

Bradshaw, J.E., and Bonierbale, M. (2010). Potatoes. In: Bradshaw, J.E. (ed). *Root and Tuber Crops, Handbook of Plant Breeding*. Springer, New York, pp. 1–52.

Bradshaw, J.E., Dale, M.F.B., and Mackay, G.R. (2003). Use of mid-parent values and progeny tests to increase the efficiency of potato breeding for combined processing quality and disease and pest resistance. *Theoretical and Applied Genetics*, **107**:36–42.

Bradshaw, J.E., Dale, M.F.B., and Mackay, G.R. (2009). Improving the yield, processing quality and disease and pest resistance of potatoes by genotypic recurrent selection. *Euphytica*, **170**:215–227.

Glendinning, D.R. (1976). Neo-Tuberosum: new potato breeding material. 4. The breeding system of Neo-Tuberosum, and the structure and composition of the Neo-Tuberosum gene pool. *Potato Research*, **19**:27–36.

Goodrich, C.E. (1863). The origination and test culture of seedling potatoes. *Transactions of New York State Agricultural Society*, **23**:89–134.

Hawkes, J.G., and Francisco-Ortega, J. (1993). The early history of the potato in Europe. *Euphytica*, **70**:1–7.

Hijmans, R.J. (2001). Global distribution of the potato crop. *American Journal of Potato Research*, **78**:403–412.

Huaman, Z., Golmirzaie, A., and Amoros, W. (1997). The Potato, in *Biodiversity in Trust: Conservation and Use of Plant Genetic Resources in CGIAR Centres* (eds D. Fuccillo, L. Sears, and P. Stapleton). Cambridge University Press, Cambridge, UK, pp. 21–28.

Huaman, Z., Hoekstra, R., and Bamberg, J.B. (2000). The inter-genebank potato database and the dimensions of available wild potato germplasm. *American Journal of Potato Research*, **77**:353–362.

Jansky, S. (2006). Overcoming hybridization barriers in potato. *Plant Breeding*, **125**:1–12.

Knight, T.A. (1807). On raising of new and early varieties of the potato (*Solanum tuberosum*). *Transactions of the Horticultural Society of London*, **1**:57–59.

Mackay, G.R. (2005). Propagation by traditional breeding methods, in *Genetic improvement of Solanaceous crops, Vol I: Potato* (eds M.K. Razdan and A.K. Mattoo). Science Publishers, Inc., Enfield, NH, USA, pp. 65–81.

Ortiz, R. (2001). The state of the use of potato genetic diversity, in *Broadening the Genetic Base of Crop Production* (eds H.D. Cooper, C. Spillane, and T. Hodgkin). CABI Publishing, Wallingford, UK, pp. 181–200.

Pieterse, L., and Hils, U. (2009). *World Catalogue of Potato Varieties*. Agrimedia GmbH, Clenze.

Spooner, D.M., McLean, K., Ramsay, G., Waugh, R., and Bryan, G.J. (2005). A single domestication for potato based on multilocus amplified fragment length polymorphism genotyping. *Proceedings of the National Academy of Sciences USA*, **102**:14694–14699.

Sun, G., Wang-Pruski, G., Mayich, M., and Jong, H. (2003). RAPD and pedigree-based genetic diversity estimates in cultivated diploid hybrids. *Theoretical and Applied Genetics*, **107**:110–115.

protected by scales. When potatoes sprout, the sprouts are lateral branches with several buds. A section across a tuber reveals a pithy central core with branches leading to each of the eyes.

37.8 Cultivars

Four cultivars account for about 75% of the potato acreage in the United States – Russet Burbank, Katahdin, Kennebeck, and Red Pontiac. These major varieties represent the most major shapes of potato – the long cylindrical and russet skin of the Russet Burbank, the red and short rounded shape of Truimph.

Many new modern cultivars have been developed for various markets – baking, frying, cooking, canning, creaming, dehydrating, and chipping. These cultivars have certain specific characteristics that make them suitable for their specific uses. The round smooth-skinned white eastern cultivars are used for chipping (potato chips) and cooking (boiling), while the mutated western types are used for baking and frozen products (mainly French fries). The russet potato varieties have higher dry matter content than the eastern types. Dry matter is measured by the specific gravity of the tuber. High dry matter (1.085 specific gravity or higher) is desired for baking and processing. For frying, boiling, or mashing, the tuber should have a specific gravity of 1.080 or higher and at least 19.8% solids plus 14% starch.

37.9 Reproductive biology

37.9.1 Floral biology

Potato has a terminal inflorescence consisting of 1–30 (but usually 7–15) flowers, depending on the cultivar. The five petals give an open flower a star shape. A flower also has a stigma that protrudes above a cluster of five large, bright-yellow anthers. The corolla color varies from white to a complex range of blue, red, and purple. Flowers open, starting with those nearest the base of the inflorescence and proceeding upwards at the rate of about 2–3 each day. At the peak bloom, there are usually 5–10 open flowers. Flowers stay open for only 2–4 days; the receptivity of the stigma and duration of pollen production is about two days.

37.9.2 Pollination

Potato is predominantly self-pollinated. The peak time of pollination is early morning. Pollen may be collected ahead of the time of use and kept in a cool dry place (e.g., in a desiccator for longevity). Pollination is most successful when temperatures are not high. Some breeders collect the desirable flowers to be used a day ahead and laid out to dry. The pollen is then shaken out over a sieve. The pollen is collected in tubes for use.

37.10 Common breeding methods

Potato has a wide array of wild germplasm that easily crosses with cultivated types. Hybridization is the principal procedure for gene transfer. Selection is conducted in the F_1 because the parents are also widely used in modern potato improvement. Protoplast fusion techniques may be used to fuse monoploid (1x) to form dihaploids (2x). A cross of 4x × 2x using a particular accession of *S. phureja* as male is a technique for generating haploids at a high frequency. Hybridization can be used to increase the frequency of tetraallelic loci and, thereby, increase the intra- and interlocus interactions for increased vigor. The techniques of unilateral sexual polyploidization (4x × 2x) or bilateral sexual polyploidization (2x × 2x) may be used in potato breeding. They are practical procedures because many diploid potatoes hybridize among themselves or with tetraploid species.

Genetic engineering procedures have been used to achieve the development of *Bt*-resistance to Colorado potato beetle and viral coat protein-based resistance to several viral disease (e.g., the potato leaf roll virus – PLRV). The cloning and use of the *AGPase* gene has enabled cultivars with high solids to be developed.

37.11 Establishing a breeding nursery

37.11.1 Field nursery

Breeders commonly cross potatoes in the greenhouse because pollination in the field does not yield good seed. Instead, some breeders produce the pollen source in the field and cut stems containing large inflorescences for use in the greenhouse. Parents for pollen sources should be free from virus infection and should be properly managed, protecting against insect pests, and fertilizing and irrigating the plants for healthy flowers to be produced.

37.11.2 Greenhouse nursery

Potato is a long-day plant. Hence, at least 16 hours of sunlight (or 20 klux of artificial light) are needed to successfully grow potato to flowering and maturity. The greenhouse temperature should be maintained at about 19 °C. Potato may be planted with seed potato or stem cuttings. Plants may be raised in a ground bed or in pots placed on raised benches. Crossing is often done in the winter. The vegetative growth is controlled by pruning and staking the plants to make the flowers more accessible.

37.12 Artificial pollination for hybridization

37.12.1 Materials and equipment

A mechanical vibrator may be used to aid pollen collection to pollinate a large number of plants. Pollen that is shed is collected in test tubes. A blunt scalpel may be used to scrape pollen from the anthers for direct deposit on the anthers.

37.12.2 Emasculation

Mature unopened buds are selected for emasculation. At this stage, the petals appear ready to open. Emasculation may be done in the afternoon for pollination the next morning. Greenhouse grown plants usually fail to set fruit, unless hand-pollinated. Consequently, in the absence of air current (e.g., from ventilators, open doors) that may agitate the flowers to cause pollen shed, emasculation is not necessary. However, it should be noted that there is always a possibility of some selfing occurring under such circumstances.

37.12.3 Pollination

The best time to pollinate is in the morning, soon after the flowers are fully open. This is the time when pollen is most abundant. Flowers with plump, bright-yellow anthers capped with a brownish tips give the best quality pollen. Because flowers in inflorescence do not all open at once, pollinating three days per week will allow most of the suitable flowers to be pollinated. It is best to pollinate flowers in the same inflorescence with the same pollen source, to reduce contamination. A blunt scalpel may be used to deposit pollen on the stigma, or the stigma may be dipped into the tube containing pollen. To authenticate hybridity, anthocyanin pigment markers may be incorporated into the breeding program.

37.13 Natural pollination

Potato has a significant amount of self-pollination. Crossing blocks may be used if one of the parents to be crossed is self-incompatible. This will make all fruits produced on the plant to be hybrids.

37.14 Seed development

Successful crosses develop into small fruits. It is recommended that each set of fruits is inserted in a paper bag about four weeks after pollination, to prevent loosing fruits (and seed) to fruit drop. Softened fruits are picked for seed extraction. The seeds are squeezed into a beaker of water and then strained through a cheese cloth. The seeds of tetraploid species have seed dormancy ranging from about six months to about two years.

37.15 Breeding objectives

Some of the major breeding objectives in potato breeding are discussed next. The order of presentation is arbitrary.

37.15.1 Tuber yield

Increased tuber yield is the primary objective of potato breeding. Tuber yield and shape are influenced by photoperiod. Responsive to photoperiod is quantitatively inherited.

37.15.2 Adaptation

- **Heat tolerance.** Temperature variation is critical in potato production. While germination and growth are favored by warm temperatures, tuberization is favored by cool temperatures. Tuberization is inhibited at temperatures above 29 °C. Heat tolerance desirable for tuberization when unseasonable weather occurs during the production season.
- **Frost resistance.** This trait is desirable for areas where fall potatoes are grown.

- **Drought resistance.** This trait is necessary for production under rainfed conditions.
- **Disease resistance.** Some of the key diseases of economic importance to potato production include:
  - **Late blight.** Caused by *Phytophthora infestans*, this fungal disease is the most economically costly in potato production, causing both foliar and tuber decay. Breeding for resistance is complicated by the fact that foliage resistance and tuber resistance may differ in the same plant. Resistance conditioned by major genes (designated R_1, R_2, R_3, etc.) has been discovered.
  - **Charcoal rot.** Caused by *Macrophomina phaseoli*, this disease occurs when temperatures are high. It is a common storage disease.
  - **Viral disease.** Several viral diseases occur in potato fields, the most economic important one being the virus X. Plant response to this virus is varied including resistance to infection, hypersensitivity, and immunity. Viral coat protein-based resistance to PLRV has been developed.
- **Root-knot nematode.** The root knot nematode (*Meloidogyne incognita*), along with other nematodes, cause economic damage to potato tubers.

37.15.3 Insect resistance

- Aphids – Pubescent cultivars have resistance to aphids.
- Colorado potato beetle (*Leptinotarsa decemlineata*) can cause over 50% loss of a potato crop by feeding on the leaves. *Bt* resistance to this insect has been developed.

37.15.4 Potato tuber quality improvement

Potatoes are sold for the fresh market (for baking, cooking) or processed (chip, frozen, starch, alcohol). The quality standards differ according to the product or end use. Long-term storage is a major aspect of marketing potatoes for chip processing. The reducing sugar content of the tuber should be low to avoid the browning of chips (caramelization) during preparation. Many cultivars have a low temperature sweetening potential, making them unsuitable for processing to chips. Cultivars are being developed that will not accumulate reducing sugars at colder storage. High solids are also desired in potato breeding. Cultivars with high solids have been developed by genetic engineering procedures, using the *AGPase* gene.

Key references and suggested reading

Jewell, L.D. (1988). *Agricultural Statistics 1987*. US Department of Agriculture, Washington, DC.

Page, O.T. (1982). *Utilization of the Genetic Resources of the Potato III*. International Potato Center, Lima, Peru.

Ross, H. (1986). Potato Breeding – Problems and Perspectives. *Journal of Plant Breeding* (Supplement 13, Advances in Plant Breeding), 132 pp.

Talburt, W.F. (1987). History of potato processing, in *Potato Processing*, 4th edn (eds W.F. Talburt and O. Smith). Van Nostrand, New York, pp. 1–10.

Internet resources

http://oregonstate.edu/potatoes/potliv.html – Potato links (accessed March 24, 2012).

http://www.umaine.edu/paa/Breeding/B&Gsec31802.htm – Web site of the Potato Association of America (accessed March, 24 2012).

Outcomes assessment

Part A

Please answer the following questions true or false.

1 Potato originates in India.
2 Potato is cool season crop.
3 Cultivated potatoes are tetraploids.
4 High reducing sugars are desired for chipping potato.

Part B

Please answer the following questions.

1 Exposure of potato tubers to light promotes the formation of a toxic alkaloid called . . . .
2 Potato bears fruits called . . . .
3 Give three major diseases and insect pests of potato.

Part C

Please write a brief essay on each of the following topics.

1 Discuss the artificial pollination of potato.
2 Discuss the importance of temperature adaptation in potato production.
3 Discuss the common potato breeding methods.

38

Breeding cotton

Taxonomy

Family	Malvaceae
Genus	*Gossypium* L
Species	*Gossypium hirsutum* L.

38.1 Economic importance

Cotton is the most important natural fiber in the world for textile manufacture, accounting for about 50% of all fibers used in the textile industry. It is more important than the various synthetic fibers, even though its use is gradually reducing. It is grown all over the world in about 80 countries. The United States is the second largest producer of cotton after China, producing 16.52 million bales on 14.6 million acres in 1999. Most of the cotton is produced in the Cotton Belt. Most of the production currently occurs west of the Mississippi River, thanks to the spread of the devastating boll weevil attack in the eastern states of the United States. The leading production states in 2000 were Texas, Mississippi, Arkansas, and California, in this order of decreasing importance. Other important cotton producing states are Alabama, Arizona, Tennessee, Georgia, Louisiana, South Carolina, North Carolina, Missouri, Oklahoma, and Minnesota.

On the world scene, 88.7 million bales of cotton were harvested from 32.2 million hectares in 2000. The leading producing countries in the world are China (Mainland), the United States, India, Pakistan, Uzbekistan, and Turkey. China and the United States produce nearly 50% of the total world production. Other major producers are Mexico, Brazil, Egypt, Greece, and Columbia.

38.2 Origin and history

The exact origin of cotton is not conclusive. However, two general centers of origin appear to have been identified – Indo-China and tropical Africa in the Old World (Mohenjo Daro, Indus Valley of Asia in 2500 BC), and South and Central America in the New World (Huaca Prieta, Peru in 8000 BC). Cotton was cultivated in India and Pakistan, and in Mexico and Peru about 5000 years ago. Cotton was grown in the Mediterranean countries in the fourteenth century and shipped from there to mills in the Netherlands in Western Europe for spinning and weaving. Wool manufacturers resisted its introduction into Europe until the law banning the manufacturing of cotton was repealed in 1736. Similarly, the English resisted the introduction of cotton mills into the United States until Samuel Slater, who previously had worked in a cotton mill in England, built one in 1790. Three types of cotton are grown in the United States – Sea Island, American–Egyptian, and Upland. These types probably originated in the Americas, the first two are believed to have come from South America. Upland cotton may have descended from Mexican cotton or from crosses of Mexican and South American species.

Principles of Plant Genetics and Breeding, Second Edition. George Acquaah.
© 2012 John Wiley & Sons, Ltd. Published 2012 by John Wiley & Sons, Ltd.

World production of cotton occurs between latitudes 45 °N and 30 °S, where the average temperature in summer is at least 25 °C. Cotton requires a frost-free production (175–225 days), availability of moisture and abundant sunshine for good plant growth, development, and ripening

38.3 Germplasm resources

Germplasm collections are held in banks in the United States and other parts of the world. In the United States, obsolete cultivars of *Gossypium hirsutum* is maintained at the US Cotton Physiology and Genetics Laboratory, Stoneville, Mississippi, whereas the US Cotton Research Laboratory, Phoenix, Arizona, maintains accessions of *G. barbadense*. Texas A&M University, College Station, maintains a collection of diploid cottons and race stocks of *G. hirsutum*, whereas the National Seed Storage Facility, Fort Collins, Colorado, maintains seeds of various cotton types.

38.4 Cytogenetics

Seven genomes of *Gossypium* species, designated *A, B, C, D, E, F,* and *G,* have been identified according to chromosomal size and affinity at meiosis. The basic chromosome number of *Gossypium* is 13. Most (45) of the 60 known species are diploids ($2n = 2x = 26$). Cotton may be divided into two major groups:

(i) **Old world cotton** ($2n = 26$). The diploids in this group have *A, B, E,* or *F* genomes. The cultivated types have the AA genome and comprise *Gossypium herbaceum,* which has five races that originated in Africa and Asia, and *G. arboreum,* which has six races of tree cotton and is found in India.

(ii) **New world cotton** ($2n = 52$). These are tetraploids with the genome *AADD* (13 pairs of each of large and small chromosomes). The dominant species are *Gossypium barbadense* (Sea Island and Egyptian cotton) and *G. hirsutum* (upland cotton).

Sixty two translocations that identify chromosomes 1 through 25 of the 26 chromosomes of the diploid cotton are known. They have been used to locate genes on chromosomes and the production of anueploids, among other uses.

38.5 Genetics

A genetic linkage map of *G. barbadense* and *G. hirsutum* is available. Homoeologous gene pairs related to several homoeologous chromosomes have been discovered. For example, there is homoeology between the loci for anthocyanin pigmentation (R_1R_2) and cluster fruiting habit (cl_1cl_2) of linkage groups II and III. Isogenic lines of a number of allotetraploid mutants (e.g., glandlessness, plant hairiness, okraleaf) have been developed.

Anthocyanin pigmentation is controlled by a multiple allelic series, while monopodial and sympodial branching habits are influenced by several major genes and minor genes. Flower color is controlled by two duplicate genes, Y_1Y_2. Y_1 conditions yellow petals in all allotetraploids (except *G. darwinii*) in which Y_2 conditions petal color. Yellow petal color in *G. barbadense* is controlled by Y_1. *G. hirsutum* has mostly cream petals ($Y_1Y_1y_2y_2$). However, both yellow and cream petals occur in wild forms. Eleven male sterility loci have been identified, 10 of which are located on *G. hirsutum* and one on *G. barbadense*.

Because of high cross-pollination in cotton, a cultivar can have a high level of heterozygosity that is uncharacteristic of self-pollinated cultivars. Further, the genetic constitution of a cultivar can change from year to year. Both genetic and CMS have been identified in tetraploid cotton. The Ms_4 dominant gene confers complete sterility on plants.

38.6 Cultivars

The varieties of cotton in cultivation are derived from four species that produce seed fibers (lint) that is of economic value.

38.6.1 G. hirsutum

About 87% of all cultivated cotton is derived from this species. They are grown in America, Africa, Asia, and Australia. About 99% of all US cotton is this type, which is also classified as upland cotton. The varieties in this type produce fiber of variable length and fineness. The plant can attain a height of two meters.

38.6.2 G. barbadense

This type accounts for 8% of the world's cotton production and is grown in America, Africa, and Asia.

The plant can reach a height of 2.5 meters and has yellow flowers and small bolls. It is also classified as Egyptian cotton and has long, fine, and strong fibers. They are used for manufacturing sewing threads. These are also called "Pima" cotton. New Pima cultivars are identified by a number (e.g., Pima S-1 and Pima S-6).

38.6.3 G. aboreum

This species constitutes about 5% of the total production and is grown mainly in East Africa and Southeast Asia. The plant can attain a height of two meters and has red flowers. The varieties from this species are also classified as Asiatic cotton.

38.6.4 G. herbaceum

This species is also classified as Asiatic cotton. The fibers produced are short (less than 1 inch) and poor quality. This type of cotton fiber is used for manufacturing surgical supplies.

Another classification of cotton based on fibers, from longest to shortest, is Sea Island > Egyptian > American Upland long staple >, American Upland short staple >, and Asiatic.

38.7 American Upland cotton

Four types of *G. hirsutum* (called American Upland cotton) are grown in the United States.

(i) **Eastern Upland.** This type has medium-sized open boll, and medium length staple. It is resistant to fusarium wilt.
(ii) **Delta highland.** This type is grown mainly in the Mississippi Delta, Texas, and Arizona. It has small to medium open boll and medium staple.
(iii) **Plains or storm-proof.** This type of cotton is so-called because the bolls are closed or only partially open at maturity, and hence the fiber resists being blown away by the wind. The bolls are large with short staple and medium fiber length. It is grown in Oklahoma and Texas.
(iv) **Acala.** Grown widely in California, this type of cotton has medium to large bolls, medium to long staple, strong fiber, and long fruiting period.

The Pima cottons are also grown in the United States in places like Arizona, New Mexico, Texas, and California, and make up only about 1% of the total US cotton acreage.

38.8 General botany

Cotton (*Gossypium* spp.) belongs to the family Malvaceae, the Mallow family. There are about 40 species in this genus, but only four species are grown for their economic importance as fiber plants. It is considered an annual plant but it grows as a perennial in tropical areas where the average temperature for the coldest months stays above 18 °C. The plant has a central stem that attains a height of 2–5 feet. It has a deep taproot system. The leaves are arranged spirally around the stem. They are petioled and lobed (3–7 lobes). The stem and leaves are pubescent.

The flowers have five separate petals with the stamens fused into a column surrounding the style. Three large leaf-like bracts occur at the base of the flower. The ovary develops into a capsule or boll (the fruit). The fruit bud (young fruit) is called a **square**. When dry, the capsule splits open along the four or five lines. Bolls average about 1.5–2 inches long. Only about 45% of the bolls produced are retained and develop to maturity. The plant is predominantly self-pollinated. When pollinated by foreign pollen, the phenomenon of **xenia** causes a reduction in fiber length.

An open, mature cotton boll reveals the economic product, a fluffy mass of fibers surrounding the seeds. Each fiber is a single-cell hair that grows from the epidermis of the seed coat. The long hair is called **lint** and the short hair, **fuzz.** Most wild species of cotton do not have lint.

38.9 Reproductive biology

38.9.1 Floral biology

Generally, cotton has perfect flowers. It takes 21–25 days for the square to reach anthesis. While in the bud form, a flower is enclosed by three bractioles (occasionally, four) forming the epicalyx. The base of the flower is occupied by a five-lobed calyx. The cotton flower is large, the petal length reaching 9 cm in some cultivars. In most species, the corolla opens widely, but *G. barbadense* and *G. raimondii* have tubular flowers. The corolla is commonly white (in *G. hirsutum* and *G. arboretum*), while it is yellow (and

Industry highlights
Cotton breeding

Don L. Keim

Delta and Pine Land Company, One Cotton Row, PO Box 157, Scott, MS 38772, USA

Modern upland cotton, *Gossypium hirsutum*, is an allotetraploid which though primarily self-pollinated is readily cross-pollinated by insects. Cotton varieties are developed primarily by pure line breeding techniques. But many varieties have traditionally been mixtures of closely related genotypes as a result of cross-pollination, cotton morphology and breeding procedures. Uniquely, the primary economic value is not the seed itself, but in the fiber produced as an extension of the seed coat cells.

Historically, a major factor contributing to the high cost of cotton production has been in the area of insect management. Insects most affecting US production have been the bollworm/budworm complex, *Heliothis virescens* and *Helicoverpa zea*, and the boll weevil, *Anthomonas grandis*. Recent adoption of new practices has dramatically reduced control costs associated with these two pests. The government-sponsored Boll Weevil Eradication Program has practically eliminated boll weevil as a pest in the major areas of the US Cotton Belt. Additionally, in 1996 farmers began adopting the use of transgenic cotton which contained a gene that confers resistance to the bollworm/budworm complex. This gene was developed by Monsanto and was given the trade name Bollgard® in cotton.

Cotton breeding procedures uniquely are affected by two key factors. Firstly, the fiber must be removed from the seed (ginning). Secondly, the remaining short fibers (seed fuzz) must also be removed with acid (delinting) so that the seed is "flowable" and can be planted with modern equipment. These processes, ginning and delinting, are costly in resources (time, effort, money). Additionally, cotton breeding is expensive due to the need for much hand harvesting (hand picking), the limited number of seed per boll, and the need for hand selfing.

These factors, along with the need for expensive fiber quality evaluations, make the resource cost per unit of genetic gain much higher for cotton when compared to several other crops (Figure B38.1). Given fixed resources, the genetic gain will generally be less with cotton.

Until recently, cotton could have been considered to be relatively "unbred" when compared to the state-of-the-art in crops like corn, soybeans and wheat. In the past, resource limits had forced cotton breeders to have fewer populations with small populations sizes and to advance limited numbers of relatively unselected strains.

The recent widespread adoption and use of transgenic varieties has greatly increased the value of the seed. Seed companies reaping this value have been able to place more resources in the hands of breeders. The recent gains in lint yield and fiber quality have been a direct result of these increased resources, as well as from efficiencies gained from the modernization of breeding procedures.

Breeding program

A major effort in my program has been to maximize the gain enabled by the added resources made available. Several technique and procedural changes have been adopted to increase the efficiency of the utilized resources. This has resulted in an increased number of populations, larger population sizes, heavy selection pressure, and the early identification of superior strains.

Such changes have included: extensive utilization of winter nurseries (time savings); modified single seed descent procedures (cost and labor savings); single boll to $F_{3:4}$ progeny row advancement (cost, labor, and time savings); $F_{4:5}$ progeny row yield testing (higher selection pressure); and yield trial technique changes (higher selection pressure). Additionally, changes in harvest mechanization and data collection have enabled better utilization of resources in the advanced stages, where developments costs are much higher. The breeding procedure utilized in my program is outlined in Table B38.1.

A key aspect addressed, is the degree of outcrossing due to pollinator activity. At Scott, the main breeding station, very little outcrossing occurs. In those nurseries the plants are treated as self-pollinated. In Costa Rica, a high degree of outcrossing occurs; therefore, selfing is required in the F_3 generation. A single selfed boll is harvested from each plant.

For each population, two opportunities are given for intensive selection, in the $F_{3:4}$ single boll-derived rows and in the $F_{4:5}$ plant-derived progeny rows. In the $F_{3:4}$, intensive visual selection is practiced, along with post-harvest selection based on lint percentage (a component of lint yield) and fiber quality. The $F_{4:5}$ progeny are planted in single rows which are replicated twice. Visual pre-selection at harvest is followed by harvest for lint yield, lint percentage and fiber quality. Values for each row are compared against the average values of four surrounding check rows. Selected lines are advanced to replicated tests and breeder increase.

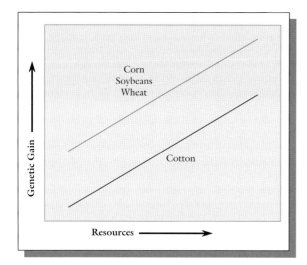

Figure B38.1 A comparison of resource cost per unit genetic gain between cotton and selected field crops. Figure courtesy of Don L. Keim.

The testing advancement schedule includes breeder tests within region for two years and an advanced testing program which involves uniform testing across all regions. Experimental lines advanced to the fourth year of testing (Stage 4) are, additionally, entered into independently conducted trials managed by the Technical Services Department.

Transgene introgression program

At the beginning the second year of testing, conventional lines are crossed to transgene donor parents to begin backcross introgression. Typically, three backcrosses are made prior to deriving lines homozygous for the desired transgene(s). Once $F_{2:3}$ rows are grown in transgene isolation, the breeder of the recurrent parent becomes responsible for selection, evaluation and advancement of lines. In many instances, lines showing close similarity are combined to form a bulk for testing and increase.

In all phases, of transgenic development the plants or seeds are evaluated for transgene presence, purity and zygosity. All seed lots through to commercialization are evaluated for transgene purity.

Key transgenic introgression lines or bulks are entered directly into the Advanced Testing Program as prior performance history exists for the conventional recurrent parents. Lines must go through gene equivalency evaluation prior to release approval by the technology provider. This is done to assure the transgene is effective in the specific genotype.

Table B38.1 Cotton breeding program.

Year	Location	Generation	Procedure/nursery	Harvest	Selection
1	Scott, MS	F_0	Crosses made		None
1	Costa Rica winter	F_1	Single row	Bulk	None
2	Scott, MS	F_2	Bulk population, F_2 yield trial	Modified bulk (single lock from 300+ plants)	Light visual, some populations discarded
2	Costa Rica	F_3	Bulk population	Single selfed boll from 200 plants	Light visual
3	Scott, MS	$F_{3:4}$	Progeny rows (7 ft)	Two plants per selected row	Heavy visual, fiber quality
4	Scott, MS	$F_{4:5}$	Progeny rows (27 ft) two replicates	Bulk row	Heavy visual, lint yield, fiber quality
5	Scott and Winterville, MS	$F_{4:6}$	Replicated tests	Bulk harvest, breeder increase	Lint yield, fiber quality, yield consistency
6	Three midsouth locations	$F_{4:7}$	Replicated tests, begin transgenic integration	Bulk harvest, breeder increase	Lint yield, fiber quality, yield consistency
7	Advanced Testing Program	$F_{4:8}$	Replicated tests	Bulk harvest, breeder increase	Lint yield, fiber quality, yield consistency

Program successes

One of the dilemmas faced in cotton improvement is the negative association of lint yield and fiber quality. In the early 1990s, as new higher yielding varieties were released, the fiber fineness (higher micronaire) and length tended to decrease. As fiber quality is determined and reported on each bale, growers suffered penalties in the marketplace. Early on, the higher yield overcame the lower price, thus increasing the growers' income. In recent years the penalties for less than desirable fiber have become more severe, negating many advantages of higher yielding varieties. Thus, a major goal of my program has been to develop higher yielding varieties with improved fiber traits.

One result has been the release of DP 491, a conventional, high yielding variety with the outstanding fiber quality traits of lower micronaire, very long fiber and high strength (Table B38.2). DP 491 was developed from a cross of DP 5415 × DP 2156. Apparently, transgressive segregation was captured in DP 491 for micronaire, length and strength. The fiber characteristics of the DP 5415 parent are average with a tendency to have high micronaire. The other parent, DP 2156, is a stripper type variety with very poor fiber characteristics.

DP 491 was converted to a "stacked gene" variety (containing the Bollgard[R] and Roundup Ready[R] transgenes) by backcrossing. The resulting recent release, DP 488 BG/RR, has shown to have similar advantages in lint yield and fiber quality over DP 458 BG/RR, the DP 5415 backcross derived, stacked gene variety (Table B38.3).

Another recent release from this program was DP 432 RR, a high yielding early Roundup Ready[R] variety. DP 432 RR has shown in tests distinct yield and fiber quality advantages over the popular ST 4793 RR, a backcross derivative of ST 474 (Table B38.4). DP 432 RR was developed from a straight cross of ST 474 × DP 5415 RR, rather than from the more typical backcross introgression of transgenes.

The examples presented demonstrate that the yield–quality barrier is not insurmountable. In fact, numerous experimental lines, developed by Delta and Pine Land Company breeders, exist which demonstrate the ability to combine high yield and high fiber quality.

The future

The future holds great promise for genetic improvement of cotton. In addition to continued lint yield and fiber quality improvement through intensive conventional breeding, opportunities exist with new transgenes, marker assisted selection, and in broadening the useful diversity in the germplasm base.

Table B38.2 Comparison between new release DP 491 and DP 5415.

Head to Head Comparisons - Beltwide

Variety	Value $/A	Lint Yield lb/A	Lint Percent	Micronaire	Length 2.5% span	Strength g/Tex
DP 491	725	1096	40.4	4.3	1.17	31.1
DP 5415	621	948	37.7	4.6	1.12	29.5
No. tests	96	105	105	101	100	96
% wins	86%	87%				
Significance[1]	***	***	***	***	***	***

[1] paired t-test across tests

Table B38.3 Comparison between new release DP 488 BG/RR and DP 458 B/RR.

Head to Head Comparisons - Beltwide

Variety	Value $/A	Lint Yield lb/A	Lint Percent	Micronaire	Length 2.5% span	Strength g/Tex
DP 488 BG/RR	808	1201	37.6	4.3	1.16	31.4
DP 458 B/RR	732	1094	36.3	4.4	1.11	30.7
No. tests	87	90	90	87	87	87
% wins	76%	76%				
Significance	***	***	***	***	***	***

Table B38.4 Comparison between new release DP 432 RR and ST 4793 RR.

Head to Head Comparisons - Beltwide

Variety	Value $/A	Lint Yield lb/A	Lint Percent	Micronaire	Length 2.5% span	Strength g/Tex
DP 432 RR	805	1208	38.1	4.4	1.13	31.2
ST 4793 RR	766	1164	38.8	4.6	1.10	30.5
No. tests	121	123	123	121	121	121
% wins	74%	65%				
Significance	***	***	***	***	***	***

Several new transgenes are being evaluated in upland cotton. Monsanto has developed a second Bt gene to be used in combination with the initial Bt gene (Bollgard II®) in order to increase effectiveness against Lepidopteron pests and to prevent or greatly delay the advent of insect resistance. Initial varieties were released in 2003.

Roundup Ready® Flex is a new glyphosate resistance gene developed by Monsanto which is designed to give a longer window of application than the original Roundup Ready® trait in cotton. Regulatory approval has been granted, and initial releases were made in 2006.

Concomitant with the great success of Bollgard® and Roundup Ready® in the marketplace (almost 80% of 2004 US cotton acreage was planted to transgenic varieties), other technology providers are entering the market with other transgenes conferring herbicide or insect resistance. Liberty Link® cotton, which contains resistance to glufosinate herbicide, was approved and became commercially available in 2004. Dow has developed Widestrike® technology (lepidopteron pest resistance) in cotton and received regulatory approval for 2005 plantings. Syngenta has developed the VIP technology in cotton providing resistance to lepidopteron pests and is seeking regulatory approval.

Marker assisted selection offers the potential to enhance screening for traits that are not easily evaluated. One example may be root-knot nematode resistance screening. Screening for this pest involves expensive and time consuming greenhouse procedures. A marker or set of markers closely linked with the resistance gene(s) would greatly aid the breeder in identifying root-knot nematode resistant genotypes. Initial work in identifying markers associated with similarly useful traits is a major focus of the Delta and Pine Land Company molecular breeding program. Once established, marker assisted selection for specific traits will be integrated into the commercial breeding programs.

My crossing program has extensively involved the use of diverse upland varieties from all over the world. A key criteria for selecting parental lines is that they be high yielding in the regions for which they were developed. These lines were crossed with germplasm adapted primarily to the US mid-south. In several instances, three-way crosses were made utilizing an additional mid-south adapted parent.

However, upland cotton germplasm derives from a very narrow base relative to other crops. Marker studies have indicated little polymorphism (5–17%). A strong need exists to reach beyond this pool into the diverse sources, such as upland race stocks and related species. Short-day sensitivity is a major barrier to the utilization of race stocks of upland cotton (modern upland varieties are day-length insensitive). Barriers brought about by chromosomal and genomic incompatibilities has limited the broad use of related species crosses. Identification of desired traits and associated markers will enable the rapid and focused utilization of exotic germplasm.

Further reading

Bowman, D.T. (2000). Attributes of Public and Private Cotton Breeding Programs. *Journal of Cotton Science*, **4**:130–136.

Cotton Varieties Planted (2004). Crop. USDA AMS – Cotton Program, Memphis, TN.

Dugger, P., and Richter, D.A. (eds) (published annually) *Proceedings of Beltwide Cotton Conference*. National Cotton Council, Memphis, TN. (http://www.cotton.org/beltwide/proceedings.cfm).

Kohel, R.J., and Lewis, C.F. (eds) (1984). *Cotton*. Agronomy Series No. 24, American Society of Agronomy, Madison, WI.

various shades of yellow) in *G. barbadense* and *G. herbaceum*.

Each flower has an anther column that may bear about 100 anthers in *G. hirsutum*. The tip of the stigma is exserted above the column. The pistil is compound and has 3–5 carpels. The ovary develops into a three- to five-loculed capsule (boll) containing 7–9 seeds that are covered with lint.

38.9.2 Pollination

Cotton is predominantly self-pollinated but up to about 30% and sometimes even higher cross-pollination occurs. Once in bloom, the flower is usually receptive to pollination for no more than eight hours. Pollination is predominantly by insects.

38.10 Common breeding methods

Like other self-pollinated species cotton improvement follows three general approaches – introduction, selection, and hybridization. The most common breeding procedure used in cotton breeding is hybridization. It is used for generating recombinants, followed commonly by pedigree selection to identify superior genotypes. The goal of breeding cultivars in cotton is to achieve sufficient uniformity for major traits (e.g., plant type, fiber properties, disease resistance) while retaining some heterozygosity for vigor. Pure line selection is rarely if ever practiced in cotton breeding because of the floral biology. Some breeders use recurrent selection to concentrate genes of quantitatively inherited traits. Introgressive hybridization to incorporate many economically important traits has been achieved in cotton. This includes genes for disease resistance, insect resistance, and fiber quality.

Hybrid cotton seed is produced by hand emasculation and pollination, or hand pollination with male sterility, in places where labor is inexpensive, such in India. Where labor is expensive, the A-B-R male sterility system is used to produce hybrid seed.

38.11 Establishing a breeding nursery

Spacing in rows may be 3–8 inches in 40-inch rows. A final stand of 2–3 plants per foot of row is optimum. Cotton responds to moderate amounts of balanced fertilization. Cotton breeders commonly use the greenhouse for crossing and selfing. The advantage of this practice is the control the breeder has over the environment to ensure proper flowering and fruiting, especially, in photoperiod sensitive cottons.

38.12 Artificial crossing

38.12.1 Materials and equipment

Some of the materials and equipment used by breeders are fine-tipped tweezers, soda straw, and wired paper tags.

38.12.2 Emasculation

Flowers are emasculated at the whitebud stage. Usually, emasculation is done in the afternoon before anthesis. The flower is large enough to allow emasculation using bare fingers to be possible (but requires skill). Commonly, tweezers are used to strip off the anthers. Cultivars of the *G. barbadense* species do not tolerate loss of petals during emasculation. A drop of gibberelic acid solution (100 ppm) may be applied at the base of the anther to increase the chance of fruit set. A section of soda straw may be used to protect the stigma prior to pollination.

38.12.3 Pollination

The day before use, the flower to be used as the pollen source should be protected from contamination from the visitation of insects by sealing the corolla by tying it with copper wire or securing it with a paper clip (or some other techniques). It is advisable to inspect the flower for intrusion by insects before using the flower. The stigma is fully receptive on the morning of anthesis. Pollen is deposited directly on the stigma using the anthers. Pollen remains viable in the field for only a few hours after anthesis. Some breeders use a soda straw to scoop up the pollen and slip it over the stigma.

To authenticate hybridity, various markers may be incorporated into the breeding program. For example, the dominant R_1 allele in *G. hirsutum* conditions a wine-red plant color. Pollinated flowers are tagged with a wired tag.

38.13 Natural pollination

As previously indicated, cotton is capable of cross-pollination to a great extent. Genetic male sterility is used to aid mass crossing in the field. The crossing block may have to be isolated for this purpose.

38.14 Seed development

Date to maturity varies among species and cultivars. Small crossing blocks are harvested by hand and delinted using a micro-gin. The fuzz is removed by chemical treatment with sulfuric acid and then dried at 43 °C before storage.

38.15 Breeding objectives

Some of the major breeding objectives in cotton are:

- **Lint fiber yield.** The primary components of fiber yield are number of bolls per plant, size of bolls, and percentage lint, of which the number of bolls per plant is most important. Plant breeders select plants that are prolific. Bolls with five locks yield higher than those with four locks. Boll number and size are negatively correlated, making it difficult to improve both traits simultaneously. Recurrent selection may be used to break this undesirable association.
- **Agromorphological traits.** Harvesting is mechanized in cotton production in the United States. Traits that facilitate mechanized harvesting include lodging resistance, bolls set high on the plant, bolls borne singly and natural leaf shedding at maturity. Early maturing and rapid fruiting are also desirable traits.
- **Adapation.** Cotton is produced in dry regions of the world. Consequently, drought resistance is an important breeding objective for this crop.
- **Disease resistance.** Some of the major diseases of cotton are:

  - **Seedling diseases.** Some of the major diseases of cotton are caused by *Fusarium* spp., *Pythium* spp., and *Rhizoctonia solani*. These soil-borne fungi cause diseases to seedlings in wet soil, including damping-off and seed rot. The consequence of this disease is reduced crop stand.
  - **Nematodes.** The root-knot nematode, *Meloidogyne incognita* is a destructive pest in some growing regions. Resistance to the disease is quantitatively inherited.
  - **Verticillium wilt.** This disease occurs widely in the United States. Some genotypes with resistance occur in the *G. barbadense* species.
  - **Bacterial blight.** Also called angular leaf spot, bacterial blight (*Xanthomonas malvaceaum*) is widespread throughout all cotton production regions. Resistance to the disease is conditioned by two or more major genes with minor or modifier genes.
- **Insect resistance.** The major insect economic insect pests of cotton are the boll weevil and bollworm.
  - **Boll weevil.** Caused by *Anthonomus grandis*, this insect pest causes cotton squares to drop. Early maturing cultivars tend to escape the pest.
  - **Cotton bollworms.** Various lepidopteran insects belong to this group of devastating insects pest of cotton. These include the cotton bollworm (*Helicoverpa zea*), tobacco budworm (*Heliothis virescens*), and pink bollworm (*Pectinophora gossypiella*). Genetic engineering procedures have been used to address the attack by these pests through the development of *Bt* cultivars.
- **Fiber quality.** The quality traits of importance in the cotton industry include fiber length, fiber strength, and fiber fineness. Breeding objectives include improving fiber length and uniformity for improved spinning performance and utility of cotton. The fiber strength determines yarn strength, while fiber fineness affects the texture of feel of the fiber. Fiber strength is important for the current technology of open-end (rotor) spinning, which demands stronger fibers.
- **Seed quality.** In addition to lint, cotton is also grown for its seed oil. A major goal in breeding seed quality in cotton is to reduce the pigmentation that discolors the seed oil. The use of glandless cotton cultivars (glands produce gossypol, a terpenoid compound responsible for the discoloration of the seed oil) helps to improve seed oil quality. However, glandless cultivars are more susceptible to insect attack.

Key references and suggested reading

Basu, A.K. and Narayanan, S.S. (1997). Cotton, in *Genetics, cytogenetics, and breeding of crop plants, Vol. 2, Cereal and commercial crops* (eds P.N. Bahl, P.M. Salimath, and A.K. Mandal). Science Publishers, Inc., Enfield, NH.

Jiang, C.X., Zhang, T.Z., Pan, J.J., and Kohel, R.J. (1998). Polyploid formation created unique avenues for response to selection in *Gossypium* cotton. *Proceedings of the National Academy of Sciences USA*, **95**:4419–4424.

Meredith, W.R. Jr. (2005). Minimum number of genes controlling cotton fiber strength in backcross population. *Crop Science*, **45**:1114–1119.

North Carolina Cooperative Extension Service (2002). *Cotton Information*. North Carolina State University, Raleigh, NC.

Rahman, M., Hussain, D., and Zafar, Y. (2002). Estimation of genetic divergence among elite cotton cultivars – genotypes by DNA fingerprinting technology. *Crop Science*, **42**:2137–2144.

Stiller, W.N., Reed, J.J. Constable, G.A., and Reid, P.E. (2005). Selection for water use efficiency traits in a cotton breeding program: Cultivar differences. *Crop Science*, **45**:1107–1113.

Outcomes assessment

Part A

Please answer the following questions true or false.

1 The fruit bud of the cotton plant is called a boll.
2 New world cotton species have a genomic formula of AA.
3 The long fiber of cotton seed is called the lint.
4 The economic important species of cotton is *G. barbadense*.

Part B

Please answer the following questions

1 Give the four species of cotton the produce economic seed fibers.
2 Distinguish between old and new world cotton groups.
3 Discuss the importance of *G. hirsutum* in cotton production.

Part C

Please write a brief essay on each of the topics.

1 Discuss the American Upland cotton types grown in the United States.
2 Discuss pollination of cotton in breeding.
3 Discuss the key breeding objectives of cotton relating to lint yield and quality.

39

Breeding tomato

Taxonomy

Family	Solanaceae
Genus	*Solanum*
Species	*lycopersicon*

Tomato belongs to the family Solanaceae, which includes more than 3000 species. Solanum section Lycopersicon includes the cultivated tomato, Solanum lycopersicum, the only domesticated species, as well as a dozen other wild relatives. The genetic diversity in wild tomatoes, especially the self-incompatible species such as S. chilense *and* S. peruvianum, *is extensive.*

39.1 Economic importance

Tomato is one of the most widely grown and eaten food crops in the world, with an annual global production of about 50 million metric tons. It is one of the most popular vegetable garden crops. Leading producers are the United States (over 7 million metric tons annually), Russia, Italy, China, Turkey, Egypt, Spain, Greece, Brazil and Romania. Tomato is marketed and used either fresh or processed. These two uses have resulted in two distinct industries for the crop, each with its distinct set of cultivars and crop production and management systems. Fresh tomato is produced either in the field or in greenhouses (it is the leading greenhouse produced vegetable in the world). It is available year round and abundant in summer months in warmer regions of the world. In Western Europe and other colder regions, greenhouses are used to extend the availability of this crop. The surge in the popularity of tomato is attributable largely to the growth in the fast food industry and increased popularity of tomato-based foods such as a pizza.

39.2 Origin and history

The center of origin of the crop appears to be the Andes mountain regions, where wild relatives occur in places including Peru, Bolivia and Ecuador. However, the name "tomato" is Mexican, where the crop is believed to have been first domesticated and cultivated. It spread to North America in the seventeenth century.

Principles of Plant Genetics and Breeding, Second Edition. George Acquaah.
© 2012 John Wiley & Sons, Ltd. Published 2012 by John Wiley & Sons, Ltd.

39.3 Commercial market classes

The two general commercial classes of tomato – fresh or processed – also form the basis of breeding programs for the crop.

39.3.1 Processing tomato

This class of tomato is grown on large acreages in the field under mechanized production systems. Cultivars developed for this class have certain desired botanical features and chemical composition. They are determinate, uniform and early maturing, firm and crack-resistant fruits, and the fruit separates easily from the vine. In terms for fruit quality, processing tomato cultivars generally have high soluble solids content. Additional qualities are influenced by the processed product for which the crop will be used. These include pH, total solids, viscosity, acidity, and fruit color. Because the fruits are harvested vine-ripened, cultivars for processed tomato production should resist the major ripe-fruit rots.

39.3.2 Fresh market tomato

Desirable characteristics of fresh market tomato include both appearance and nutritional factors. Fast food consumers prefer large and round fruits. Fresh fruits should have uniform size, well-developed color, uniform shape, and be free of blemishes. Uniformly red fruits are preferred to other colors. From the perspectives of producers, the fruits should have a long shelf-life, so they can stay firm during transportation to the market. Flavor (determined primarily by the content of free sugar, sugar:acid ratio, organics acids) is of importance to consumer of fresh tomato fruits.

39.4 Tomato types

Tomato may also be classified into types mainly by shape, color and size.

- **Beefsteak tomatoes.** This type of tomato is characterized by large, kidney-shaped fruits that are more pronouncedly ribbed on the sides. They are full-flavored and have thinner skin. When used in sandwiches, they are usually sliced from top to bottom (because of the ribbing) rather than horizontally. The slices hold together better than other varieties because of the small cells for seeds. Shelf life for this type of tomato is short.
- **Globe tomatoes.** Globe tomatoes have a round shape and smooth skin; they are the most commercially popular tomatoes. They are most flavorful when vine-ripened, otherwise their flavor is bland. They are used in salads and are easy to blanch and peel, as well as grill.
- **Plum tomatoes.** Plum tomatoes have an oval shape with thicker walls than other types. They also have less seeds and juice than others. However, they have more flesh and can maintain their shape when canned. They are bred to have a high solid content. They are ideal for slicing, dicing or chopping. Also, they are ideal for making tomato paste or sauce. The popular "Roma" tomatoes are plum tomatoes.
- **Small tomatoes.** There are several small tomato types, so-called because of their small fruit sizes. Apart from the fruit size, small tomatoes come in different shapes (e.g., round, pear, plum) and colors. Two common small tomatoes are **cherry** (round and often sweet) and **grape** (smaller and oblong). They are generally eaten whole (e.g., in salads). Larger cherry tomatoes are often cut in half before eating.
- **Heirloom tomatoes.** Heirloom tomatoes are unique in shape and color. They are open-pollinated (non-hybrid) and have a wide variety of flavors, shapes and colors (they are often used in ways that showcase these characteristics).

39.5 Germplasm

Tomato is a classic example of how researchers have successfully used interspecific hybridization (wide crosses) or introgression of genes from wide relatives to improve the characteristics of a cultivated species. Rich sources of diversity occur in the germplasm of related wild species. Remarkably, the continuing introgression of this diversity of genes for physical and chemical characteristics for the production of new cultivars has not resulted in food safety consequences. Remarkable collections of thousands of accessions of wild species of tomato are maintained at the Tomato Genetics Resource Center (TGRC) in Davis. Significant collections in the other parts of the world include the Botanical and Experimental Garden in the Netherlands. In addition to wild accessions, a large collection of mutant lines exists in locations like Cornell University.

39.6 Cytogenetics

Tomato is diploid with twelve pairs of chromosomes that are clearly differentiated and distinguishable at pachytene. Its relatively small genome has minimum internal duplication. It displays a very low tolerance of chromosomal imbalance.

39.7 Genetics

The cultivated tomato is extremely genetically poor, having suffered severe genetic bottleneck as the crop was transported from its center of origin and on its path of domestication through Central America to Europe. One estimate indicates that modern tomato contains less than 5% of the genetic variation of its relatives. Molecular genetic studies show a paucity of polymorphism in cultivated tomato.

Tomato genetics is quite advanced. Qualitative genes and quantitative trait loci (QTLs) for the domestication syndrome traits (growth habit and fruits traits) have been identified. One of the most dramatic changes in tomato through domestication is fruit size. Wild tomato has tiny berries while modern tomato cultivars are large and succulent. Mutations in about six QTLs appear to be responsible for this remarkable transformation in size. For example the QTL *fruit weight 2-2 (fw2-2)* accounts for about 30% of the changes in fruit weight. The locus codes for a negative repressor of cell division and mutations in the promoter sequences are responsible for the changes from small to large berries. Several loci have been identified for fruit shape. The *ovate* gene is responsible for the transformation from round to elongated or pear fruit shape, while the *sun* and *fs8-1* loci are responsible for elongated and square fruit shapes. QTLs for seed weigh have also been identified. In tomato, the International Solanaceae Genomics Project (SOL) was initiated in 2003 with sequencing the tomato genome as it first cornerstone

39.8 General botany

Modern tomato has two dominant leaf types, regular leaf and potato leaf. Variations exist within these fundamental types, such as rugose (darker green with rough surface) and angora (fuzzy, hairy version of the regular leaf). The regular leaf is the most typical leaf type; it has serrated edges. Variations of this leaf type include different leaf sizes, some of which are very narrow, as well variation in leaf color, some leaves having shades of green or green-blue tint. The potato leaf type has fewer cuts or serration.

The plants have two main forms, indeterminate and determinate. Indeterminate plants usually require some form of mechanical support (staking), growing to several feet (6–12 feet). They bloom, set new fruit and ripen at all at the same time during a production season. Determinate tomato plants, also called "bush" tomato, have a compact plant form, reach a height of about 3–4 ft and usually require no mechanical support in production. The plant stops growing when fruit sets on the top bud. Also, the fruits on a plant ripen together over a short period of about one or two weeks.

Tomato demonstrates wide variation in fruit size and shape. While wild tomato fruit is round, modern cultivars may be round, oblate, pear-, torpedo- or bell-shaped. Fruit color varies from yellow to colorless, while the color of the flesh varies from green to red.

In a predominantly inbreeding species, genetic variation tends to decrease, even without selection. As a consequence, genetic drift is a major process that reduces genetic variation.

39.9 Brief history of tomato breeding

Tomato is characterized by a wide variety of fruit shapes, sizes, colors and flavors. At the end of the nineteenth century, most of tomato farmers used open pollinated landraces and heirlooms. At the beginning of the twentieth century, tomato benefited from modern breeding initiated mostly in the public sector, resulting in improved cultivars for farmers. The cultivars shifted from open-pollinated to hybrids when the private sector became involved in tomato breeding. The first hybrids were single crosses, the first to be released for production occurring in 1946. The modern tomato industry is dominated by hybrids.

39.10 Breeding objectives

Breeding objectives for tomatoes have changed over the years. In the 1970s breeders focused on yield improvement; shelf-life was a key objective in the 1980s. In the 1990s, fruit quality, especially taste and nutrition, became very important to breeders. Locations differ in breeding objectives of importance to production and the market. The general breeding goals of tomato are fruit yield, fruit quality, and resistance to diseases and pests.

- **Fruit yield.** Producers use hybrid cultivars to take advantage of the phenomenon of heterosis for high yield.
- **Fruit quality.** This breeding objective includes physical characteristics like size, shape, and color, as well as chemical factors like soluble solids, acidity, taste, and sensory factors. Fresh market producers are interested in the fruit ripening shelf-life of fruits. Fruit ripening affects other quality traits like color, flavor, and soluble solids content. Similarly, during ripening some of the chemical processes negatively impact fruit storage. Genes for *polygalacturonase* and *ethylene synthase*, both compounds associated with fruit ripening, have been cloned by researchers, as well as the mutant genes, *nor-ripening (nor)* and *ripening inhibitor (rin)*, which have been used to breed long shelf-life. Fruit flavor is a complex trait that is influenced by numerous volatile compounds and sugars and acids interacting together. Genetic manipulation of sugars and acids has resulted in the improvement of fruit flavor. Manipulating volatile compounds to impact fruit flavor is more challenging.

- **Resistance to diseases and pests.** Resistance to some pests and pathogens has been transferred from the wild into cultivated species. For example, the resistance to *Cladosporium fulvum* was obtained from *S. pimpinellifolium*. Other diseases of interest to breeders include late blight resistances, Fusarium wilt, and tomato spotted wilt virus.
- **Resistance to abiotic stresses.** Breeding for resistance to abiotic stresses has focused on environmental factors such high or low temperatures, excessive water or drought, and soil salinity or alkalinity.

39.11 Common breeding methods

39.11.1 Use of wild germplasm

The use of wild germplasm in the tomato breeding is said to have been pioneered by Charlie Rick of University of California, Davis, in around 1940. Complete genetic maps of tomato, based on DNA markers, have been generated for increasing the efficiency of breeding programs. Introgression lines involving several wild tomato species are available at the Tomato Genetics Resources Center at the University of California, Davis. These include *S. pennellii*, *S. habrochaites*, *S. lycopersicoides* and *S. sitiens*. Selection of a horticultural crop like tomato is usually done on a single plant basis and with small numbers of selected plants.

Industry highlights
The breeding of tomato

Yuling Bai

Wageningen UR Plant Breeding, Droevendaalsesteeg 1, 6708 PB Wageningen, The Netherlands

Tomato is one of the most economically important vegetables, cultivated both for the fresh market and the processing industry. The global tomato production is approximately 1.4×10^8 t/year produced on about 4.8 million hectares (FAOSTAT data, 2008, http://faostat.fao.org/default.aspx). The ten leading fruit-producing countries are, in descending order, China, USA, India, Turkey, Egypt, Italy, Iran, Spain, Brazil and Mexico (FAOSTAT data, 2008). The major production per hectare is reported in the Netherlands (473 t/ha/year) where tomatoes are exclusively grown in greenhouse conditions (FAOSTAT data, 2008).

Taxonomy, origin and domestication

Tomato belongs to the botanic family of Solanaceae, also called the nightshade family. More than 3000 species belongs to the Solanaceae family, which include tuber- or fruit-bearing crops such as tomato (*Solanum lycopersicum*), potato (*Solanum tuberosum*), pepper (*Capsicum* spp.) and eggplant (*Solanum melongena*) as well as plants of horticultural (petunia *(Petunia ssp.)*) and medicinal (tobacco (*Nicotiana tabacum*)) importance. Members of the family occupy all habitats from very dry to almost aquatic, range in size from tiny annual herbs to large forest trees, and show markedly different phenotypes. The cultivated tomato was formerly referred to as *Lycopersicum esculentum*. Recent molecular phylogenetic studies, based on molecular data, reintroduced the tomato species into the genus *Solanum*, resulting in a new name: *Solanum lycopersicum*. *Solanum* section *Lycopersicon* includes the cultivated tomato (*S. lycopersicum*) and 12 additional wild relatives (Table B3.1).

Tomato originated from Andean region, which includes Colombia, Ecuador, Peru, Bolivia and Chile. It is presumed that Mexico is the most probable region of tomato domestication and Peru the centre of diversity of tomato wild relatives (Table B3.1). Domestication refers to the process of genetically adapting an animal or plant to better suit the needs of human beings, whereby the animal or plant often becomes dependent on human provision and control. *S. lycopersicum cerasiforme* is thought to be the ancestor of cultivated tomato based on its wide presence in Central America and the presence of a shorted style length of the flower. Tomato had reached a fairly advanced stage of domestication before being taken to Europe in the fifteenth century and further domestication on a much more intense level occurred throughout Europe in the eighteenth and nineteenth centuries.

Domestication has triggered a wide range of morphological and physiological traits that distinguish domesticated crops from their wild ancestors. An obvious feature of tomato domestication is the massive increase in fruit size. Wild tomato species have tiny fruits made to propagate the species and not to feed human beings. Domestication has transformed the once small wild tomato into the present-day cultivars. Because domestication occurred in prehistoric times, the evolution pathway related to the transition in tomato fruit size is unknown. Most likely, mutations associated with larger fruit were selected and accumulated during early humans selection. In tomato, domestication syndrome traits have been studied for growth habit (self-pruning, plant height and earliness) and fruit traits (set, size, shape, color and morphology).

Breeding

Heirloom and hybrid

The first achievements of tomato domestication were larger fruit size, loss of dormancy and higher rate of self-pollination. This last trait has changed tomato from partial allogamy to strict autogamy via the transition from exerted to inserted stigmas. Such a change favors homozygosity, making the offspring nearly identical to the parent. Due to this property, earlier tomato cultivars (named heirloom) were selected and inherited in a family or community. Heirloom tomato varieties are open-pollinated and are unique in size, shape and color (Watson, 1996). Available in varieties with diverse characteristics, heirloom tomatoes tend to be prized for their distinctive flavors.

At the beginning of the twentieth century, public institutes (mainly in the USA) became more involved in tomato breeding. Meanwhile, private companies were formed and commercial breeding shifted from homozygous cultivars to heterozygous hybrid cultivars due to the discovery of heterosis. This is a phenomenon whereby the F_1 hybrid obtained by crossing two genetically dissimilar parents out performs both parents for one or more traits. Hybrids combine good traits from both parents that will segregate in the progeny, discouraging seed propagation by growers. The advantages of hybrid varieties over true breeding varieties were so great that the growers would buy hybrid seeds at higher prices. In 1946 the first hybrid tomato cultivar, "Single Cross", was released (Nederlandse beschrijvende rassenlijst voor groentegewassen, 1946). Eventually, nearly all tomato cultivars for the fresh market and an increasing number of cultivars for processing became hybrids.

Hybrid breeding

The art of tomato breeding is identifying and combining the specific traits of breeding lines for each market. Breeders are continuously improving their breeding lines by applying the following strategies:

- Making new intraspecific crosses between their own breeding lines or with the cultivars of their competitors, which is allowed by breeders' law (UPOV breeders rights, 1961). The aim is to generate a segregating population and to use a pedigree method to select breeding lines with favorable traits resulting from naturally occurring recombination.
- Making interspecific crosses for introgressing characters from wild relatives. In most of cases, interspecific crosses are made aiming to transfer resistance genes from wild species to cultivated tomatoes. Via backcrosses and often with marker assisted selection (MAS), new traits are introduced from wild germplasm. Usually it takes many generations to remove the deleterious genes that go along with the introduced genes due to linkage drag. Interspecific crosses have been used extensively in tomato due to its narrow genetic basis. Embryo rescue is sometimes needed for crosses between tomato and wild species, such as *S. peruvianum* and *S. chilense* (Table B3.1).

When the parental lines are more fixed (F_4 to F_6), crosses are made to produce test hybrids. After several generations of testing at the breeders' sites and, eventually, at the farmers' sites, the best hybrids are selected for commercial use. Hybrids of tomato show some heterosis, but this is only selected for at the last stage of the breeding program, when test hybrids are generated. In earlier generations the parent lines are selected on a single plant basis but not for combining abilities or heterosis. So, recurrent selection programs to select parents with the best combining abilities, like that used in field crops, is not a common practice in tomato breeding. In breeding for processing tomatoes, recurrent selection has been employed in the development of many cultivars but not generally hybrids. Double haploid technology is a powerful alternative to classic breeding strategies and has been applied in many crops of agricultural interest. However, the extreme recalcitrancy of tomato has prevented the application of this technique in tomato breeding.

Hybrid seed can be produced by hand emasculation and hand pollination as well as by the use of male sterility followed by hand pollination. The first method is similar to the one performed in potato. As to male sterility, two gene series are mainly applied in producing hybrid seeds, the *ms* gene series and the *ps* (positional sterile) gene. Despite the fact that it is the most time consuming method, hand emasculation and hand pollination has been until now the main method in tomato hybrid seed production.

Classification of tomato cultivars

The tomato has experienced great diversification due to its adaption to different consumptions and cropping systems. According to their use, tomatoes can be divided into two groups: for fresh consumption and for processing. Cultivars for fresh consumption are grown in greenhouses (Figure B39.1) and in the open air, while cultivars for processing are only cultivated in the open air.

Tomato cultivars can also be distinguished on the basis of the indeterminate or determinate growing habit. Cultivars for processing purposes are determinate growing and plants have a compact growth habit with grouped fruits ripening at a single moment, which are suitable for a mechanical harvest. In addition, fruits for processing should have certain characteristics that are related to processing quality, such as high viscosity, dry extract, pH value and high value of total soluble acids, and so on. Indeterminate habit is typical of fresh market cultivars in a greenhouse. Important characteristics for fresh market cultivars are, for example, long shelf-life, external quality of fruits (like shape and color), and internal quality fruits (like flavor, sweetness, and juiciness).

Wild tomatoes bear fruits that are almost invariably round, while cultivars available today show a wide variety of shapes: round, oblate, pear-, torpedo- and bell-shaped (Figure B39.2). Commercially, tomato cultivars are classified mostly based on the shape and size of the fruit. Round tomatoes are most popular and have an average fruit weight of 70–100 g; cherry tomatoes are usually round-shaped, but fruits have a weight between 10 and 20 g; beef tomatoes are large tomatoes (180–250 g) often used for sandwiches or salads; plum tomatoes are oblong-shaped and characterized by a high solid content, which makes them ideal to be used by the tomato processing industry; vine or truss tomatoes may have any of the shapes mentioned above and are sold still attached to the fruiting stem.

Breeding goals

Though the goals of public and private tomato breeding programs vary widely depending on location, need and resources, breeding goals in tomato have changed over time. In general, breeding goals in tomato have gone through four phases: breeding for yield in the 1970s, for shelf life in the 1980s, for taste in the 1990s, and for nutritional quality currently.

- Yield increase has been one of the most important objectives of tomato breeding programs. The exploitation of heterosis and the development of hybrids play a critical role in yield increases. Breeders prefer to develop F_1 hybrids, not only for heterosis, but also for their uniformity and the protection against illegal reproduction.
- Tomato cultivars with long shelf-life have been obtained mainly by use of mutants *non-ripening* (*nor*) and *ripening inhibitor* (*rin*); for example, the "Daniela" hybrid that was released more than 20 years ago by the BonTom tomato breeding group (Faculty of Agriculture, the Hebrew University of Jerusalem, Israel). The genetic make-up of Daniela combines the *rin* gene with some selected polygenes for firmness and slow ripening, together with other genes generating high yields of large, quality fruit.
- Flavor is the sum of the interaction between sugars, acids and a set of approximately 30 volatile compounds. Some prediction of flavor can be made by measuring the acidity and refraction index, which is equivalent to the soluble solids content. Although flavor is a complicated trait, it has been shown that significant improvement in

Figure B39.1 Tomato cultivars for fresh consumption growing in a modern glass greenhouse. Figure courtesy of Yuling Bai.

Figure B39.2 The diversified fruit colors and shapes of modern tomato cultivars. Figure courtesy of Yuling Bai.

    tomato flavor can be attained by increasing the sugar and acid contents in tomato fruits by genetic manipulation. However, breeding for volatiles has not yet been performed intensively, as little is known about the relations between flavor, aroma and volatiles. Recently via genomics and targeted metabolite approaches, many QTLs have been identified that reproducibly alter the composition of volatile and non-volatile chemicals important to tomato flavor.

- The nutritional quality of tomato is mainly determined by its lycopene and vitamin C and E contents. Most researchers at present focus on the increasing the level of lycopene, and transgenic approaches have been used in breeding with positive results. Wild tomato accessions rich in lycopene haven been identified, making them a promising resource in tomato breeding programs. For example, the average level of lycopene in *S. pimpinellifolium* is five times higher than that in cultivated tomatoes.

    One of the common breeding objectives in tomato is breeding for resistance to the most destructive pests and pathogens. Tomato hosts more than 200 species of a wide variety of pests and pathogens that can cause significant economic losses. Nature has provided a great wealth of resistances that are available in the wild species. Many of the resistances are simply inherited, and remarkable successes have been made in transferring disease resistance genes via introgression breeding into cultivated tomato. One of the first examples was the exploitation of *Cladosporium fulvum* resistance from *S. pimpinellifolium* in 1934. The current problem with breeding for resistance to diseases is that for some diseases no monogenic genes have been found in wild relatives and/or monogenic genes are not durable. Polygenic resistance has been less used due to the difficulty of managing it in breeding programs. With the release of tomato genome sequences, a significant advance in exploitation of polygenic resistance could be obtained if molecular markers to all the known genes for resistance are available.

Since tomato is subtropical in origin, tomato production is suboptimal over large parts of the tomato crop growing areas, due to unfavorable environmental conditions caused by abiotic factors, including high or low temperatures, excessive water or drought, and soil salinity or alkalinity. In tomato breeding germplasm much genetic variation for such stress tolerance exists; this is awaiting exploitation in breeding programs to develop cultivars adapted to abiotic stresses.

Exploitation of wild tomato relatives in breeding

The collection, description, propagation and distribution of genetic materials are of utmost importance in tomato breeding. In the second half of the twentieth century, thousands of accessions of the wild *Solanum* species were collected and are maintained at the Tomato Genetics Resource Center in Davis, California (TGRC, http://tgrc.ucdavis.edu/). Other major germplasm collections, mainly focused on *S. lycopersicum*, are maintained at the United States Department of Agriculture (USDA, http://www.ars-grin.gov/npgs/) and at the Asian Vegetable Research and Development Center (AVDRC) in Taiwan (http://www.avrdc.org/germplasm.html). In The Netherlands, the Botanical and Experimental Garden (http://www.bgard.science.ru.nl/) pays special attention to the Solanaceae germplasm collection and maintains the most extensive *ex situ* plant collection of non-tuberous Solanaceae species in the world. A large tomato mutant collection, including thousands of mutant phenotypes, has been catalogued and is searchable at the Solanaceae Genome Network's web site (http://zamir.sgn.cornell.edu/mutants).

Since 1940, when plant breeder Charles Rick (University of California, Davis) observed that crosses between wild and cultivated species generated a wide array of genetic variation in the offspring, breeding has been exploiting wild species via interspecific crosses. This has led to the transfer of many favorable attributes to the cultivated tomato. Crosses between *S. lycopersicum* and wild tomato species can be easy but sometimes require strategies such as embryo rescue. Accessions of nearly all wild tomato species have been successfully used to introduce valuable traits for crop improvement, especially monogenic sources conferring resistance to fungal, nematode, bacterial and viral diseases. To introgress the wild favorable allele into cultivated tomato, MAS plays an important role and the map positions and markers linked to the genes and QTLs provided a basis for a breeder to design optimal breeding strategies. In tomato, introgression lines (ILs), which carry a single introgressed region from wild relatives and are otherwise identical for the rest of their genome to cultivated tomatoes, have been developed for several wild accessions, for example, ILs of *S. pennellii*, *S. habrochaites*, *S. lycopersicoides* and *S. sitiens*. These sets of ILs are maintained in Tomato Genetics Resource Centre in Davis, California (http://tgrc.ucdavis.edu). Together with other ILs resulting from different scientific researches, these ILs (also called pre-breds) offer tomato breeders a powerful tool to optimize the uses of the genetic variation in nature by bring together in one genotype alleles that maximize yield, resistance to biotic and abiotic stress, and so on.

Tomato, a model plant for genetics and genomics

Tomato is a favorite model species for classical genetic studies, due to the possibility of self-pollination, easy crossability, and its diploid state. It has a chromosome number of $2n = 2x = 24$. Each of the 12 different tomato chromosomes can be identified in early prophase (pachytene) on the basis of the position of the centromere, the pattern of euchromatin/heterochromatin, and the pattern of the heterochromatic knobs. Nowadays, tomato genomics has generated a huge amount of data resulting in the need for accessible databases, such as the SOL Genomics Network (SGN; http://sgn.cornell.edu). SGN gives free access to DNA sequences of the whole tomato genome as well as genetic and physical tomato maps, expressed sequence tag (EST), bacterial artificial chromosome (BAC), molecular markers, and tools for exploring available data.

With the advance of tomato genome sequences and other genomic tools, the genetic basis of plant traits will be better understood. Identifying candidate genes for important traits and studying allele variation will give knowledge to be used by breeders. This will make it possible to create novel genotypes by introgressing and pyramiding favorite unused natural alleles and/or even by shuffling and re-organization of genomic sequences. The advanced technologies that help breeders extend their germplasm with new sources or genes are applicable in tomato; these include:

- Marker assisted selection (MAS) in breeding links of phenotypic traits to DNA sequence. This will certainly be facilitated by newly made SNP (single nucleotide polymorphism) arrays that can detect SNPs genome wide at their physical locations.
- *In vitro* techniques to regenerate tomato plants from tissue or single cells, embryo rescue. For example, the future success in microspore culture for haploid embryogenesis will facilitate the application of the double haploid method in tomato hybrid breeding.
- Cell fusion techniques to combine cells of non-crossable species and transformation technologies to introduce genes from other organisms.

- Functional studies such as TILLING (targeting induced local lesions in genomes) populations that can be used to identify mutants at specific genes and study the consequences of these mutations for the plants and VIGS (virus induced gene silencing) for a rapid and transient silencing of genes of interest.
- Transformation methods to introduce single genes in the tomato genome by either *Agrobacterium tumefaciens* or particle bombardment.
- Fluorescence *in situ* Hybridization (FISH) as a diagnostic tool to detect chromosome rearrangements among *Solanum* species. FISH is a powerful techniques to reveal chromosome rearrangements, like inversions and duplications within related species (Tang *et al.*, 2008). By using a five-colour cross-species BAC FISH painting, chromosomal rearrangements among *Solanum* species have been revealed (Szinay, 2010), which has a significant implication on introgression breeding. Inversion describes a chromosome breaking and the segment inverting within the same chromosome. Assuming the gene of interest is located in a chromosomal region where an inversion is present in related wild tomato species, paring between homologous chromosomes during meiosis between cultivated tomato and such a related species will be disturbed, giving rise to meiotic abnormalities or segregation distortions and/or recombination suppression. Consequently, the inverted chromosomal region will be likely introgressed as a whole piece, resulting in linkage drag that cannot be removed via backcrossing. Thus, introgression breeding can be hampered by chromosomal rearrangements between related species when interspecific crosses are used and FISH can be applied as a diagnostic tool to investigate chromosomal rearrangements in genetic mapping and introgression breeding.

References

Szinay, D. (2010) The development of FISH tools for genetic, phylogenetic and breeding studies in tomato (*Solanum lycopersicum*). PhD thesis. Wageningen University, Wageningen, The Netherlands.

Tang, X., Szinay, D., Lang, D. C, *et al.* (2008) Cross-species bacterial artificial chromosome-fluorescence in situ hybridization painting of the tomato and potato chromosome 6 reveals undescribed chromosomal rearrangements. *Genetics*, **180**:1319–1328.

Watson, B. (1996) *Taylors Guide to Heirloom Vegetables*. Houghton Mifflin Co., New York, pp. 343.

39.11.2 Transgenic tomato breeding (the Flavr Savr tomato)

Since tomato is a relatively simple system for transformation with *Agrobacterium tumefaciens*. The first commercially produced genetically modified food crop was tomato, called the Flavr Savr.

An important breeding goal of traditional tomato breeding is enhancing shelf life of ripe fruits, a trait that is linked with the pectin-degrading enzyme polygalacturonase gene (PG). PG is associated with fruit ripening and rotting. Slow fruit ripening is desired to extend the shelf-life of fruits, especially in fruits that require transportation over long distances to markets after harvesting. Also, tomatoes have delicate skins that are easily bruised during handling and transportation. The common method of circumventing these transportation and handling challenges is for tomatoes to be harvested green (not vine-ripened) and transported to the market places where the green fruits are forced-ripened (by using carbon dioxide gas). Forced-ripened tomatoes are less flavorful. On the other hand, vine-ripened tomatoes, though more flavorful, are more prone to rotting.

Several naturally occurring mutants with decreased PG levels are available. However, using them in tomato breeding via conventional methods has been unsuccessful because these reduced ripening genes are tightly linked to undesirable genes or have undesirable pleiotropic effects.

To genetically engineer a tomato with reduced PG, the PG gene was cloned from the conventional tomato "Caligrande" in 1987. Two constructs, pCGN1416 and pCGN1436 antisense polygalacturonase binary plasmid vectors containing the FLAVR SAVR^{TM} gene in reverse orientation to the CaMV 35S constitutive promoter were constructed. They included the kanamycin resistance gene (kan^r) as selection marker. The pCGN1436 construct was delivered into tomato using *Agrobacterium* mediation. Because the FLAVR SAVR gene does not alter the structure of the PG enzyme produced in the transgenic plant (it only reduces the amount produced), the PG of both transformed and non-transformed tomatoes are identical.

The Flavr Savr tomato became the first genetically engineered tomato. It was developed by theh Calgene company in 1992, after 10 years of breeding work. The fruits were slow to rot but skin softening was not altered, making it still susceptible to bruising and bursting during transportation and handling. Another problem with the new product was that the tomato variety used in the breeding project was not very tasty to start with, making the transgenic tomato bland and less acceptable to consumers than the conventional types. Crossing the GM tomato with a tastier non-GM tomato may have corrected the problem. Embroiled in legal issues with competitor Monsanto and facing mounting losses from business missteps, the Flavr Savr tomato could not be saved and was discontinued. Calgene was eventually taken over by Monsanto.

39.11.3 Breeding hybrid tomato

Flowers are emasculated between 55 and 65 days after planting. Flower buds from the second cluster which would open within2–3 days are selected for emasculation. Pollen is collected before it is shed. The anther cones are removed from the flowers and placed in suitable containers (e.g., glassine, cellophane, or paper bags). These cones are dried, (e.g., sun-dry or use artificial light – 100 watt for 24 hours). Pollen may be stored in a freezer or refrigerator but must be warmed to room temperature before use.

Pollinate flowers after two days of emasculation by touching stigma with the tip of the index finger dipped in the pollen pool. The stigma is ready for pollination when the corolla of the emasculated flower turns bright yellow. Repeat the pollination2–3 times a week for 3–5 weeks. Usually, successful pollinations are visible within one week (fruit starts to enlarge). Thirty fruits should be allowed to develop per plant for large-fruited cultivars; the number of fruits may be up to 50 for small-fruited cultivars.

Key reference and suggested reading

Opeña, R.T., Chen, J.T., Kalb, T., and Hanson, P. (2001). Hybrid seed production in tomato. International Cooperators' guide, AVRDC pub # 01-527, Asian Vegetable Research and Development Center, Tainan, Taiwan (http://www.avrdc.org/LC/tomato/seedhybrid.pdf; accessed March 24, 2012).

Outcomes assessment

Part A

Please answer the following questions true or false.

1 Tomato is a berry.
2 Tomato originates from Africa.
3 Modern tomato industry is dominated by hybrid cultivars.
4 The "Roma" is an example of globe tomato type.

Part B

Please answer the following questions.

1 Give the key enzymes involved in the ripening of tomato.
2 What company developed the FlavrSavr tomato?
3 Describe the Beefsteak class of tomato.

Part C

Please write a brief essay on each of the following topics.

1 Discuss the use of wild germplasm in tomato breeding.
2 Describe how the FlavrSavr tomato was developed.
3 Discuss the components of fruit quality that are important in breeding of tomato.
4 Distinguish between processing tomato and fresh market tomato.

40

Breeding cucumber

Taxonomy

Family	Cucurbitaceae
Genus	*Cucumis*
Species	*Sativus*

40.1 Economic importance

Cucurbits or vine-crop family (the Cucurbitaceae) comprises a number of very popular crops, including watermelon (*Citrullus lunatus*), cucumber (*Cucumis sativus*),and melon (*Cucumis melo*). Others are the pumpkin, squash and gourds. About 80% of the production occurs in Asia, with China leading the world in production. Other major producers are Turkey, Iran, the United States and Japan. Cucumber is among the top 10 most widely cultivated crops in the world. It is the third most commonly grown vegetable crop in the world.

40.2 Origin and history

Cucumber is believed to have originated in India, with secondary centers of origin in China and the Near East. It was domesticated in Asia and later introduced into Europe.

40.3 Commercial market classes

There are two key market classes of cucumber – pickling cucumbers and slicing cucumbers. Breeding, production, and marketing are designed around these classes.

40.3.1 Pickling cucumbers

Pickling cucumbers are produced for processing or pickling. The varieties in this class are characterized by thin skins, short and block fruit, and skin color gradient that starts with dark green color at the stem to light green at the blossom end. The part of the fruit that touches the soil in field production is white or lighter green.

40.3.2 Slicing cucumber

Slicing cucumbers are grown to be sliced and eaten fresh, commonly in salads. They have longer fruits with thick skins and dark green color.

40.3.3 Varieties

- **English cucumbers.** These may have fruits as long as 24 inches. They are mostly parthenocarpic and usually grown in greenhouses.
- **East Asian cucumbers.** The fruits are slender with deep green skin color. They are ridged and bumpy on the surface. They are grown for slicing or pickling.
- **Lebanese cucumbers.** These have small fruits that are smooth. They are almost seedless.

Principles of Plant Genetics and Breeding, Second Edition. George Acquaah.
© 2012 John Wiley & Sons, Ltd. Published 2012 by John Wiley & Sons, Ltd.

- **Persian cucumbers.** These cucumbers are also called mini seedless cucumbers. The vines are parthenocarpic and require no pollinators for fruit setting.
- Other varieties include apple, Dosakai, Kekiri, and Armenian.

40.4 Cytogenetics and genetics

The genus *Cucumis* has been partitioned into two subgenera, *Cucumis* (x = 7) and *Melo* (x = 12). Cucumber (*Cucumis sativus* L., $2n = 2x = 14$) is the only species with a haploid chromosome number. Over one hundred genes have been identified in cucumber. This includes seedling mutants (e.g., non-lethal color mutants like virescent (*v*), and non-color traits like bitter-free (*bi*)); stem mutants (e.g., bush (*bu*), compact (*cp*) and determinate (*de*)); leaf mutants (e.g., blunt leaf apex (*bla*), crinkled leaf (*cr*)); flower mutants (e.g., sex expression genes and modifiers (*F, a, m*), male sterility genes); fruit mutants (e.g., bitter fruit genes (*Bt, bi*), skin texture genes (*Tu, te, P, I, H*), spine density (*f*), fruit length (*Fl, fl*)); fruit and spine color (e.g., red (*R*), black or brown spines(*B*), heavy netting (*H*)). Many genes have insect resistance and disease resistance, for example, resistance to cucumber mosaic virus (*Cmv*), scab, bacterial wilt, target leaf spot, downy mildew have been identified.

The narrow genetic base of cucumber limits the wide use of marker technology in crop improvement. The condition impedes the construction of a saturated molecular genetic map. The cucumber genome has been sequenced, thereby affording researchers insight into traits such as sex expression, disease resistance, and the biosynthesis of cucurbitacin.

40.5 General botany

Though most cucurbits are monoecious (have separate pistillate and staminate flowers), there are andromonoecious and gynoecious annuals. They are mostly indeterminate trailing vines, even though determinate types have been developed for home gardens. Prior to 1980, most cucumber cultivars were monoecious in which staminate flowers are produced first. Currently, cultivars of both pickling and slicing cucumber used in the United States, for example, are gynoecious hybrids. When produced in the fields, gynoecious cultivars are blended with a monoecious hybrid (10–15% of the total plants) or inbred to provide adequate amounts of pollen for fruit set. Cucumbers are naturally outcrossing and have a potential for interspecific hybridization.

40.6 Common breeding methods

Cucumber has a narrow genetic base, thereby limiting the possibility of crop improvement via cross breeding. Little heterosis has been observed in cucumber crosses. Both conventional and molecular methods have been used to successfully improve various traits of cucumber. The use of molecular markers is rather limited because of the challenge of using this approach when the species has a narrow genetic base and the level of polymorphism is low. The sequenced cucumber genome affords breeders insight into traits such as its sex expression, disease resistance, biosynthesis of cucurbitacin, and "fresh green" odor.

Pedigree and backcross methods have been used to incorporate simply-inherited traits like gynocey, disease resistance, and fruit into elite inbreds for release to producers as new cultivars or for use as parents in producing single cross hybrids. Recurrent selection is preferred for quantitative traits such as yield, early maturity, and fruit quality. Some reports indicate the successful use of recurrent selection for breeding increase fruit yield, disease resistance as well as herbicide resistance in the crop. S_1, half-sib and full-sib family selection methods have also been used to improve yield in cucumber.

40.7 Breeding objectives

Plant breeders have developed varieties for specific production systems and markets. For the food service industry, the major goal is to develop cultivars that are clean and spineless. Breeders approach this objective by introducing glabrousness, a recessive (*gl*) trait that was induced via mutagenesis. For greenhouse production, modern cultivars are gynoecious (all pistillate or female flowers), parthenocarpic fruits (seedless – no need for pollination), and high quality (lack of bitter principle – burpless cucumber).

Breeders are also interested in cultivars that are resistant to major diseases, including antrachnose, powdery mildew, downy mildew, angular leaf spot, cucumber mosaic virus, fusarium wilt and scab.

Key reference and suggested reading

Pierce, L.K. and Wehner, T.C. (1990). Review of genes and linkage groups in cucumber. *HortScience*, **25**:605–615.

Internet resource

http://www.brighthub.com/science/genetics/articles/27236.aspx#ixzz10Xs0SZVH – History of the genetically engineered tomato (accessed March 24, 2012).

Outcomes assessment

Part A

Please answer the following questions true or false.

1 Cucumber originated in Africa.
2 The United States is the leading producer of cucumber.

Part B

Please answer the following questions.

1 Give the scientific name of cucumber.
2 Give the basic market types of cucumber.
3 What is a parthenocarpic fruit?
4 What is the importance of cucurbitacin to cucumber?

Part C

Please write a brief essay on each of the following topics.

1 Describe important traits included in the breeding of greenhouse cucumbers.
2 Distinguish between pickling and slicing cucumbers.
3 Discuss the economic importance of cucumber.

41

Breeding roses

Taxonomy

Family	Rosaceae
Genus	*Rosa*
Species	*rosa* (over 100 species)

41.1 Economic importance

The rose flower is one of the most versatile plants in the world. Roses belong to the third-largest family of plants, Rosaceae. They are related to popular species such as raspberries, blackberries, cherries, apples, and numerous ornamental plants. Rose is the flower of choice for numerous occasions – funerals, Valentine's Day, graduation, and so on. An estimated 130 billion rose stems are sold annually worldwide. It is used as a garden plant or potted plant for the home. It is one of the most popular plants in the cut floral industry, accounting for about two-thirds of all selections worldwide. Most of the world supply of roses comes from Africa, Central America, and South America where labor is cheap. The rose plant is important in the perfume industry where varieties like "Bulgarian Rose" and highly fragrant types from Persia and Africa are coveted. Rose essential oils are used in soaps, lotion, and shampoos, and numerous other cosmetic products.

41.2 Origin and history

There are two geographical groupings of roses – the European/Mediterranean group and the Oriental group of species and hybrids roses. The European/ Mediterranean group comprises mainly of the Albas, Damasks, Damask Perpertuals, Centifolias, Gallicas, and Mosses commercial classes, which have one season of bloom per year. The Orientals are mainly the Chinas and Teas, and have repeat blooms or bloom continuously. The once distinct groups have been cross-breed to create hybrids.

41.3 Classification of roses

Rose classifications vary between the organizations that recognize them. As of 2010, the American Rose Society recognized 56 variations, while the British Association of Rose Breeders recognized 30. The World Federation of Rose Societies recognized 37 classes.

Two broad classes of rose may be recognized: **Old Garden Roses** and **Modern Roses.** The Old Garden Roses are those classified prior to 1867. This group of roses does not require extensive care and suits amateur rose enthusiast. They often bloom in summer. Examples of this group are Hybrid Perpetual, Tea,

Principles of Plant Genetics and Breeding, Second Edition. George Acquaah
© 2012 John Wiley & Sons, Ltd. Published 2012 by John Wiley & Sons, Ltd.

Damask, Noisette, China, Bourbon, and Moss roses. The Modern Roses include those classified after 1867 and have received considerable breeding attention to be full-bodied, vivacious and with delicate colors. They bloom more than once each season. Examples are the hybrid tea, grandiflora, and floribunda. Roses are placed into these general classes and then further categorized according size, ancestry, color, blooming characteristics, growth habit, date of introduction and scent.

Another general classification that may be used is:

- **Shrub roses.** These roses have spiked stems, scented flowers and large leaflets. The plants have a spreadinghabit with flowers that grow in clusters. Subclasses include the musk, rugosa, and shrub.
- **Tea/hybrid teas/climbing hybrid teas.** These roses have large flowers and typically one flower per stem. The plants are continuouslyblooming. The petals of the tea roses are typically pointed.
- **Grandiflora and floribunda.** The flowers of these roses are similar to the tea roses but smaller. The floribundas have a wide variety in blossom colors and blossom in clusters. Some grandiflora varieties grow on walls and trellises.
- **Mini flora.** These are miniature roses with small flowers that grow in clusters of 3–100. They bloom for long periods and lack notable fragrance.

41.4 Commercial classes

There are over a dozen classes of roses that are distinguished by various morphological features.

Alba roses

The Alba roses probably originated from a *R. canina* × *R. damascene* cross, or *R. corymbifera* x *T. gallica* cross. They have large, fragrant, white or light pink flowers with usually few flowers in a cluster. They may grow upright and be under five feet or become large arching shrubs.

Burbon roses

Their origin is in dispute. However, some believe they originated on the Ile Bourbon island (hence Burbon roses) in the Indian Ocean, from a cross between China "Parsons' Pink" and the red "Tous-les-Mois" (a Damask Perpetual). Key traits imparted from these ancestors include the arching growth, lush flowers and fragrance of the Damask, and the tendency to re-bloom and flower from from the China lineage. There is hardly a typical Burbon, as this class consists of a broad range of plant forms, including dwarfs and arching types, and petal colors, including deep reds, pinks, and white.

Centifolia roses

The origin of the Centifolias is also in dispute, some reports indicating that wild relatives occur in Europe and Asia. These leafy plants attain heights between four and five feet and bear lush, fragrant and pink blossoms (other colors include white, red and even stripped or spotted). The weight of these blossoms causes the branches to nod.

China roses

The China roses (*R. chinensis*) are decidedly of Chinese origin. These roses are usually bushy and have the tendency to bloom continuously. They are less cold-hardy. Common colors include deep red, pink, white and maroon.

Climbing roses

Climbing roses are so defined because they are structurally tall and wiry, and can be trained as climbers. This group of roses comprises a wide variety of unrelated varieties, the key common characteristic being the climbing ability. Many of these climbers are once-blooming, producing their bloom on wood made in the previous year. Consequently, pruning them in the fall or winter is disastrous, resulting in no bloom in the spring!

Damask roses

Damask roses (also called the oil roses) derive their name from Damascus from where they were reintroduced to Europe. These roses are believed to be descended from hybrids between *R. gallica* and *R. Phoenicia* which originate in Asia Minor. Damask roses are characterized by upright stature and arching canes, with large fragrant flowers that occur in clusters of few flowers. The color of the petals ranges from white to deep pink. They are used in the perfumery and cosmetic industry.

English roses

English roses are not considered Old Garden Roses. They derive from various varieties such as Portland Damasks, Teas, Hybrid Teas and Noisettes. They are noted for their fragrance and very big shrubs, hence they arenot suited for growing in places where space is limited. They also tend to be adapted to specific climates.

Gallica roses

Gallica roses are among the oldest roses and are a very heterogeneous group. They derive from the "French rose", *R. gallica*. They are characterized by stocky stature and open blossom that exposes the stamens. They are once-blooming, the duration of bloom lasting from 3–6 weeks, depending on the variety. The blooms tend to have very good fragrances, may be near singles to full doubles, and have colors that may be solid, stripped or spotted. Common colors include bluish pinks to rose-red.

Hybrid bracteates

These are mostly climbers that are vigorous and offer repeat bloom in the autumn. They have glossy foliage with very sharp thorns. They make good security fences.

Hybrid musks

This class of roses has a mixed heritage that includes Noisettes and Teas. They are resistant to blackspot disease that plagues many rose types. The hybrid musks are shrubs and are wider than they are tall. Some are climbers. The blooms are characterized by soft buff and honey hues.

Hybrid perpetuals

This class of roses derives from crosses of hybrid Chinas, Noisettes, and Burbons with the Damask Perpetuals that are known for their hardiness and re-blooming characteristics. Hybrid Perpetuals are characterized by big cabbage-shaped fragrant blossoms borne on arching canes. They tend to be susceptible to fungal diseases.

Hybrid teas

Hybrid Teas derive from crosses between Hybrid Perpetuals and Teas. These hybridizations result in hybrids that are bushier, leafier, and bloomed longer with more beautifully shaped flowers. However, the color range is dull and included milky whites, creamy pinks, pale coral pinks and other dull colors. They are also susceptible to diseases.

Teas

Tea roses get their name from the characteristic scent of freshly-opened packet of tea. They are thought to have been the product of *R. chineneis* x *T. gigantea*. Also, they are believed to have the most exquisite form and floral color among all roses. Of the wide color range, Teas are noted for their tones of gold, warm pink and delicate tints. Crosses with Bourbons have been most successful in producing attractive large flowers and vigorous plants.

Miniature roses

Miniature roses are so called because of their diminutive stature. The early types were discovered in China and Switzerland. They are grown in pots, gardens, and also as cut flowers. They are available in a variety of colors. Extensive breeding work on the miniatures was done in the United States by Ralph Moore. Modern miniatures are believed to have been derived mostly from breeding three varieties, "Oakington Ruby", "Tom Thumb", and "R. Rouetii" with other larger roses.

Moss roses

Moss roses arose from a mutation from the Centifolia roses. They get their name from the effect of the mutation that resulted in glands on the floral stem and sepals, making it appear as if greenish or reddish-brown moss were growing on those parts. Moss roses occur in various plant forms, depending on the background in which the mutation is placed, some varieties being tall and used as climbers.

Noisettes

The Noisette roses are bushy, multiflowered plants with double flowers. This rose type was popularized in France. Noisettes have been crossed with Hybrid Teas to develop Hybrid Musk roses.

Rugosa roses

Rugosa roses are derived from the Japanese rose, R. Rugosa, a thorny rose plant. They usually grow to heights of over five feet, but shorter varieties occur. The leaves are glossy, while the flowers are large, ranging in colors from white to pink, red and purple.

41.5 Stages of bloom

The preferred stage of harvesting roses for sale depends on the culture. A full-bud stage is preferred in the USA, while some countries prefer an open-rose stage. The commercial stages of roses are:

- **Tight green bud.** The bud is small with tightly closed sepals; the petal color is not visible. When picked at this stage, the bud will not continue to open.
- **Color break bud.** The petals are slightly visible; the buds may progress to open fully.
- **Full bud.** The bud is fully exposed and will likely open after picking.
- **Open flower.** The bud is opened to expose the stamens and pistils.
- **Hip.** The fertilized flower has developed an achene with seeds, called a hip.

41.6 Cytogenetics

Most roses are diploid ($2n = 2x = 14$). However, even octoploid species ($2n = 8x = 56$) exist. Most modern roses are tetraploids, while older classes of hybrid Chinas and hybrid multiforas are predominantly diploid. Triploids are not popular because of their reduced fertility. Breeders often artificially double the chromosome number of diploid species prior to crossing with tetraploids in breeding.

41.7 Genetics

Color inheritance in roses is very complex. Flower color may be due to one pigment gene in all cells or one pigment overlying another (bicolor) as in tomato (dark red tomato is red flesh with red skin, while bright red fruit has red flesh overlain with yellow skin). Further, a clear yellow fruit color may change to salmon with age. Color type notwithstanding, at least two factors determine flower color in roses – an absence of color factor and a color factor. Either condition may cause a white rose color to arise. Ploidy of the species influences color determination. It is estimated that at least six factors are involved in yellow flower color determination. In a tetraploid plant (7×4) where quadrivalents occur, each of these six factors could be represented four times to the exclusion of the others. The yellow color genetic system includes a gene for each of the following traits or roles – yellow color, repressing yellow, red-blue color, intensifying the color, deciding the strength of the red-blue color, and one for increasing it. In addition to these factors, at least one gene for ivory color and another for intensifying yellow color are also part of the yellow color system. Yet another complication in rose color genetics is the fact some colors fade rapidly with time to reveal the underlying colors. Red flowers may fade to magenta.

Red, yellow, and scarlet flower colors in roses are under recessive gene control while magenta color is under dominant gene control. White rose color is recessive to cream, buff or light yellow. Further, many of the so-called white roses are actually cream roses and are dominant to true whites.

When breeding for color in roses, the breeder should take note of the common linkages involving major market traits such as growth habit and health. For example, gold petal color is linked to globular bud shape. In breeding red color, breeders face the challenge of retaining the color and perfume while eliminating the undesirable weak neck trait. Of the six types of red roses, the pinky red is the easiest to breed. However, it is has short petals and poor scent. In red roses, some flowers display a clear inner side and dull outside red color, with maroon edge petals. Some flowers also have bright red color with a golden base.

Another important market trait in roses is the number of petals. The double flower trait is under quantitative genetic control. A DDDD = very double (80 petals or more), DDDd = 30–50 petals, and dddd = single flower (five petals). The long urn-shape of the flower bud is controlled by dominant genes.

Rose breeding is facilitated by the availability of genetic linkage maps, QTLs, associated with major traits such as disease resistance, flower scent, vigor, doubleness, and recurrent blooming. A larger number of segregating populations is needed for breeding polyploids than diploids.

41.8 General botany

Most rose species have a single row of five petals. However, most modern cultivars have double flowers (multiple rows of petals). These additional petals (petalloids) are derived from conversions of either stamens or pistils. The American Rose Society defines different blooms and their petals as follows: single (4–11 petals); semi-double (12–16 petals); double (17–25 petals); full (26–40 petals); and very full (more than 41 petals).

Apart from numbers, roses petals are also described by their shape or arrangement:

- **Cupped.** petals are moderately curved inward to give the appearance of a cup when viewed from the side.
- **Globular.** flower looks globular because the petals are more strongly curved inward.
- **High-centered.** the inner petals are arranged in a cone and appear higher than the outer petals.
- **Pompon.** short petals are arranged in a dome or round shape.
- **Reflexed.** the outer petals are curved downward.
- **Rosette.** short petals are arranged to create a flat, low-centered appearance.
- **Saucer-shaped.** petals are slightly curved inward to create a saucer shape.

In terms of blooming, roses are of two kinds – once-blooming, in which all flowers are produced at once in a spectacular fashion called a "flush" that lasts for several weeks, or repeat-flowering in which the plant produces two or more flushes of bloom each year.

41.9 Reproductive biology

Fertility is highly variable among rose species. Modern roses are significantly less fertile than the older varieties and, further, males are less fertile than females. Consequently, breeders tend to limit parents used in breeding programs to a few that are known to be amenable, resulting in significant inbreeding in modern roses classes. Gametophytic self-incompatibility is common in roses, especially the diploid species. Tetraploids exhibit some self-fertility, thereby necessitating routine emasculation when being bred. Rose plants produce fruits called hips, which bear achenes that are extracted for planting.

41.10 Pollination

Pollen may be used fresh or collected and stored for use at a later date.

41.11 Common breeding methods

Breeders may use recurrent selection to improve quantitative traits. However, inbreeding methods (e. g., backcross, selfing, sib-mating) along with progeny selection are effective for qualitative traits.

41.12 Common breeding objectives

The major breeding objectives in roses include:

- Vigor (or radiance)
- Long pointed or urn-shaped buds
- Glossy or semi-glossy, hard leathery foliage
- High degree of mildew and rust resistance
- Everblooming, rapid breaking habit
- Long cutting stems
- Strong neck
- Scent (fragrance)
- Absence of blue fading reaction
- High-centered open flower with 25–35 petals

41.13 The creation of the blue rose

True blue flowers and, for that matter, blue roses are non-existent in nature. To create this innovative and priceless rose plant, plant breeders have resorted to genetic engineering. Rose is one flowering plant in which color variation is astonishing. The natural colors of roses (red, yellow, pink, orange) are actually variations of the same gene. When this gene is silenced, the result is a white rose flower. The technique was used successfully to create the pro-vitamin A rice (Golden rice). For rose, the success story occurred in 2004, when an Australian-based company, Florigene, and the Japanese Suntory company collaborated to genetically engineer a true blue rose. First, they turned off (silenced) the production of the red pigment and then inserted a gene for blue pigment, resulting in the production of blue pigment.

The dihydroflavonol reductase (DFR) gene encodes the dihydroflavonol reductase that initiates pigment synthesis in flowers and produces the red pigment in roses. When this gene is silenced, the red pigment is not produced. The blue color of pansy and viola is conditioned by the delphinidin gene which produces the delphinidin pigment that produces the blue colors of pansy and viola. The DFR gene from iris is also known to favor blue pigment synthesis. In spite of the presence of a true blue gene, the transgenic rose created did not have true blue color but rather a pale violet rose flower. As it turns out, high acidity inhibits the blue pigment from manifesting. To overcome this hurdle, research efforts have focused on the search for genes that modify petal acidity or color, so that the true blue color created would be intensified.

Internet resources

http://www.ehow.com/about_6579024_rose-plant-classification.html#ixzz1bzhdG06V – Rose plant classification (accessed March 25, 2012).

http://wiki.answers.com/Q/What_is_the_scientific_classification_of_a_rose#ixzz1bzgDPvmj – Classification of a rose (accessed March 25, 2012).

http://www.agron.iastate.edu/faculty/fehr/FehrBVC.aspx – Breeding vegetable crops (accessed March 25, 2012).

Outcomes assessment

Part A

Please answer the following questions true or false.

1 Tea is a commercial class of rose.
2 Albas belong to the Oriental group of roses.
3 Moss roses arose from a mutation event.
4 The Rugosa roses originate from Italy.

Part B

Please answer the following questions.

1 Describe the color break bud blooming stage in rose production.
2 Give some of the major uses of roses in society.
3 Describe the Damask class of roses.

Part C

Please write a brief essay on each of the following topics.

1 Distinguish between the Old Garden roses and Modern roses.
2 Describe the commercial blooming stages of roses in production.
3 Distinguish between China roses and English roses.
4 Briefly discuss the origin of roses.

Supplementary chapters: review of genetic statistical principles

1

Plant cellular organization and genetic structure: an overview

S1.1 The unit of organization of living things

The **cell** is the fundamental unit of organization of living things. Some organisms consist entirely of one cell (called **unicellular**) whereas others consist of numerous cells working together (called **multicellular**). Except for a bacterium, which lacks cellular compartmentalization into discrete functional units called **organelles,** and is called a **prokaryote**, all other cells have a membrane-bound nucleus and several other membrane-enclosed organelles and are called **eukaryotes**.

The cell can be a unit for selection in breeding if, for example, molecular tools are used. The technology of genetic engineering targets single cells for manipulation. After successfully transferring foreign genes into the cell, it is isolated and nurtured into a full plant. On the other hand, when conventional tools are used, the whole plant is the unit of selection. It should be pointed out that when plants are manipulated by molecular techniques, they eventually have

to be evaluated via conventional selection process in the field using whole plants as the unit of selection.

S1.2 Levels of eukaryotic organization

An eukaryote may also be structurally organized at various levels of complexity: whole organism, organs, tissues, cells, organelles, and molecules, in this order of descending complexity. Plant breeding of sexually reproducing species by conventional tools is usually conducted at the whole plant level by crossing selected parents. Flowers are the units for crossing. The progeny of the cross is evaluated to select those with the desired combination of parental traits. The use of molecular tools allows plant breeders to directly manipulate the DNA, the hereditary material, and thereby circumvent the sexual process. Also, other biotechnological tools (e.g., tissue culture, cell culture, protoplast culture) enable genetic manipulation to be made below the whole plant level.

Principles of Plant Genetics and Breeding, Second Edition. George Acquaah.
© 2012 John Wiley & Sons, Ltd. Published 2012 by John Wiley & Sons, Ltd.

S1.3 The plant cell and tissue

The plant **cell** consists of several **organelles** and structures with distinct as well as interrelated functions (Table S1.1). Some organelles occur only in plants while others occur only in animals. The nucleus is the most prominent organelle in the cell. The extranuclear region is called the **cytoplasm**. For the plant breeder, the organelles of special interest are those directly associated with plant heredity, as discussed next.

There are three basic cell and tissue types – **parenchyma, collenchyma,** and **sclerenchyma** – with increasing thickness in the cell wall. Cells aggregate to form tissues of varying complexity and functions. Parenchyma cells have thin walls and occur in actively growing parts of the plant and extensively in herbaceous plants. The fleshy and succulent parts of fruits and other swollen parts of plants (e.g., tubers, roots) contain parenchyma cells. Collenchyma cells have a thick primary wall and play a role in the plant mechanical support system by forming strengthening tissues. Like parenchyma cells, collenchyma cells occur in regions where active growth occurs, so as to provide the plant some protection from damage. Sclerenchyma cells have both primary and secondary cell walls. The short types are called sclereids and the long cells fibers. Sclerenchyma occurs abundantly in plants that yield fiber (e.g., cotton, kenaf, flax, hemp).

S1.4 Plant genome

A **genome** may be defined as the set of chromosomes (or genes) within a gamete of a species. DNA is the hereditary material of organisms. Most of the DNA (hence most of the genes) in plants occurs in the nucleus in linear structures called **chromosomes**. The nuclear genes are subject to **Mendelian inheritance** (are transmitted according to the laws of Mendel through the processes of nuclear division) (Section S1.5). In addition to the nucleus, DNA occurs in some **plastids** (organelles that are capable of dividing, growing, and differentiating into different forms). These plastids are chloroplasts. DNA also occurs in the mitochondria. The DNA in these organelles is not subject to Mendelian inheritance but follows what is called **cytoplasmic (**or extrachromosomal or extranuclear**) inheritance**. Their distribution into gametes following nuclear division is unpredictable and not equitable. Molecular techniques may be used to separate nuclear DNA from non-nuclear DNA during DNA extraction from a tissue, for independent analysis. Some extranuclear genes are of special importance to plant breeding. Some male sterility genes are located in the mitochondria. As will be described later, cytoplasmic male sterility (CMS) is used in the breeding of corn and many other species. It is used to eliminate the need for emasculation (a time consuming and tedious operation to prepare

Table S1.1 A summary of the structures of plant cells and their functions.

Plasma membrane	This differentially permeable cell boundary delimits the cell from its immediate external environment. The surface of the may contain specific receptor molecules and may elicit an immune response.
Nucleus	Contains DNA and proteins that are condensed in strands called chromosomes (called chromatin when uncoiled).
Cytoplasm	The part of the cell excluding the nucleus and enclosed by the plasma membrane. It is made up of a colloidal material called cytosol and contains various organelles.
Endomplasmic reticulum	A membranous structure of two kinds – smooth (no ribosomes) and rough (has ribosomes). It increases the surface area for biochemical synthesis.
Ribosomes	Organelles that contain RNA and are the sites of protein sysnthesis.
Mitochondria	Organelles that are the sites of respiration; they contain DNA.
Chloroplasts	Contain DNA and chlorophyll; they are sites of photosynthesis.
Cell wall	A rigid boundary outside the plasma membrane.
Golgi apparatus	Also called dictyosomes; has a role in cell wall formation.
Vacuoles	These are storage regions of the cell for undesirable compounds. They help to regulate water pressure in cell and maintain cell rigidity.

plants for crossing by removing the anthers). Also, because genes occur in the cytoplasm but pollen grains (plant male sex units) lack cytoplasm, it is important in a hybrid program which of the two parents is used as female (provides both nuclear genes and cytoplasmic genes) or as male (provides only nuclear genes). Genes carried in the maternal cytoplasm may influence the hybrid phenotype, an effect called the **maternal effect** (Figure S1.1). When uncertain about the presence of any special beneficial genes in the cytoplasm, some breeders conduct reciprocal crossing, in which the parents take turns in being used as the female parent.

S1.5 Chromosomes and nuclear division

Genes (DNA sequences) are arranged in linear fashion in chromosomes, which may be visible as strands in the condensed stage as the cell prepares for nuclear division. Each species is characterized by a set of chromosomes per cell (Table S1.2). On the basis of the number of chromosomes, there are two kinds of cells in a sexually reproducing plant. Cells in the **gametes** (**gametic cells**) of the plant (pollen grains, eggs) contain half the set of chromosomes in the cells in other parts of the body (**somatic cells**). The somatic chromosome number is called the **diploid number (2n)**, while the gametic cells contain the **haploid number**

(**n**). Further, the somatic chromosomes can be arranged in pairs called **homologous chromosomes**, based on morphological features (size, length, centromere position). In sexually reproducing plants, one member of each pair is derived from the maternal parent (through the egg) and the other from the paternal parent (through the pollen). This occurrence is called **biparental inheritance** and, as a result, each diploid cell contains two forms of each gene (called **alleles**). At various stages in the plant life cycle, a cell nucleus may divide according to one of two processes – **mitosis** and **meiosis**.

S1.5.1 Mitosis

Mitosis occurs only in somatic cells; it is characterized by a division of the nucleus (karyokinesis) into two such that each daughter nucleus contains the same number of chromosomes as the mother cell (Figure S1.2). The cytoplasm divides (cytokinesis) so that the mitotic products are genetically identical (equational division). This conservative process produces new cells for growth and maintenance of the plant. Cells in tissue culture divide mitotically. Through the application of appropriate chemicals and other suitable environmental conditions, plant cells can be made to proliferate into an amorphous mass called **callus**. Callus is an undifferentiated mass of cells (cells with no assigned functions). It is a material

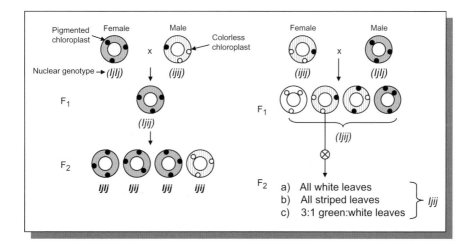

Figure S1.1 Maternal inheritance of the *iojap (ij)* gene in maize. The wild type gene is *Ij*. The green color of the leaf is caused by the chloroplasts, which are maternally inherited. The appearance of the leaf color is determined solely by the maternal phenotype. (Adapted and modified after Klug and Cummings, 1997.)

Table S1.2 Number of chromosomes per cell possessed by a variety of plant species.

Species	Scientific Name	Chromosome number (2n)
Broad bean	*Vicia faba*	24
Potato	*Solanum tuberosum*	48
Corn	*Zea mays*	20
Bean	*Phaseolus vulgaris*	22
Cucumber	*Cucumis sativus*	28
Wheat	*Triticum aestivum*	42
Rice	*Oryza sativa*	24
Tobacco	*Nicotiana tabaccum*	48
Soybean	*Gycine max*	40
Peanut	*Arachis hypogeae*	40
Cotton	*Gossypium hirsitum*	52
Alfalfa	*Midicago sativa*	32
Sugar beet	*Beta vulgaris*	18
Sunflower	*Helianthus annus*	34
Bermudagrass	*Cynodon dactylon*	18, 36

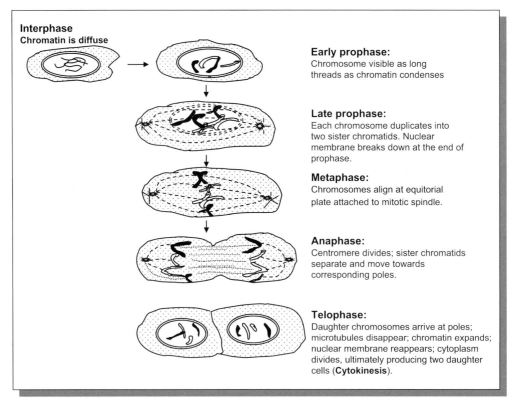

Interphase
Chromatin is diffuse

Early prophase:
Chromosome visible as long threads as chromatin condenses

Late prophase:
Each chromosome duplicates into two sister chromatids. Nuclear membrane breaks down at the end of prophase.

Metaphase:
Chromosomes align at equitorial plate attached to mitotic spindle.

Anaphase:
Centromere divides; sister chromatids separate and move towards corresponding poles.

Telophase:
Daughter chromosomes arrive at poles; microtubules disappear; chromatin expands; nuclear membrane reappears; cytoplasm divides, ultimately producing two daughter cells (**Cytokinesis**).

Figure S1.2 Diagrammatic presentation of mitosis in a cell with a diploid number of four. The male and female chromosomes are presented in different colors. Mitosis produces genetically identical daughter cells.

used in genetic engineering to receive and incorporate foreign DNA into cells.

The nuclear division process may be disrupted (e.g., using a chemical called colchicine) on purpose by scientists, by interfering with the **spindle fibers** (the structures that pull chromosomes to opposite poles of the cell). The consequence of this action is that the chromosomes fail to separate properly into the daughter cells. Instead, a mitotic product may contain a duplication of all or some of the original set of chromosomes (ploidy modification; see Chapter 24).

S1.5.2 Meiosis

Meiosis occurs only in specialized tissues in flowers of plants and produces daughter cells that contain the haploid number of chromosomes (Figure S1.3). This nuclear division is responsible for producing gametes or spores. A meiotic event called **crossing over** occurs in the diplonema stage, resulting in genetic exchange between non-sister chromatids. This event is a major source of genetic variability in flowering plants. It is responsible for the formation of new combinations of genetic material (**recombinants**) for use by plant breeders. Closely linked genes may also undergo recombination to separate them. Hence, plant breeders sometimes take advantage of this phenomenon of recombination to attempt to break undesirable genetic linkages through repeated crossing and, more importantly, to forge desirable linkage blocks. Meiosis is also critical in the lifecycle of flowering species as it pertains to the maintenance of the ploidy level of the species. By reducing the diploid number to a haploid number before fertilization, the diploid number is restored thereafter.

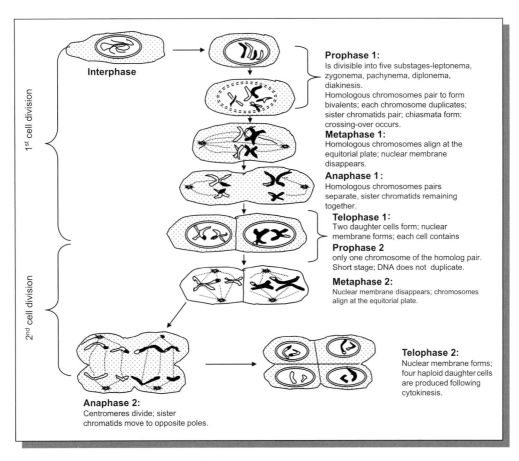

Figure S1.3 Diagrammatic presentation of meiosis in a cell with a diploid number of four. The process has two distinct cell divisions. Prophase I consists of five distinguishable stages; the most genetically significant event of crossing over occurs in the fourth stage, diplonema.

S1.6 Genetic linkage and its implications

First reported in sweet pea (*Lathrytus adoratus*) by Cambridge University geneticists, **genetic linkage** is the phenomenon whereby certain genes tend to be inherited together. Because chromosomes are allocated to gametes during nuclear division, the genes they contain tend to be inherited together, an event that violates Mendel's postulate of independent assortment of genes. Genes within a single chromosome constitute a **linkage group**. Consequently, the number of genetic linkage groups in a species corresponds to the haploid number of chromosomes.

Genes on separate chromosomes as well as genes on the same chromosomes can assort independently. When genes on the same chromosome do not assort independently, they are said to be linked (Figure S1.4). In this example, genes *A* and *B* are transmitted as one

gene (a gene block). The consequence of this linkage (called **complete linkage**) is that, instead of nine different genotypes (as would be expected with Mendelian inheritance), only three different genotypes are produced in the F$_2$, in the ratio of 1 *(AABB)*:2 *(AaBb)*:1 *(aabb)*. The meiotic products are either parental or non-crossover gametes. In the example in Figure S1.5, the phenomenon of crossing over that occurs in meiosis has caused some alteration in linkage (called **incomplete linkage**). In the absence of linkage, the testcross products would segregate in the genotypic ratio of 1:1:1:1 for the four products, *AaBb*, *Aabb*, *aaBb*, and *aabb*. However, in this example, the presence of linkage allowed most gametes to inherit parental genotypes (*AaBb*, *aabb*), as a result of normal gamete formation. Crossing over created new genotypes (*Aabb*, *aaBb*; non-parental), called **recombinants** (because they are products of recombination). When

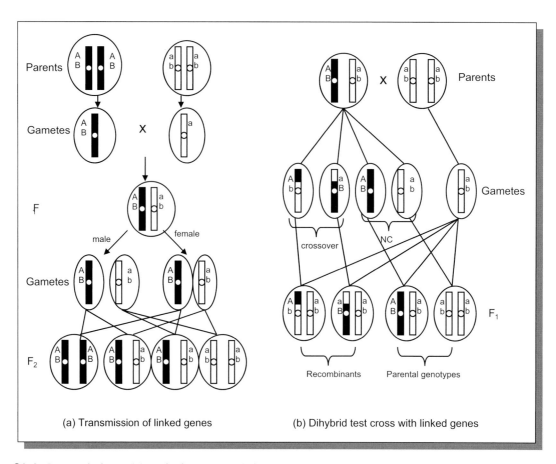

(a) Transmission of linked genes   (b) Dihybrid test cross with linked genes

Figure S1.4 Genetic linkage. (a) Linked genes *AB/ab* are transmitted intact from one generation to the next. (b) Genetic linkage may be broken by the process of recombination. A test cross may be used to reveal the occurrence of recombination. Recombinants are the individuals that are derived from gametes with crossovers.

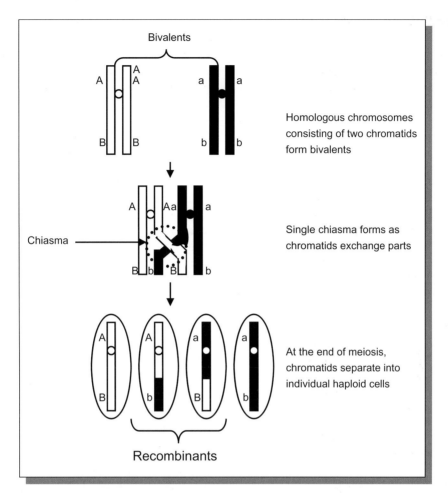

Figure S1.5 Crossover is preceded by the formation of bivalents, the pairing of homologous chromosomes. Adjacent chromatids physically exchange parts during the formation of a characteristic x-configuration, called the chiasma.

the genes of interest are arranged in a homolog such that one chromosome has both dominant alleles (in this example) while the other has both recessive alleles (*AB/ab*), the condition is described as linkage in the **coupling phase.** However, when the arrangement is *Ab/aB*, the linkage is in the **repulsion phase**.

Again, in this example, the numbers (frequency) of parental gametes were roughly equal, and so were the numbers for the recombinants. The proportion of recombinant gametes produced in meiosis in the multiple hybrid is called the **recombination frequency** (RF). If two genes are completely linked, RF = 0. Detection of linkage is accomplished by using the chi square test (Supplementary Chapter 1):

Genotype	Observed frequency (O)	Expected frequency (E)	$(O - E)^2/E$
A–B–	284	214.3	22.67
A–bb	21	71.4	35.58
aaB–	21	71.4	35.58
aabb	5	23.8	40.9
	381	380.9	134.72

Degrees of freedom (df) = 3; chi square at $\alpha = 0.05$ is 7.82. Since the calculated χ^2 is greater than tabulated, we reject the null hypothesis and declare the presence of linkage.

When a cross involves three gene pairs (a **trihybrid cross**), *ABC*, there may be recombination between *A* and *B*, *A* and *C*, and *B* and *C*. This cross is called a **three-point cross**. The most common genetic types are the parental types, and the least common, the double crossovers. A testcross should reveal eight genotypes in the progeny. The order of the genes can be deduced from a three-point cross because one gene in the middle will be the one that apparently changes places in going from the parental to the double crossover type. For example:

Recombinational events	Gametes	Testcross data
No crossover	*ABC*	401
	abc	409
Crossover in *AB* region	*Abc*	32
	aBC	28
Crossover in *BC* region	*ABc*	61
	abC	64
Crossover in both regions	*AbC*	2
(double crossovers)	*aBc*	3
		1000

Recombination between *A* and *B* is calculated for:

- parental types (*ABC*, *ABc*, *abC*, *abc*) = (401 + 61) + (64 + 409) = 935 = 93.5%
- recombinant types (*AbC*, *Abc*, *aBC*, *aBc*) = (2 + 32) + (28 + 3) = 65 = 6.5%

Recombination between *B* and *C* can be similarly calculated.

Fortunately, for the plant breeder, genes in a chromosome are not completely linked. If this were so, the lifeblood of plant breeding, genetic variation, would be very limited. However, during meiosis, as was previously indicated, the phenomenon of crossing over causes recombination or shuffling of linked genes to occur, thereby producing gametes that are unlike the mother cell. Genetic recombination is the most common source of variation in flowering species. Along with an independent assortment of genes, these two phenomena ensure that all offspring contain a diverse mixture of both maternal and paternal alleles.

Whereas breaking linkages is desirable for the creation of the much-needed variation, plant breeders would sometimes rather have certain linkages left intact. This is the case when several desirable genes are tightly linked. On the other hand, there are some occasions on which a desirable gene is linked to an undesirable gene, in which case breeders would like

to break the association. The probability of breaking a linkage depends on how close the genes are in the group or block. A tight linkage (close association) is more difficult to break than a loose linkage. An opportunity for crossover occurs whenever meiosis occurs.

S1.7 Chromosome mapping

Plant breeders develop and use "biological maps" to guide them in their work. The two basic types of maps are physical and genetic maps. Genetic maps are constructed based on the linkage relationship between genes. The degree of crossing over between any two genes or loci on a single chromosome is proportional to the distance between them. This correlation information is used to construct chromosome maps.

Chromosome maps provide information about gene locations, gene order, and the relative position of various genes, according to genetic distances. Linkage maps may be used by plant breeders to aid the selection process. If a desired gene is closely linked with a genetic marker, the breeder may use the marker to indirectly select for the desired gene. In the example of a dihybrid cross, it is possible to calculate the genetic distance between the two genes (or markers), but one cannot tell the order of the genes (i.e., whether *A* comes before *B* or *B* before *A*). A trihybrid cross is needed for this determination.

The distance between two genes is defined as the recombination frequency between them. The unit of measure is the **map unit** or **centimorgan** (cM), which is defined as 1% of crossover. In a dihybrid cross, the percentage crossover (e.g., between genes *A* and *B*) is calculated as the percentage of recombinant offspring produced in a cross. For example, for 50 recombinants out of 400 offspring, it is calculated as (50/400) × 100 = 12.5% = 12.5 map units. If the crossover between *A* and *C* is calculated as 7.5% and between *B* and *C* as 19.8%, then the gene order is *BAC*.

```
B . . . . . . . . . . . . . . A . . . . . . . . . . C
I< . . . . . . . . 12.5 . . . . . >I< . . . . . 7.5 . . >I
I< . . . . . . . . . . . 19.8 . . . . . . . . . . . . >I
```

A low frequency of double crossover between *B* and *C* will give the parental genotype, so that the crossover units will be less than the sum of those between *B* and *A*, and *A* and *C* combined. Further, genes that are separated by 50 or more crossover units are essentially non-linked and will assort independently.

Physical maps are constructed based on nucleotides, the building blocks of DNA. Genetic distance on a linkage map expressed in centimorgans is not directly correlated with the physical distance expressed in nucleotides.

S1.8 Penetrance and expressivity

It has previously been said that the environment in which a gene occurs influences how it is expressed. The source of this environmental effect could be as close and intimate as the immediate cellular environments, or as remote as the general plant external environment. In plants, breeders may transfer genes from one genetic background into another through hybridization. Sometimes, they encounter a situation in which the gene may be successfully transferred, but the desired effect is not observed. A researcher can quantitatively study the degree of expression of a trait. For example, in one case, a disease resistance gene may offer resistance in one plant but fail to do the same in another plant from the same population. This phenomenon is described as **variable gene penetrance** and measures the percentage of individuals that show some degree of expression of the genotype of interest. If 20% of plants show the desired resistance trait, the resistance gene is said to have 80% penetrance (Figure S1.6). Sometimes changes in the plant environment may cause the same plant to produce different phenotypes or degrees of expression of a trait under these different conditions. For example,

the hibiscus plant normally produces single flowers (a flower with one set of petals). A double-flowered mutant (additional petals added to the primary set) has been developed. However, the number of petals of the double flower is influenced by temperature. When grown at between 1.5 and 10 °C (35 and 50 °F), double flower characteristics are lost or diminished and the flowers produce fewer petals. This gene interaction is called **variable gene expressivity,** and describes the *range* of expression of the genotype of interest.

The effect of genetic background on the expression of a phenotype is often difficult to assess. The expression of other genes in a genome may affect the phenotype observed, a phenomenon called **genetic suppression**. Suppression of genes is known to modify the effect of primary genes. Sometimes, relocation of a gene in the genome can influence the expression of the gene, a phenomenon called **position effect**. This may occur when chromosomal mutations such as translocations and inversions occur (a region of the chromosome is relocated to another part of the chromosome).

S1.9 Nucleic acids: structure and function

Nucleic acids are polymers of **nucleotides**. There are two kinds of nucleic acid: **deoxyribonucleic acid (DNA)** and **ribonucleic acid (RNA)**. A nucleotide consists of three basic components: pentose sugar, nitrogenous base, and a phosphate group. The sugar is a cyclic five-carbon sugar; in RNA it is ribose and in DNA it is deoxyribose. Similarly, there are two kinds

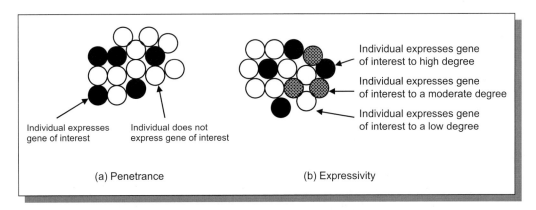

Figure S1.6 A diagrammatic presentation of (a) penetrance and (b) expressivity.

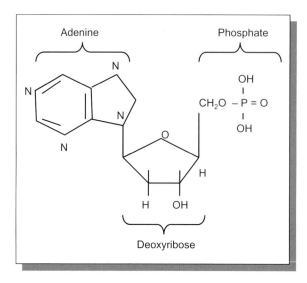

Figure S1.7 The basic chemical structure of a nucleotide molecule, showing its three constituents – a sugar, a nitrogeneous base, and a phosphate group.

of base: **purines** and **pyrimidines.** There are two purines – adenine (*A*) and guanine (*G*) – and three pyrimidines – cytosine (*C*), thymine (*T*), and uracil (*U*). Thymine occurs only in DNA, while uracil occurs only in RNA. The letters *A, C, T, G,* are casually referred to as the alphabets of life.

When a base is linked to a sugar, the product is called a **nucleoside.** A nucleoside linked to a phosphate forms a **nucleotide** (Figure S1.7). Two nucleotides may be linked by a phosphodiester group to form a dinucleotide. Shorter chains (consisting of less than 20 nucleotides) are called **oligonucleotides** while longer chains are called **polynucleotides**. A single nucleoside is also called nucleoside monophosphate (NMP), while two nucleosides form a nucleoside diphosphate (NDP). Triphosphates are important in cellular bioenergetics, especially **adenosine triphosphate (ATP)** and **guanosine triphosphate (GTP)**. When these compounds are hydrolyzed, inorganic phosphate is produced, accompanied by the release of energy (e.g., $ATP \rightarrow ADP + energy$).

S1.9.1 Structure of DNA

DNA is the universal, hereditary material (except in certain viruses – RNA viruses). The most powerful direct evidence for DNA being the hereditary material is currently provided by the cutting edge technology of **recombinant DNA (rDNA).** The structure of the DNA molecule is a **double helix** (Figure S1.8). The key features about the DNA molecule are:

- It consists of two polynucleotide chains coiled around a central axis in a spiral fashion. The most common natural form of DNA is a right-handed double helix of diameter 2.0 nm, called the **B-DNA**. A left-handed form (Z-DNA) and an A-form of DNA also occur under certain conditions.
- The polynucleotide chains are **antiparallel;** one chain runs in the 5′ to 3′ orientation and the other 3′ to 5′ (carbon atoms of a sugar are conventionally numbered from the end closest to the aldehyde or ketone).
- The two bases in each base pair lie in the same plane. Each plane is perpendicular to the axis of the helix. There are 10 base pairs per helical turn.
- The helix has two kinds of alternating external grooves; a deep groove (called the major groove) and a shallow groove (called the minor groove).
- The nitrogenous bases on one strand pair with those on the other strand in **complementary** fashion (*A* always pairs with *T*, while *G* pairs with *C*).

In addition to these features, certain implications deserve emphasis.

- Complementary base pairing means that the replicate of each strand is given the base sequence of its complementary strand when DNA replicates.

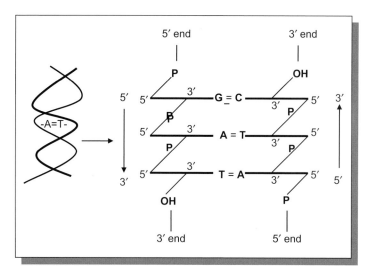

Figure S1.8 The DNA molecule has a double helix structure comprising a sugar–phosphate backbone and horizontal rungs of nitrogeneous bases. The two chains are antiparallel. The helix has minor grooves alternating with major grooves.

- Because the strands are **antiparallel**, when two nucleotides are paired the sugar portions of these molecules lie in opposite directions (one upward and the other downward along the chain).
- Because the strands are antiparallel, the convention for writing the sequence of bases in a strand is to start from the 5′ – P terminus at the left (e.g., GAC refers to a trinucleotide 5′-P′-GAC-3′-OH).
- The conventional way of expressing the base composition of an organism is by the percentage of [G] + [C]. This value is approximately 50% for most eukaryotes with only minor variations among species. In simpler organisms, there are significant variations (e.g., 27% for *Clostridium*, 50% for *E. coli*, and 76% for *Sarcina*, all of these organisms being bacteria).
- The chains of the double helix are held together by hydrogen bonds between base pairs in opposite strands. The bond between A and T is a double bond, while the bond between G and C is a triple hydrogen bond.

S1.9.2 Structure of RNA

Ribonucleic acid (RNA) is similar in structure to DNA. However, there are significant differences, the key ones being:

- RNA consists of ribose sugar (in place of deoxyribose) and uracil in place of thymine.
- Most RNA is predominantly single stranded (except in some viruses). Sometimes, the molecule folds back on itself to form double-stranded regions.

- Certain animal and plant viruses use RNA as their genetic material.
- A typical cell contains about ten times more RNA than DNA.
- Whereas DNA stores genetic information, RNA most often functions in the expression of the genetic information.
- There are three major classes of RNA known to be involved in gene expression: ribosomal RNA (rRNA), messenger RNA (mRNA), and transfer RNA (tRNA). The site of protein synthesis, the ribosome, contains rRNA.

Messenger RNA structure

Messenger RNA (mRNA) is the molecular carrier of genetic information from the DNA to ribosomes, where this DNA transcript or template is translated (the genetic information of DNA transcript is expressed) into proteins. Because genes vary in size (number of nucleotides) the mRNA species are variable in length.

Transfer RNA (tRNA) structure

The structure of transfer RNA (tRNA) is very unique among the three key RNA molecules in the cell. These molecules are small in size and very stable. tRNA molecules range in size from 75–90 nucleotides. A single-stranded tRNA molecule is able to fold back onto itself and undergo complementary base pairing in short stretches to form double strands. This

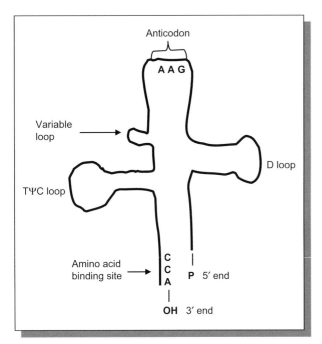

Figure S1.9 A tRNA molecule has a clover leaf shape. Two parts of special interest are the anticodon and the amino acid binding sites that are critical in polypeptide or protein synthesis.

folding also creates four characteristic loops and a cloverleaf 2-D structure (Figure S1.9). Of the four loops, three are involved in translating the message of the mRNA. The anticodon loop (or simply anticodon) consists of a sequence of three bases that are complementary to the sequence of a codon on the mRNA. The stop codons do not have tRNA with anticodons for them. Another feature of the tRNA molecule is the occurrence of the sequence pCpCpA-3′ at the 3′ end. The terminal adenine residue is the point of attachment for an amino acid and hence is called the amino acid attachment (or binding) site. During protein synthesis, the amino acid corresponding to a particular mRNA codon that base pairs with the tRNA anticodon is attached to this terminal and transported to the appropriate segment of the mRNA.

Ribosomal structure

Ribosomes are the sites ("factories") of **polypeptide synthesis** (or **protein synthesis**). A bacterial cell may contain about 1000 ribosomes. A ribosome consists of two subunits, which together form the monosome. The ribosomal particles are classified according to

their sedimentation coefficient or rate (S). Monosomes of bacteria are 70S (70S ribosomes) whereas eukaryotic monosomes are about 80S. Because sedimentation coefficients are not additive, a 70S monosome in actuality comprises two subunits that are 50S and 30S, while an 80S monosome consists of 60S and 40S subunits. A ribosome subunit consists of molecules of rRNA and proteins. For example, the 50S subunit contains one 55 rRNA, one 235 rRNA molecule, and 32 different ribosomal proteins.

S1.10 Central dogma of molecular biology

The genetic information of the DNA is changed into biological material principally through proteins, according to the **central dogma of molecular biology**. The dogma states that genetic information flow is generally unidirectional from DNA to proteins, except in special cases (Figure S1.10). This flow, mediated by **transcription** (copying of the DNA template by synthesizing the RNA molecule) and **translation** (synthesis of a polypeptide using the genetic information encoded in an mRNA molecule) and preceded by **replication** (the process of DNA synthesis), can now be reversed *in vitro* (in the test tube) by scientists. Thus, once a protein is known, the nucleotide sequence in the prescribing DNA strand can be determined and synthesized (the product is called a **complementary DNA** or **cDNA**). Production of cDNA is a technique used in genetic engineering.

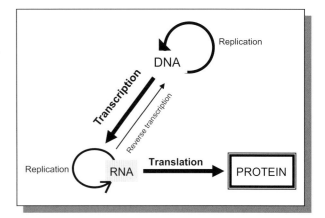

Figure S1.10 The central dogma of molecular genetics, showing the information flow involving DNA, RNA and proteins within a cell. Simply stated, DNA makes RNA, which in turn makes proteins.

S1.11 Expression of genetic information

A key question in genetics is how the information of the DNA is interpreted to produce protein. The expression of genetic information involves complex molecular events, as summarized next.

S1.11.1 The genetic code

The sequence of bases in the polynucleotide chain holds the key to DNA function. The sequence is critical because it represents the **genetic code** (the set of rules giving the correspondence between mRNA and amino acids in a protein) for the synthesis of corresponding amino acids that constitute proteins. DNA does not code for adult traits directly, there being no genes for adult traits as such. Instead, genes code for various developmental processes. The variety of protein products in a cell undertake catalytic and structural activities that eventually result in an adult phenotype.

There are about twenty commonly occurring amino acids. According to the prescribed sequence (based on the genetic code), amino acids are joined together by **peptide bonds** to form polypeptide chains (Figure S1.11). The genetic code is a triplet code. Three adjacent bases form a code for an amino acid. Each trinucleotide sequence is called a **codon** (Figure S1.12). The genetic code is read from a fixed starting point of the DNA strand.

The genetic code is said to be degenerate because nearly all amino acids are specified by at least two codons. Some (serine, arginine, leucine) are encoded by six different codons. Only tryptophan and methionine are encoded by single codons. Further, for a set of codons encoding the same amino acid, the first two letters in the figure is the same, with only the third being different (called the **wobble hypothesis**). Consequently, at least 30 different tRNA species are required to account for the 61 different triplets in the coding dictionary in just 17 (the three remaining triplets include termination codons or signals – UAG, UAA, UGA).

S1.11.2 Transcription: RNA synthesis

The genetic information of the DNA template is copied by the process of **transcription** (or RNA synthesis) to produce an RNA sequence (mRNA). The DNA strand that is transcribed is called the **template strand**. The process starts with a recognition of a special DNA sequence (called a **promoter**) and binding

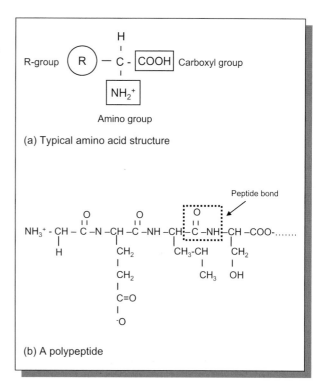

(a) Typical amino acid structure

(b) A polypeptide

Figure S1.11 (a) The basic structure of an amino acid consists of three units – an amino group, a carboxyl group, and a side chain (R) that distinguishes among the different amino acids. (b) A polypeptide chain is formed by linking many amino acids together; adjacent amino acids are linked by a peptide bond.

to it by an enzyme, a process called **template binding**. The RNA chain then grows (chain elongation) in the 3′ direction. The first product of transcription in eukaryotes is called a **pre-mRNA**, part of a group of molecules called **heterogeneous nuclear DNA (hmRNA)**. This molecule undergoes severe alterations to remove non-coding parts (**introns**) of the sequence, leaving the coding parts (**exons**) to produce the mRNA, which is typically about 25% of the original length of the pre-mRNA. After removing the introns, the splicing or linking of the pieces results in different types of mRNA (called **alternative splicing**). Consequently, different kinds of proteins may be encoded by the same gene (Figure S1.13). The mRNA is transported to the ribosomes.

S1.11.3 Translation: protein synthesis

Protein synthesis consists of three steps – **initiation, elongation, and termination**. Translation starts with

	Second base				
First base	U	C	A	G	Third base
U	UUU *Phe*	UCU *Ser*	UAU *Tyr*	UGU *Cys*	U
	UUC "	UCC "	UAC "	UGC "	C
	UUA *Leu*	UCA "	[UAA] stop	[UGA] stop	A
	UUG "	UCG "	[UAG] stop	UGG *Trp*	G
C	CUU *Leu*	CCU *Pro*	CAU *His*	CGU *Arg*	U
	CUC "	CCC "	CAC "	CGC "	C
	CUA "	CCA "	CAA *Gin*	CGA "	A
	CUG "	CCG "	CAG "	CGG "	G
A	AUU *Ile*	ACU *Thr*	AAU *Asn*	AGU *Ser*	U
	AUC "	ACC "	AAC "	AGC "	C
	AUA "	ACA "	AAA *Lys*	AGA *Arg*	A
	AUG *Met*	ACG "	AAG "	AGG "	G
G	GLU *Val*	GCU Ala	GAU *Asp*	GGU *Gly*	U
	GUC "	GCC "	GAC "	GGC "	C
	GUA "	GCA "	GAA *Glu*	GGA "	A
	GUG "	GCG "	GAG "	GGG "	G

Figure S1.12 The genetic code may be likened to a coding dictionary for constructing polypeptide chains. The triplets AUG, UAA, and UGA are termination signals and do not code for amino acids. Of the remaining codes, all amino acids are encoded by at least two codes (up to six in some), except for tryptophan.

the formation of an initiation complex that includes initiation factors that bind to the small rRNA subunit and then to the mRNA. The next step is to set the reading frame for accurate translation. The AUG triplet is usually the initiation point. The large subunit binds to the complex. The sequence of the next triplet determines what charged tRNA (has an amino acid attached) will be attached. The process is repeated until the whole mRNA is translated, adjacent amino acids being linked by peptide bonds. The termination of translation occurs when the elongation process encounters a **stop codon** or termination codon. The interval between the start and stop codons that encodes an amino acid for insertion into a polypeptide chain is called the **open reading frame (ORF)**.

Each gene codes for one polypeptide. Some proteins comprise more than one polypeptide (have multiple subunits). All genes do not code for proteins and, further, all genes in a cell are not actively transcribing mRNA all of the time. Also, most enzymes are proteins, but all proteins are not enzymes.

S1.12 Protein structure

Polypeptides are precursors of proteins. Once produced, they fold to assume three-dimentional forms, the functional stage that becomes proteins. There are four basic levels of protein structure – **primary, secondary, tertiary,** and **quaternary**. The primary structure of proteins is the sequence of the amino acids in

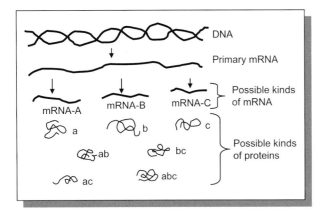

Figure S1.13 Alternative splicing of the mRNA in eukaryotes to remove introns and joining exons results in the production of different mature mRNAs and consequently different protein products.

the linear backbone of the polypeptide. The next fold (exemplified by the DNA molecule) is an alpha helix, a spiral chain of amino acids stabilized by hydrogen bonds. The secondary structure describes the arrangement of amino acids within certain areas of the polypeptide chain. The tertiary structure is a three-dimensional conformation of the entire chain in space. Proteins with more than one polypeptide chain may exhibit the quaternary protein structure through aggregations of the polypeptides.

S1.13 Regulation of gene expression

Gene regulation is a critical activity performed by plants for proper growth and development. It is not important for a gene just to be expressed but its expression must be regulated such that it is expressed at the right time only and to the desired extent. Regulation entails the "turning on" and "turning off" of genes. It is through regulation of gene expression that cellular adaptation, variation, differentiation, and development occur. Some genes are turned on all the time (called **constitutive expression**), while others are turned on only some of the time (called **differential expression**).

The underlying principle of gene regulation is that there are regulatory molecules that interact with nucleic acid sequences to control the rate of transcription or translation. Six potential levels for regulation of gene expression exist in eukaryotes, regulation of: (i) transcription, (ii) RNA processing, (iii) mRNA transport, (iv) mRNA stability, (v) translation, and (vi) protein activity. Transcription is temporarily and spatially separated from translation in eukaryotes.

A typical eukaryotic gene is shown in Figure S1.14. Unlike that of a **monocistronic** gene (lacks introns; has one transcriptional unit and one translational unit) as occurs in bacteria, eukaryotic genes are **polycistronic** (split genes with introns). Genes that encode the primary structures of proteins required by all cells for enzymatic or structural functions are called **structural genes**. In prokaryotes, these genes are organized into clusters that are transcribed as a single unit (coordinately controlled). The mRNA is called polycistronic mRNA, coding for multiple proteins involved in the same regulatory pathway (e.g., the *lac* operon).

There are two basic categories of gene regulation – negative and positive (Figure S1.15). In negative regulation, an inhibitor that is bound to a DNA (gene) must be removed in order for transcription to occur. In positive regulation, gene transcription occurs when an activator binds to the DNA. One of the main ways in which scientists genetically engineer organisms is by manipulating the gene expression.

S1.14 Synteny and plant breeding

Gene order in chromosomes is conserved over wide evolutionary distances. In some comparative studies, scientists discovered that large segments of chromosomes, or even entire chromosomes in some cases,

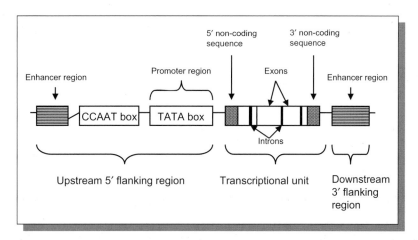

Figure S1.14 A diagrammatic presentation of a typical eukaryotic gene showing the three basic regions – the upstream 5′ flanking regions, the transcriptional unit, and the downstream 3′ flanking region – and their constitution.

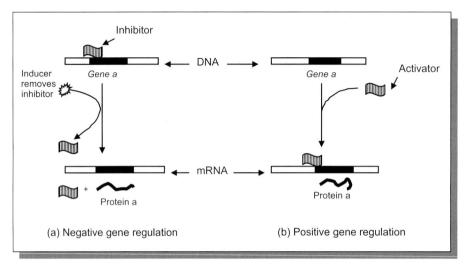

Figure S1.15 Schematic representation of the regulation of gene expression showing (a) negative gene regulation and (b) positive gene regulation.

had the same order of genes. However, the spacing between the mapped genes was not always proportional. The term colinearity is used to refer to the conservation of the gene order within a chromosomal segment between different species. The term **synteny** is technically used to refer to the presence of two or more loci on the same chromosome that may or may not be linked. The modern definition of the term has been broadened to include the homoeology (homoeologous chromosomes are located in different species or in different genomes in polyploid species and originate from a common ancestral chromosomes) of originally completely homologous chromosomes. Whole genome comparative maps have been developed for many species, but are most advanced in the Gramineae family (Poaceae). Some researchers have attempted to clone a gene in one plant species based on the detailed and sequence information (microsynteny) in a homoeologous region of another genus.

Key References and Suggested Reading

Crick, F.H.C. (1962). The genetic code. *Scientific American*, **207**: 66–77.

Clark, A.G. (2002) Limits to the prediction of phenotypes from knowledge of genotypes. *Evolutionary Biology*, **32**: 205–224.

Felsenfeld, G. (1985) *DNA. Scientific American*, **253**: 58–78.

Holland, J.B. (2001) Epistasis and plant breeding. *Plant Breeding Reviews*, **21**: 27–92.

Klug, W.S. and Cummings, M.R. (1996) *Essentials of genetics*, 2nd edn. Prentice Hall. Upper Saddle, NJ.

Klug, W.S. and Cummings, M.R. (1997) *Concepts of genetics*, 5th edn. Prentice Hall, Upper Saddle, NJ.

Stahl, F.W. (1979) Genetic recombination. *Scientific American*, **256**: 91–101.

Weinberg, R.A. (1985) The molecules of life. *Scientific American*, **253**: 48–57.

Internet resource

http://www-scf.usc.edu/~chem203/resources/DNA/doublehelix.html – DNA structure (accessed March 26, 2012).

2

Common statistical methods in plant breeding

S2.1 Role of statistics in plant breeding

The development of **statistics** arose out of a need to assist researchers in those areas where the laws of cause and effect are not apparent to the observer, and where an objective approach is needed. Plant breeders use statistics to design studies, analyze results, and draw sound conclusions. The role of statistics in plant breeding may be summarized in three key applications as follows:

(i) **To obtain a descriptive summary of the sample.** Research data are often numerous, requiring some mathematical reduction to expose hidden trends for easy interpretation. The values obtained from such descriptive analysis are sometimes called **summary statistics**.

(ii) **To provide a means of statistical inference.** The key purpose of collecting data in research is to enable the researcher to draw some kind of inference about a certain characteristic of the population from which the data were drawn. To do this, the values obtained about the sample are used.

(iii) **Comparison.** Often, the researcher has multiple sets of experimental data and needs to know whether they represent significantly different populations of measurement. Another way of putting this is that the objectives of statistics are the estimation of population parameters and testing of hypothesis about the parameters.

Statistical methods used in plant breeding can range from the simple and straight forward, such as arithmetic averages, to the more complex multivariate analysis. Computers are required for complex analysis but, sometimes, the breeder may have a small amount of data and might want to use a hand-held calculator for quick results. Hence, there is the need to know the computational basis of the commonly used statistical methods.

S2.2 Population versus sample

A statistical **population** is the totality of the units (individuals) of interest to the researcher. It follows then that, depending on the researcher's objectives, a population may be small or infinitely large. A small population can be measured in its entirety. Plant breeders often handle large populations. Obtaining measurement from the entire population is often impractical. Instead, researchers obtain measurements from a subset of the population, called a **sample**. The scores from the sample are used to infer or estimate the scores we would expect to find if it were possible to measure the entire population.

In order to draw accurate conclusions about the population, the sample must be representative of the population. To obtain a representative sample, the statistical technique of **random sampling** (in which all possible scores in the population have an equal chance of being selected for a sample). There are other methods of drawing samples from a population for a variety of purposes. These include quota, convenience, and stratified sampling methods. A number that describes a characteristic of a population is called a **sample statistic,** or simply a **statistic**. A number that describes a population is called a population parameter, or simply a **parameter**.

S2.3 The issue of causality

Scientific conclusions are drawn from the preponderance of the evidence obtained from properly conducted research. Cause and effect is implicit in the logic of researchers. However, it is difficulty to definitely prove that variable X causes variable Y. There is always the possibility that some unknown variable is actually responsible for the effect observed (change in scores). No statistical procedure will prove that one variable causes another variable to change (i.e., statistics does not prove anything!). An experiment provides evidence to argue for a certain point of view, not prove it.

S2.4 Statistical hypothesis

An **hypothesis** is an informed conjecture (educated guess) about a phenomenon. It is arrived at after taking into account pertinent scientific knowledge and personal experience. Researchers often have preconceived ideas about the phenomenon that they seek to investigate. However, they are willing to approach an investigation with an open mind. A hypothesis declares the prediction of the researcher concerning the relationship between two or more variables associated with the study. An experiment is designed to test this relationship.

In plant breeding, a breeder ends up with about a dozen promising genotypes from which one would eventually be selected for release to farmers for use in cultivation. The breeder conducts field tests or trials (over locations, years) to help in the decision making process. He or she suspects or predicts differences among these genotypes. The predicted difference represents a true phenomenon. To avoid any biases, the hypothesis is formulated in the opposite direction to the predicted or suspected outcome. That is, the researcher would state that no real differences exist among the genotypes (i.e., any differences are due to chance). This is the **null hypothesis (Ho)** or the hypothesis of no difference. The **alternative hypothesis (H$_1$)** would indicate a real difference exists. There is a standard way of mathematically stating a hypothesis. If four genotypes were being evaluated, a hypothesis could be formulated as follows:

Null hypothesis Ho: $\mu_1 = \mu_2 = \mu_3 = \mu_4$ (i.e., all genotype means are equal)
Alternate hypothesis H$_1$: $\mu_1 \neq \mu_2 \neq \mu_3 \neq \mu_4$ (i.e., all genotype means are not equal)

The Ho is accepted (i.e., automatically reject H$_1$) or rejected (accept H$_0$) at a chosen level of **statistical significance, α,** (e.g., $\alpha = 0.01$ or 0.05; acknowledging that 1% or 5% of the time you could be mistaken in your conclusion). In other words, the research does not prove anything outright, as previously pointed out. Rejecting Ho when it is true (i.e., you are saying a difference exists when in fact none really does) is called a **Type I error**. On the other hand, failure to find a true difference when it exists (accepting a false Ho) is called a **Type II error**.

The goal of a plant breeder is to conduct research in such a way that true differences, when they exist, are observed. This depends on the adoption of sound experimental procedures, often called field plot techniques and is discussed later in this chapter.

S2.5 Concept of statistical error

Error in statistics does not imply a mistake. As previously stated, statistics is not used to prove anything. Experimental conditions are seldom, if ever, perfect. If five samples of a uniform cultivar (e.g., pure line) are planted under identical conditions, it would be expected that a measurement of a trait (e.g., height) would be identical for all samples. In practice, differences, albeit minor, would be observed. This variation that cannot be accounted for is called **experimental error** (simply **error**). No effort can completely eliminate experimental error. However, efforts can be made to reduce it such that true differences in a study are not obscured. Laboratory or controlled environment research often allows the researcher more effective control over variation in the experimental environment. Field studies are subject to significant variation from the soil as well as the above-ground environment. Other sources of undesirable variation are competition among plants and operator (human) error. Plant breeders need to understand the principles of experimental design. A large error would not permit small real differences in the experiment to be detached.

Errors may be random or systematic, the former being responsible for inflated error estimates. Practical ways of reducing error include the use of proper plot size and shape. Within limits, rectangular plots and larger plots tend to reduce variation per plot. Also, the use of experimental designs that include local control of variation (e.g., randomized complete block design) helps to reduce error.

S2.6 Principles of experimental design

The subject is treated in detail in Chapter 26. It is introduced here only to further explain the concept of error. The unit to which a treatment is applied is called the **experimental unit**. In plant breeding common examples of treatment are genotypes (to be evaluated), locations (where genotypes will be evaluated), and years and seasons (over which evaluations are conducted). An experimental unit may be a plant or groups of plants (in a pot).

Experimental designs are statistical procedures for arranging experimental units (experimental design) such that experimental error is minimized. Three tactics or techniques are used in experimental designs for this purpose. These are **replication, randomization,** and **local control**.

(i) **Replication.** Replication is the number of times a treatment is repeated in a study. It is important in experimental design for several reasons, two key ones being the following:
   - **Estimation of statistical error.** To establish that experimental units treated alike vary in their response requires at least two of the same units that have been treated alike.
   - **To reduce the size of statistical error.** A measure of the consistency in a data set (standard error) is presented later in this chapter. Calculated as $\sigma/\sqrt{(\text{number of replications})}$, it is obvious that the larger the number of replications, the smaller the error (σ = standard deviation).

Another pertinent question is the number of replications to use in a study. It should be noted that the more replications used, the more expensive the experiment will be to conduct. In plant breeding, breeders commonly use fewer replications (e.g., two) for preliminary field trials, which often contain hundreds of lines, and more replications (e.g., four) for advanced trials that contain about 10–20 entries.

(ii) **Randomization.** This is the principle of equal opportunity whereby treatment allocation to experimental units is made without bias. To make the statistical test of significance valid, errors should be independent of treatment effect. Randomization may be completely random or may have restrictions to accommodate a specific factor in the experiment.

(iii) **Local control.** This is an additional tactic used by researchers to "contain" variation within an experiment through grouping of experimental units on the basis of homogeneity. Variation within groups is kept to a minimum, while enhancing variation between groups. Statistical procedures are then used to extract this group-based variation from the error estimate. Blocking is recommended if a distinct variation occurs in the experimental field. For example, where a field has a slope, there will be a fertility gradient. Completely random allocation of treatments may place all the replications of one treatment in one fertility zone. Use of the blocking techniques will allow one replication of each treatment to be represented in each distinct fertility zone by placing a restriction on randomization. The blocks should be laid across the fertility gradient.

S2.7 Probability

Statistical probability is a procedure for predicting the outcome of events. Probabilities range from 0 (an event is certain not to occur) to 1.0 (an event is certain to occur). There are two basic laws of probability – product and sum laws. The probability of two or more outcomes occurring simultaneously is equal to the product of their individual probabilities. Two events are said to be independent of one another if the outcome of each one does not affect the outcome of the other. Genetic ratios may be expressed as probabilities. Consider a heterozygous plant (Rr). The probability that a gamete will carry the R allele is ½. In a cross, $Rr \times Rr$ (selfing), the probability of a homozygous recessive (rr offspring) is ½ × ½ = ¼. The probability that one or another of several mutually exclusive outcomes will occur is the sum of their individual probabilities. Using the cross $Rr \times Rr$, the F_2 will produce RR:Rr:rr in the ratio ¼:½:¼. The probability that a progeny will exhibit a dominant phenotype (RR, Rr) = ¼ + ½ = ¾.

In using probabilities for prediction, it is important to note that a large population size is needed for accurate prediction. For example, in a dihybrid cross, the F_2 progeny will have 9:3:3:1 phenotypic ratio, indicating 9/16 will have the dominant phenotype. However, in a sample of exactly 16 plants, it is unlikely that exactly 9 plants will have the dominant phenotype. A larger sample is needed.

S2.8 Measures of central tendency

The distribution of a set of phenotypic values tends to cluster around a central value. The most common measure of this clustering is the arithmetic mean. Plant breeders use this statistical procedure very frequently in their work. The formulas for calculating means are:

(a) **Sample mean.**
$$\bar{X} = \Sigma X/n$$

(b) **Population mean.**

$$\mu = \Sigma X / N$$

where X is the measured value of the item, $\bar{X}$ is the sample mean, μ is the population mean, n is the sample size, and N is the population sample.

The sample mean is calculated as:

$$\bar{X} = \sum_{i=1}^{n} X_i/n \quad \text{(for ungrouped data; Table S2.1)}$$

or

$$\bar{X} = \Sigma X_i f_i/n \quad \text{(for grouped data; Table S2.2)}$$

where X_i is the value of the ith unit included in the sample, f_i is the frequency of the ith class and n is the Σf_i.

The sample mean of seed size of navy beans is:

$$\bar{X} = \sum_{i=1}^{n} X_i/n$$
$$= [17.2 + 18.1 + \ldots + 19.7]/10$$
$$= 190.9/10$$
$$= 19.01 \text{ g per 100 seed}$$

Using data in Table S2.2, the mean of the F_2 can be obtained as:

$$\bar{X} = \Sigma X_i f_i/n$$
$$= 2105/193$$
$$= 10.91 \text{ cm}$$

S2.9 Measures of dispersion

Measures of dispersion or variability concern the degree to which values of a data set differ from their

Table S2.1 Data for distribution of plant seedling height.

	5	6	7	8	9	10	11	12	13	14	15	16	17	Total
Distribution of seedling height (cm)														
F_1				5	14	16	12	3						50
F_2	4	10	13	17	20	28	25	18	17	13	11	10	7	193

Table S2.2 Grouped data for frequency calculation.

f_1	x_i	f_i	f_1x_i	$f_1x_i^2$	f_1x_i	$f_1x_i^2$
	5	4	20	100		
	6	10	60	360		
	7	13	91	637		
5	8	17	136	1088	40	320
14	9	20	180	1620	126	1134
16	10	28	280	2800	160	1600
12	11	25	275	3029	132	1452
3	12	18	216	2592	36	432
	13	17	221	2873	494	4938
	14	13	182	2548		
	15	11	165	2475		
	16	10	160	2560		
	17	7	119	2023		
		n	Σf_1x_i	$\Sigma f_1x_i^2$		
		193	2105	24705		

computed mean. The most commonly used measure of dispersion is the mean square deviation or **variance**. The **population variance** is given by:

$$\sigma^2 = [\Sigma(X_1 - \mu)^2]/N$$

where σ^2 is the population variance, X_1 is the value of observations in the population, μ is the mean of the population, and N is the total number of observations in the population.

The sample variance is given by:

$$s^2 = [\Sigma(X - \bar{X})^2]/(n-1)$$

where s^2 is the sample variance, X is the value of the observation in the sample, X is the mean of the sample, and n is the total number of observations in the sample.

The computational formula is:

$$s^2 = [\Sigma X^2 - (\Sigma \bar{X})^2/n]/(n-1)$$

Using the data below for number of leaves per plant:

Number of leaves										Total
X	7	6	7	8	10	7	9	8	7	10	$79 = \Sigma X$
X^2	49	36	49	64	100	49	81	64	49	100	$641 = \Sigma X^2$

$(\Sigma X)^2/n = 79^2/10 = 6241/10 = 624.1$
$s^2 = (641 - 642.1)/9 = 16.9/9 = 1.88$

Variance may also be calculated from grouped data. Using the data in Table S2.2, variance may be calculated as follows:

$$s^2 = [n\Sigma fx^2 - (\Sigma fx)^2]/n(n-1)$$
$$= [193(24705) - (2105)^2]/193(193-1)$$
$$= (4\,768065 - 4\,431025)/37056$$
$$= 337040/37056$$
$$= 9.10 \text{ (for the F}_2\text{, the most variable}$$
$$\text{generation following a cross)}$$

Variance for the F_1 is 1.67.

S2.10 Standard deviation

The **standard deviation** (SD) measures the variability that indicates by how much the value in a distribution typically deviates from the mean. It is the positive square root of the population variance. The larger the value of the standard deviation, the more the observations (data) are spread about the mean, and vice versa.

The standard deviation of the sample is simply:

$$s = \sqrt{s^2}$$

For the number of leaves per plant example:

$$\begin{aligned} s &= \sqrt{1.88} \\ &= 1.37 \end{aligned}$$

Similarly, for the seedling height data:

$$\begin{aligned} \text{SD of the } F_2 &= \sqrt{9.1} = 3.02 \\ \text{SD of the } F_1 &= \sqrt{1.67} = 1.29 \end{aligned}$$

S2.11 The normal distribution

One of the most important examples of continuous probability distribution is the **normal distribution** or the **normal curve**. It is important because it approximates many kinds of natural phenomena. If the population is normally distributed, the mean = 0.0 and the variance = 1.0. Further, a range of ±1 standard deviation from the mean will include 68.26% of the observations, whereas a range of ±2 standard deviations from the mean will capture most of the observations (95.45%) (Figures S2.1). The shape of the curve varies depending on the nature of the population.

S2.12 Coefficient of variation

The **coefficient of variation** is a measure of the relative variability of given populations. Variance estimates have units attached to them. Consequently, it is not possible to compare population measurements of different units (i.e., comparing apples with oranges). For example, one population may be measured in kilograms (e.g., yield), while another is measured in feet (e.g., plant height). A common

application of variance is the test to find out if one biological sample is more variable for one trait than for another (e.g., is plant height in soybean more variable than the number of pods per plant?). Larger organisms usually vary more than smaller ones. Similarly, traits with larger means tend to vary more than those with smaller means. For example, grain yield per hectare of a cultivar (in kg/ha) will vary more than its 100 seed weight (in grams). For these and other enquiries, the coefficient of variation facilitates the comparison because it is unit free.

The coefficient of variation (CV) is calculated as:

$$CV = (s/\bar{X}) \times 100$$

For the number of leaves per plant example:

$$\begin{aligned} CV &= 1.37/7.9 = 0.173 \\ &= 17.3\% \end{aligned}$$

A CV of 10% or less is generally desirable in biological experiments.

S2.13 Standard error of the mean

The **standard error** measures the amount of variability among individual units in a population. If several samples are taken from one population, the individuals will vary within samples as well as among samples. The **standard error of the mean (SE)** measures the variability among different sample means taken from a population. It is computed as:

$$s_x = s/\sqrt{n}$$

For the number of leaves per plant example:

$$\begin{aligned} s_x &= 1.37/\sqrt{10} \\ &= 0.433 \end{aligned}$$

The standard error of the mean indicates how precisely the population parameter has been estimated. It may be attached to the mean in presentation of results in a publication (e.g., for the leaves per plant example, it will be 7.9 ± 0.43).

S2.14 Simple linear correlation

Plant breeders are not only interested in variability as regards a single characteristic of a population, but

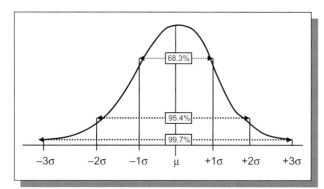

Figure S2.1 Standard normal distribution curve.

often they are interested in how multiple characteristics of the units of a population associate. If there is no association, covariance will be zero or close to zero. The magnitude of covariance is often related to the size of the variables themselves, and also depends on the scale of measurement.

The **simple linear correlation** measures the linear relationship between two variables. It measures a joint property of two variables. Plant breeding is facilitated when desirable genes are strongly associated on the chromosome. The relationship of interest in correlation is not based on cause and effect. The degree (closeness or strength) of linear association between variables is measured by the **correlation coefficient**. The correlation coefficient is free of scale and measurement and has values that lie between $+1$ and -1 (i.e., correlation can be positive or negative). If there is no linear association between variables, the correlation is zero. However, a lack of significant linear correlation does not mean no association (the association could be non-linear or curvilinear).

The population correlation coefficient (ρ) is given by:

$$\rho = \sigma^2_{XY}/\sqrt{[\sigma^2_X \times \sigma^2_Y]}$$

where σ^2_X is the variance of X, σ^2_Y is the variance of Y, and σ^2_{XY} is the covariance of X and Y. The sample covariance is called **Pearson correlation coefficient** (**r**) and is calculated as:

$$r = s^2_{XY}/\sqrt{[s^2_X \times s^2_Y]}$$

where s^2_{XY} is the sample covariance of X and Y, s^2_X is the sample variance of X, and s^2_Y is the sample variance of Y.

The computational formula is:

$$r = [N(\Sigma XY) - (\Sigma X)(\Sigma Y)]/\sqrt{[N(\Sigma X^2)} \\ - (\Sigma X)^2][N(\Sigma Y^2) - (\Sigma Y)^2]$$

The data in Table S2.3 show the seed oil and protein content of 10 soybean cultivars. The calculation yields the following results:

Covariance $= -2.45$
Correlation $= -0.757$
Intercept $= 59.34$
The slope $= -1.087$
Standard error $= 0.332$
Students' t-value $= 3.273$; probability $= 0.010$

Table S2.3 Data for oil and protein content of soybean seed.

Oil content (%)	Protein content (%)
20.1	35.7
21.2	35.1
19.5	33.2
18.3	40.6
19.0	37.5
21.3	36.1
19.8	39.5
22.6	34.8
17.5	39.1
19.9	40.2
19.92	37.68

The results indicate a significant negative association between seed oil and protein content. The breeding implication is that as one selects for high seed oil, seed protein will decrease.

S2.15 Simple linear regression

Unlike simple linear correlation, **simple linear regression** is a relationship between two variables that involves cause and effect. There is a **dependent variable** (Y) and an **independent variable** (X). For example, grain yield depends on seed size, number of seeds per pod, and so on. The changes in the dependent variable (effect) are brought about by the changes in the independent variable (cause). Another way of looking at it is that regression is a study of the relationship between variables with the objective of identifying, estimating, and validating the relationship.

Simple linear regression has the mathematical form of the equation of a straight line:

$$Y = a + bX$$

where Y is the dependent variable, X is the independent variable, b is the slope of the regression line, and a is the intercept on the Y axis.

The regression coefficient is calculated as:

$$b = [N(\Sigma XY) - (\Sigma X)(\Sigma Y)]/N(\Sigma X^2) - (\Sigma X)^2$$

The data in Table S2.4 represent the yield of soybean corresponding to various days to maturity

Table S2.4 Data on plant yield and maturity of soybean.

	Yield (bushels)	Days to maturity
	44	138
	40	136
	38	125
	35	118
	33	115
	32	111
	30	110
	28	109
	24	98
	18	93
Mean	32.2	115.3

of the crop. The results of a regression analysis are as follows:

Covariance = 110.04
Correlation = 0.976
Intercept = −27.03
Slope (b) = 0.514
Standard error = 0.040
Students' t-value = 12.794; probability = 0.000

The prediction equation is hence (Figures S2.2):

$$\hat{Y}_i = 127.03 + (0.514)X_i$$

By plugging in values for X_i, corresponding Y values can be predicted. The results indicate that the regression line will be a good predictor of an unknown value of the independent variable.

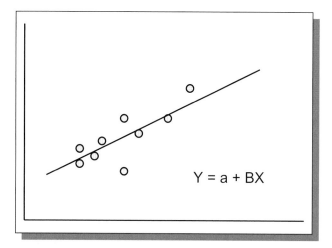

Figure S2.2 The linear regression line.

S2.16 Chi-square test

The **chi-square** (χ^2) test is used by plant breeders to test hypothesis related to categorical data such as would be collected from inheritance studies. The statistic measures the deviations of the observed frequencies of each class from that of expected frequencies. Its values can be zero or positive but not negative. As the number of degrees of freedom increases, the chi-square distribution approaches a normal distribution. It is defined mathematically as follows:

$$\chi^2 = \Sigma[(f_o - f_e)^2]/f_e$$

where f_o is an observed sample frequency and f_e is an expected frequency of the **null hypothesis (H₀)**, the hypothesis to be "disproved".

(a) **Chi-square test of goodness-of-fit.** Suppose a breeder is studying the inheritance of a trait. A cross is made and the following outcomes are recorded:

Character	Observed frequency	Expected frequency
Green cotyledon	78	75
Yellow cotyledon	22	25

If we assume that the trait is controlled by a single gene pair exhibiting dominance, we expect to find a phenotypic ratio of 3:1 in the F_2 (or 1:1 ratio in a test cross). This is the null hypothesis (Ho). The expected frequencies based on the 3:1 ratio are presented in the data above. Chi-square is calculated as follows:

$$
\begin{aligned}
\chi^2 &= \Sigma[(f_o - f_e)^2]/f_e \\
&= (78 - 75)^2/75 + (22 - 25)^2/25 \\
&= 0.12 + 0.36 \\
&= 0.48
\end{aligned}
$$

The degrees of freedom (df) = 2 − 1 = 1; the tabulated t-value is 3.81 at a probability of 0.05. The calculated chi-square value is less than the tabulated value; therefore, the discrepancy observed in the table is purely a chance event. The null hypothesis is hence accepted, and the cotyledon color is declared to be controlled by a single gene pair with complete dominance.

(b) **Chi-square test of independence.** Also called a **contingency chi test**, the chi-square test of independence may be applied to different situations. For example, it is applicable where a breeder has made one set of observations obtained under a particular set of conditions and wishes to compare it with a similar set of observations under a different set of conditions. The question being asked in contingency chi square is whether the experimental results are dependent (contingent upon) or independent of the conditions under which they were observed. In general, whenever two or more systems of classification are used, one can check for independence of the system.

There is a cross-classification when one individual is classified in multiple ways. For example, a cultivar may be classified according to species and also according to resistance to a disease. The question then is whether the classification of one individual according to one system is independent of its classification by the other system. More specifically, if there is independence in this species-infection classification, then the breeder would interpret the results to mean that there is no difference in infection rate between species.

The short-cut method for solving contingency chi-square problems is as follows:

Categories of observation

	I	II	Total
A	a	b	$a \pm b$
B	c	d	$c \pm d$
Total	$a+c$	$b+d$	$a+b+c+d$

$$\chi^2 = [(ad - bc)^2]n/[(a+b)(c+d)(d+c)(b+d)]$$

(a, b, c, d are observed frequencies)

This is called a 2×2 contingency chi-square, but can be extended to more complex problems (2×4, 4×6, etc.).

S2.17 *t*-test

The *t*-test is used to make inferences about population means. A breeder may wish to compare the yields of two cultivars, for example. Assuming the sample observations are drawn at random, the two population variances are equal, and the populations from which the samples were drawn follow the normal distributions (these are assumptions made in order to use this test), the hypothesis to be tested is:

Ho: $\mu_1 = \mu_2$ (no difference between the two means)
The alternative hypothesis to the null is:
H1: $\mu_1 \neq \mu_2$ (the two populations are not equal)

This may be tested as follows (for a small sample size):

$$t = [\bar{X}_1 - \bar{X}_2]/s_p\sqrt{[1/n_1 + 1/n_2]}$$

where $s_p = \sqrt{\{[(n_1 - 1)s_1^2 + (n_2 - 1)s_2^2]/n_1 + n_2 - 2\}}$, s_p is the pooled variance, and X_1 and X_2 are the means of samples 1 and 2, respectively.

Example

A plant breeder wishes to compare the seed size of two navy bean cultivars, A and B. Samples are drawn and the 100-seed weight obtained. The following data were compiled:

Ho: $\mu_1 = \mu_2$
H1: $\mu_1 \neq \mu_2$

	Cultivar	
	A	**B**
n	10	8
$\bar{X}$	21.2	19.5 (g/100 seeds)
s	1.3	1.1 (g/100 seeds)

$$
\begin{aligned}
t &= [\bar{X}_1 - \bar{X}_2]/s_p\sqrt{[1/n_1 + 1/n_2]} \\
s_p &= [(10 - 1)(1.3)^2 + (8 - 1)(1.1)^2]/(10 + 8 - 2) \\
&= [9(1.69) + 7(1.21)]/16 \\
&= [15.21 + 8.47]/16 \\
&= 23.68/16 \\
&= 1.48 \\
t &= 10 - 8 \\
&= 2/(1.48 \times 0.47) \\
&= 0.70 \\
&= 2/0.70
\end{aligned}
$$

t (calculated) $= 2.857$

at $\alpha_{0.05}$:

df $= 10 + 8 - 2 = 16$
t (tabulated) $= 1.746$

Since calculated t exceeds tabulated, we declare a significant difference between the two cultivars for seed size (measured as 100-seed weight).

S2.18 Analysis of variance

Frequently, the breeder needs to compare more than two cultivars. In yield trails, several advanced genotypes are evaluated at different locations and in different years. The t-test is not applicable in this circumstance but its extension, the **analysis of variance (ANOVA)**, is used instead. ANOVA allows the breeder to analyze measurements that depend on several kinds of effects, and which operate simultaneously, in order to decide which kinds of effects are important and to estimate these effects. As a statistical technique, ANOVA is used to obtain and partition the total variation in a data set according to the sources of variation present and then to determine which ones are important. The test of significance of an effect is accomplished by the F test. The results of an analysis of variance are presented in the ANOVA table, the simplest from being as follows:

Source of variation	Degrees of freedom (df)	Sum of squares (SS)	Mean sum of squares (MSS)	F
Treatment	$k-1$	SS_{Tr}	$MS_{Tr} = SS_{Tr}/k-1$	MS_{Tr}/MS_E
Error	$N-k$	SS_E	$MS_E = SS_E/N-k$	
Total	$N-1$	SS_T		

Treatment = the most important source of variance caused by the applied treatments (e.g., different cultivars, locations, years, etc.). The error = unaccounted variation.

In more detailed analysis, interaction effects between treatments are accounted for in the analysis. Examples of ANOVA for G × E interaction analysis are provided in Chapter 26. ANOVA is usually done on the computer using software such as MSTAT and SAS. As previously stated, ANOVA permits the breeder to handle more than two genotypes (or variables) in one analysis. The t-test is not efficient for comparing more than two means. Commonly used tests to separate means under such conditions include the LSD and DMRT.

S2.19 Multivariate statistics in plant breeding

Multivariate analysis is the branch of statistics concerned with analyzing multiple measurements that have been made on one or several samples of individuals. Because these variables are interdependent among themselves, they are best considered together. Unfortunately, handling data with multicolinearity can be unwieldy, hence some meaningful summarization is needed.

The multivariate techniques in use may be summarized into two groups:

(a) **Interdependence models.** for example, principal component analysis, factor analysis.
(b) **Dependence models.** for example, multivariate analysis of variance, classification functions, discriminant functions, multiple correlation, canonical correlation.

W.W. Cooley and P.R. Jones further classified multivariate procedures into four categories according to the number of populations and number of variables as follows:

(i) One set of variables; one population – for example, principal components, factor analysis.
(ii) One set of variables; two or more populations – for example, multivariate analysis of variance, discriminant functions, classification functions.
(iii) Two or more sets of variables; one population – for example, polynomials fit, multiple correlation, canonical correlations, multiple partial correlation.
(iv) Two or more sets of variables; two or more sets of populations – for example, multivariate covariance.

Multivariate analyses are done on computers because of their complexity. An overview of the common procedures is given in the next sections.

S2.19.1 Factor analysis

A variable can be explained to the extent that its variance can be attributed to an identifiable source. **Factor analysis** may be used to find ways of identifying fundamental and meaningful dimensions of a multivariate domain. It is a decision-making model for extracting subsets of covarying variables. To do this, natural or observed intercorrelated variables are reformulated into a new set (usually fewer in number) of independent variables, such that the latter set has certain desired properties specified by the analyst. Naming factors is only a mnemonic convenience. It should be done thoughtfully so as to convey information to both the analyst and the audience. For example, a large set of morphological traits may be reduced to several conceptual factors such as "architectural factor" (loaded by variables such as internode length, number of internodes, etc.), whereas a "seed size factor" may be loaded by traits such as seed length, and seed width.

S2.19.2 Principal component analysis

Principal component analysis (PCA) reduces the dimensions of multivariate data by removing intercorrelations among the traits being studied, thereby enabling multidimentional relationships to be plotted on

---

Industry highlights

Multivariate analyses procedures: applications in plant breeding, genetics and agronomy

A.A. Jaradat

Agricultural Research Service, USDA, 803 Iowa Ave., Morris 56267 MN, USA

Introduction

Plant breeders, geneticists and agronomists are increasingly faced with theoretical and practical questions of multivariate nature. With increases in germplasm sizes, numbers of plant and crop variables, and evaluation and characterization data on molecular, biochemical, morphological and agronomic traits, multivariate statistical analysis (MVA) methods are receiving increasing interest and assuming considerable significance. Some MVAs (e.g., multivariate analysis of variance, MANOVA and covariance, MANCOVA) are extensions of uni- and bivariate statistical methods appropriate for significance tests of statistical hypotheses. However, most MVAs are used for data exploration, extraction of fundamental components of large data sets, discovery of latent structural relationships, visualization and description of biological patterns. This review focuses on the salient features and applications of MVAs in multivariate data analyses of plant breeding, genetics and agronomy data. These include MANOVA, MANCOVA, data reduction methods (factor, principal components, principal coordinates, perceptual mapping and correspondence analyses), and data classification methods (discriminant analysis, clustering and additive trees).

Crop improvement programs through breeding, selection and agronomic evaluation rely on available genetic diversity for specific trait(s) in the primary and, if needed, in the secondary gene pool of a particular crop species. Classical univariate analysis procedures, limited to estimation and hypotheses testing, are not capable of detecting patterns and exploring multivariate data structures in genetic resources, breeding lines or cultivars. Therefore, MVA methods to classify and order large numbers of breeding material, trait combinations and genetic variation are gaining considerable importance and assuming considerable significance.

MANOVA and MANCOVA

MANOVA and MANCOVA perform a multivariate analysis of variance or covariance when multiple dependent variables are specified. MANOVA tests whether mean differences among groups for a combination of dependent variables are likely to have occurred by chance. A new dependent variable that maximizes group differences is created from the set of dependent variables. The new dependent variable is a linear combination of measured dependent variables, combined so as to separate the groups as much as possible. ANOVA is then performed on the newly created dependent variable. MANCOVA asks if there are statistically reliable mean differences among groups after adjusting the newly created dependent variable for differences on one or more covariates. In this case, variance associated with the covariate(s) is removed from error variance; smaller error variance provides a more powerful test of mean differences among groups.

MANOVA was used in the analysis of growth patterns and biomass partitioning of crop plants as a prerequisite for interpreting results of field experiments and in developing crop simulation models. Royo and Blanco 1999 utilized MANOVA to compare non-linear regression growth curves in spring and winter triticale and identified variables responsible for the differences between these curves. Results of these studies are partially presented in Table BS2.1, along with the smallest set of variables required to characterize the growth curves. Wilk's Lambda is the criteria for statistical inference and is estimated as the pooled ratio of error variance to effect variance plus error variance. In this example, all Wilk's Lambda and F approximation estimates are significant. For example, the differences within each growth habit (Table BS2.1) were non-significant but differences between growth habits were significant. Thermal time needed to reach the maximum LAI was the variable responsible for these differences.

Table BS2.1 Summary of the significant effect ($P < 0.05$) for leaf area index (LAI) and dry weight of stems per plant in MANOVA and determination of the smallest set of variables.

Variable	Significant source of variation	Wilk's Lambda	F approximation	Final set
LAI	Year	0.182	40.45***	y_{max}†
	Genotype	0.175	4.73*	x_{max}
	Growth habit	0.479	26.85***	
Dry weight of stems	Year	0.552	7.30*	x_{inf}
	Genotype	0.067	13.99**	x_{inf}
	Growth habit	0.480	70.17***	
	Within winter types	0.105	13.12***	

† y_{max}, maximum value of the variable; x_{max}, time in growing-degree days from sowing to y_{max}; x_{inf}, time in growing degree-days from sowing to reach maximum rate of growth. *, $P < 0.05$; **, $P < 0.01$; ***, $P < 0.001$.

Variance Components Analysis (VCA)

Experimentation is sometimes mistakenly thought to involve only the manipulation of levels of the independent variables and the observation of subsequent responses on the dependent variables. Independent variables whose levels are determined or set by the experimenter are said to have fixed effects. A second class of effects, random effects, is classification effects where the levels of the effects are assumed to be randomly selected from an infinite population of possible levels. Many independent variables of research interest are not fully amenable to experimental manipulation but, nevertheless, can be studied by considering them to have random effects.

Factor (FA) and Principal Components Analysis (PCA)

The primary purpose of FA and PCA is to define the underlying structure in a data matrix. As data reduction or exploratory methods, these procedures are used to reduce the number of variables and to detect structure in the relationships between these variables. FA reproduces the correlation matrix among variables with a few orthogonal factors; however, contrary to PCA most forms of FA are not unique. PCA is a procedure for finding hypothetical variables (components) which account for as much of the variance in a multidimensional data as possible. PCA is a unique mathematical solution; it performs simple reduction of the data set to a few components, for plotting and clustering purposes, and can be used to hypothesize that the most important components are correlated with some other underlying variables

In PCA one can obtain a "biplot" in which the objects and the variables are superimposed on the same plot so that one can study their inter-relationships (Figure BS2.1). In PCA one judges proximities among the objects using Euclidean

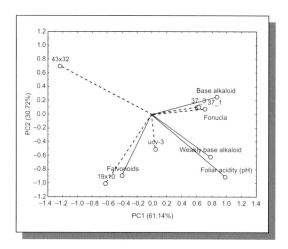

Figure BS2.1 A graph based on PCA of six sesame genotypes as operational taxonomic units (dashed lines) and three secondary metabolites in leaves and foliar acidity as variables (solid lines). Figure courtesy of A.A. Jaradat.

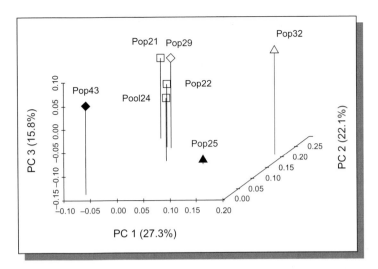

Figure BS2.2 PCoA plot of seven tropical maize populations based on modified Roger's distance. PC1, PC2, and PC3 are the first, second and third principal coordinates, respectively. Heterotic group A (Pop21, Pop22 and Pool24), Heterotic Group B (Pop25, Pop32), and populations not yet assigned to heterotic groups (Pop29, Pop34) are shown. Figure courtesy of A.A. Jaradat.

distances and among the variables using covariance or correlation matrix. PCA was utilized in determining the phyto-chemical relationship of six sesame genotypes and their resistance to whitefly (Laurentin *et al.*, 2003). Foliar acidity and flavonoides dominated PC1 and PC2, respectively. The six sesame genotypes were separated according to their phyto-chemical characteristics. A close relationship was found between secondary metabolites and foliar acidity, on the one hand, and incidence of whitefly on sesame, on the other, thus demonstrating the importance of foliar acidity values of sesame genotypes as a resistance mechanism against whitefly.

Principal Coordinates Analysis (PCoA)

PCoA focuses on samples rather than variables and is based on a matrix containing the distances between all data points. A typical usage of PCoA is reduction and interpretation of large multivariate data sets with some underlying linear structure. PCoA was instrumental in delineating relationships among tropical maize populations based on simple sequence repeats for breeding purposes (Reif *et al.*, 2003). PCoA revealed very clear association among populations within certain heterotic groups (Figure BS2.2). Reif *et al.* 2003 succeeded in identifying genetically similar germplasm based on molecular markers and concluded that PCoA provides a more economic and solid approach for making important breeding decisions early in the breeding program.

Perceptual Mapping (Biplot and GE)

Success in evaluating germplasm, breeding lines and cultivars in multiple environments and for complex traits to identify superior genotypes with specific or wide adaptation can be achieved if the genotypic (G), environmental (E) effects and their interaction (GE) are precisely estimated (Yan *et al.*, 2000). The GGE biplot procedure has been used by breeders and agronomists for dissecting GE interactions and is being used to analyze data from genotype x trait, genotype x marker, environment x QTL, and diallel cross data. The BIPLOT allows a readily visualized display of similarity and differences among environments in their differentiation of the genotypes, the similarity and differences among the genotypes in their response to locations, and the nature and magnitude of the interaction between any genotype and any location.

BIPLOT was used to compare performance of wheat cultivars under several environments in the Ontario wheat per-formance trials (Figure BS2.3) and to estimate relative variance components and their level of significance. Results of BIPLOT analysis have several implications for future breeding and cultivar evaluation. Test for optimal adaptation can be achieved through the deployment of different cultivars for mega-environments, and the unpredictable genotype x loca-tion interaction can be avoided or minimized through cultivar evaluation and selection focusing on genotype main effects.

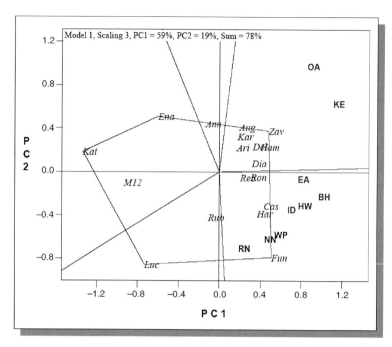

Figure BS2.3 Biplot showing performance of different wheat cultivars (in italics) in different environments (in capital letters) as a selection method to identify superior cultivars for a target environment. Figure courtesy of A. A. Jaradat.

Multiple Correspondence Analysis (MCA)

MCA is a recently developed interdependence MVA procedure that facilitates both dimensional reduction of object ratings on a set of attributes and the perceptual mapping of objects relative to these attributes. MCA helps researchers quantify the qualitative data found in nominal variables and has the ability to accommodate both non-metric data and non-linear relationships. To facilitate the use of common bean landraces in genetic improvement, Bebee *et al.* (2000) used MCA to study the structure of genetic diversity, based on RAPD, among common bean landraces of Middle American origin for breeding purposes. MCA results (Figure BS2.4) indicated that the Middle American bean germplasm is more complex than previously thought, with certain regions holding important genetic diversity which has yet to be properly explored for breeding purposes. The first dimension (Figure BS2.4a) discriminated between lowland and highland races. The second dimension discriminated among highland races, whereas the third dimension (Figure BS2.4b) divided the highland races according to their growth habit, geographical distribution and seed type. Results of MCA can be used to orient plant breeders in their search for distinct genes that can be recombined, thus contributing to higher genetic gain.

Canonical Discriminant Analysis (CDA)

CDA is used to study the variation among two or more groups (samples) of crop cultivars relative to the average variation found within the groups. Linear combinations of the original variables that account for as much as possible of total variation in the data set are constructed using PCA, then canonical correlation is used to determine a linear association between predictor variables identified in PCA and criterion measures. In CDA, more distinct differentiation of cultivars is achieved as compared with univariate analysis, since all independent variables (e.g., traits) are considered simultaneously in the process. CDA can separate "among population" effects from "within population" effects, thus maximizing the overall heritability estimates of canonical variates by placing very large weight on traits with low levels of environmental variability. CDA uses Mahalanobis distance to differentiate between cultivars or populations. The higher the canonical loadings (measures of the simple linear correlation between an original independent variable and the canonical variate) of traits of particular significance, the higher the genetic variation as compared with traits having low canonical loadings. Plant breeders can use this information to focus on particular trait(s) for genetic improvement of a particular

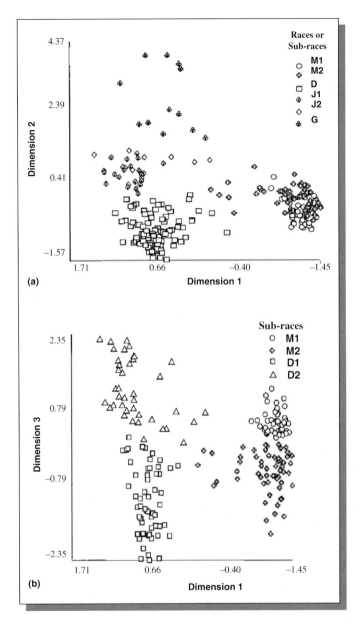

Figure BS2.4 Plots of (a) 250 Middle American bean genotypes in dimension 1 and 2 and (b) of 206 genotypes of two races in dimensions 1 and 3 of MCA based on RAPD data. Figure courtesy of A.A. Jaradat.

crop. Vaylay and van Santen (2002) employed CDA in the assessment of genetic variation in tall fescue (Figure BS2.5). They found that the genetic composition of four tall fescue cultivars differ mainly, in decreasing order, in maturity, cell wall content, flag leaf length, tiller number, and dry matter yield. Therefore, tall fescue breeders can concentrate on the most important traits of this perennial pasture crop knowing that the genetic composition of its cultivars changes with time.

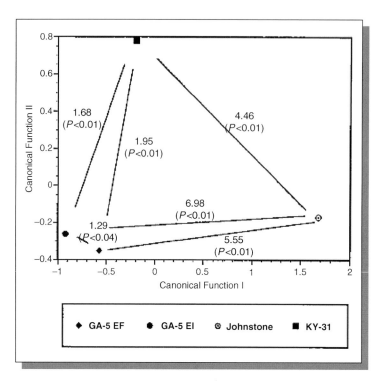

Figure BS2.5 Scatterplot of centroid values of four tall fescue cultivars on two canonical discriminant functions. Mahalanobis distances and their probability values in parenthesis measure the extent of genetic diversity between the four cultivars. Figure courtesy of A.A. Jaradat.

Cluster Analysis (CA)

CA is an analytical MVA procedure for developing meaningful subgroups of objects. It classifies a sample of objects into a small number of mutually exclusive groups based on the similarities among the objects. Stepwise clustering involves a combination or division of objects into clusters. Hierarchical CA starts with each case in a separate cluster and then combines the clusters sequentially, reducing the number of clusters at each step until only one cluster is left. The divisive clustering method begins with all objects in a single cluster, which is then divided at each step into two clusters that contain the most dissimilar objects. Additive trees, as an extension of clustering, are based on a dissimilarity distance matrix among all possible pairs of objects in order to retain the original distances among all pairs of these objects. Unlike other clustering algorithms that are based on the rigorous ultra-metric relationships between objects, the additive tree precisely reflects distances among the objects.

Cluster analysis was used as a tool to optimize and accelerate barley breeding. Karsai *et al.* (2000) evaluated barley cultivars for five physiological and agronomic traits that have significant effects on heading date and winter hardness. Cluster analysis helped identify groups of cultivars representing different adaptational types. The wide level of diversity identified in the germplasm set was valuable in studying the genetics of adaptation to certain environments. It was possible to identify (numbered 1 through 7 in Figure BS2.6) winter and spring groups, groups of cultivars with no vernilization response that had the lowest earliness *per se*, and other groups of cultivars that were the least sensitive to changes in photoperiod but with a strong vernilization response. A breeding scheme was designed on the basis of the clustering results (Figure BS2.6) and was aimed at developing new cultivars better adapted to a given environment.

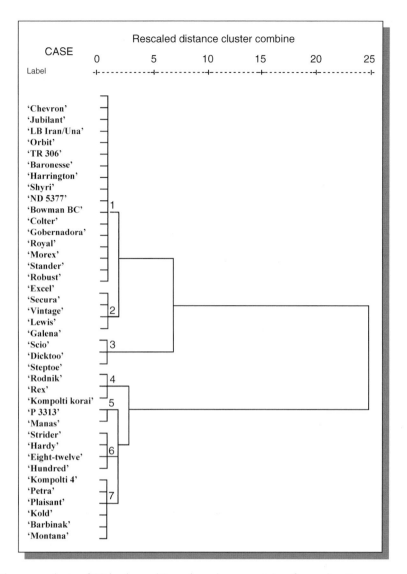

Figure BS2.6 Cluster analysis of 39 barley cultivars based on a matrix of vernalization response, photoperiod sensitivity, earliness *per se*, frost tolerance at $-10\,^{\circ}\mathrm{C}$ and $-13\,^{\circ}\mathrm{C}$, and heading dates under different photoperiod regimes. The dendrogram was created using Ward minimum variance method. Groups (1–7) were characterized by having specific levels of one or more traits. Figure courtesy of A.A. Jaradat.

References

Beebe, S., Skroch, P.W., Tohme, J., Duque, M.C., Pedraza, F., and Nienhuis, J. (2000). Structure of genetic diversity among common bean landraces of middle American origin based on correspondence analysis of RAPD. *Crop Science*, **40**:264–273.

Karsai, I., Meszaros, K., Lang, L., Hayes, P.M., and Bedo, Z. (2000). Multivariate analysis of traits determining adaptation in cultivated barley. *Plant Breeding*, **120**:21–222.

Laurentin, H., Pereira, C., and Sanabria, M. (2003). Phytochemical characterization of six sesame (*Sesamum indicum* L.) genotypes and their relationships with resistance against the sweetpotato whitefly *Bemisia tabaci* Gennadius. *Agronomy Journal*, **95**:1577–1582.

Reif, J.C., Melchinger, A.E. Xia, X.C. *et al.* (2003). Genetic distance based on simple sequence repeats and heterosis in tropical maize populations. *Crop Science*, **43**:1275–1282.

Royo, C. and Blanco, R. (1999). Growth analysis of five spring and five winter triticale genotypes. *Agronomy Journal*, **91**:305–311.

Vaylay, R. and van Santen, E. (2002). Application of canonical discriminant analysis for the assessment of genetic variation in tall fescue. *Crop Science*, **42**:534–539.

Yan, W., Hunt, L.A., Sheng, Q., and Szlavincs, Z. (2000). Cultivar evaluation and mega-environment investigation based on the GGE biplot. *Crop Science*, **40**:597–605.

two or three principal axes. PCA reduces the number of variables to be used for prediction and description. By examining a set of 15 quality traits, researchers at Michigan State University Bean Breeding Program were able to ascertain that certain quality traits (dry characteristics, soaking characteristics, and cooking characteristics) of dry beans were independent. This prompted the researchers to suggest a tandem selection procedure to be followed by the construction of selection indices for their breeding program.

S2.19.3 Discriminant analysis

Discriminant analysis assumes a population is made up of subpopulations and that it is possible to find a linear function of certain measures and attributes of the population that will allow the researcher to discriminate between the subpopulations. Consequently, discriminant procedures are not designed for seeking population groupings (that is what cluster analysis does) because the population has already been grouped. Discriminant analysis may be used in conjunction with the D^2 statistic (Mahalanobis D^2) to indicate the biological distance between separated groups.

S2.19.4 Cluster analysis

Genetic assessment of germplasm is commonly undertaken by plant breeders to understand genetic variation in the germplasm and to discover patterns of genetic diversity. **Cluster analysis**, unlike discriminant analysis, groups genetically similar genotypes. Clustering can be done on morphological or molecular basis (e.g., using DNA markers). Analysis of genetic diversity levels in germplasm helps plant breeders to make proper choices of parents to use in breeding programs.

S2.19.5 Canonical correlation analysis

The **canonical correlation** analysis is a generalization of the multiple correlation procedure. The technique is used to analyze the relationship between two sets of variables drawn from the same subjects. An assumption is made that there are unobserved variables dependent on a known set of variables X, and determining another known set, Y. The intermediating unobserved variables are used to canalize the influence of set X on set Y.

S2.20 Path analysis

Path analysis is a technique for decomposing correlations into different pieces for the interpretation of effects. The procedure is closely related to multiple regression analysis. Path analysis allows the researcher to test theoretical propositions about cause and effect without manipulating variables. Variables may be assumed to be causally related and propositions about them tested. However, it should be cautioned that, should such propositions be supported by the test, one cannot conclude that the causal assumptions are necessarily correct. A breeder may want to understand the relative contributions of yield components and morpho-phenological traits to grain yield.

The general display of a path analysis is shown in Figure S2.3. Arrows are used to indicate assumed causal relations. A single-headed arrow points from the assumed cause to its effect. If an arrow is double-

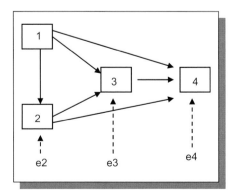

Figure S2.3 The basic concept of path analysis.

headed, only correlation is present (no causal relations are assumed). Variables to which arrows are pointed are called endogenous variables or dependent variables (Y). Exogenous variables have no arrows pointing to them. They are independent (X) variables. The direct effect of a variable assumed to be a cause on another variable assumed to be an effect, is called a **path coefficient**. Path coefficients are standardized partial regression coefficients.

Key references and suggested reading

Akroda, M.O. (1983). Principal components analysis and metroglyph of variation among Nigerian yellow yams. *Euphytica*, **32**:565–573.

Cooley, W.W. and Lohnes, P.R. (1971). *Multivariate data analysis*. John Wiley and Sons, Inc., New York, NY.

Denis, J.C. and Adams, M.W. (1978). A factor analysis of plant variables related to yield in dry beans. I. Morphological traits. *Crops Science*, **18**:74–78.

Kendall, M.G. (1965). *A course in multivariate analysis*. Charles Griffin and Co., London.

Snedecor, G.W. and Cochran, W.G. (1967). *Statistical methods*, 6th edn. Iowa State University, Ames, IA.

Outcomes Assessment

Part A

Please answer the following questions true or false.

1 Statistics do not prove anything.
2 Lack of significant linear regression does not mean a lack of relationship.
3 A t-test is used for separating more than three means.
4 Discriminant analysis is used for seeking population groupings.
5 Chi-square analysis is used for testing hypothesis involving continuous data.

Part B

Please answer the following questions.

1 What is a statistic?
2 Distinguish between standard error and standard error of the mean.
3 Distinguish between simple linear correlation and linear regression.
4 What is contingency chi-square analysis?

Part C

Please write a brief essay on each of the following.

1 Discuss the role of statistics in plant breeding.
2 Discuss the issue of causality in research.
3 Discuss the application of multivariate statistics in plant breeding.
4 Discuss the procedure of path analysis and its application in breeding.

Glossary of terms

Accession: A distinct, uniquely identified sample of seeds, plants, or other germplasm materials that is maintained as an integral part of a germplasm collection.

Adapatedness: The degree or capacity of an individual to survive in a local environment and to transmit its genotype to the next generation.

Additive gene effect: The effect of an allele expected after it has replaced another allele at a locus.

Agrobacterium: A type of soil-inhabiting bacteria that is capable of introducing DNA from plasmids in the bacteria into the genome of plant cells. Often used in the genetic transformation of plants.

Allele: One of several alternate forms (DNA sequences) that resides at the same locus on the chromosome and controls the same phenotype (although with potentially differing effects).

Allogamy: Alternative term for cross-pollination.

Alloploid (or allopolyploid): An individual with somatic cells that contain more than two sets of chromosomes, each of which derives from a different species.

Amino acid: A building block of proteins. Each protein consists of a specific sequence of amino acids (with the sequence of amino acids determined by the sequence of the underlying DNA). There are 20 types of amino acid molecules that make up proteins.

Amphidiploid (or amphiploid): An alloploid with the complete chromosome complements of two diploid species.

Aneuploid: An individual with a chromosome number that is not the exact multiple of the basic number for the species.

Antisense: The complementary strand of a coding sequence (gene); often an expressed copy of an antisense sequence is transformed into a cell or organism to shut off the expression of the corresponding gene.

Apomixis: Asexual reproduction in plants through the formation of seeds without fertilization (agamospermy).

Asexual reproduction: The reproduction process that does not involve the union of gametes.

Autoploid (or autopolyploid): An individual with more than two complete sets of the basic number of chromosomes for the species.

Average effect of a gene: The change in mean value of the population produced by combining a gene with a random sample of gametes from the original population.

Backcross: A cross of an F_1 to either parent used to generate it.

Base collection: A comprehensive collection of germplasm accessions held for the purpose of long-term conservation.

Base pair (bp): Two nitrogenous bases (adenine and thymine or guanine and cytosine) held together by weak bonds. Two strands of DNA are held together in the shape of a double helix by the bonds between base pairs.

Bioinformatics: A broad term to describe applications of computer technology and information science to organize, interpret, and predict biological structure and function. Bioinformatics is usually applied in the context of analyzing DNA sequence data.

Biopharming: The use of genetically transformed crop plants and livestock animals to produce valuable compounds, especially pharmaceuticals. Also called pharming.

Bioremediation: The use of biological organisms to render hazardous wastes non-hazardous or less hazardous.

Biotechnology: A set of biological techniques developed through basic research and now applied to research and product development.

Breeding: The science and art of manipulating the heredity of an organism for a specific purpose.

Breeding line: A genetic group that has been selected and bred for its special combinations of traits.

Breeding value: The mean genotypic value or the progeny of an individual expressed as a deviation from the population mean.

Principles of Plant Genetics and Breeding, Second Edition. George Acquaah.
© 2012 John Wiley & Sons, Ltd. Published 2012 by John Wiley & Sons, Ltd.

Bt (Bacillus thuringiensis): A naturally occurring bacterium that produces a protein toxic to certain lepidopteran insects.

Callus: A cluster of undifferentiated plant cells that have the capacity to regenerate a whole plant in some species.

Cell: The fundamental level of structural organization in complex organisms. Cells contain a nucleus (with chromosomes) and cytoplasm with the protein synthesis machinery, bounded by a membrane.

Cell culture: A technique for growing cells under laboratory conditions.

Cell fusion: The formation of a hybrid cell produced by fusing two different cells.

Centimorgan (cM): A unit of measure of recombination frequency. One centimorgan is equal to a 1% chance that a marker at one genetic locus will be separated from a marker at a second locus due to crossing over in a single generation.

Central dogma: The underlying model for describing gene structure and function. It states that genes are transcribed in the nucleus into messenger RNA molecules, which are then translated into proteins on ribosomes.

Certified seed: The progeny or increase from a breeder or foundation seed and approved by a certifying agency.

Chimera: An individual consisting of cells of two or more types.

Chromosome: A condensed structure found in the cell nucleus that contains the genes of that cell.

Clonal propagation: The reproduction of plants through asexual means, such as cuttings, grafts, or tissue culture.

Cloning: Asexually producing multiple copies of genetically identical cells or organisms descended from a common ancestor.

Codon: A triplet of nucleotides in a DNA or RNA molecule that codes for one of the 20 amino acids in proteins, or for a signal to start or stop protein production. Each gene that codes for protein is a series of codons that gives the instructions for building that protein.

Combining ability: The performance of a line with others in a cross.

Complementary: The opposite or "mirror" image of a DNA sequence. A complementary DNA sequence has an A for every T, and a C for every G. Two complementary strands of single-stranded DNA will join to form a double-stranded molecule.

Complementary DNA (cDNA): A single-stranded DNA molecule that is complementary to a specific RNA molecule and synthesized from it. Complementary DNAs are important laboratory tools as DNA probes and for isolating and studying individual genes.

Conserved sequence: A base sequence in a DNA molecule (or an amino acid sequence in a protein) that has remained essentially unchanged throughout evolution.

Crossing over: The breaking during meiosis of one maternal and one paternal chromosome, the exchange of corresponding sections of DNA, and the rejoining of the chromosomes.

Cultivar: A product of plant breeding that is released for access to producers.

Deoxyribonucleic acid (DNA): The molecule that encodes genetic information. DNA is a double-stranded molecule held together by weak bonds between base pairs of nucleotides. The four nucleotides in DNA contain the bases: adenine (A), guanine (G), cytosine (C), and thymine (T). In nature, base pairs form only between A and T and between G and C; thus the base sequence of each single strand can be deduced from that of its partner.

Diploid: A full set of genetic material consisting of paired chromosomes, one chromosome from each parental set.

DNA chip: A high density array of short DNA molecules bound to a solid surface for use in probing a biological sample to determine gene expression, marker pattern, or nucleotide sequence of DNA/RNA. See also **Microarray**.

DNA probe: A single-stranded DNA molecule used in laboratory experiments to detect the presence of a complementary sequence among a mixture of other single-stranded DNA molecules. Also called gene probe.

DNA profile: The distinctive pattern of DNA restriction fragments or PCR products that can be used to identify, with great certainty, any person, biological sample from a person, or organism from the environment.

DNA replication: The use of existing DNA as a template for the synthesis of new DNA strands. In humans and other eukaryotes, replication occurs in the cell nucleus.

DNA sequencing: Determining the order of nucleotides in a specific DNA molecule.

Domestication: The process of bringing wild plants under cultivation to produce crops under the supervision of humans.

Dominant: A phenotype that is expressed in an organism whose genotype may be either homozygous or heterozygous for the corresponding allele.

Double helix: The shape that two linear strands of DNA assume when bonded together.

Doubled haploid: An individual that is produced by doubling its gametic (n) chromosome number into 2n.

Electrophoresis: A method of separating substances, such as DNA fragments, by using an electric field to make them move through a "gel" at rates that correspond to their electric charge and size.

Embryo rescue: The removal and culture of an immature embryo to produce a plant, often following a wide cross.

Enhancement: The process of improving a germplasm accession by breeding while retaining the important genetic contributions of the accession.

Enzyme: A protein that acts as a catalyst, speeding the rate at which a biochemical reaction proceeds but not altering the direction or nature of the reaction.

Epistasis: The interaction of genes at different loci; the situation in which one gene affects the expression of another.

Eukaryote: Cell or organism with a membrane-bound, structurally discrete nucleus and other well-developed subcellular compartments.

Functional genomics: The field of study that attempts to determine the function of all genes (and gene products), largely based on knowing the entire DNA sequence of an organism.

Gamete: Mature male or female reproductive cell (sperm or ovum) with a haploid set of chromosomes.

Gene: The fundamental unit of heredity; a bundle of information for a specific biological structure or function.

Gene cloning: Isolating a gene and making many copies of it by inserting the DNA sequence into a vector, then into a cell, and allowing the cell to reproduce and make many copies of the gene.

Gene expression: The process in which a cell produces the protein, and thus the characteristic that is specified by a gene's nucleotide sequence.

Gene library: A collection of DNA fragments (carried on vector molecules) that, taken together, represents the total DNA of a certain cell type or organism.

Gene regulation: The process of controlling the synthesis or suppression of gene products in specific cells or tissues.

Gene splicing: Joining pieces of DNA from different sources using recombinant DNA technology.

Genetic code: The language in which DNA's instructions are written. The code consists of triplets of nucleotides (codons), with each triplet corresponding to one amino acid in a protein structure or to a signal to start or stop protein production.

Genetic engineering: The manipulation of genes, composed of DNA, to create heritable changes in biological organisms and products that are useful to people, living things, or the environment.

Genetic erosion: The loss of genetic diversity caused by either natural or man-made processes.

Genetic marker: A genetic factor that can be identified and thus acts to determine the presence of genes or traits linked with it but not easily identified.

Genetic stocks: Accessions that typically possess one or more special genetic traits that make them of interest for research.

Genetic vulnerability: The condition that results when a crop or a plant species is genetically and uniformly susceptible to a pest, pathogen, or environmental hazard.

Genetically modified (GM) organism: An organism whose genetic makeup has been changed by any method, including natural processes, genetic engineering, cloning, mutagenesis, or others.

Genetics: Study of the patterns of inheritance of specific traits.

Genome: The complete set of chromosomes found in each cell nucleus of an individual.

Genomics: The field of study that seeks to understand the structure and function of all genes in an organism based on knowing the organism's entire DNA sequence, with an extensive reliance on powerful computer technologies.

Genotype: The specific combination of alleles present at a single locus in the genome.

Germ cells: The sex cell(s) of an organism (sperm or egg, pollen or ovum). They differ from other cells (somatic) in that they contain only half the usual number of chromosomes. Germ cells fuse during fertilization to begin the next generation.

Germplasm: The sum total of all hereditary material in a single (interbreeding) species.

Green Revolution: An aggressive effort between 1950 and 1975 where agricultural scientists applied modern principles of genetics and breeding to improve crops grown primarily in less developed countries.

Haploid: A cell or organism with a single genome.

Heterozygosity: The presence of different alleles at one or more loci on homologous chromosomes.

Heterozygous: Situation where the two alleles at a specific genetic locus are not the same.

Homologous: Stretches of DNA that are very similar in sequence, so similar that they tend to stick together in hybridization experiments. Homologous can also be used to indicate related genes in separate organisms controlling similar phenotypes.

Homologous chromosomes: A pair of chromosomes containing the same linear gene sequences, each derived from one parent.

Homozygous: Situation where the two alleles at a specific genetic locus are identical to one another.

Hybrid: The progeny of a cross between two different species, races, cultivars, or breeding lines.

Hybridization (or crossing): The process of pollen transfer from the anther of the flower of one plant to the stigma of the flower of a different plant for the purpose of gene transfer to produce an offspring (hybrid) with a mixed parental genotype.

Hybridization: Bringing complementary single strands of nucleic acids together so that they stick and form a double strand. Hybridization is used in conjunction with DNA and RNA probes to detect the presence or absence of specific complementary nucleic acid sequences.

In vitro: Performed in a test tube or other laboratory apparatus.

In vivo: In the living organism.

Inbreeding: The breeding of individuals that are related.

Isoenzyme (isozyme): Different chemical forms of the same enzyme that can generally be distinguished from one another by electrophoresis.

Landrace: A population of plants, typically genetically heterogeneous, commonly developed in traditional agriculture from many years of farmer-directed selection, and which is specifically adapted to local conditions.

Linkage: The proximity of two or more markers (e.g., genes, RFLP markers) on a chromosome.

Linkage map: A map of the relative positions of genetic loci on a chromosome, determined on the basis of how often the loci are inherited together. Distance is measured in centimorgans (cM).

Locus: The position on a chromosome where the gene for a particular trait resides; a locus may be occupied by any one of several alleles (variants) for a given gene.

Meiosis: The process of two consecutive cell divisions in the diploid progenitors of sex cells. Meiosis results in four rather than two daughter cells, each with a haploid set of chromosomes.

Messenger RNA (mRNA): The ribonucleic acid molecule that transmits genetic information from the nucleus to the cytoplasm, where it directs protein synthesis.

Microarray: A large set of cloned DNA molecules spotted onto a solid matrix (such as a microscope slide) for use in probing a biological sample to determine the gene expression, marker pattern, or nucleotide sequence of DNA/RNA.

Microsatellite: A repeated motif of nucleotides, usually only two or three bases in length, where the number of repeats frequently differs between different members of a species.

Mitosis: The process of nuclear division in cells which produces daughter cells that are genetically identical to each other and to the parent cell.

Molecular marker: An identifiable physical location on a chromosome (e.g., restriction enzyme cutting site, gene) whose inheritance can be monitored.

Multiline: A mixture of isolines, each of which is different for a single gene conditioning different forms of the same trait.

Mutagen: A substance that induces mutations.

Mutation: A permanent change in the genetic material involving either a physical alteration in the chromosome or a biochemical change in the underlying DNA molecule.

Nitrogenous base: A nitrogen-containing molecule having the chemical properties of a base.

Nucleic acid: A large molecule composed of nucleotide subunits.

Nucleotide: A subunit of DNA or RNA consisting of a nitrogenous base (adenine, guanine, thymine, or cytosine in DNA).

Nucleus: Membrane-bound structure in the cell that contains the chromosomes (genetic material). The nucleus divides whenever the cell divides.

Pathogen: A specific biological causative agent of disease in plants or animals.

Pedigree: A record of the ancestry of an individual of family.

Phenotype: A biological characteristic or trait possessed by an organism that results from the expression of a specific gene.

Physical map: A map of the locations of identifiable landmarks on DNA (e.g., restriction enzyme cutting sites, genes), regardless of inheritance. Distance is measured in base pairs.

Plasmid: A small, self-replicating molecule of DNA that is separate from the main chromosome. Because plasmids are easily moved from cell to cell or to the test tube, scientists often cleave them with restriction enzymes and insert foreign DNA, and then transfer the recombinant DNA plasmid molecule (as a vector) into other cells.

Pollination: The transfer of pollen from the anthers to the stigma of a flower.

Polymerase chain reaction (PCR): A technique to amplify a specific DNA sequence in vitro using a DNA replicating enzyme, specific oligonucleotide primers, and repeated cycles of heating and cooling. PCR often amplifies the starting material many thousands or millions of times.

Polymorphism: The simultaneous occurrence of two or more distinct forms in a population in a frequency that cannot be accounted for by the balance of mutation and selection.

Polyploidy: An individual with more than two sets of chromosomes characteristic of the species.

Primer: Short pre-existing polynucleotide chain to which new deoxyribonucleotides can be added by DNA polymerase.

Probe: Single-stranded DNA or RNA molecules of a specific base sequence, labeled either radioactively or immunologically, that are used to detect the complementary base sequence by hybridization.

Prokaryotes: Organisms whose genetic material is not enclosed by a nucleus.

Promoter: A DNA sequence preceding a gene that contains regulatory sequences controlling the rate of RNA transcription of that gene. In effect, promoters control when and in which cells a given gene will be expressed.

Protein: A molecule composed of amino acids arranged in a special order determined by the genetic code. Proteins are required for the structure and function of all living organisms.

Pure line: The progeny of a single homozygous individual produced by repeated selfing.

Recessive: A phenotype that is expressed in organisms only if it is homozygous for the corresponding allele.

Recombinant DNA: A hybrid DNA molecule produced in the laboratory by joining pieces of DNA from different sources.

Recombinant DNA technologies: Procedures used to join together DNA segments in a cell-free system (an environment outside a cell or organism). Under appropriate conditions, a recombinant DNA molecule can enter a cell and replicate there, either autonomously or after it has become integrated into a cellular chromosome.

Recombination: The process by which progeny derive a combination of genes different from that of either parent. In higher organisms, this can occur by crossing over.

Recurrent selection: A breeding method whereby plants are repeatedly selected and intercrossed to increase the frequency of favorable alleles.

Regeneration: The process of growing an entire plant from a single cell or group of cells.

Reporter gene: A gene sequence that is easily observed when it is expressed in a given tissue or at a certain stage of development.

Restriction enzyme: An enzyme that recognizes a specific nucleotide base sequence (usually four to six base pairs in length) in a double-stranded DNA molecule and cuts both strands of the DNA molecule at every place where this sequence occurs.

Restriction fragment length polymorphism (RFLP): The presence of two or more variants in the size of DNA fragments produced by a restriction enzyme. These different sized fragments result from an inherited variation in the presence of a restriction enzyme's target sequence. RFLPs are used for gene mapping and DNA profiling.

Ribonucleic acid (RNA): A molecule that translates the instructions encoded in DNA to build proteins.

Ribosomes: Small cellular components composed of specialized ribosomal RNA and protein; site of protein synthesis.

Selection (field): The process of discriminating among genetic variability to advance a fraction to the next generation or breeding cycle.

Selection (*in vitro*): A method to retain specific cells (or clones of cells) expressing a specific trait, such as antibiotic or herbicide resistance, while killing off all other cells that do not express that trait.

Somatic cell: Cells in the body that are not involved in sexual reproduction (that is, not germ cells).

Southern blotting: Transfer by absorption of DNA fragments separated in electrophoretic gels to membrane filters for the detection of specific base sequences by radiolabeled complementary probes.

Tissue culture: Growing cells, tissues, or tissue fragments (from complex, multicellular organisms) on a nutrient medium in a dish, test tube, or flask.

Totipotent: A cell that is capable of regenerating an entire adult organism by itself.

Trait: A distinguishing characteristic or quality of an organism.

Transcription: The transfer of information from specific sequences in a DNA molecule to produce new strands of messenger RNA, which then carry this information from the nucleus to the cytoplasm (where the messenger RNA is translated into protein).

Transformation: Introduction of an exogenous DNA molecule into a cell, causing it to acquire a new phenotype (trait).

Transgenic: An organism that has been transformed with a foreign DNA sequence.

Translation: Synthesis of protein using information contained in a messenger RNA molecule.

Vector: A type of DNA molecule, usually a plasmid or virus, that is used to move recombinant DNA molecules from one cell to another.

Appendix 1

Conversion rates

Imperial unit	Metric conversion
Volume/Capacity	
Cubic inch	16.39 cubic centimeters
Bushel	0.036 cubic meters
Pint	0.57 liters
Quart	1.14 liters
Gallon	4.55 liters
Area	
Acre	4046.86 square meters
Acre	0.4 hectares
Square yard	0.8 square meters
Square feet	0.09 square meters
Mass	
Ounce (avoirdupois)	28.35 grams
Pound	0.45 kilogram
Hundredweight	50.80 kilograms
Ton	1.02 tonnes
Length	
Inch	2.54 centimeters
Foot	0.31 meter
Yard	0.91 meter
Mile	1.61 kilometers

Principles of Plant Genetics and Breeding, Second Edition. George Acquaah
© 2012 John Wiley & Sons, Ltd. Published 2012 by John Wiley & Sons, Ltd.

Index